International Table of Atomic Weights (1977)

Based on relative atomic mass of $^{12}C = 12$.

The following values apply to elements as they exist in materials of terrestrial origin and to certain artificial elements. When used with due regard to footnotes, they are reliable to ± 1 in the last digit, or ± 3 when followed by an asterisk (*). Value in parentheses is the mass number of the isotope of longest half-life.

	Symbol	Atomic number	Atomic weight		Symbol	Atomic number	Atomic weight		Symbol	Atomic number	Atomic weight
Actinium	Ac	89	227.0278	Hafnium	Hf	72	178.49*	Promethium	Pm	61	(145)
Aluminum	Al	13	26.98154[a]	Helium	He	2	4.00260[b,c,g]	Protactinium	Pa	91	231.0359[f]
Americium	Am	95	(243)	Holmium	Ho	67	164.9304[a]	Radium	Ra	88	226.0254[f,g]
Antimony	Sb	51	121.75*	Hydrogen	H	1	1.0079[b,d]	Radon	Rn	86	(222)
Argon	Ar	18	39.948[b,c,d,g]*	Indium	In	49	114.82[g]	Rhenium	Re	75	186.207[c]
Arsenic	As	33	74.9216[a]	Iodine	I	53	126.9045[a]	Rhodium	Rh	45	102.9055[a]
Astatine	At	85	(210)	Iridium	Ir	77	192.22*	Rubidium	Rb	37	85.4678*[c,g]
Barium	Ba	56	137.33[g]	Iron	Fe	26	55.847*	Ruthenium	Ru	44	101.07*[g]
Berkelium	Bk	97	(247)	Krypton	Kr	36	83.80[c,d]	Samarium	Sm	62	150.4[g]
Beryllium	Be	4	9.01218[a]	Lanthanum	La	57	138.9055*[b,g]	Scandium	Sc	21	44.9559[a]
Bismuth	Bi	83	208.9804[a]	Lawrencium	Lr	103	(260)	Selenium	Se	34	78.96*
Boron	B	5	10.81[c,d,e]	Lead	Pb	82	207.2[d,g]	Silicon	Si	14	28.0855*
Bromine	Br	35	79.904[c]	Lithium	Li	3	6.941*[c,d,e,g]	Silver	Ag	47	107.868[c,g]
Cadmium	Cd	48	112.41[g]	Lutetium	Lu	71	174.967*	Sodium	Na	11	22.98977
Calcium	Ca	20	40.08[g]	Magnesium	Mg	12	24.305[c,g]	Strontium	Sr	38	87.62[g]
Californium	Cf	98	(251)	Manganese	Mn	25	54.9380[a]	Sulfur	S	16	32.06[d]
Carbon	C	6	12.011[b,d]	Mendelevium	Md	101	(258)	Tantalum	Ta	73	180.9479*[b]
Cerium	Ce	58	140.12	Mercury	Hg	80	200.59*	Technetium	Tc	43	(98)
Cesium	Cs	55	132.9054[a]	Molybdenum	Mo	42	95.94	Tellurium	Te	52	127.60*[g]
Chlorine	Cl	17	35.453[c]	Neodymium	Nd	60	144.24*[g]	Terbium	Tb	65	158.9254[a]
Chromium	Cr	24	51.996[c]	Neon	Ne	10	20.179*[c,e]	Thallium	Tl	81	204.37*
Cobalt	Co	27	58.9332[a]	Neptunium	Np	93	237.0482[f]	Thorium	Th	90	232.0381[f,g]
Copper	Cu	29	63.546*[c,d]	Nickel	Ni	28	58.70	Thulium	Tm	69	168.9342[a]
Curium	Cm	96	(247)	Niobium	Nb	41	92.9064[a]	Tin	Sn	50	118.69*
Dysprosium	Dy	66	162.50	Nitrogen	N	7	14.0067[b,c]	Titanium	Ti	22	47.90*
Einsteinium	Es	99	(252)	Nobelium	No	102	(259)	Tungsten	W	74	183.85*
Erbium	Er	68	167.26*	Osmium	Os	76	190.2[g]	Unnilpentium	Unp	105	(262)
Europium	Eu	63	151.96[g]	Oxygen	O	8	15.9994[b,c,d]	Unnilhexium	Unh	106	(263)
Fermium	Fm	100	(257)	Palladium	Pd	46	106.4[g]	Unnilquadium	Unq	104	(261)
Fluorine	F	9	18.998403[a]	Phosphorus	P	15	30.97376[a]	Uranium	U	92	238.029[b,c,e,g]
Francium	Fr	87	(223)	Platinum	Pt	78	195.09*	Vanadium	V	23	50.9415*[b,c]
Gadolinium	Gd	64	157.25*[g]	Plutonium	Pu	94	(244)	Xenon	Xe	54	131.30[a,e,g]
Gallium	Ga	31	69.72	Polonium	Po	84	(209)	Ytterbium	Yb	70	173.04*
Germanium	Ge	32	72.59*	Potassium	K	19	39.0983*	Yttrium	Y	39	88.9059[a]
Gold	Au	79	196.9665[g]	Praseodymium	Pr	59	140.9077[a]	Zinc	Zn	30	65.38
								Zirconium	Zr	40	91.22[g]

[a]Elements with only one stable nuclide.
[b]Element with one predominant isotope (about 99 to 100% abundance).
[c]Element for which the atomic weight is based on calibrated measurements.
[d]Element for which known variation in isotopic abundance in terrestrial samples limits the precision of the atomic weight given.
[e]Element for which users are cautioned against the possibility of large variations in atomic weight due to inadvertent or undisclosed artificial isotopic separation in commercially available materials.
[f]Most commonly available long-lived isotope.
[g]In some geological specimens this element has an anomalous isotopic composition, corresponding to an atomic weight significantly different from that given.

General Chemistry

Second Edition

Saunders College Publishing
Complete Package for Teaching with
General Chemistry, 2nd edition
and
General Chemistry with Qualitative Analysis, 2nd edition

by Whitten and Gailey

Davis **Study Guide to Accompany General Chemistry**
Lippincott, Meek, Gailey, & Whitten **Experimental General Chemistry**
Whitten & Gailey **Lecture Outline to Accompany General Chemistry**
DeKorte **Solutions Manual to Accompany General Chemistry**
Whitten & Gailey **Problem Solving in General Chemistry, 2nd edition**
Davis **Instructor's Manual to Accompany General Chemistry and General Chemistry with Qualitative Analysis**
Overhead Transparencies
Test Bank
Computerized Test Bank
Wilkie **Computer Tutorial for General Chemistry**
Shakhashiri, Schreiner, & Meyer **General Chemistry Audio-Tape Lessons, 2nd edition**
Shakhashiri, Schreiner, & Meyer **Workbook for General Chemistry Audio-Tape Lessons, 2nd edition**

General Chemistry

Second Edition

Kenneth W. Whitten
University of Georgia, Athens
Kenneth D. Gailey
University of Georgia, Athens

SAUNDERS GOLDEN SUNBURST SERIES
Saunders College Publishing
Philadelphia New York Chicago
San Francisco Montreal Toronto
London Sydney Tokyo Mexico City
Rio de Janeiro Madrid

Address orders to:
383 Madison Avenue
New York, NY 10017

Address editorial correspondence to:
West Washington Square
Philadelphia, PA 19105

Text Typeface: Trump
Compositor: York Graphic Services, Inc.
Acquisitions Editor: John Vondeling
Developmental Editor: Jay Freedman
Project Editor: Sally Kusch
Copyeditor: Cate Barnett Rzasa
Managing Editor & Art Director: Richard L. Moore
Art/Design Assistant: Virginia A. Bollard
Text Design: Caliber Design Planning, Inc.
Cover Design: Lawrence R. Didona
Text Artwork: Philatek
Production Manager: Tim Frelick
Assistant Production Manager: Maureen Iannuzzi

Cover credit: Atomic fluorescence from sodium in an inductively coupled plasma, using a hollow cathode lamp as an excitation source. Photograph courtesy of Baird Corporation, Bedford, Massachusetts.

GENERAL CHEMISTRY 2/e ISBN 0-03-635-675

3456 032 987654321

CBS COLLEGE PUBLISHING
Saunders College Publishing
Holt, Rinehart and Winston
The Dryden Press

To Betty, Andy and Kathryn
Kathy, Kristen and Karen

To the Professor

In revising *General Chemistry* and *General Chemistry with Qualitative Analysis,* we have incorporated many helpful suggestions that we received from professors who used the first editions. A major effort has been made to improve clarity and accuracy throughout. The number of end-of-chapter exercises has been increased significantly, and more challenging exercises have been included. The second editions are shorter than the first editions; larger type and a more open format have increased the page count.

Chapter 2 has been re-ordered to cover composition stoichiometry first, then chemical equations and reaction stoichiometry. The mole method is used throughout basic stoichiometry. Solution stoichiometry is presented in terms of the mole/molarity method. In later chapters both acid-base and oxidation-reduction reactions are treated by the mole/molarity method, with *optional* sections on the equivalent/normality method to provide greater flexibility to professors.

Properties of metals and nonmetals have been included in Chapter 4, Chemical Periodicity, so that the idea of metallic character can be used earlier. A more thorough description of metallic bonding is presented in Chapter 9, Liquids and Solids.

We have rewritten a number of sections. Chapter 7, Chemical Reactions: A Systematic Study, has been completely reorganized and rewritten to provide a better introduction to chemical reactions. The solubility rules are presented in Chapter 7 so that students can use them as they learn to write molecular, total ionic, and net ionic equations for chemical reactions.

We have kept in mind the fact that chemistry is an experimental science and have emphasized the important role of theory in science. We have presented more of the classical experiments followed by interpretations and explanations of these milestones in the development of scientific thought.

We have defined each new term as accurately as possible and illustrated its meaning as early as practical. We begin each chapter at a very fundamental level and then progress through a series of carefully graded steps to a reasonable level of sophistication. *Numerous* illustrative examples are provided throughout the text. The first ones in each section are quite simple, the last considerably more complex. The unit-factor method has been used throughout.

We have used a blend of SI and the more traditional metric units, because many students are planning careers in areas in which SI units are not yet widely used. The health-care fields, the biological sciences, home economics, and agriculture are typical examples. We have used the joule rather than the calorie in most energy calculations.

There are thirty chapters in *General Chemistry*, and *General Chemistry with Qualitative Analysis* includes nine additional chapters.

We have chosen to present stoichiometry (Chapter 2) before atomic structure and bonding (Chapters 3–6) to establish a sound foundation for a laboratory program as early as possible. However, Chapters 3–6 are as nearly self-contained as possible in order to provide flexibility for those who wish to cover structure and bonding before stoichiometry.

Since much of chemistry involves chemical reactions, we have introduced chemical reactions in a simplified, systematic way early in the text (Chapter 7). A logical, orderly introduction to molecular, total ionic, and net ionic equations is included so that this information can be used throughout the remainder of the text. There are many references to this material in later chapters.

Because many students have difficulty in *systematizing* and *using* information, we have done our utmost to assist them. The basic ideas on chemical periodicity are introduced early (Chapter 4) and are used throughout the text. A detailed discussion of inorganic nomenclature is included at the end of Chapter 5. A simplified classification of acids and bases is introduced in Chapter 7 and then expanded in Chapter 11. References are made to the classification of acids and bases and to the solubility rules throughout the text to emphasize the importance of systematizing and using previously covered information. At many points throughout the text we summarize the results of recent discussions or illustrative examples in tabular form to help students see the "big picture."

After our excursion through Gases and the Kinetic Molecular Theory, Liquids and Solids, Solutions, and Acids, Bases and Salts, we present a short chapter on Oxidation-Reduction Reactions (Chapter 12), because most of our students need this background information. At this point we have covered sufficient material that students have appropriate grounding for a wide variety of laboratory experiments.

Comprehensive chapters are presented on Chemical Thermodynamics (Chapter 13) and Chemical Kinetics (Chapter 14). We have now developed the necessary background for a strong introduction to Chemical Equilibrium in Chapter 15, followed by three chapters on Equilibria in Aqueous Solutions. A chapter on Electrochemistry (Chapter 19) completes the "common core" of the text except for Nuclear Chemistry, Chapter 28, which is self-contained and may be studied at any point in the course.

A group of basically descriptive chapters follows. However, we have been careful to include appropriate applications of the principles that have been evolved in the first part of the text to explain descriptive chemistry. Chapters 20, The Metals and Metallurgy (which includes an interesting section on trace elements in biology), 21, The Representative Metals, and 26, The Transition Metals, give broad coverage to the chemistry of the metals. Chapter 27, Coordination Compounds, is a sound introduction to that field.

Chapters 22–25 give a comprehensive introduction to the chemistry of the nonmetals. Again, care has been taken to explain descriptive chemistry in terms of the principles that have been developed earlier.

Chapter 29 provides a detailed introduction to the hydrocarbons. Chapter 30 gives a similar introduction to some important organic functional groups.

Nine additional chapters are included in *General Chemistry with Qualitative Analysis.* In Chapter 31, the important properties of the metals of the five cation groups are tabulated, their properties are discussed, the sources of the elements are listed, their metallurgies are described, and a few uses of each metal are given.

Chapter 32 is a detailed introduction to the laboratory procedures used in semimicro qualitative analysis.

Chapters 33–37 cover the analysis of the five groups of cations. Each chapter includes a discussion of the important oxidation states of the metals, an introduction to the analytical procedures, and comprehensive discussions of the "chemistry" of each cation group. Detailed laboratory directions, set off in color, follow. Students are alerted to pitfalls in advance, and alternate confirmatory tests and "clean-up" procedures are described for troublesome cations. A set of exercises accompanies each chapter.

Chapter 38 contains a discussion of some of the more sophisticated ionic equilibria of qualitative analysis. The material is presented in a central location for the convenience of the instructor.

Chapter 39 gives identification tests for thirteen common anions. Directions for the analysis of simple mixtures that are soluble in water, hydrochloric acid, or nitric acid are also presented. The chemistry of each test is discussed.

We have included throughout the text some interesting historical notes. Marginal notes have been used to point out historical facts, to provide additional bits of information, to further emphasize important points, to relate information to ideas developed earlier, and to note the "relevancy" of various discussions.

We welcome suggestions for improvements in future editions.

Supplements

A number of ancillary materials have been prepared to assist the student in his or her study of General Chemistry and to aid the instructor in presenting the course. Each supplement can be used with either version of this text.

1. *Lecture Outline for General Chemistry,* 2nd Ed., Kenneth W. Whitten and Kenneth D. Gailey. A comprehensive lecture outline that allows professors to use valuable classroom time more effectively. It provides great flexibility for the professor and makes available more time for special topics, increased drill, or whatever the professor chooses to do.
2. *Solutions Manual for General Chemistry,* 2nd Ed., Professor John M. DeKorte, Northern Arizona University. A pace-setter! It includes detailed answers, solutions, and *explanations* for *all even-numbered* end-of-chapter exercises. In-depth answers are given for discussion questions, and helpful comments that reinforce basic concepts are included, as well as references to illustrative examples and appropriate sections of chapters in the text.
3. *Study Guide for General Chemistry,* 2nd Ed., Professor R. E. Davis, The University of Texas at Austin. It includes brief summaries of important ideas in each chapter, study goals with references to text sections and exercises, and simple preliminary tests (averaging more than 80 questions per chapter, all with answers) that reinforce basic skills and vocabulary and encourage students to think about important ideas.
4. *Instructor's Manual to Accompany General Chemistry,* 2nd Ed., Professor R. E. Davis, The University of Texas at Austin. Also includes solutions to *odd-numbered* end-of-chapter exercises and may be made available to students, if the professor chooses.
5. *Experimental General Chemistry,* W. T. Lippincott (The University of Arizona), D. W. Meek (Ohio State University), K. D. Gailey, and K. W. Whitten. A modern laboratory manual with excellent variety that includes descriptive, quantitative, and instrumental experiments. Designed for mainstream courses for science majors.
6. *Problem-Solving in General Chemistry,* 2nd Ed., K. W. Whitten and K. D. Gailey. Covers the common core of general chemistry courses for science majors.
7. *Computer Tutorial for General Chemistry,* Professor Charles A. Wilkie, Marquette University. Comprehensive review and drill in core topics. On diskettes for Apple II+ and Apple IIe computers.
8. *Test Bank for General Chemistry.*
9. *Overhead Transparencies.* One hundred figures from the text.
10. *Computerized Test Bank.*
11. *Workbook for General Chemistry Audio-Tape Lessons,* 2nd Ed., B. Shakhashiri, R. Schreiner, and P. A. Meyer (all of The University of Wisconsin, Madison).
12. *General Chemistry Audio-Tape Lessons,* Shakhashiri, Schreiner, and Meyer. Adopters of the workbook or of the text will receive up to three free copies of these tapes along with unlimited duplication rights for student use; transcripts of the tapes are also available.
13. *Modern Descriptive Chemistry,* Eugene G. Rochow, Harvard University. A 250-page paperback for those who desire more descriptive chemistry.

Acknowledgments

The list of individuals who contributed to the evolution of this book is long indeed. First, we would like to express our appreciation to the professors who contributed so much to our scientific education: Professors Arnold Gilbert, M. L. Bryant, W. N. Pirkle, the late Alta Sproull, C. N. Jones, S. F. Clark, R. S. Drago (KWW), and C. R. Russ, R. D. Dunlap, H. H. Patterson, R. L. Wells, A. L. Crumbliss, P. Smith, D. B. Chesnut, R. A. Palmer, and B. E. Douglas (KDG).

The following individuals have reviewed the manuscripts for the second editions of *General Chemistry* and *General Chemistry with Qualitative Analysis.* Their suggestions have improved the texts significantly.

Raymond E. Davis, The University of Texas
John M. DeKorte, Northern Arizona University
Darrell Eyman, The University of Iowa
Henry Heikkinen, The University of Maryland
Wilbert Hutton, Iowa State University
Marlene Kolz, The University of Illinois, Chicago Circle
Robert Kowerski, College of San Mateo

The following individuals reviewed the first editions of *General Chemistry* and *General Chemistry with Qualitative Analysis.* Their helpful suggestions contributed to the wide acceptance of the first editions, and their contributions continue in the second editions: Professors Edwin Abbott, David R. Adams, George Atkinson, Jerry Atwood, William G. Bailey, J. M. Bellama, George Bodner, Clark Bricker, Robert Broman, Thomas Cassen, Lawrence Conroy, Lawrence Epstein, Richard Gaver, Robert Hanrahan, Henry Heikkinen, Forrest C. Hentz, R. K. Hill, Arthur Hufnagel, M. D. Joesten, Philip Kinsey, Norman Kulevsky, Patricia Lee, William Litchman, Ronald Marks, William Masterton, Clinton Medbery, Randal Remmel, Ronald O. Ragsdale, Gary Riley, Eugene Rochow, John Ruff, Curtis Sears, William Scroggins, Mahesh Sharma, C. H. Stammer, Janice Turner, W. H. Waggoner, Susan Weiner, and Steve Zumdahl, and Dr. Douglas Vaughan.

The staff at Saunders College Publishing has contributed immeasurably to the evolution of this book. Our developmental editor, Jay Freedman, has done a superb job. We are convinced that he has no peer. Richard Moore and Tom Mallon have given us high quality design and artwork, respectively, that contribute to the appearance and the substance of the book. Additionally, we have drawn freely from the excellent artwork in other Saunders College texts. Our project editor, Sally Kusch, has handled innumerable details with skill and aplomb. Tricia Manning and Michelle Glazer, Assistants to John Vondeling, have facilitated communications and the flow of paper cheerfully and efficiently. We express our deepest appreciation to our editor and friend, John Vondeling, the best editor in the business. John has guided us at every step in the development of this book, and our respect and admiration for him have grown with each passing day.

Our secretary, Martha Dove, has been patient and skillful through the many revisions. We are indeed grateful for her patience, her skill, and her dedication.

Finally, we are deeply indebted to our families, Betty, Andy, and Kathryn Whitten and Kathy, Kristen, and Karen Gailey, who have supported us during the three years required to complete this project. Their understanding, encouragement, and moral support have "kept us going."

KENNETH W. WHITTEN

KENNETH D. GAILEY

To the Student

We have written this text to assist you as you study chemistry. Chemistry is a fundamental science—some call it the central science. As you and your classmates pursue diverse career goals you will find that the vocabulary and ideas presented in this text will be useful in more places and in more ways than you may imagine now.

We begin with the most basic vocabulary and ideas and then carefully evolve increasingly sophisticated ideas that are necessary and useful in all the other physical sciences, the biological sciences, and the applied sciences such as medicine, dentistry, engineering, agriculture, and home economics.

We have exerted a great effort to make many of the early chapters as nearly self-contained as possible, so that the material can be presented in the order considered most appropriate by your professor. Some professors will omit some chapters completely—the text was designed to accommodate this.

Early in each section we have attempted to provide the experimental basis for the ideas we evolve. By *experimental basis* we mean the observations and experiments on the phenomena that have been most important in developing concepts. We then present an explanation of the experimental observations.

Chemistry is an experimental science. We know what we know because we (literally thousands of scientists) have observed it to be true. Theories have been evolved to explain experimental observations (facts). Successful theories explain observations fully and accurately, but more importantly they enable us to predict the results of experiments that have not yet been performed. Thus, we should always keep in mind the fact that experiment and theory go hand-in-hand. They are intimately related parts of our attempt to understand and explain natural phenomena.

"What is the best way to study chemistry?" is a question we are asked often by our students. While there is no single answer to this question, the following suggestions may be helpful. Your professor may provide additional suggestions. A number of supplementary materials accompany this text. All are designed to assist you as you study chemistry. Your professor may suggest that you use some of them.

You should always read the assigned material before it is covered in class so that you are generally familiar with the ideas as your professor discusses them. Take careful class notes. *At the first opportunity,* your class notes should be recopied. As you do this, try to work the illustrative examples that your professor solved in class, without looking at the solution in your notes. If you must look at the solution, look at only one line (step), and then try to figure out the next step. Read the assigned material again. Reading should be much more informative the second time.

Review the "key terms" at the end of the

chapter to be sure that you know the exact meaning of each. Work the illustrative examples in the text, while covering the solutions with a sheet of paper. If you find it necessary to look at the solutions, look at only one line at a time and try to figure out the next step. Work the assigned exercises at the end of the chapter. The Appendices contain much useful information. You should become familiar with them and their contents so that you may use them whenever necessary. Answers to all even-numbered numerical exercises are given at the end of the text so that you may check your work.

We heartily recommend the *Study Guide to General Chemistry* by Professor Raymond E. Davis and the *Solutions Manual* by Professor John M. DeKorte, which were written to accompany this text. The *Study Guide* provides an overview of each chapter and emphasizes the threads of continuity that run through chemistry. It lists study goals, tells you which ideas are most important and why they are important, and provides many forward and backward references. Additionally, the *Study Guide* contains many easy-to-moderately difficult questions that enable you to gauge your progress. These short questions provide excellent practice in preparing for examinations. Answers are provided for all questions, and many have explanations or references to appropriate sections in the text.

The *Solutions Manual* contains detailed solutions and answers to all even-numbered end-of-chapter exercises. It also has many helpful references to appropriate sections and illustrative examples in the text.

If you have suggestions for improving this text, please write to us and tell us about them.

KWW

KDG

Contents Overview

Contents

5 Covalent Bonding and Inorganic Nomenclature 158

6 Molecular Orbitals in Chemical Bonding 206

22 The Nonmetallic Elements, Part I: The Noble Gases and Group VIIA 695

23 The Nonmetallic Elements, Part II: Group VIA Elements 716

1 The Foundations of Chemistry

Chemistry *is the science that describes matter, its chemical and physical properties, the chemical and physical changes that matter undergoes, and the energy changes that accompany these processes.* Consider for a moment the extraordinary breadth of this definition. Matter, after all, includes everything that is tangible, from our bodies and the stuff of our everyday lives to the grandest objects in the universe. Chemistry thus touches almost every aspect of our lives, our culture, and our environment. Its scope encompasses the air we breathe, the food we eat, the fluids we drink, our clothing, our transportation and fuel supplies, our dwellings, and our fellow creatures.

Some call chemistry the central science. It rests on the foundation of mathematics and physics, and in turn underlies the sciences of life—biology and medicine. Before we fully understand living systems, we must understand the chemical reactions and chemical influences that operate within them. In fact, the chemicals of our bodies profoundly affect even the personal world of our thoughts and emotions.

We understand simple chemical systems well; they lie near chemistry's fuzzy boundary with physics and can often be described exactly by the physicist's laws. We fare less well with more complicated systems. Even where our understanding is fairly complete, we must make approximations. And often our knowledge is far from complete. Each year, however, researchers provide new insights into the nature of matter and its interactions, and chemists find answers to old questions as they learn to ask new ones. Someone once described our scientific knowledge as an expanding sphere, which, as it grows, encounters an ever-enlarging frontier.

Despite its universal importance, chemistry, like all other sciences, begins simply—with its own vocabulary and set of fundamental concepts. This chapter lays the necessary groundwork. We shall be concerned above all with characterizing matter, which is of course the main subject of chemistry. We shall define matter, see how it is related to energy, consider what it is made of and what forms it may take, and describe some of the changes it may undergo. We shall conclude the chapter with a close look at measurements—the heart of any exact science.

1–1 Matter and Energy

Matter is anything that has mass and occupies space. Mass is a measure of the quantity of matter in a sample of any material. The more massive an object is, the more force is required to put it in motion. Since all bodies in the universe conform to the definition of matter, they all consist of matter. Our senses of sight and touch usually tell us that an object occupies space, but in the case of colorless, odorless, tasteless gases (such as air), our senses sometimes fail us. We might say that we can "touch" air when it blows in our faces, but we depend on other evidence to show that a still body of air fits our definition of matter.

Energy is commonly defined to be *the capacity to do work or to transfer heat.* We are all familiar with many forms of energy in everyday life, including mechanical energy, electrical energy, heat energy, and light energy. We know that light energy from the sun is used by plants as they grow and produce food, electrical energy allows us to light a room by flicking a switch, and heat energy cooks our food and warms our homes. As a matter of convenience, energy can be classified into two principal types: *potential energy* and *kinetic energy.*

Nuclear energy is an important kind of potential energy. Nuclear energy is widely used in the production of electricity.

Potential energy is the energy an object possesses because of its position or composition. Coal, for example, possesses chemical energy, a form of potential energy, because of its composition. Many electrical generating plants burn coal, producing heat and subsequently electrical energy. A boulder located atop a mountain possesses potential energy because of its height. It can roll down the mountainside and convert its potential energy into kinetic energy, although perhaps not often a useful sort.

The term comes from the Greek word *kinein* meaning "to move," from which the word *cinema* also is derived.

A body in motion, such as a rolling boulder, possesses energy because of its motion. Such energy is called **kinetic energy.** Kinetic energy represents the capacity for doing work directly, and is easily transferred between objects.

We discuss different forms of energy because all chemical processes are accompanied by energy changes. Many chemical processes are **exothermic,**

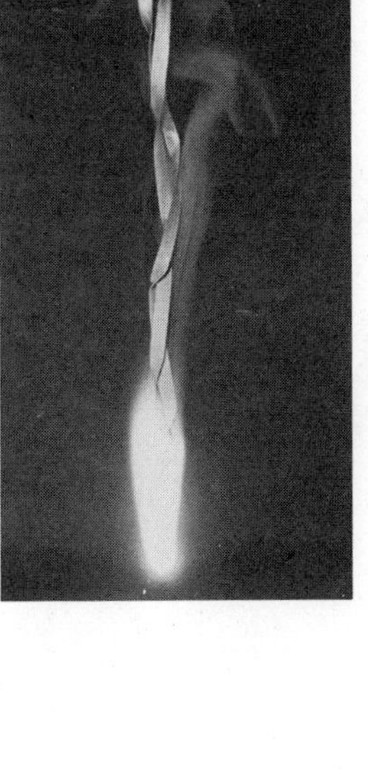

FIGURE 1–1 When magnesium burns, it combines with oxygen from the air to produce magnesium oxide, in a reaction releasing large amounts of light and heat energy. This reaction is used in photographic flashbulbs.

meaning that as reactions occur, *energy is released to the surroundings,* usually as heat energy. However, some chemical processes are **endothermic,** i.e., *they absorb energy from their surroundings.*

1 The Law of Conservation of Matter

If we burn a sample of metallic magnesium in the air, the magnesium will combine with oxygen from the air to form magnesium oxide, a white powder (Figure 1–1). This chemical reaction is accompanied by the release of large amounts of heat and light. If we then weigh the product of the reaction, magnesium oxide, we inevitably find that it is heavier than the original piece of magnesium. The increase in mass is due to the combination of the oxygen with magnesium. Numerous experiments have shown that the mass of the product of the reaction is exactly the sum of the masses of the magnesium and the oxygen that combined. Similar statements can be made for all chemical reactions. These observations are summarized in the **Law of Conservation of Matter:** *There is no observable change in the quantity of matter during an ordinary chemical reaction.* This statement is an example of a **scientific (natural) law,** a general statement based on the observed behavior of matter, to which no exceptions are known. Scientific laws cannot be rigorously proved.

2 The Law of Conservation of Energy

In exothermic chemical reactions, *chemical energy* usually is converted into *heat energy,* but some exothermic processes involve other kinds of energy changes. For example, some liberate light energy without heat, and others may produce electrical energy without heat or light. In *endothermic* reactions, heat energy, light energy, or electrical energy is converted into chemical energy. While chemical changes always involve energy changes, some energy transformations do not involve chemical changes at all. For example, heat energy may be converted into electrical energy or into mechanical energy without any simultaneous chemical changes. Electricity is produced in hydroelectric plants by converting mechanical energy (from flowing water) into electrical energy. Whatever energy changes we may choose to consider, numerous experiments have demonstrated that all of the energy involved in any change appears in some form after the change. These observations are summarized in the **Law of Conservation of Energy:** *Energy cannot be created or destroyed; it may only be converted from one form to another.*

3 The Law of Conservation of Matter and Energy

The world became aware that matter can be converted into energy in 1945, when two atomic bombs were exploded over Japan. In nuclear reactions (Chapter 28) matter is transformed into energy. The relationship between matter and energy is given by Albert Einstein's now famous equation

Einstein formulated this equation in 1905. Its validity was demonstrated in 1939.

$$E = mc^2$$

This equation, postulated as a part of the Theory of Relativity, tells us that the amount of energy released when matter is transformed into energy is the product of the mass of matter converted and the speed of light squared. At the present time, man has not (knowingly) observed the transformation of *energy into matter* on a significant scale. It does, however, happen every day on an extremely small scale in "atom smashers" or particle accelerators used to induce nuclear reactions. Now that the equivalence of matter and energy has been demonstrated, the **Laws of Conservation of Matter and Energy** are combined into a single statement: *The total amount of matter and energy available in the universe is fixed.*

1–2 States of Matter

Matter is conveniently classified into three states, although everyone can think of examples that do not fit neatly into any of the three categories. In the **solid state,** substances are rigid and have definite shapes, and their volumes do not vary much with changes in temperature and pressure. In some solids, called crystalline solids, the individual particles that make up the solid occupy definite positions in the crystal structure, and the strengths of interaction between the individual particles determine how hard and how strong the solid is. In the **liquid state,** the individual particles are confined to a given volume, but may otherwise flow and assume the shapes of their containers up to the volume of the liquid. Although the volumes of most liquids vary more with pressure changes than do the volumes of most solids, liquids are still only very slightly compressible. **Gases** fill completely any vessel in which they are confined, and assume the shape of their containers. Gases are capable of infinite expansion and are compressed easily. Because gases are easily compressed, we conclude that they consist primarily of empty space, i.e., the individual particles are quite far apart. The behavior of gases, liquids, and solids will be considered more fully in Chapters 8 and 9.

1–3 Chemical and Physical Properties

In order to distinguish samples of different kinds of matter, we determine and compare their properties. The properties of a person include height, weight, sex, and skin and hair coloration, and the many subtle features that constitute that person's general appearance. We recognize different kinds of matter by their properties, which are broadly classified into chemical properties and physical properties.

Chemical properties *are properties that matter exhibits as it undergoes changes in composition.* These properties of substances are related to the kinds of chemical changes that the substances undergo. For instance, we have already described the combination of metallic magnesium with gaseous oxygen to form magnesium oxide, a white powder. A chemical property of magnesium is that it can combine with oxygen, releasing energy in the process. Conversely, a chemical property of oxygen is that it can combine with magnesium. Attempts to make magnesium oxide combine with either magnesium or oxygen show that magnesium oxide has neither of these chemical properties.

All substances exhibit **physical properties** as well, which can be observed in the *absence* of any change in composition. Color, density, hardness, melting

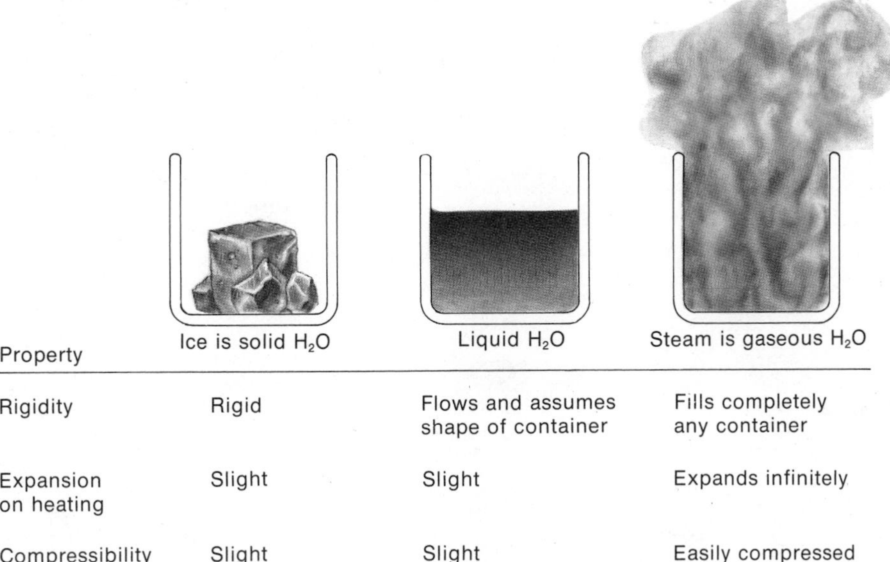

Property	Ice is solid H_2O	Liquid H_2O	Steam is gaseous H_2O
Rigidity	Rigid	Flows and assumes shape of container	Fills completely any container
Expansion on heating	Slight	Slight	Expands infinitely
Compressibility	Slight	Slight	Easily compressed

FIGURE 1–2 A comparison of some physical properties of the three states of matter (for water).

point, boiling point, and electrical and thermal conductivities are physical properties. Some physical properties of a single sample depend on the conditions, such as temperature and pressure, under which they are measured. For instance, water is a solid (ice) at low temperatures, but is a liquid at higher temperatures. At still higher temperatures, it is a gas (steam). As the sample of water is converted from one state to another, its composition is constant; therefore, its chemical properties do not change. On the other hand, the physical properties of ice, liquid water, and steam are quite dissimilar (Figure 1–2).

The physical properties of matter can be further classified as *extensive properties* or *intensive properties*. **Extensive properties** *depend on the amount of material examined.* The volume and the mass of a sample are extensive properties because they depend on, and are directly proportional to, the amount of matter in the sample examined. **Intensive properties** *do not depend on the amount of material examined.* The color and the melting point of a substance, for example, are the same for a small sample as for a large one. All chemical properties are intensive properties.

Since no two substances have identical sets of chemical and physical properties under the same conditions, we are able to identify and distinguish among different substances. For instance, water is the only clear, colorless substance that freezes at 0°C, boils at 100°C at one atmosphere of pressure, dissolves relatively large amounts of ordinary salt, and reacts violently with sodium (Figure 1–3). Table 1–1 compares several (intensive) physical properties of a few substances; a sample of any one of them can be distinguished from the others by measurements of its properties.

One atmosphere of pressure is the average atmospheric pressure at sea level.

1–4 Chemical and Physical Changes

Earlier we described the reaction of magnesium as it burns in the oxygen of the air, the reaction that occurs in photographic flashbulbs. This reaction is a *chemical change.* In any **chemical change,** (1) one or more substances are used

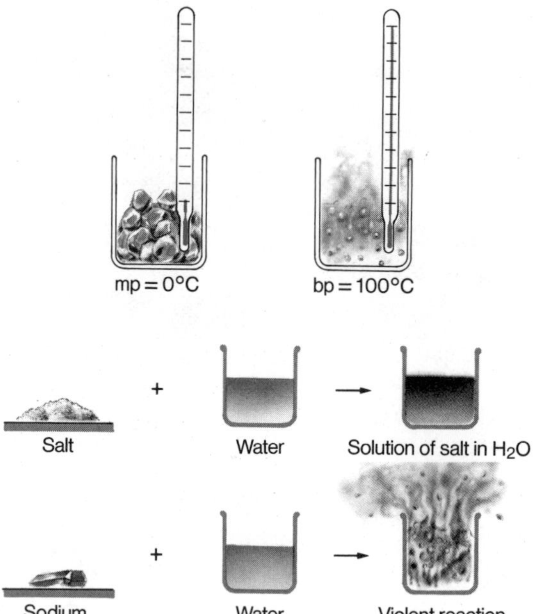

FIGURE 1–3 Some physical and chemical properties of water. Physical: (a) It melts at 0°C; (b) it boils at 100°C (at normal atmospheric pressure); (c) it dissolves salt. Chemical: (d) It reacts with sodium to form hydrogen gas and a solution of sodium hydroxide.

up (at least partially), (2) one or more new substances are formed, and (3) energy is absorbed or released. As substances undergo chemical changes they exhibit chemical properties. A **physical change,** on the other hand, occurs with *no change in chemical composition.* Physical properties may be altered as matter undergoes physical changes.

Both chemical and physical changes are always accompanied by either the absorption or the liberation of energy. Energy is required to melt ice and energy is required to boil water. Conversely, the condensation of steam to form liquid water always liberates energy, as does the freezing of liquid water to form ice. The changes in energy accompanying these physical changes are shown in Figure 1–4. All changes are at a pressure of one atmosphere. At this pressure, ice always melts at the same temperature (0°C) and pure water always boils at the same temperature (100°C). The units of energy used in the

TABLE 1–1 Physical Properties of a Few Common Substances (at atmospheric pressure)

Substance	Melting Pt (°C)	Boiling Pt (°C)	Solubility at 25°C (g/100 g) In Water	In Ethyl Alcohol	Density (g/cm^3)
Acetic acid	16.6	118.1	infinite	infinite	1.05
Benzene	5.5	80.1	0.07	infinite	0.879
Bromine	−7.1	58.8	3.51	infinite	3.12
Iron	1530	3000	insoluble	insoluble	7.86
Methane	−182.5	−161.5	0.0022	0.033	6.67×10^{-4}
Oxygen	−218.8	−183.0	0.0040	0.037	1.33×10^{-3}
Sodium chloride	801	1473	36.5	0.065	2.16
Water	0	100	—	infinite	1.00

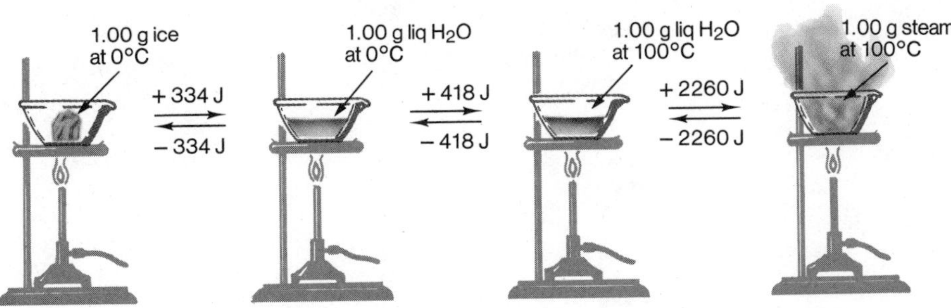

FIGURE 1–4 Changes in energy that accompany some physical changes (for water).

figure, joules (J), will be defined in Section 1–13. The positive signs preceding joules (J) above the arrows indicate that the changes in the indicated direction are endothermic. Negative signs below the arrows signify exothermic processes.

1–5 Substances, Compounds, Elements, and Mixtures

All samples of pure ethyl alcohol contain exactly the same proportions of carbon, 52.14%, hydrogen, 13.13%, and oxygen, 34.73%, by mass, regardless of the sources of the samples. All samples have the same melting point, −117.3°C, boiling point, 78.3°C, and other physical properties. Comparable statements can be made, with different values of course, for any substance. A **substance** *is any kind of matter all samples of which have identical composition and, under identical conditions, identical properties.* A substance can be either a compound or an element.

Compounds are substances consisting of two or more different elements in a fixed ratio. All compounds can be decomposed into simpler substances, either elements or simpler compounds. For example, the compound water can be decomposed by electricity into its constituent elements, hydrogen and oxygen, always in the same ratio, as shown in Figure 1–5. The compound calcium carbonate can be decomposed into simpler compounds, calcium oxide and carbon dioxide, by heating. These compounds can, in turn, be decomposed further into their constituent elements, calcium oxide into calcium and oxygen, and carbon dioxide into carbon and oxygen.

A more formal statement of this idea is the Law of Definite Proportions, which will be discussed more fully in Section 2–1.

The physical and chemical properties of a compound are different from the properties of its constituent elements. The compound sodium chloride, a white solid that we ordinarily use as table salt, is produced by the combination of sodium (a soft silvery white metal that reacts violently with water) and chlorine (a pale green corrosive gas).

If we wished to resolve sodium chloride into its elements, we could do so by electrolysis of melted (molten) sodium chloride.

Elements *are substances that cannot be decomposed into simpler substances by chemical changes.* Nitrogen, silver, aluminum, copper, gold, and sulfur are other examples of elements.

A set of **symbols** represents the known elements. Symbols are used as a matter of convenience because they can be written more quickly than names, and they occupy less space. The symbols for the first 103 elements consist of either a capital letter *or* a capital letter followed by a lower case letter, such as C (carbon) or Ca (calcium). Symbols for elements beyond number 103 consist of three letters. A list of the known elements and their symbols is included in the table inside the front cover.

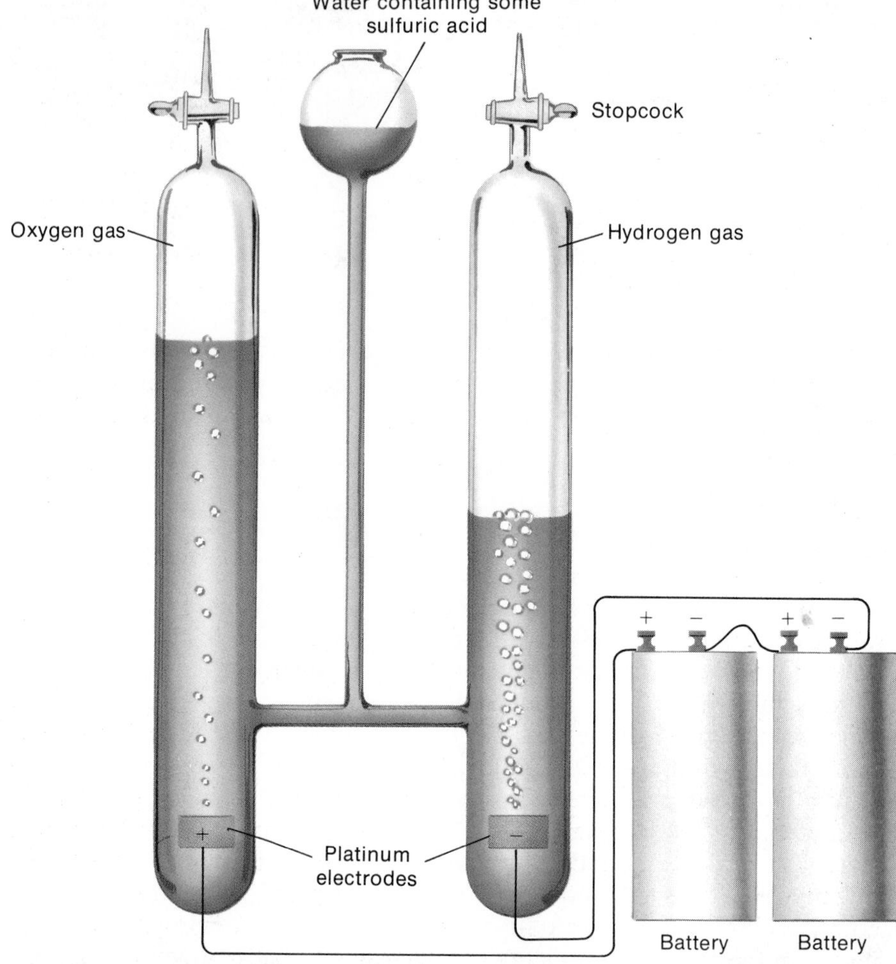

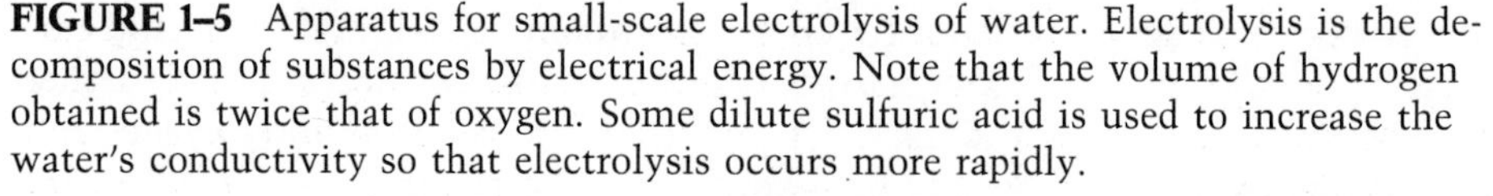
FIGURE 1–5 Apparatus for small-scale electrolysis of water. Electrolysis is the decomposition of substances by electrical energy. Note that the volume of hydrogen obtained is twice that of oxygen. Some dilute sulfuric acid is used to increase the water's conductivity so that electrolysis occurs more rapidly.

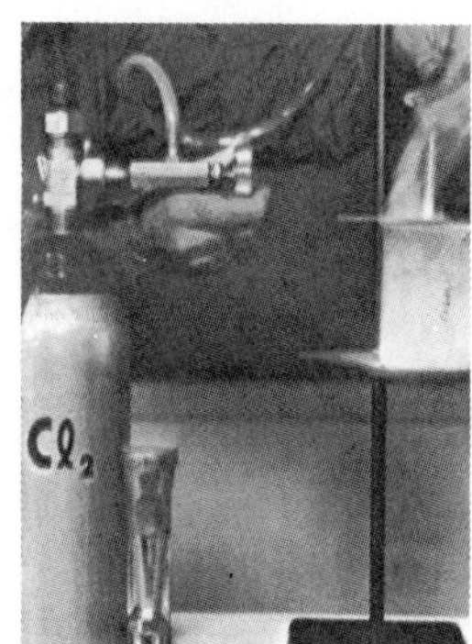

FIGURE 1–6 The reaction of sodium and chlorine to produce table salt, sodium chloride.

A short list of symbols of common elements is given in Table 1–2. Learning this list will be helpful. Many symbols consist of the first one or two letters of the element's English name. Some are derived from the element's Latin name (indicated in parentheses), and one, W for tungsten, is from the German *Wolfram*. Names and symbols for additional elements should be learned as they are encountered.

Most of the earth's crust is made up of a relatively small number of elements. For example, only 10 of the 88 naturally occurring elements make up 99+% by mass of the earth's crust, oceans, and atmosphere (Table 1–3). Oxygen accounts for roughly half, and silicon approximately one fourth of the whole.

The other known elements have been made artificially in laboratories, as described in Chapter 28. One atom of element number 109 has been reported. Element 108 has not been reported at this time, June 1983.

Relatively few elements, approximately one fourth of the naturally occurring ones, occur in nature as free elements. The rest are always found chemically combined with other elements.

TABLE 1–2 Common Elements and Their Symbols

Symbol	Element	Symbol	Element
Ag	silver (argentum)	K	potassium (kalium)
Al	aluminum	Li	lithium
Au	gold (aurum)	Mg	magnesium
Br	bromine	N	nitrogen
C	carbon	Na	sodium (natrium)
Ca	calcium	Ne	neon
Cl	chlorine	Ni	nickel
Cu	copper (cuprum)	O	oxygen
F	fluorine	P	phosphorus
Fe	iron (ferrum)	Pb	lead (plumbum)
H	hydrogen	S	sulfur
He	helium	Si	silicon
Hg	mercury (hydrargyrum)	W	tungsten (wolfram)
I	iodine	Zn	zinc

Mixtures are combinations of two or more substances in which each substance retains its own composition and properties. The compositions of mixtures can be varied widely. We can make an infinite number of different mixtures of salt and sugar by varying the relative amounts of the two substances used. Solutions of salt dissolved in water are mixtures whose composition may vary over a wide range. Air is a mixture of gases that consists primarily of nitrogen, oxygen, argon, carbon dioxide, and water vapor. There are only trace amounts of other substances in the atmosphere.

By "composition of a mixture" we mean both the identities of the substances present and their relative amounts in the mixture.

Mixtures can be classified as either homogeneous or heterogeneous. A **homogeneous mixture,** also called a **solution,** *has uniform composition and properties throughout.* Examples include air (free of particulate matter or mists), salt water, and some alloys, which are homogeneous mixtures of metals in the solid state. Mixtures that are not uniform throughout are said to be **heterogeneous.** Examples include mixtures of salt and charcoal (in which one component can be distinguished readily from the other by sight), foggy air (which includes a suspended mist of water droplets), and vegetable soup.

Mixtures can be separated by physical means because each component retains its properties (see Figures 1–7 and 1–9). For example, a mixture of salt and water can be separated by evaporating the water and leaving the solid salt

TABLE 1–3 Abundance of Elements (Earth's Crust, Oceans, and the Atmosphere)

Oxygen	O	49.5%	99.2%	Chlorine	Cl	0.19%	0.7%
Silicon	Si	25.7		Phosphorus	P	0.12	
Aluminum	Al	7.5		Manganese	Mn	0.09	
Iron	Fe	4.7		Carbon	C	0.08	
Calcium	Ca	3.4		Sulfur	S	0.06	
Sodium	Na	2.6		Barium	Ba	0.04	
Potassium	K	2.4		Chromium	Cr	0.033	
Magnesium	Mg	1.9		Nitrogen	N	0.030	
Hydrogen	H	0.87		Fluorine	F	0.027	
Titanium	Ti	0.58		Zirconium	Zr	0.023	
				All others			<0.1%

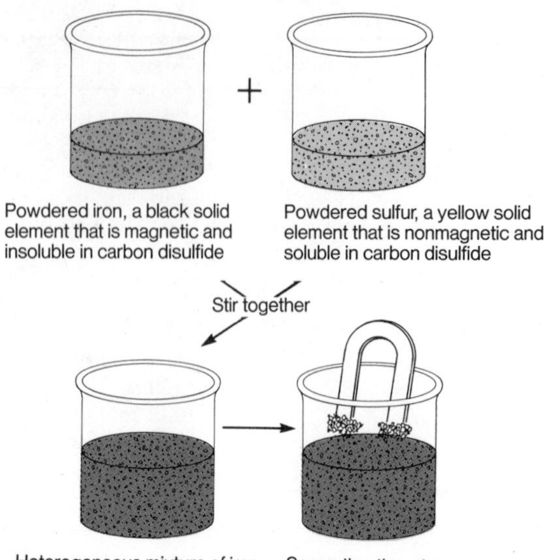
Powdered iron, a black solid element that is magnetic and insoluble in carbon disulfide
Powdered sulfur, a yellow solid element that is nonmagnetic and soluble in carbon disulfide
Stir together
Heterogeneous mixture of iron and sulfur
Separating the mixture by removing the magnetic iron

FIGURE 1–7 A mixture of iron and sulfur is a *heterogeneous* mixture. Like all mixtures, it can be separated by physical means, such as removing the iron with a magnet.

behind. A mixture of sand and salt can be separated by dissolving the salt in water, collecting the sand by filtration, and then evaporating the water to obtain the solid salt. Very fine iron powder can be mixed with powdered sulfur to give what appears to the naked eye to be a homogeneous mixture of the two. However, separation of the components of this mixture is easy. The iron may be removed by a magnet, or the sulfur may be dissolved in carbon disulfide, which does not dissolve iron (Figure 1–7).

The important characteristics of mixtures are that (1) their compositions can be varied, and (2) each component of the mixture retains its own properties. Figure 1–9 summarizes the classification of matter and the methods by which separations can be achieved.

1–6 Atoms and Molecules

An **atom** is the smallest particle of an element. Historically, an atom has been defined as the smallest particle of an element that can enter into a chemical combination. As we shall see, this historical definition leaves something to be

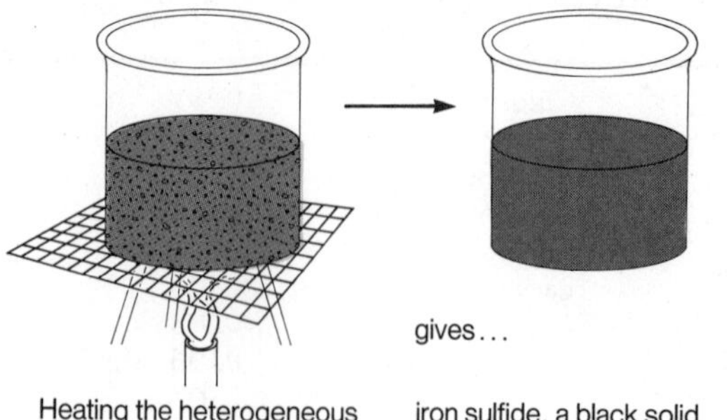
gives...
Heating the heterogeneous mixture of iron and sulfur...
iron sulfide, a black solid compound that is nonmagnetic and insoluble in carbon disulfide

FIGURE 1–8 A chemical reaction occurs when the heterogeneous mixture of iron and sulfur is heated. The result is a compound.

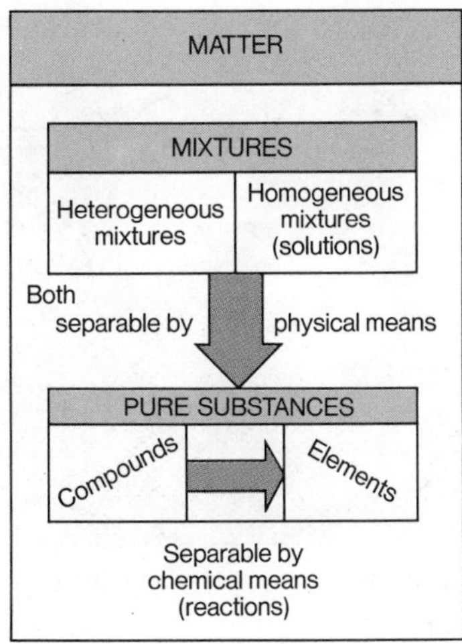

FIGURE 1–9 One classification of matter and some means by which the classes of matter can be separated.

FIGURE 1–10 Relative sizes of atoms of the noble gases.

desired, because no chemical reactions of the lighter noble gases (helium, neon, and argon) are known. A **molecule** is the smallest particle of an element or compound that can have a stable independent existence. You may now be wondering "What is the difference between an atom and a molecule of an element?" Actually, an atom and a molecule are the same for certain elements (especially the noble gases) which are capable of existence as single atoms. A molecule that consists of one atom is called a *monatomic* molecule. See Figure 1–10.

An atom of oxygen, on the other hand, cannot exist alone at room temperature and atmospheric pressure. An atom of oxygen can have a stable existence only if it combines with another atom or molecule. The oxygen we are all familiar with is made up of two atoms of oxygen; it is a *diatomic* molecule. Hydrogen, nitrogen, fluorine, chlorine, bromine, and iodine are also examples of diatomic molecules. See Figure 1–11.

Other elements exist as more complex molecules. Phosphorus exists as molecules consisting of four atoms, while sulfur exists as eight-atom molecules at ordinary temperatures and pressures. These molecules, composed of more than two atoms, are called *polyatomic* molecules. See Figure 1–12.

Methane is the principal component of natural gas.

Molecules of compounds are composed of more than one kind of atom. A water molecule, for example, consists of two atoms of hydrogen and one atom of oxygen. A molecule of methane consists of one carbon atom and four hydro-

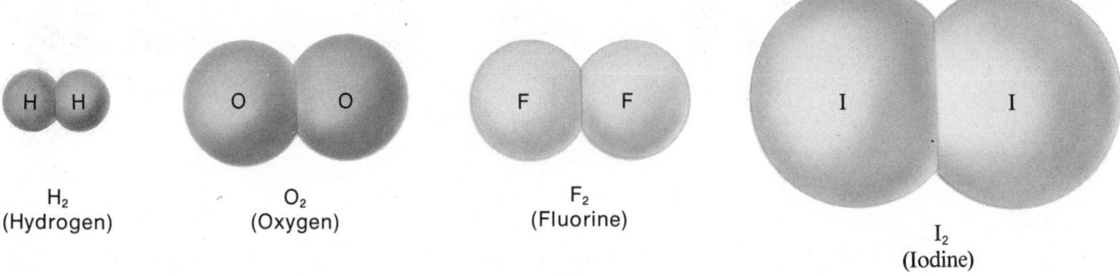

FIGURE 1–11 Models of diatomic molecules of some elements.

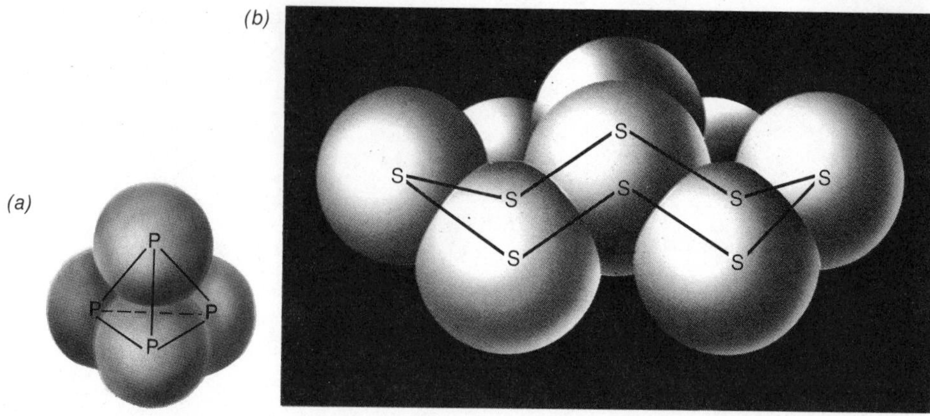

FIGURE 1–12 (a) A model of the P_4 molecule of white phosphorus; (b) a model of S_8 ring found in rhombic sulfur.

gen atoms. The shapes of these and a few other molecules are shown in Figure 1–13.

You should now be aware that atoms are the components of molecules, and that molecules are the components of elements or compounds (the compound water is made up of water molecules, which are made up of hydrogen and oxygen atoms). We are able to see samples of compounds and elements consisting of large numbers of atoms and molecules, but individual atoms and molecules are too small to be seen even with the most powerful optical microscope. To give you an idea of the size of a water molecule, it would take 100 million of them to make a row one inch long.

1–7 Measurements in Chemistry

Measurements in the scientific world are usually expressed in the units of the metric system or in its modernized and expanded successor, the International System of Units (SI), which was adopted by the National Bureau of Standards in 1964. The SI is based on the seven fundamental units listed in Table 1–4, and all other units of measurement are derived from them.

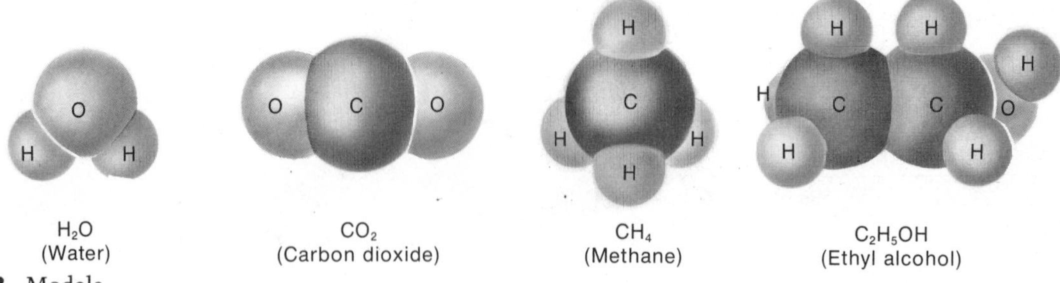

FIGURE 1–13 Models of molecules of some compounds.

TABLE 1–4 The Seven Fundamental Units of Measurement

Physical Property	Name of Unit	Symbol
Length	Meter	m
Mass	Kilogram	kg
Time	Second	s
Electric current	Ampere	A
Temperature	Kelvin	K
Luminous intensity	Candela	cd
Amount of substance	Mole	mol

Just as the general public of the United States has been reluctant to move from traditional units of distance such as the mile and the yard to the "less familiar" meter and kilometer of the metric system, the scientific community has been reluctant to dispense with its own traditional and comfortable metric units and switch completely to SI units.

In many applied sciences such as the health sciences, there has been little if any motion in the direction of using SI units. Although the metric system was endorsed by Congress in 1866, such things as tire gauges and gauges on cylinders of compressed gases are still calibrated in pounds per square inch rather than the metric units, atmospheres or millimeters of mercury of pressure. Only recently has the weather bureau begun to report atmospheric pressure in millimeters of mercury as well as inches of mercury. Thus, we must become familiar with several units of measurement for the same quantity.

In this text we shall use both metric units and SI units. As we shall see later in this chapter, conversions between non-SI and SI units are usually quite straightforward. In many instances, when calculations are done in more traditional units, we shall also specify the result in SI units parenthetically.

Appendix C lists some important units of measurement and their relationship to each other. We'll refer to it often. Appendix D lists several important physical constants for your convenience.

The metric and SI systems are *decimal systems, in which prefixes are used to indicate fractions and multiples of ten.* The same prefixes are used with all units of measurement. The distances and masses in Table 1–5 illustrate the use of some common prefixes and the relationships among them.

In the next three sections, we shall introduce the primary standards chosen for basic units of measurement. These standards were selected because they allow us to make precise measurements, and they are reproducible and unchanging. The values assigned to fundamental units are arbitrary.* In the United States, all units of measure are set, by law, by the National Bureau of Standards. Like many other governmental agencies over the world, the Bureau of Standards accepts the internationally agreed-upon values for fundamental units.

* Prior to the establishment of the U.S. Bureau of Standards in 1901, at least 50 different distances had been used as "1 foot" in measuring land within New York City. Thus, the size of a 100-ft by 200-ft lot in New York depended on the generosity of the seller and did not necessarily represent the expected dimensions.

The prefixes used in the SI and metric systems may be thought of as *multipliers*, e.g., the prefix *kilo-* indicates multiplication by 1000, while *milli-* indicates multiplication by 0.001 or by 10^{-3}.

TABLE 1–5 Common Prefixes Used in the SI and Metric Systems

Prefix	Abbreviation	Meaning	Example
mega-	M	10^6	1 megameter (Mm) = 1×10^6 m
kilo-	k	10^3	1 kilometer (km) = 1×10^3 m
deci-	d	10^{-1}	1 decimeter (dm) = 0.1 m
centi-	c	10^{-2}	1 centimeter (cm) = 0.01 m
milli-	m	10^{-3}	1 milligram (mg) = 0.001 g
micro-	μ*	10^{-6}	1 microgram (μg) = 1×10^{-6} g
nano-	n	10^{-9}	1 nanogram (ng) = 1×10^{-9} g
pico-	p	10^{-12}	1 picogram (pg) = 1×10^{-12} g

* This is the Greek letter μ (pronounced "mew").

1–8 Units of Measurement

1 Mass and Weight

We must distinguish between mass and weight. You may recall from Section 1–1 that **mass** *is a measure of the quantity of matter a body contains.* The force required to give a sample of matter a certain acceleration is a measure of its mass. The mass of a body does not vary as its position changes. On the other hand, the **weight** *of a body is a measure of the gravitational attraction of the earth for the body,* and this varies with distance from the center of the earth. An object weighs ever so slightly less high up in an airplane than at the bottom of a deep valley. Nearly everyone has seen pictures of astronauts as they experienced weightlessness during space travel. In stable orbits, they no longer feel the attraction of earth's gravity. Because the mass of a body does not vary with its position, whereas its weight does, the mass of a body is a more fundamental property than its weight. However, we have become accustomed to using the term weight when we mean mass, just because weighing is one way of measuring mass. Since we usually discuss chemical reactions at constant gravity, weight relationships are just as valid as mass relationships, but we must keep in mind that *the two are not identical.*

The basic unit of mass in the SI system is the **kilogram** (Table 1–6). The kilogram is arbitrarily defined as the mass of a platinum–iridium cylinder stored in a vault in Sèvres, near Paris, France. (See Figure 1–14.) One kilogram weighs 2.205 pounds, and a one-pound object has a mass of 0.4536 kilogram. The metric system also uses the kilogram and all of its denominations, but the basic mass unit in the metric system is the gram.

2 Length

The **meter** is the standard of length (distance) in both SI and metric systems. It is approximately 39.37 inches, or a bit more than 3 feet. In situations where the English system would use inches, the metric centimeter ($\frac{1}{100}$ meter)

TABLE 1–6 Some SI Units of Mass

*kilo*gram, kg	base unit
gram, g	1000 g = 1 kg
*milli*gram, mg	1000 mg = 1 g
*micro*gram, μg	1,000,000 μg = 1 g

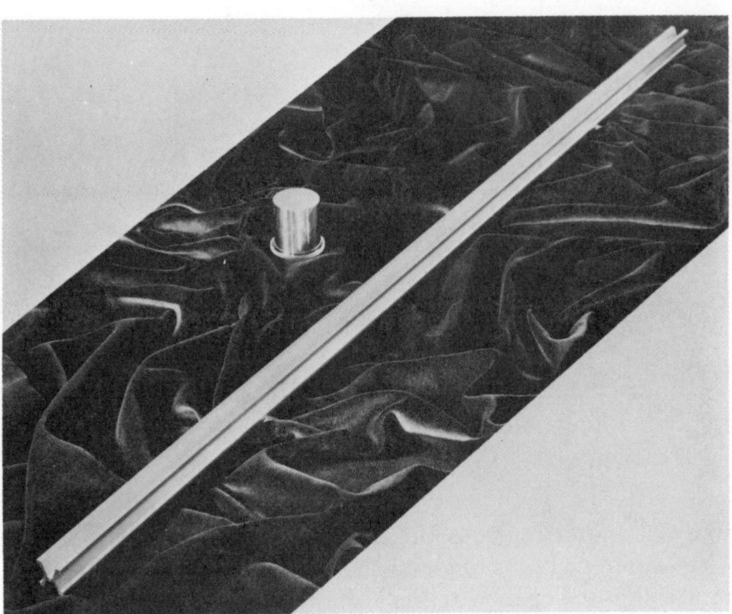

FIGURE 1–14 A duplicate of the meter bar, now replaced as the SI standard of length by a wavelength emitted by the element krypton; and a duplicate of the SI kilogram standard of mass.

is convenient. The relationship between inches and centimeters is shown in Figure 1–16. The meter was originally defined in 1790 as one ten-millionth of the distance from the North Pole to the equator. In 1875 it was redefined as the distance between two lines on the platinum–iridium bar stored in Sèvres (Figure 1–14). It is now defined as 1,650,763.73 times the wavelength of the orange-red light emitted by the element krypton under specified conditions. This is an unvarying standard that is reproducible anywhere with an instrument called a spectrointerferometer. For usual measurements in the laboratory, we still rely on the ordinary meter stick.

3 Volume

Volumes are measured in liters or milliliters in the metric system. One liter (1 L) is one cubic decimeter (1 dm^3), or 1000 cubic centimeters (1000 cm^3). One milliliter (1 mL) is 1 cm^3. In the SI the cubic meter is the basic volume

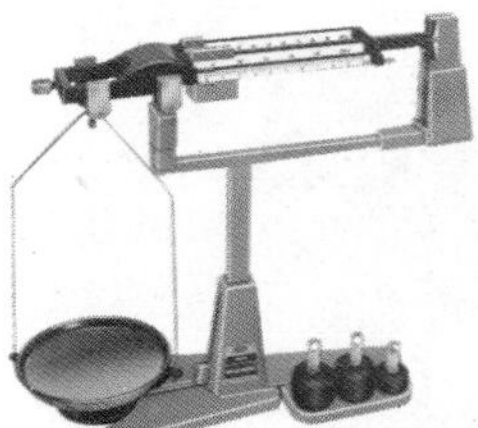

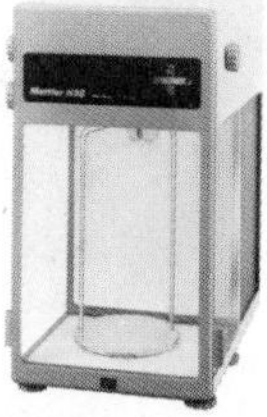

FIGURE 1–15 Three types of laboratory balances. (a) A triple beam balance used for determining mass to about ±0.01 g. (b) A modern electronic top loading balance that gives a direct readout of mass to ±0.01 g. (c) A modern analytical balance that can be used to determine mass to ±0.0001 g. Analytical balances are used when masses must be determined as accurately as possible. When less accuracy is required, other kinds of balances can be used.

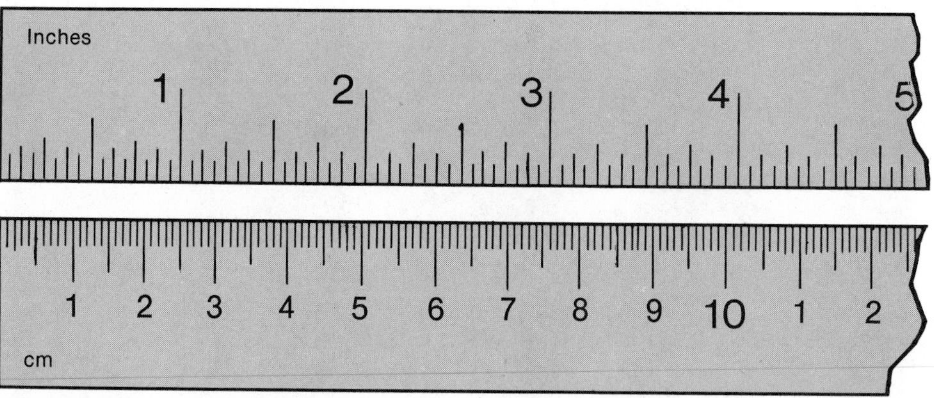

FIGURE 1–16 The relationship between centimeters and inches: 2.54 cm are exactly equal to 1 inch.

unit, and the cubic decimeter replaces the metric unit, liter (Figure 1–17).

There are many different kinds of glassware used to measure the volume of liquids. The one we choose depends on the accuracy we desire. For example, the volume of a liquid dispensed can be measured more accurately with a buret than with a small graduated cylinder (Figure 1–18).

Equivalences between common English units and metric units are summarized in Table 1–7.

1–9 Significant Figures

There are two kinds of numbers. **Exact numbers** *are numbers that are known to be absolutely accurate.* For example, the exact number of people seated in an orderly arrangement in a closed room can be counted, and there is no doubt about the number of people in the room. A dozen eggs is exactly 12 eggs, no more, no fewer.

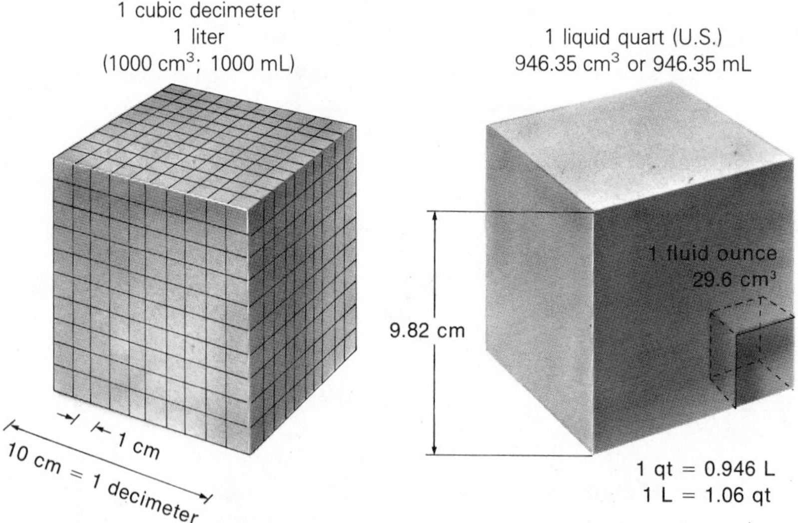

FIGURE 1–17 A comparison of common measures of volume.

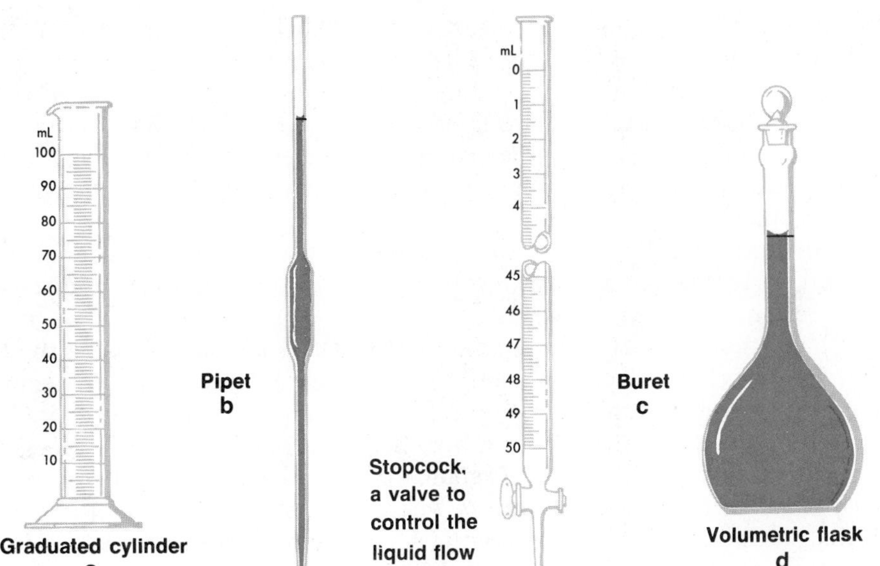

FIGURE 1–18 Laboratory apparatus used to measure volumes of liquids.

Numbers obtained from measurements are not exact. When measurements are made, some estimation is involved. For example, suppose you are asked to measure the length of this page to the nearest 0.1 mm. How do you do it? The smallest divisions (calibration lines) on an ordinary meter stick are 1 mm apart (Figure 1–16). An attempt to measure to 0.1 mm involves estimation. If you measure the length of the page to 0.1 mm three different times, will you get the same answer each time? Probably not. We deal with this problem by using significant figures.

There is some uncertainty in all measurements.

Significant figures are digits believed to be correct by the person who makes a measurement. We assume that the person is competent to use the

Significant figures indicate the *uncertainty* in measurements.

TABLE 1–7 Conversion Factors Relating Length, Volume, and Mass (Weight) Units

Metric		English		Metric–English Equivalents	
Length					
1 km	$= 10^3$ m	1 ft	= 12 in	1 in	= 2.54 cm
1 cm	$= 10^{-2}$ m	1 yd	= 3 ft	1 m	= 39.37 in*
1 mm	$= 10^{-3}$ m	1 mile	= 5280 ft	1 mile	= 1.609 km*
1 nm	$= 10^{-9}$ m				
1 Å	$= 10^{-10}$ m				
Volume					
1 mL	$= 1\ cm^3 = 10^{-3}$ liter	1 gal	= 4 qt = 8 pt	1 liter	= 1.057 qt*
1 m^3	$= 10^6\ cm^3 = 10^3$ liter	1 qt	$= 57.75\ in^3$*	1 ft^3	= 28.32 liter*
Mass					
1 kg	$= 10^3$ g			1 lb	= 453.6 g*
1 mg	$= 10^{-3}$ g	1 lb	= 16 oz	1 g	= 0.03527 oz*
1 metric tonne	$= 10^3$ kg	1 short ton	= 2000 lb	1 metric tonne	= 1.102 short ton

* These conversion factors, unlike the others listed, are inexact. They are quoted to four significant figures, which will ordinarily be more than sufficient.

measuring device. Suppose one measures a distance with a meter stick and reports the distance as 343.5 mm. What does this number mean? In this person's judgment, the distance is greater than 343.4 mm but less than 343.6 mm, and the best estimate is 343.5 mm. The number 343.5 mm contains four significant figures—the last digit, 5, represents a *best estimate* and is therefore doubtful, but it is considered to be a significant figure. In reporting numbers obtained from measurements, *we report one estimated digit, and no more.* Since the person making the measurement is not certain that the 5 is correct, clearly it would be silly, meaningless, and wrong to report the distance as 343.53 mm.

To see more clearly the part significant figures play in reporting the results of measurements, consider Figure 1–19, which shows two pieces of equipment used to measure small volumes of liquid. Graduated cylinders are used to measure volumes of liquids when a high degree of accuracy is not necessary. The calibration lines on a 50 mL graduated cylinder represent 1-mL increments. Estimation of the volume of liquid in a 50-mL cylinder to within 0.2 mL ($\frac{1}{5}$ of the calibration increments) with reasonable certainty is possible. We might measure a volume of liquid in such a cylinder and report the volume as 39.4 mL, i.e., to three significant figures. Clearly, we should not report the volume as 39.42 mL, because it is impossible to read the calibration lines to hundredths of a milliliter, and there is no basis for reporting four significant figures.

Burets are used to measure volumes of liquids when high accuracy is required. The calibration lines on a 50-mL buret represent 0.1-mL increments, allowing us to make estimates to within 0.02 mL ($\frac{1}{5}$ of the calibration increments) with reasonable certainty. Experienced individuals estimate volumes in 50-mL burets to 0.01 mL with considerable success. For example, using a 50-mL buret, we can measure out 36.95 mL (four significant figures) of liquid with reasonable accuracy. A measured volume of 36.955 mL cannot be dispensed from such a buret; there is no basis for reporting five significant figures, since this would imply accuracy of 0.002 mL ($\frac{1}{50}$ of the marked increments).

Accuracy refers to how closely a measured value agrees with the correct value. **Precision** refers to how closely individual measurements agree with each other. Ideally, all measurements should be both accurate and precise. Measurements may be quite precise, yet quite inaccurate, because of some *systematic error,* which is an error repeated in each measurement. (A faulty balance, for example, might produce a systematic error.) Very accurate measurements are seldom imprecise.

Measurements are frequently repeated in an attempt to improve accuracy and precision. Average values obtained from several measurements are usually considered more reliable than individual measurements. Significant figures indicate how accurately measurements have been made (assuming the person who made the measurements was competent).

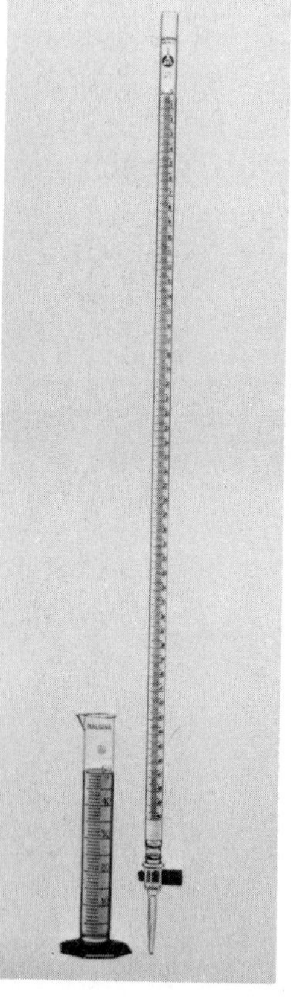

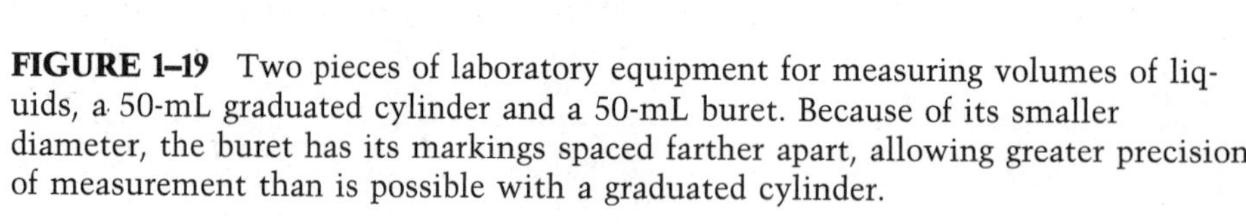

FIGURE 1–19 Two pieces of laboratory equipment for measuring volumes of liquids, a 50-mL graduated cylinder and a 50-mL buret. Because of its smaller diameter, the buret has its markings spaced farther apart, allowing greater precision of measurement than is possible with a graduated cylinder.

Some simple rules or conventions govern the use of significant figures in calculations.

Zeroes used just to position the decimal point are not significant figures.

For example, the number 0.0234 g contains only three significant figures, because the two zeroes are used to place the decimal point. The number could be reported equally well as 2.34×10^{-2} g in scientific notation (Appendix A). When zeros precede the decimal point, but come after other digits, we may have some difficulty in deciding whether the zeroes are significant figures or not. How many significant figures does the number 23,000 contain? We are given insufficient information to answer the question. If all three of the zeroes are used simply to place the decimal point, the number should be written as 2.3×10^4 (two significant figures). If only two of the zeroes are being used to place the decimal point, it is 2.30×10^4 (three significant figures), while if only one zero is being used to place the decimal point, the number should be written 2.300×10^4 (four significant figures). In the unlikely event that the number is actually known to be $23{,}000 \pm 1$, it should be written as 2.3000×10^4 (five significant figures) so that the number of significant figures is obvious.

Life would be much simpler if every person who used numbers did so correctly. As everyone knows, such is not the case. In this text we shall exert a great effort to use numbers correctly so that you will know exactly what every number means.

In multiplication and division, an answer contains no more significant figures than the least number of significant figures used in the operation.

We must, therefore, distinguish between numbers generated by electronic calculators and the result of an arithmetic operation. Electronic calculators "assume" that all numbers entered are exact numbers. Clearly, such a ridiculous assumption must produce some equally ridiculous answers.

EXAMPLE 1–1

What is the area of a rectangle 1.23 cm wide and 12.34 cm long? The area of a rectangle is its length times its width.

Solution

$A = l \times w = (12.34\ \text{cm})(1.23\ \text{cm}) = \underline{15.2\ \text{cm}^2}$

calculator result = (15.1782)

The answer should contain only three significant figures, the smallest number of significant figures used in the information given. The number generated by an electronic calculator (15.1782 shown in parentheses) is wrong; the result cannot be more accurate than the information that led to it. Since the calculator has no judgment, you must exercise yours.

$$\begin{array}{r} 12.3\underline{4}\ \text{cm} \\ \times \quad 1.2\underline{3}\ \text{cm} \\ \hline \underline{37}\ \underline{02} \\ 2\ \underline{46}\ \underline{8} \\ 12\ 3\underline{4} \\ \hline 15.\underline{17}\ \underline{82}\ \text{cm}^2 = 15.2\ \text{cm}^2 \end{array}$$

The step-by-step calculation in the margin demonstrates why the area of the rectangle should be reported as 15.2 cm^2, rather than as 15.1782 cm^2. The length, 12.34 cm, contains four significant figures while the width, 1.23 cm, contains only three. If we underline each uncertain figure, as well as each figure obtained from an uncertain figure, the step-by-step multiplication gives the result reported in Example 1–1. Thus we see that there are only two certain figures (15) in the result. We may report the first doubtful figure (.2), but no more.

Division is just the reverse of multiplication, and the same rules apply. Stated succinctly: In multiplication and division, the answer contains no more significant figures than the least number of significant figures given in the information.

In addition and subtraction, the last digit retained in the sum or difference is determined by the first doubtful digit.

The first doubtful digit is underlined in the next few examples.

EXAMPLE 1–2

a. Add 37.24 mL and 10.3 mL.

$$\begin{array}{r} 37.2\underline{4}\ \text{mL} \\ +10.\underline{3}\ \ \text{mL} \\ \hline 47.\underline{54}\ \text{mL} \end{array}$$

47.54 mL is reported as 47.5 mL (calculator gives 47.54)

b. Subtract 21.2342 g from 27.87 g

$$\begin{array}{r} 27.8\underline{7}\ \ \ \ \ \text{g} \\ -21.234\underline{2}\ \text{g} \\ \hline 6.6\underline{358}\ \text{g} \end{array}$$

6.6358 g is reported as 6.64 g (calculator gives 6.6358)

Note that in the three simple arithmetic operations we have performed, the number combination generated by an electronic calculator is not the "answer" in a single case! However, the correct result of each calculation can be obtained by "rounding off," and the rules of significant figures tell us where to round off.

The even–odd rules are intended to reduce the accumulation of errors in chains of calculations.

In rounding off, certain conventions have been adopted. When the number to be dropped is less than 5, it is just dropped (i.e., 7.34 rounds off to 7.3). When it is more than 5, the preceding number is increased by 1 (i.e., 7.37 rounds off to 7.4). When the number to be dropped is 5, the preceding number is not changed when it (the preceding number) is even (i.e., 7.45 rounds off to 7.4). When the preceding number is odd, it is increased by 1 (i.e., 7.35 rounds off to 7.4). In fairness, we must note that the even–odd rules for rounding terminal 5's are sometimes ignored; instead, the preceding number is increased by 1 when a 5 is dropped.

1–10 Dimensional Analysis (The Unit Factor Method)

Many chemical and physical processes can be described by numerical relationships. In fact, many of the most useful ideas in science may be treated mathematically. Let us, then, devote a little time to reviewing problem-solving skills. First, multiplication by unity (by one) does not change the value of an expression. However, when the units of the number "one" are chosen appropriately, many calculations can be done by just "multiplying by one." This method of performing calculations is known as **dimensional analysis,** or the **factor-label method,** or the **unit factor method.** Regardless of the name chosen, it is a very powerful mathematical tool, and is almost foolproof, as we shall demonstrate. We'll illustrate the method with some simple conversions.

Unit factors *may be constructed from any two terms that describe the same or equivalent "amounts" of whatever we may consider.* For example, 1 foot is equal to exactly 12 inches, by definition, and we may write an equation to describe this equality

$$1 \text{ ft} = 12 \text{ in}$$

Dividing both sides of the equation by 1 ft gives

$$\frac{1 \cancel{\text{ft}}}{1 \cancel{\text{ft}}} = \frac{12 \text{ in}}{1 \text{ ft}} \qquad \text{or} \qquad 1 = \frac{12 \text{ in}}{1 \text{ ft}}$$

The factor (fraction) 12 in/1 ft is a unit factor because the numerator and denominator describe the same distance. We could have divided both sides of the original equation by 12 in, and obtained 1 = 1 ft/12 in, a second unit factor, that is the reciprocal of the first. It should be obvious that the reciprocal of any unit factor is also a unit factor. Stated differently, division of an amount by the same amount always yields one!

From our knowledge of the English system of measurements we can write numerous unit factors, such as

$$\frac{1 \text{ yd}}{3 \text{ ft}}, \quad \frac{1 \text{ yd}}{36 \text{ in}}, \quad \frac{1 \text{ mile}}{5280 \text{ ft}}, \quad \frac{4 \text{ qt}}{1 \text{ gal}}, \quad \frac{2000 \text{ lb}}{1 \text{ ton}}$$

Unless otherwise indicated, a "ton" refers to a "short ton," 2000 lb. There are also the "long ton," which is 2240 lb, and the metric tonne, which is 1000 kg.

The reciprocal of each of these is also a unit factor. Items in retail stores are frequently priced with unit factors such as 39¢/lb and $3.98/gal.

Nearly all numbers have units. What does 12 mean? Usually we must supply appropriate units, such as 12 eggs or 12 people. In the unit factor method the units guide us through calculations in a step-by-step process, because all units except those in the desired result cancel.

We shall express answers to some problems in exponential form to indicate the appropriate number of significant figures. In scientific (exponential) notation, we place one nonzero digit to the left of the decimal as in the following examples.

$$4{,}300{,}000. = 4.3 \times 10^6$$

6 places to the left, ∴ exponent of 10 is 6

$$0.000348 = 3.48 \times 10^{-4}$$

4 places to the right, ∴ exponent of 10 is −4

The reverse process converts numbers from exponential to decimal form. See Appendix A for more detail, if necessary.

EXAMPLE 1–3

Express 1.47 miles in inches.

Solution

First we write down the units of what we wish to know preceded by a question mark, and then equate it to whatever we are given,

$\underline{?}$ in = 1.47 miles

Then we choose unit factors to convert the given units (miles) to the desired units (inches)

$$\underline{?}\text{ in} = 1.47\ \cancel{\text{miles}} \times \frac{5280\ \cancel{\text{ft}}}{1\ \cancel{\text{mile}}} \times \frac{12\text{ in}}{1\ \cancel{\text{ft}}} = \underline{9.31 \times 10^4\text{ in}}\ \text{(calculator gives 93139.2)}$$

The factors 5280 ft/mile and 12 in/ft contain only exact numbers, which are considered to contain an infinite number of significant figures.

Note that both miles and feet cancel, leaving only inches, the desired unit. If either of the unit factors had been written in inverted form, such as 1 mile/5280 ft, the units would not have cancelled properly. Thus there is no ambiguity as to how they should be written. The answer contains three significant figures because there are three significant figures in 1.47 miles.

Cancellation of units gives solved problems a cluttered appearance. Therefore, we have not cancelled units in the remainder of the illustrative examples in this text. You may find it useful to cancel units with a colored pencil or pen as you work through the illustrative examples.

EXAMPLE 1–4

Express 3.934 cubic yards in cubic inches.

Solution

We write down

$$\underline{?}\text{ in}^3 = 3.934\text{ yd}^3$$

and then multiply by the unit factors that convert yd^3 to in^3.

$$\underline{?}\text{ in}^3 = 3.934\text{ yd}^3 \times \underbrace{\left(\frac{3\text{ ft}}{1\text{ yd}}\right)^3 \times \left(\frac{12\text{ in}}{1\text{ ft}}\right)^3} = 3.934\text{ yd}^3 \times \left(\frac{27\text{ ft}^3}{1\text{ yd}^3}\right) \times \left(\frac{1728\text{ in}^3}{1\text{ ft}^3}\right)$$

$$= \underline{1.835 \times 10^5\text{ in}^3}\ \text{(calculator gives 183544.7)}$$

We could also use $\left(\frac{36\text{ in}}{1\text{ yd}}\right)^3$ instead of the terms with a brace beneath them.

The answer contains four significant figures because 3.934 yd^3 has four. The other numbers we used are exact.

Conversions within the SI (metric) system are easy, because measurements of a particular kind are related to each other by powers of ten.

EXAMPLE 1–5

The Ångstrom (Å) is a unit of length, 1×10^{-10} meter, that was defined to provide a convenient scale on which to express the radii of atoms (which are extremely small). The radius of a phosphorus atom is 1.10 Å. What is this distance expressed in centimeters and nanometers?

Solution

$$\underline{?}\text{ cm} = 1.10\text{ Å} \times \frac{1 \times 10^{-10}\text{ m}}{1\text{ Å}} \times \frac{1\text{ cm}}{1 \times 10^{-2}\text{ m}} = \underline{1.10 \times 10^{-8}\text{ cm}}$$

$$\underline{?}\text{ nm} = 1.10\text{ Å} \times \frac{1.0 \times 10^{-10}\text{ m}}{1\text{ Å}} \times \frac{1\text{ nm}}{1 \times 10^{-9}\text{ m}} = \underline{0.110\text{ nm}}$$

All the unit factors used in this example contain only exact numbers.

EXAMPLE 1–6

Assuming a phosphorus atom is spherical, calculate its volume in $Å^3$, cm^3, and nm^3. The volume of a sphere is $V = (\frac{4}{3})\pi r^3$. Refer to Example 1–5.

Solution

$\underline{?}\ Å^3 = (\frac{4}{3})\pi(1.10\ Å)^3 = \underline{5.58\ Å^3}$

$\underline{?}\ cm^3 = (\frac{4}{3})\pi(1.10 \times 10^{-8}\ cm)^3 = \underline{5.58 \times 10^{-24}\ cm^3}$

$\underline{?}\ nm^3 = (\frac{4}{3})\pi(1.10 \times 10^{-1}\ nm)^3 = \underline{5.58 \times 10^{-3}\ nm^3}$

$1\ Å = 10^{-10}\ m = 10^{-8}\ cm.$

EXAMPLE 1–7

How many millimeters are there in 1.39×10^4 meters?

Solution

$$\underline{?}\ mm = 1.39 \times 10^4\ m \times \frac{1000\ mm}{1\ m} = \underline{1.39 \times 10^7\ mm}$$

Note that the unit factor 1000 mm/1 m contains two exact numbers.

EXAMPLE 1–8

A sample of gold has a mass of 0.234 mg. What is its mass in g? in cg?

Solution

$$\underline{?}\ g = 0.234\ mg \times \frac{1\ g}{1000\ mg} = \underline{2.34 \times 10^{-4}\ g}$$

$$\underline{?}\ cg = 0.234\ mg \times \frac{1\ cg}{10\ mg} = \underline{0.0234\ cg} \quad \text{or} \quad \underline{2.34 \times 10^{-2}\ cg}$$

Again, we have used unit factors that contain only exact numbers.

EXAMPLE 1–9

How many square decimeters are there in 215 square centimeters?

Solution

$$\underline{?}\ dm^2 = 215\ cm^2 \times \left(\frac{1\ dm}{10\ cm}\right)^2 = 215\ cm^2 \times \left(\frac{1\ dm^2}{100\ cm^2}\right) = \underline{2.15\ dm^2}$$

Example 1–9 shows that a unit factor *squared* is still a unit factor, i.e.,

$$\left(\frac{1\ dm}{10\ cm}\right)^2 = \frac{1\ dm^2}{100\ cm^2} = 1$$

EXAMPLE 1–10

How many cubic centimeters are there in 8.34×10^5 cubic decimeters?

Solution

$$\underline{?}\ \text{cm}^3 = 8.34 \times 10^5\ \text{dm}^3 \times \left(\frac{10\ \text{cm}}{1\ \text{dm}}\right)^3 = 8.34 \times 10^5\ \text{dm}^3 \times \left(\frac{1000\ \text{cm}^3}{1\ \text{dm}^3}\right)$$

$$= \underline{8.34 \times 10^8\ \text{cm}^3}$$

Example 1–10 shows that a unit factor *cubed* is still a unit factor, that is

$$\left(\frac{10\ \text{cm}}{1\ \text{dm}}\right)^3 = \frac{1000\ \text{cm}^3}{1\ \text{dm}^3} = 1$$

We may generalize and say that a unit factor raised to any power is still a unit factor.

EXAMPLE 1–11

A common unit of energy is the erg. Convert 3.74×10^{-2} erg to the SI units of energy, joules, and kilojoules. One erg is exactly 1×10^{-7} joule.

Solution

$$\underline{?}\ \text{J} = 3.74 \times 10^{-2}\ \text{erg} \times \frac{1 \times 10^{-7}\ \text{J}}{1\ \text{erg}} = \underline{3.74 \times 10^{-9}\ \text{J}}$$

$$\underline{?}\ \text{kJ} = 3.74 \times 10^{-9}\ \text{J} \times \frac{1 \times 10^{-3}\ \text{kJ}}{\text{J}} = \underline{3.74 \times 10^{-12}\ \text{kJ}}$$

Conversions between the English and SI (metric) systems are conveniently done by the unit factor method. Several conversion factors were listed in Table 1–7. It may be helpful to remember one each for:

length	1 in = 2.54 cm (exact)
mass and weight	1 lb = 454 g (near sea level)
volume	1 qt = 0.946 L
or	1 L = 1.06 qt

EXAMPLE 1–12

Express 1.0 mL in gallons.

Solution

We write $\underline{?}$ gal = 1.0 mL and multiply by the appropriate factors

$$\underline{?}\ \text{gal} = 1.0\ \text{mL} \times \frac{1\ \text{L}}{1000\ \text{mL}} \times \frac{1.06\ \text{qt}}{1\ \text{L}} \times \frac{1\ \text{gal}}{4\ \text{qt}}$$

$$= \underline{2.6 \times 10^{-4}\ \text{gal}} \quad \text{(calculator gives 0.000265)}$$

The fact that all other units cancel to give the desired unit, gallons, shows that we have used the correct unit factors. The factors, 1L/1000 mL and 1 gal/4 qt, contain only exact numbers. The factor 1.06 qt/L contains three significant figures, while 1.0 mL contains only two; therefore the answer can contain only two significant figures.

Examples 1–1 through 1–12 show that multiplication by one or more unit factors changes the units and the number of units, but not the amount of whatever we are concerned with.

1–11 Density and Specific Gravity

The **density** of a sample of matter is defined as the mass per unit volume,

$$\text{density} = \frac{\text{mass}}{\text{volume}} \qquad \text{or} \qquad D = \frac{M}{V}$$

Densities may be used to distinguish between two substances or to assist in identifying a particular substance. They are usually expressed as g/cm^3 or g/mL for liquids and solids, and as g/L for gases. These units can also be expressed as $g \cdot cm^{-3}$, $g \cdot mL^{-1}$, and $g \cdot L^{-1}$, respectively.

Density is an *intensive property*, i.e., it does not depend upon the size of the sample. This is because density is constant for a given substance at a given temperature and pressure. The volume of 2 kilograms of ethanol at a given temperature and pressure is twice the volume of 1 kilogram of ethanol at the same temperature and pressure. Both samples of ethanol have the same density. Densities of several substances are listed in Table 1–8.

Ethanol is often called ethyl alcohol or grain alcohol.

TABLE 1–8 Densities of Common Substances*

Substance	Density, g/cm^3
Hydrogen (gas)	0.000089
Carbon dioxide (gas)	0.0019
Cork*	0.21
Oak wood*	0.71
Ethyl alcohol	0.79
Water	1.00
Magnesium	1.74
Table salt	2.16
Sand*	2.32
Aluminum	2.70
Iron	7.86
Copper	8.92
Lead	11.34
Mercury	13.59
Gold	19.3

* Cork, oak wood, and sand are common materials that have been included to provide familiar reference points. They are *not* pure elements or compounds as are the other substances listed here. High grade white sand is mostly silicon dioxide, a compound.

Densities are given at room temperature and *one atmosphere* pressure, average atmospheric pressure at sea level. Densities of solids and liquids change only slightly, but densities of gases change greatly, with changes in temperature and pressure.

EXAMPLE 1–13

A 47.3-mL sample of liquid has a mass of 53.74 g. What is its density?

Solution

$$D = \frac{M}{V} = \frac{53.74\ g}{47.3\ mL} = 1.14\ g/mL$$

EXAMPLE 1–14

If 100 g of the liquid described in Example 1–13 is needed for a chemical reaction, what volume of liquid would you use?

Solution

Recall that the density of the liquid is 1.14 g/mL

$$D = \frac{M}{V}, \qquad \text{so} \qquad V = \frac{M}{D} = \frac{100\text{ g}}{1.14\text{ g/mL}} = \underline{87.7\text{ mL}}$$

Alternatively, we can use the unit factor method to solve the problem

$$\underline{?}\text{ mL} = 100\text{ g} \times \frac{1\text{ mL}}{1.14\text{ g}} = \underline{87.7\text{ mL}}$$

The first method for solving the problem requires that one remember the formula, $D = M/V$. The second method doesn't.

The **specific gravity** (Sp. Gr.) of a substance is the ratio of its density to the density of water.

$$\text{Sp. Gr.} = \frac{D_{\text{substance}}}{D_{\text{water}}}$$

The density of water is 1.000 g/mL at 3.98°C, the temperature at which the density of water is greatest. However, variations in the density of water with changes in temperature are small enough that we may use 1.00 g/mL up to 25°C without introducing significant errors into our calculations.

Specific gravities are dimensionless numbers, as the following example demonstrates.

EXAMPLE 1–15

The density of table salt is 2.16 g/mL at 20°C. What is its specific gravity?

Solution

$$\text{Sp. Gr.} = \frac{D_{\text{salt}}}{D_{\text{water}}} = \frac{2.16\text{ g/mL}}{1.00\text{ g/mL}} = \underline{2.16}$$

The density and specific gravity of a substance are numerically equal near room temperature if density is expressed in g/mL (g/cm^3), as Example 1–15 demonstrated.

Labels on commercial solutions of acids and bases give specific gravities and the percentage by mass of the acid or base present in the solution. From this information, the amount of acid or base present in a given volume of the solution can be calculated.

At this point you need not be concerned if you do not know the chemical characteristics of an acid or a base. They will be described in Chapters 7, 11, and 16.

EXAMPLE 1–16

Battery acid is 40.0% sulfuric acid, H_2SO_4, and 60.0% water by mass. Its specific gravity is 1.31. Calculate the mass of pure sulfuric acid, H_2SO_4, in 100.0 mL of battery acid.

Solution

We have demonstrated that density and specific gravity are numerically equal at 20°C because the density of water is 1.00 g/mL. Therefore, we may write

Density = 1.31 g/mL

The solution is 40.0% H_2SO_4 and 60.0% H_2O by mass. From this information we may construct the desired unit factor

$$\frac{40.0 \text{ g } H_2SO_4}{100.0 \text{ g soln}} \leftarrow \left[\begin{array}{l}\text{because 100.0 g of solution} \\ \text{contains 40.0 g of } H_2SO_4\end{array}\right.$$

We can now solve the problem

$$\underline{?} \text{ g } H_2SO_4 = 100.0 \text{ mL soln} \times \frac{1.31 \text{ g soln}}{1 \text{ mL soln}} \times \frac{40.0 \text{ g } H_2SO_4}{100.0 \text{ g soln}} = \underline{52.4 \text{ g } H_2SO_4}$$

We first used the density as a unit factor to convert the given volume of solution to mass of solution, and then used the percentage by mass to convert the mass of solution to mass of acid.

1–12 Heat and Temperature

In Section 1–1 you learned that heat is simply one form of energy. You also learned that the many different forms of energy can be interconverted, and that in chemical processes, chemical energy is converted to heat energy or vice versa. The amount of heat a reaction uses (endothermic) or gives off (exothermic) can tell us a great deal about that reaction (this will be discussed in Chapters 9 and 13). For this reason it is important for us to be able to measure intensity of heat.

Temperature *measures the intensity of heat,* the "hotness" or "coldness" of a body. A piece of metal at 100°C feels hot to the touch, while an ice cube feels cold. Why? Because the temperature of the metal is higher, and that of the ice cube lower, than body temperature. As touching a hot surface and a cold surface demonstrate, *heat always flows spontaneously from a hotter body to a colder body*–never in the reverse direction.

Temperatures are commonly measured with mercury-in-glass thermometers. Mercury, more than most other substances, expands as its temperature rises. A mercury thermometer consists of a relatively large reservoir of mercury at the base of a glass tube, open to a very thin (capillary) column extending upward. As mercury expands in the reservoir, its movement up into the thin column is clearly visible.

Anders Celsius, a Swedish astronomer, developed the Celsius temperature scale, formerly called the centigrade temperature scale. When a well-made Celsius thermometer is placed in a beaker of crushed ice stirred with water, the mercury level stands at exactly 0°C, the lower reference point. In a beaker of water boiling at one atmosphere pressure, the mercury level stands at 100°C, the higher reference point. There are 100 evenly spaced calibration marks between these two mercury levels, and they correspond to an interval of 100 degrees between the melting point of ice and the boiling point of water at one atmosphere. Figure 1–20 shows how temperature marks between the reference points are measured.

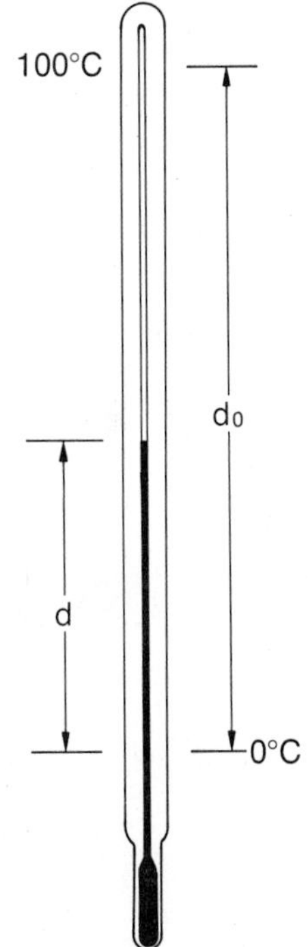

FIGURE 1–20 At 45°C, as read on a mercury-in-glass thermometer, d equals $0.45d_0$ where d_0 is the distance from the mercury level at 0°C to the level at 100°C.

Temperatures are frequently measured in the United States on the temperature scale devised by Gabriel Fahrenheit, a German instrument maker. On this scale the freezing and boiling points of water are defined as 32°F and 212°F, respectively.

In scientific work, temperatures are often expressed on the Kelvin (absolute) temperature scale. As we shall see in Section 8–5, the zero point of the Kelvin temperature scale is derived from the observed behavior of all matter. The Celsius and Fahrenheit temperature scales have no such rigorous basis.

Relationships among the three temperature scales are illustrated in Figure 1–21. The degree is the same "size" on the Celsius and Kelvin scales because there are 100 degrees between the freezing and boiling points of water in each case. Every Kelvin temperature is 273.15° above the corresponding Celsius temperature. Thus the relationship between the two scales is simply

x°C and x K refer to temperatures on the Celsius and Kelvin scales, respectively.

$$\underline{?}\text{ K} = (x°\text{C} + 273.15°\text{C})\frac{1.0\text{ K}}{1.0°\text{C}} \quad \text{or} \quad \underline{?}°\text{C} = (x\text{K} - 273.15\text{ K})\frac{1.0°\text{C}}{1.0\text{ K}}$$

In the SI system, degrees Kelvin are abbreviated simply as K rather than °K, and are called kelvins. Comparing the Celsius and Fahrenheit scales, we find that the interval between the same reference points is 100°C and 180°F, so the Celsius degree is larger than the Fahrenheit degree. That is, it takes 1.8 Fahrenheit degrees to cover the same temperature interval as 1.0 Celsius degree. From this information, we can construct the unit factors

$$\frac{1\ 8°\text{F}}{1.0°\text{C}} \quad \text{and} \quad \frac{1.0°\text{C}}{1.8°\text{F}}$$

Additionally, the starting points of the two scales are different, so we *cannot* convert a temperature on one scale to a temperature on the other by just multiplying by the unit factor. To compensate for this in converting from °F to °C, 32 Fahrenheit degrees must be added to reach the zero point on the Celsius scale

$$\underline{?}°\text{F} = \left(x°\text{C} \times \frac{1.8°\text{F}}{1.0°\text{C}}\right) + 32°\text{F}$$

and

$$\underline{?}°\text{C} = \frac{1.0°\text{C}}{1.8°\text{F}}(x°\text{F} - 32°\text{F})$$

EXAMPLE 1–17

When the temperature reaches "100°F in the shade," it's hot. What is this temperature on the Celsius scale?

Solution

100°F is 38°C.

$$\underline{?}°\text{C} = \frac{1.0°\text{C}}{1.8°\text{F}}(100°\text{F} - 32°\text{F})$$

$$= \frac{1.0°\text{C}}{1.8°\text{F}}(68°\text{F}) = \underline{38°\text{C}}$$

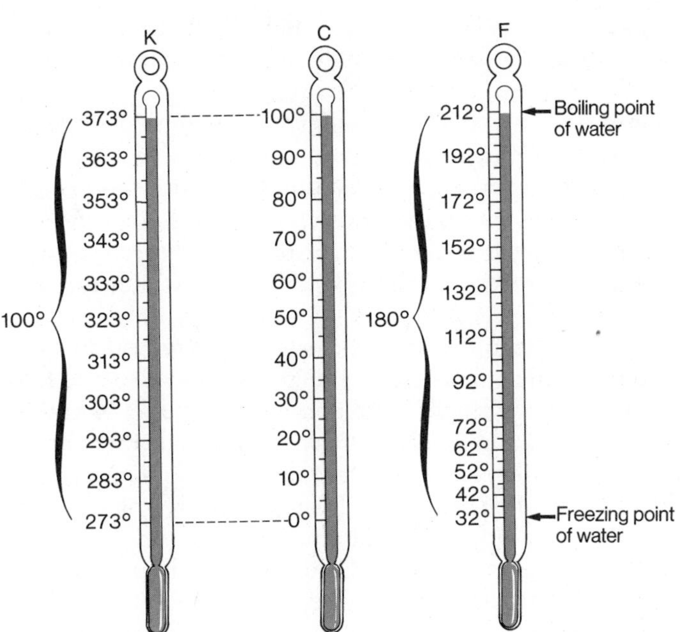

FIGURE 1–21 The relationships between the Kelvin, Celsius (centigrade), and Fahrenheit temperature scales.

EXAMPLE 1–18

When the absolute temperature is 400 K, what is the Fahrenheit temperature?

Solution

First we convert K to °C, then °C to °F.

$$\underline{?}\,°C = (400\ K - 273\ K)\frac{1.0°C}{1.0\ K} = 127°C$$

$$\underline{?}\,°F = \left(127°C \times \frac{1.8°F}{1.0°C}\right) + 32°F = \underline{261°F}$$

1–13 Heat Transfer and the Measurement of Heat

Chemical reactions and physical changes occur with either the simultaneous evolution of heat (exothermic processes) or the absorption of heat (endothermic processes). The amount of heat transferred in a process is usually expressed in calories or in the SI unit of joules. Historically, the **calorie** was originally defined as the amount of heat necessary to raise the temperature of one gram of water from 14.5°C to 15.5°C. One calorie is *now defined* as exactly 4.184 **joules.** The amount of heat necessary to raise the temperature of one gram of liquid water varies slightly with temperature, so it was necessary to specify a particular temperature increment in defining the calorie. For our purposes the variations are sufficiently small that we are justified in ignoring them. The so-called "large calorie," used to indicate energy content of foods, is really one kilocalorie, or 1000 calories. We shall do most of our calculations in joules.

The SI unit of energy and work is the **joule** (J), which is defined as $1\ kg \cdot m^2/s^2$. One joule is the amount of kinetic energy possessed by a 2 kg object moving at one meter per second. (In English units this corresponds to a 4.4 lb object moving at 197 feet per minute or 2.2 miles per hour.) You may find it more convenient to think in terms of the amount of heat required to raise the temperature of one gram of water from 14.5°C to 15.5°C, which is 4.184 joules.

The **specific heat** of a substance is the amount of heat required to raise the temperature of one gram of the substance one degree C, and also one kelvin, with no change in state. (As you might have noticed in Figure 1–4 and as we shall see again in Sections 9–8 and 9–11, changes in physical state are accompanied by the absorption or liberation of relatively large amounts of energy.) The specific heat of each substance is a physical property of the substance, and is different for the solid, liquid, and gaseous states of the substance. For example, the specific heat of ice is 2.09 J/g · °C near 0°C; for liquid water it is 4.18 J/g · °C (or 1.00 cal/g · °C), while for steam the value is 2.03 J/g · °C near 100°C. The specific heat for water is quite high compared with that for most common substances. A table of specific heats is provided in Appendix E.

The **heat capacity** of any body is the amount of heat required to raise its temperature 1°C. Clearly, the heat capacity of a body is its mass in grams times its specific heat.

EXAMPLE 1–19

How much heat, in joules, is required to raise the temperature of 500 g of water from 40°C to 90°C?

Solution

Since the specific heat of a substance is the amount of heat required to raise the temperature of 1 g of substance 1°C, or

$$\text{specific heat} = \frac{(\text{amt. of heat, J})}{(\text{mass of substance, g})(\text{temp. change,°C})}$$

then we can rearrange the equation so that

$$\begin{aligned}\text{amt. of heat} &= (\text{mass of substance})(\text{sp. ht.})(\text{temp. change}) \\ &= (500\ \text{g})(4.18\ \text{J/g}\cdot\text{°C})(50\text{°C}) = \underline{1.0 \times 10^5\ \text{J}}\end{aligned}$$

Or, we can use the unit factor approach and obtain

$$\underline{?}\ \text{J} = (500\ \text{g})(4.18\ \text{J/g}\cdot\text{°C})(50\text{°C}) = \underline{1.0 \times 10^5\ \text{J}} \text{ or } \underline{1.0 \times 10^2\ \text{kJ}}$$

Note that all units except joules cancel, and the answer contains only two significant figures because the given temperatures contain two. If 500 g of water were cooled from 90°C to 40°C, it would be necessary to remove exactly the same amount of heat, 1.0×10^2 kJ.

EXAMPLE 1–20

How much heat in calories, kilocalories, joules, and kilojoules is required to raise the temperature of 500 g of iron from 40°C to 90°C? The specific heat of iron is 0.106 cal/g·°C, or 0.444 J/g·°C.

Solution

$\underline{?}$ cal = (500 g)(0.106 cal/g·°C)(50°C) = $\underline{2.7 \times 10^3 \text{cal}}$ or $\underline{2.7 \text{ kcal}}$

$\underline{?}$ J = (500 g)(0.444 J/g·°C)(50°C) = $\underline{1.1 \times 10^4 \text{ J}}$ or $\underline{11 \text{ kJ}}$

Note that the specific heat of iron is much smaller than the specific heat of water. Therefore, much less heat is required to raise the temperature of 500 g of iron by 50°C than for 500 g of water.

EXAMPLE 1–21

If 200 g of water at 40°C is mixed with 100 g of water at 80°C in an insulated container, what will be the temperature of the mixture?

Solution

The temperature of the warm water will decrease, and the temperature of the cool water will increase. Call the new temperature t°C. It must be between 80°C and 40°C. The amount of heat gained by the cool water must equal the amount of heat lost by the warm water. Therefore, we may write

$$\text{Heat gain} = \text{Heat loss}$$
$$(200 \text{ g})(4.18 \text{ J/g}\cdot{}^\circ\text{C})(t - 40)^\circ\text{C} = (100 \text{ g})(4.18 \text{ J/g}\cdot{}^\circ\text{C})(80 - t)^\circ\text{C}$$
$$836\,t - 3.34 \times 10^4 = 3.34 \times 10^4 - 418\,t$$
$$1.25 \times 10^3\,t = 6.68 \times 10^4$$
$$\underline{t = 53}, \text{ and the final temperature is } \underline{53^\circ\text{C}}$$

Both $(t - 40)$°C and $(80 - t)$°C represent positive differences in temperature; $40 < t < 80$.

Note that when we construct an equation, *all units cancel.* This must always occur. An "equality" from which all units do not cancel is clearly not an equality!

Key Terms

Accuracy how closely a measured value agrees with the correct value.

Atom the smallest particle of an element.

Calorie the amount of heat required to raise the temperature of one gram of water from 14.5°C to 15.5°C. A unit of energy equal to 4.184 joules.

Chemical change a change in which one or more new substances are formed.

Chemical property see *Properties.*

Compound a substance composed of two or more elements in fixed proportions. Compounds can be decomposed into their constituent elements.

Density mass per unit volume, $D = M/V$.

Element a substance that cannot be decomposed into simpler substances by chemical means.

Endothermic describes processes that absorb heat energy.

Energy the capacity for doing work.

Exothermic describes processes that release heat energy.

Extensive property a property that depends upon the amount of material in a sample.

Heat a form of energy that flows between two samples of matter because of their difference in temperature.

Heat capacity the amount of heat required to raise the temperature of a body (of whatever mass) one degree Celsius.

Heterogeneous mixture a mixture that does not have uniform composition and properties throughout.

Homogeneous mixture a mixture that has uniform composition and properties throughout.

Intensive property a property that does not depend upon the amount of material in a sample.

Joule a unit of energy in the SI system. One joule is $1 \text{ kg} \cdot \text{m}^2/\text{s}^2$, which is also 0.2390 calorie.

Kinetic energy energy that matter possesses by virtue of its motion.

Law of Conservation of Energy energy cannot be created or destroyed; it may be changed from one form to another.

Law of Conservation of Matter there is no detectable change in the quantity of matter during an ordinary chemical reaction.

Law of Conservation of Matter and Energy the total amount of matter and energy available in the universe is fixed.

Mass a measure of the amount of matter in an object. Mass is usually measured in grams or kilograms.

Matter anything that has mass and occupies space.

Mixture a sample of matter composed of two or more substances, each of which retains its identity and properties.

Molecule the smallest particle of an element or compound capable of a stable, independent existence.

Physical change a change in which a substance changes from one physical state to another, but no substances with different compositions are formed.

Physical property see *Properties.*

Potential energy energy that matter possesses by virtue of its position, condition, or composition.

Precision how closely repeated measurements of the same quantity agree with each other.

Properties characteristics that describe samples of matter. Chemical properties are exhibited as matter undergoes chemical changes. Physical properties are exhibited by matter with no changes in chemical composition.

Significant figures digits that indicate the precision of measurements—digits of a measured number that have uncertainty only in the last digit.

Specific gravity the ratio of the density of a substance to the density of water.

Specific heat the amount of heat required to raise the temperature of one gram of a substance one degree Celsius.

Substance any kind of matter all specimens of which have the same chemical composition and physical properties.

Symbol a letter or group of letters that represents (identifies) an element.

Temperature a measure of the intensity of heat, i.e., the hotness or coldness of a sample or object.

Unit factor a factor in which the numerator and denominator are expressed in different units but represent the same or equivalent amounts. Multiplying by a unit factor is the same as multiplying by one.

Weight a measure of the gravitational attraction of the earth for a body.

Exercises

Basic Ideas

1. What is chemistry?
2. Define the following terms clearly and concisely. Illustrate each with a specific example.
(a) matter

(b) mass
(c) energy
(d) potential energy
(e) kinetic energy
(f) heat energy
(g) exothermic process
(h) endothermic process

3. State the following laws and provide an illustration of each.
   (a) the Law of Conservation of Matter
   (b) the Law of Conservation of Energy
   (c) the combined Law of Conservation of Matter and Energy
4. (a) Write the Einstein equation.
   (b) State the meaning of the Einstein equation in words.
5. Is the following statement true or false? Matter and energy are different forms of a single entity. Why is it true or false?
6. List the three states of matter and some characteristics of each. How are they alike? different?
7. Distinguish between the following pairs of terms and provide two specific examples to illustrate each term.
   (a) chemical properties and physical properties
   (b) chemical changes and physical changes
8. Are the following chemical properties or physical properties? Why?
   (a) the melting point of lead
   (b) hardness
   (c) color of a solid
   (d) color of a flame
   (e) ability to burn in the air
9. Are the following chemical changes or physical changes? Why?
   (a) melting of lead
   (b) burning of gasoline
   (c) rusting of iron
   (d) production of light by a glowworm
   (e) emission of light by an electric light bulb
   (f) emission of light by a whale oil lamp
10. Which of the following processes are exothermic? endothermic? Why?
    (a) combustion (d) boiling water
    (b) freezing water (e) condensing steam
    (c) melting ice
11. Define the following terms clearly and concisely. Provide two specific illustrations of each.
    (a) substance (b) mixture
    (c) element (d) compound
12. Classify each of the following as an element, a compound, or a mixture. Justify your classification.
    (a) gold (b) bronze
    (c) iron (d) steel
    (e) popcorn (f) milk
    (g) sawdust (h) sour cream
    (i) gasoline (j) natural gas
    (k) shampoo (l) table salt
13. Write symbols for the following elements.
    (a) silver (b) calcium
    (c) iron (d) iodine
    (e) neon (f) phosphorus
14. Write symbols for the following elements.
    (a) aluminum (b) chlorine
    (c) hydrogen (d) potassium
    (e) oxygen (f) lead
15. Write the name of each of the following elements.
    (a) Br (b) Cu (c) He (d) Mg (e) Na (f) S
16. Write the name of each of the following elements.
    (a) C (b) F (c) Hg (d) N (e) Ni (f) Zn
17. (a) What is the distinction between mass and weight?
    (b) Give a location in which the weight of an object that has a mass of 453.59 grams would be greater than 1.0000 pound. Justify your choice.
    (c) Give a location in which the weight of an object that has a mass of 453.59 grams would be less than 1.0000 pound. Justify your choice.

Significant Figures and Exponential Notation

18. Round off the following numbers to three significant figures. Follow the rules given in Section 1–9 exactly.
    (a) 4325 (b) 6.873×10^3
    (c) 0.17354 (d) 7.8939
    (e) 9.237×10^{-3} (f) 0.0299817
19. Which of the following are likely to be exact numbers? Why?
    (a) 15 eggs
    (b) 10 pounds of potatoes
    (c) 497 live quail

(d) 11,743 live bees
(e) 47,398 people
(f) $47,342.21
(g) 3 gnus
(h) 12 square yards of carpet

How many significant figures does each of the numbers in Exercises 20 and 21 contain?

20. (a) 0.0278 meter
(b) 1.3 centimeters
(c) 1.00 foot
(d) 8.021 yards
21. (a) 7.98×10^{-3} pound
(b) 0.2003 ton
(c) 4.69×10^{4} tons
(d) 1×10^{12} atoms
(e) 1.73×10^{24} atoms
22. Express the following numbers in proper exponential form with the indicated number of significant figures.
(a) 1000 (2 sig. fig.)
(b) 43,927 (3 sig. fig.)
(c) 0.000286 (3 sig. fig.)
(d) 0.000098765 (5 sig. fig.)
(e) 10,000 (you decide how many significant figures)
23. Express the following exponentials as ordinary numbers.
(a) 7.23×10^{4} (b) 8.193×10^{2}
(c) 1.98×10^{-3} (d) 7.51×10^{-7}
(e) 5.43×10^{0}

Perform the indicated operations and round off your answers to the proper number of significant figures in Exercises 24 and 25. Assume that all numbers were obtained from measurements.

24. (a) $18.56 + 1.233 =$
(b) $1.234 \times 0.247 =$
(c) $4.3/8.74 =$
25. (a) $8.649 - 2.8964 =$
(b) $0.06936 \times 0.384 =$
(c) $4567/2.53 =$

In Exercises 26 and 27, perform the indicated operations and round off your answers to the proper number of significant figures. Assume that all numbers were obtained from measurements. Express your answers in exponential notation.

26. (a) $(1.54 \times 10^{3}) + (2.11 \times 10^{3}) =$
(b) $(1.54 \times 10^{3}) + (2.11 \times 10^{2}) =$
(c) $(1.23 \times 10^{2})/(4.56 + 18.7) =$
(d) $(4.56 + 8.7)/(1.23 \times 10^{-2}) =$
27. (a) $(2.11 \times 10^{-3}) + (1.54 \times 10^{-3}) =$
(b) $(1.54 \times 10^{-3}) + (2.11 \times 10^{-2}) =$
(c) $(4.56 + 18.7)/(1.23 \times 10^{2}) =$
(d) $(1.23 \times 10^{-2})/(4.56 + 1.87) =$
28. Choose the appropriate word prefix to indicate the multiplier in each of the following numbers.
(a) 1×10^{3} (b) 1×10^{-3}
(c) 1×10^{6} (d) 1×10^{-1}
(e) 0.01 (f) 0.1
(g) 0.001 (h) 1×10^{-6}
29. Indicate the multiple or fraction of 10 by which a quantity is multiplied when it is preceded by the following prefixes.
(a) M (b) m
(c) c (d) d
(e) k (f) μ

In Exercises 30 and 31, choose all the units listed that could be used to express: (a) length or distance, (b) mass or weight, (c) area, (d) volume.

30. (a) km (b) mL
(c) mm^2 (d) dg
(e) m^3 (f) cm^2
31. (a) mg (b) mm
(c) dL (d) cm^3
(e) kg (f) m^2

Dimensional Analysis

32. Fill in the blanks by making the indicated conversions.
(a) 7.58 km = ______ m
(b) 758 m = ______ cm
(c) 478 g = ______ kg
(d) 9.78 kg = ______ g
(e) 1386 mL = ______ L
(f) 3.692 L = ______ mL
(g) 1126 cm^3 = ______ L
(h) 0.786 L = ______ cm^3
33. Express 1.27 feet in millimeters, centimeters, meters, and kilometers.
34. What is the length of a football field (100 yards) expressed in meters?
35. What is the distance around your waist expressed in (a) centimeters and (b) meters?

36. Express 55 miles per hour in kilometers per hour.
37. The capacity of the gasoline tank in an automobile is 12 gallons. How many liters is this?
38. Express:
    (a) 1.00 gallon in milliliters
    (b) 8.00 cubic inches in milliliters (How long is each edge of a cube of this volume?)
    (c) 4.00 cubic yards in cubic centimeters
    (d) 1.00 liter in cubic inches
39. Express:
    (a) 1.00 pint in milliliters
    (b) 1.00 square inch in square centimeters
    (c) 1.00 square foot in square centimeters
    (d) 1.00 cubic inch in cubic centimeters
    (e) 1.00 cubic foot in liters
    (f) 1.00 cubic meter in cubic feet
40. If the price of gasoline is \$0.400 per liter, what is its price per gallon?
41. If the price of gasoline is \$1.43 per gallon, what is its price per liter?
42. Express the following masses or weights in grams and kilograms (at sea level).
    (a) 1.00×10^6 milligrams
    (b) 4.25×10^5 centigrams
    (c) 3.0 pounds
    (d) 4.00 ounces
43. Express the following weights in grams.
    (a) 2.35 pounds
    (b) 1.00 ounce
    (c) 1.00 short ton
    (d) 1.00 metric tonne
44. Express:
    (a) 100 pounds in kilograms
    (b) 100 kilograms in pounds
    (c) 8.00 ounces in centigrams
    (d) 2.23 short tons in milligrams
45. If a successful professional athlete is 216 centimeters tall and has a mass of 95 kilograms, would you assume that he plays basketball or football? Why?
46. Choose the SI and traditional metric system units that would be most appropriate to indicate the following dimensions. Provide an *approximate* numerical value with each set of units. Example: A standard door is 6 feet and 8 inches high, which is approximately 2 meters high. Justify your choice of units in each case.
    (a) the length of a U.S. football field between the goal lines
    (b) the volume of the gasoline tank in a "compact" automobile
    (c) the area of the floor in your bedroom
    (d) your mass
    (e) your height
47. At a given point, the moon is 240,000 miles from the earth. How long does it take for light from a source on the earth to reach a reflector on the moon and then return to earth? The speed of light is 3.00×10^8 m/s.
48. The radius of an aluminum atom is 1.43 Å. How many aluminum atoms would have to be laid side-by-side to give a row of aluminum atoms 1.00 inch long? Assume that the atoms are spherical.
49. Cesium atoms are the largest naturally occurring atoms. The diameter of a cesium atom is 5.24 Å. If 1.00×10^9 cesium atoms were laid side-by-side to make a row, how long would the row be in inches? Assume that the atoms are spherical.

Extensive and Intensive Properties

50. Distinguish between intensive and extensive properties, and cite two specific examples of each.
51. Which of the following are extensive properties and which are intensive properties? Justify your answers.
    (a) temperature
    (b) color of copper
    (c) volume
    (d) density
    (e) melting point of tin
    (f) mass
52. Which of the following are extensive properties and which are intensive properties? Justify your answers.
    (a) color of a liquid
    (b) mass
    (c) specific gravity
    (d) boiling point of water
    (e) physical state

Density and Specific Gravity

53. Distinguish between density and specific gravity.
54. A 13.5 cubic centimeter piece of chromium has a mass of 97.2 grams. What is its density?
55. What is the mass of 13.5 cubic centimeters of mercury? Its density is 13.6 g/cm^3.
56. Refer to the two previous problems. One cm^3 of mercury is how much heavier than one cm^3 of chromium?
57. Refer to Table 1–8 and calculate the mass, at room temperature and one atmosphere pressure, of 200 mL of: (a) hydrogen, (b) carbon dioxide, (c) ethyl alcohol, (d) aluminum, and (e) gold.
58. Refer to Table 1–8. A 20.0-mL sample of which of the following has the largest mass? hydrogen, sand, gold
59. Refer to Table 1–8. A 20.0-gram sample of which of the following has the smallest volume? cork, table salt, lead
60. The specific gravity of ethyl alcohol is 0.79. What volume of ethyl alcohol has the same mass as 23 mL of water?
61. What is the specific gravity of a liquid if 300 mL of the liquid has the same mass as 400 cm^3 of water?
62. A piece of copper is placed in a graduated cylinder containing some water. The total volume increases by 7.43 mL. What is the mass of the piece of copper (density = 8.92 g/cm^3)?
63. What is the volume of 479 grams of aluminum (specific gravity = 2.70)?
64. The specific gravity of gold is 19.3. Which of the following contains the greatest mass of gold? 0.50 pound, 0.25 kilogram, or 25 milliliters
65. The specific gravity of table salt is 2.16. Which of the following occupies the largest volume: 0.50 pound, 0.25 kilogram, or 0.025 liter of salt?
66. (a) What is the volume of a bar of iron that is 4.72 centimeters long, 3.19 centimeters wide, and 0.52 centimeters thick? Its mass is 61.5 grams.
    (b) Calculate the density of iron using the data obtained in (a). Compare with the value in Table 1–8 and explain any difference.
67. The density of gold is 19.3 g/cm^3. Suppose someone offered to give you a one-gallon bucket filled with gold if you would carry the bucket of gold up a flight of stairs. Could you accept the gold?
68. (a) Given that the density of gold is 19.3 g/cm^3, calculate the volume of 100 pounds of gold in cm^3.
    (b) Assume that this sample of gold is a perfect cube. How long is each edge (in inches)?
69. An aqueous solution that is 70.0% nitric acid by mass has a specific gravity of 1.40. What volume of the solution contains (a) 100 grams of solution, (b) 100 grams of pure nitric acid, (c) 100 grams of water? What masses of (d) water and (e) pure nitric acid are contained in 100 milliliters of the solution?

Heat and Temperature

In Exercises 70 and 71 express the given temperatures on the indicated scale:

70. (a) 100.000°F as degrees Celsius
    (b) 200.000°F as degrees Celsius
    (c) 100.0 K as degrees Fahrenheit
    (d) 100.0°F as kelvins
71. (a) 0.00°F as degrees Celsius
    (b) 78.0°F as kelvins
    (c) −40.0°C as degrees Fahrenheit
    (d) 0.00 K as degrees Fahrenheit
72. The following data show the relationship between temperature and distance above the surface of the earth on a "standard day." Fill in the blanks by making the appropriate conversions.

Feet	Meters	°F	°C
(a) 1,000	______	56°	______
(b) ______	1,500	______	5°
(c) 10,000	______	______	−5°
(d) ______	4,500	5°	______
(e) 20,000	______	______	−26°
(f) ______	9,000	−47°	______
(g) 36,087	______	______	−56°

(h) Plot feet versus °F on a sheet of graph paper. Estimate the temperature at altitudes of 7500 feet and 40,000 feet.

(i) Now plot meters versus °C on another piece of graph paper. Compare the two graphs. Are they similar? Why?

Heat Transfer

73. Calculate the amount of heat required to raise the temperature of 25.0 grams of water from 10.0°C to 40.0°C. Express your answer in joules and in calories.

74. How much heat must be removed from 15.0 grams of water at 60.0°C to decrease its temperature to 10.0°C? Express your answer in kilojoules and in calories.

75. How much heat is required to raise the temperature of 25.0 grams of iron from 10.0°C to 40.0°C? Express your answer in joules. The specific heat of iron is 0.444 J/g · °C.

76. If 150 grams of liquid water at 100°C and 250 grams of water at 10.0°C are mixed in an insulated container, what is the final temperature of the mixture?

77. If 150 g of iron at 100.0°C is placed in 250 g of water at 10.0°C in an insulated container, and both are allowed to come to the same temperature, what will that temperature be? The specific heat of iron is 0.444 J/g · °C. Compare this result with the one obtained in Exercise 76.

78. (a) Calculate the amount of heat required to raise the temperature of 100 g of mercury from 10.0° to 75.0°C. The specific heat of mercury is 0.138 J/g · °C. Mercury, commonly called quicksilver, is a liquid metal at room temperature.

(b) Calculate the amount of heat required to raise the temperature of 100 g of water from 10.0°C to 75.0°C.

(c) What is the ratio of the specific heat of liquid water to that of liquid mercury?

(d) What does answer (c) tell us?

79. A 94.4-g piece of metallic iron at 152.6°C is dropped into 140 mL of octane, C_8H_{18}, at 24.6°C. The final temperature is 45.2°C. The density of octane is 0.703 g/mL. The specific heat of iron is 0.444 J/g · °C. (a) What is the specific heat of octane? (b) What is the heat capacity of octane in J/mol·°C? One mol of octane is 114 grams.

80. How much water at 80.0°C must be mixed with one liter of water in an insulated container at 15.0°C so that the temperature of the mixture will be 30.0°C?

81. How much iron at 200.0°C must be placed in 200.0 mL of water at 20.0°C so that the temperature of both will be 30.0°C? Assume no heat is lost to the surroundings.

82. If 75.0 grams of metal at 75.0°C is added to 150 grams of water at 15.0°C, the temperature of the water rises to 18.3°C. Assume no heat is lost to the surroundings. What is the specific heat of the metal?

83. The specific heat of mercury is 0.138 J/g · °C. Express this value in (a) cal/lb · °F, (b) kJ/g · °C, and (c) J/g · K.

2 Stoichiometry, Chemical Formulas, and Equations

In this chapter we shall become familiar with the language used to describe the forms of matter and the possible changes in its composition. The ideas presented here are used throughout the scientific world. For example, chemical symbols, formulas, and equations are used in such diverse areas as agriculture, home economics, engineering, geology, physics, the biological sciences, medicine, and dentistry. Many chemical terms are found in your daily newspaper or favorite magazine. It is important to learn this material well so that you can use it precisely and effectively.

The word *stoichiometry* is derived from the Greek *stoicheion,* which means "first principle or element," and *metron,* which means "measure." **Stoichiometry** describes the quantitative relationships among elements in compounds and among substances as they undergo chemical changes.

2–1 Formulas

The **formula** *for a substance shows its chemical composition.* By composition we mean the elements present as well as the ratio in which atoms occur in the substance.

The formula for a single atom is the same as the symbol for the element. Thus, Na represents a single sodium atom. It is unusual to find such isolated atoms in nature, with the exception of the noble gases (He, Ne, Ar, Kr, Xe, and Rn). A subscript following the symbol of an element indicates the number of atoms of that element in a molecule. For instance, F_2 indicates a molecule containing two fluorine atoms, and P_4 represents four phosphorus atoms in a molecule.

Some elements exist in more than one elemental form. Familiar examples include (1) oxygen, found as O_2 molecules, and ozone, found as O_3 molecules, and (2) two different crystalline forms of carbon, diamond and graphite. Different forms of the same element in the same physical state are called **allotropic modifications** or **allotropes.**

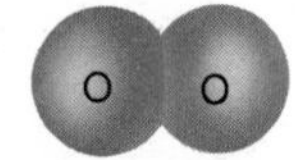

An O_2 molecule

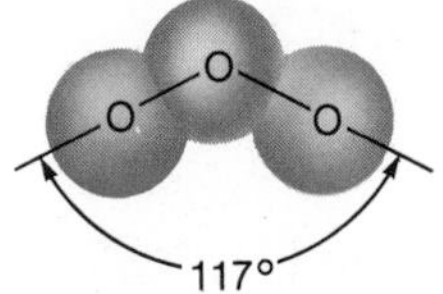

An O_3 molecule

Most compounds, we have seen, contain two or more elements in chemical combination in fixed proportions (although some do not, as explained on p. 50). Hence, each molecule of hydrogen chloride, HCl, contains one atom of hydrogen and one atom of chlorine; each molecule of carbon tetrachloride, CCl_4, contains one carbon atom and four chlorine atoms; and a molecule of propane, C_3H_8, contains three carbon atoms and eight hydrogen atoms. An aspirin molecule, $C_9H_8O_4$, contains nine carbon atoms, eight hydrogen atoms, and four oxygen atoms.

Some groups of atoms behave chemically as single entities. For instance, one nitrogen atom and two oxygen atoms may combine into a *nitro* group that can form a part of a molecule. In formulas of compounds containing two or more of the same group, the group is enclosed in parentheses to show its presence. Thus, 2,4,6-trinitrotoluene (often abbreviated TNT) contains three *nitro* groups, and its formula is $C_7H_5(NO_2)_3$. When you count up the number of atoms in this molecule, you must multiply the numbers of nitrogen and oxygen atoms in the NO_2 group by 3. There are *seven* carbon atoms, *five* hydrogen atoms, *three* nitrogen atoms, and *six* oxygen atoms in a molecule of TNT. Some examples of simple molecules are shown in Figure 2–1. A model of the TNT molecule is shown in the margin.

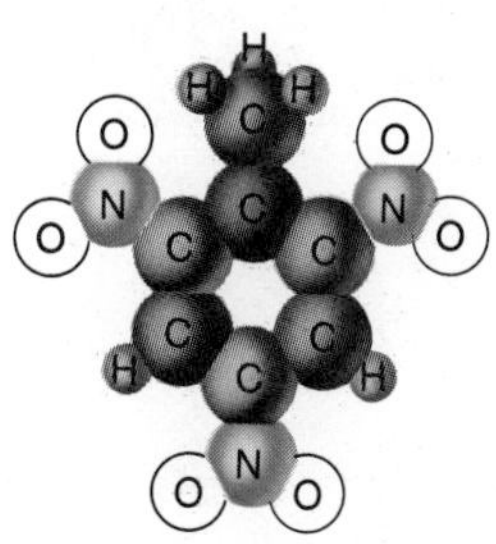
A space-filling model of a TNT, $C_7H_5(NO_2)_3$, molecule.

Compounds were first recognized as different substances because of their different physical properties, and because they could be separated from one another. Once the concept of atoms and molecules was established, the reason for these differences in properties could be understood: Two compounds differ from one another because their molecules are different. Conversely, if two molecules contain the same number of the same kinds of atoms, arranged the same way, then both are molecules of the same compound. One consequence of this reasoning is the **Law of Definite Proportions** (sometimes called the **Law of Constant Composition**): *Different pure samples of a compound always contain the same elements in the same proportions by mass.* This law was first recognized by Joseph Proust in 1799.

So far, we have cited only compounds that exist as discrete molecules. Some compounds such as sodium chloride, NaCl, consist of an infinite array of ions. An **ion** is an atom or group of atoms that carries an electrical charge. Ions that carry a positive charge are often called **cations,** and those that possess a negative charge are called **anions.** Ions will be discussed in detail in Chapter 4. One **formula unit** of a substance contains as many atoms or ions of each element as are present in the formula for the substance. One formula unit of NaCl contains one sodium ion (Na^+) and one chloride ion (Cl^-); one formula unit of

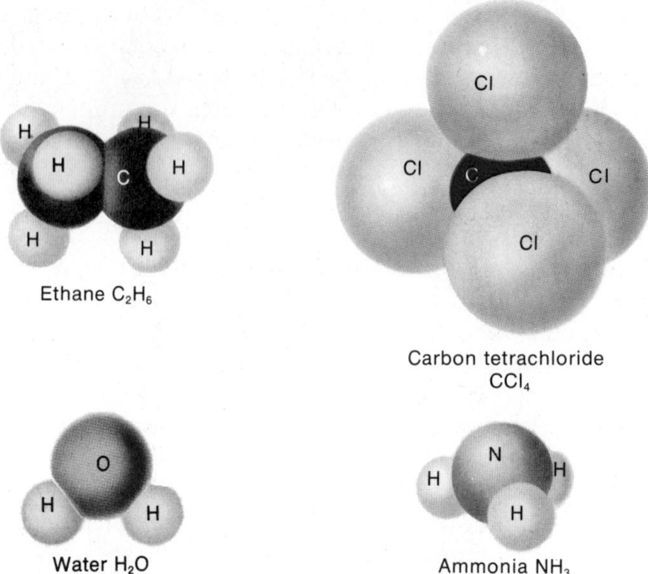

FIGURE 2–1 Representations for some simple molecules.

C_3H_8, which is the same as a C_3H_8 molecule, contains three carbon atoms and eight hydrogen atoms.

2–2 Atomic Weights

As the chemists of the eighteenth and nineteenth centuries painstakingly (and sometimes painfully!) sought information about the composition of compounds and tried to systematize their knowledge, it became apparent that each element has a characteristic mass relative to every other element. Although these early scientists did not have the experimental means to establish a mass for each kind of atom, they succeeded in defining a *relative* scale of atomic masses.

The term "atomic weight" is widely accepted because of its traditional use, although it is properly a mass rather than a weight. "Atomic mass" is sometimes used, but is not as popular.

An early observation was that carbon and oxygen have relative atomic masses, also traditionally called **atomic weights,** of approximately 12 and 16, respectively.

Thousands of experiments on the compositions of compounds have resulted in the establishment of a scale of relative atomic weights based on the **atomic mass unit (amu),** which is defined as *exactly 1/12 of the mass of an atom of a particular isotope of carbon, called carbon-12.*

On this scale, the atomic weight of hydrogen (H) is 1.0079 amu, that of sodium (Na) is 22.98977 amu, and that of magnesium (Mg) is 24.305 amu. This tells us that Na atoms have nearly 23 times the mass of H atoms, while Mg atoms are about 24 times heavier than H atoms.

2–3 The Mole

Attempts to measure the masses of individual atoms showed that these are very small indeed. Atoms are far too small to be seen or weighed individually;

the smallest bit of matter that can be reliably measured contains an enormous number of atoms. Therefore, we must deal with large numbers of atoms in any real situation, and some unit for describing conveniently a large number of atoms is desirable. The idea of a unit to describe a particular number of objects has been around for a long time. We are familiar with the dozen (12 items) and the gross (144 items); some items are grouped in tens or hundreds (think of decades, centuries, and millenia).

The unit chosen for the SI, and used universally by chemists and other scientists, is the **mole.** It is *defined* as the amount of substance that contains as many elementary entities (atoms, molecules, or other particles) as there are atoms in 0.012 kg of the pure carbon-12 isotope. As we noted in Table 1–4, the mole is a fundamental unit of the SI, and is abbreviated "mol." Many experiments, using various techniques, have refined the number described in the definition, and the currently accepted value is

$$1 \text{ mole} = 6.022045 \times 10^{23} \text{ particles}$$

This number, often rounded off to 6.022×10^{23}, is called **Avogadro's number** in honor of Amedeo Avogadro (1776–1856), whose contributions to chemistry will be discussed in Section 8–9.

Mole is derived from the Latin word *moles,* which means "a mass." Molecule is the diminutive form of this word, and means "a small mass."

According to its definition, the mole unit refers to a fixed number of "elementary entities," whose identities must be specified. Just as we may speak of a dozen eggs or a dozen automobiles, we refer to a mole of atoms or a mole of molecules (or a mole of ions, electrons, or other particles). We could even think about a mole of eggs, although the size of the required carton staggers the imagination! Helium exists as discrete He atoms, so one mole of helium consists of 6.022×10^{23} He *atoms.* Hydrogen commonly exists as diatomic (two-atom) molecules, so one mole of hydrogen contains 6.022×10^{23} H_2 *molecules* and $2(6.022 \times 10^{23})$ H atoms.

Every kind of atom, molecule, and ion has a definite characteristic mass. It follows that one mole of a given substance also has a definite mass, regardless of the source of the sample. This idea is of central importance in many calculations throughout the study of chemistry and the related sciences.

Because the mole is defined as the number of atoms in 0.012 kg (or 12 grams) of carbon-12, and the atomic mass unit is defined as 1/12 of the mass of a carbon-12 atom, the following convenient relationship is true: *The mass of one mole of atoms of a pure element in grams is numerically equal to the atomic weight of that element in amu.* For instance, if you obtain a pure sample of the metallic element titanium (Ti), whose atomic weight is 47.90 amu, and measure out 47.90 grams of it, you will have one mole or 6.022×10^{23} atoms of titanium.

The atomic weights of the elements are listed in the table inside the front cover.

The symbol for an element can (1) identify the element, (2) represent one atom of the element, or, as we shall see, (3) represent one mole of atoms of the element. The last interpretation will be extremely useful in calculations involving masses of reacting substances, introduced later in this chapter and used freely in our later work.

A quantity of a substance may be expressed in a variety of ways. For example, consider a barrel containing a dozen ten-kilogram cannon balls and a beaker containing 55.847 grams of iron filings, or one mole of iron. We can express the amount of cannon balls or iron filings present in any of several

The atomic weight of iron (Fe) is 55.847 amu.

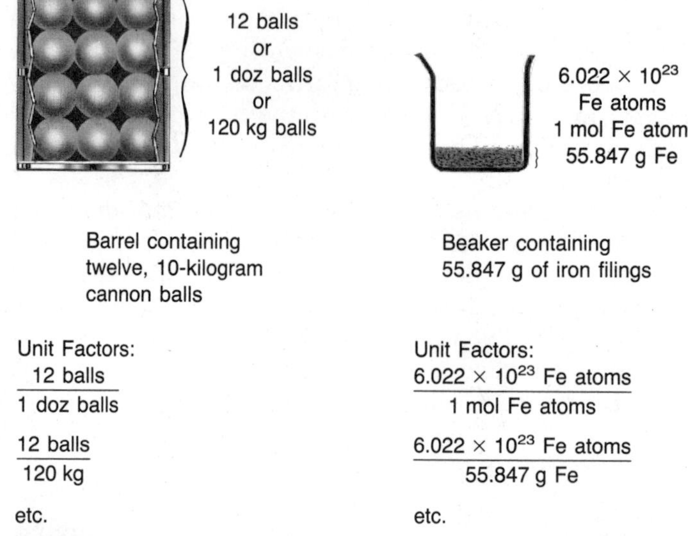

FIGURE 2–2 Representation of amounts of substances in three different ways.

different units, and we can then construct unit factors to relate an amount of the substance expressed in one kind of unit to the same amount expressed in another unit. This is illustrated in Figure 2–2.

As Table 2–1 suggests, the concept of a mole as applied to atoms is especially useful because it provides a convenient basis for comparing equal numbers of atoms of different elements.

TABLE 2–1 Mass of One Mole of Atoms of Some Common Elements

Element	A Sample with a Mass of	Contains
Carbon	12.011 g C	6.022×10^{23} C atoms or 1 mole of C atoms
Calcium	40.08 g Ca	6.022×10^{23} Ca atoms or 1 mole of Ca atoms
Titanium	47.90 g Ti	6.022×10^{23} Ti atoms or 1 mole of Ti atoms
Gold	196.9665 g Au	6.022×10^{23} Au atoms or 1 mole of Au atoms
Hydrogen	1.0079 g H_2	6.022×10^{23} H atoms or 1 mole of H atoms (3.011×10^{23} H_2 molecules or 0.5 mole of H_2 molecules)
Nitrogen	14.0067 g N_2	6.022×10^{23} N atoms or 1 mole of N atoms (3.011×10^{23} N_2 molecules or 0.5 mole of N_2 molecules)
Sulfur	32.06 g S_8	6.022×10^{23} S atoms or 1 mole of S atoms (0.7528×10^{23} S_8 molecules or 0.1250 mole of S_8 molecules)

Figure 2–3 shows what one mole of atoms of each of some common elements looks like. Each of the examples in Figure 2–3 represents 6.022×10^{23} *atoms* of the element.

The relationship between the mass of a sample of an element and the number of moles of atoms in the sample is illustrated in Example 2–1.

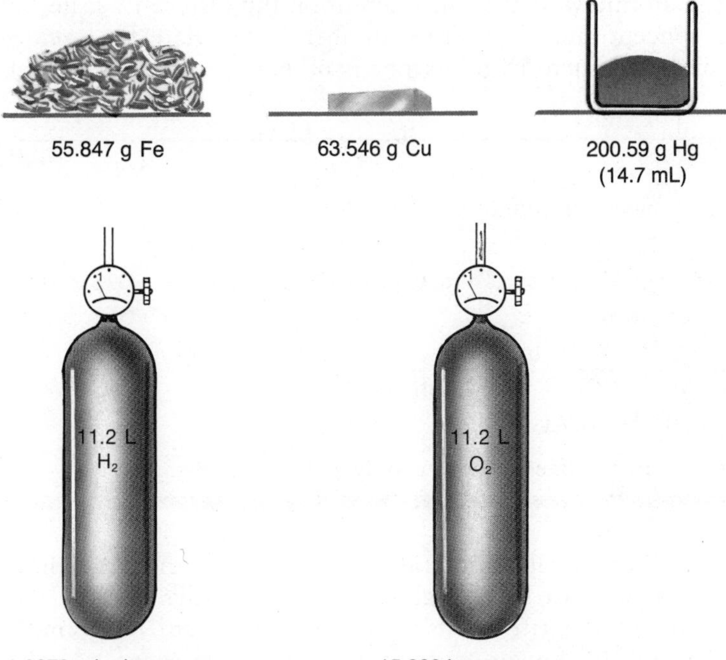

FIGURE 2–3 Representation of one mole of atoms of some common elements.

EXAMPLE 2–1

How many moles of atoms does 245.2 g of zinc metal contain?

Solution

Since the atomic weight of zinc is 65.38 amu, we know that one mole of zinc atoms is 65.38 grams of zinc.

$$\underline{?} \text{ mol Zn atoms} = 245.2 \text{ g Zn} \times \frac{1 \text{ mol Zn atoms}}{65.38 \text{ g Zn}} = \underline{3.750 \text{ mol Zn atoms}}$$

Once the number of moles of atoms of an element is known, the actual number of atoms in the sample can be calculated, as Example 2–2 illustrates.

EXAMPLE 2–2

How many atoms are contained in 0.125 mol of zinc atoms?

Solution

One mole of zinc atoms contains 6.022×10^{23} atoms

$$\underline{?} \text{ Zn atoms} = 0.125 \text{ mol Zn atoms} \times \frac{6.022 \times 10^{23} \text{ Zn atoms}}{1 \text{ mol Zn atoms}}$$

$$= \underline{7.53 \times 10^{22} \text{ Zn atoms}}$$

If we know the atomic weight of an element on the carbon-12 scale, we can use the mole concept and Avogadro's number to calculate the average mass of one atom of that element in grams (or any other mass unit we choose).

EXAMPLE 2–3

Calculate the mass of one aluminum atom in grams.

Solution

We know that one mole of Al atoms has a mass of 26.98 g and contains 6.022×10^{23} Al atoms. Hence,

$$\frac{?\text{ g Al}}{\text{Al atom}} = \frac{29.68\text{ g Al}}{1\text{ mol Al}} \times \frac{1\text{ mol Al}}{6.022 \times 10^{23}\text{ Al atoms}}$$

$$= \underline{4.480 \times 10^{-23}\text{ g Al/Al atom}}$$

Thus, we see that the mass of one Al atom is only 4.480×10^{-23} gram.

To gain some appreciation of how little this is, write 4.480×10^{-23} gram as a decimal fraction and try to name the fraction.

To illustrate just how small atoms are, let's calculate the number of nickel atoms in a micrometeorite that weighs only one billionth of a gram. The radius of a spherical micrometeorite of this mass is 3.0×10^{-4} cm. Micrometeorites of this size constantly shower down on the earth from space.

EXAMPLE 2–4

Calculate the number of atoms in one billionth of a gram of nickel metal to two significant figures.

Solution

$$\underline{?}\text{ Ni atoms} = 1.0 \times 10^{-9}\text{ g Ni} \times \frac{1\text{ mol Ni}}{58.70\text{ g Ni}} \times \frac{6.022 \times 10^{23}\text{ Ni atoms}}{1\text{ mol Ni}}$$

$$= \underline{1.0 \times 10^{13}\text{ Ni atoms}}$$

One billionth of a gram of nickel metal contains approximately 10,000,000,000,000 Ni atoms.

These two examples demonstrate how small atoms are and why we find it necessary to use large numbers of atoms in practical work.

2–4 Formula Weights, Molecular Weights, and Moles

The **formula weight** of a substance is the sum of the atomic weights of the elements in the formula, each taken the number of times the element occurs. That is, it is the mass in amu of one formula unit. Formula weights, like the atomic weights on which they are based, are relative masses expressed in amu. The formula weight (*rounded to the nearest amu*) for sodium hydroxide, NaOH, is found as follows.

	No. of Atoms of Stated Kind	Mass of One Atom
$1 \times Na =$	1	$\times$ 23 amu
$1 \times H =$	1	$\times$ 1 amu
$1 \times O =$	1	$\times$ 16 amu
	Formula weight =	40 amu

The approximate formula weight for phosphoric acid, H_3PO_4, is

	No. of Atoms of Stated Kind	Mass of One Atom	
$3 \times H =$	3	$\times$ 1 amu	= 3 amu
$1 \times P =$	1	$\times$ 31 amu	= 31 amu
$4 \times O =$	4	$\times$ 16 amu	= 64 amu
		Formula weight =	98 amu

The term **molecular weight** is used interchangeably with formula weight when reference is made to molecular (nonionic) substances, i.e., substances that exist as discrete molecules.

To illustrate how full precision arithmetic can be used to obtain a formula weight, let's calculate the formula weight for ammonium carbonate, $(NH_4)_2CO_3$. The most precisely known values for atomic weights are used and the rules for significant figures (Section 1–9) are applied.

	No. of Atoms of Stated Kind	Mass of One Atom	
$2 \times N =$	2	$\times$ 14.0067 amu	= 28.0134 amu
$8 \times H =$	8	$\times$ 1.0079 amu	= 8.0632 amu
$1 \times C =$	1	$\times$ 12.011 amu	= 12.011 amu
$3 \times O =$	3	$\times$ 15.9994 amu	= 47.9982 amu
		Formula weight =	96.086 amu

Numerous experiments have shown the mass of one mole (6.022×10^{23} formula units) of a substance, expressed in grams, is numerically equal to its formula weight, and indeed the number 6.022×10^{23} was determined just this way. One mole of sodium hydroxide is 40 g of NaOH, one mole of phosphoric acid is 98 g of H_3PO_4, and one mole of ammonium carbonate is 96 g of $(NH_4)_2CO_3$.

One mole of any molecular substance contains 6.022×10^{23} molecules of the substance, as Table 2–2 illustrates. Compare the entries for hydrogen and nitrogen with those in Table 2–1.

The physical appearance of one mole of each of some molecular substances is illustrated in Figure 2–4. Also demonstrated in this figure are two facts of general interest. First, the volumes of one mole of oxygen (O_2 molecules) and one mole of nitrogen (N_2 molecules)—and, in general, of one mole of any gas—under identical conditions are approximately the same. At 0°C and one atmosphere pressure, one mole of a gas occupies about 22.4 liters, as will be explained in Chapter 8. Second, Figure 2–4 shows two different forms of oxalic acid. The formula unit (molecule) of oxalic acid is $(COOH)_2$, with a formula weight of about 90 amu. When oxalic acid is obtained by crystalliza-

TABLE 2–2 One Mole of Some Common Molecular Substances

Substance	Mass of 1 mole	Contains
Hydrogen	2.016 g H_2	6.022×10^{23} H_2 molecules
Nitrogen	28.01 g N_2	6.022×10^{23} N_2 molecules
Oxygen	32.00 g O_2	6.022×10^{23} O_2 molecules
Methane	16.04 g CH_4	6.022×10^{23} CH_4 molecules
Oxalic acid	90.04 g $(COOH)_2$	6.022×10^{23} $(COOH)_2$ molecules
Propane	44.10 g C_3H_8	6.022×10^{23} C_3H_8 molecules
Nitroglycerine	227.1 g $C_3H_5N_3O_9$	6.022×10^{23} $C_3H_5N_3O_9$ molecules

tion from an aqueous solution, however, two molecules of water are bound to every molecule of oxalic acid and remain trapped in the crystal when it appears dry. The formula of this **hydrate** is $(COOH)_2 \cdot 2H_2O$, where the dot serves to show that the water is "attached" to the molecule. The formula weight of the hydrate is about 126 amu. The water can be driven out of the crystal by heating to leave the **anhydrous** oxalic acid. Anhydrous means "without water." This behavior occurs for many compounds.

Hydrates are solid compounds that contain H_2O molecules in their crystal structures.

Ionic compounds such as sodium chloride (Na^+Cl^-) exist as neutral combinations of ions, rather than as discrete molecules. There are no simple NaCl molecules at ordinary temperatures and pressures and, thus, it is inappropriate to refer to the "molecular weight" of NaCl. One mole of an ionic compound contains 6.02×10^{23} **formula units** of the substance. Recall that one formula unit of sodium chloride consists of one sodium ion, Na^+, and one chloride ion, Cl^-. One mole of NaCl, or 58.4 grams, contains 6.02×10^{23} Na^+ ions and 6.02×10^{23} Cl^- ions. See Table 2–3.

When fewer than four significant figures are used in calculations, Avogadro's number is rounded off to 6.02×10^{23}.

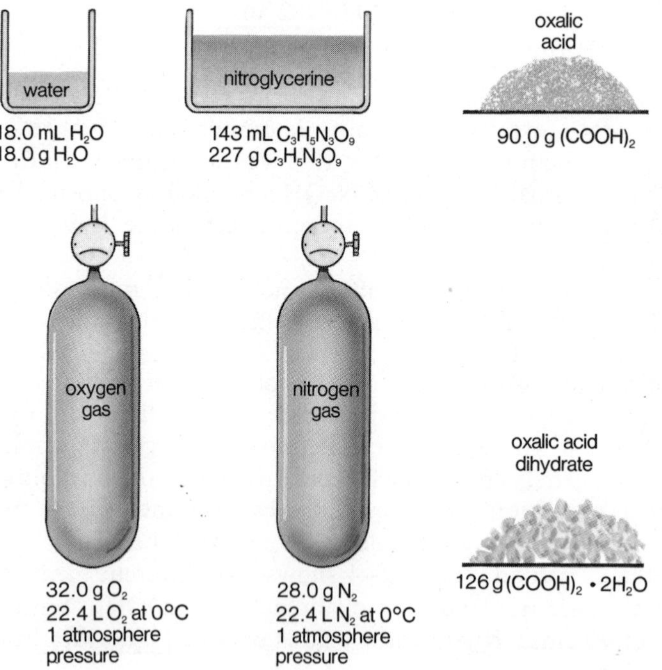

FIGURE 2–4 Representations of one mole of some molecular substances.

TABLE 2–3 One Mole of Some Ionic Compounds

Compound	Mass of 1 mole	Contains
Sodium chloride	58.4 g $NaCl$	6.02×10^{23} Na^+ ions 6.02×10^{23} Cl^- ions
Calcium chloride	111 g $CaCl_2$	6.02×10^{23} Ca^{2+} ions $2(6.02 \times 10^{23})$ Cl^- ions
Aluminum sulfate	342 g $Al_2(SO_4)_3$	$2(6.02 \times 10^{23})$ Al^{3+} ions $3(6.02 \times 10^{23})$ $SO_4{}^{2-}$ ions

The following examples show the relations between numbers of molecules, atoms, or formula units and their masses.

EXAMPLE 2–5

What is the mass in grams of 10 billion SO_2 molecules?

Solution

One mole of SO_2 contains 6.02×10^{23} SO_2 molecules and has a mass of 64 grams.

$$\underline{?}\text{ g } SO_2 = 1.0 \times 10^{10} \text{ } SO_2 \text{ molecules} \times \frac{64 \text{ g } SO_2}{6.02 \times 10^{23} \text{ } SO_2 \text{ molecules}}$$

$$= \underline{1.1 \times 10^{-12} \text{ g } SO_2}$$

It may be of interest to note that the most commonly used analytical balances are capable of weighing to ±0.0001 gram. Ten billion SO_2 molecules have a mass of only 0.0000000000011 gram.

EXAMPLE 2–6

How many (a) moles of O_2, (b) O_2 molecules, and (c) O atoms are contained in 40.0 grams of gaseous oxygen at 25°C?

Solution

One mole of O_2 contains 6.02×10^{23} O_2 molecules and its mass is 32.0 g.

$$\text{(a) } \underline{?}\text{ mol } O_2 = 40.0 \text{ g } O_2 \times \frac{1 \text{ mol } O_2}{32.0 \text{ g } O_2} = \underline{1.25 \text{ mol } O_2}$$

$$\text{(b) } \underline{?}\text{ } O_2 \text{ molecules} = 40.0 \text{ g } O_2 \times \frac{6.02 \times 10^{23} \text{ } O_2 \text{ molecules}}{32.0 \text{ g } O_2}$$

$$= \underline{7.52 \times 10^{23} \text{ molecules}}$$

$$\text{(c) } \underline{?}\text{ O atoms} = 40.0 \text{ g } O_2 \times \frac{6.02 \times 10^{23} \text{ } O_2 \text{ molecules}}{32.0 \text{ g } O_2} \times \frac{2 \text{ O atoms}}{1 \text{ } O_2 \text{ molecule}}$$

$$= \underline{1.501 \times 10^{24} \text{ O atoms}}$$

Or, we can use the number of moles of O_2 calculated in (a) to find the number of O_2 molecules

$$\underline{?}\text{ } O_2 \text{ molecules} = 1.25 \text{ mol } O_2 \times \frac{6.02 \times 10^{23} \text{ } O_2 \text{ molecules}}{1 \text{ mol } O_2}$$

$$= \underline{7.52 \times 10^{23} \text{ } O_2 \text{ molecules}}$$

EXAMPLE 2–7

Calculate the number of hydrogen atoms in 39.6 grams of ammonium sulfate, $(NH_4)_2SO_4$.

Solution

One mole of $(NH_4)_2SO_4$ is 6.02×10^{23} formula units and has a mass of 132 g.

We relate (1) g to mol, (2) mol to formula units, and (3) formula units to H atoms.

$$\underline{?}\text{ H atoms} = 39.6\text{ g }(NH_4)_2SO_4 \times \frac{1\text{ mol }(NH_4)_2SO_4}{132\text{ g }(NH_4)_2SO_4} \times \frac{6.02 \times 10^{23}\text{ formula units }(NH_4)_2SO_4}{1\text{ mol }(NH_4)_2SO_4} \times \frac{8\text{ H atoms}}{1\text{ formula unit }(NH_4)_2SO_4}$$

$$= \underline{1.44 \times 10^{24}\text{ H atoms}}$$

The term millimole (mmol) is useful in laboratory work. As the prefix indicates, one **mmol** is 1/1000 of a mole. Also, small masses are frequently expressed in milligrams (mg) rather than grams. The relation between millimoles and milligrams is the same as that between moles and grams, as illustrated in Table 2–4.

TABLE 2–4 Comparison of Moles and Millimoles

Compound	1 mole	1 millimole
NaOH	40 g	40 mg or 0.040 g
H_3PO_4	98 g	98 mg or 0.098 g
SO_2	64 g	64 mg or 0.064 g
C_3H_8	44 g	44 mg or 0.044 g

EXAMPLE 2–8

Calculate the number of millimoles of sulfuric acid in 0.147 gram of H_2SO_4.

Solution

1 mol H_2SO_4 = 98 g H_2SO_4; 1 mmol H_2SO_4 = 98 mg H_2SO_4, or 0.098 g H_2SO_4

Method 1: Convert g H_2SO_4 to mmol H_2SO_4.

$$\underline{?}\text{ mmol }H_2SO_4 = 0.147\text{ g }H_2SO_4 \times \frac{1\text{ mmol }H_2SO_4}{0.098\text{ g }H_2SO_4} = \underline{1.5\text{ mmol }H_2SO_4}$$

Or, using 0.147 g H_2SO_4 = 147 mg H_2SO_4, we have

Method 2: Convert mg H_2SO_4 to mmol H_2SO_4.

$$\underline{?}\text{ mmol }H_2SO_4 = 147\text{ mg }H_2SO_4 \times \frac{1\text{ mmol }H_2SO_4}{98\text{ mg }H_2SO_4} = \underline{1.5\text{ mmol }H_2SO_4}$$

2–5 Percent Composition and Formulas of Compounds

If the formula of a compound is known, its chemical composition can be expressed as the mass percent of each element in the compound. (As we will see in the next section, this procedure can also be reversed.) For example, one methane molecule, CH_4, contains one C atom and four H atoms. Therefore, one mole of CH_4 (16.0 g) contains one mole of carbon atoms (12.0 g) and four moles of hydrogen atoms (4.0 g). Since percentage is the part divided by the whole times 100 percent (or simply parts per 100), we can represent the percent composition of methane as:

$$\%\ C = \frac{\overset{\text{mass of one mol of C atoms}}{C}}{\underset{\text{mass of one mol of CH}_4\text{ molecules}}{CH_4}} \times 100\% = \frac{12\ g}{16\ g} \times 100\% = 75\%\ C$$

$$\%\ H = \frac{\overset{\text{mass of four mol of H atoms}}{4 \times H}}{\underset{\text{mass of one mol of CH}_4\text{ molecules}}{CH_4}} \times 100\% = \frac{4 \times 1.0\ g}{16\ g} \times 100\% = 25\%\ H$$

We have calculated that methane is 75% carbon and 25% hydrogen by mass. Strictly speaking, percentages must always add to 100%, although on occasion round-off errors may not cancel and totals such as 99.9% or 100.1% may be obtained.

EXAMPLE 2–9

Calculate the percent composition of HNO_3.

When chemists use the % notation, they mean % by mass unless otherwise specified.

Solution

The mass of one mole of HNO_3 is calculated first.

	No. of Mol of Stated Kind		Mass of One Mol	
1 × H =	1	×	1.0 g	= 1.0 g
1 × N =	1	×	14.0 g	= 14.0 g
3 × O =	3	×	16.0 g	= 48.0 g
			Mass of 1 mol of HNO_3	= 63.0 g

Now, its percent composition is

$$\%\ H = \frac{H}{HNO_3} \times 100\% = \frac{1.0\ g}{63.0\ g} \times 100\% = 1.6\%\ H$$

$$\%\ N = \frac{N}{HNO_3} \times 100\% = \frac{14.0\ g}{63.0\ g} \times 100\% = 22.2\%\ N$$

$$\%\ O = \frac{3 \times O}{HNO_3} \times 100\% = \frac{48.0\ g}{63.0\ g} \times 100\% = 76.2\%\ O$$

Total = 100.0%

We have calculated that nitric acid is 1.6% H, 22.2% N, and 76.2% O by mass. All samples of pure HNO_3 have this composition, according to the Law of Definite Proportions.

Nonstoichiometric Compounds

The Law of Definite Proportions (Constant Composition) is one of the basic tenets of chemistry, and its validity has been demonstrated for many thousands of compounds. However, there are many *solid* compounds that do not obey the "law." They are called *nonstoichiometric compounds* or "Berthollides" in honor of a French chemist, Claude Berthollet, who argued (prior to 1808) that the elemental composition of a compound could vary over wide limits depending on how it was prepared. Berthollet was refuted by Joseph Proust, who showed that nearly all compounds *do* have definite composition, and that Berthollet's "compounds" were actually mixtures. Many other compounds, however, including metal oxides, sulfides, and hydrides, are indeed nonstoichiometric compounds. Typical examples include $Ni_{0.97}O$, $TiO_{1.7-1.8}$, $FeO_{1.055}$, $Cu_{1.7}S$, $CeH_{2.69}$, and $VH_{0.56}$. Many minerals contain nonstoichiometric compounds. A number of these compounds are *superconductors*, substances that conduct electricity very efficiently at low temperatures.

As we shall see when we discuss the composition of compounds in some detail, two (and sometimes more) elements frequently form more than one compound. The **Law of Multiple Proportions** summarizes numerous observations on such compounds. It is usually stated: *When two elements, A and B, form more than one compound, the masses of element B that combine with a given mass of element A can be expressed by small whole number ratios.* Water, H_2O, and hydrogen peroxide, H_2O_2, provide an example. The ratio of masses of oxygen that combine with a given mass of hydrogen is 2:1 in H_2O_2 and H_2O. Numerous similar examples such as CO and CO_2 and SO_2 and SO_3 can be cited.

2–6 Derivation of Formulas from Elemental Composition

Each year thousands of new compounds are made in laboratories or discovered in nature. One of the first steps in characterizing a new compound is the determination of its percent composition. A *qualitative* analysis is performed to determine which elements are present in the compound. Then a *quantitative* analysis is performed to determine the percent by mass of each element.

Carbon, hydrogen, and oxygen are found in millions of compounds, and C and H analyses are conveniently performed in a C–H combustion system (Figure 2–5).

A sample of a compound of accurately known mass is burned in a furnace in a stream of oxygen so that all the carbon and hydrogen in the sample are converted to carbon dioxide and water, respectively. The resulting increases in masses in the CO_2 and H_2O absorbers can then be related to the masses and percentages of carbon and hydrogen in the original sample. Example 2–10 illustrates the procedure.

EXAMPLE 2–10

A 0.1014-gram sample of purified glucose was burned in a C–H combustion train to

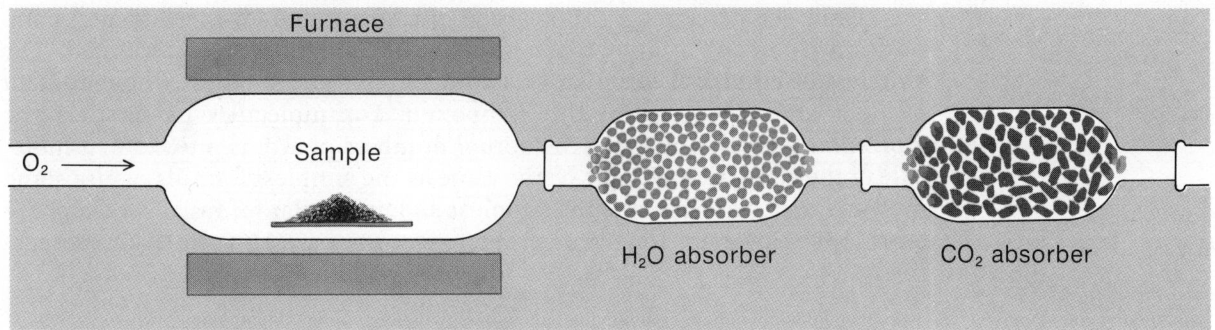

FIGURE 2–5 Combustion train used for carbon–hydrogen analysis. The absorbent for water is magnesium perchlorate, $Mg(ClO_4)_2$. Carbon dioxide is absorbed by finely divided sodium hydroxide supported on asbestos. Only a few milligrams of sample are needed for an analysis.

produce 0.1486 gram of CO_2 and 0.0609 gram of H_2O. An elemental analysis showed that glucose contains only carbon, hydrogen, and oxygen. Determine the masses of C, H, and O in the sample and the percentages of these elements in glucose.

Glucose, a simple sugar, is one of the products of carbohydrate metabolism and also the main component of intravenous feeding liquids.

Solution

We first calculate the mass of carbon that was converted to 0.1486 gram of CO_2. There is one mole of carbon atoms, 12.01 grams, in every mole of CO_2, 44.01 grams.

$$\underline{?}\ \text{g C} = 0.1486\ \text{g CO}_2 \times \frac{12.01\ \text{g C}}{44.01\ \text{g CO}_2} = \underline{0.0406\ \text{g C}}$$

Likewise, we can calculate the amount of hydrogen in the original sample from the fact that there are two moles of hydrogen atoms, 2.016 grams, per mole of H_2O, 18.02 grams.

$$\underline{?}\ \text{g H} = 0.0609\ \text{g H}_2\text{O} \times \frac{2.016\ \text{g H}}{18.02\ \text{g H}_2\text{O}} = \underline{0.00681\ \text{g H}}$$

The mass of oxygen in the sample is calculated by difference, since glucose has been shown to contain only C, H, and O.

$$\underline{?}\ \text{g O} = 0.1014\ \text{g sample} - [0.0406\ \text{g C} + 0.00681\ \text{g H}] = \underline{0.0540\ \text{g O}}$$

The percentages by mass are calculated using the relationship

$$\%\ \text{element} = \frac{\text{g element}}{\text{g sample}} \times 100\%$$

for each element in turn:

$$\%\ \text{C} = \frac{0.0406\ \text{g C}}{0.1014\ \text{g}} \times 100\% = \underline{40.0\%\ \text{C}}$$

$$\%\ \text{H} = \frac{0.00681\ \text{g H}}{0.1014\ \text{g}} \times 100\% = \underline{6.72\%\ \text{H}}$$

$$\%\ \text{O} = \frac{0.0540\ \text{g O}}{0.1014\ \text{g}} \times 100\% = \underline{53.3\%\ \text{O}}$$

Total = 100.0%

Once the percentage composition of a compound (or its elemental composition by mass) is known the simplest formula can be determined. The **simplest** or **empirical formula** for a compound is the smallest whole-number ratio of atoms present in the compound. For molecular compounds the **molecular formula** indicates the *actual* numbers of atoms present in a molecule of the compound. It may be the same as the simplest formula or else some multiple of it. For example, the simplest and molecular formulas for water are both H_2O. However, for hydrogen peroxide, they are HO and H_2O_2, respectively.

EXAMPLE 2–11

Analysis of a sample of pure compound reveals that it contains 50.1% sulfur and 49.9% oxygen by mass. What is the simplest formula?

Solution

The ratio of moles of atoms of elements in a compound is the same as the ratio of atoms in that compound, because one mole of atoms of any element is 6.02×10^{23} atoms of the element.* If we calculate the number of moles of atoms of each element in a sample of a compound, we can obtain the ratio that gives the relative number of atoms in the compound. Let's consider 100.0 grams of the compound, which must contain 50.1 grams of S and 49.9 grams of O. We shall calculate the number of moles of atoms of each.

$$\underline{?} \text{ mol S atoms} = 50.1 \text{ g S} \times \frac{1 \text{ mol S atoms}}{32.1 \text{ g S}} = \underline{1.56 \text{ mol S atoms}}$$

$$\underline{?} \text{ mol O atoms} = 49.9 \text{ g O} \times \frac{1 \text{ mol O atoms}}{16.0 \text{ g O}} = \underline{3.12 \text{ mol O atoms}}$$

A simple and useful way to obtain whole number ratios among several numbers is: (1) divide each number by the smallest, and then (2) multiply the resulting numbers by the smallest whole number that will eliminate fractions, if necessary.

Now that we know that 100.0 grams of compound contains 1.56 moles of S atoms and 3.12 moles of O atoms, let's obtain a whole number ratio between these numbers that gives the ratio of atoms in the simplest formula.

$$\frac{1.56}{1.56} = 1 \text{ S} \qquad \frac{3.12}{1.56} = 2 \text{ O} \qquad \Rightarrow \underline{SO_2}$$

* This statement is analogous to the following example. The ratio of dozens of brown eggs to dozens of white eggs in a box is the same as the ratio of brown eggs to white eggs. Suppose you have a box of eggs containing 50 dozen brown eggs and 25 dozen white eggs. The ratio of brown eggs to white eggs is

$$\frac{50 \text{ doz brown eggs}}{25 \text{ doz white eggs}} = \frac{2 \text{ brown eggs}}{1 \text{ white egg}}$$

or a simple 2:1 ratio. Similarly, if a compound contains two moles of atoms of element A and one mole of atoms of element B, the ratio of A atoms to B atoms is

$$\frac{2 \text{ moles A atoms}}{1 \text{ mole B atoms}} = \frac{2(6.02 \times 10^{23} \text{ A}) \text{ atoms}}{1(6.02 \times 10^{23} \text{ B}) \text{ atoms}} = \frac{2 \text{ A atoms}}{1 \text{ B atom}}$$

This ratio 1:2 tells us that the simplest formula for the compound is SO_2. The solution of the above problem is outlined below.

Element	Percentage	Relative number of moles of atoms	Divided by smaller number	Smallest whole number ratio of atoms
Sulfur	50.1%	$\frac{50.1}{32.1} = 1.56$	$\frac{1.56}{1.56} = 1$	1 → SO_2
Oxygen	49.9%	$\frac{49.9}{16.0} = 3.12$	$\frac{3.12}{1.56} = 2$	2 → SO_2

EXAMPLE 2–12

A 20.000-gram sample of an ionic compound is found to contain 6.072 grams of Na, 8.474 grams of S, and 6.336 grams of O. What is its simplest formula?

Solution

As before, let's calculate the number of moles of atoms of each element

$$\underline{?} \text{ mol Na atoms} = 6.072 \text{ g Na} \times \frac{1 \text{ mol Na atoms}}{23.0 \text{ g Na}} = \underline{0.264 \text{ mol Na atoms}}$$

$$\underline{?} \text{ mol S atoms} = 8.474 \text{ g S} \times \frac{1 \text{ mol S atoms}}{32.1 \text{ g S}} = \underline{0.264 \text{ mol S atoms}}$$

$$\underline{?} \text{ mol O atoms} = 6.336 \text{ g O} \times \frac{1 \text{ mol O atoms}}{16.0 \text{ g O}} = \underline{0.396 \text{ mol O atoms}}$$

We now divide each number by the smallest number to obtain the ratio of atoms in the simplest formula.

$$\frac{0.264}{0.264} = \underline{1.00 \text{ Na}}; \quad \frac{0.264}{0.264} = \underline{1.00 \text{ S}}; \quad \frac{0.396}{0.264} = \underline{1.50 \text{ O}}$$

The ratio of atoms in the simplest formula *must be a whole number ratio* (by definition). To convert this ratio, 1:1:1.5, to a whole number ratio, each number in the ratio must be multiplied by 2, which gives the simplest formula

$1.00 \text{ Na} \times 2 = 2 \text{ Na}$

$1.00 \text{ S} \times 2 = 2 \text{ S}$ —— $\underline{Na_2S_2O_3}$

$1.50 \text{ O} \times 2 = 3 \text{ O}$

In this procedure we often obtain numbers like 0.99 and 1.01 or 1.49 and 1.52. Since there is always some error in results obtained by analysis of samples (as well as round-off errors), we *assume* that 0.99 and 1.01 are equal to 1.0 and that 1.49 and 1.52 are equal to 1.5.

The general procedure for obtaining *a whole number ratio* among numbers that contain a fraction is: (1) express any decimal fraction as a common fraction (for instance, $0.5 = \frac{1}{2}$), and then (2) multiply each number in the ratio by the *denominator of the common fraction.*

For many compounds the molecular formula is a multiple of the simplest formula. Consider butane, C_4H_{10}. The simplest formula for butane is C_2H_5, but the molecular (true) formula contains twice as many atoms, i.e., $(C_2H_5)_2 = C_4H_{10}$. Benzene, C_6H_6, is another example. The simplest formula for benzene is CH, but the molecular formula contains six times as many atoms, i.e., $(CH)_6 = C_6H_6$.

Percent composition data yield only simplest formulas. In order to determine the molecular formula for a molecular compound, *both* its simplest formula and its molecular weight must be known. Methods for experimental determination of molecular weights will be introduced in Chapter 8.

EXAMPLE 2–13

In Example 2–10 we found that glucose is 40.0% C, 6.72% H, and 53.3% O. Other experiments show that its molecular weight is approximately 180. Determine the simplest formula and the molecular formula of glucose.

Solution

For simplicity we can deal with 100.0 grams of glucose, which must contain 40.0 grams of C, 6.72 grams of H, and 53.3 grams of O. We first calculate the number of moles of atoms that these masses represent.

$$\underline{?}\text{ mol C atoms} = 40.0\text{ g C} \times \frac{1\text{ mol C atoms}}{12.01\text{ g C}} = 3.33\text{ mol C atoms}$$

$$\underline{?}\text{ mol H atoms} = 6.72\text{ g H} \times \frac{1\text{ mol H atoms}}{1.008\text{ g H}} = 6.66\text{ mol H atoms}$$

$$\underline{?}\text{ mol O atoms} = 53.3\text{ g O} \times \frac{1\text{ mol O atoms}}{16.00\text{ g O}} = 3.33\text{ mol O atoms}$$

We now obtain ratios among these numbers of moles of atoms by dividing each by the smallest number.

$$\frac{3.33}{3.33} = 1.00\text{ C}, \quad \frac{6.66}{3.33} = 2.00\text{ H}, \quad \frac{3.33}{3.33} = 1.00\text{ O}$$

The simplest formula is CH_2O, which has a formula weight of 30.02 amu. Since the molecular weight of glucose is approximately 180 amu, we can determine the molecular formula by dividing the molecular weight by the simplest formula weight

$$\frac{180\text{ amu}}{30.02\text{ amu}} = 6.00$$

Traces of impurities and round-off errors often result in numbers such as 5.96, 6.02, etc., which are not integers, but are very close to integers.

The molecular weight is six times the simplest formula weight, $(CH_2O)_6 = C_6H_{12}O_6$, so the molecular formula of glucose is $C_6H_{12}O_6$.

2–7 Chemical Equations

Chemical reactions always involve changing one or more substances into one or more different substances. That is, they involve regrouping atoms or ions to form other substances.

Chemical equations *are used to describe chemical reactions,* and they show (1) *the substances that react,* called **reactants,** (2) *the substances formed,* called **products,** and (3) *the relative amounts of the substances involved.* As a typical example, let's consider the combustion (burning) of natural gas, a reaction used to heat buildings and cook foods. Natural gas is a mixture of several substances, but the principal component is methane, CH_4. The equation that describes the reaction of methane with oxygen is

$$\underbrace{CH_4 + 2O_2}_{\text{reactants}} \longrightarrow \underbrace{CO_2 + 2H_2O}_{\text{products}}$$

The arrow may be read "yields."

What does this equation tell us? In the simplest terms, it tells us that methane reacts with oxygen to produce carbon dioxide, CO_2, and water. More specifically, it says that *one* molecule of methane reacts with *two* molecules of oxygen to produce *one* molecule of carbon dioxide and *two* molecules of water. That is,

A balanced chemical equation may be interpreted on a *molecular* basis.

CH_4	+	$2O_2$	$\longrightarrow$	CO_2	+	$2H_2O$
1 molecule		2 molecules		1 molecule		2 molecules

Figure 2–6 shows the rearrangement of atoms described by this equation.

CH_4 + $2\ O_2$ → CO_2 + $2\ H_2O$

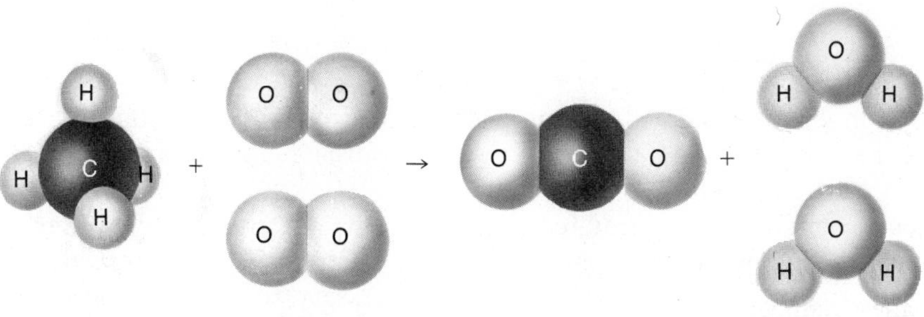

FIGURE 2–6 Representation of the reaction of methane with oxygen to form carbon dioxide and water.

As we pointed out in Section 1–1, numerous experiments have shown that *there is no detectable change in the quantity of matter during an ordinary chemical reaction.* This guiding principle, known as the **Law of Conservation of Matter,** provides the basis for "balancing" chemical equations and for calculations based on those equations. Since matter is neither created nor destroyed during a chemical reaction, a balanced chemical equation must always include the same number of each kind of atom on both sides of the equation. Since we are not dealing with a mathematical identity, an equal sign is inappropriate; instead we use an arrow (read "yields") to indicate the conversion of reactants into products.

For example, hydrogen and oxygen react to form water. This is a simple statement of fact based on numerous observations. It can be represented by the *unbalanced* "equation" that shows the formulas of the substances involved:

$$H_2 + O_2 \longrightarrow H_2O$$

Notice that substances are represented by formulas that describe them *as they exist.* We write H_2 to represent diatomic hydrogen molecules, not H, which represents hydrogen atoms. But as it now stands, the "equation" does not satisfy the Law of Conservation of Matter because there are two oxygen atoms in the O_2 molecule (left side) and only one oxygen atom in the water molecule (right side).

$$H_2 + O_2 \longrightarrow H_2O$$

Tally:

atoms of	in reactants	in products
H	2	2
O	2	1

In order to account for both oxygen atoms, *two* water molecules must be formed. *Two* H_2O molecules require two H_2 molecules to provide four hydrogen atoms. Thus, the equation is balanced by placing a 2 before the formulas of hydrogen and water. (Numbers placed before formulas are called **coefficients,** and they indicate the number of formula units of that substance involved in the reaction.) The properly balanced equation then becomes

$$2H_2 + O_2 \xrightarrow{\Delta} 2H_2O$$

Tally:

atoms of	in reactants	in products
H	4	4
O	2	2

The equation is now balanced; that is, it conforms to the Law of Conservation of Matter because there are *equal numbers* of hydrogen and oxygen atoms on both sides of the equation.

Special conditions required for some reactions are indicated by notation placed over the arrow. In the example above, the capital Greek letter delta (Δ) placed over the arrow tells us that heat is necessary to start this reaction.

Formulas must be written correctly for all reactants and products before attempting to balance an equation. Once this is done, the subscripts in the formulas cannot be changed, because doing so would violate the Law of Definite Proportions. Formulas represent experimentally determined ratios of masses of elements, and therefore ratios of numbers of atoms, and cannot be changed. Another way to look at it is that different subscripts imply different compounds, so that the equation would no longer describe the same reaction.

Let's generate the balanced equation for the reaction of aluminum metal with hydrochloric acid (hydrogen chloride dissolved in water) to produce aluminum chloride and hydrogen. The procedure is similar to that used for the hydrogen–oxygen reaction, but is slightly more complicated. The unbalanced "equation" is

The equation is *not* balanced.

$$Al + HCl \longrightarrow AlCl_3 + H_2$$

Tally:

atoms of	in reactants	in products
Al	1	1
H	1	2
Cl	1	3

Note that H atoms and Cl atoms must occur in a 1:1 ratio on the left side, while Cl atoms occur in groups of 3 and H atoms occur in groups of 2 on the right side. The least common multiple of 3 and 2 is 6. To balance the equation, we place coefficients of 2 before $AlCl_3$ and 3 before H_2 so that there will be six atoms of hydrogen and six atoms of chlorine (a 1:1 ratio) on the right side

$Al + HCl \longrightarrow 2AlCl_3 + 3H_2$

The equation is still *not* balanced.

Tally:

atoms of	in reactants	in products
Al	1	2
H	1	6
Cl	1	6

Further inspection shows that the coefficient of Al should be 2 and that of HCl should be 6 in the balanced equation

$$2Al + 6HCl \longrightarrow 2AlCl_3 + 3H_2$$

aluminum, hydrochloric acid, aluminum chloride, hydrogen

The equation is now balanced.

Tally:

atoms of	in reactants	in products
Al	2	2
H	6	6
Cl	6	6

2–8 Calculations Based on Chemical Equations

As we indicated at the beginning of the chapter, chemical equations represent a very *precise* and *versatile* language. We are now ready to use them to calculate the *amounts* of substances involved in chemical reactions. Let us again consider the chemical equation for the combustion of methane, the major component of natural gas. The balanced chemical equation for that reaction was given as

$$CH_4 + 2O_2 \xrightarrow{\Delta} CO_2 + 2H_2O$$

On a qualitative basis, the equation tells us that methane, CH_4, reacts with oxygen to form carbon dioxide and water. On a quantitative basis, at the molecular level, the equation also says

$$CH_4 + 2O_2 \longrightarrow CO_2 + 2H_2O$$

1 molecule of methane, 2 molecules of oxygen, 1 molecule of carbon dioxide, 2 molecules of water

EXAMPLE 2–14

How many O_2 molecules are required to react with 47 CH_4 molecules according to the above equation?

Solution

The *balanced* equation tells us that *one* CH_4 molecule reacts with *two* O_2 molecules. We can construct two unit factors from this fact

$$\frac{1\ CH_4\ \text{molecule}}{2\ O_2\ \text{molecules}} \quad \text{and} \quad \frac{2\ O_2\ \text{molecules}}{1\ CH_4\ \text{molecule}}$$

These are unit factors for *this* reaction because the numerator and denominator are *chemically equivalent.* In other words, the numerator and the denominator represent the same amount of reaction.

$$\underline{?}O_2\ \text{molecules} = 47\ CH_4\ \text{molecules} \times \frac{2\ O_2\ \text{molecules}}{1\ CH_4\ \text{molecule}} = \underline{94\ O_2\ \text{molecules}}$$

Chemical equations also indicate the relative amounts of each reactant and product in a given chemical reaction. We demonstrated earlier that formulas can represent moles of substances. Suppose Avogadro's number of CH_4 molecules, rather than just one CH_4 molecule, undergo this reaction. Then the equation can be written as

CH_4	+	$2O_2$	$\xrightarrow{\Delta}$	CO_2	+	$2H_2O$
6.02×10^{23} molecules		$2(6.02 \times 10^{23}$ molecules)		6.02×10^{23} molecules		$2(6.02 \times 10^{23}$ molecules)
1 mol		2 mol		1 mol		2 mol

which tells us that *one* mole of methane reacts with *two* moles of oxygen to produce *one* mole of carbon dioxide and *two* moles of water. Since we know the masses one mole of each of these substances, we can also write

We have rounded molecular weights to the nearest whole number of grams here.

CH_4 +	$2O_2$	$\longrightarrow$	CO_2 +	$2H_2O$
1 mol	2 mol		1 mol	2 mol
16 g	2 (32 g)		44 g	2 (18 g)
16 g	64 g		44 g	36 g
80 g reactants			80 g products	

The equation now tells us that 16 grams of CH_4 reacts with 64 grams of O_2 to form 44 grams of CO_2 and 36 grams of H_2O, and that the Law of Conservation of Matter is satisfied. Chemical equations describe *reaction ratios, i.e., the mole ratios of reactants and products* as well as the relative masses of reactants and products, as Example 2–15 illustrates.

EXAMPLE 2–15

What mass of oxygen is required to react with 24 grams of CH_4?

Solution

Recall the balanced equation

CH_4 +	$2O_2$	$\longrightarrow$	CO_2 +	$2H_2O$
1 mol	2 mol		1 mol	2 mol
16 g	64 g		44 g	36 g

which shows that 16 grams of CH_4 reacts with 64 grams of O_2. Since these two quantities are chemically equivalent we can construct two unit factors:

$$\frac{16\ g\ CH_4}{64\ g\ O_2} \quad \text{and} \quad \frac{64\ g\ O_2}{16\ g\ CH_4},$$

and the solution to the problem is

$$\underline{?}\text{ g } O_2 = 24 \text{ g } CH_4 \times \frac{64 \text{ g } O_2}{16 \text{ g } CH_4} = \underline{96 \text{ g } O_2}$$

Another approach to the problem we have just solved is known as the **mole method.** In this method, the number of moles of reactant or product is calculated and then converted to grams (or other desired unit). Example 2–15 asked, "What mass of oxygen is required to react with 24 grams of CH_4?" The balanced equation and the calculation by the mole method follow.

$$\begin{array}{cccc} CH_4 + & 2O_2 & \longrightarrow CO_2 + & 2H_2O \\ 1 \text{ mol} & 2 \text{ mol} & 1 \text{ mol} & 2 \text{ mol} \end{array}$$

$$\underline{?}\text{ mol } CH_4 = 24 \text{ g } CH_4 \times \frac{1 \text{ mol } CH_4}{16 \text{ g } CH_4} = \underline{1.5 \text{ mol } CH_4}$$

$$\underline{?}\text{ mol } O_2 = 1.5 \text{ mol } CH_4 \times \frac{2 \text{ mol } O_2}{1 \text{ mol } CH_4} = \underline{3.0 \text{ mol } O_2}$$

$$\underline{?}\text{ g } O_2 = 3.0 \text{ mol } O_2 \times \frac{32 \text{ g } O_2}{1 \text{ mol } O_2} = \underline{96 \text{ g } O_2}$$

A third method involves combining all the above steps into one calculation

$$\underline{?}\text{ g } O_2 = 24 \text{ g } CH_4 \times \frac{1 \text{ mol } CH_4}{16 \text{ g } CH_4} \times \frac{2 \text{ mol } O_2}{1 \text{ mol } CH_4} \times \frac{32 \text{ g } O_2}{1 \text{ mol } O_2} = \underline{96 \text{ g } O_2}$$

The same answer, 96 grams of O_2, is obtained by all three methods.

The first method by which Example 2–15 was solved is known as the unit factor method. It is a simple, direct method based on interpretation of the balanced chemical equation. The second method, the mole method, is a step-by-step method that is also based on the balanced chemical equation. The question may be reversed, as in Example 2–16.

EXAMPLE 2–16

What mass of CH_4, in grams, is required to react with 96 grams of O_2?

Solution

$$\underline{?}\text{ g } CH_4 = 96 \text{ g } O_2 \times \frac{1 \text{ mol } O_2}{32 \text{ g } O_2} \times \frac{1 \text{ mol } CH_4}{2 \text{ mol } O_2} \times \frac{16 \text{ g } CH_4}{1 \text{ mol } CH_4} = \underline{24 \text{ g } CH_4}$$

or more simply

$$\underline{?}\text{ g } CH_4 = 96 \text{ g } O_2 \times \frac{16 \text{ g } CH_4}{64 \text{ g } O_2} = \underline{24 \text{ g } CH_4}$$

Note that this is the amount of CH_4 used in Example 2–15 to react with 96 grams of O_2. Unit factors can also be constructed from moles and grams, as Example 2–17 illustrates.

EXAMPLE 2–17

Most combustion reactions occur in excess O_2, i.e., more than enough O_2 to burn the substance completely. Calculate the mass of CO_2 in grams that can be produced by burning 6.0 moles of CH_4 in excess O_2.

Solution

Recall the balanced equation

CH_4 +	$2O_2$ ⟶	CO_2 +	$2H_2O$
1 mol	2 mol	1 mol	2 mol
16 g	2(32 g)	44 g	2(18g)

which tells us that one mole of CH_4 produces one mole (44 g) of CO_2.

$$\underline{?}\text{ g } CO_2 = 6.0 \text{ mol } CH_4 \times \frac{1 \text{ mol } CO_2}{1 \text{ mol } CH_4} \times \frac{44 \text{ g } CO_2}{1 \text{ mol } CO_2} = \underline{2.6 \times 10^2 \text{ g } CO_2}$$

In a general way, the equation for the combustion of methane can be written as

CH_4 +	$2O_2$ ⟶	CO_2 +	$2H_2O$
1 mol	2 mol	1 mol	2 mol
16 g	64 g	44 g	36 g

or

16 mass units + 64 mass units ⟶ 44 mass units + 36 mass units

There are 454 g of *anything* in a pound of anything, and 2000 pounds of anything in one short ton of anything. So, the 16:64:44:36 ratio holds for any consistent set of mass or weight units.

The Law of Conservation of Matter allows us to use any weight or mass units we choose, as long as we are consistent. The equation might read

CH_4 +	$2O_2$ ⟶	CO_2 +	$2H_2O$
16 lb	64 lb	44 lb	36 lb

or

16 tons + 64 tons ⟶ 44 tons + 36 tons

EXAMPLE 2–18

What weight of O_2, in tons, is required to burn completely 7.0 tons of CH_4?

Solution

The balanced equation enables us to construct two unit factors relating tons of O_2 and CH_4: $\frac{16 \text{ tons } CH_4}{64 \text{ tons } O_2}$ and $\frac{64 \text{ tons } O_2}{16 \text{ tons } CH_4}$

$$\underline{?}\text{ tons } O_2 = 7.0 \text{ tons } CH_4 \times \frac{64 \text{ tons } O_2}{16 \text{ tons } CH_4} = \underline{28 \text{ tons } O_2}$$

From the information generated from the chemical equation for the combustion of methane, interpreted on a mole basis, we see many *chemically equivalent pairs* and numerous possibilities for unit factors relating CH_4 and O_2.

$$\frac{1 \text{ mol } CH_4}{2 \text{ mol } O_2} \qquad \frac{1 \text{ mol } CH_4}{64 \text{ g } O_2} \qquad \frac{1 \text{ mol } CH_4}{2(6.02 \times 10^{23}) \ O_2 \text{ molecules}}$$

$$\frac{16 \text{ g } CH_4}{2 \text{ mol } O_2} \qquad \frac{16 \text{ g } CH_4}{64 \text{ g } O_2} \qquad \frac{16 \text{ g } CH_4}{2(6.02 \times 10^{23}) \ O_2 \text{ molecules}}$$

$$\frac{6.02 \times 10^{23} \ CH_4 \text{ molecules}}{2 \text{ mol } O_2} \qquad \frac{6.02 \times 10^{23} \ CH_4 \text{ molecules}}{64 \text{ g } O_2} \qquad \frac{6.02 \times 10^{23} \ CH_4 \text{ molecules}}{2(6.02 \times 10^{23}) \ O_2 \text{ molecules}}$$

Note that Avogadro's number can be canceled from the last factor to leave

$$\frac{1 \ CH_4 \text{ molecule}}{2 \ O_2 \text{ molecules}},$$

which is the factor obtained when the equation is interpreted on a molecular basis. We have just written down nine unit factors relating CH_4 to O_2 for this particular reaction. The nine factors obtained by inverting each of these gives a total of 18 unit factors relating CH_4 and O_2 for *this* reaction in three kinds of units. In fact, we can write down many factors relating *any* two substances involved in any chemical reaction! Try writing down some of the others.

Please don't try to memorize unit factors for chemical reactions; rather, learn the general *method* of constructing them from balanced chemical equations!

Problem-solving usually involves interpreting a balanced chemical equation so that we can relate a *given* bit of information to the *desired* bit of information.

EXAMPLE 2–19

What mass of CH_4 must be burned to produce 3.01×10^{23} H_2O molecules?

Solution

Referring back to the balanced equation, we see that 1 mol of CH_4 produces 2 mol of H_2O.

$$\underline{?} \text{ g } CH_4 = 3.01 \times 10^{23} \ H_2O \text{ molecules} \times \frac{1 \text{ mol } H_2O}{6.02 \times 10^{23} \ H_2O \text{ molecules}}$$

$$\times \frac{1 \text{ mol } CH_4}{2 \text{ mol } H_2O} \times \frac{16 \text{ g } CH_4}{1 \text{ mol } CH_4}$$

$$= \underline{4.0 \text{ g } CH_4}$$

The possibilities for problem-solving of the type we have just done go on and on.

2–9 Percent Purity

Let us now expand our horizons by recognizing that most substances are not 100% pure. When impure substances are used, as they frequently are, account must be taken of any and all impurities. Purities (and impurities) are usually indicated by "percent purity" or some such label. For example, a sample of sugar that is 95.00% pure contains 95.00 mass units of sugar and 5.00 mass units of impurities. To illustrate the concept, suppose we have a sample that is 95.00% sugar and 5.00% salt by mass. From the percent composition we can write down several unit factors

$$\frac{95.00 \text{ g sugar}}{100.00 \text{ g mixture}}, \quad \frac{5.00 \text{ g salt}}{100.00 \text{ g mixture}}, \quad \frac{95.00 \text{ g sugar}}{5.00 \text{ g salt}}$$

and, of course, the inverse of each of these. The six unit factors may be used just as we have used other unit factors, as Examples 2–20 and 2–21 illustrate.

EXAMPLE 2–20

What mass of a mixture that contains 95.00% sugar and 5.00% salt by mass will provide 230 g of sugar?

Solution

$$\underline{?} \text{ g mixture} = 230 \text{ g sugar} \times \frac{100.00 \text{ g mixture}}{95.00 \text{ g sugar}} = \underline{242 \text{ g mixture}}$$

EXAMPLE 2–21

Calculate the masses of salt and sugar in 760 g of the mixture.

Solution

$$\underline{?} \text{ g of sugar} = 760 \text{ g mixture} \times \frac{95.00 \text{ g sugar}}{100.00 \text{ g mixture}} = \underline{722 \text{ g sugar}}$$

$$\underline{?} \text{ g salt} = 760 \text{ g mixture} \times \frac{5.00 \text{ g salt}}{100.00 \text{ g mixture}} = \underline{38 \text{ g salt}}$$

Observe the beauty of the unit-factor approach to problem-solving! When unit factors are constructed, such questions as "Do we multiply by 0.9500 or do we divide by 0.9500?" never arise. Unit factors always point toward the correct answer because we construct, and use, unit factors so that units *always* cancel until we arrive at the desired unit.

2–10 Percent Yields from Chemical Reactions

Many chemical reactions do not go to completion, i.e., the reactants are not completely converted to products. In some cases, a particular set of reactants undergoes two or more reactions simultaneously, forming undesired products as well as desired products. Reactions other than the desired one are called "side reactions." The term "percent yield" is used to indicate how much of a desired product is obtained from a particular reaction. Thus far, we have assumed that the reactions we have considered do go to completion, i.e., until one reactant has been used up. The yields or amounts of products we have calculated are known as "theoretical" or "stoichiometric" yields. The theoretical yield of a chemical reaction is the yield obtained if the reaction goes to completion. Stated differently, the **theoretical yield** is the yield calculated by *assuming* that the reaction goes to completion

$$\text{percent yield} = \frac{\text{actual yield of product}}{\text{theoretical yield of product}} \times 100\%$$

Consider the preparation of nitrobenzene, $C_6H_5NO_2$, by the reaction of excess nitric acid, HNO_3, with a limited amount of benzene, C_6H_6. The balanced equation for the reaction may be written

$$C_6H_6 + HNO_3 \longrightarrow C_6H_5NO_2 + H_2O$$

C_6H_6	HNO_3	$C_6H_5NO_2$	H_2O
1 mol	1 mol	1 mol	1 mol
78.1 g	63.0 g	123.1 g	18.0 g

EXAMPLE 2–22

A 15.6-gram sample of C_6H_6 reacts with excess HNO_3 to produce 18.0 grams of $C_6H_5NO_2$. What is the percent yield of $C_6H_5NO_2$ in this reaction?

Note that it is not necessary to know the mass of one mole of HNO_3 to solve this problem.

Solution

First, let's calculate the theoretical yield of $C_6H_5NO_2$.

$$\underline{?}\text{ g } C_6H_5NO_2 = 15.6\text{ g } C_6H_6 \times \frac{1\text{ mol } C_6H_6}{78.1\text{ g } C_6H_6} \times \frac{1\text{ mol } C_6H_5NO_2}{1\text{ mol } C_6H_6} \times \frac{123.1\text{ g } C_6H_5NO_2}{1\text{ mol } C_6H_5NO_2}$$

$$= 24.6\text{ g } C_6H_5NO_2$$

This tells us that if *all* the C_6H_6 were converted to $C_6H_5NO_2$ and isolated, we should obtain 24.6 grams of $C_6H_5NO_2$ (100 percent yield). However, the reaction produces only 18.0 grams of $C_6H_5NO_2$, which is considerably less than a 100 percent yield.

$$\text{percent yield} = \frac{\text{actual yield of product}}{\text{theoretical yield of product}} \times 100\% = \frac{18.0\text{ g}}{24.6\text{ g}} \times 100\%$$

$$= \underline{73.2\text{ percent yield}}$$

The amount of nitrobenzene obtained in this experiment is 73.2% of the amount expected if the reaction had gone to completion, if there were no side reactions, and if we could recover all of the product. Side reactions are simultaneous reactions other than the one described by the equation, and hence reactions that consume reactants (or products) without giving the desired product.

2–11 The Limiting Reagent Concept

Thus far we have worked problems in which the presence of an excess of one reactant was stated or implied. The calculations were based on the substance present in lesser amount, called the **limiting reagent.** Before we study the concept of the limiting reagent in stoichiometry, let's develop the basic idea by considering some simple but analogous nonchemical examples.

Suppose you have 20 slices of ham and 36 slices of bread and you wish to make as many ham sandwiches as possible using only one slice of ham and two slices of bread per sandwich. Obviously, you can make only 18 sandwiches, at which point you run out of bread. Therefore the bread is the "limiting reagent" and the 2 extra slices of ham are the "excess reagent."

Consider another example. Suppose you are given a box containing 93 bolts, 102 nuts, and 150 washers. How many sets consisting of one bolt, one nut, and two washers can you put together? Seventy-five sets will use up all the washers, and therefore the washers are the "limiting reagent." To be sure, 18 bolts and 27 nuts are left over. They are the "reagents present in excess."

Now let's apply the same reasoning process to chemical reactions and return to our familiar chemical equation.

EXAMPLE 2–23

What mass of CO_2 can be produced by the reaction of 8.0 grams of CH_4 with 48 grams of O_2?

Solution

Recall the balanced equation:

$$CH_4 + 2O_2 \longrightarrow CO_2 + 2H_2O$$

1 mol	2 mol	1 mol	2 mol
16 g	2(32 g)	44 g	2(18 g)

This tells us that *one* mole of CH_4 reacts with *two* moles of O_2. Since we are given masses of both CH_4 and O_2, let's calculate the number of moles of each reactant

$$\underline{?} \text{ mol } CH_4 = 8.0 \text{ g } CH_4 \times \frac{1 \text{ mol } CH_4}{16 \text{ g } CH_4} = \underline{0.50 \text{ mol } CH_4}$$

$$\underline{?} \text{ mol } O_2 = 48 \text{ g } O_2 \times \frac{1 \text{ mol } O_2}{32 \text{ g } O_2} = \underline{1.5 \text{ mol } O_2}$$

Now, we return to the balanced equation, which tells us that the ratio of reactants that actually react is

1 mole of CH_4 to 2 moles of O_2

or

0.5 mole of CH_4 to 1 mole of O_2

but we actually have

0.5 mole of CH_4 to 1.5 moles of O_2

Thus, once the 0.5 mole of CH_4 has reacted with 1 mole of the O_2, the reaction must stop for lack of CH_4, but there will be 0.5 mole of O_2 left over. The CH_4 is the limiting reagent, and we must base the calculation on it.

$$\underline{?} \text{ g } CO_2 = 8.0 \text{ g } CH_4 \times \frac{1 \text{ mol } CH_4}{16 \text{ g } CH_4} \times \frac{1 \text{ mol } CO_2}{1 \text{ mol } CH_4} \times \frac{44 \text{ g } CO_2}{1 \text{ mol } CO_2} = \underline{22 \text{ g } CO_2}$$

Thus, 22 grams of CO_2 is the maximum amount of CO_2 that can be produced from 8.0 grams of CH_4 and 48 grams of O_2. If the calculation had been based on O_2 rather than CH_4, the calculated answer would be too big and *wrong*.

Another approach to problems like Example 2–23 is to calculate the mass of one reactant required to react with the given mass of the other reactant. The calculation may be based on either moles or masses of reactants. Recall that Example 2–23 asked, "What mass of CO_2 can be produced by the reaction of 8.0 grams of CH_4 with 48 grams of O_2?"

CH_4 +	$2O_2$	$\longrightarrow$	CO_2 +	$2H_2O$
1 mol	2 mol		1 mol	2 mol
16 g	2(32 g)		44 g	2(18 g)

First, let's consider the mass of O_2 that reacts with 8.0 grams of CH_4.

$$\underline{?}\text{ g } O_2 = 8.0\text{ g } CH_4 \times \frac{1\text{ mol } CH_4}{16\text{ g } CH_4} \times \frac{2\text{ mol } O_2}{1\text{ mol } CH_4} \times \frac{32\text{ g } O_2}{1\text{ mol } O_2} = \underline{32\text{ g } O_2}$$

This calculation tells us that 8.0 grams of CH_4 reacts with 32 grams of O_2. We have 48 grams of O_2, which is more than enough (16 grams of O_2 in excess) to react with 8.0 grams of CH_4. If we choose to calculate the amount of CH_4 that reacts with 48 grams of O_2, we obtain

$$\underline{?}\text{ g } CH_4 = 48\text{ g } O_2 \times \frac{1\text{ mol } O_2}{32\text{ g } O_2} \times \frac{1\text{ mol } CH_4}{2\text{ mol } O_2} \times \frac{16\text{ g } CH_4}{1\text{ mol } CH_4} = \underline{12\text{ g } CH_4}$$

This calculation tells us that 48 grams of O_2 requires 12 grams of CH_4, and we only have 8.0 grams of CH_4. Both calculations show that CH_4 is the limiting reagent and the calculation *must* be based on CH_4.

EXAMPLE 2–24

What is the maximum mass of $Al(OH)_3$ that can be prepared by mixing 13.4 grams of $AlCl_3$ with 10.0 grams of NaOH according to the following equation?

$$AlCl_3 + 3NaOH \longrightarrow Al(OH)_3 + 3NaCl$$

Solution

Interpreting the balanced equation as usual, we have

$AlCl_3$ +	3NaOH	$\longrightarrow$	$Al(OH)_3$ +	3NaCl
1 mol	3 mol		1 mol	3 mol
134.4 g	3(40.0 g)		78.0 g	3(58.4 g)

We determine the number of moles of $AlCl_3$ and NaOH present.

$$\underline{?}\text{ mol } AlCl_3 = 13.4\text{ g } AlCl_3 \times \frac{1\text{ mol } AlCl_3}{133.4\text{ g } AlCl_3} = \underline{0.100\text{ mol } AlCl_3}$$

$$\underline{?}\text{ mol NaOH} = 10.0\text{ g NaOH} \times \frac{1\text{ mol NaOH}}{40.0\text{ g NaOH}} = \underline{0.250\text{ mol NaOH}}$$

The balanced equation tells us that the reaction ratio is:

1 mole of $AlCl_3$	to	3 moles of NaOH
	or	
0.100 mole of $AlCl_3$	to	0.300 moles of NaOH

but we have

0.100 mole of $AlCl_3$	to	0.250 mole of NaOH

Thus, NaOH is the limiting reagent because there is insufficient NaOH to react with all the $AlCl_3$. The calculation of yield must be based on NaOH

$$\underline{?} \text{ g } Al(OH)_3 = 10.0 \text{ g NaOH} \times \frac{1 \text{ mol NaOH}}{40 \text{ g NaOH}} \times \frac{1 \text{ mol } Al(OH)_3}{3 \text{ mol NaOH}} \times \frac{78.0 \text{ g } Al(OH)_3}{1 \text{ mol } Al(OH)_3}$$

$$= 6.5 \text{ g } Al(OH)_3$$

2–12 Two Reactions Occurring Simultaneously

The ideas discussed in this chapter are powerful tools that describe the quantitative relationships among substances as they undergo chemical reactions. To gain more appreciation of the versatility of these concepts, let us consider two chemical reactions occurring simultaneously to form a product common to both.

Lower case letters in parentheses indicate the physical states in which substances exist: (s) = solid, (ℓ) = liquid, (g) = gas.

EXAMPLE 2–25

A 6.940-gram sample of a mineral that contained both barium carbonate, $BaCO_3$, and calcium carbonate, $CaCO_3$, was heated until both compounds decomposed, as indicated by the following equations. After all the CO_2 had been expelled, the residue (a mixture of solid BaO and CaO) had a mass of 4.740 g. What masses and what percentages of $BaCO_3$ and $CaCO_3$ were present in the mixture?

$$BaCO_3 \text{ (s)} \xrightarrow{\Delta} BaO \text{ (s)} + CO_2 \text{ (g)}$$

$$CaCO_3 \text{ (s)} \xrightarrow{\Delta} CaO \text{ (s)} + CO_2 \text{ (s)}$$

Solution

We are able to solve this problem because CO_2 is produced in both reactions. From the statement of the problem, we can calculate the mass of CO_2 produced by *both* reactions

(6.940 g sample) − (4.740 g residue) = $\underline{2.200 \text{ g } CO_2}$

Let's look at the equations in more detail

$BaCO_3$ (s)	⟶ BaO (s)	+ CO_2 (g)
1 mol	1 mol	1 mol
197.3 g	153.3 g	44.0 g

$CaCO_3$ (s)	⟶ CaO (s)	+ CO_2 (g)
1 mol	1 mol	1 mol
100.1 g	56.1 g	44.0 g

They tell us that *197.3 g of* $BaCO_3$ produces *44.0 g of* CO_2 while *100.1 g of* $CaCO_3$ produces *44.0 g of* CO_2. But, how much of the original sample was $BaCO_3$ and how much was $CaCO_3$? To find the answer, we resort to an algebraic representation

Let

X = g $BaCO_3$ in original sample

and

$(6.940 - X)$ = g $CaCO_3$ in original sample

Now that we have represented the masses of both $BaCO_3$ and $CaCO_3$, let's convert both to grams of CO_2 because we know the total mass of CO_2 liberated

$$\underbrace{(X\text{ g BaCO}_3)\left(\frac{44.0\text{ g CO}_2}{197.3\text{ g BaCO}_3}\right)}_{\text{g CO}_2\text{ from BaCO}_3} + \underbrace{[(6.940 - X)\text{g CaCO}_3]\left(\frac{44.0\text{ g CO}_2}{100.1\text{ g CaCO}_3}\right)}_{\text{g CO}_2\text{ from CaCO}_3} = \underbrace{2.200\text{ g CO}_2}_{\text{Total g CO}_2}$$

This equation simplifies to

$$\frac{44.0X}{197.3} + \frac{44.0(6.940 - X)}{100.1} = 2.200$$

Clearing common fractions gives

$$0.223X + 3.050 - 0.440X = 2.200$$

or

$$-0.217X = -0.850$$

so

$$\underline{X = 3.92\text{ g BaCO}_3} \qquad \text{and} \qquad \underline{(6.940 - X) = 3.02\text{ g CaCO}_3}$$

Now that we know the masses of $BaCO_3$ and $CaCO_3$ in the sample, the percentage of each can easily be calculated

$$\%\text{ BaCO}_3 = \frac{\text{g BaCO}_3}{\text{g sample}} \times 100\% = \frac{3.92\text{ g}}{6.940\text{ g}} \times 100\% = \underline{56.5\%\text{ BaCO}_3}$$

$$\%\text{ CaCO}_3 = \frac{\text{g CaCO}_3}{\text{g sample}} \times 100\% = \frac{3.02\text{ g}}{6.940\text{ g}} \times 100\% = \underline{43.5\%\text{ CaCO}_3}$$

More sophisticated chemical calculations are possible, but they can wait until later in the course or until subsequent courses. We have demonstrated that chemical equations are *precise* and *versatile.* Throughout this chapter we have utilized two fundamental ideas: the Law of Definite Proportions *(Different samples of a pure compound always contain the same elements in the same proportions by mass),* and the Law of Conservation of Matter *(There is no detectable change in the quantity of matter during an ordinary chemical change).* These ideas have been very helpful in understanding interactions among countless different kinds of matter. As we continue our study of the behavior of matter, we shall observe their usefulness and validity many times.

2-13 Concentrations of Solutions

A **solution** is a homogeneous mixture of two or more substances. Simple solutions usually consist of one substance (the solute) dissolved in another substance (the solvent). The **solute** may be thought of as the *dissolved* substance and the **solvent** as the *dissolving* substance. The solutions used in the laboratory are usually liquids, and the solvent is often water. For example, solutions of hydrochloric acid are prepared by dissolving pure hydrogen chloride (HCl, a gas at room temperature and atmospheric pressure) in water. Solutions of sodium hydroxide are prepared by dissolving solid NaOH in water.

In some solutions, such as a nearly equal mixture of ethyl alcohol and water, the distinction between *solute* and *solvent* is arbitrary.

Concentrations of solutions are expressed in terms of the amount of solute present in a given mass or volume of *solution,* or the amount of solute dissolved in a given mass or volume of *solvent.*

1 Percent by Mass

Concentrations of solutions may be expressed in terms of percent by mass of solute, which gives the mass of solute per 100 mass units of solution. The gram is the usual mass unit.

$$\% \text{ solute} = \frac{\text{g solute}}{\text{g solution}} \times 100\%$$

Unless specified otherwise, water is the solvent in all solutions described in this text.

Thus, a solution that is 10.0% NaCl by mass contains 10.0 grams of NaCl in 100.0 grams of *solution.* Note that 100.0 grams of the solution contains 10.0 grams of NaCl in 90.0 grams of water. The density of a 10.0% solution of NaCl is 1.07 g/mL, so 100 mL of a 10.0% solution of NaCl has a mass of 107 grams. Observe that 100 grams of the solution does *not* occupy 100 mL. Unless otherwise specified, percent means percent *by mass.*

EXAMPLE 2–26

Calculate the mass of sodium chromate, Na_2CrO_4, required to prepare 200 grams of a 20.0% solution of Na_2CrO_4.

Solution

The percentage information given in the problem tells us that the solution contains 20.0 grams of Na_2CrO_4 in 100 grams of solution. The desired information is the mass of Na_2CrO_4 in 200 grams of solution. A unit factor is constructed by placing 20.0 grams of Na_2CrO_4 over 100 grams of solution. Multiplication of the mass of the solution, 200 grams, by the unit factor gives the desired information.

$$\underline{?}\text{ g } Na_2CrO_4 = 200\text{ g soln} \times \frac{20.0\text{ g } Na_2CrO_4}{100\text{ g soln}} = \underline{40.0\text{ g } Na_2CrO_4}$$

Placing 100 grams of solution over 20.0 grams of Na_2CrO_4 gives another unit factor, as illustrated in the next example.

EXAMPLE 2–27

Calculate the mass of a 20.0% solution of Na_2CrO_4 that contains 40.0 grams of Na_2CrO_4.

Solution

$$\underline{?}\text{ g soln} = 40.0\text{ g } Na_2CrO_4 \times \frac{100\text{ g soln}}{20.0\text{ g } Na_2CrO_4} = \underline{200\text{ g soln}}$$

EXAMPLE 2–28

Calculate the mass of Na_2CrO_4 contained in 200 mL of a 20.0% solution of Na_2CrO_4. The density of the solution is 1.19 g/mL.

Solution

The volume of a solution multiplied by its density gives the mass of the solution (Section 1–11). The mass of solution is then multiplied by the fraction of that mass due to Na_2CrO_4 (20.0 g Na_2CrO_4/100 g soln) to give the mass of Na_2CrO_4 in 200 mL of solution.

$$\underline{?}\text{ g } Na_2CrO_4 = \underbrace{200\text{ mL soln} \times \frac{1.19\text{ g soln}}{1.00\text{ mL soln}}}_{238\text{ g soln}} \times \frac{20.0\text{ g } Na_2CrO_4}{100\text{ g soln}}$$

$$= \underline{47.6\text{ g } Na_2CrO_4}$$

Volume × density = mass of solution.

EXAMPLE 2–29

What volume of a solution containing 15.0% iron(III) nitrate contains 30.0 g of $Fe(NO_3)_3$? The density of the solution is 1.16 g/mL.

Solution

$$\underline{?}\text{ mL soln} = \underbrace{30.0\text{ g } Fe(NO_3)_3 \times \frac{100\text{ g soln}}{15.0\text{ g } Fe(NO_3)_3}}_{200\text{ g soln}} \times \frac{1.00\text{ mL soln}}{1.16\text{ g soln}} = \underline{172\text{ mL}}$$

Note that the answer is not 200 mL, but considerably less because 1.00 mL of solution has a mass of 1.16 g. However, 172 mL of the solution has a mass of 200 g.

2 Molarity (Molar Concentration)

Molarity (M), or molar concentration, is the common unit for expressing the concentrations of solutions. **Molarity** is defined as *the number of moles of solute per liter of solution.* Note that the definition of molarity specifies the amount of solute *per unit of volume of solution,* whereas percent specifies the amount *per unit of mass.* Symbolically, molarity may be represented as

$$\text{molarity} = \frac{\text{number of moles of solute}}{\text{number of liters of solution}}$$

To prepare one liter of a one molar solution, one mole of solute is placed in a one-liter volumetric flask, enough water is added to dissolve the solute, and water is then added until the volume of the solution is exactly one liter. (See Figure 2–7.) Students sometimes make the mistake of assuming that one molar solutions contain one mole of solute in a liter of water. This is *not* the case; one liter of water *plus* one mole of solute usually has a total volume of more than one liter.

Frequently it is convenient to express the volume of a solution in milliliters rather than in liters. Likewise it may be convenient to express the amount of solute in millimoles (mmol) rather than in moles. Since one milliliter is 1/1000 of a liter and one millimole is 1/1000 of a mole, molarity may be also expressed as the number of millimoles of solute per milliliter of solution, i.e.,

$$\text{molarity} = \frac{\text{number of millimoles of solute}}{\text{number of milliliters of solution}}$$

FIGURE 2–7 Preparation of 1.00 molar sodium chloride solution.

(a) Weigh out 1.00 mol (58.4 g) of NaCl.

(b) Transfer NaCl into a 1.00 liter volumetric flask.

(c) Add enough distilled H_2O to dissolve NaCl.

(d) Carefully add distilled H_2O until liquid level stands at calibration mark on the neck of the flask. Stopper and invert several times to mix thoroughly.

EXAMPLE 2–30

Calculate the molarity (M) of a solution that contains 3.65 grams of HCl in 2.00 liters of solution.

Solution

We place 3.65 g HCl over 2.00 L of solution and then convert g HCl to mol HCl.

$$\frac{\underline{?}\ \text{mol HCl}}{\text{L}} = \frac{3.65\ \text{g HCl}}{2.00\ \text{L}} \times \frac{1\ \text{mol HCl}}{36.5\ \text{g HCl}} = \underline{0.0500\ \text{mol HCl/L}}$$

The concentration of the HCl solution is 0.0500 molar, and the solution is called 0.0500 M hydrochloric acid. One liter of the solution contains 0.0500 mole of HCl.

Observe that problems like the one above are worked by a familiar series of operations. The desired information, the molarity of the solution, is the number of moles of HCl in one liter of solution. Therefore, write down (on the left side of the equal sign) "$\underline{?}$ mol HCl/L." Next examine the problem to determine what information is given. In this case the mass of HCl, 3.65 grams, and the volume of the solution, 2.00 L, are given. In the desired answer the volume term (liters) will be in the denominator. Therefore, we place the mass of HCl, 3.65 grams, over the volume of solution in which it is contained, 2.00 L. The units are now g HCl/L, and the desired units are moles of HCl/L. Multiplication by the unit factor 1 mole HCl/36.5 g HCl gives the answer, 0.0500 mol HCl/L.

EXAMPLE 2–31

Calculate the molarity of a solution that contains 49.04 grams of H_2SO_4 in 250.0 mL of solution.

Solution

$$\frac{?\ \text{mol } H_2SO_4}{\text{L}} = \frac{49.04\ \text{g } H_2SO_4}{250.0\ \text{mL}} \times \frac{1\ \text{mol } H_2SO_4}{98.08\ \text{g } H_2SO_4} \times \frac{1000\ \text{mL}}{1\ \text{L}}$$

$$= \frac{49.04 \times 1 \times 1000}{250.0 \times 98.08 \times 1}\ \text{mol } H_2SO_4/\text{L} = \underline{2.000\ M\ H_2SO_4}$$

First we convert g H_2SO_4 to mol H_2SO_4, then we convert mL soln to L soln.

The concentration of the H_2SO_4 solution is 2.000 molar, and the solution is referred to as 2.000 M sulfuric acid. Note that Example 2–31 is worked exactly like Example 2–30, except that one additional factor is used to convert the volume of the solution from milliliters to liters.

EXAMPLE 2–32

Calculate the mass of $Ba(OH)_2$ required to prepare 2.500 liters of a 0.06000 molar solution of barium hydroxide.

Solution

$$?\ \text{g } Ba(OH)_2 = 2.500\ \text{L} \times \frac{0.06000\ \text{mol } Ba(OH)_2}{1\ \text{L}} \times \frac{171.4\ \text{g } Ba(OH)_2}{1\ \text{mol } Ba(OH)_2}$$

$$= (2.500 \times 0.06000 \times 171.4)\ \text{g } Ba(OH)_2$$

$$= \underline{25.71\ \text{g } Ba(OH)_2}$$

The volume of the solution, 2.500 L, is multiplied by the concentration, 0.06000 mol $Ba(OH)_2$/L, to give the number of moles of $Ba(OH)_2$. The number of moles of $Ba(OH)_2$ is then multiplied by the mass of $Ba(OH)_2$ in one mole, 171.4 g $Ba(OH)_2$/mol $Ba(OH)_2$, to give the mass of $Ba(OH)_2$ in the solution.

Concentrated commercial acids and other concentrated or "stock" solutions are used to prepare dilute solutions for laboratory use. The labels on commercial acids usually give the specific gravity and the percent of acid by mass in the solution. Therefore, it is necessary to calculate the molar concentrations of the concentrated acids before they are diluted.

EXAMPLE 2–33

Commercial sulfuric acid is 96.4% H_2SO_4 by mass and its specific gravity is 1.84. Calculate the molarity of commercial sulfuric acid.

Solution

The specific gravity of a solution is numerically equal to its density, i.e., the density of the solution is 1.84 g/mL. The solution is 96.4% H_2SO_4 by mass, and therefore 100 grams of solution contains 96.4 grams of *pure* H_2SO_4. From this information, the molarity of the solution can be calculated. First, we calculate the mass of one liter of solution

$$\frac{?\ \text{g soln}}{\text{L}} = \frac{1.84\ \text{g soln}}{\text{mL}} \times \frac{1000\ \text{mL}}{\text{L}} = 1840\ \text{g soln/L}$$

The solution is 96.4% H_2SO_4 by mass, so the mass of H_2SO_4 in one liter is

$$\frac{?\ \text{g } H_2SO_4}{\text{L}} = \frac{1840\ \text{g soln}}{\text{L}} \times \frac{96.4\ \text{g } H_2SO_4}{100.0\ \text{g soln}} = 1.77 \times 10^3\ \text{g } H_2SO_4/\text{L}$$

The molarity is the number of moles of H_2SO_4 per liter of solution.

$$\frac{?\ \text{mol } H_2SO_4}{\text{L}} = \frac{1.77 \times 10^3\ \text{g } H_2SO_4}{\text{L}} \times \frac{1\ \text{mol } H_2SO_4}{98.1\ \text{g } H_2SO_4} = \underline{18.0\ \text{mol } H_2SO_4/\text{L}}$$

Thus the solution is an 18.0 M H_2SO_4 solution. Note that this problem can be solved by using a series of three unit factors.

$$\frac{?\ \text{mol } H_2SO_4}{\text{L}} = \frac{1.84\ \text{g soln}}{\text{mL}} \times \frac{96.4\ \text{g } H_2SO_4}{100\ \text{g soln}} \times \frac{1\ \text{mol } H_2SO_4}{98.1\ \text{g } H_2SO_4} \times \frac{1000\ \text{mL}}{\text{L}}$$

$$= \frac{1.84 \times 96.4 \times 1 \times 1000}{100 \times 98.1}\ \text{mol } H_2SO_4/\text{L} = \underline{18.1\ M\ H_2SO_4}$$

Rounding errors account for the small difference.

2–14 Volumes of Solutions Required for Chemical Reactions and Dilution of Solutions

Whenever you plan to carry out a chemical reaction in an aqueous solution, you must calculate the volumes of solutions of known concentration required. If you know the molarity of a solution, you can calculate the amount of solute contained in a specified volume of that solution.

Recall that the definition of molarity is the number of moles of solute divided by the volume of the solution in liters

$$\text{molarity} = \frac{\text{number of moles of solute}}{\text{number of liters of solution}}$$

Multiplying both sides of this equation by the volume, we obtain

volume (in L) × molarity = number of moles of solute

Multiplication of the volume of a solution by its concentration gives the amount of solute in the solution. We may choose to express the volume in milliliters and the amount of solute in millimoles, giving

volume (in mL) × molarity = number of mmol of solute

Example 2–34 illustrates how we can calculate the amount of solute in a given amount of a solution of known molarity.

EXAMPLE 2–34

Calculate the (a) number of moles of H_2SO_4, (b) number of millimoles of H_2SO_4, and (c) mass of H_2SO_4 in 500 mL of 0.324 M H_2SO_4 solution.

Solution

(a) The volume of a solution in liters times its molarity gives the number of moles of solute, H_2SO_4 in this case.

$$\underline{?} \text{ mol } H_2SO_4 = 0.500 \text{ L} \times \frac{0.324 \text{ mol } H_2SO_4}{\text{L}} = \underline{0.162 \text{ mol } H_2SO_4}$$

500 mL is more conveniently expressed as 0.500 L in this problem.

(b) The volume of a solution in milliliters multiplied by its molarity gives the number of millimoles of solute, H_2SO_4

$$\underline{?} \text{ mmol } H_2SO_4 = 500 \text{ mL} \times \frac{0.324 \text{ mmol } H_2SO_4}{\text{mL}} = \underline{162 \text{ mmol } H_2SO_4}$$

(c) We may use the results of either (a) or (b) to calculate the mass of H_2SO_4 in the solution

$$\underline{?} \text{ g } H_2SO_4 = 0.162 \text{ mol } H_2SO_4 \times \frac{98.1 \text{ g } H_2SO_4}{1 \text{ mol } H_2SO_4} = \underline{15.9 \text{ g } H_2SO_4}$$

or

$$\underline{?} \text{ g } H_2SO_4 = 162 \text{ mmol } H_2SO_4 \times \frac{0.0981 \text{ g } H_2SO_4}{1 \text{ mmol } H_2SO_4} = \underline{15.9 \text{ g } H_2SO_4}$$

The mass of H_2SO_4 in the solution can be calculated without solving explicitly for the number of moles (or millimoles) of H_2SO_4

$$\underline{?} \text{ g } H_2SO_4 = 0.500 \text{ L} \times \frac{0.324 \text{ mol } H_2SO_4}{\text{L}} \times \frac{98.1 \text{ g } H_2SO_4}{1 \text{ mol } H_2SO_4} = \underline{15.9 \text{ g } H_2SO_4}$$

Volume in liters times molarity gives moles of H_2SO_4. The last unit factor converts moles to grams of H_2SO_4.

Example 2–35 demonstrates how we can relate the volume of a solution of known concentration to the amount of another reactant, expressed in grams.

EXAMPLE 2–35

What volume of a 0.324 *M* solution of sulfuric acid is required to react completely with 2.792 grams of Na_2CO_3 by the following reaction?

$$H_2SO_4 + Na_2CO_3 \longrightarrow Na_2SO_4 + CO_2 + H_2O$$

Solution

The balanced equation tells us that one mole of H_2SO_4 reacts with one mole of Na_2CO_3, and we can write

H_2SO_4 +	Na_2CO_3 ⟶	Na_2SO_4 +	CO_2 +	H_2O
1 mol	1 mol = 106.01 g	1 mol	1 mol	1 mol

which enables us to solve the problem.

$$\underline{?} \text{ L } H_2SO_4 = 2.792 \text{ g } Na_2CO_3 \times \frac{1 \text{ mol } Na_2CO_3}{106.01 \text{ g } Na_2CO_3} \times \frac{1 \text{ mol } H_2SO_4}{1 \text{ mol } Na_2CO_3} \times \frac{1 \text{ L } H_2SO_4}{0.324 \text{ mol } H_2SO_4}$$

$$= \underline{0.0813 \text{ L } H_2SO_4} \quad \text{or} \quad \underline{81.3 \text{ mL } H_2SO_4}$$

Note that we converted grams of Na_2CO_3 to moles of Na_2CO_3 (first factor), and then related moles of Na_2CO_3 to moles of H_2SO_4 (second factor), which was converted to liters of H_2SO_4 solution (third factor).

Often we must calculate the volume of solution of known molarity required to react with a specified volume of another solution. We always examine the balanced chemical equation for the reaction to determine the **reaction ratio,** i.e., *the relative numbers of moles (or millimoles) of reactants.*

EXAMPLE 2–36

What volume of 0.500 *M* NaOH solution is required to react with 40.0 mL of 0.500 *M* H_2SO_4 solution? The balanced equation for the reaction is

$$H_2SO_4 + 2NaOH \longrightarrow Na_2SO_4 + 2H_2O$$

Solution

The balanced equation tells us that *one* mole of H_2SO_4 reacts with *two* moles of NaOH

$$\underset{1\text{ mol}}{H_2SO_4} + \underset{2\text{ mol}}{2NaOH} \longrightarrow \underset{1\text{ mol}}{Na_2SO_4} + \underset{2\text{ mol}}{2H_2O}$$

Therefore, the *reaction ratio is 1 mole H_2SO_4 to 2 moles NaOH.* We are given the volume and the molarity of the H_2SO_4 solution, so we can calculate the number of moles of H_2SO_4.

The volume of H_2SO_4 solution is expressed as 0.0400 L rather than 40.0 mL.

$$\underline{?}\text{ mol } H_2SO_4 = 0.0400\text{ L} \times \frac{0.500\text{ mol } H_2SO_4}{\text{L}} = 0.0200\text{ mol } H_2SO_4$$

The number of moles of H_2SO_4 is related to the number of moles of NaOH by the reaction ratio, 1 mol H_2SO_4/2 mol NaOH.

$$\underline{?}\text{ mol NaOH} = 0.0200\text{ mol } H_2SO_4 \times \frac{2\text{ mol NaOH}}{1\text{ mol } H_2SO_4} = 0.0400\text{ mol NaOH}$$

Now we can calculate the volume of 0.500 *M* NaOH solution that contains 0.0400 mole of NaOH.

$$\underline{?}\text{ L NaOH soln} = 0.0400\text{ mol NaOH} \times \frac{1.00\text{ L NaOH soln}}{0.500\text{ mol NaOH}} = \underline{0.0800\text{ L NaOH soln}}$$

which we usually call <u>80.0 mL of NaOH solution</u>.

Now that we have worked through the problem stepwise, let us solve it in a single set-up.

$$\underline{?}\text{ L NaOH solution} = 0.0400\text{ L } H_2SO_4\text{ soln} \times \frac{0.500\text{ mol } H_2SO_4}{\text{L } H_2SO_4\text{ soln}} \times \frac{2\text{ mol NaOH}}{1\text{ mol } H_2SO_4} \times \frac{1.00\text{ L NaOH soln}}{0.500\text{ mol NaOH}}$$

$$= \underline{0.0800\text{ L NaOH soln}}$$

In the single set-up solution, we converted (1) liters of H_2SO_4 solution to moles of H_2SO_4, (2) moles of H_2SO_4 to moles of NaOH, and (3) moles of NaOH to liters of NaOH solution, the same steps we used in the stepwise procedure.

When a solution is diluted by mixing more solvent with a concentrated solution, *the number of moles of solute present does not change.* What does change is the *volume and the concentration* of the solution. Since the same number of moles of solute is divided by a larger number of liters of solution, the molarity decreases. Using a subscript 1 to represent the original concentrated solution and a subscript 2 to represent the dilute solution

$$\text{volume}_1 \times \text{molarity}_1 = \text{number of moles of solute} = \text{volume}_2 \times \text{molarity}_2$$

or

$$V_1 \times M_1 = V_2 \times M_2 \qquad \text{(for dilution only)}$$

Again, it doesn't matter whether the volumes are expressed in liters or milliliters (or, for that matter, in gallons or cubic miles) as long as both volumes are in the same units.

This expression can be used to calculate any one of four quantities when the other three are known. For instance, if a certain volume of dilute solution of a given molarity is required for use in the laboratory, and we know the concentration of the stock solution available, we can calculate how much stock solution must be used to make the dilute solution.

Caution Dilution of a concentrated solution, especially of a strong acid or base, frequently liberates a great deal of heat. This can vaporize drops of water as they hit the concentrated solution and can cause dangerous spattering. As a safety precaution, concentrated solutions are always poured *into* water, allowing the heat to be absorbed by the larger quantity of water. Although calculations are usually simpler to visualize by *assuming* that water is added to the concentrated solution, this is *never* done in practice.

EXAMPLE 2–37

Calculate the volume of 18.0 M H_2SO_4 required to prepare 1.00 L of a 0.900 M solution of H_2SO_4.

Solution

The volume (1.00 L) and molarity (0.900 M) of the final solution, as well as the molarity (18.0 M) of the original solution, are given. Therefore, the relation $V_1 \times M_1 = V_2 \times M_2$ can be used, with subscript 1 for the commercial acid solution and subscript 2 for the dilute solution

$$V_1 \times M_1 = V_2 \times M_2$$

$$V_1 = \frac{V_2 \times M_2}{M_1} = \frac{1.00\ \text{L} \times 0.900\ M}{18.0\ M} = 0.0500\ \text{L} = \underline{50.0\ \text{mL}}$$

Note that the dilute solution must contain 1.00 L × 0.900 M = 0.900 mol of H_2SO_4, so 0.900 mole of H_2SO_4 must be present in the original concentrated solution. Indeed, 0.0500 L × 18.0 M = 0.900 mol of H_2SO_4.

EXAMPLE 2–38

If 200 mL of a 4.00 M solution of NaOH and 300 mL of a 6.00 M solution of NaOH are mixed, what will be the molarity of the resulting solution? Assume that the volumes are additive.

Solution

We must calculate the volume of the final solution and the number of moles of sodium hydroxide it contains

$$\text{1st solution:} \quad 0.200 \text{ L} \times \frac{4.00 \text{ mol NaOH}}{\text{L}} = 0.800 \text{ mol NaOH}$$

$$\text{2nd solution:} \quad 0.300 \text{ L} \times \frac{6.00 \text{ mol NaOH}}{\text{L}} = 1.80 \text{ mol NaOH}$$

$$\text{final solution:} \quad 0.500 \text{ L} \quad \text{contains} \quad 2.60 \text{ mol NaOH}$$

Hence

$$M = \frac{\text{no. mol NaOH}}{\text{L}} = \frac{2.60 \text{ mol NaOH}}{0.500 \text{ L}} = \underline{5.20\ M \text{ NaOH}}$$

Key Terms

Actual yield amount of a specified pure product actually obtained from a given reaction. Compare with *Theoretical yield.*

Allotropic modifications (allotropes) different forms of the same element in the same physical state.

Atomic mass unit (amu) one twelfth of the mass of an atom of the carbon-12 isotope; a unit used for stating atomic and formula weights; also called dalton.

Atomic weight weighted average of the masses of the constituent isotopes of an element; the relative masses of atoms of different elements.

Avogadro's number see *Mole*

Chemical equation description of a chemical reaction by placing the formulas of reactants on the left and the formulas of the products on the right of an arrow.

Concentration amount of solute per unit volume or mass of solvent or of solution.

Dilution process of reducing the concentration of a solute in solution, usually simply by mixing with more solvent.

Empirical formula see *Simplest formula*

Formula combination of symbols that indicates the chemical composition of a substance.

Formula unit the smallest repeating unit of a substance, the molecule for nonionic substances.

Formula weight the mass of one formula unit of a substance in atomic mass units.

Hydrate a solid compound that contains a definite percentage of bound water.

Ion an atom or group of atoms that carries an electrical charge.

Law of Constant Composition see *Law of Definite Proportions*

Law of Definite Proportions statement that different samples of a pure compound always contain the same elements in the same proportions by mass; also called Law of Constant Composition.

Limiting reagent substance that stoichiometrically limits the amount of product(s) that can be formed.

Molarity (*M*) number of moles of solute per liter of solution.

Mole 6.022×10^{23} (Avogadro's number of) formula units of the substance under discussion.

Molecular (true) formula formula that indicates the actual number of atoms present in a molecule of a molecular substance. Compare with *Simplest formula.*

Molecular weight the mass of one molecule of a nonionic substance in atomic mass units.

Percent by mass 100% multiplied by the mass of a solute divided by the mass of the solution in which it is contained.

Percent composition the mass percent of each element in a compound.

Percent purity the percent of a specified compound or element in an impure sample.

Percent yield 100% times actual yield divided by theoretical yield.

Products substances produced in a chemical reaction.

Reactants substances consumed in a chemical reaction.

Simplest formula the smallest whole number ratio of atoms present in a compound; also called empirical formula. Compare with *Molecular (true) formula.*

Solute the dispersed (dissolved) phase of a solution.

Solution homogeneous mixture of two or more substances.

Solvent the dispersing medium of a solution.

Stoichiometry description of the quantitative relationships among elements and compounds as they undergo chemical changes.

Theoretical yield maximum amount of a specified product that could be obtained from specified amounts of reactants, assuming complete consumption of limiting reagent according to only one reaction and complete recovery of product. Compare with *Actual yield.*

Exercises

Formulas

1. Define and give an example of each of the following terms: (a) atom, (b) molecule, (c) ion, (d) formula unit, (e) allotropism.
2. Use words to describe the atomic composition of molecules in the following elements and compounds: (a) F_2, (b) HCl, (c) CH_3OH, (d) Ne, (e) SF_6, (f) H_3AsO_4.

Atomic Weights

3. (a) What is the atomic weight of an element?
   (b) Why can atomic weights be referred to as relative numbers?
4. (a) What is the atomic mass unit (amu)?
   (b) The atomic weight of vanadium is 50.942 amu and the atomic weight of ruthenium is 101.07 amu. What can we say about the relative masses of V and Ru atoms?

The Mole Concept

5. What is a mole? Why is a mole a convenient unit to work with in equations?
6. Complete the following table. You may refer to the periodic table on the inside cover of the book.

Element	Atomic Weight	Mass of 1 mole of atoms
(a) B	______	______
(b) ______	32.06 amu	______
(c) Fe	______	______
(d) ______	______	107.868 g

7. Complete the following table. You may refer to the periodic table.

Element	Formula	Mass of 1 mole of molecules
(a) H	H_2	______
(b) ______	F_2	______
(c) ______	P_4	______
(d) O	______	______
(e) He	______	______
(f) ______	S_8	______

8. Calculate:
   (a) the number of moles of boron atoms in 59.4 g of boron.
   (b) the number of calcium atoms in 20 g of calcium.
   (c) the number of iron atoms in 83.7 g of iron.
   (d) the number of lithium atoms in 1.00 g of lithium.

(e) the number of uranium atoms in 1.00 g of uranium.
(f) the ratio of the number of lithium atoms in 1.00 g of lithium to the number of uranium atoms in 1.00 g of uranium.

9. Distinguish among the following terms clearly by statements and by calculating the mass of each in grams.
(a) one nitrogen atom
(b) one nitrogen molecule
(c) one mole of nitrogen molecules
10. The atomic weight of carbon is 12.011 amu. Calculate the number of moles of carbon atoms in:
(a) 1.00 g of carbon, (b) 12.0 atomic mass units of carbon, (c) 5.66×10^{20} carbon atoms.
11. Which of the following contain eight atoms?
(a) one C_2H_6 molecule, (b) one mole of C_2H_6, (c) 30 g of C_2H_6, (d) 4.99×10^{-23} g of C_2H_6.
12. Define and illustrate the following terms. Which ones are frequently used interchangeably?
(a) formula, (b) formula weight, (c) molecular weight, (d) mole.
13. A sample of a pure element that had a mass of 1.0 g was found to contain 1.5×10^{22} atoms. If Avogadro's number is used as 6.0×10^{23}, what is the atomic weight of the element?

Percent Composition and Simplest Formulas

14. Calculate the percent composition of the following compounds:
(a) MgO, (b) Fe_2O_3, (c) Na_2SO_4, (d) $(NH_4)_2CO_3$, (e) $Al_2(SO_4)_3 \cdot 18H_2O$
15. Consider 100-g samples of each of the following compounds
Li_2O, CaO, CrO_3, As_4O_{10}, U_3O_8
Which sample contains:
(a) the greatest mass of oxygen?
(b) the smallest mass of oxygen?
(c) the largest total number of atoms?
(d) the smallest total number of atoms?
16. A compound contains 59.9% titanium and 40.1% oxygen by mass. What is the simplest formula for the compound?
17. A compound contains 66.6% titanium and 33.4% oxygen by mass. What is its simplest formula?
18. A 1.596-g sample of an oxide of iron was found to contain 1.116 g of iron and 0.480 g of oxygen. What is the simplest formula for this oxide?
19. A 2.317-g sample of an oxide of iron was found to contain 1.677 g of iron and 0.640 g of oxygen. What is the simplest formula for this oxide?
20. What law is illustrated by Exercises 16 to 19? State the law.
21. Four compounds were analyzed and found to have the following percentage compositions. What is the simplest formula for each compound?
(a) 65.20% As, 34.80% O
(b) 59.72% Ba, 21.72% As, 18.55% O
(c) 40.27% K, 26.78% Cr, 32.96% O
(d) 26.58% K, 35.35% Cr, 38.07% O
22. A sample of a compound contains 4.86 g of magnesium, 6.42 g of sulfur, and 9.60 g of oxygen. What is the simplest formula for the compound?
23. A sample of a compound contains 4.86 g of magnesium, 6.42 g of sulfur, and 12.8 g of oxygen. What is the simplest formula for the compound?
24. A sample of a compound contains 4.86 g of magnesium, 12.85 g of sulfur, and 9.60 g of oxygen. What is the simplest formula for the compound?
25. An unidentified organic compound X, containing only C, H, and O, was subjected to combustion analysis. When 228.4 mg of pure compound X was burned in the C—H combustion train, 627.4 mg of CO_2 and 171.2 mg of H_2O were obtained. (a) Determine the masses of C, H, and O in the sample. (b) Determine the simplest formula of compound X.
26. The substance responsible for the sharp odor of rancid butter is butanoic acid, an organic substance consisting entirely of C, H, and O. Combustion analysis of a

0.3164-g sample of butanoic acid yielded 0.6322 g CO_2 and 0.2588 g H_2O. (a) Determine the simplest formula of butanoic acid. (b) In an independent determination, the molecular weight of butanoic acid was shown to be approximately 88 g/mol. What is the molecular formula of butanoic acid?

Formulas and Formula Weights

27. What information does the formula for ethyl alcohol, C_2H_5OH, contain?
28. How many moles of methyl alcohol, CH_3OH, are there in 80 g of CH_3OH?
29. How many moles of ethyl alcohol, C_2H_5OH, are there in 230 g of ethyl alcohol?
30. Calculate the number of oxygen atoms in 160 g of SO_2.
31. How many hydrogen atoms are contained in 64 g of CH_4?
32. Calculate the number of hydrogen atoms in 69.8 grams of diamminepalladium(II) hydroxide, $Pd(NH_3)_2(OH)_2$.
33. A compound that is 17.2% sulfur by mass contains one sulfur atom per molecule. What is the molecular weight of the compound?
34. Lysine is an essential amino acid. One experiment showed that each molecule of lysine contains two nitrogen atoms. Another experiment showed that lysine contains 19.2% N, 9.64% H, 49.3% C, and 21.9% O by mass. What is the molecular formula for lysine?
35. In each of three experiments a sample of pure carbon was burned in oxygen to produce carbon dioxide. In Experiment 1, 5.00 grams of carbon produced 18.33 grams of carbon dioxide. In Experiment 2, 6.50 grams of carbon produced 23.83 grams of carbon dioxide. In Experiment 3, 9.25 grams of carbon produced 33.92 grams of carbon dioxide. Show how these experimental data are consistent with the Law of Constant Composition.
36. How many grams of oxygen reacted in each of the three experiments in Exercise 35?
37. When a sample of calcium chloride was decomposed, 21.7 grams of calcium and 38.3 grams of gaseous chlorine were produced. (a) What mass of calcium chloride decomposed? (b) What is the percent composition of calcium chloride?
38. Suppose that 30.0 grams of calcium are converted completely to calcium chloride (whose percent composition you calculated in Exercise 37) when mixed with an excess of chlorine. (a) What mass of calcium chloride is produced? (b) What mass of chlorine reacts?
39. Sulfur dioxide is 50.1% sulfur and 49.9% oxygen by mass. Suppose that 10.0 grams of sulfur are mixed with 30.0 grams of oxygen and heated until all the sulfur is converted to sulfur dioxide.
    (a) What mass of oxygen remains unreacted?
    (b) What mass of sulfur dioxide is produced?
40. Phosphorus burns in a limited amount of oxygen to produce a compound whose composition is 56.4% phosphorus and 43.6% oxygen by mass. Suppose that 10.0 grams of phosphorus are burned in 5.0 grams of oxygen and that all of one element is consumed.
    (a) Which element is consumed completely?
    (b) What mass of the other element remains unreacted?
    (c) What mass of the compound is produced?
41. Phosphorus reacts with an excess of chlorine to produce a compound that is 14.5% P and 85.5% Cl by mass. Suppose that 10.0 grams of phosphorus are mixed with 151.0 grams of chlorine and that all of one element is consumed.
    (a) Which element is consumed completely?
    (b) What mass of the other element remains unreacted?
    (c) What mass of the compound is produced?
42. Show that the reactions of Exercises 40 and 41 obey the Law of Conservation of Matter.

43. Phosphorus forms two chlorides. A 30.00-gram sample of one chloride decomposes to give 4.35 g P and 25.65 g Cl. A 30.00-gram sample of the other chloride decomposes to give 6.61 g P and 23.39 g Cl. Show that these compounds obey the Law of Multiple Proportions.

Chemical Equations

44. What fundamental law is the basis for balancing a chemical equation?
45. What is a chemical equation? What information does it contain?
46. Balance the following equations:
(a) $Sn + Cl_2 \longrightarrow SnCl_4$
(b) $Fe + Cl_2 \longrightarrow FeCl_3$
(c) $Fe + O_2 \longrightarrow Fe_2O_3$
(d) $Al + Cl_2 \longrightarrow Al_2Cl_6$
(e) $CaO + HCl \longrightarrow CaCl_2 + H_2O$
(f) $NaOH + SO_2 \longrightarrow Na_2SO_3 + H_2O$
(g) $Mg + HCl \longrightarrow MgCl_2 + H_2$
(h) $Na + H_2O \longrightarrow NaOH + H_2$
(i) $Al + H_2SO_4 \longrightarrow Al_2(SO_4)_3 + H_2$
(j) $NaOH + H_2SO_4 \longrightarrow Na_2SO_4 + H_2O$
47. Balance the following equations:
(a) $Ca(OH)_2 + HNO_3 \longrightarrow Ca(NO_3)_2 + H_2O$
(b) $KOH + H_3PO_4 \longrightarrow K_3PO_4 + H_2O$
(c) $Ca(OH)_2 + H_3PO_4 \longrightarrow Ca_3(PO_4)_2 + H_2O$
(d) $P_4O_{10} + NaOH \longrightarrow Na_3PO_4 + H_2O$
(e) $CuCl_2 + H_2S \longrightarrow CuS + HCl$
(f) $BiCl_3 + H_2S \longrightarrow Bi_2S_3 + HCl$
(g) $CuCl_2 + NH_3 + H_2O \longrightarrow Cu(OH)_2 + NH_4Cl$
(h) $Fe(NO_3)_3 + NH_3 + H_2O \longrightarrow Fe(OH)_3 + NH_4NO_3$
(i) $NaNO_3 \longrightarrow NaNO_2 + O_2$
(j) $(NH_4)_2Cr_2O_7 \longrightarrow N_2 + Cr_2O_3 + H_2O$
48. Balance the following equations:
(a) $Sn + F_2 \longrightarrow SnF_4$
(b) $As + O_2 \longrightarrow As_4O_{10}$
(c) $Al + O_2 \longrightarrow Al_2O_3$
(d) $CH_4 + O_2 \longrightarrow CO_2 + H_2O$
(e) $CS_2 + O_2 \longrightarrow CO_2 + SO_2$
(f) $C_2H_6 + O_2 \longrightarrow CO_2 + H_2O$
(g) $C_2H_6O + O_2 \longrightarrow CO_2 + H_2O$
(h) $KOH + H_2SO_4 \longrightarrow K_2SO_4 + H_2O$
(i) $NaOH + H_3AsO_4 \longrightarrow Na_3AsO_4 + H_2O$
(j) $Cr(OH)_3 + HClO_4 \longrightarrow Cr(ClO_4)_3 + H_2O$
49. Balance the following equations:
(a) $N_2O_5 + H_2O \longrightarrow HNO_3$
(b) $P_4O_{10} + H_2O \longrightarrow H_3PO_4$
(c) $P_4O_{10} + Mg(OH)_2 \longrightarrow Mg_3(PO_4)_2 + H_2O$
(d) $Cl_2O_7 + H_2O \longrightarrow HClO_4$
(e) $Cl_2O_7 + Ba(OH)_2 \longrightarrow Ba(ClO_4)_2 + H_2O$
(f) $H_2SnCl_6 + H_2S \longrightarrow SnS_2 + HCl$
(g) $HSbCl_4 + H_2S \longrightarrow Sb_2S_3 + HCl$
(h) $FeCl_2 + SnCl_4 \longrightarrow FeCl_3 + SnCl_2$
(i) $BiCl_3 + NH_3 + H_2O \longrightarrow Bi(OH)_3 + NH_4Cl$
(j) $NH_3 + O_2 \longrightarrow NO + H_2O$
50. Write balanced chemical equations to represent the reactions described by the following statements.
(a) Fluorine, F_2, combines with hydrogen, H_2, to form hydrogen fluoride, HF.
(b) Nitrogen, N_2, combines with hydrogen to form ammonia, NH_3.
(c) Sulfur dioxide, SO_2, combines with oxygen, O_2, to produce sulfur trioxide, SO_3.
(d) Ferrous chloride, $FeCl_2$, combines with chlorine, Cl_2, to form ferric chloride, $FeCl_3$.
(e) Heating calcium carbonate, $CaCO_3$, liberates carbon dioxide, CO_2, as a gas and leaves a residue of solid calcium oxide, CaO.
(f) When ammonium carbonate, $(NH_4)_2CO_3$, is heated it decomposes to form ammonia, NH_3, carbon dioxide, CO_2, and water.
(g) Ethane, C_2H_6, burns in oxygen to produce carbon dioxide and water.
(h) Octane, C_8H_{18}, burns in oxygen to produce carbon dioxide and water.
(i) Ethyl alcohol, C_2H_5OH, burns in oxygen to produce carbon dioxide and water.
(j) When solid mercury(II) oxide, HgO, is heated it decomposes to produce liquid mercury and gaseous oxygen.
(k) When solid potassium chlorate, $KClO_3$, is heated it decomposes to produce solid potassium chloride, KCl, and gaseous oxygen.

51. State in words the meaning of the following chemical equations:
(a) $Ca + Cl_2 \longrightarrow CaCl_2$
(b) $2Mg + O_2 \xrightarrow{\Delta} 2MgO$
(c) $2Fe + 3F_2 \longrightarrow 2FeF_3$
(d) $CH_4 + 2O_2 \xrightarrow{\Delta} CO_2 + 2H_2O$
(e) $2C_4H_{10} + 13O_2 \xrightarrow{\Delta} 8CO_2 + 10H_2O$

Calculations Based on Chemical Equations

52. What weight of CaO could be obtained from the thermal decomposition of 2.00 moles of $CaCO_3$?

$$CaCO_3 \xrightarrow{\Delta} CaO + CO_2$$

53. Aluminum and sulfur react at elevated temperatures to form aluminum sulfide as shown by the following equation.

$$2Al + 3S \xrightarrow{\Delta} Al_2S_3$$

Calculate the number of moles of:
(a) aluminum atoms that react with one mole of sulfur atoms. What mass of aluminum is this?
(b) sulfur atoms that react with one mole of aluminum atoms. What mass of sulfur is this?
(c) aluminum atoms that react with 1.00 g of sulfur. What mass of aluminum is this?
(d) sulfur atoms that react with 1.00 g of aluminum. What mass of sulfur is this?

54. A chemist needed 80 g of anhydrous copper(II) sulfate, $CuSO_4$, to perform a particular experiment but none was readily available. However, he did have available a large supply of copper(II) sulfate pentahydrate, $CuSO_4 \cdot 5H_2O$, which he could dehydrate by heating. How much of the hydrated compound did he heat to obtain 80 g of $CuSO_4$?

55. What mass of Na_3PO_4 can be prepared by the reaction of 4.9 g of H_3PO_4 with an excess of NaOH?

$$H_3PO_4 + 3NaOH \longrightarrow Na_3PO_4 + 3H_2O$$

56. What mass of K_3AsO_4 can be prepared by the reaction of 7.10 g of H_3AsO_4 with an excess of KOH?

$$H_3AsO_4 + 3KOH \longrightarrow K_3AsO_4 + 3H_2O$$

Percent Purity

57. A particular ore of lead, galena, is 10% lead sulfide, PbS, and 90% impurities by weight. What mass of lead is contained in 50 grams of this ore?

58. What mass of chromium is present in 150 grams of an ore of chromium that is 65.0% chromite, $FeCr_2O_4$, and 35.0% impurities by mass? If 90.0% of the chromium can be recovered from 100 grams of the ore, what mass of pure chromium is obtained?

59. Ethylene glycol, $C_2H_6O_2$, is used as antifreeze in automobile radiators. A method of producing small amounts of ethylene glycol in the laboratory is by reaction of 1,2-dichloroethane with sodium carbonate in a water solution, followed by distillation of the reaction mixture to purify the ethylene glycol.

$$C_2H_4Cl_2 + Na_2CO_3 + H_2O \longrightarrow C_2H_6O_2 + 2NaCl + CO_2$$

When 26.8 g of 1,2-dichloroethane is used in this reaction, 10.5 g of ethylene glycol is obtained. (a) Calculate the theoretical yield of ethylene glycol. (b) What is the percent yield of ethylene glycol in this process? (c) If the reaction had gone to completion, what mass of Na_2CO_3 would have been consumed?

60. Mesitylene, C_9H_{12}, is used in the synthesis of other organic compounds. It is prepared in poor yield from acetone, C_3H_6O, in the presence of sulfuric acid.

$$\underset{\text{acetone}}{3C_3H_6O} \xrightarrow{H_2SO_4} \underset{\text{mesitylene}}{C_9H_{12}} + 3H_2O$$

We obtain 12.4 g of mesitylene from 125 g of acetone, according to this reaction. What is the percent yield of mesitylene in this reaction?

61. Chlorobenzene, C_6H_5Cl, is an organic chemical that is manufactured on a large

scale and is then used in industrial processes to produce such useful compounds as aspirin, oil of wintergreen, and many dyes, insecticides, and disinfectants. This very useful chemical is produced by the following reaction, under suitable conditions:

$$\underset{\text{benzene}}{C_6H_6} + Cl_2 \longrightarrow \underset{\text{chlorobenzene}}{C_6H_5Cl} + HCl$$

Suppose that a particular industrial reactor is known to produce chlorobenzene by this reaction in 73% yield. What mass of benzene, in kg, would be required to produce 775 kg of chlorobenzene?

62. Ethylene oxide, C_2H_4O, a fumigant sometimes used by exterminators, is synthesized in 89% yield by reaction of ethylene bromohydrin with sodium hydroxide:

$$\underset{\text{ethylene bromohydrin}}{C_2H_5OBr} + NaOH \longrightarrow \underset{\text{ethylene oxide}}{C_2H_4O} + NaBr + H_2O$$

How many grams of ethylene bromohydrin would be consumed in the production of 125 g of ethylene oxide, at 89% yield?

Limiting Reagent

63. What mass of $Ca(NO_3)_2$ can be prepared by the reaction of 18.9 g of HNO_3 with 7.4 g of $Ca(OH)_2$?

$$2HNO_3 + Ca(OH)_2 \longrightarrow Ca(NO_3)_2 + 2H_2O$$

64. What is the maximum amount of $Ca_3(PO_4)_2$ that can be prepared from 7.4 g of $Ca(OH)_2$ and 9.8 g of H_3PO_4?

$$3Ca(OH)_2 + 2H_3PO_4 \longrightarrow Ca_3(PO_4)_2 + 6H_2O$$

65. Silver nitrate solution reacts with calcium chloride solution according to the equation

$$2AgNO_3 + CaCl_2 \longrightarrow Ca(NO_3)_2 + 2AgCl$$

All of the substances involved in this reaction are soluble in water, except for silver chloride, AgCl, which forms a solid (precipitate) at the bottom of the flask. Suppose we mix together a solution containing 12.6 g of $AgNO_3$ and 8.4 g of $CaCl_2$. What mass of AgCl would be formed?

Unscramble Them

66. What is the total mass of products formed when 19.0 g of carbon disulfide is burned in air? What mass of carbon disulfide would have to be burned to produce a mixture of carbon dioxide and sulfur dioxide that has a mass of 34.4 g?

$$CS_2 + 3O_2 \longrightarrow CO_2 + 2SO_2$$

67. A mixture of calcium oxide, CaO, and calcium carbonate, $CaCO_3$, that had a mass of 1.727 g was heated until all the calcium carbonate was decomposed according to the following equation. After heating, the sample weighed 1.551 g. Calculate the masses of CaO and $CaCO_3$ present in the original sample.

$$CaCO_3 \xrightarrow{\Delta} CaO + CO_2$$

68. Analysis of a sample of a compound shows carbon, hydrogen, and nitrogen present. Each molecule contains one nitrogen atom. A 150-mg sample of the compound produces 27.4 mg of NH_3 in which all of the nitrogen comes from the sample. Which of the following is the formula for the compound?
(a) CH_5N, (b) C_2H_7N, (c) $C_4H_{11}N$, (d) $C_5H_{13}N$, (e) C_6H_7N

69. Analysis of a sample of a compound revealed the presence of carbon, hydrogen, nitrogen, and oxygen. In one experiment in which all of the N in the compound was converted into NH_3, a 200-mg sample of the compound produced 44.7 mg of NH_3. In another experiment, C was converted to CO_2 and H was converted to H_2O to produce 405 mg of CO_2 and 94.7 mg of H_2O. Which of the following is the formula for the compound?

(a) $C_7H_8N_2O_2$, (b) C_6H_5NO, (c) $C_7H_7NO_2$, (d) $C_6H_4N_2O_4$, (e) $C_3H_5N_3O_6$

Two Reactions Occurring Simultaneously

70. A 1.020-g sample known to contain only magnesium carbonate, $MgCO_3$, and calcium carbonate, $CaCO_3$, was heated until the carbonates were decomposed to oxides as indicated by the following equations. After heating, the residue weighed 0.536 g. What masses of $MgCO_3$ and $CaCO_3$ were present in the original sample?

$$CaCO_3 \xrightarrow{\Delta} CaO + CO_2$$

$$MgCO_3 \xrightarrow{\Delta} MgO + CO_2$$

71. A mixture that contained 1.8391 g of NaCl and 2.3432 g of KCl was treated with excess $AgNO_3$. The precipitated AgCl was collected, washed, and dried. What mass of AgCl was produced?

$$NaCl + AgNO_3 \longrightarrow AgCl\ (s) + NaNO_3$$

$$KCl + AgNO_3 \longrightarrow AgCl\ (s) + KNO_3$$

72. A sample that was known to be 47.39% NaCl and 52.61% KCl by mass was treated with excess $AgNO_3$. The precipitated AgCl was collected, washed, and dried and found to have a mass of 4.2367 g. What was the mass of the sample?

73. A 0.7749-g sample that contained only sodium sulfate, Na_2SO_4, and potassium sulfate, K_2SO_4, was dissolved in water, and an excess of barium chloride solution was added. The following equations represent the reactions that occurred in the solution. The solid $BaSO_4$ was collected by filtration and dried. It had a mass of 1.167 g. What masses of Na_2SO_4 and K_2SO_4 were present in the original sample?

$$Na_2SO_4 + BaCl_2 \longrightarrow 2NaCl + BaSO_4$$

$$K_2SO_4 + BaCl_2 \longrightarrow 2KCl + BaSO_4$$

74. A sample of methane, CH_4, was burned in a quantity of oxygen that was insufficient for complete combustion. The combustion products were a mixture of carbon monoxide, CO, carbon dioxide, CO_2, and water that had a mass of 24.8 g. The water was separated from the CO and CO_2 and found to have a mass of 12.6 g. Calculate (a) the mass of CH_4 in the sample, (b) the mass of CO formed, (c) the mass of CO_2 formed, and (d) the mass of oxygen used in the reactions. Equation 1 represents the complete combustion of CH_4 and Equation 2 represents the partial combustion of CH_4.

(1) $CH_4 + 2O_2 \longrightarrow CO_2 + 2H_2O$

(2) $2CH_4 + 3O_2 \longrightarrow 2CO + 4H_2O$

Concentrations of Solutions—Percent by Mass

75. What mass of silver nitrate, $AgNO_3$, is required to prepare 500 g of a 3.50% solution of $AgNO_3$?

76. Calculate the masses of potassium dichromate, $K_2Cr_2O_7$, and water in 400 g of a 6.85% solution of $K_2Cr_2O_7$.

77. What mass of a 6.85% solution of potassium dichromate contains 40.0 g of $K_2Cr_2O_7$? What mass of water does this amount of solution contain?

78. Calculate the mass of an 8.30% solution of ammonium chloride, NH_4Cl, that contains 100 g of water. What mass of NH_4Cl does this amount of solution contain?

79. The density of a 7.50% solution of ammonium chloride, NH_4Cl, solution is 1.02 g/mL. What mass of NH_4Cl does 250 mL of this solution contain?

80. The density of a 7.50% solution of ammonium sulfate, $(NH_4)_2SO_4$, is 1.04 g/mL. What mass of $(NH_4)_2SO_4$ would be required to prepare 500 mL of this solution?

81. What volume of the solution of NH_4Cl described in Exercise 79 contains 50.0 g of NH_4Cl?

82. What volume of the solution of $(NH_4)_2SO_4$ described in Exercise 80 contains 50.0 g of $(NH_4)_2SO_4$?

83. A reaction requires 37.8 g of $(NH_4)_2SO_4$. What volume of the solution described in Exercise 80 would you use if you wished to use a 10.0% excess of $(NH_4)_2SO_4$?

Concentrations of Solutions—Molarity

84. What is the molarity of a solution that contains 490 g of phosphoric acid, H_3PO_4, in 2.00 L of solution?
85. What is the molarity of a solution that contains 1.37 g of sodium chloride in 25.0 mL of solution?
86. What is the molarity of a barium chloride solution prepared by dissolving 2.00 g of $BaCl_2 \cdot 2H_2O$ in enough water to make 200 mL of solution?
87. What mass of potassium sulfate, K_2SO_4, is contained in 750 mL of 2.00 molar solution?
88. What mass of calcium chloride, $CaCl_2$, is required to prepare 5.00 L of 0.450 molar solution?
89. What mass of hydrated copper(II) sulfate, $CuSO_4 \cdot 5H_2O$, is needed to prepare one liter of a 0.675 molar solution of $CuSO_4$?
90. What volume of 0.230 *M* Na_2CO_3 solution can be prepared from 12.4 g of Na_2CO_3?
91. What volume of 0.850 *M* $(COOH)_2$ solution can be prepared from 75.0 g of oxalic acid dihydrate, $(COOH)_2 \cdot 2H_2O$?
92. What volume of 0.250 molar nitric acid, HNO_3, is required to react with 3.70 g of calcium hydroxide, $Ca(OH)_2$, according to the following equation?

$$2HNO_3 + Ca(OH)_2 \longrightarrow Ca(NO_3)_2 + 2H_2O$$

93. What volume of a 2.00 molar solution of sulfuric acid, H_2SO_4, is required to react with 250 g of calcium carbonate, $CaCO_3$, according to the following equation?

$$CaCO_3 + H_2SO_4 \longrightarrow CaSO_4 + CO_2 + H_2O$$

94. What volume of a 0.200 molar solution of silver nitrate, $AgNO_3$, is required to react with 12.2 g of potassium phosphate, K_3PO_4, according to the following equation?

$$3AgNO_3 + K_3PO_4 \longrightarrow Ag_3PO_4 + 3KNO_3$$

95. What volume of 0.850 *M* $(COOH)_2$ solution would be required to react with 40.0 mL of 0.230 *M* Na_2CO_3 solution?

$$Na_2CO_3 + (COOH)_2 \longrightarrow Na_2(COO)_2 + H_2CO_3$$

96. What volume of 0.250 *M* $FeSO_4$ solution would be required to react with 10.0 mL of 0.200 *M* $KMnO_4$ (in sulfuric acid solution) according to the following equation?

$$10FeSO_4 + 2KMnO_4 + 8H_2SO_4 \longrightarrow 5Fe_2(SO_4)_3 + 2MnSO_4 + K_2SO_4 + 8H_2O$$

97. Commercially available concentrated sulfuric acid is 18.0 *M* H_2SO_4. Calculate the volume of concentrated sulfuric acid required to prepare 2.00 L of 0.200 *M* H_2SO_4 solution.
98. Commercial concentrated hydrochloric acid is 12.0 *M* HCl. What volume of concentrated hydrochloric acid is required to prepare 1.50 L of 1.20 *M* HCl solution?
99. Calculate the volume of 4.00 *M* NaOH solution required to prepare 100 mL of a 0.500 *M* solution of NaOH.
100. Calculate the volume of 0.0500 *M* $Ba(OH)_2$ solution that contains the same number of moles of $Ba(OH)_2$ as 120 mL of 0.0800 *M* $Ba(OH)_2$ solution.
101. Calculate the resulting molarity when 150 mL of 0.450 *M* NaCl solution is mixed with 100 mL of 2.50 *M* NaCl solution.
102. Calculate the resulting molarity when 200 mL of 3.00 *M* H_2SO_4 solution is mixed with 120 mL of 6.00 *M* H_2SO_4 solution.

3 Atomic Structure

Throughout recorded history, and undoubtedly well before, man has wondered about the nature of matter. The Greek philosophers were among the first to document their observations and "theories" on natural phenomena. We can easily imagine one of the philosophers walking along the seashore and observing that while the sea appears to be continuous matter, as does the beach from a distance, the beach underfoot obviously consists of discrete (discontinuous) particles of sand. He might logically have wondered if all matter that appears to be continuous might not actually consist of discrete particles. Unfortunately, the Greeks did not have available the devices we now have to allow closer examination of the composition of matter. Their minds were very active, but they were forced to restrict testing of their ideas almost exclusively to mental exercise and logic rather than experiment.

Around 400 BC, Democritus suggested that all matter is composed of tiny, discrete, indivisible particles which he called atoms. His ideas were rejected for 2000 years, but they began to make sense in the late eighteenth century. The term *atom* comes from the Greek language and means "not divided" or "indivisible." Both Plato and Aristotle rejected the notion of atoms, as did most other Greek philosophers.

In 1803, an English schoolteacher named John Dalton began the serious evolution of modern ideas about the existence and the nature of atoms. He summarized and supplemented the nebulous concepts of early philosophers and scientists. Taken together they form the core of **Dalton's Atomic Theory,** one of the highlights of scientific thought. In condensed form, Dalton's postulates may be stated this way:

The radius of a calcium atom is only 1.97×10^{-8} cm and its mass is 6.66×10^{-23} g.

Statement 3 is true for ordinary chemical reactions. However, it is not true for *nuclear* reactions (Chapter 28).

1. An element is composed of extremely small *indivisible* particles called atoms.
2. All atoms of a given element have identical properties, which differ from those of other elements.
3. Atoms cannot be created, destroyed, or transformed into atoms of another element.
4. Compounds are formed when atoms of different elements combine with each other in simple numerical ratios.
5. The relative numbers and kinds of atoms are constant in a given compound.

The Law of Conservation of Matter (Section 1–1) and the Law of Definite Proportions (Section 2–1) were known in Dalton's time and were the basis for his atomic theory. Dalton was also the first to state the Law of Multiple Proportions (Section 2–5), which is related to postulates 4 and 5.

Dalton believed that atoms were solid indivisible spheres, an idea we now reject, but he showed remarkable insight into the nature of matter and its interactions. Some of his postulates could not be verified (or refuted) experimentally at the time, but all were in accord with the experimental observations of his day. As we shall see, even with their errors, Dalton's postulates provided a framework that could be modified and expanded, and thus Dalton is considered the father of modern atomic theory.

3–1 Fundamental Particles

As knowledge of the structure of atoms has accumulated, we have been able to systematize chemical facts in ways that enhance our understanding of matter. Current atomic theory is an especially valuable tool in understanding the forces that hold atoms in chemical combination with each other. It is considerably less than complete, but nonetheless very useful. In our study of atomic theory, we look first at the basic building blocks of all atoms, which we call **fundamental particles.**

Atoms, hence all matter, consist principally of three fundamental particles,* *electrons, protons,* and *neutrons.* Knowledge of the nature and functions of these particles is essential to understanding chemical interactions. The masses and charges of the three fundamental particles are shown in Table 3–1. The mass of an electron is very small compared with the mass of either a proton or a neutron. The charge on a proton is equal in magnitude, but opposite in sign, to the charge on an electron.

We are now ready to look at these particles in more detail and to consider how they are distributed in the atom.

* Many other smaller particles such as positrons, neutrinos, pions, and muons have been discovered, but it is not necessary to study their characteristics to learn the fundamentals of atomic structure that are important in chemical reactions.

TABLE 3–1 Fundamental Particles of Matter

Particle	Mass	Charge
Electron (e^-)	0.00055 amu	1−
Proton (p or p^+)	1.0073 amu	1+
Neutron (n or n^0)	1.0087 amu	none

3–2 Electrons

The process is called chemical electrolysis.

In the early 1800s, the English chemist Humphrey Davy found that when he passed electrical current through some substances, the substances decomposed. This led him to propose that the elements of a chemical compound are held together by electrical forces. In 1832–33, Michael Faraday, Davy's protegé, determined the quantitative relationship between the amount of electricity used in electrolysis and the amount of chemical reaction that occurs (Chapter 19). In 1874 George Stoney carefully studied Faraday's work and suggested that units of electrical charge are associated with atoms. In 1891 he suggested that they be named *electrons*.

The most convincing evidence for the existence of electrons came from experiments in which an electric current at high voltage was passed through gases at very low pressures in *cathode ray tubes* (Figure 3–1). When a high voltage is applied across two electrodes sealed in a glass tube, from which enough air has been withdrawn, rays begin to emanate from the cathode (negative electrode), travel in straight lines to the anode (positive electrode), and cause the walls opposite the cathode to glow. If a small object is placed in the path of the cathode rays, it casts a shadow on a zinc sulfide screen placed behind the anode. The shadow proves that the rays travel from the cathode toward the anode, and that therefore they must be negatively charged.

Additionally, the rays are deflected by both magnetic and electrical fields in the directions expected for negatively charged particles. Numerous experiments showed that the properties of cathode rays are the same regardless of the metal used to construct the cathode, or the nature of the gas placed in the tube. In 1897, J. J. Thomson demonstrated that cathode rays are actually streams of negatively charged particles that he called **electrons,** the name Stoney had given them in 1891.

By studying the degree of deflections of cathode rays in different magnetic and electric fields, Thomson determined the charge (e) to mass (m) ratio for electrons. He found the ratio to be

$$e/m = 1.76 \times 10^8 \text{ coulomb}^* \text{ per gram}$$

for several different gases in the cathode ray tube. The clear implication of Thomson's work was that electrons are fundamental particles present in all atoms. We now know that this is true, and that all atoms contain integral numbers of electrons.

* The coulomb is the standard unit of *quantity* of electrical charge, and is defined as that amount of electricity which will deposit 0.001118 g of silver in an apparatus set up for plating silver. It corresponds to a *current of one ampere flowing for one second.*

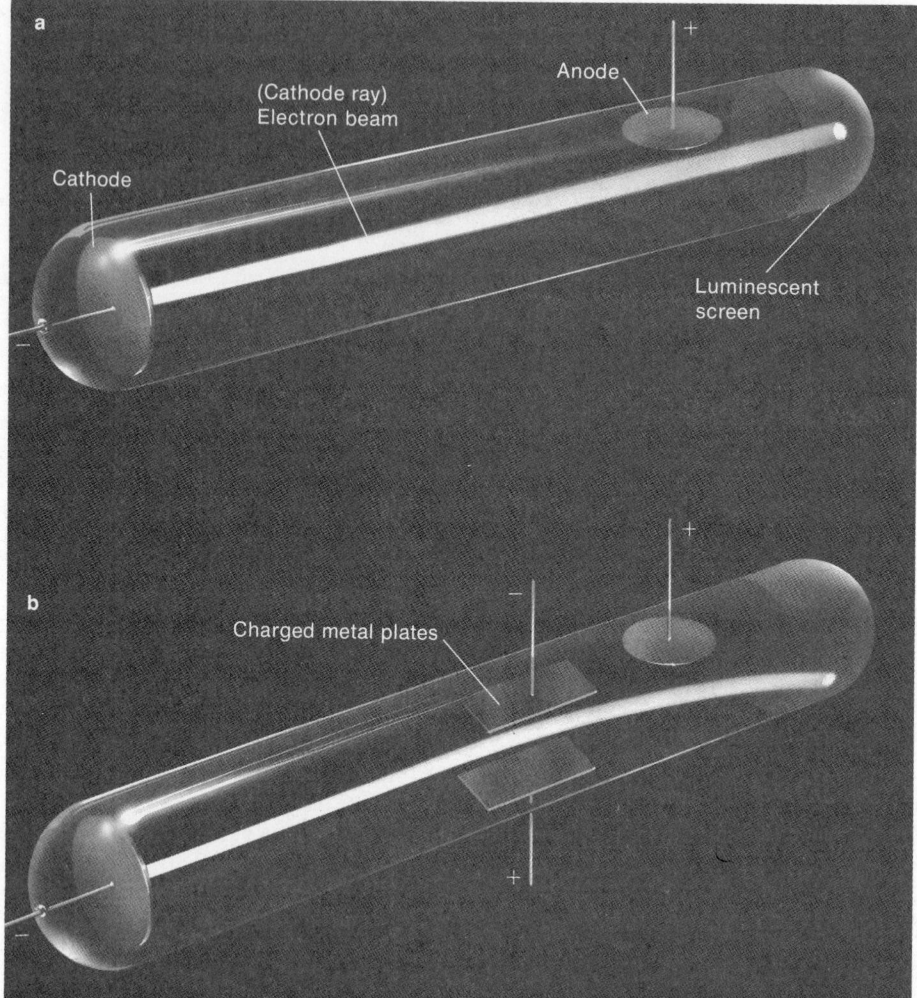

FIGURE 3–1 (a) A discharge tube showing the production of an electron beam (cathode ray). (b) Experiment showing that the electron beam possesses a negative charge; note position of the beam in an electric field. The beam is repelled by the negative plate and attracted to the positive plate. The gas pressure in the tube is very low.

Once the charge-to-mass ratio for electrons had been determined, additional experiments were necessary to determine the value of either mass or charge, so that the other could be calculated. In 1909 Robert Millikan performed his famous "oil drop" experiment and determined the charge on the electron. The apparatus is illustrated in Figure 3–2. The rates at which the tiny, uncharged droplets of oil fall in air are determined by their mass and size, both of which Millikan could measure. X-rays were then used to cause the droplets of oil to acquire electrical charges. Millikan measured the effect of an electrical field on the rates at which charged oil droplets fall under the influence of gravity. What he did was to apply an electric field sufficiently strong to keep the negatively charged droplets from falling. From these data he calcu-

X-rays are radiations of much shorter wavelength than visible light (Section 3–10), and are sufficiently energetic to knock electrons out of the atoms in the air. In Millikan's experiment, these free electrons became attached to some of the oil droplets.

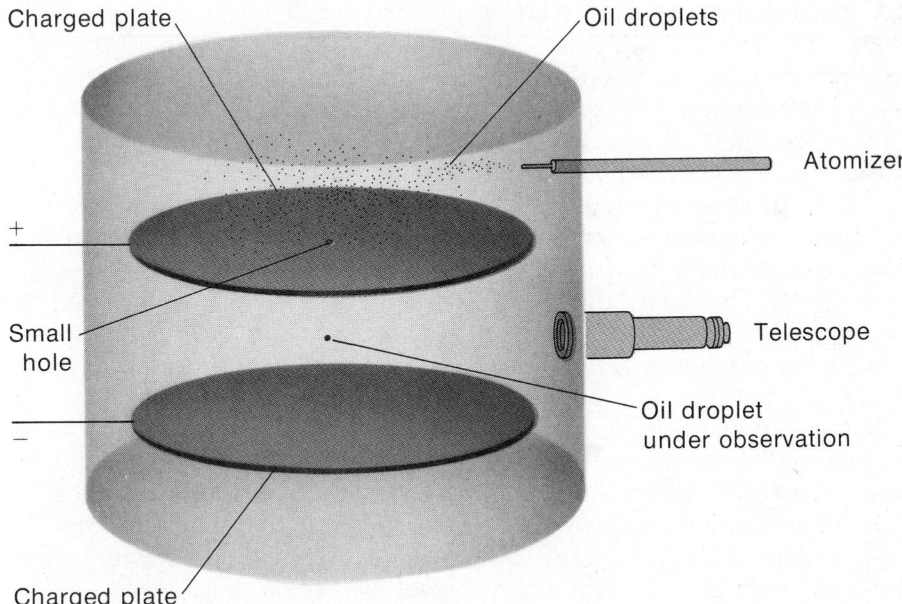

FIGURE 3–2 Millikan oil drop experiment. Tiny oil droplets are produced by an atomizer and then irradiated with x-rays so that some of the oil droplets become negatively charged. When the voltage between the plates is increased, a negatively charged drop falls more slowly, because it is attracted to the positively charged plate. At one particular voltage, the electrical and gravitational forces on the drop are exactly balanced, and the drop remains stationary. Knowing this voltage and the mass of the drop, it is possible to calculate the charge on the drop.

lated the charges on the droplets. All charges turned out to be integral multiples of the smallest observed charge. Millikan assumed that the smallest oil droplet charge was the charge on one electron, and he calculated a value of 1.60×10^{-19} coulomb for that charge.

The charge-to-mass ratio determined earlier by Thomson,

$$e/m = 1.76 \times 10^{8} \text{ coulomb/g}$$

was then used in inverse form to calculate the mass of the electron

$$m = \frac{1.00 \text{ g}}{1.76 \times 10^{8} \text{ coulomb}} \times 1.60 \times 10^{-19} \text{ coulomb} = 9.09 \times 10^{-28} \text{ g}$$

The mass of a hydrogen atom, the lightest of all atoms, is nearly 2000 times greater than the mass of an electron. Millikan's oil drop experiment stands as one of the most fundamental of all classic scientific experiments; it was the first experiment to suggest that all atoms contain integral numbers of electrons, a fact we now know to be true.

The value of e/m obtained by Thomson and the value of m obtained by Millikan differ slightly from the modern values given in this text. Scientists who work at the forefront of a research area frequently build their own apparatus, and the values they obtain may not be as accurate as desired. Important experiments are repeated many times and values are refined as equipment and techniques are improved.

3–3 Canal Rays and Protons

Careful observations of the cathode ray tube indicated that it also generates a stream of positively charged particles that moves from the anode to the cathode, first observed by Eugen Goldstein in 1886. These were termed **canal rays** because they were observed occasionally to pass through a channel or "canal" drilled in the negative electrode (Figure 3–3). We now know that these *positive rays,* or *positive ions,* are created when cathode rays knock electrons from the gaseous atoms in the tube. Different elements give positive ions with different e/m ratios. We also know that the positive charge resides in the **proton,** *a fundamental particle with a charge equal in magnitude but opposite in sign to the charge on the electron, and with a mass nearly 2000 times that of the electron.*

By the first decade of this century, therefore, it was clear that each atom contained regions of both positive and negative charge. The question was, how are these charges distributed? The dominant view was that the positive charge was distributed homogeneously throughout the atom and that the negative charge was localized on electrons, which were scattered about the atom like raisins in a raisin loaf. This picture, however, was to be short lived.

3–4 Rutherford and the Nuclear Atom

Alpha particles are now known to be helium atoms minus their two electrons, or helium nuclei, which have 2+ charges (see Chapter 28).

Radioactivity is contrary to the Daltonian idea of the indivisibility of atoms.

By the time Millikan determined the charge on the electron, Ernest Rutherford had established the fact that alpha (α) particles are positively charged particles that can be emitted by radioactive atoms, i.e., atoms that undergo spontaneous disintegration. In 1910 Rutherford's research group carried out a series of experiments that had enormous impact on the scientific world. They bombarded a very thin gold foil with α-particles from a radioactive source. A fluorescent zinc sulfide screen was used behind the foil to observe the extent to which the α-particles were scattered by the gold foil (Figure 3–4). Scintillations, or flashes, on the screen caused by the individual α-particles were counted to determine the relative numbers of α-particles deflected at various angles. Alpha particles were known to be extremely dense, much more so than even the heavy metal, gold. Further, they were known to be emitted at high kinetic energies. In light of the dominant picture of the atom, therefore, it was expected that nearly all of the α-particles would pass through the foil with little

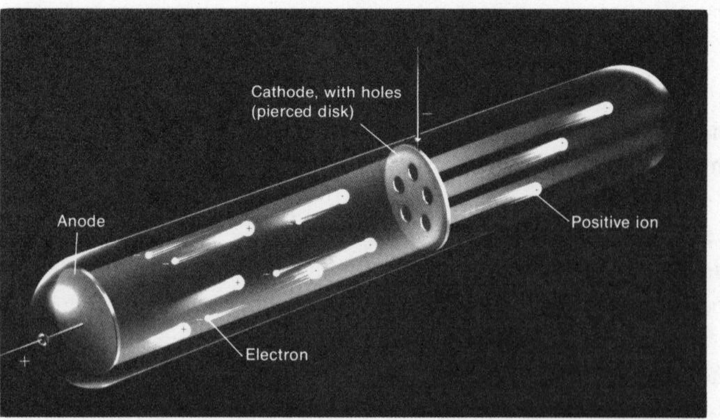

FIGURE 3–3 A cathode ray tube with a different design and with a perforated cathode was used to demonstrate that positive rays travel toward the cathode. Like cathode rays, these positive rays are deflected by magnetic or electric fields, but in opposite directions. When different elements are in the tube, positive ions with different e/m ratios are observed.

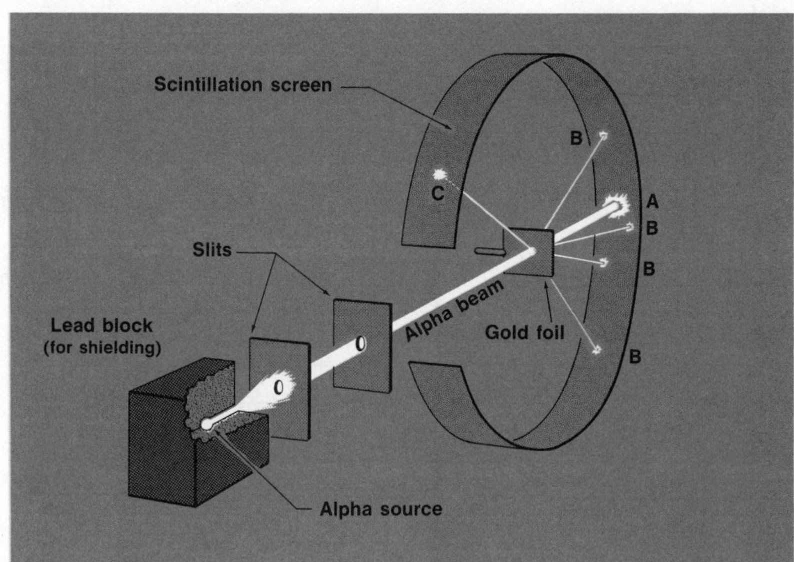

FIGURE 3–4 Rutherford scattering experiment. A narrow beam of alpha particles (helium atoms stripped of their electrons) from a radioactive source was directed at a very thin gold foil. Most of the particles passed right through the foil, striking the screen at point A. Many of the particles were deflected through moderate angles, striking the screen at points such as those labeled B. The larger deflections were surprises, but the 0.001% of the total that were reflected at acute angles, C, were totally unexpected. Similar results were observed using foils of other metals.

or no deflection. However, a few were observed to be deflected through large angles. A few α-particles even bounced back from the gold foil in the direction from which they had come. Rutherford was astounded! In his own words,

> It was quite the most incredible event that has ever happened to me in my life. It was almost as if you fired a 15-inch shell into a piece of tissue paper and it came back and hit you.

Rutherford performed a mathematical analysis which showed that the α-particle scattering was caused by repulsion from the positively charged centers in the gold foil. He concluded that the mass of one of these centers is very nearly equal to that of a gold atom, but that the diameter is no more than 1/10,000 that of an atom. After many experiments with foils of different metals yielded similar results, Rutherford concluded that each atom contains a tiny, positively charged, massive center which he called an **atomic nucleus.** Most α-particles pass through the metal foils undeflected because atoms are primarily empty space, a fact that was unknown prior to Rutherford's experiments. The few particles that are deflected are the ones that come close to or collide with the heavy, highly charged metal nuclei (Figure 3–5).

3–5 Neutrons

The third fundamental particle, the *neutron,* eluded discovery until 1932, when James Chadwick correctly interpreted experiments on the bombardment of beryllium with high-energy alpha particles. Later experiments showed that nearly all elements up to potassium, element 19, produce neutrons when they are bombarded with high-energy alpha particles. The **neutron** is an uncharged particle with a mass slightly greater than that of the proton. With its discovery, the picture of the nuclear atom was complete: Atoms consist of very small, very dense nuclei (composed of protons and neutrons) surrounded by clouds of electrons at relatively great distances from the nuclei. Nuclear diameters are about 10^{-5} nanometers; atomic diameters are about 10^{-1} nanometers.

This does not mean that elements above number 19 do not have neutrons, only that neutrons are not generally knocked out of atoms of higher atomic number by alpha-particle bombardment.

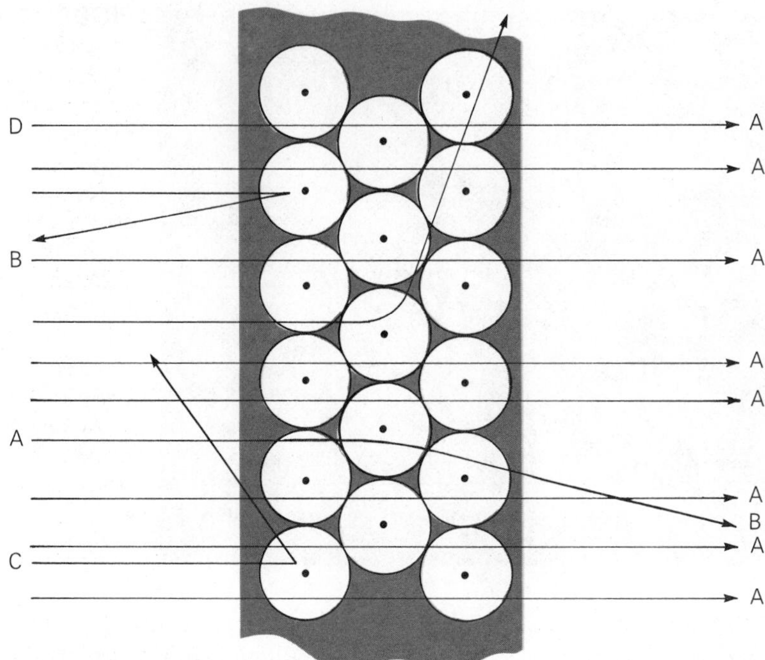

FIGURE 3–5 Interpretation of the Rutherford scattering experiment. The atom is pictured as consisting mostly of open space. At the center is a tiny and extremely dense nucleus that contains all of the atom's positive charge and nearly all of the mass. The electrons are thinly distributed throughout the open space. Most of the positively charged alpha particles (A) pass through the open space undeflected, not coming near any gold nuclei. Those that pass fairly close to a nucleus (B) are repelled by electrostatic force and thereby deflected. The few particles that are on a "collision course" with gold nuclei are repelled backward at acute angles (C). Calculations based on the results of the experiment indicated that the diameter of the open-space portion of the atom is from 10,000 to 100,000 times greater than the diameter of the nucleus.

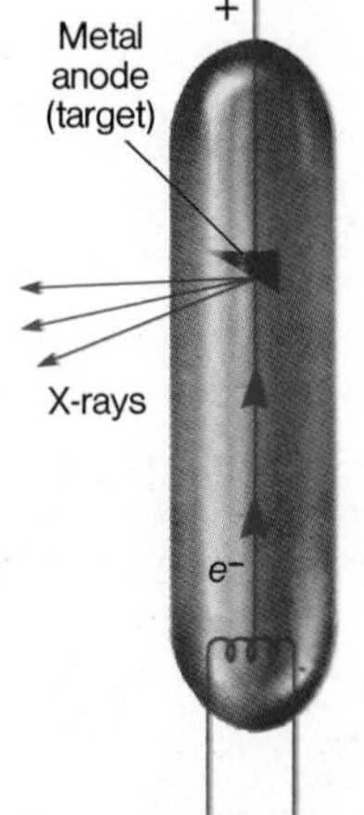

FIGURE 3–6 A simplified representation of the production of x-rays by bombardment of a solid target with a high energy beam of electrons.

3–6 Atomic Number and Mass Number

Only a few years after Rutherford's experiments showed that atomic nuclei are very heavy, positively charged centers of atoms, H. G. J. Moseley developed a technique to determine the magnitude of the positive charge. Max von Laue had demonstrated that x-rays could be diffracted by certain crystals into a spectrum in much the same way that visible light can be separated into its component colors. Moseley generated x-rays by aiming a beam of high-energy electrons at a target made of a block of a single pure element (Fig. 3–6).

When the spectra of x-rays produced by targets of different elements were recorded photographically, each was observed to consist of a series of lines representing x-rays at various wavelengths. Comparison of the spectra of different elements revealed that corresponding lines within a given series were displaced regularly toward shorter wavelengths as atomic weights of the target materials increased (with three exceptions). Moseley subjected his data to mathematical analysis and concluded that *each element differs from the preceding element by having one more positive charge in its nucleus.* For the first time it was possible to arrange all known elements in order of increasing nuclear charge. A plot summarizing this interpretation of Moseley's data appears in Figure 3–7.

We now know that every nucleus contains an integral number of protons, which is exactly equal to the number of electrons in a neutral atom of the element. The proton was actually discovered by Rutherford and Chadwick in 1919 as a particle that is emitted by bombardment of certain atoms with alpha particles. Every hydrogen atom contains one proton, every helium atom contains two protons, and every lithium atom contains three protons. The number of protons in a nucleus determines its identity and is known as its **atomic number.** The atomic number of an element is *defined* as the number of protons or positive charges in its nucleus.

Although the existence of neutrons was predicted in 1920 by Rutherford, Chadwick, and W. D. Harkins, 12 years passed before the prediction was verified in 1932. Nuclei of all atoms except the common form of hydrogen contain neutrons. However, for most elements, different nuclei of the same element can contain different numbers of neutrons.

For example, there are three distinctly different kinds of hydrogen atoms, commonly called hydrogen, deuterium, and tritium. All hydrogen atoms contain one proton in their nuclei; the common form of hydrogen contains no neutrons, but deuterium atoms contain one neutron and tritium atoms contain two neutrons in their nuclei. A simplified representation of the three kinds of hydrogen atoms is shown in Figure 3–8.

Atoms of an element that differ in mass are called *isotopes.* **Isotopes** are defined as atoms of the same element that contain different numbers of neutrons in their nuclei. Isotopes of a given element contain the same numbers of protons and electrons because they are atoms of the same element.

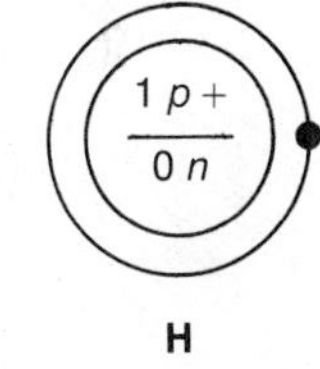

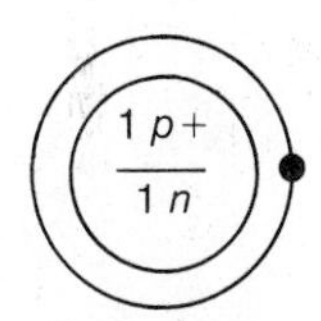

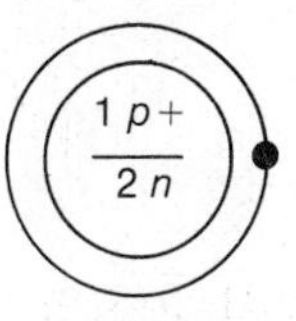

FIGURE 3–8 Isotopes of hydrogen (simplified representation).

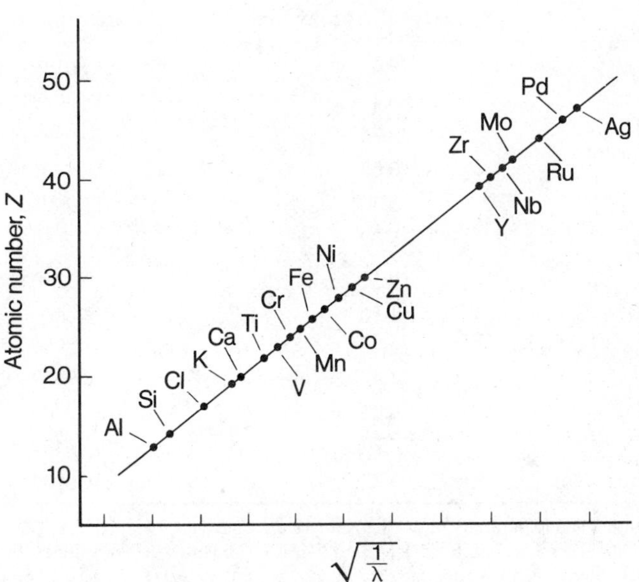

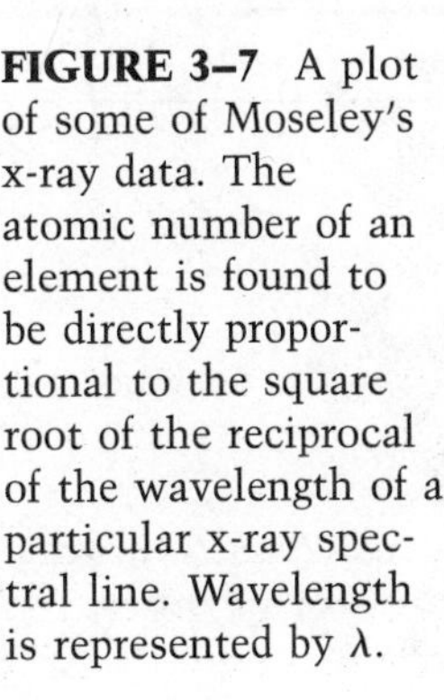

FIGURE 3–7 A plot of some of Moseley's x-ray data. The atomic number of an element is found to be directly proportional to the square root of the reciprocal of the wavelength of a particular x-ray spectral line. Wavelength is represented by λ.

The **mass number** of an atom is the sum of the number of protons and the number of neutrons in its nucleus, i.e.,

$$\text{mass number} = \text{number of protons} + \text{number of neutrons}$$
$$= \text{atomic number} + \text{neutron number}$$

Mass numbers are always integral numbers. The mass number for normal hydrogen atoms is 1, for deuterium 2, and for tritium 3. The composition of a nucleus is indicated by its **nuclide symbol.** This consists of the symbol for the element (E), with the atomic number (Z) written as a subscript at the lower left and the mass number (A) as superscript at the upper left, $^{A}_{Z}E$. By this system, the three isotopes of hydrogen are designated as $^{1}_{1}H$, $^{2}_{1}H$, and $^{3}_{1}H$.

Although some elements (such as fluorine and iodine) exist in only one form, most elements occur in nature as isotopic mixtures. Some examples of natural isotopic abundances are given in Table 3–2. The percentages are based on the numbers of naturally occurring atoms of each isotope, not their masses.

Isotopic abundances are determined in mass spectrometers, instruments that measure the mass-to-charge ratio of charged particles (Figure 3–9). A gas sample at very low pressure is bombarded with high-energy electrons. This bombardment causes electrons to be ejected from a few of the gas molecules, and creates positive ions, which are then focused into a very narrow beam and accelerated by an electric field toward a magnetic field. The magnetic field deflects the ions from their straight-line path. The extent to which the beam of ions is deflected depends upon four factors:

1. Magnitude of the accelerating voltage. The range varies from about 500 to about 2000 volts. Higher voltages result in beams of more rapidly moving particles that are deflected less than the beams of slower-moving particles produced by lower voltages.

TABLE 3–2 Naturally Occurring Isotopic Abundances for Some Common Elements

Element	Isotope	% Natural Abundance	Mass (amu)
Boron	$^{10}_{5}B$	19.6	10.01294
	$^{11}_{5}B$	80.4	11.00931
Oxygen	$^{16}_{8}O$	99.759	15.99491
	$^{17}_{8}O$	0.037	16.99914
	$^{18}_{8}O$	0.204	17.99916
Magnesium	$^{24}_{12}Mg$	78.70	23.98504
	$^{25}_{12}Mg$	10.13	24.98584
	$^{26}_{12}Mg$	11.17	25.98259
Chlorine	$^{35}_{17}Cl$	75.53	34.96885
	$^{37}_{17}Cl$	24.47	36.9659
Uranium	$^{234}_{92}U$	0.0057	234.0409
	$^{235}_{92}U$	0.72	235.0439
	$^{238}_{92}U$	99.27	238.0508

The twenty elements that have no other stable naturally occurring isotopes are: $^{9}_{4}Be$, $^{19}_{9}F$, $^{23}_{11}Na$, $^{27}_{13}Al$, $^{31}_{15}P$, $^{45}_{21}Sc$, $^{55}_{25}Mn$, $^{59}_{27}Co$, $^{75}_{33}As$, $^{89}_{39}Y$, $^{93}_{41}Nb$, $^{103}_{45}Rh$, $^{127}_{53}I$, $^{133}_{55}Cs$, $^{141}_{59}Pr$, $^{159}_{65}Tb$, $^{165}_{67}Ho$, $^{169}_{69}Tm$, $^{197}_{79}Au$, $^{209}_{83}Bi$. However, there are artificially produced isotopes of these elements.

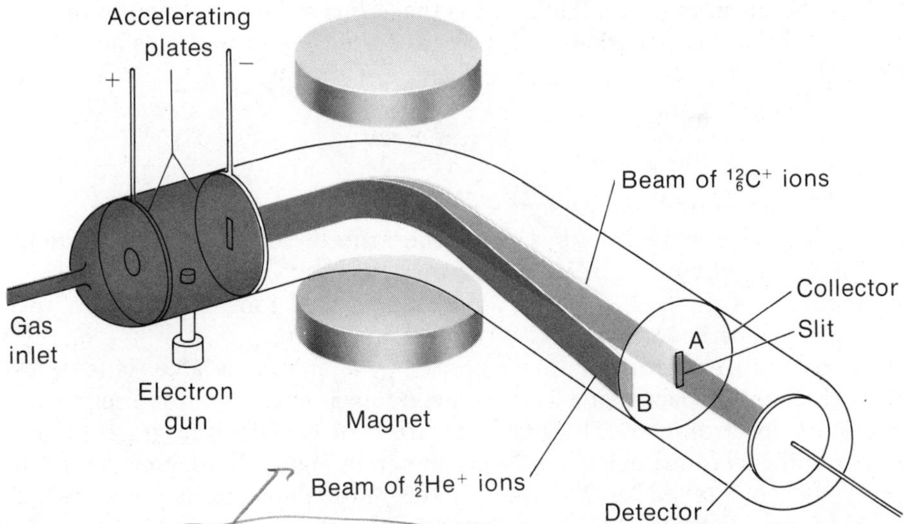

FIGURE 3–9 The mass spectrometer. In the mass spectrometer, gas molecules at low pressure are ionized and accelerated by an electric field. The ion beam is then passed through a magnetic field. In that field the beam is resolved into components, each containing particles of equal mass. Lighter particles are deflected more strongly than heavy ones. In a beam containing $^{12}_{6}C^{+}$ and $^{4}_{2}He^{+}$ ions, the lighter $^{4}_{2}He^{+}$ ions would be deflected more than the heavier $^{12}_{6}C^{+}$ ions. The spectrometer shown is adjusted to defect the $^{12}_{6}C^{+}$ ions. By changing the magnitude of the magnetic or electric field, one could move the beam of $^{4}_{2}He^{+}$ ions striking the collector from B to A where it would be detected.

2. Strengths of the electric and magnetic fields. Stronger fields deflect a given beam more than weaker fields.
3. Masses of the particles. Because of their inertia, heavier particles are deflected less than lighter particles that carry the same charge.
4. Charges on the particles. Highly charged particles interact more strongly with magnetic and electric fields and are thus deflected more than particles of equal mass with smaller charges.

The mass spectrometer is used to determine atomic weights of elements as well as isotopic abundances, which are necessary to determine the atomic weights of elements that occur in nature as two or more isotopes (as we will see in the next section).

Helium occurs in nature almost exclusively as $^{4}_{2}He$. Let's see how its atomic mass is measured. To simplify the picture, we'll assume that only ions with 1+ charge are formed in the experiment illustrated in Figure 3–9.

A carefully measured, fixed accelerating voltage is applied and a sample of vaporized carbon-12, $^{12}_{6}C$, is fed into the mass spectrometer. The magnetic field strength that causes $^{12}_{6}C^{+}$ ions to arrive at point *A* is measured. A sample of helium is then fed into the mass spectrometer with exactly the same accelerating voltage, and the lighter $^{4}_{2}He^{+}$ ions are deflected more than the heavier $^{12}_{6}C^{+}$ ions to some point *B*. The magnetic field strength is then decreased slowly until the beam of lighter ions falls at point *A*. Now that the field strengths required to focus the two kinds of ions at the same point are known, the mass of the lighter ion can be calculated. The mass of the heavier ion is

12.00 amu because, as we shall see, ${}^{12}_{6}C$ is the (arbitrary) reference point on the atomic weight scale. The relationship between masses of particles and magnetic field strength, $\mathcal{H}$, is:

$$\frac{\text{mass } {}^{4}_{2}He^{+}}{\text{mass } {}^{12}_{6}C^{+}} = \left(\frac{\mathcal{H} \text{ for } {}^{4}_{2}He^{+}}{\mathcal{H} \text{ for } {}^{12}_{6}C^{+}}\right)^{2}$$

The $\mathcal{H}$'s are magnetic field strengths required to focus the beams of ions at point A. When the experiment is done, the ratio of the two magnetic field strengths is measured as 0.5776, so the mass of ${}^{4}_{2}He^{+}$ is found to be 0.3336 times the mass of the ${}^{12}_{6}C^{+}$ ion. Since the mass of ${}^{12}_{6}C^{+}$ ions is 12.00 amu, the mass of ${}^{4}_{2}He^{+}$ ions is: 0.3336 × 12.00 amu = 4.003 amu.

$(0.5776)^2 = 0.3336$

Neon occurs in nature as three isotopes: ${}^{20}_{10}Ne$, ${}^{21}_{10}Ne$ and ${}^{22}_{10}Ne$. This statement is based on the fact that a beam of neon ions is split into three segments in the mass spectrometer. The mass spectrum of Ne^{+} ions (a graph of the relative numbers of ions of each mass) is shown in Figure 3–10. Note that the isotope ${}^{20}_{10}Ne$, of mass 19.99244 amu, is the most abundant isotope (tallest peak) and accounts for 90.9% of the atoms, while ${}^{22}_{10}Ne$ accounts for 8.8%, and ${}^{21}_{10}Ne$ only 0.3% of the atoms.

3–7 The Atomic Weight Scale and Atomic Weights

As a result of action taken by the International Union of Pure and Applied Chemistry in 1962, the **atomic weight scale** is based on the carbon-12 isotope, whose mass is assigned a value of exactly 12 atomic mass units, amu. By definition *one amu is exactly* $\frac{1}{12}$ *the mass of a carbon-12 atom* and is approximately the mass of one atom of ${}^{1}_{1}H$, the lightest isotope of the lightest weight element.

In Section 2–3, we said that one mole of atoms contains 6.02×10^{23} atoms, and that the mass of one mole of atoms of any element, in grams, is numerically equal to the atomic weight of the element. Since the mass of one carbon-12 atom is exactly 12 amu, the mass of one mole of carbon-12 atoms is exactly 12 grams.

Let us now determine the relationship between atomic mass units and grams. To do this we calculate the number of atomic mass units in one gram of

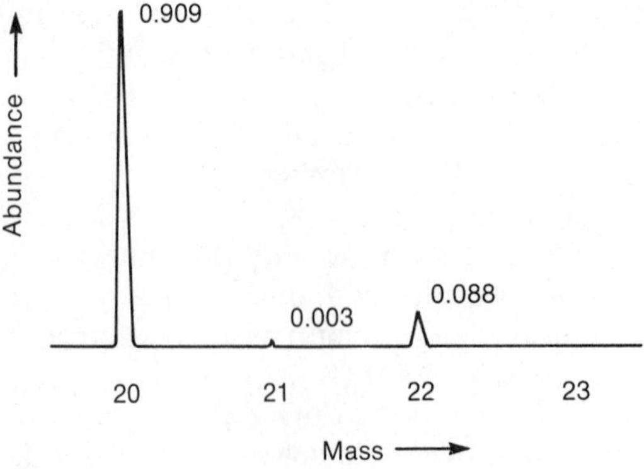

FIGURE 3–10 Mass spectrum of neon (1 + ions only). Neon contains three isotopes, of which neon-20 is by far the most abundant. The mass of that isotope, to five decimal places, is 19.99244 amu on the carbon-12 scale.

any element or isotope, say ${}^{12}_{6}C$

$$\underline{?} \text{ amu } {}^{12}_{6}C = \underbrace{\underbrace{1.00 \text{ g } {}^{12}_{6}C \times \frac{6.02 \times 10^{23} \ {}^{12}_{6}C \text{ atoms}}{12 \text{ g } {}^{12}_{6}C}}_{\text{no. of } {}^{12}_{6}C \text{ atoms in } 1.00 \text{ g of } {}^{12}_{6}C} \times \frac{12 \text{ amu } {}^{12}_{6}C}{1 \ {}^{12}_{6}C \text{ atom}}}_{\text{no. of amu } {}^{12}_{6}C \text{ in } 1.00 \text{ g of } {}^{12}_{6}C}$$

$$= 6.02 \times 10^{23} \text{ amu } {}^{12}_{6}C$$

Therefore, 1.00 g (${}^{12}_{6}C$ or anything else) = 6.02×10^{23} amu (${}^{12}_{6}C$ or anything else) or, dividing both sides by 6.02×10^{23}

1.66×10^{-24} g = 1.00 amu

You may wish to verify that the same result is obtained regardless of the element or isotope chosen.

The **atomic weight** of an element is defined as the *weighted average of the masses of its constituent isotopes.* Please note that atomic weight is NOT the sum of an element's protons and neutrons. That is the mass number. Most atomic weights are fractional numbers, not integers.

EXAMPLE 3–1

Three isotopes of magnesium occur in nature. Their abundances and masses are listed below. Use this information to calculate the atomic weight of magnesium.

Isotope	% Abundance	Mass (amu)
${}^{24}Mg$	78.70	23.98504
${}^{25}Mg$	10.13	24.98584
${}^{26}Mg$	11.17	25.98259

Solution

We multiply the fraction of each isotope by its mass and add these numbers to obtain the atomic weight of magnesium.

$$\begin{aligned} \text{at. wt.} &= 0.7870\ (23.98504 \text{ amu}) + 0.1013\ (24.98584 \text{ amu}) \\ &= 18.88 \text{ amu} + 2.531 \text{ amu} + 2.902 \text{ amu} \\ &= 24.31 \text{ amu (to four significant figures)} \end{aligned}$$

The two heavier isotopes make small contributions to the atomic weight of magnesium, since most magnesium atoms are the lightest isotope.

Example 3–2 shows how the process can be reversed so that percent abundances can be calculated from isotopic masses and from the atomic weight of an element that occurs in nature as a mixture of two isotopes.

EXAMPLE 3–2

The atomic weight of gallium is 69.72 amu. The masses of the naturally occurring isotopes are 68.9257 amu for ${}^{69}Ga$ and 70.9249 amu for ${}^{71}Ga$. Calculate the percent abundance of each isotope.

Solution

We can represent the fraction of each isotope algebraically.

Let

x = fraction of ^{69}Ga

then

$(1 - x)$ = fraction of ^{71}Ga

When a quantity is represented by fractions, the sum of the fractions must always be unity. In the present case, $x + (1 - x) = 1$.

Atomic weight is defined as the weighted average of the masses of the constituent isotopes. Therefore, the fraction of each isotope is multiplied by its mass and the sum of the results is equal to the atomic weight.

$$x(68.9257 \text{ amu}) + (1 - x)(70.9249 \text{ amu}) = 69.72 \text{ amu}$$
$$68.9257x + 70.9249 - 70.9249x = 69.72$$
$$-1.9992x = -1.20$$
$$x = 0.600$$

$$x = 0.600 = \text{fraction of } ^{69}\text{Ga} \therefore \underline{60.0\% \ ^{69}\text{Ga}}$$
$$(1 - x) = 0.400 = \text{fraction of } ^{71}\text{Ga} \therefore \underline{40.0\% \ ^{71}\text{Ga}}$$

3–8 Nuclear Stability and Binding Energy

It is an experimentally observed fact that the mass of an atom is always *less* than the sum of the masses of its constituent particles. We now know why this **mass defect** occurs, and we also know that the mass deficiency is in the nucleus of the atom and has nothing to do with the electrons. However, since tables of masses of isotopes include the electrons, we shall also include them.

The mass defect for a nucleus is the difference between the sum of the masses of electrons, protons, and neutrons in the atom, sometimes called calculated mass, and the actual measured mass of the atom.

$$\text{mass defect} = (\text{sum of masses of all } e^-, p^+, \text{ and } n^0) - (\text{actual mass of atom})$$

For most naturally occurring isotopes, the mass defect is only about 0.15% or less of the calculated mass of an atom.

EXAMPLE 3–3

Calculate the mass defect for chlorine-35 atoms. The actual mass of a chlorine-35 atom is 34.9689 amu.

Solution

The mass number of this isotope is 35. Because all Cl atoms contain 17 protons, there are 35 − 17 = 18 neutrons. Because there are 17 protons, there must be 17 electrons in a neutral atom. First, we sum the masses of the constituent particles, 17 protons, 17 electrons, and 18 neutrons

$$17 \times 1.0073 \text{ amu} = 17.124 \text{ amu} \qquad (\text{masses from Table 3–1})$$
$$17 \times 0.00055 \text{ amu} = 0.0094 \text{ amu}$$
$$18 \times 1.0087 \text{ amu} = 18.157 \text{ amu}$$
$$\text{Sum} = 35.290 \text{ amu} \longleftarrow \text{Calculated mass}$$

Then we subtract the actual mass from the "calculated" mass to obtain the mass defect.

mass defect = 35.290 amu − 34.9689 amu = 0.321 amu

We have calculated the mass defect in amu/atom. Recall from Section 3–7 that one gram is 6.02×10^{23} amu. We can easily show that a number expressed in amu/atom is equal to the same number of g/mol of atoms.

$$\frac{?\ \text{g}}{\text{mol}} = \frac{0.321\ \text{amu}}{\text{atom}} \times \frac{1\ \text{g}}{6.02 \times 10^{23}\ \text{amu}} \times \frac{6.02 \times 10^{23}\ \text{atoms}}{1\ \text{mol}\ ^{35}\text{Cl atoms}}$$

$$= 0.321\ \text{g/mol of}\ ^{35}\text{Cl atoms}$$

(mass defect, *not* the mass of a mole of Cl atoms)

What has happened to the mass represented by the mass defect? In 1905, Einstein postulated the Theory of Relativity, in which he stated that matter and energy are equivalent. This was the basis of our combination of the Laws of Conservation of Matter and Energy in Chapter 1. An obvious corollary is that matter can be transformed into energy and energy into matter. The first transformation occurs in the sun and other stars, and was first accomplished on earth when controlled nuclear fission was achieved in 1939 (Section 28–12). The reverse transformation, energy into matter, has not yet been accomplished on a large scale.

Einstein was specific. He stated that the equivalence between matter and energy could be related by a simple mathematical equation, which we first encountered in Chapter 1

$$E = mc^2$$

where E represents the energy released, m represents the mass of matter transformed into energy, and c is the speed of light in a vacuum, 2.997925×10^8 m/s, which is usually rounded off to 3.00×10^8 m/s.

A mass defect represents the amount of matter that would be converted into energy and released if a nucleus were formed from initially separated protons and neutrons. This energy is the **nuclear binding energy.** Specifically, if one mole of ^{35}Cl nuclei were to be formed from 17 moles of protons and 18 moles of neutrons, the resulting mole of nuclei would weigh 0.321 g less than the original protons and neutrons (Example 3–3).

Stated differently, the nuclear binding energy for ^{35}Cl is the amount of energy (theoretically) required to separate one mole of ^{35}Cl nuclei into 17 moles of protons and 18 moles of neutrons. This has never been done. In Example 3–3 we calculated the mass defect for ^{35}Cl atoms. Let's use this value to calculate their nuclear binding energy.

EXAMPLE 3–4

The mass defect for ^{35}Cl is 0.321 g/mol of atoms. Calculate the nuclear binding energy of ^{35}Cl in joules, where 1 joule = 1 $\text{kg} \cdot \text{m}^2/\text{s}^2$.

Solution

We substitute numbers into the Einstein equation to obtain the nuclear binding energy. If we express the mass defect as 3.21×10^{-4} kg/mol, we obtain the binding energy in J/mol of ^{35}Cl atoms.

$$E = mc^2 = (3.21 \times 10^{-4}\ \text{kg/mol})(3.00 \times 10^8\ \text{m/s})^2 = 2.89 \times 10^{13}\ \frac{\text{kg}\cdot\text{m}^2/\text{s}^2}{\text{mol}}$$

$$= 2.89 \times 10^{13}\ \text{J/mol}$$

The nuclear binding energy of a mole of ^{35}Cl nuclei, 2.89×10^{13} J/mol, is an enormous amount of energy, enough in fact to heat 6.9×10^7 kg (~76,000 tons) of water from 0°C to 100°C!

Nuclear binding energies are frequently expressed in kJ/g of nuclei to give a better idea of nuclear stabilities. Figure 3–11 is a plot of average binding energy per gram of nuclei versus mass number. It shows that nuclear binding energies (per gram) increase rapidly with increasing mass number, reach a maximum around mass number 50, and then decrease slowly. The nuclei with the highest binding energies (mass numbers 40 to 150) are the most stable, because so much energy would be required to separate these nuclei into their component neutrons and protons. Even though these nuclei are the most stable ones, *all* nuclei are stable with respect to complete decomposition into protons and neutrons because all nuclei have mass defects.* In other words, all nuclei lost some energy when they were formed from protons and neutrons, and a loss of energy means greater stability.

We have presented the important ideas about the fundamental particles of matter, electrons, protons, and neutrons. We have described atomic nuclei. Let us now turn our attention to the electrons in atoms, which determine the chemical properties of elements.

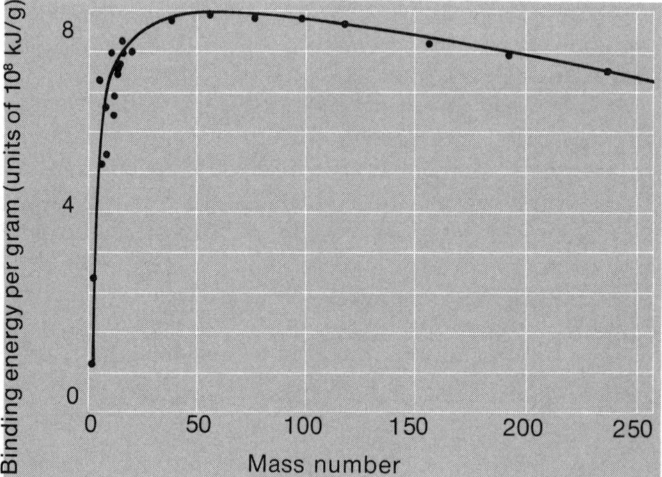

FIGURE 3–11 Plot of binding energy per gram versus mass number. Very light and very heavy nuclei are relatively unstable.

*However, some unstable radioactive nuclei do emit a single proton, a single neutron, or other subatomic particles as they decay in the direction of greater stability. (This is discussed in detail in Sections 28–3 through 28–6.) None decomposes utterly into elementary particles.

3–9 The Dual Nature of the Electron

As we noted in Section 3–2, the prevailing view at the beginning of this century was that electrons were *particles* with a definite mass and charge. Light, and indeed all electromagnetic radiation, was viewed as a *wave* phenomenon. There were, however, some experimental observations that contradicted these neat classifications. One in particular, the photoelectric effect, could not be explained by the current theories.

The apparatus for the photoelectric effect is shown in Figure 3–12. The negative electrode in the evacuated tube is made of a pure metal such as cesium. When light of a sufficiently high energy is allowed to strike the metal, electrons are knocked off its surface; they then travel to the positive electrode and form a current flowing through the circuit.

One application of the photoelectric effect is in the photoelectric sensors that open some supermarket and elevator doors when the shadow of a person interrupts the light beam.

The interesting points are these:

1. No electrons are ejected by light having energy below a certain minimum threshold energy, which is different for different metals, no matter how long the irradiation occurs.
2. The current (the number of electrons emitted per second) increases with increasing *intensity* of light but does not depend upon the *energy* of the light, as long as the energy is at or above the threshold energy.

The intensity of light is the brightness of the light.

Classical theory said that even "low" energy light should cause current to flow if the metal is irradiated long enough, because electrons should accumulate energy and be released when they have enough energy to escape from the metal atoms. Additionally, according to the old theory, if the light is made more energetic, then the current should increase even though the light intensity remains the same.

The answer to the puzzle was provided by Albert Einstein in 1905, extending Max Planck's idea that light behaves as though it were composed of *particles,* called **photons,** each with a particular amount (a quantum) of energy. According to Einstein, each photon can transfer its energy to a single electron during a collision. When we say that the intensity of light is increased, we mean that the number of photons striking a given area per second is increased.

The intensity of light is related to the amplitude of its waves.

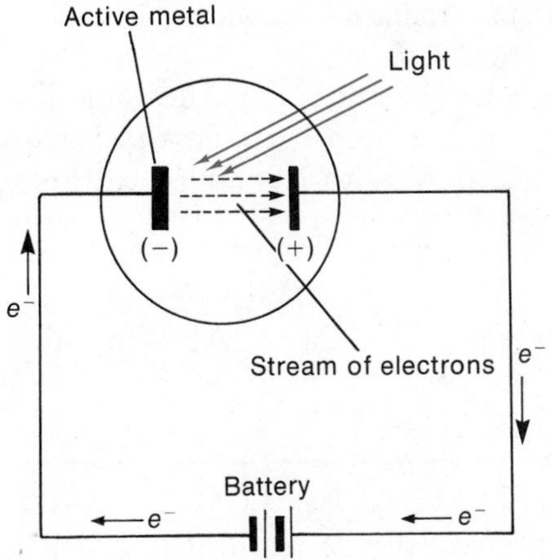

FIGURE 3–12 The photoelectric effect. When electromagnetic radiation of sufficient minimum energy strikes the surface of a metal (negative electrode) inside an evacuated tube, electrons are stripped off the metal to create an electric current. The current increases with increasing radiation intensity.

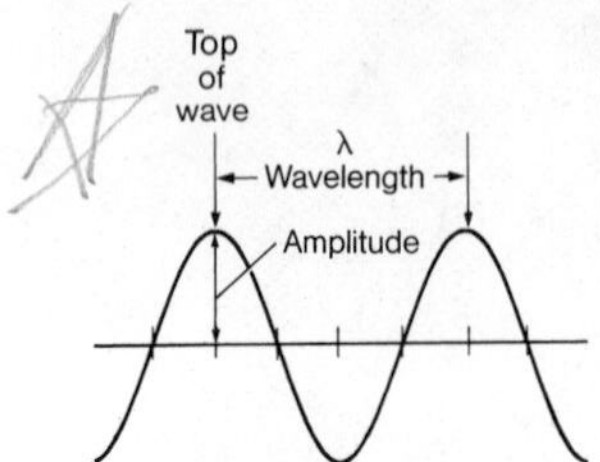

FIGURE 3–13
Illustrating the wavelength of the wave set up in a vibrating string. The distance between any two crests, the tops of the waves, is called the wavelength.

The picture is now one of a particle of light striking an electron near the surface of the metal and giving up its energy to the electron; if that energy is equal to or greater than the amount needed to liberate the electron, it can escape to join the photoelectric current. For this explanation, Einstein received the 1921 Nobel prize in physics.

The fact that light can exhibit both wave properties and particle properties suggested to Louis de Broglie the idea that very small particles, such as electrons, might also display wave properties under the proper circumstances. In his doctoral thesis in 1925, de Broglie predicted that a particle with a mass m and velocity v should have a wavelength (see Figure 3–13) associated with it, and that the numerical value of the wavelength is

$$\lambda = h/mv$$

where $h = 6.63 \times 10^{-34}$ J·s (called Planck's constant).

Two years after de Broglie's prediction, C. Davisson and L. H. Germer at the Bell Telephone Laboratory demonstrated diffraction of electrons by a crystal of nickel. This sort of behavior is possible only for waves, and shows conclusively that electrons do have wave properties. In fact, Davisson and Germer found that the wavelength associated with electrons of known energy is exactly that predicted by de Broglie. Similar diffraction experiments have been successfully performed with other particles, such as neutrons.

3–10 Electromagnetic Radiation

Our ideas about the arrangements of electrons in atoms have evolved slowly. Most of the information has been derived from **atomic emission spectra,** which are the lines or bands produced on photographic film by radiation that has passed through a refracting glass prism after being emitted from electrically or thermally excited atoms. To help us understand the nature of atomic spectra, let us first describe electromagnetic radiation in general.

All types of electromagnetic radiation, or radiant energy, are described in terms of frequency and wavelength (which determine the color of visible light). Wavelength, λ, is the distance between any two consecutive crests of a wave (Figure 3–13). Frequency, ν, is the number of wave crests passing a given point per unit time; it is usually expressed in cycles/s, or more commonly simply as 1/s or s^{-1}. One cycle per second is also called one hertz (Hz), after Rudolf Hertz, who in 1896 discovered electromagnetic radiation outside the visible range and measured its speed and wavelengths.

In a vacuum, the speed of electromagnetic radiation, c, is the same for all wavelengths, 2.9979249×10^8 m/s. The relationship between the wavelength and frequency of electromagnetic radiation, with c rounded to three significant figures, is

$$\lambda \nu = c = 3.00 \times 10^8 \text{ m/s}$$

Thus wavelength and frequency are inversely proportional to each other; the shorter the wavelength, the higher the frequency.

EXAMPLE 3–5

The frequency of violet light is 7.31×10^{14} s^{-1}, and that of red light is 4.57×10^{14} s^{-1}. Calculate the wavelength of each color.

Solution
Because frequency and wavelength are inversely proportional to each other, $\lambda = c/\nu$, we can substitute the frequencies into the relationship and calculate wavelengths

$$\text{(violet light)}\ \lambda = \frac{c}{\nu} = \frac{3.00 \times 10^8\ \text{m/s}}{7.31 \times 10^{14}\ \text{s}^{-1}} = 4.10 \times 10^{-7}\ \text{m}\ (4.01 \times 10^2\ \text{nm})$$

$$\text{(red light)}\ \lambda = \frac{c}{\nu} = \frac{3.00 \times 10^8\ \text{m/s}}{4.57 \times 10^{14}\ \text{s}^{-1}} = 6.56 \times 10^{-7}\ \text{m}\ (6.56 \times 10^2\ \text{nm})$$

Regions of the Electromagnetic Spectrum

Name	λ, m (approx.)
Radio, TV	10^{-1}–10^{6}
Microwave	10^{-3}–10^{-1}
Infrared	10^{-6}–10^{-3}
Visible	10^{-7}–10^{-3}
Ultraviolet	10^{-9}–10^{-7}
X-rays	10^{-12}–10^{-8}
Gamma Rays	10^{-16}–10^{-11}

Visible colors	λ, Å
Red	6500–7000
Orange	6000–6500
Yellow	5500–6000
Green	5000–5500
Blue	4500–5000
Violet	4000–4500

Isaac Newton first separated sunlight, which consists of waves of electromagnetic radiation, into its component colors by allowing it to pass through a prism. Since sunlight (white light) contains all wavelengths of visible light, it gives the *continuous spectrum* observed in the rainbow. But visible light represents only a tiny segment of the electromagnetic radiation spectrum. In addition to all wavelengths of visible light, sunlight also contains shorter wavelength (ultraviolet) radiation as well as longer wavelength (infrared) radiation. Neither of these can be detected by the human eye, but they may be detected and recorded photographically or by spectrophotometers designed for that purpose.

3–11 Atomic Spectra and the Bohr Atom

The energy of a photon of light is given by Planck's equation

$$E = h\nu$$

where h is Planck's constant, and ν is the frequency of the photon. Thus, energy is directly proportional to frequency. Reference to the previous example shows that violet light has higher energy than the longer wavelength, lower frequency red light.

Under high pressures, incandescent ("red hot" or "white hot") solids, liquids, and gases give continuous spectra. However, when an electric current is passed through a gas in a vacuum tube at very low pressures, the gas emits light that is dispersed by a prism into distinct lines. Such emission spectra are called **bright line spectra.** The lines can be recorded photographically, and the wavelength of light that produced each line can be calculated from the position of each line on the photograph.

EXAMPLE 3–6

A green line of wavelength 4.86×10^{-7} m is observed in the spectrum of hydrogen. Calculate the energy of one photon of this green light.

Solution
We know the wavelength of the photon, and we calculate its frequency so that we can then calculate its energy

$$\lambda\nu = c$$

$$\nu = \frac{c}{\lambda} = \frac{3.00 \times 10^8\ \text{m/s}}{4.86 \times 10^{-7}\ \text{m}} = 6.17 \times 10^{14}\ \text{s}^{-1}$$

$$E = h\nu = (6.63 \times 10^{-34}\ \text{J}\cdot\text{s})(6.17 \times 10^{14}\ \text{s}^{-1})$$
$$= 4.09 \times 10^{-19}\ \text{J/atom}$$

To gain a better appreciation of the amount of energy involved, let's calculate the energy, in kilojoules, emitted by one mole of atoms.

$$\frac{?\ \text{kJ}}{\text{mol}} = 4.09 \times 10^{-19} \frac{\text{J}}{\text{atom}} \times \frac{1\ \text{kJ}}{1 \times 10^{3}\ \text{J}} \times \frac{6.02 \times 10^{23}\ \text{atoms}}{\text{mol}}$$
$$= 2.46 \times 10^{2}\ \text{kJ/mol}$$

This calculation shows that when each atom in one mole of hydrogen atoms emits a photon of wavelength 4.86×10^{-7} m, the mole of atoms loses 246 kJ of energy as green light.

Careful studies on hydrogen showed that several series of lines in the spectrum of hydrogen are produced when an electric current is passed through the gas at very low pressures. These lines were studied intensely by many scientists. J. R. Rydberg discovered in the late nineteenth century that the wavelengths of the various lines in the hydrogen spectrum can be related by a mathematical equation

$$\frac{1}{\lambda} = R\left(\frac{1}{{n_1}^2} - \frac{1}{{n_2}^2}\right)$$

where R is $1.097 \times 10^{7}\ \text{m}^{-1}$ and is known as the Rydberg constant. The n's are positive integers, and n_1 is smaller than n_2. We now know that the n's describe *energy levels* within atoms. The Rydberg equation was derived from numerous observations, not theory, and is thus an empirical equation.

Neils Bohr, a Danish physicist, provided an explanation for Rydberg's observations in 1913. He believed that electrons can occupy only certain discrete energy levels in atoms, and that electrons absorb or emit energy in discrete amounts as they move from one energy level to another. When an electron is promoted from a lower energy level to a higher one, it absorbs a definite (or *quantized*) amount of energy. When the electron falls back to the original energy level, it emits exactly the same amount of energy it absorbed in moving from the lower to the higher energy level. Figure 3–14 illustrates these transitions schematically.

We now accept the fact that electrons do indeed occupy only certain energy levels in atoms. In most atoms, some of the energy differences between various levels correspond to the energy of light in the visible region of the spectrum, and thus colors associated with electronic transitions in such elements can be observed by the human eye. The light produced by neon in neon signs and the lightning flashes produced in electrical storms are two familiar examples.

Bohr assumed that electrons revolve around the nucleus of an atom in circular orbits, like planets around the sun. This model was modified in 1916 by Sommerfeld, who assumed elliptical rather than circular orbits. Although the Bohr theory satisfactorily explained the spectrum of hydrogen, and those of other species containing one electron, it could not explain the observed spectra of more complex species. Let us now approach the big question: How are elec-

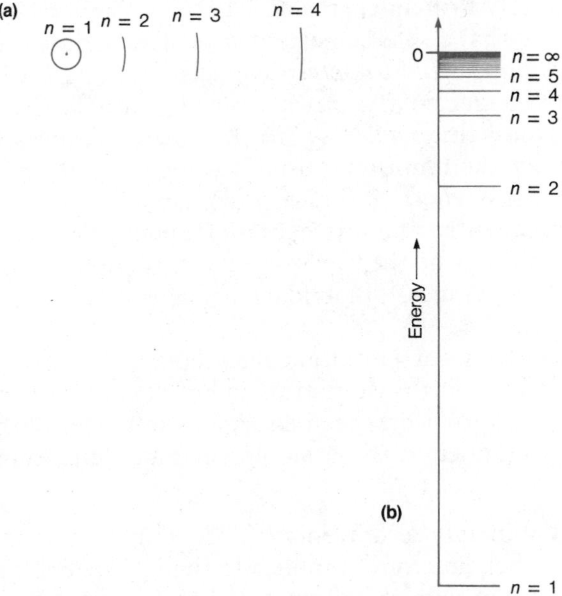

FIGURE 3–14 (a) A representation of the first four Bohr orbits for a hydrogen atom. The dot at the center represents the nucleus. The radius is proportional to the square of the value of n. (b) Relative values for the energies associated with the various energy levels in a hydrogen atom. Note that the energies become closer together as n increases. They are so close together for large values of n that they form a continuum. Also note that the energies of electrons in atoms are always negative.

trons arranged in more complicated atoms, and how do they behave? The modern picture of the atom is somewhat different from Bohr's picture, but we should not minimize Bohr's contributions, any more than those of Dalton, Rutherford, and other pioneers in the study of the atom.

3–12 The Quantum Mechanical Picture of Atoms

Through the work of de Broglie, Davisson and Germer, and others, we now know that electrons in atoms can be treated as waves more effectively than as small compact particles traveling in circular or elliptical orbits. Very small particles such as electrons, atoms, and molecules do not obey the laws of classical mechanics (Isaac Newton's laws) as do much larger objects such as golf balls and moving automobiles. A different kind of mechanics, called **quantum mechanics,** describes the behavior of very small particles much better because it is based on the idea of quantization of energy. It may be of interest to note that classical Newtonian mechanics is a special case of quantum mechanics, and Newtonian mechanics is, in fact, valid for all cases except very small particles.

One of the underlying principles of quantum mechanics is that we cannot determine precisely the paths that electrons follow as they move about atomic

Any mechanism for determining either the position or velocity of an electron of necessity changes both.

nuclei. The **Heisenberg Uncertainty Principle,** stated in 1927 by Werner Heisenberg, is a theoretical statement that also is consistent with all experimental observations. It states that *it is impossible to determine accurately both the momentum and the position of an electron (or any other very small particle) simultaneously.* Momentum is mass times velocity, *mv*. Because electrons are so small, and they move so rapidly, their motion is usually detected by electromagnetic radiation. Photons possess energies that approximate the energies associated with electrons. Consequently, the interaction of a photon with an electron severely disturbs the motion of the electron. The situation is not unlike locating the position of a moving automobile by driving another automobile into it.

Since it is not possible to determine simultaneously both the position and the velocity of an electron, we resort to a statistical approach and speak of the *probability* of finding an electron within specified regions in space. With this idea in mind, we can now enumerate the basic postulates of quantum mechanics.

1. Atoms and molecules can exist in only certain energy states that are characterized by definite energies. When an atom or molecule changes its energy state, it must absorb or emit just enough energy to bring it to the new energy state (the quantum condition).

Atoms and molecules possess various forms of energy, but let's focus our attention on the motions of electrons and the corresponding *electronic energies.*

2. When atoms or molecules absorb or emit radiation as they change their energies, the frequency of the light is related to the energy change by a simple equation

$$\Delta E = h\nu$$

Recall that $\lambda\nu = c$, so that the above equation can be written as $\Delta E = hc/\lambda$ to give a relationship between the energy change, ΔE, and the wavelength, λ, of radiation absorbed or emitted. The energy gained (or lost) by an atom as it goes from lower to higher (or higher to lower) energy states is equal to the energy of the photon absorbed (or emitted) during the transition.

3. The allowed energy states of atoms and molecules can be described by sets of numbers called *quantum numbers.*

The quantum mechanical treatment of atoms and molecules is highly mathematical, and the important point is that the solution of the Schrödinger wave equation (described in the following box) gives a set of three kinds of numbers, **quantum numbers,** that describe the energies of electrons in atoms. These numbers are in accord with those deduced from experiment and from empirical equations such as the Rydberg equation. Solutions of the Schrödinger equation also give information about the shapes and orientations of the statistical probability distributions of the electrons. (Recall that, in light of the Heisenberg Principle, this is how we describe the positions of the electrons.) These *atomic orbitals* (which we will describe in Section 3–14), deduced from the solutions of the Schrödinger equation, are directly related to the quantum numbers. In 1928, Paul A. M. Dirac reformulated electron quantum mechanics to take into account the effects of relativity. This gave rise to a fourth quantum number.

The Schrödinger Equation

In 1926 Erwin Schrödinger modified an existing equation that described a three-dimensional standing wave by imposing wavelength restrictions suggested by de Broglie's experiments. This equation allowed him to calculate the energy levels in the hydrogen atom. It is a second order differential equation, and it certainly need not be memorized or even understood to read this book. A knowledge of differential calculus would be necessary.

$$\frac{\partial^2\psi}{\partial x^2} + \frac{\partial^2\psi}{\partial y^2} + \frac{\partial^2\psi}{\partial z^2} + \frac{8\pi^2 m}{h^2}(E - V)\psi = 0$$

Hydrogen is the only atom for which it has been solved exactly. Numerous simplifying assumptions are necessary to solve the equation for more complex atoms and molecules. However, chemists and physicists have used their intuition and ingenuity (and modern computers) to apply what they have learned about hydrogen atoms to more complex systems, to the general benefit of science.

3–13 Quantum Numbers

The solutions of the Schrödinger and Dirac equations for hydrogen atoms give four quantum numbers that describe the various states available to hydrogen's single electron. We can use these quantum numbers to describe the electronic arrangements in all atoms, their so-called **electronic configurations.** The roles these quantum numbers play in describing the energy levels of electrons, as well as the shapes of the orbitals that define distributions of electrons in space, will not become completely clear until we begin to discuss atomic orbitals in the following section. For now, let's simply understand an **atomic orbital** to be *a region of space in which the probability of finding an electron is high.* We will content ourselves with defining each quantum number and describing the range of values it may take.

1. The **principal quantum number,** n, describes the energy level an electron occupies. It may take any integral value

   $n = 1, 2, 3, 4 \cdots$

2. The **subsidiary** (or **azimuthal**) **quantum number,** ℓ, designates the geometric shape of the region in space an electron occupies. Within each energy level (defined by the value of n, the principal quantum number), ℓ may take integral values from 0 up to and including $(n - 1)$

   $\ell = 0, 1, 2, \cdots (n - 1)$

   The maximum value of ℓ thus is $(n - 1)$. The subsidiary quantum number designates a **sublevel,** or a specific *kind* of atomic orbital, that an electron may occupy. For each value of ℓ, a letter has become associated, each corresponding to a different kind of atomic orbital

   $\ell = 0, 1, 2, 3, \cdots (n - 1)$
   s p d f :kind of sublevel

The *s, p, d, f* designations arise from the characteristics of spectral emission lines produced by electrons occupying the orbitals: *s* (sharp), *p* (principal), *d* (diffuse), and *f* (fundamental).

In the first energy level, the maximum value of ℓ is zero, which tells us that there is only an s sublevel, and no p sublevel. In the second energy level, the permissible values of ℓ are 0 and 1, which tells us that there is an s sublevel and a p sublevel, but no d sublevel, in this energy level.

3. The **magnetic quantum number,** m_ℓ, designates the spatial orientation of an atomic orbital. Within each sublevel, m_ℓ takes all integral values from $-\ell$ through zero up to and including $+\ell$

   $$m_\ell = (-\ell) \cdots 0 \cdots (+\ell)$$

   Clearly, the maximum value of m_ℓ depends on the value of ℓ. For example, when $\ell = 1$, which designates the p sublevel, there are three permissible values of m_ℓ, $-1, 0, +1$. There are thus three distinct regions of space, called atomic orbitals, associated with a p sublevel. We refer to these orbitals as the p_x, p_y, and p_z orbitals (next section).
4. The **spin quantum number,** m_s, refers to the spin of an electron *and* the orientation of the magnetic field produced by the motion of an electron. For every set of n, ℓ, and m_ℓ values, m_s can take the value $+\frac{1}{2}$ or $-\frac{1}{2}$

   $$m_s = \pm\tfrac{1}{2}$$

The values of n, ℓ, and m_ℓ describe a particular atomic orbital, and each atomic orbital can accommodate no more than two electrons, one with $m_s = +\frac{1}{2}$ and another with $m_s = -\frac{1}{2}$. The **Pauli Exclusion Principle** states that *no two electrons in an atom may have identical sets of four quantum numbers.* No exceptions are known. Table 3–3 summarizes permissible values for the four

TABLE 3–3 Permissible Values of the Quantum Numbers

n	ℓ	m_ℓ	m_s	Electron Capacity of Sublevel	Electron Capacity of Energy Level
1 (K)	0 (s)	0	$+\frac{1}{2}, -\frac{1}{2}$	2	2
2 (L)	0 (s)	0	$+\frac{1}{2}, -\frac{1}{2}$	2	8
	1 (p)	$-1, 0, +1$	$\pm\frac{1}{2}$ for each value of m_ℓ	6	
3 (M)	0 (s)	0	$+\frac{1}{2}, -\frac{1}{2}$	2	18
	1 (p)	$-1, 0, +1$	$\pm\frac{1}{2}$ for each value of m_ℓ	6	
	2 (d)	$-2, -1, 0, +1, +2$	$\pm\frac{1}{2}$ for each value of m_ℓ	10	
4 (N)	0 (s)	0	$+\frac{1}{2}, -\frac{1}{2}$	2	32
	1 (p)	$-1, 0, +1$	$\pm\frac{1}{2}$ for each value of m_ℓ	6	
	2 (d)	$-2, -1, 0, +1, +2$	$\pm\frac{1}{2}$ for each value of m_ℓ	10	
	3 (f)	$-3, -2, -1, 0, +1, +2, +3$	$\pm\frac{1}{2}$ for each value of m_ℓ	14	
5 (O)	0 (s)	0	$+\frac{1}{2}, -\frac{1}{2}$	2	50
	1 (p)	$-1, 0, +1$	$\pm\frac{1}{2}$ for each value of m_ℓ	6	
	2 (d)	$-2, -1, 0, +1, +2$	$\pm\frac{1}{2}$ for each value of m_ℓ	10	
	3 (f)	$-3, -2, -1, 0, +1, +2, +3$	$\pm\frac{1}{2}$ for each value of m_ℓ	14	
	4 (g)	$-4, -3, -2, -1, 0, +1, +2, +3, +4$	$\pm\frac{1}{2}$ for each value of m_ℓ	18	

quantum numbers. The quantum mechanical predictions about the number of atomic orbitals in each energy level are in agreement with the spectroscopic evidence.

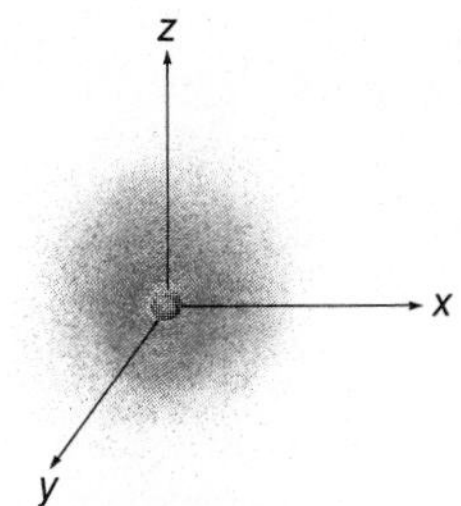

FIGURE 3–15 Electron cloud surrounding an atomic nucleus. The electron density drops off rapidly but smoothly as distance from the nucleus increases.

3–14 Atomic Orbitals

Shapes

Let us now describe the distributions of electrons in atoms. For each neutral atom, we must account for a number of electrons equal to the number of protons in the nucleus, i.e., the atomic number of the atom. Each electron is said to occupy an atomic orbital (defined by the quantum numbers n, ℓ, and m_ℓ), which, as we have seen, is simply the region in which there is a high probability of finding the electron. Within each atom, these atomic orbitals, taken together, can be represented as a diffuse, roughly spherical cloud of electrons (Figure 3–15).

The energy level of each atomic orbital in an atom is indicated by the principal quantum number n (from the Schrödinger equation). As we have seen, the principal quantum number takes integral values: $n = 1, 2, 3, 4, \cdots$, where $n = 1$ describes the first or lowest energy level. In the past, these energy levels have been referred to as electron shells designated K, L, M, N, $\cdots$ shells (Table 3–3). The correspondence between the two notations is

The letter notation comes from early experiments in spectroscopy and with x-rays.

principal quantum number n = 1, 2, 3, 4 $\cdots$
shell K, L, M, N $\cdots$
$\xrightarrow{\text{increasing radius}}$

Successive energy levels are at increasingly greater distances from the nucleus. For example, the L shell, or $n = 2$ level, has a larger radius than the K shell, or $n = 1$ level. The electron capacity of each energy level is indicated in the right-hand column of Table 3–3. (For a given n, the capacity is $2n^2$).

In accordance with the rules we set in Section 3–13, each energy level has one s sublevel (defined by $\ell = 0$) consisting of one s atomic orbital (defined by $m_\ell = 0$). Each s orbital is spherically symmetrical with respect to the nucleus; i.e., it is "round" like a basketball (Figure 3–16). Distinction between s orbitals in different energy levels is indicated by using the principal quantum number as a coefficient; $1s$ indicates the s orbital in the first energy level, $2s$ the s orbital in the second energy level, and so on.

From quantum mechanical calculations, it can be shown that the probability density of an electron at any point in an atom varies with the square of the Schrödinger wave function, ψ^2. The electron density at any point is, in turn, directly proportional to the probability density of an electron at that point.

In the upper part of Figure 3–17 the probability density of an electron at a given distance from the nucleus is plotted against distance from the nucleus. The electron clouds (electron densities) associated with the $1s$, $2s$, and $3s$ atomic orbitals are shown just below the plots. Keep in mind that the electron clouds are three-dimensional, and only cross-sections are shown here. The regions shown in some figures (Figures 3–16, 3–18, 3–19, 3–20, and 3–21) appear to have surfaces or skins only because they are arbitrarily "cut off" so that there is a 90% probability of finding an electron occupying the orbital somewhere within the surfaces.

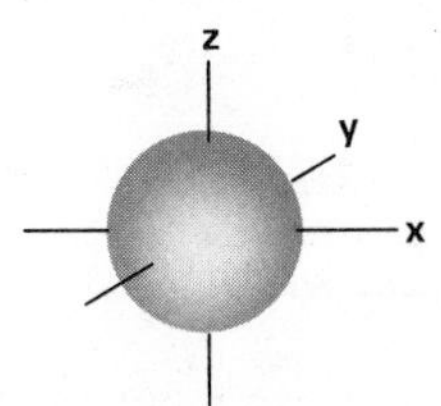

FIGURE 3–16 Atomic s orbital.

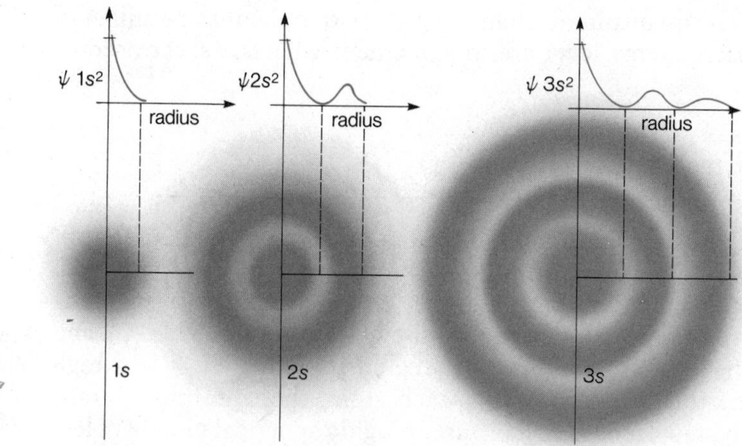

FIGURE 3–17 Electron clouds associated with *s* orbitals. Sketches show cross-section in plane of atomic nucleus.

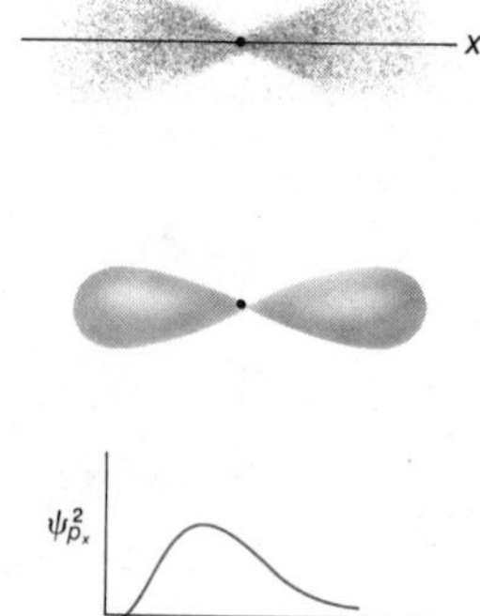

FIGURE 3–18 The shape of the $2p$ orbital—three representations.

Beginning with the second energy level, each level contains a *p* sublevel, defined by $\ell = 1$. Each of these sublevels consists of a set of *three p* atomic orbitals, corresponding to the three allowed values of m_ℓ (-1, 0, and $+1$) when $\ell = 1$. The sets are referred to as $2p$, $3p$, $4p$, $5p$, . . . orbitals to indicate the energy level. Each set of atomic *p* orbitals resembles three mutually perpendicular equal-arm dumbbells (Figure 3–18). The nucleus defines the origin of a set of Cartesian coordinates with the usual *x*, *y*, and *z* axes (Figure 3–19a). Subscripts *x*, *y*, and *z* indicate the axis along which each of the three two-lobed orbitals is directed. Atomic orbitals are seen to be *regions* in space. A set of three *p* atomic orbitals may be represented as in Figure 3–19b.

Beginning at the third energy level, each level contains a third sublevel ($\ell = 2$) composed of a set of *five d* atomic orbitals ($m_\ell = -2, -1, 0, +1, +2$). They are designated $3d$, $4d$, $5d$, . . . to indicate the energy level in which they are found. The shapes of the members of a set are indicated in Figure 3–20.

In each of the fourth and higher energy levels, there is also a fourth sublevel, containing a set of *seven f* atomic orbitals ($\ell = 3$, $m_\ell = -3, -2, -1, 0, +1, +2, +3$). These are shown in Figure 3–21.

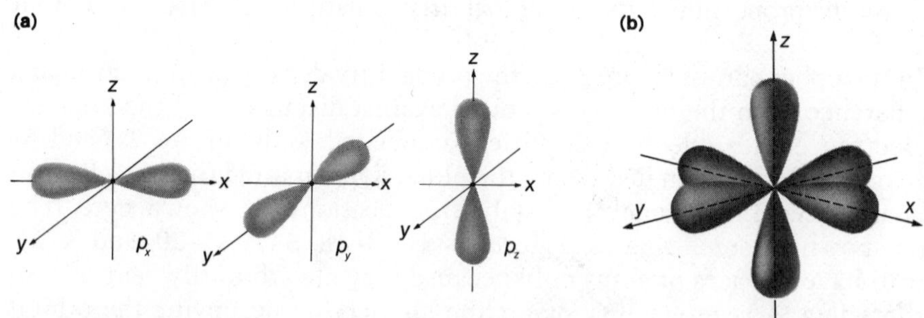

FIGURE 3–19 (a) Relative directional character of a set of atomic *p* orbitals. (b) A model of three *p* orbitals (p_x, p_y, and p_z) of a single set of orbitals. The nucleus is at the center. (The lobes are actually more diffuse than depicted.)

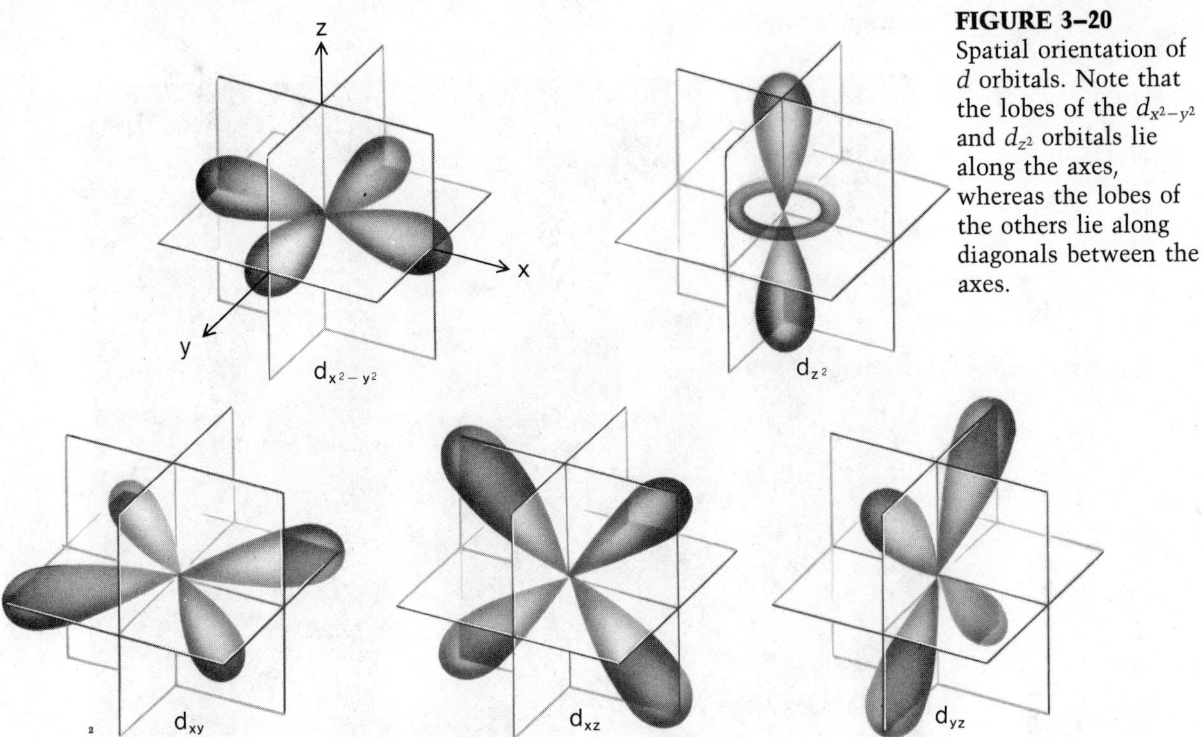

FIGURE 3–20 Spatial orientation of *d* orbitals. Note that the lobes of the $d_{x^2-y^2}$ and d_{z^2} orbitals lie along the axes, whereas the lobes of the others lie along diagonals between the axes.

Thus we see that the first energy level contains only the 1*s* orbital; the second energy level contains the 2*s* and three 2*p* orbitals; the third energy level contains the 3*s*, three 3*p*, and five 3*d* orbitals; and fourth energy level consists of a 4*s*, three 4*p*, five 4*d*, and seven 4*f* orbitals. All subsequent energy levels contain *s*, *p*, *d*, and *f* sublevels, as well as others that are not occupied in any presently known elements in the **ground states.** The term ground state is usually applied to the lowest-energy state, or unexcited state, of an atom.

Let us summarize in tabular form some of the information we have developed to this point. The principal quantum number *n* indicates the energy level. The number of sublevels per energy level is equal to *n*, the number of atomic orbitals per energy level is n^2, and the maximum number of electrons per energy level is $2n^2$, since each atomic orbital can hold two electrons.

Energy Level n	Number of Sublevels per Energy Level n	Number of Atomic Orbitals n^2	Maximum Number of Electrons $2n^2$
1	1	1 (1*s*)	2
2	2	4 (2*s*, $2p_x$, $2p_y$, $2p_z$)	8
3	3	9 (3*s*, three 3*p*'s, five 3*d*'s)	18
4	4	16	32
5	5	25	50
⋮	⋮	⋮	⋮

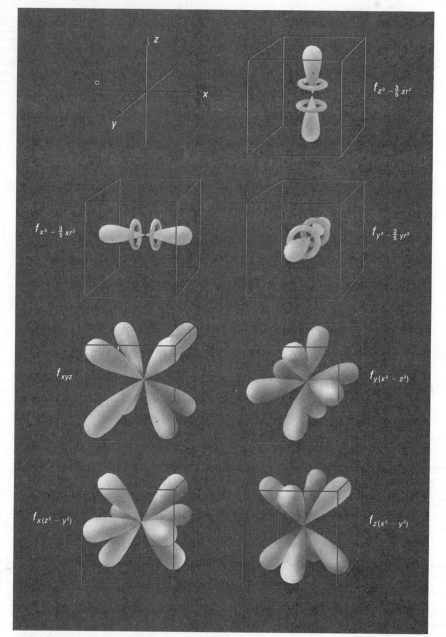

FIGURE 3–21 Relative directional character of atomic f orbitals. The seven orbitals are shown within cubes as an aid to visualization.

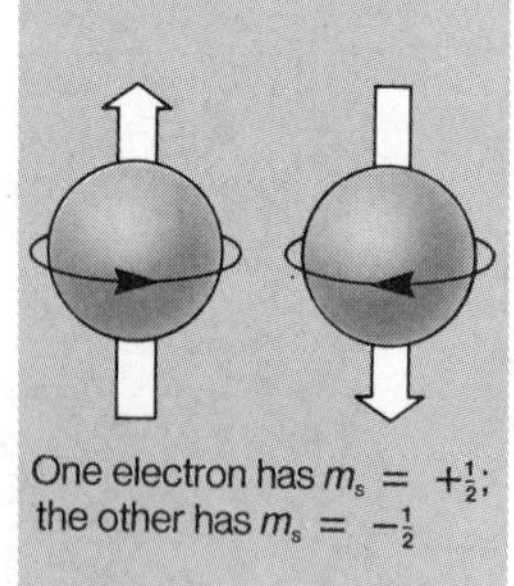

One electron has $m_s = +\frac{1}{2}$;
the other has $m_s = -\frac{1}{2}$

FIGURE 3–22
Electron spin. Electrons act as though they spin about an axis through their centers. Because there are two directions in which an electron may spin, the spin quantum number has two possible values, $+\frac{1}{2}$ and $-\frac{1}{2}$. This electron spin produces a magnetic field.

In this section, we haven't yet mentioned the fourth quantum number, the spin quantum number m_s. Since m_s has two possible values, $+\frac{1}{2}$, and $-\frac{1}{2}$, each atomic orbital, defined by the values of n, ℓ, and m_ℓ, has a capacity of two electrons. Since electrons are negatively charged, and they behave as though they were spinning about axes through their centers, they act like tiny magnets. The motions of electrons produce magnetic fields, and these (like all other magnetic fields) may interact so that they attract or repel each other. Two electrons in the same orbital having opposite m_s values, and thus interacting attractively, are said to be **spin-paired** (Figure 3–22).

3–15 The Electronic Structure of Atoms

Let us now examine the structures of atoms of different elements in some detail. For simplicity we'll indicate atomic orbitals as __ and show an unpaired electron as $\underline{\uparrow}$ and spin-paired electrons as $\underline{\uparrow\downarrow}$. By "unpaired electron" we mean an electron that occupies an orbital singly. By "unpaired electrons" we mean two or more that occupy different orbitals of a given sublevel singly, and hence electrons whose magnetic fields are not balanced out.

We shall consider the elements in order of increasing atomic number, using as our guide the **periodic table** inside the front cover of this text. We'll have much more to say about the periodic table in the next chapter.

In building up ground state electron configurations, we shall make use of the **Aufbau Principle.** This states that *the electron that distinguishes an element from the previous element (with respect to atomic number) must enter the lowest energy atomic orbital available.* In this regard, the orbitals increase in energy (in general) with increasing value of the quantum number n. For a given value of n, energy increases with increasing value of ℓ. In other words, within a particular major energy level, the s sublevel is lowest in energy, the p sublevel is next, then the d, then the f, and so on.

The German verb *aufbauen* means "to build up."

The first energy level contains only one atomic orbital, 1s, and can hold a maximum of two electrons. Hydrogen, as we have already noted, contains just one electron. Helium is a noble gas with a filled first energy level (two electrons), and the atom is so stable that no chemical reactions whatever of helium are known at the present time.

Helium's electrons can be displaced only by electrical forces, as in excitation by high-voltage discharge.

	1s	Simplified Notation
$_1$H	$\underline{\uparrow}$	$1s^1$
$_2$He	$\underline{\uparrow\downarrow}$	$1s^2$

We indicate with superscripts the number of electrons in each sublevel.

Elements of atomic numbers 3 through 10 occupy the second period or horizontal row in the periodic table.

The second energy level is filled completely in neon atoms. Neon is a noble gas, extremely stable, and no reactions of neon are known.

However, as with helium, neon's electrons can be displaced by high-voltage electrical discharge, as is observed in neon signs.

In writing electronic structures of atoms, we frequently use abbreviated notations. In the preceding tabulation, the simplified notation [He] indicates

	1s	2s	2p	Simplified Notation		
$_3$Li	$\underline{\uparrow\downarrow}$	$\underline{\uparrow}$		$1s^22s^1$	*or*	[He] $2s^1$
$_4$Be	$\underline{\uparrow\downarrow}$	$\underline{\uparrow\downarrow}$		$1s^22s^2$		[He] $2s^2$
$_5$B	$\underline{\uparrow\downarrow}$	$\underline{\uparrow\downarrow}$	$\underline{\uparrow}$ __ __	$1s^22s^22p^1$		[He] $2s^22p^1$
$_6$C	$\underline{\uparrow\downarrow}$	$\underline{\uparrow\downarrow}$	$\underline{\uparrow}$ $\underline{\uparrow}$ __	$1s^22s^22p^2$		[He] $2s^22p^2$
$_7$N	$\underline{\uparrow\downarrow}$	$\underline{\uparrow\downarrow}$	$\underline{\uparrow}$ $\underline{\uparrow}$ $\underline{\uparrow}$	$1s^22s^22p^3$		[He] $2s^22p^3$
$_8$O	$\underline{\uparrow\downarrow}$	$\underline{\uparrow\downarrow}$	$\underline{\uparrow\downarrow}$ $\underline{\uparrow}$ $\underline{\uparrow}$	$1s^22s^22p^4$		[He] $2s^22p^4$
$_9$F	$\underline{\uparrow\downarrow}$	$\underline{\uparrow\downarrow}$	$\underline{\uparrow\downarrow}$ $\underline{\uparrow\downarrow}$ $\underline{\uparrow}$	$1s^22s^22p^5$		[He] $2s^22p^5$
$_{10}$Ne	$\underline{\uparrow\downarrow}$	$\underline{\uparrow\downarrow}$	$\underline{\uparrow\downarrow}$ $\underline{\uparrow\downarrow}$ $\underline{\uparrow\downarrow}$	$1s^22s^22p^6$		[He] $2s^22p^6$

that the 1*s* orbital is completely filled, as in helium. In the tabulation on page 115, [Ne] indicates that the 1*s*, 2*s*, and 2*p* sublevels of an atom are filled with ten electrons, as in neon.

Notice that some atoms have unpaired electrons in the same set of energetically equivalent, or **degenerate,** orbitals. As we have already seen, two electrons can occupy a given atomic orbital (with the same values of n, ℓ, and m_ℓ) *only* if their spins are paired (opposite values of m_s), i.e., if their magnetic fields interact attractively. Even with pairing of spins, however, two electrons that are in the same orbital repel each other more strongly than do two electrons in different (but equal-energy) orbitals. Thus, both theory and experimental observations (see below) lead to **Hund's Rule:** *electrons must occupy all the orbitals of a given sublevel singly before pairing begins.* Thus carbon has two unpaired electrons in its 2*p* orbitals.

Substances that contain unpaired electrons are *attracted* weakly into magnetic fields and are said to be **paramagnetic.** By contrast, those in which all

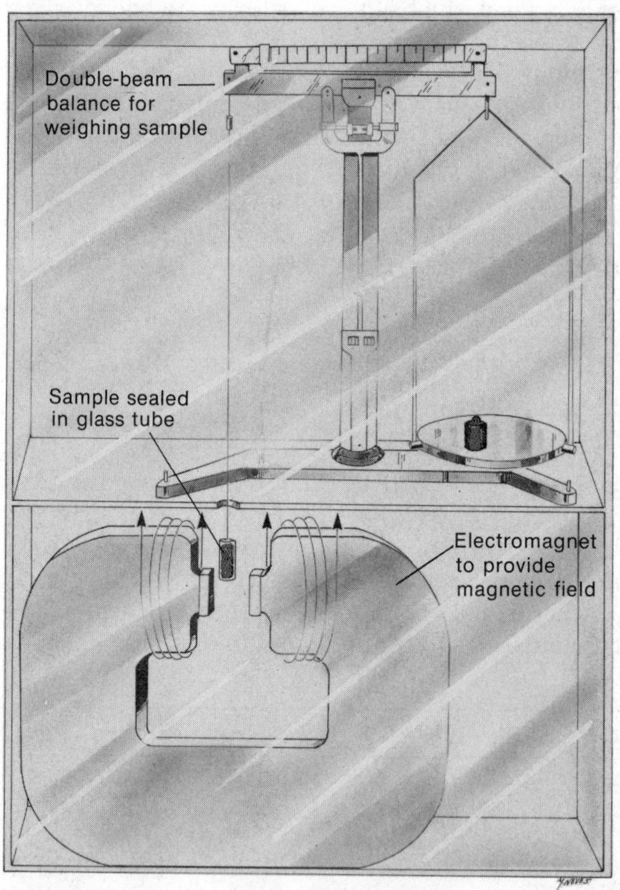

FIGURE 3–23 A magnetic balance used to measure the magnetic properties of a compound. The sample is weighed first with the electromagnet turned off, and then a current is sent through the electromagnet, providing a magnetic field of known intensity. The balance then measures the extent to which the sample is drawn into the magnetic field.

electrons are paired are repelled much more weakly by magnetic fields and are called **diamagnetic.** The paramagnetic effect can be measured by hanging a test tube full of a paramagnetic substance (such as copper sulfate) by a long thread, and bringing it near a strong magnet; the test tube will swing toward the magnet. Alternatively, the test tube can be hung just outside the gap of an electromagnet; when the current is switched on, the test tube will swing toward the magnet. If an electromagnet is placed below a sample suspended from the beam of a balance (Figure 3–23), the paramagnetic attraction per mole of substance can be measured by weighing the sample before and after energizing the magnet. The paramagnetism per mole increases with increasing number of unpaired electrons per formula unit.

Both paramagnetism and diamagnetism are hundreds to thousands of times weaker than *ferromagnetism,* the effect seen in iron bar magnets.

The next element beyond neon is sodium, and we begin to add electrons to the third energy level. Elements 11 through 18 occupy the third period in the periodic table.

	3*s*	3*p*	Simplified Notation
$_{11}Na$	[Ne] ↑		[Ne] $3s^1$
$_{12}Mg$	[Ne] ⇅		[Ne] $3s^2$
$_{13}Al$	[Ne] ⇅	↑ _ _	[Ne] $3s^23p^1$
$_{14}Si$	[Ne] ⇅	↑ ↑ _	[Ne] $3s^23p^2$
$_{15}P$	[Ne] ⇅	↑ ↑ ↑	[Ne] $3s^23p^3$
$_{16}S$	[Ne] ⇅	⇅ ↑ ↑	[Ne] $3s^23p^4$
$_{17}Cl$	[Ne] ⇅	⇅ ⇅ ↑	[Ne] $3s^23p^5$
$_{18}Ar$	[Ne] ⇅	⇅ ⇅ ⇅	[Ne] $3s^23p^6$

Although the third energy level is not yet filled (the *d* orbitals are still empty), argon is a noble gas. All noble gases, except helium in which the first energy level is filled with two electrons, have ns^2np^6 electronic configurations (where *n* indicates the highest occupied energy level), and all noble gases are extremely stable.

The third energy level contains five *d* atomic orbitals, in addition to the 3*s* and 3*p* orbitals that are filled in argon and all subsequent elements. These 3*d* orbitals can hold ten electrons. However, in most atoms the 4*s* orbital has slightly lower energy than the 3*d* orbitals, and it is an experimentally observed fact that *an electron enters the available orbital of lowest energy.* Therefore, the 4*s* atomic orbital is filled before electrons enter the 3*d* orbitals.

When the 3*d* sublevel is full, the 4*p* orbitals fill next, followed by the 5*s* orbital and then the five 4*d* orbitals. The 5*p* orbitals fill next to take us to xenon, a noble gas. The energy level diagrams in Figures 3–24 and 3–25 indicate the relative energies of the various atomic orbitals and the order of filling. Both figures are consistent with the Aufbau Principle.

Let us now examine the electronic structures of the 18 elements in the fourth period in some detail. For the first time we encounter elements that have electrons in *d* orbitals.

		$3d$	$4s$	$4p$	Simplified Notation
$_{19}$K	[Ar]		↑		[Ar] $4s^1$
$_{20}$Ca	[Ar]		⇅		[Ar] $4s^2$
$_{21}$Sc	[Ar]	↑ _ _ _ _	⇅		[Ar] $3d^14s^2$
$_{22}$Ti	[Ar]	↑ ↑ _ _ _	⇅		[Ar] $3d^24s^2$
$_{23}$V	[Ar]	↑ ↑ ↑ _ _	⇅		[Ar] $3d^34s^2$
$_{24}$Cr	[Ar]	↑ ↑ ↑ ↑ ↑	↑		[Ar] $3d^54s^1$
$_{25}$Mn	[Ar]	↑ ↑ ↑ ↑ ↑	⇅		[Ar] $3d^54s^2$
$_{26}$Fe	[Ar]	⇅ ↑ ↑ ↑ ↑	⇅		[Ar] $3d^64s^2$
$_{27}$Co	[Ar]	⇅ ⇅ ↑ ↑ ↑	⇅		[Ar] $3d^74s^2$
$_{28}$Ni	[Ar]	⇅ ⇅ ⇅ ↑ ↑	⇅		[Ar] $3d^84s^2$
$_{29}$Cu	[Ar]	⇅ ⇅ ⇅ ⇅ ⇅	↑		[Ar] $3d^{10}4s^1$
$_{30}$Zn	[Ar]	⇅ ⇅ ⇅ ⇅ ⇅	⇅		[Ar] $3d^{10}4s^2$
$_{31}$Ga	[Ar]	⇅ ⇅ ⇅ ⇅ ⇅	⇅	↑ _ _	[Ar] $3d^{10}4s^24p^1$
$_{32}$Ge	[Ar]	⇅ ⇅ ⇅ ⇅ ⇅	⇅	↑ ↑ _	[Ar] $3d^{10}4s^24p^2$
$_{33}$As	[Ar]	⇅ ⇅ ⇅ ⇅ ⇅	⇅	↑ ↑ ↑	[Ar] $3d^{10}4s^24p^3$
$_{34}$Se	[Ar]	⇅ ⇅ ⇅ ⇅ ⇅	⇅	⇅ ↑ ↑	[Ar] $3d^{10}4s^24p^4$
$_{35}$Br	[Ar]	⇅ ⇅ ⇅ ⇅ ⇅	⇅	⇅ ⇅ ↑	[Ar] $3d^{10}4s^24p^5$
$_{36}$Kr	[Ar]	⇅ ⇅ ⇅ ⇅ ⇅	⇅	⇅ ⇅ ⇅	[Ar] $3d^{10}4s^24p^6$

As you study these electronic configurations, you should be able to see how most of them are predicted from the Aufbau Principle, keeping in mind the order of filling atomic orbitals. However, as we proceed to fill the $3d$ set of orbitals, from $_{21}$Sc to $_{30}$Zn, we see that these orbitals are not filled quite regularly. Some sets of orbitals are quite close in energy (e.g., $4s$ and $3d$), so that minor changes in their relative energies may occasionally change the order of filling. Chemical and spectroscopic evidence indicate that the configurations of Cr and Cu have only one electron in the $4s$ orbital, and that their $3d$ sets are half-filled and filled, respectively, in the ground state. Calculations from the quantum mechanical equations also indicate that *stability is associated with half-filled and filled sets of equivalent orbitals.* Apparently this increased stability is sufficient to lower the $3d$ orbitals below the $4s$ orbital in energy, so that, in $_{24}$Cr, for example, the total energy of

[Ar] $3d$ ↑ ↑ ↑ ↑ ↑ $4s$ ↑

is lower than that of

[Ar] $3d$ ↑ ↑ ↑ ↑ _ $4s$ ⇅

Similar reasoning helps us rationalize the apparent exception of the configuration of $_{29}$Cu from that predicted by the Aufbau Principle. It might be asked

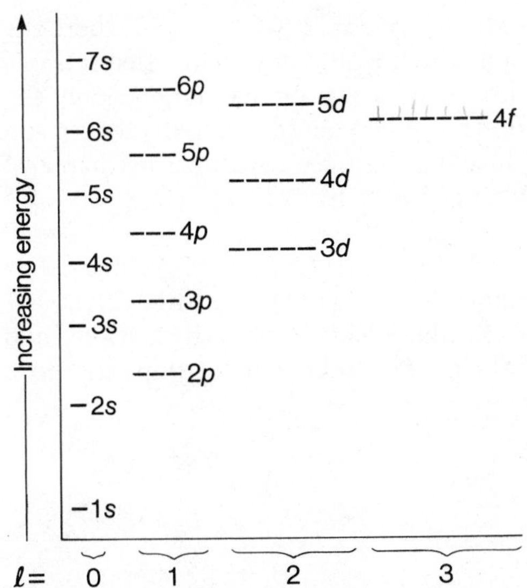

FIGURE 3–24 The usual order of energies of the orbitals of an atom. This is sometimes referred to as the **Aufbau** order. The energy scale varies for different elements, but the following main features should be noted: (1) The energies of orbitals are generally closer together at higher energies. (2) The largest energy gap is between 1*s* and 2*s* orbitals. (3) The gap between *np* and $(n + 1)s$, e.g., between 2*p* and 3*s*, or between 3*p* and 4*s*, is usually fairly large. (4) The gap between $(n - 1)df$ and *ns*, e.g., between 3*d* and 4*s*, is quite small. (5) The gap between $(n - 2)f$ and *ns*, e.g., between 4*f* and 6*s*, is even smaller.

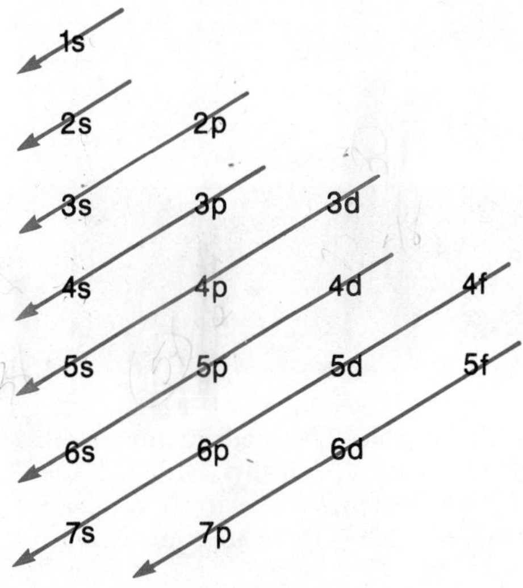

FIGURE 3–25 The order in which atomic orbitals are filled. Write all sublevels in the same major energy level on the same horizontal line. Write all like sublevels in the same column. Draw parallel arrows diagonally from upper right to lower left. The arrows are read from top to bottom, and tail to had. The order is 1*s*, 2*s*, 2*p*, 3*s*, 3*p*, 4*s*, 3*d*, 4*p*, 5*s*, 4*d*, . . . and so on.

why a similar exception does not occur in, for example, $_{32}$Ge or $_{14}$Si, where the possibility exists of an s^1p^3 configuration which would have half-filled sets of *s* and *p* orbitals. The reason that this does not occur is the very large energy gap between *ns* and *np* orbitals, in contrast to the very small gap between *ns* and $(n - 1)d$ orbitals, e.g., 4*s* and 3*d* orbitals. We shall see evidence in Chapter 4 that does, however, illustrate the enhanced stability of half-filled sets of *p* orbitals.

The periodic tables in this text are divided into "A" and "B" groups. The A groups contain elements in which *s* and *p* orbitals are being filled. Elements within any particular A group have similar electronic configurations and chemical properties, as we shall see in the next chapter. The B groups are those

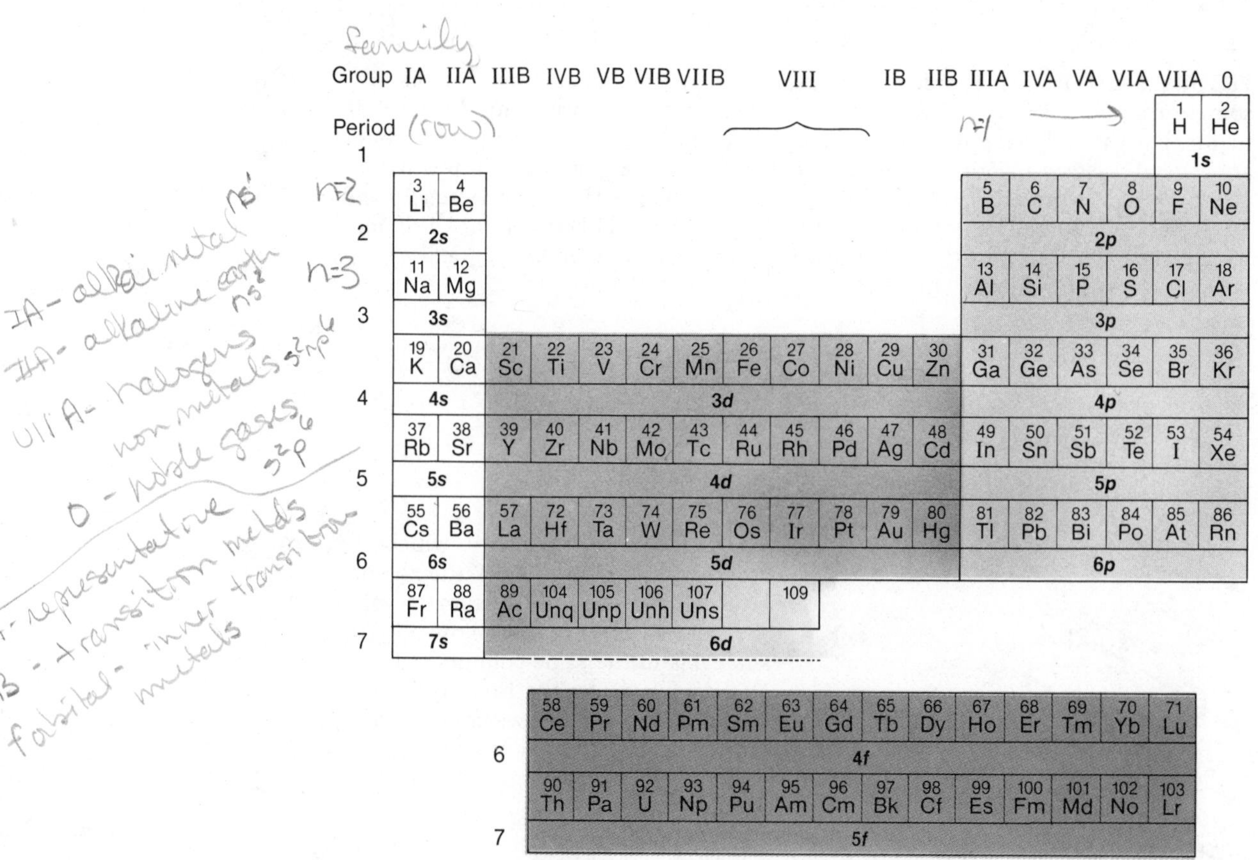

FIGURE 3–26 A periodic table shaded to show the kinds of atomic orbitals (sublevels) being filled in different parts of the periodic table. The kinds of atomic orbitals being filled are shown below the symbols of blocks of elements. Although the electronic structures of the A Group and 0 Group elements are perfectly regular and can be predicted from their positions in the periodic table, there are some exceptions in the *d* and *f* blocks.

in which there are one or two electrons in the *s* orbital of the highest occupied energy level, and *d* orbitals one energy level lower are being filled.

Observe that lithium, sodium, and potassium, elements of the leftmost column of the periodic table (Group IA), have a single electron in their outermost *s* orbital (ns^1). Beryllium and magnesium, of Group IIA, have two electrons in their highest energy level, ns^2, while boron and aluminum (Group IIIA) have three electrons in their highest energy level, $ns^2\ np^1$. Similar observations can be made for each A group.

H is shown in the 1*s* block in Figure 3–26. It is usually shown in group IA.

The electronic configurations of the A group elements and the noble gases can be predicted accurately from Figures 3–24 and 3–25. However, there are some irregularities in the B groups below the fourth period that are not easily predictable. In the heavier B group elements, the higher-energy sublevels in different principal energy levels have energies that are very nearly equal (Figure 3–24). It is easy for an electron to jump from one orbital to another of nearly the same energy in a different set because the orbitals are *perturbed* (their energies change slightly) as an extra electron is added in going from one element to the next.

Figure 3–26 displays electron configurations in abbreviated form, and a more complete picture is given inside the back cover.

Let us now write the quantum numbers to describe each electron in an atom. Keep in mind the fact that Hund's Rule must be obeyed.

Note that there is thus only one (unpaired) electron in each 2*p* orbital.

EXAMPLE 3–7

Write an acceptable set of four quantum numbers for each of the electrons of a nitrogen atom.

Solution

Nitrogen has seven electrons, which occupy the lowest-energy orbitals available. Two electrons can occupy the first energy level, $n = 1$, in which there is only one *s* orbital; when $n = 1$, then ℓ must be zero and therefore $m_\ell = 0$. The two electrons differ only in spin quantum number, m_s. The next five electrons can all fit into the second energy level, for which $n = 2$ and ℓ may be either 0 or 1. The $\ell = 0$ (*s*) sublevel fills first, and the $\ell = 1$ (*p*) sublevel is occupied next.

Electrons are indistinguishable. We have numbered them 1, 2, 3, and so on as an aid to counting them.

Electron	n	ℓ	m_ℓ	m_s	e^- Configuration
1 and 2	1	0	0	$+\frac{1}{2}$	$1s^2$
	1	0	0	$-\frac{1}{2}$	
3 and 4	2	0	0	$+\frac{1}{2}$	$2s^2$
	2	0	0	$-\frac{1}{2}$	
5, 6, 7	2	1	−1	$+\frac{1}{2}$ or $-\frac{1}{2}$	$2p_x^{\ 1}$
	2	1	0	$+\frac{1}{2}$ or $-\frac{1}{2}$	$2p_y^{\ 1}$ or $2p^3$
	2	1	+1	$+\frac{1}{2}$ or $-\frac{1}{2}$	$2p_z^{\ 1}$

EXAMPLE 3–8

Write an acceptable set of four quantum numbers that describe each electron in a chlorine atom.

Solution

Chlorine is element number 17. Its first seven electrons have the same quantum numbers as those of nitrogen in Example 3–7. Electrons 8, 9, and 10 complete the filling of the $2p$ sublevel ($n = 2$, $\ell = 1$) and therefore also the second energy level. Electrons 11 through 17 fill the $3s$ sublevel ($n = 3$, $\ell = 0$) and partially fill the $3p$ sublevel ($n = 3$, $\ell = 1$).

Electron	n	ℓ	m_ℓ	m_s	e^- Configuration
1, 2	1	0	0	$\pm\frac{1}{2}$	$1s^2$
3, 4	2	0	0	$\pm\frac{1}{2}$	$2s^2$
5–10	2	1	−1	$\pm\frac{1}{2}$	$2p^6$
	2	1	0	$\pm\frac{1}{2}$	
	2	1	+1	$\pm\frac{1}{2}$	
11, 12	3	0	0	$\pm\frac{1}{2}$	$3s^2$
13–17	3	1	−1	$\pm\frac{1}{2}$	$3p^5$
	3	1	0	$\pm\frac{1}{2}$	
	3	1	+1	$+\frac{1}{2}$ or $-\frac{1}{2}$*	

* The $3p$ orbital with only a single electron can be any one of the set, not *necessarily* the one with $m_\ell = +1$.

Acceptable sets of quantum numbers for electrons in any atom may be written using the ideas developed to this point.

Key Terms

Absorption spectrum spectrum associated with absorption of electromagnetic radiation by atoms (or other species) resulting from transitions from lower to higher energy states.

Alpha (α) particle helium ion with 2+ charge; an assembly of two protons and two neutrons.

Anode in a cathode ray tube, the positive electrode.

Atomic mass unit arbitrary mass unit defined to be exactly one-twelfth the mass of the carbon-12 isotope.

Atomic number integral number of protons in the nucleus; defines the identity of an element.

Atomic orbital region or volume in space in which the probability of finding electrons is highest.

Aufbau ("building up") Principle describes the order in which electrons fill orbitals in atoms.

Canal ray stream of positively charged particles (cations) that moves from positive electrode toward negative electrode in cathode ray tubes; observed to pass through canals in the negative electrode.

Cathode in a cathode ray tube, the negative electrode.

Cathode ray beam of electrons going from negative electrode toward positive electrode in cathode ray tube.

Cathode ray tube closed glass tube containing a gas under low pressure, with electrodes near the ends and a luminescent screen at the end near the positive electrode; produces cathode rays when high voltage is applied; also called gas discharge tube.

Continuous spectrum spectrum that contains all wavelengths in a specified region of the electromagnetic spectrum.

Degenerate of the same energy.

Diamagnetism *weak* repulsion by a magnetic field.

***d* orbitals** beginning in the third energy level, a set of five degenerate orbitals per energy level, higher in energy than *s* and *p* orbitals of the same energy level.

Electromagnetic radiation energy that is propagated by means of electric and magnetic fields that oscillate in directions perpendicular to the direction of travel of the energy.

Electron a subatomic particle having a mass of 0.00055 amu and a charge of 1−.

Electron configuration specific distribution of electrons in atomic orbitals of atoms or ions.

Electronic transition the transfer of an electron from one energy level to another.

Emission spectrum spectrum associated with emission of electromagnetic radiation by atoms (or other species) resulting from electronic transitions from higher to lower energy states.

Excited state any state other than the ground state of an atom or molecule.

***f* orbitals** beginning in the fourth energy level, a set of seven degenerate orbitals per energy level, higher in energy than *s*, *p*, and *d* orbitals of the same energy level.

Frequency the number of repeating corresponding points on a wave that pass a given observation point per unit time.

Ground state the lowest energy state or most stable state of an atom, molecule or ion.

Group a vertical column in the periodic table; also called a family.

Heisenberg Uncertainty Principle it is impossible to determine accurately both the momentum and position of an electron simultaneously.

Hund's Rule all orbitals of a given sublevel must be occupied by single electrons before pairing begins. see *Aufbau Principle*

Isotopes two or more forms of atoms of the same element with different masses; atoms containing the same number of protons but different numbers of neutrons.

Line spectrum an atomic emission or absorption spectrum.

Magnetic quantum number (m_ℓ) quantum mechanical solution to a wave equation that designates the particular orbital within a given set (*s*, *p*, *d*, *f*) in which an electron resides.

Mass defect amount of matter that would be converted into energy if an atom were formed from constituent particles.

Mass number the sum of the numbers of protons and neutrons in an atom; an integer.

Mass spectrometer an instrument that measures the mass-to-charge ratio of charged particles.

Natural radioactivity spontaneous decomposition of an atom.

Neutron a neutral subatomic nuclear particle having a mass of 1.0087 amu.

Nuclear binding energy energy equivalent of the mass defect; energy released in the formation of an atom from subatomic particles.

Nucleus the very small, very dense, positively charged center of an atom containing protons and neutrons, as well as other subatomic particles.

Nuclide symbol symbol for an atom, ${}^{A}_{Z}E$, in which E is the symbol for an element, Z is its atomic number, and A is its mass number.

Pairing of electrons, attractive (rather than repulsive) interaction of the spin-induced magnetic fields of two electrons in the same orbital.

Paramagnetism attraction toward a magnetic field, stronger than diamagnetism, but still weak, compared to ferromagnetism.

Pauli Exclusion Principle no two electrons in the same atom may have identical sets of four quantum numbers.

Period a horizontal row in the periodic table.

Photoelectric effect emission of an electron from the surface of a metal caused by impinging electromagnetic radiation of certain minimum energy; current increases with increasing intensity of radiation.

Photon a "packet" of light or electromagnetic radiation; also called a quantum of light.

***p* orbitals** beginning with the second energy level, a set of three mutually perpendicular, equal-arm, dumbbell-shaped atomic orbitals per energy level.

Principal quantum number (n) quantum mechanical solution to a wave equation that designates the major energy level or shell in which an electron resides.

Proton a subatomic particle having a mass of 1.0073 amu and a charge of +1, found in the nuclei of atoms.

Quantum a "packet" of energy.

Quantum mechanics mathematical method of treating particles on the basis of quantum theory, which assumes that energy (of small particles) is not infinitely divisible.

Quantum numbers numbers that describe the energies of electrons in atoms; derived from quantum mechanical treatment.

Radiant energy see *Electromagnetic radiation*

Rydberg equation empirical equation that relates wavelengths in the hydrogen emission spectrum to integers.

***s* orbital** a spherically symmetrical atomic orbital; one per energy level.

Spectral line any of a number of lines corresponding to definite wavelengths in an atomic emission or absorption spectrum; represents the energy difference between two energy levels.

Spectrophotometer an instrument that measures the absorption or emission of electromagnetic radiation as a function of wavelength.

Spectrum display of component wavelengths (colors) of electromagnetic radiation.

Spin quantum number (m_s) quantum mechanical solution to a wave equation that indicates the relative spins of electrons.

Subsidiary quantum number (ℓ) quantum mechanical solution to a wave equation that designates the sublevel or set of orbitals (*s*, *p*, *d*, *f*) within a given major energy level in which an electron resides.

Wavelength the distance between two corresponding points of a wave.

Exercises

Dalton's Atomic Theory

1. List the basic postulates of Dalton's atomic theory and explain the meaning of each.
2. State the Law of Conservation of Matter. How is it explained by Dalton's atomic theory?
3. State the Law of Definite Proportions. How is it explained by Dalton's atomic theory?
4. What parts of Dalton's atomic theory were in error?

Subatomic Particles and the Nuclear Atom

5. What evidence from experiments shows (a) that electrons are negatively charged and (b) that electrons have mass?
6. What evidence do we have that all atoms contain electrons?
7. Describe how Thomson determined the charge-to-mass ratio for the electron.
8. Describe Millikan's oil drop experiment. What is its significance?
9. How are canal rays produced? Are they due to subatomic particles? How do we know?
10. Describe the "plum-pudding" or "raisin loaf" model of the atom.
11. Outline Rutherford's contribution to our understanding of the nature of atoms.
12. Why were Rutherford and his co-workers so surprised that some of the alpha particles were actually scattered backwards in the gold-foil experiment?
13. Summarize Moseley's contribution to our knowledge of the structure of atoms.

14. How do the masses of protons and neutrons compare with those of electrons?
15. Compare the properties of cathode rays, protons, neutrons, electrons, and alpha particles.
16. Estimate the percentage of the total mass of a $^{109}_{47}Ag$ atom that is due to (a) electrons, (b) protons, and (c) neutrons by *assuming* that the mass of the atom is simply the sum of the masses of the appropriate numbers of subatomic particles. (The mass defect would introduce a small correction.)
17. The radius of a silver atom, Ag, is 1.44 Å. Using the results of Exercise 16 for $^{109}_{47}Ag$ and the estimate that the diameter of the nucleus of the $^{109}_{47}Ag$ atom is 1/10,000 that of the entire atom, estimate the density of the nucleus of a $^{109}_{47}Ag$ atom. Express your answer in g/cm^3. Comment on the magnitude of this calculated density.
18. Refer to Exercise 17. Estimate the density of an entire silver atom.
19. The actual density of a macroscopic sample of silver is 10.5 g/cm^3. Can the difference between this and the estimated density in Exercise 18 be due only to the assumption of Exercise 16? How can the difference in values be explained? There are two other factors. The *lesser* of them involves isotopic distributions. (If you can't think of any adequate explanation you may look ahead to Section 9–16, part 4, on metallic solids.)
20. Define and illustrate the following terms clearly and concisely: (a) atomic number, (b) mass number, (c) isotope, (d) nuclear charge.
21. How are isotopic abundances determined experimentally?
22. Complete the chart at the bottom of the page for the indicated isotopes.

The Atomic Weight Scale

23. What is the basis for the atomic weight scale?
24. The arbitrary standard for the atomic weight scale is the exact number 12 for the mass of the carbon-12 isotope. Why is the atomic weight of carbon listed as 12.011 amu?
25. Naturally occurring carbon consists of two isotopes, ^{12}C (98.89%) and ^{13}C (1.11%). There are also minute quantities of ^{14}C, but these are too small for us to consider at this point. Using these data, calculate the atomic weight of naturally occurring carbon. The mass of the ^{12}C isotope is 12.00000 amu and the mass of the ^{13}C isotope is 13.00335 amu.
26. Prior to 1962, the atomic weight scale was based on the assignment of an atomic weight of exactly 16 amu to *naturally occurring* oxygen. The atomic weight of bromine is 79.904 amu on the carbon-12 scale. What would it be on the older scale?

Kind of Atom	Atomic Number	Mass Number	Isotope	Number of Protons	Number of Electrons	Number of Neutrons
nitrogen		14				
fluorine						10
			$^{27}_{13}Al$			
sulfur						16
	24	52				
	38					50
		73		32		
					50	69
		132			54	

27. Naturally occurring silicon consists of three isotopes with the abundances indicated below. From the masses and relative abundances of these isotopes, calculate the atomic weight of naturally occurring silicon.

Isotope	Isotopic Mass (amu)	% Natural Abundance
^{28}Si	27.97693	92.21
^{29}Si	28.97649	4.70
^{30}Si	29.97376	3.09

28. The atomic weight of lithium is 6.941 amu. The two naturally occurring isotopes of lithium have the following masses: ^{6}Li, 6.01512 amu and ^{7}Li, 7.01600 amu. Calculate the percent of ^{6}Li in naturally occurring lithium.
29. Rubidium consists of two naturally occurring isotopes: ^{85}Rb, which has a mass of 84.9117 amu, and ^{87}Rb, which has a mass of 86.9085 amu. The atomic weight of rubidium is 85.4678 amu. Determine the percent abundance of each isotope in naturally occurring rubidium.
30. Refer to Table 3–2 *only* and estimate the atomic weights of boron and chlorine to one decimal place. Justify your estimates.
31. The following is a mass spectrum of the 1+ charged ions of an element. Calculate the atomic weight of the element. What is the element?

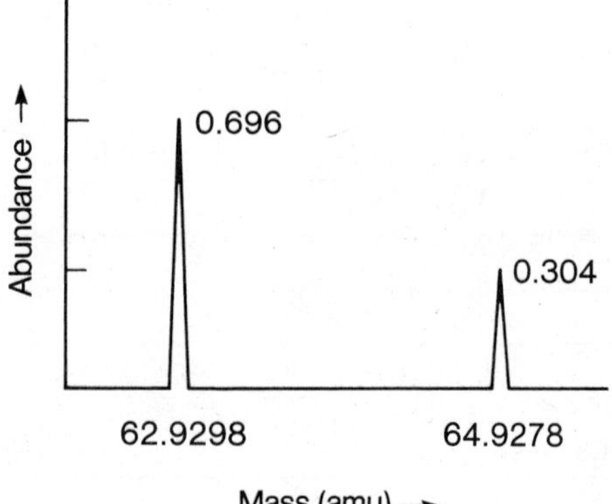

32. Consider the ions $^{14}_{7}N^{+}$, $^{15}_{7}N^{+}$, $^{14}_{7}N^{2+}$, and $^{15}_{7}N^{2+}$ produced in a mass spectrometer. Which ion's path would be deflected (a) the most and (b) the least by a magnetic field? Which one would travel (c) the most rapidly and (d) the least rapidly under the influence of a particular accelerating voltage? Justify your answers.

Mass Defect and Binding Energy

33. What is the equation that relates the equivalence of matter and energy? What does each term in this equation represent?
34. What is mass defect?
35. What is binding energy?
36. The actual mass of a ^{80}Se atom is 79.9165 amu per atom. Calculate the mass defect in amu/atom and in g/mol for this isotope.
37. What is the nuclear binding energy in kJ/mol and in kcal/mol for the isotope described in Exercise 36?
38. The actual mass of a ^{47}Ti atom is 46.9518 amu. (a) Calculate the mass defect in amu/atom and g/mol for this isotope. (b) Calculate the binding energy in kJ/mol.
39. How do nuclear binding energies vary with mass number, i.e., which elements have high and which elements have low nuclear binding energies?

Electromagnetic Radiation

40. Describe the following terms clearly and concisely in relation to electromagnetic radiation: (a) wavelength, (b) frequency, (c) amplitude, (d) color.
41. Calculate the wavelengths, in meters, of radiation of the following frequencies: (a) $5.00 \times 10^{15}\ s^{-1}$, (b) $1.36 \times 10^{14}\ s^{-1}$, (c) $7.26 \times 10^{12}\ s^{-1}$.
42. Calculate the frequencies of radiation of the following wavelengths: (a) 10,000 Å (b) 520 nm (c) 5.30×10^{-5} cm (d) 86.2 cm (e) 2.00×10^{-7} cm.
43. What is the energy of a photon of each of the radiations in Exercise 42? Express your answer in joules per photon.
44. In which regions of the electromagnetic spectrum do the radiations in Exercise 42 fall?

45. Radio station WSB-AM in Atlanta broadcasts at a frequency of 750 kHz. What is the wavelength of its signal in meters?
46. Alpha Centauri is the star closest to our solar system. It is about 2.5×10^{13} miles away. How many light years is this? A light year is the distance that light travels (in a vacuum) in one year. Assume that space is essentially a vacuum.

The Photoelectric Effect

47. What evidence supports the idea that electromagnetic radiation is (a) wave-like, (b) particle-like?
48. Describe the influence of frequency and intensity of electromagnetic radiation on the current in the photoelectric effect.
49. Cesium is often used in "electric eyes" for self-opening doors in an application of the photoelectric effect. The amount of energy required to ionize (remove an electron from) a cesium atom is 3.89 electron-volts. (One electron-volt is 1.60×10^{-19} J.) What would be the wavelength, in nanometers, of light with just sufficient energy to ionize a cesium atom?

Atomic Spectra and the Bohr Theory

50. Distinguish between a continuous spectrum and an atomic spectrum.
51. Distinguish between an atomic emission spectrum and an atomic absorption spectrum.
52. What is the Rydberg equation? Why is it called an empirical equation?
53. The Lyman series is the name given to the series of lines in the ultraviolet portion of the emission spectrum of hydrogen. These lines correspond to transitions of electrons from higher energy states to the lowest ($n = 1$) energy state. Use the Rydberg equation to calculate the wavelengths of the three lowest energy lines in the Lyman series.
54. The Rydberg equation also can be written in terms of energies rather than wavelengths,

$$\Delta E = R'\left(\frac{1}{n_1^2} - \frac{1}{n_2^2}\right)$$

in which R' is a constant. Using the Rydberg constant, $R = 1.097 \times 10^7\ \text{m}^{-1}$, determine the value and units of R'. Use the unit of joules for energy.
55. The following are prominent lines in the visible region of the emission spectra of the elements listed. The lines can be used to identify the elements. What color is the light responsible for each line? (a) sodium, 5460 Å, (b) neon, 6160 Å, (c) mercury, 4540 Å.
56. Outline the Bohr model of the hydrogen atom.
57. Why is the Bohr model for the hydrogen atom referred to as the solar system model?
58. If each atom in one mole of atoms emits a photon of wavelength 5000 Å, how much energy is lost? Express the answer in J/mol and kcal/mol. As a reference point, burning 1 mole (16 g) of CH_4 produces 819 kJ of heat.
59. Hydrogen atoms have an absorption line at 1216 Å. What is the frequency of the photons absorbed, and what is the energy difference, in joules, between the ground state and this excited state of the atom?

The Wave-Particle View of the Electron

60. What evidence supports the idea that electrons are (a) particle-like, (b) wave-like?
61. What was de Broglie's contribution to our understanding of the nature of atoms?
62. Using the de Broglie equation, show that the wavelength of a 0.13-kg baseball traveling at a velocity of 40 m/s (90 miles/hr) has such a short wavelength as to be unobservable. How does this wavelength compare with the diameter of a typical atom (see atomic radii in Figure 4–2)?

The Quantum Mechanical Picture of Atoms and Atomic Orbitals

63. State the Heisenberg Uncertainty Principle. What does it mean, and what is its significance?
64. State the basic postulates of quantum mechanics and explain the meaning and significance of each.
65. What are atomic orbitals?

66. What is the relationship among the number of electrons in a neutral atom, its atomic number, and the number of protons in its nucleus?
67. Without giving the ranges of possible values of the four quantum numbers, n, ℓ, m_ℓ and m_s, describe briefly what information each one gives.
68. How are the possible values for the subsidiary quantum number for a given electron restricted by the value of n?
69. How are the values of m_ℓ for a particular electron restricted by the value of ℓ?
70. What is the origin of the designation of atomic orbitals as s, p, d, and f orbitals?
71. What are the values of n and ℓ for the following sublevels? (a) $2s$ (b) $3p$ (c) $4s$ (d) $5d$ (e) $4f$.
72. Compare the relative energies of the s, p, d and f sublevels in a particular major energy level.
73. How many individual orbitals are there in the fourth major energy level? Write out n, ℓ and m_ℓ quantum numbers for each one and label each set as describing s, p, d or f orbitals.
74. What is Hund's Rule? What does it mean?
75. State the Pauli Exclusion Principle. What does it mean?
76. Consider the following sets of quantum numbers. Which sets represent impossible combinations? Indicate why they are impossible.

	n	ℓ	m_ℓ	m_s
(a)	1	0	1	$+\frac{1}{2}$
(b)	2	1	1	$-\frac{1}{2}$
(c)	3	2	-2	$-\frac{1}{2}$
(d)	3	3	0	$+\frac{1}{2}$
(e)	3	1	-2	$-\frac{1}{2}$
(f)	2	0	0	$-\frac{1}{2}$

77. What do we mean when we say that the highest energy electron of a lithium atom in its ground state is in the 3s orbital?
78. What is the meaning of the following statement? The probability of finding an electron at any point is directly proportional to the electron density at that point.
79. Complete the following table for the first five energy levels. What does each column indicate?

n	n^2	$2n^2$
1	1	2

80. What do we mean when we refer to the "spin of an electron"?
81. What are spin-paired electrons?

Electronic Configurations

82. What is the Aufbau Principle? What evidence do we have that it is valid?
83. Draw representations of ground state electronic configurations using ⇅ notation for the following elements: (a) He, (b) B, (c) Mg (d) Ar (e) Fe (f) Sb.
84. Give the ground state electronic configurations for the elements of Exercise 83 using shorthand notation, that is, $1s^22s^22p^6$, etc.
85. Draw representations of ground state electronic configurations using ⇅ notation for the following elements: (a) Li, (b) N, (c) Ca, (d) V, (e) Zn, (f) Mo, (g) Pb.
86. Give the ground state electronic configurations for the elements of Exercise 85 using shorthand notation, that is, $1s^22s^22p^6$, etc.
87. What do the electronic structures of elements 24 and 29 illustrate?
88. Why do electrons occupy the $4s$ orbital before they occupy the $3d$ orbitals in elements 19 and 20?
89. Distinguish between the terms diamagnetic and paramagnetic, and provide an example that illustrates the meaning of each.
90. How is paramagnetism measured experimentally?
91. Which of the elements in Exercise 83 are paramagnetic?
92. Which of the elements in Exercise 85 are paramagnetic?
93. Indicate whether each of the following elements is paramagnetic in its atomic ground state? (a) Be, (b) C, (c) Na, (d) Cl, (e) Kr, (f) Mn, (g) Cu, (h) Zn.
94. Which one of the elements in Exercise 93 exhibits the highest degree of paramagnetism in its ground state?

95. Construct a table in which you list a possible set of values for the four quantum numbers for each electron in the following atoms in their ground states: (a) Be, (b) Al, and (c) Mn.
96. Construct a table in which you list a possible set of values for the four quantum numbers for each electron in the following atoms in their ground states: (a) N, (b) Cl and (c) Kr.
97. List the n, ℓ and m_ℓ quantum numbers for the highest energy electron (or one of the highest energy electrons if there are more than one) in the following atoms in their ground state: (a) Li, (b) S, (c) Co, (d) Sr.
98. List n, ℓ and m_ℓ quantum numbers for the highest energy electron (or one of the highest energy electrons if there are more than one) in the following atoms in their ground states: (a) C, (b) Ar, (c) Rh and (d) Ce.
99. What is the distinction between the A and B groups in the periodic table?
100. How do the periodic group numbers relate to the outermost electronic configurations of the elements in the A Groups of the periodic table?
101. Draw general electronic structures for the A Group elements using the $\underline{\uparrow\downarrow}$ notation, where n is the principal quantum number for the highest occupied energy level.

	ns	*np*
IA	__	__ __ __
IIA	__	__ __ __
etc.		

102. Repeat Exercise 101 using ns^xnp^y notation.

Equation Balancing

103. Balance the following equations by inspection (to keep your skills sharp).
(a) $Sb + F_2 \longrightarrow SbF_5$
(b) $PBr_3 + Br_2 \longrightarrow PBr_5$
(c) $SnS_2 + HCl \longrightarrow H_2SnCl_6 + H_2S$
(d) $RbO_2 + H_2O \longrightarrow RbOH + O_2$
(e) $PF_3Br_2 \longrightarrow PF_5 + PBr_5$
104. Balance the following equations by inspection.
(a) $Al + O_2 \xrightarrow{\Delta} Al_2O_3$
(b) $S_8 + O_2 \xrightarrow{\Delta} SO_2$
(c) $Cl_2O_7 + H_2O \longrightarrow HClO_4$
(d) $PCl_3 + H_2O \longrightarrow H_3PO_3 + HCl$
(e) $NiSO_4 + NH_3 + H_2O \longrightarrow Ni(NH_3)_6(OH)_2 + (NH_4)_2SO_4$

4 Chemical Periodicity and Ionic Bonding

As we look around us we see that some substances are gases, others are liquids, while still others are solids. Some solids are soft (paraffin), others are quite hard (diamond), and others are quite strong (steel). Clearly, there must be significant differences among the attractive forces that hold atoms together in different substances.

The formation of chemical bonds between atoms of different elements results in the formation of compounds. When the atoms are separated, chemical bonds are broken and the original compounds cease to exist. The attractive forces between atoms are electrical in nature, and chemical reactions between atoms involve changes in their electronic structures.

Chemical bonding theories are inadequate to explain all chemical behavior, but as we shall see, they are useful because they assist in systematizing much information.

In Chapter 3 we described the structures of atoms in some detail. An element's position in the periodic table is determined by its electronic configuration, which in turn plays a role in determining the kinds of chemical bonds the element forms. Let us now turn our attention to the evolution of the periodic table, and to **chemical periodicity,** *the variation in properties of elements with their positions in the periodic table.* Then we shall introduce the two extremes in chemical bonding, ionic bonding and, in Chapter 5, covalent bonding.

4–1 The Periodic Table

In 1869 the Russian chemist Dimitri Mendeleev and Lothar Meyer, a German, independently published arrangements of known elements that closely resemble the periodic table in use today. Mendeleev's classification was based primarily on chemical properties of the elements, while Meyer's classification was based largely on physical properties. The tabulations were surprisingly similar and both emphasized the **periodicity,** or regular periodic repetition, of properties with increasing atomic weight.

Pronounced "men-del-lay'-ev."

Mendeleev arranged the known elements in order of increasing atomic weight in successive sequences so that elements with similar chemical properties fell in the same column. *He noted that both physical and chemical properties of the elements vary in a periodic fashion.* This statement is known as the **periodic law.** His periodic table of 1872, which contained the 62 elements known then, is shown in Figure 4–1.

Consider H, Li, Na, and K, all of which appear in "Gruppe I" of Mendeleev's table. All were known to combine with F, Cl, Br, and I of "Gruppe VII" to produce compounds that have similar formulas such as HF, LiCl, NaBr, and KI. All these compounds dissolve in water to produce solutions that conduct electricity. The "Gruppe II" elements form compounds such as $BeCl_2$, $MgBr_2$, and $CaCl_2$, as well as compounds with O and S from "Gruppe VI" such as MgO, CaO, MgS, and CaS.

One of the advantages of Mendeleev's periodic table was that it provided for elements that had not yet been discovered at the time he constructed the table. When he encountered "missing" elements, he left blank spaces. Some appreciation of Mendeleev's genius in constructing the table as he did can be gained by examining a comparison of predicted (1871) and observed properties

TABELLE II

REIHEN	GRUPPE I. — R^2O	GRUPPE II. — RO	GRUPPE III. — R^2O^3	GRUPPE IV. RH^4 RO^2	GRUPPE V. RH^3 R^2O^5	GRUPPE VI. RH^2 RO^3	GRUPPE VII. RH R^2O^7	GRUPPE VIII. — RO^4
1	H = 1							
2	Li = 7	Be = 9,4	B = 11	C = 12	N = 14	O = 16	F = 19	
3	Na = 23	Mg = 24	Al = 27,3	Si = 28	P = 31	S = 32	Cl = 35,5	
4	K = 39	Ca = 40	— = 44	Ti = 48	V = 51	Cr = 52	Mn = 55	Fe = 56, Co = 59, Ni = 59, Cu = 63.
5	(Cu = 63)	Zn = 65	— = 68	— = 72	As = 75	Se = 78	Br = 80	
6	Rb = 85	Sr = 87	?Yt = 88	Zr = 90	Nb = 94	Mo = 96	— = 100	Ru = 104, Rh = 104, Pd = 106, Ag = 108.
7	(Ag = 108)	Cd = 112	In = 113	Sn = 118	Sb = 122	Te = 125	J = 127	
8	Cs = 133	Ba = 137	?Di = 138	?Ce = 140	—	—	—	— — — —
9	(—)	—	—	—	—	—	—	
10	—	—	?Er = 178	?La = 180	Ta = 182	W = 184	—	Os = 195, Ir = 197, Pt = 198, Au = 199.
11	(Au = 199)	Hg = 200	Tl = 204	Pb = 207	Bi = 208	—	—	
12	—	—	—	Th = 231	—	U = 240	—	— — — —

FIGURE 4–1 Mendeleev's early periodic table, published in 1872. (J is the old German symbol for iodine.)

of germanium, which was not discovered until 1886. Mendeleev called the undiscovered element eka-silicon because it fell below silicon in his table. He was familiar with the properties of germanium's neighboring elements, and they served as the basis for his predictions of properties of germanium (Table 4–1).

Note that some of the modern values of properties of germanium differ significantly from those reported in 1886. In fairness to Mendeleev, much of the data upon which he based his predictions were as inaccurate as the 1886 values for Ge.

At this point we might observe that in many areas of human endeavor progress is slow and faltering. However, in most areas there is an occasional individual who is able to develop concepts and techniques that clarify previously confused situations. Mendeleev was such an individual. Unfortunately, although his work was of fundamental importance, it did not clarify entirely the concept of periodicity. Because Mendeleev's arrangement of the elements was based on increasing *atomic weights,* several elements appeared to be out of place in the table. Mendeleev put the controversial elements (Te and I, Co and Ni, and after the noble gases were discovered, Ar and K) in locations consistent with their properties, and attributed the apparent reversal of atomic weights to inaccurate values for those weights. Careful re-determination of the atomic weights showed that these values were correct in the first place. Indeed, resolution of the problem of these out-of-place elements had to await the development and clarification of the concept of *atomic number,* after which the periodic law was evolved in essentially its present form.

TABLE 4–1 Predicted and Observed Properties of Germanium

Property	Eka-Silicon Predicted, 1871	Germanium Reported, 1886	Modern Values
Atomic weight	72	72.32	72.59
Atomic volume	13 cm^3	13.22 cm^3	13.5 cm^3
Specific gravity	5.5	5.47	5.35
Specific heat	0.073 cal/g°C	0.076 cal/g°C	0.074 cal/g°C
Maximum valence*	4	4	4
Color	Dark gray	Grayish white	Grayish white
Reaction with water	Will decompose steam with difficulty	Does not decompose water	Does not decompose water
Reactions with acids and alkalis	Slight with acids; more pronounced with alkalis	Not attacked by HCl or dilute aqueous NaOH; reacts vigorously with molten NaOH	Not dissolved by HCl or H_2SO_4 or dilute NaOH; dissolved by concentrated NaOH
Formula of oxide	EsO_2	GeO_2	GeO_2
Specific gravity of oxide	4.7	4.703	4.228
Specific gravity of tetrachloride	1.9 at 0°C	1.887 at 18°C	1.8443 at 30°C
Boiling point of tetrachloride	100°C	86°C	84°C
Boiling point of tetraethyl derivative	160°C	160°C	186°C

* Valence refers to the number of atoms bonded to a specific atom.

As the *periodic law* is now stated, *the properties of the elements are periodic functions of their atomic numbers.* This statement tells us that if we arrange the elements in order of increasing atomic number, periodically we encounter elements that have similar chemical and physical properties. The presently used "long form" of the periodic table is such an arrangement. Table 4–2 is a modern long form of the periodic table. The vertical columns are referred to as **groups** or **families** and the horizontal rows are referred to as **periods.** Elements in a *group* have similar chemical and physical properties, while those within a *period* have properties that change progressively across the table. Several groups of elements have common names that are used so frequently they should be learned. The Group IA elements, except H, are referred to as **alkali metals,** and the Group IIA elements are called the **alkaline earth metals.** The Group VIIA elements are called **halogens,** which means "salt formers," and the Group 0 elements are called **noble (or rare) gases.**

Alkaline means basic. The character of basic compounds will be described in Section 7–1.4.

Differences Between A and B Groups

The A and B designations for groups of elements in the periodic table are arbitrary designations, and they are reversed in some periodic tables. The system used in this text is the one commonly used in the United States. Elements with the same group numbers, but with different letter designations, have relatively few similar properties. The origin of the IA, IB, etc., designations is the fact that some compounds of elements with the same group numbers, but different letter designations, have similar formulas, e.g., NaCl (IA) and AgCl (IB), $MgCl_2$ (IIA) and $ZnCl_2$ (IIB). As we shall see, variations in the properties of the B group elements across a row are not nearly as dramatic as the variations observed across a row of A group elements because electrons are being added to the $(n - 1)d$ orbitals, where n represents the highest energy level that contains electrons. Since the *outermost* electrons have the greatest influence on the properties of elements, adding an electron to an *inner d* orbital results in less striking changes in properties than adding an electron to an *outer s* or *p* sublevel.

Elements may be classified in a number of ways. One of the most useful classifications follows.

Noble gases For many years the Group 0 elements were called *inert* gases because no chemical reactions involving these elements were known, but we now know that the heavier members do form compounds, mostly with fluorine and oxygen. With the exception of helium (in which the first energy level is filled with two electrons), these elements have eight electrons in the highest energy level. Their structures may be represented as . . . ns^2np^6.

Representative elements The A group elements in the periodic table are known as representative elements. They are characterized by unfilled highest occupied energy levels, in which their "last" electron is added to an s or p orbital. These elements generally show distinct and fairly regular variations in their properties with atomic number.

d-Transition elements Elements in the B groups (except IIB) in the periodic table are known as d-transition elements, or more simply as transition elements or transition metals. They were considered as transitions between the alkaline elements (base formers) on the left and the acid formers on the right. All are metals and are characterized by electrons being added to d orbitals. Stated differently, the d-transition elements are building an inner (next to highest occupied) energy level from 8 to 18 electrons. They are referred to as

1st Transition Series: $_{21}$Sc through $_{29}$Cu
2nd Transition Series: $_{39}$Y through $_{47}$Ag
3rd Transition Series: $_{57}$La and $_{72}$Hf through $_{79}$Au
4th Transition Series: (not complete) $_{89}$Ac and elements 104 through 111

Strictly speaking, the Group IIB elements, zinc, cadmium, and mercury, are not d-transition metals since their "last" electrons go into s orbitals, but they are usually discussed with d-transition metals because their chemical properties are similar.

Inner transition elements These elements are sometimes known as *f-transition elements.* They are elements in which electrons are being added to f orbitals, or those in which the second from the highest occupied energy level is building from 18 to 32 electrons. All are metals. The inner transition elements are located between Groups IIIB and IVB in the periodic table.

1st Inner Transition Series (lanthanides): $_{58}$Ce through $_{71}$Lu
2nd Inner Transition Series (actinides): $_{90}$Th through $_{103}$Lr

TABLE 4–2 The s, p, d, and f Blocks of the Periodic Table*

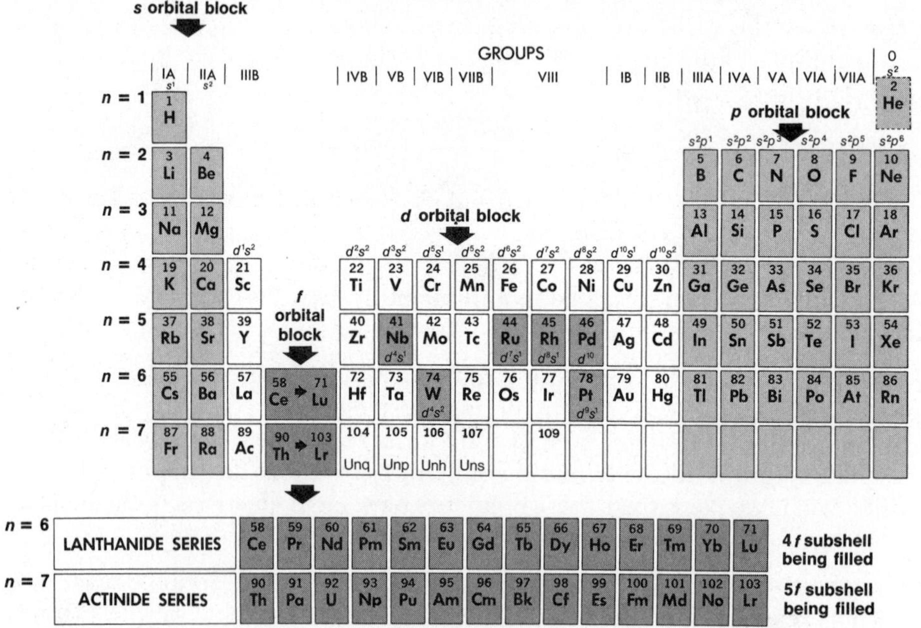

*n is the principal quantum number. The d^1s^2, d^2s^2, . . . designations represent **known** configurations. They refer to $(n-1)d$ and ns orbitals. Several exceptions to expected configurations (those indicated above each group) are highlighted.

Table 4–2 shows the divisions of the periodic table according to the kinds of atomic orbitals occupied by the "last added" electron.

4–2 Lewis Dot Representations of Elements

Table 4–3 gives **Lewis dot representations** of the representative elements. Only the electrons in the outermost occupied *s* and *p* orbitals are shown as dots. Paired and unpaired electrons are also indicated. Note that each element in a given group has the same configuration in its outer shell. These representations are useful in discussing chemical bonding, which usually involves only the outermost electrons of atoms, also called **valence electrons.**

TABLE 4–3 Electron Dot Formulas of the Representative Elements

Group	IA	IIA	IIIA	IVA	VA	VIA	VIIA	0
Number of Electrons in Outer Shell	1	2	3	4	5	6	7	8 (except He)
	H·							He:
	Li·	Be:	B·	C·	·N·	·O:	·F:	:Ne:
	Na·	Mg:	Al·	Si·	·P·	·S:	·Cl:	:Ar:
	K·	Ca:	Ga·	Ge·	·As·	·Se:	·Br:	:Kr:
	Rb·	Sr:	In·	Sn·	·Sb·	·Te:	·I:	:Xe:
	Cs·	Ba:	Tl·	Pb·	·Bi·	·Po:	·At:	:Rn:
	Fr·	Ra:						

Periodic Properties

We have found that physical and chemical properties of the elements recur periodically. Let us now investigate the nature of this periodicity in some detail, since some knowledge of periodicity is valuable in understanding bonding in simple compounds. Many physical properties, such as melting points, boiling points, and atomic volumes, show periodic variations, but for the moment let's concern ourselves with the variations that are most useful in predicting chemical properties. The variations in these properties depend upon electronic configurations, especially the configurations in the outermost occupied shell, and on how far away that shell is from the nucleus.

4–3 Atomic Radii

In a real sense, it is impossible to describe an atom as having an invariant size because the size of an atom is determined by its immediate environment, i.e., its interaction with surrounding atoms. Additionally, we cannot isolate an individual atom and measure its diameter the way we can measure the diameter of a golf ball. An indirect approach is required. By analogy, suppose we arrange a specified number of golf balls in a box in an orderly array. If we know how the balls are positioned, the number of balls, and the dimensions of the box, it is possible to calculate the diameter of an individual ball. In determining the "sizes" of atoms, the picture is complicated by the diffuse nature of the electron cloud surrounding an atom and by variations in the size of this cloud with environmental factors. Recognizing, then, that for all practical purposes, the size of an individual atom is a quantity that cannot be defined accurately, we can make measurements on samples of chemically uncombined (i.e., pure) elements. The data obtained from such measurements indicate the *relative* sizes of individual atoms. Figure 4–2 displays the relative sizes of atoms of the representative elements and the noble gases. It clearly indicates periodicity in atomic radii. [Atomic radii are often stated in **angstroms** (1 Å = 10^{-10} m). They may also be given in the SI units **nanometers** (nm, 10^{-9} m) or **picometers** (pm, 10^{-12} m).]

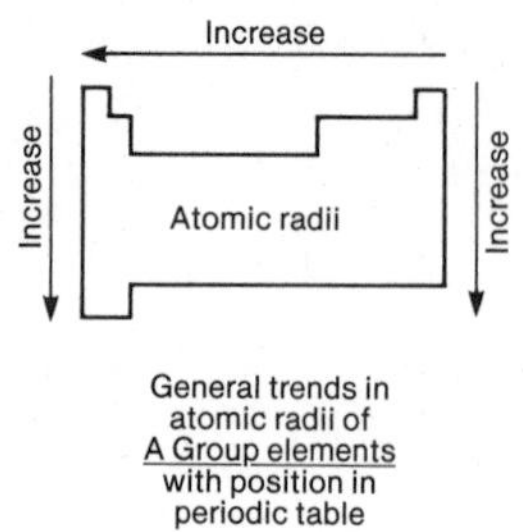

General trends in atomic radii of A Group elements with position in periodic table

As we move from left to right across a *period* in the periodic table, there is a regular decrease in atomic radii of the representative elements as electrons are added to a particular energy level. As the nuclear charge increases and electrons are added to the *same principal energy level,* the increased nuclear charge draws the electron cloud closer and closer to the nucleus. As we move down a group, atomic radii increase as electrons are added to larger orbitals in higher energy levels. For the transition elements, the variations are not so

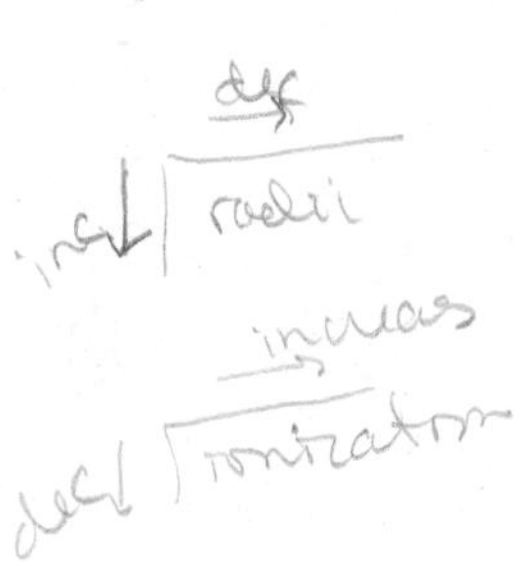

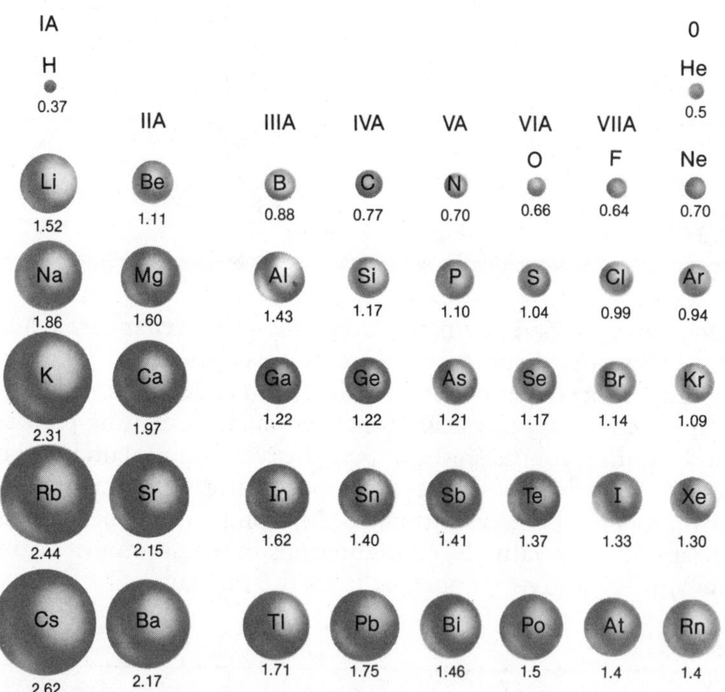

FIGURE 4–2 Atomic radii of the A group (representative) elements and the noble gases in angstroms. Atomic radii *increase* as a *group is descended* because electrons are being added to shells further from the nucleus. Atomic radii *decrease from left to right within a given row* due to increasing effective nuclear charge. Hydrogen atoms are the smallest and cesium atoms are the largest. 1Å = 1×10^{-10} *m*.

obviously regular because the electrons are being added to an inner shell, but all transition elements have radii smaller than those of the preceding Group IA and Group IIA elements in the same period.

4–4 Ionization Energy

The **first ionization energy** *is the minimum amount of energy required to remove the most loosely bound electron from an isolated gaseous atom to form an ion with a 1+ charge.* For calcium, for example, the first ionization energy, IE_1, is 590 kJ/mol

$$\text{Ca (g)} + 590 \text{ kJ} \longrightarrow \text{Ca}^+ \text{ (g)} + e^-$$

The second ionization energy is the amount of energy required to remove the second electron, which may be represented for calcium as

$$\text{Ca}^+ \text{ (g)} + 1145 \text{ kJ} \longrightarrow \text{Ca}^{2+} \text{ (g)} + e^-$$

For a given element, *IE_2 is always greater than IE_1* because it is always more difficult to remove an electron from an ion with a positive charge than from the corresponding neutral atom. Table 4–4 is a tabulation of first ionization energies.

Ionization energies measure how tightly electrons are bound to atoms. Note that ionization requires energy to remove an electron from the attractive force of the nucleus. Low ionization energies indicate ease of removal of electrons, and hence ease of positive ion (cation) formation. Figure 4–3 shows a plot of first ionization energy versus atomic number for several elements. Note that in general, *first ionization energies increase from left to right and bottom to top of the periodic table, but there are some exceptions.*

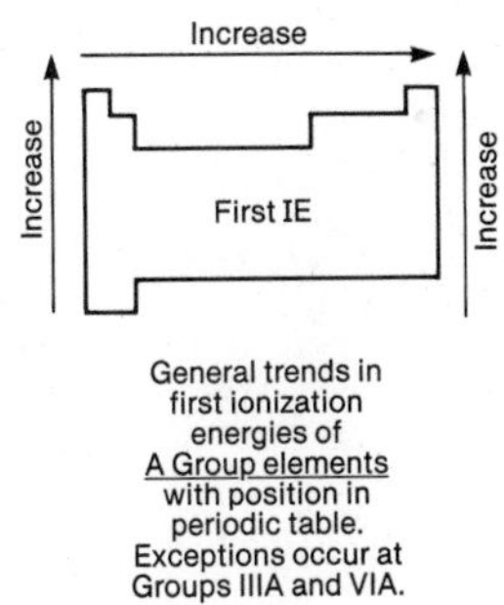

General trends in first ionization energies of A Group elements with position in periodic table. Exceptions occur at Groups IIIA and VIA.

Note also that in each periodic segment of Figure 4–3, the noble gases have the highest first ionization energies. This should not be surprising, since the noble gases are known to be very stable elements, and only the heavier members of the family show any tendency to combine with other elements to form compounds. It requires more energy to remove an electron from a helium atom (slightly less than 4.0×10^{-18} J/atom or 2372 kJ/mol) than from an atom of any other element

$$\text{He (g)} + 2372 \text{ kJ} \longrightarrow \text{He}^+ \text{ (g)} + 1\ e^-$$

TABLE 4–4 First Ionization Energies (kJ/mol of atoms) of Some Elements

H 1312																	He 2372
Li 520	Be 899											B 801	C 1086	N 1402	O 1314	F 1681	Ne 2081
Na 497	Mg 738											Al 578	Si 786	P 1012	S 1000	Cl 1251	Ar 1521
K 419	Ca 590	Sc 631	Ti 658	V 650	Cr 653	Mn 717	Fe 759	Co 758	Ni 737	Cu 745	Zn 906	Ga 579	Ge 762	As 947	Se 941	Br 1140	Kr 1351
Rb 403	Sr 549	Y 616	Zr 660	Nb 664	Mo 685	Tc 702	Ru 711	Rh 720	Pd 805	Ag 731	Cd 868	In 558	Sn 709	Sb 834	Te 869	I 1008	Xe 1170
Cs 376	Ba 503	La 538	Hf 675	Ta 761	W 770	Re 760	Os 840	Ir 878	Pt 870	Au 890	Hg 1007	Ti 589	Pb 716	Bi 703	Po 812	At 920	Rn 1037

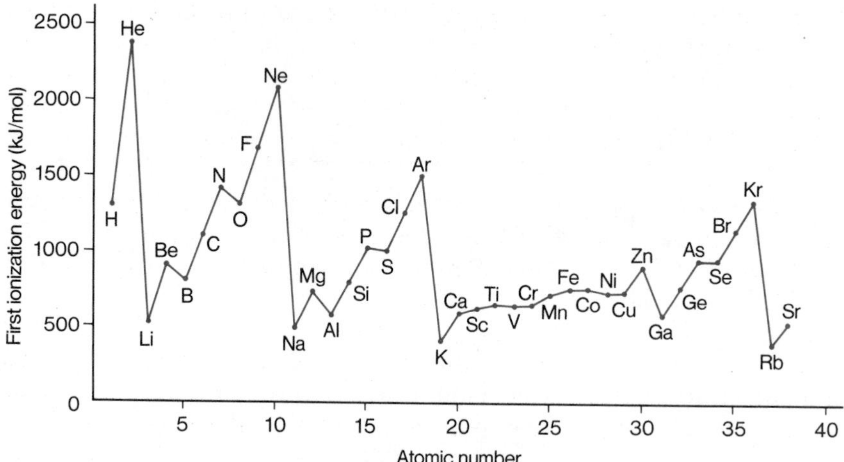

FIGURE 4–3 A plot of first ionization energies for the first 38 elements versus atomic number. Note that the noble gases have very high first ionization energies and the IA metals have low first ionization energies. Note also the similarities in the variations for elements 3 through 10 to those for elements 11 through 18, as well as the later A Group elements. Variations in B Group elements are not nearly so pronounced as those for A Group elements.

The Group IA metals (Li, Na, K, Rb, Cs) have very low first ionization energies, as is expected since these elements have only one electron in their highest energy levels (. . . ns^1). The first electron added to a principal energy level is easily removed. Note also that as we move down the group the first ionization energies become smaller. The decrease is due to the fact that filled sets of inner orbitals "shield" the nucleus from the outermost electrons which, therefore, are not very strongly held. Additionally, the force of attraction of the positively charged nucleus for electrons decreases as the square of the separation between them increases. In other words, as atomic radii increase in a given group, first ionization energies decrease because the valence electrons are further from the nucleus.

The first ionization energies for the Group IIA elements (Be, Mg, Ca, Sr, Ba) are significantly higher than those of the Group IA elements in the same periods. This is because the Group IIA elements have smaller atomic radii and higher effective nuclear charges. Thus, their valence electrons are held more tightly than those of the preceding IA metals of the same periods. Considerably more energy is required to remove an electron from the filled outermost *s* orbitals of the Group IIA elements than to remove the single electron from the half-filled outermost *s* orbitals of the Group IA elements.

The first ionization energies for the Group IIIA elements (B, Al, Ga, In, Tl) are *lower* than those of the IIA elements in the same periods because the Group IIIA elements have only a single electron in their outermost *p* orbitals. It requires less energy to remove a *p* electron than an *s* electron from the same principal energy level because an *np* orbital is higher in energy than an *ns* orbital.

The second peak on the segment of the ionization energy curve associated with each period occurs at the Group VA elements (N, P, As, Sb, Bi). These elements are characterized by three unpaired electrons in the three outermost p orbitals, that is, . . . $ns^2np_x^{\,1}np_y^{\,1}np_z^{\,1}$, a half-filled set of p orbitals. The Group VIA elements (O, S, Se, Te, Po) have slightly *lower* first ionization energies than the Group VA elements in the same periods. This tells us that it takes slightly less energy to remove a paired electron from a Group VIA element than to remove an unpaired p electron from a Group VA element in the same period. This is due to the greater repulsion between the two paired p electrons of the VIA elements, and to the relative stability of the half-filled set of p orbitals that results from removal of one electron from the VIA elements.

We have seen other consequences of the special stability of half-filled sets of equivalent orbitals, i.e., "exceptions" to the Aufbau Principle (Section 3-15).

Accurate measurements of the ionization energies (as well as atomic weights and other properties) of some elements, especially the actinides, have not yet been made. This is due to two factors: (1) some of these elements exist only as radioactive isotopes and (2) some are extremely rare and difficult to produce in large enough quantities for accurate measurements.

As we shall see, *knowledge of the relative values of ionization energies assists us in predicting which elements are likely to form ionic or molecular (covalent) compounds.* Elements with low ionization energies form ionic compounds by losing electrons to form positively charged ions (**cations**). Elements with intermediate ionization energies generally form molecular compounds by sharing electrons with other elements. Elements with very high ionization energies, such as those of Group VIA and VIIA, often gain electrons to form negatively charged ions (**anions**).

Here is one reason why trends in ionization energies are important.

The "driving force" for an atom of a **representative** element to form a monatomic ion in a compound usually lies in the formation of a stable noble gas configuration. Energy considerations are consistent with this observation. For example, one mole of Li from Group IA forms one mole Li^+ ions as it absorbs 520 kJ per mole of Li atoms. The second ionization energy is 14 times greater, 7297 kJ/mol, and is prohibitively large for the formation of Li^{2+} ions under ordinary conditions. If Li^{2+} ions were to form, an electron would have to be removed from the filled first energy level, and we recognize that this is unlikely to occur. The same behavior is seen for the other alkali metals, for the same reason.

Noble gas configurations are only stable for ions in **compounds.** In fact, Li^+ (g) is less stable than Li (g) by 520 kJ/mol.

Likewise, the first two ionization energies of Be are 899 and 1757 kJ/mol, but the third ionization energy is more than eight times larger, 14,840 kJ/mol, so Be forms Be^{2+} ions, but not Be^{3+} ions. The other alkaline earth metals, Mg, Ca, Sr, Ba, and Ra, behave in a similar way. Due to the high energy requirements, simple monatomic ions with positive charges greater than 3+ do not form under ordinary circumstances, and only the lower members of Group IIIA, beginning with Al, form 3+ ions (as do Bi and some d- and f-transition metals). Thus we see that the magnitudes of successive ionization energies support our ideas of electronic configurations as discussed in Chapter 3.

As we shall see, ionization energies are also important factors in determining *crystal lattice energies of ionic solids,* which measure the stabilities of ionic crystals (Section 13-9).

4-5 Electron Affinity

The **electron affinity** (EA) of an element is defined as *the amount of energy involved in the process in which an electron is added to an isolated gaseous atom to form an ion with a 1− charge.* The current convention is to assign a negative value when energy is released, as it usually is, and a positive value when energy is absorbed. This is consistent with thermodynamic convention. To provide specific examples, we can represent the electron affinities of beryllium and chlorine as

$$\text{Be (g)} + e^- + 241 \text{ kJ} \longrightarrow \text{Be}^- \text{ (g)} \qquad \text{EA} = 241 \text{ kJ/mol}$$

$$\text{Cl (g)} + e^- \longrightarrow \text{Cl}^- \text{ (g)} + 348 \text{ kJ} \qquad \text{EA} = -348 \text{ kJ/mol}$$

This value of EA for Cl can also be represented as -3.61 eV/atom or as -5.78×10^{-19} J/atom.

The first equation tells us that when one mole of gaseous beryllium atoms gain one electron each to form a mole of gaseous Be^- ions, 241 kJ/mol of ions is *absorbed.* The second equation tells us that when one mole of gaseous chlorine atoms gain one mole of electrons to form one mole of gaseous chloride ions, 348 kJ of energy is *liberated.* Figure 4–4 shows a plot of electron affinity versus atomic number for several elements.

The processes described by the above equations for electron affinities of Be and Cl each involve the *addition* of an electron to a neutral gaseous atom, rather than the *removal* of an electron as shown in the equation that defines the first ionization energy for a neutral atom. Thus, the electron affinity and the ionization energy for an element are related. However, these two processes are *not* simply the reverse of one another. The reverse of the ionization process for the hypothetical element X

$$\text{X}^+ \text{ (g)} + e^- \longrightarrow \text{X (g)} + \text{IE}_1$$

is not the same as

$$\text{X (g)} + e^- \longrightarrow \text{X}^- \text{ (g)} - \text{EA}$$

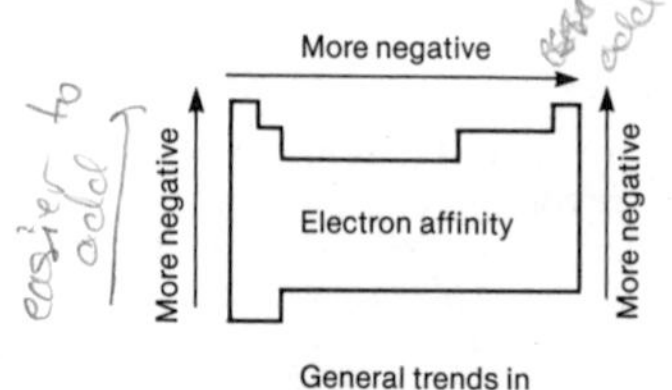

General trends in electron affinities of A Group elements with position in periodic table. Exceptions occur at Groups IIA and VA.

because the first one begins with a positive ion, while the second begins with a neutral atom. Thus, IE_1 and EA are *not* simply equal in value with the signs reversed.

Elements with very negative electron affinities gain electrons readily to form negative ions (anions). We see from Figure 4–4 that, generally speaking, electron affinities become more negative, i.e., the elements show a greater attraction for an extra electron, from left to right across a row in the periodic table, excluding the noble gases. Note that the halogens, which have the outer

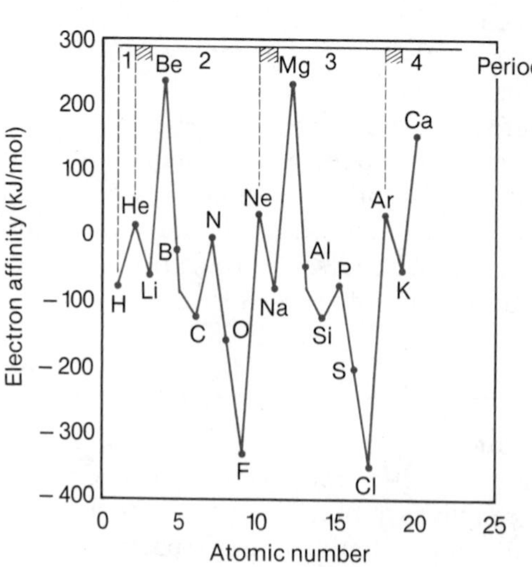

FIGURE 4–4 Plot of electron affinity versus atomic number for the first 22 elements. The general horizontal trend is that electron affinities become more negative (more energy is released as an extra electron is added) from Group IA through Group VIIA for a given period. Exceptions occur at the IIA and VA elements.

electronic configuration . . . ns^2np^5, have the most negative electron affinities. They form stable anions with noble gas configurations, . . . ns^2np^6, by gaining one electron.

Electron affinity is a precise and quantitative term, like ionization energy, but it is difficult to measure. Table 4–5 shows electron affinities for the representative elements.

For many reasons, the variations in electron affinities are not regular across a period. Exceptions to the general trend that electron affinities of the elements become more negative from left to right in each period are most noticeable for the elements of Group IIA (Be, Mg, Ca, . . .) and Group VA (N, P, As, . . .), which have less negative (more positive) values than the trends suggest (Figure 4–4). The electron affinity of a IIA metal is very positive because it involves the addition of an electron to an atom that already has only completely filled *ns* and empty *np* orbitals. The values for the VA elements are slightly less negative than expected because they apply to the addition of an electron to a relatively stable half-filled set of *np* orbitals ($ns^2np^3 \rightarrow ns^2np^4$).

The addition of a second electron to an ion with a 1− charge is usually endothermic, so electron affinities of anions are usually positive.

4–6 Ionic Radii

Many of the elements on the left side of the periodic table react with other elements by *losing* electrons to form positively charged ions. The Group IA elements (Li, Na, K, Rb, Cs) have only one electron in their highest energy level (electronic configuration . . . ns^1). They react with other elements by losing one electron to attain a noble gas configuration, and form the ions Li^+, Na^+, K^+, Rb^+, and Cs^+. The nuclear charge (number of protons) remains constant when an ion is formed by an atom losing one or more electrons. A neutral lithium atom, Li, contains three protons in its nucleus and three electrons in the region about the nucleus. However, a lithium ion, Li^+, contains three protons in its nucleus but only two electrons (helium configuration) in the region about the nucleus. The 3+ nuclear charge draws the two remaining electrons

TABLE 4–5 Electron Affinity Values (kJ/mol) of Some Elements*

H −72								He (21)
Li −60	Be (241)		B −23	C −122	N 0	O −142	F −322	Ne (29)
Na −53	Mg (231)	Cu −123	Al −44	Si −119	P −74	S −200	Cl −348	Ar (35)
K −48	Ca (156)	Ag −125	Ga (−36)	Ge −116	As −77	Se −194	Br −323	Kr (39)
Rb −47	Sr (119)	Au −222	In (−34)	Sn −120	Sb −101	Te −190	I −295	Xe (40)
Cs −45	Ba (52)		Tl (−48)	Pb −101	Bi −101	Po (−173)	At (−270)	Rn (40)
Fr (−44)								

*Estimated values in parentheses.

much closer to itself, and so a lithium ion (radius = 0.60Å) is much smaller than a neutral lithium atom (radius = 1.52 Å). The same line of reasoning convinces us that a sodium atom, Na, should be larger than a sodium ion, Na^+. The relative sizes of atoms and common ions of some representative elements are shown in Figure 4–5.

To convert from Å to nm, move the decimal point left one place, i.e., the atomic radius of Li is 1.52 Å or 0.152 nm ($1 Å = 1 \times 10^{-10}$ m = 0.1 nm).

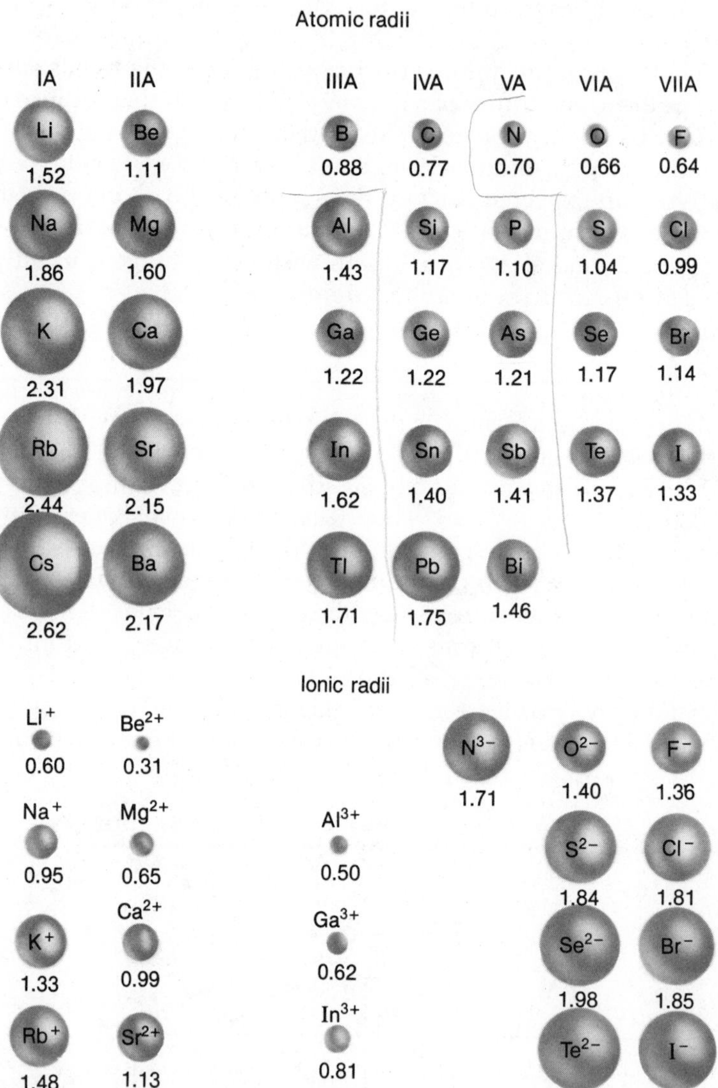

FIGURE 4–5 Sizes of atoms and ions of the A group elements in angstroms. Positive ions (cations) are always *smaller* than the netural atoms from which they are formed, whereas negative ions (anions) are always *larger* than the neutral atoms from which they are formed.

We see that the ions formed by the Group IIA elements (Be^{2+}, Mg^{2+}, Ca^{2+}, Sr^{2+}, Ba^{2+}) are significantly smaller than the **isoelectronic** ions formed by the Group IA elements in the same period. For instance, Be^{2+} (radius = 0.31 Å) is much smaller than Li^{+} (radius = 0.60 Å). *Isoelectronic* species have the same number of electrons. Again, this is just what we might expect. A beryllium ion, Be^{2+}, is formed when a beryllium atom, Be, loses two electrons while the 4+ nuclear charge remains constant. We expect the 4+ nuclear charge in Be^{2+} to draw the remaining two electrons much closer to itself. Comparison of the ionic radii for the IIA elements with the corresponding atomic radii indicates the validity of our reasoning. Similar reasoning leads us to conclude that the ions formed by the Group IIIA metals (Al^{3+}, Ga^{3+}, In^{3+}, Tl^{3+}) should be even smaller than ions formed by Group IA and Group IIA elements in the same periods, and much smaller than the neutral atoms from which they are formed.

Let us now focus our attention on the Group VIIA elements (F, Cl, Br, I), which have the outermost electronic configuration . . . ns^2np^5. These elements can fill their outermost *p* orbitals completely by *gaining* one electron to attain noble gas configurations. Thus, if a fluorine atom (atomic number = 9) with seven electrons in its highest energy level gains one electron, it becomes a fluoride ion, F^{-}, which contains ten electrons. Again, the nuclear charge of the atom remains constant, and the 9+ nuclear charge is unable to hold the electron cloud of ten electrons in the fluoride ion as tightly as it did the nine electrons in the fluorine atom. The electron cloud expands, and the fluoride ion (radius = 1.36 Å) is significantly larger than the neutral fluorine atom (radius = 0.64 Å). Similar reasoning indicates that a chloride ion, Cl^{-}, should be larger than a neutral chlorine atom, Cl. Reference to Figure 4–5 verifies our logic.

If we compare the sizes of an oxygen atom (Group VIA) and oxide ion, O^{2-}, we find that the negatively charged ion is larger than the neutral atom from which it is formed by gaining two electrons. The oxide ion is also larger than the fluoride ion because the oxide ion contains ten electrons held by a nuclear charge of only 8+, while the fluoride ion also has ten electrons held by a nuclear charge of 9+.

We may now generalize and say that simple negatively charged ions (anions) are always larger than the neutral atoms from which they are derived. As the magnitude of the negative charge on isoelectronic ions increases, so do their sizes. Conversely, positively charged ions (cations) are always smaller than the neutral atoms from which they are formed. As the magnitude of the positive charge on isoelectronic ions increases, their sizes decrease.

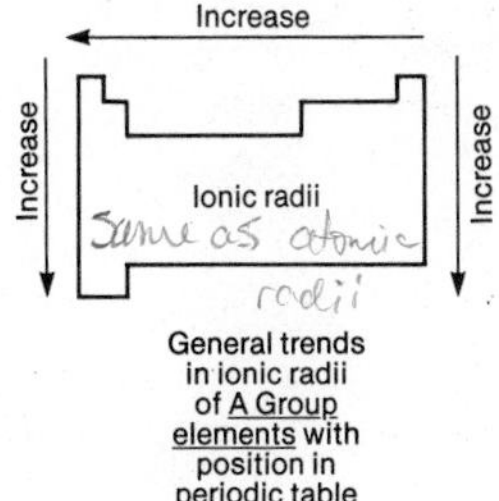

General trends in ionic radii of A Group elements with position in periodic table

Note that differences between atomic and ionic radii are nearly equal *within* a given A group.

4–7 Electronegativity

The **electronegativity** of an element *is a measure of the relative tendency of an atom to attract electrons to itself when it is chemically combined with another atom*. Electronegativities of the elements have been calculated and are expressed in Table 4–6 on a somewhat arbitrary scale, called the Pauling scale, with a maximum value of 4.0.

TABLE 4–6 Electronegativity Values of the Elements*

Key: <1.0; 1.0–1.4; 1.5–1.9; 2.0–2.4; 2.5–2.9; 3.0–4.0

1 H 2.1																
3 Li 1.0	4 Be 1.5											5 B 2.0	6 C 2.5	7 N 3.0	8 O 3.5	9 F 4.0
11 Na 1.0	12 Mg 1.2											13 Al 1.5	14 Si 1.8	15 P 2.1	16 S 2.5	17 Cl 3.0
19 K 0.9	20 Ca 1.0	21 Sc 1.3	22 Ti 1.4	23 V 1.5	24 Cr 1.6	25 Mn 1.6	26 Fe 1.7	27 Co 1.7	28 Ni 1.8	29 Cu 1.8	30 Zn 1.6	31 Ga 1.7	32 Ge 1.9	33 As 2.1	34 Se 2.4	35 Br 2.8
37 Rb 0.9	38 Sr 1.0	39 Y 1.2	40 Zr 1.3	41 Nb 1.5	42 Mo 1.6	43 Tc 1.7	44 Ru 1.8	45 Rh 1.8	46 Pd 1.8	47 Ag 1.6	48 Cd 1.6	49 In 1.6	50 Sn 1.8	51 Sb 1.9	52 Te 2.1	53 I 2.5
55 Cs 0.8	56 Ba 1.0	57 La 1.1	72 Hf 1.3	73 Ta 1.4	74 W 1.5	75 Re 1.7	76 Os 1.9	77 Ir 1.9	78 Pt 1.8	79 Au 1.9	80 Hg 1.7	81 Tl 1.6	82 Pb 1.7	83 Bi 1.8	84 Po 1.9	85 At 2.1
87 Fr 0.8	88 Ra 1.0	89 Ac 1.1														

58 Ce 1.1	59 Pr 1.1	60 Nd 1.1	61 Pm 1.1	62 Sm 1.1	63 Eu 1.1	64 Gd 1.1	65 Tb 1.1	66 Dy 1.1	67 Ho 1.1	68 Er 1.1	69 Tm 1.1	70 Yb 1.0	71 Lu 1.2
90 Th 1.2	91 Pa 1.3	92 U 1.5	93 Np 1.3	94 Pu 1.3	95 Am 1.3	96 Cm 1.3	97 Bk 1.3	98 Cf 1.3	99 Es 1.3	100 Fm 1.3	101 Md 1.3	102 No 1.3	103 Lr 1.5

*The shading of the block indicates the EN value of that element according to the key in the figure.

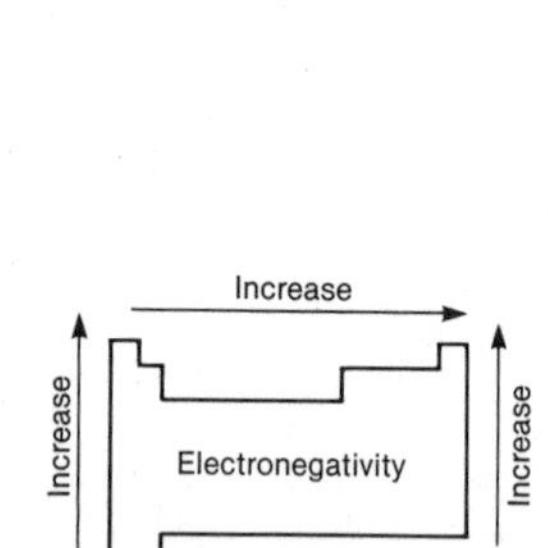

General trends in electronegativities of A Group elements with position in periodic table

Note that the electronegativity of fluorine (4.0) is higher than that of any other element. This tells us that when fluorine is chemically bonded to other elements, it has a greater tendency to attract electron density to itself than does any other element. Oxygen is the second most electronegative element and has a greater tendency to attract electron density than any element except fluorine.

Since the noble gases form few compounds, they are not included in this discussion.

For the representative elements, *electronegativities usually increase from left to right across periods and from bottom to top within groups.* Variations among the transition elements aren't as regular. Generally both ionization energies and electronegativities are low for elements at the lower left of the periodic table and high for those at the upper right.

Although the electronegativity scale is somewhat arbitrary, it can be used to make predictions about bonding with a reasonable degree of certainty. Elements with large differences in electronegativity tend to react with each other to form ionic bonds; the less electronegative element gives up its electron(s) to the more electronegative element. Elements with low electronegativity differences tend to form covalent bonds with each other, i.e., share their electrons in such a way that the more electronegative element attains a greater share. This will be discussed in detail after the next section and in Chapter 5.

4–8 Metals, Nonmetals, and Metalloids

Early in this chapter we classified elements in several ways based on their positions in the periodic table. In a broader and more general classification scheme, the elements are frequently divided into three classes: **metals, nonmetals,** and **metalloids.** The periodic tables inside the front cover of the text

About 80% of the elements are metals.

TABLE 4–7 Periodic Table, the Long-period Form*

PERIODS	IA	IIA	IIIB		IVB	VB	VIB	VIIB	VIII	VIII	VIII	IB	IIB	IIIA	IVA	VA	VIA	VIIA	0
1	1 H 1766																	1 H	2 He 1895
2	3 Li 1817	4 Be 1798												5 B 1808	6 C	7 N 1772	8 O 1772	9 F 1886	10 Ne 1898
3	11 Na 1807	12 Mg 1755												13 Al 1827	14 Si 1823	15 P 1669	16 S	17 Cl 1774	18 Ar 1894
4	19 K 1807	20 Ca 1808	21 Sc 1879		22 Ti 1791	23 V 1830	24 Cr 1797	25 Mn 1774	26 Fe	27 Co 1735	28 Ni 1751	29 Cu	30 Zn 1746	31 Ga 1875	32 Ge 1886	33 As	34 Se 1817	35 Br 1826	36 Kr 1898
5	37 Rb 1861	38 Sr 1790	39 Y 1794		40 Zr 1789	41 Nb 1801	42 Mo 1778	43 Tc 1937	44 Ru 1844	45 Rh 1803	46 Pd 1803	47 Ag	48 Cd 1817	49 In 1863	50 Sn	51 Sb	52 Te 1782	53 I 1811	54 Xe 1898
6	55 Cs 1860	56 Ba 1808	57 La 1839	58 Ce ➜ 71 Lu	72 Hf 1923	73 Ta 1802	74 W 1781	75 Re 1925	76 Os 1803	77 Ir 1803	78 Pt 1735	79 Au	80 Hg	81 Tl 1861	82 Pb	83 Bi	84 Po 1898	85 At 1940	86 Rn 1900
7	87 Fr 1939	88 Ra 1898	89 Ac 1899	90 Th ➜ 103 Lr	104 Unq 1965	105 Unp 1970	106 Unh 1974	107 Uns 1976	108	109 1982	110	111	112	113	114	115	116	117	118
8			121																

REPRESENTATIVE ELEMENTS (IA, IIA; IIIA–0) — TRANSITION ELEMENTS (IIIB–IIB) — GROUPS

6 LANTHANIDE SERIES	58 Ce 1803	59 Pr 1885	60 Nd 1843	61 Pm 1947	62 Sm 1879	63 Eu 1896	64 Gd 1880	65 Tb 1843	66 Dy 1886	67 Ho 1879	68 Er 1843	69 Tm 1879	70 Yb 1878	71 Lu 1907
7 ACTINIDE SERIES	90 Th 1828	91 Pa 1917	92 U 1789	93 Np 1940	94 Pu 1940	95 Am 1945	96 Cm 1944	97 Bk 1950	98 Cf 1950	99 Es 1952	100 Fm 1953	101 Md 1955	102 No 1958	103 Lr 1961
8	122	123	124	125	126	127	128	129	130	131	132	133	134	135

* The heavy "doorstep" line approximately separates the metallic elements, to the left, from the nonmetallic elements, to the right. Besides the atomic numbers, the dates of discovery of the elements are given. The elements with no given dates were known to the ancients. Positions of yet undiscovered elements up to 135 are shown.

and in Table 4–7 show this classification. The elements to the left of those touching the heavy stairstep line are *metals* (except hydrogen) while those to the right are *nonmetals.* Such a classification is somewhat arbitrary, and several elements do not fit neatly into either class. The elements adjacent to the heavy line are often called *metalloids* (or *semimetals*), because they show some properties that are characteristic of both metals and nonmetals.

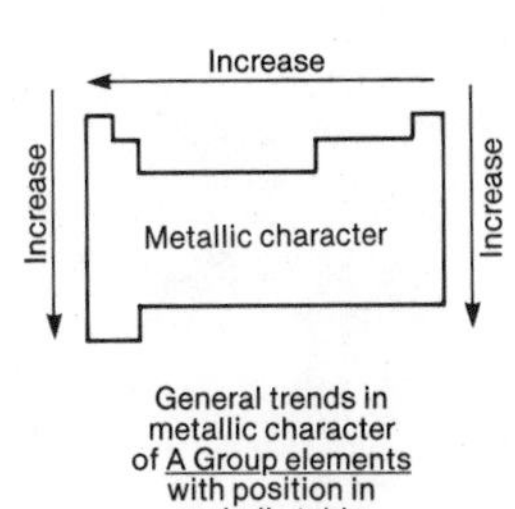

General trends in metallic character of A Group elements with position in periodic table

The physical properties that distinguish metals from nonmetals are summarized in Table 4–8. The most important thing to notice is that the general properties of metals and nonmetals are opposite. Not all metals and nonmetals possess all these properties, but they all share most of them to varying degrees. The physical properties of metals can be explained on the basis of metallic bonding within the solids, which will be discussed in Section 9–17. The strength of metallic bonding itself is influenced by the number of electrons, especially the unpaired electrons, beyond the "last" noble gas core.

As we pointed out earlier, *metalloids* show some properties that are characteristic of both metals and nonmetals. Many of the metalloids, such as silicon, germanium, and antimony, act as semiconductors, which are important

TABLE 4–8 Physical Properties of Metals and Nonmetals

Metals	Nonmetals
1. High electrical conductivity	1. Poor electrical conductivity
2. High thermal conductivity	2. Good heat insulators
3. Metallic gray or silver luster*	3. No metallic luster
4. Almost all are solids†	4. Solids, liquids, or gases
5. Malleable (can be hammered into sheets)	5. Brittle in solid state
6. Ductile (can be drawn into wires)	6. Nonductile
7. Solid state characterized by metallic bonding	7. Covalently bonded molecules; noble gases are monatomic

* Except copper and gold.

† Except mercury; cesium and gallium melt in **protected** hand.

in solid state electronic circuits. **Semiconductors** *do not conduct electrical current at room temperature but do so at higher temperatures* (Section 9–17).

Aluminum is an example of a metalloid. It is metallic in appearance, and an excellent conductor of electricity, but its electrical conductivity *increases with increasing temperature,* whereas the conductivities of metals usually decrease with increasing temperature.

Table 4–9 summarizes some of the contrasting chemical properties of metals and nonmetals.

TABLE 4–9 Some Chemical Properties of Metals and Nonmetals

Metals	Nonmetals
1. Outer shells contain few electrons; usually three or fewer	1. Outer shells contain four or more electrons
2. Low ionization energies	2. High ionization energies
3. Slightly negative or positive electron affinities	3. Very negative electron affinities
4. Low electronegativities	4. High electronegativities
5. Form cations by losing e^-	5. Form anions by gaining e^-*
6. Form ionic compounds with nonmetals	6. Form ionic compounds with metals* and molecular (covalent) compounds with other nonmetals

* Except the noble gases.

Chemical Bonding

4–9 Kinds of Chemical Bonds

The attractive forces that hold atoms together in compounds are referred to as **chemical bonds.** As a matter of convenience, we usually divide chemical bonds into classes; stated differently, we say that compounds that exhibit certain kinds of properties have one kind of chemical bond, while compounds with different properties have other kinds of bonds. There are two major classes of bonding: (1) **ionic bonding,** which results from electrostatic interactions among ions, which can be formed by the *transfer* of one or more electrons from one atom or group of atoms to another, and (2) **covalent bonding,** which results from *sharing* one or more electron pairs between two atoms. These represent two extremes, and all bonds have at least some degree of both ionic and covalent character. Compounds containing predominantly ionic bonding are called **ionic compounds,** and those containing predominantly covalent bonds are called **covalent compounds.**

Some of the properties associated with many simple ionic and covalent compounds in the extreme cases are summarized below. The differences in properties can be accounted for by the differences in bonding between the atoms or ions.

Ionic Compounds	**Covalent Compounds**
1. High melting solids	1. Gases, liquids, or low melting solids
2. Many are soluble in polar solvents such as water	2. Many are insoluble in polar solvents
3. Most are insoluble in nonpolar solvents such as hexane, C_6H_{14}	3. Most are soluble in nonpolar solvents, such as hexane, C_6H_{14}
4. Molten compounds conduct electricity well because they contain charged particles (ions)	4. Liquid or molten compounds do not conduct electricity
5. Aqueous solutions conduct electricity well because they contain charged particles (ions)	5. Aqueous solutions are *usually* poor conductors of electricity because most do not contain charged particles

The distinction between polar and nonpolar molecules will be made in Section 5–2.

As we shall see in Section 7–1, aqueous solutions of some covalent compounds do conduct electricity.

The strong electrostatic forces of attraction among ions in ionic compounds account for their relatively high melting points. According to Coulomb's Law, the force of attraction, *F*, between two oppositely charged particles of charge magnitude *q* is directly proportional to the product of the charges and inversely proportional to the square of the distance separating them, *d*.

$$F \propto \frac{q^+q^-}{d^2}$$

$\propto$ means "proportional to."

Thus, the greater the charges on the ions, and the closer the ions are to each other, the stronger is the resulting ionic bonding. Of course, like-charged ions repel each other, so the distances separating ions in solids are those at which the attractions exceed the repulsions by the greatest amount. The structure of common table salt, sodium chloride (NaCl), is shown in Figure 4–6. Typical of simple ionic compounds, NaCl exists in a regular, extended array of positive and negative ions, Na^+ and Cl^-.

Distinct molecules of ionic substances do not exist, so we must refer to **formula units** instead of molecules. While each individual interaction between two ions in an ionic compound is relatively weaker than a typical interaction between two atoms covalently bonded together, the sum of the forces of all the interactions in an ionic solid is substantial. In predominantly covalent compounds the bonds between atoms within a molecule (*intra*molecular bonds) are relatively strong but the forces of attraction between molecules (*inter*molecular forces) are relatively weak, accounting for their lower melting and boiling

Remember that a formula unit shows the simplest whole number ratio of atoms (or ions) in a compound.

Figure 4–6 has been expanded to show the spatial arrangement of ions. Adjacent ions actually are in contact with each other. The lines do *not* represent formal chemical bonds. They have been drawn to emphasize the spatial arrangement of ions.

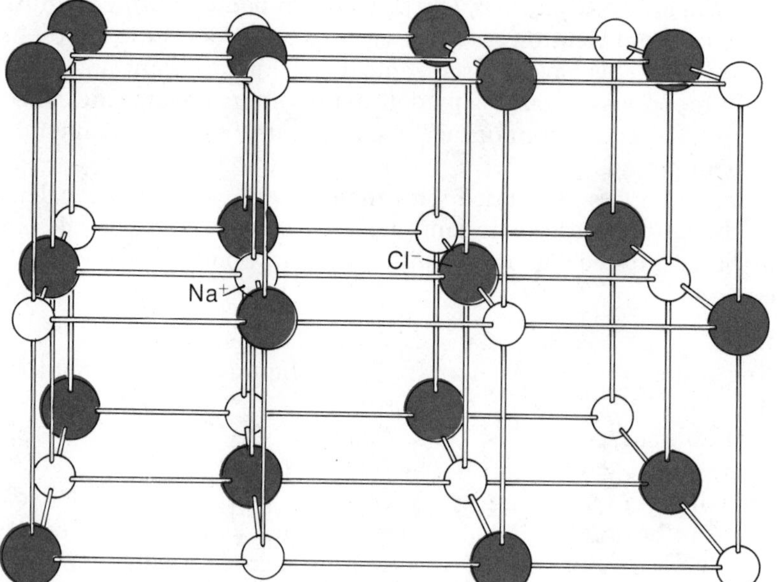

FIGURE 4–6 Crystal structure of NaCl, expanded for clarity. Like the Cl^- ion in the center of the cube, all chloride ions are surrounded by six sodium ions. Each Na^+ ion is simply surrounded by six Cl^- ions. The crystal includes billions of ions in the pattern shown. Compare with Figure 9–21, a space-filling drawing of the NaCl structure.

points. It must be emphasized, however, that both covalent and ionic bonding can be strong or weak, depending on the circumstances.

4–10 Ionic Bonding

The first kind of chemical bond we shall describe is the **ionic** or **electrovalent bond.** An ionic bond is a chemical bond that can be formed by the *transfer of one or more electrons from one atom or group of atoms to another.* As our previous discussions of ionization energy, electronegativity, and electron affinity might lead us to speculate, ionic bonding occurs most easily when elements with low ionization energies (metals) react with elements with high electronegativities and high electron affinities (nonmetals). Many metals lose electrons easily, and many nonmetals gain electrons readily.

Consider the reaction of sodium (a Group IA metal) with chlorine (a Group VIIA nonmetal). Sodium is a soft silvery metal (m.p. 98°C), and chlorine is a yellowish green corrosive gas at room temperature. Both sodium and chlorine react with water, sodium vigorously. By contrast, sodium chloride is a white solid (m.p. 801°C) that is soluble in water with the absorption of just a little heat. We can represent the reaction for its formation as

$$2Na\ (s) + Cl_2\ (g) \longrightarrow 2NaCl\ (s)$$

sodium chlorine sodium chloride

If we wish to examine this reaction in more detail, we can show electronic configurations for all species. We deal with chlorine as individual atoms rather than molecules, for simplicity.

		3s	3p				3s	3p	
$_{11}Na$	$1s^2 2s^2 2p^6$	↑	__ __ __	⟶	Na^+	$1s^2 2s^2 2p^6$	__	__ __ __	1e^- lost
$_{17}Cl$	$1s^2 2s^2 2p^6$	⥮	⥮ ⥮ ↑		Cl^-	$1s^2 2s^2 2p^6$	⥮	⥮ ⥮ ⥮	1e^- gained

We can also use Lewis dot formulas (Section 4–2) to represent the reaction.

$$Na\cdot + :\ddot{\underset{\cdot\cdot}{Cl}}\cdot \longrightarrow Na^+[:\ddot{\underset{\cdot\cdot}{Cl}}:]^-$$

In the above reaction, sodium atoms lose one electron each to form sodium ions, Na^+, which contain only ten electrons, the same number as the *preceding* noble gas, neon. We say that sodium ions have the neon electronic structure. In contrast, chlorine atoms gain one electron each to form chloride ions, Cl^-, which contain 18 electrons, the same number as the *next* noble gas, argon. Similar observations will be repeated for most ionic compounds formed by reactions between *representative metals and representative nonmetals.*

The sodium ion, Na^+, is said to be **isoelectronic** with neon, i.e., both species contain the same number of electrons, ten. The chloride ion, Cl^-, is isoelectronic with argon (18 e^-).

The formula for sodium chloride, NaCl, indicates that the compound contains Na and Cl in a 1 : 1 ratio. This is the formula we might predict based on the fact that sodium atoms contain only one electron in their highest energy level and chlorine atoms need only one electron to fill completely their outermost *p* orbitals.

The chemical formula NaCl does not explicitly indicate the ionic nature of the compound, only the ratio of atoms. So we must learn to recognize, from positions of elements in the periodic table and the known trends in electronegativity, when the difference in electronegativity is large enough to favor ionic bonding. In general, *the further apart across the periodic table two elements are, the more likely they are to form an ionic compound.*

The noble gases are excluded from this generalization.

All the Group IA metals (Li, Na, K, Rb, Cs) will react with the Group VIIA elements (F, Cl, Br, I) to form compounds of the same general composition. When Group IA metals react with Group VIIA nonmetals to form compounds of the general formula MX, the resulting ions invariably have noble gas configurations. We can generalize and write the reaction of the IA metals with the VIIA elements as:

$$2M\ (s) + X_2 \longrightarrow 2MX\ (s) \qquad \begin{array}{l} M = \text{Li, Na, K, Rb, Cs} \\ X = \text{F, Cl, Br, I} \end{array}$$

The Lewis dot representation for the generalized reaction is

$$M\cdot + :\ddot{\underset{\cdot\cdot}{X}}\cdot \longrightarrow M^+[:\ddot{\underset{\cdot\cdot}{X}}:]^-$$

Next, consider the reaction of lithium (Group IA) with oxygen (Group VIA) to form lithium oxide, a white solid ionic compound (m.p. > 1700°C). We may represent the reaction as:

$$\underset{\text{lithium}}{4Li\ (s)} + \underset{\text{oxygen}}{O_2\ (g)} \longrightarrow \underset{\text{lithium oxide}}{2Li_2O\ (s)}$$

The formula for lithium oxide, Li_2O, indicates that two atoms of lithium combine with one atom of oxygen. Again, if we examine the structures of the atoms before reaction we should predict that two lithium atoms react with one oxygen atom.

Each Li atom has one e^- in its valence shell. Each O atom has 6 e^- in its valence shell and needs 2 e^- to give it a noble gas configuration.

	1s	2s	2p				→		1s	2s	2p			
$_3$Li	⇅	↑						Li^+	⇅	—				1 e^- lost
$_3$Li	⇅	↑						Li^+	⇅	—				1 e^- lost
$_8$O	⇅	⇅	⇅	↑	↑			O^{2-}	⇅	⇅	⇅	⇅	⇅	2 e^- gained

The Lewis dot formulas for the atoms and ions are:

$$2Li\cdot + :\ddot{\underset{\cdot}{O}}\cdot \longrightarrow 2Li^+[:\ddot{\underset{\cdot\cdot}{O}}:]^{2-}$$

Note that lithium ions, Li^+, are isoelectronic with helium atoms while oxide ions, O^{2-}, are isoelectronic with neon atoms (10 e^-).

Calcium (Group IIA) reacts with oxygen (Group VIA) to form calcium oxide, a white solid ionic compound with a very high melting point, 2580°C.

$$\underset{\text{calcium}}{2Ca\ (s)} + \underset{\text{oxygen}}{O_2\ (g)} \longrightarrow \underset{\text{calcium oxide}}{2CaO\ (s)}$$

Again we write out the electronic structures of the atoms and ions, this time representing the inner electrons by the symbol of the preceding rare gas (in brackets).

In writing equations in which electrons in different energy levels in different atoms are involved, atomic orbitals are more conveniently labeled *under* the lines that represent them.

						→							
$_{20}$Ca [Ar]	⇅ (4s)						Ca^{2+} [Ar]	— (4s)					2 e^- lost
$_8$O [He]	⇅ (2s)	⇅ (2p)	↑	↑			O^{2-} [He]	⇅ (2s)	⇅ (2p)	⇅	⇅		2 e^- gained

The Lewis dot notation for the atoms and ions is

$$Ca: + :\ddot{\underset{\cdot}{O}}\cdot \longrightarrow Ca^{2+}[:\ddot{\underset{\cdot\cdot}{O}}:]^{2-}$$

Calcium ions, Ca^{2+}, are isoelectronic with argon (18 e^-), the preceding noble gas. Oxide ions, O^{2-}, are isoelectronic with neon (10 e^-), the following noble gas.

Calcium oxide melts at a much higher temperature than sodium chloride (2580°C as opposed to 801°C) because it has two charges per ion rather than one, and the ionic bonding is considerably stronger in CaO.

As our final example of ionic bonding, consider the reaction of magnesium (Group IIA) with nitrogen (Group VA) at elevated temperatures to form magnesium nitride, Mg_3N_2, a white solid ionic compound. We can represent the reaction as:

$$\underset{\text{magnesium}}{3Mg\ (s)} + \underset{\text{nitrogen}}{N_2\ (g)} \xrightarrow{\Delta} \underset{\text{magnesium nitride}}{Mg_3N_2\ (s)}$$

The formula Mg_3N_2 indicates that magnesium and nitrogen combine in a 3:2 ratio. We could predict this from the fact that magnesium atoms contain two electrons in their highest energy level, while nitrogen atoms need three electrons to fill theirs completely. Thus, three magnesium atoms lose two electrons each for a total of 6 e^-, while two nitrogen atoms gain three electrons each, also for a total of 6 $e-$. As before, examination of the structures makes the picture clearer.

$$\left.\begin{array}{l} {}_{12}\text{Mg [Ne]} \ \underset{3s}{\underline{\uparrow\downarrow}} \\ {}_{12}\text{Mg [Ne]} \ \underset{3s}{\underline{\uparrow\downarrow}} \\ {}_{12}\text{Mg [Ne]} \ \underset{3s}{\underline{\uparrow\downarrow}} \\ {}_{7}\text{N [He]} \ \underset{2s}{\underline{\uparrow\downarrow}} \ \underset{2p}{\underline{\uparrow}\ \underline{\uparrow}\ \underline{\uparrow}} \\ {}_{7}\text{N [He]} \ \underset{2s}{\underline{\uparrow\downarrow}} \ \underset{2p}{\underline{\uparrow}\ \underline{\uparrow}\ \underline{\uparrow}} \end{array}\right\} \xrightarrow{\Delta} \left\{\begin{array}{lr} \text{Mg}^{2+} \text{ [Ne]} \ \underset{3s}{\underline{\quad}} & 2\,e^- \text{ lost} \\ \text{Mg}^{2+} \text{ [Ne]} \ \underset{3s}{\underline{\quad}} & 2\,e^- \text{ lost} \\ \text{Mg}^{2+} \text{ [Ne]} \ \underset{3s}{\underline{\quad}} & 2\,e^- \text{ lost} \\ \text{N}^{3-} \text{ [He]} \ \underset{2s}{\underline{\uparrow\downarrow}} \ \underset{2p}{\underline{\uparrow\downarrow}\ \underline{\uparrow\downarrow}\ \underline{\uparrow\downarrow}} & 3\,e^- \text{ gained} \\ \text{N}^{3-} \text{ [He]} \ \underset{2s}{\underline{\uparrow\downarrow}} \ \underset{2p}{\underline{\uparrow\downarrow}\ \underline{\uparrow\downarrow}\ \underline{\uparrow\downarrow}} & 3\,e^- \text{ gained} \end{array}\right.$$

Using Lewis dot formulas for the atoms and ions we have

$$3\text{Mg}\!: + 2\cdot\ddot{\underset{\cdot}{\text{N}}}\cdot \longrightarrow 3\text{Mg}^{2+},\ 2[:\ddot{\underset{\cdot\cdot}{\text{N}}}:]^{3-}$$

In this case both magnesium ions, Mg^{2+}, and nitride ions, N^{3-}, are isoelectronic with neon (10 e^-).

Table 4–10 summarizes the general formulas of binary ionic compounds formed by the representative elements. M represents metals and X represents nonmetals from the indicated groups.

In the examples of ionic bonding examined thus far, we note that metal atoms have lost one, two, or three electrons each and nonmetals atoms have gained one, two, or three electrons. At this point we should observe that simple (monatomic) ions rarely have charges greater than 3+ or 3−. Ions with greater charges generally interact so strongly with the electron clouds of other ions in compounds that electron clouds are distorted severely, and considerable covalent character results in the bonds. This point will be developed later.

The distortion of the electron cloud of an anion by a small highly charged cation is called *polarization*.

The *d*- and *f*-transition elements form many compounds that are essentially ionic in character. Examination of the electron configurations of the transition metals and their simple ions shows that the ions usually do not have noble gas configurations.

TABLE 4–10 Simple Binary Ionic Compounds

Metal		Nonmetal		General Formula	Ions Present	Example
IA*	+	VIIA	⟶	MX	(M^+, X^-)	LiBr
IIA	+	VIIA	⟶	MX_2	$(M^{2+}, 2X^-)$	$MgCl_2$
IIIA	+	VIIA	⟶	MX_3	$(M^{3+}, 3X^-)$	GaF_3
IA*	+	VIA	⟶	M_2X	$(2M^+, X^{2-})$	Li_2O
IIA	+	VIA	⟶	MX	(M^{2+}, X^{2-})	CaO
IIIA	+	VIA	⟶	M_2X_3	$(2M^{3+}, 3X^{2-})$	Al_2O_3
IA*	+	VA	⟶	M_3X	$(3M^+, X^{3-})$	Li_3N
IIA	+	VA	⟶	M_3X_2	$(3M^{2+}, 2X^{3-})$	Ca_3P_2
IIIA	+	VA	⟶	MX	(M^{3+}, X^{3-})	AlP

* Hydrogen is considered a nonmetal, and all binary compounds of hydrogen are covalent except certain metal hydrides such as NaH and CaH_2, which contain hydride, H^-, ions.

4–11 Experimental Determination of Ionic Charges

In many cases the gross properties of a compound, such as melting point, solubility in polar or nonpolar solvents, and physical state, indicate whether the compound is ionic or covalent. For example, a pure liquid that does not conduct electricity and is miscible with benzene, a common but toxic nonpolar liquid solvent, is likely to be a covalent compound. A high melting crystalline solid that dissolves easily in water, but not in benzene, is likely to be ionic.

Inquisitive students often wonder how the actual charges on ions are determined. There are several ways. In many simple cases they may be deduced by determination of the empirical formula of a compound containing the ions of interest. Another method depends on the fact, presented in Section 4–6, that the sizes of ions vary with their charges. In some cases ionic charges can be inferred from the ionic sizes, which are found during the determination of the structure of an ionic solid by a technique called **x-ray diffraction** (described in the box preceding Section 9–16).

As we shall see, ions of the same element but with different charges, such as Cu^{+} and Cu^{2+}, and Fe^{2+} and Fe^{3+}, have very different chemical properties.

A much simpler and more common method involves determining the **electrical conductivity** of an aqueous solution of known concentration. If it is assumed that ionic compounds dissociate completely when dissolved in water, then the extent to which the resulting solutions conduct electrical current is determined by the magnitudes of the charges on the ions and their mobilities. For example, a 0.010 *M* aqueous solution of chromium(III) chloride, $CrCl_3$, should conduct electrical current better than a 0.010 *M* solution of calcium chloride, $CaCl_2$, which should, in turn, conduct better than a 0.010 *M* solution of sodium chloride, NaCl. The cations involved are Cr^{3+}, Ca^{2+}, and Na^{+}. The fact that the solutions contain Cl^{-} in a 3:2:1 ratio also contributes to the effect and must be considered. Likewise, a solution that is 0.010 *M* in sulfate ions, SO_4^{2-}, should conduct better than one that is 0.010 *M* in chloride ions, Cl^{-}, if the same cation is present in both cases. As we shall see in Section 10–16, no ionic compounds are really 100% dissociated into separate ions in every solution, although many are nearly so. However, measurement of comparative conductivities of known concentrations of dissolved compounds can provide useful information as to whether a compound is ionic, and tell something about the charges on its ions. Determination of ionic charge by electrolysis is also quite common, and is discussed in Section 19–7.

4–12 Oxidation Numbers

The **oxidation number** or oxidation state of an element in a simple binary ionic compound is the number of electrons gained or lost by an atom of that element when it forms the compound. In the cases of single-atom ions, it corresponds to the actual charge on the ion. In covalent compounds, oxidation numbers do not have the same physical significance they have in ionic compounds. However, they can be useful as mechanical aids in writing formulas and in balancing equations. In such cases positive and negative oxidation numbers indicate shifts (not transfers) of electron density from one atom toward another. The more electronegative element is assigned a negative oxidation number, while the less electronegative element is assigned a positive oxidation number. We distinguish between actual charges on ions and oxidation numbers by representing the former as $n+$ or $n-$ and the latter as $+n$ or $-n$.

TABLE 4–11 Common Oxidation Numbers for the Representative Elements

Elements in group	$e-$ Configuration ns np	Most Common (Nonzero) Oxidation Numbers: Ionic Compounds	Covalent Compounds
IA	↑ _ _ _	+1	
IIA	⇅ _ _ _	+2	+2
IIIA	⇅ ↑ _ _	+3, +1	+3
IVA	⇅ ↑ ↑ _	rare	−4, −3, −2, −1, +1, +2, +3, +4
VA	⇅ ↑ ↑ ↑	−3	−3, −1, +1, +3, +5
VIA	⇅ ⇅ ↑ ↑	−2	−2, +2, +4, +6
VIIA	⇅ ⇅ ⇅ ↑	−1	−1, +1, +3, +5, +7

The most common oxidation states of the representative elements by periodic groups are shown in Table 4–11.

The general rules for assigning oxidation numbers are given below. These rules are not comprehensive, but they cover most cases and are very useful.

A. The oxidation number of any free, uncombined element is zero. This includes multiatomic elements such as H_2, O_3, and S_8.
B. The charge on a simple (monatomic) ion is the oxidation number of the element in that ion. In a polyatomic ion, the sum of the oxidation numbers of the constituent elements is equal to the charge on the ion.
C. In compounds (whether ionic or covalent), the sum of the oxidation numbers of all elements in the compound is zero.

It follows from these rules, and the electronic configurations of the elements, that we may expect the following oxidation numbers for the representative elements:

1. The Group IA metals exhibit the +1 oxidation number in *all* their compounds. Hydrogen exhibits the +1 oxidation number in all its compounds except the metal hydrides such as NaH and CaH_2, in which hydrogen is bonded to elements less electronegative than itself, and in which it exhibits the −1 oxidation number.
2. The Group IIA metals exhibit the +2 oxidation number in *all* their compounds.
3. The Group IIIA elements exhibit the +3 oxidation number in most of their *common* compounds.
4. The Group VIIA elements exhibit the −1 oxidation number in all their binary compounds with metals (and with NH_4^+). The lower members of the family (Cl, Br, I) also exhibit +1, +3, +5, and +7 oxidation numbers in covalently bonded species that contain more electronegative elements, such as ClO^-, ClO_2^-, ClO_3^-, and ClO_4^-.
5. The Group VIA elements exhibit the −2 oxidation number in binary compounds with metals (and with NH_4^+). Oxygen exhibits the −2 oxidation number in all its compounds except OF_2 (+2), the peroxides such as H_2O_2

and Na_2O_2 (−1), and the rare superoxides such as KO_2 and RbO_2 ($-\frac{1}{2}$). The lower members of the Group VIA elements also commonly exhibit +4 and +6 oxidation numbers in covalently bonded species such as SO_2, SO_3, and SF_6.

6. The Group VA elements form few binary compounds with metals, but they exhibit the −3 oxidation number in these compounds and in species such as NH_3 and NH_4^+. The VA elements exhibit the +3 oxidation number in species such as NO_2^- and PCl_3, and the +5 oxidation number in species such as NO_3^-, PO_4^{3-}, and AsO_4^{3-} as well as in covalent compounds such as PCl_5, PF_5, and P_4O_{10}.
7. The IVA elements exhibit the +2 and +4 oxidation states in many of their compounds. They also exhibit a variety of other states.

By convention, *oxidation numbers* are represented as $+n$ and $-n$ while ionic charges are represented as $n+$ and $n-$. Both oxidation numbers and ionic charges are treated the same way algebraically.

EXAMPLE 4–1

Determine the oxidation numbers of nitrogen in the following species: (a) N_2O_4, (b) NH_3, (c) KNO_3, (d) NO_3^-, (e) N_2.

Solution

(a) Since the oxidation number of oxygen is −2 and the sum of the oxidation numbers of 2N and 4O must be zero, the oxidation number of N is +4.

ox. no. x −2
N_2O_4

$$\text{sum of ox. no.} = 2x + 4(-2) = 0$$
$$2x - 8 = 0$$
$$x = +4$$

(b) The oxidation number of H is +1 (Rule 1) and therefore that of N must be −3.

ox. no. x +1
NH_3

$$\text{sum of ox. no.} = x + 3(+1) = 0 \qquad x = -3$$

(c) The oxidation numbers of K and O are +1 and −2, respectively, and therefore that of N must be +5.

ox. no. +1 x −2
KNO_3

$$\text{sum of ox. no.} = +1 + x + 3(-2) = 0 \qquad x = +5$$

(d) The charge on the ion is 1−, and therefore the oxidation number of N is +5. This is the same as in KNO_3, which contains the NO_3^- ion.

ox. no. x −2
NO_3^-

$$\text{sum of ox. no.} = x + 3(-2) = -1 \qquad x = +5$$

(e) The oxidation number of any free element is zero.

Key Terms

Actinides elements 90 to 103 (after actinium).
Alkali metals metals of periodic Group IA.
Alkaline earths elements of periodic Group IIA.
Anion a negative ion; an atom or group of atoms that has gained one or more electrons.
Atomic radius radius of an atom.
Binary compound compound consisting of two elements.
Cation a positive ion; an atom or group of atoms that has lost one or more electrons.
Chemical bonds the attractive forces that hold

atoms together in elements and compounds.

Covalent bond chemical bond formed by the sharing of one or more electron pairs between two atoms.

Covalent compounds compounds containing predominantly covalent bonds.

***d*-Transition elements (metals)** B Group elements except IIB in the periodic table; sometimes simply transition elements.

Electrical conductivity ability to conduct electricity.

Electron affinity the amount of energy involved in the process in which an electron is added to a neutral isolated gaseous atom to form a gaseous ion with a 1− charge; has a negative value if energy is released.

Electronegativity a measure of the relative tendency of an atom to attract electrons to itself when chemically combined with another atom.

***f*-Transition elements** see *Inner transition elements*

Halogens elements of periodic Group VIIA.

Inner transition elements elements 58 to 71 and 90 to 103; also called *f*-transition elements.

Ionic bonding chemical bonding resulting from the transfer of one or more electrons from one atom or group of atoms to another.

Ionic compounds compounds containing predominantly ionic bonds.

Ionic radius radius of an ion.

Ionization energy the minimum amount of energy required to remove the most loosely held electron of an isolated gaseous atom or ion.

Isoelectronic having the same electron configurations.

Lanthanides elements 58 to 71 (after lanthanum).

Lewis dot formula representation of a molecule, ion, or formula unit by showing atomic symbols and only outer shell electrons; does not show shape.

Metal an element below and to the left of the stepwise division (metalloids) in the upper right corner of the periodic table; about 80% of all known elements are metals.

Metalloids elements with properties intermediates between metals and nonmetals: B, Al, Si, Ge, As, Sb, Te, Po, and At.

Noble gases elements of periodic Group 0; also called rare gases; formerly called inert gases.

Noble gas configuration the stable electronic configuration of a noble gas.

Nonmetals elements above and to the right of the metalloids in the periodic table.

Nuclear shielding see *Shielding effect*

Oxidation numbers arbitrary numbers that can be used as mechanical aids in writing formulas and balancing equations; for single-atom ions they correspond to the charge on the ion; more electronegative atoms are assigned negative oxidation numbers.

Periodicity regular periodic variations of properties of elements with atomic number (and position in the periodic table).

Periodic law the properties of the elements are periodic functions of their atomic numbers.

Periodic table an arrangement of elements in order of increasing atomic number that also emphasizes periodicity.

Rare earths inner transition elements.

Rare gases see *Noble gases*

Representative elements A Group elements in the periodic table.

Shielding effect electrons in filled sets of *s* and *p* orbitals between the nucleus and outer shell electrons fairly effectively shield the outer shell electrons from the effect of an equal number of protons in the nucleus; also called screening effect.

Semiconductor a substance that does not conduct electricity at room temperature but does so at higher temperatures.

Valence number of electrons that can be made available for sharing, gained, or lost by an atom of an element.

Valence electrons outermost electrons of atoms; usually those involved in bonding.

Exercises

The Periodic Table

1. In your own words, what does the word periodicity mean?
2. What was Mendeleev's contribution to the construction of the modern periodic table?
3. State the periodic law. What does it mean?
4. Mendeleev's periodic table was based on increasing atomic *weight* while the modern periodic table is based on increasing atomic *number.* In the modern table, argon comes before potassium, yet it has higher atomic weight. Explain how this can be.
5. Using Tables 21–1, 21–3, 21–5, 21–6, 22–3, 22–5, 23–1, and 24–1 look up melting points of the elements of periods 2 and 3. Show that melting point is a property that varies periodically for these elements.
6. Which of the following is a better example of deductive logic based on experimental observations, (a) or (b)? Why?
   (a) Noble gas electron configurations must be stable *because* noble gases are quite unreactive.
   (b) Noble gases must be unreactive *because* they have only filled electronic sublevels.
7. Distinguish between the following terms clearly and concisely, and provide specific examples of each: groups or families of elements and periods of elements. Write symbols for the alkali metals, the Group IIIA elements, and the Group IVB elements.
8. How do properties of elements vary within a group? Cite some specific examples.
9. How do properties of elements vary within a period? Cite some specific examples.
10. Define and illustrate the following terms clearly and concisely:
    (a) metals
    (b) nonmetals
    (c) noble gases
    (d) representative elements
    (e) *d*-transition elements
    (f) inner transition elements
11. Explain why period 1 contains two elements, period 2 contains eight, period 3 contains eight, and period 4 contains eighteen elements.
12. The third major energy level ($n = 3$) has *s*, *p*, and *d* sublevels. Why does period 3 contain only eight elements?
13. How many elements comprise period 6? Account for this number of elements.
14. What do Lewis dot formulas for atoms show? Draw Lewis dot representations for the following atoms: H, Li, N, Na, S, Ar, Ba, Sn, Se, At.
15. Draw Lewis dot representations for the following atoms: He, C, F, Al, P, Ca, As, Sb, Xe.
16. How do the physical properties of metals differ from those of nonmetals?
17. What are metalloids? List four examples.
18. Compare the metals and nonmetals with respect to (a) number of outer shell electrons, (b) ionization energies, (c) electron affinities and (d) electronegativities.
19. Of the Group VA elements, one is distinctly metallic and two are distinctly nonmetallic. Identify them and explain why.
20. By referring only to a periodic table, arrange the following in order of increasing metallic character: S, Be, Cl, Ge, Rb.

Atomic Radii

21. What is the justification for the following statement? It is impossible to describe an atom as having an invariant radius.
22. Why do atomic radii increase from top to bottom within a group of elements in the periodic table?
23. Why do atomic radii decrease from left to right within a period in the periodic table?
24. Arrange the members of each of the following groups of atoms in order of increasing atomic radii: (a) the Group IA elements, (b) the halogens, (c) the noble gases, (d) the elements in the third period, (e) Be, F, Na, I, Te.
25. Why are variations in the radii of the transition elements not as pronounced as those of the representative elements?

Ionization Energy

26. Define: (a) first ionization energy, (b) second ionization energy.

27. For a given element, why is the second ionization energy always greater than the first ionization energy?
28. Arrange the members of each of the following groups of elements in order of increasing first ionization energies: (a) the alkaline earth metals, (b) the Group VIA elements, (c) the elements in the third period, (d) Li, Rb, Be, Sr, F, I.
29. Explain why there is a general increase in first ionization energy across each period.
30. Explain the trend in first ionization energy upon descending a periodic group.
31. In a plot of first ionization energy versus atomic number for periods 2 and 3, "dips" occur at the IIIA and VIA elements. Account for these dips.
32. What is the general relationship between the sizes of the atoms of period 2 and their first ionization energies. Rationalize the relationship.
33. Why must a zinc atom absorb more energy than a calcium atom to ionize a 4*s* electron?
34. Would you expect compounds containing the Mg^{3+} ion to be stable? Why or why not? How about those containing Al^{3+}?
35. How much energy, in kJ, must be absorbed by 1.00 mol of gaseous sodium atoms to convert them all to gaseous Na^+ ions?

Electron Affinity

36. What is electron affinity?
37. Arrange the members of each of the following sets of elements in order of increasingly negative electron affinities: (a) the Group IA metals, (b) the Group VIIA elements, (c) the elements in the second period, (d) Li, K, C, F, I.
38. The electron affinities of the halogen atoms are more negative than those of the elements of Group VIA. Why is this expected?

Ionic Radii

39. Compare the sizes of cations and the neutral atoms from which they are formed by citing three specific examples.
40. Arrange the members of each of the following sets of cations in order of increasing ionic radii:
    (a) Li^+, Rb^+, Cs^+, Na^+, K^+
    (b) Na^+, Mg^{2+}, Al^{3+}
    (c) Na^+, Cs^+, Be^{2+}, Ga^{3+}
41. Compare the sizes of anions and the neutral atoms from which they are formed by citing three specific examples.
42. Arrange the following groups of anions in order of increasing ionic radii:
    (a) F^-, I^-, Cl^-, Br^-
    (b) O^{2-},N^{3-}, F^-
    (c) Se^{2-}, I^-, P^{3-}, O^{2-}
43. Compare and explain the relative sizes of H^+, H, and H^-.
44. Most transition metals can form more than one simple positive ion. For example, iron forms both Fe^{2+} and Fe^{3+} ions and copper forms both Cu^+ and Cu^{2+} ions. Which is the smaller ion of each pair and why?

Electronegativity

45. What is electronegativity?
46. Arrange the members of each of the following sets of elements in order of increasing electronegativities:
    (a) Li, F, O, Be
    (b) Be, Ba, Ca, Mg
    (c) O, Se, Te, S
    (d) Ca, Cs, O, Cl, C
47. Which of the following statements is better? Why?
    (a) The electronegativity of fluorine is high *because* fluorine has a strong attraction for electrons in a chemical bond.
    (b) Fluorine has a strong attraction for electrons in a chemical bond *because* it has a high electronegativity.
48. Comment on the validity of the following statement. Sodium has a low electronegativity because it forms sodium ions, Na^+, readily.

Chemical Bonds—Basic Ideas

49. What are chemical bonds?
50. Distinguish between ionic and covalent bonding. Why are covalent bonds called directional bonds, while ionic bonding is called nondirectional?
51. Construct a table in which four properties of ionic and covalent compounds are compared and contrasted.

52. Look up the properties of HCl and $MgCl_2$ in a handbook of chemistry. Why is HCl classified as a covalent compound and $MgCl_2$ classified as an ionic compound?
53. The following properties can be found in a handbook of chemistry:
potassium bromide (KBr)—colorless cubic crystals; specific gravity 2.75 at 20°C; melting point 730°C, boiling point 1435°C; solubility in cold water 53.48 g KCl/100 mL H_2O.
carbon tetrachloride (CCl_4)—colorless liquid; specific gravity 1.5867 at 20°C, melting point −23°C; boiling point 76.8°C; insoluble in cold water.
Would you classify these compounds as ionic or covalent? Why?
54. Based on the positions of the following pairs of elements in the periodic table, predict whether bonding between the two would be primarily ionic or covalent. Justify your answers. (a) O and Cl, (b) K and O, (c) B and O, (d) N and P, (e) Si and Si, (f) Ca and Br.
55. Predict whether the bonding between the following pairs of elements would be ionic or covalent. Justify your answers. (a) Ba and Br, (b) C and S, (c) Na and S, (d) P and O, (e) Ca and Se, (f) As and I.
56. Classify the following compounds as ionic or covalent: (a) $SiBr_4$, (b) $Ca(NO_3)_2$, (c) NO_2Cl, (d) SeO_2, (e) Na_2SO_4 (f) H_3PO_4, (g) Br_2, (h) LiF, (i) BF_3.

Ionic Bonding

57. Write electronic configurations for the following ions: (a) Ca^{2+}, (b) Cl^-, (c) Al^{3+}, (d) O^{2-}, (e) Bi^{3+}.
58. When a *d*-transition metal undergoes ionization it loses its outer *s* electrons before it loses any *d* electrons. Using [noble gas]$(n-1)d^x$ representations, write the outer electron configurations for the following ions: (a) Fe^{2+}, (b) Cu^+, (c) Cu^{2+}, (d) Mn^{3+}, (e) Cr^{3+}, (f) Ag^+, (g) Cd^{2+}.
59. Which of the following do not accurately represent stable binary ionic compounds? Why? RbF, Cs_2S, Mg_2O_3, AlF_2, Li_2O, Ba_3N, $SrCl_2$
60. Which of the following do not accurately represent stable binary ionic compounds? Why? $CaBr_2$, RbO, $InCl_3$, NaF, Ca_3S_2, MgI_3, BaO
61. Write chemical equations for reactions between the following pairs of elements. Draw electronic structures of the atoms before reaction as well as electronic structures of the ions formed in the reactions, using both the $\underline{\uparrow\downarrow}$ and ns^xnp^y notations, where n refers to the outermost occupied shell.
(a) lithium and chlorine
(b) potassium and sulfur
(c) calcium and fluorine
(d) magnesium and bromine
(e) barium and phosphorus
(f) lithium and nitrogen
62. Draw Lewis dot formulas for all atoms and ions in Exercise 61.
63. What are isoelectronic species? Cite three examples.
64. All but one of the following species are isoelectronic. Which one is not? O^{2-}, F^-, Ne, Na, Mg^{2+}, Al^{3+}
65. All but one of the following species are isoelectronic. Which one is not? P^{3+}, S^{2-}, Cl^-, Ar, K^+, Ca^{2+}
66. Write formulas for two cations and two anions that are isoelectronic with neon.
67. Write formulas for two cations and two anions that are isoelectronic with argon.
68. Write formulas for two cations that have the following electronic configurations in their highest occupied energy level. (a) . . . $4s^24p^6$, (b) . . . $5s^25p^6$
69. Write formulas for two anions that have the electronic configurations listed in Exercise 68.
70. Write chemical equations for the following reactions. Draw electronic structures of the atoms before reaction as well as electronic structures of the ions formed in these reactions, using both the $\underline{\uparrow\downarrow}$ and ns^xnp^y notations.
(a) reaction of lithium with sulfur
(b) reaction of magnesium with chlorine
(c) reaction of aluminum with fluorine
71. Draw Lewis dot formulas for all atoms and ions in Exercise 70.
72. Using M as the general symbol for metals and X as the general symbol for nonmetals, write general equations for the reactions of

the following pairs of elements. Don't forget to include charges on ions.

(a) A Group IA metal with a Group VIIA element.

(b) A Group IIA metal with a Group VIIA element.

(c) A Group IIA metal with a Group VA element.

73. Draw Lewis dot formulas for all atoms and ions in Exercise 72.

74. All compounds have some degree of both ionic and covalent character. Arrange the following compounds in order of increasing ionic character. Justify the order you choose: LiF, CsF, BeI_2, BI_3.

Oxidation Numbers

75. What are oxidation numbers?

76. List the common oxidation numbers for the A Group elements in binary compounds.

77. What oxidation numbers are the following elements expected to exhibit in simple ionic compounds? Li, Mg, O, F, Cs, Al, N

78. What oxidation numbers are the following elements expected to exhibit in simple ionic compounds? K, Ca, S, Br, Cs, I

79. Determine the oxidation number of chlorine in each of the following species: (a) $MgCl_2$, (b) $HClO_3$, (c) HClO, (d) $NaClO_2$, (e) ClO_4^-, (f) Cl_2O_7.

80. Determine the oxidation number of phosphorus in each of the following species: (a) PH_3, (b) P_4, (c) P_4O_6, (d) H_3PO_2, (e) H_3PO_4, (f) NaH_2PO_4, (g) $Ca_2P_2O_7$.

81. Determine the oxidation number of manganese in each of the following species: (a) $MnCl_3$, (b) $KMnO_4$, (c) Mn_2O_7, (d) K_2MnO_4, (e) Mn_2O_3, (f) MnO_2, (g) MnO.

5 Covalent Bonding and Inorganic Nomenclature

In Chapter 4 we divided chemical bonding into two major classes—ionic bonding and covalent bonding—and devoted the last few sections of the chapter to a study of ionic bonding. In this chapter we will examine the covalent bond and then show how certain theories, derived from our knowledge of these bonds, are used to predict shapes of covalently bonded molecules.

As you may recall, *a covalent bond is formed when two atoms share one or more pairs of electrons.* Most covalent bonds involve two, four, or six electrons or one, two, or three *pairs* of electrons. Two atoms form a **single covalent bond** when they share one pair of electrons, a **double covalent bond** when they share two electron pairs, and a **triple covalent bond** when they share three electron pairs. Covalent bonds involving one and three electrons are known, but these are relatively rare.

These are usually called simply single, double, and triple bonds.

Almost all bonds have both ionic and covalent character. By experimental means, a given bond can usually be identified as being "closer" to one or the other extreme type. We find it useful and convenient to use the labels for the major classes of bonds to describe simple substances, while keeping in mind the fact that they represent ranges of behavior.

In Chapter 4 we cited several examples of the occurrence of ionic bonding when metals react with nonmetals. Most bonds resulting from the reaction of a metal with a nonmetal are primarily ionic. However, as we shall see presently, reactions between two nonmetals result in covalent bonds.

The Covalent Bond

The principal attractive forces between atoms in covalent compounds are electrostatic in nature. Quantum mechanical calculations show that the interactions between two atoms vary with the distance between the atoms. Figure 5–1 shows a plot of energy versus internuclear distance for two hydrogen atoms. Note that the minimum energy −435 kJ/mol, corresponding to the most stable arrangement, occurs at an internuclear distance of 0.74 Å, the actual distance between two hydrogen nuclei in an H_2 molecule. At this internuclear separation, the two atoms exert the maximum attraction for each other.

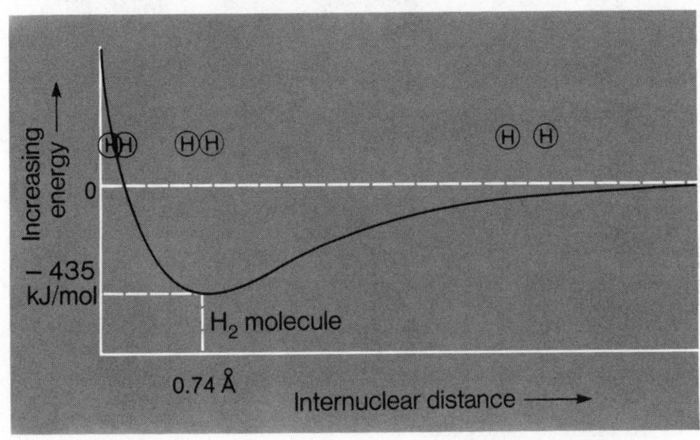

FIGURE 5–1 The potential energy of the H_2 molecule as a function of the distance between the two nuclei. The lowest point in the curve, −435 kJ/mol, corresponds to the internuclear distance actually observed in the H_2 molecule, 0.74 Å. (The minimum potential energy, −435 kJ/mol, corresponds to a value of -7.23×10^{-19} joule per H_2 molecule.) Energy is compared to that of two separated hydrogen atoms.

The high electron density in the region between the two nuclei shields the two positive charges from each other and provides the "glue" that holds the two hydrogen atoms together. At an internuclear separation of 0.74 Å, the attractive forces between the electron of each atom and the positive nucleus of the other exceed by the greatest margin the repulsive forces between like-charged particles in the molecule. At larger internuclear separations, the repulsive forces diminish, but the attractive forces decrease even faster; at smaller separations, repulsive forces grow more rapidly than attractive forces.

This picture of the covalent bond can be generalized by postulating that bonds form by the **overlap** of two atomic orbitals, each containing a single unpaired electron. This is the essence of the **Valence Bond Theory,** which we will look at in more detail later. In the case of H_2, a covalent bond forms when two 1*s* atomic orbitals overlap, thus allowing the two electrons to be shared by the two atoms.

5–1 Polar and Nonpolar Covalent Bonds

H:H
or
H—H

Covalent bonds are described as being either *polar* or *nonpolar* bonds. In **nonpolar bonds** such as that in the hydrogen molecule, H_2, the electron pair is *shared equally* between the two hydrogen nuclei. This means that the shared electrons are equally attracted to both hydrogen nuclei and therefore spend equal amounts of time near each nucleus (Figure 5–2). Stated differently, in nonpolar covalent bonds, the **electron density** is symmetrical about a plane that is perpendicular to a line between the two nuclei. This holds true for all diatomic molecules of identical atoms like H_2, O_2, N_2, F_2, and Cl_2 because the two atoms exert identical attractions for the shared electron pairs. Thus, we can generalize and say that the covalent bonds in all **homonuclear diatomic molecules** must be nonpolar.

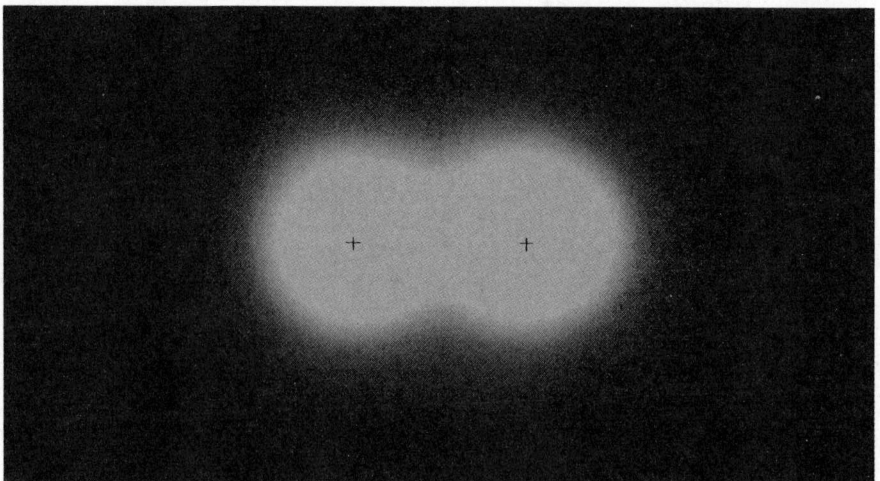

FIGURE 5–2 Electron density in H_2. The depth of shading is proportional to the probability of finding an electron in a particular region. In chemical bonds there tends to be a concentration of electronic charge between the nuclei.

When we consider **heteronuclear diatomic molecules,** the situation isn't so simple. As an example, consider the fact that hydrogen fluoride, HF, is a gaseous substance at room temperature. This tells us that it is a covalent compound (remember from Section 4–9 that ionic compounds are solids at room temperature). We also know that this covalent bond will have some degree of polarity because H and F are not identical atoms and therefore will not attract the electrons equally. But, how polar will this bond be? You may recall (Section 4–7) that we defined electronegativity as the tendency of an atom to attract electrons to itself in a chemical bond. According to Table 4–6, the electronegativity of hydrogen is 2.1 and that of fluorine is 4.0.

Clearly, the fluorine atom, with its higher electronegativity, attracts the shared electron pair much more strongly than hydrogen. We represent the structure of HF as shown below. Notice the unsymmetrical distribution of electron density; the electron density is distorted in the direction of the more electronegative fluorine atom.

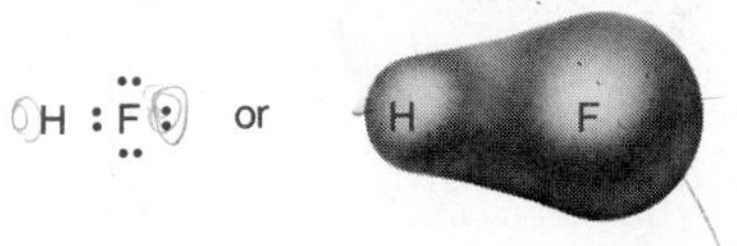

Covalent bonds, such as the one in HF, in which the *electron pairs are shared unequally* are referred to as **polar covalent bonds.** We use two kinds of notation to indicate polar bonds

$\overset{\delta+}{H}—\overset{\delta-}{F}$ *or* $\overset{+\!\!\longrightarrow}{H—F}$

The δ+ over the H atom indicates a "partial positive charge." This means that the hydrogen end of the molecule is positive *with respect* to the fluorine end of the molecule. The δ− over the F atom indicates a "partial negative charge," or that the fluorine end of the molecule is negative *with respect* to the hydrogen end. Please note that we are *not* saying that hydrogen has a charge of 1+ or that fluorine has a charge of 1−! A second way to indicate the polarity is to draw an arrow so that the head points toward the negative end (F) of the molecule and the crossed tail indicates the positive end (H).

Polar covalent bonds may be thought of as being intermediate between pure (nonpolar) covalent bonds, in which electron pairs are equally shared, and pure ionic bonds, in which there has been complete transfer of electrons from one atom to another. In fact, bond polarity is sometimes described in terms of *partial ionic character,* which usually increases with increasing difference in electronegativity between bonded atoms. Calculations based on the measured dipole moment (see the next section) of gaseous HCl indicate approximately 17% "ionic character."

The separation of charge in a polar covalent bond creates a **dipole.** The word dipole means "two poles," and refers to the positive and negative poles that result from the separation of charge within the molecule. If we consider the covalent molecules HF, HCl, HBr, and HI, we should expect their dipoles to be different because F, Cl, Br, and I have different electronegativities and therefore different tendencies to attract electron pairs shared with hydrogen. This is indeed the case. Qualitatively, we can indicate this difference as shown

The values of electronegativity are obtained from Table 4–6.

below, where Δ(EN) is the difference in electronegativity between two atoms that are bonded together.

	H—F	H—Cl	H—Br	H—I
EN:	2.1 4.0	2.1 3.0	2.1 2.8	2.1 2.5
Δ(EN)	1.9	0.9	0.7	0.4

The longest arrow indicates the largest dipole or largest separation of electron density in the molecule.

5–2 Dipole Moments

As with other concepts, it is convenient to express differences in polarities of bonds on a numerical scale. We indicate the polarity of a molecule by its *dipole moment, μ, which measures the separation of charge within a molecule.* To measure a dipole moment, a sample of the substance is placed between two electrically charged plates. Polar molecules such as HF, HCl, HBr, and HI orient themselves in the electric field, causing the measured voltage between the plates to decrease. By contrast, nonpolar molecules do not line up, and the voltage between the plates does not change (Figure 5–3).

Generally, as electronegativity differences increase in diatomic molecules, the measured dipole moments increase. This can be seen clearly in the following data for the hydrogen halides (Table 5–1).

The **dipole moment, μ,** is defined as *the product of the distance separating charges of equal magnitude and opposite sign, times the magnitude of the charge.* Unfortunately, the dipole moments of *individual bonds* can be determined only in simple diatomic molecules, because *entire molecules* rather than pairs of atoms must be subjected to measurement. Thus, tabulated values of dipole moments of polyatomic molecules refer to the average effects of all the bond dipoles in the molecules to which they apply, and reflect the *overall* polarities of the molecules. In subsequent sections we shall see that structural features, such as molecular geometry and the presence of lone (unshared) pairs

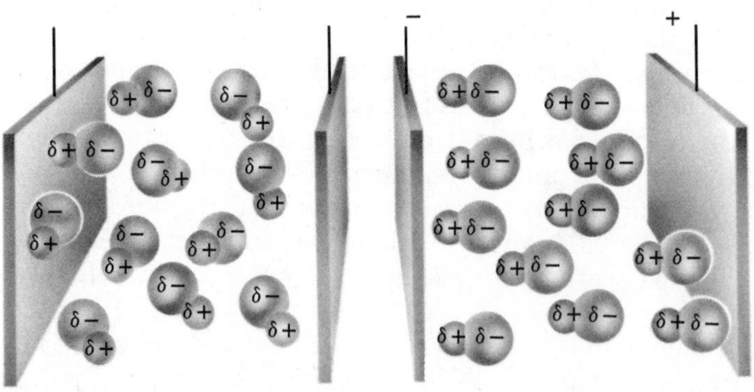

FIGURE 5–3 If polar molecules, such as HF, are subjected to an electric field, they tend to line up in a direction opposite to that of the field. This minimizes the electrostatic energy of the molecules. Nonpolar molecules are not oriented by an electric field.

TABLE 5–1 Dipole Moments and Δ(EN) Values for Hydrogen and the Pure (Gaseous) Hydrogen Halides

Hydrogen and the Hydrogen Halides	Dipole Moment*(μ)	Δ(EN)
HF	1.91 D	1.9
HCl	1.03 D	0.9
HBr	0.79 D	0.7
HI	0.38 D	0.4
H—H	0 D	0

*Dipole moments are usually expressed in Debye units, D.

of electrons, affect the polarity of a molecule. Many nonpolar molecules (those with dipole moments of zero) contain polar covalent bonds whose effects cancel one another.

In order for a molecule to be polar, *both* of the following conditions must be met:

1. There *must* be at least one polar bond present.
2. The polar bonds, if there are more than one, and lone pairs *must not* be so symmetrically arranged that their bond polarities cancel.

Putting this another way, if there are no polar bonds or lone pairs on the central atom, the molecule *cannot* be polar. Even if these are present, they may be arranged so that their polarities cancel one another, resulting in a nonpolar molecule.

5–3 Lewis Dot Formulas for Molecules and Polyatomic Ions

In Chapter 4 we drew Lewis dot formulas for atoms and monatomic ions. Lewis dot formulas can also be used to represent atoms covalently bonded in a molecule or a polyatomic ion (an ion that contains more than one atom). A water molecule is shown below. A water molecule can be represented by either of the following diagrams.

```
  ..            ..
H:O:          H—O:      two shared electron
  ..            |       pairs; two single
  H             H       covalent bonds
```

Lewis dot formula   dash formula

The O atom contributes six valence electrons and each H atom contributes one.

In "dash formulas" a shared pair of electrons is indicated by a dash. There are two *double* bonds in carbon dioxide

```
 ..    ..       ..   ..     four shared
:O::C::O:      :O=C=O:      electron pairs;
                            two double bonds
```

The C atom contributes four valence electrons and each O contributes six.

The covalent bonds in a polyatomic ion can be represented in the same way. The ammonium ion, NH_4^+, is shown below. Note that the Lewis dot formula shows only eight electrons, even though the N atom has five electrons in its valence shell and each H atom has one, for a total of 5 + 4(1) = 9. The reason is that the ion, with a charge of 1+, has one less electron than the original atoms.

The dot formula for the (uncharged) NH_3 molecule, which, like the NH_4^+ ion, has eight valence electrons about the N atom is

$$\mathrm{H:\ddot{N}:H} \atop \mathrm{\ddot{}H}$$

$$\begin{matrix} & \mathrm{H} & + \\ \mathrm{H:} & \mathrm{\ddot{N}} & \mathrm{:H} \\ & \mathrm{\ddot{}H} & \end{matrix} \qquad \begin{matrix} & \mathrm{H} & + \\ & | & \\ \mathrm{H—} & \mathrm{N} & \mathrm{—H} \\ & | & \\ & \mathrm{H} & \end{matrix}$$

Lewis dot formula dash formula

It is important to realize that Lewis dot formulas only show the number of valence electrons and the number and kinds of bonds, *and they are not intended to depict the three-dimensional shapes of molecules and ions.*

5–4 The Octet Rule

Representative elements *usually* attain stable noble gas electron configurations when they share electrons. In the water molecule (p. 163) the oxygen has a share in eight outer shell electrons, giving it the neon configuration; hydrogen shares two electrons, attaining the helium configuration. Likewise, the carbon and oxygen of CO_2 and the nitrogen of NH_3 each have a share in eight electrons in their outer shells, the neon configuration. The hydrogen atoms in H_2O and NH_3 each share two electrons, the helium configuration. We write Lewis dot formulas based on the fact that the *representative elements achieve a noble gas configuration in most of their compounds.* This rule is usually called the **octet rule.** For the present, we shall restrict our discussion to compounds of the representative elements. The octet rule, alone, does not enable us to write Lewis formulas. We still need to know how to place the electrons around the bonded atoms—that is, how many of the available valence electrons are bonding (shared) electrons and how many are associated with only a single atom. The latter kind are called unshared electrons (or lone pairs). A simple mathematical relationship is helpful here:

The method does not work for some cases in which the central atom does not achieve a noble gas configuration. Exceptions to the octet rule will be discussed later in the chapter.

$$S = N - A$$

In the equation, *S* is the total number of electrons *shared* in the molecule or polyatomic ion. *N* represents the number of valence shell electrons *needed by all the atoms in the molecule or ion to achieve noble gas configurations* (in other words, $N = 8 \times$ number of atoms *not* including H, plus $2 \times$ number of H atoms). The number *A* is the number of electrons *available* in the valence shells of all of the (representative) atoms, which is simply the sum of their periodic group numbers. (*A* is taken to be 8 for a noble gas.) For example, in CCl_4, *A* for Cl is 7 and *A* for C is 4, so *A* for CCl_4 is 4 + 4(7) = 32.

The following examples illustrate the use of this relationship.

EXAMPLE 5–1

Write the Lewis dot and dash formula for the nitrogen molecule, N_2.

Solution

Each nitrogen atom *needs* 8 e^-: $N = 2 \times 8 = 16$

Each nitrogen atom has 5 e^- *available:* $A = 2 \times 5 = 10$

Thus,

$S = N - A$
$= 16 - 10 = 6\ e^-$ shared between two N atoms

:N:::N: or   :N≡N: (a triple bond)

Only 10 e^- are available because each N atom has only five valence electrons. The only way these two N atoms can achieve a noble gas configuration (neon) is to share six electrons. Thus a triple bond is formed, leaving four unshared electrons—one pair associated with each nitrogen atom.

We were able to write the formula for N_2, a homonuclear diatomic molecule, with the two rules given thus far. Suppose, however, we were asked to write the formula for carbon disulfide, CS_2. Do we arrange the atoms as C—S—S or S—C—S? The general rules are (1) the element needing the largest number of electrons to fill its octet is the central element, and (2) the most symmetrical skeleton is usually the correct one.

EXAMPLE 5–2

Write the Lewis dot formula and dash formula for carbon disulfide, CS_2, an ill-smelling liquid.

Solution

$N = 8$ (for C) $+ 2 \times 8$ (for S) $= 24$ $S = N - A$

$A = 4$ (for C) $+ 2 \times 6$ (for S) $= 16$ $= 24 - 16 = 8\ e^-$ shared

:S̈::C::S̈:   or   :S̈=C=S̈:;   16 e^- used, of which 8 e^- are shared

(Two double bonds similar to CO_2; S is below O in Group VIA.)

Carbon is the central element, or the element in the middle of the molecule, because it needs four more electrons to acquire an octet while each sulfur atom needs only two more electrons.

EXAMPLE 5–3

Construct a dot and dash formula for sulfuric acid, H_2SO_4, a ternary acid. A **ternary acid** is a compound of hydrogen, oxygen, and another nonmetal. In such compounds, in general, the O atoms are bonded to the other nonmetal and *not* to each other, and the H atoms are almost always bonded to the O atoms.

Solution

The skeleton is

```
      O
H  O  S  O  H
      O
```

$N = 2 \times 2$ (for H) $+ 1 \times 8$ (for S) $+ 4 \times 8$ (for O) $= 44$

$A = 2 \times 1$ (for H) $+ 1 \times 6$ (for S) $+ 4 \times 6$ (for O) $= 32$

$S = N - A = 44 - 32 = 12\ e^-$ shared

Remember that H needs a share of only 2 e^- to attain the He configuration.

```
   ..                    ..
  :O:                   :O:
 .. .. ..                |
H:O:S:O:H     or    H—O—S—O—H
 .. .. ..             ..   ..
  :O:                    |
   ..                   :O:
                         ..
```

A total of 32 e^- are used (A), 12 of which are shared (S) as six single bonds.

When constructing the dot formula for a polyatomic ion, we must account for the charge on the ion by adding electrons to A for each extra negative charge or subtracting an electron for each extra positive charge. The next example illustrates this for the carbonate ion, which has a charge of 2−.

EXAMPLE 5–4

Construct dot and dash formulas for the carbonate ion, CO_3^{2-}.

Solution

The skeleton is

```
  O
O C O
```

$N = 1 \times 8$ (for C) $+ 3 \times 8$ (for O)
$= 8 + 24 = 32\ e^-$ needed by all atoms

$A = 1 \times 4$ (for C) $+ 3 \times 6$ (for O) $+ 2$ (for the 2− charge)
$= 4 + 18 + 2 = 24\ e^-$ available

$S = N - A = 32\ e^- - 24\ e^- = 8\ e^-$ (four pairs) shared

```
   ..                    ..
  :O: 2-                :O:  2-
 .. .. ..                |
:O:C::O:     or     :O—C=O:
 ..                  ..
```

Check: 24 e^- (12 pairs) have been used. At this point it doesn't matter which O is doubly bonded.

5–5 Resonance

Two other dot formulas for the CO_3^{2-} ion, in addition to the one shown in Example 5–4, are equally acceptable. In these formulas, 4 e^- could be shared between the carbon atom and either of the other two oxygen atoms

```
   ..              ..                ..
  :O:   2-        :O:   2-          :O:   2-
   |               ‖                 |
:O—C=O:  ⟷   :O—C—O:   ⟷   O=C—O:
 ..            ..    ..         ..   ..
```

A molecule or ion for which two or more equally acceptable dot formulas are available to describe the bonding for the same arrangement of atoms is said to exhibit **resonance.** The three structures above are **resonance structures** of the carbonate ion, CO_3^{2-}. The relationship among them is indicated by the double-headed arrows, ↔. This symbol should not be taken to imply that the ion flips back and forth among the three illustrated structures, but rather that the true structure is like an average of the three.

The C—O bonds in CO_3^{2-} are really *neither* double nor single bonds, but are intermediate in bond length (and strength). This has been verified experimentally. The average C—O single bond distance (based on measurements in many compounds) is 1.43 Å and the average C═O double bond distance is 1.22 Å while the C—O bond distance for each bond in the CO_3^{2-} ion is intermediate at 1.29 Å. Another way to represent this situation is by **delocalization** of bonding electrons

```
       O   2−
       ⁞⁞
  O ═══ C ═══ O        (lone pairs on O
                        atoms not shown)
```

When electrons are shared among more than two atoms, the electrons are said to be *delocalized.*

where the dashed lines indicate that the electrons shared between carbon and oxygen atoms are *delocalized* among all four atoms; that is, four pairs of electrons are equally distributed among three C—O bonds.

EXAMPLE 5–5

Draw two resonance structures for the sulfur dioxide molecule, SO_2.

Solution

$$\begin{aligned} N &= \overset{S}{1}(8) + \overset{O}{2}(8) = 24\ e^- \\ A &= 1(6) + 2(6) = 18\ e^- \\ S &= N - A = 6\ e^- \text{ shared} \end{aligned}$$

The resonance structures are

$:\ddot{O}::\ddot{S}:\ddot{O}: \longleftrightarrow :\ddot{O}:\ddot{S}::\ddot{O}:$

or, using dash formulas

$:\ddot{O}{=}\ddot{S}{-}\ddot{O}: \longleftrightarrow :\ddot{O}{-}\ddot{S}{=}\ddot{O}:$

We could show delocalization of electrons as follows

$O \equiv \ddot{S} \equiv O$ (unshared electrons on O not shown)

Remember that dot and dash formulas *do not necessarily show shapes.* SO_2 molecules are angular, not linear, for reasons we shall see beginning in Section 5–8.

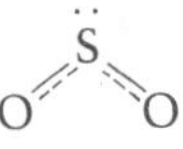

5–6 Limitations of the Octet Rule for Dot Formulas

You may recall that we stated that representative elements achieve a noble gas electronic configuration in *most* of their compounds (Section 5–4). For cases to which the octet rule is not applicable, of course, the relationship $S = N - A$ is not valid. There are five general cases for which the relationship *may not be applied* in constructing Lewis formulas. They are:

1. most covalent compounds of beryllium
2. most covalent compounds of the elements of Group IIIA "electron deficient"
3. compounds in which the central element must have a share in more than eight valence shell electrons to accommodate all attached atoms SF_6

4. compounds containing *d*- or *f*-transition metals
5. species containing an odd number of electrons (for example, NO, with 11 valence electrons; 15 total)

Let us illustrate some of these limitations below and show how dot formulas can be constructed in such cases.

EXAMPLE 5–6

Draw the Lewis dot formula for gaseous beryllium chloride, $BeCl_2$.

Solution

According to Case 1, above, we cannot apply the relationship $S = N - A$ because Be has only two electrons in its valence shell and therefore can form a maximum of two ordinary covalent bonds with two other atoms. Thus the dot formula for $BeCl_2$ must be

:C̤̈l:Be:C̤̈l:

The chlorine atoms achieve the Ar configuration, but the beryllium atom has a share of only four electrons.

One might logically predict a similar situation for compounds of the other IIA metals, Mg, Ca, Sr, Ba, and Ra. However, these elements usually form ionic compounds by losing two electrons to achieve noble gas configurations.

EXAMPLE 5–7

Draw the Lewis dot formula for boron trichloride, BCl_3, a covalent compound.

Solution

Since boron is in Group IIIA, this compound provides an example of limitation 2 listed above. The situation is similar to that in Example 5–6 in that boron has only three valence shell electrons and therefore can form only three ordinary covalent bonds with three other atoms. So the dot formula must be

```
   ..
  :Cl:
   ..
    B
 ..   ..
:Cl: :Cl:
 ..   ..
```

in which the chlorine atoms attain the Ar configuration but the boron atoms share only six valence shell electrons.

EXAMPLE 5–8

Write the Lewis formula for the covalent compound arsenic pentafluoride, AsF_5.

Solution

This is an example of limitation 3. Suppose we applied the relationship

$S = N - A$

$N = 1 \times 8 \text{ (for As)} + 5 \times 8 \text{ (for F)} = 48$

$A = 1 \times 5$ (for As) $+ 5 \times 7$ (for F) $= 40$

$S = N - A = 48 - 40 = 8\ e^-$ shared **(not correct)**

We would conclude that eight electrons should be shared. But, since arsenic is bonded to five fluorine atoms, *at least* ten electrons must be shared to account for five ordinary covalent bonds, and therefore the dot and dash formulas must be

:F: :F: As :F: :F: :F: and :F:—As(—F:)(—F:)(—F:)(—F:)

in which the arsenic atom has a share of ten electrons in its outer shell and each fluorine has attained the Ne configuration.

As the previous examples demonstrate, substituents attached to the central atom nearly always attain noble gas configurations even when the central atom does not.

We shall now apply the ideas presented thus far in this chapter to the *three-dimensional structures* of molecules.

5–7 Basic Notions of Bonding Theory

Understanding the nature of the covalent bond is one of the central challenges of chemistry. Bonding is the key to molecular structure, and structure is intimately related to the physical and chemical properties of a compound. Current bonding theories allow us to predict structures and properties that are usually accurate (although they are not always entirely satisfactory). As we begin the discussion of covalent compounds, we should keep in mind the truism that whatever we propose *must* be consistent with experimentally determined facts. When there is disagreement between facts and theory, theory must be modified to accommodate *all* known facts. In this chapter, we must account for a large body of knowledge about molecular structure, all of it based on reliable experiments.

We shall discuss two theories. The first is the Valence Shell Electron Pair Repulsion (VSEPR) Theory, which assumes that electron pairs are arranged around the central element of a compound in such a way that there is maximum *separation* (and therefore, minimum repulsion) among electron pairs. Although this statement may appear to be obvious (it is), the idea will prove to be remarkably useful in predicting the geometries of molecules and ions. The second theory is the Valence Bond (VB) Theory, which we discussed briefly in the introduction to Section 5–1. These two theories go hand-in-hand in any discussion of covalent molecules. The VSEPR theory is the key to understanding the spatial arrangement of atoms in a molecule; the VB theory explains the bonding in terms of overlapping atomic orbitals.

We shall also introduce one additional concept that will allow us to "mix" the atomic orbitals discussed in Chapter 3 to form new orbitals with different spatial orientations. This process of mixing, called **hybridization,** is often necessary to explain experimental results or to achieve the structures predicted by the VSEPR theory. The VB theory applies to hybrid atomic orbitals just as it does to "pure" atomic orbitals.

Throughout the discussion of covalent bonding we shall focus attention on the electrons in the outer shell, or valence shell, of the atoms because these are the electrons involved in bonding. To make things easier for ourselves, we shall draw Lewis dot formulas for each molecule we discuss. Recall that dot formulas show only the valence shell electrons, which may be thought of as those that were not present in the preceding noble gas, ignoring filled sets of *d* and *f* orbitals.

Covalent Bonding and Molecular Structure

5–8 Valence Shell Electron Pair Repulsion (VSEPR) Theory

The qualitative idea that geometries of molecules and polyatomic ions can be predicted by considering the number of regions of high electron density about the central atom is very useful. This theory is known as the **Valence Shell Electron Pair Repulsion Theory** or **VSEPR Theory.** A single bond, a double bond, a triple bond, or an unshared pair of electrons are each considered to be *a single region of high electron density.* The concept is based on two assumptions:

1. Each valence shell pair of electrons on an atom is significant.
2. Repulsions among valence shell pairs of electrons determine the shapes of molecules.

The basic ideas can be summarized as follows: Valence shell pairs of electrons are arranged about the *central atom* so that repulsions among them are minimized, or so that there is maximum separation among the regions of high electron density about the atom. For instance, two regions of high electron density would be most stable on opposite sides of the central atom (the linear arrangement), while three regions would be most stable when they are arranged at the corners of an equilateral triangle (the trigonal planar arrangement). The resulting *arrangement of these regions* is referred to as the **electronic geometry** of the central atom.

Table 5–2 shows the relationship between the common numbers of regions of high electron density and the corresponding electronic geometries. Consideration of the number of regions of high electron density that connect the central atom to other atoms can then lead to a prediction (or explanation) of the arrangement of the other atoms about the central atom.

5–9 Valence Bond (VB) Theory

In the opening pages of this chapter, we explained covalent bonding as the result of *orbital overlap*—the overlap of atomic orbitals from two atoms. This is the central tenet of the Valence Bond theory, which describes *how* bonding occurs. The Valence Bond theory is complementary to the VSEPR theory,

TABLE 5–2 Number of Regions of High Electron Density about A Central Atom

Number of Regions of High Electron Density	Electronic Geometry*	Angles†
2	linear	180°
3	trigonal planar	120°
4	tetrahedral	109°28′
5	trigonal bipyramidal	90°, 120°, 180°
6	octahedral	90°, 180°

* Electronic geometries are illustrated here using only single pairs of electrons as regions of high electron density.

† Angles made by imaginary lines through the nucleus and the centers of regions of high electron density.

Sets (pairs) of electrons are represented by ⊙ in Table 5–2.

which describes *where* bonding occurs, as well as *where* the lone unshared pairs of valence shell electrons on the central atom are directed. In each of the examples below, we will interpret each covalent bond in this light. We shall also assume that each lone pair occupies a separate atomic orbital. To provide the necessary atomic orbitals, or to explain the geometry (as observed experimentally or as predicted by the VSEPR theory), it will usually be necessary to invoke the concept of hybridization.

The number of regions of high electron density about a central atom in a molecule or polyatomic ion suggests the kind of hybridization, or mixing, of that atom's valence atomic orbitals that occurs (see Table 5–3). The designa-

TABLE 5–3 Common Kinds of Hybridization Associated with Particular Electronic Geometries

Number of Regions of High Electron Density About a Central Atom	Electronic Geometry	Common Hybridization	Atomic Orbitals Mixed from Valence Shell of Central Atom
2	linear	*sp*	one *s*, one *p*
3	trigonal planar	sp^2	one *s*, two *p*'s
4	tetrahedral	sp^3	one *s*, three *p*'s
5	trigonal bipyramidal	sp^3d	one *s*, three *p*'s, one *d*
6	octahedral	sp^3d^2	one *s*, three *p*'s, two *d*'s

tions given to sets of hybridized orbitals reflect the number and kind of atomic orbitals that hybridize or mix to produce the set.

We are now ready to study the structures of some simple molecules, and we shall start with the simplest examples first. For each type of molecule we shall first give the known (experimentally determined) facts about polarity and shape; then, after drawing the Lewis dot formula, we will explain the facts in terms of the VSEPR and VB theories. In each example, the simpler VSEPR theory will be presented first, to explain (or predict) the geometry of the molecule. We will then discuss the VB theory to gain a more detailed picture of the bonds in the molecule.

5–10 Linear AB_2 Molecules with No Unshared Pairs of Electrons on A

There are several linear molecules that consist of a central atom plus two atoms of another element, abbreviated as AB_2. These compounds include $BeCl_2$, $BeBr_2$, and BeI_2, as well as CdX_2 and HgX_2 where X = Cl, Br, I. All of these are known to be linear (bond angle = 180°), nonpolar, covalent compounds, although the individual bonds are polar.

Let's focus on gaseous $BeCl_2$ molecules for the moment. The electronic structures of Be and Cl *atoms* in their ground states are

	1*s*	2*s*	2*p*	3*s*	3*p*
$_4Be$	⇅	⇅			
$_{17}Cl$	⇅	⇅	⇅ ⇅ ⇅	⇅	⇅ ⇅ ↑

We drew the Lewis dot formula for $BeCl_2$ in Example 5–6. It shows two single covalent bonds, with Be and Cl each contributing one electron to each bond

$$:\ddot{\underset{..}{Cl}}:Be:\ddot{\underset{..}{Cl}}: \quad \text{or} \quad :\ddot{\underset{..}{Cl}}—Be—\ddot{\underset{..}{Cl}}:$$

1 VSEPR Theory

Valence Shell Electron Pair Repulsion theory, which assumes that regions of high electron density (electron pairs) will be as far from one another as possible, places the two electron pairs on Be 180° apart. Thus VSEPR predicts a *linear* structure for $BeCl_2$.

$$\overset{180^\circ}{:\ddot{\underset{..}{Cl}}—Be—\ddot{\underset{..}{Cl}}:}$$

Examining the bond dipoles, we see that the electronegativity difference (see Table 4–6) is large (1.5 units) and the bonds are quite polar

	Cl	Be	Cl
EN	3.0	1.5	3.0
Δ(EN)	1.5		1.5

$$\overset{\longleftarrow\!+ \quad +\!\longrightarrow}{:\ddot{\underset{..}{Cl}}—Be—\ddot{\underset{..}{Cl}}:}$$

However, the bond dipoles are identical in magnitude and opposite in direction, and therefore they cancel to give *nonpolar* molecules. We shall observe that bond dipoles always cancel in symmetrical molecules, so all symmetrical molecules are nonpolar. It is important to realize that we must distinguish between *nonpolar bonds* and *nonpolar molecules*.

The difference in electronegativity between Be and Cl is large enough that we might expect ionic bonding. However, the radius of Be^{2+} is so small (0.31 Å) and the charge density (ratio of charge-to-size) is so high that most simple beryllium compounds are covalent rather than ionic. The high charge density of Be^{2+} causes it to attract and distort the electron cloud of monatomic anions of all but the most electronegative elements to such an extent that electrons are shared rather than being localized on ions. Two exceptions are BeF_2 and BeO which are ionic compounds that contain very electronegative elements.

2 Valence Bond Theory

Let us consider the electronic configuration for Be. There are two electrons in the 1*s* orbital of Be, the central atom. These are nonvalence (inner) electrons and are not involved in the bonding in compounds of Be. There are two more electrons *paired* in the 2*s* orbital. How, then, will two Cl atoms bond to Be? The Be atom must somehow make available one orbital for each (bonding) Cl electron (the unpaired *p* electrons). Recall that the electron configuration shown below for Be is the *ground state* configuration for an isolated Be atom. Another configuration may be more stable in a bonding environment. Suppose that the Be atom *promotes* one of the paired 2*s* electrons to one of the 2*p* orbitals, the next highest energy orbitals.

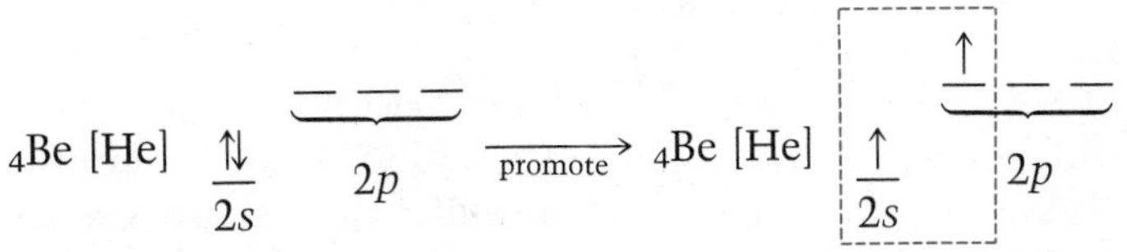

$$_4\text{Be [He]} \ \underset{2s}{\underline{\uparrow\downarrow}} \ \ \underbrace{\underline{\ \ }\ \underline{\ \ }\ \underline{\ \ }}_{2p} \ \xrightarrow{\text{promote}} \ _4\text{Be [He]} \ \underset{2s}{\underline{\uparrow}} \ \ \underbrace{\underline{\uparrow}\ \underline{\ \ }\ \underline{\ \ }}_{2p}$$

Now there would be two Be orbitals available for bonding, but we find a discrepancy with experimental fact. This "promoted pure atomic" arrangement would predict two *nonequivalent* Be—Cl bonds because the Be 2*s* and 2*p* orbitals are not expected to overlap a Cl 3*p* orbital with equal effectiveness. Yet we observe experimentally that both Be—Cl bonds are *equivalent* in bond length and bond strength. So we reject the idea of simple "promotion" as an explanation.

For these two orbitals on beryllium to become equivalent, they must **hybridize,** or combine to give two orbitals intermediate between the *s* and *p* orbitals. These are called ***sp* hybrid orbitals.** Consistent with Hund's Rule (Section 3–15), the two valence electrons of Be will occupy each of these orbitals individually

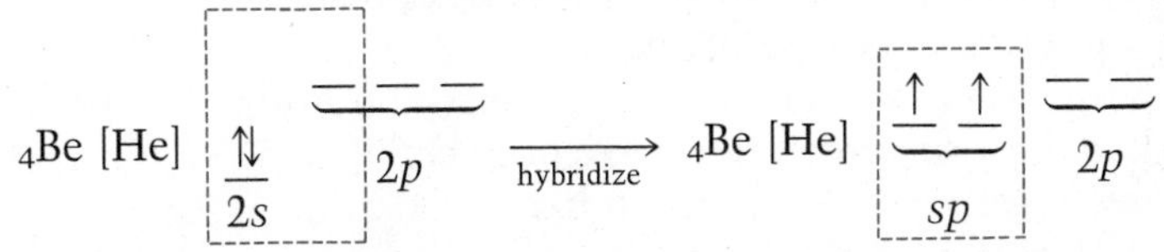

$$_4\text{Be [He]} \ \underset{2s}{\underline{\uparrow\downarrow}} \ \ \underbrace{\underline{\ \ }\ \underline{\ \ }\ \underline{\ \ }}_{2p} \ \xrightarrow[\text{hybridize}]{} \ _4\text{Be [He]} \ \underbrace{\underline{\uparrow}\ \underline{\uparrow}}_{sp} \ \ \underbrace{\underline{\ \ }\ \underline{\ \ }}_{2p}$$

The *sp* hybrid orbitals are described as *linear orbitals, and this is consistent with the experimental observation that the molecule is linear;* the Be and two

Cl nuclei lie on a straight line. In general, **sp hybridization** occurs at the central atom of a molecule, or a polyatomic ion, whenever there are two regions of high electron density around the central atom.

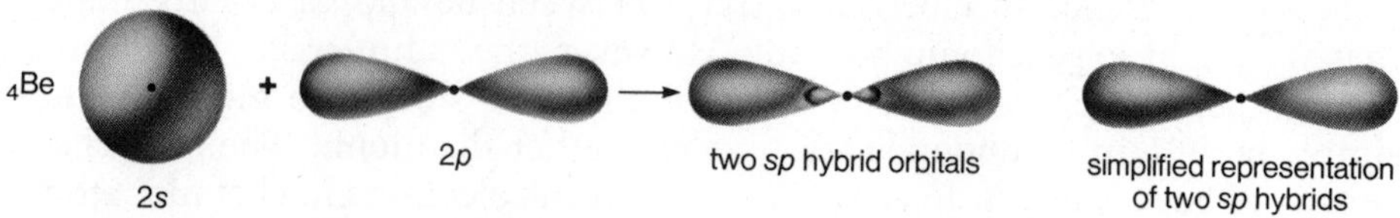

The dot in each orbital above represents the Be nucleus. Recall that each Cl atom has a 3*p* orbital that contains only one electron, so that overlap with the *sp* hybrids of Be is possible. We picture the bonding in $BeCl_2$ in the following diagram, in which only the bonding electrons are represented.

Lone pairs of e^- on Cl atoms are not shown.

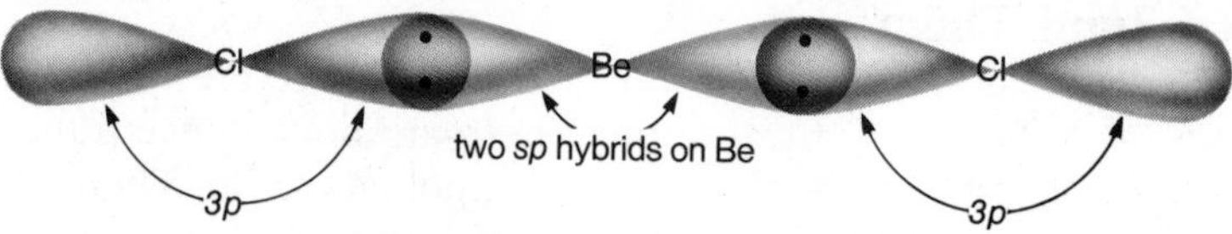

One additional idea about hybridization is worth special emphasis: The number of hybrid orbitals is always the sum of the number of atomic orbitals hybridized. Hybrid orbitals are named by indicating the *number and kind* of atomic orbitals hybridized. As we have seen, hybridization of *one 2s* orbital and *one 2p* orbital gives *two sp hybrid orbitals.* We shall see presently that hybridization of *one ns* orbital and *two np* orbitals gives *three* sp^2 hybrid orbitals, while hybridization of *one ns* orbital and *three np* orbitals gives *four* sp^3 *hybrids,* and so on.

Br and I utilize 4*p* and 5*p* orbitals, respectively, in bonding in these compounds, whereas Cl utilizes a 3*p* orbital.

The structures of beryllium bromide, $BeBr_2$, and beryllium iodide, BeI_2, are similar to that of $BeCl_2$. The chlorides, bromides, and iodides of cadmium, CdX_2, and mercury, HgX_2, are also linear, covalent molecules (where X = Cl, Br, or I). Cadmium has two electrons in its 5*s* orbital, and its 5*p* orbitals are vacant. Similarly, mercury has two electrons in its 6*s* orbital, and its 6*p* orbitals are vacant. Thus, the possibility of *sp* hybridization exists in each metal and CdX_2 and HgX_2 (where both X's are identical) are additional examples of this kind of covalent bonding.

5–11 Trigonal Planar AB_3 Molecules with No Unshared Pairs of Electrons on A

Boron is a Group IIIA element that forms many covalent compounds by bonding to three other atoms. Typical examples include boron trifluoride, BF_3, boron trichloride, BCl_3, boron tribromide, BBr_3, and boron triiodide, BI_3. All are trigonal planar (that is, flat molecules in which all three bond angles are 120°), and all are nonpolar molecules because of symmetry.

The Lewis dot formula for boron trifluoride is derived from the facts that (a) each boron atom has three electrons in its valence shell and (b) each boron atom is bonded to three fluorine (or Cl, Br, I) atoms. In Example 5–7 we drew the Lewis dot formula for BCl_3. Since F and Cl are both members of Group VIIA, the dot formula for BF_3 should be similar.

$$\begin{array}{ccc} & :\ddot{F}: & \\ & B & \\ :\ddot{F}. & & .\ddot{F}: \end{array}$$

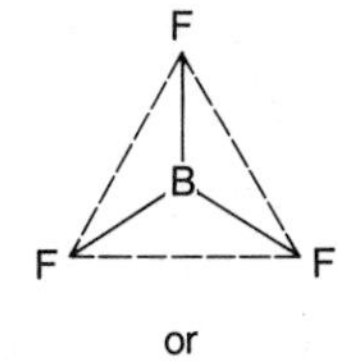

or

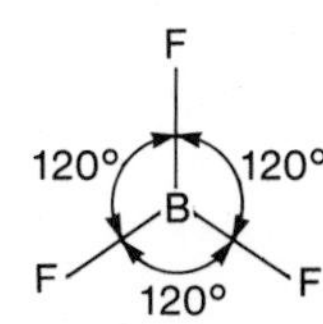

The solid lines represent bonds between B and F atoms. The dashed lines emphasize the shape of the molecules.

Note that BF_3 and other members of this class of compounds also involve central elements that do not attain a noble gas configuration by sharing electrons. Boron shares only six electrons.

1 VSEPR Theory

VSEPR theory predicts a **trigonal planar structure** for molecules such as BF_3 because this structure gives maximum separation among the three bonding electron pairs. There are no lone pairs of electrons associated with the boron atom. The maximum separation of any three items (electron pairs or F atoms) around a fourth item (B atom) is 120° angles in a single plane. The structures of BCl_3, BBr_3, and BI_3 are similar.

Examination of the bond dipoles of BF_3 shows that the electronegativity difference (Table 4–6) is very large (2.0 units) and that the bonds are very polar.

$$\begin{array}{rl} & \text{B—F} \\ \text{EN} = & \underbrace{2.0 \quad 4.0} \\ \Delta(\text{EN}) = & \quad 2.0 \end{array}$$

The B^{3+} ion is so small (radius = 0.20 Å) that boron does not form simple ionic compounds.

However, the three bond dipoles, which are really electronic vectors, cancel to give nonpolar molecules.

2 Valence Bond Theory

To be consistent with experimental findings and the predictions of VSEPR theory, the VB theory must explain three **equivalent** B—F bonds. Again, we invoke hybridization. In this case, the 2*s* orbital and two of the 2*p* orbitals of B must hybridize to form a set of three degenerate ***sp*² hybrid** orbitals.

Degenerate, in this context, refers to orbitals of the same energy.

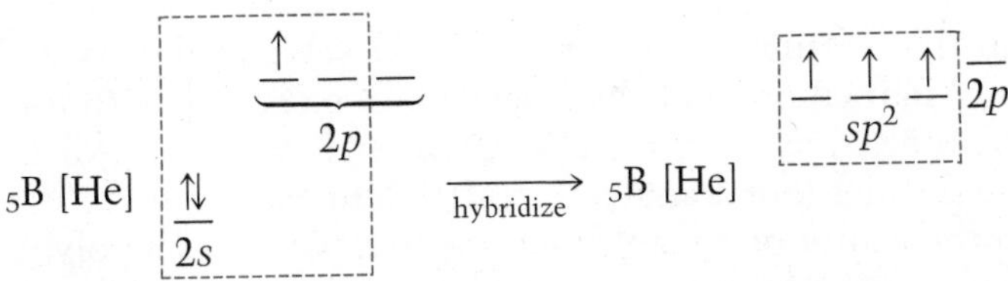

The three sp^2 hybrid orbitals are directed toward the corners of an equilateral triangle.

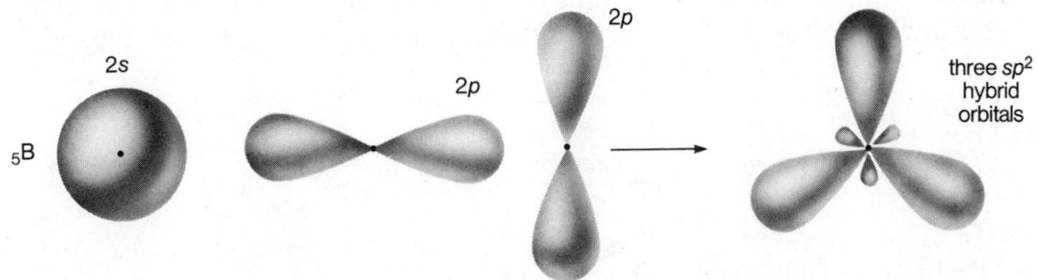

Lone pairs of e^- are not shown for F atoms.

The dots in the orbitals at the bottom of p. 175 represent the boron nucleus. Each of the three F atoms has one unpaired $2p$ electron, which can overlap the three sp^2 hybrid orbitals on boron. Three electron pairs are shared between one boron and three fluorine atoms

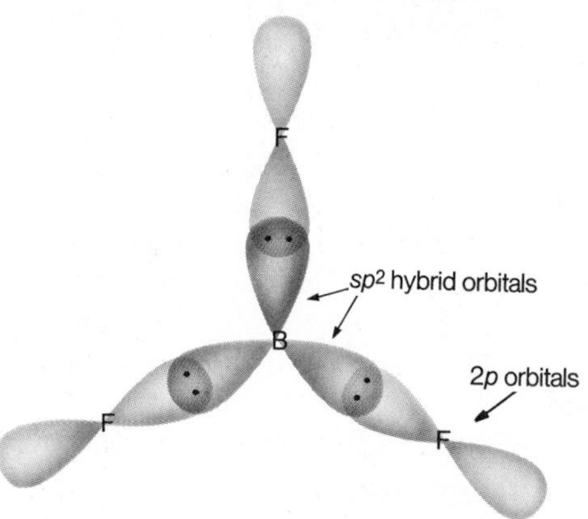

To generalize, *sp^2 hybridization* occurs at the central atom whenever there are *three* regions of high electron density around the central atom. As we shall see in Section 11–12, many molecules that have fewer than eight electrons in the valence shell of the central element frequently react as **Lewis acids.** *Lewis acids* are substances that can share in pairs of electrons contributed wholly by other species. Both beryllium chloride, $BeCl_2$, and boron trifluoride, BF_3, react as Lewis acids. The fact that both compounds so readily take a share of additional pairs of electrons provides further evidence for our earlier assumptions that beryllium and boron atoms do not acquire octets of electrons in gaseous $BeCl_2$ and BF_3.

5–12 Tetrahedral AB_4 Molecules with No Unshared Pairs of Electrons on A

The Group IVA elements have four electrons in their highest energy levels, and they form numerous covalent compounds by sharing four electrons with four other atoms. Typical examples include CH_4, CF_4, CCl_4, SiH_4, and SiF_4. All are tetrahedral, nonpolar molecules (bond angles = 109°28′). In each of them, the IVA atom is located in the center of a regular tetrahedron, which is a polyhedron with four equal-sized equilateral triangular faces. The other four atoms are located at the four corners of the tetrahedron.

Remember that in Lewis dot formulas, the emphasis is on pairs of valence electrons and *not* on the geometries of molecules. The fact that these structures have been drawn as planar is a matter of convenience and indicates nothing about their geometries, which are tetrahedral.

The Group IVA elements contribute four electrons in tetrahedral AB_4 molecules, and the other four atoms contribute one electron each. The Lewis dot formulas for methane, CH_4, and carbon tetrafluoride, CF_4, are typical

```
                     ..
   H                :F:
   ..             .. .. ..
 H:C:H           :F:C:F:
   ..             .. .. ..
   H                :F:
                     ..
```

CH_4 methane

CF_4 carbon tetrafluoride

1 VSEPR Theory

VSEPR theory predicts tetrahedral structures for AB_4 molecules that have no unshared electrons on A, because that shape affords the maximum separation for four electron pairs around one atom. Thus, the four electron pairs are directed toward the corners of a regular tetrahedron.

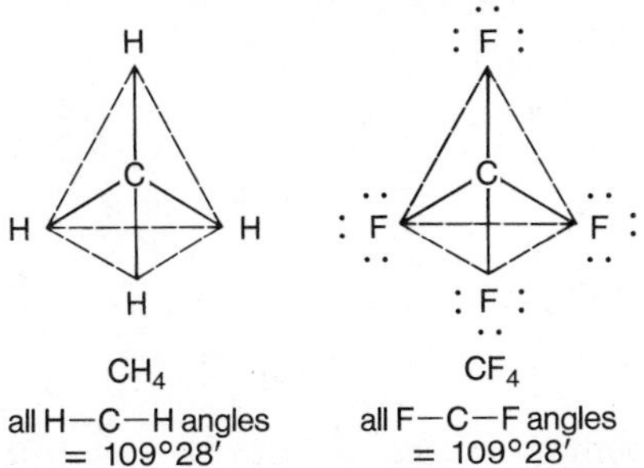

You may wonder whether square planar AB_4 molecules exist. They do, in compounds of the transition metals in which there are unshared pairs of electrons on the metal atom; but there are no *simple* square planar AB_4 molecules with no unshared electron pairs on A. The bond angles in square planar molecules are only 90°, and nearly all AB_4 molecules are tetrahedral with larger bond angles (109°28′) and therefore greater separation of valence electron pairs around A.

Examination of bond dipoles shows that in CH_4 the individual bonds are only slightly polar, while in CF_4 the bonds are quite polar.

	CH_4		CF_4
	C—H		C—F
EN =	2.5 2.1	EN =	2.5 4.0
Δ(EN) =	0.4	Δ(EN) =	1.5

In CH_4, the bond dipoles are directed toward carbon, while in CF_4 they are directed away from carbon. Because both molecules are completely symmetrical, the bond dipoles cancel and both molecules are nonpolar. Similar statements can be made about *all tetrahedral* AB_4 molecules in which there are *no unshared electron pairs on the central element.*

2 Valence Bond Theory

According to Valence Bond theory, each Group IVA atom (carbon in our example) must make available for bonding four equivalent orbitals. To do this, carbon forms four **sp^3 hybrid orbitals** in this bonding environment by mixing the *s* and all three *p* orbitals in its outer shell. This results in four unpaired electrons

$$_6\text{C [He]} \quad \underset{2s}{\underline{\uparrow\downarrow}} \quad \underbrace{\underline{\uparrow}\ \underline{\uparrow}\ \underline{\ \ }}_{2p} \quad \xrightarrow[\text{hybridize}]{} \quad _6\text{C [He]} \quad \underbrace{\underline{\uparrow}\ \underline{\uparrow}\ \underline{\uparrow}\ \underline{\uparrow}}_{sp^3}$$

These orbitals are directed toward the corners of a regular tetrahedron, which has 109°28′ angles from (any) corner to center to (any) corner.

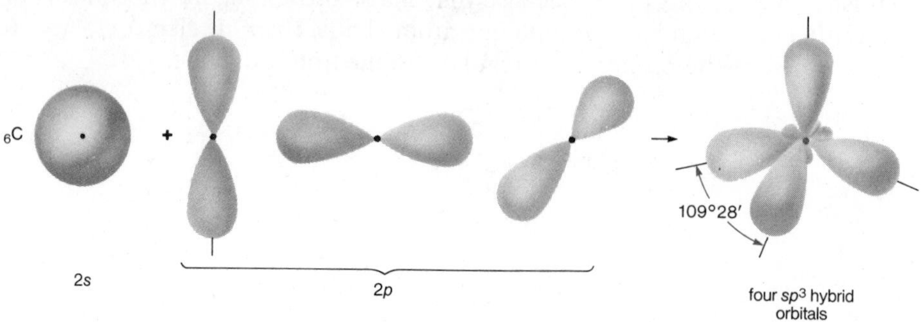

Each of the four atoms that bond to carbon possesses a half-filled atomic orbital that can overlap the half-filled sp^3 hybrids, as illustrated below for methane and carbon tetrafluoride.

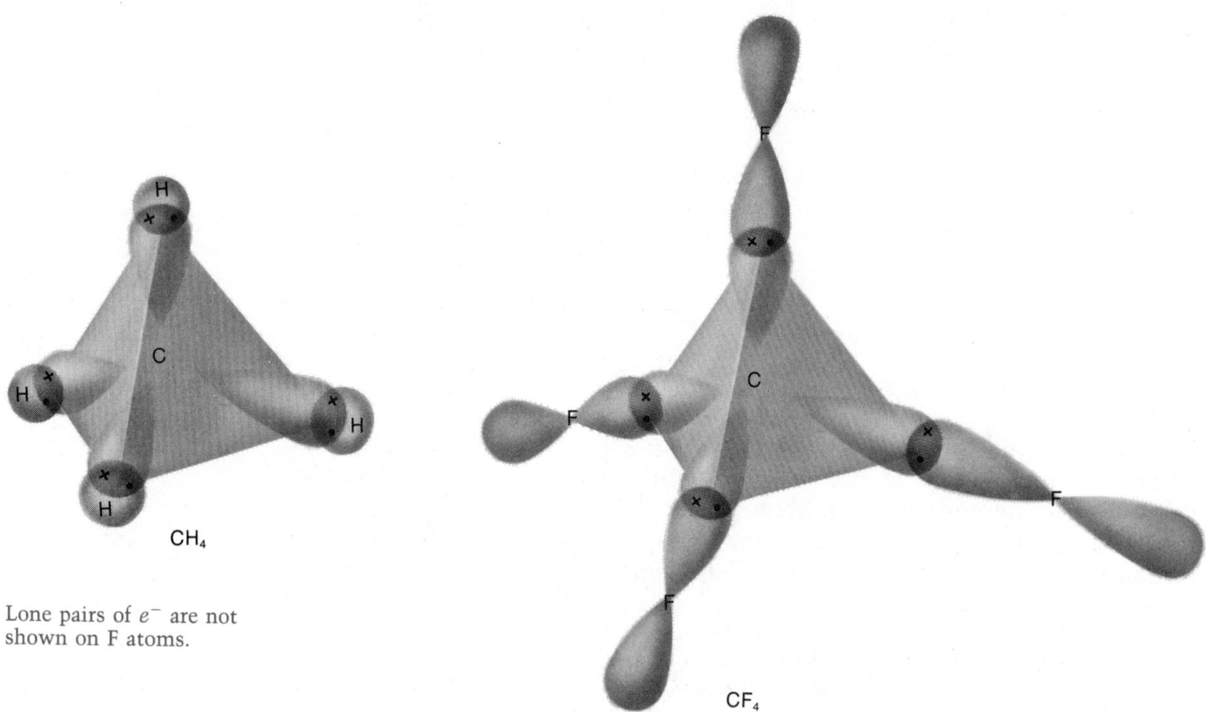

Lone pairs of e^- are not shown on F atoms.

In general, *sp^3 hybridization* occurs at the central atom of a molecule when there are *four* regions of high electron density around that atom.

5–13 Pyramidal AB_3 Molecules with One Unshared Pair of Electrons on A

Some Group VA elements also form covalent compounds by sharing all five valence electrons, as we shall see in Section 5–16.

The Group VA elements have five electrons in their valence shell, and they form some covalent compounds by sharing three electrons with three other atoms. For the moment, we'll restrict our discussion to two examples, ammonia, NH_3, and nitrogen trifluoride, NF_3. Both are pyramidal, polar molecules with an unshared electron pair on the nitrogen atom.

Since each nitrogen atom has five electrons in its highest energy level and is bonded to three other atoms in both NH_3 and NF_3 the Lewis dot formulas are

```
        ..                  .. .. ..
NH3  H:N:H        NF3   :F:N:F:
        ..                  .. .. ..
        H                     :F:
                               ..
```

1 VSEPR Theory

As in the previous examples, VSEPR theory predicts that the four electron pairs will be directed toward the corners of a tetrahedron, since this gives maximum separation. Thus both molecules have tetrahedral **electronic geometry.**

At this point we must be very careful to distinguish between **electronic geometry** and **molecular geometry.** *Electronic geometry* refers to *the geometric arrangement of valence shell electrons around the central atom. Molecular geometry* refers to *the arrangement of atoms (that is, nuclei), not just pairs of electrons, around the central atom.* For example, CH_4, CF_4, NH_3, and NF_3 all have tetrahedral electronic geometry, while only CH_4 and CF_4 have tetrahedral molecular geometry; NH_3 and NF_3 have pyramidal molecular geometry.

As we have seen, the term lone pair refers to a pair of electrons that is associated with (and move under the influence of) only one nucleus. The known geometries of numerous molecules and polyatomic ions, based on measurements of bond angles, indicate clearly that *lone pairs of electrons occupy more space than bonding pairs.* This is due to the fact that lone pairs have no other bonded atoms exerting strong attractive forces on them and they reside closer to the nucleus than do bonding electrons. These observations indicate that the relative magnitudes of the repulsive forces between pairs of electrons on an atom are as follows:

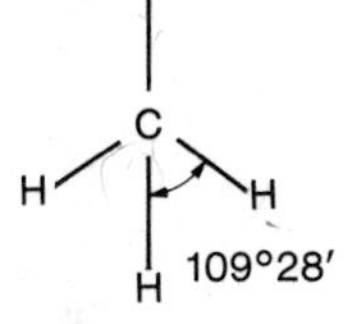

lp/lp >> lp/bp > bp/bp

where *lp* refers to lone pairs and *bp* refers to bonding pairs of valence shell electrons. From a geometric viewpoint, we are most concerned with the repulsions involving the electrons in the valence shell of the *central atom* of a molecule or polyatomic ion. The angles at which repulsive forces among valence shell electron pairs are exactly balanced are the angles at which the nuclei (and therefore bonding pairs and lone pairs) are found in covalently bonded molecules and polyatomic ions. Thus the bond angles in NH_3 and NF_3 are *less* than the angles of 109°28′ we observed in CH_4 and CF_4 molecules.

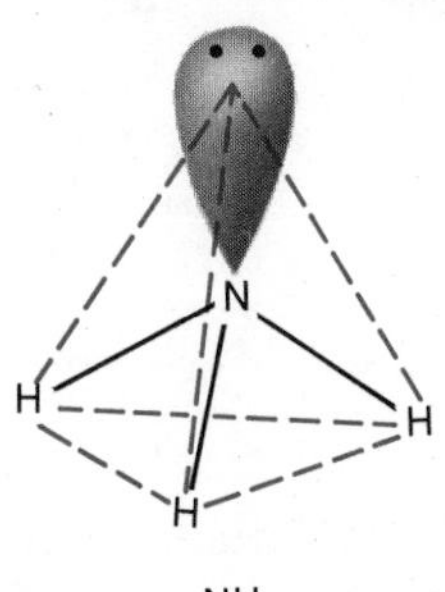

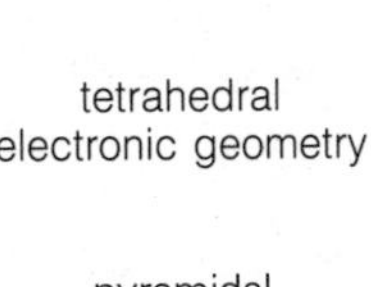

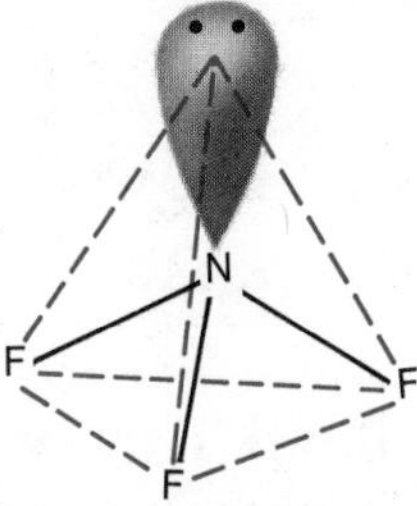

The heavy lines indicate chemical bonds while the dotted lines emphasize electronic geometry.

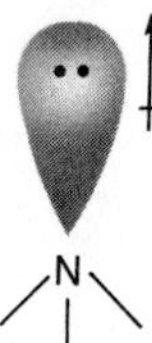

The formulas are frequently written as $:NH_3$ and $:NF_3$ to emphasize the lone pairs of electrons, which must be considered as the polarities of these molecules are examined. They are an extremely important factor! The contribution of each lone pair can be depicted as shown in the margin.

The electronegativity differences in NH_3 and NF_3 are nearly equal, *but* the resulting nearly equal bond polarities are in opposite directions.

	NH_3			NF_3	
	N—H			N—F	
EN	3.0 2.1	$\overset{\leftarrow\!\!+}{\text{N—H}}$	EN	3.0 4.0	$\overset{+\!\!\rightarrow}{\text{N—F}}$
$\Delta(\text{EN}) =$	0.9		$\Delta(\text{EN}) =$	1.0	

Thus, we have:

In NH_3 the bond dipoles **re-enforce** the effect of the unshared pair, so NH_3 is very polar ($\mu = 1.5$ D). In NF_3 the bond dipoles **oppose** the effect of the unshared pair, so NF_3 is only slightly polar ($\mu = 0.2$ D).

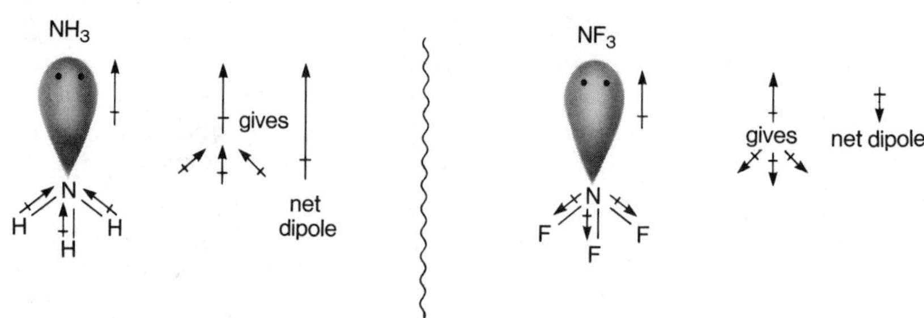

We can now use this information to explain the bond angles observed in NF_3 and NH_3. Because of the direction of the bond dipoles in NH_3 the electron-rich end of each N—H bond is at the central atom, N. In NF_3, on the other hand, the fluorine end of each bond is the electron-rich end. As a result, the lone pair can more closely approach the N in NF_3 than in NH_3. Therefore, the lone pair exerts greater repulsion toward the bonded pairs in NF_3 than in NH_3. The net effect is that the lone pair is more successful in reducing the bond angles in NF_3 than in NH_3. We might loosely represent the situation as follows

We might expect the larger fluorine atoms ($r = 0.64$ Å) to repel each other more strongly than hydrogen atoms ($r = 0.30$ Å), leading to larger bond angles in NF_3 than in NH_3. However, this is not the case.

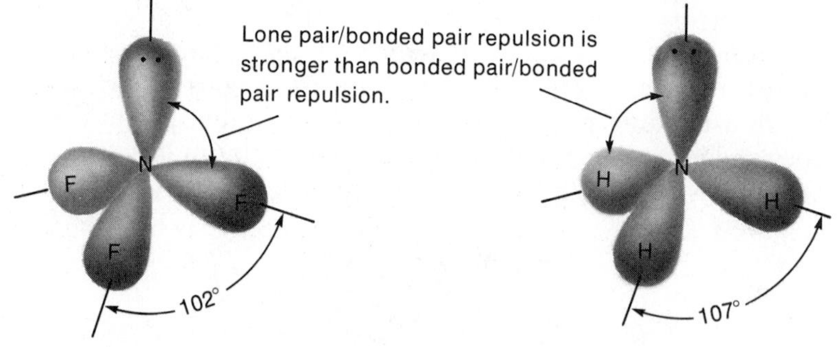

2 Valence Bond Theory

Since experimental results suggest four nearly equivalent orbitals (three involved in bonding, a fourth to accommodate the lone pair), we again need four ***sp*^{3} hybrid orbitals**

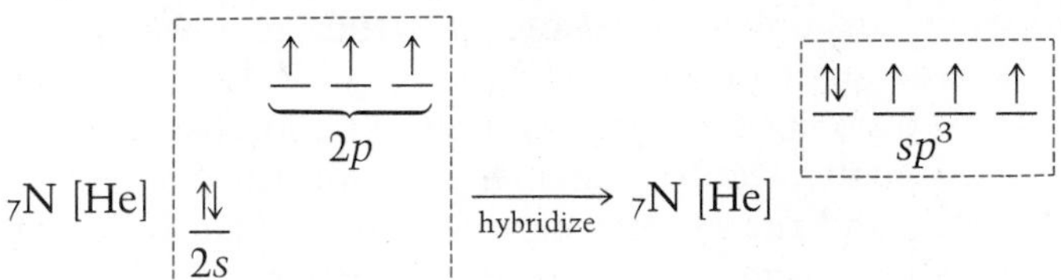

In both NH_3 and NF_3, the unshared pair of electrons occupies one of the sp^3 hybrid orbitals. Each of the other three sp^3 orbitals participates in electron sharing with another atom. They overlap with half-filled H 1*s* orbitals or F 2*p* orbitals in NH_3 and NF_3, respectively.

As we shall see in Section 11–12, many compounds that have an unshared electron pair on the central element are called **Lewis bases.** *Lewis bases react by making available an electron pair that can be shared by other species.*

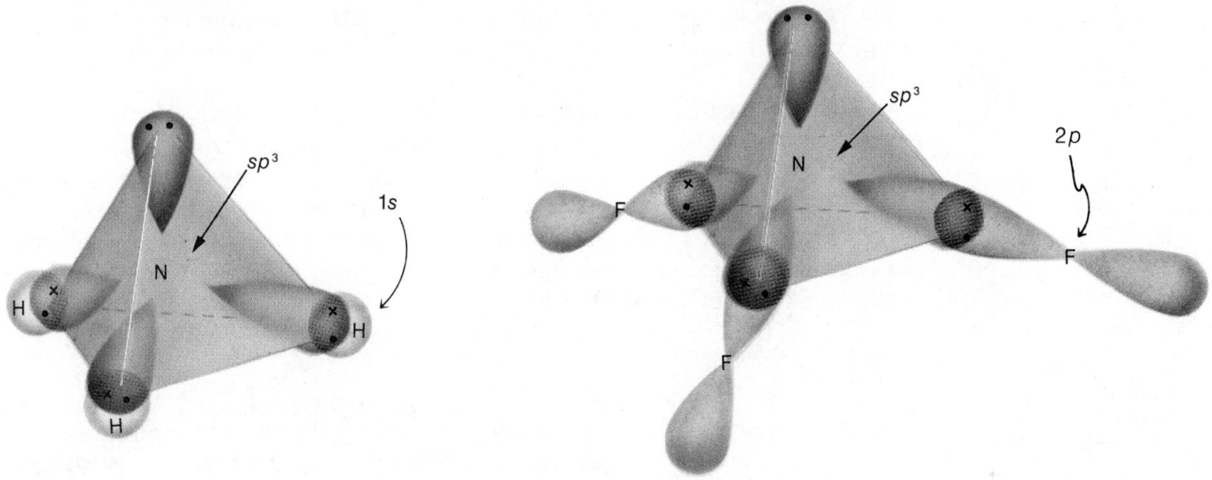

5–14 Angular AB_2 Molecules with Two Unshared Pairs of Electrons on A

The Group VIA elements have six electrons in their highest energy levels, and they form many covalent compounds by acquiring a share in two additional electrons from two other atoms. Typical examples include H_2O, H_2S, N_2O, and Cl_2O. All are angular, polar molecules. Let's consider the structure of water in detail. The bond angle in water is 104.5°, and the molecule is very polar.

Let us draw Lewis dot formulas for H_2O and H_2S.

Remember that the "shape" of a dot formula does not indicate geometry.

```
   ..       ..
 H:O:     H:S:
   ..       ..
   H        H
```

1 VSEPR Theory

VSEPR theory predicts that the four electron pairs (six electrons from oxygen and two from hydrogens) around the oxygen atom in H_2O should be 109°28′ apart in a tetrahedral arrangement. When increased repulsions between unshared pairs and bonding electron pairs are taken into consideration,

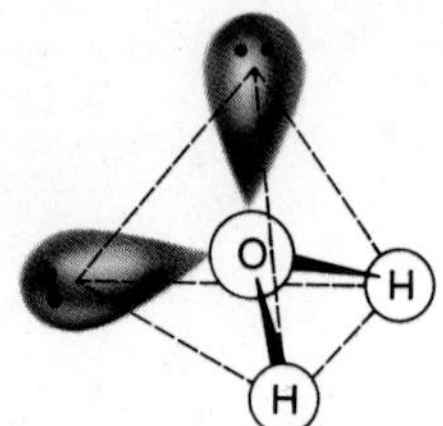
O
H
H

this theory satisfactorily explains the observed angular structure of water molecules and the observed bond angle of only 104.5°.

The electronegativity difference is large (1.4 units) and therefore the bonds are quite polar. Additionally, the bond dipoles reinforce the effect of the two unshared electron pairs, and the H_2O molecule is very polar. Its dipole moment is 1.7 D. As we shall see, water is an unusual substance, and its abnormal properties can be explained in part by its high polarity.

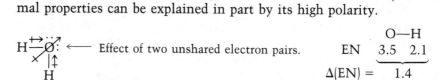

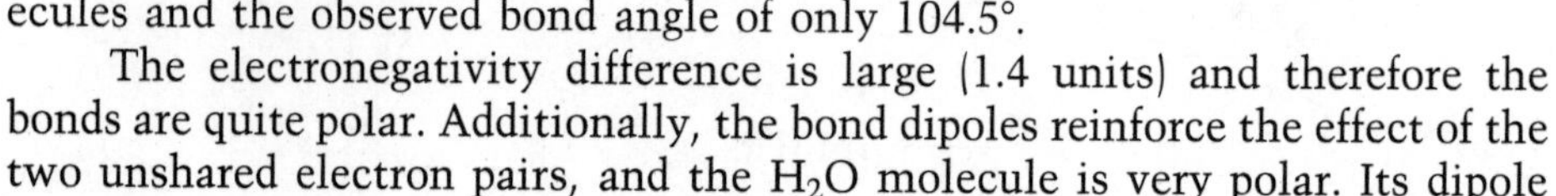

H—Ö: ⟵ Effect of two unshared electron pairs.
H

	O—H	
EN	3.5	2.1
Δ(EN) =	1.4	

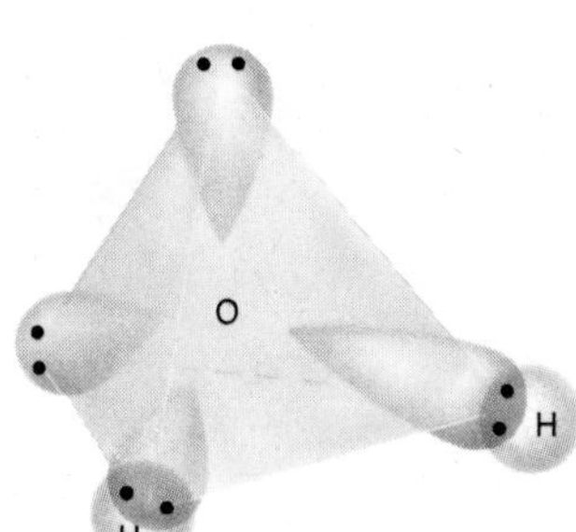
O
H
H

2 Valence Bond Theory

In light of the experimental evidence on the bond angle in H_2O, the Valence Bond theory postulates four ***sp*³ hybrid orbitals** centered on the oxygen atom: two to participate in bonding, and two to accommodate the two lone pairs.

Again, we can explain the observed bond angle of 104.5° easily. The expected bond angle for sp^3 hybridization (tetrahedral electronic geometry) is 109°28′. However, the two unshared pairs repel each other and the bonding pairs of electrons strongly. These repulsions force the bonding pairs closer together and result in the decreased bond angle. Since oxygen is very electronegative, the shared electron pairs are shifted toward the O ends of the O—H bonds, and the decrease in the H—O—H bond angle (from 109°28′ to 104.5°) is greater than the corresponding decrease in the H—N—H bond angles in ammonia.

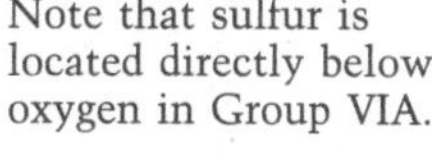

Note that sulfur is located directly below oxygen in Group VIA.

Hydrogen sulfide, H_2S, is also an angular molecule, but the H—S—H bond angle is only 92°. Therefore, we need not postulate hybrid orbitals. The two hydrogen atoms are able to exist at approximately right angles to each other (close to the 90° angles between two unhybridized $3p$ orbitals of sulfur) when they are bonded to the larger sulfur atom.

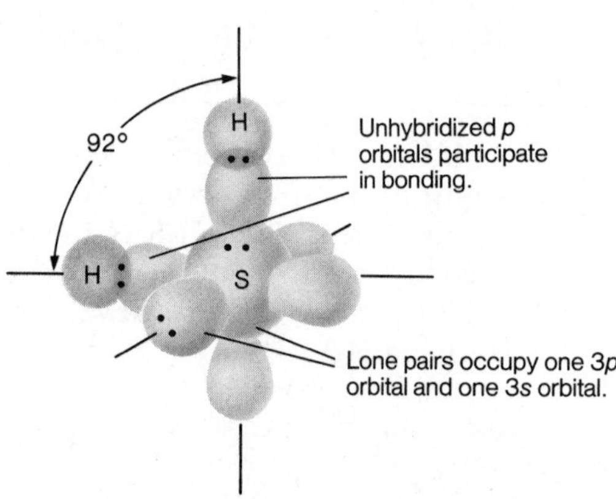
92°
H
H
S
Unhybridized p orbitals participate in bonding.
Lone pairs occupy one 3p orbital and one 3s orbital.

5–15 Linear AB Molecules with Three Unshared Pairs of Electrons on A

The Group VIIA elements have seven electrons in their highest energy levels. They form covalent compounds such as H—F, H—Cl, H—Br, and H—I by sharing one electron with another atom which contributes one electron. The Lewis dot formula for HF is

H:F:

All diatomic molecules are of necessity linear. Neither VSEPR theory nor Valence Bond theory adds anything to what we already know about the molecular geometry of HF, HCl, HBr, and HI: Overlap of any orbitals on *two atoms* produces *linear* molecules.

5–16 Trigonal Bipyramidal AB_5 Molecules with No Unshared Pairs of Electrons on A

In Section 5–13 we pointed out the fact that the Group VA elements have five electrons in their outermost shell and form some covalent molecules by sharing only three of these electrons with other atoms (for example, NH_3, NF_3). Other Group VA elements (such as P, As, and Sb) form some covalent compounds by sharing all five of their valence electrons with five other atoms. Phosphorus pentafluoride, PF_5, is such a compound. Each phosphorus atom has five valence electrons to share with five fluorine atoms, and the Lewis dot formula for PF_5, shown in the margin, is analogous to that of AsF_5, which we drew in Example 5–8. Phosphorus pentafluoride molecules are trigonal bipyramidal, nonpolar molecules. A *trigonal bipyramid* is a six-sided polyhedron consisting of two pyramids joined at a common triangular (trigonal) base.

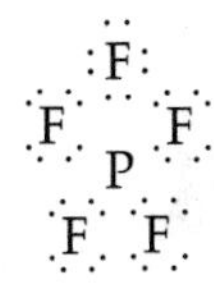

1 VSEPR Theory

VSEPR theory predicts that the five electron pairs around the phosphorus atom in PF_5 should be as far apart as possible. Maximum separation of five items (F atoms) around a sixth item (P atom) is achieved when the five items are placed at the corners and the sixth item in the center of a trigonal bipyramid. This is an agreement with experimental observation.

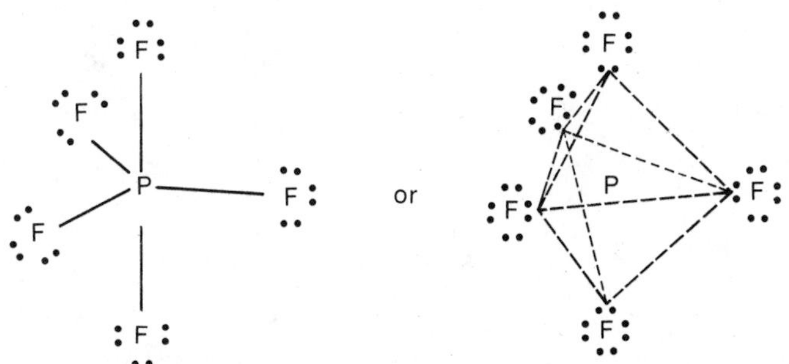

Note that the three F atoms at the corners of the equilateral triangle lie in the same plane as the P atom. These are called *equatorial* F atoms. The other two F atoms, one above the plane and one below the plane, are called *axial* F atoms; the axial P—F bonds are longer than the equatorial P—F bonds. Examination of the trigonal bipyramidal molecule indicates that the F—P—F bond angles are 90° (axial to equatorial), 120° (equatorial to equatorial), and 180° (axial to axial).

The large difference in electronegativity (1.9) indicates very polar bonds. Let's consider the bond dipoles in two groups, because there are two different kinds of P—F bonds in PF_5 molcules as shown by the different P—F bond lengths

As an exercise in geometry, how many ways can five fluorine atoms be arranged symmetrically around a phosphorus atom? Compare the hypothetical bond angles in such arrangements with those in a trigonal bipyramid.

	P—F
EN	2.1 4.0
Δ(EN) =	1.9

Axial bonds: F—P—F, 180°

Equatorial bonds: three F atoms around P, 120°

The two axial bond dipoles cancel each other, and the three equatorial bond dipoles cancel, so the PF_5 molecules are nonpolar.

2 Valence Bond Theory

Since phosphorus is the central element in PF_5 molecules, it must have available five half-filled orbitals to form bonds with five fluorine atoms. Hybridization is again the answer, this time using one *d* orbital from the vacant set of 3*d* orbitals in addition to the 3*s* and 3*p* orbitals of the phosphorus atom.

$_{15}$P [Ne] $\frac{\uparrow\downarrow}{3s}$ $\frac{\uparrow}{}\frac{\uparrow}{}\frac{\uparrow}{}$ (3*p*) _ _ _ _ _ (3*d*) $\xrightarrow{\text{hybridize}}$ $_{15}$P [Ne] $\frac{\uparrow}{}\frac{\uparrow}{}\frac{\uparrow}{}\frac{\uparrow}{}\frac{\uparrow}{}$ (sp^3d) _ _ _ _ (3*d*)

The five ***sp*^{3}*d* hybrid orbitals** are directed toward the corners of a trigonal bipyramid. Each is overlapped by the (only) 2*p* orbital of a fluorine atom that contains a single electron. The resulting pairing of P and F electrons forms a total of five covalent bonds.

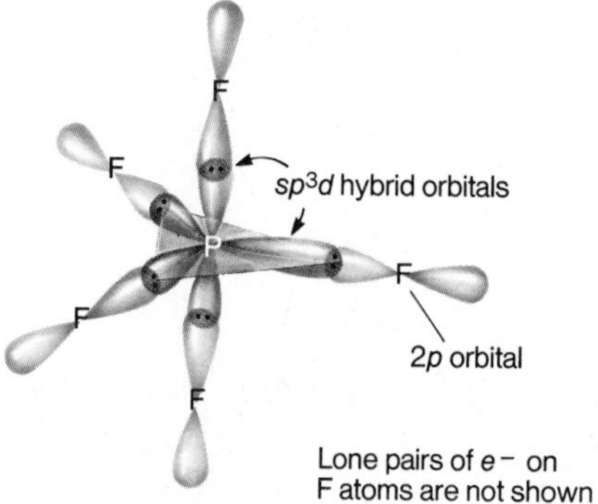

We should note that sp^3d hybridization involves utilization of an available *d* orbital in the outermost shell of the central atom, P. We mentioned at the beginning of this section that the heavier Group VA elements, P, As, and Sb, can form *five-coordinate compounds* utilizing this hybridization. But it is *not* possible for nitrogen (also in Group VA) to form such five-coordinate compounds. Two facts are important in understanding why: (1) nitrogen is too small to accommodate five (even very small) substituents without excessive crowding which causes instability, and (2) it has no low-energy *d* orbitals. In fact, we can generalize: no elements of the second period can be central elements in five-coordinate molecules (or higher-coordinate ones) because they have no low-energy *d* orbitals available for hybridization, and because they are too small. It is important to realize that the set of *s* and *p* orbitals in a given energy level, and therefore any set of hybrids composed only of them, can accommodate a *maximum* of eight electrons and participate in a *maximum* of four covalent bonds.

5–17 Octahedral AB_6 Molecules with No Unshared Pairs of Electrons on A

The heavier Group VIA elements form some covalent compounds of the AB_6 type by sharing their six electrons with six other atoms. Sulfur hexafluoride, SF_6, an unreactive gas, is an example. Sulfur hexafluoride molecules are nonpolar, octahedral molecules. An octahedron has eight faces, each of which is an equilateral triangle.

There are six shared electron pairs around the sulfur atom in SF_6 molecules, and the Lewis dot formula for SF_6 is shown in the margin.

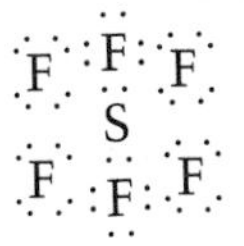

1 VSEPR Theory

In SF_6 molecules we have six electron pairs and six F atoms surrounding one S atom. Because there are no lone pairs in the valence shell of sulfur, the electronic and molecular geometries will be identical. The maximum separation possible for six F atoms surrounding one S atom is achieved when the F atoms are situated at the corners and the S atom in the center of a regular octahedron. Thus, VSEPR theory is consistent with the observation that SF_6 molecules are octahedral.

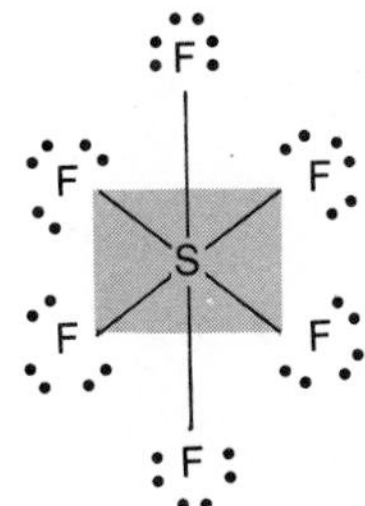

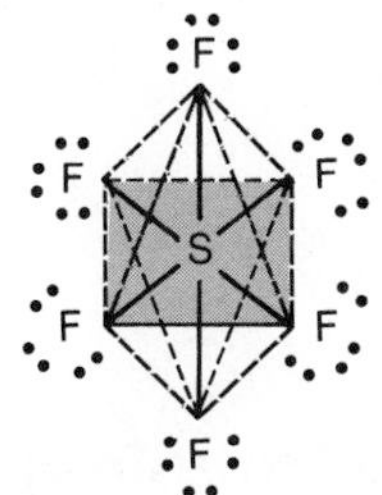

Examination of the octahedral molecule reveals that the F—S—F bond angles are 90° and 180° and the bonds are quite polar. Because the molecule is symmetrical, the large bond dipoles cancel, and the molecule is nonpolar. To see this, separate the bond dipoles into two groups as we did in the trigonal bipyramid. Notice, though, that all six bonds in such an octahedral molecule are *equivalent;* if you turn the molecule so that an "equatorial" fluorine becomes "axial," the molecule still looks exactly the same.

	S—F		
EN	2.5	4.0	
$\Delta(\text{EN}) =$	1.5		

F
$\updownarrow$
S
$\updownarrow$
F

axial bonds

F F
S
F F

equatorial bonds

2 Valence Bond Theory

The sulfur atoms can form six hybrid orbitals to accommodate six electron pairs, and in Valence Bond theory terminology, we have

$$_{16}\text{S [Ne]} \ \underset{3s}{\underline{\uparrow\downarrow}} \ \underset{3p}{\underline{\uparrow\downarrow}\ \underline{\uparrow}\ \underline{\uparrow}} \ \underset{3d}{\underline{\ }\ \underline{\ }\ \underline{\ }\ \underline{\ }\ \underline{\ }} \xrightarrow{\text{hybridize}} {}_{16}\text{S [Ne]} \ \underset{sp^3d^2}{\underline{\uparrow}\ \underline{\uparrow}\ \underline{\uparrow}\ \underline{\uparrow}\ \underline{\uparrow}\ \underline{\uparrow}} \ \underset{3d}{\underline{\ }\ \underline{\ }\ \underline{\ }}$$

The six sp^3d^2 **hybrid orbitals** are directed toward the corners of a regular octahedron. Each sp^3d^2 hybrid orbital is overlapped by a half-filled $2p$ orbital from fluorine to form a total of six covalent bonds.

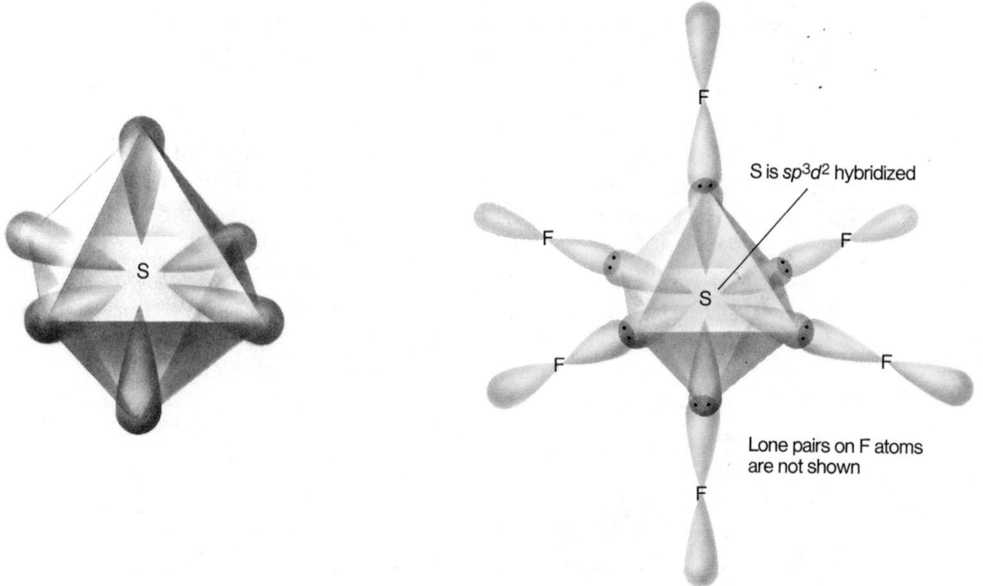

5-18 Compounds Containing Double Bonds

Early in this chapter we constructed dot formulas for some molecules and ions containing double and triple bonds, but we have not yet considered bonding and shapes for such species. Let us consider ethene (or ethylene), C_2H_4, as a specific example. Its dot formula is

$$S = N - A$$
$$24 - 12 = 12\ e^- \text{ shared}$$

H:C::C:H (each C bonded to two H) $H_2C{=}CH_2$

Because there are three regions of high electron density around each carbon atom, VSEPR theory tells us that each carbon is at the center of a trigonal plane.

It follows that Valence Bond theory tells us that each doubly bonded carbon atom is sp^2 hybridized with one electron in each sp^2 hybrid orbital and one electron in the unhybridized $2p$ orbital, which is perpendicular to the plane of the three sp^2 hybrid orbitals

$$
\begin{array}{lccccccccccc}
 & 1s & 2s & & 2p & & & 1s & & sp^2 & & 2p_z \\
{}_6\mathrm{C} & \underline{\uparrow\downarrow} & \underline{\uparrow\downarrow} & \underline{\uparrow} & \underline{\uparrow} & \underline{\;\;} & \longrightarrow & \underline{\uparrow\downarrow} & \underline{\uparrow} & \underline{\uparrow} & \underline{\uparrow} & \underline{\uparrow}
\end{array}
$$

Relative energies of atomic orbitals and hybridized orbitals are not indicated.

Recall that sp^2 hybrid orbitals are directed toward the corners of an equilateral triangle. Figure 5–4 shows top and side views of these hybrid orbitals.

The two carbon atoms interact by head-on (*end-to-end*) overlap of sp^2 hybrids pointing toward each other to form a **sigma** (σ) **bond** and by side-on overlap of the unhybridized $2p$ orbitals to form a **pi** (π) **bond.** The sigma and pi bonds together make a double bond. The $1s$ orbitals (with 1 e^- each) of four hydrogen atoms overlap the remaining four sp^2 orbitals (with 1 e^- each) on the carbon atoms to form four C—H sigma bonds.

A **sigma bond** *is a bond resulting from head-on overlap of atomic orbitals, in which the region of electron sharing is along and cylindrically around the imaginary line connecting the bonded atoms.* Most single bonds are sigma bonds, and we have seen that many kinds of pure atomic orbitals and hybridized orbitals can be involved in sigma bond formation. A **pi bond** *is a bond resulting from side-on overlap of atomic orbitals in which the regions of electron sharing are above and below the imaginary line connecting the bonded atoms, and parallel to this line.* A pi bond can form *only* if there is *also* a sigma bond between the same two atoms. A double bond consists of one sigma and one pi bond. A triple bond consists of one sigma and two pi bonds (Fig. 5–5).

As a consequence of the sp^2 hybridization of carbon atoms in carbon—carbon double bonds, each carbon atom is at the center of a trigonal plane. The fact that the p orbitals that overlap to form the π bond must be parallel to each other for effective overlap to occur adds the further restriction that these trigonal planes (sharing a common corner) must also be **coplanar.** Thus, all four atoms attached to the doubly bonded carbon atoms lie in the same plane (Figure 5–6). There are many other important examples of organic compounds containing carbon—carbon double bonds. Several are described in Chapter 29.

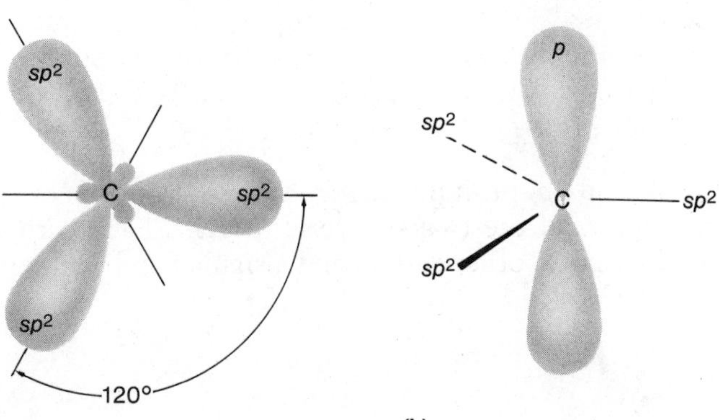

FIGURE 5–4 (a) A top view of three sp^2 hybrid orbitals. (b) A side view of a carbon atom in a trigonal planar environment, showing the remaining p orbital perpendicular to the plane of the three sp^2 hybrid orbitals.

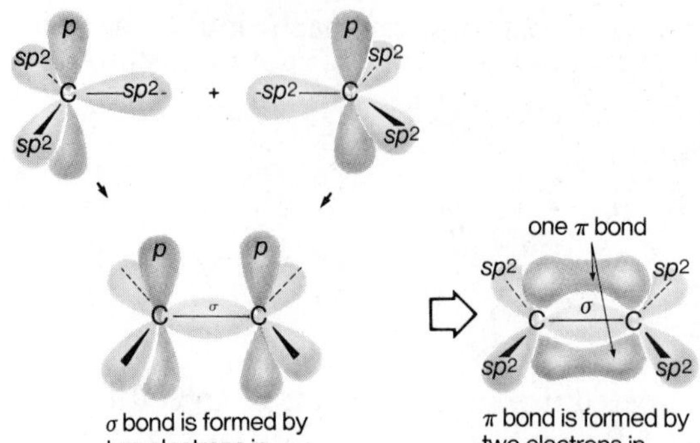

FIGURE 5–5 Schematic representation of the formation of a carbon—carbon double bond. Two sp^2 hybridized carbon atoms form a σ bond by overlap of two sp^2 orbitals and a pi (π) bond by overlap of properly aligned *p* orbitals.

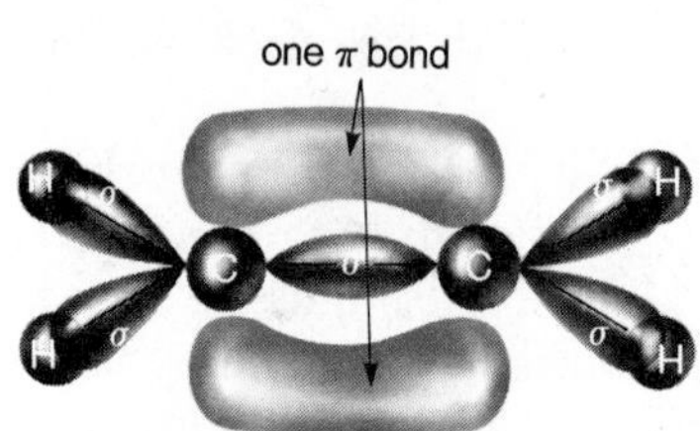

FIGURE 5–6 Four C—H σ bonds, one C—C σ bond, and one C—C π bond in the planar C_2H_4 molecule.

5–19 Compounds Containing Triple Bonds

A common compound containing a triple bond is ethyne (or acetylene), C_2H_2, which has the dot formula

$S = N - A$
$20 - 10 = 10\ e^-$ shared

H:C⋮⋮C:H H—C≡C—H

VSEPR theory requires that the two regions of high electron density around each carbon atom be 180° apart.

Valence Bond theory postulates that each triply bonded carbon atom is *sp*-hybridized (see Section 5–10) because each has two regions of high electron density. If we designate the p_x orbitals as the ones involved in hybridization, there is one carbon electron in each *sp* hybrid orbital and one electron each in the $2p_y$ and $2p_z$ orbitals (before bonding is considered). See Figure 5–7.

The three *p* orbitals in a set are indistinguishable. We label the one involved in hybridization as "p_x" to help us visualize the orientations of the two unhybridized *p* orbitals on carbon.

	1s	2s	2p				1s	sp		$2p_y$	$2p_z$
$_6C$	⇅	⇅	↑	↑	__	⟶	⇅	↑	↑	↑	↑

The unhybridized atomic $2p_y$ and $2p_z$ orbitals are *perpendicular* to each other and to a line through the centers of the two *sp* hybrid orbitals. Each carbon atom forms one sigma bond with the other carbon and a sigma bond with one hydrogen atom.

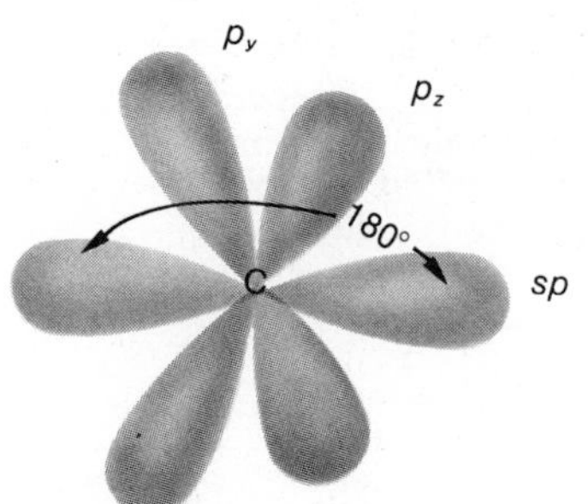

FIGURE 5–7 Diagram of the two linear hybridized *sp* orbitals of the carbon atom (in color) which lie in a straight line, and the two unhybridized *p* orbitals (grey).

Additionally, the two half-filled *p* orbitals on different carbon atoms overlap side-to-side, p_y with p_y and p_z with p_z, to form two pi bonds.

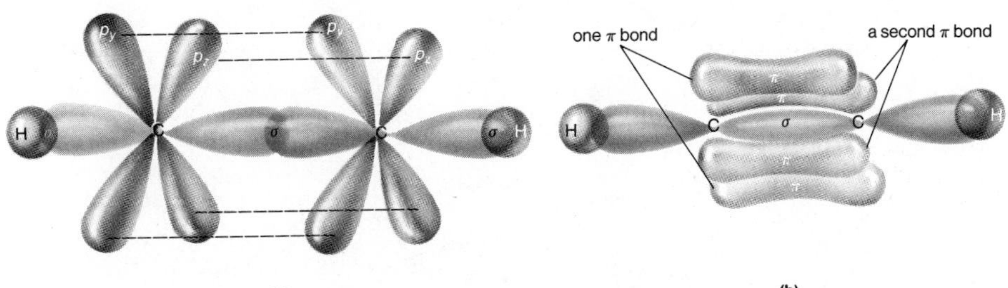

FIGURE 5–8 The acetylene molecule. (a) The overlap diagram of two *sp* hybridized carbon atoms and two *s* orbitals from two hydrogen atoms. There are two σ C—H bonds, one σ C—C bond, and two π C—C bonds (making the net carbon–carbon bond a triple bond). The dashed lines, each connecting two lobes, indicate the side-by-side overlap of the four unhybridized *p* orbitals. The hybridized *sp* orbitals are shown in color, and the unhybridized *p* orbitals are shown in grey. (b) The overall outline of the bonding orbitals in acetylene. The π bonding orbitals are positioned with one above and below the line of the σ bonds and the other behind and in front of the line of the σ bonds.

Since the *sp* hybrids are 180° apart and since *sp* hybrids on each carbon atom overlap head-on, the entire molecule must be linear.

Other molecules containing triply bonded atoms are nitrogen, :N≡N:, propyne, CH_3—C≡C—H, (the carbon at the left is sp^3 hybridized and at the center of a tetrahedron), and hydrogen cyanide, H—C≡N:. In each case both atoms involved in the triple bonds are *sp* hybridized and participate in one sigma and two pi bonds (in the triple bond). Since the carbon atom in carbon dioxide, :O=C=O:, must participate in two pi bonds (to two different oxygen atoms) in addition to two sigma bonds, it is also *sp* hybridized and the molecule is linear.

5–20 A Summary of Molecular and Electronic Geometries

We have discussed several common types of covalent molecules and provided a reasonable explanation for the observed structures and polarities of these mol-

TABLE 5-4 Electronic and Molecular Geometries

Number of Regions of High Electron Density*	Electronic Geometry	Hybridization at Central Atoms (A)	Hybridized Orbital Orientation	Examples	Molecular Geometry
2	linear	*sp* (180°)		$BeCl_2$ $HgBr_2$ CdI_2 CO_2† C_2H_2‡	linear linear linear linear linear
3	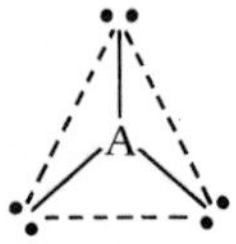 trigonal planar	*sp*^{2} (120°)	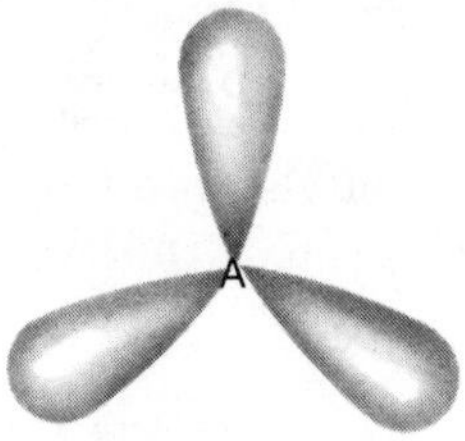	BF_3 BCl_3 SO_2§, ‖ C_2H_4**	trigonal planar trigonal planar angular planar (trig. planar at each C)
4	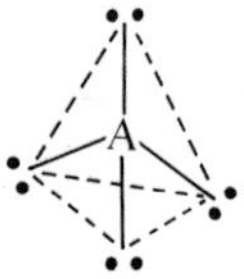 tetrahedral	*sp*^{3} (109°28′)	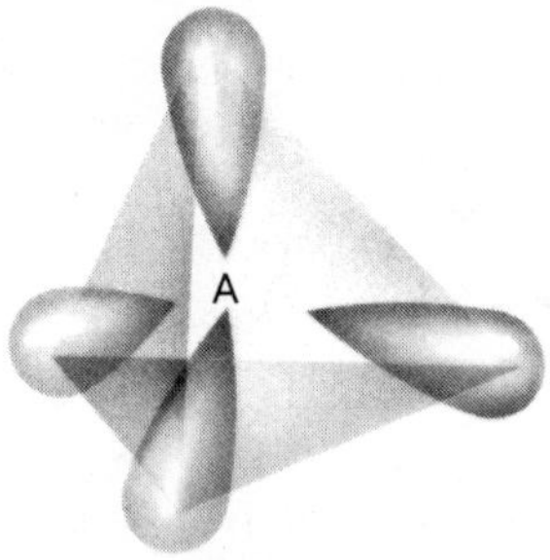	CH_4 CCl_4 $CHCl_3$ NH_3§ PF_3§ H_2O§	tetrahedral tetrahedral distorted tet. pyramidal pyramidal angular
5	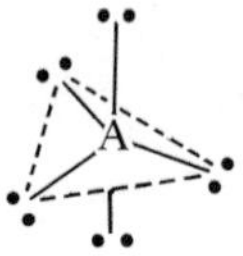 trigonal bipyramidal	*sp*^{3}*d* or *dsp*^{3} (90°, 120°, 180°)	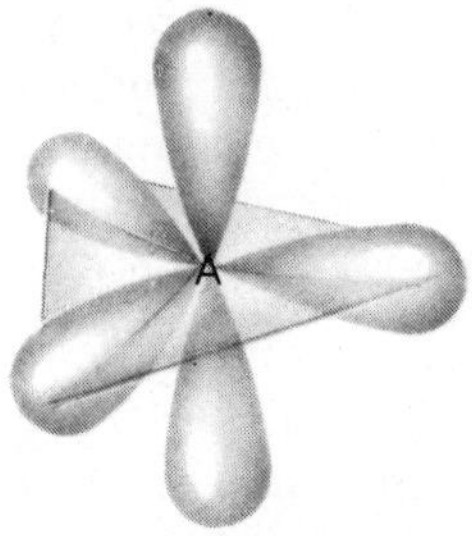	PF_5 $AsCl_5$ ICl_3§	trigonal bipyramidal trigonal bipyramidal T-shaped
6	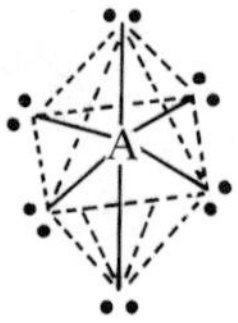 octahedral	*sp*^{3}*d*^{2} or *d*^{2}*sp*^{3} (90°, 180°)	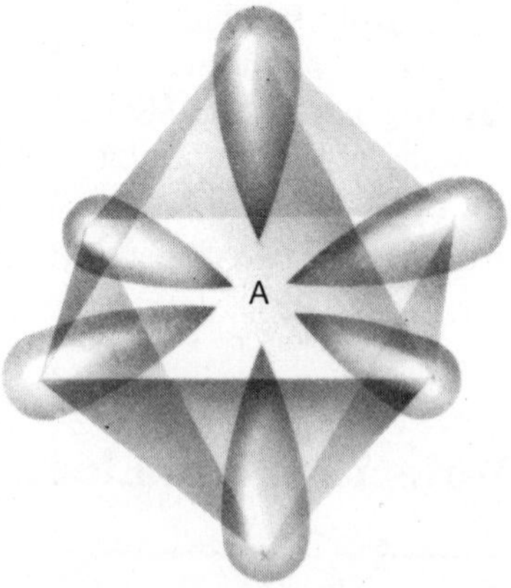	SF_6 SeF_6 XeF_4§	octahedral octahedral square planar

* The number of locations of high electron density around the central atom. A region of high electron density may be a single bond, a double bond, a triple bond, or a lone pair. These determine the electronic geometry, and thus hybridization at the central element. † Contains two double bonds. ‡ Contains a triple bond. § Central atom in molecule has lone pair(s) of electrons. ‖ Contains a resonant double bond. ** Contains one double bond.

ecules. We pointed out earlier that present knowledge of chemical bonding is far from what we would like it to be. However, as we have just learned, present theories are quite useful even if they are less than perfect. Table 5–4 includes a summation of the points developed thus far.

5–21 Geometries of Polyatomic Ions

In Section 5–4 we showed that Lewis dot formulas of ions could be constructed in the same way as dot formulas for neutral molecules, as long as the "extra" electrons of anions and "missing" electrons of cations are considered. Once the dot formula of an ion is known then we can use VSEPR and Valence Bond theories to deduce its electronic geometry, hybridization, and shape in the same way as for neutral molecules. Table 5–5 summarizes the structures of a few common ions.

TABLE 5–5 Electronic and Ionic Geometrics

Ion	Number of Regions of High Electron Density	Lewis Formula and Electronic Geometry	Hybridization of Central Atom	Hybridized Orbital Orientation	Ionic Geometry
nitrite, NO_2^- $S = N - A$ $= 24 - 18$ $= 6\ e^-$ shared	3	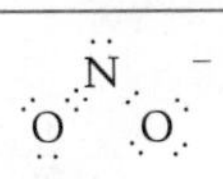 (2 resonance forms) trigonal planar 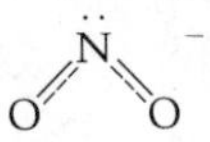	sp^2	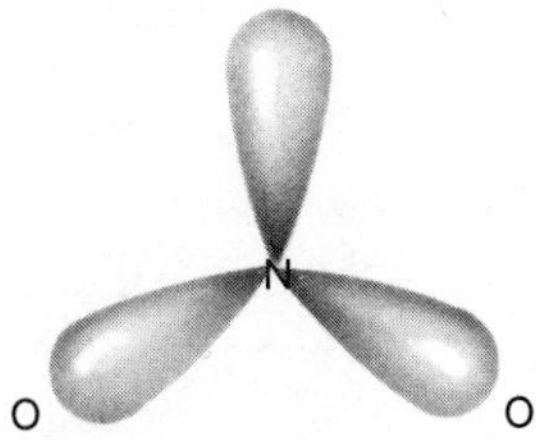 O N O	angular
nitrate, NO_3^- $S = N - A$ $= 32 - 24$ $= 8\ e^-$ shared	3	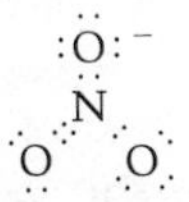 (3 resonance forms) trigonal planar 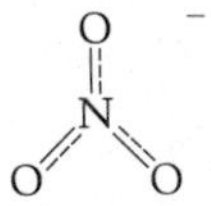	sp^2	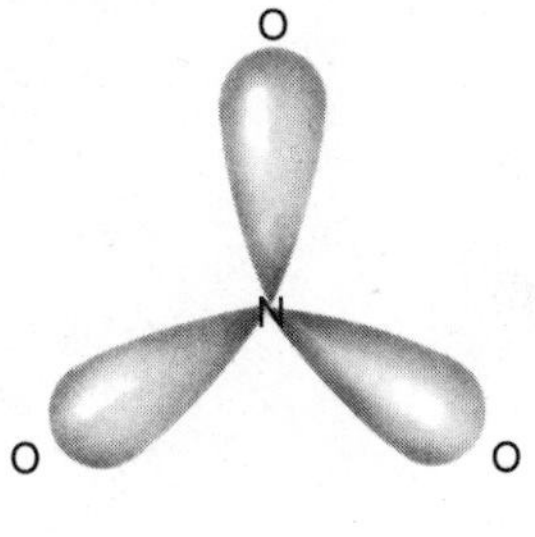 O N O O	trigonal planar
ammonium, NH_4^+ $S = N - A$ $= 16 - 8$ $= 8\ e^-$ shared	4	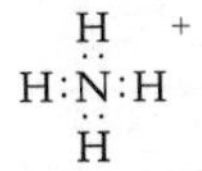 H:N:H with H above and below (+) tetrahedral H—N—H with H above and below (+) 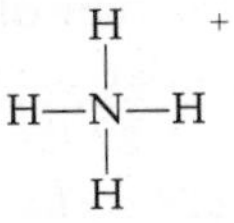	sp^3	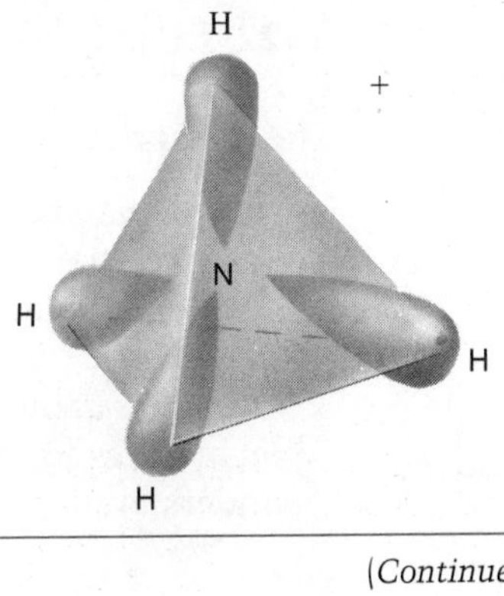 H N H H H +	tetrahedral

(Continued on next page.)

TABLE 5–5 Electronic and Ionic Geometrics *(continued)*

Ion	Number of Regions of High Electron Density	Lewis Formula and Electronic Geometry	Hybridization of Central Atom	Hybridized Orbital Orientation	Ionic Geometry
perchlorate, ClO_4^- $S = N - A$ $= 40 - 32$ $= 8\ e^-$ shared	4	$[:\ddot{O}:\ :\ddot{O}:Cl:\ddot{O}:\ :\ddot{O}:]^-$ tetrahedral $[:\ddot{O}:\ :\ddot{O}—Cl—\ddot{O}:\ :\ddot{O}:]^-$ (Cl bonded to four O)	sp^3	O, Cl, O, O, O; −	tetrahedral
sulfate, SO_4^{2-} $S = N - A$ $= 40 - 32$ $= 8\ e^-$ shared	4	$[:\ddot{O}:\ :\ddot{O}:S:\ddot{O}:\ :\ddot{O}:]^{2-}$ tetrahedral $[:\ddot{O}:\ :\ddot{O}—S—\ddot{O}:\ :\ddot{O}:]^{2-}$ (S bonded to four O)	sp^3	O, S, O, O, O; 2−	tetrahedral
sulfite, SO_3^{2-} $S = N - A$ $= 32 - 26$ $= 6\ e^-$ shared	4	$[:\ddot{O}:\ddot{S}:\ddot{O}:\ :\ddot{O}:]^{2-}$ tetrahedral $[:\ddot{O}—\ddot{S}—\ddot{O}:\ :\ddot{O}:]^{2-}$ (S bonded to three O)	sp^3	S, O, O, O; 2−	pyramidal

5–22 Final Words

The discussion of covalent bonding illustrates three important points.

1. Molecules and ions containing more than one atom have definite shapes.
2. The properties of molecules and ions are determined to a large extent by their shapes and *especially by the presence of incompletely filled electron shells or unshared pairs of electrons on the central element.*
3. Our ideas about chemical bonding have been developed over many years. As techniques for determining the *structures* of molecules have improved, our understanding of chemical bonding has improved also. Experimental obser-

vations on molecular geometry provide support for our ideas about chemical bonding. The ultimate test for any theory is: Can it predict correctly the results of experiments before they are done? When the answer is yes, we have confidence in the theory. When the answer is no, clearly the theory must be modified. Current theories of chemical bonding enable us to make predictions which are *usually* accurate.

There are other kinds of molecules and other theories of chemical bonding. They will be introduced in Chapters 6, 9, and 27.

Naming Inorganic Compounds

The rules for naming inorganic compounds were set down in 1957 by the Committee on Inorganic Nomenclature of the International Union of Pure and Applied Chemistry (IUPAC). To make naming easier we classify simple compounds in two major categories. These are **binary compounds,** those consisting of two elements, and **ternary compounds,** those consisting of three elements.

5–23 Binary Compounds

Binary compounds may be either ionic or covalent. In both cases the general rule is to name the less electronegative element first and the more electronegative element second. The more electronegative element is named by adding an "-ide" suffix to the element's characteristic (unambiguous) stem which is derived from the name of the element. Stems for the *nonmetals* are given below:

Because millions of compounds are known, it is important to be able to associate names and formulas in a systematic way.

IIIA		IVA		VA		VIA		VIIA	
								H	hydr
B	bor	C	carb	N	nitr	O	ox	F	fluor
		Si	silic	P	phosph	S	sulf	Cl	chlor
				As	arsen	Se	selen	Br	brom
				Sb	antimon	Te	tellur	I	iod

Binary ionic compounds contain metal cations and nonmetal anions. The cation is named first and the anion second according to the rule above. Some examples are:

Formula	Name	Formula	Name
KBr	potassium bromide	Rb_2S	rubidium sulfide
$CaCl_2$	calcium chloride	Al_2Se_3	aluminum selenide
NaH	sodium hydride	SrO	strontium oxide

The above method is sufficient for naming binary ionic compounds containing metals that exhibit only one oxidation number, other than zero (Section 4–12). However, most transition elements, and a few of the more electronegative representative metals, exhibit more than one oxidation number. These metals can form two or more binary compounds with the same nonmetal. In order to distinguish among all possibilities the oxidation number of the metal is indicated by a Roman numeral in parentheses following its name. Some typical examples follow.

Formula	Oxidation Number of Metal	Name	Formula	Oxidation Number of Metal	Name
Cu_2O	+1	copper(I) oxide	$SnCl_2$	+2	tin(II) chloride
CuF_2	+2	copper(II) fluoride	$SnCl_4$	+4	tin(IV) chloride
FeS	+2	iron(II) sulfide	SnS_2	+4	tin(IV) sulfide
Fe_2O_3	+3	iron(III) oxide	PbO	+2	lead(II) oxide
Hg_2Br_2	+1	mercury(I) bromide	PbO_2	+4	lead(IV) oxide
$HgBr_2$	+2	mercury(II) bromide			

Roman numerals are *not* necessary for metals that commonly exhibit only one oxidation number.

An older method, still in use but not recommended by the IUPAC, involves the use of "-ous" and "-ic" suffixes to indicate lower and higher oxidation numbers, respectively. This system can distinguish between only two different oxidation numbers for a metal and, therefore, is not as useful as the Roman numeral system. However, the older system is still widely used in many scientific, engineering, and medical fields, and familiarity with the older system is a necessity for future practitioners. The advantage of the IUPAC system is that if you know the formula you can write the exact and unambiguous name, and if you are given the name you can write the formula at once.

Formula	Oxidation Number of Metal	Name	Formula	Oxidation Number of Metal	Name
CuCl	+1	cuprous chloride	SnF_2	+2	stannous fluoride
$CuCl_2$	+2	cupric chloride	SnF_4	+4	stannic fluoride
FeO	+2	ferrous oxide	Hg_2Cl_2	+1	mercurous chloride
$FeBr_3$	+3	ferric bromide	$HgCl_2$	+2	mercuric chloride

Pseudobinary ionic compounds contain more than two elements. In these compounds one or more of the ions consist of more than one element but behave as if they were simple ions. The most common examples of such anions are the hydroxide ion, OH^-, and the cyanide ion, CN^-. As before, the name of the anion ends in *-ide.* The ammonium ion, NH_4^+, is the most common cation that behaves like a simple metal cation. Some examples are

Formula	Name	Formula	Name
NH_4I	ammonium iodide	NH_4CN	ammonium cyanide
$Ca(CN)_2$	calcium cyanide	$Cu(OH)_2$	cupric hydroxide or copper(II) hydroxide
NaOH	sodium hydroxide	$Fe(OH)_3$	ferric hydroxide or iron(III) hydroxide

Nearly all **binary covalent compounds** involve two *nonmetals* bonded together. Although many nonmetals can exhibit different oxidation numbers, their oxidation numbers properly are *not* indicated by Roman numerals or suffixes. Instead, elemental proportions in binary covalent compounds are indicated by using a prefix system for both elements. The Greek or Latin prefixes used are mono, di, tri, tetra, penta, hexa, hepta, octa, nona, deca, undeca, and dodeca. The prefix mono is omitted except in the common (trivial) name for CO, carbon monoxide. The minimum number of prefixes required to name a compound unambiguously is used.

Formula	Name	Formula	Name
SO_2	sulfur dioxide	Cl_2O_7	dichlorine heptoxide
SO_3	sulfur trioxide	CS_2	carbon disulfide
N_2O_4	dinitrogen tetroxide	As_4O_6	tetraarsenic hexoxide

Chemists sometimes name binary covalent compounds that contain two nonmetals by the system used to name compounds of metals that show variable oxidation states, i.e., the oxidation state of the less electronegative element is indicated by a Roman numeral in parentheses. However, we do not recommend this procedure because it is incapable of naming compounds *unambiguously,* which is the principal requirement for a system for naming compounds. For example, both NO_2 and N_2O_4 are called nitrogen(IV) oxide by this system, and the names do not distinguish between the two compounds. The compound P_4O_{10} is tetraphosphorus decoxide, which indicates clearly its composition. Using the Roman numeral system, it would be called phosphorus(V) oxide, which could lead to the incorrect formula, P_2O_5. The simplest formula for P_4O_{10} is P_2O_5, but the name for a covalent compound must indicate clearly the composition of its molecules, not just its simplest formula.

Binary acids are compounds of hydrogen bonded to the more electronegative nonmetals. These compounds act as acids when dissolved in water. The pure compounds are named as typical binary compounds. Their aqueous solutions are named by modifying the characteristic stem of the nonmetal with the prefix "hydro-" and the suffix "-ic" followed by the word "acid." The stem for sulfur in this instance is "sulfur" rather than "sulf." Some typical binary acids are:

Formula	Name of Compound	Name of Aqueous Solution
HCl	hydrogen chloride	hydrochloric acid, HCl(aq)
HF	hydrogen fluoride	hydrofluoric acid, HF(aq)
H_2S	hydrogen sulfide	hydrosulfuric acid, H_2S(aq)
HCN	hydrogen cyanide	hydrocyanic acid, HCN(aq)

5-24 Ternary Compounds

Ternary acids (oxyacids) are compounds of hydrogen, oxygen, and a nonmetal. A nonmetal that can exhibit more than one oxidation state can form more than one ternary acid. These ternary acids differ in the number of oxygen atoms they contain (the higher the oxidation state of the central element, the greater the number of oxygen atoms). As with binary compounds, the suffixes "-ous" and "-ic" indicate lower and higher oxidation states, respectively. These follow the stem name of the central element. One common ternary acid of each nonmetal is (somewhat arbitrarily) designated as the "*-ic acid.*" That is, it is named "stem-ic acid." The common ternary "-ic acids" are shown below.

Periodic Group of Central Elements	IIIA	IVA	VA	VIA	VIIA
	+3 H_3BO_3 boric acid	+4 H_2CO_3 carbonic acid	+5 HNO_3 nitric acid		
		+4 H_4SiO_4 silicic acid	+5 H_3PO_4 phosphoric acid	+6 H_2SO_4 sulfuric acid	+5 $HClO_3$ chloric acid
			+5 H_3AsO_4 arsenic acid	+6 H_2SeO_4 selenic acid	+5 $HBrO_3$ bromic acid
				+6 H_6TeO_6 telluric acid	+5 HIO_3 iodic acid

Note the oxidation state of the central atom is equal to its periodic group number, except for those of the halogens.

There are no common -ic ternary acids for the omitted nonmetals. It is important to learn the names and formulas of these acids, since the names of all other ternary acids and salts are derived from them.

Acids containing *one fewer oxygen* atom per central atom are named in the same way except that the "-ic" suffix is changed to "-ous" as the following examples show. Notice that the central element has a lower oxidation number in the "-ous" acid than in the "-ic" acid.

Formula	Oxidation Number	Name	Formula	Oxidation Number	Name
H_2SO_3	+4	sulfur*ous* acid	H_2SO_4	+6	sulfur*ic* acid
HNO_2	+3	nitr*ous* acid	HNO_3	+5	nitr*ic* acid
H_2SeO_3	+4	selen*ous* acid	H_2SeO_4	+6	selen*ic* acid
$HBrO_2$	+3	brom*ous* acid	$HBrO_3$	+5	brom*ic* acid

Ternary acids that have one less O atom than the "-ous" acids and two less O atoms than the "-ic" acids are named using the prefix "hypo-" and the suffix "-ous". These correspond to acids in which the oxidation state of the central nonmetal is even lower than that in the "-ous acids."

Formula	Oxidation Number	Name
$HClO$	+1	*hypochlorous* acid
H_3PO_2	+1	*hypophosphorous* acid
HIO	+1	*hypoiodous* acid
$H_2N_2O_2$	+1	*hyponitrous* acid

Notice that $H_2N_2O_2$ has a 1:1 ratio of nitrogen to oxygen, as would the hypothetical HNO.

Acids containing *one more oxygen atom* per central nonmetal atom than the normal "-ic acid" are named in the same way as the "-ic acid" but with a "per-" prefix.

Formula	Oxidation Number	Name
$HClO_4$	+7	*perchloric* acid
$HBrO_4$	+7	*perbromic* acid
HIO_4	+7	*periodic* acid

Let's recap the system for naming ternary acids, using the oxyacids of chlorine as examples.

Formula	Oxidation Number	Name
$HClO$	+1	*hypo*chlor*ous* acid
$HClO_2$	+3	chlor*ous* acid
$HClO_3$	+5	chlor*ic* acid
$HClO_4$	+7	*per*chlor*ic* acid

Ternary salts are compounds that result from replacing the hydrogen in a ternary acid with another ion. They usually contain metal cations or the ammonium ion. As with binary compounds, the cation is named first. The name of the anion is based on the name of the ternary acid from which it is derived.

An anion derived from a ternary acid with an "-ic" ending is named by dropping the "-ic acid" and replacing it with "-ate." An anion derived from an "-ous acid" is named by replacing the suffix "-ous acid" with "-ite." The "per-" and "hypo-" prefixes are retained.

Formula	Name
$(NH_4)_2SO_4$	ammonium sulfate (SO_4^{2-} derived from H_2SO_4)
KNO_3	potassium nitrate (NO_3^- derived from HNO_3)
$Ca(NO_2)_2$	calcium nitrite (NO_2^- derived from HNO_2)
$LiClO_4$	lithium perchlorate (ClO_4^- derived from $HClO_4$)
$NaBrO_2$	sodium bromite (BrO_2^- derived from $HBrO_2$)
$FePO_4$	iron(III) phosphate (PO_4^{3-} derived from H_3PO_4)
$NaClO$	sodium hypochlorite (ClO^- derived from $HClO$)

Acid salts are salts containing anions derived from ternary acids in which one or more acidic hydrogen atoms remain. These salts are named in the same way as they would be if they were the usual type of ternary salt, except that the word "hydrogen," or "dihydrogen," is inserted after the name of the metal cation to indicate the number of acidic hydrogen atoms

Formula	Name	Formula	Name
$NaHSO_4$	sodium hydrogen sulfate	KH_2PO_4	potassium dihydrogen phosphate
$NaHSO_3$	sodium hydrogen sulfite	K_2HPO_4	potassium hydrogen phosphate
		$NaHCO_3$	sodium hydrogen carbonate

An older, commonly used method (which is not recommended by the IUPAC) involves the use of the prefix "bi-" attached to the name of the anion to indicate the presence of an acidic hydrogen. According to this system, $NaHSO_4$ is called sodium bisulfate and $NaHCO_3$ is named sodium bicarbonate.

A list of common cations and anions is given in Table 5–6.

TABLE 5–6 Formulas, Ionic Charges, and Names for Some Common Ions

A. Common Cations

Formula	Charge	Name
Li^+	1+	lithium ion
Na^+	1+	sodium ion
K^+	1+	potassium ion
NH_4^+	1+	ammonium ion
Ag^+	1+	silver ion
Mg^{2+}	2+	magnesium ion
Ca^{2+}	2+	calcium ion
Ba^{2+}	2+	barium ion
Cd^{2+}	2+	cadmium ion
Zn^{2+}	2+	zinc ion
Cu^{2+}	2+	cupric ion or copper(II) ion
Hg_2^{2+}	2+	mercurous ion or mercury(I) ion
Hg^{2+}	2+	mercuric ion or mercury(II) ion
Mn^{2+}	2+	manganous ion or manganese(II) ion
Co^{2+}	2+	cobaltous ion or cobalt(II) ion
Ni^{2+}	2+	nickelous ion or nickel(II) ion
Pb^{2+}	2+	plumbous ion or lead(II) ion
Sn^{2+}	2+	stannous ion or tin(II) ion
Fe^{2+}	2+	ferrous ion or iron(II) ion
Fe^{3+}	3+	ferric ion or iron(III) ion
Al^{3+}	3+	aluminum ion
Cr^{3+}	3+	chromic ion or chromium(III) ion

B. Common Anions

Formula	Charge	Name
F^-	1−	fluoride ion
Cl^-	1−	chloride ion
Br^-	1−	bromide ion
I^-	1−	iodide ion
OH^-	1−	hydroxide ion
CN^-	1−	cyanide ion
ClO^-	1−	hypochlorite ion
ClO_2^-	1−	chlorite ion
ClO_3^-	1−	chlorate ion
ClO_4^-	1−	perchlorate ion
CH_3COO^-	1−	acetate ion
MnO_4^-	1−	permanganate ion
NO_2^-	1−	nitrite ion
NO_3^-	1−	nitrate ion
SCN^-	1−	thiocyanate ion
O^{2-}	2−	oxide ion
S^{2-}	2−	sulfide ion
HSO_3^-	1−	bisulfite ion or hydrogen sulfite ion
SO_3^{2-}	2−	sulfite ion
HSO_4^-	1−	bisulfate ion or hydrogen sulfate ion
SO_4^{2-}	2−	sulfate ion
HCO_3^-	1−	bicarbonate ion or hydrogen carbonate ion
CO_3^{2-}	2−	carbonate ion
CrO_4^{2-}	2−	chromate ion
$Cr_2O_7^{2-}$	2−	dichromate ion
PO_4^{3-}	3−	phosphate ion
AsO_4^{3-}	3−	arsenate ion

Key Terms

Bonding pair pair of electrons involved in a covalent bond.

Central atom an atom in a molecule or polyatomic ion that is bonded to more than one other atom.

Debye the unit used to express dipole moments.

Delocalization of electrons; refers to bonding electrons that are distributed among more than two atoms that are bonded together; occurs in species that exhibit resonance.

Dipole refers to the separation of charge between two covalently bonded atoms.

Dipole moment (μ) the product of the distance separating opposite charges of equal magnitude times the distance separating them; a measure of the polarity of a bond or molecule; measured dipole moments refer to the dipole moment of entire molecules.

Double bond covalent bond resulting from the sharing of four electrons (two pairs) between two atoms.

Electronic geometry refers to the geometric arrangement of orbitals containing the shared and unshared electron pairs surrounding the central atom of a molecule or polyatomic ion.

Hybridization mixing of a set of atomic orbitals to form a new set of orbitals with the same total electron capacity and with properties and energies intermediate between those of the original unhybridized orbitals.

Ionic geometry refers to the arrangement of atoms, and not lone pairs of electrons, about the central atom of a polyatomic ion.

Lewis acid substance that accepts a share in a pair of electrons from another species.

Lewis base substance that makes available a share in an electron pair.

Lewis dot formula method of representing a molecule or formula unit by showing atoms and only outer shell electrons; does not show shape.

Lone pair pair of electrons residing on one atom and not shared by other atoms; unshared pair.

Molecular geometry refers to the arrangement of atoms (nuclei), and *not* unshared electron pairs (lone pairs), around a central atom of a molecule or polyatomic ion.

Nonpolar bond covalent bond in which electron density is symmetrically distributed.

Octahedron a polyhedron with eight equal-sized, equilateral triangular faces and six apices (corners).

Octet rule many representative elements attain at least a share of eight electrons in their valence shells when they form molecular or ionic compounds; there are limitations.

Overlap of orbitals, the interaction of orbitals on different atoms in the same region of space.

Pi (π) bond a bond resulting from the side-on overlap of atomic orbitals, in which the regions of electron sharing are above and below the imaginary line connecting the bonded atoms, and parallel to this line.

Polar bond covalent bond in which there is an unsymmetrical distribution of electron density.

Pseudobinary ionic compounds compounds that contain more than two elements but are named like binary compounds.

Resonance concept in which two or more equivalent dot formulas for the same arrangement of atoms (resonance structures) are necessary to describe the bonding in a molecule or ion.

Sigma (σ) bond a bond resulting from the head-on overlap of atomic orbitals, in which the region of electron sharing is along and (cylinderically) symmetrical to the imaginary line connecting the bonded atoms.

Single bond covalent bond resulting from the sharing of two electrons (one pair) between two atoms.

Square planar a term used to describe molecules and polyatomic ions that have one atom in the center and four atoms at the corners of a square.

Tetrahedron a polyhedron with four equal-sized, equilateral triangular faces and four apices.

Trigonal bipyramid a six-sided polyhedron with five apices, consisting of two pyramids sharing a common triangular base.

Trigonal planar a term used to describe molecules and polyatomic ions that have one atom in the center and three atoms at the corners of an equilateral triangle.

Triple bond covalent bond resulting from the sharing of six electrons (three pairs) between two atoms.

Unshared pair see *Lone pair*

Valence Bond (VB) theory assumes that covalent bonds are formed when atomic orbitals on different atoms overlap and electrons are shared.

Valence Shell Electron Pair Repulsion (VSEPR) theory assumes that electron pairs are arranged around the central element of a molecule or polyatomic ion so that there is maximum separation (and minimum repulsion) among electron pairs.

Exercises

Basic Ideas

1. Distinguish between ionic and covalent bonds. Why are covalent bonds called directional bonds while ionic bonding is called nondirectional?
2. What is the nature of the attractive forces between atoms that are covalently bonded together? Describe these forces.
3. What does Figure 5–1 tell us about the attractive and repulsive forces in a hydrogen molecule?
4. Distinguish between heteronuclear and homonuclear diatomic molecules. Can homonuclear diatomic molecules be polar molecules? Why?
5. Distinguish among single, double, and triple covalent bonds.

Lewis Dot Formulas for Molecules and Polyatomic Ions

6. (a) What are Lewis dot formulas?
   (b) What are polyatomic ions?
7. What is the octet rule? Is it generally applicable to compounds of the transition metals? Why?
8. (a) What is the simple mathematical relationship that is useful in writing Lewis dot formulas?
   (b) What does each term in the relationship represent?
9. Write Lewis dot formulas for the following elements and compounds: H_2, N_2, Br_2, HCl, HI
10. Write Lewis dot formulas for the following molecules and polyatomic ions: (a) H_2O and H_2S, (b) NH_3 and NH_4^+, (c) PH_3 and PH_4^+
11. What do we mean when we refer to resonance?
12. Draw resonance structures for (a) HCO_3^-, (b) NO_3^-, (c) SO_2, and (d) SO_3. In (a), the H atom is bonded to an O atom and all O atoms are bonded to the carbon atom.
13. (a) What are the limitations of the $S = N - A$ relationship?
    (b) Cite examples of three species to which the relationship does not apply, and explain why.
14. All of the following are covalently bonded. Which ones violate the octet rule? (a) PF_3, (b) ClF_3, (c) HCN, (d) BeI_2
15. All of the following are covalently bonded. Which ones violate the octet rule? (a) CO_2, (b) BCl_3, (c) SF_4, (d) H_3PO_4
16. Draw the Lewis dot formula for the sulfur molecule, S_8, which is an eight-membered ring.
17. Draw Lewis dot formulas for the following anions: (a) ClO^-, (b) ClO_2^-, (c) ClO_3^-, (d) ClO_4^-.
18. Draw Lewis dot formulas for the following molecules: (a) CCl_4, (b) SiH_4, (c) SO_3, (d) H_2SO_3, (e) NOCl, and (f) XeF_2.
19. Draw Lewis dot formulas for the following molecules: (a) Si_2H_6, (b) AsF_5, (c) HCN, (d) $HClO_4$, (e) C_2Cl_4, (f) OF_2.
20. El is the general symbol for a representative element. In each case, state the periodic group in which the element is located. Justify your answers and cite specific examples.

(a) H—O—El—O: (with :O: bonded above El)   (b) :Br—El—Br:

(c) :Cl—El—Cl: (with O: double-bonded above El)

21. El is the general symbol for a representative element. In each case, state the periodic group in which the element is located. Justify your answers and cite specific examples.

```
          ..
         :O:
          |
     ..   |    ..
(a) :O—El—O:2−        (b) :O—El—O:−
     ..   |    ..          ..  |   ..
         :O:                  :O:
          ..                   ..

     ..  ..
(c) :O—El:−
     ..  |
        :O:
         ..
```

22. Many common stains, such as chocolate, and those of other fatty foods can be removed by dry-cleaning solvents such as tetrachloroethylene, C_2Cl_4. Is C_2Cl_4 ionic or covalent? Draw its Lewis dot formula.
23. Draw Lewis dot formulas for the following: (a) H_2O_2, (b) ClF_3, (c) N_2H_4, (d) IF_7, (e) $AlCl_4^-$, (f) CN^-.
24. Draw acceptable dot formulas for the following common air pollutants: (a) SO_2, (b) NO_2, (c) CO, (d) O_3 (ozone), (e) SO_3, (f) $(NH_4)_2SO_4$. Which one is a solid? Which ones exhibit resonance?

Electronic and Molecular (or Ionic) Geometry

25. State in your own words the basic idea of the VSEPR theory.
26. (a) Distinguish between lone pairs and bonding pairs of electrons. (b) Which has the greater spatial requirement? How do we know this? (c) Indicate the order of increasing repulsions among lone pairs and bonding pairs of electrons.
27. As an exercise in geometry that will be very useful as molecular and electronic geometries are studied, determine how many ways the following number of items (×) may be arranged around a single item (O) symmetrically: (a) two, (b) three, (c) four, (d) five, (e) six. In those cases in which more than one symmetrical arrangement is possible, which gives maximum separation among × 's?
28. Draw a Lewis dot formula for each of the following molecules and indicate the number of regions of high electron density about the central atom, as well as their electronic and molecular geometries: (a) NH_3, (b) CF_4, (c) PF_5, (d) H_2O.
29. (a) What would be the ideal bond angles in the molecules in Exercise 28, ignoring lone pair effects? (b) How do these differ, if at all, from the actual values? Why?
30. Draw a Lewis dot formula for each of the following species and indicate the number of regions of high electron density about the central atom, as well as their electronic and molecular or ionic geometries: (a) BF_3, (b) SeF_6, (c) ClO_4^-, (d) NO_2^+.
31. What are the bond angles in the species in Exercise 30?
32. Draw a Lewis dot formula for the following molecules and indicate the number of lone pairs about the central atom, the number of regions of high electron density about the central atom, and the electronic geometry: (a) $BeBr_2$ (b) H_2Se (c) AsF_3 (d) GaI_3.
33. What is the molecular geometry of each molecule in Exercise 32?
34. Draw a Lewis dot formula for each of the following polyatomic ions and indicate the number of lone pairs about the central atom, the number of regions of high electron density about the central atom, and the electronic geometry: (a) PH_4^+, (b) AlH_4^-, (c) BrO_3^-, (d) ClO_2^-.
35. What is the ionic geometry of each ion in Exercise 34?
36. Pick the member of each pair that you would expect to have the smaller bond angles and explain why: (a) SF_2 and SO_2, (b) BF_3 and BCl_3, (c) CF_4 and SF_4, (d) NH_3 and H_2O.
37. Draw a dot formula, sketch the three-dimensional shape, and name the electronic geometry and molecular geometry for the following molecules: (a) PF_3, (b) AsF_5, (c) $GeCl_2$, (d) IF_3.
38. Draw a dot formula, sketch the three-dimensional shape, and name the electronic geometry and molecular geometry for the following polyatomic ions: (a) H_3O^+, (b) $AsCl_4^-$, (c) SiF_6^{2-}, (d) ICl_2^-.
39. Draw a dot formula, sketch the three-dimensional shape, and name the electronic

geometry and molecular geometry for the following molecules: (a) CO_2, (b) H_2CO_3, (c) $COCl_2$ (Cl's bonded only to C), (d) NOCl.

40. Draw a dot formula, sketch the three-dimensional shape, and name the electronic geometry and ionic geometry for the following polyatomic ions: (a) SO_4^{2-}, (b) SO_3^{2-}, (c) CN^-, (d) $SnCl_3^-$.

41. Draw three resonance structures for each of the following species and name the molecular or ionic geometries: (a) SO_3, (b) CO_3^{2-}.

42. Draw three-dimensional representations of both of the species of Exercise 41 showing delocalization in each case.

43. Draw two resonance structures for the ozone molecule, O_3, and two for the SO_2 molecule. Then draw a single structure showing delocalization for each and describe their molecular geometries. Is the similarity surprising? Why or why not?

44. Draw dot formulas and three-dimensional structures for the following molecules that contain the noble gas, xenon. Name the molecular geometry of each: (a) XeF_2, (b) XeF_4, (c) XeO_3, (d) XeO_4.

45. As the name implies, the interhalogens are compounds that contain two halogens. Draw dot formulas and three-dimensional structures for the following bromine fluorides. Name the electronic and molecular geometry of each: (a) BrF, (b) BrF_3, (c) BrF_5.

46. A number of ions derived from the interhalogens are known. Draw dot formulas and three-dimensional structures for the following ions of interhalogens. Name the electronic and ionic geometry of each: (a) IF_2^-, (b) ICl_4^-, (c) BrF_2^+.

47. Chlorine forms four oxyanions, ClO^-, ClO_2^-, ClO_3^-, and ClO_4^-. Draw the dot formula for each and then sketch and compare the electronic and ionic geometries of these oxyanions.

48. Nitrogen dioxide, NO_2, contains one unpaired electron per molecule. Draw a dot formula for the molecule. At low temperatures, NO_2 reacts with another NO_2 molecule (dimerizes) to form N_2O_4 molecules. Suggest a reason. Draw a dot formula for N_2O_4 and describe the molecular geometry with respect to the nitrogen atoms, which are bonded to each other.

49. The pyrophosphate ion, $P_2O_7^{4-}$, contains one oxygen atom that is bonded to both phosphorus atoms. Draw a dot formula and sketch the three-dimensional shape of the ion. How would you describe the ionic geometry with respect to the central oxygen atom and with respect to each phosphorus atom?

50. Carbon forms two common oxides, CO and CO_2, both of which are linear. It forms a third (very uncommon) oxide, carbon suboxide, C_3O_2, which is also linear. The structure has terminal oxygen atoms on both ends. Draw dot and dash formulas for C_3O_2. How many regions of high electron density are there about each of the three carbon atoms?

51. (a) List three important points that were illustrated by the discussion of covalent bonding. (b) What is the significance of each?

Polarities of Bonds and of Molecules

52. Why is an HCl molecule polar while a Cl_2 molecule is nonpolar?

53. Distinguish between polar and nonpolar covalent bonds. Cite three specific examples of each kind.

54. Explain the observed polarities of HF, HCl, HBr, and HI molecules.

55. What do we mean when we refer to the dipole moment of a molecule?

56. Is there a relationship between the polarity of a diatomic molecule and its dipole moment? If so, what is it?

57. Use Table 4–6 to list the following pairs of elements in order of increasing polarities of the *bonds* between the elements: (a) S—Cl, (b) Cl—O, (c) Be—I, (d) N—F, (e) As—F.

58. Which of the following molecules are polar? Why? (a) PCl_3, (b) BCl_3, (c) H_2O, (d) HgI_2, (e) SiF_4, (f) SF_4.

59. Which of the following molecules are polar? Why? (a) CH_4, (b) CH_3Br, (c) CH_2Br_2, (d) $CHBr_3$, (e) CBr_4.

60. Which one of the following statements is a better example of deductive logic? Why? (a) Because the dot formula for SO_2 shows

that there are three regions of high electron density about the sulfur atom, SO_2 molecules must be angular, and therefore polar.

(b) Because SO_2 molecules are polar, and therefore angular, the dot formula must show three regions of high electron density about the sulfur atom.

61. Which of the following molecules have dipole moments of zero? Justify your answer. (a) SO_3, (b) BrF_3, (c) CS_2, (d) O_3, (e) S_8 (puckered eight-membered ring), (f) $AsCl_3$.
62. The PF_3Cl_2 molecule has a dipole moment of zero. Use this information to sketch its three-dimensional shape. Justify your choice.
63. Why are the bond angles in NF_3 molecules less than those in the more polar NH_3 molecule?

Valence Bond Theory—Molecules and Ions having Only Single Bonds

64. Outline the Valence Bond theory.
65. What are hybrid orbitals?
66. What angles are associated with orbitals in the following hybridized sets of orbitals? (a) *sp*, (b) sp^2, (c) sp^3, (d) sp^3d, (e) sp^3d^2.
67. Describe the electronic geometries of atoms that have undergone the kinds of hybridization listed in Exercise 66.
68. Describe the bonding in each of the following molecules with a Lewis formula and a Valence Bond structure showing all orbital overlaps. Show the orbital orientations and label the orbitals: (a) $BeCl_2$, (b) CF_4, (c) H_2S, (d) SF_6.
69. Describe the bonding in each of the following species with a Lewis formula and a Valence Bond structure showing all orbital overlaps. Show the orbital orientations and label the orbitals: (a) BF_3, (b) ClO_4^-, (c) NF_3, (d) $(CH_3)_2O$, (e) AsF_5.
70. Describe the bonding in each of the following molecules or ions with a Lewis formula and a Valence Bond structure showing all orbital overlaps. Show the orbital orientations and label the orbitals: (a) CH_3Cl, (b) ClO_3^-, (c) H_3O^+, (d) $B(OH)_3$ (boric acid).
71. Describe the bonding in each of the following molecules or ions with a Lewis formula and a Valence Bond structure showing all orbital overlaps. Show the orbital orientations and label the orbitals: (a) $SnCl_4$, (b) $SnCl_6^{2-}$, (c) OF_2, (d) $Te(OH)_6$ (telluric acid).
72. Describe the change in hybridization at the central atom of the reactant at the left that occurs in each of the following reactions.
(a) $BF_3 + F^- \longrightarrow BF_4^-$
(b) $AsCl_3 + Cl_2 \longrightarrow AsCl_5$
(c) $SF_4 + F_2 \longrightarrow SF_6$
73. What changes in hybridization occur in the following reaction?

$$NH_3 + BF_3 \longrightarrow H_3N{:}BF_3$$

74. The industrial manufacture of sulfuric acid, H_2SO_4, (Section 23–11) involves the conversion of elemental sulfur, S_8, to SO_2, then to SO_3, and finally to H_2SO_4. The S—S—S bond angles in S_8 are all about 108°. H is bonded to O in H_2SO_4. Describe the changes in hybridization that accompany these conversions.
75. Give a reason why phosphorus forms both PF_3 and PF_5 but nitrogen forms only NF_3.
76. Draw a Lewis formula and Valence Bond structure showing all orbital overlaps of each of the following molecules that contain the noble gas xenon. Label the orbitals and indicate (approximate) bond angles: (a) XeF_2, (b) XeF_4, (c) XeO_3, (d) XeO_4.
77. Draw a Lewis formula and Valence Bond structure showing all orbital overlaps of each of the following molecules and ions of the interhalogens. Label the orbitals and indicate (approximate) bond angles. (a) ClF_3, (b) BrF_5, (c) ICl_4^+, (d) ICl_4^-.

Valence Bond Theory—Polycentered Molecules and Multiple Bonding

78. For each of the molecules shown below, predict the hybridization at each carbon atom.
(a) acetone:

```
   H       H
   |       |
H—C—C—C—H
   |   ||  |
   H  :O: H
```

(b) methanol or methyl alcohol (wood alcohol):

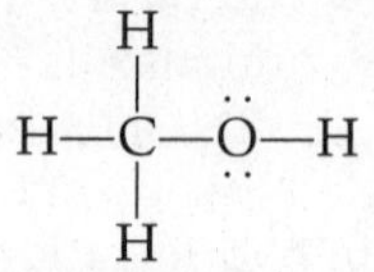

H—C—Ö—H (with H above and below C)

(c) acetic acid (vinegar is about a 5% solution of acetic acid in water):

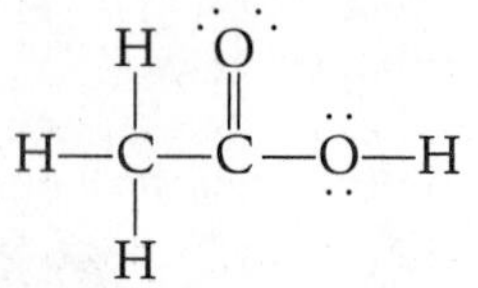

(d) tetrachloroethene (a dry-cleaning solvent):

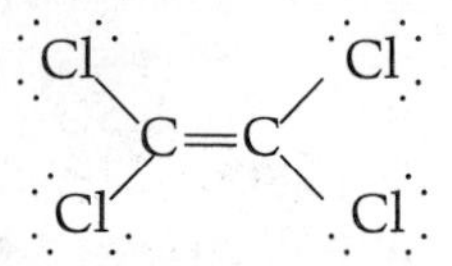

79. For each of the molecules shown below, predict the hybridization at the starred (*) atoms and the approximate bond angles at those atoms.
(a) methylamine:

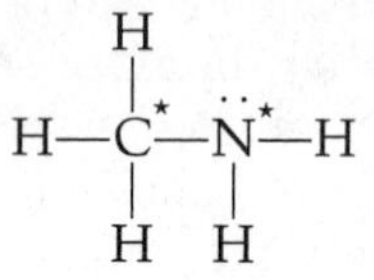

(b) urea (a product of metabolism of proteins):

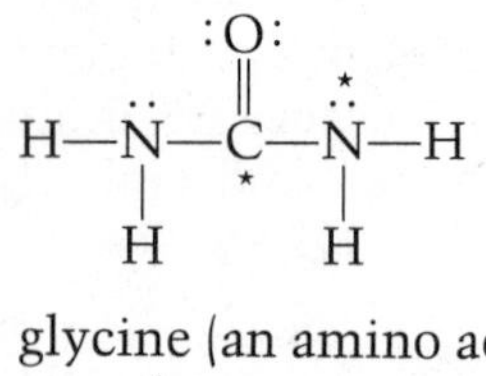

(c) glycine (an amino acid that is a constituent of many proteins)

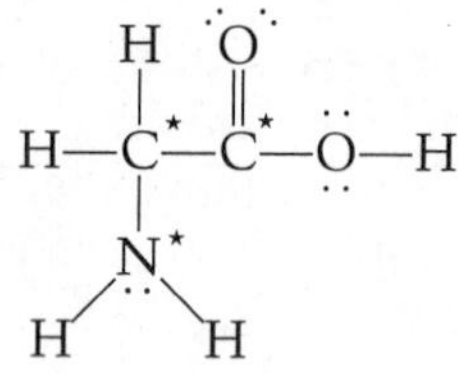

(d) asparagine (another amino acid):

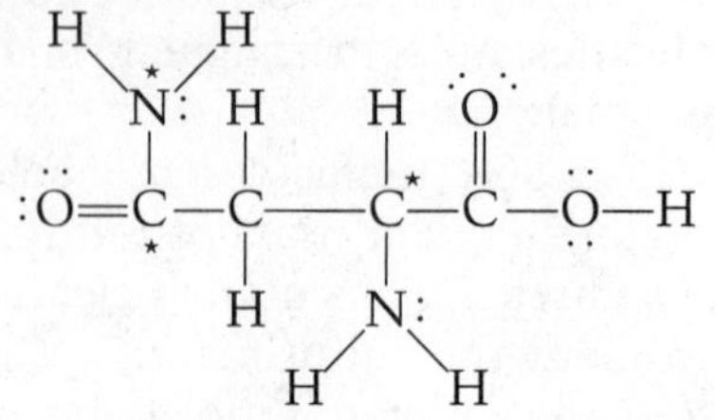

80. Sketch three-dimensional ball-and-stick models of the following ions and indicate the hybridization and approximate bond angles at the central atoms: (a) nitrate ion, NO_3^-, (b) nitrite ion, NO_2^-, (c) sulfate ion, SO_4^{2-}, (d) sulfite ion, SO_3^{2-}.
81. Sodium acetylide, $Na^+C_2H^-$, is an ionic compound. The two C atoms are bonded together. Draw a Lewis formula and a ball-and-stick model of the acetylide anion showing bond angles at each carbon atom.
82. The pyrosulfate ion, $S_2O_7^{2-}$, contains one oxygen atom bonded to both sulfur atoms, called a "bridging oxygen." Draw a dash formula and a ball-and-stick model of the ion. What are the approximate bond angles at these atoms?
83. Describe the bonding in the N_2 molecule with a three-dimensional VB structure. Show the orbital overlap and label the orbitals.
84. Draw a dash formula and a ball-and-stick model for each of the following polycentered molecules. Indicate hybridizations and bond angles at each carbon atom: (a) butane, C_4H_{10}, (b) propene, CH_3CHCH_2, (c) propyne, CH_3CCH, (d) acetaldehyde, CH_3CHO.
85. How many sigma bonds and how many pi bonds are there in each of the compounds of Exercise 84?

Inorganic Nomenclature

86. Write names and formulas for the ternary "-ic" acid of each of the following nonmetals: C, N, P, S, and Cl.

87. Write formulas for the compounds that are formed by the following pairs of ions. On a separate sheet, write the name for each compound.

	A	B	C	D	E	F
	OH^-	Cl^-	NO_3^-	SO_4^{2-}	CO_3^{2-}	PO_4^{3-}
1. Na^+						
2. NH_4^+						
3. Ca^{2+}						
4. Fe^{2+}						
5. Cu^{2+}						
6. Zn^{2+}						
7. Cr^{3+}						
8. Fe^{3+}						

88. Write formulas for the following compounds: (a) magnesium sulfide, (b) calcium hydroxide, (c) lithium acetate, (d) hydroiodic acid, (e) aluminum nitrate, (f) copper(II) sulfate, (g) barium phosphate, (h) ferric bromide.

89. Write formulas for the following compounds: (a) copper(I) cyanide, (b) ferrous sulfate, (c) ammonium phosphate, (d) nickel(II) nitrite, (e) magnesium acetate, (f) perbromic acid, (g) chloric acid, (h) mercuric chloride.

90. Write names for the following compounds: (a) FeI_2, (b) $CrCl_3$, (c) $NaNO_3$, (d) $CaSO_4$, (e) BaS, (f) $Al(CN)_3$, (g) $KClO_4$, (h) $Al(OH)_3$, (i) $Sr(MnO_4)_2$, (j) $Na_2Cr_2O_7$.

91. Write names for the following compounds: (a) $Ba(NO_3)_2$, (b) $Cu(SCN)_2$, (c) K_2CrO_4, (d) CrO_3, (e) $Cr(CH_3COO)_3$, (f) Li_2S, (g) $NaClO_3$, (h) NaClO.

92. Write names for the following compounds: (a) Hg_2Br_2, (b) $AgNO_3$, (c) P_4O_6, (d) MnO_2, (e) $Zn(ClO_4)_2$, (f) $Ca_3(AsO_4)_2$, (g) K_2CO_3, (h) $Co(HCO_3)_2$.

93. Write names for the following compounds: (a) N_2O_3, (b) $CrAsO_4$, (c) $(NH_4)_2CO_3$, (d) $CoBr_3$, (e) PCl_5, (f) $FeCl_2$, (g) Na_2CrO_4, (h) $CuSO_3$, (i) Li_2SO_4, (j) KNO_2.

94. Write formulas for the following compounds: (a) hydrofluoric acid, (b) chromic hydroxide, (c) silver sulfide, (d) barium hydrogen sulfate, (e) zinc permanganate, (f) calcium iodate, (g) magnesium phosphate, (h) sulfurous acid.

6 Molecular Orbitals in Chemical Bonding

So far we have considered bonding and molecular geometry in terms of Valence Bond theory, which assumes that electrons are located in orbitals localized on or between individual atoms, and in terms of Valence Shell Elecron Pair Repulsion theory, which considers only the effects of electrostatics on geometry.

Valence Bond theory postulates that bonds result from the sharing of electrons in overlapping orbitals of different atoms. These orbitals may be *pure atomic orbitals* or *hybridized atomic orbitals* of *individual* atoms. Electrons in overlapping orbitals of different atoms are thought of as being localized in the bonds between the two atoms involved, rather than delocalized over the entire molecule. Hybridization is invoked when it is necessary to account for either the geometry of a molecule or the number of unpaired electrons that must be available for bonding.

Molecular Orbital theory, on the other hand, postulates the *combination of atomic orbitals of different atoms to form* **molecular orbitals** (MO's), *so that electrons in them belong to the molecule as a whole.* The Valence Bond and Molecular Orbital approaches have strengths and weaknesses that are complementary. They actually represent alternative descriptions of chemical bonding, and, in the limit, produce the same results. Valence Bond theory is descriptively attractive and it lends itself well to visualization. Molecular Orbital (MO) theory gives somewhat better descriptions of electron cloud distributions, bond energies and magnetic properties, but it is harder to visualize.

In some polyatomic molecules, a molecular orbital may extend over only a fragment of the molecule. Molecular orbitals also exist for polyatomic ions such as CO_3^{2-}, SO_4^{2-}, and NH_4^+.

The Valence Bond picture of the bonding in the oxygen molecule, O_2, involves a double bond and no unpaired electrons. It predicts sp^2 hybridization at each oxygen because there are three sets of valence electrons on each oxygen atom.

$\ddot{:O}::\ddot{O}:$ (wrong!)

However, experiments show that O_2 is actually paramagnetic (p. 213) and therefore has unpaired electrons. Molecular Orbital theory *predicts* that O_2 has two unpaired electrons, and it was the ability of MO theory to account for the paramagnetism of O_2 that brought it to the forefront as a major theory in bonding. In the following sections, we shall develop some of the notions of MO theory and then apply them to some relatively simple molecules and ions.

In fact, since O_2 has an even number of valence electrons, at least two must be unpaired.

6–1 Molecular Orbitals

The mathematical pictures of hybrid orbitals in Valence Bond theory can be generated by combining the equations for electron waves that describe two or more atomic orbitals on a single atom. Similarly, mathematical descriptions of molecular orbitals can be generated by combining equations for electron waves that describe atomic orbitals on separate atoms.

The electron waves that describe atomic orbitals have positive (upward) and negative (downward) phases or amplitudes. When waves are combined, they may interact either constructively or destructively. If the two identical waves shown at the left are added they interfere constructively to produce the wave at the right. Conversely, if they are subtracted, it is as if the phases of one wave were reversed and added to the first wave. This causes destructive interference, resulting in zero amplitude in the wave at the right.

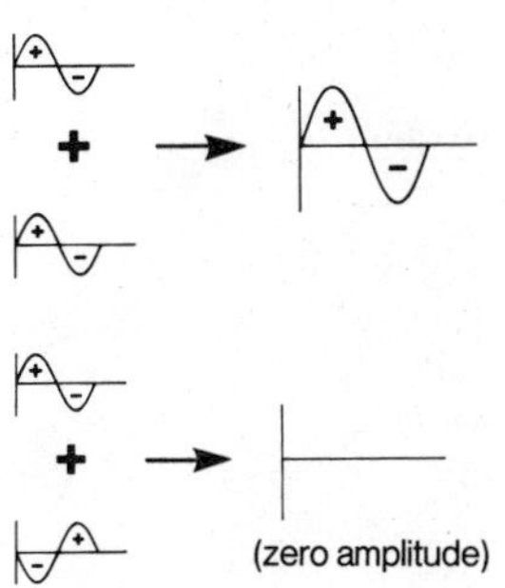

Likewise, when two atomic orbitals overlap they can be in-phase (added) or out-of-phrase (subtracted). When they overlap in-phase, constructive interference occurs in the region between the nuclei and a **bonding orbital** is produced. The energy of the bonding orbital is always lower (more stable) than the energies of the combining orbitals. When they overlap out-of-phase, destructive interference reduces the probability of finding an electron in the region between the nuclei and an **antibonding orbital** is produced. The energy of an antibonding orbital is higher (less stable) than the energies of the combining orbitals.

We can illustrate this basic principle by considering the combination of two 1*s* atomic orbitals (Figure 6–1). (Don't confuse the + and − designations, which indicate amplitudes of electron waves, with electrical charges.) When

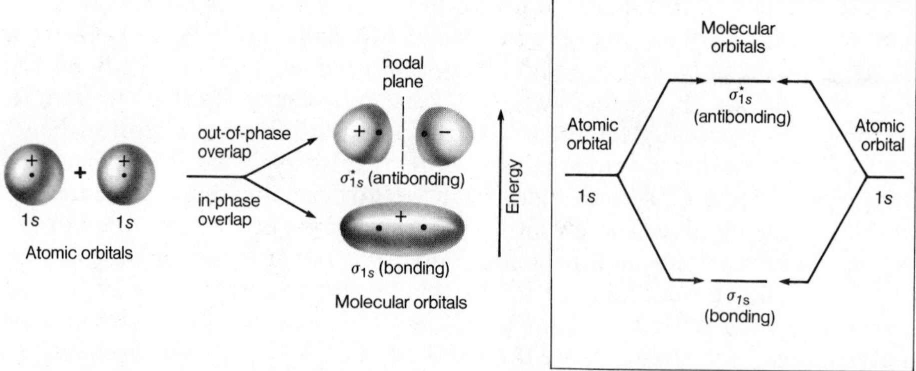

FIGURE 6–1 Molecular orbital (MO) diagram for the combination of two 1*s* atomic orbitals (at the left), to form two molecular orbitals. One is a *bonding* orbital, σ_{1s}, resulting from addition of the wave functions of the 1*s* orbitals. The other is an *antibonding* orbital, $\sigma_{1s}^{\star}$, at higher energy resulting from subtraction of the waves that describe the combining 1*s* orbitals. In all σ-type molecular orbitals, the electron density is symmetrical about an imaginary line connecting the two nuclei. The terms "subtraction of waves," "out-of-phase," and "destructive interference in the region between the nuclei" all refer to the formation of an antibonding molecular orbital. Nuclei are represented by dots.

these orbitals are occupied by electrons, the shapes of the orbitals are really plots of electron density that represent regions of space in which the probabilities of finding electrons in molecules are greatest.

In the bonding orbital, the two 1*s* orbitals have combined their densities in the region between the two nuclei by in-phase overlap, or addition of the electron waves that describe them. In the antibonding orbital, they have cancelled one another in this region by out-of-phase overlap or subtraction of the electron waves. We designate both molecular orbitals as sigma (σ) orbitals (which indicates that they are symmetrical about a line drawn through the two nuclei, the internuclear axis), and we indicate with subscripts the atomic orbitals that have been combined. The asterisk denotes an antibonding orbital. All sigma antibonding orbitals have nodal planes bisecting the internuclear axis. A **node** or **nodal plane** *is a region in which the probability of finding electrons is zero.* Thus, two 1*s* orbitals produce a σ_{1s} (read "sigma-1*s*") bonding orbital and a $\sigma_{1s}^{\star}$ (read "sigma-1*s*-star") antibonding orbital. The right-hand side of Figure 6–1 shows the relative energy levels of these orbitals.

Placing electrons in bonding molecular orbitals leads to a more stable (lower energy) system than individual atoms. Placing electrons in antibonding orbitals, which requires adding energy to the molecule, leads to destabilization of the molecule relative to individual atoms.

For any two sets of *p* orbitals on two different atoms, corresponding orbitals such as p_x orbitals can overlap head-on, giving σ_p and $\sigma_p^{\star}$ orbitals, as shown in Figure 6–2 for the head-on overlap of $2p_x$ orbitals on two atoms.

If the remaining *p* orbitals overlap (p_y with p_y and p_z with p_z), they must do so *sideways or side-on,* forming *pi (π) molecular orbitals.* Depending on whether all *p* orbitals overlap, there can be a total of two π_p and two $\pi_p^{\star}$ orbitals. See Figure 6–3, which illustrates the overlap of two corresponding 2*p*

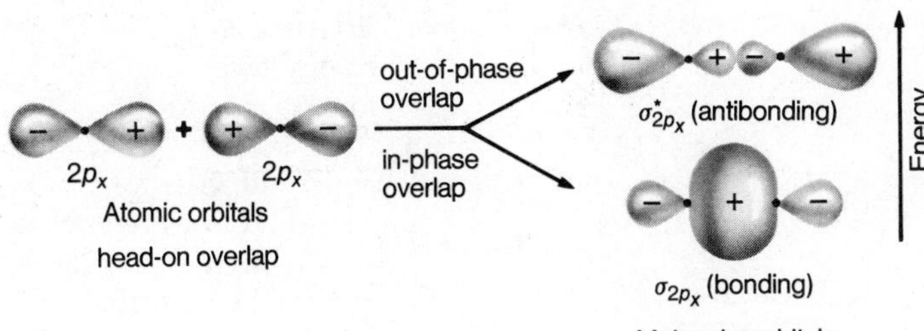

FIGURE 6–2 Production of σ_{2p_x} and $\sigma^{\star}_{2p_x}$ molecular orbitals by overlap of $2p_x$ orbitals on two atoms.

If we had chosen the z-axis as the axis of head-on overlap of the $2p$ orbitals in Figure 6–2, side-on overlap of the $2p_x - 2p_x$ and $2p_y - 2p_y$ orbitals would form the π-type molecular orbitals.

orbitals on two atoms to form π_{2p} and $\pi^{\star}_{2p}$ molecular orbitals. Note that there is a nodal plane along the internuclear axis for all pi molecular orbitals.

Note that if one views a sigma molecular orbital along the internuclear axis, it appears to be symmetrical around that axis like a pure s atomic orbital. A similar cross-sectional view of a pi molecular orbital looks like a pure p atomic orbital, with a node along the internuclear axis.

This would involve rotating Figures 6–1, 6–2, and 6–3 by 90° so that the internuclear axes are perpendicular to the plane of the pages.

6–2 Molecular Orbital Energy-Level Diagrams

Using what we know about how atomic orbitals combine to produce molecular orbitals, we can now draw molecular orbital energy-level diagrams for simple molecules. The simplest example, shown in Figure 6–4, is for most homonuclear diatomic molecules of elements in the first and second periods of the periodic table. It is simply an extension of the right-hand diagram in Figure 6–1, to which we have added molecular orbitals formed from $2s$ and $2p$ atomic orbitals.

To use the diagram in Figure 6–4, we simply follow the same rules established for atomic energy-level diagrams in Chapter 3. Count the number of electrons in the molecule, and then fill the molecular orbitals, lowest energy orbitals first, while observing the Aufbau Principle (Section 3–15), Hund's Rule (Section 3–15), and the Pauli Exclusion Principle (Section 3–13).

For the cases shown in Figure 6–4, the two π_{2p} orbitals are lower in energy than the σ_{2p} orbital. However, spectroscopic data suggests that for O_2, F_2, and Ne_2 (hypothetical) molecules, the σ_{2p} orbital is lower in energy than the π_{2p} orbitals (see Figure 6–5).

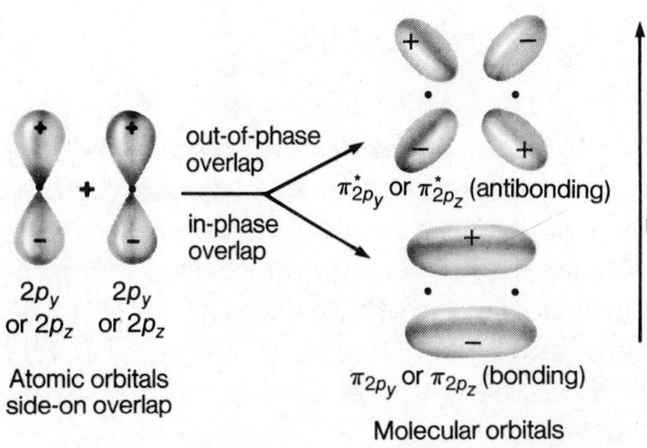

FIGURE 6–3 The π_{2p} and $\pi^{\star}_{2p}$ molecular orbitals from overlap of one pair of $2p$ atomic orbitals (for instance, $2p_y$ orbitals). There can be an identical pair of molecular orbitals at right angles to these, formed by another pair of p orbitals on the same two atoms (in this case, $2p_z$ orbitals).

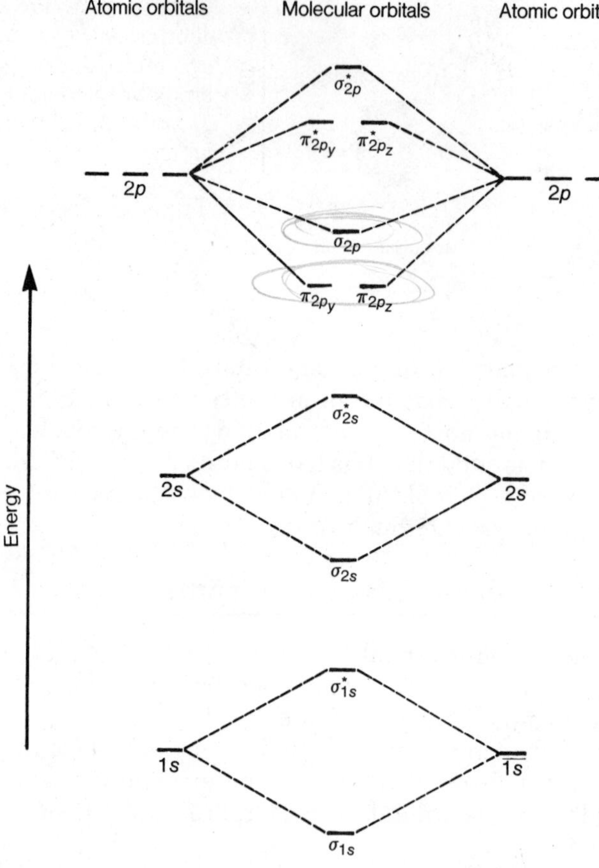

FIGURE 6–4 The energy level diagram for H_2, He_2, Li_2, Be_2, B_2, C_2, and N_2 molecules and their ions. The solid lines represent the relative energies of the indicated atomic and molecular orbitals.

2 different elements

6–3 Bond Order and Bond Stability

Now all that is needed is a criterion by which to judge the stability of a molecule, once its energy-level diagram has been filled with the appropriate number of electrons. This criterion is the **bond order,** *which is defined as the number of electrons in bonding orbitals minus the number of electrons in antibonding orbitals all divided by two.*

$$\text{bond order} = \frac{\begin{pmatrix}\text{number of } e^- \\ \text{in bonding MOs}\end{pmatrix} - \begin{pmatrix}\text{number of } e^- \\ \text{in antibonding MOs}\end{pmatrix}}{2}$$

Generally, the bond order corresponds to the number of bonds established by the Valence Bond theory. Fractional bond orders exist in species that contain an odd number of electrons, such as the nitrogen oxide molecule, NO. The greater the bond order of a diatomic molecule or ion, the more stable we predict it to be. Likewise, for a bond between two given atoms, the greater the bond order, the shorter the bond length, and the greater the bond energy. The **bond energy** *is the amount of energy necessary to break a mole of bonds;* it is therefore a measure of bond strength.

The NO molecule contains a total of 15 electrons.

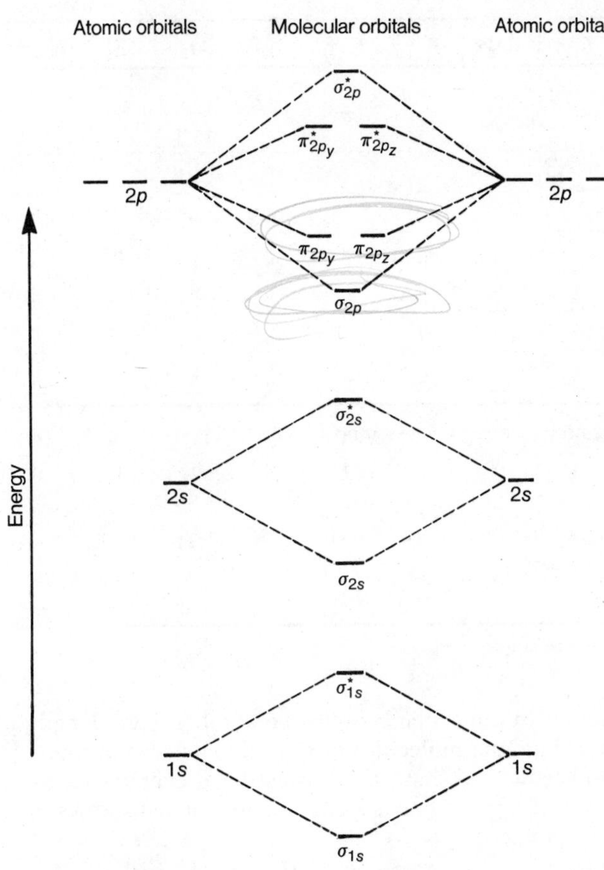

FIGURE 6–5 The energy level diagram for O_2, F_2, and Ne_2 molecules and their ions. The solid lines represent the relative energies of the indicated atomic and molecular orbitals.

6–4 Homonuclear Diatomic Molecules

The electron distributions for the homonuclear diatomic molecules of the first and second periods are shown in Figure 6–6, together with the bond order, bond length, and bond energy for each molecule. We shall now look at each in turn.

Homonuclear: Containing only atoms of the same element.

1 The Hydrogen Molecule, H_2

The overlap of the 1*s* orbitals of two hydrogen atoms produces σ_{1s} and σ^*_{1s} molecular orbitals. The two electrons of the molecule occupy the lower energy σ_{1s} orbital as shown in Figure 6–7.

Since there is one pair of electrons in a bonding orbital and none in an antibonding orbital of an H_2 molecule, the bond order is one. We conclude that the H_2 molecule is stable, and of course it is. The energy associated with two electrons in the H_2 molecule is less than that associated with the same two electrons in 1*s* atomic orbitals. The lower the energy of a system is, the more stable it is.

2 The Helium Molecule (Hypothetical), He_2

The energy level diagram for He_2 is similar to that of H_2 except that it has two more electrons, which occupy the antibonding σ^*_{1s} orbital (see Figures 6–4,

INCREASING ENERGY (not to scale) →	H_2	He_2††	Li_2†	Be_2†	B_2†	C_2†	N_2	O_2	F_2	Ne_2††
$\sigma_{2p}^{\star}$	—	—	—	—	—	—	—	—	—	⇅
$\pi_{2p_y}^{\star}, \pi_{2p_z}^{\star}$	— —	— —	— —	— —	— —	— —	— —	↑ ↑	⇅ ⇅	⇅ ⇅
σ_{2p} or	—	—	—	—	—	—	⇅	⇅ ⇅	⇅ ⇅	⇅ ⇅
π_{2p_y}, π_{2p_z}	— —	— —	— —	— —	↑ ↑	⇅ ⇅	⇅ ⇅	⇅	⇅	⇅
$\sigma_{2s}^{\star}$	—	—	—	⇅	⇅	⇅	⇅	⇅	⇅	⇅
σ_{2s}	—	—	⇅	⇅	⇅	⇅	⇅	⇅	⇅	⇅
$\sigma_{1s}^{\star}$	—	⇅	⇅	⇅	⇅	⇅	⇅	⇅	⇅	⇅
σ_{1s}	⇅	⇅	⇅	⇅	⇅	⇅	⇅	⇅	⇅	⇅
Paramagnetic?	no	no	no	no	yes	no	no	yes	no	no
Bond Order	1	0	1	0	1	2	3	2	1	0
Bond Length (Å)	0.74	—	2.67	—	1.59	1.31	1.09	1.21	1.43	—
Bond Energy (kJ/mol)	435	—	110	—	~270	602	946	498	159	—

†Exists only in vapor state at elevated temperatures.
††Unstable, i.e., unknown species.

FIGURE 6–6 Electron distribution in molecular orbitals, bond order, bond length, and bond energy of homonuclear diatomic molecules of the first and second row elements. Note that nitrogen molecules, N_2, have the highest bond energies listed because they have bond order of three (triple bonds). The doubly bonded species, C_2 and O_2, have the next highest bond energies.

6–6, and 6–7). Thus its bond order is zero, and we conclude that the molecule is not stable. In fact, He_2 is not known; helium exists only as *monatomic* molecules.

3 The Lithium Molecule, Li_2

The electron configuration for Li_2 has a total of six electrons, two each in σ_{1s}, $\sigma_{1s}^{\star}$, and σ_{2s} molecular orbitals. In shorthand notation this can be represented as $\sigma_{1s}{}^2\ \sigma_{1s}^{\star}{}^2\ \sigma_{2s}{}^2$. There are two more electrons in bonding orbitals than in antibonding orbitals, and the bond order is one. We conclude that the molecule is stable, and it is indeed known to exist in the vapor phase.

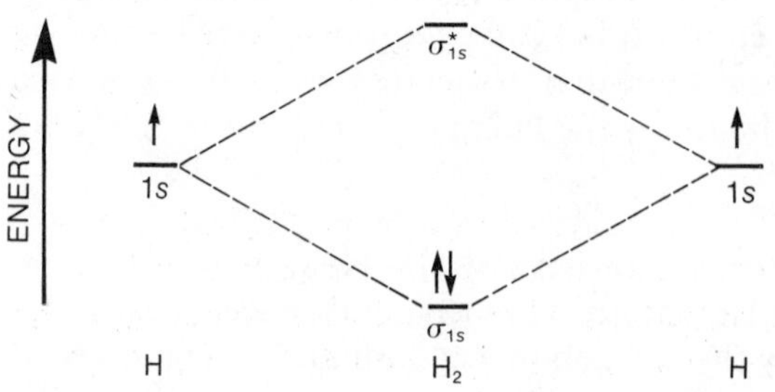

FIGURE 6–7 Molecular orbital diagram for H_2.

4 The Beryllium Molecule, Be_2

The situation for the Be_2 molecule is similar to that for He_2. Bonding and antibonding orbitals are equally populated and the bond order is zero. The electronic configuration is $\sigma_{1s}{}^2\ \sigma^{\star}_{1s}{}^2\ \sigma_{2s}{}^2\ \sigma^{\star}_{2s}{}^2$. The Be_2 molecule is known but is very unstable.

5 The Boron Molecule, B_2

The boron atom has the configuration of $1s^2 2s^2 2p^1$ and is the first element with p electrons to participate in bonding. Reference to Figures 6–4 and 6–6 suggests that the π_{p_y} and π_{p_z} molecular orbitals are lower in energy than the σ_{2p} for B_2. Thus its electron configuration is $\sigma_{1s}{}^2\ \sigma^{\star}_{1s}{}^2\ \sigma_{2s}{}^2\ \sigma^{\star}_{2s}{}^2\ \pi_{2p_y}{}^1\ \pi_{2p_z}{}^1$. This illustrates the validity of Hund's Rule in dealing with molecular orbital theory. The π_{2p_y} and π_{2p_z} orbitals are equal in energy and contain a total of two electrons. As a result, one electron occupies each orbital. The bond order is one and experiments verify not only that the molecule exists in the vapor state but that it is paramagnetic with two unpaired electrons.

Orbitals of equal energy are called **degenerate** orbitals (see Figures 6–4 and 6–6).

6 The Carbon Molecule, C_2

The carbon atom has the electronic configuration $1s^2 2s^2 2p^2$. The C_2 molecule does exist *in the gaseous state* and has the configuration $\sigma_{1s}{}^2\ \sigma^{\star}_{1s}{}^2\ \sigma_{2s}{}^2\ \sigma^{\star}_{2s}{}^2\ \pi_{2p_y}{}^2\ \pi_{2p_z}{}^2$. Its bond order is two, and it is diamagnetic because all its electrons are paired (Figure 6–6).

7 The Nitrogen Molecule, N_2

Experimental thermodynamic data show that the N_2 molecule is stable and has a very high bond energy, 946 kJ/mol, which is consistent with Molecular Orbital theory. Each nitrogen atom has seven electrons and the diamagnetic N_2 molecule has 14 electrons, distributed as follows:

$$\sigma_{1s}{}^2\quad \sigma^{\star}_{1s}{}^2\quad \sigma_{2s}{}^2\quad \sigma^{\star}_{2s}{}^2\quad \pi_{2p_y}{}^2\quad \pi_{2p_z}{}^2\quad \sigma_{2p}{}^2.$$

Since there are six more electrons in bonding orbitals than in antibonding orbitals, the bond order is three (triple bond). We also note from Figure 6–6 that N_2 has a very short bond length, only 1.09 Å the shortest of any diatomic species except H_2 (which contains only $2\ e^-$).

8 The Oxygen Molecule, O_2

Among homonuclear diatomic molecules, only N_2 and the very small H_2 have shorter bond lengths than O_2 (1.21 Å). As mentioned earlier, Valence Bond theory predicts that O_2 would be diamagnetic. However, experiments show that it is paramagnetic, with two unpaired electrons. A structure consistent with this observation is predicted by MO theory. Spectroscopic evidence suggests that for O_2 (as well as F_2 and unstable Ne_2), the σ_{2p} orbital is lower in energy than the π_{2p_y} and π_{2p_z} orbitals. Each oxygen atom has eight electrons and the O_2 molecule has 16, distributed as follows:

$$\sigma_{1s}{}^2\quad \sigma^{\star}_{1s}{}^2\quad \sigma_{2s}{}^2\quad \sigma^{\star}_{2s}{}^2\quad \sigma_{2p}{}^2\quad \pi_{2p_y}{}^2\quad \pi_{2p_z}{}^2\quad \pi^{\star}_{2p_y}{}^1\quad \pi^{\star}_{2p_z}{}^1.$$

The two unpaired electrons reside in the *degenerate* antibonding orbitals, $\pi^{\star}_{2p_y}$ and $\pi^{\star}_{2p_z}$. Since there are four more electrons in bonding orbitals than in antibonding orbitals, the bond order is two (double bond) and we conclude that the molecule should be very stable, as it is (Figures 6–5 and 6–6).

EXAMPLE 6–1

Predict the stabilities and bond orders of the O_2^+ and O_2^- ions.

Solution

The O_2^+ ion is obtained by removing one electron from the O_2 molecule. Since the electrons that are withdrawn most easily are those in the highest energy orbitals, one of the $\pi_{2p}^\star$ electrons is lost. Thus the configuration of O_2^+ is

$$\sigma_{1s}^{\ 2} \quad \sigma_{1s}^{\star\,2} \quad \sigma_{2s}^{\ 2} \quad \sigma_{2s}^{\star\,2} \quad \sigma_{2p}^{\ 2} \quad \pi_{2p_y}^{\ 2} \quad \pi_{2p_z}^{\ 2} \quad \pi_{2p_y}^{\star\,1}.$$

There are five more electrons in bonding orbitals than in antibonding orbitals, and the bond order is 2.5. We conclude that the ion would be reasonably stable relative to other diatomic ions, and it does exist. In fact, the unusual ionic compound $[O_2^+][PtF_6^-]$ played an important role in the discovery of the first noble gas compound, $XePtF_6$ (Section 22–4).

The superoxide ion, O_2^-, results from adding an electron to one of the $\pi_{2p}^\star$ orbitals of O_2. The configuration of O_2^- is

$$\sigma_{1s}^{\ 2} \quad \sigma_{1s}^{\star\,2} \quad \sigma_{2s}^{\ 2} \quad \sigma_{2s}^{\star\,2} \quad \sigma_{2p}^{\ 2} \quad \pi_{2p_y}^{\ 2} \quad \pi_{2p_z}^{\ 2} \quad \pi_{2p_y}^{\star\,2} \quad \pi_{2p_z}^{\star\,1}.$$

The bond order is 1.5 because there are three more bonding electrons than antibonding electrons. Thus we can conclude that the ion would be less stable than O_2.

Electrons in bonding orbitals are often called **bonding electrons;** electrons in antibonding orbitals, **antibonding electrons.**

It is interesting to note that the superoxides of the heavier Group IA elements, KO_2, RbO_2, and CsO_2, which contain O_2^-, all exist. These compounds are formed by combination of the free metals with oxygen (Section 7–9.2).

9 The Fluorine Molecule, F_2

Each fluorine atom has the $1s^22s^22p^5$ configuration. The 18 electrons of F_2 are distributed as follows:

$$\sigma_{1s}^{\ 2} \quad \sigma_{1s}^{\star\,2} \quad \sigma_{2s}^{\ 2} \quad \sigma_{2s}^{\star\,2} \quad \sigma_{2p}^{\ 2} \quad \pi_{2p_y}^{\ 2} \quad \pi_{2p_z}^{\ 2} \quad \pi_{2p_y}^{\star\,2} \quad \pi_{2p_z}^{\star\,2}.$$

The bond order is one. Experiments show that F_2 is stable, though less so than N_2 or O_2. The F—F bond distance is longer (1.43 Å) than the bond distance for the larger O_2 (1.21 Å) and N_2 (1.09 Å) and even C_2 (1.31 Å) molecules. The bond energy of the F_2 molecules is quite low (159 kJ/mol), and F_2 molecules are very reactive.

10 The Neon Molecule (Hypothetical), Ne_2

Neon atoms have completely filled first and second shells: $1s^22s^22p^6$. The electron distribution of the unstable and *unknown* Ne_2 molecule would involve equal numbers of bonding and antibonding electrons and a bond order of zero:

$$\sigma_{1s}^{\ 2} \quad \sigma_{1s}^{\star\,2} \quad \sigma_{2s}^{\ 2} \quad \sigma_{2s}^{\star\,2} \quad \sigma_{2p}^{\ 2} \quad \pi_{2p_y}^{\ 2} \quad \pi_{2p_z}^{\ 2} \quad \pi_{2p_y}^{\star\,2} \quad \pi_{2p_z}^{\star\,2} \quad \sigma_{2p}^{\star\,2}.$$

Like helium and the other noble gases, neon exists only as *monatomic* molecules.

11 Heavier Homonuclear Diatomic Molecules

On the surface, it would seem reasonable to use the same types of molecular orbital diagrams to predict the stability or existence of homonuclear *di-*

atomic molecules of the third and subsequent periods. However, other than the heavier halogens, Cl_2, Br_2, and I_2, containing only single (sigma) bonds, no others exist at room temperature. We would predict from both Molecular Orbital theory and Valence Bond theory that the other (nonhalogen) homonuclear diatomic molecules from below the second period would exhibit multiple bonding, and therefore pi bonding. Some heavier elements exist as diatomic species, such as S_2, in the vapor phase at elevated temperatures, but most are neither prominent nor very stable. The instability appears to be directly related to the inability of the heavier elements to form strong pi bonds with each other. This is because the degree of overlap of atomic *p* orbitals on different atoms, and therefore the strength of pi bonds, decreases quite rapidly with increasing atomic size, and the accompanying increase in sigma bond length. For example, N_2 is much more stable than P_2 because the 3*p* orbitals on one phosphorus atom do not overlap side-by-side in a pi bonding manner with corresponding 3*p* orbitals on another phosphorus atom nearly as effectively as the corresponding 2*p* orbitals on the smaller nitrogen atoms do. Note that multiple bonding is not predicted for Cl_2, Br_2, and I_2, all of which have a bond order of one.

6–5 Heteronuclear Diatomic Molecules

1 Heteronuclear Diatomic Molecules of Second Period Elements

The corresponding atomic orbitals of two different elements, such as the 2*s* orbitals of carbon and oxygen atoms, have different energies because the nuclei of the atoms have different numbers of protons. As a result, molecular orbital diagrams such as Figures 6–4 and 6–5 are inappropriate for *heteronuclear* diatomic molecules. However, if the two elements are similar (as in CO, NO, or CN molecules, for example), we can simply modify the diagram of Figure 6–4 by skewing it slightly. Accordingly, Figure 6–8 shows the energy-level diagram and electron configuration for carbon monoxide, CO. The atomic orbitals of the *more electronegative element* (oxygen in this case) are *lower* in energy than the corresponding orbitals of the less electronegative element.

Note: CN is an unstable molecule, not a cyanide ion, CN^-.

The closer in energy a molecular orbital is to the energy of one of the atomic orbitals of which it is composed, the more of the character of that atomic orbital it shows. Thus, in the case of the CO molecule, the bonding molecular orbitals have more oxygen-like atomic orbital character and the antibonding orbitals have more carbon-like atomic orbital character.

In general, the energy differences ΔE_1, ΔE_2, and ΔE_3 reflect the difference in electronegativities between the two combining atoms. The greater these energy differences are, the more polar is the bond joining the atoms and the greater is its ionic character. On the other hand, the energy differences shown in color reflect the degree of overlap between atomic orbitals; the greater these differences are, the greater is the covalent character of the bond.

Notice that CO has a total of 14 electrons, making it isoelectronic with the stable N_2 molecule. Therefore, the distribution of electrons is the same in CO as in N_2, although we expect the energy levels of the molecular orbitals to be somewhat different. Bearing out our predictions, carbon monoxide is a very

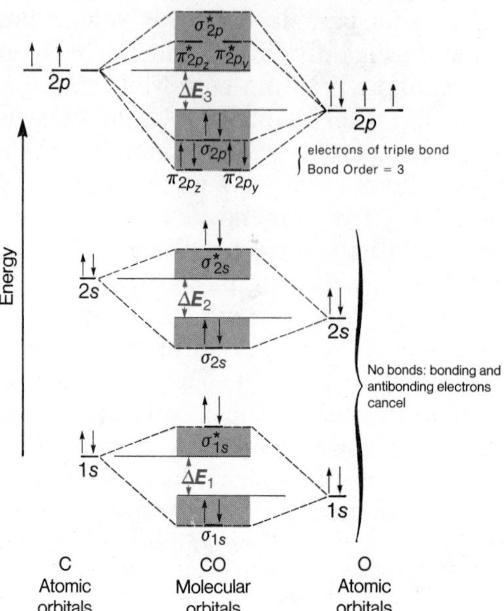

FIGURE 6–8 Energy level diagram for carbon monoxide, CO, a heteronuclear diatomic molecule. Note that the atomic orbitals of oxygen, the more electronegative element, are lower in energy than the corresponding atomic orbitals of carbon, the less electronegative element.

stable molecule, with a bond order of three and with a short carbon—oxygen (triple) bond length of 1.13 Å and a very high bond energy of 1071 kJ/mol.

2 The Hydrogen Fluoride Molecule, HF

The hydrogen fluoride molecule contains a very polar bond because the electronegativity (EN) difference between hydrogen (EN = 2.1) and fluorine (EN = 4.0) is very large (ΔEN = 1.9). Recall that the bond in HF involves the 1*s* electron of hydrogen and an unpaired electron from a fluorine 2*p* orbital. Figure 6–9 shows the overlap of the 1*s* orbital of hydrogen with a 2*p* orbital of fluorine to form σ_{sp} and $\sigma^{\star}_{sp}$ molecular orbitals. Since the remaining two fluorine 2*p* orbitals have no net overlap with hydrogen orbitals, they are called **nonbonding** orbitals. The same is true of the fluorine 2*s* and 1*s* orbitals. These nonbonding orbitals retain the characteristics of the fluorine atomic orbitals from which they are formed. The MO diagram of HF is shown in Figure 6–10.

6–6 Delocalization and the Shapes of Molecular Orbitals

In Section 5–5, we described resonance formulas for molecules and ions. Reso-

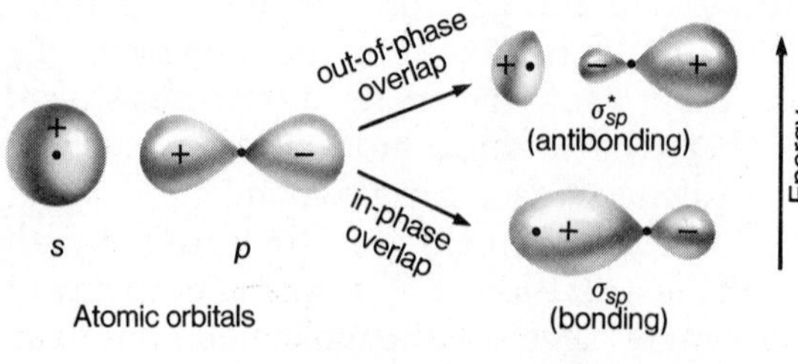

FIGURE 6–9 Formation of σ_{sp} and $\sigma^{\star}_{sp}$ molecular orbitals in HF by overlap of the 1*s* orbital of H with a 2*p* orbital of fluorine.

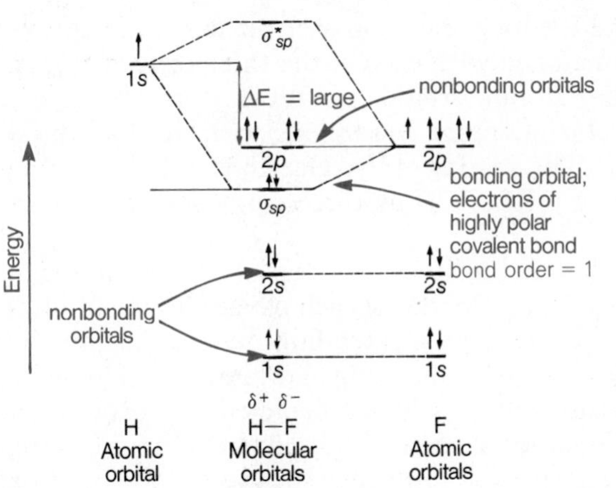

FIGURE 6–10 MO diagram for hydrogen fluoride, HF, a very polar molecule (μ = 1.9 D).

nance is said to exist when two or more equivalent Lewis dot formulas can be drawn for the same species and a single such formula does not account for the properties of a substance. A more appropriate term than resonance within the context of molecular orbital descriptions of bonding is **delocalization** of electrons. The shapes of molecular orbitals for species in which electron delocalization occurs can be determined by averaging all the contributing atomic orbitals.

1 The Carbonate Ion, CO_3^{2-}

Consider the trigonal planar carbonate ion, CO_3^{2-}, as an example. Valence Bond theory describes the ion in terms of three contributing resonance structures, shown at the top of Figure 6–11. Remember that no one of the three resonance forms adequately describes the bonding, since all the carbon—oxygen bonds in the ion have equal bond length and equal bond energy, intermediate between those of typical C—O and C═O bonds.

The bond order of a carbon—oxygen bond is $1\frac{1}{3}$ in the CO_3^{2-} ion.

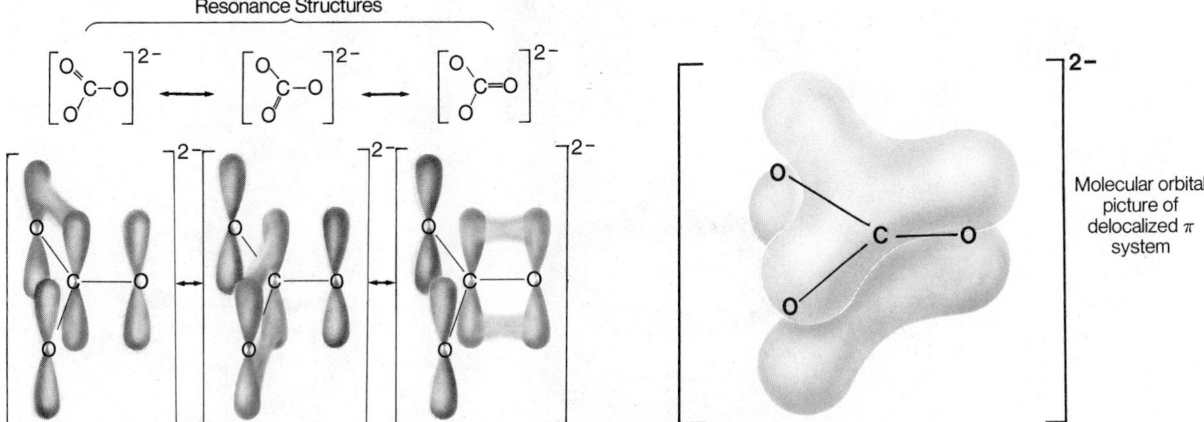

FIGURE 6–11 Resonance structures and the delocalized π MO system for the carbonate ion. A multicenter π system extends over more than two atoms (four atoms in CO_3^{2-}).

According to Valence Bond theory the carbon atom is described as sp^2-hybridized and it forms one sigma bond with each of the three oxygen atoms. This leaves one unhybridized $2p$ atomic orbital, say the $2p_z$ orbital, on the carbon atom, which is capable of overlapping and mixing with the $2p_z$ orbital of any of the three oxygen atoms. The sharing of two electrons in the resulting localized pi orbital would form a pi bond. Thus three equivalent resonance structures can be drawn.

The MO description of the pi bonding involves the overlap and mixing of the $2p_z$ orbital of the carbon with the $2p_z$ orbital of each oxygen atom simultaneously, to form a bonding pi molecular orbital extending above and below the plane of the sigma system, as well as an antibonding pi orbital. Two electrons are said to occupy the entire bonding pi MO, which is depicted at the bottom of Figure 6–11. Note that it has the same shape as we would obtain by averaging each of the three contributing Valence Bond resonance structures. The bonding

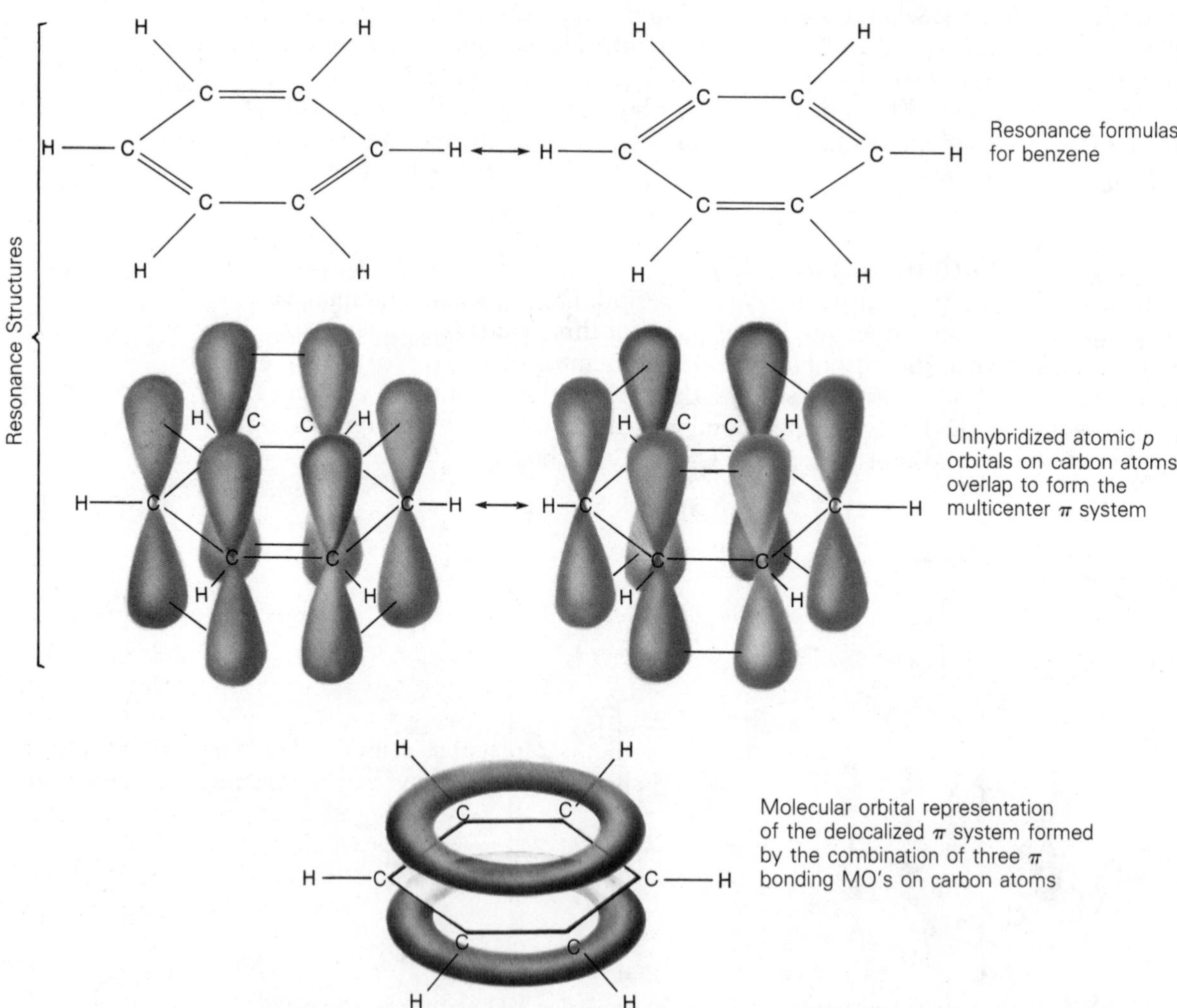

Resonance Structures
Resonance formulas for benzene
Unhybridized atomic p orbitals on carbon atoms overlap to form the multicenter π system
Molecular orbital representation of the delocalized π system formed by the combination of three π bonding MO's on carbon atoms

FIGURE 6–12 Resonance structures and molecular orbital representation of delocalized π system of benzene, C_6H_6.

in such species as nitrate ion, NO_3^-, and sulfur dioxide, SO_2, can be described in comparable terms.

2 The Benzene Molecule, C_6H_6

Now let us consider the benzene molecule, C_6H_6, whose two Valence Bond resonance forms are shown at the top of Figure 6–12. The Valence Bond description involves sp^2 hybridization at each carbon atom. Not only is each carbon atom at the center of a trigonal plane, but the entire molecule is known to be planar. There are sigma bonds from each carbon atom to the two adjacent carbon atoms and to one hydrogen atom. This leaves one unhybridized $2p_z$ orbital on each carbon atom and one remaining valence electron for each. According to Valence Bond theory, adjacent pairs of $2p_z$ orbitals, and the six remaining electrons occupy the regions of overlap to form a total of three pi bonds in either one of the two ways shown in the middle of Figure 6–12.

The actual structure of the C_6H_6 molecule indicates that it does not contain alternating single and double carbon—carbon bonds. The usual C—C single bond length is 1.54 Å and the usual C=C double bond length is 1.34 Å. Experimentally we find that all six of the carbon—carbon bonds in benzene have the same length, 1.39 Å, intermediate between those of single and double bonds. This is well explained by the MO theory which produces the prediction that the six $2p_z$ orbitals of the carbon atoms overlap and mix to form three pi bonding and three pi antibonding molecular orbitals. The bottom of Figure 6–12 shows the MO representation of the delocalized pi system due to the combination of the three pi-bonding MOs. It is this extended system of MOs that contains the six pi elecrons, which are distributed throughout the molecule as a whole, above and below the plane of the sigma-bonded framework. This results in identical character for all carbon—carbon bonds in benzene. Each carbon—carbon bond has a bond order of $1\frac{1}{2}$. Note that, as in the CO_3^{2-} ion, the MO representation of the extended pi system is the same as one would obtain by averaging the two contributing Valence Bond resonance structures.

Key Terms

Antibonding orbital a molecular orbital higher in energy than any of the atomic orbitals from which it is derived; lends instability to a molecule or ion when populated with electrons; denoted with an asterisk (*) superscript on symbol.

Bond energy the amount of energy necessary to break one mole of bonds of a given kind (in gas phase).

Bond order half the number of electrons in bonding orbitals minus half the number of electrons in antibonding orbitals.

Bonding orbital a molecular orbital lower in energy than any of the atomic orbitals from which it is derived; lends stability to a molecule or ion when populated with electrons.

Molecular orbital (MO) an orbital resulting from overlap and mixing of atomic orbitals on different atoms that belongs to the molecule as a whole.

Molecular orbital theory a theory of chemical bonding based upon the postulated existence of molecular orbitals.

Nodal plane (node) a region in which the probability of finding an electron is zero.

Nonbonding orbital a molecular orbital derived only from an atomic orbital of one atom;

lends neither stability nor instability to a molecule or ion when populated with electrons.

Pi bond bond resulting from electron occupation of a pi molecular orbital.

Pi (π) orbital molecular orbital resulting from side-on overlap of atomic orbitals.

Sigma bond bond resulting from electron occupation of a sigma molecular orbital.

Sigma (σ) orbital molecular orbital resulting from head-on overlap of two atomic orbitals.

Exercises

1. Two of the following three theories attempt to account for *how* chemical bonds are formed. The other one does not: (1) VSEPR, (2) VB, (3) MO. Which one does not? Justify your answer. What *does* this theory address?
2. What differences and similarities exist for atomic orbitals, localized hybridized atomic orbitals according to Valence Bond theory, and molecular orbitals?
3. What is the relationship between the maximum number of electrons that can be accommodated by a set of molecular orbitals and the maximum number that can be accommodated by the atomic orbitals from which the MOs are formed? What is the maximum number of electrons that one MO can hold?
4. Answer Exercise 3 after replacing "molecular orbitals" with "localized hybridized atomic orbitals."
5. Sketch and describe the molecular orbitals resulting from the following overlap of atomic orbitals: (a) two *s* orbitals (b) head-on overlap of two *p* orbitals (c) side-on overlap of two *p* orbitals.
6. State Hund's Rule. Describe the application of Hund's Rule to electron occupancy of molecular orbitals.
7. How are the electron occupancies of bonding, nonbonding, and antibonding orbitals related to the stability of a molecule or ion?
8. From memory, draw the energy-level diagram of molecular orbitals produced by the overlap of orbitals of two identical atoms from the second period. (Show the π_{2p_y} and π_{2p_z} orbitals below the σ_{2p} in energy.)
9. Utilizing sets of MOs such as you drew in Exercise 8, determine the electron configurations of the following molecules and ions: Li_2, Li_2^+, C_2, C_2^-, C_2^{2-}, N_2^-, and B_2^+.
10. (a) What is the bond order of each of the species in Exercise 9? (b) Which are diamagnetic (D) and which are paramagnetic (P)? (c) Apply MO theory to predict relative stabilities of these species. (d) Comment on the validity of the predictions from (c). What else *must* be considered in addition to electron occupancy of MOs?
11. How does the effectiveness of overlap of atomic orbitals on two atoms vary with the relative energies of the overlapping orbitals?
12. How does the polar or nonpolar character of a bond formed by overlap of atomic orbitals on two atoms depend upon the relative energies of the overlapping orbitals? Assume that two electrons occupy the resulting MO.
13. Which homonuclear diatomic molecules or ions of the second period have the following electron distributions in MOs? In other words, identify X in each case below.

    (a) X_2 $\sigma_{1s}^2\ \sigma_{1s}^{\star 2}\ \sigma_{2s}^2\ \sigma_{2s}^{\star 2}\ \pi_{2p_y}^1\ \pi_{2p_z}^1$

    (b) X_2^- $\sigma_{1s}^2\ \sigma_{1s}^{\star 2}\ \sigma_{2s}^2\ \sigma_{2s}^{\star 2}\ \sigma_{2p}^2\ \pi_{2p_y}^2\ \pi_{2p_z}^2\ \pi_{2p_y}^{\star 2}\ \pi_{2p_z}^{\star 1}$

    (c) X_2^+ $\sigma_{1s}^2\ \sigma_{1s}^{\star 2}\ \sigma_{2s}^2\ \sigma_{2s}^{\star 2}\ \sigma_{2p}^2\ \pi_{2p_y}^2\ \pi_{2p_z}^2\ \pi_{2p_y}^{\star 1}$
14. What is the bond order of each of the species of Exercise 13?
15. Assuming that the σ_{2p} MO is lower in energy than the π_{2p_y} and π_{2p_z} MOs for the following species, write out electronic configurations for all of them: O_2, O_2^-, O_2^{2-}, F_2, F_2^+, F_2^-, Ne_2, and Ne_2^+.
16. (a) What is the bond order of each of the species in Exercise 15? (b) Are they diamagnetic or paramagnetic? (c) What would the appli-

cation of MO theory predict regarding the stabilities of these species?

17. Is it possible for a molecule or complex ion in its ground state to have a negative bond order? Why?
18. From memory draw the energy level diagram of molecular orbitals for a heteronuclear diatomic molecule, XY, in which both X and Y are from period 2 and Y is the more electronegative element.
19. Draw and label the complete MO energy level diagrams for the following species: (a) Li_2, (b) Be_2^+, (c) C_2, (d) F_2, (e) Cl_2, (f) F_2^-, (g) HeH, (h) HO, (i) HCl.
20. Determine the bond order and predict the stabilities of the species in Exercise 19 based only on electron occupancy of MOs.
21. Draw MO energy-level diagrams, write out electronic configurations, determine bond orders, indicate predicted diamagnetism or paramagnetism, and predict relative stabilities of the following molecules and ions on the basis of MO theory: OF, OF^+, NF, NF^-, BC, and BC^-. Assume that the σ_{2p} MO is lower in energy than the π_{2p_y} and π_{2p_z} MOs for all except BC and BC^-.
22. Rationalize the following observations in terms of the stabilities of sigma and pi bonds. (a) The most common form of nitrogen is N_2, whereas the most common form of phosphorus is P_4 (see Figure 1–12). (b) The most common forms of oxygen are O_2 and (less common) O_3, whereas the most common form of sulfur is S_8 (see Figure 1–12).
23. Considering the shapes of MO energy level diagrams for nonpolar covalent and polar covalent molecules, what would you predict about MO diagrams, and therefore overlap of atomic orbitals, for ionic compounds?
24. Draw Lewis dot formulas depicting the resonance structures of the following species from the Valence Bond point of view, and then draw MOs for the delocalized π systems: (a) SO_2, (b) HCO_3^-, hydrogen carbonate ion, (c) SO_3, sulfur trioxide, (d) O_3, ozone, (e) C_2H_4, ethene, (f) C_2H_2, ethyne.

7 Chemical Reactions: A Systematic Study

In this chapter we shall examine chemical reactions in some detail. Since millions of reactions are known, it is useful to group them into classes or types so that we can deal with such massive amounts of information systematically. We shall classify them as (1) combination reactions, (2) displacement reactions, (3) decomposition reactions, (4) metathesis reactions, and (5) oxidation-reduction reactions. We shall also distinguish among (1) molecular, (2) total ionic, and (3) net ionic equations for chemical reactions and indicate the advantages and disadvantages of these methods for representing chemical reactions.

The classification scheme will then be illustrated by typical reactions of hydrogen, oxygen, and their compounds. These reactions will also serve to illustrate further periodic relationships with respect to chemical properties. It should be emphasized that our system is not an attempt to transform nature so that it fits into small categories; rather it is an attempt to take observations of nature and give them some order. Additionally, you may observe that many oxidation-reduction reactions also fit into one of the first three categories, and that some (relatively few) reactions do not fit neatly into any of these categories.

As we study different kinds of chemical reactions, we shall observe that we can use this knowledge to predict the products of other similar reactions.

7–1 Aqueous Solutions—An Introduction

In Chapter 2 we introduced solution terminology and methods of expressing concentrations of solutions. Many chemical reactions occur in aqueous (water) solutions. Thus it is useful to know the kinds of substances that are soluble in water, and the forms in which dissolved substances exist, before we begin our systematic study of reactions.

1 Electrolytes and Extent of Ionization

Solutes that are water soluble can be classified as either electrolytes or nonelectrolytes. **Electrolytes** are *substances whose aqueous solutions conduct electrical current.* **Strong electrolytes** are *substances that conduct electricity well in dilute aqueous solution,* while **weak electrolytes** *conduct electricity poorly in dilute aqueous solution.* Aqueous solutions of **nonelectrolytes** do not conduct electricity. Since electrical current is carried through aqueous solution by the movement of ions, the strength of an electrolyte depends upon both its tendency to ionize, or dissociate into ions, and the charges on the ions (see Figure 7–1).

Three major classes of solutes are strong electrolytes: *(1) Strong acids, (2) strong soluble bases, and (3) most soluble salts are completely or nearly completely ionized in dilute aqueous solutions, and therefore are strong electrolytes.* An **acid** can be defined as *a substance that produces hydrogen ions, H^+, in aqueous solutions,* and a **base** as *a substance that produces hydroxide ions, OH^-, in aqueous solutions.* A **salt** is *a compound that contains a cation other than H^+ and an anion other than hydroxide ion, OH^-, or oxide ion, O^{2-}.* As we shall see later in this chapter, salts are formed when acids react with bases. Solutions of salts with highly charged ions, like iron(II) sulfate, $FeSO_4$, are better conductors than equally concentrated solutions of salts with less highly charged ions like potassium bromide, KBr.

Acids and bases are further identified in Subsections 2, 3, and 4 below.

Aqueous solutions of $FeSO_4$ contain equal concentrations of Fe^{2+} and SO_4^{2-} ions; aqueous solutions of KBr contain equal concentrations of K^+ and Br^- ions.

$$FeSO_4\ (s) \xrightarrow{H_2O} Fe^{2+}\ (aq) + SO_4^{2-}\ (aq)$$

$$KBr\ (s) + \xrightarrow{H_2O} K^+\ (aq) + Br^-\ (aq)$$

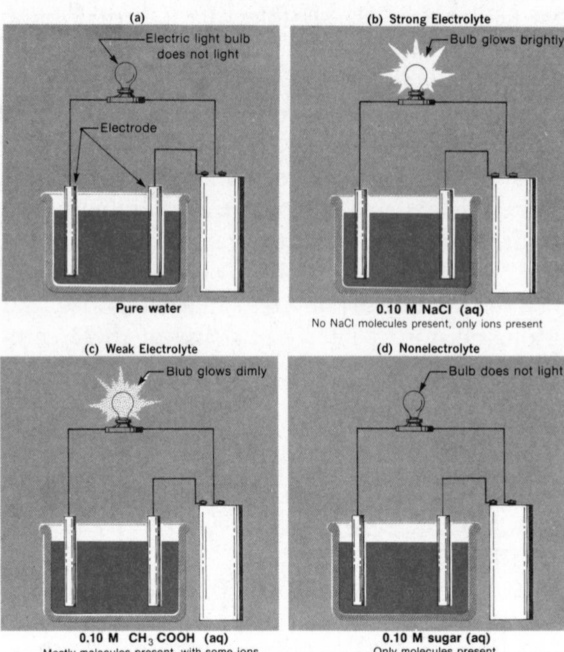

FIGURE 7–1 Representation of an experiment to determine the presence of ions in solution.

2 Strong and Weak Acids

Many properties of aqueous solutions of acids are due to H^+ ions. These are described in Section 11–4.

As a matter of convenience we place acids into two classes: strong acids and weak acids. **Strong acids** *ionize (separate into ions) completely, or very nearly completely, in dilute aqueous solution.* The seven common strong acids and their anions are listed in Table 7–1.

Since, by definition, strong acids ionize completely or very nearly completely in dilute solutions, their solutions contain (predominantly) the ions of the acid rather than acid molecules. The equation for the ionization of hydrochloric acid illustrates the point. Pure hydrogen chloride, HCl, is a polar covalent compound that is a gas at room temperature and atmospheric pressure. It dissolves in water to produce a solution that contains equal concentrations of hydrogen ions and chloride ions

The notation (aq) following ions in aqueous solution indicates that they are hydrated, that is, they interact with water molecules in solution.

$$HCl\ (g) \longrightarrow H^+\ (aq) + Cl^-\ (aq)$$

TABLE 7–1 Strong Acids and Their Anions

Common Strong Acids		Anions Derived from Common Strong Acids	
Formula	**Name**	**Formula**	**Name**
HCl	hydrochloric acid	Cl^-	chloride ion
HBr	hydrobromic acid	Br^-	bromide ion
HI	hydroiodic acid	I^-	iodide ion
HNO_3	nitric acid	NO_3^-	nitrate ion
$HClO_4$	perchloric acid	ClO_4^-	perchlorate ion
$HClO_3$	chloric acid	ClO_3^-	chlorate ion
H_2SO_4	sulfuric acid	HSO_4^-	hydrogen sulfate ion
		SO_4^{2-}	sulfate ion

Similar equations can be written for all strong acids.

The species we have shown in H^+ (aq) is sometimes shown as H_3O^+ or $[H(H_2O)^+]$ to emphasize hydration. However, it really exists in varying degrees of hydration, such as $H(H_2O)^+$, $H(H_2O)_2{}^+$ and $H(H_2O)_3{}^+$.

Weak acids *ionize only slightly in dilute aqueous solution;* the list of weak acids is very long. Many common ones are listed in Appendix F and a few of them and their anions are given in Table 7–2. We usually write the formulas of inorganic acids with hydrogen first. Organic acids can often be recognized by the presence of the COOH group in the formula.

The equation for the ionization of acetic acid, CH_3COOH, in water is typical of weak acids

$$CH_3COOH\ (\ell) \rightleftharpoons H^+\ (aq) + CH_3COO^-\ (aq)$$

The double arrow ($\rightleftharpoons$) generally signifies that the reaction occurs in both directions, and the forward reaction does not go to completion. All of us are familiar with solutions of acetic acid. Vinegar is 5% acetic acid by mass. Our use of oil and vinegar as a salad dressing tells us that acetic acid is a weak acid. We wouldn't think of using oil and 5% hydrochloric acid (which is "stomach acid") as a salad dressing! To be specific, acetic acid is 1.3% ionized (and 98.7% nonionized) in 0.10 molar solution.

Our stomachs have linings that are much more resistant to attack by acids than our other tissues.

Acetic acid is an example of an organic acid, most of which are weak acids. A multitude of organic acids occur in living systems. Organic acids contain carbon, specifically the carboxylate grouping of atoms, —COOH or $-C{\overset{\ddot{O}:}{\lessgtr}}_{\ddot{O}-H}$

They can ionize slightly by breakage of the O—H bond

$$H_3C-C(=\ddot{O}:)-\ddot{O}-H \rightleftharpoons H_3C-C(=\ddot{O}:)-\ddot{O}:^-\ (aq) + H^+\ (aq)$$

TABLE 7–2 Common Weak Acids and Their Anions

Common Weak Acids		Anions Derived from Common Weak Acids	
Formula	**Name**	**Formula**	**Name**
HF*	hydrofluoric acid	F^-	fluoride ion
CH_3COOH	acetic acid	CH_3COO^-	acetate ion
HCN	hydrocyanic acid	CN^-	cyanide ion
HNO_2†	nitrous acid	$NO_2{}^-$	nitrite ion
H_2CO_3†	carbonic acid	$HCO_3{}^-$ $CO_3{}^{2-}$	hydrogen carbonate ion carbonate ion
H_2SO_3†	sulfurous acid	$HSO_3{}^-$ $SO_3{}^{2-}$	hydrogen sulfite ion sulfite ion
H_3PO_4	phosphoric acid	$H_2PO_4{}^-$ $HPO_4{}^{2-}$ $PO_4{}^{3-}$	dihydrogen phosphate ion hydrogen phosphate ion phosphate ion
$(COOH)_2$	oxalic acid	$H(COO)_2{}^-$ $(COO)_2{}^{2-}$	hydrogen oxalate ion oxalate ion

*Note that HF is a weak acid while HCl, HBr, and HI are strong acids.

†Free acid molecules exist only in dilute aqueous solution, or not at all. However, many salts of these acids are common, stable compounds.

Inorganic acids may be strong or weak; most organic acids are weak.

Organic acids are discussed as a class in Chapter 30. Carbonic acid, H_2CO_3, and hydrocyanic acid, HCN (aq), are the common acids that contain carbon but that are considered *inorganic* acids. Inorganic acids are often called **mineral acids** because they are obtained primarily from mineral (nonliving) sources.

3 Reversible Reactions

The double arrow ($\rightleftharpoons$) also indicates that the ionization reaction for acetic acid is reversible. Reactions that can occur in both directions are **reversible reactions.** Let us illustrate the fundamental difference between reactions that go to completion and reversible reactions. We have seen that the ionization of HCl in water is nearly complete. Suppose we dissolve some table salt, NaCl, in water and then add some dilute nitric acid to it. The resulting solution contains hydrogen ions and chloride ions, the products of the ionization of HCl (as well as sodium ions and nitrate ions). The H^+ and Cl^- ions do *not* react significantly to form nonionized molecules of HCl, the reverse of the ionization of HCl. In contrast, when a sample of sodium acetate, $NaCH_3COO$, is dissolved and mixed with nitric acid, the resulting solution contains hydrogen ions and acetate ions (as well as sodium ions and nitrate ions). Most of the H^+ and CH_3COO^- ions *do* combine to produce nonionized molecules of acetic acid, which is the reverse of the ionization of the acid. Thus, the ionization of acetic acid, and that of any other weak electrolyte, is reversible

$$H^+ \text{ (aq)} + CH_3COO^- \text{ (aq)} \rightleftharpoons CH_3COOH \text{ (aq)}$$

The complete ionization of a strong electrolyte is indicated by a single arrow ($\longrightarrow$).

4 Strong Soluble Bases, Insoluble Bases, and Weak Bases

Solutions of bases have a set of common properties due to the OH^- ion. These are described in Section 11–4.

Most common bases are metal hydroxides, and most are insoluble in water. **Strong soluble bases** are *soluble in water and are dissociated completely in dilute aqueous solution.* The strong soluble bases are listed in Table 7–3. They are the hydroxides of the Group IA metals and the heavier members of Group IIA.

TABLE 7–3 Strong Soluble Bases

LiOH	lithium hydroxide		
NaOH	sodium hydroxide		
KOH	potassium hydroxide	$Ca(OH)_2$	calcium hydroxide
RbOH	rubidium hydroxide	$Sr(OH)_2$	strontium hydroxide
CsOH	cesium hydroxide	$Ba(OH)_2$	barium hydroxide

Strong soluble bases are ionic compounds in the solid state.

The equation for the dissociation of sodium hydroxide in water is typical and similar equations can be written for all strong soluble bases

$$NaOH \text{ (s)} \xrightarrow{H_2O} Na^+ \text{ (aq)} + OH^- \text{ (aq)}$$

The weak bases are *covalent* substances that ionize only slightly in water; they are sometimes called **molecular bases.**

Other metals form ionic hydroxides but these are so sparingly soluble in water that they cannot produce strongly basic solutions and are called **insoluble bases. Weak bases** such as ammonia, NH_3, are very soluble in water but ionize only slightly in solution.

$$NH_3 \text{ (g)} + H_2O \text{ } (\ell) \rightleftharpoons NH_4^+ \text{ (aq)} + OH^- \text{ (aq)}$$

5 Solubility Rules for Compounds in Aqueous Solutions

Solubility is a complex phenomenon, and it is not possible to state simple rules that cover all cases. Although the following rules for solutes in aqueous solutions are not comprehensive, they will be of great value for nearly all acids, bases, and salts encountered in general chemistry. Compounds that dissolve in water to the extent of approximately 0.02 mole per liter are usually classified as "soluble" compounds, while those that are less soluble are classified as "insoluble" compounds. No gaseous or solid substances are infinitely soluble in water.

We shall consider some of the factors that influence solubility in Chapter 10.

1. The common inorganic acids are soluble in water. Low molecular weight organic acids are soluble.
2. The common compounds of the Group IA metals (Na, K, etc.) and the ammonium ion, NH_4^+, are soluble in water.
3. The common nitrates, NO_3^-, acetates, CH_3COO^-, chlorates, ClO_3^-, and perchlorates, ClO_4^-, are soluble in water.
4. a. The common chlorides, Cl^-, are soluble in water except AgCl, Hg_2Cl_2, and $PbCl_2$.
   b. The common bromides, Br^-, and iodides, I^-, show approximately the same solubility behavior as chlorides, but there are some exceptions. As the halides ions (Cl^-, Br^-, I^-) increase in size, the solubilities of their slightly soluble compounds decrease. Although $HgCl_2$ is readily soluble in water, $HgBr_2$ is only slightly soluble and HgI_2 is even less soluble.
   c. The solubilities of the pseudohalides, CN^- (cyanide) and SCN^- (thiocyanate), are quite similar to those of the corresponding iodides. Additionally, both CN^- and SCN^- show strong tendencies to form soluble compounds containing complex ions.
5. The common sulfates, SO_4^{2-}, are soluble in water except $PbSO_4$, $BaSO_4$ and $HgSO_4$; $CaSO_4$ and Ag_2SO_4 are sparingly soluble.
6. The common metal hydroxides, OH^-, are *insoluble* in water except those of the Group IA metals and the heavier members of the Group IIA metals, beginning with $Ca(OH)_2$.
7. The common carbonates, CO_3^{2-}, phosphates, PO_4^{3-}, and arsenates, AsO_4^{3-}, are *insoluble* in water except those of the Group IA metals and NH_4^+. $MgCO_3$ is fairly soluble.
8. The common sulfides, S^{2-}, are *insoluble* in water except those of the Group IA and Group IIA metals and the ammonium ion.

They are called pseudohalides because their reactions and properties are somewhat similar to those of halide ions.

strong bases

Now that we have distinguished between strong and weak electrolytes and soluble and insoluble compounds, let us take a systematic look at chemical reactions. The ideas we have just developed become especially important when we discuss the distinction between molecular and ionic equations, beginning in Section 7-3.

The fact that a substance is insoluble in water does *not* mean that it can't take part in a reaction in contact with water.

Classification of Chemical Reactions

7-2 Combination Reactions

Reactions in which two or more substances combine to form a compound are called **combination reactions.** They may involve (1) the combination of two elements to form a compound, (2) the combination of an element and a compound to form a single new compound, or (3) the combination of two compounds to form a single new compound. Let's examine some of these reactions.

1 Combination of Two Elements to Form a Compound

Most metals react with most nonmetals to form binary ionic compounds. The Group IA metals combine with the Group VIIA nonmetals to form binary *ionic* compounds with the general formula MX (Section 4–10)

$$2M + X_2 \longrightarrow 2(M^+X^-) \quad \text{alkali metal halides} \qquad M = Li, Na, K, Rb, Cs;\ X = F, Cl, Br, I$$

TABLE 7–4 Alkali Metal Halides—Compounds Formed by Group IA and VIIA Elements*

LiF	LiCl	LiBr	LiI
NaF	NaCl	NaBr	NaI
KF	KCl	KBr	KI
RbF	RbCl	RbBr	RbI
CsF	CsCl	CsBr	CsI

* Compounds of francium (Fr) and astatine (At) are omitted from this tabulation because both are artifical, radioactive elements. Their compounds are similar to those listed.

The general equation thus represents the 20 combination reactions that form the ionic compounds listed in Table 7–4.

Specifically, sodium, a silvery white metal, combines with chlorine, a pale green gas, to form sodium chloride, or ordinary table salt

$$2Na\ (s) + Cl_2\ (g) \longrightarrow 2NaCl\ (s) \qquad \text{sodium chloride}$$

All members of both families undergo similar reactions. For example, potassium combines with fluorine to form potassium fluoride

$$2K\ (s) + F_2\ (g) \longrightarrow 2KF\ (s) \qquad \text{potassium fluoride}$$

As we might expect, the Group IIA metals also combine with the Group VIIA nonmetals to form binary ionic compounds (except $BeCl_2$, $BeBr_2$, and BeI_2 which are covalent). In general terms these combination reactions may be represented as

$$M\ (s) + X_2 \longrightarrow MX_2\ (s) \quad \text{alkaline earth metal halides} \qquad M = Be, Mg, Ca, Sr, Ba;\ X = F, Cl, Br, I$$

As a specific example of this kind of combination, consider the reaction of magnesium with fluorine to form magnesium fluoride

$$Mg\ (s) + F_2\ (g) \longrightarrow MgF_2\ (s) \qquad \text{magnesium fluoride}$$

Since all the IIA and VIIA elements undergo similar reactions, the general equation, written above, represents 20 reactions if we omit radium and astatine, the rare and highly radioactive members of the families.

When two nonmetals combine with each other they form binary *covalent* compounds. In such reactions, the oxidation number of the less electronegative element is often variable, depending upon reaction conditions. For example, phosphorus combines with a limited amount of chlorine to form phosphorus trichloride, in which phosphorus exhibits the +3 oxidation state.

$$P_4\ (s) + 6Cl_2\ (g) \longrightarrow 4\overset{+3\text{ ox. state}}{P}Cl_3\ (\ell) \qquad \text{(with limited } Cl_2)$$

With an excess of chlorine, the product is phosphorus pentachloride, which contains phosphorus in the +5 oxidation state

$$P_4\ (s) + 10Cl_2\ (g) \longrightarrow 4\overset{+5\text{ ox. state}}{P}Cl_5\ (s) \qquad \text{(with excess } Cl_2)$$

In general, *the higher oxidation state of a nonmetal is formed when it reacts with an excess of another nonmetal.* Let us illustrate further with the reactions of sulfur with a limited amount and with an excess of fluorine.

Nonmetals in odd-numbered periodic groups tend to favor odd oxidation states, while those in even-numbered groups favor even oxidation numbers in their compounds. P is in Group VA and S is in Group VIA. The *maximum* oxidation number for a representative element is equal to its periodic group number.

$$S_8\ (s) + 16F_2\ (g) \longrightarrow 8\ \overset{+4 \text{ ox. state}}{S}F_4\ (g) \qquad \text{(with limited } F_2)$$

$$S_8\ (s) + 24F_2\ (g) \longrightarrow 8\ \overset{+6 \text{ ox. state}}{S}F_6\ (g) \qquad \text{(with excess } F_2)$$

Sulfur is in the +4 and +6 oxidation states, respectively, in SF_4 and SF_6.

Many more reactions in which two elements combine to form a compound could be cited. Let us now consider combination reactions in which an element combines with a compound to form a new compound.

2 Combination of An Element and a Compound to Form a New Compound

Phosphorus in the +3 oxidation state in PCl_3 molecules can be converted to the +5 state in PCl_5 by combination with chlorine

$$PCl_3\ (g) + Cl_2\ (g) \longrightarrow PCl_5\ (s)$$

Likewise, sulfur in the +4 state is converted to the +6 state when SF_4 reacts with fluorine to form SF_6

$$SF_4\ (g) + F_2\ (g) \longrightarrow SF_6\ (g)$$

3 Combination of Two Compounds to Form a New Compound

An example of reactions in this category is the combination of calcium oxide with carbon dioxide to produce calcium carbonate

$$CaO\ (s) + CO_2\ (g) \longrightarrow CaCO_3\ (s)$$

Another is the formation of sulfuric acid by addition of pyrosulfuric acid to water, a reaction used in the manufacture of H_2SO_4

$$H_2S_2O_7\ (\ell) + H_2O\ (\ell) \longrightarrow 2H_2SO_4\ (\ell)$$

7–3 Displacement Reactions

Reactions in which one element displaces another from a compound are called **displacement reactions.** *Active metals* displace less active metals or hydrogen from their compounds in aqueous solution and also in many solids. Active metals are those with low ionization energies that readily lose electrons to form cations (see Table 7–5).

When metallic copper is added to a solution of (colorless) silver nitrate, the more active metal, copper, displaces silver ions from the solution. The resulting solution contains blue copper(II) nitrate, and metallic silver falls to the bottom of the container as a finely divided solid.

Since we did not study periodic trends in properties of transition metals in Chapter 4, it would be difficult for you to predict that Cu is more active than Ag. The fact that this reaction occurs indicates that it is.

$$2AgNO_3\ (aq) + Cu\ (s) \longrightarrow 2Ag\ (s) + Cu(NO_3)_2\ (aq)$$

The above equation is called a **molecular equation** because formulas are written as though all species existed as molecules. They don't. Both silver nitrate

TABLE 7–5
Activity Series of the Metals

Li
K
Ca
Na
Mg
Al
Mn
Zn
Cr
Fe
Cd
Co
Ni
Sn
Pb
H (a nonmetal)
Sb (a metalloid)
Cu
Hg
Ag
Pt
Au

(Bracketed ranges in the table, from the top of the series: Displace hydrogen from acids (Li through Pb); Displace hydrogen from steam; Displace hydrogen from cold water.)

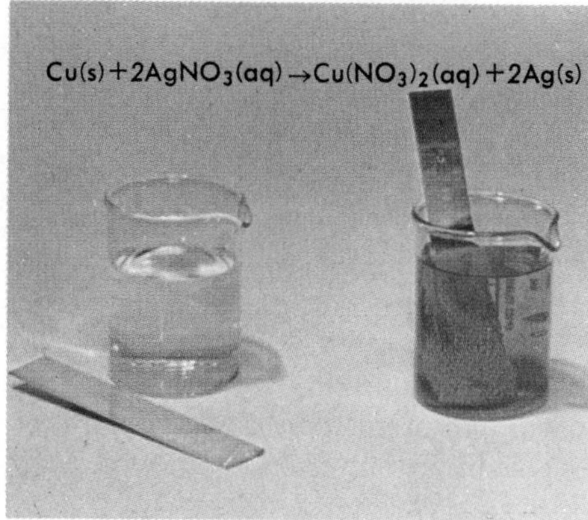

FIGURE 7–2 A displacement reaction. Copper atoms displace silver ions from an aqueous solution of silver nitrate.

and copper(II) nitrate are soluble ionic compounds (solubility rules, Section 7–1, part 5). In **total ionic equations,** formulas are written to show the (predominant) form in which each substance exists in, or in contact with, aqueous solution. The total ionic equation for this reaction is

$$[Ag^+ (aq) + NO_3^- (aq)] + Cu (s) \longrightarrow 2Ag (s) + [Cu^{2+} (aq) + 2NO_3^- (aq)]$$

Examination of the total ionic equation reveals that the nitrate ions, NO_3^-, do not participate in the reaction. They are sometimes called "spectator" ions. Elimination of species that are not actually involved in the reaction gives the **net ionic equation,** which shows only the species that react.

Brackets are not used in net ionic equations.

$$2Ag^+ (aq) + Cu (s) \longrightarrow 2Ag (s) + Cu^{2+} (aq)$$

A common method for the preparation of small amounts of hydrogen involves the reaction of active metals with acids, such as hydrochloric acid and sulfuric acid. For example, when zinc is dissolved in sulfuric acid, the reaction produces zinc sulfate; hydrogen is displaced from the acid, and it bubbles off as gaseous H_2. The molecular equation for this reaction is

$$\underset{\text{strong acid}}{Zn (s) + H_2SO_4 (aq)} \longrightarrow \underset{\text{soluble salt}}{ZnSO_4 (aq)} + H_2 (g)$$

Both sulfuric acid (in dilute solution) and zinc sulfate exist primarily as ions, so the total ionic equation is

$$Zn (s) + [2H^+ (aq) + SO_4^{2-} (aq)] \longrightarrow [Zn^{2+} (aq) + SO_4^{2-} (aq)] + H_2 (g)$$

Elimination of unreacting species common to both sides of the total ionic equation gives the net ionic equation

Nitric acid is the only common oxidizing acid. Other common inorganic acids are nonoxidizing in dilute aqueous solution. When HNO_3 reacts with metals, no H_2 is formed, but oxides of nitrogen are produced.

$$Zn (s) + 2H^+ (aq) \longrightarrow Zn^{2+} (aq) + H_2 (g)$$

Table 7–5 lists the **activity series.** When any metal listed above hydrogen in this series is added to solutions of nonoxidizing acids such as hydrochloric acid, HCl, and sulfuric acid, H_2SO_4, the metal dissolves to produce hydrogen and a salt is formed.

Very active metals can even displace hydrogen from water. However, the reactions of very active metals of Group IA are dangerous because they gener-

FIGURE 7–3 Potassium, a Group I metal, like other members of this group, reacts vigorously with water. The light for this photograph was produced by dropping a small amount of potassium into a beaker of water.

ate enough heat to cause explosive ignition of the hydrogen. The reaction of potassium, or any other metal of Group IA, with water is also a displacement reaction

$$2K\ (s) + 2H_2O\ (\ell) \longrightarrow 2KOH\ (aq) + H_2\ (g)$$

Many *nonmetals* displace less active nonmetals from combination with a metal or other cation. For example, when chlorine is bubbled through a solution containing bromide ions (derived from a soluble ionic salt such as sodium bromide, NaBr), chlorine displaces bromide ions to form elemental bromine and chloride ions (as aqueous sodium chloride):

An active nonmetal is one that forms a monatomic anion readily.

$$\underset{\text{chlorine}}{Cl_2\ (g)} + \underset{\text{bromide ions}}{2Br^-\ (aq)} \longrightarrow \underset{\text{chloride ions}}{2Cl^-\ (aq)} + \underset{\text{bromine}}{Br_2\ (\ell)}$$

Similarly, when bromine is added to a solution containing iodide ions, the iodide ions are displaced by bromine to form iodine and bromide ions

$$\underset{\text{bromine}}{Br_2\ (\ell)} + \underset{\text{iodide ions}}{2I^-\ (aq)} \longrightarrow \underset{\text{bromide ions}}{2Br^-\ (aq)} + \underset{\text{iodine}}{I_2\ (s)}$$

Each halogen will displace less electronegative (heavier) halogens from their binary salts. Conversely, a halogen will *not* displace more electronegative members from their salts

Electronegativity of the halogens decreases as the group is descended.

$$I_2\ (s) + 2F^-\ (aq) \longrightarrow \text{No Reaction}$$

> Net ionic equations allow us to focus on the *essence* of a chemical reaction in aqueous solution, that is, the changes that *really* occur. On the other hand, if we are dealing with stoichiometric calculations we frequently must deal with formula weights, and therefore with the *complete* formulas of all species. In such cases, molecular equations are more useful. Total ionic equations provide the bridge between the two.

Here is why it is important to know how and when to construct net ionic equations from molecular equations.

In general, you must answer two questions about a substance to determine whether or not it should be written in ionic form in an ionic equation:

1. Is it soluble in water?
2. If it is soluble, is it highly ionized or dissociated in water?

If *both* answers are yes, the substance is a soluble strong electrolyte and its formula is written in ionic form. If *either* answer is no, its formula is written as if it exists primarily as molecules. To answer these questions it is necessary to know the lists of strong acids (Section 7–1, part 2) and strong soluble bases (Section 7–1, part 4). These acids and bases are completely, or almost completely, ionized in dilute aqueous solutions. Other common acids and bases are either insoluble or only slightly ionized. In addition, the solubility rules (Section 7–1, part 5) allow you to determine which salts are soluble in water. Most salts that are soluble in water are also strong electrolytes. Exceptions such as lead acetate, $Pb(CH_3COO)_2$, which is soluble but predominantly nonionized, will be noted as they are encountered.

> The only common substances that can be written in ionized or dissociated form in ionic equations are (1) strong acids, (2) strong soluble bases, and (3) soluble ionic salts.

7–4 Decomposition Reactions

Decomposition reactions are those in which a compound decomposes to produce: (1) two elements, (2) one or more elements *and* one or more compounds, or (3) two or more compounds. Let us consider an example of each.

1 Decomposition of a Compound into Two Elements

The electrolysis of water (Section 1–5) produces two elements by the decomposition of a compound. A compound that ionizes, such as NaOH or H_2SO_4, is added to increase the conductivity of water and the rate of the reaction (Figure 1–5), but it does not participate in the reaction

$$2H_2O\ (\ell) \xrightarrow{\text{electrolysis}} 2H_2\ (g) + O_2\ (g)$$

Small amounts of oxygen usually are prepared by the thermal decomposition of certain oxygen-containing compounds. Some metal oxides, such as mercury(II) oxide, HgO, decompose on heating to produce oxygen as the following equation indicates:

$$\underset{\text{mercury(II) oxide}}{2HgO\ (s)} \xrightarrow{\Delta} 2Hg\ (\ell) + O_2\ (g)$$

2 Decomposition of a Compound into an Element and One or More Compounds

Manganese dioxide, MnO_2, is used as a **catalyst,** a substance that speeds up (or occasionally slows down) a chemical reaction but is not consumed. Here it allows the decomposition to proceed at a lower temperature.

The alkali metal chlorates, such as $KClO_3$, decompose when heated to produce the corresponding chlorides and liberate oxygen. Potassium chlorate is a common laboratory source of small amounts of oxygen

$$\underset{\text{potassium chlorate}}{2KClO_3\ (s)} \xrightarrow[MnO_2]{\Delta} \underset{\text{potassium chloride}}{2KCl\ (s)} + 3O_2\ (g)$$

3 Decomposition of a Compound into Two or More Compounds

The thermal decomposition of calcium carbonate (limestone) and other carbonates produces two compounds, a metal oxide and carbon dioxide

$$CaCO_3\ (s) \xrightarrow{\Delta} CaO\ (s) + CO_2\ (g)$$

7–5 Metathesis Reactions

Metathesis reactions are those in which two compounds react to form two new compounds and *no changes occur in oxidation number.* (Recall the rules for assigning oxidation numbers, given in Section 4–12.) They are frequently described as reactions in which the ions of two compounds simply "change partners."

Among the more common types of metathesis reactions are **precipitation reactions,** those in which a solid, or **precipitate,** forms, which separates from solution. An example is the formation of insoluble lead(II) iodide as a result of mixing solutions of the soluble, ionic compounds, lead(II) nitrate and potassium iodide (Figure 7–4)

$$Pb(NO_3)_2\ (aq) + 2KI\ (aq) \longrightarrow \underset{\text{yellow precipitate}}{PbI_2\ (s)} + 2KNO_3\ (aq) \qquad \text{molecular}$$

$$[Pb^{2+}\ (aq) + 2NO_3^-\ (aq)] + 2[K^+\ (aq) + I^-\ (aq)] \longrightarrow PbI_2\ (s) + 2[K^+\ (aq) + NO_3^-\ (aq)] \qquad \text{total ionic}$$

$$Pb^{2+}\ (aq) + 2I^-\ (aq) \longrightarrow PbI_2\ (s) \qquad \text{net ionic}$$

The reaction of an acid with a metal hydroxide base produces a salt and water and is another common example of a metathesis reaction. Consider, for

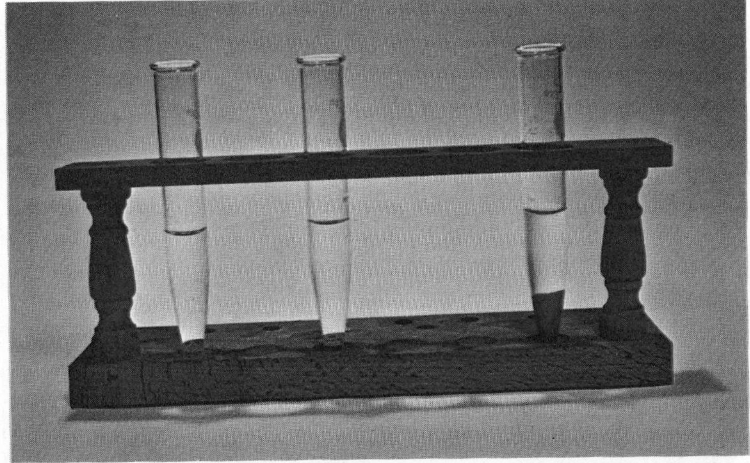

FIGURE 7–4 An ionic precipitation reaction. When aqueous potassium iodide is added to an aqueous solution of lead(II) nitrate, lead(II) iodide is precipitated. The resulting solution contains potassium nitrate.

example, the reaction of hydrochloric acid with aqueous sodium hydroxide. The salt produced, sodium chloride, contains the cation of the base and the anion of the acid

$HCl\ (aq) + NaOH\ (aq) \longrightarrow NaCl\ (aq) + H_2O\ (\ell)$ molecular

$[H^+\ (aq) + Cl^-\ (aq)] + [Na^+\ (aq) + OH^-\ (aq)] \longrightarrow [Na^+\ (aq) + Cl^-\ (aq)] + H_2O$ total ionic

$H^+\ (aq) + OH^-\ (aq) \longrightarrow H_2O\ (\ell)$ net ionic

This is the net ionic equation for all reactions of a strong acid with a strong soluble base to form a soluble salt and water. It is called a **neutralization reaction** because the corrosive properties of H^+ ions and OH^- ions are neutralized as they form water molecules.

EXAMPLE 7–1

Predict the products of the reactions involving the reactants shown. Write balanced molecular, total ionic, and net ionic equations for each one.

a. $HI\ (aq) + Ca(OH)_2\ (aq) \longrightarrow$
b. $Li\ (s) + H_2O\ (\ell)\ (excess) \longrightarrow$
c. $Na_3PO_4\ (aq) + Ca(NO_3)_2\ (aq) \longrightarrow$

Solution

a. This is an acid–base neutralization (metathesis) reaction; therefore the products are water and the salt containing the cation of the base, Ca^{2+}, and the anion of the acid, I^-

$2HI\ (aq) + Ca(OH)_2\ (aq) \longrightarrow CaI_2\ (aq) + 2H_2O$ molecular

HI is a strong acid (Table 7–1), $Ca(OH)_2$ a strong soluble base (Table 7–3), and CaI_2 is a soluble ionic salt (Section 7–1, part 5), so these are all written in ionic form in ionic equations

Ca^{2+} (aq) and I^- (aq) ions are spectator ions.

$2[H^+\ (aq) + I^-\ (aq)] + [Ca^{2+}\ (aq) + 2OH^-\ (aq)] \longrightarrow [Ca^{2+}\ (aq) + 2I^-\ (aq)] + 2H_2O$ total ionic

$2H^+\ (aq) + 2OH^-\ (aq) \longrightarrow 2H_2O\ (\ell)$

$H^+\ (aq) + OH^-\ (aq) \longrightarrow H_2O\ (\ell)$ net ionic

b. This is a displacement reaction in which the active Group IA metal, lithium, displaces hydrogen from water to form lithium hydroxide and gaseous hydrogen. LiOH is a strong soluble base and is written in ionic form in ionic equations

$2Li\ (s) + 2H_2O\ (\ell) \longrightarrow 2LiOH\ (aq) + H_2\ (g)$ molecular

$2Li\ (s) + 2H_2O\ (\ell) \longrightarrow 2[Li^+\ (aq) + OH^-\ (aq)] + H_2\ (g)$ total ionic

There are no spectator ions.

$2Li\ (s) + 2H_2O\ (\ell) \longrightarrow 2Li^+\ (aq) + 2OH^-\ (aq) + H_2\ (g)$ net ionic

c. Both reactants are soluble ionic salts, but according to the solubility rules, the calcium ions, Ca^{2+} (aq), and phosphate ions, PO_4^{3-} (aq), can form an insoluble compound, calcium phosphate, $Ca_3(PO_4)_2$. Thus, this is a precipitation (metathesis) reaction. The other reaction product, sodium nitrate, $NaNO_3$, is a soluble ionic salt. All species except the precipitate are shown in ionic form in ionic equations for this reaction

$2Na_3PO_4\ (aq) + 3Ca(NO_3)_2\ (aq) \longrightarrow Ca_3(PO_4)_2\ (s) + 6NaNO_3\ (aq)$ molecular

$$2[3Na^+ (aq) + PO_4^{3-} (aq)] + 3[Ca^{2+} (aq) + 2NO_3^- (aq)] \longrightarrow Ca_3(PO_4)_2 (s) + 6[Na^+ (aq) + NO_3^- (aq)] \quad \text{total ionic}$$

$$2PO_4^{3-} (aq) + 3Ca^{2+} (aq) \longrightarrow Ca_3(PO_4)_2 (s) \quad \text{net ionic}$$

Na^+ (aq) and NO_3^- (aq) ions are spectator ions.

7–6 The Removal of Ions from Aqueous Solution

Many reactions occurring in aqueous solution result in the removal of ions from solution. This is often accomplished in one of three ways: (1) formation of predominantly nonionized molecules (weak or nonelectrolytes) in solution from ions, (2) formation of a precipitate from its ions, and (3) formation of a gas which bubbles off. You have seen examples of the first two in acid–base neutralization and precipitation reactions.

Let us illustrate the third case. When an acid, for example, hydrochloric acid, is added to aqueous sodium carbonate, a metathesis reaction occurs in which carbonic acid, a weak acid, is produced:

$$Na_2CO_3 (aq) + 2HCl (aq) \longrightarrow 2NaCl (aq) + H_2CO_3 (aq)$$

or

$$CO_3^{2-} (aq) + 2H^+ (aq) \longrightarrow H_2CO_3 (aq)$$

The heat generated in the reaction causes thermal decomposition of carbonic acid to gaseous carbon dioxide and water

$$H_2CO_3 (aq) \longrightarrow CO_2 (g) + H_2O (\ell)$$

Most of the CO_2 bubbles off and the reaction goes to completion (with respect to the limiting reagent). The net effect is the conversion of ionic species into nonionized molecules of a gas and water.

7–7 Oxidation-Reduction Reactions—An Introduction 

Reactions in which substances undergo changes in oxidation number are called **oxidation-reduction reactions,** or simply **redox** reactions. Several redox reactions were discussed earlier in this chapter, although they were not identified as such. All displacement reactions and those combination and decomposition reactions involving free elements as either reactants or products are *always* redox reactions. Metathesis reactions are *never* redox reactions. We shall define the following important terms and illustrate their meanings.

Oxidation—an algebraic increase in oxidation number; also, a process in which electrons are (or appear to be) lost.

Reduction—an algebraic decrease in oxidation number; also, a process in which electrons are (or appear to be) gained.

Oxidizing agents—substances that undergo a decrease in oxidation number (gain electrons) and oxidize other substances. Oxidizing agents are always reduced.

Reducing agents—substances that undergo an increase in oxidation number (lose electrons) and reduce other substances. Reducing agents are always oxidized.

Oxidation and reduction always occur simultaneously, and the total increases and decreases in oxidation number are equal in all redox reactions. This is physically reasonable, since the electrons that cause the reduction of one substance must come from somewhere (by the Law of Conservation of Matter), and that "somewhere" is the substance that is simultaneously oxidized.

The combination of hydrogen and fluorine to form hydrogen fluoride is a redox reaction as well as a combination reaction. The *unbalanced* equation is

$$H_2\ (g) + F_2\ (g) \longrightarrow HF\ (g) \qquad \text{unbalanced}$$

The oxidation number of hydrogen increases from zero in H_2 to $+1$ in HF. (You may wish to refer to the rules for assigning oxidation numbers in Section 4–12.) Hydrogen is therefore oxidized, and it is the reducing agent. The oxidation number of fluorine decreases from zero in F_2 to -1 in HF. Fluorine is therefore reduced, and it is the oxidizing agent.

Oxidation numbers are useful in balancing equations for redox reactions using the "change-in-oxidation number" method, which will be introduced here and covered in more detail in Chapter 12. The changes in oxidation numbers may be summarized by "helping equations" for this purpose

$$H^0 \xrightarrow{1\uparrow} H^{+1}$$

$$F^0 \xrightarrow{1\downarrow} F^{-1}$$

in which the elements that are oxidized and reduced are shown with their oxidation numbers. The changes in oxidation number are indicated over the arrows by $1\uparrow$ and $1\downarrow$, which indicate an increase and a decrease in oxidation number by one, respectively. Since the oxidation number of each H atom increases by one while that of each F atom decreases by one, the reaction ratio is 1 : 1. There are two H atoms per H_2 molecule and two F atoms per F_2 molecule, and therefore the equation is balanced by placing a 2 before HF. The reactions of hydrogen with the other halogens, such as Cl_2 and Br_2, are similar.

$$H_2\ (g) + F_2\ (g) \longrightarrow 2HF\ (g) \qquad \text{balanced}$$

The dissolution of nitrogen dioxide in water to form nitric acid and nitrogen oxide is a redox reaction. The *unbalanced* equation is

$$NO_2\ (g) + H_2O\ (\ell) \longrightarrow HNO_3\ (\ell) + NO\ (g) \qquad \text{unbalanced}$$

Three of the compounds in this equation contain nitrogen, and it is the only element that undergoes a change in oxidation number. The hydrogen and oxygen atoms are in the $+1$ and -2 oxidation states, respectively, on both sides of the equation (see Section 4–12).

Compound	**Ox. No. of Nitrogen**
NO_2	$+4$
HNO_3	$+5$
NO	$+2$

The "helping" equations are

$N^{+4} \xrightarrow{1\uparrow} N^{+5}$

$N^{+4} \xrightarrow{2\downarrow} N^{+2}$

To balance the changes in oxidation numbers, we multiply the first equation by two and the second equation by one

$2(N^{+4} \xrightarrow{1\uparrow} N^{+5})$ total increase = 2↑

$1(N^{+4} \xrightarrow{2\downarrow} N^{+2})$ total decrease = 2↓

This tells us that the number of nitrogen atoms must total three on each side. Furthermore, two of the nitrogen atoms on the right side must be in the +5 oxidation state, that is in HNO_3

$3NO_2 + H_2O \longrightarrow 2HNO_3 + 1NO$ balanced

Inspection shows that the hydrogen and oxygen atoms also are balanced.

In the above reaction, some molecules of nitrogen dioxide are oxidized and other molecules of nitrogen dioxide are reduced. Therefore, NO_2 is *both* the oxidizing agent and the reducing agent. Such reactions are called auto-oxidation-reduction reactions, or preferably **disproportionation reactions.**

Many oxidation-reduction equations are quite complex and difficult to balance by inspection. The ideas introduced in this section will be expanded in Chapter 12 to allow us to balance more complicated oxidation-reduction equations.

EXAMPLE 7–2

Classify each of the following reactions as one or two of the following kinds of reactions, as appropriate: (1) combination, (2) displacement, (3) decomposition, (4) metathesis, or (5) oxidation-reduction.

a. NH_3 (g) + HCl (g) ⟶ NH_4Cl (s)
b. 2HBr (aq) + $Ba(OH)_2$ (aq) ⟶ $BaBr_2$ (aq) + $2H_2O$ (ℓ)
c. Mg (s) + 2HCl (aq) ⟶ $MgCl_2$ (aq) + H_2 (g)
d. 3Cu (s) + $8HNO_3$ (aq) ⟶ $3Cu(NO_3)_2$ (aq) + 2NO (g) + $4H_2O$ (ℓ)
e. $2N_2O$ (g) $\xrightarrow{\Delta}$ $2N_2$ (g) + O_2 (g)

Solution

a. combination, b. metathesis, c. displacement and redox, d. redox, e. decomposition and redox.

Some combination reactions and decomposition reactions, all displacement reactions, and no metathesis reactions are also oxidation-reduction reactions.

Chemical Reactions and Periodicity

Now we shall further illustrate the classification of reactions and periodicity of chemical properties by considering some characteristic reactions of hydrogen, oxygen, and their compounds.

7–8 Hydrogen and Hydrides

1 Hydrogen

The name means "acid former."

Elemental hydrogen is a colorless, odorless, tasteless, diatomic gas with the lowest atomic weight and density of any known substance. Discovery of the element is attributed to the Englishman Henry Cavendish, who prepared it in 1766 by passing steam through a red-hot gun barrel (mostly iron) and by the reaction of acids on various metals. In each case, H_2 is liberated by a displacement (and redox) reaction, of the kind described in detail in Section 7–3. (See also the activity series, Table 7–5.) The latter is still the method commonly used for the preparation of small amounts of H_2 in the laboratory

Mg + 2HCl → H₂↑ + MgCl₂

$$3Fe\ (s) + 4H_2O\ (g) \xrightarrow{\Delta} Fe_3O_4\ (s) + 4H_2\ (g)$$ displacement and redox

steam

Can you write the net ionic equation for the reaction of Zn with HCl (aq)?

$$Zn\ (s) + 2HCl\ (aq) \longrightarrow ZnCl_2\ (aq) + H_2\ (g)$$ displacement and redox

Hydrogen also can be prepared by electrolysis of water (Section 1–5)

$$2H_2O\ (\ell) + \text{Energy} \xrightarrow{\text{Elect.}} 2H_2\ (g) + O_2\ (g)$$ decomposition and redox

If it becomes economical to convert solar energy into electrical energy that can be used to electrolyze water, H_2 could become an important fuel in the future (although the dangers of storage and transportation would have to be overcome). The *combustion* of H_2 liberates a great deal of heat. **Combustion** is the highly exothermic combination of a substance with oxygen, usually with a visible flame. (See Section 7–9, part 3.)

Note that this is the reverse of the decomposition of H_2O above.

$$2H_2\ (g) + O_2\ (g) \xrightarrow[\text{or } \Delta]{\text{spark}} 2H_2O\ (\ell) + \text{Energy}$$ combination and redox

Hydrogen is no longer used in blimps and dirigibles; it has been replaced by slightly denser, but much safer and nonflammable helium.

Hydrogen is very flammable, and was responsible for the Hindenberg airship disaster in 1937. A spark is enough to initiate the **combustion reaction,** which is exothermic enough to provide the heat necessary to sustain the reaction.

Vast quantities of hydrogen are produced commercially each year by the reaction of methane with steam at 830°C in the presence of a nickel catalyst

$$CH_4\ (g) + H_2O\ (g) \xrightarrow[\text{Ni}]{\Delta} CO\ (g) + 3H_2\ (g)$$ redox

Hydrogen is also prepared by the "water gas reaction," which results from the passage of steam over white-hot coke (impure carbon, a nonmetal) at 1500°C. "Water gas" is used industrially as a fuel since both components, CO and H_2, undergo combustion.

$$\underset{\text{in coke}}{C\ (s)} + \underset{\text{steam}}{H_2O\ (g)} \longrightarrow \underbrace{CO\ (g) + H_2\ (g)}_{\text{"water gas"}}$$ displacement and redox

Hydrogen is prepared in the petroleum industry by the thermal cracking of hydrocarbons by **decomposition reactions. Thermal cracking** refers to heating a substance in the *presence* of a catalyst in the *absence* of air

methane $$CH_4\ (g) \xrightarrow[\text{catalyst}]{\Delta} C\ (s) + 2H_2\ (g)$$ decomposition and redox butane C_4H_{10} (g) $\xrightarrow[\text{catalyst}]{600°C}$ $2C_2H_2$ (g) + $3H_2$ (g) decomposition and redox

2 Reactions of Hydrogen and the Hydrides

Now let us turn to some of the combination reactions of hydrogen with metals and other nonmetals to form binary compounds called **hydrides,** and also to characteristic reactions of the hydrides. Atomic hydrogen has the $1s^1$ electronic configuration and it forms (1) **ionic hydrides,** containing hydride ions, H^-, by gaining one electron per atom from an active metal, or (2) **covalent hydrides** by sharing its electrons with an atom of another nonmetal to form a single covalent bond. The H^- ion has the stable He configuration, $1s^2$. In covalent bonds the hydrogen atom attains a *share* of two outer shell electrons, so it again has the He configuration.

The use of the term *hydride* does not necessarily imply the presence of the hydride ion, H^-.

The ionic or covalent character of the binary compounds of hydrogen depends upon the position of the other element in the periodic table (Figure 7–5.) The combination (and redox) reactions of H_2 with the *alkali* (IA) and the heavier (more active) *alkaline earth* (IIA) *metals* result in solid *ionic hydrides,* often called saline or salt-like hydrides. The reaction with the IA metals may be represented in general terms as

$$2M \text{ (molten)} + H_2 \text{ (g)} \longrightarrow 2(M^+, H^-) \text{ (s)} \qquad M = \text{Li, Na, K, Rb, Cs}$$

Thus, hydrogen combines with lithium to form lithium hydride and with sodium to form sodium hydride

The ionic hydrides are named by naming the metal first, followed by the stem for *hydr*ogen with the *-ide* ending.

$$2Li \text{ (molten)} + H_2 \text{ (g)} \longrightarrow 2LiH \text{ (s)} \qquad \text{lithium hydride}$$

$$2Na \text{ (molten)} + H_2 \text{ (g)} \longrightarrow 2NaH \text{ (s)} \qquad \text{sodium hydride}$$

In general terms, the reactions of the heavier (most active) IIA metals may be represented as

$$M \text{ (molten)} + H_2 \text{ (g)} \longrightarrow (M^{2+}, 2H^-) \text{ (s)} \qquad M = \text{Ca, Sr, Ba}$$

Thus, calcium combines with hydrogen to form calcium hydride

$$Ca \text{ (s)} + H_2 \text{ (g)} \longrightarrow CaH_2 \text{ (s)} \qquad \text{calcium hydride}$$

These *ionic hydrides are all basic* because hydride ions reduce water to form hydroxide ions and hydrogen. When water is added by drops to lithium hy-

Thus ionic hydrides can serve as sources of hydrogen. However, they must be carefully stored in environments free of moisture and O_2.

IA	IIA	IIIA	IVA	VA	VIA	VIIA
LiH	BeH_2	B_2H_6	CH_4	NH_3	H_2O	HF
NaH	MgH_2	$(AlH_3)_x$	SnH_4	PH_3	H_2S	HCl
KH	CaH_2	Ga_2H_6	GeH_4	AsH_3	H_2Se	HBr
RbH	SrH_2	InH_3	SnH_4	SbH_3	H_2Te	HI
CsH	BaH_2	TlH	PbH_4	BiH_3	H_2Po	HAt

FIGURE 7–5 Common hydrides of the representative elements. The ionic hydrides are shaded heavily (▨), covalent hydrides are unshaded, and those of intermediate character are shaded lightly (⁘).

dride, for example, lithium hydroxide and hydrogen are produced. The reaction of calcium hydride is similar

We show LiOH and $Ca(OH)_2$ as solids here (rather than as aqueous LiOH or $Ca(OH)_2$) because not enough water is available to act as a solvent.

$$\left.\begin{array}{l} LiH\ (s) + H_2O\ (\ell) \longrightarrow LiOH\ (s) + H_2\ (g) \\ CaH_2\ (s) + 2H_2O\ (\ell) \longrightarrow Ca(OH)_2\ (s) + 2H_2\ (g) \end{array}\right\} \text{displacement and redox}$$

Hydrogen reacts with *nonmetals* to form binary *covalent hydrides.* For example, H_2 combines with the halogens to form colorless, gaseous, hydrogen halides

$$H_2\ (g) + X_2 \longrightarrow \underset{\text{hydrogen halides}}{2HX\ (g)} \qquad X = F, Cl, Br, I$$

The hydrogen halides are named by the word hydrogen followed by the stem for the *hal*ogen with an *-ide* ending. Recall that the names of all binary compounds have *-ide* endings (Section 5–23).

Specifically, hydrogen reacts with fluorine to form hydrogen fluoride and with chlorine to form hydrogen chloride

$$H_2\ (g) + F_2\ (g) \longrightarrow 2HF\ (g) \qquad \text{hydrogen fluoride}$$

$$H_2\ (g) + Cl_2\ (g) \longrightarrow 2HCl\ (g) \qquad \text{hydrogen chloride}$$

Hydrogen also combines with the Group VIA elements to form covalent compounds. You already know that hydrogen and oxygen combine at elevated temperatures to form water

These compounds are named hydrogen oxide (H_2O), hydrogen sulfide (H_2S), hydrogen selenide (H_2Se), and hydrogen telluride (H_2Te).

$$2H_2\ (g) + O_2\ (g) \xrightarrow{\Delta} 2H_2O\ (g)$$

The heavier members of this family also combine with hydrogen to form binary covalent compounds, which are gases even at room temperature, whose formulas resemble that of water.

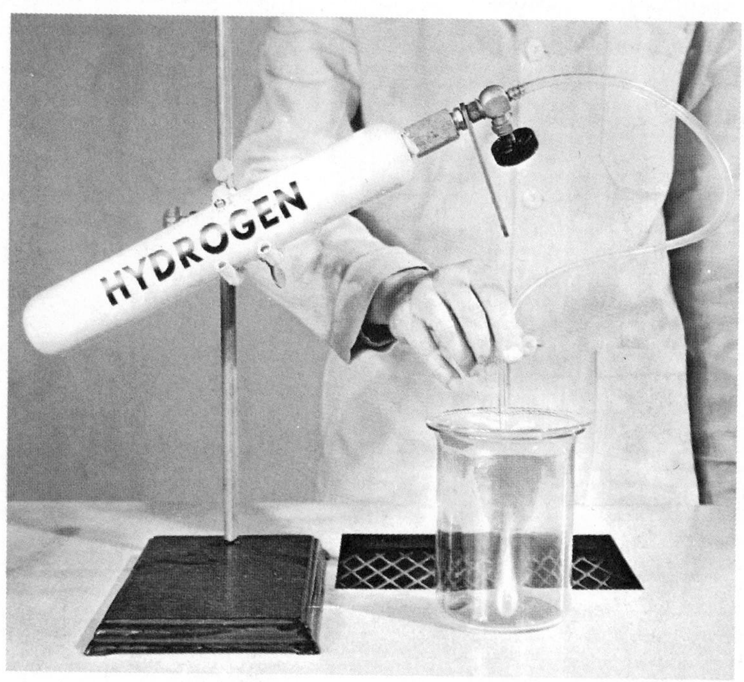

FIGURE 7–6 Hydrogen burns in an atmosphere of bromine vapor. Hydrogen bromide is formed.

FIGURE 7–7 When used as a fertilizer, ammonia may be applied directly to the soil.

The primary industrial use of H_2 is in the synthesis of ammonia, a covalent hydride, by the Haber process (Section 15–5). Most of the NH_3 is used as liquid ammonia as a fertilizer or to make other fertilizers, such as ammonium nitrate, NH_4NO_3, and ammonium sulfate, $(NH_4)_2SO_4$

$$N_2\ (g) + 3H_2\ (g) \xrightarrow[\Delta,\ \text{high pressure}]{\text{catalysts}} 2NH_3\ (g)$$

Many of the covalent (nonmetal) hydrides are acidic; their aqueous solutions produce hydrogen ions. These include, HF, HCl, HBr, HI, H_2S, H_2Se, and H_2Te.

As we shall see in Chapter 11, even H_2O is weakly acidic.

Many transition metals react with H_2 to form **interstitial hydrides,** substances in which the small H_2 molecules merely fit into spaces (interstices) in the crystal lattices of the metals. It is questionable whether these should be called compounds, because there is little evidence for chemical reaction when they are formed, and the properties of the interstitial hydrides closely resemble those of the metals themselves. Some transition metals are able to absorb several hundred times their own volume of H_2 under high pressure of H_2. This H_2 often can be released by reducing the pressure or by heating. For this reason, some transition metals may become very useful for storing hydrogen in a safe, economical, easily transportable way, especially if hydrogen ever becomes an important fuel.

EXAMPLE 7–3

Predict the products of the reactions involving the reactants shown. Write a balanced molecular equation for each one.

a. $H_2\ (g) + I_2\ (g) \xrightarrow{\Delta}$
b. $K\ (\ell) + H_2\ (g) \xrightarrow{\Delta}$
c. $NaH\ (s) + H_2O\ (\ell)\ (excess) \longrightarrow$
d. $S_8\ (s) + H_2\ (g) \xrightarrow{\Delta}$

Solution

Remember that hydride ions, H^-, react with water to produce OH^- ions and H_2 (g).

a. $H_2\ (g) + I_2\ (g) \xrightarrow{\Delta} 2HI\ (g)$
b. $2K\ (\ell) + H_2\ (g) \xrightarrow{\Delta} 2KH\ (s)$
c. $NaH\ (s) + H_2O\ (\ell) \longrightarrow NaOH\ (aq) + H_2\ (g)$
d. $S_8\ (s) + 8H_2\ (g) \xrightarrow{\Delta} 8H_2S\ (g)$

EXAMPLE 7–4

Predict the ionic or covalent character of the products of the reactions of Example 7–3.

Solution

Reactions a and d are combination reactions involving hydrogen and another nonmetal. The products, HI and H_2S, must be covalent. Reaction b is the combination of hydrogen with an active Group IA metal. Thus KH must be ionic. The products of reaction c are covalent H_2 (g) and the strong soluble base, NaOH, which is ionic.

7–9 Oxygen and Oxides

1 Oxygen and Ozone

Oxygen was discovered in 1774 by an English minister and scientist, Joseph Priestley, who observed the thermal decomposition of mercury(II) oxide, a red powder

$$2HgO\ (s) \xrightarrow{\Delta} 2Hg\ (\ell) + O_2\ (g) \qquad \text{decomposition and redox}$$

That part of the earth we see, comprising land, water, and air, is approximately 50% oxygen by mass. About two-thirds of the mass of the human body is due to oxygen. Elemental oxygen, O_2, is an odorless and nearly colorless gas that makes up about 21% by volume of dry air. In the liquid and solid states it is pale blue. Oxygen also exists in a second allotropic form, ozone, O_3, also a gas. **Allotropes** are different forms of the same element in the same physical state.

Oxygen molecules are only very slightly soluble in water; only about 0.004 gram dissolves in 100 grams of water at 25°C. Yet this is sufficient to sustain fish and other marine organisms. The greatest single industrial use of O_2 is for oxygen-enrichment in steel blast furnaces (Section 20–6) for the conversion of pig iron to steel.

Oxygen is obtained commercially by the fractional distillation of liquid air. In this process, air is first liquefied under high pressures by cooling it below the boiling points of its various components. The components of liquid air have different boiling points, and they can be vaporized and collected individually by slowly increasing the temperature. Liquid oxygen boils at −183°C under one atmosphere pressure.

Liquid O_2 is used as an oxidizer for rocket fuels. O_2 also is used in health areas for oxygen-enriched air.

Like hydrogen, very pure oxygen is obtained by electrolytic decomposition of water. Small amounts of very pure oxygen are obtained by immersing a platinized nickel foil catalyst in a 30% solution of hydrogen peroxide and heating gently

$$2H_2O_2\ (\ell) \xrightarrow[Pt]{\Delta} 2H_2O\ (\ell) + O_2\ (g) \qquad \text{decomposition and redox}$$

Oxygen can also be obtained by the reaction of sodium peroxide, Na_2O_2, with water. Other peroxides undergo similar reactions

$$2Na_2O_2\ (s) + 2H_2O\ (\ell) \longrightarrow 4NaOH\ (aq) + O_2\ (g) \qquad \text{redox}$$

A common laboratory method for preparing oxygen is the thermal decomposition of potassium chlorate, $KClO_3$, in the presence of a catalyst, MnO_2

$$2KClO_3\ (s) \xrightarrow[\Delta]{MnO_2} 2KCl\ (s) + 3O_2\ (g) \qquad \text{decomposition and redox}$$

Ozone: its properties and presence in the upper atmosphere Ozone, O_3, is an unstable, pale blue substance (a gas at room temperature) that is formed by passing an electrical discharge through gaseous oxygen. It has a unique pungent odor that is often noticed during electrical storms and in the vicinity of electrical equipment. Not surprisingly, its density is about $1\frac{1}{2}$ times that of O_2. At −112°C it condenses to a deep blue liquid. It is a very strong oxidizing agent. As a concentrated gas or a liquid, ozone can easily decompose explosively

$$2O_3\ (g) \longrightarrow 3O_2\ (g)$$

Oxygen atoms, or **radicals,** are intermediates in this exothermic decomposition of O_3 to O_2. These act as strong oxidizing agents in such applications as destroying bacteria in water purification.

A **radical** is a species containing one or more unpaired electrons; many radicals are very reactive.

The ozone molecule is angular and diamagnetic. Both oxygen—oxygen bond lengths (1.28 Å) are identical, and are intermediate between typical single and double bond lengths. Ozone is represented by the following resonance formulas:

The advantage of using ozone instead of chlorine for water purification is that ozone leaves no residual taste.

O O O ⟷ O O O

O O O 117° (Lone pairs on end O's are not shown.)

Ozone is formed in the upper atmosphere as O_2 molecules absorb high-energy electromagnetic radiation. Its concentration in the stratosphere is about 10 ppm, whereas it is only about 0.04 ppm in the troposphere. The ozone layer is responsible for absorbing some of the ultraviolet light from the sun which, if it reached the surface of the earth in higher intensity, could cause damage to plants and animals (including humans). It has been predicted that the incidence of skin cancer would increase by 2% for every 1% decrease in the concentration of ozone in the stratosphere. Although it decomposes rapidly in the upper atmosphere, the ozone supply is constantly replenished.

The concentration unit *ppm* stands for *parts per million,* that is, molecules of ozone per million molecules of all components of air.

Small amounts of O_3 at the surface of the earth decompose rubber and plastic products by oxidation.

FIGURE 7–8 This satellite, launched in 1979, is used to measure the amount of ozone at various latitudes from a circular orbit nearly 400 miles above the earth.

2 Reactions of Oxygen and Oxides

Oxygen combines directly with almost all other elements—except the noble gases and noble (unreactive) metals (Au, Pd, Pt)—to form **oxides,** binary compounds that contain oxygen. While such reactions are generally very exothermic, many proceed quite slowly and require heating to supply the energy necessary to break the strong bonds in O_2 molecules. Once these reactions are initiated, most release more than enough energy to be self-sustaining and sometimes result in incandescence.

In general, metallic oxides (and peroxides and superoxides) are ionic solids. The Group IA metals combine with oxygen to form three kinds of solid ionic products called oxides, peroxides, and superoxides. Lithium combines with oxygen to form lithium oxide

$$4Li\ (s) + O_2\ (g) \longrightarrow 2Li_2O\ (s) \qquad \text{lithium oxide}$$

By contrast, sodium reacts with an excess of oxygen to form sodium peroxide, Na_2O_2, rather than sodium oxide, Na_2O, as the *major* product

$$2Na\ (s) + O_2\ (g) \longrightarrow Na_2O_2\ (s) \qquad \text{sodium peroxide}$$

Peroxides contain the $:\ddot{\underset{..}{O}}-\ddot{\underset{..}{O}}:^{2-}$ group, in which the oxidation number of oxygen is -1, whereas *normal oxides* such as lithium oxide, Li_2O, contain oxide ions, $:\ddot{\underset{..}{O}}:^{2-}$. The heavier members of the family (K,Rb,Cs) react with an excess of oxygen to form **superoxides,** which contain the superoxide ion, O_2^-, in which the oxidation number of oxygen is $-\frac{1}{2}$. The reaction with potassium is

$$K\ (s) + O_2\ (g) \longrightarrow KO_2\ (s) \qquad \text{potassium superoxide}$$

Thus, the tendency of the Group IA metals to form oxygen-rich compounds increases upon descending the group, i.e., with increasing cation radii and decreasing charge density on the metal ion. A similar trend is observed in the reactions of the Group IIA metals with oxygen.

With the exception of Be, the Group IIA metals react with oxygen at normal temperatures to form normal ionic oxides, MO, and at high pressures of oxygen the heavier ones form ionic peroxides, MO_2 (Table 7–6).

Beryllium reacts with oxygen only at elevated temperatures and forms only the normal oxide, BeO.

$$2M\ (s) + O_2\ (g) \longrightarrow 2(M^{2+}, O^{2-})\ (s) \qquad M = \text{Be, Mg, Ca, Sr, Ba}$$

$$M\ (s) + O_2\ (g) \longrightarrow (M^{2+}, O_2^{2-})\ (s) \qquad M = \text{Ca, Sr, Ba}$$

For example, the equations for the reactions of calcium and oxygen are

$$2Ca\ (s) + O_2\ (g) \longrightarrow 2CaO\ (s) \qquad \text{calcium oxide}$$
$$Ca\ (s) + O_2\ (g) \longrightarrow CaO_2\ (s) \qquad \text{calcium peroxide}$$

TABLE 7–6 Oxygen Compounds of the IA and IIA Metals*

	IA					IIA				
	Li	Na	K	Rb	Cs	Be	Mg	Ca	Sr	Ba
Oxide	Li_2O	Na_2O	K_2O	Rb_2O	Cs_2O	BeO	MgO	CaO	SrO	BaO
Peroxide	Li_2O_2	Na_2O_2	K_2O_2	Rb_2O_2	Cs_2O_2			CaO_2	SrO_2	BaO_2
Superoxide		NaO_2	KO_2	RbO_2	CsO_2					

* The compounds underlined represent the principal product of the direct reaction of the metal with oxygen.

The other metals, with the exceptions noted above (Au, Pd, and Pt), react with oxygen to form solid metal oxides. Since many metals to the right of Group IIA show variable oxidation states, several oxides may be formed. For example, iron combines with oxygen to form the following oxides in a series of reactions

$$2Fe\ (s) + O_2\ (g) \xrightarrow{\Delta} 2FeO\ (s) \qquad \text{iron(II) oxide or ferrous oxide}$$

$$6FeO\ (s) + O_2\ (g) \xrightarrow{\Delta} 2Fe_3O_4\ (s) \qquad \text{magnetic iron oxide (a mixed oxide)}$$

$$4Fe_3O_4\ (s) + O_2\ (g) \xrightarrow{\Delta} 6Fe_2O_3\ (s) \qquad \text{iron(III) oxide or ferric oxide}$$

Fe_3O_4 is a mixed oxide containing FeO and Fe_2O_3 in a 1:1 ratio. Ordinary iron rust is primarily Fe_2O_3.

Copper reacts with a limited amount of oxygen to form red Cu_2O

$$4Cu\ (s) + O_2\ (g) \xrightarrow{\Delta} 2Cu_2O\ (s) \qquad \text{copper(I) oxide or cuprous oxide}$$

and with excess oxygen to form black CuO

$$2Cu\ (s) + O_2\ (g) \xrightarrow{\Delta} 2CuO\ (s) \qquad \text{copper(II) oxide or cupric oxide}$$

Metals that exhibit variable oxidation states (Section 4–12) react with a limited amount of oxygen to give lower oxidation state oxides (such as FeO and Cu_2O), while reaction with an excess of oxygen gives higher oxidation state oxides (such as Fe_2O_3 and CuO).

Oxides of metals are called **basic anhydrides** because many of them combine with water to form bases *with no change in oxidation state of the metal. Anhydride* means "without water"; in a sense, the metal oxide is a hydroxide base with the water "removed." Metal oxides that are soluble in water react to

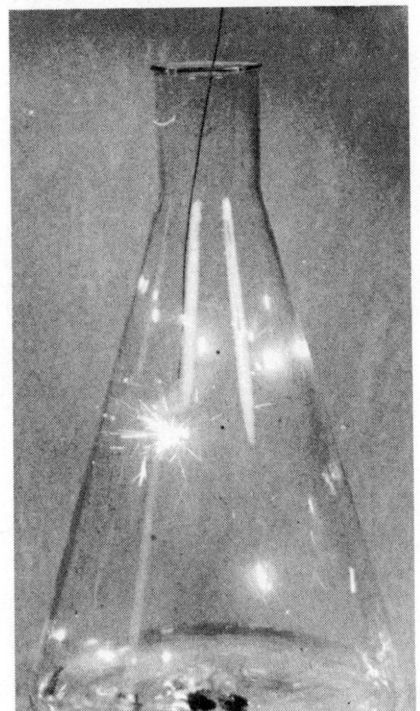

FIGURE 7–9 Iron burns brilliantly in pure oxygen to form the iron oxide Fe_3O_4.

produce the corresponding hydroxides. For example, sodium oxide dissolves in water to produce sodium hydroxide in a combination reaction

Although Na_2O_2 is the common product of the reaction of Na with O_2, it is possible to prepare Na_2O.

$$\overset{+1}{Na_2O}\ (s) + H_2O\ (\ell) \longrightarrow 2\overset{+1}{Na}OH\ (aq) \qquad \text{sodium hydroxide}$$

Barium oxide dissolves in water to produce barium hydroxide solution

$$\overset{+2}{BaO}\ (s) + H_2O\ (\ell) \longrightarrow \overset{+2}{Ba}(OH)_2\ (aq) \qquad \text{barium hydroxide}$$

As we saw in Section 7–1, part 4, both NaOH and $Ba(OH)_2$ are ionic solids which are nearly completely dissociated in aqueous solution.

The oxides of the Group IA metals and the heavier Group IIA metals dissolve in water to give solutions of strong soluble bases. Most other metals oxides are insoluble in water.

Oxygen combines with many nonmetals to form covalent oxides. For example, carbon burns in oxygen to form carbon monoxide or carbon dioxide, depending on the relative amounts of carbon and oxygen, as the following equations show

$$2C\ (s) + O_2\ (g) \longrightarrow \underset{\text{carbon monoxide}}{2\overset{+2 \text{ ox. no.}}{C}O\ (s)} \qquad \text{(excess C and limited } O_2\text{)}$$

$$C\ (s) + O_2\ (g) \longrightarrow \underset{\text{carbon dioxide}}{\overset{+4 \text{ ox. no.}}{C}O_2\ (g)} \qquad \text{(excess } O_2 \text{ and limited C)}$$

Carbon monoxide is also produced by the incomplete combustion of carbon-containing compounds such as gasoline and diesel fuel. It is a poisonous material, and each year newspapers carry many reports of carbon monoxide asphyxiation. The Lewis dot formula for carbon monoxide, :C:::O:, shows one unshared pair of electrons on the carbon atom. This pair of electrons can be shared with the iron atom in hemoglobin in blood. The resulting bond is stronger than the bond that oxygen molecules form with the iron atom in hemoglobin. Attachment of the CO molecule to the iron atom destroys the ability of hemoglobin to pick up oxygen in the lungs and carry it to the brain and muscle tissues. Carbon monoxide poisoning is particularly insidious because the gas has no odor, and because the victim first becomes drowsy.

Unlike carbon monoxide, carbon dioxide is not toxic. It is one of the products of the respiratory process. It is used to make carbonated beverages, which are mostly saturated solutions of carbon dioxide in water. A small amount of the carbon dioxide combines with the water to form carbonic acid, H_2CO_3, which is a very weak acid.

A limited amount of oxygen reacts with phosphorus to form tetraphosphorus hexoxide, P_4O_6

$$P_4\ (s) + 3O_2\ (g) \longrightarrow \overset{+3 \text{ ox. no.}}{P}_4O_6\ (s) \qquad \text{tetraphosphorus hexoxide}$$

while an excess of oxygen reacts with phosphorus to form tetraphosphorus decoxide, P_4O_{10}

$$P_4\ (s) + 5O_2\ (g) \longrightarrow \overset{+5 \text{ ox. no.}}{P}_4O_{10} \qquad \text{tetraphosphorus decoxide}$$

Unlike carbon and phosphorus, sulfur burns in oxygen to form primarily sulfur dioxide and small amounts of sulfur trioxide.

The production of SO_3 at a reasonable rate requires the presence of a catalyst.

$$S_8\ (s) + 8O_2\ (g) \longrightarrow 8\overset{+4\text{ ox. no.}}{S}O_2\ (g) \qquad \text{sulfur dioxide}$$

$$S_8\ (s) + 12O_2\ (g) \longrightarrow 8\overset{+6\text{ ox. no.}}{S}O_3\ (g) \qquad \text{sulfur trioxide}$$

Nonmetals exhibit more than one oxidation state in their compounds. In general, the *most* common oxidation states of a nonmetal are (1) its periodic group number, (2) its periodic group number minus two, and (3) its periodic group number minus eight. The reactions of nonmetals with a limited amount of oxygen usually give products that contain the nonmetals (other than oxygen) in lower oxidation states, usually case (2), while reaction with an excess of oxygen gives products in which the nonmetals exhibit higher oxidation states, case (1). The examples we have cited are CO and CO_2, P_4O_6 and P_4O_{10}, and SO_2 and SO_3. In some instances, the *true* formulas of the oxides are not always easily predictable but the *simplest* formulas are. For example, the two most common oxidation states of phosphorus in covalent compounds are +3 and +5. The simplest formulas for the corresponding phosphorus oxides therefore are P_2O_3 and P_2O_5, respectively. The true formulas are twice these, P_4O_6 and P_4O_{10}.

Nonmetal oxides are called **acid anhydrides** because many of them dissolve in water to form an acid *with no change in oxidation state of the nonmetal.* Several ternary acids can be prepared by reaction of the appropriate nonmetal oxides with water. Consider the combination of carbon dioxide with water to form carbonic acid

$$\overset{+4}{C}O_2\ (g) + H_2O\ (\ell) \longrightarrow H_2\overset{+4}{C}O_3\ (aq) \qquad \text{carbonic acid}$$

Similarly, the dissolution of sulfur dioxide in water produces sulfurous acid

$$\overset{+4}{S}O_2\ (g) + H_2O\ (\ell) \longrightarrow H_2\overset{+4}{S}O_3\ (aq) \qquad \text{sulfurous acid}$$

and the dissolution of sulfur trioxide in water produces sulfuric acid.

$$\overset{+6}{S}O_3\ (g) + H_2O\ (\ell) \longrightarrow H_2\overset{+6}{S}O_4\ (aq) \qquad \text{sulfuric acid}$$

Nitric acid may be produced by the dissolution of dinitrogen pentoxide in water.

$$\overset{+5}{N}_2O_5\ (s) + H_2O\ (\ell) \longrightarrow 2H\overset{+5}{N}O_3\ (aq) \qquad \text{nitric acid}$$

Phosphoric acid solutions are prepared by dissolving tetraphosphorus decoxide in water

$$\overset{+5}{P}_4O_{10}\ (s) + 6H_2O\ (\ell) \longrightarrow 4H_3\overset{+5}{P}O_4\ (aq) \qquad \text{phosphoric acid}$$

FIGURE 7–10 Sulfur burns in oxygen to form sulfur dioxide gas.

An amphoteric oxide is one that shows some acidic and some basic properties (Section 11–8).

Increasing acidic character →

Increasing base character ↓

IA	IIA	IIIA	IVA	VA	VIA	VIIA
Li_2O	BeO	B_2O_3	CO_2	N_2O_5		F_2O
Na_2O	MgO	Al_2O_3	SiO_2	P_4O_{10}	SO_3	Cl_2O_7
K_2O	CaO	Ga_2O_3	GeO_2	As_2O_5	SeO_3	Br_2O_7
Rb_2O	SrO	In_2O_3	SnO_2	Sb_2O_5	TeO_3	I_2O_7
Cs_2O	BaO	Tl_2O_3	PbO_2	Bi_2O_5	PoO_3	At_2O_7

FIGURE 7–11 The normal oxides of the representative elements in their maximum oxidation states. Acidic oxides are shaded, amphoteric oxides are lightly shaded, and basic oxides are not shaded.

Except for the oxides of boron and silicon, which are insoluble, nearly all oxides of nonmetals dissolve in water to give solutions of ternary acids.

Another common kind of combination reaction is the *combination of metal oxides (basic anhydrides) with nonmetal oxides (acid anhydrides), with no change in oxidation states, to form salts.* Consider, for example, the combination reaction of calcium oxide, CaO, with sulfur trioxide, SO_3, to form calcium sulfate, $CaSO_4$.

$$\overset{+2}{Ca}O\ (s) + \overset{+6}{S}O_3\ (g) \longrightarrow \overset{+2}{Ca}\overset{+6}{S}O_4\ (s) \qquad \text{calcium sulfate}$$

Calcium oxide is the anhydride of calcium hydroxide, $Ca(OH)_2$, and sulfur trioxide is the anhydride of sulfuric acid, H_2SO_4.

The salts produced in these reactions are the same ones produced in the neutralization reactions of the ternary acids and bases related to the oxides themselves.

Carbon dioxide, the anhydride of carbonic acid, H_2CO_3, reacts with magnesium oxide, the anhydride of magnesium hydroxide, $Mg(OH)_2$, to produce magnesium carbonate

$$\overset{+4}{C}O_2\ (g) + \overset{+2}{Mg}O\ (s) \longrightarrow \overset{+2}{Mg}\overset{+4}{C}O_3\ (s) \qquad \text{magnesium carbonate}$$

As a third example, consider the reaction of tetraphosphorus decoxide, P_4O_{10}, with sodium oxide, Na_2O, to form sodium phosphate, Na_3PO_4

$$\overset{+5}{P_4}O_{10}\ (s) + 6\overset{+1}{Na_2}O\ (s) \longrightarrow 4\overset{+1}{Na_3}\overset{+5}{P}O_4\ (s) \qquad \text{sodium phosphate}$$

EXAMPLE 7–5

Arrange the following oxides in order of increasing covalent character: SO_3, Cl_2O_7, CaO, and PbO_2.

Solution

Covalent character of an oxide increases as nonmetallic character of the element combined with oxygen increases

Increasing Nonmetallic Character →

	Ca <	Pb <	S <	Cl
Periodic Group:	IIA	IVA	VIA	VIIA

Thus the order is

$$\xrightarrow{\text{Increasing Covalent Character}}$$
$$CaO < PbO_2 < SO_3 < Cl_2O_7$$

EXAMPLE 7–6

Arrange the oxides in Example 7–5 in order of increasing basicity.

Solution
The greater the covalent character of an oxide, the more acidic it is. Thus the most basic oxides have the least covalent (or most ionic) character

$$\xrightarrow{\text{Increasing Basic Character}}$$
$$Cl_2O_7 < SO_3 < PbO_2 < CaO$$

most covalent most ionic

EXAMPLE 7–7

Predict the products of the reactions involving the following reactants. Write a balanced molecular equation for each one.

a. $Cl_2O_7\ (\ell) + H_2O \longrightarrow$

b. $As_4\ (s) + O_2\ (g)$ (excess) $\xrightarrow{\Delta}$

c. $Mg\ (s) + O_2\ (g)$ (low pressure) $\xrightarrow{\Delta}$

Solution

a. This is the reaction of a nonmetal oxide (acid anhydride) with water to form a ternary acid in which the nonmetal (Cl) has the same oxidation state (+7) as in the oxide. Thus the acid is perchloric acid, $HClO_4$

$$Cl_2O_7\ (\ell) + H_2O\ (\ell) \longrightarrow 2HClO_4\ (aq)$$

b. Arsenic, a nonmetal of Group VA, exhibits common oxidation states of +5 and $+5 - 2 = +3$. Reaction of arsenic with *excess* oxygen produces the higher oxidation state oxide, As_2O_5. By analogy with the oxide of phosphorus in the +5 oxidation state, P_4O_{10}, one might have predicted a formula of As_4O_{10}, but the oxide is usually represented as As_2O_5 because its structure is unknown

$$As_4\ (s) + 5O_2\ (g) \xrightarrow{\Delta} 2As_2O_5\ (s)$$

c. The reaction of a Group IIA metal with oxygen (at low pressures) produces the normal metal oxide, MgO in this case

$$2Mg\ (s) + O_2\ (g) \xrightarrow{\Delta} 2MgO\ (s)$$

EXAMPLE 7–8

Predict the products of the following pairs of reactants. Write a balanced molecular equation for each reaction.

a. $Sn\ (s) + O_2\ (g)$ (limited amount) $\xrightarrow{\Delta}$

b. $CaO\ (s) + H_2O\ (\ell) \xrightarrow{\Delta}$

c. $Li_2O\ (s) + SO_3\ (g) \xrightarrow{\Delta}$

Solution

a. Tin (a metal of Group IVA) exhibits two oxidation states, +2 and +4. Its reaction with a *limited* amount of O_2 produces tin(II) oxide, SnO

$$2Sn\ (s) + O_2\ (g) \xrightarrow{\Delta} 2SnO\ (s)$$

b. The reaction of a metal oxide with water produces the metal hydroxide

$$CaO\ (s) + H_2O\ (\ell) \xrightarrow{\Delta} Ca(OH)_2\ (aq)$$

c. The reaction of a metal oxide with a nonmetal oxide produces a salt containing the cation of the metal oxide and the anion of the acid for which the nonmetal oxide is the anhydride. SO_3 (+6 ox. state) is the acid anhydride of sulfuric acid, H_2SO_4 (+6 ox. state)

$$Li_2O\ (s) + SO_3\ (g) \xrightarrow{\Delta} Li_2SO_4\ (s)$$

3 Combustion Reactions

Combustion, or burning, is simply an oxidation-reduction reaction in which oxygen combines rapidly with oxidizable materials in highly exothermic reactions, with a visible flame (see Section 7–8, part 2). The complete combustion of **hydrocarbons,** in fossil fuels for example, produces carbon dioxide and water (steam) as the major products

Hydrocarbons are compounds that contain only hydrogen and carbon.

$$\overset{-4+1}{CH_4}\ (g) + \underset{\text{excess}}{\overset{0}{2O_2}}\ (g) \xrightarrow{\Delta} \overset{+4-2}{CO_2}\ (g) + \overset{+1-2}{2H_2O}\ (g) + \text{Heat}$$

$$\underset{\text{cyclohexane}}{\overset{-2+1}{C_6H_{12}}}\ (g) + \underset{\text{excess}}{\overset{0}{9O_2}}\ (g) \xrightarrow{\Delta} \overset{+4-2}{6CO_2}\ (g) + \overset{+1-2}{6H_2O}\ (g) + \text{Heat}$$

As we have seen, the origin of the term *oxidation* lies in just such reactions, in which oxygen "oxidizes" another species.

4 Combustion of Fossil Fuels and Air Pollution

Fossil fuels are mixtures of variable composition that consist primarily of hydrocarbons. They are burned because they release energy, rather than to obtain chemical products. The incomplete combustion of hydrocarbons yields undesirable products, carbon monoxide (Section 25–3) and elemental carbon (soot), which pollute the air. Unfortunately, all fossil fuels—natural gas, coal, gasoline, kerosene, oil, etc.—also have undesirable nonhydrocarbon impurities that undergo combustion to produce oxides that act as additional air pollutants. At this time it is not economically feasible to remove all of these impurities.

Carbon or soot is one of many kinds of *particulate matter* in polluted air.

Fossil fuels result from the decay of animal and vegetable matter, and since all living matter contains some sulfur and nitrogen, fossil fuels also con-

TABLE 7–7 Some Typical Coal Compositions in Percent (Dry, Ash-Free)

	C	H	O	N	S
Lignite	70.59	4.47	23.13	1.04	0.74
Subbituminous	77.2	5.01	15.92	1.30	0.51
Bituminous	80.2	5.80	7.53	1.39	5.11
Anthracite	92.7	2.80	2.70	1.00	0.90

tain sulfur and nitrogen impurities to varying degrees. Table 7–7 gives composition data for some common kinds of coal.

Combustion of sulfur produces sulfur dioxide, SO_2, probably the single most harmful pollutant

$$\overset{0}{S_8}\ (s) + 8O_2\ (g) \xrightarrow{\Delta} 8\overset{+4}{S}O_2\ (g)$$

Sulfur dioxide is corrosive; it damages plants, structural materials, and humans. It is a nasal, throat, and lung irritant. Sulfur dioxide is slowly oxidized to sulfur trioxide, SO_3, by oxygen in air

$$2\overset{+4}{S}O_2\ (g) + O_2\ (g) \longrightarrow 2\overset{+6}{S}O_3\ (\ell)$$

Sulfur trioxide combines with moisture in the air to form the strong, corrosive acid, sulfuric acid

SO_3 is an *acid anhydride.* No changes in oxidation numbers accompany this kind of reaction.

$$\overset{+6}{S}O_3 + H_2O\ (\ell) \longrightarrow H_2\overset{+6}{S}O_4\ (\ell)$$

FIGURE 7–12 The luxuriant growth of vegetation during the carboniferous age is the source of our coal deposits today.

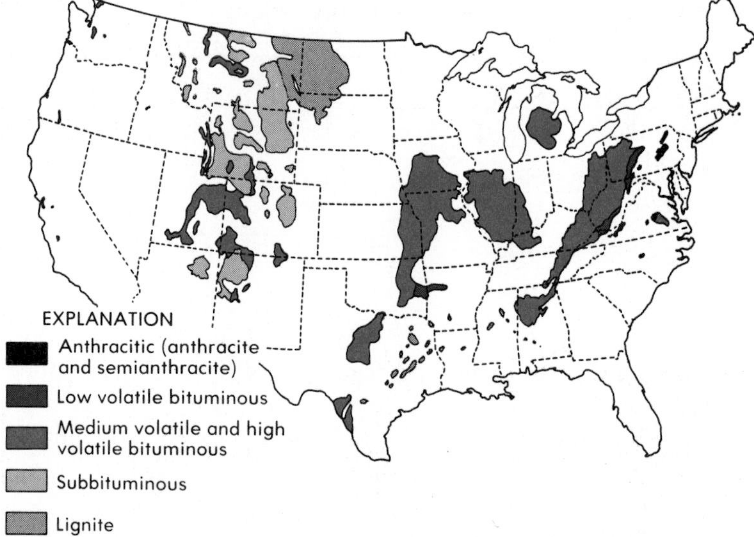

FIGURE 7–13 The various forms of coal are widely distributed in the United States.

FIGURE 7–14 A marble statue, the victim of "acid rain." When fossil fuels are burned, oxides of nitrogen and (usually) sulfur are released into the atmosphere where they dissolve in moisture to form acidic solutions. These acids react with marble, a form of calcium carbonate, and slowly corrode it.

It is of interest to note that the oxides of sulfur result from roasting the ores of metals that occur in nature as sulfides. This is done in the process of extracting the free (elemental) metals. **Roasting** involves heating an ore in the presence of air. For many metal sulfides this produces a metal oxide and SO_2. The metal ions are subsequently reduced to the free metals. Consider lead sulfide, PbS, as an example

$$2PbS\ (s) + 3O_2\ (g) \longrightarrow 2PbO\ (s) + 2SO_2\ (g)$$

Compounds of nitrogen are also impurities in fossil fuels and they undergo combustion to form nitric oxide, NO. However, most of the nitrogen in the NO in exhaust gases from furnaces, automobiles, airplanes, etc., comes from the air that is mixed with the fuel

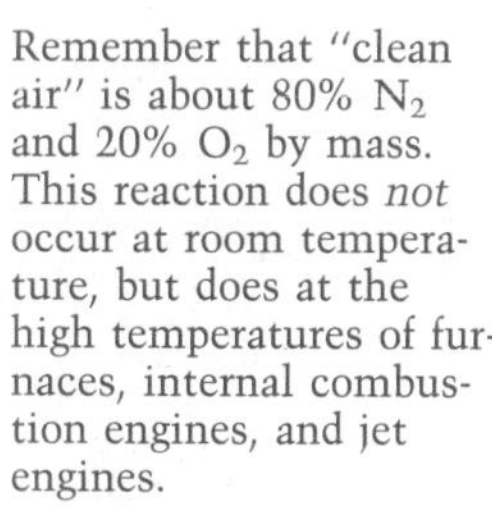

Remember that "clean air" is about 80% N_2 and 20% O_2 by mass. This reaction does *not* occur at room temperature, but does at the high temperatures of furnaces, internal combustion engines, and jet engines.

$$\overset{0}{N_2}\ (g) + O_2\ (g) \longrightarrow 2\overset{+2}{N}O\ (g)$$

NO can be oxidized further by oxygen to nitrogen dioxide, NO_2; this reaction is enhanced in the presence of ultraviolet light from the sun

$$2\overset{+2}{N}O\ (g) + O_2\ (g) \xrightarrow[\text{light}]{\text{uv}} 2\overset{+4}{N}O_2\ (g) \qquad \text{reddish-brown gas}$$

NO_2 is responsible for the reddish-brown haze that hangs over many cities in the afternoons of sunny days, and probably for most of the respiratory problems associated with this kind of air pollution. It can react to produce other oxides of nitrogen and other secondary pollutants, as described in more detail in Section 24–13.

FIGURE 7–15 Photochemical smog (a brown haze) enveloping the city of Los Angeles. (Los Angeles County Air Pollution Control District)

In addition to being a pollutant itself, nitrogen dioxide reacts with water in the air to form nitric acid, which is a major contributor to "acid rain"

$$3\overset{+4}{N}O_2\ (g) + H_2O\ (\ell) \longrightarrow 2H\overset{+5}{N}O_3\ (\ell) + \overset{+2}{N}O\ (g)$$

Federal and state regulations regarding permissible limits of pollutants in exhaust gases have been enacted over the last several years, and this has resulted in the lowering, to some extent, of the concentrations of pollutants in the air.

Key Terms

Acid a substance that produces H^+ (aq) ions in aqueous solution. Strong acids are completely or almost completely ionized in dilute aqueous solution. Weak acids are only slightly ionized.

Acid anhydride the oxide of a nonmetal that reacts with water to form an acid.

Active metal metal with low ionization energy that loses electrons readily to form cations.

Activity series a listing of metals (and hydrogen) in order of decreasingly active metals.

Amphoterism the ability to react with both acids and bases.

Base a substance that produces OH^- (aq) ions in aqueous solution. Strong, soluble bases are soluble in water and are completely dissociated.

Basic anhydride the oxide of a metal that reacts with water to form a base.

Catalyst a substance that speeds up a chemical reaction without being consumed itself in the reaction.

Combination reaction reaction in which two substances (elements or compounds) combine to form one compound.

Combustion reaction reaction of a substance with oxygen in a highly exothermic reaction, usually with a visible flame.

Displacement reactions reactions in which one element displaces another from a compound.

Disproportionation reactions redox reactions in which the oxidizing agent and the reducing agent are the same species.

Electrolyte a substance whose aqueous solutions conduct electricity.

Hydride a binary compound of hydrogen.

Metathesis reactions reactions in which two compounds react to form two new compounds, with no changes in oxidation number, sometimes called reactions in which the ions of two compounds exchange partners.

Molecular equation equation for a chemical reaction in which all formulas are written *as if* all substances existed as molecules; only complete formulas are used.

Net ionic equation equation that results from cancelling spectator ions and eliminating brackets from a total ionic equation.

Neutralization the reaction of an acid with a base to form a salt and water; the reaction of hydrogen ions with hydroxide ions to form water molecules.

Nonelectrolyte a substance whose aqueous solutions do not conduct electricity.

Normal oxide a compound containing oxide ion, O^{2-}, or oxygen in the -2 oxidation state.

Oxidation an algebraic increase in oxidation number; may correspond to a loss of electrons.

Oxidation-reduction reactions reactions in which oxidation and reduction occur; also called redox reactions.

Oxide a binary compound of oxygen.

Oxidizing agent the substance that oxidizes another substance and is reduced.

Peroxide a compound containing an O_2^{2-} ion, or oxygen in the -1 oxidation state.

Precipitate an insoluble solid that forms and separates from a solution.

Precipitation reactions reactions in which a precipitate forms.

Radical a species containing one or more unpaired electrons; many radicals are very reactive.

Reducing agent the substance that reduces another substance and is oxidized.

Reduction an algebraic decrease in oxidation number; may correspond to a gain of electrons.

Reversible reactions reactions that occur in both directions; indicated by double arrows ($\rightleftharpoons$).

Roasting heating an ore of an element in the presence of air.

Salt a compound that contains a cation other than H^+ and an anion other than OH^- or O^{2-}.

Spectator ions ions in solution that do not participate in a chemical reaction.

Strong electrolyte a substance that conducts electricity well in dilute aqueous solution.

Superoxide a compound containing an O_2^- ion, or oxygen in the $-\frac{1}{2}$ oxidation state.

Ternary acid an acid containing three elements, H, O, and (usually) another nonmetal.

Thermal cracking decomposition by heating a substance in the presence of a catalyst and in the absence of air.

Total ionic equation equation for a chemical reaction written to show the predominant form of all species in, or in contact with, aqueous solution.

Weak electrolyte a substance that conducts electricity poorly in dilute aqueous solution.

Exercises

Aqueous Solutions—An Introduction

1. Define acids, bases and neutralization.
2. How are the ions of which a salt is composed related to an acid and a base?
3. Define and distinguish among (a) strong electrolytes, (b) weak electrolytes, and (c) nonelectrolytes.
4. Distinguish between strong and weak acids.

5. Classify each of the following as a strong or weak acid and name each one: (a) HNO_3 (b) HNO_2 (c) HCl (d) HF (e) H_2SO_4 (f) CH_3COOH
6. Classify each of the following as a strong or weak acid and name each one: (a) H_3PO_4 (b) HI (c) $HClO_4$ (d) $(COOH)_2$ (e) H_2SO_3
7. Write equations for the ionization of the following acids in water: (a) HBr (b) CH_3COOH (c) dilute H_2SO_4 (d) $HClO_3$ (e) HClO
8. What (functional) group of atoms characterizes an organic acid?
9. Distinguish among strong soluble bases, insoluble bases, and weak bases. Give an example of each.
10. Which of the following are strong soluble bases? (a) $Al(OH)_3$ (b) KOH (c) $Mg(OH)_2$ (d) $Ca(OH)_2$ (e) $Cu(OH)_2$
11. Which of the following are insoluble bases? (a) $Fe(OH)_2$ (b) $Fe(OH)_3$ (c) NaOH (d) $Cu(OH)_2$ (e) $Be(OH)_2$ (f) NH_3
12. Which of the following are classified as water-soluble salts? (a) $NaClO_4$ (b) $CaCl_2$ (c) $(NH_4)_2S$ (d) AgI (e) $Ca_3(PO_4)_2$ (f) K_2CO_3
13. Which of the following are classified as water-soluble salts? (a) $PbSO_4$ (b) $CaCO_3$ (c) NaBr (d) FeS (e) BaS (f) $Mg_3(AsO_4)_2$
14. Which of the following are classified as water-*insoluble* salts? (a) $NiSO_4$ (b) Cu_2S (c) $Fe_2(CO_3)_3$ (d) $BaSO_4$ (e) $Hg(CH_3COO)_2$ (f) Hg_2Cl_2
15. Which of the following are classified as water-*insoluble* salts? (a) $Cr(NO_3)_3$ (b) $Fe(ClO_3)_2$ (c) KF (d) NH_4CN (e) AgBr (f) Li_2SO_3
16. Which of the following substances are strong electrolytes? (a) HClO (b) LiOH (c) $KClO_3$ (d) HBr (e) $MgCO_3$ (f) HNO_2
17. Which of the following substances are strong electrolytes? (a) $Sr(OH)_2$ (b) HNO_3 (c) $PbCO_3$ (d) HCN (e) NaOH (f) HNO_2
18. Which of the following are strong electrolytes? (a) Na_2SO_4 (b) HF (c) NH_4F (d) $HClO_3$ (e) $NaNO_3$ (f) $CuCl_2$

Identifying Reaction Types

The following reactions apply to Exercises 19–26.

a. $H_2SO_4\ (aq) + 2KOH\ (aq) \longrightarrow K_2SO_4\ (aq) + 2H_2O\ (\ell)$

b. $2Rb\ (s) + Br_2\ (\ell) \xrightarrow{\Delta} 2RbBr\ (s)$

c. $2KI\ (aq) + F_2\ (g) \longrightarrow 2KF\ (aq) + I_2\ (s)$

d. $CaO\ (s) + SiO_2\ (s) \xrightarrow{\Delta} CaSiO_3\ (s)$

e. $S\ (s) + O_2\ (g) \xrightarrow{\Delta} SO_2\ (g)$

f. $BaCO_3\ (s) \xrightarrow{\Delta} BaO\ (s) + CO_2\ (g)$

g. $HgS\ (s) + O_2\ (g) \xrightarrow{\Delta} Hg\ (\ell) + SO_2\ (g)$

h. $AgNO_3\ (aq) + HCl\ (aq) \longrightarrow AgCl\ (s) + HNO_3\ (aq)$

i. $Pb\ (s) + 2HBr\ (aq) \longrightarrow PbBr_2\ (s) + H_2\ (g)$

j. $2HI\ (aq) + H_2O_2\ (aq) \longrightarrow I_2\ (s) + 2H_2O\ (\ell)$

k. $RbOH\ (aq) + HNO_3\ (aq) \longrightarrow RbNO_3\ (aq) + H_2O\ (\ell)$

l. $N_2O_5\ (s) + H_2O\ (\ell) \longrightarrow 2HNO_3\ (aq)$

m. $H_2O\ (g) + CO\ (g) \xrightarrow{\Delta} H_2\ (g) + CO_2\ (g)$

n. $MgO\ (s) + H_2O\ (\ell) \longrightarrow Mg(OH)_2\ (s)$

o. $PbSO_4\ (s) + PbS\ (s) \xrightarrow{\Delta} 2Pb\ (s) + 2SO_2\ (g)$

19. Identify the metathesis reactions.
20. Identify the metathesis reactions that are also precipitation reactions.
21. Identify the metathesis reactions that are also classical acid-base reactions.
22. Identify the oxidation-reduction reactions.
23. Identify the oxidizing agent and reducing agent for each oxidation-reduction reaction.
24. Identify the oxidation-reduction reactions that are also displacement reactions.
25. Identify the decomposition reactions.
26. Identify the combination reactions.

Molecular, Total Ionic, and Net Ionic Equations for Metathesis Reactions and Identification of Products

27. Write balanced molecular equations for the following reactions occurring in water, or in contact with water. For acid-base reactions, assume complete neutralization.
(a) nitric acid + calcium hydroxide ⟶
(b) silver nitrate + ammonium bromide ⟶
(c) hydrocyanic acid + sodium hydroxide ⟶
(d) dilute sulfuric acid + barium chloride ⟶
28. Write balanced total ionic equations for the reactions of Exercise 27.

29. Write balanced net ionic equations for the reactions of Exercise 27.
30. Write balanced molecular equations for the following reactions in water, or in contact with water. For acid–base reactions, assume complete neutralization.
    (a) copper(II) hydroxide + hydrochloric acid $\longrightarrow$
    (b) sulfurous acid + potassium hydroxide $\longrightarrow$
    (c) aqueous ammonia + perchloric acid $\longrightarrow$
    (d) lithium phosphate + calcium chloride $\longrightarrow$
31. Write balanced total ionic equations for the reactions of Exercise 30.
32. Write balanced net ionic equations for the reactions of Exercise 30.
33. Write balanced molecular equations for the preparation of the following salts from acids and bases in aqueous solution or in contact with water.
    (a) magnesium sulfate
    (b) lithium carbonate
    (c) calcium nitrate
    (d) copper(II) sulfide
    (e) barium acetate
34. Write balanced total ionic equations for the reactions of Exercise 33.
35. Write balanced net ionic equations for the reactions of Exercise 33.
36. Which of the following metal hydroxides are amphoteric? (a) $Fe(OH)_3$ (b) KOH (c) $Zn(OH)_2$ (d) $Fe(OH)_2$ (e) $Pb(OH)_2$
37. Write balanced molecular, total ionic and net ionic equations for the reactions of beryllium hydroxide, $Be(OH)_2$, with (a) hydrochloric acid and (b) nitrous acid.
38. Repeat Exercise 37 for chromium(III) hydroxide, $Cr(OH)_3$, instead of $Be(OH)_2$.
39. Describe how each of the following reactions involves the removal of ions from solution:
    (a) $K_2SO_3\ (aq) + 2HBr\ (aq) \longrightarrow 2KBr\ (aq) + H_2SO_3\ (aq) \xrightarrow{\Delta} SO_2\ (g) + H_2O\ (\ell)$
    (b) $Ba(OH)_2\ (aq) + 2HCl\ (aq) \longrightarrow BaCl_2\ (aq) + 2H_2O\ (\ell)$
    (c) $BaCl_2\ (aq) + H_2SO_4\ (aq) \longrightarrow BaSO_4\ (s) + 2HCl\ (aq)$
    (d) $2NaOH\ (aq) + NiCl_2\ (aq) \longrightarrow Ni(OH)_2\ (s) + 2NaCl\ (aq)$

Oxidation-Reduction Reactions

40. Define (a) oxidation, (b) reduction, (c) oxidizing agent, and (d) reducing agent.
41. Write balanced molecular equations for the following redox reactions:
    (a) aluminum with sulfuric acid, H_2SO_4, to produce aluminum sulfate, $Al_2(SO_4)_3$, and hydrogen
    (b) nitrogen, N_2, with hydrogen, H_2, to form ammonia, NH_3
    (c) zinc sulfide, ZnS, with oxygen, O_2, to form zinc oxide, ZnO, and sulfur dioxide, SO_2
    (d) carbon with nitric acid, HNO_3, to produce nitrogen dioxide, NO_2, carbon dioxide, CO_2, and water
    (e) sulfuric acid with hydrogen iodide, HI, to produce sulfur dioxide, SO_2, iodine, I_2, and water
42. Identify the oxidizing agents and reducing agents in the oxidation-reduction reactions given in Exercise 41.
43. Write total ionic and net ionic equations for the following redox reactions occurring in aqueous solution or in contact with water:
    (a) $Fe + 2HCl \longrightarrow FeCl_2 + H_2$
    (b) $2KMnO_4 + 16HCl \longrightarrow 2MnCl_2 + 2KCl + 5Cl_2 + 8H_2O$ (Note: $MnCl_2$ is water-soluble.)
    (c) $4Zn + 10HNO_3 \longrightarrow 4Zn(NO_3)_2 + NH_4NO_3 + 3H_2O$.
44. Of the possible displacement reactions below, which one(s) will occur?
    (a) $2Cl^-\ (aq) + Br_2\ (\ell) \longrightarrow 2Br^-\ (aq) + Cl_2\ (g)$
    (b) $2Br^-\ (aq) + F_2\ (g) \longrightarrow 2F^-\ (aq) + Br_2\ (\ell)$
    (c) $2I^-\ (aq) + Cl_2\ (g) \longrightarrow 2Cl^-\ (aq) + I_2\ (s)$
    (d) $2Br^-\ (aq) + Cl_2\ (g) \longrightarrow 2Cl^-\ (aq) + Br_2\ (\ell)$
45. Which of the metals listed below will displace hydrogen from water or steam? (a) K (b) Ca (c) Cr (d) Ni (e) Hg (f) Pt
46. Refer to the metals of Exercise 45. Which one(s) will *not* displace hydrogen from nonoxidizing acids? (Assume that any oxide coatings on the metals have been brushed off.)
47. Which of the reactions below are disproportionation reactions?

(a) $KClO_4 + H_2SO_4 \longrightarrow KHSO_4 + HClO_4$
(b) $CH_4 + 2O_2 \longrightarrow CO_2 + 2H_2O$
(c) $3NaClO \longrightarrow NaClO_3 + 2NaCl$
(d) $2NO_2 \longrightarrow N_2O_4$
(e) $3NaOH + P_4 + 3H_2O \longrightarrow 3NaH_2PO_2 + PH_3$

Hydrogen and the Hydrides

48. Summarize the physical properties of hydrogen.
49. Write balanced molecular equations for (a) the reaction of iron with steam, (b) the reaction of calcium with hydrochloric acid, (c) the electrolysis of water, (d) the "water gas" reaction, and (e) the thermal cracking of a hydrocarbon such as propane, C_3H_8 (g).
50. Write a balanced molecular equation for the preparation of (a) an ionic hydride and (b) a covalent hydride.
51. Classify the following hydrides as covalent, ionic, or interstitial: (a) NaH (b) H_2S (c) BaH_2 (d) $TiH_{1.7}$, (e) NH_3
52. Explain how a formula like $TiH_{1.7}$ can represent a hydride.
53. Write molecular equations for the reactions of (a) NaH and (b) BaH_2 with a limited amount of water.
54. Name the following compounds: (a) H_2S (b) HF (c) KH (d) NH_3 (e) H_2Se (f) MgH_2

Oxygen and the Oxides

55. Draw Lewis dot formulas for O_2 and O_3. Why is that for O_2 inadequate?
56. Briefly compare and contrast the properties of oxygen with those of hydrogen.
57. Write molecular equations to show how oxygen can be prepared from (a) mercury(II) oxide, HgO, (b) hydrogen peroxide, H_2O_2, and (c) potassium chlorate, $KClO_3$.
58. Which of the following elements form normal oxides as the *major* products of reactions with oxygen? (a) Li (b) Na (c) Rb (d) Mg (e) Zn (exhibits only one common oxidation state) (f) Al
59. Write molecular equations for the primary reactions of oxygen with the following elements: (a) Li (b) Na (c) K (d) Ca
60. Write molecular equations for the reactions of the following elements with a *limited* amount of oxygen: (a) Sr (b) Fe (c) Mn (d) Cu
61. Write molecular equations for the reactions of the following elements with an *excess* of oxygen: (a) Sr (b) Fe (c) Mn (d) Cu
62. Write molecular equations for the reactions of the following elements with a *limited* amount of oxygen: (a) C (b) As_4 (c) Ge
63. Write molecular equations for the reactions of the following elements with an *excess* of oxygen: (a) C (b) As_4 (c) Ge
64. Distinguish among normal oxides, peroxides, and superoxides. What is the oxidation state of oxygen in each case?
65. Which of the following can be classified as basic anhydrides? (a) SO_2 (b) Li_2O (c) SeO_3 (d) CaO (e) N_2O_5
66. Which of the following can be classified as acid anhydrides? (a) CuO (b) CO_2 (c) Cl_2O_7 (d) Na_2O_2 (e) As_2O_5
67. Write balanced molecular equations for the following reactions and name the products:
(a) sulfur dioxide, SO_2, with water
(b) sulfur trioxide, SO_3, with water
(c) selenium trioxide, SeO_3, with water
(d) dinitrogen pentoxide, N_2O_5, with water
(e) dichlorine heptoxide, Cl_2O_7, with water
68. Write balanced molecular equations for the following reactions and name the products.
(a) sodium oxide, Na_2O, with water
(b) calcium oxide, CaO, with water
(c) lithium oxide, Li_2O, with water
(d) magnesium oxide, MgO, with sulfur dioxide, SO_2
(e) barium oxide, BaO, with carbon dioxide, CO_2
69. Identify the acid anhydride of the following ternary acids: (a) H_2SO_4 (b) H_2CO_3 (c) H_2SO_3 (d) H_3AsO_4 (e) HNO_2
70. Identify the basic anhydride of the following metal hydroxides: (a) NaOH (b) $Mg(OH)_2$ (c) $Fe(OH)_2$ (d) $Al(OH)_3$
71. Define combustion. Why must all combustion reactions also be redox reactions as well?
72. Write equations for the complete combustion of the following compounds: (a) ethane, C_2H_6 (g) (b) propane, C_3H_8 (g) (c) ethanol, C_2H_5OH (ℓ)
73. Write equations for the incomplete combustion of the following compounds to produce carbon monoxide: (a) ethane, C_2H_6 (g), and (b) propane, C_3H_8 (g)

74. Write equations for the complete combustion of the following compounds. Assume sulfur is converted to SO_2 and nitrogen is converted to NO. (a) C_6H_5N (ℓ) (b) C_2H_5SH (ℓ) (c) $C_7H_{10}NO_2S$ (ℓ)

75. Describe the formation of the reddish-brown haze of some cities experiencing this kind of air pollution.

76. Account for the occurrence of "acid rain."

8 Gases and the Kinetic-Molecular Theory

Matter is generally regarded as existing in three physical states: solids, liquids, and gases. Some samples do not fit neatly into any of these categories, but this classification enables us to deal with large amounts of information systematically. In the solid state water is known as ice, in the liquid state it is called water, while in the gaseous state it is known as steam or water vapor. Most, but not all, substances can exist in all three states. Most solids change to liquids and most liquids change to gases as they are heated.

8–1 Comparison of Solids, Liquids, and Gases

Liquids and gases are known as **fluids** because they flow freely. Solids and liquids are referred to as **condensed states** because they have much higher

TABLE 8–1 Comparison of Densities (g/mL) of Substances in Different States at Atmospheric Pressure

Substance	Solid	Liquid (20°C)	Gas (100°C)
Water (H_2O)	0.917 (0°C)	0.998	0.000588
Benzene (C_6H_6)	0.899 (0°C)	0.879	0.00255
Carbon tetrachloride (CCl_4)	1.7 (−25°C)	1.59	0.00503

The density of ice is less than the density of liquid water, which is quite unusual. Most substances are more dense in the solid state.

densities than gases. Table 8–1 displays the densities of a few common substances in different states.

As the data in Table 8–1 indicate, the densities of solids and liquids are many times greater than those of gases. The molecules must be very far apart in gases and much closer together in liquids and solids. For example, the volume of one mole of liquid water is slightly more than 18 milliliters, whereas one mole of steam occupies 30,600 milliliters at 100°C at atmospheric pressure. The possibilities for interaction among gaseous molecules would be minimal (because they are so far apart) were it not for their rapid motion.

A comparison of some additional properties of the three states of matter will be made in Chapter 9. Because the molecules are so far apart, and because interactions among molecules are minimal in the gaseous state, gases fill completely any container in which they are placed and are easily compressible.

All substances that are gases at room temperature may be liquified by cooling and compressing them. Volatile liquids, those of low boiling point, are easily converted to gases at room temperature or slightly above. The term **vapor** describes gases formed by boiling or evaporating liquids.

8–2 Composition of the Atmosphere and Some Common Properties of Gases

Many important chemical substances are gases. The atmosphere we breathe is composed of a mixture of gases (Table 8–2). The major components are oxygen, O_2, (bp, −182.98°C) and nitrogen, N_2 (bp, −195.79°C), with smaller amounts of other gases. All gases are miscible; that is, they mix completely in all proportions *unless* they react with each other.

Several scientists, notably Torricelli (1643), Boyle (1660), Charles (1787), and Graham (1831) laid an experimental foundation upon which our present understanding of gases is based. For example, their investigations showed that

compression

1. Gases can be compressed into smaller volumes; that is, their densities can be increased by applying increased pressure.
2. Gases exert pressure on their surroundings, and in turn pressure must be exerted to confine gases.
3. Gases expand without limit so that gas samples completely and uniformly occupy the volume of any container.
4. Gases diffuse into each other (mix) so that two samples of gas placed in the same container mix completely almost immediately. Conversely, different gases in a mixture like air do not separate on standing.

TABLE 8–2 Volume Percentage Composition of Dry Air

Gas		Percent by Volume
N_2, nitrogen		78.09
O_2, oxygen		20.94
Ar, argon		0.93
CO_2, carbon dioxide		0.03 (variable)
He, Ne, Kr, Xe (noble gases)		0.002
CH_4, methane		0.00015 (variable)
H_2, hydrogen		0.00005
All others*	less than	0.00004

*Atmospheric moisture (expressed as relative humidity) varies.

5. The amounts and properties of gases are described fully in terms of temperature, pressure, the volume they occupy, and the number of molecules present. For example, a sample of hot gas occupies a much larger volume than it does cold at the same pressure, but the number of molecules does not change.

8–3 Pressure

Pressure is defined as force per unit area (lb/in^2, commonly known as *psi*, for example). Pressure may be expressed in many different units as we shall see. The mercury **barometer** is a simple device for measuring atmospheric pressures. A glass tube (about 800 mm long) is sealed at one end, filled with mercury, and then carefully inverted into a dish of mercury without allowing air to enter. The mercury in the tube falls to the level where the pressure of the air on the surface of the mercury in the dish supports the column of mercury in the tube. The air pressure is measured in terms of the height of the mercury column, i.e., the vertical distance between the surface of the mercury in the open dish and that inside the closed tube. The pressure exerted by the atmosphere is equal to the pressure exerted by the column of mercury. Figure 8–1 illustrates the "heart" of a mercury barometer. A commercial model is shown in the margin. Since mercury barometers are simple and well known, gas pressures are frequently expressed in terms of millimeters of mercury (mm Hg, or just mm). In recent years the unit **torr** has been used to indicate mm Hg pressure, that is, 1 mm Hg = 1 torr.

A commercial barometer.

The unit torr was named for Evangelista Torricelli (1608–1647), who invented the mercury barometer.

Mercury **manometers** consist of a glass tube partially filled with mercury. One arm is open to the atmosphere and the other is connected to a container of gas. The pressure exerted by the gas in the container is equal to the atmospheric pressure plus or minus the difference in mercury levels, Δh, as shown in Figure 8–2.

Atmospheric pressure varies with atmospheric conditions and distance above sea level. It is less at high elevations because the air at low levels is compressed by the air above it. For example, one half of the quantity of matter in the atmosphere is below 20,000 feet above sea level, and therefore atmospheric pressure is only one half as great at 20,000 feet as it is at sea level. Mountain climbers and pilots use small portable barometers to determine their altitudes.

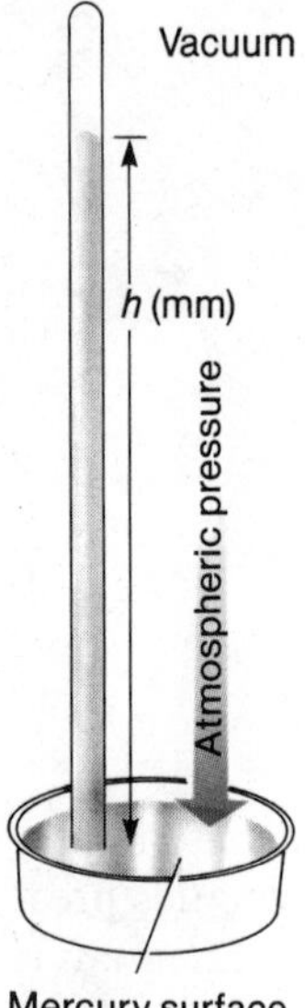

FIGURE 8–1 The barometer. At the level of the lower mercury surface, the pressure both inside and outside the tube must be that of the atmosphere. Inside the tube, the pressure is exerted by the mercury column h mm high. Hence, the atmospheric pressure must equal h mm Hg.

At sea level, at a latitude of 45°, the average atmospheric pressure supports a column of mercury 760 mm high in a simple mercury barometer when the mercury is at 0°C. This average sea level pressure of 760 mm Hg is called **one atmosphere of pressure**

1 atmosphere (atm) = 760 mm Hg = 760 torr = 1 bar

Acceleration is the change in velocity (m/s) per unit time (s), m/s^2.

The SI unit of pressure is the pascal (Pa). A **pascal** is defined as the pressure exerted by a force of one newton acting on an area of one square meter. By definition, one newton (N) is the force required to give a mass of one kilogram an acceleration of one meter per second per second. Symbolically we represent one newton as

$$1\ \text{N} = \frac{1\ \text{kg} \cdot \text{m}}{\text{s}^2}$$

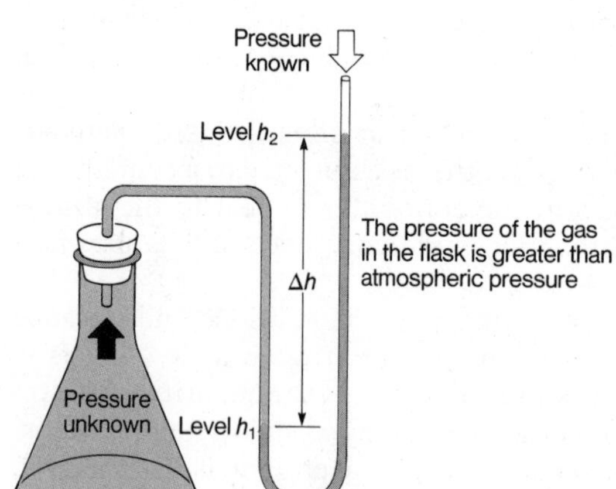

FIGURE 8–2 The two-arm mercury barometer is called a manometer. At the level h_1, the total pressure on the mercury in the left arm must equal the total pressure on the mercury in the right arm.

while one pascal is represented as

$$1 \text{ Pa} = \frac{1 \text{ N}}{\text{m}^2} = \frac{1 \text{ kg}}{\text{m} \cdot \text{s}^2}$$

One atmosphere of pressure is 1.013×10^5 pascals or 101.3 kilopascals

$$1 \text{ atm} = 1.013 \times 10^5 \text{ Pa} = 101.3 \text{ kPa}$$

8–4 Boyle's Law: The Relation of Volume to Pressure at Constant Temperature

In 1662 Robert Boyle summarized the results of experiments on various samples of gases. He found that *at constant temperature, the volume occupied by a definite mass of a gas is inversely proportional to the applied pressure.* In a typical experiment (Figure 8–3) a sample of a gas was trapped in a U-tube and allowed to come to constant temperature, and its volume and the difference in heights of the two mercury columns were recorded. This difference in heights plus the pressure of the atmosphere represents the pressure on the gas. The results of several such experiments are given in Table 8–3.

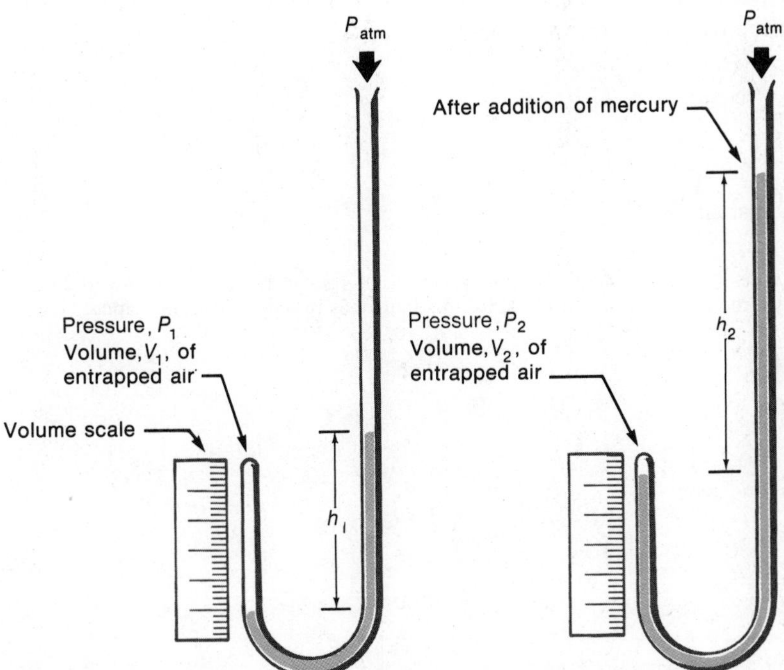

FIGURE 8–3 An artist's representation of Boyle's experiment. A sample of air is entrapped in a tube, such that the pressure on the air can be changed and the volume of the air measured. P_{atm} is the atmospheric pressure, measured with a barometer. $P_1 = h_1 + P_{atm}$, $P_2 = h_2 + P_{atm}$.

TABLE 8–3

Pressure	Volume	Pressure × Volume*	$1/P$
5.0	40.0	200	0.20
10.0	20.0	200	0.10
15.0	13.3	200	0.067
17.0	11.8	201	0.059
20.0	10.0	200	0.050
22.0	9.10	200	0.045
30.0	6.70	201	0.033

*Both pressure and volume are in arbitrary units in this example.

When the pressure of a gas is plotted against its volume, the resulting curve is one branch of a hyperbola (Figure 8–4a).* This is a graphic illustration of an inverse relationship. If volume is plotted *versus* the reciprocal of the pressure, a straight line results (Figure 8–4b). **Boyle's Law** states: *at a given temperature, the product of pressure and volume of a definite mass of gas is constant* (200 in the example above).

Boyle's observations show that for a definite mass of a gas at constant temperature, volume varies inversely with applied pressure. This may be stated symbolically as

n is the number of moles of gas in the sample.

$$V \propto \frac{1}{P} \qquad \text{(constant } n, T)$$

where the symbol $\propto$ reads "is proportional to." A proportionality is converted into an equality by introducing a proportionality constant, k.

The units of "k" are determined by the units used to express the volume, V, and pressure, P.

$$V = k\left(\frac{1}{P}\right) \qquad \text{(constant } n, T)$$

Rearranging the equation gives

$$PV = k \qquad \text{(constant } n, T)$$

* Since neither the volume nor the pressure of a sample of gas can be less than zero, the other branch of the hyperbola, which lies in the third quadrant, has no physical significance.

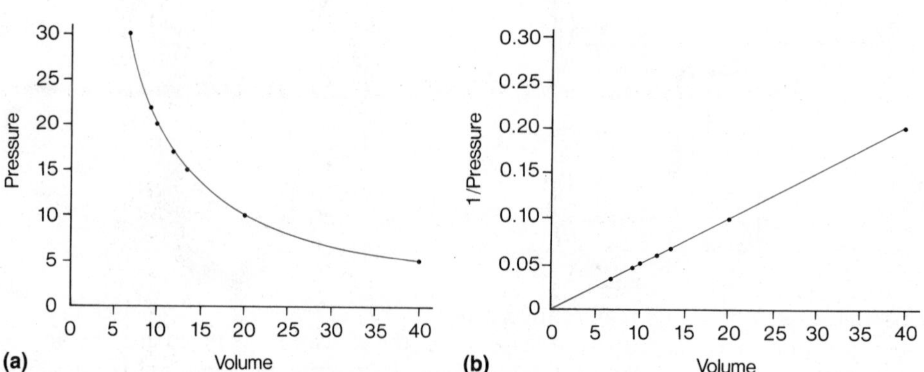

FIGURE 8–4 A graphical illustration of Boyle's Law, using the data of Table 8–3.

which is an expression of Boyle's law. If we denote the first pressure and volume in Table 8–3 as P_1 and V_1, we have

$$P_1V_1 = k_1 \qquad \text{(constant } n, T)$$

Changing the pressure (temperature is constant) gives new values for P and V, the second values in Table 8–3, and we have

$$P_2V_2 = k_2 \qquad \text{(constant } n, T)$$

The data in Table 8–3 show that $k_1 = k_2$, and since "things equal to the same thing are equal to each other," we may write the useful form of Boyle's Law

$$P_1V_1 = P_2V_2 \qquad \text{(for a given mass of a gas at constant temperature)}$$

At normal temperatures and pressures, most gases obey Boyle's Law rather well. We call this *ideal behavior.* Deviations will be discussed in Section 8–15. The following illustrative examples demonstrate the utility of Boyle's Law.

EXAMPLE 8–1

A sample of gas occupies 10 liters under a pressure of 1 atmosphere. What will its volume be if the pressure is increased to 2 atmospheres? Assume the temperature of the gas sample does not change.

Solution

Boyle's Law tells us that $P_1V_1 = P_2V_2$ and solving for V_2, gives

$$V_2 = \frac{P_1V_1}{P_2},$$

and substitution yields

$$V_2 = \frac{(1\text{ atm})(10\text{ L})}{(2\text{ atm})} = \underline{5\text{ L}}$$

Pressure and volume are inversely proportional and doubling the pressure halves the volume of a sample of gas at constant temperature.

Another approach to the problem is to multiply the original volume by a "Boyle's Law Factor." The pressure increases from one atmosphere to two atmospheres, and therefore the volume *decreases* by the factor (1 atm/2 atm). The solution then becomes

$\underline{?}$ L = 10 L × (Boyle's Law factor that represents a decrease in volume)

$$= 10\text{ L} \times \left(\frac{1\text{ atm}}{2\text{ atm}}\right) = \underline{5\text{ L}}$$

EXAMPLE 8–2

A sample of oxygen occupies 10.0 liters under a pressure of 790 torr (105 kPa). At what pressure will it occupy 13.4 liters if the temperature does not change?

Solution

We know that $P_1V_1 = P_2V_2$ and solving for P_2 gives

$$P_2 = \frac{P_1V_1}{V_2} = \frac{(790\text{ torr})(10.0\text{ L})}{13.4\text{ L}} = \underline{590\text{ torr}} \qquad (78.6\text{ kPa})$$

Note that the factors used in Examples 8–1 and 8–2 are *correction factors*, not unit factors.

The statement of the problem tells us that volume *increases* from 10.0 liters to 13.4 liters (at constant temperature), and therefore the pressure must *decrease*. We can also solve the problem by multiplying the original pressure by the volume factor less than unity, that is, 10.0 L/13.4 L.

$$\underline{?} \text{ torr} = 790 \text{ torr} \times \frac{10.0 \text{ L}}{13.4 \text{ L}} = \underline{590 \text{ torr}} \qquad (78.6 \text{ kPa})$$

8–5 The Kelvin (Absolute) Temperature Scale

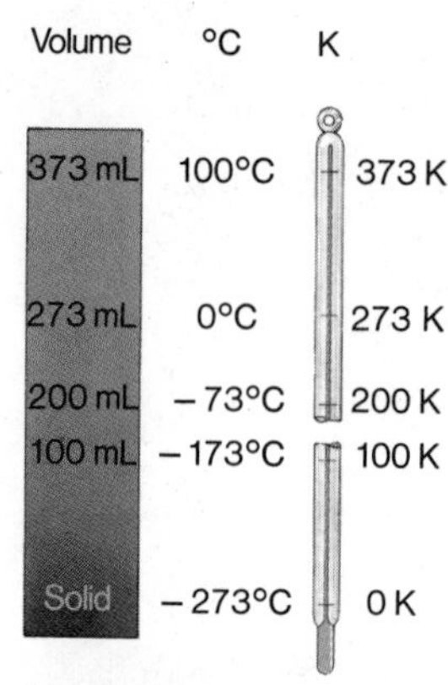

FIGURE 8–5
Volume–temperature relationship for a sample of gas at constant pressure.

The absolute temperature scale was discussed in Section 1–12, and the fact that this temperature scale is based on the observed behavior of gases was emphasized. Let us consider the origin and utility of the scale in more detail.

In his pressure–volume studies on gases, Robert Boyle noticed that heating a sample of gas caused some volume change but he failed to follow up on this observation. About 1800, two French scientists, Jacques Charles and Joseph Gay-Lussac, pioneer balloonists at the time, began studying the expansion of gases with increasing temperature. Their studies led them to conclude that the rate of expansion with increased (Celsius scale) temperature was constant and the same for all the gases they studied as long as pressure remained constant. The implications of their discovery were not fully realized until nearly a century later when scientists recognized that this behavior of gases would become the basis of a new temperature scale, the absolute temperature scale.

Experiments have shown that when a 273-mL sample of gas at 0°C is heated to 1°C its volume increases by 1 mL to 274 mL. At 10°C the volume increases (by 10 mL) to 283mL if the pressure remains constant in both cases. If a 273-mL sample of gas at 0°C is cooled to −1°C, its volume decreases to 272 mL, while at −10°C the volume decreases to 263 mL if the pressure remains constant. For each °C *increase* in temperature, the volume of a sample of gas *increases* by the fraction 1/273 of its volume at 0°C. For each °C *decrease* in temperature, the volume of a sample of gas *decreases* by the fraction 1/273 of its volume at 0°C. Thus, if a 273-mL sample of gas at 0°C is heated to 273°C at constant pressure, its volume doubles to 546 mL. On the other hand, if the temperature of a 273-mL sample of gas at 0°C should be lowered from 0°C to −273°C, the gas "should" have no volume at all because the volume should decrease at the rate of 1/273 of its volume at 0°C for each degree by which temperature is decreased, and so become zero. However, all gases liquefy, and most solidify, before the temperature reaches −273°C. This temperature, −273°C (to be more exact, −273.15°C), is *absolute zero* and is taken as the zero point on the Kelvin temperature scale. As pointed out in Section 1–12, the relationship between the Celsius and Kelvin temperature scales is

$$\text{K} = {}^\circ\text{C} + 273^\circ$$

Recall that degrees Kelvin is called simply kelvins and is represented by K, not °K.

Figure 8–5 shows the relationship between temperature and volume for a 273-mL sample of gas originally at 0°C (273 K) at constant pressure.

8–6 Charles' Law: The Relation of Volume to Temperature at Constant Pressure

A summary of observations on the effect of temperature changes on the volumes of samples of gases at constant pressure is known as **Charles' Law.** *At constant pressure, the volume occupied by a definite mass of a gas is directly proportional to its absolute temperature.* The temperature–volume relationship, at constant pressure, is illustrated visually and graphically in Figures 8–6a and 8–6b.

The solid line portions of Figure 8–6b illustrate the behavior observed by Charles and Gay-Lussac. The lines represent the same sample under different pressures. Lord Kelvin, a British physicist, noticed that an extension of the different temperature–volume lines back to zero volume (dashed line) yields a common intercept. This intercept is −273.15°C on the temperature axis, and Kelvin named this temperature **absolute zero.** Since the degrees are the same size over the entire scale, 0°C becomes 273.15 degrees above absolute zero. In honor of Lord Kelvin's work, this temperature scale is called the Kelvin temperature scale.

Absolute zero may be thought of as the limit of thermal contraction for an ideal gas.

The data in Figure 8–5 illustrate the validity of Charles' Law. The law was, of course, derived from similar observations. Charles' Law tells us that at constant pressure, the volume occupied by a definite mass of a gas is directly proportional to the absolute temperature, and we may write

$V \propto T \qquad \text{(constant } n, P)$

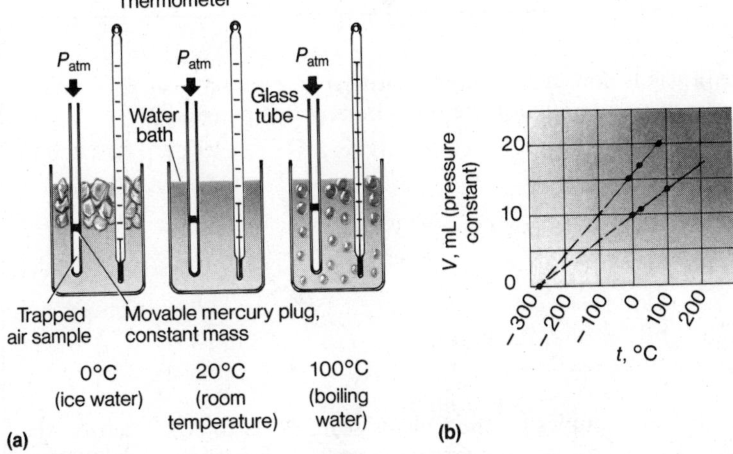

FIGURE 8–6 (*a*) An illustration showing that the volume of a gas increases as the temperature is increased at constant pressure. A mercury plug of constant weight plus atmospheric pressure (P_{atm}) maintains a constant pressure on the trapped air. (*b*) A plot (graph) of volume versus temperature shows that gases expand when heated at constant pressure. The two lines represent the same weight of the same gas at different pressures.

A proportionality constant (k) is introduced to convert the proportionality into an equality

$V = kT$ (for a definite mass of gas at constant pressure)

Rearranging the expression gives $V/T = k$, a concise statement of Charles' Law. If the temperature increases, the volume must also increase proportionally.

If the volume and temperature of a sample of gas are measured, a numerical value can be calculated for k. Call it k_1.

$$\frac{V_1}{T_1} = k_1 \qquad (\text{constant } n, P)$$

Changing the temperature causes the volume to change, and new values are obtained for V and T

$$\frac{V_2}{T_2} = k_2 \qquad (\text{constant } n, P)$$

Experiments show that for a given sample of gas at constant pressure, $k_1 = k_2$. In Figure 8–5, $k_1 = k_2 = 1$, because the initial volume was chosen so that the V/T ratio was unity. Since things equal to the same thing are equal to each other, we may write

$$k_1 = k_2 \quad \text{or} \quad \frac{V_1}{T_1} = \frac{V_2}{T_2} \qquad (\text{constant } n \text{ and } P)$$

which is the useful form of Charles' Law. The following examples illustrate the use of Charles' Law. This relationship is valid *only* when temperature, T, is expressed on an absolute (usually the Kelvin) scale.

EXAMPLE 8–3

A 250-mL sample of gas is confined under one atmosphere pressure at 25°C. What volume will the gas occupy at 50°C if the pressure remains constant?

Solution

$V_1 = 250$ mL $\qquad V_2 = \underline{?}$

$T_1 = 25°\text{C} + 273° = 298$ K $\qquad T_2 = 50°\text{C} + 273° = 323$ K

$$\frac{V_1}{T_1} = \frac{V_2}{T_2} \quad \text{and} \quad V_2 = \frac{T_2 V_1}{T_1}$$

$$V_2 = \frac{(323\text{ K})(250\text{ mL})}{(298\text{ K})} = \underline{271\text{ mL}}$$

Note that the temperature doubles on the Celsius scale (25°C to 50°C), but on the Kelvin scale the fractional absolute temperature increase is much smaller (298 K to 323 K). Another approach to the problem is to multiply the original volume by a Charles' Law factor. The temperature *increases* from 298 K to 323 K which causes the volume to *increase* by the factor, 323 K/298 K

$$\underline{?}\text{ mL} = 250\text{ mL} \times \frac{323\text{ K}}{298\text{ K}} = \underline{271\text{ mL}}$$

EXAMPLE 8–4

A sample of nitrogen occupies 400 milliliters at 100°C. At what temperature will it occupy 200 milliliters if the pressure does not change?

Solution

$V_1 = 400$ mL $\qquad V_2 = 200$ mL

$T_1 = 100°C + 273° = 373$ K $\qquad T_2 = \underline{?}$

$$\frac{V_1}{T_1} = \frac{V_2}{T_2} \quad \text{so} \quad T_2 = \frac{V_2 T_1}{V_1}$$

$$T_2 = \frac{(200 \text{ mL})(373 \text{ K})}{400 \text{ mL}} = 186 \text{ K}$$

°C = 186 K − 273° = $\underline{-87°C}$

We see that the temperature must be reduced to −87°C to reduce the pressure to its original value. Recall that according to Boyle's Law, if the temperature had remained constant, the volume decrease would have caused an increase in pressure. As in Example 8–3, this problem can also be solved using a Charles' Law factor. A decrease in volume causes the temperature to decrease at constant pressure so the initial temperature must be multiplied by the volume factor less than one, 200 mL/400 mL

$$\underline{?}\text{ K} = 373 \text{ K} \times \frac{200 \text{ mL}}{400 \text{ mL}} = 186 \text{ K, which is } \underline{-87°\text{C}}$$

8–7 Standard Temperature and Pressure

We have demonstrated that both temperature and pressure affect the volumes (and therefore densities) of gases. It is convenient to choose some "standard" temperature and pressure as a reference point in discussing gases. **Standard conditions of temperature and pressure** (STP or SC) are, by international agreement, 0°C (273 K) and one atmosphere of pressure (760 torr).

8–8 The Combined Gas Laws

Boyle's Law relates the pressure and volume of a sample of gas at constant temperature, while Charles' Law relates the volume and temperature at constant pressure, that is, $P_1V_1 = P_2V_2$ and $V_1/T_1 = V_2/T_2$. Combination of the essence of Boyle's Law and Charles' Law into a single expression gives the **combined gas law equation**

$$\frac{P_1V_1}{T_1} = \frac{P_2V_2}{T_2} \qquad (\text{constant } n)$$

The following examples illustrate applications of the combined gas laws.

EXAMPLE 8–5

A sample of neon occupies 100 liters at 27°C under a pressure of 1000 torr. What volume would it occupy at standard conditions?

Solution

$V_1 = 100\ \text{L}$ $V_2 = \underline{?}$

$P_1 = 1000\ \text{torr}$ $P_2 = 760\ \text{torr}$

$T_1 = 27°\text{C} + 273° = 300\ \text{K}$ $T_2 = 273\ \text{K}$

$$\frac{P_1V_1}{T_1} = \frac{P_2V_2}{T_2}$$

Solving the combined gas law expression for V_2 gives

$$V_2 = \frac{P_1V_1T_2}{P_2T_1} = \frac{(1000\ \text{torr})(100\ \text{L})(273\ \text{K})}{(760\ \text{torr})(300\ \text{K})} = \underline{120\ \text{L}}$$

Alternatively, we can multiply the original volume by a Boyle's Law factor and a Charles' Law factor. The pressure decreases from 1000 torr to 760 torr (volume will increase) so the Boyle's Law factor is 1000 torr/760 torr. The temperature decreases from 300 K to 273 K (volume will decrease) and the Charles' Law factor is 273 K/300 K. Multiplication of the original volume by these factors gives the same result obtained above

$$\underline{?}\ \text{L} = 100\ \text{L} \times \frac{1000\ \text{torr}}{760\ \text{torr}} \times \frac{273\ \text{K}}{300\ \text{K}} = \underline{120\ \text{L}}$$

Note that the pressure decrease (from 1000 torr to 760 torr) alone would result in a significant *increase* in the volume of neon. The temperature decrease (from 300 K to 273 K) alone would give only a small *decrease* in the volume of neon. The net result of the two changes is that the volume increases from 100 liters to 120 liters.

As a reference, one atm is 101.3 kPa.

EXAMPLE 8–6

A sample of gas occupies 10.0 liters at 240°C under a pressure of 80.0 kPa. At what temperature will the gas occupy 20.0 liters if the pressure is increased to 107 kPa?

Solution

$V_1 = 10.0\ \text{L}$ $V_2 = 20.0\ \text{L}$

$P_1 = 80.0\ \text{kPa}$ $P_2 = 107\ \text{kPa}$

$T_1 = 240°\text{C} + 273° = 513\ \text{K}$ $T_2 = \underline{?}$

Solving the combined gas law expression for T_2 gives

$$\frac{P_1V_1}{T_1} = \frac{P_2V_2}{T_2} \quad \text{so} \quad T_2 = \frac{P_2V_2T_1}{P_1V_1}$$

$$T_2 = \frac{(107\ \text{kPa})(20.0\ \text{L})(513\ \text{K})}{(80.0\ \text{kPa})(10.0\ \text{L})} = \underline{1.37 \times 10^3\ \text{K}}$$

Since $\text{K} = °\text{C} + 273°$, $°\text{C} = 1.37 \times 10^3\ \text{K} - 273° = \underline{1.10 \times 10^3\ °\text{C}}$

The combined gas law equation is useful when both the temperature and pressure of a sample of gas are changed. Or, when any five of the variables in the equation are known, the sixth variable can be calculated.

The combined gas law equation is frequently used to correct volumes of gases to *standard conditions* (STP) as Example 8–7 illustrates.

EXAMPLE 8–7

A sample of dry oxygen prepared in the laboratory occupies 750 milliliters at room temperature, 21°C, on a day when the barometric pressure is 745 torr. What volume will it occupy at standard conditions?

Solution

$V_1 = 750$ mL $\qquad V_2 = \underline{?}$

$P_1 = 745$ torr $\qquad P_2 = 760$ torr

$T_1 = 21°C + 273° = 294$ K $\qquad T_2 = 273$ K

The combined gas law expression is

$$\frac{P_1V_1}{T_1} = \frac{P_2V_2}{T_2}$$

and solving for V_2 gives

$$V_2 = \frac{P_1V_1T_2}{P_2T_1} = \frac{(745\text{ torr})(750\text{ mL})(273\text{ K})}{(760\text{ torr})(294\text{ K})} = \underline{683\text{ mL}}$$

Both temperature and pressure changes decrease the volume slightly and so the volume at STP is significantly less. Alternatively, the volume of oxygen could be multiplied by a Boyle's Law factor, 745 torr/760 torr, and by a Charles' Law factor, 273 K /294 K, to give the same result

$$\underline{?}\text{ mL} = 750\text{ mL} \times \frac{745\text{ torr}}{760\text{ torr}} \times \frac{273\text{ K}}{294\text{ K}} = \underline{683\text{ mL}}$$

8–9 Gas Densities and the Standard Molar Volume

In 1811 Amedeo Avogadro postulated that *at the same temperature and pressure, equal volumes of all gases contain the same number of molecules.* Numerous experiments have demonstrated that Avogadro's hypothesis was correct to within about ±2%, and the statement is now known as **Avogadro's Law.** Thus, the number of moles of gas in a sample does not change with temperature or pressure.

Gas densities depend on pressure and temperature. Recall that density is defined as mass per unit volume, or density = mass/volume. Since pressure changes affect volumes of gases according to Boyle's Law, while temperature changes affect volumes of gases according to Charles' Law, gas densities measured at various temperatures and pressures can be converted to *standard temperature and pressure* by application of Boyle's and Charles' Laws. Table 8–4 gives the densities of several gases at standard conditions. The volume occupied by a mole of gas at STP is very nearly a constant for all gases and is referred to as the standard molar volume. For an ideal gas the **standard molar volume** is taken to be 22.4 liters per mole at STP.

TABLE 8–4 Densities and Standard Molar Volumes for Some Common Gases

Gas	Formula	Mol. Wt. (g/mol)	Density at STP (g/L)	Standard Molar Volume (L/mol)
Hydrogen	H_2	2.02	0.090	22.428
Helium	He	4.003	0.178	22.426
Neon	Ne	20.18	0.900	22.425
Nitrogen	N_2	28.01	1.250	22.404
Oxygen	O_2	32.00	1.429	22.394
Argon	Ar	39.95	1.784	22.393
Carbon dioxide	CO_2	44.01	1.977	22.256
Ammonia	NH_3	17.03	0.771	22.094
Chlorine	Cl_2	70.91	3.214	22.063

The following examples illustrate the usefulness of the standard molar volume idea.

EXAMPLE 8–8

If 1.00 mole of a gas occupies 22.4 liters at STP, what volume will 2.65 moles of gas occupy at STP?

Solution

Since the standard molar volume is about 22.4 L (± about 2%) for all gases, we can use the unit factor, 22.4 L_{STP}/mol of gas.

$$\underline{?}\ L_{STP} = 2.65 \text{ mol} \times \frac{22.4\ L_{STP}}{1 \text{ mol gas}}$$

$$= \underline{59.4 \text{ L at STP}}$$

EXAMPLE 8–9

A mole of a given gas occupies 27.0 liters and its density is 1.41 g/L at a particular temperature and pressure. What is its molecular weight? What is the density of the gas at STP?

Solution

We multiply the density under the original conditions by the unit factor constructed from the equality 27.0 L = 1 mol, in order to generate the appropriate units, g/mol.

$$\frac{\underline{?}\ g}{mol} = \frac{1.41 \text{ g}}{L} \times \frac{27.0 \text{ L}}{mol} = \underline{38.1 \text{ g/mol}}$$

Hence the molecular weight is <u>38.1 amu</u>. At STP, one mole of the gas, 38.1 g, occupies only 22.4 L and its density is

$$\text{Density} = \frac{38.1 \text{ g}}{22.4 \text{ L}} = \underline{1.70 \text{ g/L}} \text{ at STP.}$$

8–10 A Summary of the Gas Laws—The Ideal Gas Equation

Let us summarize what we have learned about gases. Gases can be described in terms of their pressure, temperature (Kelvin), volume, and the number of moles, n, present. Any three of these variables determine the fourth. Most gases behave quite similarly so that the behavior of an ideal gas is illustrative of gases in general. An **ideal gas** is one that obeys the gas laws exactly. Many real gases show slight deviations from ideality, but at normal temperatures and pressures the deviations are small enough to be ignored in most cases. We'll do so for the present, and discuss deviations later.

Summarizing the behavior of gases in a general way, we have

Boyle's Law	$V \propto \frac{1}{P}$	(at constant T and n)
Charles' Law	$V \propto T$	(at constant P and n)
Avogadro's Law	$V \propto n$	(at constant T and P)
and therefore,	$V \propto nT/P$	
or, rearranging,	$PV \propto nT.$	

As before, the proportionality, $PV \propto nT$, can be converted into an equality by introducing a proportionality constant (in this case we'll use the symbol R as the proportionality constant).

$$PV = nRT$$

This relationship, $PV = nRT$, is called the **ideal gas equation** or the **ideal gas law.** The numerical value of R, the proportionality constant, depends upon the choices of the units for P, V, and T. Recall that 1.00 mole of an ideal gas occupies 22.4 liters at 1.00 atmosphere and 273 K (STP). Solving the ideal gas law for R gives

$$R = \frac{PV}{nT}$$

Substituting the above values for P, V, n, and T yields

$$R = \frac{(1.00 \text{ atm})(22.4 \text{ L})}{(1.00 \text{ mol})(273 \text{ K})} = \frac{0.0821 \text{ L} \cdot \text{atm}}{\text{mol} \cdot \text{K}}$$

The proportionality constant, R, is called the **universal gas constant,** and its more exact value is 0.08206 L · atm/mol · K. It is a constant of nature and may be expressed in a variety of units.

R .08206 l atm / mol · deg (K)

EXAMPLE 8–10

R can have any *energy* units per mole per Kelvin. Calculate R in terms of joules per mole per degree Kelvin and in SI units of kPa · dm³/mol · K.

Solution

Appendix C shows that 1 L · atm = 101.32 joules.

$$R = \frac{0.08206 \text{ L} \cdot \text{atm}}{\text{mol} \cdot \text{K}} \times \frac{101.32 \text{ J}}{1 \text{ L} \cdot \text{atm}} = 8.314 \text{ J/mol} \cdot \text{K}$$

Recall that 1 dm^3 = 1 L.

Let us now evaluate R in SI units. One atmosphere pressure is 101.3 kilopascals and the molar volume at STP is 22.4 dm^3

$$R = \frac{PV}{nT} = \frac{101.3\text{ kPa} \times 22.4\text{ dm}^3}{1\text{ mol} \times 273\text{ K}} = 8.31\,\frac{\text{kPa}\cdot\text{dm}^3}{\text{mol}\cdot\text{K}}$$

We have now evaluated R, the universal gas constant, in three different sets of units, and so we can write

$$R = 0.08206\,\frac{\text{L}\cdot\text{atm}}{\text{mol}\cdot\text{K}} = \frac{8.314\text{ J}}{\text{mol}\cdot\text{K}} = 8.31\,\frac{\text{kPa}\cdot\text{dm}^3}{\text{mol}\cdot\text{K}}$$

Often, Boyle's, Charles', and Avogadro's Laws are sufficient to calculate a change for a sample of gas. The usefulness of the ideal gas equation lies in the fact that it relates the four variables, P, V, n, and T, that describe a sample of gas at a given set of conditions. If three variables are known, the fourth can be calculated as the following examples show.

EXAMPLE 8–11

What is the volume of a gas balloon filled with 4.00 moles of helium when the atmospheric pressure is 748 torr and the temperature is 30°C?

Solution

We always look at the variables first and arrange them in some order with the proper units

$$\underline{?}\text{ atm} = 748\text{ torr} \times \frac{1\text{ atm}}{760\text{ torr}} = 0.984\text{ atm }(P) \qquad \underline{?}\text{ mol} = 4.00\text{ mol }(n)$$

$$\underline{?}\text{ K} = 30°\text{C} + 273° = 303\text{ K }(T)$$

Solving the ideal gas equation for V and substituting these values gives

$$PV = nRT, \qquad V = \frac{nRT}{P}$$

$$V = \frac{(4.00\text{ mol})\left(0.0821\,\frac{\text{L}\cdot\text{atm}}{\text{mol}\cdot\text{K}}\right)(303\text{ K})}{0.984\text{ atm}} = 101\text{ L}$$

You may wonder whether pressures are given in torr or mm Hg and temperatures in °C just to confuse the issue. This is not the case. Pressures are usually measured with Torricellian (mercury) barometers, while temperatures are measured with Celsius thermometers.

EXAMPLE 8–12

How many moles of hydrogen are present in a 5.00-liter container at a pressure of 400 torr at 26°C?

Solution

$$P = 400\text{ torr} \times \frac{1\text{ atm}}{760\text{ torr}} = 0.526\text{ atm} \qquad V = 5.00\text{ L}$$

$$T = 26°\text{C} + 273° = 299\text{ K}$$

Solving $PV = nRT$ for n, and substituting gives

$$n = \frac{PV}{RT} = \frac{(0.526\ \text{atm})(5.00\ \text{L})}{\left(0.0821\dfrac{\text{L}\cdot\text{atm}}{\text{mol}\cdot\text{K}}\right)(299\ \text{K})} = \underline{0.107\ \text{mol}}$$

EXAMPLE 8–13

What pressure in kilopascals is exerted by 54.0 grams of xenon, Xe, in a 1.00-liter flask at 20°C?

Solution

$V = 1.00\ \text{L} = 1.00\ \text{dm}^3 \qquad n = 54.0\ \text{g Xe} \times \dfrac{1\ \text{mol}}{131.3\ \text{g Xe}} = 0.411\ \text{mol}$

$T = 20°\text{C} + 273° = 293\ \text{K}$

Solving $PV = nRT$ for P and substituting gives

$$P = \frac{nRT}{V} = \frac{(0.411\ \text{mol})\left(\dfrac{8.31\ \text{kPa}\cdot\text{dm}^3}{\text{mol}\cdot\text{K}}\right)(293\ \text{K})}{1.00\ \text{dm}^3}$$

$$= \underline{1.00 \times 10^3\ \text{kPa}} \qquad (9.87\ \text{atm})$$

8–11 Determination of Molecular Weights and True Formulas of Gaseous Compounds

In Section 2–6, we distinguished between the simplest formula and the true formula of a compound and illustrated how simplest formulas can be calculated from the percent compositions of compounds. The formula (molecular) weight must be known to determine the true (molecular) formula for a compound. For compounds that are gases at reasonable temperatures and pressures, the ideal gas law provides a basis for calculating actual molecular weights, as Example 8–14 illustrates.

EXAMPLE 8–14

A 0.109-gram sample of gas occupies 112 milliliters at 100°C and 750 torr. What is the molecular weight of the gas?

Solution

We can use the ideal gas law, $PV = nRT$, to calculate the number of moles of gas

$V = 0.112\ \text{L} \qquad T = 100°\text{C} + 273 = 373\ \text{K}$

$P = 750\ \text{torr} \times \dfrac{1\ \text{atm}}{760\ \text{torr}} = 0.987\ \text{atm}$

$$n = \frac{PV}{RT} = \frac{(0.987\ \text{atm})(0.112\ \text{L})}{\left(0.0821\dfrac{\text{L}\cdot\text{atm}}{\text{mol}\cdot\text{K}}\right)(373\ \text{K})} = 0.00361\ \text{mol}$$

We now know that 0.109 g of the gas is 0.00361 mole, and so we can calculate the mass of one mole, the molecular weight of the gas.

$$\frac{?\ g}{mol} = \frac{0.109\ g}{0.00361\ mol} = \underline{30.2\ g/mol}$$

The molecular weight of the gas is 30.2 amu.

EXAMPLE 8–15

Analysis of a sample of a gaseous compound shows that it contains 85.7% carbon and 14.3% hydrogen by mass. At standard conditions, 100 milliliters of the compound has a mass of 0.188 gram. What is the true (molecular) formula for the compound?

Solution

Let us determine the empirical formula for the compound as we did in Section 2–6

$$?\ \text{mol C atoms} = 85.7\ \text{g C} \times \frac{1\ \text{mol C atoms}}{12.0\ \text{g C}} = \underline{7.14\ \text{mol C atoms}}$$

$$?\ \text{mol H atoms} = 14.3\ \text{g H} \times \frac{1\ \text{mol H atoms}}{1.01\ \text{g H}} = \underline{14.2\ \text{mol H atoms}}$$

The ratio of moles of carbon atoms to moles of hydrogen atoms is

$$\frac{7.14}{7.14} = 1\ \text{C} \qquad \text{simplest formula} = CH_2$$

and

$$\frac{14.2}{7.14} = 2\ \text{H} \qquad \text{formula weight} = 14\ \text{amu}$$

The simplest formula is CH_2 (formula weight = 14 amu) and the true formula of the compound is some multiple of CH_2. Let us calculate the number of moles in the sample of gas by substituting into the ideal gas law, $PV = nRT$

$V = 0.100$ L, $P = 1.00$ atm, $T = 273$ K

$$n = \frac{PV}{RT} = \frac{(1.00\ \text{atm})(0.100\ \text{L})}{\left(0.0821\ \frac{\text{L} \cdot \text{atm}}{\text{mol} \cdot \text{K}}\right)(273\ \text{K})} = 0.00446\ \text{mol}$$

This result tells us that 0.00446 mole has a mass of 0.188 g, which enables us to calculate the molecular weight

$$\frac{?\ g}{mol} = \frac{0.188\ g}{0.00446\ mol} = \underline{42.2\ g/mol}$$

We know that one mole of the compound has a mass of 42.2 grams, and we can determine its true formula. We divide the molecular weight of the compound (which corresponds to its true formula) by the simplest formula weight to obtain an integer

$$\frac{\text{molecular weight}}{\text{simplest formula weight}} = \frac{42.2\ \text{amu}}{14\ \text{amu}} = 3$$

This tells us that the true formula is three times the simplest formula. Therefore, the true formula is $(CH_2)_3$ = $\underline{C_3H_6}$, and the gas is either propene or cyclopropane, both of which have the formula C_3H_6.

The kind of calculation illustrated in Example 8–15 can be extended to volatile liquids. In the **Dumas method,** a sample of volatile liquid is placed in a previously weighed flask, the flask is placed in a boiling water bath and the liquid is allowed to evaporate, and the excess vapor escapes into the air. As the last bit of liquid evaporates, the container is filled completely with vapor of the volatile liquid at 100°C. At this point the container is removed from the boiling water bath, cooled quickly, and weighed again. Rapid cooling condenses the vapor that filled the flask at 100°C, and the flask is filled with the air again. Recall that the volume of a liquid is extremely small in comparison to the volume occupied by the same mass of the gaseous compound at atmospheric pressure. The difference between the mass of the flask containing the condensed liquid and the empty flask is the mass of the condensed liquid and therefore, the mass of the vapor that filled the flask at 100°C and the atmospheric pressure (Figure 8–7). The "empty" flask contains air, of course, but it also contained air when weighed previously, so the mass of air remains constant and does not affect the result. Example 8–16 illustrates the determination of the molecular weight of a volatile liquid by the Dumas method. Although the compound is a liquid at room temperature, it is a gas at 100°C, and therefore can be treated by the appropriate gas laws. If the water surrounding the flask is replaced by a liquid with a higher boiling point, the Dumas method can be applied to liquids that vaporize at or below the boiling point of the liquid surrounding the flask (Figure 8–7).

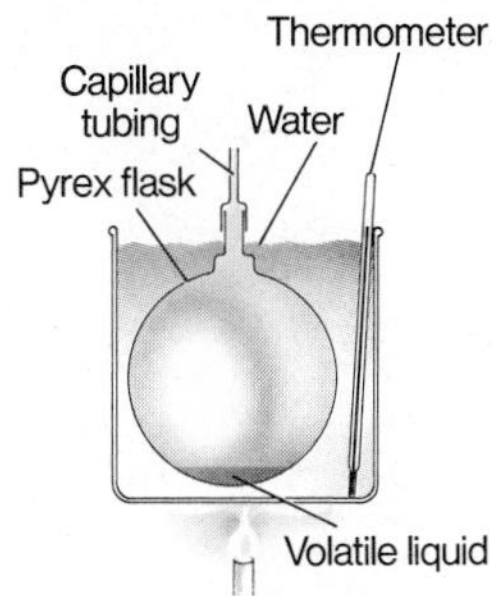

FIGURE 8–7 Determination of molecular weight by the Dumas method. A volatile liquid is vaporized in the boiling water bath, and excess vapor is driven from the flask. The remaining vapor, at known *P*, *T*, and *V*, is condensed and weighed.

EXAMPLE 8–16

The molecular weight of a volatile liquid was determined by the Dumas method. A 120-milliliter flask contained 0.345 gram of vapor at 100°C and one atmosphere pressure. What is the molecular weight of the compound?

Solution

We substitute into the ideal gas law, $PV = nRT$, to determine the number of moles of vapor that filled the flask

$V = 0.120\text{ L},\ P = 1.00\text{ atm},\ T = 100°\text{C} + 273 = 373\text{ K}$

$$n = \frac{PV}{RT} = \frac{(1.00\text{ atm})(0.120\text{ L})}{\left(0.0821\,\frac{\text{L}\cdot\text{atm}}{\text{mol}\cdot\text{K}}\right)(373\text{ K})} = 0.00392\text{ mol}$$

The mass of 0.00392 mole of vapor is 0.345 g, and so we can calculate the mass of one mole

$$\frac{?\text{ g}}{\text{mol}} = \frac{0.345\text{ g}}{0.00392\text{ mol}} = 88.0\text{ g/mol}$$

The molecular weight of the volatile liquid is 88.0 amu.

Let's carry the calculation one step further in Example 8–17.

EXAMPLE 8–17

Analysis of the volatile liquid in Example 8–16 showed that it contained 54.5% carbon, 9.1% hydrogen, and 36.4% oxygen by mass. What is the true formula for the volatile liquid?

Solution

Let's calculate the simplest formula for the compound:

$$\underline{?} \text{ mol C atoms} = 54.5 \text{ g C} \times \frac{1 \text{ mol C atoms}}{12.0 \text{ g C}} = 4.54 \text{ mol C atoms}$$

$$\underline{?} \text{ mol H atoms} = 9.1 \text{ g H} \times \frac{1 \text{ mol H atoms}}{1.0 \text{ g H}} = 9.1 \text{ mol H atoms}$$

$$\underline{?} \text{ mol O atoms} = 36.4 \text{ g O} \times \frac{1 \text{ mol O atoms}}{16.0 \text{ g O}} = 2.28 \text{ mol O atoms}$$

$$\frac{4.54}{2.28} = 2 \text{ C}$$

$$\frac{9.1}{2.28} = 4 \text{ H}$$

$$\frac{2.28}{2.28} = 1 \text{ O}$$

Simplest formula is C_2H_4O
Formula weight = 44 amu

Division of the molecular weight by the simplest formula weight gives

$$\frac{\text{molecular weight}}{\text{simplest formula weight}} = \frac{88 \text{ amu}}{44 \text{ amu}} = 2$$

Therefore, the true formula is $(C_2H_4O)_2 = \underline{C_4H_8O_2}$.

One compound that has this composition is ethyl acetate, a common solvent used in nail polishes and polish removers.

8–12 Dalton's Law of Partial Pressures

Many gases, including our atmosphere, are mixtures of gases that consist of different kinds of gas molecules. Since the number of molecules of each type in a mixture can be expressed as a number of moles, the total number of moles in the gas mixture is given by

$$n_{\text{total}} = n_A + n_B + n_C + \cdots + n_x$$

where n_A, n_B, etc., represent the number of moles of each of the x types of gas present.

Rearrangement of the ideal gas equation and solving for n gives

$$n = \frac{PV}{RT}$$

All the gas molecules in a mixture occupy the same container at the same temperature and pressure. Therefore, we can write an ideal gas law expression for the number of moles of each component in a mixture of gases, that is,

$$n_A = \left(\frac{V}{RT}\right) P_A, \qquad n_B = \left(\frac{V}{RT}\right) P_B, \qquad n_C = \left(\frac{V}{RT}\right) P_C$$

The pressure exerted by the molecules of gas A is proportional to n_A, the number of moles of A, and so on. The pressure term P_A is called the **partial pressure** of gas A.

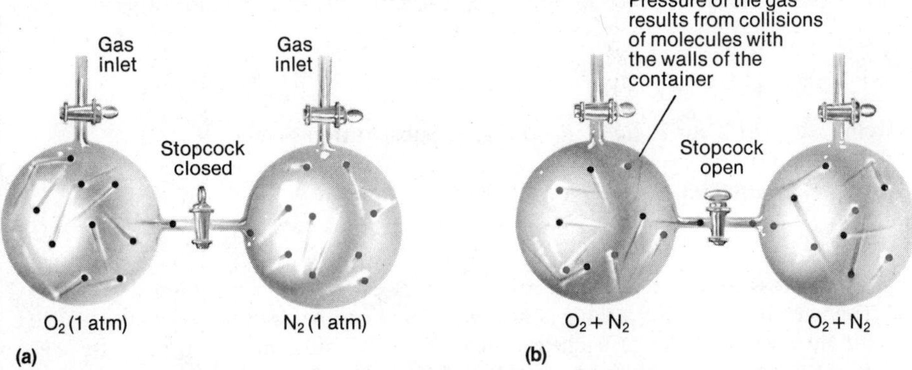

FIGURE 8–8 Representation of diffusion of gases. The space between the molecules allows for ease of admission of one gas into another. Collisions of molecules with the walls of the container are responsible for the pressure of the gas.

Substituting these equations into the equation for n_{total} gives an interesting result

$$n_{total} = \frac{V}{RT}(P_{total}) = \frac{V}{RT}(P_A) + \frac{V}{RT}(P_B) + \frac{V}{RT}(P_C) + \cdots$$

$$\frac{V}{RT}(P_{total}) = \frac{V}{RT}(P_A + P_B + P_C + \cdots)$$

Cancelling common terms gives the total pressure of a mixture of gases in terms of the partial pressures of the individual components.

$$P_{total} = P_A + P_B + P_C + \cdots$$

The total pressure exerted by a mixture of gases is the sum of the partial pressures of those gases. This is known as **Dalton's Law of Partial Pressures,** since John Dalton was the first to notice this effect in 1807 while studying the composition of moist and dry air.

Two gases at one atmosphere pressure in vessels of equal volume connected by a closed stopcock are depicted in Figure 8–8. If the stopcock is opened, both gases diffuse to occupy both vessels equally; each exhibits a partial pressure of one-half atmosphere and the pressure is one atmosphere after the gases are mixed.

Dalton's Law is useful in describing gaseous mixtures because it allows us to relate total measured pressures to the composition of mixtures. Consider the following example.

EXAMPLE 8–18

A 10.0-liter flask contains 0.200 mole of methane, 0.300 mole of hydrogen, and 0.400 mole of nitrogen at 25°C. (a) What is the pressure in atmospheres inside the flask? (b) What is the partial pressure of each component of the mixture of gases?

Solution

(a) We are given the numbers of moles of each component, and the ideal gas law is used to calculate the total pressure

$n = 0.200 \text{ mol } CH_4 + 0.300 \text{ mol } H_2 + 0.400 \text{ mol } N_2 = 0.900 \text{ mol of gas}$

$V = 10.0 \text{ L}$

$T = 25°C + 273° = 298 \text{ K}$

Solving $PV = nRT$ for P gives, $P = nRT/V$. Substitution gives

$$P = \frac{(0.900 \text{ mol})\left(0.0821 \dfrac{\text{L}\cdot\text{atm}}{\text{mol}\cdot\text{K}}\right)(298 \text{ K})}{10.0 \text{ L}} = \underline{2.20 \text{ atm}}$$

The total pressure exerted by the mixture of gases is 2.20 atmospheres.

(b) The partial pressure of each gas in the mixture can be calculated by substituting the number of moles of each gas into $PV = nRT$ individually, that is, for CH_4, $n = 0.200$ mole and the value for V and T are the same as above

$$P_{CH_4} = \frac{(n_{CH_4})RT}{V} = \frac{(0.200 \text{ mol})\left(0.0821 \dfrac{\text{L}\cdot\text{atm}}{\text{mol}\cdot\text{K}}\right)(298 \text{ K})}{10.0 \text{ L}}$$

$P_{CH_4} = \underline{0.489 \text{ atm}}$

Similar calculations for the partial pressures of hydrogen and nitrogen give

$P_{H_2} = \underline{0.734 \text{ atm}}$ and $P_{N_2} = \underline{0.979 \text{ atm}}$

Dalton's Law states, "The total pressure of a mixture of gases is the sum of the partial pressures of the component gases." Addition of the partial pressures of this mixture of gases should give the total pressure

$P_{CH_4} + P_{H_2} + P_{N_2} = P_{total}$,

and substitution gives

$0.489 \text{ atm} + 0.734 \text{ atm} + 0.979 \text{ atm} = \underline{2.20 \text{ atm}}$

Frequently gases, or mixtures of gases, are conveniently collected over water. Figure 8–9 illustrates the collection of a sample of hydrogen over water. A gas produced in a reaction will displace the more dense water from the inverted water-filled jars. Gases that are soluble in water or that react with water are not collected by this method. The pressure on the gas inside the collection vessel can be made equal to atmospheric pressure by raising or low-

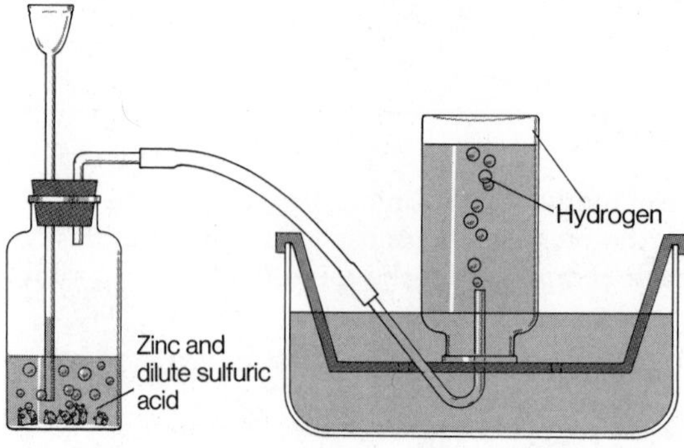

FIGURE 8–9 An apparatus for preparing hydrogen from zinc and dilute sulfuric acid. ($Zn\ (s) + 2H^+\ (aq) \longrightarrow Zn^{2+}\ (aq) + H_2\ (g)$) The hydrogen is collected over water.

TABLE 8–5 Vapor Pressure of Water Near Room Temperature

Temperature (°C)	Vapor Pressure of Water (torr)	Temperature (°C)	Vapor Pressure of Water (torr)
15	12.79	23	21.07
16	13.63	24	22.38
17	14.53	25	23.76
18	15.48	26	25.21
19	16.48	27	26.74
20	17.54	28	28.35
21	18.65	29	30.04
22	19.83	30	31.82

ering the vessel until the water level inside the vessel is the same as that outside. The atmospheric pressure is measured by an ordinary mercury barometer.

However, one complication arises. A gas in contact with water soon becomes saturated with water vapor, i.e., the pressure inside the vessel is the sum of the partial pressure of the gas itself *plus* the partial pressure exerted by the water molecules in the gas (that is, the vapor pressure of water). Every liquid shows characteristic vapor pressures that vary with temperature. Table 8–5 displays vapor pressures of water near room temperature. The relevant point here is that a gas collected over water is saturated with water vapor, and the pressure of the mixture is equal to the applied pressure. If a gas is collected at atmospheric pressure, we can write

$$P_{atm} = P_{gas} + P_{H_2O} \quad \text{or} \quad P_{gas} = P_{atm} - P_{H_2O}$$

Example 8–19 provides a detailed illustration.

EXAMPLE 8–19

A 300-milliliter sample of hydrogen was collected over water at 21°C on a day the atmospheric pressure was 748 torr. (a) What volume will the dry hydrogen occupy at STP? (b) What is the mass of the sample of dry hydrogen?

Solution

(a) We can correct the volume of hydrogen to STP as in Section 8–8, but we must also correct for the vapor pressure of water at 21°C (see Table 8–5).

$V_1 = 300$ mL $\qquad V_2 = \underline{?}$

$P_1 = P_{atm} - P_{H_2O} = 748$ torr $- 19$ torr $\qquad P_2 = 760$ torr

$P_1 = 729$ torr

$T_1 = 21°C + 273° = 294$ K $\qquad T_2 = 273$ K

$$\underline{?}\text{ L} = 300\text{ mL} \times \frac{729\text{ torr}}{760\text{ torr}} \times \frac{273\text{ K}}{294\text{ K}} = \underline{267\text{ mL of } H_2 \text{ at STP}}$$

Now that we know the volume of dry hydrogen at STP, we can calculate its mass. One mole of dry hydrogen occupies 22,400 mL at STP and has a mass of 2.02 g.

$$\text{(b) } \underline{?}\text{ g } H_2 = 267\text{ mL} \times \frac{2.02\text{ g } H_2}{22{,}400\text{ mL } H_2} = \underline{0.0241\text{ g } H_2}$$

A 300-mL sample of hydrogen collected over water at 21°C, when the atmospheric pressure is 748 torr, has a mass of 0.0241 gram. The dry hydrogen occupies 267 mL at STP.

8–13 The Kinetic-Molecular Theory

As early as 1738, Johann Bernoulli envisioned gaseous molecules in ceaseless motion striking the walls of their container and thereby exerting pressure. In 1857, Rudolf Clausius published a theory that attempted to explain various experimental observations that had been summarized by Boyle's, Dalton's Charles', and Avogadro's Laws. The basic assumptions of his **kinetic-molecular theory** are:

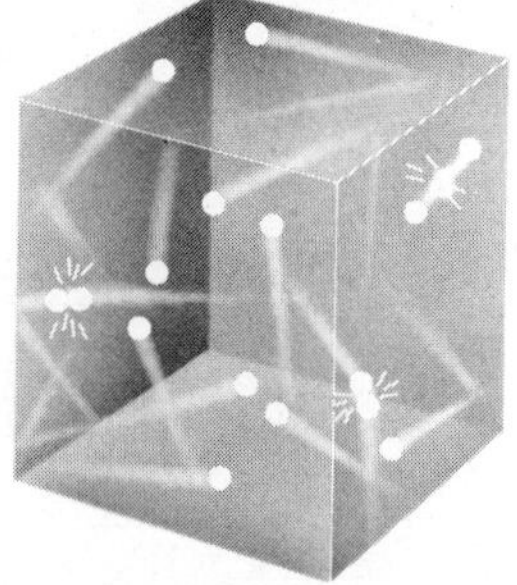

FIGURE 8–10 Visual demonstration of molecular motion.

1. Gases consist of discrete molecules. The individual molecules are relatively far apart, and they exert very little attraction for each other except near the liquefaction point (the point at which a gas liquefies). The volume occupied by the gas molecules themselves is insignificant compared to the volume occupied by the gas at ordinary temperatures and pressures (see Section 8–1).
2. Gaseous molecules are in continuous random, straight-line motion with varying velocities (see Figure 8–10). Collisions between gas molecules and with the walls of the container are elastic, i.e., there is no net gain or loss of energy in these collisions. At any given instant only a small number of molecules are involved in collisions relative to the number present.
3. The average kinetic energies of molecules of different gases are equal at a given temperature. Kinetic energies of gases increase with increasing temperature and decrease as temperature decreases. Kinetic energy is the energy a body possesses by virtue of its motion. It is $\frac{1}{2}mv^2$, where m, the body's mass, can be expressed in grams, and v, its velocity, can be expressed in meters per second, m/s. Thus, small, light-weight molecules like hydrogen and helium have much higher average velocities than heavier molecules like carbon dioxide and sulfur dioxide at the same temperature. All gases possess the same average kinetic energy at a given temperature. We may summarize: *The average kinetic energy of gaseous molecules is proportional to the absolute temperature.* Figure 8–11 shows the distribution of velocities of gaseous molecules at two temperatures.

The utility of the kinetic-molecular theory is that it satisfactorily explains most of the observed behavior of gases. Numerous experiments have indicated its basic validity. Let's look at the gas laws in light of the kinetic-molecular theory.

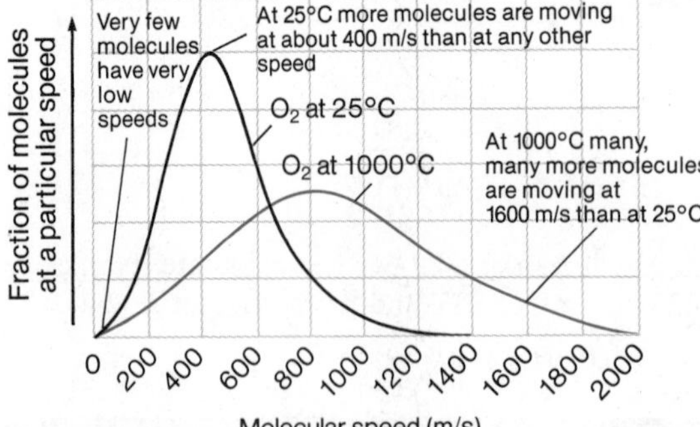

FIGURE 8–11 The Maxwellian distribution function for molecular speeds. This graph shows the relative numbers of O_2 molecules having any given speed at 25°C and at 1000°C. At 25°C most O_2 molecules have speeds between 200 and 600 m/s.

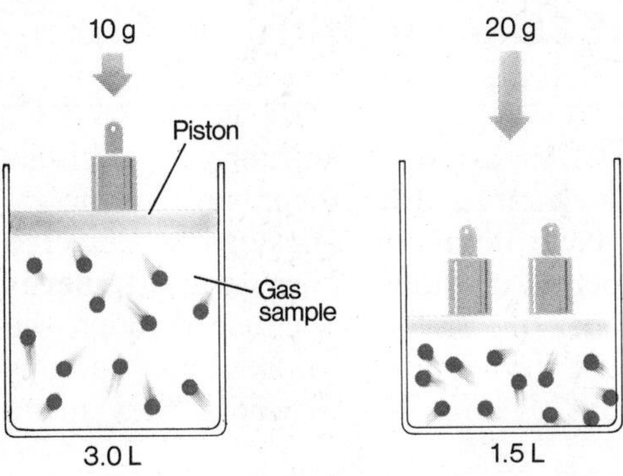

FIGURE 8–12 An illustration of the change in volume of a gas with changes in pressure (temperature constant). The entire apparatus is enclosed in a vacuum.

Boyle's and Dalton's Laws

The pressure exerted by a gas upon the walls of its container is caused by gas molecules striking the walls. Clearly, pressure depends upon two factors: (1) the number of molecules striking the walls per unit time, and (2) how vigorously the molecules strike the walls. Halving the volume of a given sample of gas doubles the pressure at constant temperature because twice as many molecules strike a given area on the container walls per unit time. Likewise, doubling the volume of a sample of gas halves the pressure because only half as many gas molecules strike a given area on the container walls per unit time (Figure 8–12).

Charles' Law

Recall that kinetic energy is directly proportional to the absolute temperature. Doubling the *absolute* temperature of a sample of gas doubles the average kinetic energy of the gaseous molecules, and the increased vigor of collision doubles the volume at constant pressure. Similarly, halving the absolute temperature (at constant pressure) decreases kinetic energy to one half its original value and the volume decreases by one half because of reduced vigor of collision with the container walls (Figure 8–13).

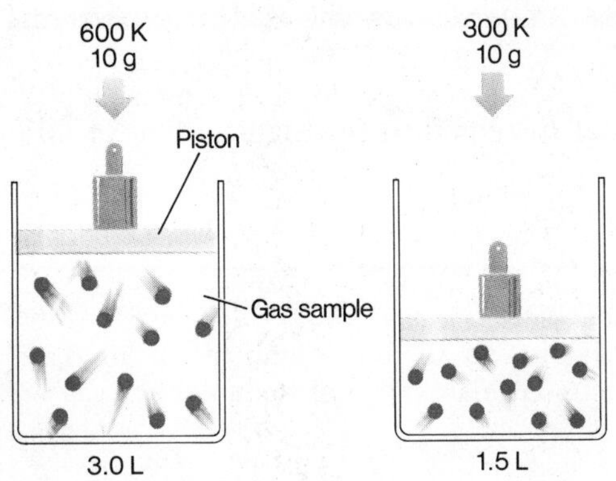

FIGURE 8–13 An illustration of the change in volume of a gas with changes in temperature (pressure constant). The entire apparatus is enclosed in a vacuum.

8–14 Graham's Law: Rates of Effusion (Diffusion) of Gases

Strictly speaking, chemists use the word *effusion* to describe the escape of a gas through a tiny hole, and the word *diffusion* to describe movement of a gas into a space or the mixing of one gas with another. The distinction that chemists make is somewhat sharper than Webster's distinction between the meanings of the two words.

Because gas molecules are in constant, rapid, random motion, they diffuse quickly throughout any container. For example, if hydrogen sulfide (the essence of rotten eggs) is released in a large room, the odor can be detected throughout the room in a very short period of time. If a mixture of gases is placed in a container with porous walls, the molecules effuse through the walls. Lighter gas molecules always effuse through the tiny openings of porous materials faster than heavier molecules (Figure 8–14) because they move faster.

The rates of effusion of different gases were studied by Thomas Graham in 1832, and he demonstrated that *the rates of effusion of gases are inversely proportional to the square roots of their molecular weights* (or densities). This statement is known as Graham's Law. Symbolically, **Graham's Law** may be represented

$$\frac{\text{Rate of effusion of gas A}}{\text{Rate of effusion of gas B}} = \sqrt{\frac{\text{Molecular weight of gas B}}{\text{Molecular weight of gas A}}}$$

Or, to use simpler notation

$$\frac{\text{Rate}_A}{\text{Rate}_B} = \sqrt{\frac{M_B}{M_A}}$$ where: Rate refers to rates of effusion and M refers to molecular weights

Example 8–20 demonstrates Graham's Law for two common gases, methane CH_4, and sulfur dioxide, SO_2.

EXAMPLE 8–20

Calculate the ratio of the rate of effusion of methane to that of sulfur dioxide, that is, $\text{Rate}_{CH_4}/\text{Rate}_{SO_2}$.

Solution

Substitution into the Graham's Law equation gives

$$\frac{\text{Rate}_{CH_4}}{\text{Rate}_{SO_2}} = \sqrt{\frac{M_{SO_2}}{M_{CH_4}}} = \sqrt{\frac{64\text{ amu}}{16\text{ amu}}} = \sqrt{\frac{4}{1}} = \frac{2}{1}$$

At a given temperature the average velocity of CH_4 molecules is twice that of SO_2 molecules.

Since $\text{Rate}_{CH_4}/\text{Rate}_{SO_2} = 2/1$, we can write $\text{Rate}_{CH_4} = 2\ \text{Rate}_{SO_2}$, which tells us that the rate of effusion of CH_4 is twice that of SO_2.

Example 8–21 provides additional insight into the significance of Graham's Law.

EXAMPLE 8–21

A 100-milliliter sample of hydrogen effuses through a porous container four times as rapidly as an unknown gas. Calculate the molecular weight of the unknown gas.

Solution

The Graham's Law equation

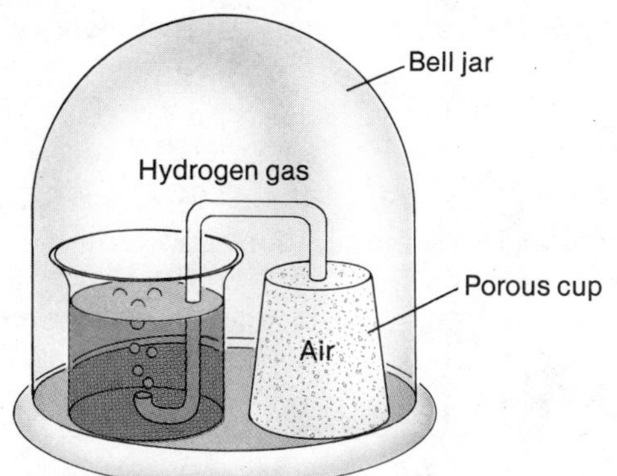

FIGURE 8–14 Effusion of gases. If a bell jar full of hydrogen is brought down over a porous cup full of air, hydrogen will diffuse into the cup faster than the oxygen and nitrogen in the air can effuse out of the cup. This causes an increase in pressure in the cup sufficient to produce bubbles in the water in the beaker.

$$\frac{\text{Rate}_{H_2}}{\text{Rate}_{\text{Unk}}} = \sqrt{\frac{M_{\text{Unk}}}{M_{H_2}}}$$

can be solved for M_{Unk}. Let's square both sides of the equation, solve the resulting expression for M_{Unk}, and substitute the known values into the expression

$$\left(\frac{\text{Rate}_{H_2}}{\text{Rate}_{\text{Unk}}}\right)^2 = \frac{M_{\text{Unk}}}{M_{H_2}} \quad \text{and} \quad M_{\text{Unk}} = \left(\frac{\text{Rate}_{H_2}}{\text{Rate}_{\text{Unk}}}\right)^2 M_{H_2}$$

$$M_{\text{Unk}} = \left(\frac{4}{1}\right)^2 (2.0 \text{ amu}) = \left(\frac{16}{1}\right) 2.0 \text{ amu} = \underline{32 \text{ amu}}$$

The unknown gas is probably oxygen.

Interestingly, Graham's Law can be derived from the part of the kinetic-molecular theory which says that *at a given temperature the average kinetic energies of molecules of different gases are equal.* If we represent the average kinetic energy of gas molecules A as $KE_A = \frac{1}{2}m_A v_A^2$, and the average kinetic energy of gas molecules B as $KE_B = \frac{1}{2}m_B v_B^2$, we can write

$$KE_A = KE_B \quad \text{or} \quad \frac{1}{2}m_A v_A^2 = \frac{1}{2}m_B v_B^2$$

Multiplying the latter expression by 2 gives $m_A v_A^2 = m_B v_B^2$, and, rearranging the expression, we obtain

$$\frac{v_A^2}{v_B^2} = \frac{m_B}{m_A}$$

Extracting the square root of both sides of this equation gives

$$\frac{v_A}{v_B} = \sqrt{\frac{m_B}{m_A}}$$

which says that the ratio of the velocity of molecule A to that of molecule B is the square root of the ratio of the mass of molecule B to the mass of molecule A. Rates of effusion are directly proportional to velocities and molecular weights are numerically equal to masses of molecules. Therefore, we may write

$$\frac{\text{Rate}_A}{\text{Rate}_B} = \sqrt{\frac{M_B}{M_A}}$$

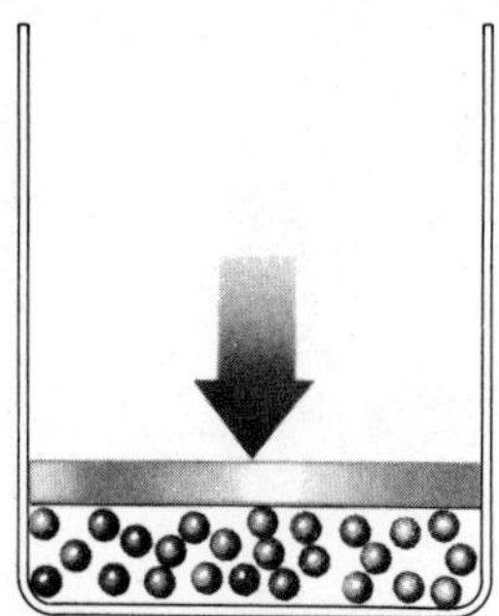

FIGURE 8–15 A sample of gas under high pressure. The molecules are quite close together.

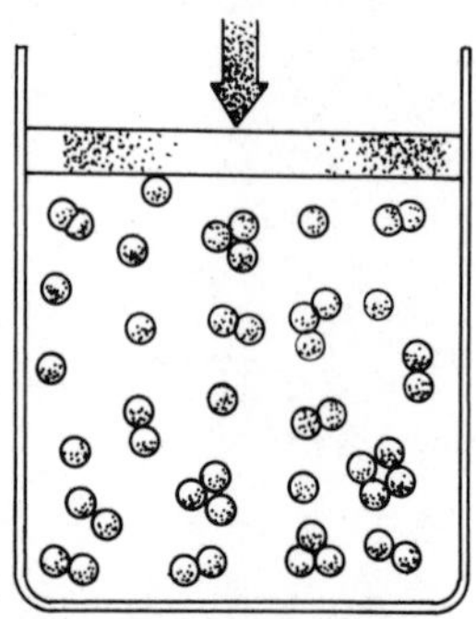

FIGURE 8–16 A sample of gas at a low temperature. Each sphere represents a molecule.

The van der Waals equation is known as an *equation of state,* i.e., an equation that describes a state of matter.

which is the usual form of Graham's Law. We can also show that the above equation is equivalent to

$$\frac{\text{Rate}_A}{\text{Rate}_B} = \sqrt{\frac{D_B}{D_A}}$$

where D refers to gas densities measured at the same pressure and temperature.

8–15 Real Gases—Deviations from Ideality

Under ordinary conditions most real gases behave like ideal gases, in that they obey the postulates of the kinetic-molecular theory and the ideal gas law. Recall that according to the kinetic-molecular model most of the volume of a gas sample is empty space and gases are very compressible. Also, the molecules of the *ideal* gases exhibit no forces of attraction for each other because they are so far apart and moving so rapidly.

But, at *low temperatures* and/or *high pressures,* near the liquefaction point, a real gas deviates significantly from ideality because the postulates do not describe the behavior of matter accurately. Under high pressures a gas is compressed to the point that the volume of the molecules themselves becomes a significant fraction of the total volume occupied by the gas. At low temperatures the molecules travel more slowly than at higher temperatures and forces of attraction can overcome the motion of some of the molecules. Both of these factors result in a greater tendency for molecules to "stick together," at least briefly, as they collide (Figures 8–15 and 8–16). In 1867, after studying deviations of real gases from ideal behavior, Johannes van der Waals empirically adjusted the ideal gas equation

$$(P)(V) = nRT$$

to take into account two complicating factors. The van der Waals equation is

$$\left(P + \frac{n^2a}{V^2}\right)(V - nb) = nRT$$

in which P, V, T, and n represent the same variables as in the ideal gas law, but a and b are experimentally measured constants that differ for different gases. The a term corrects for the fact that molecules do exert attractive forces upon each other. Large values of a indicate strong attractive forces between molecules. The b factor corrects for the volume occupied by the molecules themselves; larger molecules have larger values of b. Note that if both a and b are zero, the van der Waals equation reduces to the ideal gas equation. Table 8–6 gives van der Waals constants for some common gases.

The tendency of the molecules of a gas to "stick together" when they collide depends upon the forces of attraction between them. Note that a for helium is very small. This is the case for all noble gases and many other nonpolar molecules, since only very weak attractive forces, called London forces, exist among them. London forces result from short-lived electrical dipoles produced by the attraction of one atom's nucleus for an adjacent atom's electrons. These forces exist for all molecules but are especially important for nonpolar molecules which would never liquify if these forces did not exist. Polar mole-

TABLE 8–6 van der Waals Constants

Gas	a ($L^2 \cdot atm/mol^2$)	b (L/mol)
H_2	0.244	0.0266
He	0.034	0.0237
N_2	1.39	0.0391
NH_3	4.17	0.0371
CO_2	3.59	0.0427
CH_4	2.25	0.0428

cules, like ammonia, NH_3, have permanent charge separations (dipoles) and therefore exhibit greater forces of attraction for each other. Note the high value of a for ammonia. London forces and permanent dipole forces of attraction are discussed in more detail in Chapter 9.

The following example illustrates the deviation of methane, CH_4, from ideal gas behavior under a typical set of conditions.

EXAMPLE 8–22

Calculate the pressure exerted by 1.00 mole of methane, CH_4, in a 10.0-liter vessel at 25°C assuming (a) ideal behavior and (b) nonideal behavior.

Solution

(a) Ideal gases obey the ideal gas law and we'll assume that methane does

$$PV = nRT$$

$$P = \frac{nRT}{V} = \frac{(1.00\ \text{mol})\left(0.0821\ \frac{\text{L}\cdot\text{atm}}{\text{mol}\cdot\text{K}}\right)(298\ \text{K})}{10.0\ \text{L}} = \underline{2.45\ \text{atm}}$$

(b) As a real gas it obeys the van der Waals equation

$$\left(P + \frac{n^2a}{V^2}\right)(V - nb) = nRT$$

For CH_4, $a = 2.25\ L^2 \cdot atm/mol^2$ and $b = 0.0428$ L/mol (Table 8–6).

$$\left[P + \frac{(1.00\ \text{mol})^2(2.25\ \text{L}^2\cdot\text{atm/mol}^2)}{(10.0\ \text{L})^2}\right]\left[10.0\ \text{L} - (1.00\ \text{mol})\left(0.0428\ \frac{\text{L}}{\text{mol}}\right)\right]$$

$$= (1.00\ \text{mol})\left(0.0821\ \frac{\text{L}\cdot\text{atm}}{\text{mol}\cdot\text{K}}\right)(298\ \text{K})$$

Combining terms and canceling units we get

$$[P + 0.0225\ \text{atm}][9.957\ \text{L}] = 24.47\ \text{L}\cdot\text{atm}$$

$$P + 0.0225\ \text{atm} = 2.46\ \text{atm}$$

$$P = \underline{2.44\ \text{atm}}$$

This value is 0.01 atmosphere (0.4%) less than the pressure calculated from the ideal gas law. We see that almost no error is introduced by assuming that methane behaves as an ideal gas under these conditions.

8–16 Gay-Lussac's Law: The Law of Combining Volumes

Gases react in simple, definite proportions by volume. For example, *one* volume of hydrogen always combines (reacts) with *one* volume of chlorine to form *two* volumes of hydrogen chloride. It is understood that all volumes are measured at the same temperature and pressure

$$H_2\text{ (gas)} + Cl_2\text{ (gas)} \longrightarrow 2HCl\text{ (gas)}$$

$$1\text{ volume} + 1\text{ volume} \longrightarrow 2\text{ volumes}$$

Volumes may be expressed in any units as long as the same unit is used for all. Gay-Lussac (1788–1850) summarized several experimental observations on combining volumes of gases. The summary is known as **Gay-Lussac's Law** or the **Law of Combining Volumes:** *At constant temperature and pressure, the volumes of reacting gases can be expressed as a ratio of simple whole numbers.* The ratio is obtained from the coefficients in the balanced equation for the reaction. Clearly, the law applies only to *gaseous* substances at the same temperature and pressure. No generalizations can be made about the volumes of solids and liquids as they undergo chemical reactions. Consider the following examples, based on experimental observations at constant temperature and pressure. Hundreds more could be cited.

1. One volume of nitrogen reacts with three volumes of hydrogen to form two volumes of ammonia

   $$N_2\text{ (g)} + 3H_2\text{ (g)} \longrightarrow 2NH_3\text{ (g)}$$

   $$1\text{ volume} + 3\text{ volumes} \qquad 2\text{ volumes}$$

2. One volume of methane reacts with (burns in) two volumes of oxygen to give one volume of carbon dioxide and two volumes of steam

   $$CH_4\text{ (g)} + 2O_2\text{ (g)} \longrightarrow CO_2\text{ (g)} + 2H_2O\text{ (g)}$$

   $$1\text{ volume} + 2\text{ volumes} \qquad 1\text{ volume} + 2\text{ volumes}$$

3. Sulfur (a solid) reacts with one volume of oxygen to form one volume of sulfur dioxide

   $$S\text{ (s)} + O_2\text{ (g)} \longrightarrow SO_2\text{ (g)}$$

   $$1\text{ volume} \qquad 1\text{ volume}$$

4. Four volumes of ammonia burn in five volumes of oxygen to produce four volumes of nitric oxide and six volumes of steam

   $$4NH_3\text{ (g)} + 5O_2\text{ (g)} \longrightarrow 4NO\text{ (g)} + 6H_2O\text{ (g)}$$

   $$4\text{ volumes} \quad 5\text{ volumes} \qquad 4\text{ volumes} \quad 6\text{ volumes}$$

Avogadro's Law (Section 8–9) provides an explanation of Gay–Lussac's Law. Consider the reaction of fluorine and hydrogen to produce hydrogen fluoride. All volumes are measured at the same temperature and pressure. Experiments show that

$$1\text{ volume of fluorine} + 1\text{ volume of hydrogen} \longrightarrow 2\text{ volumes of hydrogen fluoride}$$

Figure 8–17 illustrates the volumetric relationships in the combination of fluorine and hydrogen to form hydrogen fluoride. Avogadro's Law tells us that equal volumes of fluorine, hydrogen, and hydrogen fluoride contain equal numbers of molecules.

The experimentally observed facts and deductions based on these facts are:

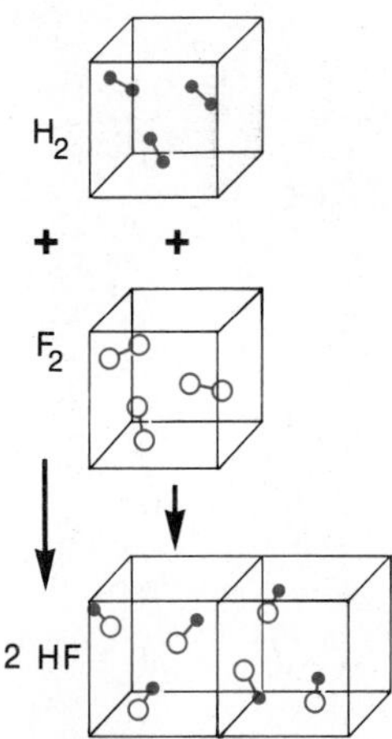

FIGURE 8–17 A representation of the combination of fluorine and hydrogen.

1. Two molecules of hydrogen fluoride are formed from one molecule of fluorine and one molecule of hydrogen.
2. Each hydrogen fluoride molecule must contain *at least* one fluorine atom and one hydrogen atom.
3. Therefore, two hydrogen fluoride molecules contain at least two fluorine atoms and two hydrogen atoms, which must have been present in one fluorine molecule and one hydrogen molecule

   1 volume of F_2 + 1 volume of H_2 $\longrightarrow$ 2 volumes of HF

   or

   1 molecule 1 molecule 2 molecules

4. There are no known reactions in which one molecule of fluorine or one molecule of hydrogen contains enough atoms to form more than two molecules of product. Therefore, we may safely assume that each fluorine molecule and each hydrogen molecule contains exactly two atoms, i.e., that the formulas for the molecules are F_2 and H_2.

Similar experiments and reasoning show that the other common diatomic elements (two atoms per molecule) are oxygen, O_2; nitrogen, N_2; chlorine, Cl_2; bromine, Br_2; and iodine, I_2.

Other reactions involving gases can be explained by Avogadro's and Gay-Lussac's Laws. It is interesting to note that Dalton, in the first decade of the nineteenth century, considered, but then quite properly rejected, the notion that equal volumes of gases contain the same numbers of *atoms*. The idea of the existence of diatomic (and more complex) molecules had not yet evolved, and it didn't occur to Dalton.

Application of Gay-Lussac's Law to gas phase reactions accurately predicts the volumes of gases involved.

EXAMPLE 8–23

What volume of hydrogen will combine with 40 liters of oxygen to form steam at 750°C and atmospheric pressure?

Solution

The balanced equation for the reaction

$2H_2$ (gas) + O_2 (gas) $\longrightarrow$ $2H_2O$ (gas)

tells us that two hydrogen molecules combine with one oxygen molecule to form two water molecules. Or, applying Gay-Lussac's Law, we can write

$2H_2$	+ O_2	$\longrightarrow$ $2H_2O$
2 volumes	1 volume	2 volumes

We can use any units we choose to describe the volumes of H_2 and O_2 as long as we use the same unit for both.

and we have

$$\underline{?}\ L\ H_2 = 40\ L\ O_2 \times \frac{2\ \text{volumes}\ H_2}{1\ \text{volume}\ O_2} = \underline{80\ L\ H_2}$$

8–17 Mass–Volume Relationships in Reactions Involving Gases

Small amounts of oxygen are produced in the laboratory by heating potassium chlorate, $KClO_3$, in the presence of a small amount of a catalyst, manganese dioxide, MnO_2. Potassium chloride, KCl, and oxygen are produced as the following equation shows

$$2KClO_3\ (s) \xrightarrow[\Delta]{MnO_2} 2KCl\ (s) + 3O_2\ (g)$$

As in Section 2–8, we can represent the stoichiometry of the reaction as

$$2KClO_3\ (s) \longrightarrow 2KCl\ (s) + 3O_2\ (g)$$

$2KClO_3$ (s)	2KCl (s)	$3O_2$ (g)
2 mol	2 mol	3 mol
2(122.6 g)	2(74.6 g)	3(32.0 g)

But, we now know that one mole of gas, measured at STP, occupies 22.4 liters, and this information can be utilized. We can write the above equation as

$$2KClO_3\ (s) \longrightarrow 2KCl\ (s) + 3O_2\ (g)$$

$2KClO_3$ (s)	2KCl (s)	$3O_2$ (g)
2 mol	2 mol	3 mol
245.2 g	149.2 g	96.0 g
		$3(22.4\ L_{STP}) = 67.2\ L\ STP$

Unit factors can be constructed using any two of these quantities just as in Section 2–8.

EXAMPLE 8–24

What volume of O_2 (STP) can be produced by heating 100.0 grams of $KClO_3$?

Solution

Reference to the above equation shows that two moles of $KClO_3$ produce three moles of O_2

$$\underline{?}\ L_{STP}\ O_2 = 100.0\ g\ KClO_3 \times \frac{1\ mol\ KClO_3}{122.6\ g\ KClO_3} \times \frac{3\ mol\ O_2}{2\ mol\ KClO_3} \times \frac{22.4\ L_{STP}\ O_2}{1\ mol\ O_2}$$

$$= \underline{27.4\ L_{STP}\ O_2}$$

Alternatively, we could have solved this problem using a single unit factor, $67.2\ L_{STP}\ O_2/245.2\ g\ KClO_3$, read directly from the equation

$$\underline{?}\ L_{STP}\ O_2 = 100\ g\ KClO_3 \times \frac{67.2\ L_{STP}\ O_2}{245.2\ g\ KClO_3} = \underline{27.4\ L_{STP}\ O_2}$$

This calculation shows that the thermal decomposition of 100 grams of $KClO_3$ produces 27.4 liters of oxygen measured at standard conditions.

EXAMPLE 8–25

A 1.80-gram mixture of potassium chlorate, $KClO_3$, and potassium chloride, KCl, is heated until all the $KClO_3$ has decomposed. The liberated oxygen occupies 400 milliliters collected over water at 25°C when the barometric pressure is 750 torr. What percentage of the mixture is $KClO_3$?

Solution

Note that heating KCl produces no oxygen, and therefore we are concerned with the same reaction as before, that is,

$$2KClO_3\ (s) \longrightarrow 2KCl\ (s) + 3O_2\ (g)$$

The O_2 is collected over water at 25°C and 750 torr. Therefore, we must subtract out the vapor pressure of water (Table 8–5) to obtain the pressure exerted by O_2,

$$P_{O_2} = 750 \text{ torr} - 24 \text{ torr} = 726 \text{ torr}$$

We can now use the ideal gas law to calculate the number of moles of O_2 produced

$$V = 0.400 \text{ L}$$

$$P = 726 \text{ torr} \times \frac{1 \text{ atm}}{760 \text{ torr}} = 0.955 \text{ atm} \qquad T = 25°\text{C} + 273 = 298 \text{ K}$$

$$n = \frac{PV}{RT} = \frac{(0.955 \text{ atm})(0.400 \text{ L})}{\left(0.0821 \frac{\text{L}\cdot\text{atm}}{\text{mol}\cdot\text{K}}\right)(298 \text{ K})} = 0.0156 \text{ mol } O_2$$

We now know that 0.0156 mole of O_2 was produced, so we can calculate the mass of $KClO_3$ that decomposed to produce it.

$$\underline{?}\text{ g } KClO_3 = 0.0156 \text{ mol } O_2 \times \frac{2 \text{ mol } KClO_3}{3 \text{ mol } O_2} \times \frac{122.6 \text{ g } KClO_3}{1 \text{ mol } KClO_3}$$

$$= \underline{1.28 \text{ g } KClO_3}$$

The sample contained 1.28 grams of $KClO_3$. The percent of $KClO_3$ in the sample is

$$\% \ KClO_3 = \frac{\text{g } KClO_3}{\text{g sample}} \times 100\% = \frac{1.28 \text{ g}}{1.80 \text{ g}} \times 100\% = \underline{71.1\% \ KClO_3}$$

The sample contains 71.1% $KClO_3$ and 28.9% KCl (obtained by difference).

Key Terms

Absolute zero zero degrees on the absolute temperature scale; −273.15°C or 0 K; theoretically, the temperature at which molecular motion ceases.

Atmosphere a unit of pressure; the pressure that will support a column of mercury 760 mm high at 0°C.

Avogadro's Law at the same temperature and pressure, equal volumes of all gases contain the same number of molecules.

Barometer a simple device for measuring pressure. See Figure 8–1.

Boyle's Law at constant temperature the volume occupied by a definite mass of a gas is inversely proportional to the applied pressure.

Charles' Law at constant pressure the volume occupied by a definite mass of a gas is directly proportional to the absolute temperature.

Condensed states the solid and liquid states.

Dalton's Law see *Law of Partial Pressures*

Dumas method a method used to determine the molecular weights of volatile liquids. See Figure 8–7.

Equation of state an equation that describes the behavior of matter in a given state; the van der Waals equation describes the behavior of the gaseous state.

Fluids substances that flow freely; gases and liquids.

Gay-Lussac's Law see *Law of Combining Volumes*

Graham's Law The rates of effusion of gases are inversely proportional to the square roots of their molecular weights (or densities).

Ideal gas a hypothetical gas that obeys exactly all postulates of the kinetic-molecular theory.

Ideal Gas Law the product of the pressure and volume of an ideal gas is directly proportional to the number of moles of the gas and the absolute temperature.

Kinetic-Molecular Theory a theory, originally published by Clausius in 1857, that attempts to explain macroscopic observations on gases in microscopic or molecular terms.

Law of Combining Volumes at constant temperature and pressure, the volumes of reacting gases (and any gaseous products) can be expressed as ratios of small whole numbers; also known as Gay-Lussac's Law.

Law of Partial Pressures the total pressure exerted by a mixture of gases is the sum of the partial pressures of the individual gases; also called Dalton's Law.

Manometer a two-armed barometer. See Figure 8–2.

Partial pressure the pressure exerted by one gas in a mixture of gases.

Pressure force per unit area.

Real gases gases that deviate from ideal gas behavior; all real gases.

Standard conditions (STP or SC) standard temperature, 0°C, and standard pressure, one atmosphere, are standard conditions for gases.

Standard molar volume the volume occupied by one mole of an ideal gas under standard conditions; 22.4 liters.

Universal gas constant R, the proportionality constant in the ideal gas equation, $PV = nRT$.

Vapor a gas formed by boiling or evaporating a liquid.

Vapor pressure the pressure exerted by a vapor at the surface of its mother liquid.

van der Waals' Equation an equation of state that extends the ideal gas law to real gases by inclusion of two empirically determined parameters, which are different for different gases.

Exercises

General Ideas

1. What are the three states of matter? Compare and contrast them.
2. What evidence is there for the statement that the individual molecules are quite far apart in gases?
3. What is the composition of the atmosphere?
4. On what experimental foundation does our current understanding of gases rest?

Pressure

5. (a) What is pressure?
   (b) Describe the mercurial barometer. How does it work?
6. What is a manometer? How does it work?
7. List some units of pressure. Express one atmosphere of pressure in several different units.
8. Does a tire gauge measure the absolute pressure inside a tire? Why?
9. What does the statement that an automobile tire has 28 pounds of air in it mean?

Boyle's Law: The Pressure–Volume Relationship

10. On what kinds of observations (measurements) is Boyle's Law based? State the law.
11. Could the words "a fixed number of moles" be substituted for "a definite mass of" in the statement of Boyle's Law? Why?

12. Why is a plot of pressure versus volume, at constant temperature, (Fig. 8–4) one branch of a hyperbola?
13. Use the statement of Boyle's Law to derive a simple mathematical expression for Boyle's Law.
14. A sample of nitrogen occupies 11.2 liters under a pressure of 580 torr at 32°C. What volume would it occupy at 32°C if the pressure were increased to 840 torr?
15. A sample of oxygen occupies 47.2 liters under a pressure of 730 torr at 25°C.
    (a) What volume would it occupy at 25°C if the pressure were increased to 1240 torr?
    (b) By what percent does the pressure increase?
    (c) By what percent does the volume decrease?
    (d) Is answer (b) the same as answer (c)? Why?
16. Under what pressure would the sample of oxygen described in Exercise 15 occupy 100 liters at 25°C? Express the answer in torr, atm, Pa, and kPa.
17. If the pressure on a 100-liter sample of hydrogen is doubled at constant temperature, by what fraction and by what percent do the volume change?
18. If the pressure on a 100-liter sample of hydrogen is halved at constant temperature, by what multiple and by what percent do the volume change?

The Kelvin Temperature Scale

19. Describe the experiments that led to the evolution of the absolute temperature scale. What is the relationship between the Celsius and Kelvin temperature scales?
    (a) What does "absolute temperature scale" mean?
    (b) What does "absolute zero" mean?
20. (a) Can an absolute temperature scale based on "Fahrenheit" rather than "Celsius" degrees be evolved? Why?
    (b) Can an absolute temperature scale that is based on a "degree" twice as large as a Celsius degree be developed? Why?

Charles' Law

21. On what kind of observations (measurements) is Charles' Law based? State the law.
22. Why is a plot of volume versus temperature at constant pressure (Fig. 8–6b) a straight line?
23. Use the statement of Charles' Law to derive a simple mathematical expression for Charles' Law.
24. A 400-mL sample of hydrogen is confined under a pressure of 0.500 atm at 50.0°C. What volume would it occupy at 100°C under the same pressure?
    (a) By what fraction and by what percent does the temperature increase on the Celsius scale? On the Kelvin scale?
    (b) By what fraction and by what percent does the volume increase? Why?
25. Refer to Exercise 24. If the pressure were expressed as 380 torr or 50.65 kilopascals would the same answer be obtained? Why?
26. A sample of methane, CH_4, occupies 800 mL at 150°C. At what temperature will it occupy 400 mL if the pressure does not change?
27. Which of the following statements are true? Which are false? Why is each true or false? *Assume constant pressure* in each case.
    (a) If a sample of gas is heated from 100°C to 200°C the volume will double.
    (b) If a sample of gas is heated from 0°C to 273°C the volume will double.
    (c) If a sample of gas is cooled from 400°C to 200°C the volume will decrease by a factor of two.
    (d) If a sample of gas is cooled from 1000°C to 200°C the volume will decrease by a factor of five.
    (e) If a sample of gas is heated from 200°C to 2000°C the volume will increase by a factor of ten.

Standard Conditions

28. (a) What are standard conditions?
    (b) Express standard temperature on three scales.
    (c) Express standard pressure in torr, mm Hg, atm, Pa, and kPa.
29. Why do we have a standard temperature and a standard pressure?

The Combined Gas Laws

30. (a) To what does "the combined gas laws" refer?

(b) What is the mathematical expression for the combined gas laws?
(c) Outline the logic used to obtain the mathematical expression for the combined gas laws.

31. A sample of nitrogen occupies a volume of 300 mL under a pressure of 380 torr at 177°C. What volume would the gas occupy at STP?
32. A sample of gas occupies 250 mL at 100°C under a pressure of one atmosphere. If the pressure were increased to 1900 torr, to what temperature would the sample have to be heated to occupy a volume of 500 mL?
33. A sample of gas occupies 300 mL at STP. Under what pressure would this sample occupy 150 mL if the temperature were increased to 546°C?
34. A sample of hydrogen occupies 250 mL at STP. If the temperature were increased to 546°C, how much would the pressure have to increase in order to keep the volume constant at 250 mL?
35. A sample of nitrogen occupies 400 mL at 400°C under a pressure of 4.00 atm. If the pressure were increased to 12.00 atm, how much would the temperature have to increase to keep the volume constant at 400 mL?

Gas Densities and the Standard Molar Volume

36. What is Avogadro's Law? What does it mean?
37. What does standard molar volume mean?
38. A mole of oxygen occupies 22.4 L at STP. What is its density in g/L and g/mL at STP?
39. Refer to Exercise 38. What is the volume of 3.85 moles of oxygen at STP?
40. (a) How many moles of oxygen are contained in 83.7 liters at STP?
    (b) What is the mass of this sample of oxygen?
41. At a given temperature and pressure one mole of a gas occupies 38.3 liters. Its density is 1.37 g/L under these conditions. What is the mass of one mole of the gas?
42. If 4.00 g of a gas occupies a volume of 4.48 liters at STP, what is the molecular weight of the gas?
43. Calculate the mass of a 6.72-liter sample of carbon monoxide, CO, measured at standard conditions.
44. If 4.00 grams of a gas occupies 1.12 liters at STP, what is the mass of 4 moles of the gas?

The Ideal Gas Equation

45. What is an "ideal gas?"
46. (a) What is the ideal gas equation?
    (b) Outline the logic used to obtain the ideal gas equation.
    (c) What is R? How is it obtained?
47. Calculate R in $L \cdot atm/mol \cdot K$, $kPa \cdot dm^3/mol \cdot K$, $J/mol \cdot K$, and $kJ/mol \cdot K$.
48. Calculate the volume occupied by 64.0 g of CH_4 at 127°C under a pressure of 1520 torr.
49. Calculate the pressure exerted by 60.0 g of C_2H_6 in a 30.0 liter vessel at 27.0°C.
50. How many moles of nitrogen are contained in 328 mL of the gas under a pressure of 3040 torr at 527°C? How many nitrogen atoms does the sample contain?
51. What is the mass of 8.21 liters of CH_4 at 227°C under a pressure of 1520 torr?
52. What volume would 90.0 grams of C_2H_6 occupy at 27°C under a pressure of 380 torr?
53. What pressure would a mixture of 3.2 g of O_2, 6.4 g of CH_4, and 6.4 g of SO_2 exert if the gases were placed in a 40.0 L container at 127°C?
54. A student was given a container of ethane, C_2H_6, that had been sealed at STP. By making appropriate measurements, the student found that the mass of the sample of ethane was 0.218 gram and the volume of the container was 165 mL. Use the student's data to calculate the molecular weight of ethane. What percent error is obtained? Explain some possible sources of the error.
55. A student was given a sample of an unknown gas that occupied 41.5 mL at 25°C and 754 torr. The mass of the sample of gas was 0.0761 g. The student was asked to calculate the molecular weight of the gas. What value did she obtain? The student was then told that the sample of gas was propane, C_3H_8, but that the sample was contaminated with either ethane, C_2H_6, or *n*-butane, C_4H_{10}. What was the contaminant? Justify your answer.

Molecular Weights and True Formulas of Gaseous Compounds

56. Distinguish between simplest (empirical) formulas and true (molecular) formulas.
57. A compound containing only carbon and hydrogen is 80.0% C and 20.0% H by mass. At STP, 1.12 liters of the gas has a mass of 1.50 g. What is the molecular (true) formula for the compound?
58. A 0.580-g sample of a compound containing only carbon and hydrogen contains 0.480 g of carbon and 0.100 g of hydrogen. At STP 33.6 mL of the gas has a mass of 0.087 g. What is the molecular (true) formula for the compound?
59. Analysis of a volatile liquid shows that it contains 62.04% carbon, 10.41% hydrogen, and 27.54% oxygen by mass. At 150°C and 1.00 atm, 500 mL of the vapor has a mass of 1.673 g.
    (a) What is the molecular weight of the compound?
    (b) What is its true formula?

Dalton's Law

60. (a) State Dalton's Law. Express it symbolically.
    (b) What are partial pressures of gases?
61. One-liter samples of N_2, H_2, He, and O_2 are collected in four different vessels at the same temperature under a pressure of 1 atmosphere. If the four gases are then forced into one of the containers at the original temperature, what will the resulting pressure be?
62. If 2.00 liters of nitrogen under a pressure of 2.00 atmospheres at 127°C and 3.00 liters of nitrogen under a pressure of 1.00 atmosphere at 127°C are forced into a 4.00 liter container at 127°C, what will the final pressure be?
63. Suppose that one-half mole of CO_2 at STP and one-half mole of CO at STP are forced into a container which has a volume of 11.2 liters at a pressure of one atmosphere. What will the temperature of the mixture be in the new container?
64. A 5.42-liter sample of a gas was collected over water on a day when the temperature was 24°C and the barometric pressure was 706 torr. The dry sample of gas had a mass of 5.60 g. What is the mass of three moles of the dry gas? At 24°C the vapor pressure of water is 22 torr.
65. A 296-mL sample of oxygen is collected over water at 23°C on a day when the barometric pressure is 753 torr. What volume would the dry oxygen occupy at 487°C under a pressure of one atmosphere?
66. Calculate the mass of dry hydrogen in 750 mL of moist hydrogen collected over water at 25.0°C and 755 mm Hg. (The density of hydrogen is 0.08987 g/L at STP.) The vapor pressure of water is 24 torr at 25°C.

The Kinetic-Molecular Theory

67. Outline the kinetic-molecular theory.
68. How do average velocities of gaseous molecules vary with temperature?
69. How does the kinetic-molecular theory explain
    (a) Boyle's Law?
    (b) Dalton's Law?
    (c) Charles' Law?

Graham's Law: Effusion of Gases

70. State Graham's Law. What does it mean?
71. Show how the kinetic-molecular theory leads to the mathematical expression for Graham's Law.
72. (a) Explain how the hydrogen bubbler works. (Figure 8–14)
    (b) Could sulfur hexafluoride, SF_6, be used in this bubbler? Why?
73. At a given temperature and pressure, the average velocity of O_2 molecules would be ______ times as great as the average velocity of HI molecules.
74. At 500°C the average velocity of HI molecules should be ______ times the average velocity of CH_3OH molecules.
75. Calculate the ratio of the rate of effusion of CH_4 to that of SO_2.
76. If the average velocity of helium atoms is 0.707 mile per second at room temperature, what will be the average velocity of oxygen molecules at the same temperature?

Real Gases and Deviations from Ideality

77. How do "real" and "ideal" gases differ?

78. Under what kind of conditions are deviations from ideality most important? Why?
79. What is the van der Waals equation? How does it differ from the ideal gas equation?
80. Suppose you have samples of each of the gases listed in Table 8–6 under a pressure of 100 atmospheres, each at a temperature 10°C above its liquefaction point (at 100 atm). For which gas would you expect deviation from ideal behavior to be (a) greatest, and (b) least? Justify your answers.
81. (a) How do the van der Waals constants (Table 8–6) for He and H_2 differ? Why?
    (b) Compare the van der Waals constants for H_2 and NH_3. Explain why those for NH_3 would be expected to be larger.
82. Use both the ideal gas law and the van der Waals equation to calculate the pressure exerted by a 150-gram sample of ammonia in a 60.0-liter container at 1000°C. By what percent do the two results differ?
83. (a) Use both the ideal gas law and the van der Waals equation to calculate the pressure exerted by a 150-gram sample of ammonia in a 600-liter container at 1000°C. By how much do these results differ?
    (b) Explain the difference between the results obtained in Exercises 82 and 83a.

Gay-Lussac's Law and Mass-Volume Relationships in Reactions

84. (a) What is Gay-Lussac's Law?
    (b) What does it mean?
    (c) What is the Law of Combining Volumes?
85. What volume of chlorine, Cl_2, is necessary to prepare 38 liters of hydrogen chloride, HCl, if an excess of hydrogen, H_2, is used? All gases are measured at 100°C at 1 atm.
86. (a) What volume of ammonia, NH_3, can be prepared from 50 liters of nitrogen, N_2, and excess hydrogen, H_2, if all gases are measured at the same temperature and pressure?
    (b) What volume of hydrogen would be required in (a)?
87. What volume of hydrogen (STP) would be required to react with 0.100 mole of nitrogen?

$$N_2 + 3H_2 \longrightarrow 2NH_3$$

88. What volume of oxygen (STP) would be produced by the thermal decomposition of 0.600 mole of potassium nitrate?

$$2KNO_3\ (s) \longrightarrow 2KNO_2\ (s) + O_2\ (g)$$

89. What mass of $KClO_3$ would have to be decomposed to produce 13.4 liters of oxygen gas measured at STP?

$$2KClO_3\ (s) \longrightarrow 2KCl\ (s) + 3O_2\ (g)$$

90. An impure sample of $KClO_3$ that had a mass of 30.0 g was heated until all the $KClO_3$ had decomposed. The liberated oxygen occupied 6.72 liters at STP. What percent of the sample was $KClO_3$? Refer to Exercise 89 for the balanced equation.
91. What volume of dry NO (gas) at STP could be prepared by reacting 6.35 grams of Cu with an excess of HNO_3?

$$3Cu\ (s) + 8HNO_3\ (aq) \longrightarrow 3Cu(NO_3)_2\ (aq) + 2NO\ (g) + 4H_2O\ (\ell)$$

92. If sufficient acid is used to react completely with 4.862 g of magnesium, what volume of hydrogen (STP) would be produced? What volume would the gas occupy if it were collected over water on a day when the temperature was 75°F and the barometric pressure was 740 torr?

$$Mg\ (s) + 2HCl\ (aq) \longrightarrow MgCl_2\ (aq) + H_2\ (g)$$

93. If sufficient acid is used to react completely with 13.5 g of aluminum, what volume of hydrogen (STP) would be produced? What volume would the hydrogen occupy if it were collected over water on a day when the temperature was 80°F and the barometric pressure was 750 torr?

$$2Al\ (s) + 6HCl\ (aq) \longrightarrow 2AlCl_3\ (aq) + 3H_2\ (g)$$

94. What volume of hydrogen (STP) would be required to produce 0.400 mole of HCl?

$$H_2\ (g) + Cl_2\ (g) \longrightarrow 2HCl\ (g)$$

95. If 0.500 mole of carbon disulfide reacts with oxygen completely according to the following equation, what volume (*total*) would the products occupy if they were measured at STP?

$$CS_2\ (\ell) + 3O_2\ (g) \longrightarrow CO_2\ (g) + 2SO_2\ (g)$$

96. Based on the following equation, what volume of hydrogen (STP) would be required to react with 10.4 g of C_2H_2? If a chemist obtained 18.0 g of C_2H_6 by this reaction, what total volume (STP) of hydrogen and C_2H_2 did he use?

$$2H_2\ (g) + C_2H_2\ (g) \xrightarrow{\text{cat.}} C_2H_6\ (g)$$

9 Liquids and Solids

The molecules of most gases are so widely separated at ordinary temperatures and pressures that they do not interact with each other significantly. Consequently, the physical properties of gases are reasonably well described by the relatively simple relationships described in Chapter 8. In liquids and solids, the so-called **condensed phases,** the particles are closely spaced and interact strongly. Although the properties of liquids and solids can be described, they cannot be adequately explained by simple mathematical relationships. Figure 9–1 and Table 9–1 summarize some of the distinguishing characteristics of gases, liquids, and solids.

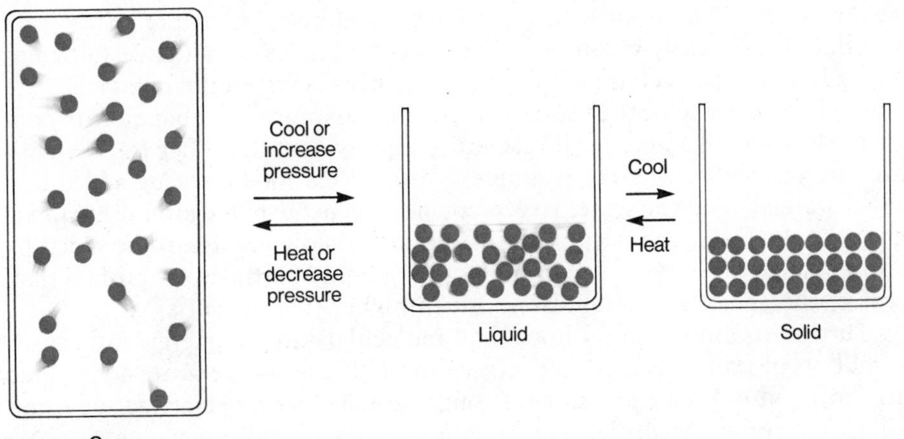

FIGURE 9–1 A representation of the three states of matter at the molecular level.

9–1 Kinetic-Molecular Description of Liquids and Solids

The properties listed in Table 9–1 can be qualitatively explained in terms of the kinetic-molecular theory of Chapter 8. We saw in Section 8–13 that the average kinetic energy of a collection of gas molecules decreases as the temperature is lowered. As the temperature falls, the rapid, random motion of gaseous molecules decreases. If a sample of gas is cooled and compressed sufficiently, the molecules approach each other closely enough for intermolecular attractions to overcome the kinetic energies. At this point condensation, or liquefaction, occurs. Since different kinds of molecules have different attractive forces, the temperatures and pressures required for condensation vary from gas to gas.

*Inter*molecular attractions refer to attractions between different molecules. *Intra*molecular refer to attractions between atoms within a single molecule.

In the liquid state, the forces of attraction among particles are great enough that disordered clustering occurs. The particles are so close together

TABLE 9–1 Some Common Characteristics of Gases, Liquids, and Solids

Gases	Liquids	Solids
1. No definite shape (fill containers completely)	1. No definite shape (assume shapes of containers)	1. Definite shape (resist deformation)
2. Compressible	2. Only very slightly compressible.	2. Nearly incompressible
3. Low density	3. Intermediate density	3. High density
4. Fluid	4. Fluid	4. Not fluid
5. Diffuse rapidly	5. Diffuse through other liquids	5. Diffuse only very slowly through solids
6. Extremely disordered particles; much empty space; rapid, random motion in three dimensions	6. Disordered clusters of particles; quite close together; random motion in three dimensions	6. Ordered arrangement of particles; vibrational motion only; particles very close together

that very little of the total volume occupied by a liquid is empty space. As a result liquids are nearly incompressible. Liquid particles do not have sufficient energy of motion to overcome the attractive forces between them, but they are able to slide past each other so that the liquid assumes the shape of its container. They are also able to diffuse quite rapidly through other liquids with which they are *miscible.* For example, a drop of red food coloring added to a glass of water causes the water to become homogeneously red after diffusion is complete. However, the natural diffusion rate is slow enough to be aided by swirling. Since the average separations among liquid particles are far less than in gases, the densities of liquids are much higher than densities of gases.

The miscibility of two liquids refers to their ability to mix and produce a homogeneous solution.

Further cooling a liquid lowers its molecular kinetic energy and causes the motion of its molecules to decrease more. If the temperature is lowered sufficiently, at ordinary pressures, stronger but shorter-range attractive interactions overcome the decreasing kinetic energies of the molecules to cause solidification. The temperature required for crystallization at a given pressure is dependent upon the nature of short-range interactions among the particles and is characteristic of each substance.

Most solids are characterized by ordered arrangements of particles with very restricted motion. Particles in the solid state are unable to move freely past one another, and only vibrate about fixed positions. Consequently solids have definite shapes. Because the particles are so close together, solids are nearly incompressible, and are very dense relative to gases. Solid particles do not diffuse readily into other solids. However, analysis of two blocks of different solids, such as copper and lead, after being pressed together for a period of years, shows that each block contains some atoms of the other element. This demonstrates that solids do diffuse, but very slowly (Figure 9–2).

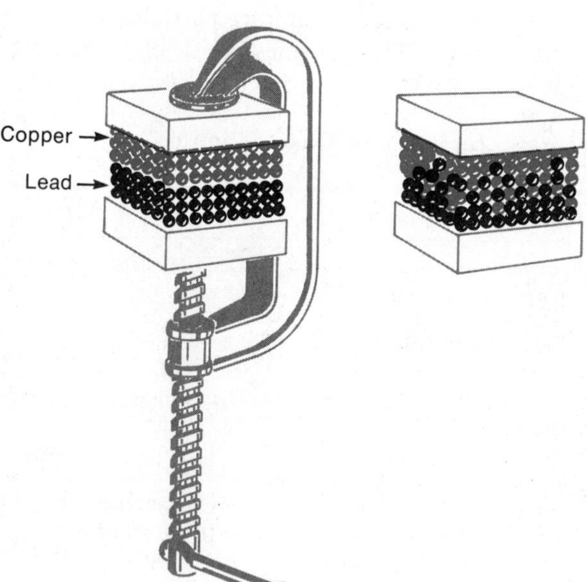

FIGURE 9–2 A representation of diffusion in solids. When blocks of two different metals are clamped together for a long time, a few atoms of each metal diffuse into the other metal.

The Liquid State

In the next several sections, we shall briefly describe several properties of the liquid state. These properties vary markedly among various liquids. Such variations depend on the nature and strength of the attractive forces between the particles (atoms, molecules, ions) making up the liquid. In Section 9–9, we shall discuss the nature of the intermolecular forces in more detail.

9–2 Viscosity

Viscosity is the resistance to flow of a liquid and is related to the intermolecular forces of attraction (Section 9–9) and to the sizes and shapes of the constituent particles. Molasses has a high viscosity at room temperature and freely flowing diethyl ether has a low viscosity. Viscosity can be measured by using a viscometer such as the one depicted in Figure 9–3, in which the time it takes for a known volume of a liquid to flow through a small neck of known size is measured. Intermolecular attractions, and therefore viscosity, decrease with increasing temperature, so long as no change in composition occurs.

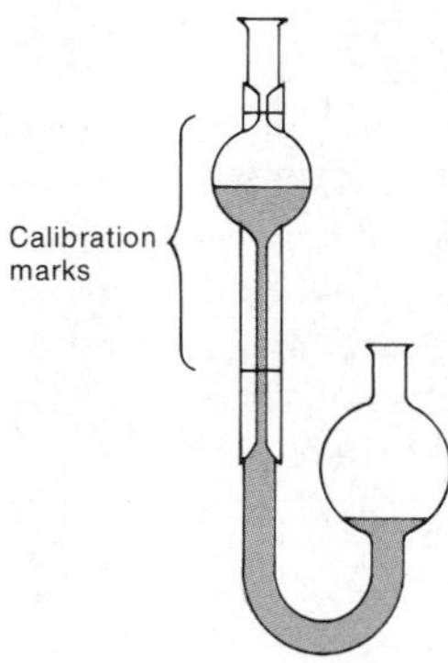

FIGURE 9–3 The Ostwald viscometer, a device used to measure viscosity.

9–3 Surface Tension

Molecules below the surface of a liquid are influenced by intermolecular attractions from all directions, but those on the surface are attracted only toward the interior, as shown in Figure 9–4. The attractions pull the surface layer toward the center. From an energetic point of view, the most stable situation is one in which the surface area is minimal. This is the reason that drops of liquid tend to assume spherical shapes. **Surface tension** is a measure of the inward forces that must be overcome in order to expand the surface area of a liquid.

Fire polishing removes the rough edges of laboratory glass tubing because of the tendency of the heated (liquid) glass to form a rounded surface. Surface tension is also responsible for the ability of water to support the mass of some insects that are more dense than water itself. Perhaps you've heard of floating a needle in a glass of water.

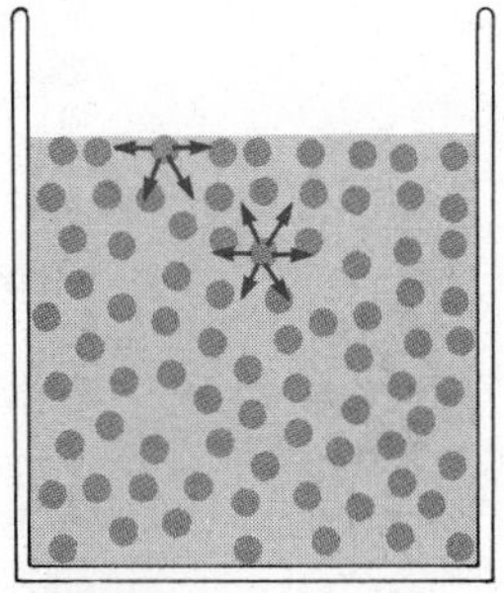

FIGURE 9–4 Molecular level view of the attractive forces experienced by molecules at and below the surface of a liquid.

9–4 Capillary Action

All the forces holding a liquid together are called **cohesive forces.** The forces of attraction between a liquid and another surface are **adhesive forces.** Water wets glass and increases its surface area as it creeps up the side of a glass tube due to the strong adhesive forces between water and glass, which have similar structures. The surface of the water, its **meniscus,** takes on a concave shape (Figure 9–5). On the other hand, mercury does not wet glass because its cohesive forces are much stronger than its attraction to glass. Thus, its meniscus is convex. **Capillary action** occurs when one end of a capillary tube, a glass tube with small bore (diameter), is immersed in a liquid. If adhesive forces exceed cohesive forces, the liquid continually creeps up the sides of the tube until a balance is reached between adhesive forces and the weight of liquid. The smaller the bore, the higher the liquid climbs. Plant roots take up water and

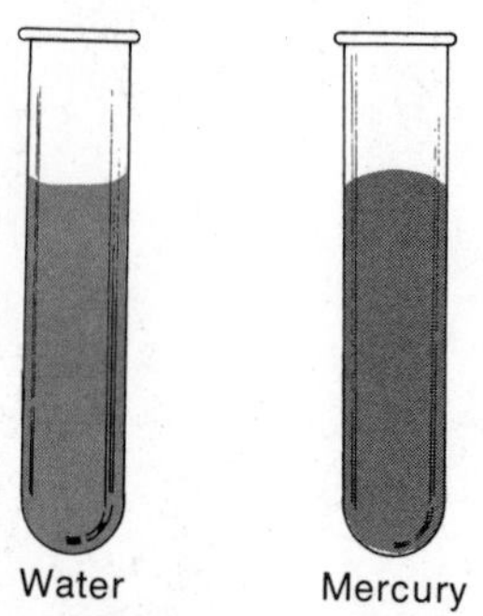

FIGURE 9–5 The meniscus, as observed in glass tubes with both water and mercury.

dissolved nutrients from the soil and transmit them up the stems by capillary action. The roots, like glass, exhibit strong adhesive forces for water.

9–5 Evaporation

Kinetic energies of liquids show the same type of temperature dependence as those of gases. The distribution of kinetic energies among liquid molecules at two different temperatures is shown in Figure 9–6. **Evaporation** or **vaporization** is the process by which molecules on the surface of the liquid break away and go into the gas phase (Figure 9–7). In order to break away, the molecules must possess a minimum kinetic energy, one half the product of mass times the square of the escape velocity. Figure 9–6 shows that the higher the temperature, the greater the fraction of molecules possessing at least that minimum energy. The rate of evaporation increases as temperature increases.

Further, it is only the higher energy molecules that can escape from the liquid phase. The average molecular kinetic energy of the molecules remaining in the liquid state is thereby lowered, so that a lower temperature in the liquid results. Since the liquid would then be cooler than its surroundings, it absorbs heat from its surroundings. You are familiar with a common example of the cooling effect of evaporation on the surroundings of a liquid, the cooling of the body by evaporation of perspiration.

A molecule in the vapor may collide with a molecule of the air, bounce back to the liquid surface, and be captured there. This process is the reverse of evaporation and is called **condensation.** As evaporation proceeds in a closed container, the volume of liquid decreases and the number of gas molecules above the surface increases. Since more collisions will then occur in the gas phase, the rate of condensation increases. The system composed of the liquid and gas molecules (of the same substance) eventually achieves a **dynamic equi-**

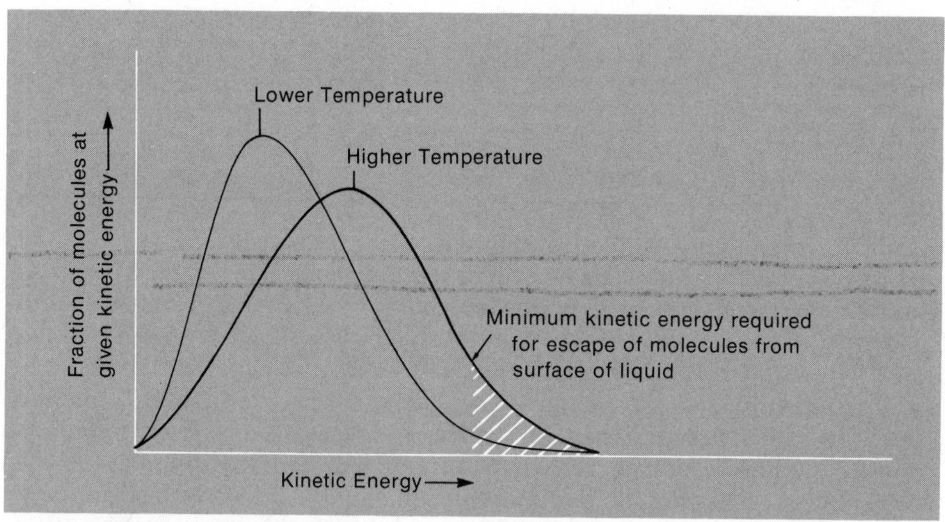

FIGURE 9–6 Distribution of kinetic energies of molecules in a liquid at different temperatures. Note that at the lower temperature, a smaller fraction of the molecules has the energy required to escape from the liquid, so evaporation is slower at the lower temperature.

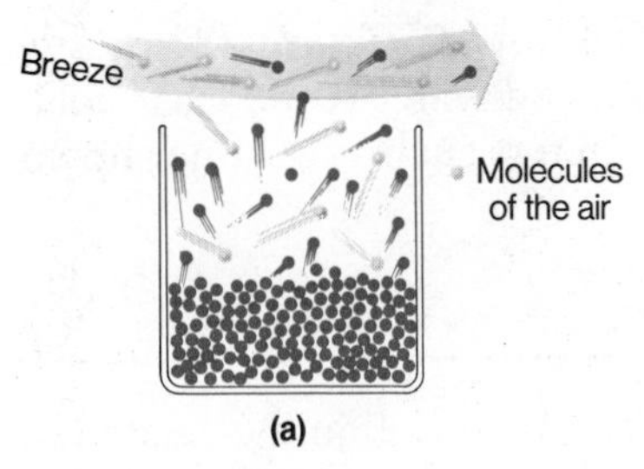

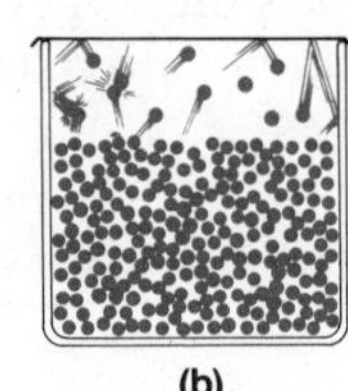

FIGURE 9–7 (*a*) Liquid continually evaporates from an open vessel; (*b*) equilibrium between liquid and vapor is established in a closed container in which molecules return to the liquid at the same rate as they leave it.

librium in which the rate of evaporation equals the rate of condensation in the closed container

$$\text{Liquid} \underset{\text{condensation}}{\overset{\text{evaporation}}{\rightleftharpoons}} \text{Vapor}$$

The two opposing rates counterbalance each other and are not equal to zero, hence the use of the term "dynamic" rather than "static" equilibrium.

However, if the vessel were left open to the air this equilibrium could not be established, since molecules would diffuse away and the slightest air currents would also sweep some gas molecules away from the liquid surface, allowing more evaporation to occur in order to replace the lost vapor molecules. Consequently, a liquid eventually evaporates entirely if it is left uncovered. This situation illustrates **LeChatelier's Principle** which states that a system at equilibrium, or striving to attain equilibrium, responds in the way that best accommodates or "undoes" any stress placed upon it. In this example, the stress is the removal of molecules in the vapor phase, and the response is the continued evaporation of the liquid.

This is one of the guiding principles that allows us to understand chemical equilibrium. It is discussed further in Chapter 15.

9–6 Vapor Pressure

Vapor molecules cannot escape when vaporization of a liquid occurs in a closed container. As more molecules leave the liquid, more collisions between gaseous molecules and the walls of the container occur and more condensation occurs. This is responsible for the formation of liquid droplets that adhere to the sides of the vessel above a liquid surface and for the eventual establishment of equilibrium between liquid and vapor.

The partial pressure of vapor molecules above the surface of a liquid at equilibrium is the **vapor pressure** of the liquid. Since, the rate of evaporation increases and the rate of condensation decreases with increasing temperature, vapor pressures of liquids always increase as temperature increases. At any given temperature the vapor pressures of different liquids are different (Table 9–2) because their cohesive forces are different. Easily vaporized liquids are

TABLE 9–2 Vapor Pressures (in torr) of Some Liquids at Different Temperatures

	0°C	25°C	50°C	75°C	100°C
Water	4.6	23.8	92.5	300	760
Benzene	27.1	94.4	271	644	1360
Methyl alcohol	29.7	122	404	1126	
Diethyl ether	185	470	1325	2680	4859

called **volatile** liquids, and they have relatively high vapor pressures. The most volatile liquid given in Table 9–2 is diethyl ether, and water is the least volatile. Vapor pressures can be measured with manometers such as the one shown in Figure 9–8.

9–7 Boiling Points and Distillation

Heating a liquid always increases its vapor pressure. When a liquid is heated to a sufficiently high temperature under a given applied (atmospheric) pressure, bubbles of vapor begin to form below the surface. They rise to the surface and burst, releasing the vapor into the air. This process is called boiling, and is distinctly different from evaporation.* If the vapor pressure inside the bubbles is less than the applied pressure on the surface of the liquid, the bubbles collapse as soon as they form and no boiling occurs. The **boiling point** of a liquid under a given pressure is the temperature at which its vapor pressure is just equal to the applied pressure. The **normal boiling point** is the temperature at which the vapor pressure of a liquid is equal to exactly one atmosphere (760 torr) pressure. The vapor pressure of water is 760 torr at 100°C, its normal boiling point. If the applied pressure is reduced below 760 torr, say on the top of a mountain, well above sea level, water boils below 100°C. It takes longer to cook food in boiling water at high altitudes than at or below sea level. A pressure cooker is efficient in cooking food rapidly because water boils at higher

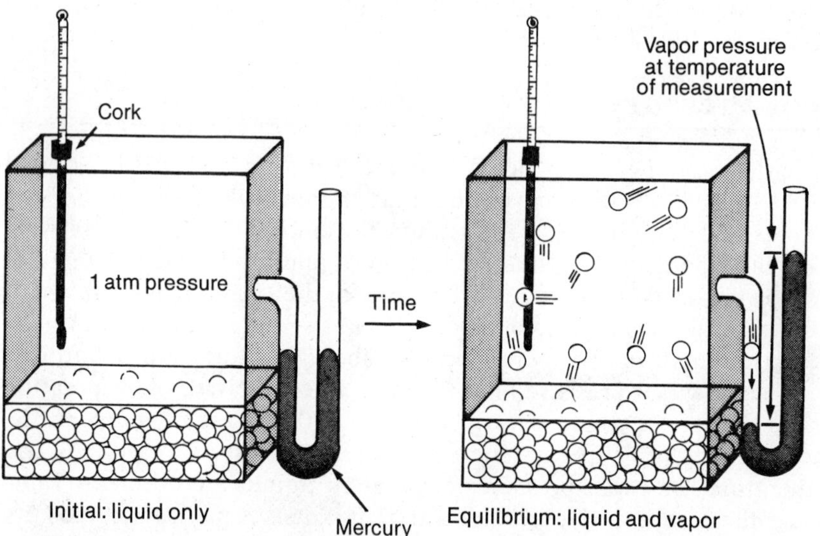

FIGURE 9–8 A simplified representation of the measurement of vapor pressure of a liquid at a given temperature. At the instant the liquid is added to the container, the space above the liquid is occupied by air only. As time passes some of the liquid vaporizes until equilibrium is established. The difference between the heights of the mercury columns is a measure of the vapor pressue of the liquid at that temperature.

* When water is heated, but before it boils, small bubbles may begin to adhere to the container, and then rise to the surface and burst. This is not boiling of water, but rather the formation of bubbles of previously dissolved gases such as carbon dioxide or oxygen, whose solubilities in water decrease with increasing temperature.

temperatures under high pressures. The increased heat content of the boiling water allows the foods to absorb heat from the water more rapidly than at 100°C.

Since different liquids have different cohesive forces, they have different vapor pressures, and boil at different temperatures. Hence a mixture of liquids with different enough boiling points can often be separated into its components by **distillation.** In this process the mixture is heated slowly, until the temperature reaches the point at which the most volatile component boils off. If this component is a liquid under ordinary conditions, it is subsequently recondensed in a water-cooled condensing column (Figure 9–9) and collected as a distillate. After enough heat has been added to vaporize all of the most volatile liquid, the temperature again rises slowly until the boiling point of the next substance is reached and the process continues. Any nonvolatile substances dissolved in the liquid do not boil, but remain in the distilling flask. Impure water can be purified and separated from its dissolved salts by distillation. Compounds with similar boiling points, especially those that interact very strongly with each other, are not well separated by simple distillation but require a modification called fractional distillation which is discussed in Section 10–12.

9–8 Heat Transfer Involving Liquids

Heat must be added to a liquid in order to raise its temperature (Section 1–13). The **specific heat** (J/g · °C) or **molar heat capacity** (J/mol · °C) of a liquid is the amount of heat that must be added to the stated mass of liquid to raise its temperature by one degree Celsius. If heat is supplied to a liquid at a constant rate, the temperature rises at a constant rate until its boiling point is reached. At that point the temperature remains constant until enough heat has been added to boil away all the liquid. The **molar heat of vaporization** (ΔH_{vap}) of a liquid is the amount of heat that must be added to one mole of the liquid at its boiling point to convert it to vapor with no change in temperature. Heats of vaporization can also be expressed in joules per gram. The heat of vaporization for water is 2.26×10^3 J/g which corresponds to 40.7 kJ/mol, as shown below.

Because molar heats of vaporization are larger than molar heat capacities, the former are usually expressed kilojoules rather than joules.

$$\frac{?\ \text{kJ}}{\text{mol}} = \frac{2.26 \times 10^3\ \text{J}}{\text{g}} \times \frac{1\ \text{kJ}}{1000\ \text{J}} \times \frac{18.0\ \text{g}}{\text{mol}} = 40.7\ \frac{\text{kJ}}{\text{mol}}$$

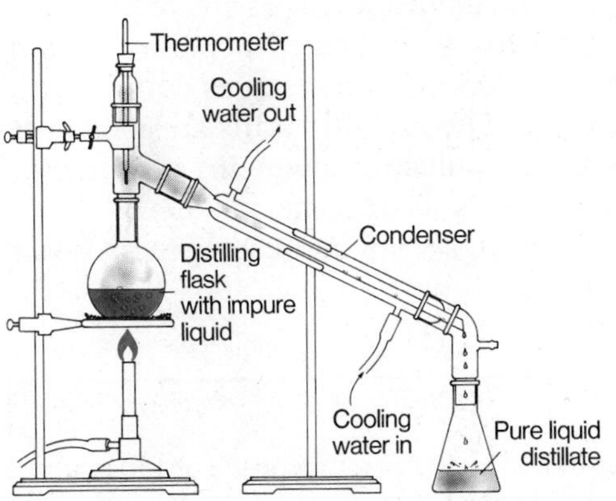

FIGURE 9–9 Laboratory set-up for distillation. During distillation of an impure liquid, nonvolatile substances remain in the distilling flask. The liquid is vaporized and condensed before being collected in the receiving flask.

TABLE 9–3 Heats of Vaporization, Boiling Points, and Vapor Pressures of Some Common Liquids

Liquid	Boiling Point at 1 atm (°C)	Vapor Pressure (torr at 20°C)	Heat of Vaporization at Boiling Point	
			(J/g)	(kJ/mol)
Water	100	17.5	2.26×10^3	40.7
Ethyl alcohol	78.3	43.9	858	39.3
Benzene	80	74.6	395	30.8
Diethyl ether	34.6	442	351	26.0

As with many other properties of liquids, their heats of vaporization reflect the strengths of intermolecular forces. Heats of vaporization generally increase as boiling points and intermolecular forces increase and vapor pressures decrease. Table 9–3 illustrates this.

The unusually high heat of vaporization of water, and to a lesser extent, that of ethyl alcohol, are due mainly to the strong hydrogen bonding interactions in these liquids (Section 9–9). The high value for water makes it very effective and efficient as a coolant, and, in the form of steam, as a source of heat.

Many large buildings are heated by steam; the condensation of steam liberates the heat that warms the buildings.

Liquids can be converted to vapors even below their boiling points by evaporation. The water in perspiration is a good coolant for our bodies because as it evaporates, each gram absorbs 2.41 kJ of heat from the body and carries it into the air in the form of water vapor. We feel even cooler in a breeze because evaporation of perspiration is facilitated (Section 9–5) and heat is removed more rapidly.

The heat of vaporization of water is higher at 37°C (normal body temperature) than at 100°C (2.41 kJ/g as opposed to 2.26 kJ/g).

Condensation is the reverse of evaporation. The heat that must be removed from a vapor to condense it (without change in temperature) is called the **heat of condensation.**

$$\text{Liquid} + \text{Heat} \underset{\text{condensation}}{\overset{\text{evaporation}}{\rightleftharpoons}} \text{Vapor}$$

The heat of condensation of a liquid is equal in magnitude, but opposite in sign, to the heat of vaporization, and is released by the vapor during condensation.

Since 2.26 kJ/g must be absorbed in order to vaporize water at 100°C, it follows that 2.26 kJ/g must be released to the environment as steam at 100°C condenses to form liquid water at 100°C. In steam-heated radiators, steam condenses and releases 2.26 kJ of heat per gram as its molecules collide with the cooler radiator walls and condense there. The metallic walls are good heat conductors, and they transfer the heat to air molecules bouncing off the outside walls of the radiator. Other substances such as benzene and diethyl ether are much less effective as heating and cooling agents because of their lower heats of vaporization and condensation. (See Table 9–3.)

Burns caused by steam at 100°C are much more severe than burns caused by liquid water at 100°C because of the large amount of heat released by the steam as it condenses.

EXAMPLE 9–1

Calculate the amount of heat in joules required to raise the temperature of 180 grams of water from 10.0°C to 40.0°C. The specific heat of water is 4.18 J/g · °C.

Solution

$\underline{?}\ J = 180\ g \times \frac{4.18\ J}{g \cdot °C} \times (40.0°C - 10.0°C) = \underline{2.26 \times 10^4\ J}$

Note that all units except joules *cancel*.

EXAMPLE 9–2

Calculate the amount of heat in joules required to convert 180 grams of water at 10.0°C to steam at 105.0°C.

Solution

The total amount of heat absorbed is the sum of the amounts required to (a) raise the temperature of the liquid water from 10.0°C to 100.0°C, (b) convert the liquid water to steam at 100°C, and (c) raise the temperature of the steam from 100.0°C to 105.0°C. Steps 1 and 3 involve the specific heats of water and steam, 4.18 J/g · °C and 2.03 J/g · °C, respectively (see Appendix E), while step 2 involves the heat of vaporization of water (2.26×10^3 J/g).

(a) $\underline{?}\ J = 180\ g \times \frac{4.18\ J}{g \cdot °C} \times (100.0°C - 10.0°) = 6.77 \times 10^4\ J \quad = 0.677 \times 10^5\ J$

(b) $\underline{?}\ J = 180\ g \times \frac{2.26 \times 10^3\ J}{g} \quad = 4.07 \times 10^5\ J$

(c) $\underline{?}\ J = 180\ g \times \frac{2.03\ J}{g \cdot °C} \times (105.0°C - 100.0°C) = 1.83 \times 10^3\ J = 0.0183 \times 10^5\ J$

Heat absorbed = $\underline{4.77 \times 10^5\ J}$

EXAMPLE 9–3

Compare the amount of "cooling" experienced by an individual who drinks 400 mL of ice water (0°C) with the amount of "cooling" experienced by an individual who "sweats out" 400 mL of water.

Solution

The density of water is very nearly 1.00 g/mL at both 0°C and 37°C, average body temperature. The heat of vaporization of water is 2.41 kJ/g at 37°C. The amount of heat required to raise the temperature of 400 g of water from 0°C to 37°C is

$\underline{?}\ J = (400\ g)(4.18\ J/g \cdot °C)(37°C) = 6.19 \times 10^4\ J$ or $\underline{61.9\ kJ}$

Evaporating, i.e., "sweating out," 400 mL of water at 37°C requires

$\underline{?}\ J = (400\ g)(2.41 \times 10^3\ J/g) = 9.64 \times 10^5\ J$ or $\underline{964\ kJ}$

Thus, we see that "sweating out" 400 mL of water removes 964 kJ of heat from one's body, while drinking 400 mL of ice water absorbs only 61.9 kJ. Stated differently, "sweating" is 15.6 times (964/61.9) more efficient than drinking ice water!

Absorbs means "removes from the body."

9–9 Intermolecular Forces of Attraction and Phase Changes

We have referred to the influence of the strengths of intermolecular forces of attraction on properties of liquids such as boiling point, vapor pressure, and heat of vaporization. Such forces are also directly related to the properties of solids, such as melting point and heat of fusion (page 313). *Inter*molecular forces refer to the forces between individual particles (atoms, molecules, ions) of a substance. Intermolecular forces are quite weak relative to *intra*molecular forces, i.e., covalent and ionic bonds within compounds. For example, 920 kJ of heat are required to decompose one mole of water vapor into hydrogen and oxygen atoms. This reflects the strength of intramolecular forces (chemical bonds). But it requires only 40.7 kJ, ΔH_{vap}, to convert one mole of liquid water into steam at 100°C. This reflects the strength of the intermolecular forces of attraction between the water molecules, mainly *hydrogen bonding,* one of the stronger types of intermolecular forces.

If it were not for the existence of intermolecular forces, condensed phases (liquids and solids) could not exist. These are the forces that hold the particles together. As we shall see, the effects of these forces on melting points of solids parallel those on boiling points of liquids. High boiling points are associated with compounds exhibiting strong intermolecular attractions. Let us consider the general types of forces that exist among ionic, covalent, and monatomic species and their effects on boiling points.

1 Ion–Ion Interactions

According to Coulomb's Law, the *force of attraction* between two oppositely charged ions is directly proportional to the charges on the ions, q^+ and q^-, and inversely proportional to the square of the distance between them, d

$$F \propto \frac{q^+q^-}{d^2}$$

Energy has the units of force × distance, $F \times d$, so the *energy of attraction* between two oppositely charged ions is directly proportional to the charges on the ions and inversely proportional to the distance of separation, rather than the square of the distance

$$E \propto \frac{q^+q^-}{d}$$

Since ionic compounds such as NaCl, $CaBr_2$, and K_2SO_4 exist as extended arrays of discrete ions in the solid state, ionic bonding may be thought of as both *inter-* and *intramolecular* bonding. Most ionic bonding is strong, and as a result, most ionic compounds have high melting points. At high enough temperatures ionic solids will melt as the added heat energy overcomes the energy associated with the attraction of oppositely charged ions that holds the crystal together. The ions in the resulting *molten* samples are free to move about more or less randomly, accounting for the ability of molten ionic compounds to act as very strong conductors of electricity. Melting of solids also produces greater average separations among the ions, which means that the forces (and energies) of attraction among the ions are less than in the solid state because d is greater in the melt. Such energies of attraction, though, are generally considerably greater than the energies of attraction among the molecules of covalent or monatomic liquids. This accounts for the very high boiling points of molten ionic substances.

Additionally, since the product q^+q^- increases as the charges on ions increase, ionic substances containing multiply charged ions such as Al^{3+},

Mg^{2+}, O^{2-}, and S^{2-} ions usually have higher melting and boiling points than salts containing only univalent ions such as Na^+, K^+, F^-, and Cl^-.

2 Dipole–Dipole Interactions

Permanent dipole–dipole interactions occur between polar covalent molecules, because of the attraction of the δ+ atoms of one molecule to the δ− atoms of another molecule (Section 5–2). Electrostatic forces between two ions decrease by the factor $1/d^2$ as their separation, *d*, increases. But dipole–dipole forces vary as $1/d^4$ and, in order to be at all effective, must operate over only very short distances. In addition, such forces are weaker than in the ion–ion cases because q^+ and q^- are less due to the presence of only "partial charges." Average dipole–dipole interaction energies are approximately 4 kJ per mole of bonds, in contrast to average values for typical ionic and covalent bond energies, which are approximately 400 kJ per mole of bonds. Substances in which permanent dipole–dipole interaction affects physical properties include bromine fluoride, BrF, and sulfur dioxide, SO_2. Dipole–dipole interactions are illustrated in Figure 9–10. All dipole–dipole interactions, including hydrogen bonding (discussed below), are somewhat directional. An increase in temperature causes an increase in translational, rotational, and vibrational motion of molecules. This produces more randomness of orientation of molecules relative to each other, and a consequent decrease in the strength of dipole–dipole interactions. All these factors make compounds having dipole–dipole interactions more volatile than ionic compounds.

> Scientists often must change their thinking to accommodate new ideas and especially new experimental findings. The material in the above paragraph is an example. Until quite recently chemists believed that the forces of attraction between permanent dipoles varied as $1/d^7$, where *d* represents the distance between dipoles in different molecules. Theoretical chemists who study molecular mechanics now interpret permanent dipole–dipole forces as varying with $1/d^4$. The mathematics associated with molecular mechanics is beyond the scope of this text. In addition, we emphasize again that energy can be expressed as force times distance, $F \times d$, and that equations that show the *energies* associated with different kinds of molecular and ionic interactions have d^n in their denominators. In energy terms, *n* is one less than the value of the exponent in equations that describe *forces.*

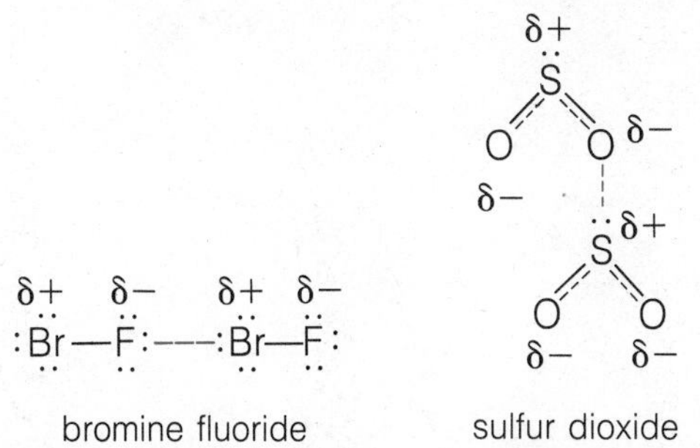

FIGURE 9–10 Dipole-dipole interactions (----) in bromine fluoride and sulfur dioxide.

3 Hydrogen Bonding

Ordinarily, we restrict usage of the term hydrogen bonding only to compounds in which H is directly bonded to F, O, or N.

Hydrogen bonds are not really chemical bonds in the formal sense, but are rather a very strong, special case of dipole–dipole interaction that occurs among polar covalent molecules containing hydrogen and one of the small, highly electronegative elements, such as fluorine, oxygen, or nitrogen. As with ordinary dipole–dipole interactions, they result from the attractions between $\delta+$ atoms of one molecule, in this case hydrogen atoms, and the $\delta-$ atoms of another molecule. The small sizes of the fluorine, oxygen, and nitrogen atoms, combined with their extremely high electronegativities, serve to concentrate the electrons of these molecules around the $\delta-$ atoms. This effect forces the connected hydrogen atom to behave, in some ways, as a bare proton. This, in turn, has a dramatic effect on the hydrogen atom's attraction to the $\delta-$ pole of adjacent molecules.

Hydrogen bonding is responsible for the abnormally high melting and boiling points of compounds such as water, ethyl alcohol, and ammonia relative to other compounds of similar molecular weight and molecular geometry (refer to Figure 9–13). Hydrogen bonding in hydrogen fluoride is discussed and illustrated in Section 22–11. Hydrogen bonding between carboxylate groups ($—CO_2—$) and amino groups ($—NH_2$) of amino acid subunits is very important in establishing the so-called tertiary (three-dimensional) structures of proteins.

4 London Forces

London forces are named after the German-born physicist, Fritz London, who initially postulated their existence on the basis of quantum mechanical theory in 1930.

London forces are weak attractive forces that are important only over *extremely* short distances since they vary as $1/d^7$. They exist for all types of molecules in condensed phases but are very weak for small molecules. Nevertheless, London forces are the only kind of intermolecular forces present

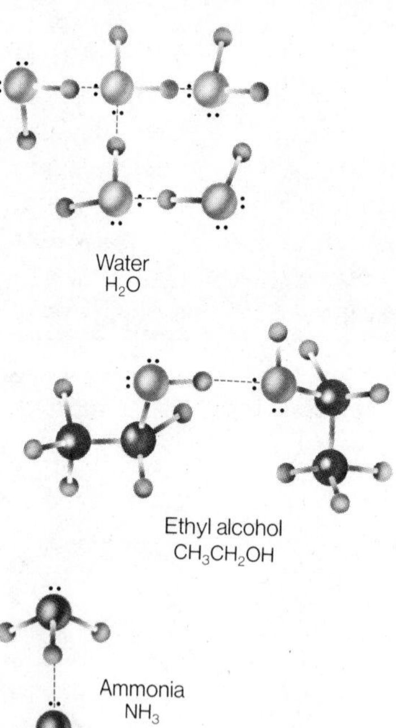

FIGURE 9–11 Hydrogen bonding ( ) in water, ethyl alcohol, and ammonia. This is a special case of very strong dipole-dipole interaction.

among symmetrical nonpolar substances such as SO_3, CO_2, O_2, N_2, Br_2, H_2, and monatomic species such as the noble gases. Thus London forces are responsible for the liquefaction of these substances which in some instances occurs only at very low temperatures and/or high pressures. Although the term *van der Waals forces* usually refers to all intermolecular attractions, it is also often used interchangeably with London forces, as are the terms *dispersion forces* and *dipole-induced dipole forces.*

London forces result from the attraction of the positively charged nucleus of one atom for the electron cloud of an adjacent atom in another molecule. The resulting transient dipoles induced in the atoms or molecules increase in strength as the polarizability of their electron clouds increases. Polarizability increases with increasing atomic or molecular size, because as electron clouds become larger and more diffuse they are attracted less strongly by their own (positively charged) nuclei, and thus they are more easily distorted, or polarized, by adjacent nuclei. London forces are depicted in Figure 9–12 for monatomic argon. They exist in polar covalent substances as well as in nonpolar substances.

Figure 9–13 shows that polar covalent compounds with hydrogen bonding (H_2O, HF, NH_3) boil at higher temperatures than analogous, polar compounds without hydrogen bonding (H_2S, HCl, PH_3) which, in turn, boil at higher temperatures than symmetrical, nonpolar compounds (CH_4, SiH_4) of comparable molecular weight. It also shows that in the absence of hydrogen bonding, boiling points of analogous substances (CH_4, SiH_4, GeH_4, SnH_4) increase fairly regularly with increasing molecular size, and therefore, with increasing molecular weight, due to increasing effectiveness of London forces of attraction. This is true even in the case of polar covalent molecules. The increasing effectiveness of London forces accounts for the increase in boiling points in the sequences HCl < HBr < HI and $H_2S < H_2Se < H_2Te$, which involve nonhydrogen-bonded polar covalent molecules. Since the differences in electronegativities between hydrogen and the other nonmetals *decrease* in these sequences the effectiveness of *permanent* dipole-dipole interactions de-

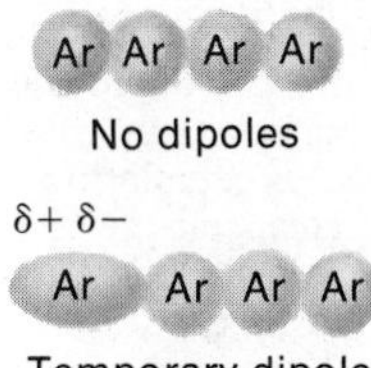

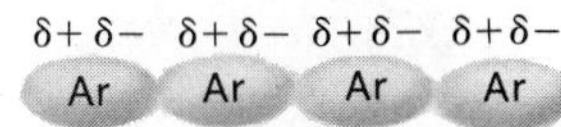

Induced dipoles in other atoms

FIGURE 9–12 Simplified picture of the generation of London forces in symmetrical (nonpolar) argon atoms. An instantaneous nonspherical electronic distribution which gives a dipole in one atom induces a temporary dipole in an adjacent atom. The resulting pair of transient dipoles causes a weak attraction between the two atoms.

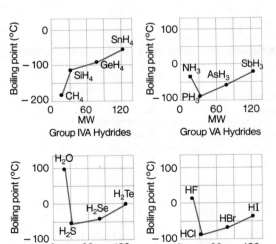

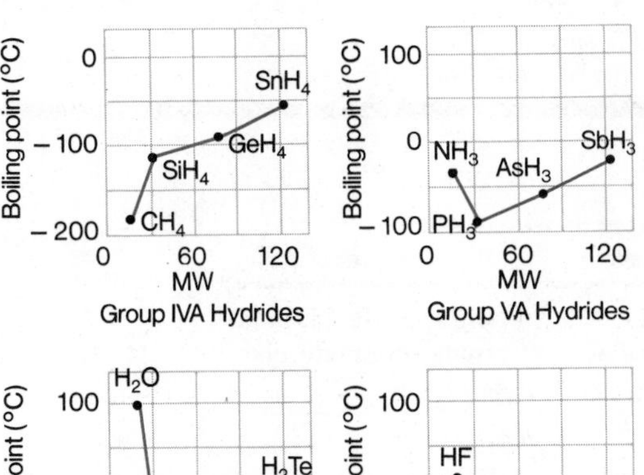

FIGURE 9–13 Boiling points of some hydrides as a function of molecular weight. The high boiling points of NH_3, H_2O, and HF are due to hydrogen bonding. The electronegativity difference between H and C is so small that CH_4 is not hydrogen bonded.

The exact lack of correlation of total energy with boiling points for Ar and CO is probably due to approximations made in calculating total energies, and the fact that some kinds of interaction that make very small contributions have been omitted.

TABLE 9–4 Approximate Contributions to the Total Energy of Interaction Between Molecules as kJ/mol

Molecule	Permanent Dipole Moment (D)	Permanent Dipole–Dipole Energy	London Energy	Total Energy	Boiling Point (K)
Ar	0	0	8.5	8.5	87
CO	0.1	0.0	8.7	8.7	81
HCl	1.03	3.3	17.8	21	188
NH_3	1.5	13*	16.3	29	240
H_2O	1.8	36*	10.9	47	373

* Hydrogen bonded.

creases. This effect alone would lead to *decreasing* boiling points. Since the boiling points actually *increase* in the order of increasing molecular size, this tells us that the increasing London forces actually override the decreasing permanent dipole–dipole forces in these cases.

It is interesting to compare the magnitudes of the various contributions to the total energy of interactions in a group of simple molecules. Table 9–4 shows the permanent dipole moments and the energy contributions for five simple molecules. Note that the contribution from London forces is large in all cases. The variations of these total energies of interaction are closely related to boiling points, as we might expect.

EXAMPLE 9–4

Predict the order of increasing boiling points for the following: H_2S, H_2O, CH_4, Ne, and KBr.

Solution

Since KBr is ionic, it boils at the highest temperature. Water exhibits hydrogen bonding and boils at the next highest temperature. Hydrogen sulfide is the only other polar covalent substance so it boils below H_2O. Nonpolar CH_4 boils at a higher temperature than the noble gas Ne.

$$\underrightarrow{Ne < CH_4 < H_2S < H_2O < KBr}$$
increasing boiling points

TABLE 9–5 General Effects of Intermolecular Attractions on Physical Properties

Property	Volatile Liquids Weak Intermolecular Attractions	Nonvolatile Liquids Strong Intermolecular Attractions
Cohesive forces	Low	High
Viscosity	Low	High
Surface tension	Low	High
Specific heat	Low	High
Vapor pressure	High	Low
Rate of evaporation	High	Low
Boiling point	Low	High
Heat of vaporization	Low	High

The general effects of intermolecular attractions on the physical properties of liquids are summarized in Table 9–5. Please keep in mind the fact that high and low are relative terms. Table 9–5 is included to show only very general trends.

The Solid State

9–10 Melting Point

The **melting point** of a solid, which is the same as the **freezing point** of its liquid, is the temperature at which the rate of melting of a solid is the same as the rate of freezing of its liquid under a given applied pressure. That is, it is *the temperature at which solid and liquid exist in equilibrium*

$$\text{Liquid} \underset{\text{melting}}{\overset{\text{freezing}}{\rightleftharpoons}} \text{Solid}$$

The normal melting point of a substance is its melting point at one atmosphere pressure. Variations in melting and boiling points of substances are generally parallel since intermolecular forces are similar.

9–11 Heat Transfer Involving Solids

When heat is added to a solid below its melting point, its temperature rises in accordance with its specific heat. After enough heat has been added to reach the melting point, the addition of more heat is not accompanied by a temperature rise because the solid begins to melt. Only after all the solid has melted will the continued addition of heat result in an increase in the temperature of the liquid. This is illustrated graphically on the left side of Figure 9–14, a typical heating curve.

The amount of heat required to melt one gram of a solid at its melting point is its **heat of fusion.** The **molar heat of fusion** (ΔH_{fus}; kJ/mol) *is the amount of heat that must be absorbed to melt one mole of a solid at its melting point.* The heat of fusion is a measure of the *inter*molecular forces of attraction in the solid state. These are the forces of attraction that "hold the molecules together" as a solid. Heats of fusion usually increase as melting points increase. Melting points and heats of fusion of some common substances are shown in Table 9–6.

Notice that melting is always endothermic. The term fusion literally means melting.

TABLE 9–6 Melting Points and Heats of Fusion of Some Common Substances

Substance	Melting Point (°C)	Heat of Fusion (J/g)	Heat of Fusion (kJ/mol)
Methane	−182	58.6	0.92
Ethyl alcohol	−117	109	5.02
Water	0	334	6.02
Aluminum	658	395	10.6
Sodium chloride	801	519	30.3

Specific heats and molar heat capacities have units of energy/g°C and energy/mol °C, respectively, but heats of transition have no temperature units because no temperature changes accompany the processes to which they apply.

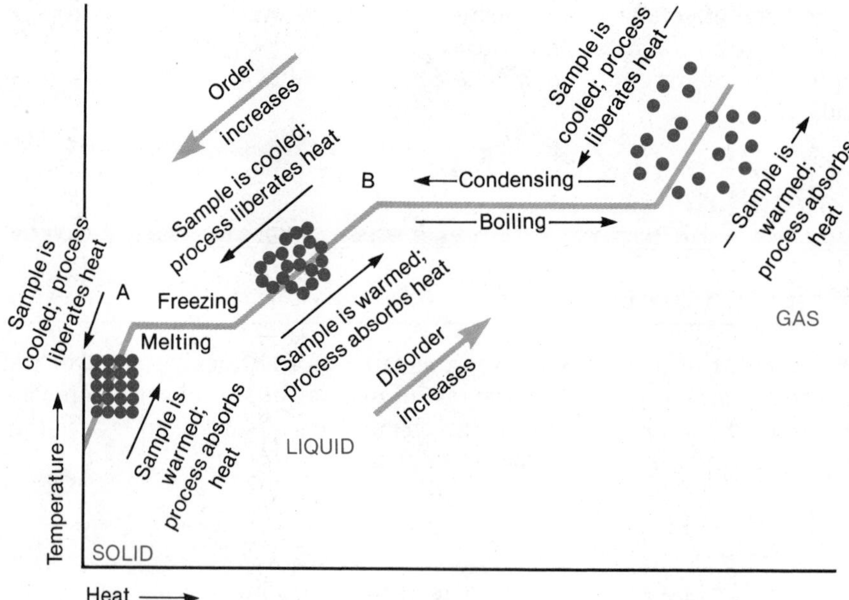

FIGURE 9–14 When heat energy is added to a solid below its melting point, the temperature of the solid rises until its melting point is reached (point A). If the solid is heated at its melting point, its temperature remains constant until the solid has melted, because the melting process requires energy. When all of the solid has melted, heating the liquid raises its temperature until its boiling point is reached (point B). If heat is added to the liquid at its boiling point, the added heat energy is absorbed as the liquid boils. When all the liquid has been converted to a gas (vapor), the addition of more heat raises the temperature of the gas. Each step in the process can be reversed by removing heat.

The **heat of solidification,** ΔH_{sol}, *of a liquid is equal in magnitude, but opposite in sign, to the heat of fusion,* ΔH_{fus}

$$\Delta H_{\text{solidification}} = -\Delta H_{\text{fusion}}$$

It represents the amount of heat that must be *removed* from a given amount (1 g or 1 mol) of liquid to solidify the liquid at its freezing point. For water

$$\text{Water} \underset{+334\ \text{J/g}}{\overset{-334\ \text{J/g}}{\rightleftharpoons}} \text{Ice} \qquad (\text{at } 0°\text{C})$$

The heat of fusion, like other heats of transition, is an *extensive property* of a substance, because the amount of heat absorbed depends upon the amount of solid that melts.

EXAMPLE 9–5

The molar heat of fusion, ΔH_{fus}, of sodium at its melting point, 97.5°C, is 2.6 kJ/mol. How much heat must be absorbed by 5.0 grams of solid sodium at 97.5°C to convert it to molten sodium?

Solution

The molar heat of fusion tells us that every mole of Na, 23 grams, will absorb 2.6 kJ of heat at 97.5°C during the melting process. We are interested in the amount of heat that 5.0 grams would absorb. Thus we choose the appropriate unit factors, constructed from the atomic weight and ΔH_{fus}, to calculate the amount of heat absorbed by 5.0 g of Na

$$\underline{?}\ \text{kJ} = 5.0\ \text{g Na} \times \frac{1\ \text{mol Na}}{23\ \text{g Na}} \times \frac{2.6\ \text{kJ}}{1\ \text{mol Na}} = \underline{0.57\ \text{kJ}}$$

The two unit factors could be combined into a single unit factor, 2.6 kJ/23 g Na.

EXAMPLE 9–6

Calculate the amount of heat that must be absorbed by 50.0 grams of ice at −12.0°C to convert it to water at 20.0°C.

Solution

We must determine the heat absorbed during three processes: (a) raising the temperature of 50.0 g of ice from −12.0°C to its melting point, 0.0°C, which is governed by the specific heat of ice, 2.09 J/g·°C; (b) melting the ice with no change in temperature, which is governed by the heat of fusion of ice at 0.0°C, 334 J/g; and (c) raising the temperature of the resulting water from 0.0°C to 20.0°C, which is governed by the specific heat of water, 4.18 J/g·°C.

$$\text{(a)}\ 50.0\ \text{g} \times \frac{2.09\ \text{J}}{\text{g}\cdot{}^\circ\text{C}} \times [0.0 - (-12.0)]^\circ\text{C} = 1.25 \times 10^3\ \text{J} = 0.125 \times 10^4\ \text{J}$$

$$\text{(b)}\ 50.0\ \text{g} \times \frac{334\ \text{J}}{\text{g}} = 1.67 \times 10^4\ \text{J}$$

$$\text{(c)}\ 50.0\ \text{g} \times \frac{4.18\ \text{J}}{\text{g}\cdot{}^\circ\text{C}} \times (20.0 - 0.0)^\circ\text{C} = 4.18 \times 10^3\ \text{J} = 0.418 \times 10^4\ \text{J}$$

$$\text{Total heat absorbed} = \underline{2.21 \times 10^4\ \text{J} = 22.1\ \text{kJ}}$$

Note that most of the heat absorbed was required in step 2, melting the ice.

9–12 Sublimation and the Vapor Pressure of Solids

Some solids, such as iodine and Dry Ice, vaporize without passing through the liquid state at atmospheric pressure. We say they **sublime.** Solids exhibit vapor pressures just as liquids do, but generally have much lower vapor pressures. Solids with high vapor pressures sublime easily. The characteristic odors of the common household solids, naphthalene (moth balls) and paradichlorobenzene (bathroom deodorizer) are due to sublimation. The reverse process, by which a vapor solidifies without passing through the liquid phase, is called **deposition**

Solid carbon dioxide is called Dry Ice.

$$\text{Solid} \underset{\text{deposition}}{\overset{\text{sublimation}}{\rightleftharpoons}} \text{Gas}$$

Some impure solids can be purified by sublimation and subsequent deposition of the vapor (as a solid) onto the surface of a cooler object. A sublimation apparatus is illustrated in Figure 9–15. Iodine is commonly purified by sublimation.

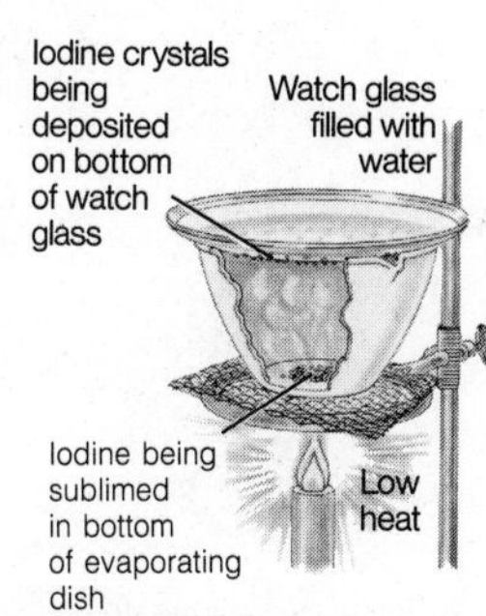

FIGURE 9–15 Apparatus for the recrystallization of iodine by sublimation.

9–13 Phase Diagrams

After discussing the general properties of the three phases of matter we can now describe **phase diagrams,** which show the equilibrium pressure–temperature relationships between the different phases of a given substance. Figure 9–16 shows a portion of the phase diagrams for water and carbon dioxide. As one can see from a quick glance at the axes, the curves are not drawn to scale. The shapes of the curves are exaggerated to allow us to describe the changes of state accompanying pressure or temperature changes using one diagram rather than several.

The curved line from *A* to *B* in Figure 9–16a is a vapor pressure curve obtained experimentally by measuring the vapor pressures of water at various temperatures (see Table 9–7). The points along this curve represent the temperature–pressure combinations for which liquid and gas (vapor) coexist in equilibrium. At points directly above *AB*, the stable form of water is liquid; below the curve it is vapor.

TABLE 9–7 Points on the Vapor Pressure Curve for Water

Temperature (°C)	−10	0	20	30	50	70	90	95	100	101
Vapor pressure (torr)	2.1	4.6	17.5	31.8	92.5	234	526	634	760	788

Line *AC* represents the liquid–solid equilibrium conditions. Note that it has a negative slope. Water is one of the very few substances for which this is the case. The slope up and to the left indicates that increasing the pressure sufficiently on the surface of ice causes it to melt. This is because ice is *less dense* than liquid water in the vicinity of the liquid–solid equilibrium. This causes ice to float in liquid water and is due to the fact that the network of hydrogen bonding in ice, which is more extensive than that in liquid water,

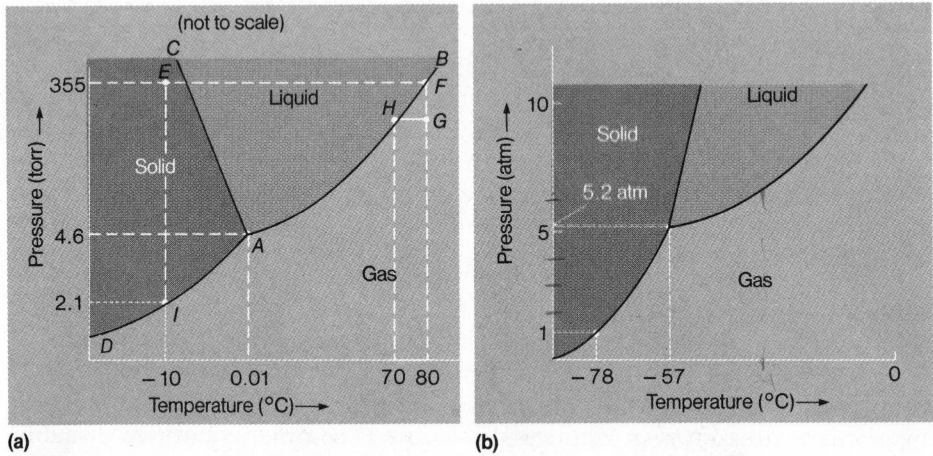

FIGURE 9–16 Phase diagrams. (*a*) Diagram for water (not to scale). For the few such substances for which the solid is less dense than the liquid, the solid-liquid equilibrium line (AC) has negative slope. (*b*) Diagram for carbon dioxide (not to scale), a substance for which the solid is more dense that the liquid. Note that the solid-liquid equilibrium line slopes to the right, i.e., has positive slope. This is true for most substances.

requires a greater separation of H_2O molecules than in the liquid. It is this property that helps make ice skating possible. Frictional heating and the pressure exerted by the skate blade on the ice melt the ice just below it, thus producing a very slick surface which refreezes after the skater has passed. One could not skate on most other solids, even at temperatures and pressures near the solid–liquid equilibrium because most solids are more compact than their liquids. Most substances, therefore, have positive slopes associated with line *AC*.

The stable form of water at points to the left of *AC* is solid (ice). Thus *AC* is called a melting curve. At pressures and temperatures along *AD*, the sublimation curve, solid and vapor exist in equilibrium. There is only one point, *A*, at which all three phases—ice, liquid water, and water vapor—can coexist at equilibrium. This is called the **triple point,** and for water it occurs at 4.6 torr and 0.01°C.

Note that at pressures below the triple point pressure, the liquid phase cannot exist; rather the substance goes directly from solid to gas (sublimes) or the reverse (crystals are formed directly from the gas). Consider CO_2 (Figure 9–16b). The triple point is at 5.2 atmospheres and 57°C. Since this pressure is *above* normal atmospheric pressure, liquid CO_2 cannot exist at atmospheric pressure. Dry Ice (solid CO_2) sublimes, and does not melt at atmospheric pressure.

To illustrate the use of a phase diagram in determining the physical state or states of a system under different sets of pressures and temperatures, let's consider a sample of water at point *E* in Figure 9–16a (355 torr and −10°C). At this point all the water is in the form of ice. Suppose we hold the pressure constant and gradually increase the temperature; in other words, trace a path from left to right along *EF*. When we reach the temperature at which *EF* intersects *AC*, the melting curve, some of the ice melts and if we stopped here equilibrium between ice and liquid water would eventually be established. That is, both phases would be present. If we continue to add heat, all the ice melts with no temperature change. Recall that all phase changes of pure substances occur at constant temperature. Once the ice is completely melted additional heat causes the temperature to rise again. Eventually at point *F* (355 torr and 80°C) some of the liquid begins to boil; liquid and vapor coexist at equilibrium. Adding more heat at constant pressure vaporizes the rest of the water with no temperature change. Complete vaporization also occurs if, at point *F* and before all the liquid had vaporized, the temperature were held constant and the pressure were decreased to, say, 234 torr at point *G*. If we then held the pressure at 234 torr and desired to condense some of the vapor it would be necessary to cool the vapor to 70°C, point *H*, at which we again arrive at the vapor pressure curve, *AB*. To state this in another way, the vapor pressure of water at 70°C is 234 torr.

Compare the description here with that accompanying Figure 9–14. Each horizontal line on a phase diagram contains the information in one heating curve, obtained at a particular pressure. Phase diagrams are obtained by combining the results of heating curves measured experimentally at different pressures.

Suppose we move back to ice at point *E* (355 torr and −10°C) again. If we now hold temperature at −10°C and reduce the pressure we move vertically down along *EI*. At a pressure of 2.1 torr we reach the sublimation curve, at which point some of the solid passes directly into the gas phase (sublimes). Decreasing the pressure further vaporizes all the ice. An important application of this phenomenon is in the freezing-drying of foods. In this process a water-containing food is cooled below the freezing point of water to form ice, which is then removed as a vapor by decreasing the pressure.

9–14 Amorphous Solids and Crystalline Solids

We have already seen that solids have definite shapes and volumes, are not very compressible, are dense, diffuse only very slowly into other solids, and are generally characterized by compact ordered arrangements of particles that vibrate about fixed positions in their crystal structures.

However a small class of noncrystalline solids, *amorphous solids,* has no well-defined, ordered structure. Examples are rubber, some kinds of plastics, and amorphous sulfur.

Glasses are sometimes called amorphous solids and sometimes called **supercooled liquids.** The justification for calling them supercooled liquids is that, like liquids, they flow, although very slowly. The irregular structures of glasses are really intermediate between those of freely flowing liquids and crystalline solids; there is only short-range order. Unlike crystalline solids, glasses and other amorphous solids do not exhibit sharp melting points. They soften over a temperature range, while crystalline solids like ice and sodium chloride all have well defined, sharp melting temperatures. Since particles in amorphous solids are irregularly arranged, intermolecular forces among particles vary in strength within the sample. Melting occurs at different temperatures for various portions of the same sample as the intermolecular forces are overcome.

One test for the purity of a crystalline solid is the sharpness of its melting point. Impurities disrupt the intermolecular forces and cause melting to occur over a considerable temperature range.

The shattering of a crystalline solid produces fragments having the same interfacial angles and structural characteristics as the original sample. The shattering of a large cube of rock salt produces several small cubes of rock salt. This symmetrical shattering of crystals occurs preferentially along crystal lattice planes between which the interionic or intermolecular forces of attraction are weakest. Amorphous solids with irregular structures, like fused glasses, shatter irregularly to yield pieces with jagged edges and irregular angles.

The lattice planes are planes within the crystal containing ordered arrangements of particles.

X-Ray Diffraction

Atoms, molecules, and ions are much too small to be seen with the eye. The arrangements of particles in crystalline solids are determined indirectly by x-ray diffraction (scattering). In 1912 the German physicist Max von Laue showed mathematically that any crystal could serve as a three-dimensional diffraction grating for incident electromagnetic radiation with wavelengths approximating the internuclear separations ($\sim$1 Å; 1 Ångstrom = 10^{-10} m) of atoms in the crystal. Such radiation is in the x-ray region of the electromagnetic spectrum. Using an apparatus similar to that shown in Figure 9–17, a monochromatic (single wavelength) x-ray beam is defined by a system of slits and directed onto the surface of a slowly rotated crystal so as to vary the angle of incidence θ. At various angles, strong beams of deflected x-rays impinge upon a photographic plate which, upon development, shows a central spot due to the primary beam and a set of symmetrically disposed spots due to deflected x-rays. Different kinds of crystals produce different arrangements of spots.

In 1913, the English scientists William and Lawrence Bragg found that the Laue photographs are more easily interpreted by treating the crystal as a reflection grating rather than a diffraction grating. The analysis of the spots is really quite complicated, but an experienced crystallographer can determine the separations between atoms within identical layers and the distances between layers of atoms. Moreover, it is possible not only to determine the relative positions of particles but also to determine unambiguously the identities of individual atoms if they are sufficiently different in atomic weight. Heavier atoms (with more electrons) deflect x-rays more strongly than light atoms.

Figure 9–18 illustrates the determination of spacings between layers of atoms. The x-ray beam strikes the surface of the crystal at an angle θ. Those rays colliding with atoms in the first layer are reflected at the same angle θ. Those passing through the first layer may be reflected from the second layer, third layer, and so forth. A strong reflected beam results only if all rays are in phase. The ray FCH travels farther than the ray EAG by a distance $BC + CD$. The two rays will be in phase only if the difference in their path lengths (the distance $BC + CD$) equals an integral number of wavelengths of the x-rays. Or,

$$BC + CD = n\lambda \qquad (n = \text{whole number})$$

Since AB is perpendicular to BC, angle ABC is 90°. Since the sum of the angles in any triangle is 180° then

$$\angle BAC + \angle BCA = 90°$$

Since the angle ACI is 90° and angle BCI is θ then

$$\angle ACB = 90° - \theta$$

Therefore, $\angle CAB + (90° - \theta) = 90°$, and $\angle CAB = \theta$

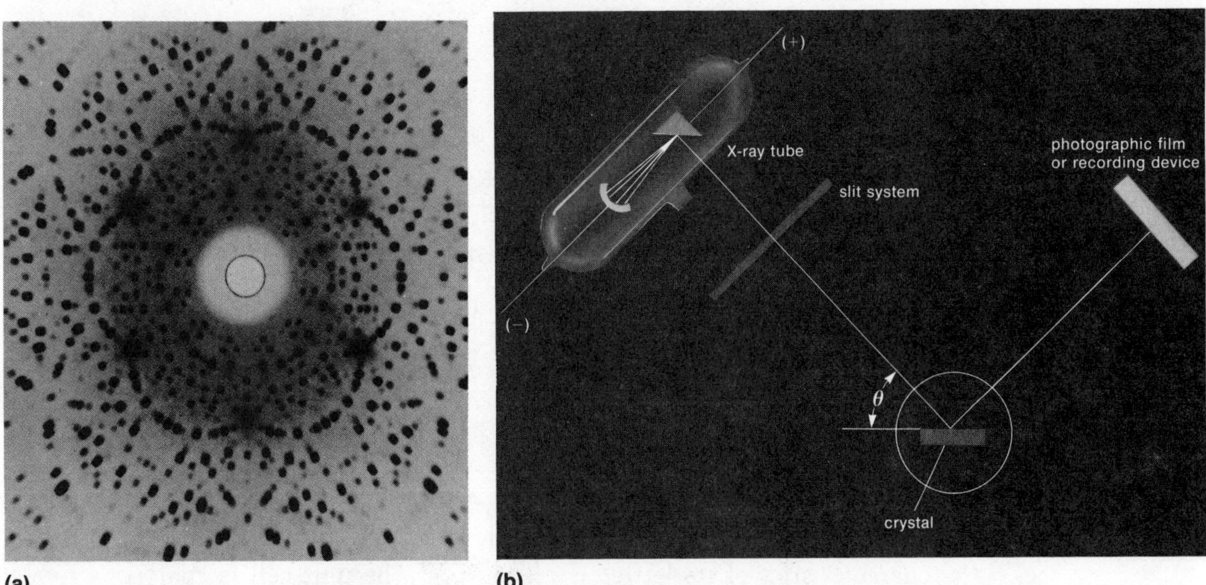

FIGURE 9–17 (*a*) Laue photograph of a crystal. The beam of x-rays was focused perpendicular to one of the lattice planes. A copper target x-ray tube was used, with an exposure time of 10 hours and a crystal-to-film distance of 2–3 cm. Courtesy of Eastman Kodak. (*b*) X-ray diffraction of crystals (schematic).

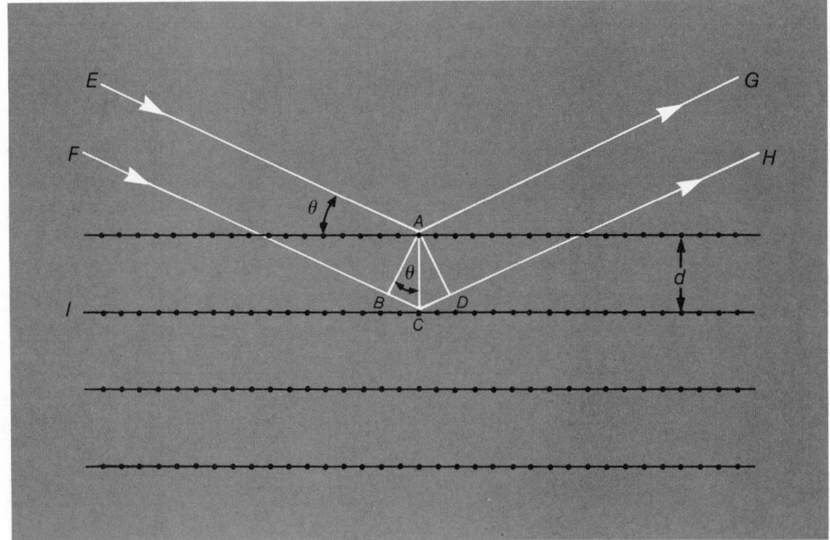

FIGURE 9–18 Reflection of a monochromatic beam of x-rays by two planes of a crystal.

The sine of θ ($\angle CAB$) is BC/AC where AC is the distance between layers in the crystal, d. Thus

$$\sin \theta = \frac{BC}{d}$$

on rearranging, $BC = d \sin \theta$

Also $CD = d \sin \theta$, and adding the last two equations,

$BC + CD = 2d \sin \theta$

In order for the waves to be in phase (interact constructively), as we have seen, the following condition must be met.

$$n\lambda = 2d \sin \theta \qquad \text{or} \qquad \sin \theta = \frac{n\lambda}{2d}$$

This is known as the Bragg equation. It tells us that for x-rays of a given wavelength, λ, atoms in planes separated by distances, d, give rise to intense reflections for angles of incidence, θ. The intensities of reflections decrease and angles θ increase with increasing order, $n = 1, 2, 3$, and so forth.

9–15 Structures of Crystals

All crystals are polyhedra consisting of regularly repeating arrays of atoms, molecules, or ions. The smallest unit of volume of a crystal showing all the characteristics of its lattice is a **unit cell.** The unit cell is analogous to the repeating unit of a wallpaper pattern, but in three dimensions. Once we identify the pattern of a wallpaper we can repeat it in two dimensions to cover a wall. Unit cells are stacked in three dimensions to build crystals. It is found that a unit cell fits into one of the seven crystal systems (Table 9–8), distin-

TABLE 9–8 The Unit Cell Dimensions for the Seven Crystal Systems

System	Axis Lengths	Angles	Example (Common Name)
Cubic	$a = b = c$	$\alpha = \beta = \gamma = 90°$	NaCl (rock salt)
Tetragonal	$a = b \neq c$	$\alpha = \beta = \gamma = 90°$	TiO_2 (rutile)
Orthorhombic	$a \neq b \neq c$	$\alpha = \beta = \gamma = 90°$	$MgSO_4 \cdot 7H_2O$ (epsomite)
Monoclinic	$a \neq b \neq c$	$\alpha = \gamma = 90°;\ \beta \neq 90°$	$CaSO_4 \cdot 2H_2O$ (gypsum)
Triclinic	$a \neq b \neq c$	$\alpha \neq \beta \neq \gamma \neq 90°$	$K_2Cr_2O_7$ (potassium dichromate)
Hexagonal	$a = b \neq c$	$\alpha = \beta = 90°;\ \gamma = 120°$	SiO_2 (silica)
Rhombohedral	$a = b = c$	$\alpha = \beta = \gamma \neq 90°$	$CaCO_3$ (calcite)

guished by the relative lengths of their axes, *a*, *b*, *c* (which are related to the spacings between layers, *d*), and the angles between the axes, α, β, and γ (Figure 9–19), and the symmetry of the resulting three-dimensional patterns.

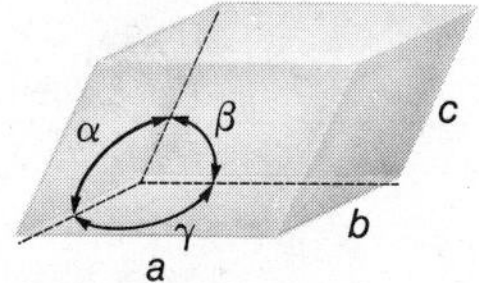

FIGURE 9–19 Representation of a unit cell.

Crystals have the same symmetry as their constituent unit cells since all crystals are repetitive multiples of such cells. The corners of unit cells of simple compounds are occupied by atoms, molecules, or ions. Equivalent sets of particles may occupy certain other positions, resulting in different modifications of the same unit cell, giving rise to a total of fourteen crystal lattices whose unit cells are shown in Figure 9–20. Different substances which crystallize in the same type of lattice with the same atomic arrangement are said to be **isomorphous.** When a single substance can crystallize in more than one form, it is said to be **polymorphous.** In the simple cubic lattice, equivalent particles occupy only the eight corners of the cubic unit cell. In the body-centered cubic (bcc) lattice an additional particle is located at the center of the unit cell. The face-centered cubic (fcc) structure involves eight particles at the corners and an additional six particles, one in the middle of each of the six square faces of the cube.

In certain types of crystals, particles other than those forming the framework of the unit cell may occupy extra positions within the unit cell. For example, sodium chloride crystallizes in a face-centered cubic lattice. The unit cell can be visualized as consisting of chloride ions at the corners and middles of the faces and sodium ions occupying spaces on the edges between the chloride ions and a space at the center (Figure 9–21a). Or by simple translation of the unit cell by half a unit cell in any direction within the lattice, we could visualize the unit cell in which sodium and chloride ions have exchanged positions. Such an exchange is not possible for most types of crystals. Most particles of a unit cell are not entirely within the unit cell. Those at corners are shared by the eight unit cells that meet at the center of the particle. Those lying on edges, but not at corners, are shared by four unit cells and those on faces are shared by two unit cells (Figure 9–22). While a unit cell of NaCl *may appear* to consist of 14 chloride ions and 12 sodium ions (or the reverse), it really consists of $8\left(\frac{1}{8}\right) + 6\left(\frac{1}{2}\right) = 4$ chloride ions and $12\left(\frac{1}{4}\right) + 1 = 4$ sodium ions (or the reverse). The unit cell reflects the symmetry of the crystal and the stoichiometric proportions of its chemical formula.

We have discussed the structure of NaCl in some detail because it is quite easy to visualize. However, it should be noted that many compounds with complicated molecules or complicated formula units crystallize in structures with unit cells that are much more difficult to visualize. Experimental deter-

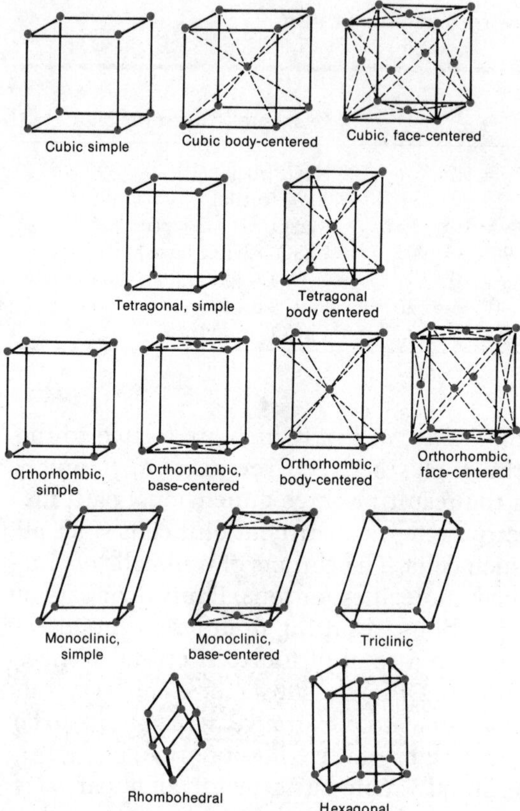

FIGURE 9–20 The fourteen possible crystal lattices.

minations of the crystal structures of such solids are usually correspondingly more complex as well.

9–16 Bonding in Solids

We can now classify crystalline solids in four categories depending upon the types of particles making up the lattice and the bonding or interactions among

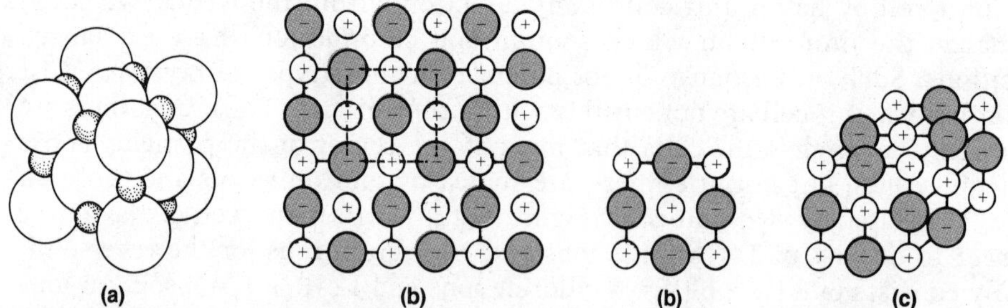

FIGURE 9–21 (*a*) The crystal structure of sodium chloride. (*b*) A cross-section of the space lattice of NaCl, showing the repeating pattern of its unit cell at the right. The dashed lines outline an alternative choice of the unit cell. Note that the entire pattern is generated by repeating either unit cell (and its contents) in all three directions. Several such choices of unit cells are usually possible. (*c*) The three-dimensional representation of (*b*).

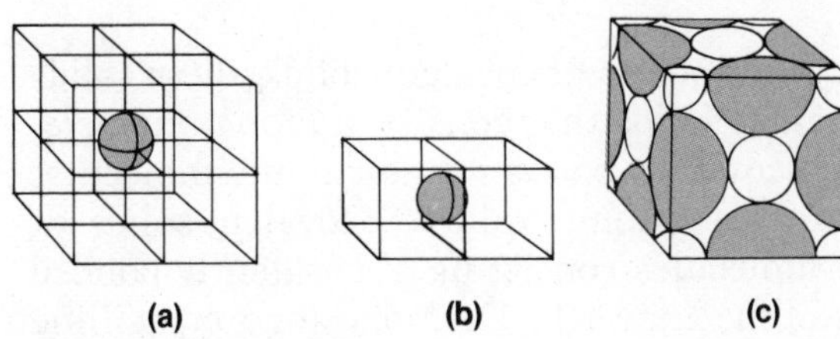

FIGURE 9–22 (*a*) The sharing of a corner atom by eight unit cells. (*b*) The sharing of a face-centered atom by two unit cells. (*c*) A representation of the unit cell of sodium chloride that indicates the relative sizes of the Na^+ and Cl^- ions as well as how ions are shared between unit cells. Particles at the corners and faces of unit cells are shared by other unit cells.

them. The four categories are (1) molecular solids, (2) covalent solids, (3) ionic solids, and (4) metallic solids. Table 9–9 summarizes some of their important characteristics.

1 Molecular Solids

The lattice positions that describe unit cells of molecular solids are occupied by molecules or monatomic elements (sometimes referred to as monatomic molecules). While the bonds *within* the molecules are covalent, the forces of attraction *between* molecules are quite weak. They range from hydrogen bonds and weaker dipole–dipole interactions in polar molecules like H_2O and SO_2 to very weak London forces in symmetrical, nonpolar molecules such as CH_4, CO_2, and O_2 and monatomic elements such as the noble gases. Because of the relatively weak intermolecular forces of attraction, molecules can be easily displaced and molecular solids are characteristically soft substances with low melting points. Because electrons do not move from one molecule to another under ordinary conditions, molecular solids are poor electrical conductors and good insulators.

London forces are also present among polar molecules.

TABLE 9–9 Characteristics of Types of Solids

	Molecular	Covalent	Ionic	Metallic
Particles of unit cell	Molecules (or atoms)	Atoms	Anions, cations	Metal ions in "electron gas"
Strongest interparticle forces	London, dipole–dipole, and/or hydrogen bonds	Covalent bonds	Electrostatic	Metallic bonds (electrical attraction between cations and e^-'s)
Properties	Soft; poor heat and electrical conductors; low melting points	Very hard; poor heat and electrical conductors; high melting points	Hard; brittle; poor heat and electrical conductors; high melting points	Soft to very hard; good heat and electrical conductors; wide range of melting points
Examples	CH_4 (methane) P_4, O_2, Ar, CO_2, H_2O, S_8	C (diamond) SiO_2 (quartz)	NaCl, $CaBr_2$, K_2SO_4 (typical salts)	Li, K, Ca, Cu, Cr, Ni (metals)

2 Covalent Solids

Although atoms occupy the lattice points of covalent solids, these solids should not be confused with molecular solids that consist of *monatomic elements,* such as argon. There are no covalent bonds in monatomic molecular solids (sometimes classified separately as "atomic solids"). Covalent solids (or "network solids") are really giant molecules consisting of covalently bonded atoms in an extended, rigid crystalline network. Diamond (one crystalline form of carbon, Section 25–1), and quartz, SiO_2 (Section 25–9), are examples of covalent solids. Because of their rigid, strongly bonded structures, most covalent solids are very hard and melt at high temperatures. Since the valence electrons are localized in covalent bonds they are not freely mobile so covalent solids are usually poor thermal and electrical conductors at ordinary temperatures.

3 Ionic Solids

Most salts crystallize as ionic solids with ions occupying lattice sites. Sodium chloride has been cited as an example. Many other salts crystallize in the sodium chloride (face-centered cubic) lattice. Examples are the halides, of Li^+, K^+, and Rb^+, and $M^{2+}X^{2-}$ oxides and sulfides like MgO, CaO, CaS, and MnO. Such isomorphous solids, with ions of the same charge, may substitute in each other's lattice to some extent. Two other common ionic structures are the cesium chloride, CsCl (simple cubic lattice), and zinc blende, or ZnS (face-centered cubic lattice) structures, shown in Figure 9–23. Salts that are isomorphous with the CsCl structure include CsBr, CsI, NH_4Cl, TlCl, TlBr, and TlI. The sulfides of Be^{2+}, Cd^{2+}, Zn^{2+}, and Hg^{2+}, together with CuBr, CuI, AgI, and ZnO, are isomorphous with the zinc blende structure. Several other ionic structural types exist.

Since the ions in an ionic solid are not free to move, other than to vibrate about their fixed positions, ionic solids are poor electrical and thermal conductors. However, molten salts are excellent conductors because their ions are freely mobile.

Ionic radii like those in Figure 4–5 and Table 10–1 are obtained from x-ray crystallographic determinations of unit cell dimensions assuming that adjacent ions are in contact with each other. Examples 9–7 and 9–8 illustrate such calculations.

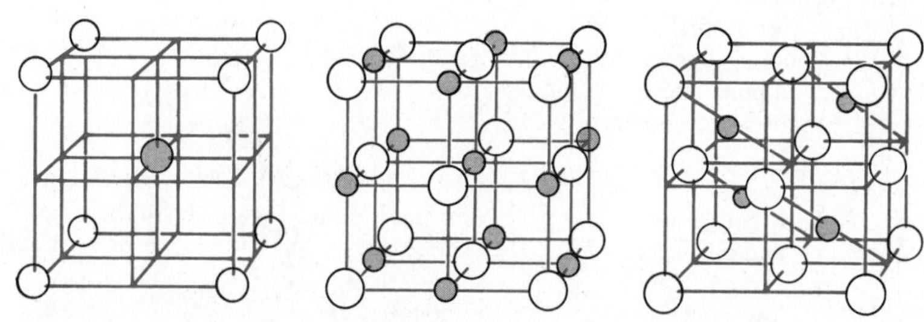

(a) Cesium chloride CsCl; ● $= Cs^+$, ○ $= Cl^-$ **(b)** Sodium chloride NaCl; ● $= Na^+$, ○ $= Cl^-$ **(c)** Zinc blende ZnS; ● $= Zn^{2+}$, ○ $= S^{2-}$

FIGURE 9–23 Crystal structures of ionic compounds of the MX type. The blue circles represent cations.

EXAMPLE 9–7

Lithium bromide, LiBr, crystallizes in the NaCl face-centered cubic lattice with a unit cell edge length of $a = b = c = 5.501$ Å. Assume that the bromide ions at the corners of the unit cell are in contact with those at the centers of the faces. Determine the ionic radius of the bromide ion. One face of the unit cell is depicted in Figure 9–24.

Solution

We may visualize the face as consisting of two right isosceles triangles sharing a common hypotenuse, h, and having sides of length $a = 5.501$ Å. The hypotenuse is equal to four times the radius of the bromide ion

$$h = 4\, r_{Br^-}$$

The hypotenuse can be calculated from the Pythagorean theorem, $h^2 = a^2 + a^2$. The length of the hypotenuse equals the square root of the sum of the squares of the sides

$$h = \sqrt{a^2 + a^2} = \sqrt{2\,a^2} = \sqrt{2(5.501\text{ Å})^2} = 7.780\text{ Å}$$

The radius of the bromide ion is one-fourth of h

$$r_{Br^-} = \frac{7.780\text{ Å}}{4} = 1.945\text{ Å}$$

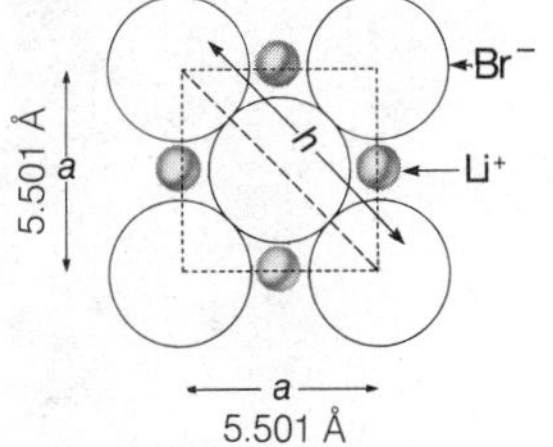

FIGURE 9–24 One face of the face-centered cubic lattice of lithium bromide.

This value agrees well with the ionic radius of the bromide ion listed in Figure 4–5. However, the tabulated value is the *average* value obtained from a number of crystal structures of compounds containing bromide ions. Calculations of ionic radii are based on the assumption that anion–anion contact exists. This may not always be true. Therefore, calculated radii vary from structure to structure. We should not place too much emphasis on a value of radius obtained from any single structure determination.

EXAMPLE 9–8

Refer to Example 9–7. Calculate the ionic radius of Li^+ in LiBr, assuming anion–cation contact along an edge of the unit cell.

Solution

The edge length, $a = 5.501$ Å, is two times the radius of the bromide ion plus two times the radius of the lithium ion. We know the radius for the bromide ion is 1.945 Å from Example 9–7

$$5.501\text{ Å} = 2\, r_{Br^-} + 2\, r_{Li^+}$$
$$2\, r_{Li^+} = 5.501\text{ Å} - 2\,(1.945\text{ Å}) = 1.611\text{ Å}$$
$$r_{Li^+} = 0.806\text{ Å}$$

The radius of Li^+ is calculated to be 0.806 Å but Figure 4–5 lists it as 0.60 Å. The discrepancy lies in the fallacy of the assumption that the lithium ion is in contact with both Br^- ions simultaneously. It is simply too small and is free to vibrate about a fixed-center position between two large Br^- ions. We now see that there is occasionally some difficulty in determining precise values of ionic radii. Similar difficulties can arise in the determination of atomic radii from molecular and covalent solids, or metallic radii from solid metals.

(a)

(b)

FIGURE 9–25 (*a*) Spheres in the same plane, packed as closely as possible; each sphere touches six others. (*b*) Spheres in two planes, packed as closely as possible. In real crystals, there are far more than two planes. Each sphere touches six others in its own layer, three others in the layer below it, and three in the layer above it, resulting in a total contact (coordination number) for each sphere of 12 other spheres.

4 Metallic Solids

Metals crystallize as metallic solids in which metal ions occupy the lattice sites and are embedded in a cloud of delocalized valence electrons. Practically all metals crystallize in one of three types of lattices: (1) body-centered cubic, (2) face-centered cubic (also called cubic close-packed), and (3) hexagonal close-packed. The latter two types are called close-packed structures because the particles (in this case metal atoms) are packed together as closely as possible. The differences between the two close-packed structures are illustrated in Figures 9–25 and 9–26. We shall let spheres of equal size represent the identical metal atoms, or any other particles, that form close-packed structures. Consider a layer of spheres packed in a plane, *A*, as closely as possible (Figure 9–26a). An identical plane of spheres, *B*, is placed in the depressions of plane *A*. If the third plane is placed with its spheres directly above those in plane *A*, the *ABA* arrangement results. This is the hexagonal close-packed structure (Figure 9–26a). The extended pattern of arrangement of planes is *ABABAB* . . . If the third layer is placed in the alternate set of depressions in the second layer so that spheres in the first and third layers are *not* directly above and below each other, the cubic close-packed (fcc) structure, *ABCABCABC* . . . , results (Figure 9–26b). In closest packed structures each sphere has a coordination number of twelve, i.e., twelve nearest neighbors. In all (ideal) close-packed structures it is found that 76% of a given volume is due to spheres and 24% is empty space. The body-centered cubic structure is less efficient in packing; each sphere has only eight nearest neighbors, and there is more empty space.

The term coordination number is used in crystallography in a somewhat different sense than it is in coordination compounds (Section 27–3).

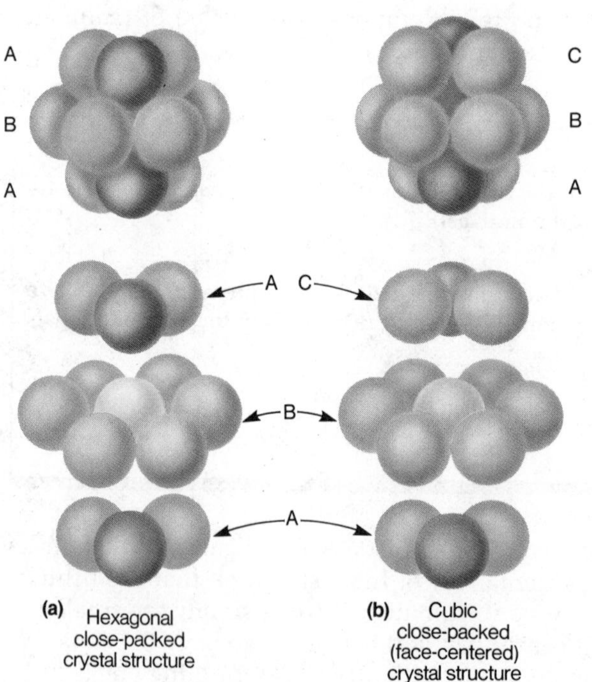

FIGURE 9–26 There are two crystal structures in which atoms are packed together as compactly as possible. The diagrams show the structures expanded to clarify the difference between them. In the hexagonal close-packed structure (left), the first and third layers are oriented in the same direction. In the cubic close-packed structure (right), the layers are oriented in opposite directions. In both cases, every atom is surrounded by twelve other atoms if the structure is extended infinitely, and so has a coordination number of twelve. Although it is not obvious from this figure, the cubic close-packed structure is also face-centered cubic. To see this we would have to include additional atoms and rotate the resulting cluster of atoms.

EXAMPLE 9–9

The unit cell of metallic silver is face-centered cubic (or cubic close-packed) with $a = b = c = 4.086$ Å. Calculate (a) the radius of a silver atom, (b) the volume of a silver atom in cm^3 given that for a sphere, $V = (\frac{4}{3})\pi r^3$, and (c) the percent of the volume of a unit cell that is occupied by silver atoms and the percent that is empty space.

Solution

(a) One face of the face-centered cubic (fcc) unit cell can be visualized as consisting of (parts of) five silver atoms describing two right triangles sharing a hypotenuse. (This can be visualized by picturing the Br^- ions in Figure 9–24 as Ag atoms and removing the Li^+ ions.) The hypotenuse, h, is four times the radius of the silver atom. By the Pythagorean theorem we can evaluate r_{Ag}

$$h = \sqrt{a^2 + a^2} = \sqrt{2\,a^2} = \sqrt{2(4.086\text{ Å})^2} = 5.778\text{ Å} = 4\,r_{Ag}$$

$$r_{Ag} = \frac{5.778\text{ Å}}{4} = \underline{1.444\text{ Å}}$$

(b) The volume of a silver atom is $(\frac{4}{3})\pi r^3{}_{Ag}$

$$V = (\tfrac{4}{3})\pi(1.444\text{ Å})^3 = 12.612\text{ Å}^3$$

We now convert Å^3 to cm^3: 1 Å = 10^{-8} cm

$$\underline{?}\text{ cm}^3 = 12.612\text{ Å}^3 \times \left(\frac{10^{-8}\text{ cm}}{\text{Å}}\right)^3 = \underline{12.612 \times 10^{-24}\text{ cm}^3}$$

(c) The entire fcc unit cell contains $8(\frac{1}{8}) + 6(\frac{1}{2}) = 4$ silver atoms. Each silver atom occupies a volume of 12.612×10^{-24} cm^3. Thus the volume of the unit cell that silver atoms occupy is four times this volume

$$V_{\text{Ag atoms}} = 4\text{ Ag atoms} \times \frac{12.612 \times 10^{-24}\text{ cm}^3}{\text{Ag atom}}$$

$$= 50.448 \times 10^{-24}\text{ cm}^3$$

Since the unit cell is cubic its entire volume is a^3

$$V_{\text{unit cell}} = (4.086\text{ Å})^3 = 68.217\text{ Å}^3$$

We now convert this to cm^3

$$\underline{?}\text{ cm}^3 = 68.217\text{ Å}^3 \times \left(\frac{10^{-8}\text{ cm}}{\text{Å}}\right)^3 = \underline{68.217 \times 10^{-24}\text{ cm}^3}$$

The percentage of the volume of the unit cell occupied by the silver atoms can now be calculated

$$\%\text{ Ag}_{\text{volume}} = \frac{V_{\text{Ag atoms}}}{V_{\text{unit cell}}} \times 100\%$$

$$= \frac{50.448 \times 10^{-24}\text{ cm}^3}{68.217 \times 10^{-24}\text{ cm}^3} \times 100\% = 74.0\%$$

$$\%\text{ empty space} = (100.0 - 74.0)\% = \underline{26.0\%}$$

This result agrees well with the observation made earlier that ideal close-packed structures are only about 24% empty space.

EXAMPLE 9–10

From data in Example 9–9, calculate the density of metallic silver. Use 6.022×10^{23} as Avogadro's number.

Solution

We first determine the mass of a unit cell, that is, the mass of four atoms of silver

$$\underline{?}\text{ g Ag} = 4\text{ Ag atoms} \times \frac{1\text{ mol Ag}}{6.022 \times 10^{23}\text{ Ag atoms}} \times \frac{107.87\text{ g Ag}}{1\text{ mol Ag}}$$
$$= 7.165 \times 10^{-22}\text{ g Ag}$$

The density of the unit cell, and therefore of silver, is its mass divided by its volume

$$\text{Density} = \frac{7.165\text{ g} \times 10^{-22}\text{ g}}{68.217 \times 10^{-24}\text{ cm}^3} = \underline{10.50\text{ g/cm}^3}$$

A handbook gives 10.5 g/cm^3 as the density of silver at 20°C.

EXAMPLE 9–11

Titanium crystallizes in a hexagonal close-packed structure with a density of 4.50 g/cm^3. Assume that the unit cell contains 24% empty space (it involves a close-packed structure) and calculate the volume and the radius of a titanium atom.

Solution

We first calculate the total volume occupied by one mole of titanium atoms from its density and atomic weight

$$V_{\text{molar}} = \frac{1\text{ cm}^3}{4.50\text{ g}} \times \frac{47.90\text{ g Ti}}{1\text{ mol Ti atoms}} = \frac{10.6\text{ cm}^3}{\text{mol Ti atoms}}$$

This calculation shows that 6.02×10^{23} titanium atoms occupy a volume of 10.6 cm^3. Since 24% of an ideal close-packed solid is empty space, 76% of it is occupied by titanium atoms. Now we can calculate the volume occupied by *only the atoms* in one mole of titanium atoms

$$\frac{\underline{?}\text{ cm}^3}{\text{mol Ti atoms}} = \frac{10.6\text{ cm}^3}{\text{mol Ti atoms}} \times 0.76 = \underline{8.1\text{ cm}^3/\text{mol Ti atoms}}$$

The volume of an individual titanium atom can now be obtained

$$V_{\text{atomic}} = \frac{8.1\text{ cm}^3}{\text{mol Ti}} \times \frac{1\text{ mol Ti atoms}}{6.02 \times 10^{23}\text{ atoms}} = \underline{1.3 \times 10^{-23}\text{ cm}^3/\text{atom}}$$

Since $V_{\text{atomic}} = (\frac{4}{3})\,\pi(r_{\text{atomic}})^3$ and V_{atomic} is known, we can calculate the atomic radius of titanium

$$r_{\text{atomic}} = \sqrt[3]{\frac{V_{\text{atomic}}}{(\frac{4}{3})\pi}} = \sqrt[3]{\frac{1.3 \times 10^{-23}\text{ cm}^3}{(\frac{4}{3})(3.14)}} = \underline{1.5 \times 10^{-8}\text{ cm or } 1.5\text{ Å}}$$

This agrees with the tabulated value of 1.46 Å for the atomic radius of titanium.

9–17 Metallic Bonding and Band Theory

As described in the previous section, the vast majority of metals crystallize in close-packed lattices. The close packing suggests strong attractive interactions among the 8 to 12 nearest neighbors. This is somewhat difficult to rationalize if we recall that each Group IA and Group IIA metal atom has only one or two valence electrons available for bonding, and this is too few to participate in bonds localized between it and each of its nearest neighbors.

Bonding in metals is called **metallic bonding,** and results from the electrical attractions among positively charged metal ions and mobile, delocalized electrons belonging to the crystal as a whole. The properties associated with metallic bonding, i.e., metallic luster, high thermal and electrical conductivity, and others, are well explained by the **band theory** of metals, which we shall now describe.

The interactions of two atomic orbitals, say the 3*s* orbitals of two sodium atoms, produces two molecular orbitals, one bonding orbital and one antibonding orbital (Chapter 6). If *N* atomic orbitals interact, *N* molecular orbitals are formed. In a single metallic crystal containing one mole of sodium atoms, for example, the interaction of 6.022×10^{23} 3*s* atomic orbitals produces 6.022×10^{23} molecular orbitals. Atoms interact more strongly with those nearby than with those farther away. As discussed in Chapter 6, the energy separating bonding and antibonding molecular orbitals resulting from two given atomic orbitals decreases as the interaction (overlap) between the atomic orbitals decreases. Thus, when all possible interactions among the mole of sodium atoms are considered, there results a series of very closely spaced molecular orbitals (formally σ_{3s} and $\sigma^{\star}_{3s}$) *that comprise a nearly continuous* **band** of orbitals belonging to the crystal as a whole. The 6.022×10^{23} orbitals in the band are half-filled since one mole of sodium atoms contributes 6.022×10^{23} valence electrons (see Figure 9–27).

The empty 3*p* atomic orbitals of the sodium atoms also interact to form a wide band of $3 \times 6.022 \times 10^{23}$ orbitals. Since the 3*s* and 3*p* atomic orbitals are quite close in energy, the two fanned-out bands of molecular orbitals actually overlap as shown in Figure 9–28. The two overlapping bands contain $4 \times 6.022 \times 10^{23}$ orbitals and only 6.022×10^{23} electrons. Since each orbital can accommodate two electrons, the resulting combination of bands is only one-eighth full.

The factor of 3 is due to the presence of three 3*p* orbitals per atom.

Each Na atom contributes four valence shell orbitals and one electron.

The ability of metallic sodium to conduct electricity is due to the ability of any of the highest energy electrons in the "3*s*" band to jump to a slightly

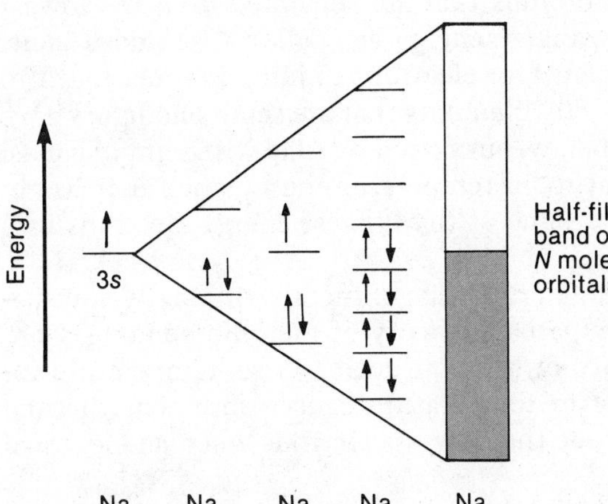

FIGURE 9–27 The band of orbitals resulting from interaction of the 3*s* orbitals in a crystal of sodium.

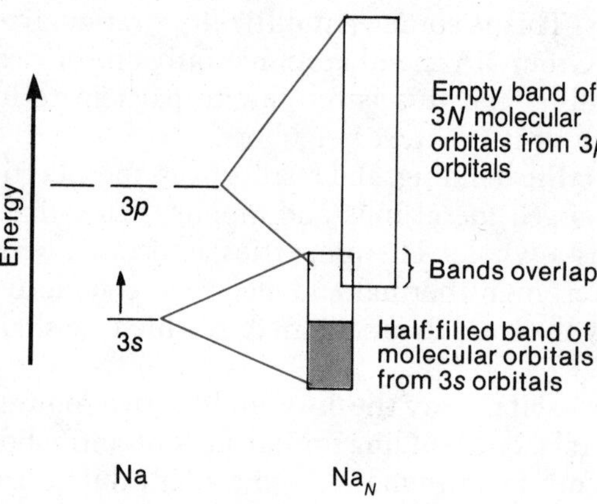

FIGURE 9–28 Overlapping of half-filled "3s" band with empty "3p" band of Na_N crystal.

higher energy vacant orbital in the same band when an electric field is applied. As this occurs, the net flow of electrons through the crystal is in the direction of the applied field.

The alkali metals are those of Group IA and the alkaline earth metals are those of Group IIA.

The fact that the "3s" and "3p" bands overlap is of no consequence in explaining the ability of sodium, or any other alkali metal, to conduct electricity since it could do so utilizing only the half filled "3s" band. However, such overlap is of great significance for the alkaline earth metals. Consider a crystal of magnesium as an example. The 3s atomic orbital of an isolated magnesium atom is filled with two electrons. Thus, in the absence of this overlap, the "3s" band in a crystal of magnesium also would be filled. Magnesium would not be able to conduct electricity at room temperature if the highest energy electrons were not able to move readily into vacant orbitals in the "3p" band (Figure 9–29).

According to band theory, the highest energy electrons of metallic crystals occupy either a partially filled band or a filled band that overlaps an empty band. A band within which or into which electrons must move to allow electrical conduction is called a **conduction band.** Crystalline nonmetals, like diamond and phosphorus, are **insulators** in that they do not conduct electricity (or heat or have metallic luster). This is because their highest energy electrons occupy filled bands of molecular orbitals that are separated from the lowest empty band (conduction band) by a large energy gap called a **forbidden zone,** which is too large an energy difference for electrons to jump in order to get to the conduction band (see Figure 9–30). Elements that are **semiconductors** have filled bands that are only slightly below, but do not overlap with, empty bands. They do not conduct electricity at room temperature but a small increase in temperature is sufficient to excite some of the highest energy electrons into the empty conduction band.

Let us now explain some of the properties of metals, originally summarized in Section 4–8, in terms of the band theory of metallic bonding. Conversely, the electrical conductivity of a metal decreases as temperature increases. Presumably, the increase in temperature causes sufficient thermal agitation of the metal ions to impede the flow of electrons when an electrical field is applied.

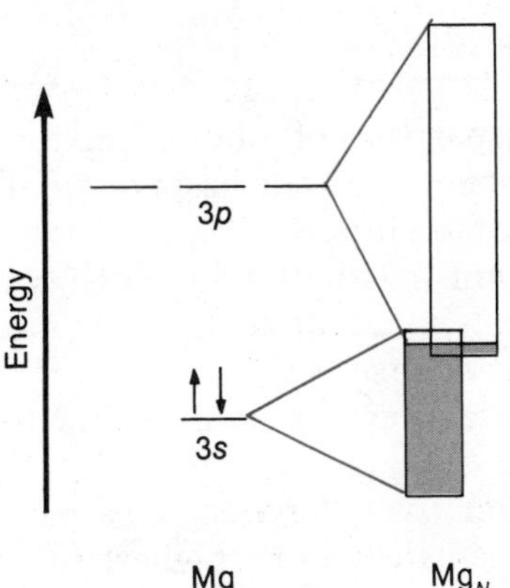

FIGURE 9–29 Overlapping of filled "3*s*" band with empty "3*p*" band of Mg_N crystal. The higher energy electrons are able to move into the "3*p*" band due to overlap.

1. We have just accounted for the *ability of metals to conduct electricity.*
2. Metals also are *good conductors of heat.* They can absorb heat as electrons become thermally excited to low-lying vacant orbitals in a conduction band. The reverse process accompanies the release of heat.
3. Metals have a *lusterous appearance* because the mobile electrons can absorb a wide range of wavelengths of radiant energy and then immediately emit photons of visible light as they first jump to higher energy levels and then fall back to lower levels within the conduction band.
4. Metals are *malleable and/or ductile.* A crystal of a metal is easily deformed when a mechanical stress is applied to it. Since all the metal ions are identical and since they are imbedded in a "sea of electrons," as bonds are broken new ones are readily formed with adjacent metal ions as soon as the deformation is complete. The features of the lattice remain unchanged and the environment of each metal ion is essentially the same as before the deformation occurred (Figure 9–31). The breakage of bonds involves the promotion of electrons to higher energy levels. The formation of bonds is accompanied by the return of the electrons to the original energy levels.

A malleable substance can be rolled or pounded into sheets; a ductile substance can be drawn into wires.

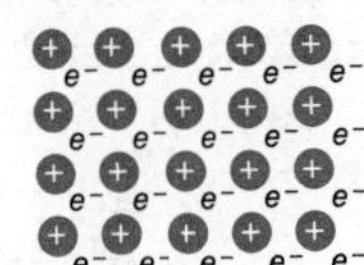

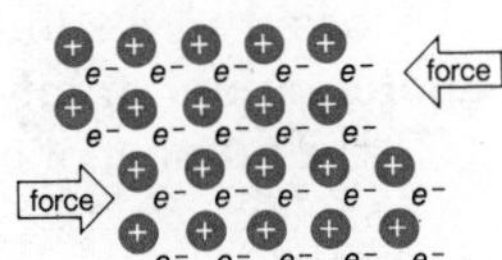

FIGURE 9-31 When a metal is distorted (e.g., rolled into sheets or drawn into wires), new metallic bonds are formed and the environment around each atom is essentially unchanged. This explains why sheet metal wires are strong.

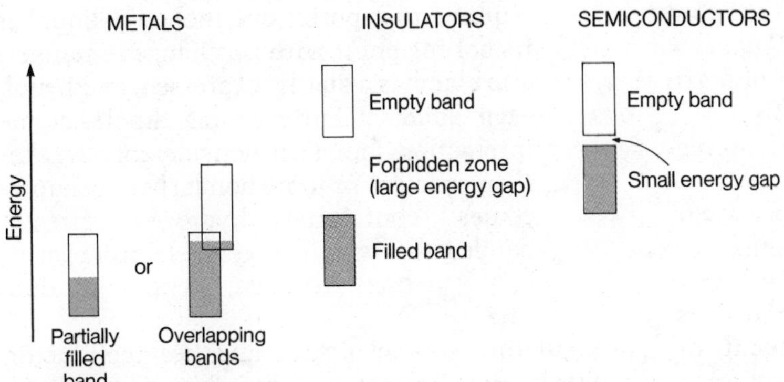

FIGURE 9–30 Distinction between metals, insulators, and semiconductors. In each case an unshaded area represents a conduction band.

Key Terms

Adhesive forces forces of attraction between a liquid and another surface.

Allotropes different forms of the same element in the same physical state.

Amorphous solid a noncrystalline solid with no well-defined ordered structure.

Band a series of very closely spaced, nearly continuous molecular orbitals that belong to the crystal as a whole.

Band theory of metals theory that accounts for the bonding and properties of metallic solids.

Boiling point the temperature at which the vapor pressure of a liquid is equal to the applied pressure; also the condensation point.

Capillary action the drawing of a liquid up the inside of a small bore tube when adhesive forces exceed cohesive forces, or the depression of the surface of the liquid when cohesive forces exceed adhesive forces.

Cohesive forces all the forces of attraction among particles of a liquid.

Condensation liquefaction of vapor.

Condensed phases the liquid and solid phases; phases in which particles interact strongly.

Conduction band a partially filled band or a band of vacant energy levels just higher in energy than a filled band; a band within which, or into which, electrons must be promoted to allow electrical conduction to occur in a solid.

Coordination number in describing crystals, the number of nearest neighbors of an atom or ion.

Crystal lattice pattern of arrangement of particles in a crystal.

Crystalline solid a solid characterized by a regular, ordered arrangement of particles.

Deposition the direct solidification of a vapor by cooling; the reverse of sublimation.

Dipole–dipole interactions attractive interactions between polar molecules, that is, between molecules with permanent dipoles.

Dipole-induced dipole interactions see *London forces*

Dispersion forces see *London forces*

Distillation the separation of a liquid mixture into its components on the basis of differences in boiling points.

Dynamic equilibrium a situation in which two (or more) processes occur at the same rate so that no net change occurs.

Evaporation vaporization of a liquid below its boiling point.

Forbidden zone a relatively large energy separation between an insulator's highest filled electron energy band and the next higher energy vacant band.

Heat of condensation the amount of heat that must be removed from one gram of a vapor at its condensation point to condense the vapor with no change in temperature.

Heat of crystallization the amount of heat that must be removed from one gram of a liquid at its freezing point to freeze it with no change in temperature.

Heat of fusion the amount of heat required to melt one gram of a solid at its melting point with no change in temperature. Usually expressed in J/g. The *molar heat of fusion* is the amount of heat required to melt one mole of a solid at its melting point with no change in temperature and is usually expressed in kJ/mol.

Heat of vaporization the amount of heat required to vaporize one gram of a liquid at its boiling point with no change in temperature. Usually expressed in J/g. The *molar heat of vaporization* is the amount of heat required to vaporize one mole of a liquid at its boiling point with no change in temperature and is usually expressed in kJ/mol.

Hydrogen bond a fairly strong dipole–dipole interaction (but still considerably weaker than covalent or ionic bonds) between molecules containing hydrogen directly bonded to a small, highly electronegative atom such as nitrogen, oxygen, or fluorine.

Insulator poor electrical and heat conductor.

Intermolecular forces forces between individual particles (atoms, molecules, ions) of a substance.

Intramolecular forces bonds between atoms (or

ions) within molecules (or formula units).

Isomorphous refers to crystals having the same atomic arrangement.

LeChatelier's Principle states that a system at equilibrium, or striving to attain equilibrium, responds in such a way as to "counteract" any stress placed upon it.

London forces very weak and very short-range attractive forces between short-lived temporary (induced) dipoles; also called dispersion forces.

Melting point the temperature at which liquid and solid coexist in equilibrium; also the freezing point.

Meniscus the shape assumed by the surface of a liquid in a cylindrical container.

Metallic bonding bonding within metals due to the electrical attraction of positively charged metal ions for mobile electrons that belong to the crystal as a whole.

Normal boiling point the temperature at which the vapor pressure of a liquid is equal to one atmosphere pressure.

Normal melting point the melting (freezing) point at one atmosphere pressure.

Phase diagram diagram that shows equilibrium temperature-pressure relationships for different phases of a substance.

Polymorphous refers to solids that can crystallize in more than one type of lattice.

Semiconductor a substance that does not conduct electricity well at room temperature but does at higher temperatures.

Sublimation the direct vaporization of a solid by heating without passing through the liquid state.

Supercooled liquids liquids which, when cooled, apparently solidify but actually continue to flow very slowly under the influence of gravity.

Surface tension a measure of the inward intermolecular forces of attraction among liquid particles that must be overcome in order to expand the surface area.

Triple point point on a phase diagram for a substance corresponding to the only pressure and temperature at which solid, liquid and gas can coexist at equilibrium.

Unit cell smallest repeating unit showing all the structural characteristics of a crystal.

Vapor pressure the partial pressure of a vapor at the surface of its parent liquid.

Viscosity the tendency of a liquid to resist flow; a measure of its fluidity.

Volatility refers to the ease with which a liquid vaporizes.

Exercises

General Concepts

1. List several characteristics that distinguish solids from liquids and liquids from gases.
2. Why are the properties of solids and liquids more difficult to describe in general mathematical relationships than those of gases?
3. Arrange the following in order of increasing strength of interaction: hydrogen bonds, covalent bonds, London forces, permanent dipole–dipole forces.
4. Which of the following substances have permanent dipole–permanent dipole forces? (a) HCl (b) Cl_2 (c) BrF (d) BrF_5 (e) SF_6 (f) CO_2 (g) CO (h) Kr (i) CCl_4 (j) $CHBr_3$
5. For which of the substances in Exercise 4 are London forces the only important ones in determining boiling points? How do London forces operate?
6. Which of the following substances exhibit strong hydrogen bonding in the liquid and solid states? (a) CH_3OH, methyl alcohol (b) PH_3 (c) CH_4 (d) CH_2Cl_2 (e) H_2S (f) NH_3 (g) SiH_4 (h) HF (i) HCl (j) CH_3NH_2

The Liquid State

7. Describe the behavior of liquids with changing temperature in terms of kinetic-molecular theory.
8. Why are liquids more dense than gases?
9. We don't see "holes" in liquids even though there are spaces between particles or clusters of particles. Why? Atoms and ions are mostly emtpy space. We don't see or feel spaces in solids. Why?
10. Give two examples of highly viscous liquids and two examples of only slightly viscous

liquids. What is responsible for high viscosity?

11. Distinguish between cohesion and adhesion. How are they related to the shape of the meniscus of a liquid in a cylindrical container?
12. Explain how plants are able to obtain nutrients from fertilizers applied as solids.
13. Distinguish between evaporation and boiling. Explain the dependence of rate of evaporation on temperature in terms of kinetic-molecular theory.
14. Support or criticize the statement that liquids with high normal boiling points have low vapor pressures. Give examples of three liquids that have relatively high vapor pressures at 25°C and three that have low vapor pressures at 25°C.
15. The normal boiling point of H_2 is −259°C and that of F_2 is −188°C. Why does H_2 have the lower boiling point?
16. Using Table 9–2, draw a graph of vapor pressure vs. temperature for the four listed liquids. From your graph estimate the normal boiling point of each liquid. Give an explanation for the fact that the normal boiling point of water is higher than for the other three liquids.
17. Consider a liquid in a beaker being heated with a Bunsen burner. Bubbles begin to form as it begins to boil. Where do they form and why? What are the bubbles?
18. Explain the effectiveness of a pressure cooker in cooking.
19. Explain how liquids can be separated from each other by distillation. Is the separation more effective if boiling points of liquids are close together or far apart? Why should the temperature of the vapor near the entrance to the condenser arm be monitored during distillation in an apparatus like that in Figure 9–9?
20. Arrange each compound in the following lists in order of increasing magnitude of heat of vaporization. (You may consult a handbook.) (a) H_2S, H_2O, H_2Te, H_2Se (b) HBr, HF, HI, HCl (c) PH_3, AsH_3, NH_3, SbH_3, BiH_3
21. Arrange the compounds in Exercise 20 in order of increasing boiling points and explain the orders. (You may consult a handbook.)
22. Arrange the following substances in order of increasing (a) melting and (b) boiling points (You may consult a handbook.) $CH_3CH_2CH_3$, Na_2S, CH_3CH_3, and H_2O. Why is this order observed?

Changes of State and Heat Transfer

23. Explain why perspiration has a cooling effect on the skin.
24. Write the equation for a reaction for which the amount of heat released or absorbed by one mole of starting material (enthalpy change) is equal to
    (a) the heat of sublimation of carbon dioxide
    (b) the heat of fusion of ice
    (c) the heat of vaporization of pentane, C_5H_{12}
    (d) the heat of condensation of sulfur, S_8

The following values will be useful in working the problems below.

Specific heat of ice	2.09 J/g·°C
Heat of fusion of ice at 0°C	333 J/g
Specific heat for liquid H_2O	4.18 J/g·°C
Heat of vaporization of liquid H_2O at 100°C	2.26×10^3 J/g
Specific heat for steam	2.03 J/g·°C

25. Calculate the amount of heat required to raise the temperature of 10.0 grams of water from 10.0°C to 50.0°C.
26. If 100 grams of liquid water at 100.0°C and 200 grams of water at 20.0°C are mixed in an insulated container, what is the final temperature?
27. Calculate the amount of heat required to convert 10.0 grams of ice at 0.0°C to liquid water at 50.0°C.
28. Calculate the amount of heat required to convert 10.0 grams of ice at 0.0°C to liquid water at 100.0°C.
29. Calculate the amount of heat required to convert 10.0 grams of ice at −20.0°C to steam at 120.0°C.
30. Calculate the amount of heat given up when 10.0 grams of steam at 100.0°C is condensed and cooled to 20.0°C.
31. If 10.0 grams of ice at −10.0°C and 10.0

grams of liquid water at 100°C are mixed in an insulated container, what will the final temperature be?

32. If 105 grams of liquid water at 0.0°C and 10.5 grams of steam at 110.0°C are mixed in an insulated container, what will the final temperature be?
33. If 10.0 grams of steam at 110.0°C are bubbled slowly into 50.0 grams of liquid water at 0.0°C in an insulated container, will all of the steam be condensed?
34. If 100 g of iron at 100.0°C is placed in 200 g of water at 20.0°C in an insulated container, what will the temperature of the iron and water be when both are the same? The specific heat of iron is 0.444 J/g·°C. Compare this result with the one obtained in Exercise 26 when 100 g of water at 100.0°C was mixed with 200 g of water at 20.0°C.
35. (a) Calculate the amount of heat required to raise the temperature of 100 g of mercury from 20.0°C to 90.0°C. The specific heat of mercury is 0.138 J/g·°C.
    (b) Calculate the amount of heat required to raise the temperature of 100 g of water from 20.0°C to 90.0°C.
    (c) What is the ratio of the specific heat of liquid water to that of liquid mercury.
    (d) What does the answer to (c) tell us?

Phase Diagrams

36. How can a substance be purified by sublimation?

Refer to the phase diagram of CO_2 in Figure 9–16b to answer Exercises 37–40.

37. What phase of CO_2 exists at 2 atm pressure and a temperature of −78°C? −50°C? 0°C?
38. What phases of CO_2 are present at (a) a temperature of −78°C and a pressure of 1 atm? (b) at −57°C and a pressure of 5.2 atm?
39. List the phases that would be observed if a sample of CO_2 at 6 atm pressure were heated from −100°C to 0°C.
40. How does the melting point of CO_2 change with pressure? What does this indicate about the relative density of solid CO_2 versus liquid CO_2?

The Solid State

41. Comment on the statement: "The only perfectly ordered state of matter is the crystalline state."
42. Ice floats in water. Why? Would you expect solid mercury to float in liquid mercury at its freezing point? Explain why or why not.
43. How do amorphous and crystalline solids differ? Give three examples of each.
44. We normally think of glasses as being solid. What is the rationale for describing glasses as supercooled liquids?
45. Since even the largest atoms are so small that they cannot be seen even under very powerful microscopes, how are the relative arrangements of atoms in solids determined?
46. Sodium chloride, NaCl, and magnesium sulfide, MgS, are isomorphous. Which has the higher melting point? Why?
47. Covalent bonding occurs in both molecular solids and covalent solids, yet molecular solids are relatively soft and have low melting points, while covalent solids are hard and have high melting points. Why is this so?
48. The molecular weights of compound A and compound B differ only slightly. The molar heat of sublimation of A is much larger than that of B. What does this tell us about the strengths of intermolecular interactions in solid A and solid B? Which would you expect to have the higher vapor pressure at room temperature? Why?
49. Classify each of the following into one of the four categories of solids: molecular, ionic, covalent (network), or metallic.

	Melting Point (°C)	Boiling Point (°C)	Electrical Conductivity	
			Solid	Liquid
Ar	−189.3	−185.6	no	no
Ag	960.8	1950	yes	yes
Sc	1541	2831	yes	yes
Ge	937	2830	poor	poor
$MgCl_2$	708	1412	no	yes
PBr_3	−40	173	no	no
OF_2	−223.8	−144.8	no	no

50. Determine the number of sodium ions and chloride ions that belong to the unit cell shown in Figure 9–21.
51. Determine the number of atoms per unit cell for the first three cells shown in Figure 9–20.
52. Determine the number of ions present in each unit cell shown in Figure 9–23.
53. Illustrate and describe the significance of the following equation to the determination of the structure of crystals

$$\sin \theta = \frac{n\lambda}{2d}$$

54. What is a unit cell? Sketch a unit cell and label the dimensions a, b, c, and angles α, β, and γ. How do cubic, tetragonal, and hexagonal crystal systems differ?
55. Distinguish among and sketch simple cubic, body-centered cubic, and face-centered cubic lattices. Use CsCl, sodium, and nickel as examples of solids existing in simple cubic, bcc, and fcc lattices, respectively.
56. What are the two types of closest-packed crystal structures? How do they differ from other types of structures?
57. Sketch the arrangement of atoms in the crystal structures of cadmium and copper, which have cubic close-packed and hexagonal close-packed structures, respectively. Represent the atoms as spheres.
58. Distinguish among and compare the characteristics of molecular, covalent, ionic, and metallic solids. Give two examples of each kind of solid.
59. Distinguish between isomorphous and polymorphous solids.

Unit Cell Data and Atomic and Ionic Sizes

60. Metallic rhodium crystallizes in a face-centered cubic lattice with a unit cell edge length of 3.803 Å. Calculate the molar volume of rhodium, i.e., the volume of one mole of rhodium (including the empty spaces).
61. Zinc selenide, ZnSe, crystallizes in a face-centered cubic unit cell and has a density of 5.267 g/cm^3. Determine the edge length of the unit cell.
62. The atomic radius of palladium is 1.375 Å. The unit cell of palladium is a face-centered cube. Calculate the density of palladium.
63. A certain metal is found to have a specific gravity of 10.200 at 25°C. It is found to crystallize in a body-centered cubic lattice with a unit cell edge length of 3.147 Å. Determine the atomic weight and identify the metal.
64. A certain metal crystallizes in the hexagonal closest-packed structure and has a density of 1.737 g/cm^3. Its atomic radius is 1.60 Å. There are four atoms per unit cell. Determine its atomic weight. What is the metal?
65. Nickel crystallizes in a face-centered cubic unit cell with an edge length of 3.524 Å. Its density is 8.902 g/cm^3 and its atomic weight is 58.70. From these data calculate Avogadro's number.
66. Barium crystallizes in a body-centered cubic unit cell with an edge length of 5.025 Å. Calculate the fraction of empty space in the body-centered cubic lattice. How does this compare with the fraction of empty space in close-packed lattices? (Hint: the atom in the center touches the atoms at the corners of the cube. Atoms along edges of the cube do not touch.)
67. Zirconium crystallizes in a hexagonal closest-packed lattice. Its density is 6.51 g/cm^3, and its atomic radius is 1.57 Å. Estimate the molar volume of zirconium and the volume occupied by a mole of zirconium atoms (no empty space).
68. The unit cell of nickel is a face-centered cube having a volume of 43.756 Å. The atom at the center of each face just touches the atoms at the corners. Determine the atomic radius and atomic volume of nickel.
69. The spacing between successive planes of platinum atoms comprising the parallel faces of face-centered unit cells is 2.256 Å. When x-radiation emitted by molybdenum strikes a crystal of platinum metal, the minimum diffraction angle of x-rays is 9.045°. What is the wavelength of the Mo radiation?
70. Gold crystallizes in an fcc lattice. When x-radiation of 1.5404 Å wavelength from copper is used for structure determination of

metallic gold, the minimum diffraction angle of x-rays by the gold is 19.14°. Calculate the spacing between parallel layers of gold atoms.

Metallic Bonding

71. Compare the temperature dependence of electrical conductivity of a metal with that of a typical metalloid. Explain the difference.

72. In general, metallic solids are ductile and malleable, whereas ionic salts are brittle and shatter readily (although they are hard). Refer to Figures 9–21 and 9–31 to explain this observation.

73. What single factor accounts for the ability of metals to conduct both heat and electricity in the solid state? Why are ionic solids poor conductors of heat and electricity even though they are composed of charged particles?

10 Solutions

Solutions are common in nature and are extremely important in all life processes and in all scientific areas. The body fluids of all forms of life are solutions, and variations in their concentrations accompany many illnesses. Analysis of body fluids, especially blood and urine, give physicians valuable clues about one's health.

A solution is defined as a *homogeneous mixture* of pure substances in which no settling occurs. True solutions consist of a solvent and one or more solutes whose proportions vary from solution to solution. By contrast, pure substances have fixed composition. As noted in Chapter 2, the *solvent* is the medium in which the *solutes* are dissolved. The fundamental units of solutes are usually ions or molecules.

Solutions may involve many different combinations in which a solid, liquid, or gas acts as either solvent or solute. The most commonly encountered kinds are those in which the solvent is a liquid. For instance, seawater is an aqueous solution of many salts and some gases (such as carbon dioxide and oxygen). Carbonated water is a saturated solution of carbon dioxide in water. However, examples of solutions in which the solvent is not a liquid also are numerous and quite common. Air is a solution of gases (with variable composition). Dental fillings are solid amalgams, or solutions of liquid mercury dissolved in solid metals. Alloys are solid solutions of one or more solid metals dissolved in another. Thus 14-karat yellow gold is a solid solution of copper in gold, and white gold is a solution of nickel, palladium, or zinc in gold. Sterling silver is a solution of 7.5% copper and 92.5% silver.

It is usually obvious which of the components of a solution is the solvent and which is/are the solute(s): the solvent is usually the most abundant species present. When sugar is dissolved in water, the sugar is clearly the solute and water is the solvent. Likewise, in a cup of instant coffee, the coffee and any added sugar are considered solutes, and the hot water is the solvent. If 10 grams of alcohol are mixed with 90 grams of water, the alcohol is the solute. If 10 grams of water are mixed with 90 grams of alcohol, the water is the solute. But what is the solute and what is the solvent in a solution that contains 50 grams of water and 50 grams of alcohol? In such cases, the decision is arbitrary and, in fact, unimportant.

The Dissolution Process

10–1 Mixing and Spontaneity of the Dissolution Process

In Section 7–1 we covered the solubility rules that apply to common solutes in water. Now we shall investigate the major factors that influence solubility in general. A substance may dissolve with or without reaction with the solvent. Metallic sodium "dissolves" in water with the evolution of bubbles of hydrogen. A chemical change occurs in which hydrogen and soluble, ionic sodium hydroxide, NaOH, are produced. The total ionic equation is

$$2Na\ (s) + 2H_2O \longrightarrow 2[Na^+\ (aq) + OH^-\ (aq)] + H_2\ (g)$$

Solid sodium chloride, NaCl, on the other hand, dissolves in water with no evidence of chemical reaction

$$NaCl\ (s) \xrightarrow{H_2O} Na^+\ (aq) + Cl^-\ (aq)$$

If the first solution is evaporated to dryness, solid sodium hydroxide, NaOH, is obtained rather than metallic sodium. This, along with the production of bubbles of hydrogen, is evidence for a reaction with the solvent. But evaporation of

Strictly speaking, even solutes that don't "react" with the solvent undergo solvation, a kind of reaction in which molecules of solvent are attached in oriented clusters to the solute particles.

the sodium chloride solution yields the original NaCl. We shall consider only dissolving of the latter type, in which no irreversible reaction occurs between components.

The ease of the dissolution process depends upon two factors: (1) the change in energy (exothermicity or endothermicity), and (2) the change in disorder (called entropy change) accompanying the process. In Chapter 13, we shall study both of these factors in detail for many kinds of physical and chemical changes. For now, suffice it to say that the spontaneity of a process is *favored* by (1) a decrease in the energy of the system, which corresponds to an *exothermic process,* and by (2) an *increase in the disorder,* or randomness, of the system.

Let us concentrate first on the factors that determine the change in heat content, factor (1) above. This change is called the **heat of solution, $\Delta H_{solution}$.** In a pure liquid to be used as a solvent, the intermolecular forces are all between like molecules; when the liquid and a solute are mixed, each molecule then experiences forces from molecules (or ions) unlike it as well as from like molecules. The relative strengths of such interactions help to determine the extent of solubility of the solute in the solvent. The interactions that must be considered to assess the heat of solution for the dissolution of a specific solute in a specific solvent are

1. solute–solute interactions
2. solvent–solvent interactions
3. solvent–solute interactions

Dissolution is favored when the first two of these interactions are relatively small and the third is relatively large, as summarized in Figure 10–1. The intermolecular or interionic attractions among solute particles in the pure solute must be overcome (step a) to dissolve the solute. This part of the process requires an *input* of energy. Likewise, separating the solvent molecules from each other (step b) to "make room" for the solute particles also requires the *input* of energy. However, energy is *released* as the solute particles and solvent molecules interact in the solution (step c). Thus, the dissolving process is endothermic (and disfavored with respect to heat of solution) if the amount of heat absorbed in (hypothetical) steps a and b is greater than the amount of heat released in step c. It is exothermic (and favored) if the amount of heat absorbed in steps a and b is less than the amount of heat released in step c.

Many solids dissolve in liquids by endothermic processes. The reason such solids are soluble in liquids is that the endothermicity is outweighed by a great increase in disorder of the solute accompanying the dissolving process. The solute particles are very highly ordered in a solid crystal, but are free to move about randomly in liquid solutions. Likewise, the solvent particles increase in their degree of disorder as the solution is formed, since they then are in a more random environment; they are surrounded by a mixture of solvent and solute particles (Figure 10–1).

All dissolution processes are accompanied by an increase in the disorder of both solvent and solute. Thus, this disorder factor is *always* favorable to solubility. The determining factor is whether the heat of solution also favors dissolution or, if it does not, whether it is small enough to be outweighed by the favorable effects of the increasing disorder. In gases, for instance, the mole-

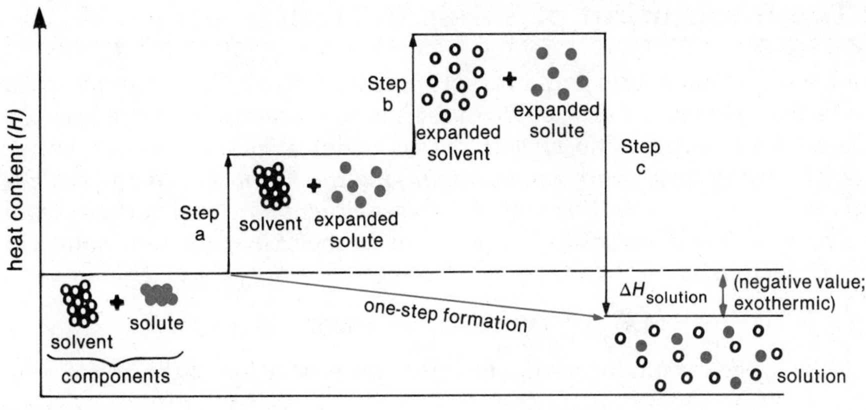

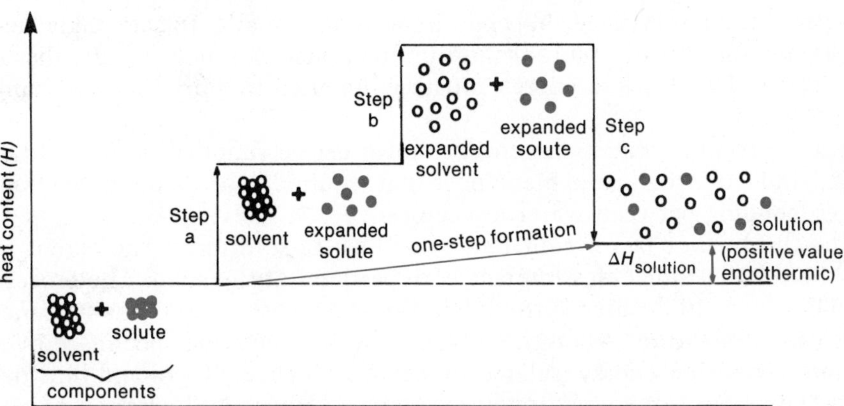

FIGURE 10–1 Diagram representing the changes in heat content associated with the hypothetical three-step sequence in a dissolution process. The figure is presented as for a solid solute dissolving in a liquid solvent. Similar considerations would apply to other combinations. (a) an exothermic process: Heat absorbed in steps a and b is *less* than heat released in step c, so heat of solution is favorable. (b) an endothermic process: Heat absorbed in steps a and b is *greater* than heat released in step c, so heat of solution is unfavorable.

cules are so far apart that intermolecular forces are quite weak. Thus, when gases are mixed, changes in the intermolecular forces are very slight, so the very favorable increase in disorder that accompanies mixing always is more important than possible changes in intermolecular attractions (energy). So, gases can always be mixed with each other in any proportion. (This statement does not apply to gases that react with each other chemically. In such cases, the gases are converted to other substances.)

In the next several sections, we shall consider in more detail the types of solutions encountered most often—those in which the solvent is a liquid.

10–2 The Dissolution of Solids in Liquids

The ease with which a solid goes into solution depends to a large extent on the crystal lattice energy, or the strength of interactions among the particles holding the lattice together. The **crystal lattice energy** is defined *as the energy change (change in heat content) accompanying the formation of one mole of formula units in the crystalline state from constituent particles in the gaseous state,* a process that is invariably exothermic. In the case of an ionic solid, the process is written as

$$M^+ \text{ (g)} + X^- \text{ (g)} \longrightarrow MX \text{ (s)} + \text{crystal lattice energy}$$

The reverse of the crystal formation reaction, the separation of the crystal into ions

$$MX \text{ (s)} + \text{crystal lattice energy} \longrightarrow M^+ \text{ (g)} + X^- \text{ (g)}$$

can be considered as the hypothetical first step (a in Figures 10–1a and b) in forming a solution. It is invariably endothermic. The smaller the crystal lattice energy (a measure of the solute–solute interactions), the more readily dissolution occurs. That is, less energy must be supplied to start the dissolving process.

If the solvent is water, the energy that must be supplied to expand the solvent (b in Figure 10–1a and b) includes that required to break up some of the hydrogen bonding between water molecules.

The third major factor contributing to heat of solution is the extent to which the solvent molecules interact with particles of the solid. The process by which solvent molecules surround and interact with solute ions or molecules is called **solvation.** When the solvent is water, the more specific term is **hydration. Hydration energy** (related to step c in Figures 10–1a and b) is defined as the amount of energy involved in the (exothermic) hydration of one mole of gaseous ions

$$M^{n+} \text{ (g)} + xH_2O \longrightarrow M(OH_2)_x{}^{n+} + \text{hydration energy of cation}$$

$$X^{y-} \text{ (g)} + rH_2O \longrightarrow X(H_2O)_r{}^{y-} + \text{hydration energy of anion}$$

Hydration is generally quite highly exothermic for ionic or polar covalent compounds, since the polar water molecules interact very strongly with ions and polar molecules (as we saw in Section 9–9). In fact, the only solutes that are appreciably soluble in water either undergo dissociation (ionization) or are able to hydrogen bond with water. Nonpolar solids, like naphthalene, $C_{10}H_8$, do not dissolve appreciably in polar solvents like water because the two substances do not attract each other significantly. This is true despite the fact that crystal lattice energies of solids consisting of nonpolar molecules are much smaller than those of ionic solids. However, naphthalene dissolves readily in nonpolar solvents like benzene because there are no strong attractive forces between solute molecules or between solvent molecules. These facts provide the basis for explaining the observation that "like dissolves like."

Let us consider what happens when a cube of sodium chloride, a typical ionic solid, is placed in water. Any Cl^- ions located at the corners of the cube (Figure 10–2) are held less strongly by the lattice than any other Cl^- ions because they have only three nearest neighbor Na^+ ions. They go into solution

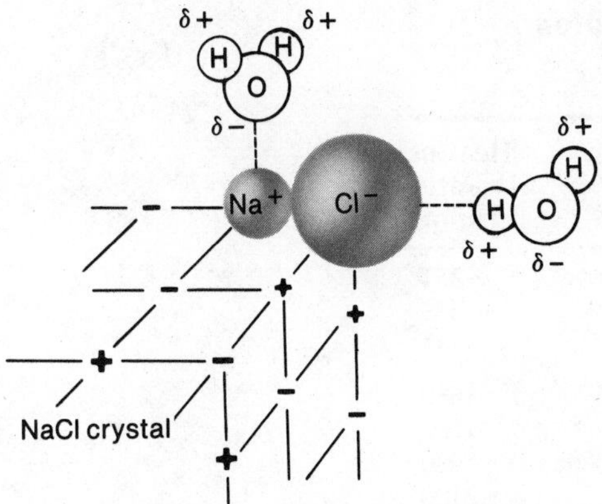

FIGURE 10–2 Electrostatic attraction in the dissolution of NaCl in water.

most readily. Ions on the edges are attached to four nearest neighbors of opposite charge. Those on faces are held by electrostatic attraction to five such nearest neighbors, while those within the cube have six nearest neighbors of opposite charge.

The δ+ ends of water molecules associate with the negative chloride ions, as shown in Figure 10–2. As chloride ions at the corners go into solution, sodium ions, now surrounded by only three nearest neighbor chloride ions, are exposed to water molecules whose δ− ends orient themselves toward the sodium ions and solvate them. This is also shown in Figure 10–3.

When we write Na^+ (aq) and Cl^- (aq) we refer to hydrated ions. The number of water molecules directly attached to an ion is different for different ions. Sodium ions are thought to be hexahydrated; that is, Na^+ (aq) probably represents $Na(OH_2)_6{}^+$. Most cations in aqueous solution are surrounded by four to nine water molecules, with six being the most common number. Generally, larger cations can accommodate more water molecules around them than smaller cations.

For simplicity we frequently omit the (aq) designations from dissolved ions. But we must remember that all ions in aqueous solution are hydrated, whether this is indicated or not.

Many solids that are appreciably soluble in water are ionic in nature. Crystal lattice energies generally increase with increasing charge and decreas-

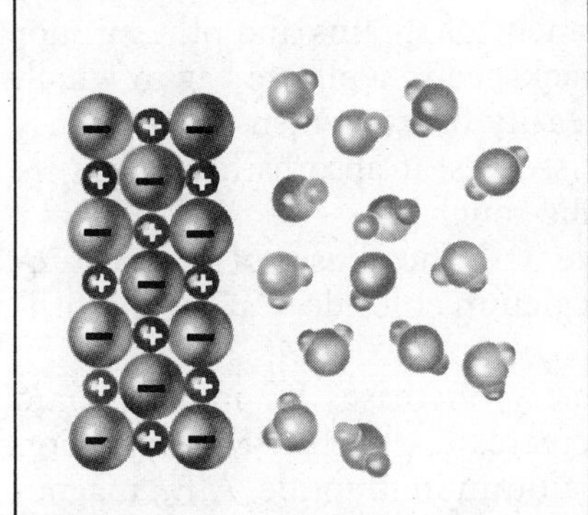

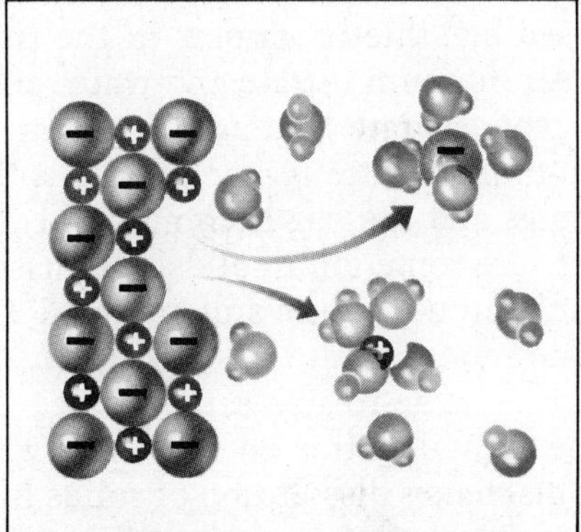

FIGURE 10–3 A two-dimensional representation of the dissolving of an ionic solid, such as NaCl, in water. The hydrated ions that form as the solid dissolves are shown on the right.

TABLE 10–1 Ionic Radii, Charge/Radius Ratios, and Heats of Hydration for Some Cations

Ion	Ionic Radius, Å	Charge/Radius Ratio*	Heat of Hydration (kJ/mol)†
K^+	1.33	0.75	−351
Na^+	0.95	1.05	−435
Li^+	0.60	1.67	−544
Ca^{2+}	0.99	2.02	−1650
Mn^{2+}	0.78	2.56	−1900
Fe^{2+}	0.76	2.63	−1980
Zn^{2+}	0.74	2.70	−2100
Co^{2+}	0.74	2.70	−2110
Cu^{2+}	0.72	2.78	−2160
Ni^{2+}	0.72	2.78	−2170
Fe^{3+}	0.64	4.69	−4340
Cr^{3+}	0.62	4.84	−4370
Al^{3+}	0.52	5.77	−4750

* The charge/radius ratio is the ionic charge divided by the ionic radius in Ångstroms. The cations are listed in order of increasing charge/radius ratio, which is a measure of the *charge density* around the ion.

† The negative values indicate that heat is *released* during hydration.

ing size of ions. That is, lattice energies increase as the ionic charge densities increase, and therefore, as electrostatic attractions within the lattice increase. Hydration energies also increase in the same order (Table 10–1). As we indicated earlier, crystal lattice energies and hydration energies are generally much smaller for molecular solids than for ionic solids.

The effects of lattice energies and hydration energies oppose each other in the dissolution process, but since lattice energies and hydration energies are usually of approximately the same magnitude for low-charge species, they tend to cancel each other. As a result, the dissolution process is slightly endothermic for many ionic substances. Ammonium nitrate, NH_4NO_3, is typical of salts that dissolve endothermically. This property is utilized in the "instant ice packs" used by athletic trainers in the treatment of sprains and other minor injuries. Ammonium nitrate and water are packaged in a plastic bag in which they are kept separate by a partition that is easily broken when squeezed. As the ammonium nitrate reaches water and dissolves, it absorbs heat from its surroundings and the bag becomes cold to the touch.

There are some ionic solids that dissolve with the release of heat. Examples are anhydrous sodium sulfate, Na_2SO_4, calcium chloride, $CaCl_2$, and lithium sulfate monohydrate, $Li_2SO_4 \cdot H_2O$.

As the charge-to-size ratio (charge density) increases for ions in ionic solids, the crystal lattice energy usually increases faster than the hydration energy. This makes dissolution of solids like aluminum fluoride, AlF_3, magnesium oxide, MgO, and chromium(III) oxide, Cr_2O_3, very endothermic. As we have already noted, high endothermicity usually results in very *low* solubility. Chromium(III) oxide is quite insoluble in water.

10–3 The Dissolution of Liquids in Liquids (Miscibility)

Miscibility refers to the ability of one liquid to mix with (dissolve in) another. The three kinds of attractive interactions listed above (solvent–solvent, solute–solute, and solute–solvent) must be considered for liquid–liquid solutions just as they were for solid–liquid solutions. Since solute–solute attractions are usually much lower for liquid solutes than for solids (as measured by crystal lattice energy), this factor is less important, so that the liquid–liquid mixing process is often exothermic for miscible liquids.

As we have seen in Section 9–9, polar liquids tend to interact strongly with and dissolve readily in polar solvents. Methanol, CH_3OH, ethanol, CH_3CH_2OH, acetonitrile, CH_3CN, and sulfuric acid, H_2SO_4, are all polar liquids that are soluble in most polar solvents such as water. The hydrogen bonding between methanol and water molecules, and the dipolar interaction between acetonitrile and water molecules, are both depicted in Figure 10–4.

In the case of aqueous sulfuric acid, H_2SO_4, the sulfate ion is able to undergo strong hydrogen bonding with the water molecules. X-ray determinations of crystal structures of many solid hydrated sulfate compounds like copper(II) sulfate pentahydrate, $CuSO_4 \cdot 5H_2O$, reveal that the sulfate anion interacts very strongly with water molecules. This is also presumed to be the case in aqueous solutions containing sulfate ions, especially in concentrated solutions. The 109°28′ O—S—O bond angles in the tetrahedral sulfate ion are the right size for strong attractive interactions with both hydrogens of the $\delta+$ end of a water molecule, which has a bond angle of about 105° (Figure 10–5).

This compound is more appropriately formulated as $[Cu(OH_2)_4]SO_4 \cdot H_2O$ and named tetraaquacopper(II) sulfate hydrate (see Section 27–4).

Because this interaction is so strong, large amounts of heat are released when concentrated sulfuric acid is diluted with water, often enough to cause the solution to boil and spatter. For this reason, *sulfuric acid* (as well as all other mineral acids) *is always diluted by adding the acid slowly and carefully to water. Water is never added to the acid.* If spattering does occur when the acid is added to water, it is mainly water that spatters, not the corrosive concentrated acid.

The potential danger when water is added to concentrated acid is due more to the spattering of the acid itself than to the steam from boiling water.

The interaction between sulfate ion and water is also enhanced by the presence of the 2− charge on the ion. The perchlorate ion, ClO_4^-, is also tetrahedral and interacts with water molecules. Dissolution and dilution of perchloric acid, $HClO_4$, in water is exothermic, but less so than for H_2SO_4; the ClO_4^- has only a 1− charge and does not interact with water nearly so strongly as does the SO_4^{2-} ion with its 2− charge.

Nonpolar liquids that do not react with the solvent generally are not very soluble in polar liquids, because of the mismatch of forces of interaction. Nonpolar liquids are, however, often quite soluble in other nonpolar liquids.

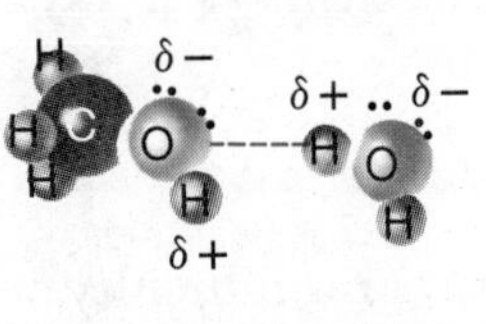

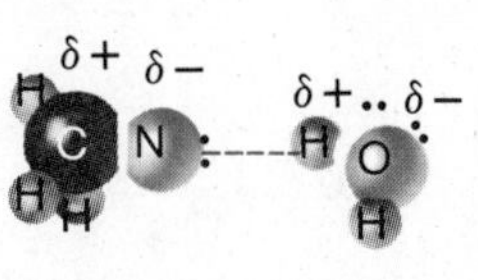

FIGURE 10–4 (a) Hydrogen bonding in methanol/water solution; (b) dipolar interaction in acetonitrile/water solution.

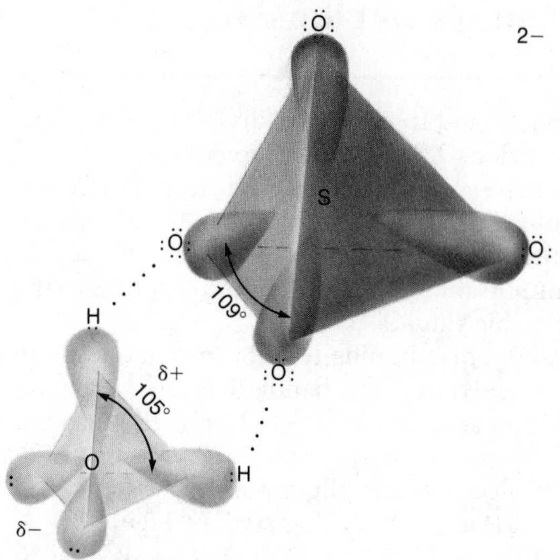

FIGURE 10–5 Interaction of a sulfate ion and a water molecule.

10–4 The Dissolution of Gases in Liquids

Based on what you have read in Section 9–9 and the foregoing discussion, it should come as no surprise that polar gases are generally most soluble in polar solvents and nonpolar gases are most soluble in nonpolar liquids. The hydrogen halides, HF, HCl, HBr, and HI are all polar covalent gases. In the gas phase the interactions among the widely separated molecules are not very strong, so solute–solute attractions are minimal and the dissolution processes are exothermic. The resulting solutions, called hydrohalic acids, contain predominantly ionized HX (X = Cl, Br, I). The ionization involves protonation of a water molecule by HX to form a hydrated hydrogen ion and halide ion X^- (which is also hydrated). HCl is used as an example below.

Protonation describes reactions in which H^+ (a bare proton) combines with another species.

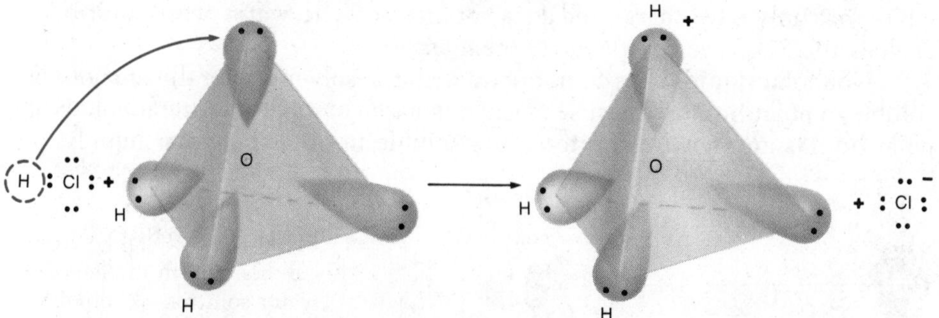

HF is only slightly ionized in aqueous solution because of the difficulty of breaking the strong covalent bond between highly electronegative fluorine and hydrogen. In addition, the more polar bond between H and the small F in HF is conducive to very strong hydrogen bonding between H_2O and the largely intact HF molecules

$$\overset{\delta+}{\text{H}}-\overset{\delta-}{\ddot{\text{O}}}\text{:---}\overset{\delta+}{\text{H}}-\overset{\delta-}{\ddot{\text{F}}}\text{:} \quad (\text{O bonded to H}^{\delta+}) \qquad \text{as well as} \qquad \overset{\delta+}{\text{H}}-\overset{\delta-}{\ddot{\text{F}}}\text{:---}\overset{\delta+}{\text{H}}-\overset{\delta-}{\ddot{\text{O}}}\text{:} \quad (\text{O bonded to H}^{\delta+})$$

Hydrogen fluoride also hydrogen bonds to itself in aqueous solutions (as well as in the gas phase), forming species that can be represented as $(HF)_n$, where n may be as large as five

$$\text{F}-\text{H}\cdots\text{F}(\angle 140^\circ)-\text{H}\cdots\text{F}-\text{H}\cdots\text{F}-\text{H}\cdots\text{F}-\text{H}$$

Although carbon dioxide, CO_2, and oxygen, O_2, are nonpolar gases, they do dissolve to limited extents in water. Carbon dioxide is somewhat more soluble because it reacts with water to form carbonic acid, H_2CO_3, which in turn ionizes slightly in two steps to give hydrogen ions, bicarbonate ions, and carbonate ions.

$$CO_2\ (g) + H_2O\ (\ell) \rightleftharpoons H_2CO_3\ (aq) \qquad \text{carbonic acid (exists only in solution)}$$

$$H_2CO_3\ (aq) \rightleftharpoons H^+\ (aq) + HCO_3^-\ (aq)$$

$$HCO_3^-\ (aq) \rightleftharpoons H^+\ (aq) + CO_3^{2-}\ (aq)$$

Carbon dioxide is called an acid anhydride or an "acid without water." As noted in Section 7-9.2, many other oxides of nonmetals, such as N_2O_3, SO_2, and P_4O_{10}, are also acid anhydrides, and most are soluble in water.

Approximately 1.45 grams of CO_2 (0.033 mole) dissolves in a liter of water at 25°C and one atmosphere pressure. But only about 0.0045 gram of O_2 (1.4×10^{-4} mole) dissolves in a liter of water under the same conditions. Yet this is sufficient to support aquatic life.

To summarize, the only gases that dissolve appreciably in water are those that are capable of hydrogen bonding (such as HF), those that ionize (such as HCl, HBr, and HI), and those that react with water (such as CO_2).

10-5 Rates of Dissolution and Saturation

At a given temperature, the rate of dissolution of a solid increases if large crystals are ground to a powder. This increases the surface area which, in turn, increases the number of solute ions or molecules in contact with the solvent. Pulverization also increases the number of corners and edges, where the ions or molecules are less firmly held. When an ionic solid is placed in water, some of its ions solvate and dissolve. The rate of this process slows as time passes because the surface area of each crystallite gets smaller and smaller. At the same time, the number of ions in solution increases and collisions between dissolved ions and solid occur more frequently. Some such collisions result in recrystallization or precipitation. After some time the rates of the two opposing processes become equal, and the solid and dissolved ions are then said to be in equilibrium with each other

$$\text{Solid} \underset{\text{crystallization}}{\overset{\text{dissolution}}{\rightleftharpoons}} \text{Dissolved ions}$$

FIGURE 10-6
A mortar and pestle for grinding a solid.

Dynamic equilibria occur in all saturated solutions; for instance, there is a continuous exchange of oxygen molecules across the surface of water in an open container. This is fortunate for fish, which "breathe" dissolved oxygen.

The double arrows ($\rightleftharpoons$) signify that both processes occur simultaneously, and the equilibrium is called a dynamic (moving) rather than a static (stationary) equilibrium. The slow "patching" of defects on the surfaces of crystals without increases in mass when imperfect crystals are placed in saturated solutions of their ions provides evidence for the dynamic character of the equilibrium. After equilibrium is established, no more solid dissolves without the simultaneous crystallization of an equal mass of dissolved ions. Such a solution is said to be **saturated.** Saturation occurs at very low concentrations of dissolved species for slightly soluble substances and at high concentrations for very soluble substances.

The solubilities of many solids increase at higher temperatures, and **supersaturated solutions,** which actually contain higher than saturation concentrations of solute, can sometimes be prepared by saturating a solution at a high temperature. The saturated solution is cooled slowly, without agitation, to a temperature at which the solute is less soluble. The resulting solution is **metastable** and produces crystals rapidly if it is slightly disturbed or if it is "seeded" with a dust particle or a crystallite. Under such conditions enough solid crystallizes to leave a solution that is just saturated (Figure 10–7).

FIGURE 10–7 Seeding a supersaturated solution of sodium acetate (top); the excess solute precipitates in the form of needle-like crystals (middle). The end result is a saturated solution of sodium acetate in equilibrium with excess solid.

10–6 Effect of Temperature on Solubility

In Section 9–6 we introduced LeChatelier's Principle, which states that *when a stress is applied to a system at equilibrium, the system responds in a way that best relieves the stress.* Recall that exothermic processes release heat and endothermic processes absorb heat

Exothermic: Reactants $\longrightarrow$ Products + Heat

Endothermic: Reactants + Heat $\longrightarrow$ Products

Since many solids dissolve by endothermic processes, their solubilities in water usually *increase* as heat is added and the temperature increases. For example, KCl dissolves endothermically

$$\text{KCl (s)} + 17.0\text{ kJ} \xrightarrow{H_2O} \text{K}^+\text{ (aq)} + \text{Cl}^-\text{ (aq)}$$

Figure 10–8 shows that the solubility of KCl increases as the temperature increases because more heat is available to drive the dissolving process.

Some solids, and many liquids and gases, dissolve by exothermic processes, and their solubilities, therefore, usually decrease as temperature increases. Anhydrous Na_2SO_4 is an example. The major reason for exothermicity of the dissolving of liquids and gases is that solute–solute interactions are much weaker than in solids because there is no crystal lattice energy to be overcome.

The solubility of O_2 in water decreases (by 22%) from 0.0045 gram per liter of water at 25°C to 0.0035 gram per liter at 50°C. The **thermal pollution** of rivers and lakes by heated waste water from industrial plants and nuclear power plants refers to the increased heat content of the water. A slight temperature increase causes a small but significant decrease in the amount of dissolved oxygen, so the water can no longer adequately support the marine life it ordinarily could.

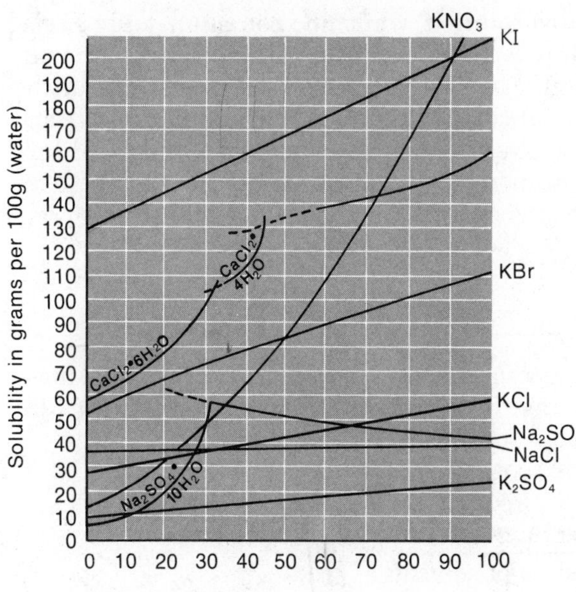

FIGURE 10–8 A graph illustrating the effect of temperature on the solubilities of some salts.

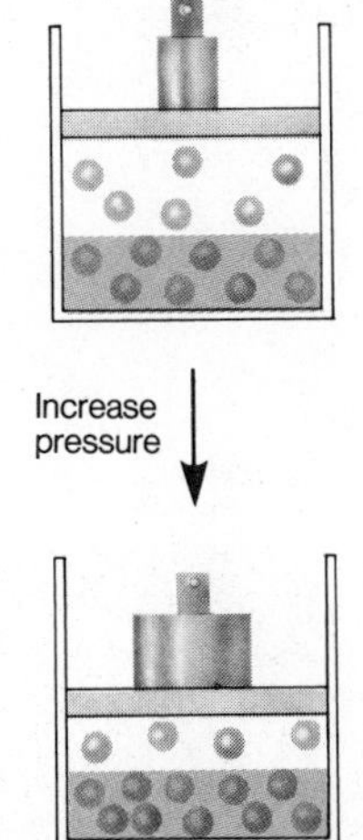

FIGURE 10–9 Illustration of Henry's Law. The solubility of a gas (that does not react with the solvent) increases with increasing pressure of the gas above the solution.

10–7 Effect of Pressure on Solubility

Changing the pressure has no appreciable effect on the solubilities of either solids or liquids in liquids. However, the solubilities of gases in all solvents increase as the partial pressures of the gases increase. Carbonated water is a saturated solution of carbon dioxide in water under pressure. When a can or bottle of a carbonated beverage is opened, the pressure on the surface of the beverage is reduced to atmospheric pressure and much of the CO_2 bubbles out of solution. If the container is left open, the beverage becomes "flat" faster than it would if tightly stoppered, because CO_2 is released and escapes faster from the open container.

The solubility of CO_2 in water at 25°C (approximately room temperature) and one atmosphere pressure is only 1.45 grams per liter (Section 10–4).

Henry's Law is applicable to gases that do not react with the solvent in which they dissolve (or in some cases gases that react to some extent, but some gas remains unreacted). It is usually stated: *the concentration of a gas in a solution is proportional to the pressure of the gas above the surface of the solution.*

Henry's Law can be represented symbolically as

$$M_{gas} = kP_{gas}$$

in which M_{gas} is the molar concentration of the gas in solution, P_{gas} is the pressure of the gas above the solution, and k is a constant for a particular gas and solvent at a particular temperature. The relationship is valid at low concentrations and low pressures (see Figure 10–9).

Concentrations of Solutions: A Review

The amount of a given solute dissolved in a specified amount of solvent or solution is its *concentration.* A solution containing a high concentration of solute is said to be a **concentrated solution,** while one containing only a relatively low concentration of solute is a **dilute solution.** There are many methods for expressing concentration. The four methods used most often in the sciences are: the percentage of solute by mass, molarity (M), normality (N), and molality (m). The first two have already been introduced in Section 2–13. In this chapter we shall expand that material as well as introduce molality. Normality will be discussed in Chapters 11 and 12.

10–8 Percent by Mass

The percent by mass of a solute in a solution is the percentage of the total mass of a sample of a given solution that is due to the solute alone (Section 2–13)

$$\% \text{ solute} = \frac{\text{mass of solute}}{\text{mass of solution}} \times 100\%$$

$$= \frac{\text{mass of solute}}{\text{mass of solute} + \text{mass of solvent}} \times 100\%$$

The percent solute or solvent always refers to percent by mass unless otherwise specified.

Clearly, a solution made up of 25.0 grams of alcohol and 75.0 grams of water is 25% alcohol by mass and 75.0% water by mass. The same is true for a solution composed of 50.0 grams of alcohol and 150 grams of water.

Consider an aqueous solution that is 14.0% ammonium chloride, NH_4Cl. Several unit factors may be constructed from this information. The following examples illustrate the usefulness of such factors.

The solvent is assumed to be water unless otherwise specified.

EXAMPLE 10–1

What mass of ammonium chloride would you use to prepare 250 grams of 14.0% solution of NH_4Cl?

Solution

$$\underline{?}\text{ g } NH_4Cl = 250 \text{ g soln} \times \frac{14.0 \text{ g } NH_4Cl}{100 \text{ g soln}} = \underline{35.0 \text{ g } NH_4Cl}$$

EXAMPLE 10–2

What volume of water would you use to prepare 250 grams of 14.0% solution of NH_4Cl?

Solution

$$\underline{?}\text{ mL } H_2O = \underbrace{\underbrace{250 \text{ g soln} \times \frac{86.0 \text{ g } H_2O}{100 \text{ g soln}}}_{\text{g } H_2O} \times \frac{1.00 \text{ mL } H_2O}{1.00 \text{ g } H_2O}}_{\text{mL } H_2O} = \underline{215 \text{ mL } H_2O}$$

A volume of 215 mL of water is required. This corresponds to 215 g H_2O *only* because the density of pure water is 1.00 g/mL near room temperature.

EXAMPLE 10–3

In carrying out a reaction, a student wishes to introduce 25.0 grams of ammonium chloride into a reaction mixture. What volume of 14.0% solution of NH_4Cl does he use? The density of the solution is 1.04 g/mL.

Solution

$$\underline{?}\text{ mL soln} = \underbrace{\underbrace{25.0\text{ g }NH_4Cl \times \frac{100\text{ g soln}}{14.0\text{ g }NH_4Cl}}_{\text{g soln}} \times \frac{1\text{ mL soln}}{1.04\text{ g soln}}}_{\text{mL soln}} = \underline{172\text{ mL soln}}$$

10–9 Molarity

The most common method of expressing concentrations of solutions is *molarity* (M). Recall that molarity is defined as the number of moles of solute dissolved per liter of solution (not solvent). Alternatively, it is also the number of millimoles of solute per milliliter of solution.

$$\text{Molarity} = \frac{\text{number of moles solute}}{\text{liter of solution}} = \frac{\text{number of millimoles solute}}{\text{milliliter of solution}}$$

EXAMPLE 10–4

The directions for an experiment call for the addition of 50.0 mL of a 0.0150 M solution of potassium permanganate, $KMnO_4$, to another solution. How many millimoles and how many grams of $KMnO_4$ are contained in this solution?

Solution

As we learned in Section 2–14, the product of the volume (mL) of a solution times its concentration (M) is the amount of solute present (mmol)

$$\text{mL} \times M = \text{number of mmol solute}$$

or

$$\text{number of mL} \times \frac{\text{number of mmol solute}}{\text{mL}} = \text{number of mmol solute}$$

$$\underline{?}\text{ mmol }KMnO_4 = 50.0\text{ mL} \times \frac{0.0150\text{ mmol }KMnO_4}{\text{mL}} = \underline{0.750\text{ mmol }KMnO_4}$$

The mass of $KMnO_4$ in 0.750 mmol can be calculated easily

$$\underline{?}\text{ g }KMnO_4 = 0.750\text{ mmol }KMnO_4 \times \frac{158\text{ mg }KMnO_4}{\text{mmol }KMnO_4} \times \frac{1.00\text{ g}}{1000\text{ mg}}$$

$$= \underline{0.118\text{ g }KMnO_4}$$

EXAMPLE 10–5

Commercial nitric acid, HNO_3, is 70.0% HNO_3 by mass, and its specific gravity is 1.42. What is its molarity?

Solution

$$\underline{?}\ \frac{\text{mol } HNO_3}{\text{L soln}} = \underbrace{\underbrace{\underbrace{\frac{1.42 \text{ g soln}}{1 \text{ mL soln}} \times \frac{1000 \text{ mL}}{1 \text{ L}}}_{\text{g soln/L soln}} \times \frac{70.0 \text{ g } HNO_3}{100 \text{ g soln}}}_{\text{g } HNO_3\text{/L soln}} \times \frac{1 \text{ mol } HNO_3}{63.0 \text{ g } HNO_3}}_{\text{mol } HNO_3\text{/L soln}}$$

$$= \underline{15.8\ M\ HNO_3}$$

EXAMPLE 10–6

Commercial nitric acid, 15.8 M HNO_3, is also known as "concentrated nitric acid." An experiment requires a student to prepare 100 mL of 3.00 M HNO_3 from concentrated HNO_3. How many milliliters of concentrated acid does he use?

Solution

Since the number of moles (or millimoles) of HNO_3 does not change during dilution, the following relationships apply [where the subscripts "1" and "2" refer to the concentrated and diluted solutions, respectively (Section 2–14)].

$$\text{number of mmol}_1\ HNO_3 = \text{number of mmol}_2\ HNO_3$$

$$\text{mL}_1 \times M_1 = \text{mL}_2 \times M_2$$

$$\text{mL}_1 = \frac{\text{mL}_2 \times M_2}{M_1}$$

Substitution of the given data gives

$$\text{mL}_1 = \frac{100 \text{ mL} \times 3.00\ M}{15.8\ M} = \underline{19.0 \text{ mL}}$$

The addition of 19.0 mL of concentrated nitric acid to sufficient water to make 100 mL of solution gives a 3.00 M solution of HNO_3.

Colligative Properties of Solutions

10–10 Molality

Physical properties of solutions that depend upon the *number*, but not the *kind*, of solute particles in a given amount of solvent are called **colligative properties.** The mathematical treatment of colligative properties is often made easier by expressing concentrations in molality units. Other concentration units are more useful in describing *chemical* properties. The **molality,** m, of a solute in solution is the number of moles of solute *per kilogram of solvent* (not solution):

$$\text{molality} = \frac{\text{number of moles solute}}{\text{number of kilograms solvent}}$$

EXAMPLE 10–7

Determine the molality of a solution containing 45.0 grams of potassium chloride in 1100 grams of water.

Solution

$$\underline{?}\ \frac{\text{mol KCl}}{\text{kg } H_2O} = \frac{45.0 \text{ g KCl}}{1.100 \text{ kg } H_2O} \times \frac{1 \text{ mol KCl}}{74.6 \text{ g KCl}} = \underline{0.548 \text{ mol KCl/kg } H_2O}$$

EXAMPLE 10–8

How many grams of H_2O must be used to dissolve 50.0 grams of sucrose to prepare a 0.100 *m* solution of sucrose, $C_{12}H_{22}O_{11}$?

Solution

$$\underline{?} \text{ mol } C_{12}H_{22}O_{11} = 50.0 \text{ g } C_{12}H_{22}O_{11} \times \frac{1 \text{ mol } C_{12}H_{22}O_{11}}{342 \text{ g } C_{12}H_{22}O_{11}} = 0.146 \text{ mol } C_{12}H_{22}O_{11}$$

Since molality of solution $= \dfrac{\text{number of mol } C_{12}H_{22}O_{11}}{\text{number of kg } H_2O}$, rearranging gives

$$\text{number of kg } H_2O = \frac{\text{number of mol } C_{12}H_{22}O_{11}}{\text{molality of solution}}$$

$$= \frac{0.146 \text{ mol } C_{12}H_{22}O_{11}}{0.100\ m}$$

$$= 1.46 \text{ kg } H_2O$$

$$= \underline{1.46 \times 10^3 \text{ g } H_2O}$$

The four most important colligative properties are lowering of the vapor pressure of a solvent, elevation of its boiling point, depression of its freezing point, and its osmotic pressure. Each will be discussed in turn.

10–11 Lowering of Vapor Pressure and Raoult's Law

Numerous experiments have shown that solutions containing *nonvolatile* liquids or solids as solutes always have lower vapor pressures than the pure solvents (Figure 10–10). The vapor pressure of a liquid depends upon the ease with which the molecules are able to escape from the surface of the liquid. When a solute is dissolved in a liquid, some of the total volume of the solution is occupied by solute molecules, and there are therefore fewer solvent molecules per unit area at the surface. As a result, solvent molecules vaporize at a slower rate than if no solute were present. This results in lowering the vapor pressure of the solvent, which is a colligative property; it is a function of the number, and not the kind, of solute particles in solution. We should point out that solutions of gases or low boiling (volatile) liquids have *higher* total vapor pressures than the pure solvent, so this discussion does not apply to them.

The lowering of vapor pressure associated with nonvolatile, nonionizing solutes is summarized by **Raoult's Law:** *The vapor pressure of a solvent in a*

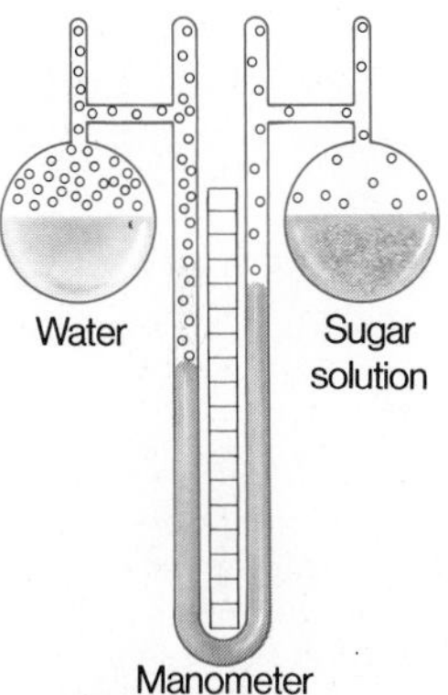

FIGURE 10–10 Lowering of vapor pressure. If no air is present in the apparatus, the pressure above the two liquids is due to water vapor. This pressure is less over the solution of water and sugar, because there are fewer water molecules on the surface to evaporate.

solution decreases as its mole fraction decreases. The relationship can be expressed mathematically as

$$P_{solvent} = X_{solvent}P^0_{solvent}$$

In a solution containing components A and B,

$$X_A = \frac{\text{mol A}}{\text{mol A} + \text{mol B}}$$

in which $X_{solvent}$ represents the mole fraction of the solvent in a solution, $P^0_{solvent}$ is the vapor pressure of the pure solvent, and $P_{solvent}$ is the vapor pressure of the solvent *in the solution.*

The *lowering* of vapor pressure, $\Delta P_{solvent}$, is defined as

$$\Delta P_{solvent} = P^0_{solvent} - P_{solvent}$$

Thus

$$\begin{aligned}\Delta P_{solvent} &= P^0_{solvent} - (X_{solvent}P^0_{solvent})\\ &= (1 - X_{solvent})P^0_{solvent}\end{aligned}$$

Since $X_{solvent} + X_{solute} = 1$, then $1 - X_{solvent} = X_{solute}$. So, we can express the lowering of vapor pressure in terms of the mole fraction of solute

$$\Delta P_{solvent} = X_{solute}P^0_{solvent}$$

Solutions that obey this relationship exactly are called **ideal solutions.**

10–12 Fractional Distillation

In Section 9–7 we described simple distillation as a process in which a liquid solution can be separated into volatile and nonvolatile components. However, 100% separation of volatile components is not possible by this method. Consider a liquid solution consisting of two volatile components. If the temperature is slowly raised, the solution will begin to boil when the sum of the vapor pressures of the components reaches the applied pressure on the surface of the solution, which is often atmospheric pressure. However, since both components exert a vapor pressure, both are carried away as a vapor. As the less volatile component condenses, heat is released, which helps the more volatile component to remain in the vapor phase. The resulting distillate is richer in the more volatile component, in proportion to the ratio of the vapor pressures. Repeated distillations of the distillate would make the distillate even richer in the more volatile component.

Distillation under vacuum lowers the applied pressure and allows boiling at lower temperatures than under atmospheric pressure. This technique allows distillation of some substances that would decompose at higher temperatures.

Successive distillations may be avoided by employing **fractional distillation,** in which a fractionating column is inserted above the solution and attached to the condenser, as shown in Figure 10–11. The column is packed with many small glass helices or beads, which provide surfaces upon which condensation can occur. Contact between the vapor and the packing favors condensation of the less volatile component. By the time the vapor reaches the top of the column, practically all of the less volatile component has condensed and fallen back into the flask. The more volatile component goes into the condenser, where it is liquefied and delivered as a pure distillate into the collection flask. The longer the column or the more packing, the better is the separation.

10–13 Boiling Point Elevation

The vapor pressure of a solvent at a given temperature is lowered by the presence of a nonvolatile solute in it, and such a solution must be heated to a

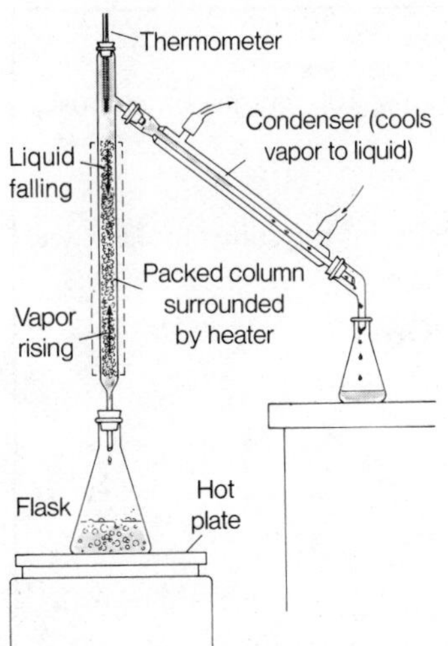

FIGURE 10–11 A fractional distillation apparatus. The vapor phase rising in the column is in equilibrium with the liquid phase condensed out and flowing slowly back down the column.

higher temperature than the pure solvent in order for its vapor pressure to equal atmospheric pressure (Figures 10–10 and 10–12). Recall that the boiling point of a liquid is the temperature at which its vapor pressure just equals the applied pressure on its surface or, for liquids in open containers, atmospheric pressure. Therefore, in accord with Raoult's Law, the elevation of the boiling point of a solvent caused by the presence of a nonvolatile, nonionized solute is proportional to the number of moles of solute dissolved in a given mass of solvent. Mathematically, this is usually expressed as follows:

$$\Delta T_b = K_b m$$

$\Delta T_b = T_{b\,(sol'n)} - T_{b\,(solv)}$ Note that ΔT_b is always positive for solutions that contain nonvolatile solutes because the boiling points of such solutions are always higher than the boiling points of the pure solvents.

The term ΔT_b represents the elevation of boiling point of the solvent, i.e., the boiling point of the solution minus the boiling point of the pure solvent. The m is the molality of the solute, and K_b is a proportionality constant called the *molal boiling point elevation constant,* which is different for different solvents and does not depend on the solute. Values of K_b for various solvents are tabulated in Table 10–2. They correspond numerically to the change in boiling points produced by one molal solutions of nonvolatile nonelectrolytes. Elevations of boiling points and depressions of freezing points, which are discussed below, are usually quite small for solutions of typical concentrations. As a result they can be measured accurately only with specially constructed (and expensive) differential thermometers that measure small temperature changes accurately to the nearest 0.001°C.

TABLE 10–2 Properties of Some Common Solvents

Solvent	Boiling Point (°C)	K_b (°C/m)	Freezing Point (°C)	K_f (°C/m)
Water	100	0.512	0	1.86
Benzene	80.1	2.53	5.48	5.12
Acetic acid	118.1	3.07	16.6	3.90
Nitrobenzene	210.88	5.24	5.7	7.00
Phenol	182	3.56	43	7.40

EXAMPLE 10–9

Calculate the boiling point of a solution that contains 100 grams of sucrose, $C_{12}H_{22}O_{11}$, in 500 g of water.

Solution

We must relate the increase of boiling point to the molality of the solution. First we determine the molality of the solution

$$\underline{?}\ \frac{\text{mol } C_{12}H_{22}O_{11}}{\text{kg } H_2O} = \frac{100 \text{ g } C_{12}H_{22}O_{11}}{0.500 \text{ kg } H_2O} \times \frac{1 \text{ mol } C_{12}H_{22}O_{11}}{342 \text{ g } C_{12}H_{22}O_{11}}$$

$$= 0.585\ m \text{ sucrose solution}$$

Now, $\Delta T_b = K_b m$

From Table 10–2, K_b for H_2O = 0.512°C/m, so

$\Delta T_b = (0.512°C/m)(0.585\ m) = \underline{0.300°C}$

The normal boiling point of pure water is exactly 100°C, so the boiling point of this solution is 100.300°C.

10–14 Freezing Point Depression

Ethylene glycol, CH_2OHCH_2OH, the major component of "permanent" antifreeze, effectively depresses the freezing point of water in an automobile radiator and raises its boiling point so that the solution remains in the liquid state over a wider temperature range than does pure water.

Molecules of most liquids approach each other more closely as the temperature is lowered, because their kinetic energies decrease and collisions become less frequent and less vigorous. The freezing point of a liquid is the temperature at which the forces of attraction among molecules are just great enough to cause a phase change from the liquid to the solid state. Strictly speaking, the freezing (melting) point of a substance is the temperature at which the liquid and solid phases exist in equilibrium. Clearly, the solvent molecules in a solution are somewhat more separated from each other (because of solute particles)

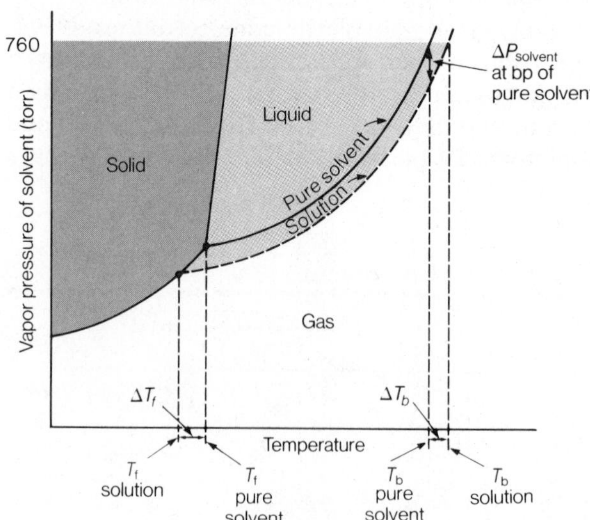

FIGURE 10–12 Because a *nonvolatile* solute lowers the vapor pressure of a solvent, the boiling point of a solution is higher and the freezing point lower than the corresponding points for the pure solvent. The magnitude of boiling point elevation, ΔT_b, is less than the magnitude of freezing point depression, ΔT_f.

than they are in the pure solvent. Consequently, the temperature of a solution must be lowered below the freezing point of the pure solvent in order to freeze it. (See Figure 10–12.)

The freezing point depression of solutions of nonelectrolytes has been found to be equal to the molality of the solute times a proportionality constant called the *molal freezing point depression constant*, K_f

$$\Delta T_f = K_f m$$

ΔT_f is the *depression* of freezing point: $\Delta T_f = T_{f\,(solvent)} - T_{f\,(sol'n)}$; it is always *positive.*

The values of K_f for different solvents were given in Table 10–2. They are numerically equal to the freezing point depression caused by dissolving one mole of a nonelectrolyte in one kilogram of solvent.

EXAMPLE 10–10

When 15.0 g of ethyl alcohol, C_2H_5OH, are dissolved in 750 g of formic acid, the freezing point of the solution is 7.20°C. The freezing point of pure formic acid is 8.40°C. Evaluate K_f for formic acid.

Solution

The relationship, $\Delta T_f = K_f m$, can be solved for K_f, and values for ΔT_f and m can be substituted into the expression

$$K_f = \frac{\Delta T_f}{m}$$

The molality and the depression of the freezing point are calculated first:

$$\underline{?}\ \frac{\text{mol } C_2H_5OH}{\text{kg formic acid}} = \frac{15.0 \text{ g } C_2H_5OH}{0.750 \text{ kg formic acid}} \times \frac{1 \text{ mol } C_2H_5OH}{46.0 \text{ g } C_2H_5OH} = \underline{0.435\ m}$$

$$\Delta T_f = (T_{f\text{ formic acid}}) - (T_{f\text{ solution}})$$
$$= 8.40°C - 7.20°C = 1.20°C \text{ depression}$$

$$\text{Then } K_f = \frac{\Delta T_f}{m} = \frac{1.20°C}{0.435\ m} = 2.76°C/m$$

K_f = $\underline{2.76°C/m}$, the molal freezing point depression constant for formic acid.

Recall that the molal freezing point depression constant is the depression for a one molal solution of a nonelectrolyte.

EXAMPLE 10–11

What is the freezing point of a solution of 23.0 g of C_2H_5OH in 600 grams of water?

Solution

For H_2O, $K_f = 1.86°C/m$. First we must calculate the molality of the solution

$$\underline{?}\ \frac{\text{mol } C_2H_5OH}{\text{kg } H_2O} = \frac{23.0 \text{ g } C_2H_5OH}{0.600 \text{ kg } H_2O} \times \frac{1 \text{ mol } C_2H_5OH}{46.0 \text{ g } C_2H_5OH} = 0.833\ m\ C_2H_5OH$$

Then $\Delta T_f = (1.86°C/m)(0.833\ m) = 1.55°C$

The temperature at which the solution freezes (one atmosphere pressure) is 1.55°C *below* the freezing point of pure water, or

$T_{f\ \text{solution}} = 0.00°\text{C} - 1.55°\text{C} = \underline{-1.55°\text{C}}$

10–15 Determination of Molecular Weight by Freezing Point Depression

The colligative properties of freezing point depression and, to a lesser extent, boiling point elevation are useful in the determination of molecular weights of solutes. The solutes must be nonvolatile in the temperature range of the investigation if boiling point elevations are to be determined. They must also be nonelectrolytes, since partial dissociation of the solute would produce more solute particles than would be anticipated if no dissociation occurred.

EXAMPLE 10–12

A sample of an unknown covalent organic compound having a mass of 1.20 grams is dissolved in 50.0 grams of benzene. The resulting solution freezes at 4.92°C. Determine the molecular weight of the compound.

Solution

From Table 10–2, the freezing point of benzene is 5.48°C and K_f for benzene is 5.12°C/m

$\Delta T_f = 5.48°\text{C} - 4.92°\text{C} = 0.56°\text{C}$

$$m = \frac{\Delta T_f}{K_f} = \frac{0.56°\text{C}}{5.12°\text{C}/m} = 0.11\ m$$

The molality is the number of moles of solute per kilogram of benzene, so the number of moles in 0.0500 kg of benzene can be calculated

$$0.11\ m = \frac{\underline{?}\ \text{mol solute}}{0.0500\ \text{kg benzene}}$$

$$\underline{?}\ \text{mol solute} = (0.11\ m)(0.0500\ \text{kg}) = 0.0055\ \text{mol solute}$$

$$\text{mass of 1.0 mol} = \frac{\text{no. of g solute}}{\text{no. of mol solute}}$$

$$= \frac{1.20\ \text{g solute}}{0.0055\ \text{mol solute}} = 2.2 \times 10^2\ \text{g/mol}$$

molecular weight = $\underline{2.2 \times 10^2\ \text{amu}}$

10–16 Apparent Percentage Dissociation of Strong Electrolytes

$\Delta T_f = K_f m$
$= (1.86°\text{C}/m)(0.100\ m)$
$= 0.186°\text{C}$

As we said earlier, colligative properties depend upon the *number* of solute particles in a given amount of solvent. A 0.100 molal *aqueous* solution of a covalent compound that does not ionize gives a freezing point depression of 0.186°C. We might predict that a 0.100 molal solution of a 1:1 strong electrolyte like KBr would have a freezing point depression of 2 × 0.186°C, or

0.372°C, since 0.100 m KBr would have an *effective* molality of 0.200 m, or 0.100 m K^+ + 0.100 m Br^-, assuming dissociation were complete. In fact, the *observed* depression is only 0.349°C. The discrepancy is due to the fact that some of the ions undergo **association** in solution. At any given instant, some K^+ and Br^- ions collide and "stick together." During the brief time that they are in contact, they behave as a single particle, and this tends to reduce the effective molality, and therefore the freezing point depression (as well as boiling point elevation and lowering of vapor pressure).

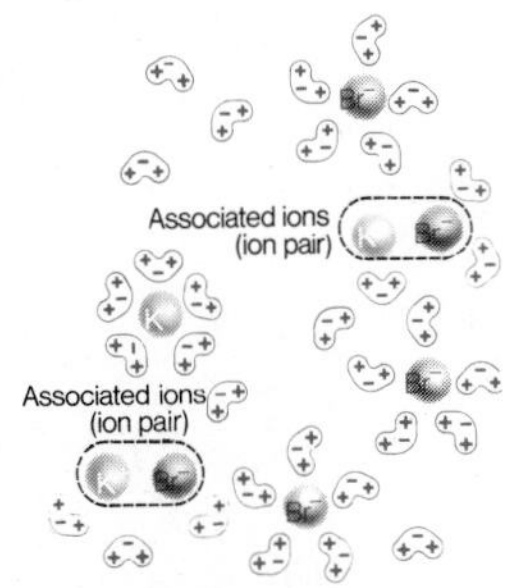

FIGURE 10–13 Diagrammatic representation of the various species thought to be present in a solution of KBr in water, which would explain unexpected values for its colligative properties, such as freezing point depression.

A (more concentrated) 1.00 m solution of KBr might be expected to have a freezing point depression of 2 × 1.86°C, but the observed depression is 3.29°C. There is a greater deviation from the depression predicted (neglecting ionic association) in the more concentrated solution. This is due to the greater frequency of collisions of ions, and consequent increase in ionic association, in the more concentrated solution. Table 10–3 lists the expected depressions of freezing point (assuming 100% dissociation) for several ionic compounds in water and the actual depressions, as well as values for nonelectrolytes.

TABLE 10–3 Expected and Observed Freezing Point Depressions for Aqueous Solutions of Nonelectrolytes and of Strong Electrolytes

Compound	Freezing Point 1.0 m Solution	Freezing Point 0.10 m Solution
Sucrose, $C_{12}H_{22}O_{11}$	−1.86°C	−0.186°C
nonelectrolytes	−1.86°C	−0.186°C
KBr	−3.29°C	−0.349°C
NaCl	−3.40°C	−0.348°C
If 2 ions/formula unit	−3.72°C	−0.372°C
K_2CO_3	−4.45°C	−0.455°C
K_2CrO_4	−3.62°C	−0.444°C
If 3 ions/formula unit	−5.58°C	−0.558°C
$K_3[Fe(CN)_6]$		−0.530°C
If 4 ions/formula unit	−7.44°C	−0.744°C

Example 10–13 involves determination of apparent percentage dissociation from a freezing point depression experiment.

EXAMPLE 10–13

A 0.0311 m solution of iron(III) chloride, $FeCl_3$, in water freezes at −0.206°C. Determine the apparent percentage dissociation of $FeCl_3$ in this solution.

Solution

If we let x = molality of the $FeCl_3$ that is apparently dissociated at any instant, then the molality of apparently undissociated $FeCl_3$ remaining is $(0.0311 - x)\ m$. The molalities of the hydrated Fe^{3+} and Cl^- ions will be $x\ m$ and $3x\ m$, respectively, since each formula unit of $FeCl_3$ that apparently dissociates produces one Fe^{3+} and three Cl^- ions

$$\begin{array}{lll} FeCl_3 & \longrightarrow Fe^{3+} & + 3Cl^- \\ (0.0311 - x)\ m & x\ m & 3x\ m \end{array}$$

The freezing point depression is caused by *all the particles in solution;* thus, we need to have an expression for the *total molality,* or the sum of the molalities of all the solute particles

$$m_{\text{total}} = m_{FeCl_3} + m_{Fe^{3+}} + m_{Cl^-}$$
$$= (0.0311 - x)\,m + x\,m + 3x\,m$$

$$m_{\text{total}} = (0.0311 + 3x)\,m$$

We can calculate m_{total} from the freezing point depression and K_f for water

$$m_{\text{total}} = \frac{\Delta T_f}{K_f} = \frac{0.206°C}{1.86°C/m} = 0.111\,m$$

Now we set the two expressions for m_{total} equal to each other and solve for x

$$0.111\,m = (0.0311 + 3x)m$$
$$0.0799 = 3x$$
$$0.0266 = x$$

The molality of $FeCl_3$ that apparently dissociates is 0.0266 m.

We are now in a position to calculate the apparent percentage dissociation of $FeCl_3$ in the solution, which is given by the expression below

$$\underline{?}\ \%\ \text{dissociation} = \frac{m\ FeCl_3\ \text{dissociated}}{m\ FeCl_3\ \text{original}} \times 100\%$$
$$= \frac{0.0266\,m}{0.0311\,m} \times 100\% = \underline{85.5\%\ \text{dissociated}}$$

10–17 Activities and Activity Coefficients

The "effective concentrations" or "apparent concentrations" of ions in solution as discussed above may be described in terms of **activities** of the ions. We can see from Example 10–13 that the concentration of Fe^{3+} ions in 0.0311 m $FeCl_3$ solution is only 85.5% of what we might expect. So its activity, a, is 0.0266 m

$$a_{Fe^{3+}} = 0.855\,(0.0311\,m) = 0.0266\,m$$

The activity of the Cl^- ions is

$$a_{Cl^-} = 0.855\,(3 \times 0.0311\,m) = 0.0798\,m$$

The 0.855 (85.5% expressed as a decimal) is called the **activity coefficient,** γ. In general, the activity of an ion equals the activity coefficient of the ion times its expected concentration assuming 100% dissociation. Activity coefficients decrease with increasing concentration

$$a = \gamma \times \text{expected concentration if 100\% dissociation (ionization)}$$

10–18 Percentage Ionization of Weak Electrolytes

Many weak electrolytes, such as weak acids and weak bases, are quite soluble in water, but they ionize only slightly. Example 10–14 illustrates how the percentage ionization of a weak electrolyte can be determined from freezing point depression data.

EXAMPLE 10–14

Lactic acid, $C_2H_4(OH)(COOH)$, is a weak acid, and therefore a weak electrolyte, found in sour milk. It is also formed in muscles during intense physical activity and is responsible for the pain felt during strenuous exercise. The freezing point of a 0.0100 m aqueous solution of lactic acid is −0.0206°C. Calculate the percentage ionization in the solution.

Solution

The equation for the ionization of lactic acid and the concentrations of the various species in the solution are (where $x = m$ of lactic acid ionized)

$$C_2H_4(OH)COOH \longrightarrow H^+ + C_2H_4(OH)COO^-$$

$$(0.0100 - x)\ m \qquad x\ m \qquad x\ m$$

The "total" molality of the solution is $(0.0100 - x + x + x)\ m$, or

$$m_{\text{total}} = (0.0100 + x)\ m$$

From $\Delta T_f = K_f m_{\text{total}}$, we can solve for m_{total}, equate the two representations for m_{total}, and then solve for x

$$m_{\text{total}} = \frac{\Delta T_f}{K_f} = \frac{0.0206°\text{C}}{1.86°\text{C}/m} = 0.0111\ m$$

$$(0.0100 + x) = 0.0111$$
$$x = 0.0011\ m$$

Hence the percentage ionization for 0.0100 m lactic acid is

$$\%\ \text{lactic acid ionized} = \frac{0.0011\ m}{0.0100\ m} \times 100\% = 11\%$$

This calculation tells us that in 0.0100 m solution, 89% of the lactic acid exists as nonionized molecules and 11% exists as H^+ and $C_2H_4(OH)COO^-$ ions.

10–19 Membrane Osmotic Pressure

Osmosis is the spontaneous process by which solvent molecules pass through a semipermeable membrane from a solution of lower concentration into a solution of higher concentration. A **semipermeable membrane** (such as cellophane) is used as a thin film or partition between two solutions. Solvent molecules may pass through the membrane in either direction, but the rate at which they pass into the more concentrated solution is found to be greater than the rate in the reverse direction. This is because more solvent molecules hit the membrane, per unit time, on the dilute side than on the concentrated side. The initial difference between the two rates is directly proportional to the difference in concentration between the two solutions. Solvent particles continue to pass through the membrane (Figure 10–14), and the column of liquid continues to rise, until the hydrostatic pressure of the column is sufficient to force solvent molecules back through the membrane at the same rate they enter from the dilute side. The pressure exerted under this condition is called the **osmotic pressure** of the solution.

On a microscopic scale, there are channels through the membrane that are wide enough for the small solvent molecules to pass through, but too narrow for the large solute molecules or solvated ions.

The osmotic pressure of a given aqueous solution can be measured with an apparatus such as that depicted in Figure 10–14. The solution of interest is

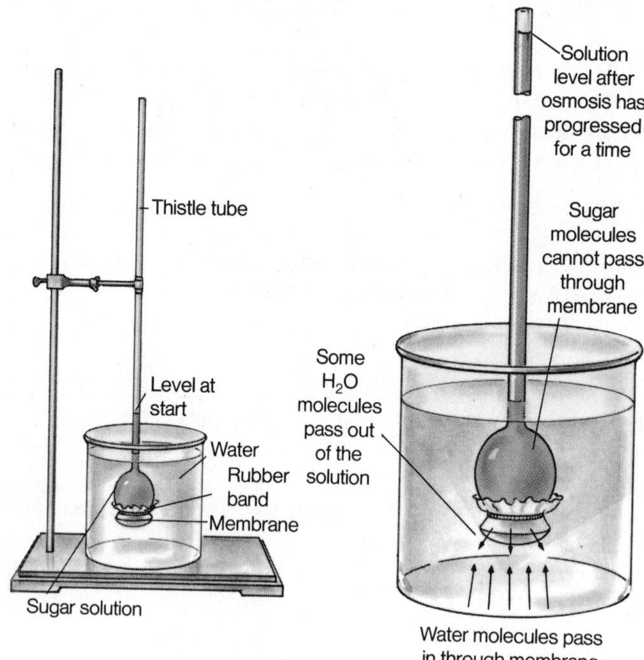

FIGURE 10–14 Laboratory apparatus for demonstrating osmosis. The picture at the right gives some details of the process.

placed inside an inverted thistle tube which has a cellophane film firmly fastened across the bottom. This part of the thistle tube and its cellophane are then immersed into a container of pure water. As time passes, the height of the solution in the neck rises until the weight of the column of solution is counterbalanced by the osmotic pressure.

Osmotic pressure is a measure of the forces that bind solvent molecules together, thereby causing molecules of the pure solvent to pass through the membrane into the solution in order to replace those that have been tied up by interactions with the solute. Osmotic pressure is therefore dependent upon the number, and not the kind, of solute particles in solution; it is therefore a colligative property.

The greater the number of solute particles, the greater height to which the column rises, and the greater the osmotic pressure.

Solute particles are quite widely separated in very dilute solutions and do not interact significantly with each other. For very dilute solutions, osmotic pressure, represented by π, is found to follow the equation

$$\pi = \frac{nRT}{V}$$

In this equation, n is the number of moles of solute in volume V (in liters) of the solution. The other quantities have the same meaning as in the ideal gas law. The term (n/V) is a concentration term. If concentration is expressed in terms of molarity (M), we obtain

$$\pi = MRT$$

Osmotic pressure is proportional to T because T affects the number of solvent–membrane collisions per unit time, and π is proportional to M because M affects the difference in the number of solvent molecules hitting the membrane from each side. For *dilute aqueous solutions,* the molarity is approximately equal to the molality (because the density is nearly 1 kg/liter), and the relationship becomes

$$\pi = mRT$$

Osmotic pressures represent very significant forces. For example, a 1.0 molal solution of a nonelectrolyte in water at 0°C produces an equilibrium osmotic pressure of approximately 22.4 atmospheres. Remember that atmospheric pressure is nearly 15 lb/in^2.

For a dilute solution of an electrolyte $\pi = m_{\text{total}}RT$.

The use of measurements of osmotic pressure for the determination of molecular weights has both advantages and disadvantages. An important advantage is that even very dilute solutions give rise to easily measurable osmotic pressures. This is in contrast to the vanishingly small freezing point depressions and boiling point elevations associated with the same solutions. This method therefore finds application in determination of the molecular weight of very expensive substances, or substances that can be prepared in only very small amounts, or substances of very high molecular weight such as polymers and many biological compounds. Because many high molecular weight biological materials are difficult, and in some cases impossible, to obtain in a high state of purity, determinations of their molecular weights are not as accurate as we might like. Nonetheless, osmotic pressures provide a very useful method of estimating molecular weights, as Example 10–15 illustrates.

EXAMPLE 10–15

Pepsin is an enzyme present in the human digestive tract. An enzyme is a protein that acts as a biological catalyst. Pepsin catalyzes the metabolic cleavage of amino acid chains (called peptide chains) in other proteins. A solution of a 0.500-gram sample of purified pepsin in 30.0 mL of benzene exhibits an osmotic pressure of 8.92 torr at 27.0°C. Estimate the molecular weight of pepsin.

Solution

$$\pi = MRT = \left(\frac{n}{V}\right)RT$$

We solve for n, the number of moles of solute (pepsin), and convert 8.92 torr to atmospheres to be consistent with the units of R

$$n = \frac{\pi V}{RT} = \frac{\left(8.92\text{ torr} \times \dfrac{1\text{ atm}}{760\text{ torr}}\right)(0.0300\text{ L})}{\left(0.0821\,\dfrac{\text{L}\cdot\text{atm}}{\text{mol}\cdot\text{K}}\right)(300\text{ K})}$$

$$n = 1.43 \times 10^{-5}\text{ mol pepsin}$$

Thus 0.500 gram of pepsin is 1.43×10^{-5} mol. We can now estimate its molecular weight

$$\underline{?}\text{ g/mol} = \frac{0.500\text{ g}}{1.43 \times 10^{-5}\text{ mol}} = \underline{3.50 \times 10^{4}\text{ g/mol}}$$

The molecular weight of pepsin is approximately 35,000 amu. This is typical for medium-size proteins.

The freezing point depression of this very dilute solution would be only about 0.003°C, which would be extremely hard to measure accurately. The osmotic pressure of 8.92 torr, on the other hand, is easily measured.

10–20 Water Purification by Reverse Osmosis

Consider a salt solution separated from pure water by a semipermeable membrane. If the saline solution is pressurized under a greater pressure than its osmotic pressure, the direction of osmosis can be reversed. That is, the net flow of water molecules will be from the saline solution through the membrane into the pure water. This process, depicted in Figure 10–15 and called **reverse osmosis,** has been used for the purification of brackish (mildly saline) water. The membrane usually consists of cellulose acetate or hollow fibers of a material structurally similar to nylon. This is only one of several methods for water purification, but it has the economic advantages of low cost, ease of apparatus construction, and simplicity of operation. Notice that it requires no heat, and so has a great advantage over distillation. However, membranes suitable for the economically feasible purification of seawater, which is much more concentrated than brackish water, have not yet been developed.

Some energy is required to run the pump.

Sarasota, Florida, is currently building the world's largest reverse osmosis plant for the purification of drinking water. It will process 4.5 million gallons of water per day from local wells. Total dissolved solids will be reduced in concentration from 1744 parts per million (0.1744% by mass) to 90 parts per million (ppm). This water will then be mixed with additional well water purified by an ion-exchange system. The final product will be 12 million gallons of water per day containing less than 500 ppm of total dissolved solids, which is the standard for drinking water set by the World Health Organization.

Colloids

At the beginning of the chapter, a solution was defined as a homogeneous mixture in which no settling occurs and in which solute particles are at the molecular or ionic state of subdivision. This represents one extreme of mixtures. The other extreme would be a suspension, a clearly heterogeneous mix-

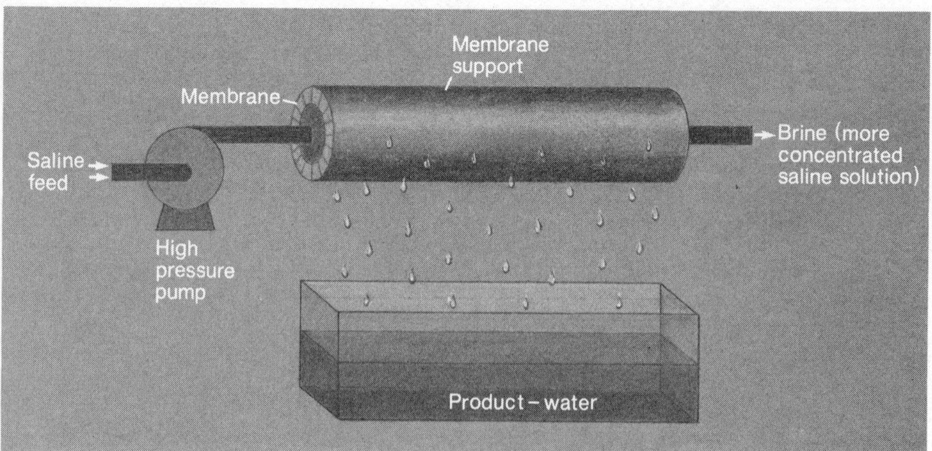

FIGURE 10–15 The reverse osmosis method of desalination.

TABLE 10–4 Types of Colloids and Examples

Dispersed (Solute-like) Phase		Dispersing (Solvent-like) Medium	Common Name	Examples
Solid	in	Solid	Solid sol	Many alloys (such as steel and duralumin), colored gems, reinforced rubber, porcelain, pigmented plastics
Liquid	in	Solid	Solid emulsion	Cheese, butter, jellies
Gas	in	Solid	Solid foam	Sponge, rubber, pumice, Styrofoam
Solid	in	Liquid	Sols and gels	Milk of magnesia, paints, mud, puddings
Liquid	in	Liquid	Emulsion	Milk, face cream, salad dressings, mayonnaise
Gas	in	Liquid	Foam	Shaving cream, whipped cream, foam on beer
Solid	in	Gas	Solid aerosol	Smoke, airborne viruses and particulate matter, auto exhaust
Liquid	in	Gas	Liquid aerosol	Fog, mist, aerosol spray, clouds

ture in which solute-like particles immediately settle out after mixing with a solvent-like phase. Such a situation results when a handful of sand is dropped into water. **Colloids, colloidal suspensions,** or **colloidal dispersions** represent an intermediate kind of mixture in which the solute-like particles, or **dispersed phase,** are suspended in the solvent-like phase, or **dispersing medium.** The particles of the dispersed phase are small enough that no settling occurs, yet large enough to make the mixture appear cloudy (and in many cases, opaque), owing to scattering of light as it passes through the colloid.

Table 10–4 indicates that all combinations of solids, liquids, and gases are candidates for colloids except mixtures of gases in gases (all of which are homogeneous and, therefore, true solutions). Whether a given mixture forms a solution, a colloidal dispersion, or a suspension depends upon the size of the solute-like particles (Table 10–5), as well as solubility and miscibility.

TABLE 10–5 Dispersed Particle Sizes

Dispersion	Example	Particle Size
Suspension	Sand in water	Larger than 10,000 Å
Colloidal dispersion	Starch in water	10–10,000 Å
Solution	Sugar in water	1–10 Å

10–21 The Tyndall Effect

The scattering of light by colloidal particles is called the **Tyndall effect** (Figure 10–16). In order to scatter visible light, a particle must have at least one di-

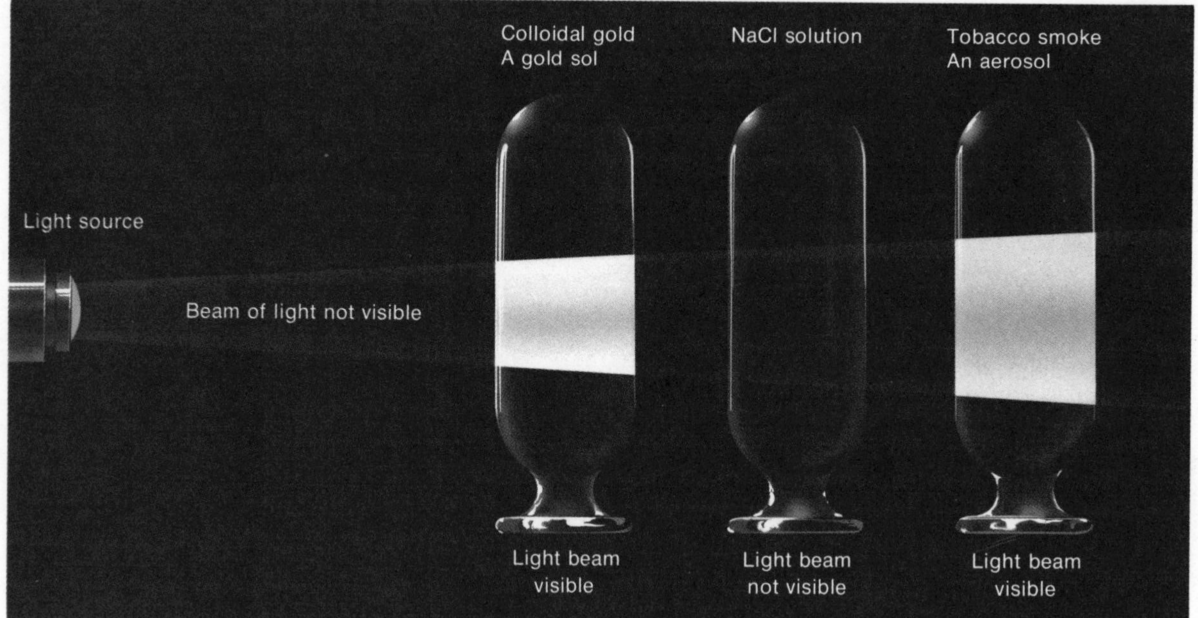

FIGURE 10–16 The dispersion of a beam of light by colloidal particles is called the Tyndall effect. The presence of colloidal particles is easily detected with the aid of a light beam.

mension of approximately 10 Å. Solute particles in solutions are below this limit. The maximum dimension of a colloidal particle is about 10,000 Å.

The scattering of light from automobile headlights by fogs and mists is an example of the Tyndall effect, as is the reflection of a light beam from a movie projector by dust particles in the air in a darkened room.

10–22 The Adsorption Phenomenon

Much of the chemistry of everyday life is the chemistry of colloids, as one can tell from a quick glance at Table 10–4. Since colloidal particles are so finely divided, they have tremendously high total surface area in relation to their volume. It is not surprising, therefore, that an understanding of colloidal behavior requires an understanding of surface phenomena.

Atoms on the surface of a colloidal particle are bonded only to other atoms of the particle on and below the surface. Since these atoms tend to bond in all directions, they have a tendency to interact with whatever comes in contact with the surface. Colloidal particles often adsorb ions or other charged particles, as well as gases and liquids. The process of **adsorption** involves adhesion of any such species onto the surfaces of particles. For example, a bright red **sol** (solid dispersed in liquid) is formed by the addition of hot water to a concentrated aqueous solution of iron(III) chloride

$$2x[Fe^{3+}(aq) + 3Cl^{-}(aq)] + x(3 + y)H_2O \rightarrow [Fe_2O_3 \cdot yH_2O]_x(s) + 6x[H^{+} + Cl^{-}]$$

yellow solution | bright red sol

Each colloidal particle of this sol consists of a cluster of many hydrated "Fe_2O_3

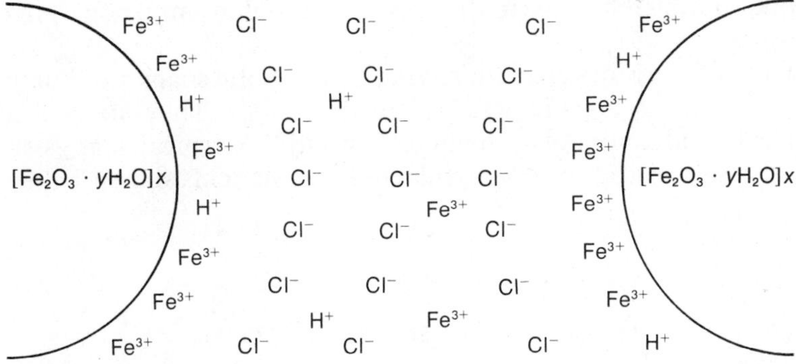

FIGURE 10–17 Stabilization of a colloid by electrostatic forces.

molecules," which tends to attract positively charged Fe^{3+} ions to its surface. Since each particle is then surrounded by a shell of positively charged ions, the particles repel each other and cannot combine to the extent necessary to cause actual precipitation (see Figure 10–17).

The surfaces of hydrated Fe_2O_3 particles tend to adsorb ions that fit readily into the crystal structure, so Fe^{3+} is preferentially adsorbed rather than Cl^-.

10–23 Hydrophilic and Hydrophobic Colloids

Colloids can be classified as **hydrophilic** ("water loving") or **hydrophobic** ("water hating") on the basis of the surface characteristics of the dispersed particles.

1 Hydrophilic Colloids

Proteins like the oxygen-carrier hemoglobin form hydrophilic sols when they are suspended in saline aqueous body fluids like blood plasma. Such proteins are macromolecules (giant molecules) that fold and twist in an aqueous environment so that polar groups are exposed to the fluid, while nonpolar groups are encased (see Figure 10–18). Protoplasm and human cells are examples of **gels,** which are special types of sols in which the solid particles (in this case mainly proteins and carbohydrates) join together in a semirigid network structure that encloses the dispersing medium. Other examples of gels are gelatin and jellies and gelatinous precipitates such as $Al(OH)_3$.

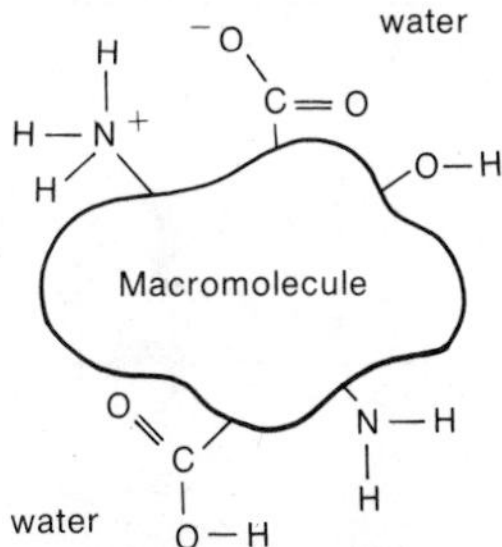

FIGURE 10–18 Examples of hydrophilic groups at the surface of a giant molecule (macromolecule) that help keep the macromolecule suspended in water.

2 Hydrophobic Colloids

Hydrophobic colloids cannot exist in polar solvents without the presence of **emulsifying agents,** or **emulsifiers,** which coat the particles of the dispersed phase to prevent their coagulation into a separate phase. Milk and mayonnaise are good examples of hydrophobic phases (milk fat in the first, vegetable oil in the second) that stay suspended with the aid of emulsifying agents (casein in milk and egg yolk in mayonnaise).

Consider the mixture resulting from vigorous shaking of oil (nonpolar) and water (polar), or salad oil and vinegar. Droplets of hydrophobic oil are temporarily suspended in the water. However, in a short time, the very polar water molecules, which attract each other strongly, squeeze out the nonpolar oil molecules, which in turn coalesce and float to the top. The oil floats on the surface of the water, since it is less dense. If we now add an emulsifying agent such as egg yolk, and shake or beat the mixture, a stable emulsion called mayonnaise results.

Sodium stearate is also a major component of most stick deodorants.

Soaps and other detergents are emulsifying agents. Solid soaps are usually sodium salts of long-chain organic acids called fatty acids. They are said to have a polar "head" and a nonpolar "hydrocarbon tail." A typical soap is sodium stearate, the salt of sodium hydroxide and stearic acid, shown below.

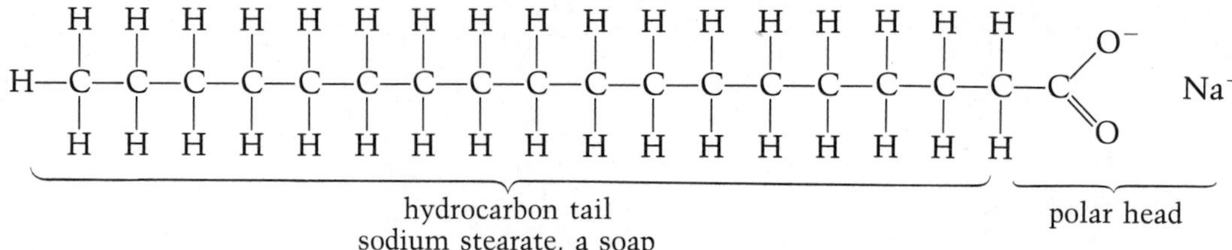

sodium stearate, a soap

If soap is added to an oil–water mixture and the mixture is vigorously shaken, a true emulsion forms. The hydrocarbon tails are attracted to the nonpolar oil, and droplets of the oil are surrounded by a shell of soap molecules with their polar heads in contact with the water. As a result, the oil droplets are unable to coalesce and they remain suspended in the water (Figure 10–19). The mechanism of the cleansing action by which soaps and other detergents remove dirt, grease, or stains is the same (Figure 10–20).

"Hard" water contains Fe^{3+}, Ca^{2+}, and/or Mg^{2+} ions, all of which displace Na^+ from soap molecules to form precipitates. This removes the soap from the water and puts an undesirable coating on the bathtub or on the fabric being laundered. **Synthetic detergents** are soap-like emulsifiers that contain sulfonate, $—SO_3^-$, sulfate, $—OSO_3^-$, or phosphate groups instead of carboxylate groups, $—COO^-$. They do not form precipitates with the ions of hard water, so they can be used in hard water as soap substitutes without forming undesirable scum.

However, the use of phosphate detergents (see Section 24–23, part 3) is now discouraged because of their tendency to cause **eutrophication** in rivers and streams that receive sewage. This is a condition (not related to colloids) in

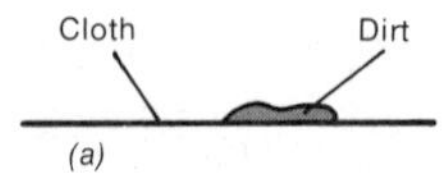

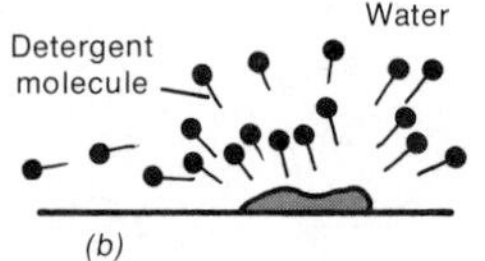

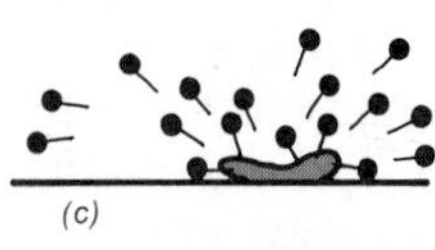

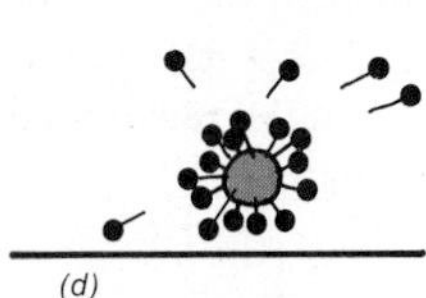

FIGURE 10–20 Representation of the method by which soaps and detergents act on dirt.

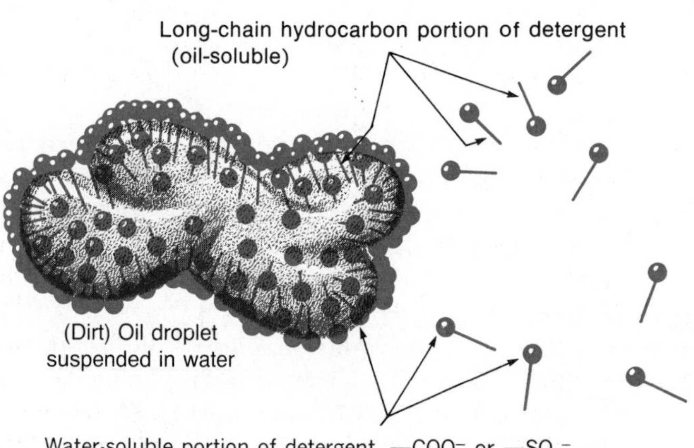

FIGURE 10–19 Attachment of soap or detergent molecules to a droplet of oily dirt.

which there is an unsightly overgrowth of vegetation caused by the high concentration of phosphorus, which is a plant nutrient. This overgrowth is accompanied by a decreased content of dissolved oxygen in the water, which causes the gradual elimination of marine life. There is also a foaming problem associated with alkylbenzenesulfonate (ABS) detergents in streams and in pipes, tanks, and pumps of sewage treatment plants. Such detergents are not **biodegradable;** that is, they cannot be broken down by bacteria. However, currently used linear-chain alkylsulfonate (LAS) detergents are biodegradable and do not cause such foaming.

A foam is a gas suspended in a liquid, like soap suds or detergent suds.

Alkyl is a term used in organic chemistry to designate noncyclic hydrocarbon structures.

ABS's have branched chain substituents.

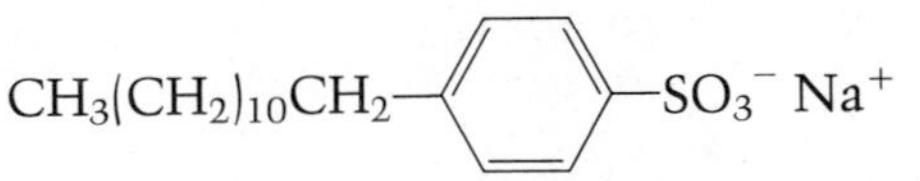

$$CH_3(CH_2)_{10}CH_2-C_6H_4-SO_3^-\ Na^+$$

sodium lauryl benzenesulfonate
a linear alkylbenzenesulfonate (LAS)
a biodegradable detergent

$$CH_3\overset{CH_3}{\overset{|}{C}}HCH_2\overset{CH_3}{\overset{|}{C}}HCH_2\overset{CH_3}{\overset{|}{C}}HCH_2\overset{CH_3}{\overset{|}{C}}H-C_6H_4-SO_3^-\ Na^+$$

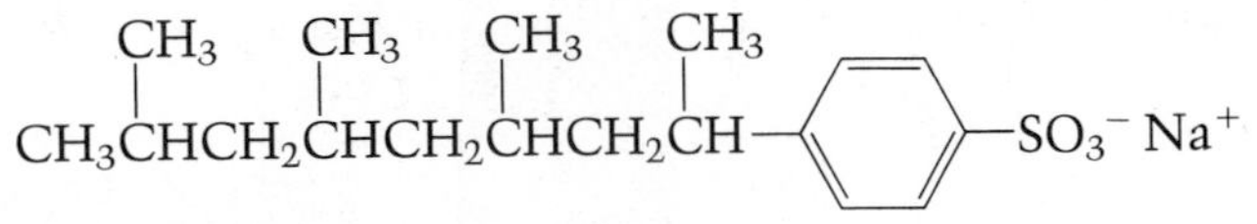

a sodium alkylbenzenesulfonate (ABS)
a nonbiodegradable detergent

10–24 Removal of Colloidal Particles

As we have seen, some colloids are desirable and necessary adjuncts to life and our culture. However, some are not, and we must often separate collidal particles from the suspending medium in order to purify water or process a product. **Coagulation,** or **flocculation,** can sometimes be brought about by heating the colloid, which increases the motion of the particles of the dispersed phase. This causes more frequent collisions and allows the particles to grow in size until precipitation occurs. In the purification of water, gelatinous substances such as $Al(OH)_3$ are often formed by the addition of the appropriate chemicals. These trap bacteria and suspended matter, but then must be removed. They may be separated from the water by filtration more readily after heating.

Adding electrolytes to a hydrophobic colloid can neutralize charged species on the surface of dispersed particles, and so cause them to coagulate and settle. Mud in river water represents a hydrophobic suspension of clay particles, whose surfaces are coated with hydroxide ions. As river water containing the suspended clay particles empties into an ocean, sodium and magnesium ions from the salt water neutralize the colloids. As the resulting neutral colloidal particles collide, they grow in size and eventually settle out to form a delta of deposited silt.

Smokes are solid **aerosols** (solids dispersed in gases). The solid particles, such as those of carbon and dust, have charged surfaces and so can be removed from the smoke or industrial stack gas by means of a Cottrell precipitator (Figure 10–21) in which the gases are passed over electrically charged plates. The particles lose their charge upon contact with the electrodes, and then precipitate as a dust or can be removed by filtration. This not only reduces air pollution but also allows the recovery of some valuable materials that would otherwise be lost in the smoke.

Dialysis is a process that resembles osmosis. Ions and other adsorbed matter can be separated from colloidal particles by passing the colloid over a huge area of semipermeable membrane with pores large enough for the ions, but not the colloidal particles, to pass through (Figure 10–22). Artificial kidney

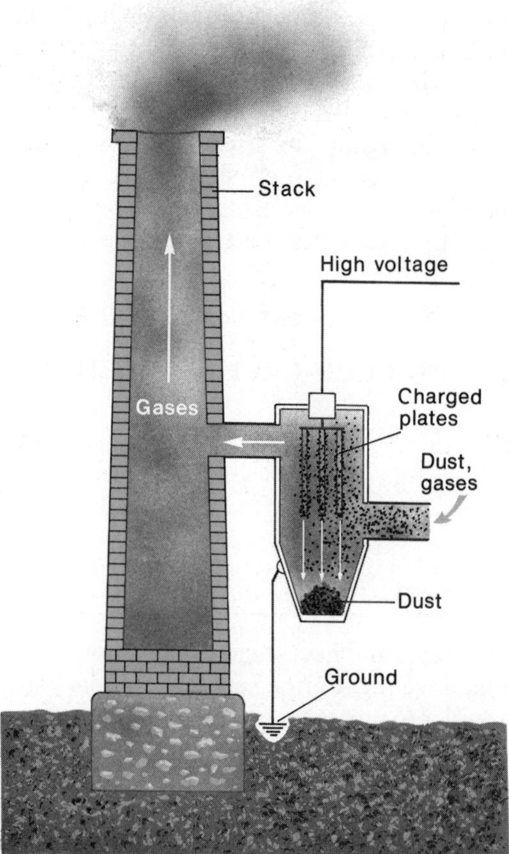

FIGURE 10–21 A Cottrell precipitator. Positively- and negatively-charged particles in smoke are precipitated as they pass over the electrically charged plates.

machines utilize this technique in the purification of blood. In addition to their function in restoring the proper ion concentrations in the blood, they allow molecules of uric acid and other toxic substances to diffuse through the membrane and to be separated from the bloodstream.

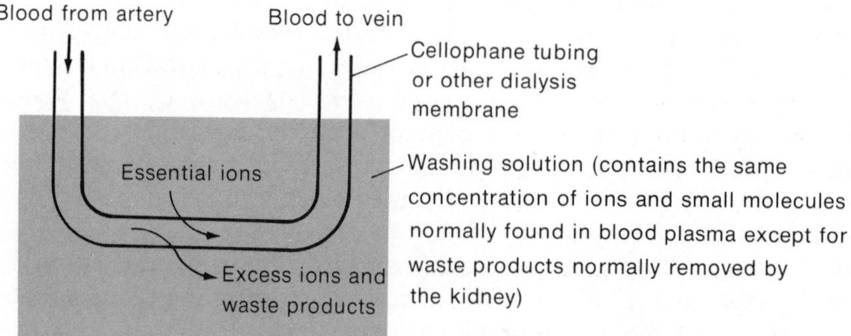

FIGURE 10–22 Dialysis; a simplified representation.

Key Terms

The following terms were listed at the end of Chapter 2: **concentration, dilution, molarity, percent by mass, solute, solution,** and **solvent.**

Activity of a dissolved species, concentration multiplied by a fraction called an activity coefficient that corrects for deviation from ideality. The "effective concentration" of a dissolved species.

Activity coefficient of a dissolved species, the decimal fraction that when multiplied by the actual concentration of a dissolved species gives its "effective" concentration.

Adsorption adhesion of species onto surfaces of particles.

Apparent percentage dissociation (ionization) of strong electrolytes, the percentage of formula units of solute that appear to be dissociated into discrete ions in a solution of a given concentration.

Associated ions short-lived species formed by the chance collision of dissolved ions of opposite charge.

Biodegradability the ability of a substance to be broken down into simpler substances by bacteria.

Boiling point elevation the increase in the boiling point of a solvent caused by dissolution of a nonvolatile solute.

Coagulation combination of colloidal particles into particles that are large enough to precipitate.

Colligative properties physical properties of solutions that depend upon the number but not the kind of solute particles present.

Colloid a heterogeneous mixture in which solute-like particles do not settle out after introduction into a solvent-like phase.

Crystal lattice energy energy released when one mole of formula units of a crystalline solid is formed from its ions, atoms, or molecules in the gas phase.

Detergent a soap-like emulsifier that contains a sulfonate, $—SO_3^-$, sulfate, $—OSO_3^-$, or phosphate group instead of a carboxylate group.

Dialysis process similar to osmosis in which ions can be separated from colloidal particles by passing through a semipermeable membrane.

Differential thermometer a very expensive thermometer used for accurate measurement of very small changes in temperature.

Dispersed phase the solute-like species in a colloid.

Dispersing medium the solvent-like phase in a colloid.

Distillation the process in which components of a mixture are separated by boiling away the more volatile liquid.

Emulsifier see *Emulsifying agent*

Emulsifying agent a substance that coats the particles of a dispersed phase and prevents coagulation of colloidal particles; an emulsifier.

Emulsion colloidal suspension of a liquid in a liquid.

Eutrophication the undesirable overgrowth of vegetation caused by high concentration of plant nutrients in bodies of water.

Flocculation see *Coagulation*

Foam colloidal suspension of a gas in a liquid.

Fractional distillation the process in which a fractionating column is used in a distillation apparatus to separate components of a liquid mixture that have different boiling points.

Freezing point depression the decrease in the freezing point of a solvent caused by the dissolution of a solute.

Gel colloidal suspension of a solid dispersed in a liquid; a semirigid sol.

Hard water water containing Fe^{3+}, Ca^{2+}, and/or Mg^{2+} ions, which form precipitates with soaps.

Heat of solution (molar) the amount of heat absorbed in the formation of a solution that contains one mole of solute; the value is positive if heat is absorbed (endothermic) and negative if heat is released (exothermic).

Henry's Law the concentration of a gas in a solution is proportional to the pressure of the gas above the surface of the solution.

Hydration the interaction (surrounding) of an ion with water molecules.

Hydration energy (molar) of an ion, the amount of energy released in the hydration of a mole of gaseous ions.

Hydrophilic colloids colloidal particles that attract water molecules.

Hydrophobic colloids colloidal particles that repel water molecules.

Ideal solution a solution that obeys Raoult's Law exactly.

Ion pair see *Associated ions*

Liquid aerosol colloidal suspension of a liquid in gas.

Molality (m) concentration expressed as number of moles of solute per kilogram of solvent.

Miscibility the ability of one liquid to mix with (dissolve in) another liquid.

Mole fraction of a component in solution, the number of moles of the component divided by the sum of the number of moles of all components.

Osmosis the process by which solvent molecules pass through a semipermeable membrane from a dilute solution into a more concentrated solution.

Osmotic pressure the hydrostatic pressure produced on the surface of a semipermeable membrane by osmosis.

Percentage ionization of weak electrolytes, the percentage of the weak electrolyte that actually ionizes in a solution of given concentration.

Raoult's Law the vapor pressure of a solvent in an ideal solution decreases as its mole fraction decreases.

Reverse osmosis forcing solvent molecules to flow through a semipermeable membrane from a concentrated solution into a dilute solution by application of greater hydrostatic pressure on the concentrated side than the osmotic pressure opposing it.

Saturated solution solution in which no more solute will dissolve.

Semipermeable membrane a thin partition between two solutions through which certain molecules can pass but others cannot.

Soap an emulsifier that can disperse nonpolar substances in water; the sodium salt of a long chain organic acid; consists of a long hydrocarbon chain attached to a carboxylate group, $—CO_2^-Na^+$.

Sol colloidal suspension of a solid dispersed in a liquid.

Solvation the process by which solvent molecules surround and interact with solute ions or molecules.

Solid aerosol colloidal suspension of a solid in a gas.

Solid emulsion colloidal suspension of a liquid dispersed in a solid.

Solid foam colloidal suspension of a gas dispersed in a solid.

Solid sol colloidal suspension of a solid dispersed in a solid.

Supersaturated solution a (metastable) solution that contains a higher than saturation concentration of solute; slight disturbance causes crystallization of excess solute.

Suspension a heterogeneous mixture in which solute-like particles settle out of a solvent-like phase some time after their introduction.

Thermal pollution introduction of heated waste water into natural waters.

Total molality the sum of the molalities of all solute particles in solution.

Tyndall effect the scattering of light by colloidal particles.

Exercises

General Concepts—The Dissolution Process

1. There are no true solutions in which the solvent is gaseous and the solute is either liquid or solid. Why?
2. Explain why (a) solute–solute, (b) solvent–solvent, and (c) solute–solvent interactions are important in determining the ease of dissolution of a solute in a solvent.
3. What is the relative importance of each factor listed in Exercise 2 when (a) solids, (b) liquids, and (c) gases dissolve in water?
4. Why do many solids dissolve in water in endothermic processes, while most miscible

liquids mix with each other in exothermic processes?

5. Define and distinguish between solvation and hydration.
6. The amount of heat released or absorbed in the dissolution process is important in determining whether or not the dissolution process is spontaneous, i.e., whether or not it can occur. What is the other important factor? How does it influence solubility?
7. Consider the following solutions. In each case predict whether the solubility of the solute should be high or low. Justify your answer. (a) KCl in H_2O (b) KCl in CCl_4 (c) NH_4Cl in pentane, C_5H_{12} (d) H_2O in CH_3OH (e) CH_3OH in H_2O (f) CCl_4 in H_2O (g) HCl in H_2O (h) HF in H_2O (i) N_2 in H_2O (j) Fe_2O_3 in H_2O
8. For those solutions that can be prepared in "reasonable" concentrations in Exercise 7, classify the solutes as nonelectrolytes, weak electrolytes, or strong electrolytes.
9. Give a reasonable explanation for the fact that the metal hydroxides of Group IA and those of the lower members of Group IIA are considered water-soluble, while other metal hydroxides are insoluble in water.
10. Refer to Figure 10–8 to answer the following questions.
    (a) A solution is prepared by dissolving 80 g of potassium iodide, KI, in 100 g of water at 60°C. Is the solution supersaturated, saturated, or unsaturated?
    (b) The solution of (a) is slowly cooled to 35.0°C with no crystallization. Is the solution supersaturated, saturated, or unsaturated?
11. Is it possible for a saturated solution to be a dilute solution also? If so, give an example.
12. Describe the preparation of a saturated aqueous solution of calcium hydroxide, $Ca(OH)_2$. Describe what happens at the molecular or ionic level as saturation is achieved.
13. Supersaturated solutions do not involve equilibrium between a solid and its dissolved ions or molecules. Why?
14. Describe and explain the effect of changing the temperature on the solubility of most solids that dissolve by (a) an exothermic and (b) an endothermic process.
15. What is the effect of raising the temperature on the solubility of most gases in water?
16. Describe the effect of increasing the pressure on the solubility of gases in liquids.

Percent by Mass

17. Write out six unit factors relating masses of solute, solvent, and/or solution that can be obtained from the information that an aqueous solution is 40.0% $Ca(NO_3)_2$ by mass. What do the unit factors tell you? How can they be used?
18. Calculate the mass of NaOH contained in 100 mL of a solution that is 20% NaOH by mass. The density of the solution is 1.6 g/mL.
19. Calculate the mass of water in 200 mL of a solution that is 15.0% KCl by mass. The specific gravity of the solution is 1.10.
20. How much solute is contained in 500 mL of a solution if the solution contains 15.0% solute by mass? The density of the solution is 1.20 g/mL.
21. How much solvent is contained in 500 mL of a solution if the solution contains 15.0% solute by mass? The density of the solution is 1.20 g/mL.

Molarity

22. Define molarity in terms of millimoles and milliliters.
23. Calculate the molarity of a solution that contains 49 g of H_3PO_4 in 500 mL of solution.
24. Calculate the molarity of a solution that contains 9.0 g of $(COOH)_2$ in 500 mL of solution.
25. Calculate the molarity of a solution that contains 12.6 g of $(COOH)_2 \cdot 2H_2O$ in 500 mL of solution.
26. Calculate the molarity of a solution that contains 288 g of $(NH_4)_2CO_3$ in 6.0 liters of solution.
27. Calculate the molarity of a solution that contains 30.8 g of $Al_2(SO_4)_3$ in 450 mL of solution.
28. Calculate the mass of NaOH required to prepare 2.5 liters of 0.50 M NaOH solution.
29. Calculate the mass of $(NH_4)_2SO_4$ required to prepare 480 mL of 1.5 M $(NH_4)_2SO_4$ solution.

30. Calculate the mass of $Mg(NO_3)_2 \cdot 6H_2O$ required to prepare 640 mL of 0.750 *M* $Mg(NO_3)_2$ solution.
31. Calculate the molarity of a solution that is 49.0% H_2SO_4 by mass. The specific gravity of the solution is 1.39.
32. Calculate the molarity of a solution that is 31.5% HNO_3 by mass. The specific gravity of the solution is 1.20.

Mole Fraction

33. A solution of ethanol and water has a mole fraction of ethanol of 0.36. What is the mole fraction of water?
34. Calculate the mole fraction of carbon tetrachloride, CCl_4, in a solution prepared by mixing 36.0 g of CCl_4 with 64.0 g of benzene, C_6H_6.
35. What is the mole fraction of ethanol, C_2H_5OH, in an alcoholic beverage that is 50% ethanol and 50% water by mass?
36. What mass of glucose, $C_6H_{12}O_6$, should be dissolved in 100 mL of water so that the mole fraction of $C_6H_{12}O_6$ is 0.100? The density of water is 1.00 g/mL.
37. What is the percent by mass of benzene, C_6H_6, and carbon tetrachloride, CCl_4, in a solution in which each has a mole fraction of 0.500?

Raoult's Law and Vapor Pressure

38. In your own words, explain briefly *why* the vapor pressure of a solvent is lowered by dissolving a nonvolatile solute in it.
39. Calculate (a) the lowering of vapor pressure and (b) the vapor pressure of a solution at 20°C prepared by dissolving 60.0 g of naphthalene, $C_{10}H_8$, (a nonvolatile nonelectrolyte) in 100.0 g of benzene, C_6H_6. Assume the solution is an ideal solution. The vapor pressure of benzene is 74.6 torr at 20°C.
40. (a) Calculate the lowering of vapor pressure associated with dissolving 75.0 grams of (solid) sucrose, $C_{12}H_{22}O_{11}$, in 300 grams of water at 25.0°C. (b) What is the vapor pressure of the solution? Assume the solution is ideal. The vapor pressure of water at 25°C is 23.76 torr.
41. What is the vapor pressure of the solution of Exercise 40 at 100°C?

Molality

42. What is the molality of a solution prepared by dissolving 6.00 g of benzoic acid, C_6H_5COOH, in 150 g of benzene, C_6H_6.
43. Calculate the molality of a solution that contains 60 g of methyl alcohol, CH_3OH, in 350 g of water.
44. Calculate the molality of a solution that contains 150 g of ethyl alcohol, C_2H_5OH, in 500 mL of benzene. The density of benzene is 0.879 g/mL.
45. Wht is the molality of ethanol, C_2H_5OH, in a solution prepared by mixing 50.0 mL of C_2H_5OH with 100 mL of water at 20°C. The density of C_2H_5OH is 0.789 g/mL at 20°C.
46. A student wishes to prepare an aqueous solution of sucrose, $C_{12}H_{22}O_{11}$, that is 0.250 *m*. What mass of sucrose must be dissolved in 300 g of water?
47. Urea, NH_2CONH_2, is a product of metabolism of proteins. An aqueous solution is 20.0% urea by mass and has a density of 1.05 g/mL. Calculate the molality of urea in the solution.

Boiling Point Elevation and Freezing Point Depression—Solutions of Nonelectrolytes

48. Refer to Table 10–2. Suppose you had 0.100 molal solutions of a nonvolatile nonelectrolyte in each of the solvents listed there. Which one would have (a) the greatest freezing point depression, (b) the lowest freezing point, (c) the greatest boiling point elevation and (d) the highest boiling point?
49. Explain qualitatively why boiling points are elevated, whereas freezing points are depressed by dissolving nonvolatile solutes in solvents.
50. What is the significance of (a) the molal freezing point depression constant, K_f, and (b) the molal boiling point elevation constant, K_b? How could their values for given solvents be determined in the laboratory?
51. Explain why automobile radiator antifreeze also protects against radiator boilover.
52. Calculate the freezing point and the boiling point of a solution that contains 30.0 g of urea, N_2H_4CO, in 250 g of water. Urea is a nonvolatile nonelectrolyte.

53. Calculate the freezing point and boiling point of a solution that contains 3.42 g of ordinary sugar (sucrose, $C_{12}H_{22}O_{11}$) in 20.0 g of water. Sucrose is a nonvolatile nonelectrolyte.
54. Using data from Table 10–2, calculate the freezing point and boiling point of a solution that contains 12.2 g of benzoic acid (C_6H_5COOH) in 100 g of benzene (C_6H_6). Benzoic acid is a nonvolatile nonelectrolyte in benzene.
55. When 2.00 g of an unknown nonelectrolyte is dissolved in 10.0 g of water, the resulting solution freezes at −3.72°C. What is the molecular weight of the unknown compound?
56. When 0.400 g of an unknown nonelectrolyte is dissolved in 40.0 g of water, the resulting solution freezes at −0.465°C. What is the molecular weight of the compound?
57. What minimum mass of ethylene glycol, CH_2OHCH_2OH (an antifreeze component), must be mixed with 6.00 gallons of water to prevent it from freezing at − 10.0°F? One gallon is 3.785 liters.
58. If ethylene glycol is treated as a nonvolatile solute, at what temperature would the solution of Exercise 57 boil?

Boiling Point Elevation and Freezing Point Depression—Solutions of Electrolytes

59. What is ion association in solution? Can you suggest why the term ion pairing is sometimes used to describe this phenomenon?
60. Why do the percentage ionization of a weak acid and the activity of a strong electrolyte increase with increasing dilution?
61. Acetic acid (CH_3COOH) dissolves in water according to

$$CH_3COOH \rightleftharpoons H^+ + CH_3COO^-$$

A 0.0100 *m* acetic acid solution freezes at −0.01938°C. Calculate the percentage ionization of CH_3COOH in this solution.
62. A 0.100 *m* acetic acid solution in water freezes at −0.1884°C. Calculate the percentage ionization of CH_3COOH in this solution.
63. A weak acid, HX, dissolves in water according to

$$HX \rightleftharpoons H^+ + X^-$$

In a 0.100 *m* solution, HX is 15.0% ionized. Calculate the freezing point of this solution.
64. NaCl dissolves in water according to

$$NaCl \longrightarrow Na^+ + Cl^-$$

A 0.100 *m* solution of NaCl freezes at −0.348°C. Calculate the apparent percentage dissociation of NaCl in this solution.
65. K_2SO_4 dissolves in water according to

$$K_2SO_4 \longrightarrow 2K^+ + SO_4^{2-}$$

A 0.100 *m* solution of K_2SO_4 freezes at −0.432°C. Calculate the apparent percentage dissociation of K_2SO_4 in this solution.
66. The complex compound $K_3[Fe(CN)_6]$ dissolves in water according to

$$K_3[Fe(CN)_6] \longrightarrow 3K^+ + [Fe(CN)_6]^{3-}$$

A 0.100 *m* solution of $K_3[Fe(CN)_6]$ freezes at −0.530°C. Calculate the apparent percentage dissociation of $K_3[Fe(CN)_6]$ in this solution.
67. A solution of 23.8 grams of mercury(II) chloride, $HgCl_2$, in 500 g of water boils at 100.085°C at one atmosphere pressure. (a) Calculate the percentage ionization for $HgCl_2$ in this solution. (b) Would you classify $HgCl_2$ as a nonelectrolyte, a weak electrolyte, or a strong electrolyte?

Osmotic Pressure

68. What are osmosis, osmotic pressure, and reverse osmosis?
69. Show how the expression

$$\pi = MRT,$$

where π is osmotic pressure, is similar to the ideal gas law. Rationalize qualitatively why this should be so.
70. Show numerically that the molality and molarity of 1.00×10^{-3} *M* aqueous sodium chloride are nearly equal. Why is this true? Would this be true if another solvent, say acetonitrile (methyl cyanide), CH_3CN, replaced water? Why? The density of CH_3CN is 0.786 g/mL at 20°C.

71. Estimate the molecular weight of a biological macromolecule if a 0.250-gram sample dissolved in 75.0 mL of benzene has an osmotic pressure of 12.17 torr at 25.0°C. The density of benzene is 0.879 g/mL.
72. Estimate the osmotic pressure associated with 30.0 g of an enzyme of molecular weight 1.2×10^5 dissolved in 1700 mL of benzene at 40.0°C.
73. Calculate the freezing point depression and boiling point elevation associated with the solution of Exercise 72.
74. A solution of 0.900 g of an unknown nonelectrolyte in 300 mL of water at 27.0°C has an osmotic pressure of 38.4 torr. What is the molecular weight of the compound?

Colloids

75. Distinguish between solutions and colloids. Give examples of each.
76. Distinguish among (a) sol, (b) gel, (c) emulsion, (d) foam, (e) solid sol, (f) solid emulsion, (g) solid foam, (h) solid aerosol, and (i) liquid aerosol. Try to give an example of each that is not already listed in Table 10–4.
77. What is the Tyndall effect and how is it caused?
78. Distinguish between hydrophilic and hydrophobic colloids.
79. What is an emulsifer?
80. What is coagulation? Coagulation does not occur readily in colloidal aqueous ferric oxide. Why?
81. Distinguish between soaps and detergents. How do they interact with hard water? Write an equation to show the interaction between a soap and hard water.
82. What is the disadvantage of alkylbenzenesulfonate (ABS) detergents compared to linear alkylbenzenesulfonate (LAS) detergents?
83. Describe three methods by which colloidal particles can be removed.

11 Acids, Bases, and Salts

In highly developed societies such as we have in the United States, acids, bases, and salts are indispensable compounds. Table 11–1 lists the 18 compounds that were included in the top 50 chemicals produced in the United States in 1981. Those marked with an asterisk are organic compounds.

It is of interest to note that the 1981 production of H_2SO_4 was about twice as great as the production of NH_3, the second chemical in the top 50 list. Sixty-five percent of the H_2SO_4 is used in the production of fertilizers.

Other than the 15 inorganic chemicals in Table 11–1, most of the top 50 are organic compounds. Yet, the total production of inorganic chemicals was more than twice that of organic compounds in the top 50.

Many acids, bases, and salts occur in nature and serve a wide variety of purposes. For instance, your "digestive juice" contains approximately 0.10 mole of hydrochloric acid per liter.

TABLE 11–1 1981 Production of Acids, Bases, and Salts in the United States

Formula	Name	Billions of Pounds	Major Uses
H_2SO_4	Sulfuric acid	81.35	Manufacture of fertilizers and other chemicals
NH_3	Ammonia	38.07	Fertilizer; manufacture of fertilizers and other chemicals
CaO, $Ca(OH)_2$	Lime (calcium oxide and calcium hydroxide)	35.99	Manufacture of other chemicals, steelmaking, water treatment
$NaOH$	Sodium hydroxide	21.30	Manufacture of other chemicals, pulp and paper, soap and detergents, aluminum, textiles
H_3PO_4	Phosphoric acid	19.83	Manufacture of fertilizers and detergents
HNO_3	Nitric acid	18.08	Manufacture of fertilizers, explosives, plastics, and lacquers
NH_4NO_3	Ammonium nitrate	17.58	Fertilizer and explosive
Na_2CO_3	Sodium carbonate (soda ash)	16.56	Manufacture of glass, other chemicals, detergents, pulp, and paper
$C_6H_4(COOH)_2$*	Terephthalic acid	6.35	Manufacture of fibers (polyesters), films, and bottles
HCl	Hydrochloric acid	4.89	Manufacture of other chemicals and rubber; metal cleaning
$(NH_4)_2SO_4$	Ammonium sulfate	4.22	Fertilizer
CH_3COOH*	Acetic acid	2.71	Manufacture of acetate esters
$Al_2(SO_4)_2$	Aluminum sulfate	2.41	Water treatment; dyeing textiles
Na_2SO_4	Sodium sulfate	2.33	Manufacture of paper, glass, and detergents
$CaCl_2$	Calcium chloride	1.83	Deicing roads in winter, controlling dust in summer; concrete additive
Na_2SiO_3	Sodium silicate	1.48	Manufacture of detergents, cleaning agents, and adhesives
$Na_5P_3O_{10}$	Sodium tripolyphosphate	1.37	Builder in detergents
$C_4H_8(COOH)_2$*	Adipic acid	1.21	Manufacture of Nylon 66

* Organic compounds.

11–1 Historical Notes: The Arrhenius Theory

In 1680 Robert Boyle noted that acids (1) dissolve many substances, (2) change the colors of some natural dyes (indicators), and (3) lose their characteristic properties when mixed with alkalies (bases). By 1814 Gay-Lussac concluded that acids *neutralize* bases, and that the two classes of substances can be defined only in terms of their reactions with each other. The latter is an extremely important idea!

In 1884 Svante Arrhenius presented his theory of electrolytic dissociation, which resulted in the Arrhenius theory of acid-base reactions. According to this view, an *acid* is defined as a substance that contains hydrogen and produces H^+ in aqueous solution. A *base* is defined as a substance that contains the OH group and produces hydroxide ions, OH^-, in aqueous solution. *Neutralization* is defined as the combination of hydrogen ions with hydroxide ions to form water molecules

It is now known that all ions are hydrated in aqueous solution.

$$H^+ (aq) + OH^- (aq) \longrightarrow H_2O (\ell) \qquad \text{(neutralization)}$$

The Arrhenius theory of acid-base behavior, which is sometimes called the water-ion concept, satisfactorily explained reactions of protonic acids (those containing acidic hydrogen atoms) with metal hydroxides (hydroxy bases), and represented a significant contribution to chemical thought and theory in the latter part of the nineteenth century. We used this theory in introducing acids and bases and discussing some of their reactions in Sections 7–1, 7–5, 7–6, and 7–9, part 2. You may wish to refer back to these sections to refresh your memory. However, as we shall see, the theory does have some serious limitations. For example, it does not explain the basic character and amphoteric behavior of species that do not contain OH groups. The Arrhenius model of acids and bases, although limited in scope, led to the development of other general theories of acid-base behavior, which will be considered in some of the sections that follow.

Amphoteric behavior, the ability of a substance to act as either an acid or a base, will be discussed in Sections 11–3 and 11–8.

11–2 The Hydrated Hydrogen Ion

Although Arrhenius described H^+ ions in water as bare ions, we now know that they are hydrated in aqueous solution and exist as $H^+(H_2O)_n$ in which n is some small integer. This is due to the attraction of the H^+ ions, or protons, for the oxygen end ($\delta-$) of water molecules. While we do not know the extent of hydration of H^+ in most solutions, we often represent the hydrated hydrogen ion as the hydronium ion, H_3O^+, or $H^+(H_2O)_n$ in which $n = 1$. The hydrated hydrogen ion is the species that gives aqueous solutions of acids their characteristic acidic properties. Whether we use the designation H^+ (aq) or H_3O^+, we are always referring to the hydrated hydrogen ion.

The most common isotope of hydrogen, ^1_1H, has no neutrons. Thus $^1_1H^+$ is a bare proton.

$$H^+ + :\ddot{\underset{\cdot\cdot}{O}}:H \longrightarrow H:\ddot{\underset{\cdot\cdot}{O}}:H^+$$
$$\quad\ \ H \qquad\qquad\quad H$$

11–3 The Brønsted-Lowry Theory

In 1923 Brønsted and Lowry independently presented logical extensions of the Arrhenius theory. Brønsted's contribution was more thorough than Lowry's, and the result is known as the Brønsted theory or the **Brønsted-Lowry theory.**

An **acid** is defined in this theory as a *proton donor,* H^+, and a **base** is defined as a *proton acceptor.* The definitions are sufficiently broad that any hydrogen-containing molecule or ion capable of releasing a proton, H^+, is an acid, while any molecule or ion that can accept a proton is a base according to the Brønsted-Lowry theory.

According to this theory *an acid-base reaction is the transfer of a proton from an acid to a base.* Thus the complete ionization of hydrogen chloride, HCl, a *strong* acid, in water is an acid-base reaction in which water acts as a base, a proton acceptor.

Step 1:	$HCl\ (g) \xrightarrow{H_2O} H^+\ (aq) + Cl^-\ (aq)$	(Arrhenius description)
Step 2:	$H^+\ (aq) + H_2O\ (\ell) \longrightarrow H_3O^+$	
Overall:	$HCl\ (g) + H_2O\ (\ell) \longrightarrow H_3O^+ + Cl^-\ (aq)$	(Brønsted-Lowry description)

$$\text{(H)}:\ddot{\underset{\cdot\cdot}{Cl}}: + H:\ddot{\underset{H}{O}}: \longrightarrow H:\overset{H}{\underset{H}{O}}:^+ + :\ddot{\underset{\cdot\cdot}{Cl}}:^-$$

acid$_1$ base$_2$ acid$_2$ base$_1$

Various measurements (electrical conductivity, freezing point depression, etc.) indicate that HF is only *slightly* ionized in water.

Likewise, the ionization of hydrogen fluoride, a *weak* acid, is similar, except that it occurs to only a slight extent, so we use double arrows to indicate that the reaction is reversible.

$$\underset{\text{acid}_1}{HF\ (g)} + \underset{\text{base}_2}{H_2O\ (\ell)} \rightleftharpoons \underset{\text{acid}_2}{H_3O^+} + \underset{\text{base}_1}{F^-\ (aq)}$$

$$\text{H}:\ddot{\underset{..}{\text{F}}}: + \text{H}:\ddot{\underset{\text{H}}{\text{O}}}: \rightleftharpoons \text{H}:\overset{\text{H}}{\ddot{\underset{\text{H}}{\text{O}}}}:^+ + :\ddot{\underset{..}{\text{F}}}:^-$$

We can describe Brønsted-Lowry acid-base reactions in terms of **conjugate acid-base pairs,** *which are species on opposite sides of an equation that differ by a proton.* In the above equation, HF ($acid_1$) and F^- ($base_1$) are one conjugate acid-base pair and H_2O ($base_2$) and H_3O^+ ($acid_2$) are the other pair. The members of each conjugate pair are designated by the same numerical subscript, although it makes no difference which pair, HF and F^- or H_2O and H_3O^+, is assigned the subscript 1 or 2.

In the forward reaction, HF and H_2O act as acid and base, respectively. In the reverse reaction, H_3O^+ functions as the acid, the proton donor, and F^- functions as the base, or proton acceptor. In an aqueous solution of HF there are far more nonionized HF molecules than H_3O^+ or F^- ions, which tells us that F^- is stronger as a base than HF is as an acid.

On the other hand, in dilute hydrochloric acid solutions H_3O^+ and Cl^- ions are the predominant species, and there are (practically) no nonionized HCl molecules. This tells us that HCl is a strong acid, but that the Cl^- ion has virtually no tendency to accept a proton to form an HCl molecule in dilute aqueous solution and is, therefore, an *extremely weak* base.

We can generalize: *The weaker an acid is, the greater is the base strength of its conjugate base. Likewise, the weaker a base is, the stronger is its conjugate acid.* Strong and weak, like many other adjectives, are used in a relative sense. For example, we do not mean to imply that the fluoride ion, F^-, is a strong base compared with species such as the hydroxide ion, OH^-. We mean that *relative to the anions of strong acids,* which are extremely weak bases, F^- is a much stronger base.

Ammonia is very soluble in water. In 0.10 *M* solution, NH_3 is only 1.3% ionized and 98.7% nonionized.

Ammonia acts as weak Brønsted-Lowry base and water acts as an acid in the ionization of aqueous ammonia.

$$\underset{\text{base}_1}{NH_3\ (g)} + \underset{\text{acid}_2}{H_2O\ (\ell)} \rightleftharpoons \underset{\text{acid}_1}{NH_4^+\ (aq)} + \underset{\text{base}_2}{OH^-\ (aq)}$$

$$\text{H}:\overset{\text{H}}{\underset{\text{H}}{\ddot{\text{N}}}}: + \text{H}:\ddot{\underset{\text{H}}{\text{O}}}: \rightleftharpoons \text{H}:\overset{\text{H}}{\underset{\text{H}}{\ddot{\text{N}}}}:\text{H}^+ + :\ddot{\underset{\text{H}}{\text{O}}}:^-$$

In three dimensions, the molecular structures are:

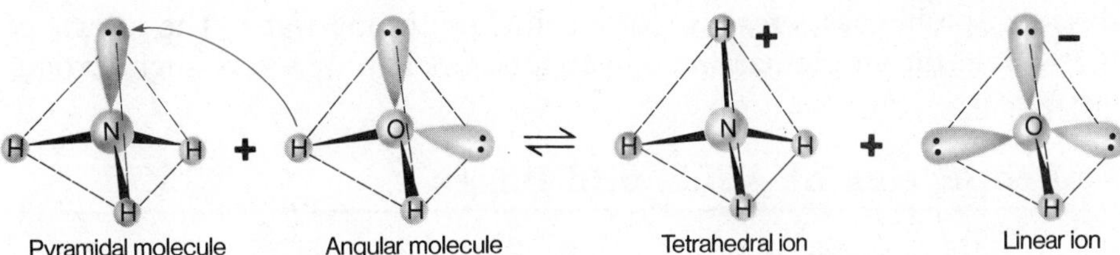

Ammonium ion, NH_4^+, is the conjugate acid of NH_3, and hydroxide ion, OH^-, is the conjugate base of water in the reverse reaction.

Note that water acts as an acid (H^+ donor) in its reaction with NH_3, whereas it acts as a base (H^+ acceptor) in its reaction with HCl and HF. Whether water acts as an acid or a base depends on its environment, i.e., the other species present.

Careful measurements have shown that pure water ionizes ever so slightly to produce equal numbers of hydrated hydrogen ions and hydroxide ions

$$\underset{\text{base}_1}{H_2O} + \underset{\text{acid}_2}{H_2O} \rightleftharpoons \underset{\text{acid}_1}{H_3O^+} + \underset{\text{base}_2}{OH^-}$$

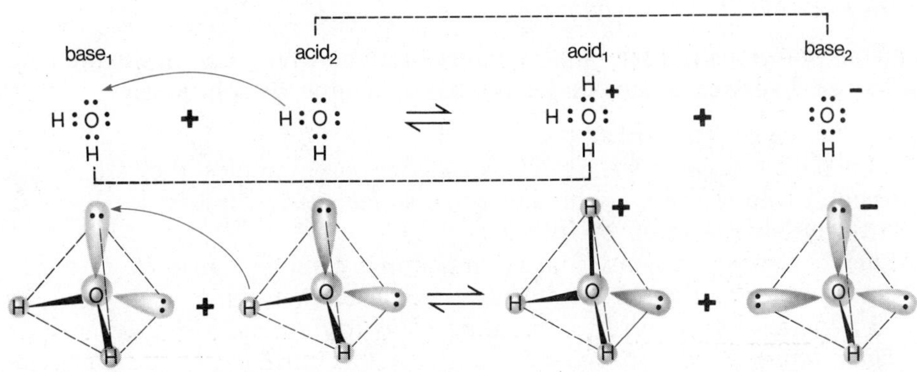

or in simplified notation

$$H_2O \rightleftharpoons H^+ \text{ (aq)} + OH^- \text{ (aq)}$$

The **autoionization** of water is an acid-base reaction according to the Brønsted-Lowry theory. One H_2O molecule (the acid) donates a proton to another H_2O molecule (the base). The H_2O molecule that donates a proton becomes an OH^- ion, which is called the conjugate base of water. The H_2O molecule that accepts a proton becomes an H_3O^+. Examination of the reverse reaction (right-to-left) shows that H_3O^+ (an acid) donates a proton to OH^- (a base) to form two H_2O molecules. Since one H_2O molecule behaves as an acid while the other acts as a base in the autoionization of water, water is said to be **amphiprotic,** that is, H_2O molecules can accept and donate protons.

The prefix amphi- means "of both kinds." Amphiprotism refers to amphoterism by accepting and donating a proton in different reactions.

As we saw in Section 7-5, the combination of H_3O^+ and OH^- ions to form nonionized water molecules is the reaction that occurs when strong acids and strong soluble bases react to form soluble salts and water. The reverse of this reaction, the autoionization of water, occurs to only a very slight extent, as expected.

11–4 Properties of Acids and Bases

Aqueous solutions of most protonic acids exhibit certain properties, which are properties of hydrated hydrogen ions.

1. *They have a sour taste.* Pickles are usually preserved in vinegar, a 5% solution of acetic acid. Many pickled condiments contain large amounts of sugar so that the taste of acetic acid isn't so pronounced. Lemons contain citric acid, which is responsible for their characteristic sour taste.
2. *They change the colors of many indicators* (highly colored dyes). Acids turn blue litmus red, and bromothymol blue changes from blue to yellow.
3. *Nonoxidizing acids react with metals above hydrogen in the activity series (Section 7–3) to liberate hydrogen, H_2.*
4. *They react with (neutralize) metal oxides and metal hydroxides to form salts and water (Section 7–5).*
5. *They react with salts of weaker or more volatile acids to form the weaker or more volatile acid and a new salt.*
6. *Aqueous solutions of protonic acids conduct an electric current because they are wholly or partly ionized.*

Aqueous solutions of bases also exhibit certain general properties, which are due to the hydrated hydroxide ions of bases in aqueous solutions:

1. *They have a bitter taste.*
2. *They have a slippery feeling.* Soaps are common examples; they are mildly basic. A solution of household bleach also feels very slippery because it is strongly basic (Section 22–14).
3. *They change the colors of many indicators:* Litmus changes from red to blue, and bromothymol blue changes from yellow to blue.
4. *They react with (neutralize) protonic acids to form salts and water.*
5. *Their aqueous solutions conduct an electric current because they are ionized or dissociated.*

11–5 Strengths of Binary Acids

The strengths of a series of *binary protonic acids* can be predicted from electronegativity differences. The ionization of a binary protonic acid involves breaking H—X bonds; as the electronegativity difference between H and element X increases, so does bond strength, making ionization more difficult. The electronegativity differences for the hydrogen halides are (see Section 5–1):

	H—F	H—Cl	H—Br	H—I
ΔEN	1.9	0.9	0.7	0.4

The electronegativity difference for HF is much greater than for HCl, and consequently the HF bond is much stronger than the HCl bond. Hydrogen fluoride ionizes only slightly in dilute aqueous solutions because the HF bond is so strong and because of hydrogen bonding between HF and H_2O (Section 22–12).

$$HF\ (g) \xrightleftharpoons{H_2O} H^+\ (aq) + F^-\ (aq)$$

However, HCl, HBr, and HI ionize completely or nearly completely in dilute aqueous solutions because the H—X bonds are much weaker

$$HX\ (g) \xrightarrow{H_2O} H^+\ (aq) + X^-\ (aq) \qquad X = Cl, Br, I$$

The order of *bond strengths* for the hydrogen halides is

$$HF \gg HCl > HBr > HI$$

while, as we shall see presently, the order of *acid strengths* is

$$HF \ll HCl < HBr < HI$$

In dilute aqueous solutions, hydrochloric, hydrobromic, and hydroiodic acids are completely ionized and all show the same apparent acid strength. Water is sufficiently basic that it does not distinguish among the acid strengths of HCl, HBr, and HI, and therefore it is referred to as a **leveling solvent.** It is not possible to determine the order of the strengths of these three acids in water because they are so nearly completely ionized.

When these compounds are dissolved in anhydrous acetic acid, or other solvents less basic than water, however, significant differences in their acid strengths are observed. The order of increasing acid strengths is the order expected based on decreasing differences in electronegativities and bond strengths

The strengths of *ternary* acids will be discussed in Section 11–8.

$$HCl < HBr < HI \qquad \text{(strongest acid)}$$

Similar statements can be made for other vertical series of similar binary acids. For those of the Group VIA elements, the order of increasing acid strength is

$$H_2O \ll H_2S < H_2Se < H_2Te \qquad \text{(strongest acid)}$$

Table 11–2 displays relative acid and base strengths of a number of conjugate acid-base pairs.

One more observation is appropriate to describe the leveling effect:

The hydrated hydrogen ion is the strongest acid that can exist in aqueous solution.

The trends in binary acid strengths *across* a period (e.g., $CH_4 < NH_3 < H_2O < HF$) are the reverse of those predicted from trends in bond energies and electronegativity differences. The correlations used for *vertical* trends *cannot* be used for *horizontal* trends because a "horizontal" series of compounds has different stoichiometries and different numbers of lone pairs of electrons on their central atoms.

Acids stronger than H^+ (aq) react with water to produce H^+ (aq) and their conjugate bases. For example, $HClO_4$ (see Table 11–2) reacts with H_2O completely to form H^+ (aq) and ClO_4^- (aq). Similar observations can be made for aqueous solutions of strong soluble bases such as NaOH and KOH. Both are completely dissociated in dilute aqueous solutions

$$Na^+OH^-\ (s) \xrightarrow{H_2O} Na^+\ (aq) + OH^-\ (aq)$$

The hydroxide ion is the strongest base that can exist in aqueous solution.

TABLE 11–2 Relative Strengths of Conjugate Acid–Base Pairs

Acid			Base	
$HClO_4$	100% ionized in dil. aq. soln. No molecules of nonionized acid.		Negligible base strength in water	ClO_4^-
HI				I^-
HBr				Br^-
HCl				Cl^-
HNO_3		$\underset{+H^+}{\overset{-H^+}{\rightleftarrows}}$		NO_3^-
H_3O^+	Equilibrium mixture of nonionized molecules of acid, conjugate base, and H^+ (aq).			H_2O
HF				F^-
CH_3COOH				CH_3COO^-
HCN				CN^-
NH_4^+				NH_3
H_2O			Reacts completely with H_2O. Cannot exist in aqueous solution.	OH^-
NH_3				NH_2^-

Acid Strength Increases (upward) — Base Strength Increases (downward)

Bases stronger than OH^- react with H_2O to produce OH^- and their conjugate acids. When metal amides such as sodium amide, $NaNH_2$, are placed in H_2O the amide ion, NH_2^-, reacts with H_2O completely as shown by the following equation

$$NH_2^- + H_2O \longrightarrow NH_3\ (aq) + OH^-\ (aq)$$

Thus, we see that H_2O is a leveling solvent for all bases stronger than OH^-.

11–6 Reactions of Acids and Bases

Common Strong Acids

Binary	Ternary
HCl	$HClO_4$
HBr	$HClO_3$
HI	HNO_3
	H_2SO_4

Strong Soluble Bases

LiOH	
NaOH	
KOH	$Ca(OH)_2$
RbOH	$Sr(OH)_2$
CsOH	$Ba(OH)_2$

In Section 7–5 we introduced classical acid-base reactions as one kind of metathesis reaction. We defined neutralization as the reaction of an acid with a base to form a salt and (in most cases) water. Most salts are ionic compounds that contain a cation other than H^+ and an anion other than OH^- or O^{2-}. The strong acids and strong soluble bases are listed in the margin. Recall that other common acids may be assumed to be weak. The other common metal hydroxides (bases) are insoluble in water.

Also in Section 7–5 we introduced the distinctions among molecular, total ionic, and net ionic equations. We pointed out that there are instances in which molecular equations are desirable (as when stoichiometric calculations are to be performed, for example), and other instances in which net ionic equations are more useful in that they focus attention on only the species that *actually participate* in the reaction. The total ionic equation provides the bridge between the two kinds of equations.

Arrhenius and Brønsted-Lowry acid-base neutralization reactions all have one thing in common, they involve the reaction of an acid with a base to form a salt that contains the cation characteristic of the base and the anion characteristic of the acid and usually water. This is indicated in the molecular equation. However, the general form of the net ionic equation, and the essence of the reaction, are different for different acid-base reactions, depending upon the solubility and extent of ionization of each reactant and product. We shall illustrate this idea in this section.

The writing of ionic equations from molecular equations requires a knowledge of the lists of strong acids and strong soluble bases, as well as of the generalizations on solubilities of inorganic compounds presented in Section 7–1, part 5. In writing ionic equations, we always write the formulas of the predominant forms of the compounds in, or in contact with, aqueous solution: ionized for compounds that are both dissolved and strong electrolytes, and nonionized for insoluble compounds or for those that are soluble and only slightly ionized. The forms of the equations differ *only* to reflect differences in the characteristics of reactants and products.

A few generalizations for common compounds are very useful as we write equations for chemical reactions.

1. Strong acids and strong soluble bases are ionized (or dissociated) completely, or very nearly completely, in dilute aqueous solutions, and are represented as ions in ionic equations.
2. Weak acids and weak bases exist primarily as nonionized molecules in dilute aqueous solutions, and they are represented as molecules, rather than as ions, in ionic equations.
3. Nearly all *soluble salts* are dissociated (ionized) extensively in dilute aqueous solutions, and they are represented as ions in ionic equations. Lead acetate, $Pb(CH_3COO)_2$, and mercury(II) chloride, $HgCl_2$, are two common exceptions; both are *soluble covalent salts.* Other exceptions will be noted as they are encountered.

Insoluble compounds are represented as undissociated in ionic equations.

In Section 7–5 we examined some reactions of strong acids with strong soluble bases to form soluble salts. Let us illustrate one additional example.

Perchloric acid, $HClO_4$, reacts with sodium hydroxide to produce sodium perchlorate, $NaClO_4$, a soluble ionic salt

$$HClO_4\ (aq) + NaOH\ (aq) \longrightarrow NaClO_4\ (aq) + H_2O\ (\ell)$$

The total ionic equation for this reaction is

$$[H^+\ (aq) + ClO_4^-\ (aq)] + [Na^+\ (aq) + OH^-\ (aq)] \longrightarrow [Na^+\ (aq) + ClO_4^-\ (aq)] + H_2O\ (\ell)$$

If we show H^+ (aq) as H_3O^+ here, we must also show $2H_2O$ (ℓ) on the right, rather than H_2O (ℓ).

The perchlorate and sodium ions are common to both sides of the equation, and eliminating them gives the net ionic equation

$$H^+\ (aq) + OH^-\ (aq) \longrightarrow H_2O\ (\ell)$$

This is the same as $H_3O^+ + OH^- \rightarrow 2H_2O$.

This is the net ionic equation for the reaction of all strong acids with strong soluble bases to form soluble salts and water.

There are many reactions of weak acids with strong soluble bases to form soluble salts and water. For example, acetic acid, CH_3COOH, reacts with sodium hydroxide, NaOH, to produce sodium acetate, $NaCH_3COO$, a soluble salt. The molecular equation for this reaction is

$$CH_3COOH\ (aq) + NaOH\ (aq) \longrightarrow NaCH_3COO\ (aq) + H_2O\ (\ell)$$

The total ionic equation for this reaction is

$$CH_3COOH\ (aq) + [Na^+\ (aq) + OH^-\ (aq)] \longrightarrow [Na^+\ (aq) + CH_3COO^-\ (aq)] + H_2O\ (\ell)$$

Elimination of Na^+ from both sides gives the net ionic equation

$$CH_3COOH\ (aq) + OH^-\ (aq) \longrightarrow CH_3COO^-\ (aq) + H_2O\ (\ell)$$

Monoprotic acids contain one, diprotic acids contain two, and triprotic acids contain three acidic (ionizable) hydrogen atoms per formula unit. Polyprotic acids (those that contain more than one ionizable hydrogen atom) will be discussed in detail in Chapter 16.

In general terms, the reaction of a weak monoprotic acid with a strong soluble base to form a soluble salt may be represented as

$$HA\ (aq) + OH^-\ (aq) \longrightarrow A^-\ (aq) + H_2O\ (aq) \qquad \text{(net ionic equation)}$$

EXAMPLE 11–1

Write (a) molecular, (b) total ionic, and (c) net ionic equations for the reaction of phosphoric acid, H_3PO_4, with calcium hydroxide, $Ca(OH)_2$. Assume complete neutralization.

Solution

See the solubility rules in Section 7–1, part 5.

(a) The salt produced is calcium phosphate, $Ca_3(PO_4)_2$, and the molecular equation is

$$2H_3PO_4\ (aq) + 3Ca(OH)_2\ (aq) \longrightarrow Ca_3(PO_4)_2\ (s) + 6H_2O\ (\ell)$$

(b) H_3PO_4 is a weak acid, $Ca(OH)_2$ is a strong soluble base, and $Ca_3(PO_4)_2$ is an insoluble salt. The total ionic equation is

$$2H_3PO_4\ (aq) + 3[Ca^{2+}\ (aq) + 2OH^-\ (aq)] \longrightarrow Ca_3(PO_4)_2\ (s) + 6H_2O\ (\ell)$$

(c) No species are common to both sides of the equation so the net ionic equation is

$$2H_3PO_4\ (aq) + 3Ca^{2+}\ (aq) + 6OH^-\ (aq) \longrightarrow Ca_3(PO_4)_2\ (s) + 6H_2O\ (\ell)$$

EXAMPLE 11–2

Write (a) molecular, (b) total ionic, and (c) net ionic equations for the neutralization of aqueous ammonia with nitric acid.

Solution

(a) The salt that is produced is ammonium nitrate, NH_4NO_3. The molecular equation is

$$HNO_3\ (aq) + NH_3\ (aq) \longrightarrow NH_4NO_3\ (aq)$$

(b) HNO_3 is a strong acid, aqueous ammonia is a weak base, and NH_4NO_3 is a soluble ionic salt. The total ionic equation is

$$[H^+\ (aq) + NO_3^-\ (aq)] + NH_3\ (aq) \longrightarrow [NH_4^+\ (aq) + NO_3^-\ (aq)]$$

(c) Cancelling the spectator ions, NO_3^-, we obtain the net ionic equation

$$H^+\ (aq) + NH_3\ (aq) \longrightarrow NH_4^+\ (aq)$$

This is the net ionic equation for the reaction of any strong acid with aqueous ammonia.

EXAMPLE 11–3

Write (a) molecular, (b) total ionic, and (c) net ionic equations for the neutralization of hydrochloric acid with copper(II) hydroxide.

Solution

(a) The salt produced is copper(II) chloride. The molecular equation is

$$2HCl\ (aq) + Cu(OH)_2\ (s) \longrightarrow CuCl_2\ (aq) + 2H_2O\ (\ell)$$

(b) HCl is a strong acid, $Cu(OH)_2$ is an insoluble base, and $CuCl_2$ is a soluble ionic salt. The total ionic equation is

$$2[H^+\ (aq) + Cl^-\ (aq)] + Cu(OH)_2\ (s) \longrightarrow [Cu^{2+}\ (aq) + 2Cl^-\ (aq)] + 2H_2O\ (\ell)$$

(c) Cancelling the spectator ions, Cl^-, we obtain the net ionic equation

$$2H^+\ (aq) + Cu(OH)_2\ (s) \longrightarrow Cu^{2+}\ (aq) + 2H_2O\ (\ell)$$

There are other kinds of reactions between acids and bases, but we have provided enough examples to illustrate how chemical equations for them are written.

11–7 Acidic Salts and Basic Salts

To this point we have examined acid-base reactions in which stoichiometric amounts of acids and bases have been mixed to form *normal salts.* As the name implies, **normal salts** contain no unreacted H^+ or OH^- ions.

If less than stoichiometric amounts of bases react with *polyprotic* acids, the resulting salts are known as **acidic salts** because they are still capable of neutralizing bases. The reaction of phosphoric acid, H_3PO_4, a weak acid, with strong bases can produce three different salts, depending on the relative amounts of acid and base used.

$$H_3PO_4\ (aq) + NaOH\ (aq) \longrightarrow NaH_2PO_4\ (aq) + H_2O\ (\ell)$$

1 mole　　1 mole　　sodium dihydrogen phosphate, an acidic salt

$$H_3PO_4\ (aq) + 2NaOH\ (aq) \longrightarrow Na_2HPO_4\ (aq) + 2H_2O\ (\ell)$$

1 mole　　2 moles　　sodium hydrogen phosphate, an acidic salt

$$H_3PO_4\ (aq) + 3NaOH\ (aq) \longrightarrow Na_3PO_4\ (aq) + 3H_2O\ (\ell)$$

1 mole　　3 moles　　sodium phosphate, a normal salt

There are many additional examples of acidic salts. Sodium hydrogen carbonate, $NaHCO_3$, is commonly called sodium bicarbonate and is properly classified as an acidic salt. However, it is the acidic salt of an extremely weak acid, carbonic acid, H_2CO_3, and solutions of sodium bicarbonate are slightly basic. In the older system for naming compounds, the prefix "bi-" indicates the presence of an acidic hydrogen atom but not the acidic or basic character of solutions of the salt. Sodium bisulfate is $NaHSO_4$, and its proper name is sodium hydrogen sulfate.

Acidic salts are capable of neutralizing bases, but their solutions are not necessarily acidic.

When *polyhydroxy bases* (bases that contain more than one OH per formula unit) react with less than stoichiometric amounts of acids they form **basic salts,** i.e., salts that contain unreacted OH groups. For example, the reaction of aluminum hydroxide with hydrochloric acid can produce three different salts.

The basic aluminum salts are common components of "roll-on" deodorants ("aluminum chlorohydrate").

$Al(OH)_3$ (s) + HCl (aq)	⟶	$Al(OH)_2Cl$ (s)	+ H_2O (ℓ)
1 mole 1 mole		aluminum dihydroxide chloride, a basic salt	
$Al(OH)_3$ (s) + $2HCl$ (aq)	⟶	$Al(OH)Cl_2$ (s)	+ $2H_2O$ (ℓ)
1 mole 2 moles		aluminum hydroxide dichloride, a basic salt	
$Al(OH)_3$ (s) + $3HCl$ (aq)	⟶	$AlCl_3$ (aq)	+ $3H_2O$ (ℓ)
1 mole 3 moles		aluminum chloride, a normal salt	

Aqueous solutions of *basic* salts can neutralize acids, but are not necessarily basic, and most basic salts are rather insoluble in water.

11–8 Strengths of Ternary Acids and Amphoterism

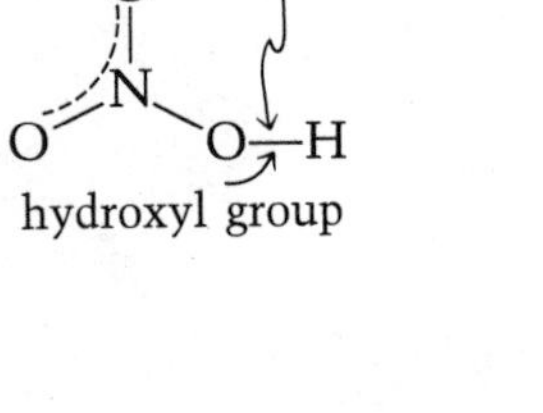

Ternary acids ionize to produce H^+ (aq), but they are *hydroxides of nonmetals.* The formula for nitric acid is commonly written HNO_3 to emphasize the presence of an acidic hydrogen atom, but it could also be written as $NO_2(OH)$ as its structure shows (see margin).

We usually reserve the term hydroxide for substances that produce basic solutions and call the other "hydroxides" acids because they ionize to produce H^+ (aq). In ternary acids the hydroxyl oxygen is bonded to a fairly electronegative element such as nitrogen. In nitric acid the nitrogen draws the electrons of the N—O (hydroxyl) bond more closely toward itself than would a less electronegative element such as sodium. The oxygen pulls the electrons of the O—H bond close enough so that the hydrogen atom ionizes as H^+, leaving NO_3^-

$$HNO_3\ (\ell) \xrightarrow{H_2O} H^+\ (aq) + NO_3^-\ (aq)$$

In contrast let us consider hydroxides of metals. Oxygen is so much more electronegative than most metals, such as sodium, that it draws the electrons of the sodium-oxygen bond in NaOH (a strong soluble base) close enough to itself the bonding is ionic, and therefore NaOH exists as Na^+ and OH^- ions, even in the solid state.

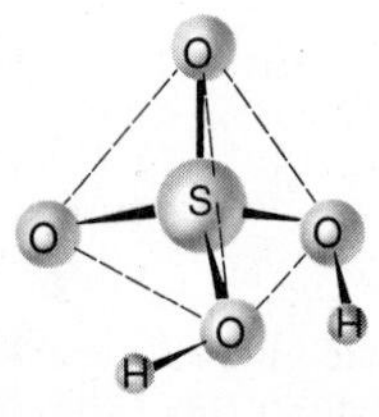

We usually write the formula for sulfuric acid as H_2SO_4 to emphasize the fact that it is an acid. However, the formula can also be written as $(HO)_2SO_2$, because the structure of sulfuric acid (margin) shows clearly that H_2SO_4 contains two —O—H groups bound to a sulfur atom. Since the O—H bonds are easier to break than the S—O bonds, sulfuric acid ionizes as an acid when it is dissolved in water.

Sulfuric acid is called a polyprotic acid because it has more than one ionizable hydrogen atom per molecule. It is the only common polyprotic acid that is also a strong acid.

1st Step: $$H_2SO_4\ (\ell) \xrightarrow{H_2O} H^+\ (aq) + HSO_4^-\ (aq)$$

2nd Step: $$HSO_4^-\ (aq) \xrightarrow{H_2O} H^+\ (aq) + SO_4^{2-}\ (aq)$$

The first step in the ionization of H_2SO_4 is complete in dilute aqueous solution. The second step is nearly complete only in very dilute solutions. As we shall see in Section 16–10, the first step in the ionization of a polyprotic acid occurs to a greater extent than the second step.

Sulfurous acid, H_2SO_3, is a polyprotic acid that contains the same elements as H_2SO_4. However, H_2SO_3 is a weak acid, which tells us that the H—O bonds in H_2SO_3 are stronger than those in H_2SO_4.

Comparison of the acid strengths of nitric acid, HNO_3, and nitrous acid, HNO_2, shows that HNO_3 is a much stronger acid than HNO_2. Similar observations show that *acid strengths of most ternary acids containing the same central element increase with the oxidation state of the central element and with increasing numbers of oxygen atoms.* The following orders of increasing strength are typical

$H_2SO_3 < H_2SO_4$
$HNO_2 < HNO_3$ (strongest acids are on the right side)
$HClO < HClO_2 < HClO_3 < HClO_4$

Acid strengths of most ternary acids containing different elements in the same oxidation state from the same group in the periodic table increase with increasing electronegativity of the central element, as in the following examples

$H_2SeO_4 < H_2SO_4$ $H_2SeO_3 < H_2SO_3$
$H_3PO_4 < HNO_3$
$HBrO_4 < HClO_4$ $HBrO_3 < HClO_3$

Contrary to what we might expect, H_3PO_3 is a stronger acid than HNO_2.

A word of caution is necessary. Care must be exercised to compare acids that have *similar structures.* For example, H_3PO_2, which contains two H atoms bonded to the P atom, is a stronger acid than H_3PO_3, which contains one H atom bonded to the P atom. H_3PO_3 is a stronger acid than H_3PO_4 which has no H atoms bonded to the P atom.

Remember that in *most* ternary inorganic acids, all H atoms are bonded to O.

As we have noted earlier, whether a particular substance behaves as an acid or as a base depends on its environment. In Section 11–3 we noted the *amphiprotic nature of water.* **Amphoterism** is the general term that describes the ability of a substance to react either as an acid or as a base. Amphiprotic behavior describes the particular case in which a substance exhibits amphoterism by accepting and by donating a proton, H^+.

Several *insoluble* metal hydroxides are amphoteric, i.e., they react with acids to form salts and water as expected, but they also dissolve in and react with excess strong soluble bases.

The energy of hydration of a small, highly charged cation is quite high. However, such a cation makes so great a contribution to the lattice energy of its hydroxide that all hydroxides containing small highly charged metal ions are insoluble in water.

Aluminum hydroxide is a typical amphoteric metal hydroxide. Its behavior as *a base* may be illustrated by its reaction with nitric acid to form a *salt*

$$Al(OH)_3\ (s) + 3HNO_3\ (aq) \longrightarrow Al(NO_3)_3\ (aq) + 3H_2O\ (\ell)$$

$$Al(OH)_3\ (s) + 3[H^+\ (aq) + NO_3^-\ (aq)] \longrightarrow [Al^{3+}\ (aq) + 3NO_3^-\ (aq)] + 3H_2O\ (\ell)$$

$$Al(OH)_3\ (s) + 3H^+\ (aq) \longrightarrow Al^{3+}\ (aq) + 3H_2O\ (\ell)$$

As seen from the net ionic equation, this is a typical neutralization, with $Al(OH)_3$ as the base.

When an excess of sodium hydroxide solution (or any other strong soluble base) is added to solid aluminum hydroxide, the aluminum hydroxide dissolves. The equation for the reaction is usually written as

The proper name of "sodium aluminate" is *sodium tetrahydroxoaluminate.* See Section 27–4.

$$\underset{\text{an acid}}{Al(OH)_3\ (s)} + \underset{\text{a base}}{NaOH\ (aq)} \longrightarrow \underset{\text{sodium aluminate, a soluble compound}}{NaAl(OH)_4\ (aq)}$$

The total ionic and net ionic equations are

$$Al(OH)_3\ (s) + [Na^+\ (aq) + OH^-\ (aq)] \longrightarrow [Na^+\ (aq) + Al(OH)_4^-\ (aq)]$$

$$Al(OH)_3\ (s) + OH^-\ (aq) \longrightarrow Al(OH)_4^-\ (aq)$$

We have indicated that ions are hydrated in aqueous solution, that is, bound to water molecules. Many solid compounds also contain water molecules bound to ions. We have written the formula for aluminum hydroxide as $Al(OH)_3$ (and many other formulas similarly) for simplicity. The formula for aluminum hydroxide is more accurately written as $Al(OH_2)_3(OH)_3$, because three water molecules are bound to the aluminum ion. Its structure may be represented as

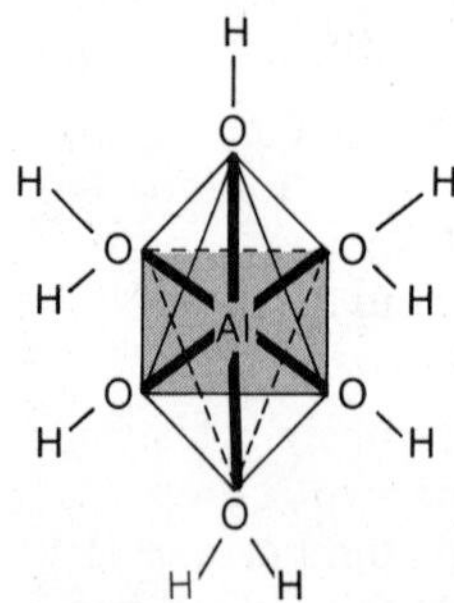

which shows that the δ− ends of polar H_2O molecules (Section 5–14) are bound to the positive aluminum ion. This sort of bonding is called **coordination.**

The H—O bonds in the water molecules are slightly weaker than in uncoordinated H_2O molecules because the positively charged aluminum ion withdraws some electron density from the O atoms. Thus, the H atoms in coordinated H_2O molecules are slightly more acidic than are H atoms in uncoordinated H_2O molecules. The hydroxide ion from a strong soluble base can attack one of the coordinated H_2O molecules and abstract (remove) a proton, which leaves another OH^- bound to the aluminum ion.

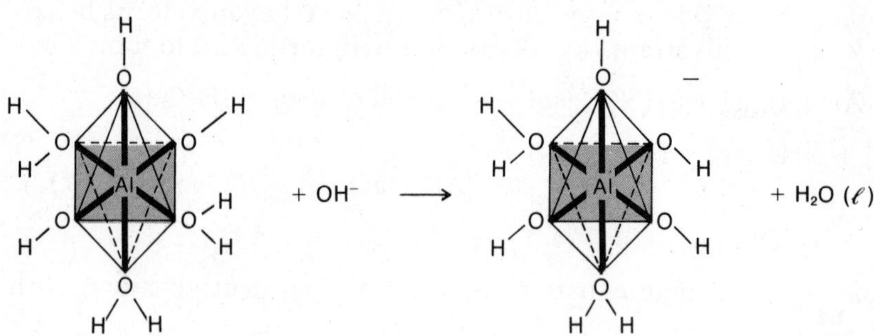

The reaction may be represented in condensed form as

$$Al(OH_2)_3(OH)_3\ (s) + OH^-\ (aq) \longrightarrow Al(OH_2)_2(OH)_4^- + H_2O\ (\ell)$$

In simplified notation, it becomes the same as the net ionic equation we saw before

$$Al(OH)_3 + OH^-\ (aq) \longrightarrow Al(OH)_4^-\ (aq)$$

Other amphoteric metal hydroxides undergo similar reactions.

Table 11–3 contains a list of the common amphoteric hydroxides. Three are hydroxides of the metalloids, the elements located along the line that divides metals and nonmetals in the periodic table.

Generally, it is elements of intermediate electronegativity that form amphoteric hydroxides; those of high and low electronegativity form acidic and basic "hydroxides," respectively.

TABLE 11–3 Amphoteric Hydroxides

Metal or Metalloid Ions	Insoluble Amphoteric Hydroxide	Complex Ion Formed in an Excess of a Strong Soluble Base
Be^{2+}	$Be(OH)_2$	$[Be(OH)_4]^{2-}$
Al^{3+}	$Al(OH)_3$	$[Al(OH)_4]^-$
Cr^{3+}	$Cr(OH)_3$	$[Cr(OH)_4]^-$
Zn^{2+}	$Zn(OH)_2$	$[Zn(OH)_4]^{2-}$
Sn^{2+}	$Sn(OH)_2$	$[Sn(OH)_3]^-$
Sn^{4+}	$Sn(OH)_4$	$[Sn(OH)_6]^{2-}$
Pb^{2+}	$Pb(OH)_2$	$[Pb(OH)_4]^{2-}$
As^{3+}*	$As(OH)_3$	$[As(OH)_4]^-$
Sb^{3+}*	$Sb(OH)_3$	$[Sb(OH)_4]^-$
Si^{4+}*	$Si(OH)_4$	SiO_4^{4-} and SiO_3^{2-}
Co^{2+}†	$Co(OH)_2$	$[Co(OH)_4]^{2-}$
Cu^{2+}†	$Cu(OH)_2$	$[Cu(OH)_4]^{2-}$

* As, Sb, and Si are metalloids.

† $Co(OH)_2$ and $Cu(OH)_2$ are only slightly amphoteric, i.e., a very large excess of strong soluble base is required to dissolve small amounts of these insoluble hydroxides.

11–9 The Preparation of Acids

Binary acids may be prepared by combination of appropriate elements with hydrogen (Section 7–8, part 2).

Small quantities of the hydrogen halides (their solutions are called hydrohalic acids), and other **volatile acids,** are usually prepared by dropping concentrated nonvolatile acids onto the appropriate salts. The reactions of concentrated (96%) sulfuric acid with solid sodium fluoride and sodium chloride evolve heat and produce gaseous hydrogen fluoride and hydrogen chloride, as the following equations show

Sulfuric and phosphoric acids are classified as nonvolatile acids because they have much higher boiling points than other common acids.

$H_2SO_4\ (\ell)$ +	$NaF\ (s)$	$\longrightarrow$	$NaHSO_4\ (s)$ +	$HF\ (g)$
sulfuric acid bp = 338°C	sodium fluoride		sodium hydrogen sulfate	hydrogen fluoride bp = 19.59°C
$H_2SO_4\ (\ell)$ +	$NaCl\ (s)$	$\longrightarrow$	$NaHSO_4\ (s)$ +	$HCl\ (g)$
	sodium chloride			hydrogen chloride bp = −84.9°C

Because concentrated sulfuric acid is a fairly strong oxidizing agent, it cannot be used to prepare hydrogen bromide and hydrogen iodide; instead, the free halogens are produced. Phosphoric acid, a nonoxidizing acid, is dropped onto solid sodium bromide and sodium iodide to produce hydrogen bromide and hydrogen iodide, as the following equations show

$$H_3PO_4\ (\ell) + NaBr\ (s) \xrightarrow{\Delta} NaH_2PO_4\ (s) + HBr\ (g)$$

phosphoric acid bp = 213°C	sodium bromide	sodium dihydrogen phosphate	hydrogen bromide bp = −67.0°C

$$H_3PO_4\ (\ell) + NaI\ (s) \xrightarrow{\Delta} NaH_2PO_4\ (s) + HI\ (g)$$

	sodium iodide		hydrogen iodide bp = −35°C

This kind of reaction may be generalized as the reaction of *a nonvolatile acid with the salt of a volatile acid to produce the volatile acid and a salt of the nonvolatile acid.*

If dilute sulfuric acid is added to substances containing sulfide ions, volatile hydrogen sulfide is evolved. Consider the reaction of solid sodium sulfide with sulfuric acid to produce sodium hydrogen sulfate and hydrogen sulfide

$$Na_2S\ (s) + 2H_2SO_4\ (\ell) \longrightarrow 2NaHSO_4\ (aq) + H_2S\ (g)$$

Some volatile *ternary* acids can also be prepared by evolution from their salts. Addition of concentrated sulfuric acid to sodium acetate produces acetic acid

As the five preceding examples show, volatile is a relative term.

$$NaCH_3COO\ (s) + H_2SO_4\ (\ell) \xrightarrow{\Delta} NaHSO_4\ (s) + CH_3COOH\ (g)$$

Acetic acid (bp = 118°C) is not sufficiently volatile to be vaporized from the reaction mixture, but can be distilled out of it. Similarly, the reaction of sulfuric acid with solid sodium nitrate produces nitric acid

$$NaNO_3\ (s) + H_2SO_4\ (\ell) \xrightarrow{\Delta} NaHSO_4\ (s) + HNO_3\ (g)$$

Although nitric acid (bp = 83°C) is not particularly volatile, it is when compared to H_2SO_4 (bp of concentrated H_2SO_4 solution = 338°C).

In Section 7–9, part 2 we pointed out the fact that many nonmetal oxides, called acid anhydrides, react with water to form **ternary acids** with no changes in oxidation numbers. Some additional examples are given here.

Carbon dioxide dissolves in water to form carbonic acid, H_2CO_3, an extremely weak acid

$$\overset{+4}{C}O_2\ (g) + H_2O\ (\ell) \rightleftharpoons H_2\overset{+4}{C}O_3\ (aq)$$

Dichlorine heptoxide, Cl_2O_7, forms perchloric acid when it is dissolved in water.

$$\overset{+7}{Cl_2}O_7\ (\ell) + H_2O\ (\ell) \longrightarrow 2H\overset{+7}{Cl}O_4\ (aq)$$

Since $HClO_4$ is a strong acid (in fact, the strongest of all inorganic acids), the equation is more accurately written as

See Section 5–24 for the rules of naming these compounds.

$$\overset{+7}{Cl_2O_7}\ (\ell) + H_2O\ (\ell) \longrightarrow 2[H^+\ (aq) + \overset{+7}{ClO_4^-}\ (aq)]$$

Some *high oxidation state transition metal oxides* are acidic oxides, that is, they dissolve in water to give solutions of ternary acids. Manganese(VII) oxide, Mn_2O_7, and chromium(VI) oxide, CrO_3, are the most common examples

$$\underset{\text{manganese(VII) oxide}}{\overset{+7}{Mn_2O_7}\ (\ell)} + H_2O\ (\ell) \longrightarrow \underset{\text{permanganic acid}}{2[H^+\ (aq) + \overset{+7}{MnO_4^-}\ (aq)]}$$

Note the similarity between this reaction and the preparation of perchloric acid.

Chromium(VI) oxide is a deep red solid that dissolves in water to give a deep red solution containing dichromic acid, $H_2Cr_2O_7$

$$\underset{\text{chromium(VI) oxide}}{2\overset{+6}{CrO_3}\ (s)} + H_2O\ (\ell) \longrightarrow \underset{\text{dichromic acid}}{[2H^+\ (aq) + \overset{+6}{Cr_2O_7^{2-}}\ (aq)]}$$

Neither permanganic acid nor dichromic acid has been isolated in pure form.

The halides and oxyhalides of some nonmetals hydrolyze (react with water) to produce two acids—a (binary) hydrohalic acid and a (ternary) oxyacid of the nonmetal, with no changes in oxidation numbers.

The dissolution of phosphorus trihalides in water produces the corresponding hydrohalic acids and phosphorous acid, a weak diprotic acid (one H atom is bonded directly to phosphorus and is not acidic)

$$\overset{+3\ -1}{PX_3} + 3H_2O\ (\ell) \longrightarrow \overset{+3}{H_3PO_3}\ (aq) + 3H\overset{-1}{X}\ (aq)$$

As a specific example, consider the reaction of phosphorus trichloride with water

$$\underset{\text{phosphorus trichloride}}{\overset{+3\ -1}{PCl_3}\ (\ell)} + 3H_2O\ (\ell) \longrightarrow \underset{\text{phosphorous acid}}{\overset{+3}{H_3PO_3}\ (aq)} + \underset{\text{hydrochloric acid}}{3H\overset{-1}{Cl}\ (aq)}$$

Both phosphorus pentachloride and phosphorus oxytrichloride (phosphoryl chloride) react with water to produce phosphoric acid and hydrochloric acid

$$\underset{\text{phosphorus pentachloride}}{\overset{+5\ -1}{PCl_5}\ (s)} + 4H_2O\ (\ell) \longrightarrow \underset{\text{phosphoric acid}}{\overset{+5}{H_3PO_4}\ (aq)} + 5H\overset{-1}{Cl}\ (aq)$$

$$\underset{\text{phosphoryl chloride}}{\overset{+5\ \ -1}{POCl_3}\ (\ell)} + 3H_2O\ (\ell) \longrightarrow \overset{+5}{H_3PO_4}\ (aq) + 3H\overset{-1}{Cl}\ (aq)$$

11–10 The Preparation of Strong Soluble Bases

Small amounts of strong soluble bases are usually prepared by dissolving the appropriate metal oxide, called a basic anhydride, in water. In Section 7–9, part 2 we presented reactions for the preparation of sodium and barium hydroxides.

Lithium oxide dissolves in water to give a solution of lithium hydroxide

$$Li_2O\ (s) + H_2O\ (\ell) \longrightarrow 2[Li^+\ (aq) + OH^-\ (aq)]$$

This reaction, and others similar to it, involve the reaction of the oxide ion, O^{2-}, with water to form hydroxide ions

$$O^{2-} + H_2O\ (\ell) \longrightarrow 2OH^-$$

Like the other soluble metal oxides, lithium oxide is an ionic compound. Recall that the hydroxide ion is the strongest base that can exist in aqueous solution (Table 11–2). The oxide ion is a much stronger base that reacts with water instantaneously and irreversibly to form hydroxide ions. All of the Group IA metal oxides as well as the heavier Group IIA metal oxides (CaO, SrO, BaO) undergo similar reactions.

Peroxides of the IA metals and the heavier IIA metals react with water to form metal hydroxides and liberate oxygen

$$\underset{\text{sodium peroxide}}{2Na_2O_2\ (s)} + 2H_2O\ (\ell) \longrightarrow 4[Na^+\ (aq) + OH^-\ (aq)] + O_2\ (g)$$

The alkali metals and heavier alkaline earth metals reduce water to produce solutions of metal hydroxides and liberate hydrogen gas

The reactions of the IA metals liberate enough energy to ignite hydrogen. Extreme caution must be exercised.

$$2Na\ (s) + 2H_2O\ (\ell) \longrightarrow 2[Na^+\ (aq) + OH^-\ (aq)] + H_2\ (g)$$

$$Ca\ (s) + 2H_2O\ (\ell) \longrightarrow [Ca^{2+}\ (aq) + 2OH^-\ (aq)] + H_2\ (g)$$

Sodium hydroxide is prepared commercially by the electrolysis of sodium chloride solutions (Section 19–4)

$$2[Na^+\ (aq) + Cl^-\ (aq)] + 2H_2O \xrightarrow{\text{electrolysis}} 2[Na^+\ (aq) + OH^-\ (aq)] + Cl_2\ (g) + H_2\ (g)$$

The process utilizes salt and water as starting materials and produces three important industrial chemicals, sodium hydroxide, chlorine, and hydrogen.

11–11 The Preparation of Insoluble Bases

Small amounts of insoluble bases are normally prepared by the reaction of a soluble salt containing the metal ion of interest with a strong soluble base for cases in which metal oxides do not react with water and are not soluble in water. In many cases aqueous ammonia, a weak base, may be used instead of a strong soluble base. When a solution of copper(II) nitrate, or some other soluble copper(II) salt, is treated with a stoichiometric amount of sodium hydroxide solution, insoluble copper(II) hydroxide precipitates from the solution

$$Cu(NO_3)_2\ (aq) + 2NaOH\ (aq) \longrightarrow Cu(OH)_2\ (s) + 2NaNO_3\ (aq)$$

$$[Cu^{2+}\ (aq) + 2NO_3^-\ (aq)] + 2[Na^+\ (aq) + OH^-\ (aq)] \longrightarrow Cu(OH)_2\ (s) + 2[Na^+\ (aq) + NO_3^-\ (aq)]$$

Cu^{2+} (aq) + $2OH^-$ (aq) $\longrightarrow$ $Cu(OH)_2$ (s)

The net ionic equation correctly suggests that most soluble copper(II) salts react with most strong soluble bases to form insoluble copper(II) hydroxide and a soluble salt.

Caution must be exercised in choosing reactants for precipitation reactions so that only one insoluble compound is formed. For example, mixing aqueous solutions of copper(II) sulfate and barium hydroxide produces insoluble copper(II) hydroxide, but barium sulfate, the other product of the reaction, is also insoluble

Such reactions are of no value for preparative purposes because they produce a mixture of two insoluble compounds that cannot be separated easily.

$CuSO_4$ (aq) + $Ba(OH)_2$ (aq) $\longrightarrow$ $Cu(OH)_2$ (s) + $BaSO_4$ (s)

Magnesium hydroxide, $Mg(OH)_2$, is insoluble in water. It may be prepared by reaction of a soluble salt, such as magnesium chloride, with a strong soluble base such as calcium hydroxide

$MgCl_2$ (aq) + $Ca(OH)_2$ (aq) $\longrightarrow$ $Mg(OH)_2$ (s) + $CaCl_2$ (aq)

$[Mg^{2+}$ (aq) + $2Cl^-$ (aq)] + $[Ca^{2+}$ (aq) + $2OH^-$ (aq)] $\longrightarrow$
$Mg(OH)_2$ (s) + $[Ca^{2+}$ (aq) + $2Cl^-$ (aq)]

Mg^{2+} (aq) + $2OH^-$ (aq) $\longrightarrow$ $Mg(OH)_2$ (s)

11–12 The Lewis Theory

In 1923 Professor G. N. Lewis presented the most comprehensive of the classical acid-base theories. In the Lewis view, an **acid** *is defined as any species that can accept a share in an electron pair.* A **base** *is any species that can make available or "donate" a share in an electron pair.* The definitions do *not* specify that an electron pair must be transferred from one atom to another—only that an electron pair, residing originally on one atom, be shared between two atoms. *Neutralization* is defined as the formation of a **coordinate covalent bond,** which is a covalent bond in which both electrons were furnished by one atom.

The reaction of boron trifluoride with ammonia is a typical Lewis acid-base reaction

BF_3 (g) + NH_3 (g) $\longrightarrow$ $F_3B:NH_3$

acid base product

```
  :F:       H           :F:  H
:F:B   +  :N:H  ⟶   :F:B : N:H
  :F:       H           :F:  H
```

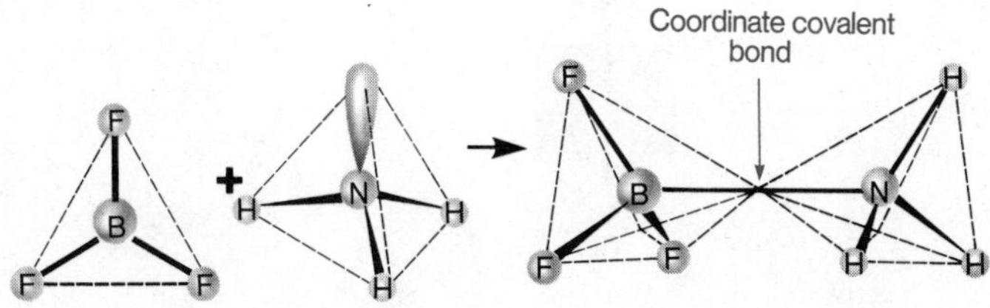

The Lewis theory is sufficiently general that it covers *all* acid-base reactions that the other theories include, plus many additional reactions such as complex formation (Chapter 27).

The autoionization of water (Section 11–3) was described in terms of the Brønsted-Lowry theory. In Lewis theory terminology, this is also an acid-base reaction. The acceptance of a proton, H^+, by a base involves the formation of a coordinate covalent bond

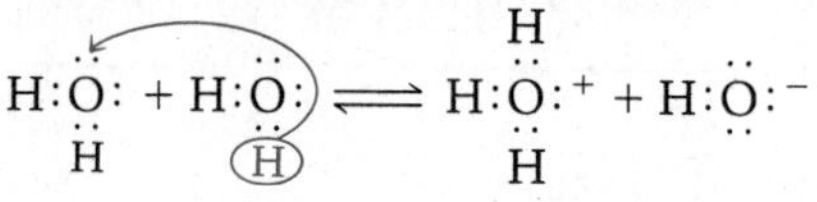

$$\mathrm{H\!:\!\ddot{\underset{\cdot\cdot}{O}}\!:\!H} + \mathrm{H\!:\!\ddot{\underset{\cdot\cdot}{O}}\!:\!H} \rightleftharpoons \left[\mathrm{H\!:\!\ddot{O}\!:\!H}\right]^+ (\text{with third H}) + \mathrm{H\!:\!\ddot{\underset{\cdot\cdot}{O}}\!:}^-$$

base acid

Theoretically, any species that contains an unshared electron pair is a base. In fact, most ions and molecules that contain unshared electron pairs undergo some reactions by sharing their electron pairs. Conversely, many Lewis acids contain only six electrons in the highest occupied energy level of the central element, and react by accepting a share in an additional pair of electrons. These species are said to have an **open sextet.** Many compounds of the Group IIIA elements are Lewis acids, as illustrated by the reaction of boron trifluoride with ammonia presented above.

Anhydrous aluminum chloride is a common Lewis acid that is used to catalyze many organic reactions. The dissolution of $AlCl_3$ in hydrochloric acid gives a solution that contains $AlCl_4^-$ ions

$$AlCl_3\ (s) + Cl^-\ (aq) \longrightarrow AlCl_4^-\ (aq)$$

acid base product

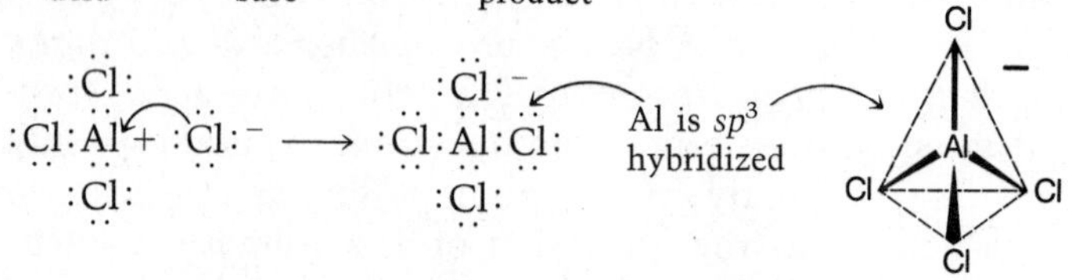

$$:\!\ddot{Cl}\!:\!Al + :\!\ddot{Cl}\!:^- \longrightarrow :\!\ddot{Cl}\!:\!\ddot{Al}\!:\!\ddot{Cl}\!:^-$$

Al is sp^3 hybridized

Cl, Al, Cl, Cl, Cl, −

Other ions and molecules behave as Lewis acids by expansion of the valence shell of the central element. Anhydrous tin(IV) chloride, often called stannic chloride, is a colorless liquid that also is frequently used as a Lewis acid catalyst. The tin atom (Group IVA) can expand its valence shell by utilizing vacant *d* orbitals and so can accept shares in two additional electron pairs, as its reaction with hydrochloric acid illustrates

$$SnCl_4\ (\ell) + 2Cl^-\ (aq) \longrightarrow SnCl_6^{2-}\ (aq)$$

acid base

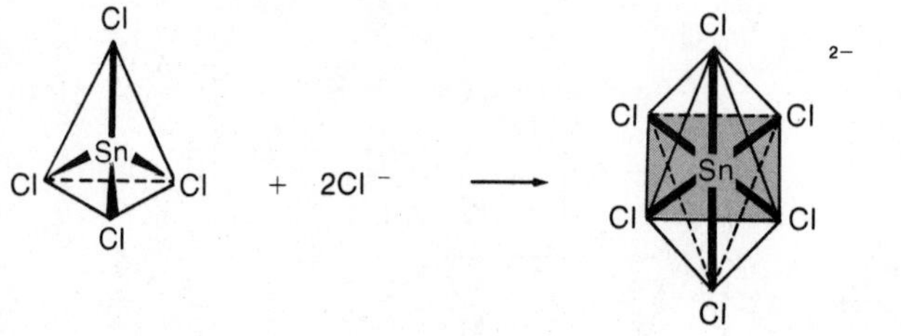

Cl, Sn, Cl, Cl, Cl + $2Cl^-$ ⟶ Cl, Cl, Cl, Sn, Cl, Cl, Cl, 2−

Sn is sp^3 hybridized

Sn is sp^3d^2 hybridized

Experienced chemists find the Lewis theory to be very useful because so many chemical reactions are covered by it. The less experienced usually find the theory less useful, but as their knowledge expands so does the utility of the Lewis theory.

11–13 The Solvent System Theory

Water is by far the most common solvent for chemical reactions. However, many other solvents have been studied extensively and some are useful media for chemical reactions.

Although ammonia is a gas at room temperature, it is easily liquefied to a clear, colorless liquid (bp = −33.4°C) that looks like water. Liquid ammonia is an excellent solvent for many reactions that don't occur in water, and it is strikingly similar to water in many instances.

The autoionization of liquid ammonia occurs to a much smaller extent than that of water, but both reactions involve the transfer of a single proton from one solvent molecule to another

pure H_2O: $$H_2O\ (\ell) + H_2O\ (\ell) \rightleftharpoons H_3O^+\ (aq) + OH^-\ (aq)$$

liq. NH_3: $$NH_3\ (\ell) + NH_3\ (\ell) \rightleftharpoons \underset{\text{ammonium ion}}{NH_4^+} + \underset{\text{amide ion}}{NH_2^-}$$

The inquisitive student may wonder if ions are solvated in liquid ammonia. The answer is yes, but the notation has been omitted from the equations for reactions in liquid ammonia.

In the solvent system theory, an **acid** is defined as any substance that produces the cation that is characteristic of the solvent, and a **base** is any substance that produces the anion that is characteristic of the solvent. Examination of the equations for the autoionization of pure water and pure ammonia given above tells us the following

In water:

acid = any substance that produces H_3O^+
base = any substance that produces OH^-

In liquid ammonia:

acid = any substance that produces NH_4^+
base = any substance that produces NH_2^-

Neutralization is defined as the formation of solvent molecules, that is,

In water:

$$H_3O^+\ (aq) + OH^-\ (aq) \longrightarrow 2H_2O\ (\ell)$$

In liquid ammonia:

$$NH_4^+ + NH_2^- \longrightarrow 2NH_3$$

Ammonium salts are acids in liquid ammonia, and in many cases their reactions are similar to reactions of protonic acids in water. The metal amides are bases in liquid ammonia, and many of their reactions are similar to reactions of metal hydroxides in water. The following examples illustrate this.

1. **Neutralization**

(H_2O)

$$HNO_3 + NaOH \longrightarrow NaNO_3 + H_2O$$

$$[H_3O^+ + NO_3^-] + [Na^+ + OH^-] \longrightarrow [Na^+ + NO_3^-] + 2H_2O$$

$$H_3O^+ + OH^- \longrightarrow 2H_2O$$

Physical states and solvation are omitted from these equations so that attention is easily focused on the reactions.

(liq. NH$_3$) $NH_4NO_3 + NaNH_2 \longrightarrow NaNO_3 + 2NH_3$

ammonium nitrate, sodium amide, sodium nitrate

$$[NH_4^+ + NO_3^-] + [Na^+ + NH_2^-] \longrightarrow [Na^+ + NO_3^-] + 2NH_3$$

$$NH_4^+ + NH_2^- \longrightarrow 2NH_3$$

2. **Reaction of Acids with Metal Oxides**

(H$_2$O) $2HNO_3 + CaO \longrightarrow Ca(NO_3)_2 + H_2O$

(liq. NH$_3$) $2NH_4NO_3 + CaO \longrightarrow Ca(NO_3)_2 + 2NH_3 + H_2O$

3. **Reaction of Salts with Strong Soluble Bases to Form Insoluble Bases**

(H$_2$O) $3NaOH + Al(NO_3)_3 \longrightarrow Al(OH)_3\ (s) + 3NaNO_3$

(liq. NH$_3$) $3NaNH_2 + Al(NO_3)_3 \longrightarrow Al(NH_2)_3\ (s) + 3NaNO_3$

sodium amide, aluminum amide

4. **Amphoteric Behavior of Bases**

(H$_2$O) excess $NaOH + Al(OH)_3\ (s) \longrightarrow NaAl(OH)_4$

(liq. NH$_3$) excess $NaNH_2 + Al(NH_2)_3\ (s) \longrightarrow NaAl(NH_2)_4$

There are many other kinds of reactions that occur in liquid ammonia and that are quite similar to those that occur in water. The four types presented are typical.

Many other solvents have been studied as reaction media. Reactions that occur best in a strongly acidic medium may be run in anhydrous sulfuric acid. Liquid hydrogen fluoride, liquid sulfur dioxide, and liquid phosgene, $COCl_2$, are typical nonaqueous solvents that have been investigated in some detail. Many organic reactions occur in organic solvents such as ethyl alcohol, diethyl ether, or benzene.

Some organic reactions will be discussed in Chapters 29 and 30.

Concentrations and Acid-Base Reactions in Aqueous Solution

11–14 Calculations Involving Molarity

In Sections 2–13, 2–14, and 10–8 and 10–9, we introduced methods for expressing concentrations of solutions. Reference to these sections may be helpful as we learn more about acid-base reactions in solutions.

In *some cases* one mole of an acid reacts with one mole of a base as indicated in the following examples

$$HCl + NaOH \longrightarrow NaCl + H_2O$$

$$HNO_3 + KOH \longrightarrow KNO_3 + H_2O$$

$$HClO_4 + NH_3 \longrightarrow NH_4ClO_4$$

Since one mole of each acid reacts with one mole of each base in these examples, *one liter of a one molar solution of any of these acids* reacts with *one liter of a one molar solution of any of these bases.* Note that the acids used in the above examples have only one acidic hydrogen per formula unit, and that the bases have one hydroxide ion per formula or (in the case of ammonia) that one formula unit of base reacts with one hydrogen ion.

EXAMPLE 11–4

If 100 mL of 1.00 *M* HCl solution and 100mL of 1.00 *M* NaOH are mixed, what is the molarity of the salt in the resulting solution? You may assume that the volumes are additive, because experiments have shown that volumes of dilute aqueous solutions are very nearly additive and no significant error is introduced by making this assumption.

Solution

The following tabulation shows that equal numbers of moles (or millimoles) of HCl and NaOH are mixed, and therefore the resulting solution contains only NaCl, the salt formed by the reaction, and water

	NaOH	+	HCl	$\longrightarrow$	NaCl + H_2O
reaction ratio:	1 mol		1 mol		1 mol 1 mol
start:	$100\text{ mL}\left(\frac{1.00\text{ mmol}}{\text{mL}}\right)$		$100\text{ mL}\left(\frac{1.00\text{ mmol}}{\text{mL}}\right)$		
	100 mmol NaOH		100 mmol HCl		
after reaction:	0 mmol		0 mmol		100 mmol NaCl

The HCl and NaOH neutralize each other exactly, and the resulting solution contains 100 mmol of NaCl in 200 mL of solution. Its molarity is:

$$\underline{?}\ \frac{\text{mmol NaCl}}{\text{mL}} = \frac{100\text{ mmol}}{200\text{ mL}} = \underline{0.500\ M\text{ NaCl}}$$

EXAMPLE 11–5

If 100 mL of 1.00 *M* HCl solution and 100 mL of 0.75 *M* NaOH are mixed, what is the molarity of the resulting solution?

Solution

The following tabulation shows that some of the HCl remains unreacted and some NaCl is produced, but no NaOH remains

	HCl	+ NaOH	$\longrightarrow$	NaCl	+ H_2O
reaction ratio:	1 mol	1 mol		1 mol	1 mol
start:	100 mmol	75 mmol			
after reaction:	25 mmol HCl	0 mmol		75 mmol NaCl	

Since two solutes are present in the solution after reaction, we must calculate the concentrations of both

$$\underline{?}\ \frac{\text{mmol HCl}}{\text{mL}} = \frac{25\text{ mmol HCl}}{200\text{ mL}} = \underline{0.12\ M\text{ HCl}}$$

$$? \frac{\text{mmol NaCl}}{\text{mL}} = \frac{75 \text{ mmol NaCl}}{200 \text{ mL}} = 0.38\ M \text{ NaCl}$$

Both HCl and NaCl are strong electrolytes, so the solution is 0.12 M in H^+ (aq), (0.12 + 0.38) M = 0.50 M in Cl^- and 0.38 M in Na^+ ions.

In many cases one mole of an acid will not neutralize one mole of a base, or one mole of a base will not neutralize one mole of an acid, as shown in the following examples

$$\begin{array}{llll} H_2SO_4 + & 2NaOH & \longrightarrow Na_2SO_4 + 2H_2O \\ 1 \text{ mol} & 2 \text{ mol} & \quad 1 \text{ mol} \end{array}$$

$$\begin{array}{llll} 2HCl + & Ca(OH)_2 & \longrightarrow CaCl_2 + 2H_2O \\ 2 \text{ mol} & 1 \text{ mol} & \quad 1 \text{ mol} \end{array}$$

The first equation shows that one mole of H_2SO_4 reacts with two moles of NaOH, and therefore *two* liters of 1 molar NaOH solution are required to neutralize one liter of 1 molar H_2SO_4 solution. The second equation shows that two moles of HCl react with one mole of $Ca(OH)_2$, and therefore *two* liters of HCl solution are required to neutralize one liter of $Ca(OH)_2$ solution of equal molarity.

EXAMPLE 11–6

What volume of 0.00300 M HCl solution will just neutralize 30.0 mL of 0.00100 M $Ca(OH)_2$ solution?

Solution

The balanced equation for the reaction is

$$\begin{array}{llll} 2HCl + & Ca(OH)_2 \longrightarrow & CaCl_2 + & 2H_2O \\ 2 \text{ mol} & 1 \text{ mol} & 1 \text{ mol} & 2 \text{ mol} \end{array}$$

We can convert (a) mL of $Ca(OH)_2$ solution to mol of $Ca(OH)_2$ using molarity as a unit factor, 0.00100 mol $Ca(OH)_2$/1000 mL $Ca(OH)_2$ solution, (b) mol of $Ca(OH)_2$ to mol of HCl using the unit factor 2 mol HCl/1 mol $Ca(OH)_2$ (from the balanced equation), and (c) mol of HCl to mL of HCl solution using the unit factor 1000 mL HCl/0.00300 mol HCl.

$$? \text{ mL HCl} = \underbrace{\underbrace{\underbrace{30.0 \text{ mL } Ca(OH)_2 \times \frac{0.00100 \text{ mol } Ca(OH)_2}{1000 \text{ mL } Ca(OH)_2}}_{\text{mol } Ca(OH)_2} \times \frac{2 \text{ mol HCl}}{1 \text{ mol } Ca(OH)_2}}_{\text{mol HCl}} \times \frac{1000 \text{ mL HCl}}{0.00300 \text{ mol HCl}}}_{\text{mL HCl}}$$

$$= \underline{20.0 \text{ mL HCl}}$$

Note that in Example 11–6 we used the unit factor 2 mol HCl/1 mol $Ca(OH)_2$ to convert moles of $Ca(OH)_2$ to moles of HCl, because the balanced equation for the reaction shows that two moles of HCl are required to neutralize one mole of $Ca(OH)_2$. We must always write balanced equations for reactions and determine the **reaction ratio,** i.e., *the relative numbers of moles of reactants,* from the balanced equation for the reaction of interest.

EXAMPLE 11–7

If 100 mL of 1.00 *M* H_2SO_4 solution is mixed with 200 mL of 1.00 *M* KOH, what salt is produced, and what is its molarity?

Solution

	H_2SO_4	+	2KOH	$\longrightarrow$	K_2SO_4	$+ 2H_2O$
reaction ratio:	1 mol		2 mol		1 mol	
start:	$100 \text{ mL} \times \frac{1.00 \text{ mmol}}{\text{mL}}$		$200 \text{ mL} \times \frac{1.00 \text{ mmol}}{\text{mL}}$			
	100 mmol		200 mmol			
change:	−100 mmol		−200 mmol		+100 mmol	
after reaction:	0 mmol H_2SO_4		0 mmol KOH		100 mmol K_2SO_4	

The reaction ratio is 1 mol H_2SO_4:2 mol KOH.

The reaction produces 100 mmol of potassium sulfate, which are contained in 300 mL of solution, and therefore the concentration is

$$\underline{?}\ \frac{\text{mmol } K_2SO_4}{\text{mL}} = \frac{100 \text{ mmol } K_2SO_4}{300 \text{ mL}} = 0.333\ M\ K_2SO_4$$

Since K_2SO_4 is a strong electrolyte, this corresponds to 0.666 *M* K^+ and 0.333 *M* SO_4^{2-}.

11–15 Standardization of Solutions and Acid-Base Titrations

Solutions of accurately known concentrations are called **standard solutions.** Standard solutions of some substances can be prepared by dissolving a carefully weighed solid sample in enough water to give an accurately known volume of solution. Other substances cannot be weighed out accurately and conveniently because they react with the atmosphere (primarily with CO_2 and/or H_2O). Solutions of such substances are prepared and then their concentrations are determined by titration.

Titration is the process in which a solution of one reactant, the titrant, is carefully added to a solution of another reactant, and the volume of titrant required for complete reaction is measured.

Standardization is the process by which one determines the concentration of a solution being standardized by measuring carefully the volume of solution required to react with an exactly known amount of a **primary standard.** The standardized solution is known as a **secondary standard** and is to be used in analysis of an unknown. The concentration of a secondary standard solution must be determined by titrating it with a primary standard.

How does one know when to stop a titration, that is, when the chemical reaction is just complete? In most cases a few drops of an *indicator* solution are added to the solution to be titrated. An **indicator** for an acid-base titration is a substance that can exist in solution in two different forms, with different colors that depend upon the concentration of hydrogen ions in the solution. At least one of these forms must be very intensely colored so that very little indicator is necessary for one to detect its presence visually.

Suppose we titrate an acid solution of unknown concentration by adding a standardized solution of sodium hydroxide dropwise from a **buret** (Figure 11–1). A common buret is graduated in large intervals of 1 mL and in smaller intervals of 0.1 mL so that it is possible to estimate the volume of a solution dispensed to within at least ±0.02 mL. Experienced individuals can often read a buret to within ±0.01 mL. The analyst tries to choose an indicator that changes color clearly at the point at which stoichiometrically equivalent amounts of acid and base have reacted, the **equivalence point.** The point at which the indicator changes color and the titration is stopped is called the **end point.** Ideally, the end point should coincide with the equivalence point. Phenolphthalein is a very common indicator for the titration of a solution of a

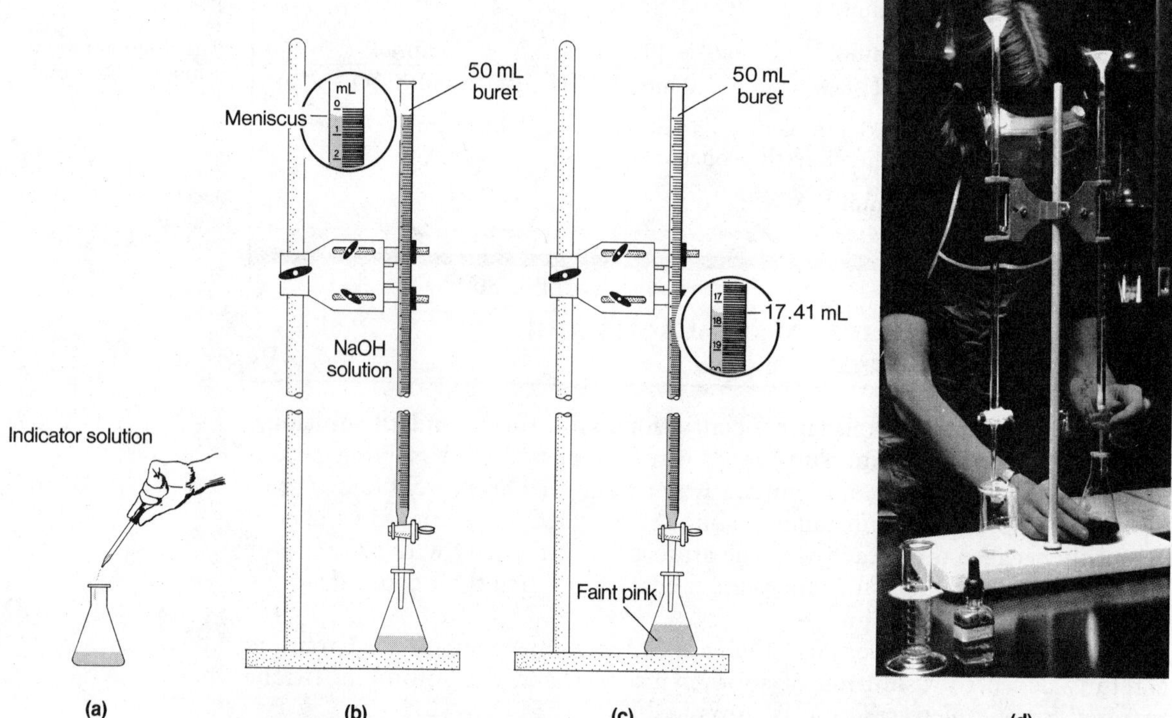

FIGURE 11–1 The titration process. (a) The solution to be titrated is placed in an Erlenmeyer flask, and a few drops of indicator are added. (b) The buret is filled with a standard solution (or the solution to be standardized). The volume of solution in the buret is read carefully. The meniscus describes the surface of the liquid in the buret. Aqueous solutions wet glass and therefore the meniscus of an aqueous solution is always concave. The position of the bottom of the meniscus is read and recorded. (c) The solution in the buret is added (dropwise near the end point) to the Erlenmeyer flask until the end point in the titration is reached. The end point is signaled by the appearance (or sometimes disappearance) of color in the solution being titrated. The volume of the liquid is read again—the difference between the final and initial buret readings is the volume of solution used in the titration. (d) A typical setup for titration in a teaching laboratory.

strong acid with a solution of a base. Phenolphthalein is colorless in acidic solution and reddish-violet in basic solution. The end point in a titration in which a base is added to an acid, and in which phenolphthalein is used as the indicator, is signaled by the first appearance of a faint pink coloration that persists for at least 15 seconds (Figure 11–1), as the solution is swirled. Indicators will be described in more detail in Section 16–5.

Now let us consider the properties of *ideal* **primary standards.** They include the following:

1. must not react with or absorb the components of the atmosphere, such as water vapor, oxygen, or carbon dioxide
2. must react according to one invariable reaction
3. have a high percentage purity
4. have a high formula weight to minimize error in weighing
5. soluble in the solvent of interest
6. nontoxic

Each of these characteristics minimizes the errors involved in analysis. An additional factor of low cost is desirable but not necessary. Because primary standards are often costly and difficult to prepare, secondary standards are often used in day-to-day work.

1 The Mole Method and Molarity

Let's now describe a few primary standards for acids and bases and illustrate how they are used. A common primary standard for solutions of acids is sodium carbonate, Na_2CO_3, a solid compound. Consider the following reaction

H_2SO_4 +	Na_2CO_3 ⟶	Na_2SO_4 +	CO_2 +	H_2O
1 mol	1 mol	1 mol	1 mol	1 mol
98.08 g	106.0 g			

1 mol Na_2CO_3 = 106.0 g and 1 mmol Na_2CO_3 = 0.1060 g

Sodium carbonate is a salt (of carbonic acid and sodium hydroxide). However, since a base can be broadly defined as a substance that reacts with hydrogen ions, Na_2CO_3 can be thought of as a base in *this* reaction. Notice that H_2SO_4 and Na_2CO_3 react in a 1:1 mole ratio.

EXAMPLE 11–8

Calculate the molarity of a solution of H_2SO_4 if 40.0 milliliters of the solution react with 0.364 gram of Na_2CO_3.

Solution

We know from the balanced equation that one mol of H_2SO_4 reacts with one mol of Na_2CO_3, 106.0 g. This provides the unit factors that convert 0.364 g of Na_2CO_3 to the corresponding number of moles of H_2SO_4.

$$\underline{?}\ \text{mol } H_2SO_4 = 0.364\ \text{g } Na_2CO_3 \times \frac{1\ \text{mol } Na_2CO_3}{106.0\ \text{g } Na_2CO_3} \times \frac{1\ \text{mol } H_2SO_4}{1\ \text{mol } Na_2CO_3}$$

$$= 0.00343\ \text{mol } H_2SO_4$$

We could have used mmol of H_2SO_4 and mL of solution rather than mol H_2SO_4 and L of solution.

Now we calculate the molarity of the H_2SO_4 solution

$$\frac{? \text{ mol } H_2SO_4}{L} = \frac{0.00343 \text{ mol } H_2SO_4}{0.0400 \text{ L}} = 0.0858 \, M \, H_2SO_4$$

Most inorganic bases are metal hydroxides, all of which are solids. However, most inorganic bases react rapidly with CO_2 (an acid anhydride) from the atmosphere, even in the solid state, and most metal hydroxides also absorb H_2O from the air. In order to obtain solutions of bases of accurately known concentration, they are commonly standardized against an acidic salt, potassium hydrogen phthalate, $KC_6H_4(COO)(COO\underline{H})$, which is produced by neutralization of one of the two ionizable hydrogens of an organic acid, phthalic acid

$$C_6H_4(COO\boxed{H})(COOH) + K\boxed{OH} \longrightarrow H_2O + KC_6H_4(COO^-K^+)(COO\underline{H})$$

phthalic acid
$C_6H_4(COO\underline{H})_2$

potassium hydrogen phthalate ($K\underline{H}P$)
$KC_6H_4(COO)(COO\underline{H})$

This acidic salt, known as KHP, has one acidic hydrogen (underlined), and reacts with bases. It is easily obtained in a high state of purity, is soluble in water, and is used as a primary standard for bases.

The P in KHP stands for the phthalate ion, $C_6H_4(COO)_2^{2-}$, not phosphorus.

EXAMPLE 11–9

A 20.00-mL sample of a solution of NaOH reacts with 0.3641 gram of KHP. Calculate the molarity of the basic solution.

Solution

$$\underset{1 \text{ mol}}{NaOH} + \underset{\substack{1 \text{ mol} \\ 204.2 \text{ g}}}{KHP} \longrightarrow \underset{1 \text{ mol}}{NaKP} + \underset{1 \text{ mol}}{H_2O}$$

We first calculate the number of moles of NaOH that react with 0.3641 gram of KHP. Note that NaOH and KHP react in a 1:1 mole ratio

$$? \text{ mol NaOH} = 0.3641 \text{ g KHP} \times \frac{1 \text{ mol KHP}}{204.2 \text{ g KHP}} \times \frac{1 \text{ mol NaOH}}{1 \text{ mol KHP}}$$

$$= 0.001783 \text{ mol NaOH}$$

Then we calculate the molarity of the NaOH solution

$$\frac{? \text{ mol NaOH}}{L} = \frac{0.001783 \text{ mol NaOH}}{0.02000 \text{ L}} = 0.08915 \, M \text{ NaOH}$$

Impure samples of acids can be titrated with standard solutions of bases, and the results can be used to determine percentage purity of the acid, as Example 11–10 illustrates.

EXAMPLE 11–10

A 0.2000-g sample of impure oxalic acid, $(COOH)_2$, required 39.82 mL of 0.1023 *M* NaOH solution for complete neutralization. No acidic impurities were present. Calculate the percentage purity of the $(COOH)_2$.

Solution

The equation for the reaction of NaOH with $(COOH)_2$

$$\underset{\text{2 mol}}{2NaOH} + \underset{\substack{\text{1 mol}\\ \text{90.04 g}}}{(COOH)_2} \longrightarrow \underset{\text{1 mol}}{Na_2(COO)_2} + \underset{\text{2 mol}}{2H_2O}$$

shows that *two* moles of NaOH are required to neutralize completely one mole of $(COOH)_2$. The number of moles of NaOH that react is the volume times the molarity of the solution

$$\underline{?} \text{ mol NaOH} = 0.03982 \text{ L} \times \frac{0.1023 \text{ mol NaOH}}{\text{L}} = 0.004074 \text{ mol NaOH}$$

Now we calculate the mass of $(COOH)_2$ that reacts with 0.004074 mol NaOH

$$\underline{?} \text{ g } (COOH)_2 = 0.004074 \text{ mol NaOH} \times \frac{1 \text{ mol } (COOH)_2}{2 \text{ mol NaOH}} \times \frac{90.04 \text{ g } (COOH)_2}{1 \text{ mol } (COOH)_2}$$

$$= 0.1834 \text{ g } (COOH)_2$$

The sample contained 0.1834 g of $(COOH)_2$, and its percentage purity is

$$\% \text{ purity} = \frac{0.1834 \text{ g } (COOH)_2}{0.2000 \text{ g sample}} \times 100\% = \underline{91.70\% \text{ pure } (COOH)_2}$$

Each molecule of $(COOH)_2$ contains two acidic H's.

$$\begin{array}{l} \text{O} \\ \| \\ \text{C—O—H} \\ | \\ \text{C—O—H} \\ \| \\ \text{O} \end{array}$$

2 Equivalent Weights and Normality

Because one mole of an acid does not necessarily neutralize one mole of a base, many chemists choose a method of expressing concentration other than molarity in order to retain a one-to-one relationship. Concentrations of solutions of acids and bases are frequently expressed as **normality** (*N*). The normality of a solution is defined as the *number of equivalent weights, or simply equivalents (eq), of solute per liter of solution.* (The term *equivalent mass* is not widely used.) Normality may be represented symbolically as

An **equivalent weight** is often referred to simply as an **equivalent (eq).**

$$\text{normality} = \frac{\text{number of equivalent weights of solute}}{\text{liter of solution}}$$

By definition there are 1000 milliequivalent weights (meq) in one equivalent weight of an acid or base. Thus, we see that normality may also be expressed as

A **milliequivalent weight** is often referred to simply as a **milliequivalent (meq).**

$$\text{normality} = \frac{\text{number of milliequivalent weights of solute}}{\text{milliliter of solution}}$$

In acid-base reactions the **equivalent weight** or **one equivalent (eq),** of an acid is defined as the mass of the acid (expressed in grams) that will furnish

6.022×10^{23} hydrogen ions (1 mole) or that will react with 6.022×10^{23} hydroxide ions (1 mole). One mole of an acid contains 6.022×10^{23} formula units of the acid. Consider hydrochloric acid as a typical monoprotic acid

$$HCl \xrightarrow{H_2O} H^+\,(aq) + Cl^-\,(aq)$$

HCl	H^+ (aq)	Cl^- (aq)
1 mol	1 mol	1 mol
36.46 g	1.008 g	35.45 g
6.022×10^{23}	6.022×10^{23}	6.022×10^{23}

This equation shows that one mole of HCl can produce 6.022×10^{23} H^+, and therefore *one mole of HCl is one equivalent.* The same is true for all monoprotic acids.

Sulfuric acid is a diprotic acid because it can furnish two H^+ ions per molecule of H_2SO_4

$$H_2SO_4 \xrightarrow{H_2O} 2H^+\,(aq) + SO_4^{2-}\,(aq)$$

H_2SO_4	$2H^+$ (aq)	SO_4^{2-} (aq)
1 mol	2 mol	1 mol
98.08 g	2(1.008 g)	96.06 g
6.022×10^{23}	$2(6.022 \times 10^{23})$	6.022×10^{23}

This equation shows that one mole of H_2SO_4 can produce $2 \times 6.022 \times 10^{23}$ H^+; therefore one mole of H_2SO_4 is two equivalent weights in all reactions in which both acidic hydrogen atoms react with a base. Or, one equivalent weight of H_2SO_4 is one half of a mole. The equivalent weight of an acid is obtained by dividing its formula weight by the number of acidic hydrogens furnished by one formula unit of the acid, or by the number of hydroxide ions with which one formula unit of the acid reacts. The equivalent weights of some common acids are tabulated in Table 11–4.

The equivalent weight of a base is defined as the mass of the base (expressed in grams) that will furnish 6.022×10^{23} hydroxide ions or the mass of the base that will react with 6.022×10^{23} hydrogen ions. The equivalent weight of a base is obtained by dividing its formula weight in grams by the number of hydroxide ions furnished by one formula unit, or by the number of

TABLE 11–4 Equivalent Weights* of Common Acids and Bases

Acids					Bases				
Symbolic Representation				**One eq**	**Symbolic Representation**				**One eq**
$\frac{HNO_3}{1}$	=	$\frac{63.02\text{ g}}{1}$	=	63.02 g HNO_3	$\frac{NaOH}{1}$	=	$\frac{40.00\text{ g}}{1}$	=	40.00 g NaOH
$\frac{CH_3COO\underline{H}}{1}$	=	$\frac{60.03\text{ g}}{1}$	=	60.03 g $CH_3COO\underline{H}$	$\frac{NH_3}{1}$	=	$\frac{17.04\text{ g}}{1}$	=	17.04 g NH_3
$\frac{KC_6H_4(COO)(COO\underline{H})}{1}$	=	$\frac{204.2\text{ g}}{1}$	=	204.2 g K(COO)(COO$\underline{H}$)	$\frac{Ca(OH)_2}{2}$	=	$\frac{74.10\text{ g}}{2}$	=	37.05 g $Ca(OH)_2$
$\frac{H_2SO_4}{2}$	=	$\frac{98.08\text{ g}}{2}$	=	49.04 g H_2SO_4	$\frac{Ba(OH)_2}{2}$	=	$\frac{171.36\text{ g}}{2}$	=	85.68 g $Ba(OH)_2$

*Complete neutralization is assumed.

hydrogen ions with which one formula unit of the base reacts. Likewise, we define one equivalent weight of a salt as that mass of salt that would be produced by the reaction of one equivalent weight of an acid or a base. The equivalent weights of some common bases also are tabulated in Table 11–4.

From the definitions of one equivalent of an acid and a base, we see that *one equivalent of any acid reacts with one equivalent of any base.* It is not true that one mole of any acid reacts with one mole of any base, but as a consequence of the definition of equivalents, 1 eq acid $\simeq$ 1 eq base. In general terms, we may write the following relationship for all acid-base reactions that go to completion

The notation $\simeq$ reads "is equivalent to."

no. eq of acid = no. eq of base

The product of the volume of a solution, in liters, and its normality is equal to the number of equivalents of solute contained in the solution. For a solution of an acid

Remember that the product of volume and concentration equals the amount of solute.

$$\text{L}_{\text{acid}} \times \frac{\text{eq acid}}{\text{L}_{\text{acid}}} = \text{eq acid} \qquad \text{or} \qquad \text{L}_{\text{acid}} \times N_{\text{acid}} = \text{eq acid}$$

Alternatively

$$\text{mL}_{\text{acid}} \times \frac{\text{meq acid}}{\text{mL}_{\text{acid}}} = \text{meq acid} \qquad \text{or} \qquad \text{mL}_{\text{acid}} \times \text{N}_{\text{acid}} = \text{meq acid}$$

Similar relationships can be written for bases. Since 1 eq of acid always reacts with 1 eq of base we may write

no. eq of acid = no. eq of base

or

$$\text{L}_{\text{acid}} \times N_{\text{acid}} = \text{L}_{\text{base}} \times N_{\text{base}}$$

or

$$\text{mL}_{\text{acid}} \times N_{\text{acid}} = \text{mL}_{\text{base}} \times N_{\text{base}}$$

three different forms of a very useful relationship.

EXAMPLE 11–11

What volume of 0.100 N HNO_3 is required to neutralize 50.0 mL of a 0.150 N solution of $Ba(OH)_2$?

Solution

Since number of meq acid = number of meq of base, we can write

$$\underline{?}\ \text{mL}_{\text{acid}} = \frac{\text{mL}_{\text{base}} \times N_{\text{base}}}{N_{\text{acid}}} = \frac{50.0 \text{ mL} \times 0.150\ N}{0.100\ N} = \underline{75.0 \text{ mL of HNO}_3 \text{ solution}}$$

EXAMPLE 11–12

Calculate the normality of a solution that contains 4.202 g of HNO_3 in 600 mL of solution.

Solution

$$N = \frac{\text{number of eq HNO}_3}{\text{L}}$$

$$? \frac{\text{eq HNO}_3}{\text{L}} = \underbrace{\frac{4.202 \text{ g HNO}_3}{0.600 \text{ L}} \times \frac{1 \text{ mol HNO}_3}{63.02 \text{ g HNO}_3}}_{M_{HNO_3}} \times \frac{1 \text{ eq HNO}_3}{\text{mol HNO}_3} = \underline{0.111\ N \text{ HNO}_3}$$

Notice that *normality* is *molarity* times the number of equivalents per mole of solute; it is always equal to or greater than molarity

$$M \times \frac{\text{number eq}}{\text{mol}} = N$$

EXAMPLE 11–13

What is the normality of a solution of 9.50 g of barium hydroxide in 2000 mL of solution? What is the molarity?

Solution

$$? \frac{\text{eq Ba(OH)}_2}{\text{L}} = \underbrace{\frac{9.50 \text{ g Ba(OH)}_2}{2.00 \text{ L}} \times \frac{1 \text{ mol Ba(OH)}_2}{171.4 \text{ g Ba(OH)}_2}}_{M_{Ba(OH)_2}} \times \frac{2 \text{ eq}}{\text{mol}} = \underline{0.0554\ N \text{ Ba(OH)}_2}$$

The $Ba(OH)_2$ solution is 0.0554 N and 0.0277 M.

Let us again solve Example 11–8 in Example 11–14, this time using normality rather than molarity. The balanced equation for the reaction of H_2SO_4 with Na_2CO_3 interpreted in equivalent weights is

H_2SO_4 +	Na_2CO_3 ⟶	Na_2SO_4 +	CO_2 +	H_2O
1 mol	1 mol	1 mol	1 mol	1 mol
2 eq	2 eq	2 eq	2 eq	2 eq
98.08 g	106.0 g			

1 eq Na_2CO_3 = 53.0 g and 1 meq Na_2CO_3 = 0.0530 g

EXAMPLE 11–14

Calculate the normality of a solution of H_2SO_4 if 40.0 mL of the solution react with 0.364 g of Na_2CO_3.

Solution

First we calculate the number of milliequivalents of Na_2CO_3 in the sample

$$\text{no. meq Na}_2\text{CO}_3 = 0.364 \text{ g Na}_2\text{CO}_3 \times \frac{1 \text{ meq Na}_2\text{CO}_3}{0.0530 \text{ g Na}_2\text{CO}_3} = 6.87 \text{ meq Na}_2\text{CO}_3$$

Because number of meq H_2SO_4 = number of meq Na_2CO_3, we may write

$$\text{mL}_{H_2SO_4} \times N_{H_2SO_4} = 6.87 \text{ meq H}_2\text{SO}_4$$

$$N_{H_2SO_4} = \frac{6.87 \text{ meq } H_2SO_4}{mL_{H_2SO_4}} = \frac{6.87 \text{ meq } H_2SO_4}{40.0 \text{ mL}}$$
$$= \underline{0.172\ N\ H_2SO_4}$$

Note that the normality of this H_2SO_4 solution is twice the molarity obtained in Example 11–8 because 1 mol of H_2SO_4 is two equivalents.

EXAMPLE 11–15

A 0.2000-gram sample of impure oxalic acid, $(COOH)_2$, required 39.82 mL of 0.1023 *N* NaOH solution for complete neutralization. No acidic impurities were present. Calculate the percentage purity of the $(COOH)_2$.

This is Example 11–10 using normality rather than molarity.

Solution

$$2NaOH + (COOH)_2 \longrightarrow Na_2(COO)_2 + 2H_2O$$

2NaOH	$(COOH)_2$
2 mol	1 mol
2 eq	2 eq
	90.04 g

The equation for the reaction of NaOH with $(COOH)_2$ shows that *one mole* of $(COOH)_2$ is *two* equivalents. Therefore one equivalent weight of $(COOH)_2$ is 90.04 g/2 = 45.02 g. The number of equivalents of NaOH that react is the volume times the normality of the solution.

$$\underline{?} \text{ eq NaOH} = 0.03982 \text{ L} \times \frac{0.1023 \text{ eq NaOH}}{\text{L}} = 0.004074 \text{ eq NaOH}$$

Now we calculate the mass of $(COOH)_2$ that reacts with 0.004074 eq NaOH

$$\underline{?} \text{ g } (COOH)_2 = 0.004074 \text{ eq NaOH} \times \frac{1 \text{ eq } (COOH)_2}{1 \text{ eq NaOH}} \times \frac{45.02 \text{ g } (COOH)_2}{1 \text{ eq } (COOH)_2}$$
$$= 0.1834 \text{ g } (COOH)_2$$

The sample contained 0.1834 g of $(COOH)_2$, and its percentage purity is

$$\% \text{ purity} = \frac{0.1834 \text{ g } (COOH)_2}{0.2000 \text{ g sample}} \times 100\% = \underline{91.70\% \text{ pure } (COOH)_2}$$

Key Terms

Acidic salt a salt containing an ionizable hydrogen atom; does not necessarily produce acidic solutions.

Amphiprotism ability of a substance to exhibit amphoterism by accepting or donating protons.

Amphoterism ability of a substance to act as either an acid or a base.

Anhydrous without water.

Autoionization an ionization reaction between identical molecules.

Basic salt a salt containing an ionizable OH group.

Brønsted-Lowry acid a proton donor.

Brønsted-Lowry base a proton acceptor.

Buret a piece of volumetric glassware, usually graduated in 0.1-mL intervals, that is used to deliver solutions to be used in titrations in a quantitative (dropwise) manner.

Conjugate acid-base pair in Brønsted-Lowry terminology, a reactant and product that differ by a proton, H^+.

Coordinate covalent bond covalent bond in which both shared electrons were furnished by the same species; bond between a Lewis acid and a Lewis base.

End point the point at which an indicator changes color and a titration is stopped.

Equivalence point the point at which chemically equivalent amounts of reactants have reacted.

Equivalent weight the mass of an acid or base that furnishes or reacts with 6.022×10^{23} H_3O^+ or OH^- ions.

Hydration the process by which water molecules bind to ions or molecules in the solid state or in solution.

Hydrolysis reaction of a substance with water.

Indicators for acid-base titrations, organic compounds that exhibit different colors in solutions of different acidities; used to determine the point at which reaction between two solutes is complete.

Ionization the breaking up of a compound into separate ions.

Leveling effect effect by which all acids stronger than the acid that is characteristic of the solvent react with solvent to produce that acid; similar statement applies to bases. The strongest acid (base) that can exist in a given solvent is the acid (base) characteristic of that solvent.

Lewis acid any species that can accept a share in an electron pair.

Lewis base any species that can make available a share in an electron pair.

Monoprotic acid acid that contains one ionizable hydrogen atom per formula unit.

Normality the number of equivalent weights (or equivalents) of solute per liter of solution.

Normal salt a salt containing no ionizable H atoms or OH groups.

Open sextet refers to species that have only six electrons in the highest energy level of the central element (many Lewis acids).

Polyprotic acid an acid that contains more than one ionizable hydrogen atom per formula unit.

Primary standard a substance of a known high degree of purity that undergoes one invariable reaction with the other reactant of interest.

Protonic acid an Arrhenius (classical) acid, or a Brønsted-Lowry acid.

Secondary standard a solution that has been titrated against a primary standard. A standard solution is a secondary standard.

Solvent system acid any substance that produces the cation characteristic of the solvent.

Solvent system base any substance that produces the anion characteristic of the solvent.

Standard solution a solution of accurately known concentration.

Standardization the process by which the concentration of a solution is accurately determined by titrating it against an accurately known amount of a primary standard.

Titration the process by which the volume of a standard solution required to react with a specific amount of a substance is determined.

Exercises

Basic Ideas—The Arrhenius Theory

1. What is the significance of the idea that acids and bases can be defined only in terms of their reactions with each other?
2. Outline Arrhenius' ideas about acids and bases. (a) How did he define the following terms: acid, base, neutralization? (b) Provide a specific example that illustrates each term.
3. Define and illustrate the following terms clearly and concisely, and provide a specific example of each:
(a) strong electrolyte
(b) weak electrolyte

(c) nonelectrolyte
(d) strong acid
(e) strong soluble base
(f) weak acid
(g) weak base

4. Distinguish between the following pairs of terms and provide a specific example of each:
   (a) strong acid and weak acid
   (b) strong soluble base and weak base
   (c) strong soluble base and insoluble base
5. Write formulas and names for:
   (a) the common strong acids
   (b) three weak acids
   (c) the common strong soluble bases
   (d) the most common weak base
   (e) three soluble ionic salts
   (f) three insoluble salts

The Hydrated Hydrogen Ion

6. (a) Describe the hydrated hydrogen ion in words and with formulas.
   (b) Why is the hydrated hydrogen ion important?
   (c) Criticize the following statement: "The hydrated hydrogen ion should always be represented as H_3O^+."

Brønsted-Lowry Theory

7. State the basic ideas of the Brønsted-Lowry theory.
8. Use Brønsted-Lowry terminology to define the following terms. Illustrate each with a specific example.
   (a) acid
   (b) conjugate base
   (c) base
   (d) conjugate acid
   (e) conjugate acid-base pair
9. Write balanced equations that describe the ionization of the following acids in dilute aqueous solution. Use a single arrow ($\rightarrow$) to represent complete, or nearly complete, ionization and a double arrow ($\rightleftharpoons$) to represent a small extent of ionization.
   (a) HNO_3 (b) CH_3COOH
   (c) HBr (d) HCN
   (e) HF (f) $HClO_4$
10. Use words and equations to describe how ammonia can act as a base in (a) aqueous solution, and (b) the pure state, i.e., as gaseous ammonia molecules when it reacts with gaseous hydrogen chloride or a similar anhydrous acid.
11. What does autoionization mean? How can the autoionization of water be described as an acid-base reaction?
12. What do we mean when we say that water is amphiprotic? (a) Can we also describe water as an amphoteric substance? Why? (b) Illustrate the amphiprotic nature of water by writing three equations for reactions in which water exhibits this property.

Acidic and Basic Character

13. List six characteristics (properties) of aqueous solutions of protonic acids.
14. List five characteristics of strong soluble bases in aqueous solution. Does aqueous ammonia exhibit these properties? Why?
15. What does "acid strength" mean?
16. What does "base strength" mean?
17. Classify the following compounds as (a) strong soluble bases, (b) insoluble bases, (c) strong acids, or (d) weak acids: HBr, HF, KOH, $Cu(OH)_2$, CH_3COOH, $Ba(OH)_2$, $HClO_4$, HNO_3, HCN, $Fe(OH)_3$.
18. (a) What are binary protonic acids?
    (b) Write names and formulas for four binary protonic acids.
19. (a) How can the order of increasing acid strength in a series of similar binary protonic acids be explained?
    (b) Illustrate your answer for the series HF, HCl, HBr, and HI.
    (c) What is the order of increasing base strength of the conjugate bases of the acids in (b)? Why?
    (d) Is your explanation applicable to the series H_2O, H_2S, H_2Se, and H_2Te? Why?
20. What does the term "leveling effect" mean? Illustrate your answer with three specific examples.
21. We say that strong acids, weak acids, and weak bases ionize when they are placed in water, while strong soluble bases dissociate when they are placed in water. What does this statement mean?

22. Distinguish between solubility in water and extent of ionization in water. Provide specific examples that illustrate the meaning of both terms.
23. Write three general statements that describe the extent to which acids, bases, and salts are ionized in dilute aqueous solutions.

Reactions of Acids and Bases

24. Why are acid-base reactions described as neutralization reactions or simply as neutralization?
25. Distinguish among (a) molecular equations, (b) total ionic equations, and (c) net ionic equations. What are the advantages and limitations of each?

For Exercises 26 through 28, write balanced (1) molecular equations, (2) total ionic equations, and (3) net ionic equations for reactions between the following acid-base pairs. Name all compounds except water. Assume complete neutralization.

26. (a) $HCl + NaOH \longrightarrow$
    (b) $H_2SO_4 + KOH \longrightarrow$
    (c) $H_3PO_4 + RbOH \longrightarrow$
    (d) $HNO_3 + Ba(OH)_2 \longrightarrow$
    (e) $CH_3COOH + Ca(OH)_2 \longrightarrow$
27. (a) $H_2CO_3 + Sr(OH)_2 \longrightarrow$
    (b) $H_2SO_4 + Ba(OH)_2 \longrightarrow$
    (c) $H_3PO_4 + Ca(OH)_2 \longrightarrow$
    (d) $H_2S + KOH \longrightarrow$
    (e) $H_3AsO_4 + KOH \longrightarrow$
28. (a) $HClO_4 + NaOH \longrightarrow$
    (b) $HBr + NH_3 \longrightarrow$
    (c) $H_2SO_4 + NH_3 \longrightarrow$
    (d) $HCl + Fe(OH)_2 \longrightarrow$
    (e) $H_2SO_4 + Al(OH)_3 \longrightarrow$
29. Although many salts may be formed by a variety of reactions, salts are usually thought of as being derived from the reaction of an acid with a base. For each of the salts listed below, choose the acid and base which react with each other to form the salt. Write the (1) molecular equation, (2) total ionic equation, and (3) net ionic equation for the formation of each salt.
    (a) KCl
    (b) $Ca(ClO_4)_2$
    (c) $Mg_3(AsO_4)_2$
    (d) lithium fluoride
30. Repeat Exercise 29 for the following salts:
    (a) sodium sulfate
    (b) magnesium bromide
    (c) ammonium sulfate
    (d) chromium(III) nitrate
31. Write the (1) molecular equation, (2) total ionic equation, and (3) net ionic equation for the formation of each salt in the following list from the appropriate acid and base.
    (a) $NaNO_3$
    (b) $(NH_4)_2CO_3$
    (c) $Al(CH_3COO)_3$
    (d) CaI_2
32. Repeat Exercise 31 for the following salts:
    (a) potassium acetate
    (b) ammonium nitrate
    (c) iron(II) chloride
    (d) copper(II) phosphate

Acidic and Basic Salts

33. What are polyprotic acids? Write names and formulas for five polyprotic acids.
34. What are acidic salts?
    (a) Write balanced equations to show how the following acidic salts can be prepared from the appropriate acid and base: $NaHSO_4$, $NaHCO_3$, NaH_2AsO_4, Na_2HAsO_4.
    (b) Indicate the molar ratio of acid and base required in each case.
35. What are polyhydroxy bases? Write names and formulas for five polyhydroxy bases.
36. What are basic salts?
    (a) Write balanced equations to show how the following basic salts can be prepared from the appropriate acid and base:
    $Ca(OH)Cl$ $Cu(OH)NO_3$
    $Fe(OH)Cl_2$ $Fe(OH)_2Cl$
    (b) Indicate the molar ratio of acid and base required in each case.

Strengths of Ternary Acids and Amphoterism

37. What are ternary acids? Write names and formulas for four of them.
38. Why can we describe nitric and sulfuric acids as "hydroxides" of nonmetals?

39. Explain the order of increasing acid strength for the following groups of acids and the order of increasing base strength for their conjugate bases.
(a) H_2SO_3, H_2SO_4
(b) HNO_2, HNO_3
(c) H_3PO_3, H_3PO_4
(d) $HClO$, $HClO_2$, $HClO_3$, $HClO_4$

40. (a) Write a generalization that describes the order of acid strengths for a series of ternary acids that contain different elements in the same oxidation state from the same group in the periodic table.
(b) Indicate the order of acid strengths for the following:
(1) HNO_3, H_3PO_4
(2) H_3PO_4, H_3AsO_4
(3) H_2SO_4, H_2SeO_4
(4) $HClO_3$, $HBrO_3$

41. What are amphoteric metal hydroxides? (a) Are they bases? (b) Write the names and formulas for five amphoteric metal hydroxides.

42. Aluminum hydroxide and zinc hydroxide are typical amphoteric hydroxides.
(a) Write the molecular equation, the total ionic equation, and the net ionic equation for the complete reaction of each hydroxide with nitric acid.
(b) Write the same kinds of equations for the reaction of each hydroxide with an excess of sodium hydroxide solution. Reference to Table 11–3 may be helpful.

43. Write the kinds of equations listed in Exercise 42 for the reaction of chromium(III) hydroxide with hydrochloric acid and with an excess of potassium hydroxide solution.

Preparation of Acids and Bases

44. Outline a method of preparing each of the following acids, and write appropriate balanced equations for each preparation: (a) HF, (b) HCl, (c) HBr, (d) HI, (e) H_2S, (f) CH_3COOH, (g) HNO_3 (2 methods).

45. Repeat Exercise 44 for (a) carbonic acid, (b) perchloric acid, (c) permanganic acid, and (d) phosphoric acid (2 methods).

46. How are strong soluble bases prepared? Write balanced equations for reactions that produce the following bases: (a) $LiOH$, (b) $NaOH$ (2 methods), (c) KOH, (d) $Ca(OH)_2$.

47. How are insoluble bases prepared? Write balanced equations for reactions that produce the following bases: (a) $Mg(OH)_2$, (b) $Fe(OH)_2$, (c) $Al(OH)_3$, (d) $Zn(OH)_2$.

The Lewis Theory

48. Define and illustrate the following terms clearly and concisely. Write an equation to illustrate the meaning of each term: (a) Lewis acid, (b) Lewis base, (c) neutralization (according to Lewis theory).

49. What are the advantages and limitations of (a) the Arrhenius theory, (b) the Brønsted-Lowry theory, and (c) the Lewis theory of acid-base behavior?

50. Draw a Lewis formula for each species in the following equations. Label the acids and bases using Lewis theory terminology.
(a) $H_2O + H_2O \rightleftharpoons H_3O^+ + OH^-$
(b) $HBr\ (g) + H_2O \longrightarrow H_3O^+ + Br^-$
(c) $NH_3\ (g) + H_2O \rightleftharpoons NH_4^+ + OH^-$
(d) $NH_3\ (g) + HBr\ (g) \longrightarrow NH_4Br\ (s)$

51. Repeat Exercise 50 for the following:
(a) $NH_3\ (g) + BF_3\ (g) \longrightarrow H_3N{:}BF_3\ (s)$
(b) $AlCl_3 + Cl^- \longrightarrow AlCl_4^-$
(c) $SnCl_2 + 2Cl^- \longrightarrow SnCl_4^{2-}$
(d) $SnCl_4 + 2Cl^- \longrightarrow SnCl_6^{2-}$

52. Repeat Exercise 50 for the following:
(a) $SO_2 + MgO \longrightarrow MgSO_3$
(b) $CO_2 + CaO \longrightarrow CaCO_3$
(c) $SO_3 + K_2O \longrightarrow K_2SO_4$

The Solvent System Theory

53. Define and illustrate the following terms clearly and concisely. Write an equation to illustrate the meaning of each term: (a) solvent system acid, (b) solvent system base, (c) neutralization (according to the solvent system theory).

54. Show that the autoionization of water and the autoionization of liquid ammonia are acid-base reactions according to the (a) Brønsted-Lowry, (b) Lewis, and (c) solvent system theories.

55. Use the solvent system theory to demonstrate the fact that neutralizations in water and in liquid ammonia are quite similar. (a) Could the Brønsted-Lowry or the Lewis theory be used equally well? (b) Why?

56. Write molecular, total ionic, and net ionic equations for reactions between the following acids and bases (or acids and basic oxides). Reactions in aqueous solution are preceded by (H_2O) while those that occur in liquid ammonia are preceded by (NH_3). The examples have been chosen so that the metal hydroxides and the metal amides, as well as the reaction products, show similar solubility behavior in water and liquid ammonia, respectively. Name all compounds except NH_3 and H_2O.
(a) (H_2O) $HNO_3 + KOH \longrightarrow$
(NH_3) $NH_4NO_3 + KNH_2 \longrightarrow$
(b) (H_2O) $HNO_3 + BaO \longrightarrow$
(NH_3) $NH_4NO_3 + BaO \longrightarrow$

57. The reactions of salts with strong soluble bases in liquid ammonia are quite similar to the analogous reactions in aqueous solutions. Write balanced molecular, total ionic, and net ionic equations for the following reactions. Name all compounds except H_2O and NH_3.
(a) (H_2O) $KOH + Al(NO_3)_3 \longrightarrow$
(NH_3) $KNH_2 + Al(NO_3)_3 \longrightarrow$
(b) (H_2O) $NaOH + Zn(NO_3)_2 \longrightarrow$
(NH_3) $NaNH_2 + Zn(NO_3)_2 \longrightarrow$

58. Many amides exhibit amphoterism in liquid ammonia that is similar to the amphoterism exhibited by hydroxides in water. Write balanced molecular, total ionic, and net ionic equations for reactions between the following amphoteric bases and an excess of strong soluble base.
(a) (H_2O) $Al(OH)_3 + NaOH \longrightarrow$
(NH_3) $Al(NH_2)_3 + NaNH_2 \longrightarrow$
(b) (H_2O) $Zn(OH)_2 + KOH \longrightarrow$
(NH_3) $Zn(NH_2)_2 + KNH_2 \longrightarrow$

Molarity

59. Calculate the molarities of solutions that contain the following masses of solute in the indicated volumes:
(a) 75 g of H_3AsO_4 in 500 mL of solution
(b) 8.3 g of $(COOH)_2$ in 600 mL of solution
(c) 13.0 g of $(COOH)_2 \cdot 2H_2O$ in 750 mL of solution

60. Calculate the molarities of solutions that contain the following:
(a) 17.5 g of $(NH_4)_2SO_4$ in 300 mL of solution
(b) 143 g of $Cr_2(SO_4)_3$ in 3.00 liters of solution
(c) 143 g of $Cr_2(SO_4)_3 \cdot 18H_2O$ in 3.00 liters of solution

61. Calculate the mass of NaOH required to prepare 2.5 liters of 1.50 *M* NaOH solution.

62. What mass of $(NH_4)_2SO_4$ is required to prepare 10.0 liters of 1.5 *M* $(NH_4)_2SO_4$ solution?

63. Calculate the mass of $Mg(NO_3)_2 \cdot 6H_2O$ required to prepare 1200 mL of 0.375 *M* $Mg(NO_3)_2$ solution.

64. Calculate the molarity of a solution that is 20.0% H_2SO_4 by mass. The specific gravity of the solution is 1.14.

65. What is the molarity of a solution that is 19.0% HNO_3 by mass? The specific gravity of the solution is 1.11.

66. If 200 mL of 2.00 *M* HCl solution is added to 400 mL of 1.00 *M* NaOH solution, the resulting solution will be ______ molar in NaCl.

67. If 400 mL of 0.400 *M* HCl solution is added to 800 mL of 0.100 *M* $Ba(OH)_2$ solution, the resulting solution will be ______ molar in $BaCl_2$.

68. If 200 mL of 2.00 *M* H_3PO_4 solution is added to 600 mL of 2.00 *M* NaOH solution, the resulting solution will be ______ molar in Na_3PO_4.

69. What volumes of 1.00 *M* KOH and 0.750 *M* HNO_3 solutions would be required to produce 4.80 g of KNO_3?

70. What volumes of 1.00 *M* NaOH and 0.500 *M* H_3PO_4 solutions would be required to form 1.00 mole of Na_3PO_4?

71. What volumes of 0.0200 *M* $Ca(OH)_2$ and 0.0300 *M* H_3AsO_4 solutions would be required to prepare 1.00 g of $Ca_3(AsO_4)_2$?

72. Ammonia reacts with sulfuric acid to produce ammonium sulfate. What volume of sulfuric acid solution (sp gr = 1.82; 90.0% H_2SO_4 by mass) would be required to prepare 3.00 pounds of ammonium sulfate?

Dilution

73. Calculate the volume of 6.00 M NaOH required to prepare 500 mL of 0.120 M NaOH solution.
74. Calculate the volume of 0.200 M H_2SO_4 required to prepare 200 mL of 0.100 M H_2SO_4 solution.
75. Calculate the volume of 3.00 M H_3PO_4 required to prepare 900 mL of 0.100 M H_3PO_4.

Standardization and Acid-Base Titrations—Mole Method

76. Define and illustrate the following terms clearly and concisely: (a) standard solution, (b) titration, (c) primary standard, (d) secondary standard.
77. Distinguish between the terms *equivalence point* and *end point* of a titration.
78. (a) What are the desirable characteristics of a primary standard? (b) What is the significance of each?
79. Why can sodium carbonate be used as a primary standard for solutions of acids?
80. (a) What is potassium hydrogen phthalate? (b) How is it prepared? (c) For what is it used?
81. What volume of 0.100 molar hydrochloric acid solution just neutralizes 30.0 mL of 0.150 molar sodium hydroxide solution?
82. What volume of 0.100 molar sodium hydroxide solution would be required to react with 20.0 mL of 0.400 molar sulfuric acid solution? Assume complete neutralization.
83. A 20-mL sample of 0.10 molar phosphoric acid solution required 120 mL of barium hydroxide solution for complete neutralization. Calculate the molarity of the barium hydroxide solution.
84. What volume of 0.10 molar barium hydroxide would be required to react with 20 mL of 0.10 molar phosphoric acid? Assume complete neutralization.
85. Calculate the molarity of a KOH solution if 20.0 mL of the KOH solution reacted with 0.408 g of KHP.
86. An impure sample of $(COOH)_2 \cdot 2H_2O$ that had a mass of 2.00 g was dissolved in water and titrated with standard NaOH solution. The titration required 40.0 mL of 0.200 M NaOH solution. Calculate the percentage of $(COOH)_2 \cdot 2H_2O$ in the sample.
87. A 50-mL sample of 0.050 M $Ca(OH)_2$ is added to 10 mL of 0.20 M HNO_3. (a) Is the resulting solution acidic or basic? (b) How many moles of excess acid or base are present? (c) How many mL of 0.050 M $Ca(OH)_2$ or 0.050 M HNO_3 would be required to neutralize the solution?
88. A solution of sodium hydroxide is standardized against potassium hydrogen phthalate. From the following data, calculate the molarity of the NaOH solution:

mass of KHP	0.8407 g
buret reading before titration	0.06 mL
buret reading after titration	44.68 mL

89. The secondary standard solution of NaOH of Exercise 88 was used to titrate a solution of unknown concentration of HCl. A 30.00-mL sample of the HCl solution required 24.21 mL of the NaOH solution for complete neutralization. What is the molarity of the HCl solution?
90. A 34.53-mL sample of a solution of sulfuric acid, H_2SO_4, is neutralized by 28.24 mL of the NaOH solution of Exercise 88. Calculate the molarity of the sulfuric acid solution.
91. Benzoic acid, C_6H_5COOH, is sometimes used as a primary standard for the standardization of solutions of bases. A 1.862-gram sample of the acid is neutralized by 31.62 mL of an NaOH solution. What is the molarity of the base solution?

$$C_6H_5COOH\ (s) + NaOH\ (aq) \longrightarrow C_6H_5COONa\ (aq) + H_2O\ (\ell)$$

92. An antacid tablet containing magnesium hydroxide as an active ingredient requires 22.6 mL of 0.597 M HCl for complete neutralization. What mass of $Mg(OH)_2$ did the tablet contain?
93. Butyric acid, whose empirical formula is C_2H_4O, is the acid that is responsible for the odor of rancid butter. The acid has one ionizable hydrogen per molecule. A 1.000-gram sample of butyric acid is neutralized by 54.42 mL of standardized 0.2088 M

NaOH solution. What are (a) the molecular weight and (b) the molecular formula of butyric acid?

94. The typical concentration of HCl in stomach acid (gastric juice) is a concentration of about 8.0×10^{-2} *M*. One experiences "acid stomach" when the stomach contents reach about 1.0×10^{-1} *M* HCl. One Rolaids® tablet (an antacid) contains 334 mg of active ingredient, $NaAl(OH)_2CO_3$. Assume that you have acid stomach and that your stomach contains 300 mL of 1.0×10^{-1} *M* HCl. What percentage of a tablet should be just sufficient to return the molarity of HCl to 8.0×10^{-2} *M*? Assume the neutralization reaction produces NaCl, $AlCl_3$, CO_2, and H_2O.

Standardization and Acid-Base Titrations—Equivalent Weight Method

In answering Exercises 95 through 99, assume that the acids and bases will be completely neutralized.

95. Calculate the normality of a solution that contains 4.9 g of H_2SO_4 in 100 mL of solution.

96. What is the normality of a solution that contains 7.08 g of H_3AsO_4 in 100 mL of solution?

97. Calculate the molarity and the normality of a solution that was prepared by dissolving 34.2 g of barium hydroxide in enough water to make 4000 mL of solution.

98. Calculate the molarity and the normality of a solution that was prepared by dissolving 19.6 g of phosphoric acid in enough water to make 600 mL of solution.

99. What are the molarity and normality of a sulfuric acid solution that is 19.6% H_2SO_4 by mass? The density of the solution is 1.14 g/mL.

100. A 20-mL sample of 0.20 normal nitric acid solution required 40 mL of barium hydroxide solution for neutralization. Calculate the molarity of the barium hydroxide solution.

101. Vinegar is an aqueous solution of acetic acid, CH_3COOH. Suppose you titrate a 25.00-mL sample of vinegar with 17.62 mL of a standardized 0.1060 *N* solution of NaOH. (a) What is the normality of acetic acid in the vinegar? (b) What is the mass of acetic acid contained in 1.000 liter of vinegar?

102. A 44.4-mL sample of sodium hydroxide solution was titrated with 40.0 mL of 0.100 *N* sulfuric acid solution. A 36.0-mL sample of hydrochloric acid required 40.0 mL of the sodium hydroxide solution for titration. What is the normality of the hydrochloric acid solution?

103. Calculate the normality and molarity of an H_2SO_4 solution if 20.0 mL of the solution reacted with 0.212 g of Na_2CO_3.

$$H_2SO_4 + Na_2CO_3 \longrightarrow Na_2SO_4 + CO_2 + H_2O$$

104. Calculate the normality and molarity of an HCl solution if 30.0 mL of the solution reacted with 0.318 g of Na_2CO_3.

$$2HCl + Na_2CO_3 \longrightarrow 2NaCl + CO_2 + H_2O$$

105. What are the normality and molarity of a $Ca(OH)_2$ solution if 50.0 mL of the $Ca(OH)_2$ solution reacted with 0.102 g of KHP?

106. An impure sample of KHP that had a mass of 1.000 g was dissolved in water and titrated with 30.00 mL of 0.1000 *N* NaOH solution. Calculate the percentage of KHP in the sample.

107. A 20-mL sample of 0.20 *M* H_2SO_4 is added to 20 mL of 0.20 *M* NaOH. (a) Is the resulting solution acidic or basic? (b) How many meq of excess acid or base are present? (c) What volume of 0.10 *N* H_2SO_4 or NaOH would be required to neutralize the solution?

12 Oxidation-Reduction Reactions

Reactions accompanied by changes in oxidation number are called **oxidation-reduction reactions,** and they occur in every area of chemistry and biochemistry. In this chapter we will learn to identify oxidizing agents and reducing agents, and to balance oxidation-reduction equations. Acquisition of such skills is a prerequisite to the study of electrochemistry in Chapter 19, which involves electron transfer between physically separated oxidizing and reducing agents, and interconversions between chemical energy and electrical energy. These skills are also fundamental to biology and biochemistry, since practically all reactions associated with metabolism are redox reactions.

12–1 Basic Concepts

Reactions in which atoms undergo changes in oxidation number are called **oxidation-reduction reactions** or **redox reactions.** They involve, or appear to involve, electron transfer. The concepts and rules of assigning oxidation numbers were treated in Section 4–12. Figure 12–1 contains a listing of some of the more common oxidation numbers of elements.

These rules should be reviewed if necessary.

The term *oxidation* originally referred to the combination of a substance with oxygen, which implied an increase in the oxidation state of an element in

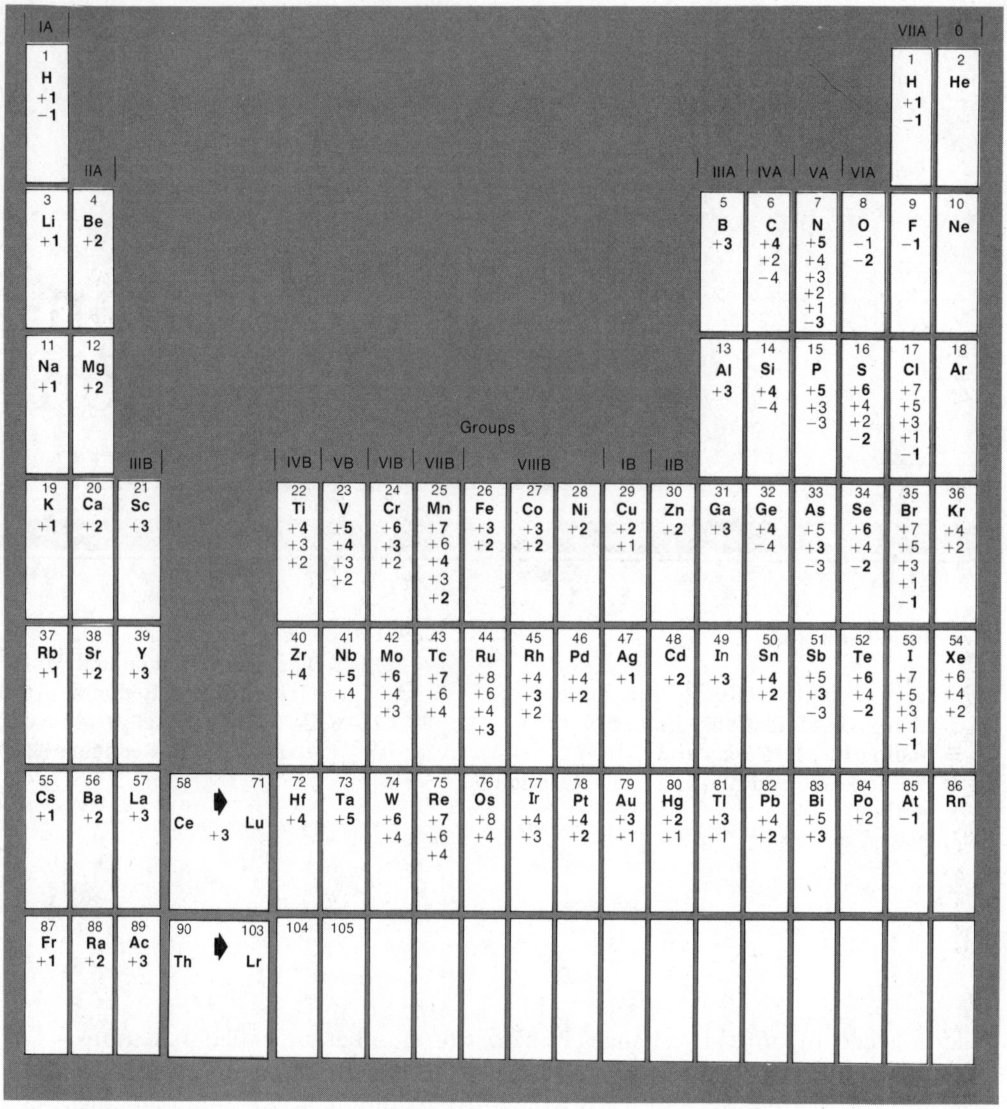

Groups

IA	IIA	IIIB		IVB	VB	VIB	VIIB	VIIIB	VIIIB	VIIIB	IB	IIB	IIIA	IVA	VA	VIA	VIIA	0
1 H +1 −1																	1 H +1 −1	2 He
3 Li +1	4 Be +2												5 B +3	6 C +4 +2 −4	7 N +5 +4 +3 +2 +1 −3	8 O −1 −2	9 F −1	10 Ne
11 Na +1	12 Mg +2												13 Al +3	14 Si +4 −4	15 P +5 +3 −3	16 S +6 +4 +2 −2	17 Cl +7 +5 +3 +1 −1	18 Ar
19 K +1	20 Ca +2	21 Sc +3		22 Ti +4 +3 +2	23 V +5 +4 +3 +2	24 Cr +6 +3 +2	25 Mn +7 +6 +4 +3 +2	26 Fe +3 +2	27 Co +3 +2	28 Ni +2	29 Cu +2 +1	30 Zn +2	31 Ga +3	32 Ge +4 −4	33 As +5 +3 −3	34 Se +6 +4 −2	35 Br +7 +5 +3 +1 −1	36 Kr +4 +2
37 Rb +1	38 Sr +2	39 Y +3		40 Zr +4	41 Nb +5 +4	42 Mo +6 +4 +3	43 Tc +7 +6 +4	44 Ru +8 +6 +4 +3	45 Rh +4 +3 +2	46 Pd +4 +2	47 Ag +1	48 Cd +2	49 In +3	50 Sn +4 +2	51 Sb +5 +3 −3	52 Te +6 +4 −2	53 I +7 +5 +3 +1 −1	54 Xe +6 +4 +2
55 Cs +1	56 Ba +2	57 La +3	58 Ce → 71 Lu +3	72 Hf +4	73 Ta +5	74 W +6 +4	75 Re +7 +6 +4	76 Os +8 +4	77 Ir +4 +3	78 Pt +4 +2	79 Au +3 +1	80 Hg +2 +1	81 Tl +3 +1	82 Pb +4 +2	83 Bi +5 +3	84 Po +2	85 At −1	86 Rn
87 Fr +1	88 Ra +2	89 Ac +3	90 Th → 103 Lr	104	105													

FIGURE 12–1 The most common nonzero oxidation states of the elements.

that substance. According to the original definition, the following reactions involve oxidation of the element or compound shown on the far left side of each equation.

Since oxidation number is a purely formal concept adopted for our convenience, the numbers are determined solely by reliance upon rules. These rules can result in a fractional oxidation number, as shown here. This does not mean that electronic charges are split.

1. The formation of rust, Fe_2O_3, iron(III) oxide: oxidation state of Fe

$$4Fe\ (s) + 3O_2\ (g) \longrightarrow 2Fe_2O_3\ (s) \qquad 0 \longrightarrow +3$$

2. Combustion reactions: oxidation state of C

$$C\ (s) + O_2\ (g) \longrightarrow CO_2\ (g) \qquad 0 \longrightarrow +4$$

$$2CO\ (g) + O_2\ (g) \longrightarrow 2CO_2\ (g) \qquad +2 \longrightarrow +4$$

$$C_3H_8\ (g) + 5O_2\ (g) \longrightarrow 3CO_2\ (g) + 4H_2O\ (g) \qquad -8/3 \longrightarrow +4$$

Originally the term *reduction* described the removal of oxygen from a compound. Ores are reduced to metals (a very real reduction in mass; Chapter 20) and oxides to elements. For example, tungsten for use in light bulb filaments can be prepared by reduction of tungsten(VI) oxide with hydrogen at 1200°C:

$$WO_3\ (s) + 3H_2\ (g) \longrightarrow W\ (s) + 3H_2O\ (g) \qquad \begin{matrix}\text{oxidation state of W}\\ +6 \longrightarrow 0\end{matrix}$$

Tungsten is reduced and its oxidation state decreases from +6 to zero.

The terms oxidation and reduction are now applied much more broadly. **Oxidation** is defined as an *algebraic increase in oxidation number* and corresponds to the *loss, or apparent loss, of electrons.* **Reduction** refers to an *algebraic decrease in oxidation number* and corresponds to a *gain, or apparent gain, of electrons.* Since electrons cannot be created or destroyed, oxidation and reduction always occur simultaneously in ordinary chemical reactions, and to the same extent. In the four equations cited above as examples of oxidation, observe that not only do the oxidation numbers of iron and carbon atoms increase as they are oxidized, but in each case oxygen is reduced as its oxidation state decreases from zero to −2. In the reduction of WO_3 according to the equation above, hydrogen also is oxidized from the zero to the +1 oxidation state.

We pointed out in Section 7–7 that **oxidizing agents** *are the species that* (1) *gain* (or appear to gain) *electrons,* (2) *are reduced,* and (3) *oxidize other substances.* **Reducing agents** *are the species that* (1) *lose* (or appear to lose) *electrons,* (2) *are oxidized,* and (3) *reduce other substances.* The following equations represent other examples of redox reactions. Oxidation numbers are shown above the formulas, and oxidizing and reducing agents are indicated for the molecular equations

$$\underset{\text{reducing agent}}{2\overset{0}{Fe}\ (s)} + \underset{\text{oxidizing agent}}{3\overset{0}{Cl_2}\ (g)} \longrightarrow 2\overset{+3}{Fe}\overset{-1}{Cl_3}\ (s)$$

$$\underset{\text{reducing agent}}{2\overset{+3}{Fe}\overset{-1}{Br_3}\ (aq)} + \underset{\text{oxidizing agent}}{3\overset{0}{Cl_2}\ (g)} \longrightarrow 2\overset{+3}{Fe}\overset{-1}{Cl_3}\ (aq) + 3\overset{0}{Br_2}\ (\ell)$$

Equations for redox reactions may be written as total ionic and net ionic equations as well as molecular equations. For example, the last equation may also be written as shown below. We shall distinguish between oxidation numbers and actual charges on ions by denoting oxidation numbers as $+n$ or $-n$ just above the symbols of the elements, and actual charges as $n+$ or $n-$ above and to the right of formulas of ions.

$$2[Fe^{3+}\ (aq) + 3Br^-\ (aq)] + 3Cl_2\ (g) \longrightarrow 2[Fe^{3+}\ (aq) + 3Cl^-\ (aq)] + 3Br_2\ (\ell) \quad \text{(total ionic equation)}$$

$$2Br^-\ (aq) + Cl_2\ (g) \longrightarrow 2Cl^-\ (aq) + Br_2\ (\ell) \quad \text{(net ionic equation)}$$

Notice that the spectator ions, Fe^{3+} ions in this case, do not participate in electron transfer. Their cancellation (Section 7–3) allows us to focus more clearly on the oxidizing agent, Cl_2 (g), and the reducing agent, Br^- (aq).

EXAMPLE 12–1

Write each of the following molecular equations as a net ionic equation if the two differ. Which ones represent redox reactions? In those that are redox reactions, identify the oxidizing agent, the reducing agent, the species oxidized, and the species reduced.

(a) $2AgNO_3\ (aq) + Cu\ (s) \longrightarrow Cu(NO_3)_2\ (aq) + 2Ag\ (s)$

(b) $2KClO_3\ (s) \xrightarrow{\Delta} 2KCl\ (s) + 3O_2\ (g)$

(c) $3AgNO_3\ (aq) + K_3PO_4\ (aq) \longrightarrow Ag_3PO_4\ (s) + 3KNO_3\ (aq)$

(d) $CaCO_3\ (s) \xrightarrow{\Delta} CaO\ (s) + CO_2\ (g)$

Solution

(a) According to the solubility rules (Section 7–1, Part 5), both silver nitrate, $AgNO_3$, and copper(II) nitrate, $Cu(NO_3)_2$, are water-soluble, ionic compounds. The total ionic equation and oxidation states (rules given in Section 4–12) are

$$2[\overset{+1}{Ag^+}\ (aq) + \overset{+5\,-2}{NO_3^-}\ (aq)] + \overset{0}{Cu}\ (s) \longrightarrow [\overset{+2}{Cu^{2+}}\ (aq) + 2\,\overset{+5\,-2}{NO_3^-}\ (aq)] + 2\overset{0}{Ag}\ (s)$$

The nitrate ions, NO_3^-, are spectator ions. Cancelling them from both sides gives the net ionic equation

$$2\overset{+1}{Ag^+}\ (aq) + \overset{0}{Cu}\ (s) \longrightarrow \overset{+2}{Cu^{2+}}\ (aq) + 2\overset{0}{Ag}\ (s)$$

This is a redox equation. The oxidation number of silver decreases from +1 to zero; silver ion is reduced, and is the oxidizing agent. The oxidation number of copper increases from zero to +2; copper is oxidized, and is the reducing agent.

(b) This reaction involves two solids and a gas, so the molecular and net ionic equations are identical. It is a redox reaction

$$2\overset{+1\,+5\,-2}{KClO_3}\ (s) \xrightarrow{\Delta} 2\overset{+1\,-1}{KCl}\ (s) + 3\overset{0}{O_2}\ (g)$$

The oxidizing agent is chlorine in the +5 oxidation state. It is reduced to the −1 oxidation state. The oxidation number of oxygen increases from −2 to zero; it is oxidized, and is the reducing agent. We might also say that $KClO_3$ is both oxidizing agent and reducing agent; this is a disproportionation reaction.

(c) The solubility rules indicate that all these compounds are soluble and ionic except for silver phosphate, Ag_3PO_4. The total ionic equation is

$$3[Ag^+\ (aq) + NO_3^-\ (aq)] + [3K^+\ (aq) + PO_4^{3-}\ (aq)] \longrightarrow Ag_3PO_4\ (s) + 3[K^+\ (aq) + NO_3^-\ (aq)]$$

Elimination of the spectator ions, NO_3^- (aq) and K^+ (aq), generates the net ionic equation. Oxidation numbers are assigned

$$3\overset{+1}{Ag^+}\ (aq) + \overset{+5\,-2}{PO_4^{3-}}\ (aq) \longrightarrow \overset{+1\,+5\,-2}{Ag_3PO_4}\ (s)$$

There are no changes in oxidation numbers; therefore, the reaction is not a redox reaction.

(d) $$\overset{+2\,+4\,-2}{CaCO_3}\ (s) \xrightarrow{\Delta} \overset{+2\,-2}{CaO}\ (s) + \overset{+4\,-2}{CO_2}\ (g)$$

As in (b), the equation describes a solid state decomposition, and the molecular and net ionic equations are identical. There are no changes in oxidation numbers and this is not a redox reaction.

Balancing Oxidation-Reduction Equations

Our rules for assigning oxidation numbers are constructed so that the *total increase in oxidation numbers must equal the total decrease in oxidation numbers in all redox reactions.* This equivalence provides the basis for balancing redox equations. Although there is no single "best method" for balancing all redox equations, two methods are particularly useful, (1) the change-in-oxidation-number method and (2) the ion-electron method.

Most redox equations can be balanced by both methods, but in some instances one may be easier to use than the other.

12–2 Change-in-Oxidation-Number Method

The next few examples illustrate this method, which is based on *equal total increases and decreases in oxidation numbers.* Although many redox equations can be balanced by simple inspection, it is important to learn the method because it can be used to balance more difficult equations. The general procedure is:

1. Write the overall unbalanced equation.
2. Assign oxidation numbers to determine which elements undergo changes in oxidation number.
3. Insert coefficients to make the total increase and decrease in oxidation numbers equal.
4. Balance the remaining atoms by inspection.

EXAMPLE 12–2

Aluminum dissolves in hydrochloric acid to form aqueous aluminum chloride and gaseous hydrogen. Balance the net ionic and molecular equations and identify the oxidizing and reducing agents.

Solution

The unbalanced molecular equation and oxidation numbers are

$$\overset{0}{\mathrm{Al}}\ (s) + \overset{+1\,-1}{\mathrm{HCl}}\ (aq) \longrightarrow \overset{+3\,-1}{\mathrm{AlCl_3}}\ (aq) + \overset{0}{\mathrm{H_2}}\ (g)$$

The oxidation number of aluminum increases from zero to +3. Therefore, aluminum is the reducing agent and is oxidized. (Recall from Section 7–7 that we represent an increase in oxidation number by $n\uparrow$ and a decrease in oxidation number by $n\downarrow$, where n is the magnitude of the change.)

$$\overset{0}{\mathrm{Al}} \xrightarrow{3\uparrow} \overset{+3}{\mathrm{Al}}$$

There is a decrease in the oxidation number of hydrogen, which is reduced and is the oxidizing agent. Since *at least* two hydrogen atoms in the zero oxidation state are produced (H_2), we must have at least two hydrogen atoms on the left at this point

$$2\overset{+1}{\mathrm{H}} \xrightarrow{2\downarrow} \overset{0}{\mathrm{H_2}}$$

The total increase in oxidation number for aluminum must equal the total decrease in oxidation number for hydrogen. The lowest common multiple of two and three is six. Therefore we multiply the aluminum "helping" equation by two and the hydrogen "helping" equation by three, so that there are six units of increase in oxidation number for every six units of decrease.

$$2(\overset{0}{\text{Al}} \xrightarrow{3\uparrow} \overset{+3}{\text{Al}}) \quad \text{or} \quad 2\overset{0}{\text{Al}} \xrightarrow{6\uparrow} 2\overset{+3}{\text{Al}}$$

$$3(2\overset{+1}{\text{H}} \xrightarrow{2\downarrow} \overset{0}{\text{H}_2}) \quad \text{or} \quad 6\overset{+1}{\text{H}} \xrightarrow{6\downarrow} 3\overset{0}{\text{H}_2}$$

From these numbers we can generate the balanced net ionic equation

$$2\text{Al (s)} + 6\text{H}^+ \text{ (aq)} \longrightarrow 2\text{Al}^{3+} \text{ (aq)} + 3\text{H}_2 \text{ (g)} \qquad \text{(balanced net ionic equation)}$$

All balanced equations must satisfy two criteria:

1. There must be mass balance. That is, the same number of atoms of each kind must be shown as reactants and products.
2. There must be a charge balance. The sums of actual charges on the left and right sides of the equation must be equal.

In this example there are 2Al and 6H on each side, so the first criterion is satisfied; the second is satisfied by charges of 6+ on each side.

The ionic equation is easily converted to a molecular equation that is also balanced with respect to atoms (2Al, 6H, 6Cl) and charge (zero) by adding $6Cl^-$ to each side.

$$2\text{Al (s)} + 6\text{HCl (aq)} \longrightarrow 2\text{AlCl}_3 \text{ (aq)} + 3\text{H}_2 \text{ (g)} \qquad \text{(balanced molecular equation)}$$

EXAMPLE 12–3

Sodium carbonate, Na_2CO_3, reacts with carbon and nitrogen at high temperatures to form sodium cyanide, NaCN, and carbon monoxide, CO. This is one industrial preparation of sodium cyanide, an important reactant in the synthesis of many commercially important products such as nylon. Balance the molecular equation for the reaction and identify the oxidizing and reducing agents. (Note: Assume that the oxidation state of carbon in the cyanide ion is +2.)

Solution

The unbalanced equation and oxidation numbers are

$$\overset{+1\ +4\ -2}{\text{Na}_2\text{CO}_3} \text{ (s)} + \overset{0}{\text{C}} \text{ (s)} + \overset{0}{\text{N}_2} \text{ (g)} \longrightarrow \overset{+1\ +2\ -3}{\text{NaCN}} \text{ (s)} + \overset{+2\ -2}{\text{CO}} \text{ (g)}$$

The oxidation numbers of sodium and oxygen do not change, but those of both carbon and nitrogen do change. The oxidation number of nitrogen decreases by three, so the nitrogen is reduced and is the oxidizing agent. Nitrogen atoms occur in groups of two on the left, so we write N_2 on the left

$$\overset{0}{\text{N}_2} \xrightarrow{6\downarrow} 2\overset{-3}{\text{N}}$$

The oxidation number of the carbon atoms in *sodium carbonate,* Na_2CO_3, decreases by two in forming *some* of the sodium cyanide, NaCN, and CO. Thus, these carbon atoms also function as oxidizing agents and are reduced

$$\overset{+4}{\text{C}} \xrightarrow{2\downarrow} \overset{+2}{\text{C}}$$

(Please keep in mind the fact that these designations *do not imply* that there are ions present.) The *elemental carbon* atoms are oxidized because their oxidation number increases by two. Elemental carbon is the reducing agent

$$\overset{0}{\text{C}} \xrightarrow{2\uparrow} \overset{+2}{\text{C}}$$

We now balance the changes in oxidation numbers. The total decrease is $6\downarrow + 2\downarrow = 8\downarrow$. There must be eight unit increases for every eight unit decreases.

$$(\overset{0}{N_2} \xrightarrow{6\downarrow} \overset{-3}{2N})$$

in Na_2CO_3 → $(\overset{+4}{C} \xrightarrow{2\downarrow} \overset{+2}{C})$

in elemental C → $4(\overset{0}{C} \xrightarrow{2\uparrow} \overset{+2}{C})$

(both $\overset{+2}{C}$ in NaCN and CO)

We sum these to obtain

$$\overset{0}{N_2} + \overset{+4}{C} + \overset{0}{4C} \longrightarrow \overset{-3}{2N} + \overset{+2}{5C}$$

From this information we can write the balanced molecular equation. The coefficient of 2 preceding N in the product tells us that there must be 2NaCN on the right. Since $2\overset{+2}{C}$ are a part of the 2NaCN, there must be 3CO to account for the remaining $3\overset{+2}{C}$

$$N_2\,(g) + Na_2CO_3\,(s) + 4C\,(s) \longrightarrow 2NaCN\,(s) + 3CO \qquad \text{(balanced)}$$

We now check for atom balance and find 2Na, 5C, 3O, and 2N on each side. The equation is also balanced with respect to charge, since all species are electrically neutral in all *molecular* equations.

12–3 Adding H^+, OH^-, or H_2O

Frequently it is necessary to complete the mass balance for reactions in aqueous solution by adding hydrogen atoms or oxygen atoms. For reactions in *acidic solution*, H^+ or H_2O or both may be added as needed, since both are readily available. In *basic solution*, OH^- or H_2O or both may be added as needed. This is illustrated in the next two examples.

EXAMPLE 12–4

Potassium dichromate reacts with hydroiodic acid to produce potassium iodide, chromium(III) iodide, and iodine. Write balanced net ionic and molecular equations for the reaction, and identify the oxidizing and reducing agents.

Solution

The incomplete and unbalanced molecular equation and oxidation states are

$$\overset{+1}{K_2}\overset{+6}{Cr_2}\overset{-2}{O_7} + \overset{+1}{H}\overset{-1}{I} \longrightarrow \overset{+1}{K}\overset{-1}{I} + \overset{+3}{Cr}\overset{-1}{I_3} + \overset{0}{I_2}$$

Chromium is reduced from the +6 to the +3 oxidation state. *Some* of the iodine is oxidized from the −1 to the zero oxidation state, while some remains in the −1 state in KI and CrI_3

$$\overset{+6}{Cr} \xrightarrow{3\downarrow} \overset{+3}{Cr}$$

$$2\overset{-1}{I} \xrightarrow{2\uparrow} \overset{0}{I_2}$$

We now balance the changes in oxidation numbers and add the two resulting "helping" equations

$$2(\overset{+6}{Cr} \xrightarrow{3\downarrow} \overset{+3}{Cr})$$

$$3(2\overset{-1}{I} \xrightarrow{2\uparrow} \overset{0}{I_2})$$ (Only the I^- that undergoes a change in oxidation number is converted to I_2.)

$$2\overset{+6}{Cr} + 6\overset{-1}{I} \longrightarrow 2\overset{+3}{Cr} + 3\overset{0}{I_2}$$

There are two chromium atoms per formula unit of $Cr_2O_7^{2-}$, so the coefficient of $Cr_2O_7^{2-}$ is one:

$Cr_2O_7^{2-}$ (aq) + $6I^-$ (aq) $\longrightarrow$ $2Cr^{3+}$ (aq) + $3I_2$ (s) (incomplete)

The equation is not yet balanced with respect to atoms or charge. Now we must add H_2O as suggested in the beginning of this section. There are seven atoms of oxygen in $Cr_2O_7^{2-}$ on the left, and none on the right; these oxygen atoms must appear in the product as water (since none of the other products of the reaction contain oxygen). Seven H_2O are required on the right to balance the seven oxygen atoms on the left. Fourteen H^+ can then be added to the left (since this reaction occurs in an acidic solution) to achieve atom (and charge) balance

$Cr_2O_7^{2-}$ (aq) + $14H^+$ (aq) + $6I^-$ (aq) $\longrightarrow$ $2Cr^{3+}$ (aq) + $3I_2$ (s) + $7H_2O$ (ℓ)
(balanced net ionic equation)

K⁺ and some of the I⁻ ions are spectator ions.

Two things must now be done to obtain the molecular equation from this net ionic equation. First, note that the ionic equation shows a total charge of 6+ on each side, whereas the molecular equation must be neutral on both sides; therefore, we must add species with a total charge of 6− to both sides. Second, the potassium ions in potassium dichromate and the iodide ions that counterbalance Cr^{3+} and K^+ ions must be accounted for. Thus, we add $2K^+$ and $8I^-$ to each side

$Cr_2O_7^{2-}$ (aq) + $14H^+$ (aq) + $6I^-$ (aq) $\longrightarrow$ $2Cr^{3+}$ (aq) + $3I_2$ (s) + $7H_2O$ (ℓ)
(net ionic equation)

adding spectator ions

+$2K^+$ (aq) + $8I^-$ (aq) $\longrightarrow$ $6I^-$ (aq) + $2K^+$ (aq) + $2I^-$ (aq) (spectator ions)

$K_2Cr_2O_7$ (aq) + 14HI (aq) $\longrightarrow$ $2CrI_3$ (aq) + 2KI (aq) + $3I_2$ (s) + $7H_2O$ (ℓ)
(molecular equation)

The dichromate ion, which contains chromium in the +6 oxidation state, is the oxidizing agent. Iodine in the −1 oxidation state, I^-, is the reducing agent.

EXAMPLE 12–5

In basic solution, permanganate ions oxidize iodide ions to iodine and form manganese(IV) oxide. Write and balance the net ionic equation for the reaction.

Solution

Here we have shown only the oxidation numbers for elements that change.

We first write the incomplete unbalanced equation and assign oxidation numbers. MnO_2 is insoluble in water and is written in molecular form

$$\overset{+7}{Mn}O_4^- \text{ (aq)} + \overset{-1}{I^-} \text{ (aq)} \longrightarrow \overset{0}{I_2} \text{ (s)} + \overset{+4}{Mn}O_2 \text{ (s)}$$

Manganese is reduced from the +7 to the +4 oxidation state. The oxidation number of iodine increases from −1 to zero as iodide ions are oxidized to iodine molecules, I_2

$$\overset{+7}{Mn} \xrightarrow{3\downarrow} \overset{+4}{Mn}$$

$$2\overset{-1}{I} \xrightarrow{2\uparrow} \overset{0}{I_2}$$

To balance the changes in oxidation numbers the first "helping" equation is multiplied by two, and the second by three, and then they are added

$$2(\overset{+7}{Mn} \xrightarrow{3\downarrow} \overset{+4}{Mn})$$

$$3(2\overset{-1}{I} \xrightarrow{2\uparrow} \overset{0}{I_2})$$

$$2\overset{+7}{Mn} + 6\overset{-1}{I} \longrightarrow 2\overset{+4}{Mn} + 3\overset{0}{I_2}$$

We next begin to balance the net ionic equation using these coefficients

$$2MnO_4^-\ (aq) + 6I^-\ (aq) \longrightarrow 3I_2\ (s) + 2MnO_2\ (s) \qquad \text{(unbalanced)}$$

The changes in oxidation number have been balanced, but we do not yet have mass balance or charge balance. (It is important to realize that at this point the charge *cannot* be balanced by adding or removing electrons.) Both mass and charge balance are accomplished by adding OH^- and/or H_2O since the reaction occurs in basic solution. There are eight oxygen atoms on the left and only four on the right. We balance oxygen atoms while adding equal numbers of hydrogen atoms to both sides. Note that both OH^- and H_2O contain only hydrogen and oxygen, but that water is "hydrogen-rich." Therefore we add twice as many OH^- ions to the side needing oxygen atoms as we add H_2O to the other side. In this case $8OH^-$ are added to the right and $4H_2O$ to the left to give the balanced equation

$$4H_2O\ (\ell) + 2MnO_4^-\ (aq) + 6I^-\ (aq) \longrightarrow 3I_2\ (s) + 2MnO_2\ (s) + 8OH^-\ (aq)$$

The equation now has atom balance (8H, 12O, 2Mn, 6I on each side) and charge balance (8− on each side).

12-4 The Ion-Electron Method (Half-Reaction Method)

The ion-electron method relies upon the separate and complete balancing of equations describing so-called oxidation and reduction **half-reactions.** This is followed by equalizing the numbers of electrons gained and lost in each and, finally, adding the resulting half-reactions to give the overall balanced equation. The general procedure follows.

1. Write as much of the overall unbalanced equation as possible.
2. Construct unbalanced oxidation and reduction half-reactions (these are usually incomplete as well as unbalanced).
3. Balance the atoms in each half-reaction.
4. Balance the charge in each half-reaction by adding electrons.
5. Balance the electron transfer by multiplying the balanced half-reactions by appropriate integers.
6. Add the resulting half-reactions, and eliminate any common terms to obtain the balanced equation.

Note that reactions between MnO_4^- and I^- ions in basic and acidic solutions are *not* the same. See Example 12–5.

EXAMPLE 12–6

Permanganate ions oxidize iodide ions to iodine in acidic solution, and are reduced to manganese(II) ions. Write and balance the net ionic equation for the reaction by the ion-electron method.

Solution

Applying the first step in the procedure to the statement of the problem, we can write

$$MnO_4^- \text{ (aq)} + I^- \text{ (aq)} \longrightarrow Mn^{2+} \text{ (aq)} + I_2 \text{ (s)} \qquad \text{(incomplete and unbalanced)}$$

From the information we begin to construct half-reactions for the oxidation and reduction parts of the reaction

$$2I^- \longrightarrow I_2$$

$$MnO_4^- \longrightarrow Mn^{2+}$$

The first half-reaction, $2I^- \text{ (aq)} \rightarrow I_2 \text{ (s)}$ (unbalanced), shows mass balance but not charge balance. Addition of 2 electrons to the right gives a balanced half-reaction; this must be the oxidation because the system loses electrons.

$$2I^- \text{ (aq)} \longrightarrow I_2 \text{ (s)} + 2e^- \qquad \text{(balanced } \textit{oxidation} \text{ half-reaction)}$$

The other half-reaction is balanced by:

(1) adding $4H_2O$ to the right because there are 4 oxygen atoms on the left,

$$MnO_4^- \longrightarrow Mn^{2+} + 4H_2O \qquad \text{(unbalanced)}$$

(2) adding $8H^+$ to the left to give mass balance,

$$MnO_4^- + 8H^+ \longrightarrow Mn^{2+} + 4H_2O \qquad \text{(unbalanced)}$$

(3) adding 5 electrons to the left to give charge balance and a *balanced* reduction half-reaction.

$$MnO_4^- + 8H^+ + 5e^- \longrightarrow Mn^{2+} + 4H_2O \qquad \text{(balanced } \textit{reduction} \text{ half-reaction)}$$

Examination of the balanced half-reactions

$$2I^- \text{ (aq)} \longrightarrow I_2 \text{ (s)} + 2e^-$$

$$MnO_4^- \text{ (aq)} + 8H^+ \text{ (aq)} + 5e^- \longrightarrow Mn^{2+} \text{ (aq)} + 4H_2O \text{ } (\ell)$$

indicates that the first should be multiplied by *5* and the second by *2* to balance the electron transfer. Addition of the resulting half-reactions term-by-term gives

$$5(2I^- \longrightarrow I_2 + 2e^-)$$

$$2(MnO_4^- + 8H^+ + 5e^- \longrightarrow Mn^{2+} + 4H_2O)$$

$$\overline{10I^- + 2MnO_4^- + 16H^+ + 10e^- \longrightarrow 5I_2 + 2Mn^{2+} + 8H_2O + 10e^-}$$

Elimination of terms common to both sides ($10e^-$) gives the balanced net ionic equation.

$$10I^- \text{ (aq)} + 2MnO_4^- \text{ (aq)} + 16H^+ \text{ (aq)} \longrightarrow 5I_2 \text{ (s)} + 2Mn^{2+} \text{ (aq)} + 8H_2O \text{ } (\ell)$$

EXAMPLE 12–7

The chromate ion, CrO_4^{2-}, oxidizes the stannite ion, $HSnO_2^-$, to the stannate ion, $HSnO_3^-$, and is reduced to chromite ion, CrO_2^-, in basic solution. Balance the net ionic equation for this reaction.

Solution

From the statement of the problem we may write

CrO_4^{2-} (aq) + $HSnO_2^-$ (aq) $\longrightarrow$ $HSnO_3^-$ (aq) + CrO_2^- (aq) (unbalanced)

Let us deal with oxidation of the tin-containing species first

$HSnO_2^- \longrightarrow HSnO_3^-$ (unbalanced)

The hydrogen and tin atoms are balanced, but one more oxygen is required on the left. We add two OH^- to the left and one H_2O to the right to maintain the hydrogen balance

$2OH^- + HSnO_2^- \longrightarrow HSnO_3^- + H_2O$ (unbalanced)

The net charge on the left is 3−, and on the right it is 1−, so 2 electrons are added to the right

$2OH^- + HSnO_2^- \longrightarrow HSnO_3^- + H_2O + 2e^-$ (balanced *oxidation* half-reaction)

Reduction involves conversion of chromate ions to chromite ions

$CrO_4^{2-} \longrightarrow CrO_2^-$ (unbalanced)

We add $4OH^-$ to the right and $2H_2O$ to the left to balance oxygen and hydrogen atoms

$2H_2O + CrO_4^{2-} \longrightarrow CrO_2^- + 4OH^-$ (unbalanced)

To balance the charge (2− left, 5− right), 3 electrons are added to the left

$3e^- + 2H_2O + CrO_4^{2-} \longrightarrow CrO_2^- + 4OH^-$ (balanced *reduction* half-reaction)

We now balance the electron transfer and add the two half-reactions

$3(2OH^- + HSnO_2^- \longrightarrow HSnO_3^- + H_2O + 2e^-)$ (oxidation)
$2(3e^- + 2H_2O + CrO_4^{2-} \longrightarrow CrO_2^- + 4OH^-)$ (reduction)

$$6OH^- + 3HSnO_2^- + 6e^- + 4H_2O + 2CrO_4^{2-} \longrightarrow 3HSnO_3^- + 3H_2O + 6e^- + 2CrO_2^- + 8OH^-$$

Elimination of terms common to both sides ($6e^-$, $3H_2O$, and $6OH^-$) gives the balanced net ionic equation

$$3HSnO_2^-\ (aq) + H_2O\ (\ell) + 2CrO_4^{2-}\ (aq) \longrightarrow 3HSnO_3^-\ (aq) + 2CrO_2^-\ (aq) + 2OH^-\ (aq)$$

EXAMPLE 12–8

Balance the net ionic equation for the reaction by which phosphine, PH_3, a colorless, poisonous gas with an odor like decaying fish, and the hypophosphite ion, $H_2PO_2^-$, are prepared from elemental phosphorus, P_4, in acidic solution.

Solution

P_4 (s) $\longrightarrow$ PH_3 (g) + $H_2PO_2^-$ (aq) (unbalanced)

Phosphorus is both oxidized and reduced in this reaction. Such reactions are called *disproportionation* reactions. The unbalanced reduction half-reaction is

$P_4 \longrightarrow PH_3$ (unbalanced)

Following the steps previously outlined gives the balanced reduction half-reaction

$$P_4 \longrightarrow 4PH_3$$

$$12H^+ + P_4 \longrightarrow 4PH_3$$

$$12e^- + 12H^+ + P_4 \longrightarrow 4PH_3 \quad \text{(balanced } \textit{reduction} \text{ half-reaction)}$$

The balanced oxidation half-reaction is obtained by a series of similar steps

$$P_4 \longrightarrow 4H_2PO_2^-$$

$$8H_2O + P_4 \longrightarrow 4H_2PO_2^- + 8H^+$$

$$8H_2O + P_4 \longrightarrow 4H_2PO_2^- + 8H^+ + 4e^- \quad \text{(balanced } \textit{oxidation} \text{ half-reaction)}$$

We now equalize the electron transfer and add the two half-reactions algebraically

$$(12e^- + 12H^+ + P_4 \longrightarrow 4PH_3) \quad \text{(reduction)}$$
$$3(8H_2O + P_4 \longrightarrow 4H_2PO_2^- + 8H^+ + 4e^-) \quad \text{(oxidation)}$$
$$12e^- + 12H^+ + P_4 + 24H_2O + 3P_4 \longrightarrow 4PH_3 + 12H_2PO_2^- + 24H^+ + 12e^-$$

Elimination of common terms ($12e^-$ and $12H^+$) and collection of like terms (P_4) gives

$$4P_4\ (s) + 24H_2O\ (\ell) \longrightarrow 4PH_3\ (g) + 12H_2PO_2^-\ (aq) + 12H^+\ (aq)$$

The equation is divisible by 4 to give the balanced net ionic equation

$$P_4\ (s) + 6H_2O\ (\ell) \longrightarrow PH_3\ (g) + 3H_2PO_2^-\ (aq) + 3H^+\ (aq)$$

Redox Titrations

One method of analyzing samples quantitatively for the presence of oxidizable or reducible substances is by **redox titration.** In such analyses, the concentration of a solution is determined by allowing the solution to react with a carefully measured amount of a standard solution of an oxidizing or reducing agent.

Concentrations of solutions involved in redox titrations can be expressed in terms of either molarity or normality, and amounts of solutes can be described in terms of either moles or equivalent weights. We shall illustrate redox titrations, separately, by both the mole/molarity method and the equivalent weight/normality method.

12–5 The Mole Method and Molar Solutions of Oxidizing and Reducing Agents

As in other kinds of chemical reactions, we must pay particular attention to the mole ratio, i.e., the reaction ratio, in which oxidizing agents and reducing agents react. The examples that follow will re-emphasize the importance of the reaction ratio.

Potassium permanganate, $KMnO_4$, is a strong oxidizing agent, and it has been the "workhorse" of redox titrations over the years. For example, in acidic solution, $KMnO_4$ can be used to react with iron(II) sulfate, $FeSO_4$, according to the balanced molecular and net ionic equations given below. (A strong acid, such as H_2SO_4, is used in such titrations.)

$$2KMnO_4 + 10FeSO_4 + 8H_2SO_4 \longrightarrow 2MnSO_4 + 5Fe_2(SO_4)_3 + K_2SO_4 + 8H_2O$$

$$MnO_4^-\ (aq) + 5Fe^{2+}\ (aq) + 8H^+\ (aq) \longrightarrow Mn^{2+}\ (aq) + 5Fe^{3+}\ (aq) + 4H_2O\ (\ell)$$

We have provided the balanced molecular and net ionic equations to emphasize the fact that the reaction ratio is 1 mol $KMnO_4$:5 mol $FeSO_4$ or 1 mol MnO_4^-:5 mol Fe^{2+}. (Examples 12–9 and 12–11 are based on this reaction.)

A word about conventions. The above reaction involves permanganate ions, MnO_4^-, and iron(II) ions, Fe^{2+}, in acidic solution. The source of MnO_4^- ions is the soluble ionic compound, $KMnO_4$. We often refer to "permanganate solutions." Clearly such solutions contain cations, K^+, in this case. Likewise, we often refer to "iron(II) solutions" without specifying that the anion is SO_4^{2-}.

Because it has an intense purple color, $KMnO_4$ acts as its own indicator. One drop of 0.020 *M* $KMnO_4$ solution imparts a pink color to a liter of pure water. When $KMnO_4$ solution is added to a solution of a reducing agent, the end point in the titration is taken as the point at which a pale pink color appears in the solution being titrated and persists for at least 30 seconds.

EXAMPLE 12–9

What volume of 0.0200 *M* $KMnO_4$ solution is required to oxidize 40.0 mL of 0.100 *M* $FeSO_4$ in sulfuric acid solution?

Solution

From the above discussion, we know the balanced equation and the reaction ratio

$$MnO_4^-\ (aq) + 8H^+\ (aq) + 5Fe^{2+}\ (aq) \longrightarrow 5Fe^{3+}\ (aq) + Mn^{2+}\ (aq) + 4H_2O$$

reaction ratio: 1 mol 5 mol

We first determine the number of moles of Fe^{2+} to be titrated.

$$\underline{?}\ \text{mol Fe}^{2+} = 40.0\ \text{mL} \times \frac{0.100\ \text{mol Fe}^{2+}}{1000\ \text{mL}} = 4.00 \times 10^{-3}\ \text{mol Fe}^{2+}$$

1.000 L has been expressed as 1000 mL because small volumes of solutions are used in titrations.

Now we use the balanced equation to determine the number of moles of MnO_4^- required

$$\underline{?}\ \text{mol MnO}_4^- = 4.00 \times 10^{-3}\ \text{mol Fe}^{2+} \times \frac{1\ \text{mol MnO}_4^-}{5\ \text{mol Fe}^{2+}} = 8.00 \times 10^{-4}\ \text{mol MnO}_4^-$$

Finally we calculate the volume of 0.0200 *M* MnO_4^- solution that contains 8.00×10^{-4} mol MnO_4^-

$$\underline{?}\ \text{mL MnO}_4^-\ \text{soln} = 8.00 \times 10^{-4}\ \text{mol MnO}_4^- \times \frac{1000\ \text{mL MnO}_4^-\ \text{soln}}{0.0200\ \text{mol MnO}_4^-}$$

$$= \underline{40.0\ \text{mL MnO}_4^-\ \text{soln}}$$

Potassium dichromate, $K_2Cr_2O_7$, is another frequently used oxidizing agent. However, an indicator must be used when reducing agents are titrated with dichromate solutions because $K_2Cr_2O_7$ is not intensely colored enough to serve as its own indicator, and the reduction product, Cr^{3+}, is intensely colored (green).

Consider the oxidation of sulfite ions, SO_3^{2-}, to sulfate ions, SO_4^{2-}, by $Cr_2O_7^{2-}$ ions in the presence of a strong acid such as sulfuric acid. We shall balance the equation by the ion-electron method, beginning with the reduction of $Cr_2O_7^{2-}$

$$Cr_2O_7^{2-} \longrightarrow Cr^{3+}$$

$$Cr_2O_7^{2-} \longrightarrow 2Cr^{3+}$$

$$14H^+ + Cr_2O_7^{2-} \longrightarrow 2Cr^{3+} + 7H_2O$$

$$6e^- + 14H^+ + Cr_2O_7^{2-} \longrightarrow 2Cr^{3+} + 7H_2O \quad \text{(balanced } \textit{reduction} \text{ half-reaction)}$$

Now we balance the oxidation half-reaction by the usual procedure:

$$SO_3^{2-} \longrightarrow SO_4^{2-}$$

$$SO_3^{2-} + H_2O \longrightarrow SO_4^{2-} + 2H^+$$

$$SO_3^{2-} + H_2O \longrightarrow SO_4^{2-} + 2H^+ + 2e^- \quad \text{(balanced \textit{oxidation} half-reaction)}$$

We now equalize the electron transfer, add the balanced half-reactions, and eliminate common terms:

$$(6e^- + 14H^+ + Cr_2O_7^{2-} \longrightarrow 2Cr^{3+} + 7H_2O) \quad \text{(reduction)}$$
$$3(SO_3^{2-} + H_2O \longrightarrow SO_4^{2-} + 2H^+ + 2e^-) \quad \text{(oxidation)}$$
$$8H^+ (aq) + Cr_2O_7^{2-} (aq) + 3SO_3^{2-} (aq) \longrightarrow 2Cr^{3+} (aq) + 3SO_4^{2-} (aq) + 4H_2O (\ell)$$

The balanced equation tells us that the reaction ratio is 3 mol SO_3^{2-}/mol $Cr_2O_7^{2-}$ or 1 mol $Cr_2O_7^{2-}$/3 mol SO_3^{2-}. Potassium dichromate is the usual source of $Cr_2O_7^{2-}$ ions and Na_2SO_3 is the usual source of SO_3^{2-} ions. Thus, the above reaction ratio could also be expressed as 1 mol $K_2Cr_2O_7$/3 mol Na_2SO_3.

EXAMPLE 12–10

A 20.00-mL sample of Na_2SO_3 was titrated with 36.30 mL of 0.05130 *M* $K_2Cr_2O_7$ solution in the presence of sulfuric acid. Calculate the molarity of the Na_2SO_3 solution.

Solution

From the above discussion we know the balanced equation and the reaction ratio

$$\underset{3\text{ mol}}{3SO_3^{2-}} + \underset{1\text{ mol}}{Cr_2O_7^{2-}} + 8H^+ \longrightarrow 3SO_4^{2-} + 2Cr^{3+} + 4H_2O$$

The number of moles of $Cr_2O_7^{2-}$ used is

$$\underline{?}\text{ mol } Cr_2O_7^{2-} = 0.03630\text{ L} \times \frac{0.05130\text{ mol } Cr_2O_7^{2-}}{\text{L}} = 0.001862\text{ mol } Cr_2O_7^{2-}$$

The number of moles of SO_3^{2-} that reacted with 0.001862 mole of $Cr_2O_7^{2-}$ is

$$\underline{?}\text{ mol } SO_3^{2-} = 0.001862\text{ mol } Cr_2O_7^{2-} \times \frac{3\text{ mol } SO_3^{2-}}{1\text{ mol } Cr_2O_7^{2-}} = 0.005586\text{ mol } SO_3^{2-}$$

The Na_2SO_3 solution contained 0.005586 mole of SO_3^{2-} (and of necessity 0.005586 mole of Na_2SO_3). Its molarity is

$$\underline{?}\ \frac{\text{mol } Na_2SO_3}{\text{L}} = \frac{0.005586\text{ mol } Na_2SO_3}{0.02000\text{ L}} = \underline{0.2793\text{ M } Na_2SO_3}$$

12–6 Equivalent Weights and Normal Solutions of Oxidizing and Reducing Agents

When we studied acid-base reactions, we defined the term *equivalent weight* so that one equivalent weight, or one equivalent (eq), of any acid reacts with one equivalent of any base. We now define the term as it applies to redox reactions.

In redox reactions, **one equivalent weight** (or **equivalent**) *of a substance is the mass of the oxidizing or reducing substance that gains or loses*

6.022×10^{23} electrons. This definition tells us that one equivalent of any oxidizing agent reacts exactly with one equivalent of any reducing agent.

For all redox reactions we may write

$$\text{number of eq oxidizing agent} = \text{number of eq reducing agent}$$

Using the definition of normality, the relationship may also be expressed as

$$L_O \times N_O = L_R \times N_R$$

or as

$$\text{number of meq oxidizing agent} = \text{number of meq reducing agent}$$
$$mL_O \times N_O = mL_R \times N_R$$

Note that from a *stoichiometric* point of view, the only difference between calculations on acid-base reactions and those on redox reactions is the *definition* of the equivalent.

We saw earlier that the balanced net ionic equation for the reaction of $KMnO_4$ with $FeSO_4$ in acidic solution is

$$MnO_4^- \text{ (aq)} + 8H^+ \text{ (aq)} + 5Fe^{2+} \text{ (aq)} \longrightarrow Mn^{2+} \text{ (aq)} + 5Fe^{3+} \text{ (aq)} + 4H_2O \text{ } (\ell)$$

The balanced half-reactions are

$$Fe^{2+} \text{ (aq)} \longrightarrow Fe^{3+} \text{ (aq)} + 1e^- \qquad \text{(oxidation)}$$

$$MnO_4^- \text{ (aq)} + 8H^+ \text{ (aq)} + 5e^- \longrightarrow Mn^{2+} \text{ (aq)} + 4H_2O \text{ } (\ell) \qquad \text{(reduction)}$$

Each Fe^{2+} ion loses one electron in this reaction, and one mole of $FeSO_4$ loses 6.022×10^{23} electrons. Thus, one mole of $FeSO_4$ (151.9 grams) is one equivalent *in this reaction.*

The equivalent weight of an oxidizing agent or reducing agent depends upon the specific reaction it undergoes.

Each permanganate ion gains five electrons in the above reaction, and each mole of MnO_4^- that reacts gains $5(6.022 \times 10^{23}$ electrons). *In this reaction,* one mole of $KMnO_4$ (158.0 grams) provides five equivalents.

We may summarize our discussion of this reaction as follows.

	Compound	e^- Transferred per Formula Unit	One Mole	One eq
Oxidizing agent:	$KMnO_4$	5	158.0 g	$\frac{158.0 \text{ g}}{5} = 31.60$ g
Reducing agent:	$FeSO_4$	1	151.9 g	$\frac{151.9 \text{ g}}{1} = 151.9$ g

The following tabulation is equally appropriate.

	Species	e^- Transferred per Ion	One Mole	One eq
Oxidizing agent:	MnO_4^- (aq)	5	118.9 g	$\frac{118.9 \text{ g}}{5} = 23.8$ g
Reducing agent:	Fe^{2+} (aq)	1	55.8 g	$\frac{55.8 \text{ g}}{1} = 55.8$ g

In this reaction, 31.60 grams of $KMnO_4$ would react with 151.9 grams of $FeSO_4$. In terms of ions, 23.8 grams of MnO_4^- ions react with 55.8 grams of Fe^{2+} ions.

Notice that Example 12–11 is the same as Example 12–9 except that concentrations are expressed as *normality* rather than as *molarity*.

EXAMPLE 12–11

What volume of 0.1000 *N* $KMnO_4$ is required to oxidize 40.0 mL of 0.100 *N* $FeSO_4$ in sulfuric acid solution?

Solution

From the above discussion we know that the balanced equation is

$$MnO_4^- \text{ (aq)} + 8H^+ \text{ (aq)} + 5Fe^{2+} \text{ (aq)} \longrightarrow 5Fe^{3+} \text{ (aq)} + Mn^{2+} \text{ (aq)} + 4H_2O\ (\ell)$$

Since

$$mL_{MnO_4^-} \times N_{MnO_4^-} = mL_{Fe^{2+}} \times N_{Fe^{2+}}$$

we can solve for the number of milliliters of $KMnO_4$ solution

$$\underline{?}\ mL_{MnO_4^-} = \frac{mL_{Fe^{2+}} \times N_{Fe^{2+}}}{N_{MnO_4^-}} = \frac{40.0 \text{ mL} \times 0.100\ N}{0.1000\ N} = \underline{40.0 \text{ mL } KMnO_4 \text{ solution}}$$

The equivalent weight of an oxidizing agent or reducing agent depends upon the specific reaction it undergoes. In the reaction given below, the equivalent weight of $KMnO_4$ is different from the value we calculated before, because the permanganate ion undergoes a three-electron change rather than a five-electron change. The dissolution of metallic zinc in mildly basic potassium permanganate solution produces solid manganese(IV) oxide and zinc hydroxide. In this reaction the half-reactions are

$$2(MnO_4^- + 2H_2O + 3e^- \longrightarrow MnO_2 + 4OH^-) \qquad \text{(reduction)}$$

$$3(Zn + 2OH^- \longrightarrow Zn(OH)_2 + 2e^-) \qquad \text{(oxidation)}$$

and the balanced net ionic equation is

$$2MnO_4^- \text{ (aq)} + 3Zn\text{ (s)} + 4H_2O\ (\ell) \longrightarrow 2MnO_2\text{ (s)} + 3Zn(OH)_2\text{ (s)} + 2OH^- \text{ (aq)}$$

One mole of $KMnO_4$ is three equivalents *in this reaction*. One mole of zinc is two equivalents, as the following tabulation shows.

	Substance	e^- Transferred per Formula Unit	One Mole	One eq
Oxidizing agent:	$KMnO_4$ (aq)	3	158.0 g	$\frac{158.0\text{ g}}{3} = 52.7\text{ g}$
Reducing agent:	Zn (s)	2	65.4 g	$\frac{65.4\text{ g}}{2} = 32.7\text{ g}$

One equivalent of $KMnO_4^-$ is 52.7 grams (not 31.6 grams as before), and 52.7 grams of $KMnO_4$ react with 32.7 grams of Zn.

EXAMPLE 12–12

How many grams of $KMnO_4$ are contained in 35.0 mL of 0.0500 *N* $KMnO_4$ used in the following reaction in basic solution?

$$2MnO_4^- \text{ (aq)} + 3Zn \text{ (s)} + 4H_2O \text{ (}\ell\text{)} \longrightarrow 2MnO_2 \text{ (s)} + 3Zn(OH)_2 \text{ (s)} + 2OH^- \text{ (aq)}$$

Solution
As we saw earlier, one mole of $KMnO_4$ is three equivalents in this reaction

$$\underline{?} \text{ g } KMnO_4 = 35.0 \text{ mL} \times \frac{0.0500 \text{ eq } KMnO_4}{1000 \text{ mL}} \times \frac{1 \text{ mol}}{3 \text{ eq}} \times \frac{158.0 \text{ g}}{1 \text{ mol}}$$

$$= \underline{0.0922 \text{ g } KMnO_4}$$

To illustrate further how normality is used in oxidation-reduction titrations, let us solve Example 12–10, using normality rather than molarity, as illustrated in Example 12–13(a).

EXAMPLE 12–13

A 20.00-mL sample of Na_2SO_3 was titrated with 36.30 mL of 0.3078 *N* $K_2Cr_2O_7$ solution in the presence of H_2SO_4. (a) Calculate the normality of the Na_2SO_3 solution. (b) What mass of Na_2SO_3 was present in the sample? (c) What mass of $K_2Cr_2O_7$ was used to prepare 500 mL of the $K_2Cr_2O_7$ solution?

Solution
Recall that the balanced equation for the oxidation of SO_3^{2-} ions to SO_4^{2-} ions by $Cr_2O_7^{2-}$ ions in acidic solution is

$$8H^+ + Cr_2O_7^{2-} + 3SO_3^{2-} \longrightarrow 2Cr^{3+} + 3SO_4^{2-} + 4H_2O$$

and the balanced half-reactions are

$$6e^- + 14H^+ + Cr_2O_7^{2-} \longrightarrow 2Cr^{3+} + 7H_2O$$

$$SO_3^{2-} + H_2O \longrightarrow SO_4^{2-} + 2H^+ + 2e^-$$

(a) We know the volume and normality of the $K_2Cr_2O_7$ solution as well as the volume of the Na_2SO_3 solution

$$\text{mL}_O \times N_O = \text{mL}_R \times N_R$$

$$N_R = \frac{\text{mL}_O \times N_O}{\text{mL}_R} = \frac{36.30 \text{ mL} \times 0.3078\ N}{20.00 \text{ mL}}$$

$$N_R = 0.5587\ N\ Na_2SO_3$$

This calculation tells us that one liter of Na_2SO_3 solution contains 0.5587 eq Na_2SO_3

The answer to Example 12–10 is 0.2793 *M*. Is this the same concentration?

The half-reactions (above) give us the information needed to answer (b) and (c)

$$1 \text{ mol } K_2Cr_2O_7 = 6 \text{ eq } K_2Cr_2O_7 = 294.2 \text{ g } K_2Cr_2O_7$$

$$1 \text{ mol } Na_2SO_3 = 2 \text{ eq } Na_2SO_3 = 126.0 \text{ g } Na_2SO_3$$

$$\text{(b) } \underline{?} \text{ g } Na_2SO_3 = 20.00 \text{ mL} \times \frac{0.5587 \text{ eq } Na_2SO_3}{1000 \text{ mL}} \times \frac{1 \text{ mol } Na_2SO_3}{2 \text{ eq } Na_2SO_3}$$

$$\times \frac{126.0 \text{ g } Na_2SO_3}{1 \text{ mol } Na_2SO_3} = \underline{0.7040 \text{ g } Na_2SO_3}$$

(c) $? \text{ g } K_2Cr_2O_7 = 500 \text{ mL} \times \frac{0.3078 \text{ eq } K_2Cr_2O_7}{1000 \text{ mL}} \times \frac{1 \text{ mol } K_2Cr_2O_7}{6 \text{ eq } K_2Cr_2O_7}$

$\times \frac{294.2 \text{ g } K_2Cr_2O_7}{1 \text{ mol } K_2Cr_2O_7} = 7.546 \text{ g } K_2Cr_2O_7$

Key Terms

Equivalent weight of oxidizing or reducing agent, the mass that gains (oxidizing agents) or loses (reducing agents) 6.022×10^{23} electrons in a redox reaction.

Half-reaction either the oxidation part or the reduction part of a redox reaction.

Oxidation an algebraic increase in oxidation number; may correspond to a loss of electrons.

Oxidation-reduction reactions reactions in which oxidation and reduction occur; also called redox reactions.

Oxidizing agent the substance that oxidizes another substance and is reduced.

Redox reaction an oxidation-reduction reaction.

Redox titration the quantitative analysis of the amount or concentration of an oxidizing or reducing agent in a sample by observing its reaction with a known amount or concentration of a reducing or oxidizing agent.

Reducing agent the substance that reduces another substance and is oxidized.

Reduction an algebraic decrease in oxidation number; may correspond to a gain of electrons.

Exercises

1. Define and illustrate the following terms: (a) oxidation (b) reduction (c) oxidizing agent (d) reducing agent (e) equivalent weights of oxidizing and reducing agents

Assigning Oxidation Numbers

2. Assign oxidation numbers to the element specified in each group of compounds.
   (a) N in NO, N_2O_3, N_2O_4, NH_3, N_2H_4, NH_2OH, HNO_3
   (b) C in CO, CO_2, CH_2O, CH_4O, C_2H_6O, $(COOH)_2$, Na_2CO_3
   (c) S in S_8, H_2S, SO_2, SO_3, Na_2SO_3, H_2SO_4, K_2SO_4
3. Assign oxidation numbers to the element specified in each group of compounds.
   (a) P in PCl_3, P_4O_6, P_4O_{10}, HPO_3, H_3PO_4, $POCl_3$, $H_4P_2O_7$, $Mg_3(PO_4)_2$
   (b) Cl in Cl_2, HCl, $HClO$, $HClO_2$, $KClO_3$, Cl_2O_7, $Ca(ClO_4)_2$
   (c) Mn in MnO, MnO_2, $Mn(OH)_2$, K_2MnO_4, $KMnO_4$, Mn_2O_7
   (d) O in OF_2, Na_2O, Na_2O_2, KO_2
4. Assign oxidation numbers to the element specified in each group of ions.
   (a) S in S^{2-}, SO_3^{2-}, SO_4^{2-}, $S_2O_3^{2-}$, $S_4O_6^{2-}$
   (b) Cr in CrO_2^-, $Cr(OH)_4^-$, CrO_4^{2-}, $Cr_2O_7^{2-}$
   (c) B in BO_2^-, BO_3^{3-}, $B_4O_7^{2-}$
5. Assign oxidation numbers to the element specified in each group of ions.
   (a) N in N^{3-}, NO_2^-, NO_3^-, N_3^-, NH_4^+
   (b) Br in Br^-, BrO^-, BrO_3^-, BrO_4^-

Identification of Redox Reactions

6. Determine which of the following are oxidation-reduction reactions. For those that are, identify the oxidizing and reducing agents:
   (a) $HgCl_2\ (aq) + 2KI\ (aq) \longrightarrow HgI_2\ (s) + 2KCl\ (aq)$
   (b) $4NH_3\ (g) + 3O_2\ (g) \longrightarrow 2N_2\ (g) + 6H_2O\ (g)$
   (c) $CaCO_3\ (s) + 2HNO_3\ (aq) \longrightarrow Ca(NO_3)_2\ (aq) + CO_2\ (g) + H_2O\ (\ell)$
   (d) $PCl_3\ (\ell) + 3H_2O\ (\ell) \longrightarrow 3HCl\ (aq) + H_3PO_3\ (aq)$

7. Determine which of the following are oxidation-reduction reactions. For those that are, identify the oxidizing and reducing agents.
   (a) $2H_2O_2\ (\ell) \longrightarrow 2H_2O\ (\ell) + O_2\ (g)$
   (b) $ICl\ (s) + H_2O\ (\ell) \longrightarrow HCl\ (aq) + HOI\ (aq)$
   (c) $3HCl\ (aq) + HNO_3\ (aq) \longrightarrow Cl_2\ (g) + NOCl\ (g) + 2H_2O\ (\ell)$
   (d) $Fe_2O_3\ (s) + 3CO\ (g) \xrightarrow{\Delta} 2Fe\ (s) + 3CO_2\ (g)$

Balancing Redox Equations

8. Balance the following as net ionic equations.
   (a) $MnO_4^-\ (aq) + H^+\ (aq) + Br^-\ (aq) \longrightarrow Mn^{2+}\ (aq) + Br_2\ (\ell) + H_2O\ (\ell)$
   (b) $Cr_2O_7^{2-}\ (aq) + H^+\ (aq) + I^-\ (aq) \longrightarrow Cr^{3+}\ (aq) + I_2\ (s) + H_2O\ (\ell)$
   (c) $MnO_4^-\ (aq) + SO_3^{2-}\ (aq) + H^+\ (aq) \longrightarrow Mn^{2+}\ (aq) + SO_4^{2-}\ (aq) + H_2O\ (\ell)$
   (d) $Cr_2O_7^{2-}\ (aq) + Fe^{2+}\ (aq) + H^+\ (aq) \longrightarrow Cr^{3+}\ (aq) + Fe^{3+}\ (aq) + H_2O\ (\ell)$
9. Balance the following as net ionic equations.
   (a) $Cr(OH)_4^-\ (aq) + OH^-\ (aq) + H_2O_2\ (aq) \longrightarrow CrO_4^{2-}\ (aq) + H_2O\ (\ell)$
   (b) $MnO_2\ (s) + H^+\ (aq) + NO_2^-\ (aq) \longrightarrow NO_3^-\ (aq) + Mn^{2+}\ (aq) + H_2O\ (\ell)$
   (c) $Sn(OH)_3^-\ (aq) + Bi(OH)_3\ (s) + OH^-\ (aq) \longrightarrow Sn(OH)_6^{2-}\ (aq) + Bi\ (s)$
10. Balance the following as net ionic equations.
    (a) $Al\ (s) + NO_3^-\ (aq) + OH^-\ (aq) + H_2O \longrightarrow Al(OH)_4^-\ (aq) + NH_3\ (g)$
    (b) $NO_2\ (g) + OH^-\ (aq) \longrightarrow NO_3^-\ (aq) + NO_2^-\ (aq) + H_2O\ (\ell)$
    (c) $MnO_4^-\ (aq) + H_2O\ (\ell) + NO_2^-\ (aq) \longrightarrow MnO_2\ (s) + NO_3^-\ (aq) + OH^-\ (aq)$
    (d) $I^-\ (aq) + H^+\ (aq) + NO_2^-\ (aq) \longrightarrow NO\ (g) + H_2O\ (\ell) + I_2\ (s)$
    (e) $Hg_2Cl_2\ (s) + NH_3\ (aq) \longrightarrow Hg\ (\ell) + HgNH_2Cl\ (s) + NH_4^+\ (aq) + Cl^-\ (aq)$
11. Balance the following as net ionic equations.
    (a) $CrO_4^{2-}\ (aq) + H_2O\ (\ell) + HSnO_2^-\ (aq) \longrightarrow CrO_2^-\ (aq) + OH^-\ (aq) + HSnO_3^-\ (aq)$
    (b) $C_2H_4\ (g) + MnO_4^-\ (aq) + H^+\ (aq) \longrightarrow CO_2\ (g) + Mn^{2+}\ (aq) + H_2O\ (\ell)$
    (c) $H_2S\ (aq) + H^+\ (aq) + Cr_2O_7^{2-}\ (aq) \longrightarrow Cr^{3+}\ (aq) + S\ (s) + H_2O\ (\ell)$
    (d) $ClO_3^-\ (aq) + H_2O\ (\ell) + I_2\ (s) \longrightarrow IO_3^-\ (aq) + Cl^-\ (aq) + H^+\ (aq)$
    (e) $Cu\ (s) + H^+\ (aq) + SO_4^{2-}\ (aq) \longrightarrow Cu^{2+}\ (aq) + H_2O\ (\ell) + SO_2\ (g)$
12. Balance the following as net ionic equations for reactions in acidic solution. H^+ or H_2O (but not OH^-) may be added as necessary.
    (a) $Fe^{2+}\ (aq) + MnO_4^-\ (aq) \longrightarrow Fe^{3+}\ (aq) + Mn^{2+}\ (aq)$
    (b) $Br_2\ (\ell) + SO_2\ (g) \longrightarrow Br^-\ (aq) + SO_4^{2-}\ (aq)$
    (c) $Cu\ (s) + NO_3^-\ (aq) \longrightarrow Cu^{2+}\ (aq) + NO_2\ (g)$
    (d) $PbO_2\ (s) + Cl^-\ (aq) \longrightarrow PbCl_2\ (s) + Cl_2\ (g)$
    (e) $Zn\ (s) + NO_3^-\ (aq) \longrightarrow Zn^{2+}\ (aq) + N_2\ (g)$
13. Balance the following as net ionic equations for reactions in acidic solution. H^+ or H_2O (but not OH^-) may be added, as necessary.
    (a) $P_4\ (s) + NO_3^-\ (aq) \longrightarrow H_3PO_4\ (aq) + NO\ (g)$
    (b) $H_2O_2\ (aq) + MnO_4^-\ (aq) \longrightarrow Mn^{2+}\ (aq) + O_2\ (g)$
    (c) $HgS\ (s) + Cl^-\ (aq) + NO_3^-\ (aq) \longrightarrow HgCl_4^{2-}\ (aq) + NO_2\ (g) + S\ (s)$
    (d) $HBrO\ (aq) \longrightarrow Br^-\ (aq) + O_2\ (g)$
    (e) $Cl_2\ (g) \longrightarrow ClO_3^-\ (aq) + Cl^-\ (aq)$
14. Balance the following as net ionic equations in basic solution. OH^- or H_2O (but not H^+) may be added as necessary.
    (a) $MnO_4^-\ (aq) + NO_2^-\ (aq) \longrightarrow MnO_2\ (s) + NO_3^-\ (aq)$
    (b) $Zn\ (s) + NO_3^-\ (aq) \longrightarrow NH_3\ (aq) + Zn(OH)_4^{2-}\ (aq)$
    (c) $N_2H_4\ (aq) + Cu(OH)_2\ (s) \longrightarrow N_2\ (g) + Cu\ (s)$
    (d) $Mn^{2+}\ (aq) + MnO_4^-\ (aq) \longrightarrow MnO_2\ (s)$
    (e) $Cl_2\ (g) \longrightarrow ClO_3^-\ (aq) + Cl^-\ (aq)$
15. Balance the following as net ionic equations in basic solution. OH^- or H_2O (but not H^+) may be added as necessary.
    (a) $Mn(OH)_2\ (s) + H_2O_2\ (aq) \longrightarrow MnO_2\ (s)$
    (b) $CN^-\ (aq) + MnO_4^-\ (aq) \longrightarrow CNO^-\ (aq) + MnO_2\ (s)$
    (c) $As_2S_3\ (s) + H_2O_2\ (aq) \longrightarrow AsO_4^{3-}\ (aq) + SO_4^{2-}\ (aq)$

(d) CrI_3 (aq) + H_2O_2 (aq) $\longrightarrow$ CrO_4^{2-} (aq) + IO_4^- (aq)

16. Balance the following molecular equations.
 (a) H_2SO_4 (aq) + C (s) $\longrightarrow$ CO_2 (g) + SO_2 (g) + H_2O (ℓ)
 (b) MnO_2 (s) + HCl (aq) $\longrightarrow$ Cl_2 (g) + $MnCl_2$ (aq) + H_2O (ℓ)
 (c) $KMnO_4$ (aq) + NaI (aq) + H_2SO_4 (aq) $\longrightarrow$ I_2 (s) + $MnSO_4$ (aq) + Na_2SO_4 (aq) + K_2SO_4 (aq) + H_2O (ℓ)
 (d) $K_2Cr_2O_7$ (aq) + KBr (aq) + H_2SO_4 (aq) $\longrightarrow$ Br_2 (ℓ) + K_2SO_4 (aq) + $Cr_2(SO_4)_3$ (aq) + H_2O (ℓ)
17. Balance the following molecular equations.
 (a) HNO_3 (aq) (dil) + ZnS (s) $\longrightarrow$ S (s) + NO (g) + $Zn(NO_3)_2$ (aq) + H_2O (ℓ)
 (b) HNO_3 (aq) (conc) + Cu (s) $\longrightarrow$ $Cu(NO_3)_2$ (aq) + NO_2 (g) + H_2O (ℓ)
 (c) $KMnO_4$ (aq) + H_2O_2 (aq) + H_2SO_4 (aq) $\longrightarrow$ O_2 (g) + $MnSO_4$ (aq) + K_2SO_4 (aq) + H_2O (ℓ)
 (d) $(NH_4)_2Cr_2O_7$ (s) $\xrightarrow{\Delta}$ Cr_2O_3 (s) + H_2O (g) + N_2 (g)

Redox Titrations—Mole Method

18. What volume of 0.10 *M* $KMnO_4$ would be required to oxidize 20 mL of 0.10 *M* $FeSO_4$ in acidic solution? Refer to Exercise 12(a).
19. What volume of 0.10 *M* $K_2Cr_2O_7$ would be required to oxidize 60 mL of 0.10 *M* Na_2SO_3 in acidic solution? The products include Cr^{3+} and SO_4^{2-} ions.
20. What volume of 0.10 *M* $KMnO_4$ would be required to oxidize 50 mL of 0.10 *M* KI in acidic solution?
21. What volume of 0.10 *M* $K_2Cr_2O_7$ would be required to oxidize 50 mL of 0.10 *M* KI in acidic solution?
22. A 5.026-gram sample of an ore of iron was dissolved in an acid solution. The iron was converted to the +2 oxidation state. The resulting solution was titrated by 30.68 mL of 0.06402 molar $KMnO_4$. What are the mass and the percentage of iron in the ore?
23. A 0.683-gram sample of an ore of iron is dissolved in acid and converted to the ferrous form. The sample is oxidized by 38.50 mL of 0.161 *M* ceric sulfate, $Ce(SO_4)_2$, solution during which the ceric ion, Ce^{4+}, is reduced to Ce^{3+} ion. (a) Write a balanced equation for the reaction. (b) What is the percentage of iron in the ore?
24. Limonite is an ore of iron that contains $Fe_2O_3 \cdot 1\frac{1}{2}\ H_2O$ (or $2Fe_2O_3 \cdot 3H_2O$). A 0.5166-gram sample of limonite is dissolved in acid and treated so that all the iron is converted to ferrous ion, Fe^{2+}. This sample requires 42.96 mL of 0.02130 *M* sodium dichromate solution, $Na_2Cr_2O_7$, for titration. Fe^{2+} is oxidized to Fe^{3+} and $Cr_2O_7^{2-}$ is reduced to Cr^{3+}. What is the percentage of iron in the limonite?
25. Given the following unbalanced equation:

$$KMnO_4 + Na_2(COO)_2 + H_2SO_4 \longrightarrow K_2SO_4 + MnSO_4 + Na_2SO_4 + H_2O + CO_2$$

 (a) Balance the molecular equation and write the balanced total ionic and net ionic equations for this reaction. All salts in this equation may be assumed to be ionic compounds.
 (b) Calculate the molarity of a $KMnO_4$ solution if 20.0 mL of the $KMnO_4$ solution reacts with 0.268 g of solid $Na_2(COO)_2$.
 (c) Calculate the mass of $Na_2(COO)_2$ required to prepare 12.0 liters of 0.600 *M* $Na_2(COO)_2$ solution.
 (d) Calculate the mass of solid $KMnO_4$ required to prepare 10.0 liters of 0.0400 *M* $KMnO_4$ solution.

Redox Titrations—Equivalent Weight Method

26. Calculate the molarity and normality of a solution that contains 15.8 g of $KMnO_4$ in 500 mL of solution to be used in the reaction in Exercise 16(c).
27. Calculate the molarity and normality of a solution that contains 2.94 g of $K_2Cr_2O_7$ in 100 mL of solution to be used in the reaction in Exercise 16(d).
28. Calculate the molarity and normality of a solution that contains 16.2 g of $FeSO_4$ in 200 mL of solution to be used in the reaction in Exercise 12(a).
29. Calculate the molarity and normality of a solution that contains 12.6 g of Na_2SO_3 in 1.00 liter of solution to be used in the reaction in Exercise 8(c).

30. Arrange in a tabular form (a) formula weight, (b) equivalent weight, and (c) the number of grams of solute per liter of a 0.1500 *N* solution for the following.
$KMnO_4$ (in acid solution)
$K_2Cr_2O_7$
$Na_2S_2O_3 \cdot 5H_2O$ (which will be converted to $S_4O_6^{2-}$ ion)
SO_2 (which will be converted to H_2SO_4)
HNO_2 (which will be converted to HNO_3)
31. What is the percentage of iron in an ore sample with a mass of 0.6835 gram if the sample, when dissolved in acid and converted to iron(II), requires 38.50 mL of 0.1607 *N* ceric sulfate, $Ce(SO_4)_2$, solution? The Ce^{4+} is reduced to Ce^{3+} and Fe^{2+} is oxidized to Fe^{3+}.
32. A sample containing vanadium is to be analyzed by converting the vanadium to V_2O_5, which is dissolved in acid and reduced to VO^{2+}. The VO^{2+} is then oxidized by a standardized $KMnO_4$ solution to VO_2^+ ion. Calculate the equivalent weight of V_2O_5 for this reaction.
33. A 1.164-gram sample containing some Fe_2O_3 is dissolved and all the iron is reduced to the iron(II) state. The resulting solution is titrated with 19.68 mL of a freshly standardized 0.1104 *N* solution of $KMnO_4$. Calculate the percentage of Fe_2O_3 in the original sample.
34. What mass of $KMnO_4$ would be required to prepare 1.0 liter of 0.10 *N* $KMnO_4$ to be used as an oxidizing agent in acidic solution?
35. Given the following unbalanced equation:

$$KMnO_4 + (COOH)_2 + H_2SO_4 \longrightarrow MnSO_4 + K_2SO_4 + CO_2 + H_2O$$

(a) Balance the molecular equation and write the balanced total ionic and net ionic equations for this reaction. All salts in this equation may be assumed to be ionic.
(b) What is the equivalent weight of $(COOH)_2$ in this reaction?
(c) What is the equivalent weight of $KMnO_4$ in this reaction?
(d) A standard 0.300 *N* $(COOH)_2$ solution is to be prepared by dissolving an accurately known mass of $(COOH)_2 \cdot 2H_2O$ in water. What mass of $(COOH)_2 \cdot 2H_2O$ would be required to prepare 18.0 liters of this solution?
(e) What mass of $(COOH)_2 \cdot 2H_2O$ is contained in 2.50 liters of a $(COOH)_2$ solution, if 40.0 mL of the $(COOH)_2$ solution reacts with 60.0 mL of a 0.100 *N* $KMnO_4$ solution?

36. Given the following unbalanced equation:

$$K_2Cr_2O_7 + SnSO_4 + H_2SO_4 \longrightarrow K_2SO_4 + Cr_2(SO_4)_3 + Sn(SO_4)_2 + H_2O$$

(a) Balance the molecular equation and write the balanced total ionic and net ionic equations for this reaction. All salts in this equation may be assumed to be ionic compounds.
(b) Calculate the molarity of a solution prepared by dissolving 5.88 grams of $K_2Cr_2O_7$ in enough water to make 500 mL of solution.
(c) What is the normality of the $K_2Cr_2O_7$ solution prepared in part (b)?
(d) What is the normality of a $SnSO_4$ solution that required 20.0 mL of the above $K_2Cr_2O_7$ solution to react with 30.0 mL of $SnSO_4$ solution?
(e) What is the molarity of the $SnSO_4$ solution in part (d)?
(f) What mass of $K_2Cr_2O_7$ would be required to prepare 15.0 liters of 0.0500 *N* $K_2Cr_2O_7$ solution?

13 Chemical Thermodynamics

In Chapter 9 we dealt with the heat transfer associated with physical changes. Heat is only one of many different forms of energy, such as electrical energy, mechanical energy, and chemical bond energy, all of which are interconvertible. **Thermodynamics** is the study of energy changes (or transfers) accompanying physical and chemical processes. Ultimately, we shall be able to use thermodynamic information as a tool to predict whether or not a particular reaction can occur under specified conditions. If the answer is yes, the process is said to be thermodynamically **spontaneous.** If the answer is no, the reaction is **nonspontaneous,** and *cannot* occur under the given conditions.

The fact that a process is spontaneous does not mean that it will occur at an observable rate. It may occur rapidly, at a moderate rate, or very slowly. The rate at which a spontaneous reaction occurs is addressed by kinetics, a subject we shall study in Chapter 14. The only absolute statement that can be made on the basis of thermodynamic data is that under a given set of conditions a nonspontaneous reaction will not occur.

13–1 First Law of Thermodynamics

Many spontaneous reactions proceed with the *evolution* of energy in the form of heat, and so are called *exothermic* reactions. The combustion reactions of fossil fuels are familiar examples. Hydrocarbons, including methane, the principal component of natural gas, and octane, one of the minor components of gasoline, undergo combustion to yield carbon dioxide and water with the release of energy, as shown on the product side of each equation. (The heats are shown for the reaction of one mole of methane and two moles of octane.)

A hydrocarbon is a binary compound of hydrogen and carbon, as in the examples used here. Hydrocarbons may be gaseous, liquid, or solid; all burn.

$$CH_4\ (g) + 2O_2\ (g) \longrightarrow CO_2\ (g) + 2H_2O\ (\ell) + 890\ kJ$$

$$2C_8H_{18}\ (\ell) + 25O_2\ (g) \longrightarrow 16CO_2\ (g) + 18H_2O\ (\ell) + 1.090 \times 10^4\ kJ$$

A thermochemical equation always shows the amount of heat released (or absorbed if left of the arrow) in the reaction involving the number of *moles* of reactants and products specified by the coefficients.

In such reactions the total energy of the products is lower than that of the reactants by the amount of energy released, most of which is heat.

Another common example of an exothermic reaction is the combination of hydrogen and oxygen to form water, with the liberation of heat energy

$$2H_2\ (g) + O_2\ (g) \longrightarrow 2H_2O\ (\ell) + 571.5\ kJ$$

Some initial activation by heat is needed to get the reaction started, as shown in Figure 13–1. However, an equal amount of energy *plus* 571.5 kJ are released as two moles of water molecules are formed.

Fuel cells convert the energy evolved by a chemical reaction directly into electrical energy, as described in Section 19–25. The Apollo spacecraft were powered by solar energy and hydrogen-oxygen fuel cells, which utilized this reaction to supply energy as well as drinking water.

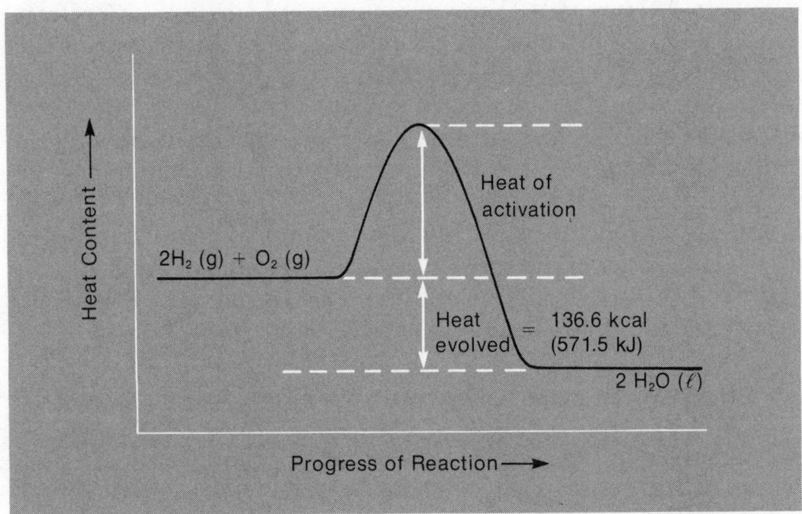

FIGURE 13–1 The difference between the heat content of the reactants (hydrogen and oxygen) and the product (water) is the amount of heat evolved in an exothermic reaction, 571.5 kJ, in this reaction. Some initial activation by heat is needed to get the reaction started. In the absence of such initiating energy, a mixture of hydrogen and oxygen can be kept at room temperature for a long time without reacting. This aspect of reaction energetics will be covered in Chapter 14.

Energy changes always accompany physical changes as well as chemical changes, as we have seen in Chapter 9. For example, the melting of one mole of ice at 0°C must be accompanied by the absorption of 6.02 kilojoules of energy

$$H_2O\ (s) + 6.02\ kJ \longrightarrow H_2O\ (\ell)$$

This tells us that the total energy of the water is raised by 6.02 kJ in the form of heat during the phase change.

The concept of energy is somewhat abstract, but it can be defined as the *capacity to do work or transfer heat.* The **First Law of Thermodynamics,** like all physical laws, is a broad, general statement summarizing numerous observations on natural phenomena; no exceptions to the law are known.

It states: *The total amount of energy in the universe is constant.* Stated in another way, it is referred to as the **Law of Conservation of Energy:** *Energy is neither created nor destroyed in ordinary chemical reactions and physical changes.*

In our discussion of atomic structure (Section 3–8), we pointed out the equivalence of matter and energy. The word "energy" is understood to include the energy equivalent of all matter in the universe. Stated differently, *the total amount of mass and energy in the universe is constant.*

Once we get them started, the combustion of hydrocarbons and the hydrogen-oxygen reaction are all exothermic and spontaneous. The energy contents of the products are lower than those of the reactants. When we say that a chemical or physical change is spontaneous, we mean only that it *can* occur, not necessarily that it will occur at an observable rate. *Spontaneous processes tend toward a state of lower energy.* However, *it is not necessary that exothermic reactions be spontaneous, or that spontaneous reactions be exothermic.* When energy is released during a chemical reaction or physical change, spontaneity is *favored* but not required. Exothermicity is but one of two considerations. Another factor, related to the disorder of reactants and products and called **entropy,** must also be considered.

As we saw in Section 10–2, the dissolution of ammonium nitrate, NH_4NO_3, in water is spontaneous; yet if one holds a beaker in which this process occurs, one's hand becomes colder. In this case the system (consisting of the water, the NH_4NO_3, and the resulting hydrated NH_4^+ and NO_3^- ions) absorbs heat from the surroundings, of which the hand is a part, in order for the endothermic process to occur. Nevertheless, the process is spontaneous because the system increases in disorder as the regularly arranged ions of crystalline ammonium nitrate become randomly distributed hydrated ions in solution. An increase in disorder in the system favors the spontaneity of a reaction and, *in this particular case,* overrides the effect of endothermicity. The effect of disorder is discussed in more detail in Section 13–11.

13–2 Some Thermodynamic Terms

The substances involved in the chemical and physical changes of interest are called the **system,** while everything else in the system's environment constitutes its **surroundings.** The **universe** is the system plus its surroundings, and the system may be thought of as the part of the universe under investigation. The first law tells us that the energy of the universe is constant, i.e., energy is neither created nor destroyed, but is merely transferred between the system and its surroundings. As exothermic reactions occur, heat energy is transferred from the system (reactants and products) to the surroundings. As endothermic reactions occur, the system absorbs heat energy from its surroundings.

The **thermodynamic state of a system** *is a set of conditions that completely specifies all of the properties of the system.* This set includes the temperature, pressure, composition (identity and number of moles of each component), and physical state (gas, liquid, or solid forms) of each part of the system. Once the state has been specified, all other properties—both physical and chemical—are fixed.

The conditions that make up the state of a system, and the variables that measure them, are called **state functions.** Their defining characteristic is that the *value* of a state function depends *only* on the state of the system, and not on the manner by which the system came to be in that state (what we might call the history of the system). A *change* in a state function describes the *differences* between two states and is independent of the process or pathway by which the change occurs.

An example of a state function is the potential energy of an object in a gravitational field. It is the product of the mass of the object, m, the gravitational attraction, g, and the height of the object above an arbitrary reference level, h.

$$E_{\text{potential}} = mgh$$

State functions are designated by capital letters.

The potential energy of an object at a particular height is the same regardless of the path followed in attaining the height. For instance, a rock perched on the edge of a cliff has a certain potential energy with respect to the ground below (the reference level); it does not matter whether the rock was lifted straight up the cliff face or carried up the back of the mountain.

Any property of a system that depends only on the values of its state functions is also a state function. For instance, the volume of a sample of matter, which depends only on temperature, pressure, composition, and physical state, is a state function. We shall encounter other state functions as we continue our study of thermodynamics.

13–3 Changes in Internal Energy, ΔE

The **internal energy,** E, of a specific amount of a substance represents all the energy contained within the substance. It includes such forms as kinetic energies of the molecules, energies of attraction and repulsion among subatomic particles, atoms, ions, or molecules, as well as other forms of energy. The internal energy of a collection of molecules is a state function. The difference between the internal energy of the products and the internal energy of the reactants of a chemical reaction or physical change, ΔE, is given by the equation

Internal energy is a state function.

$$\Delta E = E_{\text{products}} - E_{\text{reactants}} = q - w$$

The terms q and w represent heat and work, respectively. **Work** is a form of energy in which a body is moved through a distance, d, by application of a force, f; that is, $w = fd$. Since E_{products} and $E_{\text{reactants}}$ are both state functions, ΔE is also state function

The individual quantities q and w are *not* state functions.

$$\Delta E = (\text{amount of heat added to the system}) - (\text{amount of work done by system})$$

The following conventions apply to the signs of q and w

q is positive:	heat is *absorbed* by the system from the surroundings
q is negative:	heat is *released* by the system to the surroundings
w is positive:	work is done *by* system on the surroundings
w is negative:	work is done *on* system by the surroundings

When energy is released by a reacting system, ΔE is negative and energy is written as a product in the equation for the reaction. When the system absorbs energy from the surroundings, the energy of the molecules increases, ΔE is positive, and energy is written as a reactant in the equation.

For example, the combustion of methane at 25°C

At 25°C the change in internal energy for the combustion of methane is −887 kJ/mol, while (Section 13–1) the change in heat content is −890 kJ/mol. The small difference is due to work done on the system.

$$CH_4\ (g) + 2O_2\ (g) \longrightarrow CO_2\ (g) + 2H_2O\ (\ell) + 887\ kJ$$

releases energy. We could also write

$$CH_4\ (g) + 2O_2\ (g) \longrightarrow CO_2\ (g) + 2H_2O\ (\ell) \qquad \Delta E = -887\ kJ$$

(− indicates release of energy)

On the other hand, the reverse of this reaction *absorbs* energy

$$CO_2\ (g) + 2H_2O\ (\ell) + 887\ kJ \longrightarrow CH_4\ (g) + 2O_2\ (g)$$

It could also be written as

$$CO_2\ (g) + 2H_2O\ (\ell) \longrightarrow CH_4\ (g) + 2O_2\ (g) \qquad \Delta E = +887\ kJ$$

(+ indicates absorption of energy)

The latter reaction is *not* one that occurs spontaneously. But if it were forced to occur, the system would have to absorb 887 kJ of energy from its surroundings.

Other than electrical work, which will be covered in Chapter 19, the only type of work we need to be concerned with in most chemical and physical changes is pressure-volume work. From dimensional analysis we can see that the product of pressure and volume is work. Pressure is the force exerted per unit area, where area is distance squared, d^2; volume is distance cubed, d^3. Thus the product of pressure and volume is force times distance, or work:

$$\underbrace{\text{pressure}}_{\frac{f}{d^2}} \times \underbrace{\text{volume}}_{d^3} = \text{work} = fd = w$$

When a gas is produced against constant external pressure, such as laboratory conditions in an open vessel at atmospheric pressure, the gas does work as its molecules expand against the pressure of the atmosphere. This, of course, results in a decrease in the energy of the molecules of the system, if no heat is absorbed. On the other hand, when a gas is consumed in a reaction, the atmosphere does work on the reacting system, whose molecules increase in energy as the volume they occupy decreases. Compression of a gaseous system increases its potential energy.

Let us illustrate the latter case by considering the complete reaction of a 2:1 mole ratio of hydrogen and oxygen to produce steam at some constant temperature above 100°C and at one atmosphere pressure, which is depicted in Figure 13–2. Assuming that the constant temperature bath (mineral oil, for

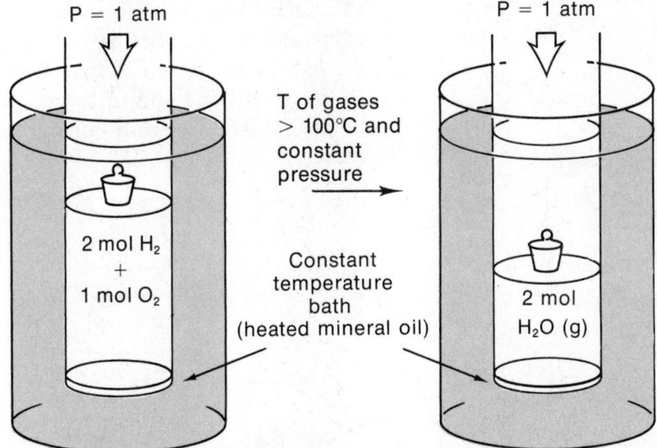

FIGURE 13–2 An illustration of the decrease in volume by a factor of one-third that accompanies the reaction (Note: The temperature is above 100°C so H_2O is a gas.)

$2H_2$ (g) + O_2 (g) $\longrightarrow$ $2H_2O$ (g)

example) surrounding the cylinder in which the reaction occurs completely absorbs all the evolved heat, the temperature of the gaseous system does not change and the volume of the system decreases by one third (3 mol gaseous reactants → 2 mol gaseous products). The surroundings, exerting a constant pressure of one atmosphere, do work on the system by compressing it, and the potential energy of the system simultaneously increases by an amount equal to the amount of work done on it.

The work done on or by a gaseous system equals the change in the product of pressure and volume, $\Delta(PV)$. When the pressure is constant, this is the same as the pressure (P) times the volume change, or $P\Delta V$. If we substitute $P\Delta V$ for w in the equation $\Delta E = q - w$, we obtain

$$\Delta E = q - P\Delta V$$

In constant-volume reactions, no work can be done. Although the pressure varies, the absence of a change in volume means that nothing "moves through a distance," so $d = 0$ and $fd = 0$.

Do not make the error of setting work equal to $V\Delta P$.

Since solids and liquids do not expand or contract significantly as pressure changes, their production or consumption involves only very little or negligible work ($\Delta V \approx 0$). In reactions in which equal numbers of moles of gases are produced and consumed at constant temperature and pressure, no work is done since, by the ideal gas law, $P\Delta V = (\Delta n)RT$ and $\Delta n = 0$, where Δn equals the number of moles of gaseous products minus the number of moles of gaseous reactants. Thus, the work term, w, has a significant value *only when there are different numbers of moles of gaseous products and reactants and the volume of the system varies.* In other cases, the change in internal energy of the system is just the amount of heat absorbed or released at constant volume, q_v; that is

$$\Delta E = q_v$$

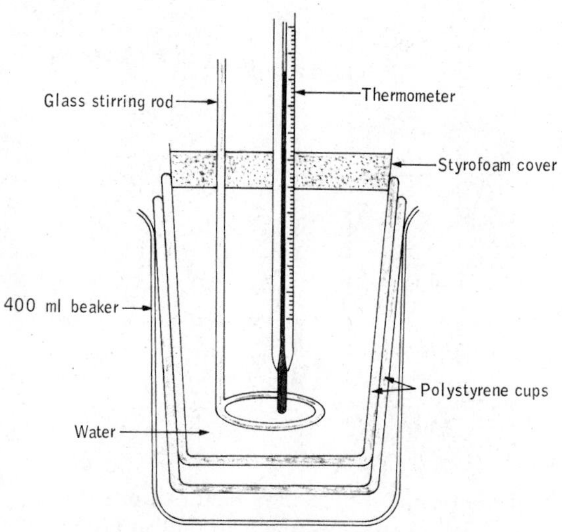

FIGURE 13–3 A coffee-cup calorimeter. The stirring rod is moved up and down to insure thorough mixing and uniform heating of the solution during reaction.

As we shall see shortly, q_p is also called the enthalpy change, ΔH.

13–4 Calorimeters

A "coffee-cup" calorimeter made of polystyrene and Styrofoam is frequently used in teaching laboratories to measure "heats of reaction" in aqueous solution at constant pressure, q_p. Reactions are chosen so that there are no gaseous reactants or products, so that all reactants and products remain in the vessel throughout the experiment. For example, a coffee-cup calorimeter such as the one shown in Figure 13–3 could be used to measure the heat evolved upon neutralization of known quantities of hydrochloric acid and potassium hydroxide

$$HCl\ (aq) + KOH\ (aq) \longrightarrow KCl\ (aq) + H_2O\ (\ell) + \text{heat}$$

for which the net ionic equation is simply

$$H^+\ (aq) + OH^-\ (aq) \longrightarrow H_2O\ (\ell) + \text{heat}$$

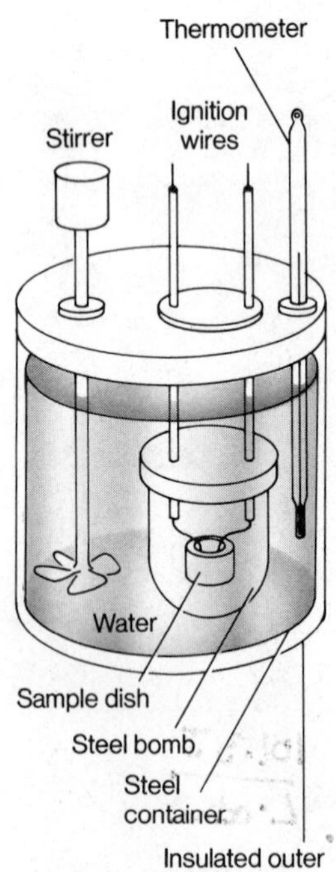

FIGURE 13–4 A bomb calorimeter measures the amount of heat evolved or absorbed by a reaction occurring at constant volume.

The double polystyrene walls and Styrofoam top provide enough insulation that only an insignificant amount of heat escapes. The heat evolved can be calculated by relating it to the rise it causes in the temperature of the known mass of solution and to the previously determined amount of heat the calorimeter itself absorbs. The heat absorbed by the calorimeter is known as its "water equivalent" and will be described below for the "bomb" calorimeter.

A bomb calorimeter is a device that measures the heat evolved or absorbed by a reaction occurring at constant volume (Figure 13–4). In this type of calorimeter a strong steel vessel (the bomb) is immersed in a large volume of water. As heat is produced or absorbed by a reaction going on inside the steel vessel, the heat is transferred to or from the large volume of water, so only very small temperature changes occur. For all practical purposes, the energy changes associated with the reactions are measured at constant volume and constant temperature. No work is done when a reaction is carried out in a "bomb" calorimeter even if gases are involved, because $\Delta V = 0$, and therefore

$$\Delta E = q_v \qquad \text{(constant volume)}$$

For exothermic reactions, we may write

heat lost by system = (heat gained by calorimeter bomb) + (heat gained by water)

In order to simplify calculations, the amount of heat absorbed by the calorimeter is usually expressed as its **water equivalent,** which refers to the amount of water that would absorb the same amount of heat as the calorimeter per degree temperature change. The water equivalent of a calorimeter is determined by burning a sample of a compound that produces a known amount of heat, and measuring the temperature rise of the calorimeter. For example, if it is known that burning 1.000 gram of a particular compound produces 9.598 kJ of heat, which raises the temperature of a calorimeter and its 3000 grams of water by 0.629°C, then

Benzoic acid, C_6H_5COOH, is usually used to determine the water equivalent. It is a solid that can be compressed into pellets, and its heat of combustion is 3227 kJ/mol or 26.46 kJ/g.

$$\begin{aligned}\text{heat gained by calorimeter} &= (\text{heat lost by sample}) - (\text{heat gained by water})\\ &= 9598\text{ J} - (3000\text{ g})(4.184\text{ J/g}\cdot{}^\circ\text{C})(0.629^\circ\text{C})\\ &= (9598 - 7895)\text{ J}\\ &= 1703\text{ J}\quad(1.703\text{ kJ})\end{aligned}$$

(temp. change; sp. ht. of H_2O; amount of water)

We can now calculate the amount of water that would absorb the same amount of heat in going through the same temperature change.

$$\underline{?}\text{ g } H_2O = \frac{1703\text{ J}}{0.629^\circ\text{C}} \times \frac{1\text{ g}^\circ\text{C}}{4.184\text{ J}} = 647\text{ g } H_2O$$

Thus, the water equivalent of the calorimeter is 647 g. This means that the amount of heat required to raise the temperature of the steel vessel 1.000°C is the same as the amount of heat required to raise the temperature of 647 grams of water 1.000°C. *Or,* since the water equivalent of the calorimeter is known, the temperature change experienced by this calorimeter is the same as the change expected for 647 grams of H_2O.

The methods and concepts of Chapter 9 are applied to calorimetric data to relate temperature changes of water to amounts of heat transferred and, therefore, to ΔE.

An alternative to measuring the water equivalent of the calorimeter is to calibrate the calorimeter prior to use by electrical heating to provide the "calorimeter constant" in joules per degree.

EXAMPLE 13–1

A 1.000-gram sample of ethanol, C_2H_5OH, was burned in the sealed bomb calorimeter described above, which was surrounded by 3000 grams of water. The temperature of the water rose from 24.284°C to 26.225°C. Determine ΔE for the reaction in joules per gram of ethanol, then in kilojoules per mole of ethanol. The specific heat of water is 4.184 J/g·°C. The combustion reaction is

$$C_2H_5OH\,(\ell) + 3O_2\,(g) \longrightarrow 2CO_2\,(g) + 3H_2O\,(\ell)$$

Solution

Heat from the sealed compartment (the system) raises the temperature of the calorimeter and water. Recall that the water equivalent of this calorimeter is 647 grams, so we have 3000 + 647 = 3647 g of water as the *effective* total. The increase is

$$\underline{?}\ ^\circ\text{C} = 26.225^\circ\text{C} - 24.284^\circ\text{C} = \underline{1.941^\circ\text{C rise}}$$

The number of joules of heat responsible for the increase in the water temperature is

$$\underline{?}\text{ J} = 1.941^\circ\text{C} \times \frac{4.184\text{ J}}{\text{g}^\circ\text{C}} \times 3647\text{ g} = \underline{2.962 \times 10^4\text{ J}} = 29.62\text{ kJ}$$

In order to measure temperature changes to the nearest thousandth of a degree, very sensitive and expensive differential thermometers are used.

One gram of C_2H_5OH liberates 29.62 kJ of energy in the form of heat. That is,

$\Delta E = q_v = -29.62\ kJ/g\ C_2H_5OH$

Now we may evaluate ΔE in kilojoules per mole by converting grams of C_2H_5OH to moles.

$$\frac{?\ kJ}{mol} = \frac{-29.62\ kJ}{g} \times \frac{46.07\ g\ C_2H_5OH}{1\ mol\ C_2H_5OH} = -1365\ kJ/mol$$

$\Delta E = -1365\ kJ/mol$

This calculation shows that for the combustion of ethanol at constant temperature and constant volume, the change in internal energy is -1365 kJ/mol. The negative sign indicates that energy is released to the surroundings.

Gasohol

The potential use of "gasohol" as an automobile fuel has been well publicized recently. Gasohol is a mixture of 90% gasoline (mainly hydrocarbons) and 10% ethanol which, as Example 13–1 showed, also liberates energy when it is burned. Let us now compare the energy yields for the combustion of one gram of octane and one gram of ethanol. The second equation in Section 13–1 shows that combustion of one mole of octane liberates 5450 kilojoules. Since one mole of octane has a mass of 114.2 grams, combustion of one gram of octane evolves 47.72 kilojoules.

$$?\ \frac{kJ}{g} = \frac{-5450\ kJ}{mol\ C_8H_{18}} \times \frac{1\ mol\ C_8H_{18}}{114.2\ g\ C_8H_{18}} = -47.72\ kJ/g$$

In Example 13–1 we found that combustion of one gram of ethanol produces only 29.62 kilojoules. Although the energy yield per gram of ethanol is less than that for octane and other hydrocarbons, it may serve as a useful fuel supplement to slow the consumption rate of hydrocarbon fuels. Both ethanol and methanol, CH_3OH, are relatively inexpensive and may be used as fuels. The alcohols burn more cleanly than hydrocarbons and cause fewer pollution problems. Methanol can be obtained by destructive distillation of wood, or from coal by reaction of carbon monoxide with hydrogen.* Ethanol is obtained by fermentation of grains or can be synthesized industrially by reaction of ethene (ethylene), C_2H_4, with water. Hydrocarbon fuels, of course, are not easily synthesized, and known reserves are decidedly limited.

Destructive distillation involves heating in the absence of air.

* The red-hot coal is first subjected to a blast of steam

$$C\ (s) + H_2O\ (g) \longrightarrow CO\ (g) + H_2\ (g)$$

and the gaseous products are combined with extra hydrogen to produce methanol

$$CO + 2H_2 \xrightarrow{ZnO} CH_3OH$$

13–5 Enthalpy Change, ΔH

It is much more convenient to measure the amount of heat released or absorbed by a reaction under constant (atmospheric) pressure, i.e., laboratory conditions, than to measure the heat transferred at constant volume and temperature. The heat change of a reaction at *constant pressure*, q_p, is also called the **heat of reaction** or **enthalpy change,** ΔH. It can be shown by using the ideal gas law, $PV = nRT$, that the enthalpy change is related to the change in internal energy by both the equations below

$$\Delta H = \Delta E + P\Delta V$$

or since $P\Delta V = (\Delta n)RT$, a work term,

$$\Delta H = \Delta E + (\Delta n)RT$$

In this equation Δn refers to the number of moles of *gaseous products minus* the number of moles of *gaseous reactants.*

Combustion of one mole of ethanol at 298 K and constant pressure releases 1367 kilojoules of heat, and therefore we can say

$$\Delta H = -1367 \frac{\text{kilojoules}}{\text{mole ethanol}}$$

In Example 13–1 we found that the change in internal energy, ΔE, for the combustion of ethanol is -1365 kilojoules per mole at 298 K. The difference, 2 kJ, is due to the work term, $P\Delta V$ or $(\Delta n)RT$. In the reaction

$$C_2H_5OH\ (\ell) + 3O_2\ (g) \longrightarrow 2CO_2\ (g) + 3H_2O\ (\ell)$$

there are fewer moles of gaseous products than gaseous reactants: $\Delta n = 2 - 3 = -1$. Thus, the atmosphere does work on the reacting system and the system is compressed. Let us evaluate the work term, $(\Delta n)RT$.

$$w = (\Delta n)RT = (-1\ \text{mol})\left(\frac{8.314\ \text{J}}{\text{mol} \cdot \text{K}}\right)(298\ \text{K}) = -2.48 \times 10^3\ \text{J}$$

The negative sign is consistent with the fact that work is done on the system.

$$w = -2.48\ \text{kJ}$$

We can now calculate ΔE for the reaction from ΔH and $(\Delta n)RT$ values:

$$\Delta E = \Delta H - (\Delta n)RT$$

$$= [-1367 - (-2.48)]\ \text{kJ}$$

$$\Delta E = \underline{-1365\ \text{kJ}}$$

This value agrees exactly with the result that we obtained in Example 13–1. Note also that the work term (-2.48 kJ) is practically insignificant in comparison to ΔH (-1367 kJ). This is true for many reactions. Of course, $\Delta H = \Delta E$ if $\Delta n = 0$. Measurements of ΔH for reactions are usually simpler to make than measurements of ΔE and are entirely adequate substitutes in most cases.

13–6 Standard Molar Enthalpies of Formation, ΔH_f^0

A temperature of 25°C is 77°F (Section 1–12). This is slightly above room temperature, which is about 20°C or 68°F.

The **thermochemical standard state** *of a substance is its most stable state under standard pressure (one atmosphere) and at some specific temperature (usually 25°C unless otherwise specified).* Examples of elements in their standard states at 25°C are hydrogen, gaseous diatomic molecules, H_2 (g); mercury, a silver-colored liquid metal, Hg (ℓ); sodium, a silvery-white solid metal, Na (s); and carbon, a grayish-black solid called graphite, C (graphite). The designation C (graphite) is used instead of C (s) to distinguish it from another allotropic modification of carbon, C (diamond). Examples of standard states of compounds include ethanol (ethyl alcohol or grain alcohol), a liquid, C_2H_5OH (ℓ); water, a liquid, H_2O (ℓ); calcium carbonate, a solid, $CaCO_3$ (s); and carbon dioxide, a gas, CO_2 (g).

A special name, **standard molar enthalpy of formation, ΔH_f^0,** *is given to the amount of heat absorbed in a reaction in which one mole of a substance in a specified state is formed from its elements in their standard states.* Standard molar enthalpy of formation is often called **standard molar heat of formation,** or more simply, **heat of formation.** The superscript zero in ΔH_f^0 signifies standard pressure, one atm. Negative values for ΔH_f^0 describe exothermic reactions, while positive values for ΔH_f^0 describe endothermic reactions.

Note that the enthalpy change for a normally balanced equation may not give directly a molar enthalpy of formation for the compound formed. For instance, the heat of reaction at standard state conditions for the exothermic reaction

$$H_2\ (g) + Br_2\ (\ell) \longrightarrow 2HBr\ (g) + 72.8\ kJ$$

is −72.8 kilojoules, or $\Delta H^0_{rxn} = -72.8$ kJ. However, *two* moles of HBr (g) are formed in the reaction as written; half as much energy, 36.4 kilojoules, is liberated when one mole of HBr (g) is produced from its constituent elements in their standard states. For HBr (g), $\Delta H_f^0 = -36.4$ kJ/mol. This can be shown by dividing all coefficients in the balanced equation by 2

The coefficients ½ preceding H_2 (g) and Br_2 (ℓ) do *not* imply half a molecule of each. In thermodynamic problems the coefficients always refer to the number of *moles* under consideration.

$$\tfrac{1}{2}H_2\ (g) + \tfrac{1}{2}Br_2\ (\ell) \longrightarrow HBr\ (g) \qquad \Delta H^0_{rxn} = \Delta H^0_{f\ HBr\ (g)} = -36.4\ kJ$$

Standard heats of formation of some common substances are tabulated in Table 13–1. Appendix K contains a larger listing. Note that ΔH_f^0 values for *elements* in their standard states are zero.

TABLE 13–1 Selected Standard Molar Enthalpies of Formation at 298 K ΔH_f^0

Substance	ΔH_f^0 kJ/mol	Substance	ΔH_f^0 kJ/mol
Br_2 (ℓ)	0	H_2 (g)	0
Br_2 (g)	30.91	HBr (g)	−36.4
C (diamond)	1.897	H_2O (ℓ)	−285.8
C (graphite)	0	H_2O (g)	−241.8
CH_4 (g)	−74.81	N_2 (g)	0
C_2H_4 (g)	52.26	N_2H_4 (ℓ)	50.63
C_6H_6 (ℓ)	49.03	NO (g)	90.25
C_2H_5OH (ℓ)	−277.7	Na (s)	0
CO (g)	−110.52	NaCl (s)	−411.0
CO_2 (g)	−393.51	O_2 (g)	0
CaO (s)	−635.5	Pb (s)	0
$CaCO_3$ (s)	−1207	PbO (s) yellow	−217.3
Cl (g)	121.7	SO_2 (g)	−296.8
Cl_2 (g)	0	SiH_4 (g)	34
Hg (ℓ)	0	$SiCl_4$ (g)	−657.0
HgS (s) red	−58.2	SiO_2 (s)	−910.9

EXAMPLE 13–2

The standard molar enthalpy of formation of benzene, C_6H_6 (ℓ), is 49.03 kJ/mol. Write the equation for the reaction for which ΔH^0_{rxn} = 49.03 kJ/mol.

Solution

The desired reaction involves the formation of one mole of liquid benzene from its constituent elements in their standard states at 25°C and one atmosphere. The standard state for carbon is graphite.

$6C$ (graphite) + $3H_2$ (g) $\longrightarrow$ C_6H_6 (ℓ) $\quad \Delta H^0_{rxn} = 49.03$ kJ/mol $\quad$ (endothermic)

We may also write: $6C$ (graphite) + $3H_2$ (g) + 49.03 kJ $\rightarrow$ C_6H_6 (ℓ)

13–7 Hess' Law

In 1840, G. H. Hess published his **law of heat summation** which he derived on the basis of numerous thermochemical calculations. It states that *the enthalpy change for a reaction is always the same whether it occurs by one step or a series of steps.* This is consistent with the fact that enthalpy is a state function, and, therefore, independent of the pathway by which a reaction occurs. We do not need to know whether the reaction *does* or even *can* occur by the series of steps used in the calculation. That is, the steps (only formally) must result in the overall reaction. The utility of Hess' Law is that we can often calculate, from tabulated values, enthalpy changes for reactions for which the quantity can be measured only with difficulty, if at all. Consider, for example, the reaction

C (graphite) + $\frac{1}{2}O_2$ (g) $\longrightarrow$ CO (g) $\quad \Delta H^0 = ?$

The enthalpy change for this reaction cannot be measured directly. Even though CO (g) is the predominant product of the reaction of graphite with a limited amount of O_2 (g), some CO_2 (g) is always produced as well. However, the following reactions do go to completion with excess O_2 (g), and therefore ΔH^0 can be measured directly for them.

C (graphite) + O_2 (g) $\longrightarrow$ CO_2 (g) $\quad \Delta H^0 = -393.5$ kJ $\quad$ (1)

CO (g) + $\frac{1}{2}O_2$ (g) $\longrightarrow$ CO_2 (g) $\quad \Delta H^0 = -283.0$ kJ $\quad$ (2)

The common allotropic forms of carbon are graphite and diamond.

If we reverse equation (2) (and change the sign of its ΔH^0 value) and then add it to equation (1), cancelling equal numbers of moles of the same species on each side, we obtain the equation for the reaction of interest. Adding the corresponding enthalpy changes gives us the enthalpy change of interest.

The reverse of an exothermic reaction is endothermic.

	ΔH^0	
C (graphite) + O_2 (g) $\longrightarrow$ CO_2 (g)	-393.5 kJ	(1)
CO_2 (g) $\longrightarrow$ CO (g) + $\frac{1}{2}O_2$ (g)	$-(-283.0$ kJ$)$	(-2)
C (graphite) + $\frac{1}{2}O_2$ (g) $\longrightarrow$ CO (g)	$\Delta H^0 = -110.5$ kJ	

In general terms, Hess' Law of heat summation may be represented as

$$\Delta H^0 = \Delta H^0_a + \Delta H^0_b + \Delta H^0_c + \cdots$$

where a, b, c, . . . refer to the equations that can be summed to give the equation for the reaction of interest. ΔH^0 is sometimes represented as ΔH^0_{rxn}.

EXAMPLE 13–3

Use the equations and ΔH^0 values given below to determine ΔH^0 at 25°C for the reaction

C (graphite) + $2H_2$ (g) ⟶ CH_4 (g)

	ΔH^0	
C (graphite) + O_2 (g) ⟶ CO_2 (g)	−393.5 kJ	(1)
H_2 (g) + $\frac{1}{2}O_2$ (g) ⟶ H_2O (ℓ)	−285.8 kJ	(2)
CO_2 (g) + $2H_2O$ (ℓ) ⟶ CH_4 (g) + $2O_2$ (g)	+890.3 kJ	(3)

Solution

If we double (i.e., multiply by 2) equation (2) and add it to equations (1) and (3) we obtain the desired result

	ΔH^0	
C (graphite) + O_2 (g) ⟶ CO_2 (g)	−393.5 kJ	(1)
$2H_2$ (g) + O_2 (g) ⟶ $2H_2O$ (ℓ)	2(−285.8 kJ)	2 × (2)
CO_2 (g) + $2H_2O$ (ℓ) ⟶ CH_4 (g) + $2O_2$ (g)	+890.3 kJ	(3)
C (graphite) + $2H_2$ (g) ⟶ CH_4 (g)	ΔH^0_{rxn} = −74.8 kJ	

Note that ΔH^0_{rxn} is ΔH^0_f for CH_4 (g). However, we should not jump to the conclusion that methane can be produced by this series of reactions on a large scale. There are many complicating factors.

EXAMPLE 13–4

Given the following equations and ΔH^0 values, calculate the heat of reaction at 298 K for the reaction

C_2H_4 (g) + H_2O (ℓ) ⟶ C_2H_5OH (ℓ)

ethylene ethanol

	ΔH^0	
C_2H_5OH (ℓ) + $3O_2$ (g) ⟶ $3H_2O$ (ℓ) + $2CO_2$ (g)	−1367 kJ	(1)
C_2H_4 (g) + $3O_2$ (g) ⟶ $2CO_2$ (g) + $2H_2O$ (ℓ)	−1411 kJ	(2)

Solution

If we reverse equation (1) to give (−1) and then add it to equation (2) as written, the equation for the reaction of interest is obtained

	ΔH^0	
$3H_2O$ (ℓ) + $2CO_2$ (g) ⟶ C_2H_5OH (ℓ) + $3O_2$ (g)	+1367 kJ	(−1)
C_2H_4 (g) + $3O_2$ (g) ⟶ $2CO_2$ (g) + $2H_2O$ (ℓ)	−1411 kJ	(2)
C_2H_4 (g) + H_2O (ℓ) ⟶ C_2H_5OH (ℓ)	ΔH^0 = −44 kJ	

Notice that when equation (1) is reversed, the sign of ΔH^0 is also changed because the reverse of an exothermic reaction is an endothermic reaction. The ΔH^0 for the reaction of interest is −44 kilojoules per mole of C_2H_5OH (ℓ) formed. Note, however, that since this reaction does not involve formation of C_2H_5OH (ℓ) from its constituent elements ΔH^0 is *not* equal to ΔH^0_f for C_2H_5OH (ℓ).

A more generally useful formulation of Hess' Law enables us to use extensive tables of ΔH^0_f values to calculate the enthalpy change accompanying the desired reaction. Let us illustrate this approach by reconsidering the reaction of Example 13–4

$$C_2H_4\ (g) + H_2O\ (\ell) \longrightarrow C_2H_5OH\ (\ell)$$

A table of ΔH^0_f values (Appendix K) gives the values $\Delta H^0_{f\ C_2H_5OH\ (\ell)} = -277.7$ kJ/mol, $\Delta H^0_{f\ C_2H_4\ (g)} = 52.3$ kJ/mol, and $\Delta H^0_{f\ H_2O\ (\ell)} = -285.8$ kJ/mol. We may express this information in the form of the thermochemical equations shown below

		$\Delta H^0 = \Delta H^0_f$	
for $C_2H_5OH\ (\ell)$:	$2C\ (graphite) + 3H_2\ (g) + \frac{1}{2}O_2\ (g) \longrightarrow C_2H_5OH\ (\ell)$	−277.7 kJ	(1)
for $C_2H_4\ (g)$:	$2C\ (graphite) + 2H_2\ (g) \longrightarrow C_2H_4\ (g)$	52.3 kJ	(2)
for $H_2O\ (\ell)$:	$H_2\ (g) + \frac{1}{2}O_2\ (g) \longrightarrow H_2O\ (\ell)$	−285.8 kJ	(3)

We may generate the equation for the desired net reaction by adding equation (1) to the reverse of equations (2) and (3). The value of ΔH^0 for the desired reaction is then simply the sum of the corresponding ΔH^0 values

	ΔH^0	
$2C\ (graphite) + 3H_2\ (g) + \frac{1}{2}O_2\ (g) \longrightarrow C_2H_5OH\ (\ell)$	−277.7 kJ	(1)
$C_2H_4\ (g) \longrightarrow 2C\ (graphite) + 2H_2\ (g)$	−52.3 kJ	(−2)
$H_2O\ (\ell) \longrightarrow H_2\ (g) + \frac{1}{2}O_2\ (g)$	+285.8 kJ	(−3)
net rxn: $C_2H_4\ (g) + H_2O\ (\ell) \longrightarrow C_2H_5OH\ (\ell)$	$\Delta H^0 =$ −44.2 kJ	

Note that ΔH^0 for this reaction is given by

$$\Delta H^0_{rxn} = \Delta H^0_{(1)} + \Delta H^0_{(-2)} + \Delta H^0_{(-3)}$$

and *also* by

Product — Reactants

$$\Delta H^0_{rxn} = \Delta H^0_{f\ C_2H_5OH\ (\ell)} - [\Delta H^0_{f\ C_2H_4\ (g)} + \Delta H^0_{f\ H_2O\ (\ell)}]$$

In general terms, Hess' Law may be stated

$$\Delta H^0_{rxn} = \Sigma\, n\Delta H^0_{f\ products} - \Sigma\, n\Delta H^0_{f\ reactants}$$

In applying this relationship one must remember that the Σn implies that the ΔH^0_f value of each product and reactant must be multiplied by the coefficient of that species, *n*, in the balanced equation for the reaction.

The capital Greek letter sigma (Σ) is read "the sum of."

This general result is a very useful form of Hess' Law: *The standard enthalpy change of a reaction is equal to the sum of the standard molar enthalpies of formation of the products, each multiplied by its coefficient, n, in the reaction of interest, minus the corresponding sum of the standard molar enthalpies of formation of the reactants.*

Let us now work some examples that further illustrate the point.

EXAMPLE 13–5

Calculate ΔH^0 at 298 K for the decomposition of calcium carbonate, $CaCO_3$ (s) using the ΔH^0_f values in Table 13–1.

$$CaCO_3\ (s) \longrightarrow CaO\ (s) + CO_2\ (g)$$

Solution

$$\Delta H^0 = \Sigma\, n\Delta H^0_{f\ \text{products}} - \Sigma\, n\Delta H^0_{f\ \text{reactants}}$$
$$= \Delta H^0_{f\ CaO\ (s)} + \Delta H^0_{f\ CO_2\ (g)} - \Delta H^0_{f\ CaCO_3\ (s)}$$
$$= (1\ \text{mol})(-635.5\ \text{kJ/mol}) + (1\ \text{mol})(-393.5\ \text{kJ/mol}) - (1\ \text{mol})(-1207\ \text{kJ/mol})$$
$$\Delta H^0 = +178\ \text{kJ (an endothermic reaction)}$$

EXAMPLE 13–6

Calculate ΔH^0 for the following reaction at 298 K.

$$SiH_4\ (g) + 2O_2\ (g) \longrightarrow SiO_2\ (s) + 2H_2O\ (\ell)$$

Solution

$$\Delta H^0 = \Sigma\, n\Delta H^0_{f\ \text{products}} - \Sigma\, n\Delta H^0_{f\ \text{reactants}}$$
$$\Delta H^0 = \Delta H^0_{f\ SiO_2\ (s)} + 2\ \Delta H^0_{f\ H_2O\ (\ell)} - [\Delta H^0_{f\ SiH_4\ (g)} + 2\ \Delta H^0_{f\ O_2\ (g)}]$$
$$= (1\ \text{mol})(-910.9\ \text{kJ/mol}) + (2\ \text{mol})(-285.8\ \text{kJ/mol})$$
$$- [(1\ \text{mol})(+34\ \text{kJ/mol}) + 2\ \text{mol}\ (0\ \text{kJ/mol})]$$

O_2 (g) is standard state of the element

$$\Delta H^0 = -1516\ \text{kJ}$$

Notice that there are two moles of water in the balanced equation, and the enthalpy of formation of water, −285.8 kJ/mol, is multiplied by 2 moles.

If we know ΔH^0 at 298 K for a reaction and all but one of the ΔH^0_f values for reactants and products, the unknown ΔH^0_f value can be calculated, as Example 13–7 shows.

EXAMPLE 13–7

Given the following information

$$\underset{\text{yellow}}{PbO\ (s)} + CO\ (g) \longrightarrow Pb\ (s) + CO_2\ (g) \qquad \Delta H^0_{rxn} = -65.69\ \text{kJ}$$

ΔH^0_f for CO_2 (g) = −393.5 kJ/mol

ΔH^0_f for CO (g) = −110.5 kJ/mol

Determine, without consulting Table 13–1 or Appendix K, ΔH^0_f for yellow PbO (s).

Solution

$$\Delta H^0_{rxn} = \Sigma\, n\Delta H^0_{f\ \text{products}} - \Sigma\, n\Delta H^0_{f\ \text{reactants}}$$
$$\Delta H^0_{rxn} = \Delta H^0_{f\ Pb\ (s)} + \Delta H^0_{f\ CO_2\ (g)} - [\Delta H^0_{f\ PbO\ (s)} + \Delta H^0_{f\ CO\ (g)}]$$
$$-65.69\ \text{kJ} = 0\ \text{kJ} + (-393.5\ \text{kJ}) - [\Delta H^0_{f\ PbO\ (s)} + (-110.5\ \text{kJ})]$$

Rearranging to solve for the unknown

$$\Delta H^0_{f\ PbO\ (s)} = 65.69\ \text{kJ} - 393.5\ \text{kJ} + 110.5\ \text{kJ}$$
$$= -217.3\ \text{kJ/mol}$$

which agrees with the value listed in Table 13–1.

13–8 Bond Energies

Chemical reactions involve the breaking and making of chemical bonds. Energy is always required to break a chemical bond. The **bond energy,** which is the same as bond enthalpy for all practical purposes, is the amount of energy necessary to dissociate a bond in a covalent substance in the gaseous state into atoms in the gaseous state. For example, for the reaction

We use the term bond *energy* rather than bond *enthalpy* only because it is common practice to do so. Tabulated values of average bond energies are actually average bond enthalpies.

$$H_2\ (g) \longrightarrow 2H\ (g) \qquad \Delta H^0_{rxn} = \Delta H_{H—H} = +435\ kJ$$

the bond energy of the hydrogen–hydrogen bond is 435 kilojoules per mole of bonds. The reaction is *endothermic* (ΔH^0_{rxn} is positive) and could be written

$$H_2\ (g) + 435\ kJ \longrightarrow 2H\ (g)$$

In Table 13–2, each bond energy refers to the covalent bond between the element at the top of the column and the one at the end of the row. Table 13–3 gives bond energies of some multiple bonds.

Let us consider more complex molecules. For the reaction

$$CH_4\ (g) \longrightarrow C\ (g) + 4H\ (g)$$

$\Delta H^0_{rxn} = 1.66 \times 10^3$ kJ (398 kcal). Since the four hydrogen atoms are identical, all the C—H bonds are identical in bond length and energy *in methane mole-*

TABLE 13–2 Some Average Single Bond Energies in kJ/mol of Bonds (kcal/mol of Bonds)

H	C	N	O	F	Si	P	S	Cl	Br	I	
435 (104)	414 (99)	389 (93)	464 (111)	569 (136)	293 (70)	318 (76)	339 (81)	431 (103)	368 (88)	297 (71)	H
	347 (83)	293 (70)	351 (84)	439 (105)	289 (69)	264 (63)	259 (62)	330 (79)	276 (66)	238 (57)	C
		159 (38)	201 (48)	272 (65)	—	209 (50)	—	201 (48)	243? (58?)	—	N
			138 (33)	184 (44)	368 (88)	351 (84)	—	205 (49)	—	201 (48)	O
				159 (38)	540 (129)	490 (117)	327 (78)	255 (61)	197? (47?)	—	F
					176 (42)	213 (51)	226 (54)	360 (86)	289 (69)	213 (51)	Si
						213 (51)	230 (55)	331 (79)	272 (65)	213 (51)	P
							213 (51)	251 (60)	213 (51)	—	S
								243 (58)	218 (52)	209 (50)	Cl
									192 (46)	180 (43)	Br
										151 (36)	I

TABLE 13–3 Some Multiple Bond Energies

N=N	418 kJ	(100 kcal)	C=C	611 kJ	(146 kcal)
N≡N	946 kJ	(226 kcal)	C≡C	837 kJ	(200 kcal)
C=N	615 kJ	(147 kcal)	C=O	741 kJ	(177 kcal)
C≡N	891 kJ	(213 kcal)	C≡O	1.07×10^2 kJ	(256 kcal)
		O=O	498 kJ	(119 kcal)	

cules. However, the energies required to break the individual C—H bonds differ for successively broken bonds, as shown below.

Since the hydrogen atoms are indistinguishable, it makes no difference which one is removed first.

$$\begin{array}{ll} CH_4\ (g) \longrightarrow CH_3\ (g) + H\ (g) & \Delta H^0 = +\ 427\ kJ/mol \\ CH_3\ (g) \longrightarrow CH_2\ (g) + H\ (g) & \Delta H^0 = +\ 439\ kJ/mol \\ CH_2\ (g) \longrightarrow CH\ (g) + H\ (g) & \Delta H^0 = +\ 452\ kJ/mol \\ CH\ (g) \longrightarrow C\ (g) + H\ (g) & \Delta H^0 = +\ 347\ kJ/mol \\ \hline CH_4\ (g) \longrightarrow C\ (g) + 4H\ (g) & \Delta H^0 = +1665\ kJ \end{array}$$

"Average" C—H bond energy = $\Delta H^0/4$ = 416 kJ (99.5 kcal)

We see that the *"average" C—H bond energy* in methane is one fourth of 1665 kJ, or 416 kJ, although no mole of single C—H bonds is actually broken by absorption of exactly that amount of energy. Average C—H bond energies differ slightly from compound to compound, as in CH_4, CH_3Cl, CH_3NO_2, etc., but are nevertheless sufficiently constant to be useful, as illustrated below.

A special case of Hess' Law involves the use of bond energies to *estimate* heats of reaction. Consider the reaction of hydrogen with gaseous bromine to form hydrogen bromide

The standard state of Br_2 at 25°C is the liquid state, but the liquid exists in equilibrium with its vapor.

$$H_2\ (g) + Br_2\ (g) \longrightarrow 2HBr\ (g) \qquad \Delta H^0_{rxn} = ?$$

The bond energies are given by ΔH^0 values for the following reactions

$$\begin{array}{lll} & \boldsymbol{\Delta H^0} & \\ HBr\ (g) \longrightarrow H\ (g) + Br\ (g) & 368\ kJ/mol\ (\Delta H_{H—Br}) & (1) \\ H_2\ (g) \longrightarrow 2H\ (g) & 435\ kJ/mol\ (\Delta H_{H—H}) & (2) \\ Br_2\ (g) \longrightarrow 2Br\ (g) & 192\ kJ/mol\ (\Delta H_{Br—Br}) & (3) \end{array}$$

If we reverse equation (1) and multiply it by two, and then add it to equations (2) and (3), we obtain the equation for the reaction of interest and ΔH^0_{rxn}.

$$\begin{array}{llr} & \boldsymbol{\Delta H^0\ (kJ)} & \\ 2H\ (g) + 2Br\ (g) \longrightarrow 2HBr\ (g) & 2(-368)\ (-2\Delta H_{H—Br}) & (-1) \\ H_2\ (g) \longrightarrow 2H\ (g) & 435\ (\Delta H_{H—H}) & (2) \\ Br_2\ (g) \longrightarrow 2Br\ (g) & 192\ (\Delta H_{Br—Br}) & (3) \\ \hline H_2\ (g) + Br_2\ (g) \longrightarrow 2HBr\ (g) & \Delta H^0_{rxn} = -109\ kJ\ (-26\ kcal) & \end{array}$$

From this result we see that

$$\Delta H^0_{rxn} = \Delta H_{H—H} + \Delta H_{Br—Br} - 2\Delta H_{H—Br}$$

You may wish to verify that if ΔH^0_{rxn} is calculated from ΔH^0_f data for H_2 (g), Br_2 (g), and HBr (g) at 298 K, the value obtained is −104 kJ. Since the tabulated bond energies represent average bond energies, the correspondence is quite good. In general terms, ΔH^0_{rxn} is related to the bond energies of the reactants and products in gas phase reactions by the following version of Hess' Law

$$\Delta H^0_{rxn} = \Sigma \text{ B.E.}_{\text{reactants}} - \Sigma \text{ B.E.}_{\text{products}} \quad \text{in gas phase reactions only}$$

Note that this involves bond energies of *reactants minus* bond energies of *products*, rather than [products] minus [reactants] as when molar enthalpies of formation are used.

The net enthalpy change of a reaction is the energy required to break all the bonds in reactant molecules *minus* the energy required to break all the bonds in product molecules. Stated in another way, the amount of energy released when a bond is formed is equal to the amount absorbed when the same bond is broken. Therefore, the heat of reaction can also be described as the energy released in forming all the bonds in the product molecules minus the energy released in forming all the bonds in the reactant molecules for gas phase reactions. The situation is depicted in Figure 13–5.

EXAMPLE 13–8

Using the bond energies listed in Table 13–2, estimate the heat of reaction at 298 K for the reaction below. All bonds are single bonds:

$Br_2\ (g) + 3F_2\ (g) \longrightarrow 2BrF_3\ (g)$

Solution

Each BrF_3 molecule contains three Br—F bonds, so two moles of BrF_3 contain six moles of Br—F bonds. Three moles of F_2 contain a total of three moles of F—F bonds and one mole of Br_2 contains one mole of Br—Br bonds. Using the bond energy form of Hess' Law

$$\begin{aligned} \Delta H^0_{rxn} &= \Delta H_{Br—Br} + 3\Delta H_{F—F} - [6\Delta H_{Br—F}] \\ &= 192 \text{ kJ} + 3(159 \text{ kJ}) - 6(197 \text{ kJ}) \\ \Delta H^0_{rxn} &= \underline{-513 \text{ kJ}} \quad (-123 \text{ kcal}) \end{aligned}$$

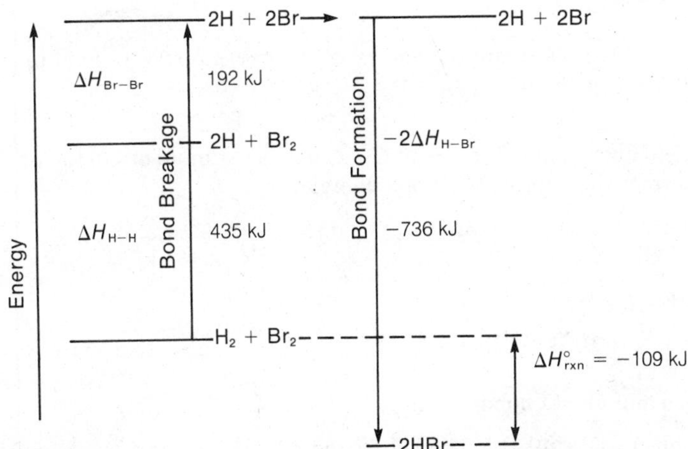

FIGURE 13–5 The heat of reaction for the formation of HBr from H_2 (g) and Br_2 (g).

EXAMPLE 13–9

Use the information from Example 13–8 and the following information

$Br_2\ (\ell) \longrightarrow Br_2\ (g)$, $\Delta H^0_{vaporization} = 31$ kJ

to determine ΔH^0_{rxn} for the following reaction at 298 K.

$Br_2\ (\ell) + 3F_2\ (g) \longrightarrow 2BrF_3\ (g)$

Solution

This problem is solved in much the same way as Example 13–8, except that we must also consider the fact that the reaction involves $Br_2\ (\ell)$ rather than Br_2 (g).

	ΔH^0
$Br_2\ (g) + 3F_2\ (g) \longrightarrow 2BrF_3\ (g)$	−513 kJ
$Br_2\ (\ell) \longrightarrow Br_2\ (g)$	+31 kJ
$Br_2\ (\ell) + 3F_2\ (g) \longrightarrow 2BrF_3\ (g)$	$\Delta H^0_{rxn} = -482$ kJ

The next example shows a variation of the bond energy method: given a reaction for which ΔH^0_{rxn} and all but one of the bond energies are known, we can calculate the missing bond energy.

EXAMPLE 13–10

Given the following equation and average bond energies

$C_3H_8\ (g) + 5O_2\ (g) \longrightarrow 3CO_2\ (g) + 4H_2O\ (g) \qquad \Delta H^0_{rxn} = -2.05 \times 10^3$ kJ

Bond	Energy
C—C	347 kJ/mol of bonds
C—H	414 kJ/mol of bonds
C═O	741 kJ/mol of bonds
O—H	464 kJ/mol of bonds

Without consulting Table 13–3, estimate the energy of the oxygen-oxygen bond in O_2 molecules. Compare this with the value listed in Table 13–3.

Solution

There are 8 moles of C—H bonds and 2 moles of C—C bonds per mole of C_3H_8. As before, we add and subtract the appropriate bond energies

$$\Delta H^0_{rxn} = 2\Delta H_{C—C} + 8\Delta H_{C—H} + 5\Delta H_{O═O} - [6\Delta H_{C═O} + 8\Delta H_{O—H}]$$
$$-2.05 \times 10^3 \text{ kJ} = 2(347 \text{ kJ}) + 8(414 \text{ kJ}) + 5\Delta H_{O═O} - [6(741 \text{ kJ}) + 8(464 \text{ kJ})]$$

Rearranging, we obtain

$$-5\Delta H_{O═O} = 694 \text{ kJ} + 3.31 \times 10^3 \text{ kJ} - 4.45 \times 10^3 \text{ kJ} - 3.71 \times 10^3 \text{ kJ} + 2.05 \times 10^3 \text{ kJ}$$
$$5\Delta H_{O═O} = 2.11 \times 10^3 \text{ kJ}$$
$$\Delta H_{O═O} = 422 \text{ kJ per mol O═O bonds}$$

The value listed in Table 13–3 is 498 kJ. The calculated value is in error by about 15%. While it is a reasonable estimate of the O═O bond energy, it does point out the limitations of using *average* bond energies.

13–9 Born-Haber Cycle

We have just dealt with average bond energies of covalent molecules. While covalent substances exist as discrete molecules, solid ionic substances exist in arrays of alternating positive and negative ions called crystal lattices. The energy holding a crystal together is called the **crystal lattice energy, ΔH^0_{xtal}**. It is the amount of energy absorbed when a mole of an ionic substance is formed from its gaseous ions, as illustrated below for sodium chloride

$$Na^+ (g) + Cl^- (g) \longrightarrow NaCl (s) \qquad \Delta H^0_{rxn} = \Delta H^0_{xtal}$$

Because energy is always released in such processes, crystal lattice energies have negative values.

Crystal lattice energies *cannot* be measured directly. However, by application of Hess' Law to a series of reactions beginning with consumption of Na (s) and Cl_2 (g) and ending with the production of NaCl (s), ΔH_{xtal} can be evaluated. Such a sequence of reactions for ionic substances is called a **Born-Haber cycle,** after Max Born and Fritz Haber, who developed the general procedure. Its use is illustrated below for sodium chloride.

The heat of formation can easily be determined:

$$Na (s) + \tfrac{1}{2}Cl_2 (g) \longrightarrow NaCl (s) \qquad \Delta H_{rxn} = \Delta H_{f\ NaCl\ (s)} = -411 \text{ kJ}$$

The following reactions and experimental data can be summed to give the same equation:

Sublimation is the process by which a solid is converted directly to a gas (Section 9–12).

1. Sublimation of sodium metal. The heat change is ΔH_f for Na (g) = ΔH_{subl}.

   $$Na (s) \longrightarrow Na (g) \qquad \Delta H_{subl} = 109 \text{ kJ}$$

2. Dissociation of one-half mole of Cl_2 (g) into gaseous chlorine atoms. The heat change is ΔH_f for Cl (g) = $\frac{1}{2}\Delta H_{diss}$ (also $\frac{1}{2}$ the bond energy for Cl_2 (g)).

   $$\tfrac{1}{2}Cl_2 (g) \longrightarrow Cl (g) \qquad \tfrac{1}{2}\Delta H_{diss} = 122 \text{ kJ}$$

3. Ionization of a mole of sodium atoms. The heat change is the first ionization energy for sodium expressed in kilojoules.

   $$Na (g) \longrightarrow Na^+ (g) + e^- \qquad \Delta H_{ie} = 496 \text{ kJ}$$

4. Addition of one mole of electrons to one mole of gaseous chlorine atoms. This heat change is the electron affinity of chlorine expressed in kilojoules.

   $$Cl (g) + e^- \longrightarrow Cl^- (g) \qquad \Delta H_{ea} = -348 \text{ kJ}$$

5. Condensation of the gaseous ions to form a mole of solid NaCl. This heat change cannot be measured directly.

   $$Na^+ (g) + Cl^- (g) \longrightarrow NaCl (s) \qquad \Delta H_{xtal} = \underline{?}$$

Summation of these five reactions and their ΔH^0 values allows us to calculate ΔH^0_{xtal}.

1. $Na (s) \longrightarrow Na (g)$	$\Delta H_{subl} = 109$ kJ
2. $\frac{1}{2}Cl_2 (g) \longrightarrow Cl (g)$	$\frac{1}{2}\Delta H_{diss} = 122$ kJ
3. $Na (g) \longrightarrow Na^+ (g) + e^-$	$\Delta H_{ie} = 496$ kJ
4. $Cl (g) + e^- \longrightarrow Cl^- (g)$	$\Delta H_{ea} = -348$ kJ
5. $Na^+ (g) + Cl^- (g) \longrightarrow NaCl (s)$	$\Delta H_{xtal} = \underline{?}$
$Na (s) + \frac{1}{2}Cl_2 (g) \longrightarrow NaCl (s)$	ΔH_{rxn} = sum of five ΔH's given above

We have just shown by Hess' Law that for a series of reactions, the overall ΔH_{rxn} value can be calculated. In this case it is:

$$\Delta H_{rxn} = \Delta H_{subl} + \tfrac{1}{2}\Delta H_{diss} + \Delta H_{ie} + \Delta H_{ea} + \Delta H_{xtal}$$
$$-411 \text{ kJ} = 109 \text{ kJ} + 122 \text{ kJ} + 496 \text{ kJ} + (-348 \text{ kJ}) + \Delta H_{xtal}$$

Solving for ΔH_{xtal},

$$\Delta H_{xtal} = (-411 - 109 - 122 - 496 + 348) \text{ kJ}$$
$$= -790 \text{ kJ per mol of NaCl (s)}$$

The entire Born-Haber cycle is summarized in Figure 13–6.

Crystal lattice energies cannot be measured directly, but they can be calculated from other known enthalpy values.

The crystal lattice energy is a measure of the stability of an ionic solid. The more negative its value, the more energy is released in the hypothetical reaction in which a mole of ionic solid is formed from its constituent ions in the gaseous state. The value of −790 kJ per mol of NaCl (s) indicates that this is a very stable solid.

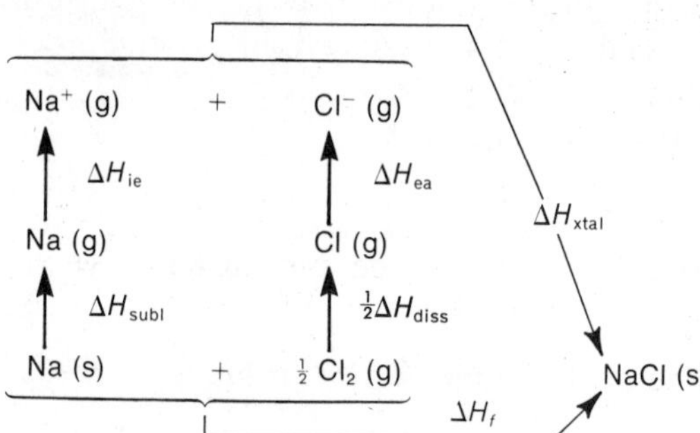

FIGURE 13–6 The Born-Haber cycle for NaCl (s).

13–10 Second Law of Thermodynamics

Early in this chapter we stated that two factors must be considered in determining whether or not a reaction is spontaneous under a given set of conditions. The effect of the first factor, the enthalpy change, is that spontaneity is *favored* (but not required) by exothermicity. Nonspontaneity is favored by endothermicity. The effect of the other factor, called *entropy,* is related to the **Second Law of Thermodynamics,** which states that *in spontaneous changes the universe tends toward a state of greater disorder.* It should be noted that the second law is based on our experiences, but it is not a provable statement.

The general applicability of this law in the macroscopic world is illustrated by a few common examples. When mirrors are dropped they shatter. When a drop of food coloring is added to a glass of water, it diffuses until a homogeneously colored solution results. When a truck drives through the wall of a building, the wall crumbles and the truck becomes more disordered. On the other hand, we would be very surprised to drop the pieces of a jigsaw puzzle to the floor and find that the puzzle had spontaneously assembled itself.

13–11 Entropy

If the degree of disorder, or **entropy,** of a system increases during a process, the spontaneity of the process is *favored* (but not required). Entropy, *S,* like enthalpy, is a state function. It is possible for the entropy of a system to decrease during a spontaneous process, or for the entropy of a system to increase during a nonspontaneous process. The second law states that the entropy of the *universe* (not the system) increases during a spontaneous process. Thus, for a spontaneous process:

$$\Delta S_{universe} = \Delta S_{system} + \Delta S_{surroundings} > 0$$

If ΔS_{system} is negative (decrease in disorder), then $\Delta S_{universe}$ may still be positive (increase in disorder) if $\Delta S_{surroundings}$ is more positive than ΔS_{system} is negative. And $\Delta S_{universe}$ still may be negative (and the process will be nonspontaneous) if ΔS_{system} is positive and $\Delta S_{surroundings}$ is even more negative. A good illustration of the first kind of process is an ordinary refrigerator, which removes heat from the inside of the box (the system) and ejects that heat, *plus* the heat generated by the refrigerator compressor, into the room (the surroundings). Although the entropy of the system decreases (because the air molecules inside the box move more slowly), the increase in the entropy of the surroundings more than makes up for it; the entropy of the universe (system plus surroundings) increases.

TABLE 13–4 Absolute Entropies for a Few Common Substances

	S^0 (J/mol K)
C (diamond)	2.38
C (g)	158.0
H_2O (ℓ)	69.91
H_2O (g)	188.7
I_2 (s)	116.1
I_2 (g)	260.6

For any particular substance, the particles in the solid state are more highly ordered than those in the liquid state which, in turn, are more highly ordered than those in the gaseous state. The more ordered a substance is, the lower is its absolute entropy, S^0. Consider, as examples, the absolute entropies at 298 K listed in Table 13–4. See Figure 13–7 also.

The entropy of a given substance increases as it goes from solid to liquid to gas. The **Third Law of Thermodynamics** states that *the entropy of a pure, perfect crystalline substance* (infinitely ordered) *is zero at absolute zero Kelvin.* Of course, no such substance exists because we are unable to attain this temperature exactly; even the particles making up the most perfectly ordered crystals are able to vibrate about fixed positions in the crystal at all temperatures even infinitesimally above absolute zero.

Enthalpies can be measured only as *differences* with respect to an arbitrary standard state. Entropies, in contrast, are defined relative to an absolute zero level.

Let's consider the entropy changes that occur when a liquid solidifies at a temperature below its freezing (melting) point. We all know that this is a spontaneous process, yet ΔS_{system} is negative, since a solid forms from its liquid. However, $\Delta S_{surroundings}$ is positive and of larger magnitude than ΔS_{system}. The reason is that a liquid releases heat to its surroundings (atmosphere) as it crystallizes, and the released heat increases the motion (randomness, disorder) of the molecules of the surroundings. As a result, $\Delta S_{universe}$ is positive. As the temperature decreases, the $\Delta S_{surroundings}$ contribution becomes more important, $\Delta S_{universe}$ becomes more positive, and the process becomes more spontaneous.

Similar arguments apply for condensing a gas at its condensation point (boiling point).

The situation is reversed when a liquid is boiled or a solid is melted. For example, at temperatures above its melting point, a solid spontaneously melts and ΔS_{system} is positive. The heat absorbed as the solid (system) melts is removed from the surroundings, and thus decreases the motion of the molecules of the surroundings. Therefore, $\Delta S_{surroundings}$ is negative (the surroundings become less disordered). However, the positive ΔS_{system} is larger than the negative $\Delta S_{surroundings}$, so that $\Delta S_{universe}$ is positive.

Above the melting point, $\Delta S_{universe}$ is positive for melting. Below the melting point, $\Delta S_{universe}$ is positive for freezing. At the melting point, $\Delta S_{surroundings}$ is equal in magnitude and opposite in sign to ΔS_{system}. Then

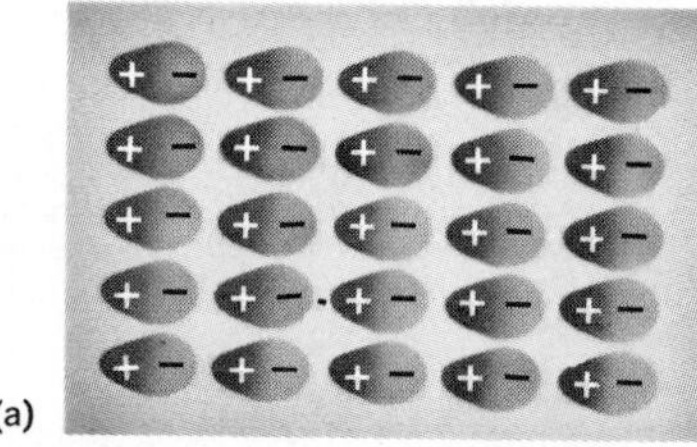

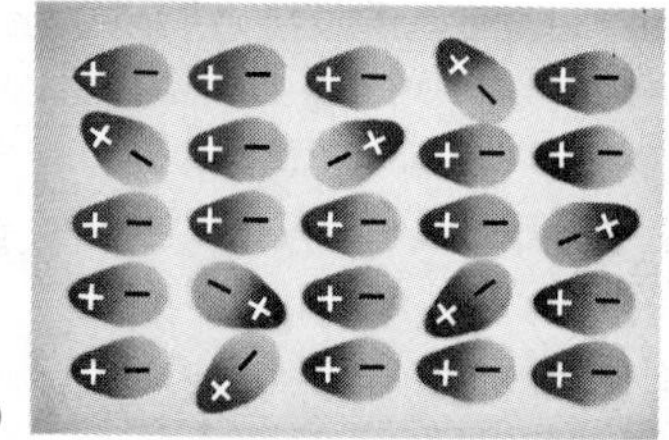

FIGURE 13–7 (a) A simplified representation of a side view of a perfect crystal of HCl at 0 K. Note the perfect alignment of the dipoles in all molecules in a "perfect" crystal, which causes its entropy to be zero at 0 K. However, there are no perfect crystals because even the purest substances that scientists have prepared to date are contaminated by traces of impurities that occupy a few of the positions in crystal lattices. Additionally, there are some vacancies in the crystal lattices of even very highly purified substances such as those used in semiconductors (Section 9–17). (b) A simplified representation of the same "perfect" crystal at a temperature slightly above 0 K. Vibrations of the individual molecules within the crystal lattice cause some dipoles to be oriented in directions other than those in a "perfect" crystal at 0 K. The entropy of such a crystalline solid is greater than zero because there is some disorder in the crystal.

Can you develop a comparable table for boiling (liquid → gas) and condensation (gas → liquid)?

$\Delta S_{universe}$ is zero for both melting and freezing, and the system is at equilibrium. Table 13–5 summarizes the entropy effects associated with changes of physical state.

TABLE 13–5 Entropy Effects Associated with Melting and Freezing

Change	Temperature	Sign of ΔS_{sys}	Sign of ΔS_{surr}	(Magnitude of ΔS_{sys}) compared with (Magnitude of ΔS_{surr})	$\Delta S_{univ} = \Delta S_{sys} + \Delta S_{surr}$	Spontaneity
1. Melting (solid → liquid)	(a) > mp	+	−	>	> 0	Spontaneous
	(b) = mp	+	−	=	= 0	Equilibrium
	(c) < mp	+	−	<	< 0	Nonspontaneous
2. Freezing (liquid → solid)	(a) > mp	−	+	>	< 0	Nonspontaneous
	(b) = mp	−	+	=	= 0	Equilibrium
	(c) < mp	−	+	<	> 0	Spontaneous

Notice that $\Delta S_{universe}$ is positive for all spontaneous processes. Unfortunately, it is not possible to make direct measurements of $\Delta S_{universe}$. Consequently, entropy changes accompanying physical and chemical changes are reported in terms of ΔS_{system}. The subscript "system" is commonly omitted and the symbol ΔS refers to the change in entropy of the reacting *system*, just as ΔH refers to the change in enthalpy of the reacting system.

The *absolute entropies* (S^0) of various substances under standard conditions are tabulated in Appendix K. The *standard entropy change* (ΔS^0) of a reaction can be determined from the absolute entropies in the same way as an enthalpy change can be determined from values of standard enthalpies of formation of reactants and products. The relationship is similar to Hess' Law:

$$\Delta S^0_{rxn} = \Sigma\, nS^0_{products} - \Sigma\, nS^0_{reactants}$$

As with the corresponding relationship describing ΔH_{rxn} in Section 13–7, one must remember that the Σn implies multiplication of each S value by the appropriate coefficient, n, from the balanced chemical equation.

Values of S^0 are generally quite small in comparison to ΔH^0_f values, and are usually tabulated in units of *J/mol·K* rather than the larger units, kJ/mol, that we used for enthalpy changes.

EXAMPLE 13–11

Using the values of standard molar entropies in Appendix K, calculate the entropy change at 25°C and one atmosphere pressure for the reaction of hydrazine with hydrogen peroxide. The mixture is explosive, and has been used for rocket propulsion. Do you think the reaction is spontaneous?

The balanced equation for the reaction is

$$N_2H_4\,(\ell) + 2H_2O_2\,(\ell) \longrightarrow N_2\,(g) + 4H_2O\,(g) + 642.2\text{ kJ}$$

Solution

$$\Delta S^0_{rxn} = [S^0_{N_2\,(g)} + 4S^0_{H_2O\,(g)}] - [S^0_{N_2H_4\,(\ell)} + 2S^0_{H_2O_2\,(\ell)}]$$
$$= 1\text{ mol }(191.5\text{ J/mol·K}) + 4\text{ mol }(188.7\text{ J/mol·K}) - 1\text{ mol }(121.2\text{ J/mol·K}) - 2\text{ mol }(109.6\text{ J/mol·K})$$
$$\Delta S^0_{rxn} = \underline{+605.9\text{ J/K}} \qquad (+0.6059\text{ kJ/K})$$

We have multiplied the absolute entropy of each species by the number of moles of that species and *included the unit moles.* It is common practice to omit the unit "mol" in such calculations for conciseness, and we shall do so in later examples.

Although it may not appear to be, +605.9 J/K is a relatively large value of ΔS (for the system). The entropy change is positive and favors spontaneity. In this case, the reaction is also exothermic, $\Delta H^0 = -642.2$ kilojoules. Therefore, the reaction *must* be spontaneous, as we shall demonstrate in the next section, because ΔS^0 is positive and ΔH^0 is negative.

13–12 Free Energy Change, ΔG

Energy has been defined as the capacity to do work. If heat is released in a chemical reaction (ΔH is negative), some of the heat may be converted into useful work, but some of it may be expended to increase the order of the system (if ΔS is negative). If a system becomes more disordered ($\Delta S > 0$), then more useful energy becomes available than indicated by the enthalpy change alone. J. Willard Gibbs, a prominent nineteenth-century American professor of mathematics and physics, formulated the relationship between enthalpy and entropy changes for a process in terms of the change in another state function, ΔG, called the **Gibbs free energy change.** The relationship, which holds at constant temperature, is known as the **Gibbs-Helmholtz equation,** and is

This is now sometimes called simply the Gibbs energy change, or the free energy change.

$$\Delta G = \Delta H - T\Delta S \qquad \text{(constant temperature)}$$

where T is the absolute temperature.

The change in Gibbs free energy is the *maximum useful energy* obtainable from a given process at constant temperature. It is also the *indicator of spontaneity of a reaction or physical change.* If there is a net release of useful energy, ΔG is negative and the process is spontaneous. Note from the equation that ΔG becomes more negative as ΔH becomes more negative (exothermic) and as ΔS becomes more positive (increase in disorder). If there is a net absorption of free energy by the system during a process, ΔG is positive and the process is nonspontaneous. This means, of course, that the reverse process is spontaneous under the given conditions. When $\Delta G = 0$, there is no net transfer of energy and both the forward and reverse processes occur at the same rate, and the system is at equilibrium. Thus, $\Delta G = 0$ describes a system at equilibrium.

The relationship between ΔG and the spontaneity of a reaction may be summarized as:

ΔG	Spontaneity of Reaction
ΔG is positive	Reaction is nonspontaneous
ΔG is zero	System is at equilibrium
ΔG is negative	Reaction is spontaneous

Values of ΔG may be determined by relating them to values of standard free energy change, ΔG^0, which in turn may be calculated from experimentally determined equilibrium constants (discussed in Chapter 15), from the electrochemical potential for a process (discussed in Chapter 19), and by other techniques. Since values of ΔG and ΔH usually can be obtained experimentally with little difficulty, ΔS values can be calculated by the Gibbs relationship. There is no way of measuring ΔS directly.

Values of standard molar free energy of formation, ΔG_f^0, for many substances are tabulated in Appendix K. The values may be used to calculate the standard free energy change of a reaction at 298 K by using the following relationship. For elements in their standard states, $\Delta G_f^0 = 0$.

$$\Delta G^0_{rxn} = \Sigma\, n\Delta G^0_{f\ products} - \Sigma\, n\Delta G^0_{f\ reactants}$$

Alternatively, ΔG^0 can be calculated by using the Gibbs-Helmholtz relationship at standard state conditions,

$$\Delta G^0 = \Delta H^0 - T\Delta S^0 \quad \text{(constant temperature)}$$

EXAMPLE 13–12

Diatomic nitrogen and oxygen molecules make up 99% of all the molecules in reasonably "unpolluted" air. Evaluate ΔG^0 for the following reaction at 298 K, using ΔG_f^0 values from Appendix K. Is the reaction spontaneous?

Solution

$N_2\,(g) + O_2\,(g) \longrightarrow 2NO\,(g)$ nitrogen oxide

$$\Delta G^0_{rxn} = 2\Delta G^0_{f\,NO\,(g)} - [\Delta G^0_{f\,N_2\,(g)} + \Delta G^0_{f\,O_2\,(g)}]$$
$$= (2(86.57) - [0 + 0])\ \text{kJ}$$

$\Delta G^0_{rxn} = +173.1$ kJ for the reaction as written

For the reverse reaction at 298 K, $\Delta G^0 = -173.1$ kJ; it is spontaneous.

Since ΔG^0 is positive, the reaction is nonspontaneous at 298 K under standard state conditions. This is fortunate, for otherwise we *might* all have to breathe nitrogen oxide, which is poisonous. (Remember that thermodynamic spontaneity does not guarantee that a process will occur at an observable rate.)

EXAMPLE 13–13

Make the same determination as in Example 13–12, using heats of formation and absolute entropies rather than free energies of formation.

Solution

First, we must evaluate ΔH^0_{rxn} and ΔS^0_{rxn}

$$\begin{aligned}
\Delta H^0_{rxn} &= \Sigma n \Delta H^0_{f\,products} - \Sigma n \Delta H^0_{f\,reactants} \\
&= 2\Delta H^0_{f\,NO\,(g)} - [\Delta H^0_{f\,N_2\,(g)} + \Delta H^0_{f\,O_2\,(g)}] \\
&= (2[90.25] - [0 + 0])\text{ kJ} \\
\Delta H^0_{rxn} &= 180.5\text{ kJ}
\end{aligned}$$

$$\begin{aligned}
\Delta S^0_{rxn} &= \Sigma n S^0_{products} - \Sigma n S^0_{reactants} \\
&\quad 2S^0_{NO\,(g)} - [S^0_{N_2\,(g)} + S^0_{O_2\,(g)}] \\
&= (2[210.7] - [191.5 + 205.0])\text{ J/K} \\
\Delta S^0_{rxn} &= 24.9\text{ J/K} = 0.0249\text{ kJ/K}
\end{aligned}$$

Now we may use the Gibbs relationship with $T = 298$ K, since we are evaluating the free energy change under standard state conditions at 298 K

$$\begin{aligned}
\Delta G^0_{rxn} &= \Delta H^0_{rxn} - T\Delta S^0_{rxn} \\
&= 180.5\text{ kJ} - (298\text{ K})(0.0249\text{ kJ/K}) \\
&= 180.5\text{ kJ} - 7.42\text{ kJ} \\
\Delta G^0_{rxn} &= \underline{+173.1\text{ kJ}}, \text{ the same value obtained in Example 13–12.}
\end{aligned}$$

The Gibbs-Helmholtz equation can be used to estimate the temperature at which physical processes are in equilibrium, as is shown in the following example.

EXAMPLE 13–14

Use the thermodynamic data in the appendices to estimate the normal boiling temperature of liquid bromine, Br_2. Assume that ΔH and ΔS do not change with temperature.

Actually, both ΔH and ΔS vary with temperature, but not enough to introduce significant errors when the temperature changes are small. The value of ΔG, on the other hand, is strongly dependent on the temperature.

Solution

The reaction of interest is

$$Br_2\,(\ell) \longrightarrow Br_2\,(g)$$

By definition, the normal boiling point of a liquid is the temperature at which liquid and gas exist in equilibrium at one atmosphere, and therefore $\Delta G = 0$. We may assume that $\Delta H_{rxn} = \Delta H^0_{rxn}$ and $\Delta S_{rxn} = \Delta S^0_{rxn}$

$$\begin{aligned}
\Delta H_{rxn} &= \Delta H^0_{f\,Br_2\,(g)} - \Delta H^0_{f\,Br_2\,(\ell)} \\
&= 30.91\text{ kJ} - 0\text{ kJ} \\
&= 30.91\text{ kJ} \qquad (7.388\text{ kcal})
\end{aligned}$$

$$\begin{aligned}
\Delta S_{rxn} &= S^0_{Br_2\,(g)} - S^0_{Br_2\,(\ell)} \\
&= (245.4 - 152.2)\text{ J/K} \\
&= 93.2\text{ J/K} = 0.0932\text{ kJ/K} \qquad (0.0223\text{ kcal/K})
\end{aligned}$$

We can now solve for the temperature at which the system is in equilibrium, i.e., the boiling point of Br_2

$$\Delta G_{rxn} = \Delta H_{rxn} - T\Delta S_{rxn} = 0$$

$$\Delta H_{rxn} = T\Delta S_{rxn}$$

$$T = \frac{\Delta H_{rxn}}{\Delta S_{rxn}} = \frac{30.91 \text{ kJ}}{0.0932 \text{ kJ/K}} = 332 \text{ K } (59°\text{C})$$

The value listed in a handbook of chemistry and physics is 58.78°C.

As we have noted before, the free energy change and spontaneity of a reaction depend upon both enthalpy and entropy changes. Since both ΔH and ΔS may be either positive or negative, we can classify reactions into four categories with respect to spontaneity.

$\Delta G = \Delta H - T\Delta S$ *(constant temperature)*

1. $\Delta H = -, \Delta S = +$ Reactions are spontaneous at all temperatures.
2. $\Delta H = -, \Delta S = -$ Reactions become spontaneous as temperature decreases.
3. $\Delta H = +, \Delta S = +$ Reactions become spontaneous as temperature increases.
4. $\Delta H = +, \Delta S = -$ Reactions are nonspontaneous at all temperatures.

Table 13–6 gives examples of reactions in each category as well as temperatures at which the changeover from spontaneous to nonspontaneous occurs, where appropriate.

TABLE 13–6 Thermodynamic Classes of Reactions*

Class	Reaction	ΔH (kJ)	ΔS (J/K)	Temperature Range of Spontaneity
1	$2H_2O_2\ (\ell) \longrightarrow 2H_2O\ (\ell) + O_2\ (g)$	−196	+126	All temperatures
	$H_2\ (g) + Br_2\ (\ell) \longrightarrow 2HBr\ (g)$	−72.8	+114	All temperatures
2	$NH_3\ (g) + HCl\ (g) \longrightarrow NH_4Cl\ (s)$	−176	−285	Lower temperatures (<619K)
	$2H_2S\ (g) + SO_2\ (g) \longrightarrow 3S\ (s) + 2H_2O\ (\ell)$	−233	−424	Lower temperatures (<550K)
3	$NH_4Cl\ (s) \longrightarrow NH_3\ (g) + HCl\ (g)$	+176	+285	Higher temperatures (>619K)
	$CCl_4\ (\ell) \longrightarrow C\ (\text{graphite}) + 2Cl_2\ (g)$	+136	+235	Higher temperatures (>517K)
4	$2H_2O\ (\ell) + O_2\ (g) \longrightarrow 2H_2O_2\ (\ell)$	+196	−126	Nonspontaneous, all temperatures
	$3O_2\ (g) \longrightarrow 2O_3\ (g)$	+285	−137	Nonspontaneous, all temperatures

* The numbering system corresponds to that given above.

The next two examples illustrate how we can estimate the temperature range over which a chemical reaction is spontaneous, by evaluation of ΔH^0_{rxn} and ΔS^0_{rxn} from tabulated data.

EXAMPLE 13–15

Mercury(II) sulfide is found in a dark red mineral called cinnabar. Metallic mercury is obtained by roasting the sulfide in a limited amount of air (Section 20–3). Estimate the temperature range in which the reaction is spontaneous.

$HgS\ (s) + O_2\ (g) \longrightarrow Hg\ (\ell) + SO_2\ (g)$

Solution

We first evaluate ΔH^0_{rxn} and ΔS^0_{rxn} and *assume* that their values are independent of temperature.

$$\Delta H^0_{rxn} = \Delta H^0_{f\,Hg\,(\ell)} + \Delta H^0_{f\,SO_2\,(g)} - [\Delta H^0_{f\,HgS\,(s)} + \Delta H^0_{f\,O_2\,(g)}]$$
$$= (0 - 296.8 + 58.2 - 0)\ \text{kJ}$$

$$\Delta H^0_{rxn} = -238.6\ \text{kJ}$$

$$\Delta S^0_{rxn} = S^0_{Hg\,(\ell)} + S^0_{SO_2\,(g)} - [S^0_{HgS\,(s)} + S^0_{O_2\,(g)}]$$
$$= (76.02 + 248.1 - 82.4 - 205.0)\ \text{J/K}$$

$$\Delta S^0_{rxn} = +36.7\ \text{J/K}$$

Since ΔH^0_{rxn} is negative and ΔS^0_{rxn} is positive, the reaction is spontaneous at all temperatures. The reverse reaction is, therefore, nonspontaneous at all temperatures. This illustrates a corollary of the Second Law of Thermodynamics: The reverse of a spontaneous reaction or process is nonspontaneous.

EXAMPLE 13–16

Estimate the temperature range for which the following reaction is spontaneous.

$$SiO_2\ (s) + 2C\ (graphite) + 2Cl_2\ (g) \longrightarrow SiCl_4\ (g) + 2CO\ (g)$$

Solution

We proceed as in Example 13–15.

$$\Delta H_{rxn} = \Delta H^0_{f\,SiCl_4\,(g)} + 2\Delta H^0_{f\,CO\,(g)} - [\Delta H^0_{f\,SiO_2\,(s)} + 2\Delta H^0_{f\,C\,(graphite)} + 2\Delta H^0_{f\,Cl_2\,(g)}]$$
$$= (-657.0 + 2(-110.5) - [(-910.9) - 2(0) - 2(0)])\ \text{kJ}$$
$$= \underline{+32.9\ \text{kJ}} \qquad (+7.86\ \text{kcal})$$

$$\Delta S_{rxn} = S^0_{SiCl_4\,(g)} + 2S^0_{CO\,(g)} - [S^0_{SiO_2\,(s)} + 2S^0_{C\,(graphite)} + 2S^0_{Cl_2\,(g)}]$$
$$= [330.6 + 2(197.6) - [41.84 + 2(5.740) + 2(223.0)]]\ \text{J/K}$$
$$= 226.5\ \text{J/K} = \underline{0.2265\ \text{kJ/K}} \qquad (54.13\ \text{cal/K})$$

When $\Delta G = 0$, neither the forward nor the reverse reaction is spontaneous. Let's calculate the temperature at which $\Delta G = 0$ and the system is at equilibrium.

$$\Delta G = \Delta H - T\Delta S = 0$$

$$\Delta H = T\Delta S$$

$$T = \frac{\Delta H}{\Delta S} = \frac{+32.9\ \text{kJ}}{+0.2265\ \text{kJ/K}}$$

$\underline{T = 145\ \text{K}}$, system is at equilibrium

At temperatures above 145 K the $T\Delta S$ term is larger ($-T\Delta S$ is more negative) than the ΔH term, which makes ΔG negative, so the reaction is spontaneous. At temperatures below 145 K the $T\Delta S$ term is smaller than the ΔH term, which makes ΔG positive, so the reaction is nonspontaneous below 145 K.

However, 145 K (−128°C) is a very low temperature. Thus, for all practical purposes, the reaction is spontaneous at all but very low temperatures. In practice, it is carried out at 800 – 1000°C, in order to attain a useful and economical rate of reaction.

Key Terms

Absolute entropy (of a substance) the increase in the entropy of a substance as it goes from a perfectly ordered crystalline form at 0 K (where its entropy is zero) to the temperature in question.

Bomb calorimeter a device used to measure the heat transfer between system and surroundings at constant volume.

Bond energy the amount of energy necessary to break one mole of bonds in a substance, dissociating the substance in the gaseous state into atoms of its elements in the gaseous state.

Born-Haber cycle a series of reactions (and accompanying enthalpy changes) which, when summed, represents the hypothetical one-step reaction by which elements in their standard states are converted into crystals of ionic compounds (and the accompanying enthalpy changes).

Calorimeter a device used to measure the heat transfer between system and surroundings.

Crystal lattice energy energy that holds a crystal together; energy released when a mole of solid is formed from its constituent atoms or ions (for ionic compounds) in the gaseous state.

Endothermicity the absorption of heat by a system as a process occurs.

Enthalpy, *H* the heat content of a specific amount of a substance; thermodynamically defined as $E - PV$.

Entropy, *S* a thermodynamic state property that measures the degree of disorder or randomness of a system; it represents a form of energy.

Equilibrium or chemical equilibrium a state of dynamic balance in which the rates of forward and reverse reactions are equal; the state of a system when neither the forward nor the reverse reaction is thermodynamically favored.

Exothermicity the release of heat by a system as a process occurs.

First Law of Thermodynamics the total amount of energy in the universe is constant; (also known as the Law of Conservation of Energy) energy is neither created nor destroyed in ordinary chemical reactions and physical changes.

Free energy, *G* (also known as Gibbs free energy) the thermodynamic state function of a system which indicates the amount of energy available for the system to do useful work.

Free energy change, ΔG the indicator of spontaneity of a process. If ΔG is negative for a process, it is spontaneous.

Gasohol a mixture of gasoline and alcohol which represents a potential fuel or fuel supplement.

Gibbs free energy see *Free energy, G.*

Hess' Law of heat summation the enthalpy change of a reaction is the same whether it occurs in one step or a series of steps.

Internal energy, *E* all forms of energy associated with a specific amount of a substance.

Pressure-volume work work done by a gas when it expands against an external pressure or work done on a system as gases are compressed or consumed in the presence of an external pressure.

Second Law of Thermodynamics the universe tends toward a state of greater disorder in spontaneous processes.

Spontaneity of a process, its property of being energetically favorable and therefore capable of proceeding in the forward direction (but not necessarily at an observable rate).

Standard entropy, S^0 (of a substance) the absolute entropy of a substance in its standard state at 298 K.

Standard molar enthalpy of formation, ΔH_f^0 the amount of heat absorbed in the formation of one mole of a substance in a specified state from its elements in their standard states.

State function a variable that defines the state of a system; a function that is independent of the pathway by which a process occurs.

Surroundings everything in the environment of the system.

System the substances of interest in a process; the part of the universe under investigation.

Thermodynamics the study of the energy transfers accompanying physical and chemical processes.

Thermodynamic state of a system a set of conditions that completely specifies all of the properties of the system.

Thermodynamic standard state of a substance its most stable state under one atmosphere pressure and at some specific temperature (usually 25°C unless otherwise specified).

Third Law of Thermodynamics the entropy of a hypothetical pure, perfect, crystalline substance at absolute zero temperature is zero.

Universe the system plus the surroundings.

Water equivalent of a calorimeter, the amount of water that would absorb the same amount of heat as the calorimeter per degree temperature increase.

Work the application of a force through a distance; for chemical reactions at constant pressure, it is the pressure times the change in volume $P(\Delta V)$.

Exercises

General Concepts

1. State precisely the meaning of each of the following terms. You may need to review Chapter 1 to refresh your memory concerning terms introduced there.
   (a) energy
   (b) kinetic energy
   (c) potential energy
   (d) calorie
   (e) joule
   (f) system
   (g) surroundings
   (h) thermodynamic state of system
   (i) work
   (j) calorimeter water equivalent
2. Distinguish between spontaneous and nonspontaneous processes.
3. What can be said about the sign of ΔH^0 for (a) a spontaneous process and (b) a nonspontaneous process?
4. Distinguish between endothermic and exothermic processes. If we know that a reaction is exothermic in one direction, what can be said about the reaction in the reverse direction?
5. What is entropy? What do we mean when we say that an increase in entropy favors the spontaneity of a process?
6. What is a state function? Would Hess' Law be a law if enthalpy were not a state function?
7. Which of the following are examples of state functions? (a) your bank balance, (b) your mass, (c) your weight, (d) the heat lost by perspiration during a climb up a mountain along a fixed path.
8. (a) Distinguish between ΔH and ΔH^0 for a reaction.
   (b) Distinguish between ΔH^0 and ΔH_f^0.
9. Can we calculate the total energy content at 25°C of one liter of gasoline of a specified composition? If so, how?
10. Can we calculate the heat of combustion at 25°C of one liter of gasoline of a specified composition? If so, how?
11. Both of the following reactions are exothermic. Which one is more exothermic and why?

$$H_2\,(g) + Br_2\,(g) \longrightarrow 2HBr\,(g)$$

$$H_2\,(g) + Br_2\,(\ell) \longrightarrow 2HBr\,(g)$$

12. According to the First Law of Thermodynamics, the total amount of energy in the universe is constant. Why, then, do we say that we are experiencing a declining supply of energy?

Internal Energy and Changes in Internal Energy

13. What is internal energy? What is ΔE? What are the sign conventions for q, the amount of heat added to or removed from a system?
14. Show that $P\Delta V$ is equal to work, and then that it is proportional to the change in number of moles of gases in a process at a particular absolute temperature.

15. A sample of a gas is heated by the addition of 4000 kJ of heat. (a) If the volume remains constant, what is the change in internal energy, ΔE, of this sample? (b) Suppose that in addition to putting heat into the sample, we also do 1000 kJ of work on the sample. What is the change in internal energy, ΔE, of the sample? (c) Suppose, instead, that as the original sample is heated, it is allowed to expand against the atmosphere so that it does 6000 kJ of work on its surroundings. What is the change in internal energy, ΔE, of the sample?
16. Calculate q, w, and ΔE for the vaporization of 1.00 g of liquid ethanol (C_2H_5OH) at 1.00 atm at 78.0°C, to form gaseous ethanol at 1.00 atm at 78.0°C. Make the following simplifying assumptions: (a) the density of liquid ethanol at 78.0°C is 0.789 g/mL and (b) gaseous ethanol is adequately described by the ideal gas equation. The heat of vaporization of ethanol is 854 J/g.
17. Assuming the gases are ideal, calculate the amount of work done (in joules) in each of the following reactions. In each case, is the work done *on* or *by* the system?
    (a) A reaction in the Mond process for purifying nickel which involves formation of the volatile gas, nickel(0) tetracarbonyl, at 50—100°C. Assume one mole of nickel is used and a constant temperature of 75°C is maintained.

    $$Ni\ (s) + 4CO\ (g) \longrightarrow Ni(CO)_4\ (g)$$

    (b) The conversion of one mole of brown nitrogen dioxide into colorless dinitrogen tetroxide at 30.0°C:

    $$2NO_2\ (g) \longrightarrow N_2O_4\ (g)$$

    (c) The decomposition of one mole of an air pollutant, nitric oxide, at 300°C:

    $$2NO\ (g) \longrightarrow N_2\ (g) + O_2\ (g)$$

Calorimetry

18. A strip of magnesium metal having a mass of 1.22 g dissolves in 100 mL of 6.02 *M* HCl, which has a specific gravity of 1.10. The hydrochloric acid is initially at 23.0°C, and the resulting solution reaches a final temperature of 45.5°C. The heat capacity of the calorimeter in which the reaction occurs is 562 J/°C. Calculate ΔH for the reaction [per mole of Mg (s) consumed] under the conditions of the experiment, assuming the specific heat of the final solution is the same as that for water, 4.18 J/g·°C.

    $$Mg\ (s) + 2HCl\ (aq) \longrightarrow MgCl_2\ (aq) + H_2\ (g)$$

19. A coffee-cup calorimeter having a heat capacity of 472 J/°C is used to measure the heat evolved when the following aqueous solutions, both initially at 22.6°C, are mixed: 100 g of solution containing 6.62 g of lead(II) nitrate, $Pb(NO_3)_2$, and 100 g of solution containing 6.00 g of sodium iodide, NaI. The final temperature is 24.2°C. Assume that the specific heat of the mixture is the same as that for water, 4.18 J/g°C. The reaction that occurs is

    $$Pb(NO_3)_2\ (aq) + 2NaI\ (aq) \longrightarrow PbI_2\ (s) + 2NaNO_3\ (aq)$$

    (a) Calculate the heat evolved in the reaction.
    (b) Calculate the ΔH for the reaction [per mole of $Pb(NO_3)_2$ (aq) consumed] under the conditions of the experiment.
20. What is a bomb calorimeter? How do bomb calorimeters give us useful data?
21. A 2.00-g sample of hydrazine, N_2H_4, is burned in a bomb calorimeter that contains 7.00×10^3 g of H_2O and the temperature increases from 25.00°C to 26.17°C. The water equivalent of the calorimeter is known to be 9.00×10^2 g. Calculate ΔE for the combustion of N_2H_4 in kJ/g and in kJ/mol.
22. The combustion of 1.048 grams of benzene, C_6H_6 (ℓ), in a calorimeter compartment surrounded by 826 grams of water raised the temperature of the water from 23.640° to 33.700°C. The calorimeter water equivalent is 216 g of H_2O.
    (a) Write the balanced equation for the combustion reaction, assuming that CO_2 (g) and H_2O (ℓ) are the only products.

(b) Use the calorimetric data to calculate ΔE for the combustion of benzene in kJ/g and in kJ/mol.

Enthalpy and Changes in Enthalpy

23. How do enthalpy and internal energy differ?
24. When 1.00 mole of ice melts at 0°C and a constant pressure of 1.00 atm, 6.02 kJ of heat are absorbed by the system. The molar volumes (the volume occupied by 1 mole) of ice and water are 0.0196 and 0.0180 liter, respectively. Calculate ΔH and ΔE. Comment on the value(s) of ΔH and ΔE. Can you make a generalization on the relative values of ΔH and ΔE for processes in which only liquids and solids are involved?
25. Our bodies convert chemical bond energy from compounds such as carbohydrates into useful heat energy by oxidation. The combustion of cane sugar or sucrose (a carbohydrate) is shown below. Combustion is rapid oxidation accompanied by the liberation of heat and light. The same amount of energy is released in slow oxidation.

$$C_{12}H_{22}O_{11}\ (s) + 12O_2\ (g) \longrightarrow 12CO_2\ (g) + 11H_2O\ (\ell)$$

(a) Calculate the work done in the oxidation of one mole of sucrose at 25°C and one atmosphere pressure.
(b) Use data from Appendix K to calculate ΔH^0_{298} for the combustion of sucrose.
(c) Calculate ΔE^0 for the oxidation of one mole of sucrose.
(d) Most hydrocarbons and carbohydrates (compounds containing C, H, and O) undergo exothermic and thermodynamically spontaneous oxidation. Since a large portion of our bodies is made up of carbohydrates, how can our bodies remain intact when they are in constant contact with oxygen?

26. (a) Use data from Appendix K to calculate ΔH^0 for the combustion of benzene, C_6H_6, at 25°C. The products are CO_2 (g) and H_2O (ℓ). (b) Compare the value of ΔH^0 with the value of ΔE calculated in Exercise 22.
27. In Exercise 25 you found that $\Delta E^0 = \Delta H^0$ for the combustion of sucrose. In Exercises 22 and 26 you found that $\Delta E^0 \neq \Delta H^0$ for the combustion of benzene. Why?

Standard Molar Enthalpy of Formation, ΔH^0_f

28. Explain the meaning of the term "thermochemical standard state of a substance."
29. Explain the significance of each word in the term "standard molar enthalpy of formation."
30. From the data in Appendix K, determine the form that represents the standard state for each of the following elements: (a) fluorine, (b) iron, (c) carbon, (d) sulfur, (e) hydrogen, (f) mercury, (g) bromine.
31. Write thermochemical equations that describe the formation of the following compounds from their elements in their standard states:
(a) PCl_5 (g)
(b) N_2O_4 (g)
(c) NH_4NO_3 (s)
32. Comment on the statement: "Since ΔH^0_f is zero for all elements in their standard states, their heat contents are all zero kilojoules per mole."

Thermochemical Equations and Hess' Law

33. State Hess' Law of heat summation.
34. Calculate the standard molar enthalpy change, ΔH^0, at 25°C for each of the following reactions:
(a) $SiO_2\ (s) + Na_2CO_3\ (s) \longrightarrow Na_2SiO_3\ (s) + CO_2\ (g)$
(b) $H_2SiF_6\ (aq) \longrightarrow 2HF\ (aq) + SiF_4\ (g)$
(c) $2MgS\ (s) + 3O_2\ (g) \longrightarrow 2MgO\ (s) + 2SO_2\ (g)$
35. Calculate ΔH^0 at 25°C for the following reactions:
(a) $2NO\ (g) + H_2\ (g) \longrightarrow N_2O\ (g) + H_2O\ (g)$
(b) $N_2O\ (g) + H_2\ (g) \longrightarrow N_2\ (g) + H_2O\ (g)$
36. A common laboratory method for generating O_2 (g) is the thermal decomposition of potassium chlorate, $KClO_3$ (s) (in the presence of a catalyst). Show that this reaction is exothermic:

$$2KClO_3\ (s) \longrightarrow 2KCl\ (s) + 3O_2\ (g)$$

37. Calculate ΔH^0 at 25°C, using data in Appendix K, for the preparation of hydrogen fluoride by the reaction of calcium fluoride with a nonvolatile acid, sulfuric acid:

$$CaF_2\ (s) + H_2SO_4\ (aq) \longrightarrow CaSO_4\ (s) + 2HF\ (g)$$

38. Calculate ΔH^0 at 25°C, using data in Appendix K, for the preparation of the World War I poison gas, phosgene, from carbon monoxide and chlorine:

$$CO\ (g) + Cl_2\ (g) \longrightarrow COCl_2\ (g)$$

39. Calculate ΔH^0 at 25°C, using data in Appendix K, for the industrial roasting of zinc sulfide ore prior to smelting (formation of the metal):

$$2ZnS\ (s) + 3O_2\ (g) \longrightarrow 2ZnO\ (s) + 2SO_2\ (g)$$

40. The equation for a reaction used to prepare ultrapure silicon for making semiconductors follows. Calculate ΔH^0 at 25°C for this reaction, using data in Appendix K:

$$SiBr_4\ (\ell) + 4Na\ (s) \longrightarrow 4NaBr\ (s) + Si\ (s)$$

41. Using data in Appendix K, calculate the standard heat of vaporization, ΔH^0_{vap}, at 298 K for tin(IV) chloride, $SnCl_4$, in kilojoules per mole and joules per gram:

$$SnCl_4\ (\ell) \longrightarrow SnCl_4\ (g)$$

42. Calculate the amount of heat released, in kilojoules, in the reaction of 14.2 g of magnesium with excess chlorine to produce magnesium chloride under standard thermodynamic conditions:

$$Mg\ (s) + Cl_2\ (g) \longrightarrow MgCl_2\ (s)$$

43. Calculate the amount of heat released, in kilojoules, in the complete oxidation of 13.2 g of aluminum at 25°C and one atmosphere pressure to form aluminum oxide, the protective coating on aluminum doors and windows.

$$4Al\ (s) + 3O_2\ (g) \longrightarrow 2Al_2O_3\ (s)$$

44. Calculate the amount of heat released in the roasting (heating in the presence of oxygen) of:
(a) 1.35 grams of iron pyrite, FeS_2
(b) 1.35 tons of iron pyrite

$$4FeS_2\ (s) + 11O_2\ (g) \longrightarrow 2Fe_2O_3\ (s) + 8SO_2\ (g)$$

45. The air pollutant sulfur dioxide can be partially removed from roaster stack gases (see Exercise 44) and converted to sulfur trioxide, the acid anhydride of commercially important sulfuric acid. Calculate the standard enthalpy change for the reaction of 25.0 grams of SO_2 with O_2 (g):

$$2SO_2\ (g) + O_2\ (g) \longrightarrow 2SO_3\ (g)$$

46. Calculate the standard enthalpy change at 25°C accompanying the reaction 39.2 g of sulfur trioxide with a stoichiometric quantity of water:

$$SO_3\ (g) + H_2O\ (\ell) \longrightarrow H_2SO_4\ (\ell)$$

47. Solid phosphorus exists in two allotropic forms, red and white. Both react with chlorine to produce phosphorus trichloride, a colorless liquid that fumes in air:

$$P_4\ (white) + 6Cl_2\ (g) \longrightarrow 4PCl_3\ (\ell)$$
$$\Delta H^0 = -1.15 \times 10^3\ kJ$$

$$P_4\ (red) + 6Cl_2\ (g) \longrightarrow 4PCl_3\ (\ell)$$
$$\Delta H^0 = -1.23 \times 10^3\ kJ$$

Calculate the standard enthalpy of reaction at 25°C for the reaction by which red phosphorus is converted to white phosphorus:

$$P_4\ (red) \longrightarrow P_4\ (white)$$

48. From the following equations and ΔH^0 values,

$$H_2\ (g) + Cl_2\ (g) \longrightarrow 2HCl\ (g)$$
$$\Delta H^0 = -185\ kJ$$

$$2H_2\ (g) + O_2\ (g) \longrightarrow 2H_2O\ (g)$$
$$\Delta H^0 = -483.7\ kJ$$

calculate ΔH^0 for the following reaction:

$$4HCl\ (g) + O_2\ (g) \longrightarrow 2Cl_2\ (g) + 2H_2O\ (g)$$

49. Given the following equations and ΔH^0 values at 25°C:

	ΔH^0
$H_3BO_3\ (aq) \longrightarrow HBO_2\ (aq) + H_2O\ (\ell)$	-0.02 kJ

$H_2B_4O_7\ (s) + H_2O\ (\ell) \longrightarrow 4HBO_2\ (aq)$ $\quad -11.3$ kJ

$H_2B_4O_7\ (s) \longrightarrow 2B_2O_3\ (s) + H_2O\ (\ell)$ $\quad 17.5$ kJ

calculate ΔH^0 for the following reaction:

$$2H_3BO_3\ (aq) \longrightarrow B_2O_3\ (s) + 3H_2O\ (\ell)$$

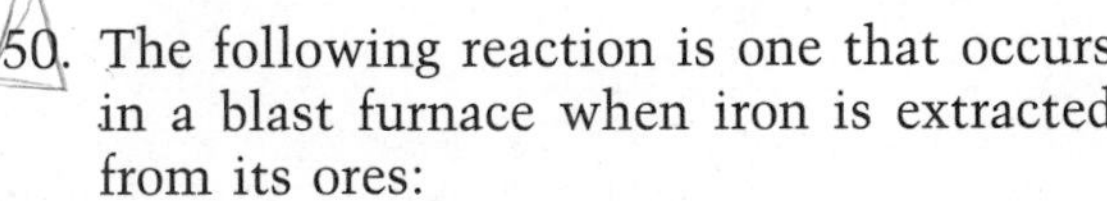

50. The following reaction is one that occurs in a blast furnace when iron is extracted from its ores:

$$Fe_2O_3\ (s) + 3CO\ (g) \longrightarrow 2Fe\ (s) + 3CO_2\ (g)$$

Evaluate ΔH^0 for this reaction at 298 K given the following enthalpy changes at 298 K:

	ΔH^0
$3Fe_2O_3\ (s) + CO\ (g) \longrightarrow 2Fe_3O_4\ (s) + CO_2\ (g)$	-46.4 kJ
$FeO\ (s) + CO\ (g) \longrightarrow Fe\ (s) + CO_2\ (g)$	9.0 kJ
$Fe_3O_4\ (s) + CO\ (g) \longrightarrow 3FeO\ (s) + CO_2\ (g)$	-41.0 kJ

51. What do the last three equations of Exercise 50 have in common?

52. The reaction that occurs during the discharge of a typical automobile battery, called a lead storage battery, is

$$Pb\ (s) + PbO_2\ (s) + 2H_2SO_4\ (\ell) \longrightarrow 2PbSO_4\ (s) + 2H_2O\ (\ell)$$

Evaluate ΔH^0 at 25°C for this reaction from the following information:

	ΔH^0
$SO_3\ (g) + H_2O\ (\ell) \longrightarrow H_2SO_4\ (\ell)$	-133 kJ
$Pb\ (s) + PbO_2\ (s) + 2SO_3\ (g) \longrightarrow 2PbSO_4\ (s)$	-775 kJ

53. "Water gas" is an important industrial fuel consisting of a mixture of carbon monoxide and hydrogen. It is produced by the action of superheated steam on red-hot coke:

$$\underset{\text{coke}}{C\ (s)} + H_2O\ (g) + 131.3\text{ kJ} \xrightarrow{1000°C} \underbrace{CO\ (g) + H_2\ (g)}_{\text{water gas}}$$

Given that ΔH_f^0 for $H_2O\ (g) = -241.8$ kJ/mol, and assuming that coke has the same molar enthalpy of formation as graphite, calculate ΔH_f^0 for CO.

54. Methanol, CH_3OH, is a possible fuel or fuel supplement of the future (in gasohol, for example). It can be prepared industrially by the reaction of carbon monoxide with hydrogen ("water gas" with excess H_2) at high temperatures and pressures in the presence of appropriate catalysts:

$$CO\ (g) + 2H_2\ (g) \xrightarrow[ZnO]{Cr_2O_3} CH_3OH\ (\ell)$$

Evaluate ΔH^0 for this reaction, given the following:

$2C\ (graphite) + O_2\ (g) \longrightarrow 2CO\ (g)$
$\Delta H^0 = -221$ kJ

$C\ (graphite) + O_2\ (g) \longrightarrow CO_2\ (g)$
$\Delta H^0 = -393.5$ kJ

$2CH_3OH\ (\ell) + 3O_2\ (g) \longrightarrow 2CO_2\ (g) + 4H_2O\ (\ell)$
$\Delta H^0 = -1453$ kJ

$2H_2\ (g) + O_2\ (g) \longrightarrow 2H_2O\ (\ell)$
$\Delta H^0 = -571.6$ kJ

55. Welders use acetylene, C_2H_2, torches to attain very high temperatures. Calculate ΔH^0 at 25°C for the complete combustion of one mole of acetylene:

$$2C_2H_2\ (g) + 5O_2\ (g) \longrightarrow 4CO_2\ (g) + 2H_2O\ (\ell)$$

given the following ΔH^0 values at 25°C:

	ΔH^0
$H_2\ (g) + \frac{1}{2}O_2\ (g) \longrightarrow H_2O\ (\ell)$	-285.8 kJ
$C\ (graphite) + O_2\ (g) \longrightarrow CO_2\ (g)$	-393.5 kJ
$2C\ (graphite) + H_2\ (g) \longrightarrow C_2H_2\ (g)$	226.7 kJ

56. From the following data, calculate the standard molar heat of formation of gaseous PCl_5:

$P_4\ (white) + 6Cl_2\ (g) \longrightarrow 4PCl_3\ (\ell)$
$\Delta H^0 = -1.15 \times 10^3$ kJ

PCl_3 (ℓ) + Cl_2 (g) ⟶ PCl_5 (g)
$\Delta H^0 = -111$ kJ

57. Given the following, at 25°C:

P_4O_{10} (s) + $4HNO_3$ (ℓ) ⟶
$4HPO_3$ (s) + $2N_2O_5$ (g)
$\Delta H^0 = -102.3$ kJ

ΔH_f^0 for P_4O_{10} (s) is −2984 kJ/mol, that for HNO_3 (ℓ) is −174.1 kJ/mol, and that for N_2O_5 (g) is 11 kJ/mol, calculate ΔH_f^0 for HPO_3 (s) at 25°C.

58. Silicon carbide or carborundum, SiC, is one of the hardest substances known and is used as an abrasive. It has the structure of diamond with half of the carbons replaced by silicon. It is prepared industrially by reduction of sand (SiO_2) with coke (C) at 3000°C in an electric furnace:

SiO_2 (s) + 3C (s) ⟶ SiC (s) + 2CO (g)

Given that ΔH^0 for this reaction is 624.6 kJ, and that ΔH_f^0 for SiO_2 (s) and CO (g) are −910.9 kJ/mol and −110.5 kJ/mol, respectively, calculate ΔH_f^0 for silicon carbide.

59. The tungsten used in filaments for light bulbs can be prepared from tungsten(VI) oxide by reduction with hydrogen at 1200°C:

114.9 kJ + WO_3 (s) + $3H_2$ (g) ⟶
W (s) + $3H_2O$ (g)

The ΔH_f^0 value for H_2O (g) is −241.8 kJ/mol. What is ΔH_f^0 for WO_3 (s)?

Bond Energies

60. How is the heat released or absorbed in a *gas phase reaction* related to bond energies of products and reactants?
61. Write equations for reactions that illustrate the definition of (a) bond energy and (b) average bond energy.
62. Suggest a reason for the fact that different amounts of energy are required for the successive removal of the three hydrogen atoms of an ammonia molecule, even though all N—H bonds in ammonia are equivalent.
63. Suggest why the N—H bonds in different compounds such as ammonia, NH_3, methylamine, CH_3NH_2, and ethylamine, $C_2H_5NH_2$, have slightly different bond energies.
64. The structural formula for benzene, C_6H_6, can be represented by two resonance hybrid forms:

```
        H                       H
        |                       |
  H     C      H          H     C      H
    \ //  \  /              \ /  \\  /
     C     C                 C     C
     |     ||      ←→        ||    |
     C     C                 C     C
    / \\  /  \              /  \ //  \
  H     C      H          H     C      H
        |                       |
        H                       H
```

How would you expect the carbon-carbon bond energies (and lengths) to compare with those of ethane and ethene or ethylene,

```
   H  H           H         H
   |  |            \       /
H—C—C—H,            C═C          ?
   |  |            /       \
   H  H           H         H
```

65. (a) Calculate the carbon-oxygen bond energies in CO_2 and CO. For CO_2 (g), $\Delta H_f^0 = -393.5$ kJ/mol; for CO (g), $\Delta H_f^0 = -110.5$ kJ/mol. For O (g) and C (g), $\Delta H_f^0 = 249.2$ kJ/mol and 716.7 kJ/mol, respectively.
(b) Compare these bond energies with those listed in Tables 13–2 and 13–3. Would you classify the bonds as single, double, or triple covalent bonds?
66. Using data in Appendix K, calculate the average N—H bond energy in NH_3.
67. Using data in Appendix K, calculate the H—F bond energy in HF (g).
68. Using data in Appendix K, calculate the average H—O bond energy in H_2O (g).
69. Hess' Law states that

$$\Delta H^0_{rxn} = \Sigma\, n\Delta H^0_{f\,products} - \Sigma\, n\Delta H^0_{f\,reactants}$$

and the relationship between ΔH^0_{rxn} and bond energies for a *gas phase reaction* is:

$$\Delta H^0_{rxn} = \Sigma\text{ Bond Energies}_{reactants} - \Sigma\text{ Bond Energies}_{products}$$

It is *not* true, in general, that ΔH_f^0 for a substance is equal to the negative of the sum of the bond energies of the substance? Why?

70. Use the bond energies in Table 13–2 to estimate the heat of reaction at 298 K for

$$ICl_3\ (g) + 2H_2\ (g) \longrightarrow HI\ (g) + 3HCl\ (g)$$

71. Use the bond energies in Tables 13–2 and 13–3 to estimate the heat of reaction at 298 K for

$$2ClF_3\ (g) + 2O_2\ (g) \longrightarrow Cl_2O\ (g) + 3OF_2\ (g)$$

72. Ethylamine undergoes an endothermic gas phase dissociation to produce ethylene (or ethene) and ammonia:

$$54.68\ kJ + H_3C{-}CH_2{-}NH_2 \xrightarrow{\Delta} H_2C{=}CH_2 + NH_3$$

Given the following average bond energies per mole of bonds: C—H = 414 kJ, C—C = 347 kJ, C=C = 611 kJ, N—H = 389 kJ. Calculate the C—N bond energy. Compare this with the value in Table 13–2.

Crystal Lattice Energy and Born-Haber Cycles

73. Determine the crystal lattice energy for LiF (s), given the following: sublimation energy for Li = 155 kJ/mol, ΔH_f^0 for F (g) = 78.99 kJ/mol, the first ionization energy of Li = 520 kJ/mol, the electron affinity of fluorine = −322 kJ/mol, and the standard molar enthalpy of formation of LiF (s) = −589.5 kJ/mol.

74. From the following information determine the crystal lattice energy of KI: sublimation energy for potassium = 90 kJ/mol, the first ionization energy for potassium = 419 kJ/mol, the sublimation energy for iodine = 62 kJ/mol, the dissociation energy for iodine = 151 kJ mol, and the electron affinity for iodine = −295 kJ/mol.

75. Construct a Born-Haber cycle for the production of calcium oxide, CaO (s) from Ca (s) and O_2 (g). It is not necessary to include the values of enthalpy changes for the various steps.

76. Calculate the crystal lattice energy for $CaBr_2$ from the following information: ΔH_{subl} for calcium = 193 kJ/mol, ΔH_{ie1} for calcium = 590 kJ/mol, ΔH_{ie2} for calcium = 1145 kJ/mol, ΔH_{vap} for Br_2 is 315 kJ/mol [Br_2 (ℓ) is standard state for bromine], ΔH_{diss} for Br_2 (g) = 193 kJ/mol, ΔH_{ea} for bromine = −324 kJ/mol, and ΔH_f^0 for $CaBr_2$ (g) = −675 kJ/mol.

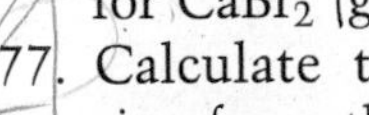

77. Calculate the electron affinity for chlorine from the following information: ΔH_f^0 for $MgCl_2$ (s) = −642 kJ/mol, ΔH_{subl} for Mg = 151 kJ/mol, the first and second ionization energies for Mg are 738 and 1451 kJ/mol, ΔH_{diss} for Cl_2 = 243 kJ/mol, and the crystal lattice energy for $MgCl_2$ (s) = −2529 kJ/mol.

78. The value of ΔH_{ie1} is *always* positive, yet the ionization of a metal atom to form a metal ion with a noble gas electron configuration, $M(g) \rightarrow M^{n+}\ (g) + ne^-$, occurs readily in the formation of many ionic solids from their elements. What is the primary factor that makes the overall process favorable?

Entropy and Entropy Changes

79. State the Second Law of Thermodynamics. We cannot use $\Delta S_{universe}$ directly as a measure of the spontaneity of a reaction. Why?

80. State the Third Law of Thermodynamics. What does it mean?

81. Explain why the term $T\Delta S$ is an energy term.

82. Why is it impossible for any substance to have an absolute entropy of zero at temperatures above 0 K?

83. Describe the characteristic of a reaction that its entropy change measures.

84. Which of the following processes are accompanied by an increase in entropy?

(a) The freezing of water.

(b) The evaporation of carbon tetrachloride, CCl_4.
(c) The sublimation of iodine,

$$I_2\ (s) \xrightarrow{\Delta} I_2\ (g)$$

(d) The precipitation of white silver chloride, AgCl, from a solution containing silver ions and chloride ions.
(e) The reaction $PCl_5\ (g) \rightarrow PCl_3\ (g) + Cl_2\ (g)$
(f) The reaction $PCl_3\ (g) + Cl_2\ (g) \rightarrow PCl_5\ (g)$
(g) Thirty-five pennies are removed from a bag and placed heads up on a table.
(h) The pennies of (g) are swept off the table and back into the bag.

85. Use data from Appendix K to calculate the value of ΔS^0_{298} for each reaction in Exercise 17. Compare the signs and magnitudes for these ΔS^0_{298} values and explain your observations.
86. List six kinds of processes that result in increases in entropy and explain why each does.
87. Explain why ΔS^0 may be referred to as a contributor to spontaneity.

Free Energy Changes and Spontaneity

88. What are the two factors that favor spontaneity of a process?
89. What is free energy? Change in free energy?
90. Most spontaneous reactions are exothermic, but some are not. Explain.
91. Explain how the signs and magnitudes of ΔH^0 and ΔS^0 are related to the spontaneity of a process and how they affect it.
92. Is it true that the change in the available energy of a system is equal to the change in the total energy of the system plus the change in energy due to the change in the order/disorder of the system? Why?
93. Under a specific set of conditions, why must the reverse of a spontaneous reaction be nonspontaneous?
94. It is not necessary to know the absolute values of free energies, *G*, and enthalpies, *H*, of substances in order to predict the spontaneity of reactions involving the substances. Why?
95. Using values of standard free energy of formation, ΔG^0_f, from Appendix K, calculate the standard free energy change for each of the following reactions.
(a) $2K\ (s) + 2H_2O\ (\ell) \longrightarrow 2KOH\ (s) + H_2\ (g)$
(b) $3Fe\ (s) + 4H_2O\ (\ell) \longrightarrow Fe_3O_4\ (s) + 4H_2\ (g)$
96. Make the same calculations as in Exercise 95, using values of standard enthalpy of formation and absolute entropy instead of values of ΔG^0_f.
97. Use Appendix K to calculate ΔG^0_f at 298 K for each of the following reactions. Which ones are spontaneous at 298 K?
(a) $CaH_2\ (s) + 2H_2O\ (\ell) \longrightarrow Ca(OH)_2\ (s) + 2H_2\ (g)$
(b) $Na_2CO_3\ (s) + 2HCl\ (g) \longrightarrow 2NaCl\ (aq) + H_2O\ (\ell) + CO_2\ (g)$
(c) $CaO\ (s) + H_2O\ (\ell) \longrightarrow Ca(OH)_2\ (aq)$
(d) $2C_2H_2\ (g) + 5O_2\ (g) \longrightarrow 4CO_2\ (g) + 2H_2O\ (\ell)$
(e) $C_2H_4\ (g) + H_2O\ (g) \longrightarrow C_2H_5OH\ (\ell)$

Temperature Range of Spontaneity

98. What are the effects of temperature changes on the spontaneity of reactions for which:
(a) ΔH is negative and ΔS is positive?
(b) ΔH is negative and ΔS is negative?
(c) ΔH is positive and ΔS is positive?
(d) ΔH is positive and ΔS is negative?
99. Determine the temperature ranges over which the following reactions are spontaneous.
(a) The reaction by which sulfuric acid droplets from polluted air convert water-insoluble limestone or marble (calcium carbonate) to slightly soluble calcium sulfate, which is slowly washed away by rain:

$$CaCO_3\ (s) + H_2SO_4\ (\ell) \longrightarrow CaSO_4\ (s) + H_2O\ (\ell) + CO_2\ (g)$$

(b) The reaction by which Antoine Lavoisier achieved the first laboratory preparation of oxygen in the late eighteenth century: the thermal decomposition of the red-orange powder, mercury(II) oxide, to oxygen and the silvery liquid metal, mercury:

$$2HgO\ (s) \longrightarrow 2Hg\ (\ell) + O_2\ (g)$$

(c) The reaction of coke (carbon) with carbon dioxide to form the reducing agent, carbon monoxide, which is used to reduce some metal ores to the metals:

$$CO_2\ (g) + C\ (s) \longrightarrow 2CO\ (g)$$

(d) The reverse of the reaction by which iron rusts:

$$2Fe_2O_3\ (s) \longrightarrow 4Fe\ (s) + 3O_2\ (g)$$

100. (a) Estimate the temperature range over which the reaction below is spontaneous:

$$2CO\ (g) + O_2\ (g) \longrightarrow 2CO_2\ (g)$$

(b) In light of your answer to (a), how do you explain the fact that carbon monoxide from auto exhausts is a major health hazard?

101. Estimate the temperature ranges over which the following reactions are spontaneous.
(a) $2Al\ (s) + 3Cl_2\ (g) \longrightarrow 2AlCl_3\ (s)$
(b) $2NOCl\ (g) \longrightarrow 2NO\ (g) + Cl_2\ (g)$
(c) $4NO\ (g) + 6H_2O\ (g) \longrightarrow 4NH_3\ (g) + 5O_2\ (g)$
(d) $2PH_3\ (g) \longrightarrow 3H_2\ (g) + 2P\ (g)$

102. (a) Estimate the boiling point of water (the temperature at which liquid and vapor are in equilibrium with each other) at one atmosphere pressure, using Appendix K.
(b) Compare the temperature obtained with the known boiling point of water. Can you explain the discrepancy?

103. Estimate the normal boiling point of titanium(IV) chloride, $TiCl_4$, using Appendix K.

104. Estimate the sublimation temperature (solid to vapor) of a dark violet solid, iodine, I_2, using the data of Appendix K. Sublimation and subsequent deposition onto a cold surface is a common method of purification of I_2 and other solids that sublime readily.

Unscramble Them

105. When is it true that $\Delta S = \dfrac{\Delta H}{T}$?

106. Dissociation reactions are those in which molecules break apart. Why do high temperatures favor the spontaneity of most dissociation reactions?

107. Determine the standard entropy changes at 298 K for each of the following reactions from ΔG_f^0 and ΔH_f^0 values:
(a) The careful thermal decomposition of ammonium nitrate, a laboratory preparation of nitrous oxide (dinitrogen oxide), also known as laughing gas, a mild anesthetic often used in dental work:

$$NH_4NO_3\ (s) \xrightarrow{\Delta} 2H_2O\ (g) + N_2O\ (g)$$

(b) Reaction (a) can become dangerous if large (bulk) quantities of ammonium nitrate are used or if the salt is heated too strongly. The reaction below was partially responsible for the disastrous explosions that occurred in the port of Texas City, Texas in 1947:

$$2NH_4NO_3\ (s) \longrightarrow 4H_2O\ (g) + 2N_2\ (g) + O_2\ (g)$$

108. Using your chemical intuition, decide which of the following reactions are spontaneous and occur at an observable rate at room temperature and atmospheric pressure.
(a) $H_2O\ (\ell) \longrightarrow H_2O\ (s)$
(b) $2O\ (g) \longrightarrow O_2\ (g)$
(c) $2H_2O\ (\ell) \longrightarrow 2H_2\ (g) + O_2\ (g)$
(d) $N_2\ (g) + O_2\ (g) \longrightarrow 2NO\ (g)$
(e) $NaOH\ (s) + \frac{1}{2}H_2\ (g) \longrightarrow Na\ (s) + H_2O\ (\ell)$

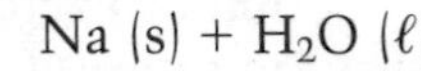

109. Without consulting Appendix K, decide whether or not the following reactions are spontaneous at room temperature and atmospheric pressure. Why must the reactions be spontaneous or nonspontaneous,

as the case may be?

(a) $(NH_4)_2Cr_2O_7$ (s) $\longrightarrow$
Cr_2O_3 (s) + N_2 (g) + $4H_2O$ (g) + 300 kJ

(b) Cs^+ (aq) + F^- (aq) + 8.4 kJ $\longrightarrow$ CsF (aq)

110. (a) Using data in Appendix K, calculate ΔG^0 at 25°C for each of the following reactions, which are utilized in the industrial production of nitric acid, HNO_3, by the oxidation of ammonia (the Ostwald Process).

$4NH_3$ (g) + $5O_2$ (g) $\longrightarrow$ 4NO (g) + $6H_2O$ (g)

2NO (g) + O_2 (g) $\longrightarrow$ $2NO_2$ (g)

$3NO_2$ (g) + H_2O (ℓ) $\longrightarrow$
$2HNO_3$ (ℓ) + NO (g) (recycled)

(b) Calculate the standard enthalpy change accompanying complete conversion of 100 grams of ammonia into nitric acid by this process

111. If half of a given sample of ammonia is used to prepare nitric acid as in Exercise 110, the remainder can be used to neutralize the nitric acid to form ammonium nitrate, an important nitrogen fertilizer.

NH_3 (g) + HNO_3 (ℓ) $\longrightarrow$ NH_4NO_3 (s)

Calculate the standard (a) enthalpy and (b) free energy changes at 25°C for the neutralization of the amount of nitric acid prepared in part (b) of Exercise 110. (c) Is the entropy change positive or negative?

112. Standard entropy changes cannot be measured directly in the laboratory. They are calculated from experimentally obtained values of ΔG^0 and ΔH^0. Calculate ΔS^0 at 298 K for each of the reactions below:

	ΔH^0	ΔG^0
(a) $3NO_2$ (g) + H_2O (ℓ) $\longrightarrow$ $2HNO_3$ (ℓ) + NO (g)	−71.71 kJ	8.28 kJ
(b) OF_2 (g) + H_2O (g) $\longrightarrow$ O_2 (g) + 2HF (g) oxygen difluoride	−323 kJ	−358.4 kJ
(c) CaC_2 (s) + $2H_2O$ (ℓ) $\longrightarrow$ $Ca(OH)_2$ (s) + C_2H_2 (g) calcium carbide; acetylene	−125.4 kJ	−145.4 kJ
(d) 2PbS (s) + $3O_2$ (g) $\longrightarrow$ 2PbO (s) (red) + $2SO_2$ (g)	−830.8 kJ	−780.9 kJ
(e) SnO_2 (s) + 2CO (g) $\longrightarrow$ $2CO_2$ (g) + Sn (s)	14.8 kJ	5.23 kJ

113. Many organic compounds containing nitro (—NO_2) and nitrate (—O—NO_2) groups are explosives. An example is nitroglycerin, which undergoes the following explosive decomposition.

$4C_3H_5(NO_3)_3$ (ℓ) $\longrightarrow$
$12CO_2$ (g) + $10H_2O$ (g) + $6N_2$ (g) + O_2 (g)

(Note the similarity to the explosion reaction of Exercise 107b.) At 298 K this reaction releases about 6.28 kJ per gram of nitroglycerin.

(a) What is ΔH^0 for the reaction in kilojoules per mole of nitroglycerin?

(b) What is ΔH^0 in kilojoules for the reaction as written?

(c) Using Appendix K, calculate ΔH_f^0 for nitroglycerin.

(d) What is suggested by the sign and magnitude of ΔH_f^0 for nitroglycerin?

(e) What can you say about the entropy change of the reaction?

(f) What can you say about the temperature range over which the explosion is spontaneous?

(g) How much work (calories) does the system do on the environment as it expands against the atmosphere if 50.0 grams of nitroglycerin explode at 25°C?

(Most of the damage caused by such an explosion is due to the shock wave resulting from the sudden expansion of the gases produced. The expansion is accelerated by the evolved heat.)

114. Carbon tetrachloride was formerly used widely as a dry cleaning solvent for removal of greasy or oily stains because of its nonpolar character. Its use has been curtailed because it is now a suspected carcinogen. It is still used (with care!) as a reactant or solvent in many organic reactions. It can be prepared by the action of chlorine on chloroform.

$$CHCl_3\ (\ell) + Cl_2\ (g) \longrightarrow CCl_4\ (\ell) + HCl\ (g)$$

(a) Evaluate ΔG^0 at 25°C for this reaction using ΔH_f^0 and S^0 values from Appendix K.
(b) Evaluate ΔG^0 for the same reaction at 500°C when all species are in the gas phase, assuming ΔH^0 and ΔS^0 have the same values as at 25°C.
(c) Evaluate ΔG^0 for the gas phase reaction at 1000°C.
(d) At which of these temperatures is the reaction thermodynamically most favorable?

14 Chemical Kinetics

Chemical kinetics refers to the study of *rates* of chemical reactions and the *mechanisms* by which they occur. The **rate** of a reaction is the *change in concentration* of a product or reactant *per unit time.* The **mechanism** of a reaction is the *pathway* (or series of steps) by which the reaction occurs.

The question of whether a reaction *can* occur is addressed by thermodynamics. The question of whether substantial reaction *will* occur in a certain period of time is addressed by kinetics. If a reaction is thermodynamically unfavorable, it will not occur appreciably under the given conditions. If a reaction is thermodynamically favorable, it can occur but not necessarily at a measurable rate.

This is the major reaction occurring in one's digestive system when an antacid, containing relatively insoluble magnesium hydroxide, neutralizes excess stomach acid. We are, of course, thankful that the reaction is not only spontaneous but also rapid.

The reactions of acids with bases are thermodynamically favorable and occur at very rapid rates. Consider, for example, the reaction of one molar hydrochloric acid with solid magnesium hydroxide. It is thermodynamically spontaneous under standard state conditions, as indicated by the negative ΔG^0 value, and it occurs rapidly

$$2HCl\ (aq) + Mg(OH)_2\ (s) \longrightarrow MgCl_2\ (aq) + 2H_2O\ (\ell) \qquad \Delta G^0 = -97\ kJ$$

The reaction of diamond, a form of solid carbon, with oxygen is also spontaneous

$$C\ (\text{diamond}) + O_2\ (g) \longrightarrow CO_2\ (g) \qquad \Delta G^0 = -396\ \text{kJ}$$

However, we know from experience that diamonds exposed to air do not disappear and form carbon dioxide even over long periods of time. Thus the reaction does not occur at an observable rate at room temperature or at body temperature. The reason for this observation is explained by kinetics, not thermodynamics.

Four factors that have been found to affect reaction rates are (1) nature of reactants, (2) concentrations of reactants, (3) temperature, and (4) catalysis. The effects of each will be discussed and illustrated after sections on the **collision theory** of reaction rates and **transition state theory.** These are complementary theories, consistent with experimental evidence, that provide general descriptions of how reactions occur.

14–1 Collision Theory of Reaction Rates

The fundamental notion of the **collision theory of reaction rates** is that in order for reaction to occur between atoms, ions, or molecules, they must first collide. Increased concentrations of reacting species result in greater numbers of collisions per unit time. However, not all collisions result in reaction, i.e., not all collisions are **effective collisions.** In order for a collision to be effective, the reacting species must (1) possess at least a certain minimum energy necessary to rearrange outer electrons in breaking bonds and forming new ones, and (2) have the proper orientations toward each other at the time of collision. Collision is a necessary, but not sufficient, condition for chemical reaction.

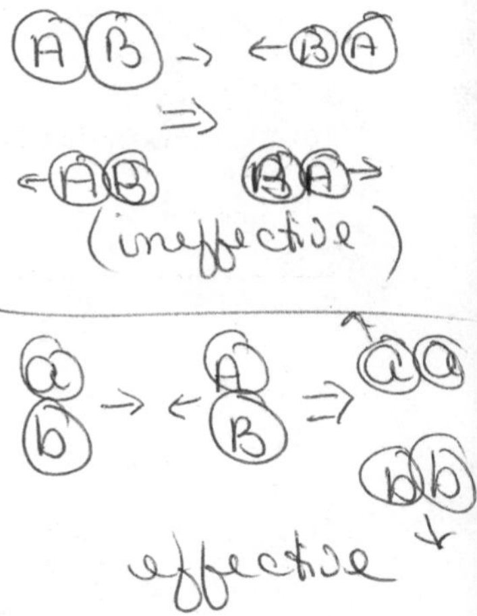

Recall from Chapter 8 that the average kinetic energy of a collection of molecules is proportional to the absolute temperature. At higher temperatures, a greater fraction of the molecules in the collection possess sufficient energy to react. We will discuss this further in Section 14–5.

If colliding molecules have improper orientations, they do not react even though they may possess sufficient energy. Figure 14–1 depicts ineffective collisions between molecules of carbon monoxide and water vapor, each possessing sufficient energy to react according to the equation

$$CO\ (g) + H_2O\ (g) \longrightarrow CO_2\ (g) + H_2\ (g)$$

This is the final reaction in an important commercial method of preparing hydrogen. The carbon monoxide is produced by reaction of methane, CH_4, or some other hydrocarbon, with steam in the presence of catalysts.

The fraction of energetically favorable collisions that have the proper orientation can be increased for many reactions by the introduction of a heterogeneous catalyst, discussed in Section 14–6.

14–2 Transition State Theory

Chemical reactions involve the making and breaking of chemical bonds by shifting electrons. The energy associated with a chemical bond is a form of potential energy, and reactions are accompanied by changes in potential energy. Many reactions occur with release of energy to the surroundings. (Recall that if the energy is released in the form of heat, the reaction is said to be exothermic.) Consider the following hypothetical, one-step exothermic reac-

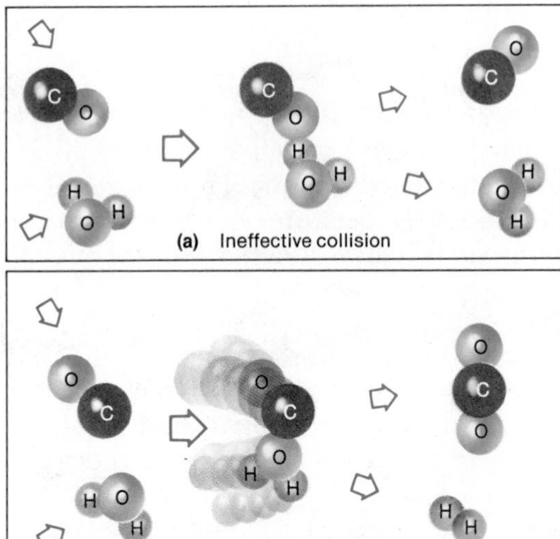

FIGURE 14–1 (a) Ineffective and (b) effective collisions between carbon monoxide and water molecules in the gas phase. The molecules must possess both sufficient energy to react and the proper orientations relative to each other.

tion at a certain temperature

$$A_2 + B_2 \longrightarrow 2AB + \text{energy}$$

Figure 14–2 shows a plot of potential energy versus progress of reaction (time). The ground state energy of the reactants, A_2 and B_2, is higher than the ground state energy of the products, 2AB. The energy released in the reaction is the difference between these two energies, ΔE, which is related to the change in enthalpy or heat content.

According to the **transition state theory,** the reactants pass through a short-lived, high-energy intermediate state, called a *transition state,* before the products are formed.

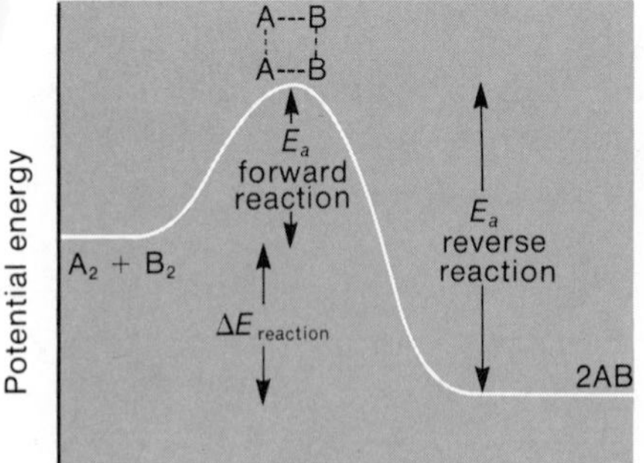

FIGURE 14–2 Potential energy diagram for a reaction that releases energy.

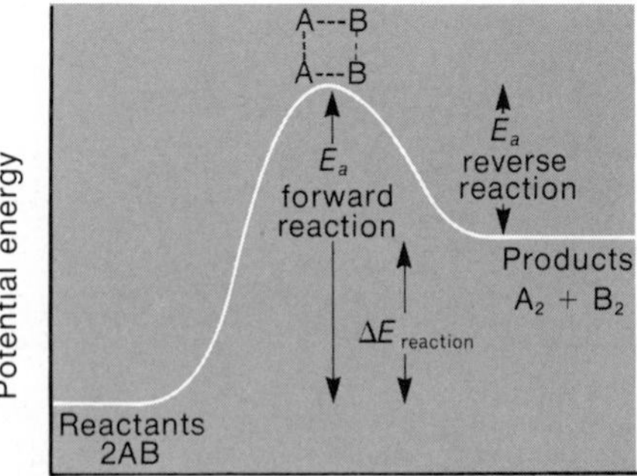

FIGURE 14–3 Potential energy diagram for a reaction that absorbs energy.

partial bonds

A B		A—B		A——B
| + |	⟶	| |	⟶	+
A B		A——B		A——B
reactants		transition state		product
$A_2 + B_2$		A_2B_2		$2AB$

The **activation energy, E_a,** is the energy that must be absorbed by the reactants in their ground states to allow them to reach the transition state. If A_2 and B_2 do not possess the necessary amount of energy, E_a, above their ground states when they collide, no reaction will occur. If they do possess sufficient energy to "climb the energy barrier" to the transition state, the reaction can proceed to completion with the release of an amount of energy equal to the energy of activation *plus* ΔE. Thus, the activation energy must be supplied to the system from its environment, but is subsequently released to the environment along with additional energy in reactions that evolve energy. The net release of energy is ΔE.

The reverse of any such reaction requires a net absorption of energy from its surroundings. (The reaction is endothermic if the energy is absorbed as heat.) A typical example, $2AB \rightarrow A_2 + B_2$, is represented by Figure 14–3. In this case ΔE is the *net* energy *absorbed* by the system from its surroundings as the reaction occurs. Not all the energy of activation is returned to the environment when the products are formed. Note that the activation energy for this reaction is greater than for the (original) forward reaction of Figure 14–2, and the magnitude, but not the sign, of ΔE is the same.

Factors Affecting Reaction Rates

14–3 Nature of Reactants

The rates at which reactions occur depend upon the chemical identities and physical states of substances mixed together. As we have seen, solutions of an acid and a base react as soon as they are mixed, as does metallic sodium when it contacts water. However, metallic calcium reacts slowly with water at room temperature.

The physical states of reacting substances are important in determining their reactivities. White phosphorus and red phosphorus are different solid forms (allotropes) of elemental phosphorus. White phosphorus ignites when exposed to oxygen in the air. By contrast, red phosphorus may be kept in open containers for long periods of time.

The state of subdivision of solids can be crucial in determining reaction rates. Large chunks of most metals do not burn, but many powdered metals with larger surface areas, and consequently more atoms exposed to the oxygen of the air, burn easily. The effects of large surface area on rates of combustion are violently exemplified by the dust explosions that can occur in grain elevators, coal mines, and chemical plants in which large amounts of powdered oxidizable substances are produced.

Samples of dry solid potassium sulfate, K_2SO_4, and dry solid barium nitrate, $Ba(NO_3)_2$, can be mixed with no appreciable reaction occurring over a period of several years. But, if solutions of the two in reasonable concentrations are mixed, a reaction occurs rapidly as evidenced by the immediate formation of a white precipitate, barium sulfate

$$Ba^{2+}\ (aq) + SO_4^{2-}\ (aq) \longrightarrow BaSO_4\ (s)$$

Additionally, it is observed that spontaneous reactions involving ions in solution are generally much more rapid than spontaneous reactions in which bonds must be broken.

14–4 Concentration of Reactants

Consider a hypothetical reaction

$$aA + bB \longrightarrow cC + dD$$

The rate at which the reaction proceeds can be measured in terms of the rate at which one of the reactants disappears, $-\Delta[A]/\Delta t$ or $-\Delta[B]/\Delta t$, **or** the rate at which one of the products appears, $\Delta[C]/\Delta t$ or $\Delta[D]/\Delta t$. The negative sign indicates a decrease in the concentration of reactants A or B in a given time interval, Δt. Of course the rate must be a positive quantity since it relates to the forward (left-to-right) reaction, which consumes A and B. Concentrations are usually expressed in moles per liter and are designated by square brackets, [ ]. Thus, the rates of change are expressed as moles per liter per unit time. The rate of a reaction may be determined by following the change in concentration of any product or reactant that can be detected conveniently by quantitative means.

In reactions involving gases, rates of reaction may be related to rates of change of partial pressures, since pressures and concentrations of gases are directly proportional.

The changes in concentration are related to each other. In order to generate an equality based on the rate of change of concentration of each species, it is necessary to divide each change by its coefficient in the balanced equation for the reaction. The coefficients of the balanced equation (above) tell us, for example, that the [A] decreases a/b times as rapidly as [B] and a/c times as fast as [C] increases.

Rates of *decrease* of conc'n. of reactants — Rates of *increase* of conc'n. of products

$$\text{Rate of reaction} = \frac{-\Delta[A]}{a\Delta t} = \frac{-\Delta[B]}{b\Delta t} = \frac{\Delta[C]}{c\Delta t} = \frac{\Delta[D]}{d\Delta t}$$

This representation gives several equalities, any one of which can be used to relate changes in observed concentrations to the rate of reaction.

Consider as a specific example the gas-phase reaction of 2.000 moles of iodine chloride and 1.000 mole of hydrogen at 230°C in a closed 1.000 liter container

$$2ICl\ (g) + H_2\ (g) \longrightarrow I_2\ (g) + 2HCl\ (g)$$

$$\text{Rate of reaction} = \frac{-\Delta[H_2]}{\Delta t} = \frac{-\Delta[ICl]}{2\Delta t} = \frac{\Delta[I_2]}{\Delta t} = \frac{\Delta[HCl]}{2\Delta t}$$

Table 14–1 lists the concentrations of reactants remaining at one second intervals, beginning with the time of mixing ($t = 0$ seconds), and the rate of reaction measured in terms of the rate of decrease of concentration of hydrogen. Verify for yourself that the rate of loss of ICl is twice that for H_2, and therefore that the rate of reaction could also be expressed as Rate $= -\Delta[\text{ICl}]/2\Delta t$. Increases in concentrations of products could be used instead. Figure 14–4 shows graphically the rates of change of concentration of all reactants and products.

Figure 14–5 is a plot of the rate of change of hydrogen concentration versus time, using the data of Table 14–1.

The rate of reaction at any time t is the negative of the slope of the tangent to the curve at time t, $-\Delta[H_2]/\Delta t$. Clearly the rate decreases with time. Lower concentrations of H_2 and ICl result in fewer collisions and less chance for reaction. The slope of the tangent to the curve at $t = 3.5$ seconds (for exam-

TABLE 14–1 Concentration and Rate Data for Reaction of 2.000 *M* ICl and 1.000 *M* H_2 at 230°C

[ICl] (mol/L)	$[H_2]$ (mol/L)	Rate $= \dfrac{-\Delta[H_2]}{\Delta t}$ $(\text{mol} \cdot \text{L}^{-1} \cdot \text{s}^{-1})$	Time (t) (seconds)
2.000	1.000		0
		0.326	
1.348	0.674		1
		0.148	
1.052	0.526		2
		0.090	
0.872	0.436		3
		0.062	
0.748	0.374		4
		0.046	
0.656	0.328		5
		0.035	
0.586	0.293		6
		0.028	
0.530	0.265		7
		0.023	
0.484	0.242		8
		0.020	
0.444	0.222		9
		0.016	
0.412	0.206		10
		0.014	
0.384	0.192		11
		0.012	
0.360	0.180		12
		0.011	
0.338	0.169		13
		0.009	
0.320	0.160		14
		0.008	
0.304	0.152		15
		0.008	
0.288	0.144		16

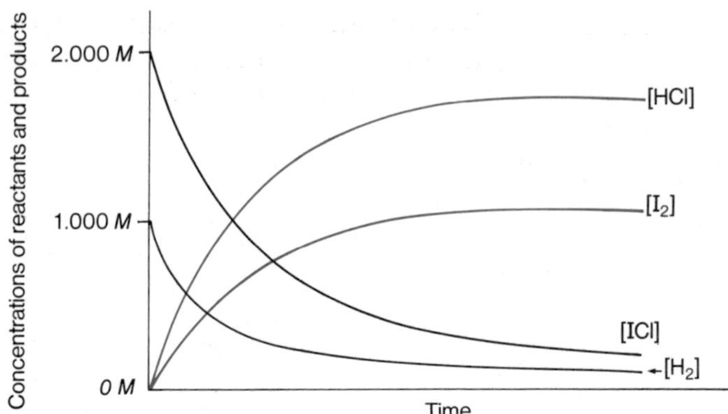

FIGURE 14–4 A plot of concentrations of all reactants and products in the reaction of 2.000 *M* ICl with 1.000 *M* H_2 versus time.

The slope of a line is the change in the quantity plotted along the vertical axis for a specified segment of the line, divided by the corresponding change along the horizontal axis. In this case the slope of the line (the tangent) is negative because the $[H_2]$ decreases as time passes (increases).

ple) can be estimated by dividing the change in $[H_2]$ in the interval between 3 and 4 seconds by 1 second (the corresponding change in time). The initial rate, or the rate at the instant of mixing the reactants, is the negative of the slope at t = zero. Had we looked at a plot of the rate of change of concentration of a product, I_2 or HCl, the rate would have been equal to the (positive) slope of the tangent.

From measurements of the initial rates of change of concentrations at constant temperature, the **rate-law expression** for a reaction may be determined. The rate-law expression for a reaction has the general form

$$\text{Rate} = k[\text{A}]^x[\text{B}]^y \ldots$$

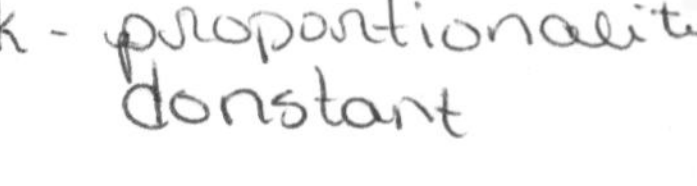

in which k is called the **specific rate constant** for the reaction at a particular temperature. The powers to which the concentrations are raised, x and y, are usually integers or zero, but are occasionally fractional. The values of x, y, and k can be determined *only experimentally,* and bear no necessary relationship

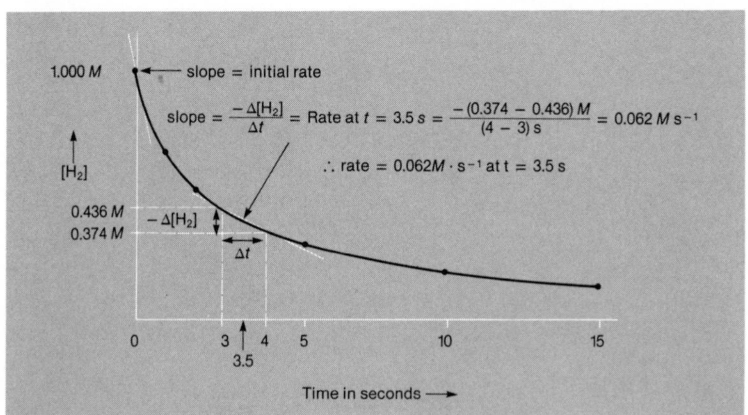

FIGURE 14–5 A plot of H_2 concentration versus time for the reaction of 2.000 *M* ICl with 1.000 *M* H_2 at 230° C from data in Table 14–1. The rate of reaction, which decreases with time, equals the negative of the slope of the tangent to the curve.

to the coefficients in the *balanced* chemical equation for the reaction. The value of x is said to be the **order** of the reaction with respect to A, while the order of the reaction with respect to B is y. The ***overall order*** of the reaction is $x + y$. Some examples are given below.

There is *no way* to predict the orders of reactants from the balanced chemical equation. They *must* be determined experimentally.

$3NO\ (g) \longrightarrow N_2O\ (g) + NO_2\ (g)$

Rate $= k[NO]^2$ second order in NO; second order overall

$2NO_2\ (g) + F_2\ (g) \longrightarrow 2NO_2F\ (g)$

Rate $= k[NO_2][F_2]$ first order in NO_2 and F_2; second order overall

$H_2O_2\ (aq) + 3I^-\ (aq) + 2H^+\ (aq) \longrightarrow 2H_2O\ (\ell) + I_3^-\ (aq)$

Rate $= k[H_2O_2][I^-]$ first order in H_2O_2 and I^-; zero order in H^+; second order overall.

Any number raised to the zero power is one. When a reaction is "zero order" in a reactant, its rate is independent of the concentration of that reactant.

The tabulated data below represent typical experimentally measured kinetic data and refer to the hypothetical reaction

$A_2 + 2B \longrightarrow 2AB$

at a specific temperature. The brackets refer to the concentrations of the reacting species *at the beginning* of the experimental runs listed in the first column, i.e., the initial concentration for each trial run.

Trial Run	Initial $[A_2]$	Initial [B]	Initial Rate of Formation of AB ($M \cdot s^{-1}$)
1	$1.0 \times 10^{-2}\ M$	$1.0 \times 10^{-2}\ M$	4.0×10^{-4}
2	$1.0 \times 10^{-2}\ M$	$2.0 \times 10^{-2}\ M$	4.0×10^{-4}
3	$2.0 \times 10^{-2}\ M$	$3.0 \times 10^{-2}\ M$	8.0×10^{-4}

Since we are describing the same reaction in each trial run, each is governed by the same rate law

Rate $= k[A_2]^x[B]^y$

We shall compare the initial rates of formation of product (reaction rates) for different trial runs to determine how changes in concentrations of reactants affect the rate of reaction, and, by so doing, evaluate x and y, and then k.

We observe that the initial concentration of A_2 in trials 1 and 2 is the same, so for these trials any change in reaction rate would be due to different initial concentrations of B. We can evaluate y by solving the ratio of the rate expressions for the two trial runs for y. Note that the operation involves dividing both sides of an equation by terms that are equal.

$$\frac{\text{Rate}_{(1)}}{\text{Rate}_{(2)}} = \frac{k[A_2]^x_{(1)}[B]^y_{(1)}}{k[A_2]^x_{(2)}[B]^y_{(2)}}$$

The value of k always cancels from such ratios, since it is constant at any particular temperature. For trials 1 and 2, the initial concentrations of A_2 are equal, so they too cancel. Thus, the expression simplifies to

$$\frac{\text{Rate}_{(1)}}{\text{Rate}_{(2)}} = \left(\frac{[\text{B}]_{(1)}}{[\text{B}]_{(2)}}\right)^y$$

The only unknown in this equation is y. Substituting the appropriate values from the data for trials 1 and 2 into the equation, we get

$$\frac{4.0 \times 10^{-4}}{4.0 \times 10^{-4}} = \left(\frac{1.0 \times 10^{-2}}{2.0 \times 10^{-2}}\right)^y$$

$n^0 = 1$ for any value of n.

$$1.0 = (0.5)^y \qquad \text{and} \qquad \underline{y = 0}$$

Since the units of $\text{Rate}_{(1)}$ and $\text{Rate}_{(2)}$ are identical, they obviously cancel. Similarly, the units of $[\text{B}]_{(1)}$ and $[\text{B}]_{(2)}$ are identical, and they too cancel. Both sets of units have been omitted to simplify the mathematical expression. Because any number raised to the zero power equals one, y must be zero. This result indicates that the initial concentration of B is raised to the zero power in the rate expression, and that the rate at which the reaction occurs is *independent* of the concentration of B. Thus far we know that the rate expression is

$$\text{Rate} = k[\text{A}_2]^x[\text{B}]^0 \qquad \text{or} \qquad \text{Rate} = k[\text{A}_2]^x$$

Next we shall evaluate x. In trial runs 1 and 3, the initial concentration of A_2 is doubled and the rate doubles. We may neglect the initial concentration of B since we have just shown that it does not affect the rate of the reaction.

$$\frac{\text{Rate}_{(3)}}{\text{Rate}_{(1)}} = \frac{k[\text{A}_2]^x_{(3)}}{k[\text{A}_2]^x_{(1)}} = \left(\frac{[\text{A}_2]_{(3)}}{[\text{A}_2]_{(1)}}\right)^x$$

$$\frac{8.0 \times 10^{-4}}{4.0 \times 10^{-4}} = \left(\frac{2.0 \times 10^{-2}}{1.0 \times 10^{-2}}\right)^x$$

$n^1 = n$ for any value of n.

$$2.0 = (2.0)^x \qquad \text{and} \qquad \underline{x = 1}$$

The power to which $[A_2]$ is raised in the rate-law expression is one, and we know that the rate law for this reaction is

$$\text{Rate} = k[\text{A}_2]^1 = \underline{k[\text{A}_2]}$$

The specific rate constant, k, can be evaluated by substituting any of the three sets of data into the rate-law expression. Substitution of the data from trial run 1 gives

$$\text{Rate}_{(1)} = k[\text{A}_2]_{(1)}$$

$$4.0 \times 10^{-4}\ M \cdot \text{s}^{-1} = k(1.0 \times 10^{-2}\ M)$$

$$k = \frac{4.0 \times 10^{-4}}{1.0 \times 10^{-2}}\ \text{s}^{-1} = \underline{4.0 \times 10^{-2}\ \text{s}^{-1}}$$

At the temperature at which the measurements were made, the rate-law expression for this reaction is

$$\text{Rate} = k[\text{A}_2], \qquad \text{or} \qquad \text{Rate} = \underline{4.0 \times 10^{-2}\ \text{s}^{-1}\ [\text{A}_2]}$$

We may check our calculations by evaluating k from one of the other sets of data. Using the data from trial number 3 in the rate-law expression gives

$$\text{Rate}_{(3)} = k[\text{A}_2]_{(3)}$$

$$8.0 \times 10^{-4}\ M \cdot \text{s}^{-1} = k(2.0 \times 10^{-2}\ M)$$

$$k = \frac{8.0 \times 10^{-4}}{2.0 \times 10^{-2}}\ \text{s}^{-1} = \underline{4.0 \times 10^{-2}\ \text{s}^{-1}}$$

This is the same value for k obtained above, and verifies the earlier calculation. The units of k vary for different reactions, and depend upon the form of the rate-law expression for the reaction under consideration.

EXAMPLE 14–1

Given the data tabulated below, determine the rate-law expression for the reaction

$2\text{A} + \text{B}_2 + \text{C} \longrightarrow \text{A}_2\text{B} + \text{BC}$

Trial Run	Initial [A]	Initial $[\text{B}_2]$	Initial [C]	Initial Rate of Formation of BC
1	0.20 M	0.20 M	0.20 M	$2.4 \times 10^{-6}\ M \cdot \text{min}^{-1}$
2	0.40 M	0.30 M	0.20 M	$9.6 \times 10^{-6}\ M \cdot \text{min}^{-1}$
3	0.20 M	0.30 M	0.20 M	$2.4 \times 10^{-6}\ M \cdot \text{min}^{-1}$
4	0.20 M	0.40 M	0.40 M	$4.8 \times 10^{-6}\ M \cdot \text{min}^{-1}$

Solution

The rate law is of the form shown below, and we must evaluate x, y, z, and k

$\text{Rate} = k[\text{A}]^x[\text{B}_2]^y[\text{C}]^z$

In trial runs 1 and 3, the initial concentrations of A and C are constant, and any change in the rate is due to the change in concentration of B_2. Thus, k and the initial concentrations of A and C cancel from the ratio $\text{Rate}_{(3)}/\text{Rate}_{(1)}$, and y can be evaluated

$$\frac{\text{Rate}_{(3)}}{\text{Rate}_{(1)}} = \frac{k[\text{A}]^x_{(3)}[\text{B}_2]^y_{(3)}[\text{C}]^z_{(3)}}{k[\text{A}]^x_{(1)}[\text{B}_2]^y_{(1)}[\text{C}]^z_{(1)}}$$

$$\frac{\text{Rate}_{(3)}}{\text{Rate}_{(1)}} = \left(\frac{[\text{B}_2]_{(3)}}{[\text{B}_2]_{(1)}}\right)^y$$

$$\frac{2.4 \times 10^{-6}\ M \cdot \text{min}^{-1}}{2.4 \times 10^{-6}\ M \cdot \text{min}^{-1}} = \left(\frac{0.30}{0.20}\right)^y$$

$$1.0 = (1.5)^y \quad \text{and} \quad \underline{y = 0}$$

We now know that the rate of reaction is independent of the initial concentration of B_2. The quantity $[\text{B}_2]$ is raised to the *zero* power in the rate expression, and we can write

$\text{Rate} = k[\text{A}]^x(1)[\text{C}]^z = k[\text{A}]^x[\text{C}]^z$

We can neglect $[\text{B}_2]$ in the following calculations. Trials 1 and 4 involve the same initial concentration of A. Let us compare these to evaluate z.

$$\frac{\text{Rate}_{(4)}}{\text{Rate}_{(1)}} = \frac{k[\text{A}]^x_{(4)}[\text{C}]^z_{(4)}}{k[\text{A}]^x_{(1)}[\text{C}]^z_{(1)}} = \left(\frac{[\text{C}]_{(4)}}{[\text{C}]_{(1)}}\right)^z$$

$$\frac{4.8 \times 10^{-6}\ M \cdot \text{min}^{-1}}{2.4 \times 10^{-6}\ M \cdot \text{min}^{-1}} = \left(\frac{0.40\ M}{0.20\ M}\right)^z$$

$$2.0 = (2.0)^z \quad \text{and} \quad \underline{z = 1}$$

Now that we know y and z, we can write the ratio for trials 1 and 2

$$\frac{\text{Rate}_{(2)}}{\text{Rate}_{(1)}} = \frac{k[\text{A}]^x_{(2)}[\text{C}]_{(2)}}{k[\text{A}]^x_{(1)}[\text{C}]_{(1)}} = \left(\frac{[\text{A}]_{(2)}}{[\text{A}]_{(1)}}\right)^x$$

$$\frac{9.6 \times 10^{-6}\, M \cdot \text{min}^{-1}}{2.4 \times 10^{-6}\, M \cdot \text{min}^{-1}} = \left(\frac{0.40\, M}{0.20\, M}\right)^x$$

$$4.0 = (2.0)^x \quad \text{and} \quad x = 2.$$

From this we can write the complete rate-law expression

$$\text{Rate} = k[\text{A}]^2[\text{B}_2]^0[\text{C}]^1, \quad \text{or} \quad \text{Rate} = k[\text{A}]^2[\text{C}]$$

We may evaluate the specific rate constant, k, by substituting any of the four sets of data into the rate-law expression we have just derived. Data from trial 2 are used here.

$$\text{Rate}_{(2)} = k[\text{A}]^2_{(2)}[\text{C}]_{(2)}$$

$$k = \frac{\text{Rate}_{(2)}}{[\text{A}]^2[\text{C}]}$$

$$k = \frac{9.6 \times 10^{-6}\, M \cdot \text{min}^{-1}}{(0.40\, M)^2(0.20\, M)} = 3.0 \times 10^{-4}\, M^{-2} \cdot \text{min}^{-1}$$

The rate-law expression, Rate $= k[\text{A}]^2[\text{C}]$, can now be written as

$$\text{Rate} = 3.0 \times 10^{-4}\, M^{-2} \cdot \text{min}^{-1}\, [\text{A}]^2[\text{C}]$$

This expression allows us to calculate the rate at which the reaction occurs with any known concentrations of A and C (so long as some B_2 is present). As we shall see presently, changes in temperature change the reaction rates, and this expression is valid *only* at the temperature at which the data were collected.

EXAMPLE 14–2

Consider the following initial rate data, and determine the rate-law expression for the reaction $3A + 2B \rightarrow 2C + D$.

Trial Run	Initial [A]	Initial [B]	Initial Rate of Formation of D
1	$1.00 \times 10^{-2}\, M$	$1.00 \times 10^{-2}\, M$	$6.00 \times 10^{-3}\, M \cdot \text{min}^{-1}$
2	$2.00 \times 10^{-2}\, M$	$3.00 \times 10^{-2}\, M$	$1.44 \times 10^{-1}\, M \cdot \text{min}^{-1}$
3	$1.00 \times 10^{-2}\, M$	$2.00 \times 10^{-2}\, M$	$1.20 \times 10^{-2}\, M \cdot \text{min}^{-1}$

Solution

Let us first compare trials 1 and 3, since [A] is the same in both cases. Doubling the initial concentration of B doubles the initial rate of formation of D, so the power to which [B] is raised in the rate expression is one.

$$\frac{\text{Rate}_{(3)}}{\text{Rate}_{(1)}} = \frac{k[\text{A}]^x_{(3)}[\text{B}]^y_{(3)}}{k[\text{A}]^x_{(1)}[\text{B}]^y_{(1)}} = \left(\frac{[\text{B}]_{(3)}}{[\text{B}]_{(1)}}\right)^y$$

$$\frac{1.20 \times 10^{-2}\, M \cdot \text{min}^{-1}}{6.00 \times 10^{-3}\, M \cdot \text{min}^{-1}} = \left(\frac{2.00 \times 10^{-2}\, M}{1.00 \times 10^{-2}\, M}\right)^y$$

$$2.00 = (2.00)^y \quad \text{and} \quad y = 1$$

$$\text{Rate} = k[\text{A}]^x[\text{B}]^1 = k[\text{A}]^x[\text{B}]$$

Next, we shall evaluate x by determining how the initial rate depends upon the [A]. There are no trial runs in which [B] is constant. Therefore, as we determine the value of x, the known effect of the change in [B] on the initial rate of reaction must be taken into account. Let us compare trials 1 and 2. Here [B] is tripled, and since [B] is raised to the first power in the rate expression, the rate will triple due to this effect *alone*. If the rate *only* tripled it would be $3(6.00 \times 10^{-3}\ M\cdot\text{min}^{-1})$, or $1.80 \times 10^{-2}\ M\cdot\text{min}^{-1}$ for trial 2. But we observe that the rate actually increases to $1.44 \times 10^{-1}\ M\cdot\text{min}^{-1}$. This increase involves multiplication of the rate in trial 1 by an additional factor of eight, that is, $1.44 \times 10^{-1}/1.80 \times 10^{-2} = 8$. This increase in rate is due to doubling [A] in trial 2. Since doubling [A] causes the rate to increase by a factor of eight, [A] is cubed in the rate expression, because $2^3 = 8$. In other words, comparing trial 1 to trial 2, tripling [B] causes the rate to triple (because [B] is raised to the first power) *and* doubling [A] causes the rate to increase by an additional factor of eight. The *overall* increase in rate is a factor of $8 \times 3 = 24$. Note that $6.00 \times 10^{-3}\ M\cdot\text{min}^{-1} \times 24 = 1.44 \times 10^{-1}\ M\cdot\text{min}^{-1}$. Thus, the rate-law expression has the form

$$\text{Rate} = k[\text{A}]^3[\text{B}]$$

The same result can be obtained by the mathematical procedure we used above, once we know that the concentration of B is raised to the first power. Let us use data from trial runs 1 and 2

$$\frac{\text{Rate}_{(1)} = k[\text{A}]^x_{(1)}[\text{B}]^1_{(1)}}{\text{Rate}_{(2)} = k[\text{A}]^x_{(2)}[\text{B}]^1_{(2)}} = \frac{[\text{A}]^x_{(1)}(1.0 \times 10^{-2})}{[\text{A}]^x_{(2)}(3.0 \times 10^{-2})}$$

$$\frac{6.00 \times 10^{-3}\ M\cdot\text{min}^{-1}}{1.44 \times 10^{-1}\ M\cdot\text{min}^{-1}} = \frac{(1.0 \times 10^{-2}\ M)^x}{(2.0 \times 10^{-2}\ M)^x} \times \frac{1}{3}$$

$$\frac{1}{24} = \frac{1}{3}\left(\frac{1}{2}\right)^x \qquad \frac{1}{8} = \left(\frac{1}{2}\right)^x \qquad \text{or} \qquad x = 3$$

Therefore, the concentration of A is cubed, and the rate expression is

$$\text{Rate} = k[\text{A}]^3[\text{B}]$$

the same expression we obtained by "reasoning" our way through the data.

In Section 14–7 we shall see how to apply experimentally determined rate-law expressions toward postulating the mechanisms by which reactions occur. Let us now consider the effects of temperature on rates of reaction.

14–5 Temperature

The average kinetic energy of a collection of molecules is proportional to the absolute temperature. At any particular temperature, T_1, a definite fraction of the molecules possesses the average kinetic energy and the others possess kinetic energies above and below the average for that temperature. Therefore, at a given temperature, a definite fraction of the reactant molecules have sufficient energy, E_a, to react to form product molecules upon collision. At a higher temperature, T_2, a greater percentage of the molecules possesses the necessary activation energy, and the reaction proceeds at a faster rate. This situation is depicted in Figure 14–6.

From experimental observations, a Swedish chemist of the late nineteenth and early twentieth centuries, Svante Arrhenius, developed the mathematical relationship among activation energy, absolute temperature, and the

This is the same Arrhenius who proposed a theory of acids and bases (Chapters 7 and 11).

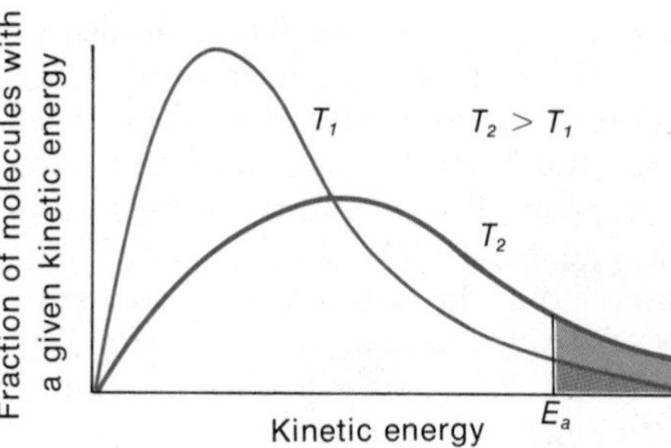

FIGURE 14–6 The effect of temperature on the number of molecules that have kinetic energies greater than E_a. Note that at T_2, a higher fraction of molecules possesses at least E_a, the activation energy.

specific rate constant of a reaction, k, at that temperature. The relationship is

$$k = Ae^{-E_a/RT}$$

or, in log form (to the base 10)

$$\log k = \log A - \frac{E_a}{2.303\ RT}$$

e ≈ 2.718 is the base of the *natural* logarithms. The number 2.303 (which is the natural logarithm of 10) is a conversion factor from natural to base-10 logarithms.

In this expression A is a constant having the same units as the rate constant, and is proportional to the frequency of collisions between reacting molecules. R is the universal gas constant, with the same energy units in its numerator as are used for E_a. The important point to note is that k decreases as E_a increases, so the rate of reaction decreases (as E_a increases).

Let's look at how the rate constant varies with temperature for a single reaction. If we assume that the activation energy and the factor A do not depend on temperature, we can write the Arrhenius equation twice for two different temperatures, subtract one equation from the other, and rearrange the result to obtain

$$\log \frac{k_2}{k_1} = \frac{E_a}{2.303\ R}\left(\frac{T_2 - T_1}{T_1 T_2}\right)$$

We can now substitute some representative values into this equation. The activation energy for many reactions near room temperature is about 50 kJ/mol (or 12 kcal/mol), and R is 8.314 J/mol K; so for an increase from 300 K to 310 K we find

$$\log \frac{k_2}{k_1} = \frac{50{,}000\ \text{J/mol}}{(2.303)(8.314\ \text{J/mol K})}\left(\frac{310\ \text{K} - 300\ \text{K}}{(300\ \text{K})(310\ \text{K})}\right) = 0.2808$$

$$\frac{k_2}{k_1} = 1.91 \approx 2$$

This is summarized by the rule of thumb that the rate of a reaction approximately doubles with a 10°C rise in temperature. Such a rule must be used with care, however, since it depends critically on the activation energy.

Although we can use the Arrhenius equation to treat activation energy in a quantitative way, as we have just seen, we shall deal with it only qualitatively in the sections that follow.

14–6 Catalysts

Catalysts are substances that can be added to reacting systems either to increase or, rarely, to decrease the rate of reaction. They allow reactions to occur via alternate pathways, and these affect reaction rates by altering activation energies. The activation energy is lowered in most catalyzed reactions, as depicted in Figures 14–7 and 14–8. However, in the case of inhibitory catalysis, the activation energy is raised as the reaction is forced to occur by a less favorable route. Although a catalyst may enter into a reaction, it does not appear in the balanced equation for the reaction. If a catalyst does react, all of it is regenerated in subsequent steps; if a species is not regenerated, it is not a catalyst but a reactant.

Catalysts may be classified in two categories: (1) homogeneous catalysts and (2) heterogeneous catalysts or contact catalysts. A **homogeneous catalyst** exists in the same phase as the reactants. Strong acids function as homogeneous catalysts in the acid-catalyzed hydrolysis of esters (a class of organic compounds). Using ethyl acetate (a component of nail polish removers) as an exam-

"Hydrolysis" means reaction with water and is discussed in detail in Chapter 17.

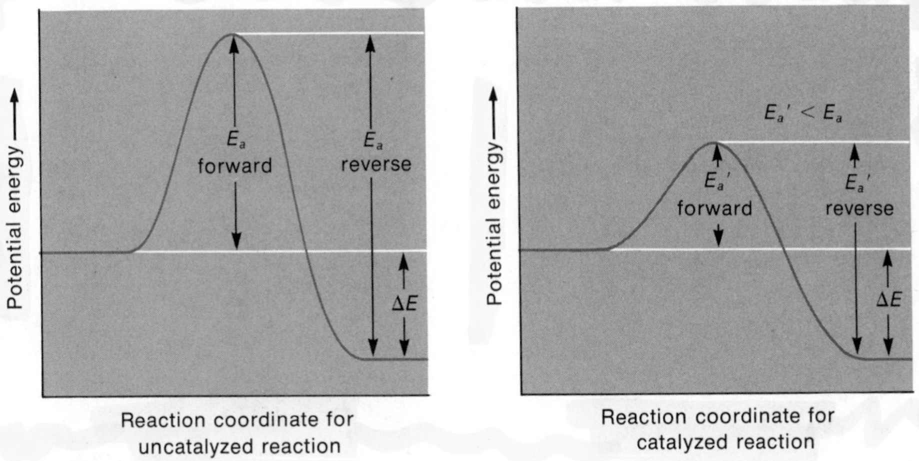

FIGURE 14–7 The effect a catalyst has on the potential energy diagram. The reaction mechanism is changed by the catalyst which provides a low-energy mechanism for the formation of the products. ΔE has the same value for each path, since the value of ΔE depends on the state of the reactants and products only.

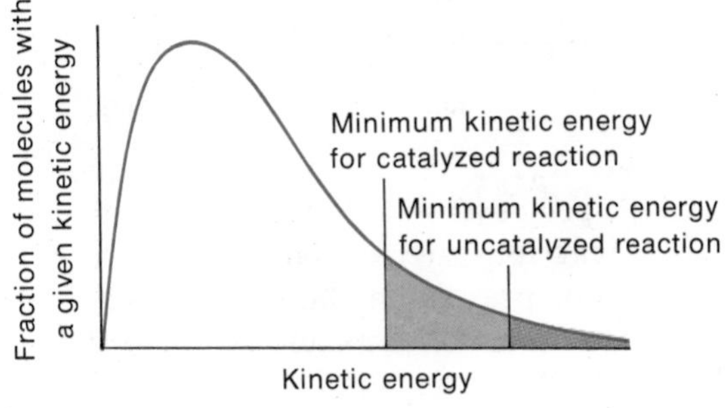

FIGURE 14–8 When a catalyst is present, more molecules possess the minimum kinetic energy necessary for reaction.

ple of an ester, the overall reaction can be written

$$\underset{\text{ethyl acetate}}{CH_3{-}\overset{\overset{\Large O}{\|}}{C}{-}OCH_2CH_3\,(aq)} + H_2O \xrightarrow{H^+} \underset{\text{acetic acid}}{CH_3{-}\overset{\overset{\Large O}{\|}}{C}{-}OH\,(aq)} + \underset{\text{ethanol}}{CH_3CH_2OH\,(aq)}$$

Some reaction intermediates are so unstable and short-lived that it is very difficult to prove experimentally that they exist. Others are so stable that they can be sold commercially.

This is a thermodynamically favorable reaction, but because of its high energy of activation it occurs only very, very slowly when no catalyst is present. However, in the presence of strong acids, the reaction occurs rapidly. In acid-catalyzed hydrolysis, different intermediates with lower activation energies are formed. The *postulated* sequence of steps is shown below.

$$CH_3{-}\overset{\overset{O}{\|}}{C}{-}OCH_2CH_3 + H^+ \longrightarrow \left[CH_3{-}\overset{\overset{\overset{+}{O}H}{\|}}{C}{-}OCH_2CH_3\right] \quad (1)$$

$$CH_3{-}\overset{\overset{\overset{+}{O}H}{\|}}{C}{-}OCH_2CH_3 + H_2O \longrightarrow \left[CH_3{-}\overset{\overset{OH}{|}}{\underset{\underset{H\diagup\ \ \diagdown H}{O^+}}{\underset{|}{C}}}{-}OCH_2CH_3\right] \quad (2)$$

$$CH_3{-}\overset{\overset{OH}{|}}{\underset{\underset{H\diagup\ \ \diagdown H}{O^+}}{\underset{|}{C}}}{-}OCH_2CH_3 \longrightarrow \left[CH_3{-}\overset{\overset{OH}{|}}{\underset{\underset{HO}{|}}{C}}{-}\underset{\underset{H}{|}}{\overset{+}{O}}{-}CH_2CH_3\right] \quad (3)$$

$$CH_3{-}\overset{\overset{OH}{|}}{\underset{\underset{OH}{|}}{C}}{-}\underset{\underset{H}{|}}{\overset{+}{O}}{-}CH_2CH_3 \longrightarrow \left[CH_3{-}\overset{\overset{^+OH}{\|}}{C}{-}OH\right] + \underset{\text{ethanol}}{HOC_2H_5} \quad (4)$$

$$CH_3{-}\overset{\overset{^+OH}{\|}}{C}{-}OH \longrightarrow \underset{\text{acetic acid}}{CH_3{-}\overset{\overset{O}{\|}}{C}{-}OH} + H^+ \quad (5)$$

$$\text{Overall:}\quad \underset{\text{ethyl acetate}}{CH_3{-}\overset{\overset{O}{\|}}{C}{-}OCH_2CH_3} + H_2O \xrightarrow{H^+} \underset{\text{acetic acid}}{CH_3{-}\overset{\overset{O}{\|}}{C}{-}OH} + \underset{\text{ethanol}}{CH_3CH_2OH}$$

Note that H^+ is a reactant in equation (1) and a product in equation (5) and is, therefore, a catalyst. The charged species (shown in brackets) produced in equations (1) through (4) are called **reaction intermediates;** they are not considered catalysts because they were not present in the original reaction mixture. Of course, ethyl acetate and water are reactants and acetic acid and ethanol are products of the overall catalyzed reaction.

Heterogeneous catalysts exist in a different phase than the reactants. They are usually solids, and they lower activation energies by providing surfaces upon which reactions can occur. One or more of the reactants may be preferentially adsorbed onto catalytic surfaces in particular orientations, resulting in more effective collisions between reactants than when a catalyst is not present. Most *contact catalysts* are more effective as finely divided powders, with large surface areas.

The reaction between hydrogen and oxygen to produce water is a thermodynamically favorable reaction

$$2H_2\ (g) + O_2\ (g) \longrightarrow 2H_2O\ (\ell) \qquad \Delta G^0 = -474\ \text{kJ}$$

but it does not proceed significantly at room temperature in the absence of energy input or a catalyst. When finely divided noble metals (unreactive metals such as platinum or palladium) are used as catalysts, the reaction can occur explosively at room temperature. The effect of this catalysis is depicted in Figure 14–9.

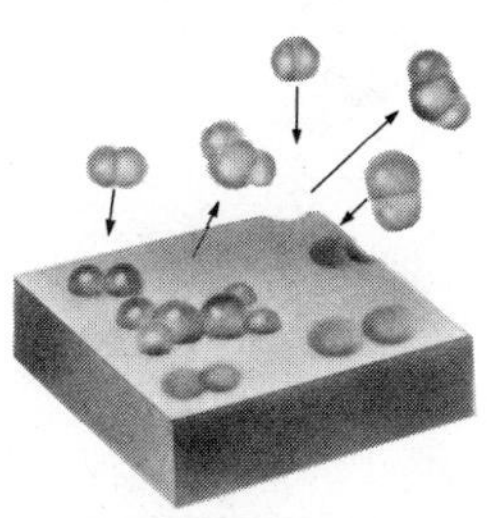

FIGURE 14–9 Catalysis of $2H_2 + O_2 \longrightarrow 2H_2O$ on metal surface.

The catalytic mufflers built into today's automobiles contain two types of heterogeneous catalysts, powdered noble metals and powdered transition metal oxides. They catalyze the oxidation of unburned fuel and of partial combustion products such as carbon monoxide (see Figure 14–10).

$$2C_8H_{18}\ (g) + 25O_2\ (g) \xrightarrow[\text{NiO}]{\text{Pt}} 16CO_2\ (g) + 18H_2O\ (g)$$

iso-octane (a component of gasoline)

$$2CO\ (g) + O_2\ (g) \xrightarrow[\text{NiO}]{\text{Pt}} 2CO_2\ (g)$$

The same catalysts also catalyze the decomposition of nitrogen oxide, NO, one of the by-products of combustion of any fuel, into harmless nitrogen and oxygen

$$2NO\ (g) \xrightarrow[\text{NiO}]{\text{Pt}} N_2\ (g) + O_2\ (g)$$

Nitrogen oxide is a serious air pollutant because it is oxidized to nitrogen dioxide, NO_2, which reacts with water to form nitric acid and with alcohols to form nitrites (eye irritants).

These three reactions, catalyzed in catalytic mufflers, are all exothermic and thermodynamically favorable. Unfortunately, other energetically favored reactions are also accelerated by the mixed catalysts. All fossil fuels contain sulfur impurities, which are oxidized to sulfur dioxide during combustion.

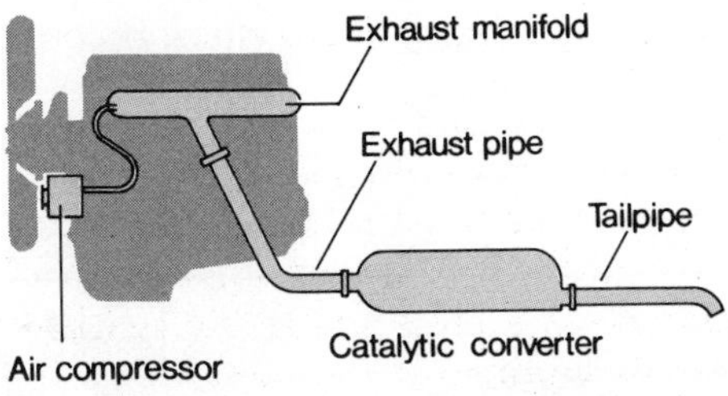

FIGURE 14–10 The arrangement of a catalytic converter.

Sulfur dioxide, itself an air pollutant, undergoes further oxidation to form sulfur trioxide as it passes through the catalytic bed:

$$2SO_2\ (g) + O_2\ (g) \xrightarrow[NiO]{Pt} 2SO_3\ (g)$$

Sulfur trioxide is probably a worse pollutant than sulfur dioxide, because SO_3 is the acid anhydride of strong, corrosive sulfuric acid. Sulfur trioxide reacts with water vapor in the air, as well as in auto exhausts, to form sulfuric acid droplets. This is a major problem that must be overcome if the current type of catalytic converter is to see continued use. These same catalysts also suffer from the problem of being "poisoned," i.e., made inactive, by lead. Leaded fuels, which contain tetraethyl lead, $Pb(C_2H_5)_4$, and tetramethyl lead, $Pb(CH_3)_4$, are not suitable for automobiles equipped with catalytic converters and are excluded by law from use in such cars.

The Haber process for the preparation of ammonia, an extremely important industrial chemical, involves the use of iron as a catalyst at 450° to 500°C and high pressures. The reaction is thermodynamically spontaneous, but very slow

$$N_2\ (g) + 3H_2\ (g) \xrightarrow{Fe} 2NH_3\ (g)$$

In the absence of a catalyst, the reaction does not occur at room temperature, and even with a catalyst is not efficient at atmospheric pressure. It is discussed in detail in Section 15–5.

Enzymes are proteins that act as catalysts in living systems for specific biochemical reactions. The reaction between nitrogen and hydrogen to form ammonia is catalyzed at room temperature and atmospheric pressure by a class of enzymes, called nitrogenases, that are present in some bacteria. Most of the essential nutrients for both plants and animals contain nitrogen. Legumes are plants that support certain bacteria, which are able to obtain nitrogen as N_2 from the atmosphere and convert it to ammonia. The process is called **nitrogen fixation.** The ammonia can then be used in the synthesis of proteins, nucleic acids, and other nitrogen-containing biological compounds.

Currently, one of the most active areas of chemical research involves attempts to discover or synthesize catalysts that are able to mimic the efficiency of naturally occurring enzymes such as nitrogenases. Such a development would be industrially important because it would eliminate the costs of high temperature and high pressure currently necessary in the Haber process, and could decrease the cost of food grown with the aid of ammonia-based fertilizer. Ultimately this would aid greatly in feeding the world's growing population.

In comparison to man-made catalysts, most enzymes are tremendously efficient under very mild conditions. If chemists and biochemists could develop catalysts with but a small fraction of the efficiency of enzymes, it could be a great boon to the world's health and economy.

14–7 Reaction Mechanisms and the Rate-Law Expression

Most reactions occur in a *series* of steps. In many mechanisms (series of steps), one step is slower than the others and is called the **rate-determining step.** The speed at which the slow step occurs determines the rate at which the overall reaction occurs. That is, a reaction can never occur faster than its slowest step. For the general overall reaction

$$aA + bB \longrightarrow cC + dD$$

the experimentally determined rate-law expression has the form: Rate = $k[A]^x[B]^y$. The values of x and y are influenced by the coefficients of the reactants in the slowest step and, in some cases, earlier steps (*not* the overall balanced equation, which is the sum of the individual steps).

Using a combination of experimental data and chemical intuition, we can *postulate* a mechanism by which a reaction occurs. There is no way to verify absolutely the validity of a proposed mechanism. All we can do is postulate a mechanism that is consistent with experimental data. If further investigations produce information inconsistent with a postulated mechanism, such as the identification of reaction intermediate species that are not part of the proposed mechanism, it must be modified to conform to all observations.

As an example, the reaction of nitrogen dioxide and carbon monoxide has been found to be second order with respect to NO_2 and zero order with respect to CO below 225°C.

$$NO_2\ (g) + CO\ (g) \longrightarrow NO\ (g) + CO_2\ (g)$$

$$\text{Rate} = k[NO_2]^2$$

The following proposed two-step mechanism is consistent with the observed rate-law expression:

$$\begin{array}{lll}
(1)\ NO_2 + NO_2 & \longrightarrow N_2O_4 & \text{(slow)} \\
(2)\ N_2O_4 + CO & \longrightarrow NO + CO_2 + NO_2 & \text{(fast)} \\
\hline
\quad NO_2 + CO & \longrightarrow NO + CO_2 & \text{overall}
\end{array}$$

The rate-determining step of this mechanism involves a *bimolecular* collision between two NO_2 molecules. This is consistent with the rate expression involving the square of the initial NO_2 concentration. Since the CO is involved only after the slow step has occurred, the reaction would be zero order in CO if this is the actual mechanism.

However, nitrogen trioxide, NO_3, has been identified from other experimental data as a transient (short-lived) reaction intermediate. The mechanism thought to be correct is shown below.

$$\begin{array}{lll}
(1)\ NO_2 + NO_2 & \longrightarrow NO_3 + NO & \text{(slow)} \\
(2)\ NO_3 + CO & \longrightarrow NO_2 + CO_2 & \text{(fast)} \\
\hline
\quad NO_2 + CO & \longrightarrow NO + CO_2 & \text{overall}
\end{array}$$

This postulated mechanism obeys both required criteria of all mechanisms: (1) it is consistent with the rate expression (in that $2NO_2$ and no CO are reactants in the slow step), and (2) the steps add to give the equation for the overall reaction. It is important to note that one cannot infer from the balanced overall equation that one molecule of NO_2 collides with one molecule of CO. Instead, in this case, it seems that two molecules of NO_2 collide to produce one molecule each of NO_3 and NO. The NO_3 then collides with one molecule of CO and reacts very rapidly to produce one molecule each of NO_2 and CO_2. Even though two NO_2 molecules are reactants in the first step, one is also produced

Note that *both* mechanisms are consistent with the observed rate law.

in step two. The net result is that only one NO_2 is a reactant in the overall equation.

One of the earliest kinetic studies undertaken involved the gas-phase reaction of hydrogen and iodine to form hydrogen iodide. The reaction was found to be first order in both hydrogen and iodine:

$$H_2\ (g) + I_2\ (g) \longrightarrow 2HI\ (g)$$

$$\text{Rate} = k[H_2][I_2]$$

The mechanism originally postulated involved collision of single molecules of H_2 and I_2 in a simple one-step reaction. However, present evidence indicates a more complex process. Most kineticists now accept the following mechanism

(1) $I_2 \longrightarrow 2I$		(fast)
(2) $I + H_2 \longrightarrow H_2I$		(fast)
(3) $H_2I + I \longrightarrow 2HI$		(slow)
$H_2 + I_2 \longrightarrow 2HI$		overall

In this case, neither of the original reactants appears in the rate-determining step, but both appear in the rate-law expression. Since each step is a reaction in itself, it follows that (according to transition state theory) each step has its own activation energy. Since step three is the slowest, its activation energy is the highest, as shown in Figure 14–11.

When several steps have similar activation energies, the analysis of experimental data is complex.

The gas-phase reaction of nitrogen oxide and bromine is known to be second order in NO and first order in Br_2

$$2NO\ (g) + Br_2\ (g) \longrightarrow 2NOBr\ (g)$$

$$\text{Rate} = k[NO]^2[Br_2]$$

A one-step, three-molecule collision involving two NO molecules and one Br_2 molecule would be consistent with the experimentally determined rate-law expression. However, the likelihood of all three molecules colliding simultaneously is far less than the likelihood that two will collide. Therefore, *more favorable routes involving only bimolecular collisions (or unimolecular decompositions) are thought to be important in reaction mechanisms.* The mechanism is believed to be that given below

(1) $NO + Br_2 \longrightarrow NOBr_2$		(fast)
(2) $NOBr_2 + NO \longrightarrow 2NOBr$		(slow)
$2NO + Br_2 \longrightarrow 2NOBr$		overall

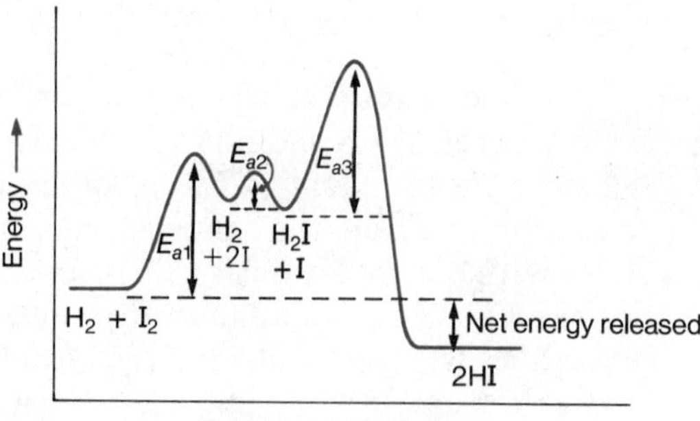

FIGURE 14–11 A graphical representation of the relative energies of activation for a postulated mechanism for the gas phase reaction $H_2\ (g) + I_2\ (g) \longrightarrow 2HI\ (g)$.

The first step involves the collision of one NO and one Br_2 to produce the intermediate species, $NOBr_2$, which then reacts relatively slowly with one NO molecule to produce two NOBr molecules. The rate-determining step involves one molecule of NO and one of $NOBr_2$. We could express the rate as

$$\text{Rate} = k'[NOBr_2][NO]$$

However, the $NOBr_2$ is a reaction intermediate, and its concentration at the beginning of the second step cannot be directly measured. Since the $NOBr_2$ is produced from one NO molecule and one Br_2 molecule, its concentration is proportional to the concentrations of both

$$[NOBr_2] = k''[NO][Br_2]$$

Thus, if we substitute the right side of the equation above for $[NOBr_2]$ in the rate expression we arrive at the original experimentally determined expression

$$\text{Rate} = k'k''[NO][Br_2][NO]$$

$$\text{Rate} = k[NO]^2[Br_2]$$

Similar situations apply to most other overall third or higher order reactions, as well as some lower order reactions, such as that of H_2 and I_2 above.

We see now that the experimentally determined reaction orders of reactants indicate the number of molecules of those reactants involved in (1) the slow step only or (2) the slow step *and* the steps preceding the slow step.

The mechanism by which nitrogen oxide from auto exhaust reacts with oxygen to form another pollutant, NO_2, is thought to be a two-step mechanism. An intermediate, N_2O_2, is formed in the first step. This then reacts with O_2 to form two NO_2 molecules, as depicted in Figure 14–12.

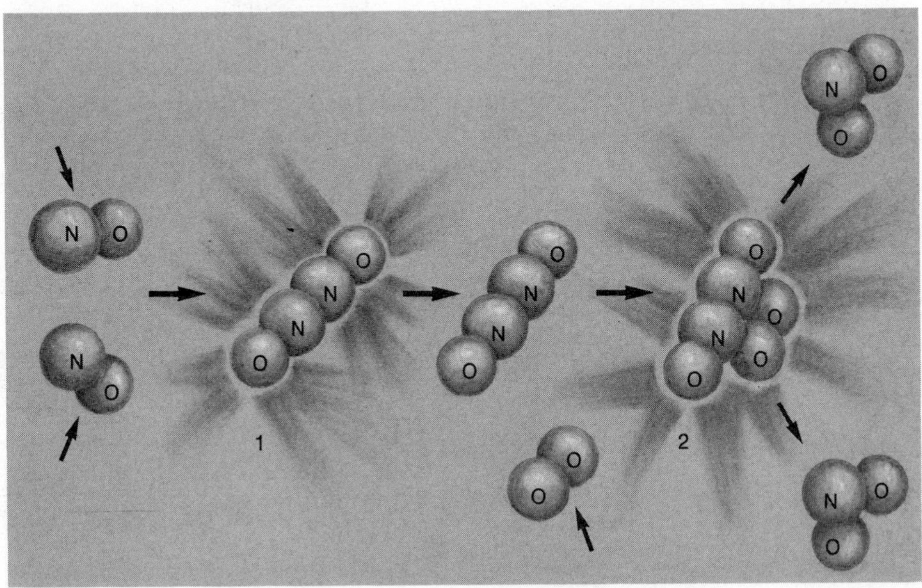

FIGURE 14–12 The two steps in the reaction 2NO (g) + O_2 (g) $\longrightarrow$ $2NO_2$ (g). The first step is fast, and the second is slow.

(1) $2NO\ (g) \longrightarrow N_2O_2\ (g)$	(fast)	
(2) $N_2O_2\ (g) + O_2\ (g) \longrightarrow 2NO_2\ (g)$	(slow)	
$2NO\ (g) + O_2\ (g) \longrightarrow 2NO_2\ (g)$	overall	Rate $= k[NO]^2[O_2]$

14–8 Half-Life of a Reactant

When given amounts of two reacting species are brought into contact with each other, they react at a rate that becomes slower and slower as time passes, since fewer and fewer reactant molecules are available to collide with each other. The **half-life, $t_{1/2}$,** of a reactant is the time it takes for half of it to be converted into product. The half-life is directly related to the specific rate constant of a reaction. We shall look at the relationships between half-lives and specific rate constants for reactions that are overall first order, or are second order with respect to a particular reactant and second order overall.

For reactions that are **first order in a particular reactant, A, and first order overall,** the following relationship has been found to be valid

For interested students with experience in calculus, we point out that this equation can be derived by integrating a differential form of the rate-law expression.

$$\log\left(\frac{[A]_0}{[A]}\right) = \frac{kt}{2.303}$$

The concentration of A at some time, t, after the reaction begins is [A], and $[A]_0$ is the initial concentration of A. If we solve this relationship for t we get:

$$t = \log\left(\frac{[A]_0}{[A]}\right) \times \frac{2.303}{k}$$

By definition, $[A] = \frac{1}{2}[A]_0$ at $t = t_{1/2}$. Thus

$$t_{1/2} = \log\left(\frac{[A]_0}{\frac{1}{2}[A]_0}\right) \times \frac{2.303}{k}$$

$$= \log(2) \times \frac{2.303}{k}$$

$$= 0.301 \times \frac{2.303}{k}$$

$$t_{1/2} = \frac{0.693}{k}$$

This is the relationship between the half-life of a reactant in an overall first-order reaction and its rate constant k. Note that in such reactions the half-life is independent of the initial concentration of A (but not in all types of reactions, as we shall see).

EXAMPLE 14–3

Compound A decomposes to form B and C in a reaction that is first order with respect to A and first order overall. At 25°C the specific rate constant for the reaction is $0.0450\ s^{-1}$ What is the half-life of A at 25°C?

$A \longrightarrow B + C$

Solution

$$t_{1/2} = \frac{0.693}{k} = \frac{0.693}{0.0450\ s^{-1}} = \underline{15.4\ s}$$

After 15.4 seconds of reaction, half of the original reactant remains.

EXAMPLE 14–4

The reaction $N_2O_5\ (g) \rightarrow N_2O_4\ (g) + \frac{1}{2}O_2\ (g)$ obeys the rate law Rate $= k[N_2O_5]$ in which the specific rate constant is $1.68 \times 10^{-2}\ s^{-1}$ at a certain temperature. If 2.50 moles of N_2O_5 (g) are placed in a 5.00-liter container at that temperature, how many moles of N_2O_5 would remain after 1.00 minute? How much O_2 would have been produced?

Solution

We may apply the relationship

$$\log\left(\frac{[N_2O_5]_0}{[N_2O_5]}\right) = \frac{kt}{2.303}$$

First we must determine the original concentration of N_2O_5:

$$[N_2O_5]_0 = \frac{2.50\ mol}{5.00\ L} = 0.500\ M$$

The only unknown is $[N_2O_5]$ after 1.00 minute. Let us solve for the unknown. Since $\log x/y = \log x - \log y$,

$$\log [N_2O_5]_0 - \log [N_2O_5] = \frac{kt}{2.303}$$

$$\log [N_2O_5] = \log [N_2O_5]_0 - \frac{kt}{2.303}$$

$$= \log (0.500) - \frac{(1.68 \times 10^{-2}\ s^{-1})(60.0\ s)}{2.303}$$

$$= -0.301 - 0.438$$

$$\log [N_2O_5] = -0.739$$

Taking the antilogarithm of both sides

$$[N_2O_5] = 10^{-0.739} = 10^{+0.261 - 1.000} = 10^{+0.261} \times 10^{-1.000}$$

$$[N_2O_5] = 1.82 \times 10^{-1}\ M$$

Thus, after 1.00 minute of reaction, the concentration of the reactant is $0.182\ M\ N_2O_5$. Now we convert this to moles of unreacted N_2O_5 and compute the amount of O_2 produced.

unreacted: $\ ?\ mol\ N_2O_5 = 5.00\ L \times \frac{1.82 \times 10^{-1}\ mol}{L} = \underline{0.910\ mol\ N_2O_5}$

reacted: $\ ?\ mol\ N_2O_5 = 2.50\ mol - 0.91\ mol = 1.59\ mol\ N_2O_5$

produced: $\ ?\ mol\ O_2 = 1.59\ mol\ N_2O_5 \times \frac{0.5\ mol\ O_2}{1\ mol\ N_2O_5} = \underline{0.795\ mol\ O_2}$

For a reaction that is **second order with respect to A and second order overall,** the relationship is

$$\frac{1}{[A]} - \frac{1}{[A]_0} = kt$$

For $t = t_{1/2}$, we have $[A] = \frac{1}{2}[A]_0$ and

$$\frac{1}{\frac{1}{2}[A]_0} - \frac{1}{[A]_0} = kt_{1/2}$$

Multiplying both numerator and denominator of the term $1/[A]_0$ by $\frac{1}{2}$ we get

$$\frac{1}{\frac{1}{2}[A]_0} - \frac{\frac{1}{2}}{\frac{1}{2}[A]_0} = kt_{1/2}$$

$$\frac{1 - \frac{1}{2}}{\frac{1}{2}[A]_0} = kt_{1/2}$$

Rearranging, we obtain the relationship between the rate constant and $t_{1/2}$ for a reaction that is *overall second order and second order in A*

$$t_{1/2} = \frac{1}{k[A]_0}$$

Note that, in this case, $t_{1/2}$ is *dependent upon the initial concentration of A.* This equation may also be used for reactions that are second order overall and first order with respect to each of two reactants initially present in equal concentrations.

EXAMPLE 14–5

Compounds A and B react to form C and D in a reaction that was found to be second order overall and second order in B. The rate constant at 30°C is 0.622 liter per mole per minute. What is the half-life of B if 4.10×10^{-2} *M* B reacts with excess A?

$$A + B \longrightarrow C + D$$

Unless excess A is present, the reaction stops before B is consumed.

Solution

As long as A is present in stoichiometric excess, only the concentration of B affects the rate. The half-life is given by the equation

$$t_{1/2} = \frac{1}{k[B]_0} = \frac{1}{(0.622\ M^{-1} \cdot \text{min}^{-1})(4.10 \times 10^{-2}\ M)} = \underline{39.2\ \text{min}}$$

EXAMPLE 14–6

What is the half-life of B in the reaction of Example 14–5 when an initial 3.50×10^{-4} *M* concentration of B reacts with excess A?

Solution

$$t_{1/2} = \frac{1}{k[B]_0} = \frac{1}{(0.622\ M^{-1} \cdot \text{min}^{-1})(3.50 \times 10^{-4}\ M)} = \underline{4.59 \times 10^{3}\ \text{min}}$$

We see that for a second-order reaction the half-life is dependent upon the initial concentration of reactant. The half-life of this reaction is 4.59×10^{3} minutes, or 3.19 days.

EXAMPLE 14–7

Consider the reaction of Example 14–5 at 30°C. How many minutes does it take to use up 7.0×10^{-3} *M* of B if 4.00×10^{-2} *M* of B is mixed with excess A?

Solution

We first determine how much B remains after 7.0×10^{-3} *M* of B is used up

$$\underline{?}\ M\ \text{B remaining} = 0.040\ M - 0.0070\ M = 0.033\ M = [\text{B}]$$

We may apply the relationship $\dfrac{1}{[\text{B}]} - \dfrac{1}{[\text{B}]_0} = kt$

and solving for t gives $t = \dfrac{1}{k}\left(\dfrac{1}{[\text{B}]} - \dfrac{1}{[\text{B}]_0}\right) = \dfrac{1}{k}\left(\dfrac{[\text{B}]_0 - [\text{B}]}{[\text{B}][\text{B}]_0}\right)$

$$= \frac{1}{0.622\ M^{-1}\cdot\text{min}^{-1}}\left(\frac{7.0 \times 10^{-3}\ M}{0.00132\ M^2}\right)$$

$$t = \underline{8.5\ \text{minutes}}$$

EXAMPLE 14–8

Consider the reaction of Example 14–5 at 30°C. What concentration of B will remain after 10.0 minutes of reaction, starting with 2.40×10^{-2} *M* of B and excess A?

Solution

Again we start with the expression $\dfrac{1}{[\text{B}]} - \dfrac{1}{[\text{B}]_0} = kt$

We must solve for [B]: $\dfrac{1}{[\text{B}]} = kt + \dfrac{1}{[\text{B}]_0} = \dfrac{[\text{B}]_0kt + 1}{[\text{B}]_0}$

$$[\text{B}] = \frac{[\text{B}]_0}{[\text{B}]_0\,kt + 1}$$

$$= \frac{0.0240\ M}{(0.0240\ M)(0.622\ M^{-1}\cdot\text{min}^{-1})(10.0\ \text{min}) + 1}$$

$$[\text{B}] = 0.0209\ M = \underline{2.09 \times 10^{-2}\ M}$$

Thus, only about 13% of the original concentration of B reacts within the first ten minutes. This is reasonable since, as you may easily verify, the reaction has a half-life of 67.0 minutes.

Key Terms

Activation energy energy that must be absorbed by reactants in their ground states in order to reach the transition state so that a reaction can occur.

Catalyst a substance that alters (usually increases) the rate at which a reaction occurs.

Chemical kinetics the study of rates and mechanisms of chemical reactions.

Collision theory theory of reaction rates that states that effective collisions between reactant molecules must occur in order for reaction to occur.

Contact catalyst see *Heterogeneous catalyst*

Effective collision collision between molecules resulting in reaction; one in which molecules collide with proper relative orientations and sufficient energy to react.

Enzyme a protein that acts as a catalyst in biological systems.

Half-life of a reactant, the time required for half of a given reactant to be converted into product(s).

Heterogeneous catalyst a catalyst that exists in a different phase (solid, liquid, or gas) from the reactants; a contact catalyst.

Homogeneous catalyst a catalyst that exists in the same phase (solid, liquid, or gas) as the reactants.

Inhibitory catalyst inhibitor, a catalyst that decreases the rate of a reaction.

Mechanism the sequence of steps by which reactants are converted into products.

Order of a reactant, the power to which the reactant's concentration is raised in the rate-law expression; of a reaction, the sum of the powers to which all concentrations are raised in the rate-law expression (also called overall order).

Rate-determining step the slowest step in a mechanism; the step that determines the overall rate of reaction.

Rate-law expression equation relating the rate of a reaction to the concentrations of the reactants and the specific rate constant.

Rate of reaction change in concentration of a reactant or product per unit time.

Reaction intermediate a usually short-lived species produced and consumed during a reaction.

Specific rate constant an experimentally determined (proportionality) constant, which is different for different reactions and which changes only with temperature; k in the rate-law expression: Rate = $k[A]^x[B]^y$.

Thermodynamically favorable (spontaneous) reaction reaction that occurs with a net release of free energy, G; reaction for which ΔG is negative (see Section 13–12).

Transition state relatively high-energy state in which bonds in reactant molecules are partially broken and new ones are partially formed.

Transition state theory theory of reaction rates that states that reactants pass through high-energy transition states before forming products.

Exercises

Basic Concepts

1. Summarize briefly the effects of each of the four factors affecting rates of reactions.
2. Describe the basic features of collision theory and transition state theory.
3. What is a rate-law expression? Describe how it is determined for a particular reaction.
4. Distinguish between reactions that are thermodynamically favorable and reactions that are kinetically favorable. What can be said about relationships between the two?
5. Draw typical reaction-energy diagrams for one-step reactions that release energy and that absorb energy. Distinguish between the net energy change for each kind of reaction and the activation energy. Indicate potential energies of products and reactants for both kinds of reactions.
6. Describe and illustrate graphically how the presence of a catalyst can affect the rate of a reaction.
7. What is meant by the order of a reaction?
8. Define reaction mechanism. Why do we assume that only bimolecular collisions and unimolecular decompositions are important in most reaction mechanisms?
9. What, if anything, can be said about the relationship between the coefficients of the balanced *overall* equation for a reaction and the powers to which concentrations are raised in the rate-law expression? To what are these powers related?

10. How do homogeneous catalysts and heterogeneous catalysts differ? What is an inhibitor?

Rate-Law Expression

11. If tripling the initial concentration of a reactant triples the initial rate of reaction, what is the order of the reaction with respect to the reactant? If the rate increases by a factor of nine, what is the order? If the rate remains the same, what is the order?
12. What are the units of the rate constant for reactions that are overall: (a) first order? (b) second order? (c) third order? (d) of order $1\frac{1}{2}$?
13. The following rate data were obtained at 25°C for the following reaction. What is the rate-law expression for this reaction?

$2A + B \longrightarrow 3C$

Experiment	[A] [mol/L]	[B] [mol/L]	Initial Rate of Formation of C
1	0.10	0.10	4.0×10^{-4} *M*/min
2	0.30	0.30	1.2×10^{-3} *M*/min
3	0.10	0.30	4.0×10^{-4} *M*/min
4	0.20	0.40	8.0×10^{-4} *M*/min

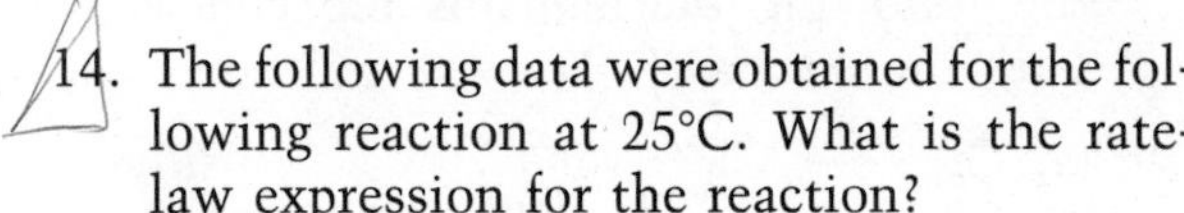

14. The following data were obtained for the following reaction at 25°C. What is the rate-law expression for the reaction?

$2A + B + 2C \longrightarrow 3D$

Experiment	Initial [A]	Initial [B]	Initial [C]	Initial Rate of Formation of D
1	0.20 *M*	0.10 *M*	0.10 *M*	4.0×10^{-4} *M*/min
2	0.20 *M*	0.30 *M*	0.20 *M*	1.2×10^{-3} *M*/min
3	0.20 *M*	0.10 *M*	0.30 *M*	4.0×10^{-4} *M*/min
4	0.60 *M*	0.30 *M*	0.40 *M*	3.6×10^{-3} *M*/min

15. The following data were collected for the following reaction at a particular temperature. What is the rate-law expression for this reaction?

$A + B \longrightarrow C$

Experiment	Initial [A]	Initial [B]	Initial Rate of Formation of C
1	0.10 *M*	0.10 *M*	4.0×10^{-4} *M*/min
2	0.20 *M*	0.20 *M*	3.2×10^{-3} *M*/min
3	0.10 *M*	0.20 *M*	1.6×10^{-3} *M*/min

16. The rate-law expression for the following reaction is found to be of the form: Rate = $k[N_2O_5]$. What is the overall reaction order?

$2N_2O_5\ (g) \longrightarrow 4NO_2\ (g) + O_2\ (g)$

17. We know that the rate expression for the following reaction at a certain temperature is Rate = $k[NO]^2[O_2]$. If two experiments involving this reaction are carried out at the same temperature, but if in the second experiment the initial concentration of NO is doubled while the initial concentration of O_2 is halved, the initial rate in the second experiment will be ______ times that of the first.

$2NO + O_2 \longrightarrow 2NO_2$

18. We know that the rate expression for the following reaction is Rate = $k[A][B_2]^2$. According to this expression, if during a reaction the concentrations of both A and B_2 are suddenly tripled, the rate of the reaction will ______ by a factor of ______.

$A + B_2 \longrightarrow \text{Products}$

19. Nitric oxide reacts with hydrogen to produce nitrogen and water vapor according to the following equation

$2NO\ (g) + 2H_2\ (g) \longrightarrow N_2\ (g) + 2H_2O\ (g)$

This reaction is thought to proceed by the following two-step mechanism.

$2NO + H_2 \longrightarrow N_2O + H_2O$ (slow)
$N_2O + H_2 \longrightarrow N_2 + H_2O$ (fast)

(a) According to this mechanism, what is the rate-law expression for this reaction?
(b) What is the overall reaction order?

20. Given the following data for the reaction $A + B \rightarrow C$, write the rate-law expression.

Experiment	Initial [A]	Initial [B]	Initial Rate of Formation of C
1	0.20 *M*	0.10 *M*	5.0×10^{-6} *M*/s
2	0.30 *M*	0.10 *M*	7.5×10^{-6} *M*/s
3	0.40 *M*	0.20 *M*	4.0×10^{-6} *M*/s

21. Given the following data for the reaction A + B → C, write the rate-law expression.

Experiment	Initial [A]	Initial [B]	Initial Rate of Formation of C
1	0.10 *M*	0.10 *M*	8.0×10^{-5} *M*/s
2	0.20 *M*	0.10 *M*	3.2×10^{-4} *M*/s
3	0.20 *M*	0.20 *M*	1.28×10^{-3} *M*/s

22. Given the following data for the reaction A + B → C, write the rate-law expression.

Experiment	Initial [A]	Initial [B]	Initial Rate of Formation of C
1	0.10 *M*	0.10 *M*	2.0×10^{-4} *M*/s
2	0.20 *M*	0.10 *M*	8.0×10^{-4} *M*/s
3	0.40 *M*	0.20 *M*	2.56×10^{-2} *M*/s

23. Consider a chemical reaction between compounds A and B that is first order in A and first order in B. From the information given below, fill in the blanks.

Experiment	Rate ($M \cdot s^{-1}$)	[A]	[B]
1	0.10	0.20 *M*	0.050 *M*
2	0.40	____ *M*	0.050 *M*
3	0.80	0.40 *M*	____ *M*

24. Consider a chemical reaction of compounds A and B that was found to be first order in A and second order in B. From the information given below, fill in the blanks.

Experiment	Rate ($M \cdot s^{-1}$)	[A]	[B]
1	0.10	1.0 *M*	0.20 *M*
2	____	2.0 *M*	0.20 *M*
3	____	2.0 *M*	0.40 *M*

Activation Energy and Temperature

25. Consider the reaction below, for which the activation energy of the forward reaction is 69.9 kJ and that of the reverse reaction is 82.0 kJ at 25°C.

$$A + 2B \longrightarrow C + D$$

Sketch a potential energy versus progress of reaction diagram for the reaction. Does the reaction release energy or absorb energy?

26. In the chapter, it was calculated that the rate constant, k, for a reaction whose activation energy is 50 kJ/mol approximately doubles (factor of 1.9) when the temperature is increased from 300 K to 310 K. Over this same temperature range, determine the factor by which k increases for a reaction whose activation energy is 100 kJ/mol. Can you rationalize this result?

27. Consider the forward reaction of Exercise 25. If its second order rate constant, k, is $6.0 \times 10^{-5}\ M^{-1}\ s^{-1}$ at 25°C, what is its rate constant at 50°C?

Reaction Mechanisms

28. The rate equation for the reaction

$$Cl_2\ (aq) + H_2S\ (aq) \longrightarrow S\ (s) + 2HCl\ (aq)$$

is found to be Rate = $k[Cl_2][H_2S]$. Which of the following mechanisms are consistent with the rate law?

(a)
$Cl_2 \longrightarrow Cl^+ + Cl^-$ slow
$Cl^- + H_2S \longrightarrow HCl + HS^-$ fast
$Cl^+ + HS^- \longrightarrow HCl + S$ fast
$Cl_2 + H_2S \longrightarrow S + 2HCl$ overall

(b)
$Cl_2 + H_2S \longrightarrow HCl + Cl^+ + HS^-$ slow
$Cl^+ + HS^- \longrightarrow HCl + S$ fast
$Cl_2 + H_2S \longrightarrow S + 2HCl$ overall

(c)
$Cl_2 \longrightarrow Cl + Cl$ fast
$Cl + H_2S \longrightarrow HCl + HS$ fast
$HS + Cl \longrightarrow HCl + S$ slow
$Cl_2 + H_2S \longrightarrow S + 2HCl$ overall

29. The ozone, O_3, of the stratosphere can be decomposed by reaction with nitrogen oxide (or, more commonly, nitric oxide), NO, from high-flying jet aircraft.

$$O_3\ (g) + NO\ (g) \longrightarrow NO_2\ (g) + O_2\ (g)$$

The rate expression is Rate = $k[O_3][NO]$. Which of the following mechanisms are consistent with the observed rate expression?

(a)

$NO + O_3 \longrightarrow NO_3 + O$	slow
$NO_3 + O \longrightarrow NO_2 + O_2$	fast
$O_3 + NO \longrightarrow NO_2 + O_2$	overall

(b)

$NO + O_3 \longrightarrow NO_2 + O_2$	slow (one step)

(c)

$O_3 \longrightarrow O_2 + O$	slow
$O + NO \longrightarrow NO_2$	fast
$O_3 + NO \longrightarrow NO_2 + O_2$	overall

(d)

$NO \longrightarrow N + O$	slow
$O + O_3 \longrightarrow 2O_2$	fast
$O_2 + N \longrightarrow NO_2$	fast
$O_3 + NO \longrightarrow NO_2 + O_2$	overall

(e)

$NO \longrightarrow N + O$	fast
$O + O_3 \longrightarrow 2O_2$	slow
$O_2 + N \longrightarrow NO_2$	fast
$O_3 + NO \longrightarrow NO_2 + O_2$	overall

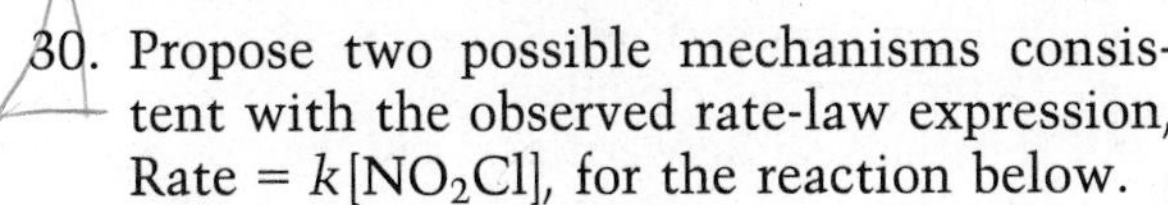

30. Propose two possible mechanisms consistent with the observed rate-law expression, Rate = $k[NO_2Cl]$, for the reaction below.

$$2NO_2Cl \longrightarrow 2NO_2 + Cl_2$$

31. Propose a possible mechanism for the reaction of iodide ion, I^-, with hypochlorite ion, OCl^-.

$$I^- + OCl^- \longrightarrow OI^- + Cl^-$$

The observed rate expression is Rate = $k[I^-][OCl^-]$.

32. Propose a mechanism consistent with the rate expression Rate = $k[NO]^2[O_2]$ for the reaction

$$2NO + O_2 \longrightarrow 2NO_2$$

Half-Life of a Reactant

33. What is meant by the half-life of a reactant?
34. The decomposition of carbon disulfide, CS_2, to carbon monosulfide, CS, and sulfur is first order with $k = 2.8 \times 10^{-7}\ s^{-1}$ at 1000°C.

$$CS_2 \longrightarrow CS + S$$

(a) What is the half-life of this reaction at 1000°C?
(b) How many days would pass before a 1.00-gram sample of CS_2 had decomposed to the extent that 0.60 gram of CS_2 remained?
(c) Refer to (b). How many grams of CS would be present after this length of time?
(d) How much of a 1.00-gram sample of CS_2 would remain after 35.0 days?

35. The first order rate constant for the conversion of cyclobutane to ethylene at 1000°C is $87\ s^{-1}$.

$$C_4H_8 \text{ (cyclobutane)} \longrightarrow 2\ H_2C{=}CH_2 \text{ (ethylene)}$$

cyclobutane ethylene

(a) What is the half-life of this reaction at 1000°C?
(b) If one started with 1.00 gram of cyclobutane, how long would it take to consume 0.70 gram of it?
(c) How much of an initial 1.00-gram sample of cyclobutane would remain after 1.00×10^{-3} second?

36. For the reaction

$$2NO_2 \longrightarrow 2NO + O_2$$

the rate equation is

$$\text{Rate} = 1.4 \times 10^{-10}\ M^{-1}\,s^{-1}[NO_2]^2 \text{ at } 25°C.$$

(a) If 2.50 moles of NO_2 are initially present in a sealed 1.00-liter vessel at 25°C, what is the half-life of the reaction?
(b) Refer to (a). What concentration and how many grams of NO_2 remain after 100 years?
(c) Refer to (b). What concentration of NO would have been produced during the same period of time?

15 Chemical Equilibrium

15–1 Basic Concepts

Most chemical reactions do not go to completion. That is, when reactants are mixed in stoichiometric quantities, they are not completely converted to products. Reactions that do not go to completion *and* that can occur in either direction are called *reversible reactions.*

In Section 7–1, part 3 we discussed the ionization of acetic acid in water as a specific example of a reversible reaction

$$CH_3COOH\ (aq) + H_2O\ (\ell) \rightleftharpoons H_3O^+\ (aq) + CH_3COO^-\ (aq)$$

You may wish to refresh your memory by referring to that section.

Reversible reactions may be represented in general terms as shown below, where the capital letters represent formulas and the lower case letters represent coefficients in the balanced equation. (In the equation for the ionization of acetic acid, a, b, c, and d are all ones.)

$$aA + bB \rightleftharpoons cC + dD$$

The double arrow ($\rightleftharpoons$) indicates that the reaction is reversible and that both the forward (left-to-right) and reverse (right-to-left) reactions can occur simultaneously. When A and B react to form C and D at the same rate at which C and D react to form A and B, the system is at *equilibrium.* **Chemical equilibrium** exists when two opposing reactions occur simultaneously at the same rate. Chemical equilibria are **dynamic equilibria** in the sense that individual molecules are continually reacting, even though the overall composition of the reaction mixture does not change. In a system at equilibrium, the equilibrium is said to lie toward the right if more C and D are present than A and B, and to lie toward the left if more A and B are present.

The dynamic nature of chemical equilibrium can be proved experimentally by "tagging" a small percentage of reactant molecules with radioactive atoms and following them through the reaction. Even when the initial mixture is at equilibrium, radioactive atoms will appear in product molecules.

Consider the simple case in which the coefficients of the equation for a reaction are all one, *and* the forward and reverse reactions occur by one-step mechanisms. If substances A and B are mixed, the rate of the forward reaction decreases as time passes because the concentrations of A and B decrease.

$$A + B \longrightarrow C + D \qquad (1)$$

As C and D are formed, these species react to form A and B.

$$C + D \longrightarrow A + B \qquad (2)$$

The rate of reaction between C and D increases with time because as more C and D molecules are formed, more collide and react. Eventually, the two reactions occur at the same rate and the system is at equilibrium (see Figure 15–1).

The square brackets, [ ], represent concentrations of the species enclosed within them in moles per liter.

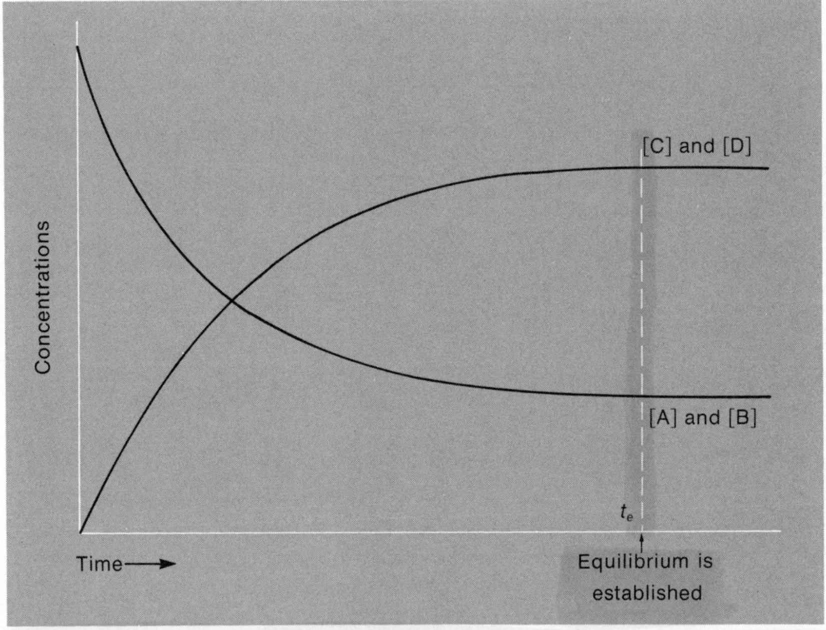

FIGURE 15–1 Variation in the concentrations of species present in the $A + B \rightleftharpoons C + D$ system as equilibrium is approached beginning with equal concentrations of A and B only. This equilibrium lies far to the right.

If a reaction begins with only C and D present, reaction (2) proceeds at a rate which decreases with time and the rate of reaction (1) increases with time until the two rates are equal.

In previous chapters we have dealt almost exclusively with reactions that were assumed to occur in only one direction. Those that go essentially to completion are commonly written with only a single arrow, signifying that when equilibrium is established only insignificantly small amounts of reactants remain.

The SO_2, O_2, SO_3 System

Let us examine the reversible reaction of sulfur dioxide with oxygen to form sulfur trioxide at 1500 K. The numbers in the following discussion were determined experimentally.

$$2SO_2\ (g) + O_2\ (g) \rightleftharpoons 2SO_3\ (g)$$

Suppose 0.400 mole of SO_2 and 0.200 mole of O_2 are injected into a closed container of 1.00-liter capacity. When equilibrium is established (at time t_e, Figure 15–2), 0.056 mole of SO_3 has formed and 0.344 mole of SO_2 and 0.172 mole of O_2 remain unreacted; *or,* 0.056 mole of SO_2 and 0.028 mole of O_2 have reacted to form 0.056 mole of SO_3. The reaction does not go to completion. But the reaction ratio is a 2:1:2 mole ratio as required by the coefficients of the balanced equation; that is, 2(0.028) mole of SO_2 and 1(0.028) mole of O_2 have been consumed and 2(0.028) mole of SO_3 has been produced. The situation is summarized below, using molarity units rather than moles. (They are numerically identical since the volume of the reaction vessel is 1.00 liter.) The *net reaction* is represented by the *changes* in concentrations. A positive (+) change represents an increase, while a negative (−) change indicates a decrease in concentration.

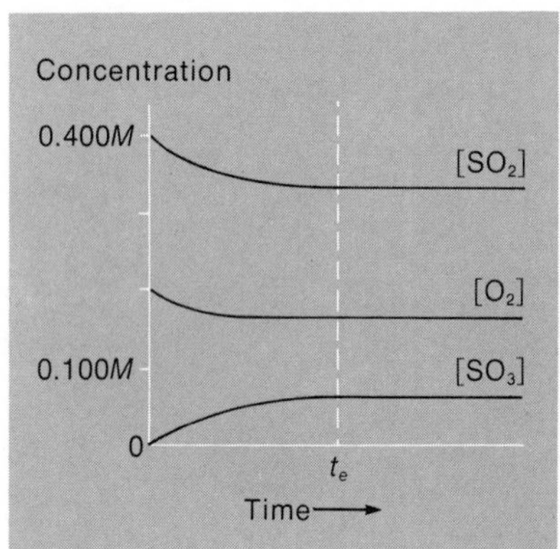

FIGURE 15–2 Establishment of equilibrium in the $2SO_2 + O_2 \rightleftharpoons 2SO_3$ system, beginning with only SO_2 and O_2.

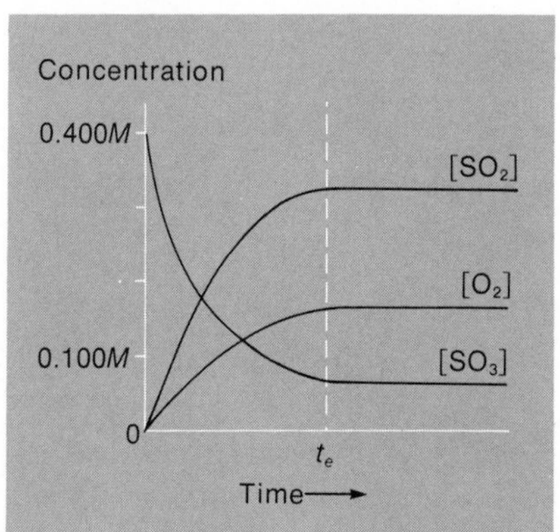

FIGURE 15–3 Establishment of equilibrium in the $2SO_2 + O_2 \rightleftharpoons 2SO_3$ system beginning with only SO_3. Greater changes in concentrations occur to establish equilibrium starting with SO_3 than starting with SO_2 and O_2. The equilibrium favors SO_2 and O_2.

	$2SO_2$ (g) +	O_2 (g)	⇌ $2SO_3$ (g)
initial conc'n	0.400 *M*	0.200 *M*	0
change in conc'n due to rxn (ratio *set* by coefficients)	−0.056 *M*	−0.028 *M*	+0.056 *M*
equilibrium conc'n	0.344 *M*	0.172 *M*	0.056 *M*

In another experiment, only 0.400 mole of SO_3 is introduced into a closed 1.00-liter container. The same total numbers of sulfur and oxygen atoms are present as in the previously described experiment. When equilibrium is established (time t_e in Figure 15–3), 0.056 mole of SO_3, 0.172 mole O_2, and 0.344 mole of SO_2 are present. These are the same amounts found at equilibrium in the previous case. Note that the net reaction proceeds from *right to left* as the equation is written, and that the changes in concentration are in the same 2:1:2 ratio as in the previous case, as required by the coefficients of the balanced equation.

	$2SO_2$ (g) +	O_2 (g)	⇌ $2SO_3$ (g)
initial conc'n	0	0	0.400 *M*
change in conc'n due to rxn (ratio *set* by coefficients)	+0.344 *M*	+0.172 *M*	−0.344 *M*
equilibrium conc'n	0.344 *M*	0.172 *M*	0.056 *M*

The data below summarize the results of these two experiments.

	Initial Concentrations			Equilibrium Concentrations		
	$[SO_2]$	$[O_2]$	$[SO_3]$	$[SO_2]$	$[O_2]$	$[SO_3]$
Experiment 1	0.400 *M*	0.200 *M*	0 *M*	0.344 *M*	0.172 *M*	0.056 *M*
Experiment 2	0 *M*	0 *M*	0.400 *M*	0.344 *M*	0.172 *M*	0.056 *M*

These results are displayed graphically in Figures 15–2 and 15–3.

15–2 The Equilibrium Constant

Assume that the reversible reaction below occurs via a *simple one-step mechanism.*

Recall that the simultaneous collision of three bodies is an unlikely occurrence from a statistical point of view (Section 14–7).

$$2A + B \rightleftharpoons A_2B$$

The rate of the forward reaction is $\text{Rate}_f = k_f\,[A]^2[B]$ while the rate of the reverse reaction is $\text{Rate}_r = k_r\,[A_2B]$. In these expressions k_f and k_r are the *specific rate constants* of the forward and reverse reactions, respectively. By definition the two rates are equal at equilibrium, and since $\text{Rate}_f = \text{Rate}_r$, we may write

$$k_f[A]^2[B] = k_r[A_2B]$$

Rearranging the expression by dividing both sides of the equation by $k_r[A]^2[B]$

gives

$$\frac{k_f}{k_r} = \frac{[A_2B]}{[A]^2[B]}$$

At any specific temperature, k_f/k_r is a constant, since both k_f and k_r are constants. This ratio is given a special name and symbol, the **equilibrium constant,** K_c or simply K, where the subscript refers to concentrations. In this reaction,

$$K_c = \frac{[A_2B]}{[A]^2[B]}$$

The square brackets, [ ], in an equilibrium constant expression indicate *equilibrium* concentrations in moles per liter.

We have dealt with a simple one-step reaction. Suppose, instead, that the reaction involves a two-step mechanism with the rate constants shown.

(1) $2A \underset{k_{1r}}{\overset{k_{1f}}{\rightleftharpoons}} A_2$

(2) $A_2 + B \underset{k_{2r}}{\overset{k_{2f}}{\rightleftharpoons}} A_2B$

The overall reaction is
$2A + B \rightleftharpoons A_2B$

An equilibrium constant expression can be written for each step:

$$K_1 = \frac{k_{1f}}{k_{1r}} = \frac{[A_2]}{[A]^2}$$

$$K_2 = \frac{k_{2f}}{k_{2r}} = \frac{[A_2B]}{[A_2][B]}$$

By multiplying K_1 by K_2 the $[A_2]$ term may be eliminated. Since K_1 and K_2 are both constants, K_c is also a constant and we obtain

$$K_1 \times K_2 = \frac{[A_2]}{[A]^2} \times \frac{[A_2B]}{[A_2][B]} = \frac{[A_2B]}{[A]^2[B]} = K_c$$

Regardless of the mechanism by which a reaction occurs, the same expression for the equilibrium constant is obtained. For a reaction in general terms,

$$\underbrace{aA + bB}_{\text{Reactants}} \rightleftharpoons \underbrace{cC + dD}_{\text{Products}}$$

the equilibrium constant expression is given by

$$K_c = \frac{[C]^c[D]^d}{[A]^a[B]^b} \quad \begin{matrix}\leftarrow \text{Products} \\ \leftarrow \text{Reactants}\end{matrix}$$

By convention we refer to the substances on the right side of the balanced equation as "products" and those on the left as "reactants," even if the reaction is initiated by "products," with no "reactants" initially present. The equilibrium constant, K_c, is defined as the product of the *equilibrium concentrations* (moles per liter) of the products, each raised to the power that corresponds to its coefficient in the balanced equation, divided by the product of the *equilibrium concentrations* of reactants, each raised to the power that corresponds to its coefficient in the balanced equation.

The equilibrium constant expressions are given for the following reversible reactions to illustrate some common types

$$N_2\ (g) + O_2\ (g) \rightleftharpoons 2NO\ (g), \qquad K_c = \frac{[NO]^2}{[N_2][O_2]}$$

$$CH_4\ (g) + Cl_2\ (g) \rightleftharpoons CH_3Cl\ (g) + HCl\ (g), \qquad K_c = \frac{[CH_3Cl][HCl]}{[CH_4][Cl_2]}$$

$$N_2\ (g) + 3H_2\ (g) \rightleftharpoons 2NH_3\ (g), \qquad K_c = \frac{[NH_3]^2}{[N_2][H_2]^3}$$

Numerical values for equilibrium constants must be determined experimentally. Let us reconsider the $SO_2/O_2/SO_3$ equilibrium described earlier in this chapter. The equilibrium concentrations were the same in the two experiments. We may use these equilibrium concentrations to calculate the value of the equilibrium constant for the reaction. The equilibrium concentrations were $[SO_3] = 0.056\ M$, $[SO_2] = 0.344\ M$, and $[O_2] = 0.172\ M$

	$2SO_2$ (g) +	O_2 (g) $\rightleftharpoons$	$2SO_3$ (g)
equil conc'n	0.344 M	0.172 M	0.056 M

Substitution of these values into the equilibrium expression allows evaluation of the equilibrium constant

$$K_c = \frac{[SO_3]^2}{[SO_2]^2[O_2]} = \frac{(0.056\ M)^2}{(0.344\ M)^2(0.172\ M)} = 1.5 \times 10^{-1}\ M^{-1}$$

For the reaction written *as it is*, K_c is 0.15 M^{-1} at 1500 K. The value and units of K_c (in this case M^{-1}) depend upon the form of the balanced equation for the reaction. The units are commonly omitted. The value of K_c, for any particular equation, varies only with the temperature, is constant for the reaction at a given temperature, and is independent of initial concentrations.

The magnitude of K_c is a measure of the extent to which reaction occurs. A very large value of K_c indicates that at equilibrium most of the species on the left are converted into the species on the right side of the balanced chemical equation. If K_c is much smaller than 1, equilibrium is established when most of the reactants (left side) remain unreacted and relatively small amounts of products (right side) are formed.

EXAMPLE 15–1

Some nitrogen and hydrogen gases are pumped into an empty 5.00-liter vessel at 500°C. When equilibrium is established, 3.00 moles of N_2, 2.10 moles of H_2, and 0.298 mole of NH_3 are present. Evaluate K_c for the reaction of nitrogen and hydrogen at 500°C.

$$N_2\ (g) + 3H_2\ (g) \rightleftharpoons 2NH_3\ (g)$$

Solution

The equilibrium concentrations are obtained by dividing the number of moles of each reactant and product by the volume, 5.00 liters

$$[N_2] = 3.00 \text{ mol}/5.00 \text{ L} = 0.600\ M$$

$$[H_2] = 2.10 \text{ mol}/5.00 \text{ L} = 0.420\ M$$

$$[NH_3] = 0.298 \text{ mol}/5.00 \text{ L} = 0.0596\ M$$

The resulting concentrations (not numbers of moles) are then substituted into the expression for the equilibrium constant:

$$K_c = \frac{[NH_3]^2}{[N_2][H_2]^3} = \frac{(0.0596)^2}{(0.600)(0.420)^3} = 0.080$$

Thus, for the reaction of hydrogen and nitrogen to form ammonia at 500°C, we can write

$$K_c = \frac{[NH_3]^2}{[N_2][H_2]^3} = \underline{0.080}$$

The small value of K_c indicates that the equilibrium lies far to the left.

Variation of K_c with the Form of the Balanced Equation

We wrote the equation for the reaction of SO_2 with O_2 to form SO_3 as

$$2SO_2\ (g) + O_2\ (g) \rightleftharpoons 2SO_3\ (g)$$

and showed the equilibrium constant to be

$$K_c = \frac{[SO_3]^2}{[SO_2]^2[O_2]} = \frac{(0.056)^2}{(0.344)^2(0.172)} = 0.15$$

Suppose we had written the equation for the same reaction in reverse,

$$2SO_3\ (g) \rightleftharpoons 2SO_2\ (g) + O_2\ (g)$$

The equilibrium constant, K_c', for the reverse reaction is

$$K_c' = \frac{[SO_2]^2[O_2]}{[SO_3]^2} = \frac{(0.344)^2(0.172)}{(0.056)^2} = 6.5$$

We see that K_c', the equilibrium constant for the reverse reaction, is *the reciprocal* of K_c, the equilibrium constant for the forward reaction. If the equation for the reaction is written as

$$SO_2\ (g) + \tfrac{1}{2}O_2\ (g) \rightleftharpoons SO_3\ (g)$$

then

$$K_c'' = \frac{[SO_3]}{[SO_2][O_2]^{1/2}} = \frac{0.056}{(0.344)(0.172)^{1/2}} = 0.39 = K_c^{1/2}$$

where K_c'' is the square root of K_c.

To generalize, if an *equation* for a reaction is *multiplied* by a positive or negative integer or fraction, n, then the *original value of* K_c is raised to the nth *power.* (Reversing an equation is the same as multiplying all coefficients by -1.)

15–3 Uses of the Equilibrium Constant

The equilibrium constant for a given reaction at a given temperature is a type of *ratio* of product concentrations to reactant concentrations, and it expresses the extent of the reaction at equilibrium. We have seen how to calculate its value from one set of equilibrium concentrations; once that value is obtained, the computational process can be turned around to calculate equilibrium *concentrations* from the equilibrium *constant.* (Remember that the equilibrium constant is a "constant" only if the temperature does not change. We will see in Section 15–11 how to compute a new equilibrium constant for a different temperature.)

As in any algebraic equation, one of the quantities involved can be calculated if values are specified for all of the others. In Example 15–1, the equation involved three concentrations and the equilibrium constant; we were given values of the three equilibrium concentrations, and we calculated the equilibrium constant. In the next example, given all the *initial* concentrations and the equilibrium constant, we shall calculate the equilibrium concentrations.

EXAMPLE 15–2

For the reaction

$$A + B \rightleftharpoons C + D$$

the equilibrium constant is 144 at a certain temperature. If 0.400 mole each of A and B are placed in a 2.00-liter container at that temperature, what concentrations of all species are present at equilibrium?

Solution

In practically all equilibrium calculations it is helpful to write down the values, or symbols for the values, of initial concentrations, changes in concentrations (due to the reaction), and the concentrations at equilibrium as shown below. Since the coefficients of the equation are all ones, the reaction ratio must be 1:1:1:1.

Let x = moles per liter of both A and B that have reacted, and then
x = moles per liter of C and D that have been formed

	A	+ B	$\rightleftharpoons$ C	+ D
initial	0.200 M	0.200 M	0 M	0 M
change	$-x$ M	$-x$ M	$+x$ M	$+x$ M
at equilibrium	$(0.200 - x)$ M	$(0.200 - x)$ M	x M	x M

Now substitute the equilibrium concentrations (*not* the initial concentrations) into the equation for the equilibrium constant

$$K_c = 144 = \frac{[C][D]}{[A][B]}$$

$$144 = \frac{(x)(x)}{(0.200 - x)(0.200 - x)} = \frac{x^2}{(0.200 - x)^2}$$

We have a quadratic equation that is also a perfect square, and it can be solved by extracting the square root of both sides of the equation and then solving for x

$$\frac{x}{0.200 - x} = 12.0$$

$$x = 2.40 - 12.0x$$
$$13.0x = 2.40$$
$$x = \frac{2.40}{13.0} = 0.185\ M$$

Now that the value of x is known, the equilibrium concentrations may be easily obtained

$[A] = [B] = (0.200 - x)\ M = \underline{0.015\ M}$

$[C] = [D] = x = \underline{0.185\ M}$

These values may be substituted into the equilibrium expression to provide a quick check on the calculation. Recall that $K_c = 144$.

$$K_c = \frac{[C][D]}{[A][B]} = \frac{(0.185)^2}{(0.015)^2} = 1.5 \times 10^2$$

Note that the value 1.5×10^2 differs somewhat from 144. However, its calculation involved squaring two numbers, each of which contains considerable round-off error, and then dividing one of these numbers into the other. You might like to play with the number combinations generated by your calculator until you are convinced that the statement is correct.

Example 15–3 is similar to Example 15–2, except that A and B are mixed in nonstoichiometric amounts.

EXAMPLE 15–3

Consider the same system as in Example 15–2 at the same temperature. If 0.600 mole of A and 0.200 mole of B are mixed in a 2.00-liter container and allowed to reach equilibrium, what are the equilibrium concentrations of all species?

Solution

Let x = concentrations of A and B that have reacted, and
x = concentrations of C and D formed

	A	+	B	$\rightleftharpoons$	C	+	D
initial	$0.300\ M$		$0.100\ M$		$0\ M$		$0\ M$
change	$-x\ M$		$-x\ M$		$+x\ M$		$+x\ M$
equilibrium	$(0.300 - x)\ M$		$(0.100 - x)\ M$		$x\ M$		$x\ M$

Note that initial concentrations are governed only by the amounts of reactants mixed together. But *changes in concentrations* due to reaction must occur in a 1:1:1:1 ratio as required by the coefficients in the balanced equation. Now

$$K_c = \frac{[C][D]}{[A][B]} = 144$$

$$\frac{(x)(x)}{(0.300 - x)(0.100 - x)} = 144$$

This equation is not a perfect square but it is a quadratic equation, and we can rearrange it into the standard form of a quadratic equation

$$\frac{x^2}{0.0300 - 0.400x + x^2} = 144$$

$$x^2 = 4.32 - 57.6x + 144x^2$$

$$143x^2 - 57.6x + 4.32 = 0$$

All quadratic equations of the form

$$ax^2 + bx + c = 0$$

can be solved by use of the quadratic formula, which is

$$x = \frac{-b \pm \sqrt{b^2 - 4ac}}{2a} \quad \text{(see Appendix A)}$$

In this case, $a = 143$, $b = -57.6$, and $c = 4.32$. Substituting in the appropriate values

$$x = \frac{-(-57.6) \pm \sqrt{(-57.6)^2 - 4(143)(4.32)}}{2(143)} = \frac{57.6 \pm \sqrt{3318 - 2471}}{286}$$

$$= \frac{57.6 \pm \sqrt{847}}{286} = \frac{57.6 \pm 29.1}{286}$$

$$x = 0.303 \text{ or } 0.0997$$

Solutions of quadratic equations always yield two roots. One root has physical meaning (the answer) and the other root is extraneous, i.e., it has no physical meaning. The maximum acceptable value of x is 0.100 M (based on the chemical equation). The value of x is defined as the number of moles of A per liter and the number of moles of B per liter that react, and no more B can be consumed than was initally present (0.100 M), as would be the case if $x = 0.303$. Thus $x = 0.0997\ M$ is the root in which we are interested, and the extraneous root is 0.303.

The two roots for quadratic equations that are also perfect squares are equal in magnitude but opposite in sign. Clearly, concentrations can never be less than zero.

The equilibrium concentrations are

$$[A] = 0.300 - x = 0.200\ M$$

$$[B] = 0.100 - x = 0.0003\ M$$

$$[C] = [D] = x = 0.0997\ M$$

The following table summarizes the last two examples. Recall that $K_c = 144$.

	Initial Concentration (M)				Equilibrium Concentration (M)			
	[A]	[B]	[C]	[D]	[A]	[B]	[C]	[D]
Example 15–2	0.200	0.200	0	0	0.015	0.015	0.185	0.185
Example 15–3	0.300	0.100	0	0	0.200	0.0003	0.0997	0.0997

The data from the table can be substituted into the equilibrium constant expression to show that even though the reaction is initiated by different relative amounts of reactants in the two cases, the equilibrium ratios of concentrations of products to reactants (each raised to the first power) are identical within round-off error. Note that B is the limiting reagent in Example 15–3,

Very large round-off errors are generated because numbers with one or two significant figures cannot be used to calculate numbers with three significant figures.

and that less B remains at equilibrium than in Example 15–2, where reactants are present in stoichiometric amounts.

15–4 Reaction Quotient

"The Law of Mass Action" is another name for LeChatelier's Principle.

We define the **mass action expression** or **reaction quotient, Q,** for the general reaction

$$aA + bB \rightleftharpoons cC + dD$$

as

$$Q = \frac{[C]^c[D]^d}{[A]^a[B]^b} \quad \text{not necessarily equilibrium concentrations}$$

The reaction quotient has the same *form* as the equilibrium constant, except that it involves specific concentrations that are *not necessarily* equilibrium concentrations. However, if they are equilibrium concentrations, then $Q = K_c$. As we shall see, the concept of reaction quotient is extremely valuable because we can compare the magnitude of Q with that of K for any particular reaction under a given set of conditions to decide whether the forward or reverse reaction will proceed to a greater extent to establish equilibrium.

If at any time during an experiment $Q < K$, the forward reaction proceeds to a greater extent than the reverse reaction until equilibrium is established. This is because when $Q < K$, the numerator of the reaction quotient is too small and the denominator is too large. The only way to reduce the denominator and increase the numerator is by A and B reacting to produce C and D. Conversely, if $Q > K$, the reverse reaction proceeds to a greater extent than the forward reaction until equilibrium is established. When $Q = K$, the system is at equilibrium and both forward and reverse reactions occur at the same rate. In summary,

$Q < K$	forward reaction occurs to a greater extent than reverse reaction until equilibrium is established
$Q = K$	system is at equilibrium
$Q > K$	reverse reaction occurs to a greater extent than forward reaction until equilibrium is established

EXAMPLE 15–4

At a very high temperature, $K_c = 1 \times 10^{-13}$ for

$$2HF\ (g) \rightleftharpoons H_2\ (g) + F_2\ (g)$$

At a certain time the following concentrations were detected. Is the system at equilibrium? If not, what must occur in order for equilibrium to be established?

$[HF] = 0.500\ M$, $[H_2] = 1.00 \times 10^{-3}\ M$, and $[F_2] = 4.00 \times 10^{-3}\ M$

Solution

We substitute these concentrations into the mass action expression

$$Q = \frac{[H_2][F_2]}{[HF]^2} = \frac{(1.00 \times 10^{-3})(4.00 \times 10^{-3})}{(0.500)^2} = 1.60 \times 10^{-5}$$

and find that $Q = 1.60 \times 10^{-5}$ and therefore $Q > K_c$. The system is *not* at equilibrium. In order for equilibrium to be established, the value of Q must become smaller until it equals K_c. This can occur only if the numerator decreases and the denominator increases. Thus, the right-to-left reaction must occur to a greater extent than the forward reaction, i.e., H_2 and F_2 react to form more HF.

15–5 Factors That Affect Equilibria

Once a reacting system has achieved equilibrium, it remains at equilibrium until it is disturbed by some change of conditions. We shall now consider the different types of changes, remembering that the *value* of an equilibrium constant changes only with temperature. The guiding principle is known as **LeChatelier's Principle,** which was introduced in Section 9–5. It states: *If a change of conditions (stress) is applied to a system at equilibrium, the system will respond in the way that best reduces the stress in reaching a new state of equilibrium.* There are four types of changes to consider: (1) concentration changes, (2) pressure changes (which can also be treated as volume changes for gas phase reactions), (3) temperature changes, and (4) introduction of catalysts.

For reactions involving gases at constant temperature, pressure changes cause volume changes and vice versa.

In this section we shall look at the effects of these types of stresses from a purely qualitative or descriptive point of view. In Section 15–6 we shall illustrate the validity of our discussion with several quantitative examples.

1 Changes in Concentration

Consider the following system at equilibrium:

$$A + B \rightleftharpoons C + D \qquad K_c = \frac{[C][D]}{[A][B]}$$

If an additional amount of any reactant or product is *added* to the system, the stress is relieved by shifting the equilibrium in the direction that consumes some of the added substance. Let us compare the mass action expressions for Q and K. If more A or more B is added, then $Q < K$ and the forward reaction proceeds to a greater extent than the reverse reaction until equilibrium is reestablished. If more C or D is added, $Q > K$ and the reverse reaction occurs more rapidly until equilibrium is re-established.

The terminology used here is not as precise as we might like. However, it is widely used. When we say the equilibrium is shifted to the left, we mean that the reaction to the left occurs to a greater extent than the reaction to the right.

If a reactant or product is *removed* from a system at equilibrium, the equilibrium shifts in the direction that produces the substance. For example, if some C is removed, then $Q < K$, and the forward reaction is favored until equilibrium is re-established. If A is removed, the equilibrium shifts to the left. To summarize

Stress	Q	Direction of Shift
Increase concentration of A or B	$Q < K$	→ right
Increase concentration of C or D	$Q > K$	left ←
Decrease concentration of A or B	$Q > K$	left ←
Decrease concentration of C or D	$Q < K$	→ right

2 Changes in Volume and Pressure

Changes in pressure do not significantly affect the concentrations of solids or liquids, since they are nearly incompressible. However, concentrations of gases are significantly altered by changes in pressure. Therefore, such changes affect the value of Q for reactions in which the number of moles of gaseous reactants differs from the number of moles of gaseous products. For an ideal gas

$$PV = nRT \quad \text{and} \quad P = (n/V)(RT)$$

The term n/V is concentration in moles per liter. At constant temperature n, R, and T are constants. Thus, if the volume occupied by a gas decreases, its partial pressure increases and its concentration (n/V) increases. If the volume increases, both its partial pressure and its concentration decrease.

Consider the following gaseous system at equilibrium

$$\text{A (g)} \rightleftharpoons \text{2D (g)} \qquad K = \frac{[\text{D}]^2}{[\text{A}]}$$

At constant temperature, an increase in pressure (decrease in volume) increases the concentrations of both A and D. In the mass action expression, the concentration of D is squared and the concentration of A is raised to the first power. As a result, the numerator of Q increases more dramatically with a given increase in pressure than does the denominator. Thus, $Q > K$ and this equilibrium shifts to the left. Conversely, a decrease in pressure (increase in volume) shifts this reaction to the right until equilibrium is re-established, since $Q < K$. Thus, we may summarize the effect of pressure (volume) changes on *this* gas phase system at equilibrium:

Stress	Q*	Direction of Shift
Pressure increase (Volume decrease)	$Q > K$	Toward smaller number of moles of gas
Pressure decrease (Volume increase)	$Q < K$	Toward larger number of moles of gas

* Refers to Q for *this* reaction *as written.*

In general an increase in pressure (decrease in volume) shifts a system at equilibrium in the direction that produces the smaller number of moles of gases, and a decrease in pressure shifts it in the opposite direction. If there is no change in number of moles of gases in a reaction, a pressure (volume) change does not affect the position of equilibrium.

It should be noted that the foregoing argument applies only when pressure changes are due to volume changes. It *does not apply* if the total pressure of a gaseous system is raised by pumping in an inert gas, which has no effect on the position of equilibrium. In such a situation, the *partial* pressure of each reacting gas remains constant.

3 Changes in Temperature

Consider the following system at equilibrium

$$\text{A + B} \rightleftharpoons \text{C + D + Heat}$$

Heat is a product of the forward (exothermic) reaction. Increasing the temperature at constant pressure and volume increases the amount of heat in the system. This drives the equilibrium to the left, consuming some of the extra heat. Lowering the temperature shifts the reaction to the right as the system replenishes some of the heat removed. Likewise, for the following system at equilibrium

$$W + X + \text{Heat} \rightleftharpoons Y + Z$$

an increase in temperature at constant pressure and volume shifts the reaction to the right, and a decrease in temperature shifts it to the left. In fact, the *values* of equilibrium constants change as temperature changes. No other stresses affect the value of K.

4 Introduction of a Catalyst

A catalyst may be added to a system to change the rate of a reaction (Section 14–6), but it *cannot* shift the equilibrium in favor of either products or reactants. Since a catalyst affects the activation energy of both forward and reverse reactions, it changes both equally, and only the time required to establish equilibrium is altered.

Not all reactions attain equilibrium; they simply occur too slowly, or else products or reactants are continually added or removed. Such is the case with most reactions in biological systems. On the other hand, some reactions, such as typical acid-base neutralizations, achieve equilibrium very rapidly.

The Haber Process

The Haber process for the industrial production of ammonia is interesting because it provides insight into both the kinetic and thermodynamic factors that influence reaction rates and the positions of equilibria. Ammonia is produced by the combination of nitrogen and hydrogen, an exothermic, reversible reaction that in fact never reaches a state of equilibrium, but moves toward it.

More than 38 billion pounds of NH_3 were produced in the USA in 1981.

$$N_2\ (g) + 3H_2\ (g) \rightleftharpoons 2NH_3\ (g) + 92.22\ \text{kJ}$$

$$K_c = \frac{[NH_3]^2}{[N_2][H_2]^3} = 4.0 \times 10^8$$

The process is diagrammed in Figure 15–4. The reaction is run at about 450°C under pressures ranging from 200 to 1000 atmospheres. Hydrogen is obtained from coal gas or petroleum refining, and nitrogen from liquefied air.

The equilibrium constant is 4.0×10^8 at 25°C, and the very large value of K_c indicates that *at equilibrium* virtually all of the nitrogen and hydrogen (mixed in a 1:3 mole ratio) are converted to ammonia. However, the reaction occurs so slowly at 25°C that no measurable amount of ammonia is produced within a reasonable period of time. Thus, the large equilibrium constant (a thermodynamic factor) indicates that the reaction proceeds toward the right almost completely, *but* it tells us nothing about how fast the reaction occurs (a kinetic factor).

Since there are four moles of gases on the left side of the balanced equation and only two moles of gas on the right side, increasing the pressure on this system shifts the equilibrium to the right and favors the production of ammo-

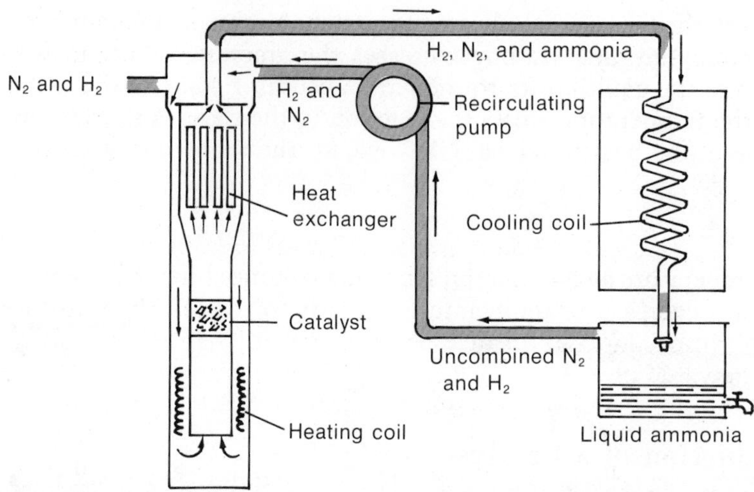

FIGURE 15–4 A simplified representation of the Haber process for synthesizing ammonia.

nia. Therefore, the Haber process is carried out at very high pressures, as high as the equipment will safely stand.

The reaction is exothermic ($\Delta H^0_{rxn} = -92.22$ kJ), so increasing the temperature shifts the equilibrium to the left and favors the decomposition of ammonia. The rates of both forward and reverse reactions increase as temperature increases. Therefore the reaction is carried out at high temperatures *to speed up* the reaction, even though the equilibrium is shifted somewhat to the left. The economic advantage of the increase in rate outweighs the disadvantage of shifting the equilibrium to the left and thus decreasing the yield of ammonia.

The addition of a catalyst consisting of finely divided iron and small amounts of selected oxides speeds up both the forward and reverse reactions because activation energies are lowered. This enables the reaction to proceed at a lower temperature, which is more favorable for yield and for life of the equipment.

TABLE 15–1 Effect of *T* and *P* on Yield of Ammonia

		Mole % NH_3 in Equilibrium Mixture		
°C	K_c	10 atm	100 atm	1000 atm
200	650	51	82	98
400	0.5	4	25	80
600	0.014	0.5	5	13

Table 15–1 shows the effects of increases in temperature and pressure on the yield of ammonia from the Haber process. Note that K_c decreases from 4.0×10^8 at 25°C to only 1.4×10^{-2} at 600°C, which tells us that the equilibrium is shifted *very far to the left* as the temperature is raised. Casual examina-

tion of the data in Table 15–1 might indicate that the reaction should be run at 200°C, since a high percentage of the nitrogen and hydrogen mixture is converted into ammonia. However, the reaction occurs so slowly, even in the presence of a catalyst, that it cannot be run economically until the temperature reaches about 450°C, at as high a pressure as possible. In practice, the mixed reactants are compressed by special pumps and injected into the heated reaction vessel.

The emerging reaction mixture is cooled down and ammonia (bp = −33.43°C) is removed periodically by liquefaction. This, of course, favors the forward reaction. The unreacted nitrogen and hydrogen are recycled. Excess nitrogen is used to shift the equilibrium to the right.

The Haber process is economically the most important process by which atmospheric nitrogen is converted to a soluble, reactive form (N_2 is extremely stable and therefore very unreactive). Many dyes, plastics, explosives, fertilizers, and synthetic fibers are nitrogen-containing compounds produced from ammonia.

15–6 Application of Stress to a System at Equilibrium

Example 15–5 illustrates how an equilibrium constant can be used to determine a set of new equilibrium concentrations resulting from addition of one or more species to a system at equilibrium.

EXAMPLE 15–5

Some hydrogen and iodine are mixed at 400°C in a 1.00-liter container, and when equilibrium is established the following concentrations are present: [HI] = 0.490 *M*, $[H_2]$ = 0.080 *M*, and $[I_2]$ = 0.060 *M*. If an additional 0.300 mole of HI is added, what concentrations will be present when the new equilibrium is established?

$$H_2\ (g) + I_2\ (g) \rightleftharpoons 2HI\ (g)$$

Solution

We can calculate the equilibrium constant from the first set of equilibrium concentrations

$$K_c = \frac{[HI]^2}{[H_2][I_2]} = \frac{(0.490)^2}{(0.080)(0.060)} = 50$$

When 0.300 mole of HI is added, the concentration of HI instantaneously increases by 0.300 *M* (the volume is 1.00 liter).

	H_2 (g)	+ I_2 (g)	$\rightleftharpoons$ 2HI (g)
equilibrium	0.080 *M*	0.060 *M*	0.490 *M*
mol/L added	0 *M*	0 *M*	+0.300 *M*
new initial conc'n	0.080 *M*	0.060 *M*	0.790 *M*

Substitution of these concentrations into the mass action expression shows that $Q > K$

$$Q = \frac{[HI]^2}{[H_2][I_2]} = \frac{(0.790)^2}{(0.080)(0.060)} = 130$$

Since $Q > K$, the equilibrium shifts to the left to establish a new equilibrium, and the new equilibrium concentrations can be determined as follows. Let x = concentration of H_2 and I_2 formed, and $2x$ = concentration of HI consumed. Then

	H_2 (g)	+	I_2 (g)	$\rightleftharpoons$	2HI(g)
new initial	0.080 M		0.060 M		0.790 M
change	$+x$ M		$+x$ M		$-2x$ M
new equilibrium	$(0.080 + x)$ M		$(0.060 + x)$ M		$(0.790 - 2x)$ M

Substitution of these values into K_c allows us to evaluate x

$$K_c = 50 = \frac{(0.790 - 2x)^2}{(0.080 + x)(0.060 + x)}$$

$$50 = \frac{0.624 - 3.16x + 4x^2}{0.0048 + 0.14x + x^2}$$

$$0.24 + 7.0x + 50x^2 = 0.624 - 3.16x + 4x^2$$
$$46x^2 + 10.2x - 0.38 = 0$$
$$x^2 + 0.22x - 0.0083 = 0$$

Solution by the quadratic formula gives $x = 0.033$ or -0.25.

Clearly $x = -0.25$ is the extraneous root, because x cannot be less than zero (nor greater than 0.790 M/2); it is impossible to consume a negative quantity of HI, since the equilibrium is shifted toward the left. Since $x = 0.033$ M is the root with physical meaning, the new equilibrium concentrations are

$$[H_2] = (0.080 + x)\ M = (0.080 + 0.033)\ M = \underline{0.113\ M}$$

$$[I_2] = (0.060 + x)\ M = (0.060 + 0.033)\ M = \underline{0.093\ M}$$

$$[HI] = (0.790 - 2x)\ M = (0.790 - 0.066)\ M = \underline{0.724\ M}$$

The data below summarize this example.

Original Equilibrium	Stress Applied	New Equilibrium
$[H_2]$ = 0.080 M $[I_2]$ = 0.060 M [HI] = 0.490 M	Add 0.300 M HI	$[H_2]$ = 0.113 M $[I_2]$ = 0.093 M [HI] = 0.724 M

Knowing the value of the equilibrium constant is crucial to the determination of the new equilibrium concentrations. We see that some of the additional HI is consumed, but not all of it. More HI remains after the new equilibrium is established than was present before the stress was imposed. However, the new equilibrium concentrations of H_2 and I_2 are substantially larger than the original equilibrium concentrations. At the new equilibrium conditions, the equilibrium constant, K_c, must still be satisfied (within round-off error).

Example 15–6 shows how equilibrium constants can be used to calculate new equilibrium concentrations resulting from decreasing the volume (increasing the pressure) of a gaseous system that was initially at equilibrium.

EXAMPLE 15–6

At 25°C the equilibrium constant, K_c, for the following reaction is 4.66×10^{-3}. (a) If 0.800 mole of N_2O_4 is injected into a closed 1.00-liter glass container at 25°C, what will the equilibrium concentrations of the two gases be? (b) What will the concentrations be at equilibrium if the volume is suddenly halved at constant temperature?

$$N_2O_4\ (g) \rightleftharpoons 2NO_2\ (g) \qquad K_c = \frac{[NO_2]^2}{[N_2O_4]} = 4.66 \times 10^{-3}$$

Solution

(a)

	N_2O_4 (g)	$\rightleftharpoons$ $2NO_2$ (g)
initial	0.800 *M*	0 *M*
change	$-x$ *M*	$+2x$ *M*
equilibrium	$(0.800 - x)$ *M*	$2x$ *M*

$$K_c = 4.66 \times 10^{-3} = \frac{(2x)^2}{0.800 - x} = \frac{4x^2}{0.800 - x}$$

$$3.73 \times 10^{-3} - 4.66 \times 10^{-3}x = 4x^2$$
$$4x^2 + 4.66 \times 10^{-3}x - 3.73 \times 10^{-3} = 0$$
$$x^2 + 1.16 \times 10^{-3}x - 9.32 \times 10^{-4} = 0$$

Solving by the quadratic formula gives $x = 3.00 \times 10^{-2}$ and $x = -3.11 \times 10^{-2}$. The value of x must be between 0 *M* and 0.800 *M*, and the acceptable value of x is

$x = 3.00 \times 10^{-2}\ M$

The original equilibrium concentrations are

$[NO_2] = 2x\ M = \underline{6.00 \times 10^{-2}\ M}$

$[N_2O_4] = (0.800 - x)\ M = (0.800 - 3.00 \times 10^{-2})\ M = \underline{0.770\ M}$

(b) When the volume of the reaction vessel is halved (the pressure is doubled), the concentrations are doubled instantaneously, so the new initial concentrations of N_2O_4 and NO_2 are $2(0.770\ M) = 1.54\ M$ and $2(6.00 \times 10^{-2}\ M) = 0.120\ M$, respectively. A decrease in volume (increase in pressure) favors the production of N_2O_4, i.e., shifts equilibrium to the left.

	N_2O_4 (g)	$\rightleftharpoons$ $2NO_2$ (g)	
new initial	1.54 *M*	0.120 *M*	$Q = \frac{(0.120)^2}{1.54} = 9.35 \times 10^{-3}$
change	$+x$ *M*	$-2x$ *M*	$Q > K$ shift left ←
new equil	$(1.54 + x)$ *M*	$(0.120 - 2x)$ *M*	

$$K_c = 4.66 \times 10^{-3} = \frac{(0.120 - 2x)^2}{1.54 + x}$$

$$7.18 \times 10^{-3} + 4.66 \times 10^{-3}x = 1.44 \times 10^{-2} - 0.480x + 4x^2$$

Rearranging into the general form of a quadratic equation

$$4x^2 - 0.485x + 7.22 \times 10^{-3} = 0$$
$$x^2 - 0.121x + 1.81 \times 10^{-3} = 0$$

Solving by the quadratic formula gives the value of x: $x = 0.104$ and 0.017.

The maximum value of x is 0.060 *M* (or 0.120 *M*/2), since $2x$ may not exceed the concentration of NO_2 initially present when the volume is halved. Thus, $x = 0.104$ is the extraneous root and $x = 0.017$ *M* is the root with physical significance. The new equilibrium concentrations in the 0.500 liter container are

$[NO_2] = (0.120 - 2x)\ M = (0.120 - 0.034)\ M = \underline{0.086\ M}$

$[N_2O_4] = (1.54 + x)\ M = (1.54 + 0.017) = \underline{1.56\ M}$

The data below summarize this example.

First Equilibrium	Stress	New Equilibrium
$[N_2O_4] = 0.770\ M$ $[NO_2] = 0.0600\ M$	Decrease volume from 1.00 L to 0.500 L	$[N_2O_4] = 1.56\ M$ $[NO_2] = 0.086\ M$

The fact that the concentrations of both N_2O_4 and NO_2 *increase* is a consequence of the large decrease in volume. But you may verify that the *number of moles* of N_2O_4 increases and the *number of moles* of NO_2 decreases, as one would predict by applying LeChatelier's Principle.

15–7 Partial Pressures and the Equilibrium Constant

It is often more convenient to measure pressures rather than concentrations of gases. Recall that if the ideal gas law is solved for pressure, the result is

$PV = nRT$

$$P = \frac{n}{V}(RT)$$

Thus, the pressure of a gas is directly proportional to its concentration (n/V). For equilibria involving gases, the equilibrium constant is often expressed in terms of partial pressures in atmospheres (K_p) rather than in terms of concentrations (K_c). For instance, for the following reversible reaction

$$N_2\ (g) + 3H_2\ (g) \rightleftharpoons 2NH_3\ (g)$$

K_p is defined as

$$K_p = \frac{(P_{NH_3})^2}{(P_{N_2})(P_{H_2})^3}$$

EXAMPLE 15–7

The following partial pressures were measured at equilibrium at 500°C: $P_{NH_3} = 0.147$ atm, $P_{N_2} = 5.00$ atm, and $P_{H_2} = 6.00$ atm. Evaluate K_p at 500°C for the reaction.

$$N_2\ (g) + 3H_2\ (g) \rightleftharpoons 2NH_3\ (g)$$

Solution

$$K_p = \frac{(P_{NH_3})^2}{(P_{N_2})(P_{H_2})^3} = \frac{(0.147\ \text{atm})^2}{(5.00\ \text{atm})(6.00\ \text{atm})^3}$$

$$K_p = 2.0 \times 10^{-5}\ \text{atm}^{-2}$$

In general, for a reaction involving gases,

$$aA\ (g) + bB\ (g) \rightleftharpoons cC\ (g) + dD\ (g)$$

K_p is defined as

$$K_p = \frac{(P_C)^c(P_D)^d}{(P_A)^a(P_B)^b}$$

15–8 Relationship between K_p and K_c

The term n/V is a concentration term (moles per liter). If the ideal gas law is rearranged, the concentration of a gas is

$$\frac{n}{V} = \frac{P}{RT} \quad \text{or} \quad [\text{conc'n}] = \frac{P}{RT}$$

Substituting P/RT for n/V into the K_c expression for the ammonia equilibrium gives the relationship between K_c and K_p

$$K_c = \frac{[NH_3]^2}{[N_2][H_2]^3} = \frac{\left(\frac{P_{NH_3}}{RT}\right)^2}{\left(\frac{P_{N_2}}{RT}\right)\left(\frac{P_{H_2}}{RT}\right)^3} = \frac{(P_{NH_3})^2}{(P_{N_2})(P_{H_2})^3} \times \frac{\left(\frac{1}{RT}\right)^2}{\left(\frac{1}{RT}\right)^4}$$

$K_c = K_p\,(RT)^2 \quad \text{and} \quad K_p = K_c\,(RT)^{-2}$

In general, the relationship between K_c and K_p is

$$K_p = K_c\,(RT)^{\Delta n}$$

where Δn is the number of moles of *gaseous* products minus the number of moles of *gaseous* reactants in the balanced equation

$$\Delta n = (n_{\text{gas prod}}) - (n_{\text{gas react}})$$

For reactions in which there are the same number of moles of gases on both sides of the balanced equation, $\Delta n = 0$ and $K_p = K_c$. For the ammonia equilibrium, $N_2\text{ (g)} + 3H_2\text{ (g)} \rightleftharpoons 2NH_3\text{ (g)}$, we find $\Delta n = -2$, so at 500°C

$$K_p = K_c\,(RT)^{\Delta n} \quad \text{and} \quad K_c = K_p\,(RT)^{-\Delta n}$$
$$K_c = (2.0 \times 10^{-5}\text{ atm}^{-2})[(0.0821\text{ L} \cdot \text{atm/mol K})(773\text{ K})]^{-(-2)}$$
$$= (2.0 \times 10^{-5})(4028)\text{ L}^2/\text{mol}^2 = 0.081\ M^{-2}$$

which agrees with the value at 500°C calculated in Example 15–1, within the range of round-off error.

15–9 Heterogeneous Equilibria

Thus far we have considered only equilibria involving species in a single phase, i.e., homogeneous equilibria. **Heterogeneous equilibria** are equilibria involving species in more than one phase. Consider the following reversible reaction at 25°C:

$$2HgO\text{ (s)} \rightleftharpoons 2Hg\text{ }(\ell) + O_2\text{ (g)}$$

When equilibrium is established for this system, a solid, a liquid, and a gas are present. Neither solids nor liquids are affected significantly by changes in pressure unless the changes are extreme. The concentrations of solids and pure liquids are directly proportional to their densities, which vary only with temperature.* Therefore, the concentrations and partial pressures of pure solids and liquids remain constant in any reaction as long as any solid or liquid is present and the temperature remains constant. As a result, we can simplify the equilibrium constant expressions for the above reaction, which we represent as

$$K_c' = \frac{[Hg]^2[O_2]}{[HgO]^2} \quad \text{and} \quad K_p' = \frac{(P_{Hg})^2(P_{O_2})}{(P_{HgO})^2},$$

by including the concentrations or partial pressures of Hg and HgO (which are constants) in the values of the equilibrium constants. We do this by multiplying both sides of the K_c' equation by $[HgO]^2/[Hg]^2$, and by multiplying both sides of the K_p' equation by $(P_{HgO})^2/(P_{Hg})^2$. The new equilibrium constants are K_c and K_p.

$$\underbrace{K_c' \frac{[HgO]^2}{[Hg]^2}}_{\text{constants}} = [O_2] = K_c \qquad \underbrace{K_p' \frac{(P_{HgO})^2}{(P_{Hg})^2}}_{\text{constants}} = P_{O_2} = K_p$$

The equilibrium constant expressions indicate that equilibrium exists at a given temperature for *one and only one* concentration and one partial pressure of oxygen in contact with liquid mercury and solid mercury(II) oxide.

By convention, tabulated values of equilibrium constants for reactions involving solids and/or pure liquids include the concentrations (partial pressures) of the solids and liquids as part of the constants. Therefore, concentration and partial pressure terms for pure liquids and solids are omitted from equilibrium constant expressions.

EXAMPLE 15–8

Write both K_c and K_p expressions for the following reversible reactions:

Solution

(a) $2SO_2\ (g) + O_2\ (g) \rightleftharpoons 2SO_3\ (g)$

$$K_c = \frac{[SO_3]^2}{[SO_2]^2[O_2]} \qquad K_p = \frac{(P_{SO_3})^2}{(P_{SO_2})^2(P_{O_2})}$$

(b) $CaCO_3\ (s) \rightleftharpoons CaO\ (s) + CO_2\ (g)$

$$K_c = [CO_2] \qquad K_p = P_{CO_2}$$

(c) $H_2O\ (\ell) + CO_2\ (g) \rightleftharpoons H_2CO_3\ (aq)$

$$K_c = \frac{[H_2CO_3]}{[CO_2]} \qquad K_p = \frac{1}{P_{CO_2}} = (P_{CO_2})^{-1}$$

(d) $2NH_3\ (g) + H_2SO_4\ (\ell) \rightleftharpoons (NH_4)_2SO_4\ (s)$

$$K_c = \frac{1}{[NH_3]^2} = [NH_3]^{-2} \qquad K_p = \frac{1}{(P_{NH_3})^2} = (P_{NH_3})^{-2}$$

Liquids acting as solvents in all but the most concentrated solutions are "nearly pure" liquids in the sense that almost all molecules in the solution are solvent molecules. These liquids are treated as if they were pure.

$$^* \frac{\text{mol}}{\text{L}} = \frac{\text{g}}{\text{mL}} \times \frac{1000\ \text{mL}}{\text{L}} \times \frac{\text{mol}}{\text{g}}$$

$$\text{conc} = \text{density} \times \text{constant} \times \frac{1}{\text{formula weight}}$$

$$= \text{constant} \times \text{constant} \times \text{constant}$$

(e) $S\ (s) + H_2SO_3\ (aq) \rightleftharpoons H_2S_2O_3\ (aq)$

$K_c = \dfrac{[H_2S_2O_3]}{[H_2SO_3]}$ K_p undefined; no gases involved

15–10 Relationship Between ΔG^0 and the Equilibrium Constant

The standard free energy change for a reaction is ΔG^0. This is the free energy change that accompanies complete conversion of reactants to products, at standard conditions (*all* reactants initially present in one molar concentrations in solution, or as pure solids or pure liquids, or as gases at partial pressures of one atmosphere converted to *all* products in the same conditions). The free energy change under any other set of conditions is ΔG (no superscript zero). The two quantities are related by the equation

$$\Delta G = \Delta G^0 + RT \ln Q$$

Demonstration of the validity of this relationship is beyond the scope of this text.

or, after converting natural log ($\log_e$ or ln) to $\log_{10}$ (common log),

$$\Delta G = \Delta G^0 + 2.303RT \log Q$$

where R is the universal gas constant, T is the absolute temperature, and Q is the reaction quotient. Recall from Section 15–4 that the reaction quotient has the same form as the equilibrium expression for a reaction, but does not (necessarily) involve concentrations at equilibrium.

When a system is at equilibrium, $\Delta G = 0$ (see Section 13–12), and

$$0 = \Delta G^0 + 2.303RT \log Q$$

or

$$\Delta G^0 = -2.303RT \log Q$$

At equilibrium $Q = K$, and therefore

$$\Delta G^0 = -2.303RT \log K$$

This equation shows the relationship between the standard free energy change and the thermodynamic equilibrium constant for a reaction. In general, the thermodynamic equilibrium constant is a sort of "hybrid" of K_p and K_e. The mass action expression to which it is related involves concentration terms for species in solution and partial pressures for gases. For reactions involving only gases, the thermodynamic equilibrium constant (related to ΔG^0) is K_p. How can we rationalize the fact that ΔG^0 is the free energy change that occurs during a reaction *beginning* with all reactants at unit molarities or partial pressures, and ending with all products at unit molarities or partial pressures, while K involves equilibrium concentrations?

When ΔG^0 is negative, the reaction is spontaneous, i.e., the forward reaction may proceed to a greater extent than the reverse reaction *until* equilibrium is established. At that instant ($\Delta G = 0$, and $Q = K$) and thereafter, equilibrium concentrations are present. Conversely, if ΔG^0 for a reaction is positive, the *reverse reaction* is spontaneous and is favored over the forward reaction until equilibrium is established. In the special case where the system is already at equilibrium (when all species are present in unit concentrations or at unit partial pressures), $\Delta G^0 = 0$ and $Q = K = 1$.

Thus, the sign of ΔG^0 indicates whether the forward or reverse reaction is spontaneous at unit concentrations or partial pressures, and the magnitude indicates how spontaneous or nonspontaneous a reaction is under these conditions. When ΔG^0 is large and negative, K is very large, indicating that *after equilibrium is established,* the ratio of products to reactants will be very large. When ΔG^0 is large and positive, K is very small, and *at equilibrium* the ratio of products to reactants will be very small. In other words, ΔG^0 tells us *how far away from equilibrium* the starting reaction mixture is and whether the reaction occurs to the right or left to achieve equilibrium. To summarize:

ΔG^0 is $-$, $K > 1$	forward reaction is spontaneous at unit concentrations or partial pressures
ΔG^0 is 0, $K = 1$	system is at equilibrium at unit concentrations or partial pressures (very rare case)
ΔG^0 is $+$, $K < 1$	reverse reaction is spontaneous at unit concentrations or partial pressures

Always keep in mind that if a reaction is spontaneous, the reverse reaction is nonspontaneous. Also, $\Delta G = 0$ when a system is at equilibrium, and ΔG^0 is usually not equal to zero.

EXAMPLE 15–9

Using the data in Appendix K, evaluate the equilibrium constant for the following reaction at 25°C.

$$2C_2H_2\ (g) + 5O_2\ (g) \rightleftharpoons 4CO_2\ (g) + 2H_2O\ (g)$$

Solution

We may evaluate ΔG^0_{rxn} for the forward reaction from ΔG^0_f values.

$$\begin{aligned}\Delta G^0_{rxn} &= 4\Delta G^0_{f\,CO_2(g)} + 2\Delta G^0_{f\,H_2O(g)} - [2\Delta G^0_{f\,C_2H_2(g)} + 5\Delta G^0_{f\,O_2(g)}]\\ &= [4(-394.4) + 2(-228.6)]\ \text{kJ} - [2(209.2) + 5(0)]\ \text{kJ}\\ &= -2.45 \times 10^3\ \text{kJ} \quad \text{or} \quad -2.45 \times 10^6\ \text{J}\end{aligned}$$

This is a gas-phase reaction, and ΔG^0_{rxn} is related to K_p by

$$\Delta G^0_{rxn} = -2.303RT \log K_p$$

$$\log K_p = \frac{-2.45 \times 10^6\ \text{J/mol}}{-2.303(8.314\ \text{J/mol}\cdot\text{K})(298\ \text{K})}$$

$$\log K_p = 429$$

$$K_p = \text{antilog } 429 = \underline{10^{429}}$$

The very large value for K_p tells us that the equilibrium lies *very* far to the right.

EXAMPLE 15–10

In Examples 13–12 and 13–13 we evaluated ΔG^0 (+173.1 kJ) for the following reaction. Calculate K_p for this reaction at 25°C.

$$N_2\ (g) + O_2\ (g) \rightleftharpoons 2NO\ (g)$$

Solution

$$\Delta G^0 = -2.303RT \log K_p$$

$$\log K_p = \frac{\Delta G^0}{-2.303RT} = \frac{1.731 \times 10^5 \text{ J/mol}}{(-2.303)(8.314 \text{ J/mol K})(298 \text{ K})}$$

$$\log K_p = -30.34$$
$$K_p = 10^{0.66} \times 10^{-31}$$
$$K_p = 4.6 \times 10^{-31}$$

Clearly, this very small number indicates that at equilibrium almost no N_2 and O_2 are converted to NO at 25°C. For all practical purposes, the reaction does not occur at 25°C.

EXAMPLE 15–11

The equilibrium constant, K_p, for the following reaction is 5.04×10^{17} atm^{-1} at 25°C. Calculate ΔG^0.

$$C_2H_4\text{ (g)} + H_2\text{ (g)} \rightleftharpoons C_2H_6\text{ (g)}$$

ethylene ethane

Solution

$$\Delta G^0 = -2.303RT \log K_p$$
$$= -2.303(8.314)(298) \log (5.04 \times 10^{17})$$
$$= -5706 \log (5.04 \times 10^{17})$$
$$= -5706(17.702)$$
$$= -1.010 \times 10^5 \text{ J}$$
$$\Delta G^0 = -101 \text{ kJ}$$

15–11 Evaluation of Equilibrium Constants at Different Temperatures

Chemists have determined equilibrium constants for thousands of reactions. It would be an impossibly large task to catalog such constants at every temperature of interest for each reaction. Fortunately, there is no need to do this. If we determine the equilibrium constant, K_{T_1}, for a reaction at one temperature, T_1, and also its ΔH^0, we can then estimate the equilibrium constant at a second temperature, T_2, using the **van't Hoff equation:**

$$\frac{\Delta H^0(T_2 - T_1)}{2.303RT_2T_1} = \log \left(\frac{K_{T_2}}{K_{T_1}}\right)$$

Derivation of the van't Hoff Equation

Let us derive this very useful equation. The following relationships apply for standard free energy changes

$$\Delta G^0 = \Delta H^0 - T\Delta S^0$$
$$\Delta G^0 = -2.303RT \log K$$

Thus,

$$\Delta H^0 - T\Delta S^0 = -2.303RT \log K$$

This is strictly true only for constant temperature and pressure. However, ΔH^0 and ΔS^0 are approximately temperature-independent, i.e., they usually vary only slightly with temperature.

Let's consider a reversible reaction at equilibrium at two temperatures, T_1 and T_2, with equilibrium constants K_{T_1} and K_{T_2}. We write the preceding equation for each temperature

$$\Delta H^0 - T_1\Delta S^0 = -2.303RT_1 \log K_{T_1} \quad (1)$$

and

$$\Delta H^0 - T_2\Delta S^0 = -2.303RT_2 \log K_{T_2} \quad (2)$$

We may rearrange each equation

$$\frac{\Delta H^0}{T_1} - \Delta S^0 = -2.303R \log K_{T_1} \quad (3)$$

and

$$\frac{\Delta H^0}{T_2} - \Delta S^0 = -2.303R \log K_{T_2} \quad (4)$$

If we subtract equation (4) from (3) we get (5)

$$\frac{\Delta H^0}{T_1} - \frac{\Delta H^0}{T_2} - \Delta S^0 - (-\Delta S^0) = -2.303R(\log K_{T_1} - \log K_{T_2}) \quad (5)$$

or

$$\Delta H^0\left(\frac{1}{T_1} - \frac{1}{T_2}\right) = -2.303R \log\left(\frac{K_{T_1}}{K_{T_2}}\right) \quad (6)$$

or

$$\Delta H^0\left(\frac{1}{T_1} - \frac{1}{T_2}\right) = 2.303R \log\left(\frac{K_{T_2}}{K_{T_1}}\right) \quad (7)$$

Multiplying $1/T_1$ by T_2/T_2 and $1/T_2$ by T_1/T_1, we get (8) (this amounts to obtaining a common denominator)

$$\Delta H^0\left(\frac{T_2 - T_1}{T_1 T_2}\right) = 2.303R \log\left(\frac{K_{T_2}}{K_{T_1}}\right) \quad (8)$$

Dividing both sides by $2.303R$ gives (9), the van't Hoff equation

$$\frac{\Delta H^0(T_2 - T_1)}{2.303RT_2T_1} = \log\left(\frac{K_{T_2}}{K_{T_1}}\right) \quad (9)$$

Thus, if we know ΔH^0 for a reaction and K (or ΔG^0, from which we can calculate K) at a given temperature (say 298 K), we can use the van't Hoff equation to calculate the value of K at any other temperature.

EXAMPLE 15–12

We found in Examples 15–10 and 13–13 that $K_p = 4.6 \times 10^{-31}$ at 25°C and $\Delta H^0 =$ 180.5 kJ for the reaction

$$N_2\ (g) + O_2\ (g) \rightleftharpoons 2NO\ (g)$$

Evaluate the equilibrium constant for this reaction at 2400 K, and then let us compare $K_{2400\,K}$ with $K_{298\,K}$.

This is a typical temperature inside the combustion chambers of automobile engines. Large quantities of N_2 and O_2 are present during gasoline combustion, since the gasoline is mixed with air.

Solution

Let $T_1 = 298$ K and $T_2 = 2400$ K. Then

$$\frac{\Delta H^0(T_2 - T_1)}{2.303RT_2T_1} = \log\left(\frac{K_{T_2}}{K_{T_1}}\right)$$

We know all of the terms except K_{T_2}. Let us solve for $\log K_{T_2}$. First we use the general relation, $\log (x/y) = \log x - \log y$

$$\frac{\Delta H^0(T_2 - T_1)}{2.303RT_2T_1} = \log K_{T_2} - \log K_{T_1}$$

Now we solve for $\log K_{T_2}$

$$\frac{\Delta H^0(T_2 - T_1)}{2.303RT_2T_1} + \log K_{T_1} = \log K_{T_2}$$

If we express R in J/mol K, then ΔH^0 must be expressed in J/mol

$$\log K_{T_2} = \frac{(1.805 \times 10^5\ \text{J/mol})(2400\ \text{K} - 298\ \text{K})}{(2.303)(8.314\ \text{J/mol K})(2400\ \text{K})(298\ \text{K})} + \log (4.6 \times 10^{-31})$$

$$\log K_{T_2} = 27.71 + \log (4.6 \times 10^{-31})$$
$$= 27.71 + (-30.34) = -2.63$$

$$K_{T_2} = 10^{-2.63} = 10^{0.37} \times 10^{-3}$$
$$K_{T_2} = 2.34 \times 10^{-3}$$

We see that K_{T_2} (K_p at 2400 K) is quite small, and that the equilibrium favors N_2 and O_2 rather than NO. However, K_{T_2} is very much larger than K_{T_1}, which is 4.6×10^{-31}. At 2400 K, significantly more NO is present at equilibrium, relative to N_2 and O_2, than at 298 K. Thus, small but significant amounts of NO are exhausted into the atmosphere by automobiles. Only small amounts of NO are necessary to cause severe air pollution problems. Catalytic converters are designed to catalyze the forward and reverse reactions (and others) involving breakdown of nitrogen oxide into nitrogen and oxygen.

$$2NO\ (g) \rightleftharpoons N_2\ (g) + O_2\ (g)$$

The forward reaction as written here is spontaneous. Remember that catalysts do not shift the position of equilibrium. That is, they favor neither the consumption nor production of nitrogen oxide. They merely allow the system to reach equilibrium more rapidly. The time factor is particularly important in this system, since the NO stays in the engine for only a tiny fraction of a second.

Key Terms

Chemical equilibrium a state of dynamic balance in which the rates of forward and reverse reactions are equal; there is no net change in concentrations of reactants or products while a system is at equilibrium.

Dynamic equilibrium an equilibrium in which

processes occur continuously, with no *net* change.

Equilibrium constant a quantity that characterizes the position of equilibrium for a reversible reaction; its magnitude is equal to the mass action expression at equilibrium. K varies with temperature.

Heterogeneous equilibria equilibria involving species in more than one phase.

Homogeneous equilibria equilibria involving only species in a single phase, i.e., all gases, all liquids, or all solids.

LeChatelier's Principle if a stress (change of conditions) is applied to a system at equilibrium, the system shifts in the direction that reduces the stress.

Mass action expression for a reversible reaction,

$$aA + bB \rightleftharpoons cC + dD$$

the product of the concentrations of the products (species on the right), each raised to the power that corresponds to its coefficient in the balanced chemical equation, divided by the product of the concentrations of the reactants (species on the left), each raised to the power that corresponds to its coefficient in the balanced chemical equation. At equilibrium the mass action expression is equal to K; at other conditions, it is Q.

$$\frac{[C]^c[D]^d}{[A]^a[B]^b} = Q\text{, or at equilibrium, } K$$

Reaction quotient, Q the mass action expression under any set of conditions (not necessarily equilibrium); its magnitude relative to K determines the direction in which reaction must occur to establish equilibrium.

Reversible reactions reactions that do not go to completion and occur in both the forward and reverse directions.

van't Hoff equation the relationship between ΔH^0 for a reaction and its equilibrium constants at two different temperatures.

Exercises

Basic Concepts

1. Define and illustrate the following terms: (a) reversible reaction, (b) chemical equilibrium, (c) equilibrium constant, K.
2. Why are chemical equilibria called dynamic equilibria?
3. Write the expression for K for each of the following reactions in terms of concentrations:
   (a) $2NO_2\ (g) \rightleftharpoons 2NO\ (g) + O_2\ (g)$
   (b) $CO\ (g) + H_2O\ (g) \rightleftharpoons CO_2\ (g) + H_2\ (g)$
   (c) $2H_2S\ (g) + 3O_2\ (g) \rightleftharpoons 2H_2O\ (g) + 2SO_2\ (g)$
   (d) $PCl_5\ (g) \rightleftharpoons PCl_3\ (g) + Cl_2\ (g)$
   (e) $MgCO_3\ (s) \rightleftharpoons MgO\ (s) + CO_2\ (g)$
   (f) $P_4\ (g) + 3O_2\ (g) \rightleftharpoons P_4O_6\ (s)$
   (g) $H_2\ (g) + I_2\ (s) \rightleftharpoons 2HI\ (g)$
4. Explain the significance of (a) a very large value of K, (b) a very small value of K, (c) a value of K of about 1.0.
5. Why may we omit concentrations of pure solids and pure liquids from equilibrium constant expressions?
6. Arrange the following in order of increasing tendency of the forward reactions to proceed toward completion at 298 K and one atmosphere pressure.
   (a) $H_2O\ (g) \rightleftharpoons H_2O\ (\ell)$ $\quad K_c = 782$
   (b) $F_2\ (g) \rightleftharpoons 2F\ (g)$ $\quad K_c = 4.9 \times 10^{-21}$
   (c) $C\ (\text{graphite}) + O_2\ (g) \rightleftharpoons CO_2\ (g)$ $\quad K_c = 1.3 \times 10^{69}$
   (d) $H_2\ (g) + C_2H_4\ (g) \rightleftharpoons C_2H_6\ (g)$ $\quad K_c = 9.8 \times 10^{18}$
   (e) $N_2O_4\ (g) \rightleftharpoons 2NO_2\ (g)$ $\quad K_c = 4.6 \times 10^{-3}$
7. Define the reaction quotient Q. Distinguish between Q and K.
8. Why is it useful to compare Q with K? What is the situation in the case of (a) $Q = K$, (b) $Q < K$, (c) $Q > K$?
9. Given the equilibrium constants for the following two reactions at a particular temperature:

$$NiO\ (s) + H_2\ (g) \rightleftharpoons Ni\ (s) + H_2O\ (g) \quad K_c = 25$$

$$NiO\ (s) + CO\ (g) \rightleftharpoons Ni\ (s) + CO_2\ (g) \quad K_c = 500$$

Calculate the value for the equilibrium constant, K_c, for the reaction

$$CO_2\ (g) + H_2\ (g) \rightleftharpoons CO\ (g) + H_2O\ (g)$$

at the same temperature.

Calculation and Uses of *K*

10. Given: $A\ (g) + B\ (g) \rightleftharpoons C\ (g) + 2D\ (g)$

    One mole of A and one mole of B are placed in a 2.00-liter container. After equilibrium has been established, 0.10 mole of C is present in the container. Calculate the equilibrium constant for the reaction.

11. For the gas phase decomposition of nitrogen oxide, NO, according to the following equation $K_c = 25$ at 25°C. If 0.050 mole of NO is admitted into a 2.00-liter container at 25°C, what will be the concentrations of NO, N_2, and O_2 when equilibrium is established?

    $$2NO\ (g) \rightleftharpoons N_2\ (g) + O_2\ (g)$$

12. Refer to Exercise 11. What is the percent dissociation of NO?

13. Refer to Exercise 11. (a) Suppose 0.050 mole of N_2 (g) and 0.050 mole O_2 (g) are initially present in a 2.00-liter container at 25°C. What will be the equilibrium concentrations of NO, N_2, and O_2 at 25°C? (b) The combination of N_2 with O_2 to form NO does not occur at an observable rate at 25°C. Can you suggest a reason why?

14. Phosgene, $COCl_2$, is a poisonous gas that decomposes into carbon monoxide and chlorine according to the following equation. At 900°C, $K_c = 0.083$. If 0.200 mole of $COCl_2$ is admitted into a 2.00-liter container at 900°C, what will be the concentrations of $COCl_2$, Cl_2, and CO at equilibrium?

    $$COCl_2\ (g) \rightleftharpoons CO\ (g) + Cl_2\ (g)$$

15. Refer to Exercise 14. What is the percent dissociation of $COCl_2$?

16. Refer to Exercise 14. If 0.200 mole of CO and 0.200 mole of Cl_2 are admitted into a 2.00-liter container at 900°C, what will be the concentrations of $COCl_2$, Cl_2, and CO at equilibrium?

17. Given: $A\ (g) + 3B\ (g) \rightleftharpoons C\ (g) + 2D\ (g)$

    One mole of A and one mole of B are placed in a 5.00-liter container. After equilibrium has been established, 0.50 mole of D is present in the container. Calculate the equilibrium constant for the reaction.

18. Antimony pentachloride decomposes in a gas phase reaction at 448°C as follows:

    $$SbCl_5\ (g) \rightleftharpoons SbCl_3\ (g) + Cl_2\ (g)$$

    An equilibrium mixture in a 5.00-liter vessel is found to contain 3.84 grams of $SbCl_5$, 9.14 grams of $SbCl_3$, and 2.84 grams of Cl_2. Evaluate K_c at 448°C.

19. If 10.0 grams of $SbCl_5$ are placed in a 5.00-liter container at 448°C and allowed to establish equilibrium as in Exercise 18, what will be the equilibrium concentrations of all species? For this reaction, $K_c = 2.51 \times 10^{-2}$.

20. If 5.00 grams of $SbCl_5$ and 5.00 grams of $SbCl_3$ are placed in a 5.00-liter container and allowed to establish equilibrium as in Exercise 18, what will the equilibrium concentrations be? ($K_c = 2.51 \times 10^{-2}$)

21. At 600°C, K_c for the reaction below is 0.542.

    $$CO_2\ (g) + H_2\ (g) \rightleftharpoons CO\ (g) + H_2O\ (g)$$

    (a) If 0.100 mole of CO_2 and 0.100 mole of H_2 are placed into a closed 2.00-liter container, what equilibrium concentrations of all species will result?
    (b) If 0.200 mole of CO_2 and 0.100 mole of H_2 are placed into a closed 2.00-liter container, what equilibrium concentrations of all species will result?

22. At a certain temperature, K_c for the reaction below is 0.300. If 0.600 mole of $POCl_3$ is placed in a closed 3.00-liter vessel at this temperature, what percentage of it will be dissociated when equilibrium is established?

    $$POCl_3\ (g) \rightleftharpoons POCl\ (g) + Cl_2\ (g)$$

23. Bromine chloride, BrCl, a reddish covalent gas with properties similar to those of Cl_2, may eventually replace Cl_2 as a water disinfectant (see Section 22–8). One mole of chlorine and one mole of bromine are enclosed in a 5.00-liter flask and allowed to reach equilibrium at a certain temperature.

    $$Cl_2\ (g) + Br_2\ (g) \rightleftharpoons 2BrCl\ (g) \qquad K_c = 4.7 \times 10^{-2}$$

(a) What percent of the chlorine has reacted at equilibrium?
(b) What mass of each species is present at equilibrium?
(c) How would a decrease in volume shift the position of equilibrium, if at all?

The Equilibrium Constant in Terms of Partial Pressures

24. Write the K_p expression for each of the reactions in Exercise 3.
25. Under what conditions are no units associated with (a) K_c and (b) K_p?
26. Under what conditions are K_c and K_p for a reaction numerically equal?
27. Given: $2Cl_2\ (g) + 2H_2O\ (g) \rightleftharpoons 4HCl\ (g) + O_2\ (g) \qquad K_p = 4.6 \times 10^{-14}$

    Write the equilibrium constant expressions for K_c and K_p and calculate their values for the reaction written in the following forms:
    (a) $4HCl\ (g) + O_2\ (g) \rightleftharpoons 2Cl_2\ (g) + 2H_2O\ (g)$
    (b) $2HCl\ (g) + \frac{1}{2}O_2\ (g) \rightleftharpoons Cl_2\ (g) + H_2O\ (g)$
    (c) $Cl_2\ (g) + H_2O\ (g) \rightleftharpoons 2HCl\ (g) + \frac{1}{2}O_2\ (g)$
    (d) $4Cl_2\ (g) + 4H_2O\ (g) \rightleftharpoons 8HCl\ (g) + 2O_2\ (g)$
    (e) $\frac{1}{2}Cl_2\ (g) + \frac{1}{2}H_2O\ (g) \rightleftharpoons HCl\ (g) + \frac{1}{4}O_2\ (g)$
28. K_p for the decomposition of N_2O_4 is 0.113 at 25°C. If the initial pressure of N_2O_4 in a 1.00-liter flask at 25°C is 0.0100 atm, what will be the partial pressures of N_2O_4 and NO_2 and the total pressure at equilibrium?

    $N_2O_4\ (g) \rightleftharpoons 2NO_2\ (g)$
29. What are the equilibrium concentrations of N_2O_4 and NO_2 in Exercise 28?
30. The dark-red solid interhalogen compound iodine chloride, ICl, decomposes to the dark violet solid, I_2, and the greenish-yellow gas, Cl_2. At 25°C, K_p for the reaction is 0.24 atm.

    $2ICl\ (s) \rightleftharpoons I_2\ (s) + Cl_2\ (g)$

    (a) If 1.0 mole of ICl is placed in a closed vessel, what will be the pressure of Cl_2 at equilibrium? Note that it is not necessary to specify the volume of the vessel in this case.
    (b) If the 1.0 mole of ICl is placed in a 1.0-liter container, what concentration of Cl_2 is present at equilibrium?
    (c) How many grams of ICl will remain at equilibrium in the 1.0-liter container?
31. Given: $PCl_5\ (g) \rightleftharpoons PCl_3\ (g) + Cl_2\ (g)$

    At 250°C, a sample of PCl_5 was placed in a 24.0-liter evacuated reaction vessel and allowed to come to equilibrium. Analysis showed that at equilibrium 0.42 mole of PCl_5, 0.64 mole of PCl_3, and 0.64 mole of Cl_2 were present in the vessel. Calculate K_c and K_p for this system at 250°C.
32. Given: $H_2\ (g) + I_2\ (g) \rightleftharpoons 2HI\ (g)$

    1.00 g of H_2 and 127 g of I_2 are injected into a 10.0-liter evacuated reaction vessel at 448°C. K_c for this reaction is 50.
    (a) What is K_p at 448°C?
    (b) What are the partial pressures of each substance in the equilibrium mixture and the total pressure in the vessel at equilibrium?
    (c) How many moles and grams of I_2 remain unreacted at equilibrium?

Factors Affecting *K* and Stresses on Equilibria

33. What would be the effect of decreasing the temperature on each of the following systems at equilibrium?
    (a) $H_2\ (g) + I_2\ (g) \rightleftharpoons 2HI\ (g) + 9.45\ kJ$
    (b) $PCl_5\ (g) + 92.5\ kJ \rightleftharpoons PCl_3\ (g) + Cl_2\ (g)$
    (c) $2SO_2\ (g) + O_2\ (g) \rightleftharpoons 2SO_3\ (g) \qquad \Delta H^0 = -198\ kJ$
    (d) $2NOCl\ (g) \rightleftharpoons 2NO\ (g) + Cl_2\ (g) \qquad \Delta H^0 = 75\ kJ$
    (e) $C\ (s) + H_2O\ (g) + 131\ kJ \rightleftharpoons CO\ (g) + H_2\ (g)$
34. What would be the effect of increasing the volume of each of the following systems at equilibrium?
    (a) $2CO\ (g) + O_2\ (g) \rightleftharpoons 2CO_2\ (g)$
    (b) $2NO\ (g) \rightleftharpoons N_2\ (g) + O_2\ (g)$
    (c) $N_2O_4\ (g) \rightleftharpoons 2NO_2\ (g)$
    (d) $Ni\ (s) + 4CO\ (g) \rightleftharpoons Ni(CO)_4\ (g)$
    (e) $N_2\ (g) + 3H_2\ (g) \rightleftharpoons 2NH_3\ (g)$
35. What would be the effect of increasing the pressure by decreasing the volume of each of the systems of Exercise 34 at equilibrium?

36. What would be the effect of each of the following concentration changes on the systems at equilibrium?
   (a) $CO (g) + Cl_2 (g) \rightleftharpoons COCl_2 (g)$
   :add Cl_2 (g)
   (b) $CO (g) + Cl_2 (g) \rightleftharpoons COCl_2 (g)$
   :remove $COCl_2$ (g)
   (c) $PCl_3 (g) + Cl_2 (g) \rightleftharpoons PCl_5 (g)$
   :remove PCl_3 (g)
   (d) $CaO (s) + SO_3 (g) \rightleftharpoons CaSO_4 (s)$
   :add SO_3 (g)
   (e) $CaO (s) + SO_3 (g) \rightleftharpoons CaSO_4 (s)$
   :add $CaSO_4$ (s)
   (f) $2PbS (s) + 3O_2 (g) \rightleftharpoons 2PbO (s) + 2SO_2 (g)$
   :add O_2 (g)
   (g) $2PbS (s) + 3O_2 (g) \rightleftharpoons 2PbO (s) + 2SO_2 (g)$
   :remove PbS (s)

37. What effect does the presence of a catalyst have on the position of equilibrium for a reversible reaction? Why?

38. Given: $PCl_5 (g) \rightleftharpoons PCl_3 (g) + Cl_2 (g)$ $\quad \Delta H^0 = 92.5$ kJ

   How will each of the following influence this system at equilibrium? (a) increased temperature, (b) increased pressure, (c) increased concentration of Cl_2, (d) increased concentration of PCl_5, (e) adding a catalyst.

39. Given: $N_2 (g) + 3H_2 (g) \rightleftharpoons 2NH_3 (g)$ $\quad \Delta H^0 = -92$ kJ

   How will each of the following influence this system at equilibrium? (a) increased temperature, (b) increased pressure, (c) increased concentration of NH_3, (d) increased concentration of N_2, (e) adding a catalyst.

40. The value of K_c is 0.063 at 300°C for the reaction shown below. If there are 0.080 mole of PCl_5, 0.30 mole of PCl_3, and 0.50 mole of Cl_2 in a 1.00-liter container at 300°C, is the system at equilibrium or must the forward or reverse reaction occur to a greater extent to bring the system to equilibrium?

   $PCl_5 (g) \rightleftharpoons PCl_3 (g) + Cl_2 (g)$

41. The value of K_c is 0.10 at 2000°C for the reaction shown below. If there are 0.050 mole of N_2, 0.050 mole of O_2, and 0.10 mole of NO in a 1.00-liter container at 2000°C, is the system at equilibrium or must the forward or reverse reaction occur to a greater extent to bring the system to equilibrium?

   $N_2 (g) + O_2 (g) \rightleftharpoons 2NO (g)$

42. Given: $A (g) + B (g) \rightleftharpoons C (g) + D (g)$
   (a) At equilibrium a 1.00-liter container was found to contain 1.6 mole of C, 1.6 mole of D, 0.20 mole of A, and 0.20 mole of B. Calculate the equilibrium constant for this reaction.
   (b) If 0.20 mole of A and 0.20 mole of B are added to this system, what will the new equilibrium concentration of A be?

43. Given: $A (g) + B (g) \rightleftharpoons C (g) + D (g)$.

   When one mole of A and one mole of B are mixed and allowed to reach equilibrium at room temperature, the mixture is found to contain $\frac{2}{3}$ mole of C and $\frac{2}{3}$ mole of D.
   (a) Calculate the equilibrium constant.
   (b) If three moles of A were mixed with one mole of B and allowed to reach equilibrium, how many moles of C would be present at equilibrium?

44. Given: $A (g) \rightleftharpoons B (g) + C (g)$
   (a) When the system is at equilibrium at 200°C, the concentrations are found to be: $[A] = 0.20\ M$, $[B] = 0.30\ M = [C]$. Calculate K_c.
   (b) If the volume of the container in which the system is at equilibrium is suddenly doubled at 200°C, what will the new equilibrium concentrations be?
   (c) Refer back to part (a). If the volume of the container is suddenly halved at 200°C, what will the new equilibrium concentrations be?

45. The equilibrium constant, K_c, for the dissociation of phosphorus pentachloride is 4.0×10^{-2} at 250°C. How many moles and grams of PCl_5 must be added to a 3.0-liter flask to obtain a Cl_2 concentration of 0.15 M? Refer to Exercise 38 for the equation.

46. (a) What is K_p for the reaction of Exercise 45 at 250°C?
   (b) What are the equilibrium partial pressures of each of the species?
   (c) What is the total pressure of the system at equilibrium?

(d) What was the partial pressure of PCl_5 before any of it decomposed?

47. Consider the following system *at equilibrium* at 1800°C.

$$CO_2\ (g) + H_2\ (g) \rightleftharpoons H_2O\ (g) + CO\ (g)$$

A 2.00-liter vessel contains 0.48 mole of CO_2, 0.48 mole of H_2, 0.96 mole of H_2O, and 0.96 mole of CO.
(a) How many moles and how many grams of H_2 must be added to bring the concentration of CO to 0.60 *M*?
(b) How many moles and how many grams of CO_2 must be added to bring the CO concentration to 0.60 *M*?
(c) How many moles of H_2O must be removed to bring the CO concentration to 0.60 *M*?

48. At 25°C, K_c is 4.66×10^{-3} for the dissociation of dinitrogen tetroxide to nitrogen dioxide.

$$N_2O_4\ (g) \rightleftharpoons 2NO_2\ (g)$$

(a) Calculate the equilibrium concentrations of both gases when 2.50 grams of N_2O_4 are placed in a 2.00-liter flask at 25°C.
(b) What will be the new equilibrium concentrations if the volume of the system is suddenly increased to 4.00 liters at 25°C?
(c) What will be the new equilibrium concentrations if the volume is decreased to 1.00 liter at 25°C?

49. Consider the equilibrium described in part (a) of Exercise 48.
(a) What will be the new equilibrium concentrations if 0.20 gram of NO_2 is removed?
(b) What will be the new equilibrium concentrations if 0.20 gram of NO_2 is added?
(c) What will be the new equilibrium concentrations if 0.20 gram of N_2O_4 is added?

Relationships Among K, ΔG^0, ΔH^0, and T

50. What must be true of the value of ΔG^0 for a reaction if (a) $K \gg 1$; (b) $K = 1$; (c) $K \ll 1$?

51. What kind of equilibrium constant can be calculated from a ΔG^0 value for a reaction involving only gases?

52. Given that K_p is 4.6×10^{-14} at 25°C for the reaction below:

$$2Cl_2\ (g) + 2H_2O\ (g) \rightleftharpoons 4HCl\ (g) + O_2\ (g) \qquad \Delta H^0 = +115 \text{ kJ}$$

Calculate K_p and K_c for the reaction at 500°C and at 1000°C.

53. A mixture of 3.00 moles of Cl_2 and 3.00 moles of CO is enclosed in a 5.00-liter flask at 600°C. At equilibrium, 3.3% of the Cl_2 has been consumed.

$$CO\ (g) + Cl_2\ (g) \rightleftharpoons COCl_2\ (g)$$

(a) Calculate K_c for the reaction at 600°C.
(b) Calculate ΔG^0 for the reaction at this temperature.

54. The following is an example of an alkylation reaction important in the production of iso-octane (2,2,4-trimethylpentane) from two components of crude oil, isobutane and isobutene. Iso-octane is an anti-knock additive for gasoline.

$$\underset{\text{isobutane}}{CH_3-\overset{\overset{CH_3}{|}}{\underset{\underset{CH_3}{|}}{C}}-H} + \underset{\text{isobutene}}{CH_3-\overset{\overset{CH_3}{|}}{C}=CH_2} \xrightleftharpoons[\text{catalyst}]{\Delta} \underset{\text{iso-octane}}{CH_3-\overset{\overset{CH_3}{|}}{\underset{\underset{CH_3}{|}}{C}}-CH_2-\overset{\overset{CH_3}{|}}{\underset{\underset{H}{|}}{C}}-CH_3}$$

The thermodynamic equilibrium constant, K, for this reaction at 25°C is 4.3×10^6, and ΔH^0 is -78.58 kJ.
(a) Calculate ΔG^0 at 25°C.
(b) Calculate K_c at 700°C.
(c) Calculate ΔG^0 at 700°C.
(d) How does the spontaneity of the forward reaction at 700°C compare with that at 25°C?
(e) Why do you think the reaction mixture is heated in the industrial preparation of iso-octane?
(f) What is the purpose of the catalyst? Does it affect the forward reaction more than the reverse reaction?

16 Equilibria in Aqueous Solutions—I

Acids and Bases

We have seen that water-soluble compounds may be classified as either electrolytes or nonelectrolytes. **Electrolytes** are compounds that ionize (or dissociate into their constituent ions) to produce aqueous solutions that conduct an electric current. **Nonelectrolytes** exist as molecules in aqueous solution, and such solutions do not conduct an electric current. Most organic compounds are nonelectrolytes.

16–1 Strong Electrolytes

Solid ionic compounds that are soluble in water dissociate into their constituent ions when they dissolve. Aqueous solutions of soluble, ionic compounds such as sodium chloride, NaCl, and potassium sulfate, K_2SO_4, conduct an electric current well. Aqueous solutions of many polar covalent compounds such as hydrogen chloride and hydrogen bromide are electrolytes because they react with water as they dissolve to produce ions

$$HCl\ (g) + H_2O\ (\ell) \longrightarrow H_3O^+ + Cl^-\ (aq)$$
$$HBr\ (g) + H_2O\ (\ell) \longrightarrow H_3O^+ + Br^-\ (aq)$$

Electrolytes are rather arbitrarily divided into two classes. **Strong electrolytes** are compounds that conduct electric current well in dilute aqueous solutions. From this observation we conclude that solutions of strong electrolytes contain high concentrations of ions. They are completely, or nearly completely, ionized in dilute aqueous solutions. **Weak electrolytes** are compounds that conduct electric current only poorly. Solutions of weak electrolytes contain low concentrations of ions, which indicates that they are only slightly ionized in dilute aqueous solutions.

Three common classes of compounds are strong electrolytes.

Common Strong Acids	
HCl	HNO_3
HBr	$HClO_4$
HI	$HClO_3$
	H_2SO_4

1. *Strong acids* are acids that ionize completely, or very nearly completely, in dilute aqueous solution. Recall that there are only seven *common* strong acids, and all are soluble in water. Refer to the margin if necessary to refresh your memory. Strong acids exist almost exclusively as ions in dilute aqueous solutions. For example, in dilute solutions of nitric acid the following reaction occurs so completely that we may accurately represent such solutions as consisting of equal numbers of H_3O^+ and NO_3^- ions

$$HNO_3\ (\ell) + H_2O\ (\ell) \longrightarrow H_3O^+ + NO_3^-\ (aq)$$

Strong Soluble Bases	
LiOH	
NaOH	
KOH	$Ca(OH)_2$
RbOH	$Sr(OH)_2$
CsOH	$Ba(OH)_2$

2. *Strong soluble bases* are soluble metal hydroxides that ionize completely, or very nearly completely, in dilute aqueous solution. Recall that the list of strong soluble bases is short (see margin), because most metal hydroxides are relatively insoluble in water. The strong soluble bases are the hydroxides of the Group IA metals and the heavier members of the Group IIA metals. These hydroxides are ionic compounds in the solid state. Thus, solid sodium hydroxide consists of equal numbers of Na^+ and OH^- ions arranged in a crystal lattice. Dissolving sodium hydroxide in water results in the hydration of the Na^+ and OH^- ions, thereby dissociating them, so aqueous solutions of sodium hydroxide conduct an electric current well. Similar statements can be made for all strong soluble bases.

The solubility rules are given in Section 7–1, part 5.

3. Most *soluble salts* are ionic compounds that dissociate into their constituent ions in aqueous solutions. Two common exceptions are lead acetate, $Pb(CH_3COO)_2$, and mercury(II) chloride, $HgCl_2$, which are soluble, but are covalent* salts and nonelectrolytes. Most soluble salts are ionic, both in the solid state and in solution. For example, aqueous solutions of sodium bromide, NaBr, contain equal numbers of Na^+ and Br^- ions. Solutions of calcium nitrate, $Ca(NO_3)_2$, contain Ca^{2+} and NO_3^- ions in a 1:2 ratio.

*It is impossible to make a general statement that enables us to predict (absolutely) whether a given salt is primarily ionic or covalent. Differences in electronegativity do not give a definite answer. For example, the electronegativity difference between magnesium and iodine in magnesium iodide is 1.3 units. Chemists agree that magnesium iodide, MgI_2, is an ionic salt. Its aqueous solutions conduct electric current well. The electronegativity difference between mercury and chlorine is 1.1 units (only slightly less than ΔEN for MgI_2), yet aqueous solutions of mercury(II) chloride, $HgCl_2$, do not conduct electric current. We conclude that some soluble salts such as $HgCl_2$ and $Pb(CH_3COO)_2$ are covalent compounds.

Calculation of the concentrations of constituent ions in solutions of *strong* electrolytes is simple, as the following examples illustrate.

EXAMPLE 16–1

Calculate the concentrations of ions in 0.010 *M* hydrochloric acid solution.

Solution

$$\underset{0.010\ \text{mol/L}}{HCl\ (g)} + H_2O\ (\ell) \longrightarrow \underset{0.010\ \text{mol/L}}{H_3O^+} + \underset{0.010\ \text{mol/L}}{Cl^-\ (aq)}$$

Recall that we use a single arrow to indicate that a reaction goes virtually to completion in the indicated direction.

From the equation for the ionization of HCl, we see that one mole of HCl produces one mole of H_3O^+ *and* one mole of Cl^-. Therefore, 0.010 mol/L of HCl produces 0.010 mol/L of H_3O^+ *and* 0.010 mol/L of Cl^-. The concentration of *both* H_3O^+ and Cl^- is 0.010 molar, i.e., $[H_3O^+] = [Cl^-] = 0.010\ M$.

EXAMPLE 16–2

Calculate the concentrations of ions in a 0.030 *M* solution of barium hydroxide.

Solution

$$\underset{0.030\ \text{mol/L}}{Ba(OH)_2\ (s)} \xrightarrow{H_2O} \underset{0.030\ \text{mol/L}}{Ba^{2+}\ (aq)} + \underset{2(0.030\ \text{mol/L})}{2OH^-\ (aq)}$$

From the equation for the dissociation of barium hydroxide, we see that *one* mole of $Ba(OH)_2$ produces *one* mole of Ba^{2+} and *two* moles of OH^-. Therefore, 0.030 mole of $Ba(OH)_2$ produces 0.030 mole of Ba^{2+} and 0.060 mole of OH^-. Thus, the concentrations of ions are $[Ba^{2+}] = 0.030\ M$ and $[OH^-] = 0.060\ M$.

EXAMPLE 16–3

Calculate the concentrations of ions in a solution that contains 0.92 gram of $MgBr_2$ in 500 milliliters of solution. Magnesium bromide is a soluble ionic salt.

Solution

First, we calculate the concentration of $MgBr_2$ in moles per liter of solution, i.e., the molarity of the solution

This does not imply the existence of "molecules" of $MgBr_2$ in solution.

$$\frac{?\ \text{mol}\ MgBr_2}{\text{L soln}} = \frac{0.92\ \text{g}\ MgBr_2}{0.500\ \text{L}} \times \frac{1\ \text{mol}\ MgBr_2}{184\ \text{g}\ MgBr_2} = 0.010\ \text{mol}\ MgBr_2/\text{L}$$

The solution is 0.010 *M* in $MgBr_2$. The equation for the dissociation of $MgBr_2$,

$$\underset{0.010\ M}{MgBr_2\ (s)} \xrightarrow{H_2O} \underset{0.010\ M}{Mg^{2+}\ (aq)} + \underset{0.020\ M}{2Br^-\ (aq)}$$

shows that *one* mole of $MgBr_2$ produces *one* mole of Mg^{2+} and *two* moles of Br^-. Therefore, 0.010 molar $MgBr_2$ solution contains 0.010 mole/liter of Mg^{2+} and 0.020 mole/liter of Br^-, i.e., $[Mg^{2+}] = 0.010\ M$ and $[Br^-] = 0.020\ M$.

16–2 The Ionization of Water

Careful experiments on its electrical conductivity have shown that pure water ionizes to a very slight extent:

$$H_2O\ (\ell) + H_2O\ (\ell) \rightleftharpoons H_3O^+ + OH^-\ (aq)$$

The equilibrium constant for this reaction is

$$K_c = \frac{[H_3O^+][OH^-]}{[H_2O]^2}$$

Since the concentration of water is constant in pure water, it can be included in the equilibrium constant

$$K_c[H_2O]^2 = [H_3O^+][OH^-]$$

In pure water the concentration of H_3O^+ is obviously equal to the concentration of OH^-, since the formation of an H_3O^+ ion by the ionization of water invariably results in the formation of an OH^- ion, so we can write $[H_3O^+] = [OH^-]$. Careful measurements show that the concentration of both ions is 1.0×10^{-7} mol/L at 25°C

$$[H_3O^+] = [OH^-] = 1.0 \times 10^{-7}\ \text{mol/L}$$

Since the concentrations of both H_3O^+ and OH^- are known for pure water at 25°C, substitution into the expression

$$K_c[H_2O]^2 = [H_3O^+][OH^-]$$

gives

$$K_c[H_2O]^2 = [H_3O^+][OH^-] = (1.0 \times 10^{-7})(1.0 \times 10^{-7}) = 1.0 \times 10^{-14}$$

or, in a more useful form,

$$[H_3O^+][OH^-] = 1.0 \times 10^{-14}$$

The quantity, 1.0×10^{-14}, is known as the **ion product** for water and is usually represented as K_w.

Solutions in which the concentration of solute is less than 1 mole/liter are usually called dilute solutions.

Although the expression $K_w = [H_3O^+][OH^-] = 1.0 \times 10^{-14}$ was obtained for pure water, *it is valid for dilute aqueous solutions at 25°C,* because the concentration of water remains nearly constant in all such solutions and is very close to that in pure water. Indeed, this is one of the most useful relationships chemists have discovered. It is useful because it gives a simple (inverse) relationship between H_3O^+ and OH^- concentrations in *all* dilute aqueous solutions.

EXAMPLE 16–4

Calculate the concentrations of H_3O^+ and OH^- in a 0.050 *M* solution of nitric acid.

Solution

The equation for the ionization of HNO_3, a strong acid,

$$\underset{0.050\,M}{HNO_3\ (\ell)} + H_2O\ (\ell) \longrightarrow \underset{0.050\,M}{H_3O^+} + \underset{0.050\,M}{NO_3^-\ (aq)}$$

shows that 0.050 mol/L of HNO_3 produces 0.050 mol/L of H_3O^+ *and* 0.050 mol/L of NO_3^-. Thus, $[H_3O^+] = [NO_3^-] = 0.050\ M$.

The concentration of OH^- is calculated from the ion product for water and the concentration of H_3O^+ in the solution.

$$[H_3O^+][OH^-] = 1.0 \times 10^{-14} \qquad [OH^-] = \frac{1.0 \times 10^{-14}}{[H_3O^+]}$$

Substitution gives

$$[OH^-] = \frac{1.0 \times 10^{-14}}{5.0 \times 10^{-2}} = 2.0 \times 10^{-13}\ M$$

In solving this problem, we have assumed that *all* of the H_3O^+ comes from the ionization of HNO_3 and have neglected the H_3O^+ formed by the ionization of water. The ionization of water produces only $2.0 \times 10^{-13}\ M$ H_3O^+ and $2.0 \times 10^{-13}\ M$ OH^- in this solution. The $2.0 \times 10^{-13}\ M$ H_3O^+ produced by the ionization of water is so small compared to the concentration of H_3O^+ formed by the ionization of HNO_3 ($5.0 \times 10^{-2}\ M$) that it is neglected. We are justified in assuming that the H_3O^+ concentration is derived solely from HNO_3.

When nitric acid is added to water, large numbers of H_3O^+ ions are produced instantaneously. The large increase in H_3O^+ concentration shifts the water equilibrium to the left sharply (LeChatelier's Principle) and the concentration of OH^- is decreased

$$H_2O\ (\ell) + H_2O\ (\ell) \rightleftharpoons H_3O^+\ (aq) + OH^-\ (aq)$$

In acidic solutions the H_3O^+ concentration is always greater than the OH^- concentration. We should not conclude that acidic solutions contain no OH^- ions, but rather that the concentration of OH^- is always less than $1.0 \times 10^{-7}\ M$ in acidic solutions. The converse is true for basic solutions, where the concentration of OH^- is always greater than $1.0 \times 10^{-7}\ M$. By definition, "neutral" aqueous solutions are solutions in which $[H_3O^+] = [OH^-] = 1.0 \times 10^{-7}\ M$.

16–3 The pH Scale

The pH scale provides a convenient way of expressing the acidity and basicity of dilute aqueous solutions. The pH of a solution is defined as

$$pH = \log \frac{1}{[H_3O^+]} \qquad \text{or} \qquad pH = -\log [H_3O^+]$$

This is the base-10 (common) logarithm, *not* the base-*e* (natural) logarithm.

Note that we use pH rather than pH_3O. At the time the pH concept was developed, it was believed that all aqueous solutions of acids contained H^+ ions. Today we represent the species that was formerly written H^+ as H_3O^+, even though it is even more highly hydrated. A number of "p" terms are currently used. For example, $pOH = -\log [OH^-]$. In general, a lower case "**p**" before a symbol is read "negative logarithm of the symbol." Thus, **pH** is the negative logarithm of H_3O^+ concentration, pOH is the negative logarithm of OH^- concentration, pAg refers to the negative logarithm of Ag^+ concentration, and pK refers to the negative logarithm of an equilibrium constant.

If H_3O^+ and OH^- concentrations are known for a particular solution, the pH and pOH can be calculated by taking the negative logarithms of these concentrations, as Examples 16–5 and 16–6 illustrate.

EXAMPLE 16–5

Calculate the pH of a solution in which the concentration of H_3O^+ is 0.010 M.

Solution

We are given $[H_3O^+] = 0.010\ M$, or more conveniently, $[H_3O^+] = 1.0 \times 10^{-2}\ M$.

$\text{pH} = -\log [H_3O^+] = -\log [1.0 \times 10^{-2}] = -[0.00 + (-2)] = \underline{2.00}$.

Note that pH = 2.00 contains two significant figures because the 2 only indicates decimal position in the number from which pH was calculated.

EXAMPLE 16–6

Calculate the pH of a solution in which the H_3O^+ concentration is 0.050 mol/L.

Solution

We are given, $[H_3O^+] = 0.050\ M$ *or* $[H_3O^+] = 5.0 \times 10^{-2}\ M$. Calculation of pH involves taking the logarithm of 5.0×10^{-2} and then multiplying by -1.

$\text{pH} = -\log [H_3O^+] = -\log [5.0 \times 10^{-2}] = -[0.699 + (-2)] = 1.301$

Since we are justified in using only two significant figures, pH = $\underline{1.30}$. The "1" in 1.30 is derived from the exponential term and is *not* a significant figure.

EXAMPLE 16–7

The pH of a solution is 3.301. What is the concentration of H_3O^+ in this solution?

Solution

Recall that $\text{pH} = -\log [H_3O^+]$. Therefore, we can write

$-\log [H_3O^+] = 3.301$

Multiplying through by -1 gives

$\log [H_3O^+] = -3.301$

Taking the antilogarithm of both sides of the equation gives

$[H_3O^+] = 10^{-3.301}$, or in general terms, $[H_3O^+] = 10^{-\text{pH}}$

Since logarithm tables contain only positive mantissas, we express the negative exponent in terms of a positive fractional exponent *and* a negative integral exponent.

$[H_3O^+] = 10^{-3.301} = 10^{0.699} \times 10^{-4}$

We look up 0.699 in the table of mantissas, which gives

$\underline{[H_3O^+] = 5.00 \times 10^{-4}\ M}$

Mantissas are fractional powers of the base 10; for example, $10^{0.699} = 5.00$.

Scientific calculators can handle negative logarithms directly.

A convenient relationship between pH and pOH can be derived easily. Recall that for *all dilute aqueous solutions,*

$$[H_3O^+][OH^-] = 1.0 \times 10^{-14}$$

Taking the logarithm of both sides of this equation gives

$$\log [H_3O^+] + \log [OH^-] = \log (1.0 \times 10^{-14})$$

Multiplying both sides of this equation by −1 gives

$$(-\log [H_3O^+]) + (-\log [OH^-]) = -\log (1.0 \times 10^{-14})$$

or, in slightly different form

$$pH + pOH = 14$$

This is an extraordinarily useful expression because it gives a simple relationship between pH and pOH in dilute aqueous solutions. We now have expressions for relating $[H_3O^+]$ and $[OH^-]$ as well as pH and pOH, i.e.,

$$[H_3O^+][OH^-] = 1.0 \times 10^{-14} \quad \text{and} \quad pH + pOH = 14$$

These two expressions are used frequently enough to warrant a small location in your memory!

From the relationship pH + pOH = 14, we see that positive values of both pH and pOH are limited to the range *0 to 14.* If either pH or pOH is greater than 14, the other is obviously negative. As a matter of convention, we do not normally express acidities and basicities on the pH and pOH scale outside the range 0 to 14. This does not mean that pH and pOH cannot have negative values or values greater than 14, but just that they are seldom used. Example 16–8 illustrates the relationships among $[H_3O^+]$, pH, $[OH^-]$, and pOH.

EXAMPLE 16–8

Calculate $[H_3O^+]$, pH, $[OH^-]$, and pOH for 0.010 *M* HNO_3 solution.

Solution

$$HNO_3 + H_2O \longrightarrow H_3O^+ + NO_3^-$$

Because nitric acid is a strong acid (it ionizes completely), we know that

$$[H_3O^+] = \underline{1.0 \times 10^{-2}\ M}$$

$$pH = -\log [H_3O^+] = -\log [1.0 \times 10^{-2}] = -[0.00 + (-2)] = \underline{2.00}$$

We also know that pH + pOH = 14. Therefore,

$$pOH = 14 - pH = 14 - 2.00 = \underline{12.00}$$

Since $[H_3O^+][OH^-] = 1.0 \times 10^{-14}$, $[OH^-]$ is easily calculated.

$$[OH^-] = \frac{1.0 \times 10^{-14}}{1.0 \times 10^{-2}} = \underline{1.0 \times 10^{-12}\ M}$$

In the remainder of this chapter, we'll discuss compounds that are soluble in water. All ions are hydrated in aqueous solution, and we'll omit the designation (ℓ), (g), (s), (aq), etc. to simplify writing equations.

Also, since pOH = 12.00, we have $-\log [OH^-] = 12$ or $[OH^-] = 1.0 \times 10^{-12}\ M$.

To develop familiarity with the pH scale, consider a series of solutions in which the H_3O^+ concentration varies from 1.0 *M* to 1.0×10^{-14} *M*. Obviously the OH^- concentration will vary from 1.0×10^{-14} *M* to 1.0 *M* in these solutions. Let us tabulate $[H_3O^+]$, $[OH^-]$, pH, and pOH for this series of solutions, so the relationships become obvious (Table 16–1).

TABLE 16–1 Relationships among $[H_3O^+]$, $[OH^-]$, pH, and pOH

$[H_3O^+]$ (*M*)	$[OH^-]$ (*M*)	pH	pOH	
1.0	1.0×10^{-14}	0	14	↑ increasing acidity
1.0×10^{-1}	1.0×10^{-13}	1	13	
1.0×10^{-2}	1.0×10^{-12}	2	12	
1.0×10^{-3}	1.0×10^{-11}	3	11	
1.0×10^{-4}	1.0×10^{-10}	4	10	
1.0×10^{-5}	1.0×10^{-9}	5	9	
1.0×10^{-6}	1.0×10^{-8}	6	8	
1.0×10^{-7}	1.0×10^{-7}	7	7	neutral
1.0×10^{-8}	1.0×10^{-6}	8	6	increasing basicity ↓
1.0×10^{-9}	1.0×10^{-5}	9	5	
1.0×10^{-10}	1.0×10^{-4}	10	4	
1.0×10^{-11}	1.0×10^{-3}	11	3	
1.0×10^{-12}	1.0×10^{-2}	12	2	
1.0×10^{-13}	1.0×10^{-1}	13	1	
1.0×10^{-14}	1.0	14	0	

Recall that in pure water at 25°C, $[H_3O^+] = [OH^-] = 1.0 \times 10^{-7}$ mol/L. Although 1.0×10^{-7} mol/L is a *low concentration* of ions, it does represent a significant number of ions, as the following illustration shows.

EXAMPLE 16–9

Calculate the number of H_3O^+ ions in one liter of pure water at 25°C.

Solution

Since we know that the concentration of H_3O^+ is 1.0×10^{-7} mol/L in pure water, we need only multiply the concentration of H_3O^+ by Avogadro's number to obtain the number of H_3O^+ ions

$$\underline{?}\ H_3O^+ = \frac{1.0 \times 10^{-7}\ \text{mol}\ H_3O^+}{\text{L}} \times \frac{(6.02 \times 10^{23})\ H_3O^+}{1\ \text{mol}\ H_3O^+}$$

$$= \underline{6.0 \times 10^{16}\ H_3O^+/\text{L}}$$

which is about 60,000,000,000,000,000 H_3O^+ ions in one liter of pure water at 25°C! The number of OH^- ions in one liter of pure water is the same.

The pH of an aqueous solution may be determined by using indicators (Section 16–5) or by using pH meters, which are electrical instruments constructed for that purpose.

These are similar to the standard solutions used for acid-base titrations.

In the indicator method, a series of solutions of known pH, called standard solutions, is prepared and the appropriate mixture of indicators is added to each so that solutions of different acidities have different colors. The same mixture of indicators is added to the unknown solution, and its color is then

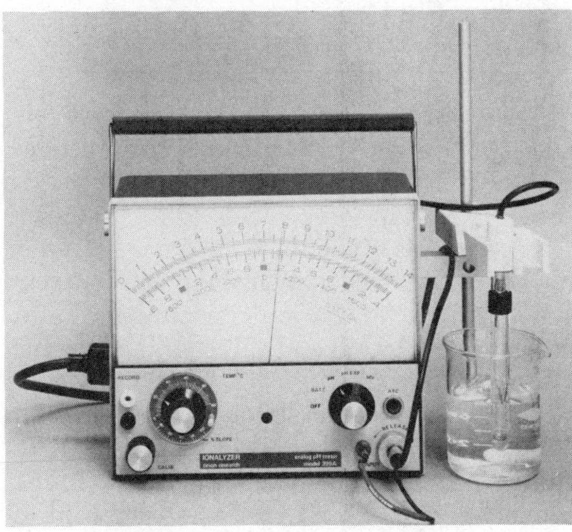

FIGURE 16–1 A pH meter is an instrument that reads the pH of a solution directly. The electrode is dipped into the solution of interest, and the meter gives the pH of the solution. Before using, a pH meter must be calibrated with a series of solutions of known pH.

compared to those of the standard solutions. Solutions that have the same pH develop the same color.

Alternatively, "universal indicator" papers may be used to determine the pH of a solution. These are pieces of paper that are impregnated with a mixture of indicators; a color chart is attached to each container of paper. A piece of paper is dipped into a solution, and its color is then compared to the color chart to establish the pH of the solution.

The pH meter (Figure 16–1) is based on the glass electrode, a sensing device that generates a voltage. This voltage is proportional to the pH of the solution in which the electrode is placed. The rest of the instrument consists of an electrical circuit to amplify the voltage from the electrode, and a meter that relates the voltage to the pH.

Like other scientific instruments, pH meters must be calibrated. This is accomplished by using a series of solutions of known pH. Once calibrated, the pH meter can then be used to determine the pH of an unknown solution.

pH Range for a Few Common Substances

Substances	pH Range
Gastric contents (human)	1.0–3.0
Limes	1.8–2.0
Soft drinks	2.0–4.0
Lemons	2.2–2.4
Vinegar	2.4–3.4
Apples	2.9–3.3
Tomatoes	4.0–4.4
Beer	4.0–5.0
Bananas	4.5–4.7
Urine (human)	4.8–8.4
Milk (cow's)	6.3–6.6
Saliva (human)	6.5–7.5
Blood plasma (human)	7.3–7.5
Egg white	7.6–8.0
Milk of magnesia	10.5
Household ammonia	11–12

16–4 Ionization Constants for Weak Monoprotic Acids and Bases

Thus far we have restricted our discussion to strong acids and strong soluble bases. As we indicated earlier, there are relatively few common strong acids and strong soluble bases.

Weak acids are much more numerous than strong acids. For this reason you were asked to learn the list of strong acids (Table 7–1). You may assume, with a high degree of certainty, that the other acids you encounter in this text will be weak acids. The names, formulas, and ionization constants for a few common weak acids are listed in Table 16–2; Appendix F contains a more comprehensive list.

Weak acids ionize only slightly in dilute aqueous solution. Note that our classification of acids as strong or weak is based only on the *extent to which they ionize in dilute aqueous solution.* Many students know that hydrofluoric acid will dissolve glass and erroneously conclude that HF is a strong acid. It isn't. This dissolution of glass in hydrofluoric acid is due to the fact that silicates react with hydrofluoric acid to produce silicon tetrafluoride, SiF_4, a very volatile compound. This particular reaction tells us exactly nothing about the acid strength of hydrofluoric acid.

Would you eat pickles preserved in 5% hydrochloric acid, a strong acid?

How do you *know* that carbonated beverages are *saturated* solutions?

Would you think of using sulfuric or nitric acids for any of these uses?

We can think of several weak acids with which we are familiar because of their common occurrence and uses. Vinegar is a 5% solution of acetic acid, CH_3COOH. Carbonated beverages are saturated solutions of carbon dioxide dissolved in water to produce carbonic acid ($CO_2 + H_2O \rightleftharpoons H_2CO_3$). Citrus fruits contain citric acid, $C_3H_5O(COOH)_3$, and many of the ointments and powders used for medicinal purposes contain boric acid, H_3BO_3. These everyday uses of weak acids indicate that there is indeed a fundamental difference between the properties of strong and weak acids. The difference is that strong acids ionize completely in dilute aqueous solution while weak acids ionize only slightly.

Let us consider the reaction that occurs when a weak acid, such as acetic acid, is dissolved in water. The equation for the ionization of acetic acid is

$$CH_3COOH + H_2O \rightleftharpoons H_3O^+ + CH_3COO^-$$

The equilibrium constant for this reaction is

$$K_c = \frac{[H_3O^+][CH_3COO^-]}{[CH_3COOH][H_2O]}$$

TABLE 16–2 Ionization Constants for Some Weak Monoprotic Acids

Acid	Ionization Reaction	K_a at 25°C
Hydrofluoric acid	$HF + H_2O \rightleftharpoons H_3O^+ + F^-$	7.2×10^{-4}
Nitrous acid	$HNO_2 + H_2O \rightleftharpoons H_3O^+ + NO_2^-$	4.5×10^{-4}
Acetic acid	$CH_3COOH + H_2O \rightleftharpoons H_3O^+ + CH_3COO^-$	1.8×10^{-5}
Hypochlorous acid	$HClO + H_2O \rightleftharpoons H_3O^+ + ClO^-$	3.5×10^{-8}
Hydrocyanic acid	$HCN + H_2O \rightleftharpoons H_3O^+ + CN^-$	4.0×10^{-10}

Note that this expression contains the concentration of water. If we restrict our discussion to *dilute* aqueous solutions, the concentration of water is very high. There are 55.6 moles of water in one liter of pure water. In dilute aqueous solutions, the concentration of water is *essentially* constant; if we assume that it *is* constant, we can rearrange the above expression to give

$$K_c[H_2O] = \frac{[H_3O^+][CH_3COO^-]}{[CH_3COOH]}$$

Since K_c is a constant (at a specified temperature), and the concentration of water is treated as a constant, we define the **ionization constant** of a weak acid as the product $K_c[H_2O]$, i.e.,

$$K_a = K_c[H_2O]$$

For acetic acid,

$$K_a = \frac{[H_3O^+][CH_3COO^-]}{[CH_3COOH]} = 1.8 \times 10^{-5}$$

This expression tells us that in dilute aqueous solutions of acetic acid, the concentration of H_3O^+ multiplied by the concentration of CH_3COO^- and then divided by the concentration of *nonionized* acetic acid is equal to 1.8×10^{-5}.

Ionization constants for weak acids (and bases) must be calculated from data that are determined experimentally. Conductivity, depression of freezing point, and measurements of pH provide data from which ionization constants can be calculated. Values for some weak acids were listed in Table 16–2, and more are given in Appendix F.

EXAMPLE 16–10

In 0.0100 *M* solution, acetic acid is 4.2% ionized. Calculate its ionization constant from this information.

Solution

The equation for the ionization of CH_3COOH is

$$CH_3COOH + H_2O \rightleftharpoons H_3O^+ + CH_3COO^-$$

The equilibrium constant expression is

$$K_a = \frac{[H_3O^+][CH_3COO^-]}{[CH_3COOH^-]}$$

The concentrations of H_3O^+ and CH_3COO^- are equal, and each is 4.2% of the total acid concentration

$$[H_3O^+] = [CH_3COO^-] = 0.042 \times 0.0100\ M = 4.2 \times 10^{-4}\ M$$

The concentration of nonionized CH_3COOH is 95.8% (100.0% − 4.2%) of the total acid concentration

$$[CH_3COOH] = 0.958 \times 0.0100\ M = 9.58 \times 10^{-3}\ M$$

Substitution of these values into the K_a expression enables us to calculate K_a for CH_3COOH.

$$K_a = \frac{[H_3O^+][CH_3COO^-]}{[CH_3COOH]} = \frac{(4.2 \times 10^{-4})(4.2 \times 10^{-4})}{9.58 \times 10^{-3}} = \underline{1.8 \times 10^{-5}}$$

EXAMPLE 16–11

The pH of a 0.100 *M* solution of a weak monoprotic acid is 1.47. Calculate K_a for the acid.

Solution

The ionization of the weak monoprotic acid (call it HA) may be represented as

$$HA + H_2O \rightleftharpoons H_3O^+ + A^-$$

and its ionization constant expression is

$$K_a = \frac{[H_3O^+][A^-]}{[HA]}$$

We can calculate $[H_3O^+]$ from the definition of pH

$$pH = -\log [H_3O^+]$$
$$[H_3O^+] = 10^{-pH} = 10^{-1.47}$$
$$[H_3O^+] = 10^{0.53} \times 10^{-2} = 3.4 \times 10^{-2}\ M$$

Recall that the ionization of pure water produces only $1.0 \times 1.0^{-7}\ M\ H_3O^+$. The concentration of H_3O^+ *from* water is even less when an acid is dissolved in water (see the discussion following Example 16–4).

The ionization equation tells us that the concentrations of H_3O^+ and A^- must be equal, since they are formed in a 1:1 ratio

$$[H_3O^+] = [A^-] = 3.4 \times 10^{-2}\ M$$

Since each molecule of HA that ionizes produces one H_3O^+ (and one A^-), the concentration of nonionized HA is

$$[HA]_{nonionized} = [HA]_{total} - [HA]_{ionized}$$
$$= [HA]_{total} - [H_3O^+]$$
$$[HA]_{nonionized} = (0.100 - 0.034)\ M = 0.066\ M$$

Now that all concentrations are known, the ionization constant can be calculated

$$K_a = \frac{[H_3O^+][A^-]}{[HA]} = \frac{(3.4 \times 10^{-2})(3.4 \times 10^{-2})}{6.6 \times 10^{-2}} = 1.8 \times 10^{-2}$$

Since ionization constants are equilibrium constants for ionization reactions, their values indicate the extents to which weak electrolytes ionize. Acids with larger ionization constants ionize to greater extents (and are therefore stronger acids) than acids with smaller ionization constants. If we look back at Table 16–2, we see that for the five weak acids listed there, the order of decreasing acid strength is

$$HF > HNO_2 > CH_3COOH > HClO > HCN$$

Conversely, we note that in terms of Brønsted-Lowry terminology, the order of increasing base strength of the anions of these acids is

$$F^- < NO_2^- < CH_3COO^- < ClO^- < CN^-$$

Knowing the value of the ionization constant for a weak acid enables us to calculate the concentrations of the various species in solutions of the weak acid of known concentrations, as Example 16–12 illustrates.

EXAMPLE 16–12

Calculate the concentrations of the various species in 0.10 *M* hypochlorous acid, for which $K_a = 3.5 \times 10^{-8}$.

Solution

The equation for the ionization of HOCl (Table 16–2) is

$$HOCl + H_2O \rightleftharpoons H_3O^+ + OCl^-$$

and the expression for the ionization constant is

$$K_a = \frac{[H_3O^+][OCl^-]}{[HOCl]} = 3.5 \times 10^{-8}$$

We have written the formula for hypochlorous acid as HOCl rather than HClO to emphasize that the structure is H—O—Cl.

Having written *both* the equation for the ionization of HOCl and the expression for the ionization constant, we are ready to attack the problem. We would like to know the concentrations of H_3O^+, OCl^-, and nonionized HOCl in the solution. An algebraic representation of concentrations is required, since there is no other obvious way to obtain the desired concentrations. If we let $x = [H_3O^+]$, we can also let $x = [OCl^-]$ because the ionization of one HOCl molecule produces one H_3O^+ and one OCl^-. If $x = [H_3O^+]$, then the equilibrium concentration of *nonionized* HOCl is $(0.10 - x)$. We have now represented the concentrations of the various species algebraically, i.e.,

We will neglect the 1.0×1.0^{-7} *M* of H_3O^+ produced by the ionization of water. Recall (Section 16–2) that the addition of an acid to water suppresses the ionization of water.

	$HOCl$	$+ H_2O \rightleftharpoons$	H_3O^+	$+ OCl^-$
Orig	0.10 *M*		~0 *M*	0 *M*
Change	$-x$ *M*		$+x$ *M*	$+x$ *M*
Equil	$(0.10 - x)$ *M*		x *M*	x *M*

Substituting the algebraic representations into the equilibrium constant expression gives

$$\frac{[H_3O^+][OCl^-]}{[HOCl]} = \frac{(x)(x)}{(0.10 - x)} = 3.5 \times 10^{-8}$$

This is a quadratic equation, but it is not necessary to solve it by the quadratic formula. If we assume that $(0.10 - x)$ is very nearly equal to 0.10 (see the tinted section that follows), the equation becomes

$$\frac{x^2}{0.10} \approx 3.5 \times 10^{-8}$$

$$x^2 \approx 3.5 \times 10^{-9}$$

$$x \approx 5.9 \times 10^{-5}$$

In our algebraic representation, we let

$$[H_3O^+] = x = \underline{5.9 \times 10^{-5}\ M}$$

$$[OCl^-] = x = \underline{5.9 \times 10^{-5}\ M}$$

$$[HOCl] = 0.10 - x = 0.10 - 0.000059 = \underline{0.10\ M}$$

Simplifying Quadratic Equations

A property of quadratic equations is: When the linear variable (x) is added to or subtracted from a much larger number, the linear variable may be disregarded if it is sufficiently small. A reasonable rule-of-thumb for determining whether the variable can be disregarded is: If the exponent of 10 in the equilibrium constant is −5 or less (−5, −6, −7, etc.), then the variable can be disregarded when it is added to or subtracted from a number greater than 0.05.

Let's examine the assumption as it applies to Example 16–12 in detail. Our quadratic equation is

$$\frac{(x)(x)}{(0.10 - x)} = 3.5 \times 10^{-8}$$

If we convert this to the standard quadratic equation form, $ax^2 + bx + c = 0$, it becomes

$$x^2 + (3.5 \times 10^{-8})x - 3.5 \times 10^{-9} = 0$$

Since x is obviously a small number, then $(3.5 \times 10^{-8})x$ is so small that it is of no significance, and we are justified in assuming that $(0.10 - x) = 0.10$. By making the simplifying assumption, our quadratic equation becomes $x^2 = 3.5 \times 10^{-9}$, which is easily solved by taking the square root of both sides. You may wish to use the quadratic formula to verify that the answer obtained this way is correct within round-off error.

Although the above argument is purely algebraic, we could use our chemical intuition to arrive at exactly the same conclusion. A small ionization constant (10^{-5} or less) tells us that the extent of ionization is very small. If the extent of ionization is small, most of the weak acid exists as nonionized molecules, and the amount that ionizes is not significant when compared to the concentration of nonionized weak acid.

From our calculations, we can make some observations. In a solution containing only a weak monoprotic acid, the concentration of H_3O^+ is equal to the concentration of the anion of the weak monoprotic acid. Unless the weak acid is *very* dilute, say less than 0.050 *M*, the concentration of nonionized acid is approximately equal to the molarity of the solution. Only if the ionization constant for the weak acid is much greater than 10^{-5} will the extent of ionization be sufficiently large so that there will be a significant difference between the concentration of nonionized acid and the molarity of the solution. Example 16–13 further illustrates these statements.

EXAMPLE 16–13

Calculate the percentage ionization of a 0.10 *M* solution of acetic acid.

Solution

The equation for the ionization of CH_3COOH is

$$CH_3COOH + H_2O \rightleftharpoons H_3O^+ + CH_3COO^-$$

Having written the appropriate equation, let us decide how to proceed. Since percentage is defined as (part/whole) × 100%, we represent the percentage ionization of acetic acid as

$$\%\ \text{ionization} = \frac{[CH_3COOH]_{\text{ionized}}}{[CH_3COOH]_{\text{total}}} \times 100\%$$

Examination of the chemical equation shows that each molecule of CH_3COOH that ionizes produces one H_3O^+. Therefore, we can represent the concentration of

CH_3COOH that ionizes by the equilibrium concentration of H_3O^+, and the above expression becomes

$$\% \text{ ionization} = \frac{[H_3O^+]}{[CH_3COOH]_{total}} \times 100\%$$

We can solve for $[H_3O^+]$ as we did in Example 16–12. Let $x = [H_3O^+] = [CH_3COO^-]$, and $[CH_3COOH] = (0.10 - x)$ at equilibrium.

	CH_3COOH	$+ H_2O \rightleftharpoons$	H_3O^+	$+ CH_3COO^-$
Orig	$0.10\,M$		$\sim 0\,M$	$0\,M$
Change	$-x\,M$		$+x\,M$	$+x\,M$
Equil	$(0.10 - x)\,M$		$x\,M$	$x\,M$

Substituting into the ionization constant expression gives

$$K_a = \frac{[H_3O^+][CH_3COO^-]}{[CH_3COOH]} = \frac{(x)(x)}{(0.10 - x)} = 1.8 \times 10^{-5} \qquad \text{(from Table 16–2)}$$

If we make the simplifying assumption as in Example 16–12, namely that $(0.10 - x)$ is approximately equal to 0.10, we have

$$\frac{x^2}{0.10} = 1.8 \times 10^{-5} \qquad \text{and} \qquad x^2 = 1.8 \times 10^{-6}$$

which gives $x = 1.3 \times 10^{-3}$ mol/L $= [H_3O^+]$ and pH = 2.89. Now that we know $[H_3O^+]$, we can calculate the percentage ionization for 0.10 *M* CH_3COOH solution.

$$\% \text{ ionization} = \frac{[CH_3COOH]_{ionized}}{[CH_3COOH]_{total}} \times 100\% = \frac{[H_3O^+]}{[CH_3COOH]_{total}} \times 100\%$$

$$\% \text{ ionization} = \frac{1.3 \times 10^{-3}}{0.10} \times 100\% = \underline{1.3\%}$$

Note that our assumption that $(0.10 - x)$ is approximately 0.10 is reasonable because $(0.10 - x) = (0.10 - 0.0013)$. If we follow the rules for significant figures, $(0.10 - 0.0013)$ is 0.10. However, if the ionization constant for a weak acid is significantly greater than 10^{-5}, this assumption would introduce considerable error into calculations.

In dilute solutions, acetic acid exists primarily as nonionized molecules (as do all weak acids), and there are relatively few hydronium and acetate ions. To be specific, in 0.10 *M* solution, CH_3COOH is 1.3% ionized; for each 1000 formula units of CH_3COOH in the solution, there are 13 H_3O^+ ions, 13 CH_3COO^- ions, and 987 nonionized CH_3COOH molecules. For weaker acids, the number of molecules of nonionized acid is even larger.

By now we should be developing some "feel" for the strength of an acid by simply looking at its ionization constant. Let's consider 0.10 *M* solutions of HCl (a strong acid), CH_3COOH (Example 16–13), and HOCl (Example 16–12). If we calculate the % ionization for 0.10 *M* HOCl (as we did for 0.10 *M* CH_3COOH in Example 16–13), we find that it is 0.059% ionized. In 0.10 *M* solution, HCl is very nearly completely ionized. The data in Table 16–3 show that the concentration of H_3O^+ in 0.10 *M* HCl is approximately 77 times greater than that in 0.10 *M* CH_3COOH, and approximately 1700 times greater than that in 0.10 *M* HOCl.

TABLE 16–3 Comparison of Ionizations of Acids

Acid Solution	Ionization Constant	$[H_3O^+]$	pH	% Ionization
0.10 *M* HCl	very large	0.10 *M*	1.00	~100
0.10 *M* CH_3COOH	1.8×10^{-5}	0.0013 *M*	2.89	1.3
0.10 *M* HClO	3.5×10^{-8}	0.000059 *M*	4.23	0.059

Thus far we have focused our attention on acids. Very few common weak bases are soluble in water. Aqueous ammonia is the most frequently encountered example. The organic derivatives of ammonia are called *amines,* and many are also soluble weak bases. Hundreds of amines are known, and some are very important in the structures of proteins. From our earlier discussion of bonding in covalent compounds (Section 5–13) we recall that there is one unshared pair of electrons on the nitrogen atom in ammonia (frequently written $:NH_3$ for emphasis), and that when ammonia dissolves in water it accepts a proton from a water molecule as it ionizes slightly (see Section 11–3). The low molecular weight amines show similar behavior

$$:NH_3 + H_2O \rightleftharpoons NH_4^+ + OH^-$$

Solutions of aqueous ammonia are basic because OH^- ions are produced as indicated by the above equation. Amines are thought of as derivatives of ammonia in which one or more hydrogen atoms have been replaced by organic groups, as the following structures indicate.

H, H—N:, H	H, H_3C—N:, H	H_3C, H_3C—N:, H	H_3C, H_3C—N:, H_3C
ammonia	methylamine	dimethylamine	trimethylamine

The unshared pair of electrons on the nitrogen atom is responsible for the fact that aqueous solutions of amines are basic. The dissolution of trimethylamine in water, for example, results in the formation of trimethylammonium ions and OH^- ions. Note the structural similarities of the ammonium and trimethylammonium ions

$$(H_3C)_3N: + H_2O \rightleftharpoons (H_3C)_3NH^+ + OH^-$$

trimethylamine → trimethylammonium ion

Let us now consider the behavior of ammonia in aqueous solutions. As indicated earlier, the reaction of ammonia with water is

$$NH_3 + H_2O \rightleftharpoons NH_4^+ + OH^-$$

and the ionization constant expression is

The subscript b indicates that the substance ionizes as a base.

$$K_b = \frac{[NH_4^+][OH^-]}{[NH_3]} = 1.8 \times 10^{-5}$$

Because the concentration of water is constant, or very nearly so, it is included in the ionization constant as it was for weak acids. The fact that the ionization constant for aqueous ammonia has the same value as the ionization constant

TABLE 16–4 Ionization Constants for Some Weak Bases

Base	Ionization Reaction	K_b at 25°C
Ammonia	$NH_3 + H_2O \rightleftharpoons NH_4^+ + OH^-$	1.8×10^{-5}
Methylamine	$(CH_3)NH_2 + H_2O \rightleftharpoons (CH_3)NH_3^+ + OH^-$	5.0×10^{-4}
Dimethylamine	$(CH_3)_2NH + H_2O \rightleftharpoons (CH_3)_2NH_2^+ + OH^-$	7.4×10^{-4}
Trimethylamine	$(CH_3)_3N + H_2O \rightleftharpoons (CH_3)_3NH^+ + OH^-$	7.4×10^{-5}
Pyridine	$C_5H_5N + H_2O \rightleftharpoons C_5H_5NH^+ + OH^-$	1.5×10^{-9}

for acetic acid is pure coincidence. However, it does tell us that in aqueous solutions of the same concentration, CH_3COOH and NH_3 are ionized to the same extent. Table 16–4 lists ionization constants for a few common weak bases. Appendix G includes a longer list.

Ionization constants for weak bases are used in the same way as ionization constants for weak acids, as the following examples illustrate.

EXAMPLE 16–14

Calculate the concentration of OH^-, the pH, and the % ionization for a 0.20 *M* solution of aqueous ammonia.

Solution

The equation for the ionization of aqueous ammonia and the algebraic representations of equilibrium concentrations are

	NH_3	$+ H_2O \rightleftharpoons$	NH_4^+	$+ OH^-$
Orig	$0.20\ M$		$0\ M$	$\sim 0\ M$
Change	$-x\ M$		$+x\ M$	$+x\ M$
Equil	$(0.20 - x)\ M$		$x\ M$	$x\ M$

Substitution into the ionization constant expression gives

$$K_b = \frac{[NH_4^+][OH^-]}{[NH_3]} = 1.8 \times 10^{-5} = \frac{(x)(x)}{(0.20 - x)}$$

If we assume that $(0.20 - x) \approx 0.20$, we have

$$\frac{x^2}{0.20} = 1.8 \times 10^{-5} \quad \text{and} \quad x^2 = 3.6 \times 10^{-6}$$

which gives $x = 1.9 \times 10^{-3}\ M = [OH^-]$.

Since $[OH^-] = 1.9 \times 10^{-3}\ M$, pOH = 2.72 and pH = 11.28.

The percentage ionization may be represented as

$$\%\ \text{ionization} = \frac{[NH_3]_{\text{ionized}}}{[NH_3]_{\text{total}}} \times 100\% = \frac{[OH^-]}{[NH_3]_{\text{total}}} \times 100\%$$

Each NH_3 molecule that ionizes produces one OH^- ion, and therefore we represent $[NH_3]_{\text{ionized}}$ by $[OH^-]$. Substitution into the above relationship gives

$$\frac{[OH^-]}{[NH_3]_{\text{total}}} \times 100\% = \frac{1.9 \times 10^{-3}}{0.20} \times 100\% = 0.95\%\ \text{ionized}$$

EXAMPLE 16–15

The pH of an aqueous solution of ammonia is 11.50. Calculate the molarity of the solution.

Solution

Let's examine the problem to determine what we are given and what we would like to know. Since pH = 11.50, we know that pOH = 2.50, and $[OH^-] = 10^{-2.50}$ or

$$[OH^-] = 10^{-2.50} = 10^{0.50} \times 10^{-3} = 3.2 \times 10^{-3}\ M$$

Now that we know $[OH^-] = 3.2 \times 10^{-3}\ M$, we also know that $[NH_4^+] = 3.2 \times 10^{-3}\ M$, and we can represent the equilibrium concentrations by letting $x\ M$ represent the original concentration of NH_3.

	NH_3	$+ H_2O \rightleftharpoons$	NH_4^+	$+$	OH^-
Orig	$x\ M$		$0\ M$		$\sim 0\ M$
Change	$-3.2 \times 10^{-3}\ M$		$+3.2 \times 10^{-3}\ M$		$+3.2 \times 10^{-3}\ M$
Equil	$(x - 3.2 \times 10^{-3})\ M$		$3.2 \times 10^{-3}\ M$		$3.2 \times 10^{-3}\ M$

Since x mol/L of NH_3 is present in the solution and 3.2×10^{-3} mol/L of NH_3 ionizes, the equilibrium concentration of NH_3 is $(x - 3.2 \times 10^{-3})$ mol/L. Substituting these values into the equilibrium expression for aqueous ammonia gives

$$K_b = \frac{[NH_4^+][OH^-]}{[NH_3]} = \frac{(3.2 \times 10^{-3})(3.2 \times 10^{-3})}{(x - 3.2 \times 10^{-3})} = 1.8 \times 10^{-5}$$

Examination of the above equation suggests that $(x - 3.2 \times 10^{-3})$ is approximately equal to x (that is, 3.2×10^{-3} is very small compared to x). Making this assumption simplifies the calculation.

$$\frac{(3.2 \times 10^{-3})(3.2 \times 10^{-3})}{x} = 1.8 \times 10^{-5}, \quad \text{and} \quad x = 0.57\ M\ NH_3$$

Therefore, the solution is 0.57 molar in ammonia. Our assumption that $(x - 3.2 \times 10^{-3}) \approx x$ was justified.

16–5 Acid-Base Indicators

In Section 11–15 we defined **titration** as the procedure in which a solution of accurately known concentration, a standard solution, is added slowly to a solution of unknown concentration until the chemical reaction between the two solutes is complete. Suppose we have a standard solution of sodium hydroxide and a solution of hydrochloric acid of unknown concentration (Figure 11–1). Clearly the acid and base react in a 1:1 mole ratio to form ordinary salt and water

$$NaOH\ (aq) + HCl\ (aq) \longrightarrow NaCl\ (aq) + H_2O$$

How can we determine the point at which the reaction is complete? Some organic compounds exhibit different colors in solutions of different acidities, and change colors when the hydrogen ion concentration in the solution reaches a definite value. They are called **indicators** and are used to determine (as nearly as possible) the point at which the reaction between the acid and base is complete. We introduced indicators in Section 11–15.

Phenolphthalein is also the active component of the laxative "ExLax," and is sometimes added to laboratory ethyl alcohol to discourage its consumption.

Phenolphthalein is a common acid-base indicator. It is colorless in aqueous solution of pH less than 8 (that is, $[H_3O^+] > 10^{-8}\ M$), and turns red as the pH approaches 10.

Most acid-base indicators are weak organic acids, HIn, or weak organic bases, InOH, where "In" represents complex organic groups. Bromthymol blue is typical of the acid-base indicators that are weak organic acids. Its ionization constant is about 1×10^{-7}. We can represent its dissociation in dilute aqueous solution as

$$\underset{\substack{\text{color 1} \\ \text{yellow}}}{HIn} + H_2O \rightleftharpoons H_3O^+ + \underset{\substack{\text{color 2} \\ \text{blue}}}{In^-} \qquad \text{(for bromthymol blue)}$$

where HIn represents nonionized acid molecules and In^- represents the anion (conjugate base) of the weak acid. The essential characteristic of acid-base indicators is that HIn and In^- *must* have different colors. The relative amounts of the two species determine the colors of solutions. The ionization constant expression for such indicators is

$$K_a = \frac{[H_3O^+][In^-]}{[HIn]}$$

Addition of an acid shifts the equilibrium to the left and gives more HIn molecules, while addition of a base shifts the equilibrium to the right and gives more In^- ions.

The ionization constant expression can be rearranged to

$$\frac{[In^-]}{[HIn]} = \frac{K_a}{[H_3O^+]}$$

which shows clearly the dependence of the $[In^-]/[HIn]$ ratio on $[H_3O^+]$ (or on pH) and the value of the ionization constant for the indicator. As a rule-of-thumb, when $[In^-]/[HIn] \geq 10$, color 2 is observed; conversely, when $[In^-]/[HIn] \leq \frac{1}{10}$, color 1 is observed. The point at which the color change for the indicator occurs in a titration, known as the **end point**, is determined by the value of the ionization constant for the indicator. Table 16–5 shows a few acid-base indicators and the pH ranges over which their colors change. Typically, color changes occur over a range of 1.5 to 2.0 pH units.

The *equivalence point* (Section 11–15) is the point at which chemically equivalent amounts of acid and base have reacted. Ideally, the end point and the equivalence point should coincide in a titration. In practice, the best we can do is to select an indicator whose range of color change spans the theoretical equivalence point of the reaction of interest. By choosing an indicator with the proper end point and using the same procedures in both standardization and analysis, any error arising from a difference between end point and equivalence point can be minimized.

Indicators: (1) intensely colored substances (2) Diff. indicators change color in diff. pH range. (3) Many are very complex weak organic acids or bases.

TABLE 16–5 Range and Color Changes of Some Common Acid-Base Indicators

Indicators	Color (lower pH)	pH range of color change	Color (higher pH)
* Methyl orange	red	3.1–4.4	yellow
Methyl red	red	4.4–6.2	yellow
Bromthymol blue	yellow	6.2–7.6	blue
Neutral red	red	6.8–8.0	yellow
* Phenolphthalein	colorless	8.0–10.0	red; colorless beyond 13.0

Methyl violet — Color (acid) yellow — pH range 0-2 — Color (base) purple
Litmus — red — 4.7-8.2 — blue

16–6 Titration Curves for Strong Acids and Strong Bases

A **titration curve** is a plot of pH versus the amount (volume, usually) of acid or base added. It displays graphically the change in pH as acid or base is added to a solution, and indicates clearly how pH changes near the equivalence point.

Titrations are usually done with 50 mL, or smaller, burets. We have chosen to use 100 mL of solution in this illustrative example to simplify the arithmetic.

Consider the titration of 100.0 mL of a 0.100 *M* solution of HCl with a 0.100 *M* solution of NaOH. We shall calculate the pH of the solution as NaOH is added. As pointed out earlier, NaOH and HCl react in a 1:1 ratio. Before any NaOH is added to the 0.100 *M* HCl solution, its pH is

$$HCl + H_2O \longrightarrow H_3O^+ + Cl^-$$
$$[H_3O^+] = 0.10\ M$$
$$pH = 1.00$$

After 20.0 mL of 0.100 *M* NaOH has been added, the pH is

	HCl	+ NaOH	$\longrightarrow$ NaCl	$+ H_2O$
Start	10.0 mmol	2.00 mmol	0 mmol	
Change	−2.00 mmol	−2.00 mmol	+2.00 mmol	
After rxn	8.00 mmol	0 mmol	2.00 mmol	

$$M_{HCl} = \frac{8.00\ \text{mmol}}{120\ \text{mL}} = 0.067\ M$$
$$[H_3O^+] = 6.7 \times 10^{-2}\ M$$
$$pH = 1.17$$

After 50.0 mL of 0.100 *M* NaOH has been added (midpoint of the titration), the pH is

	HCl	+ NaOH	$\longrightarrow$ NaCl	$+ H_2O$
Start	10.0 mmol	5.00 mmol	0 mmol	
Change	−5.00 mmol	−5.00 mmol	+5.00 mmol	
After rxn	5.00 mmol	0 mmol	5.00 mmol	

$$M_{HCl} = \frac{5.00\ \text{mmol}}{150\ \text{mL}} = 0.033\ M\ \text{HCl}$$
$$[H_3O^+] = 3.3 \times 10^{-2}\ M$$
$$pH = 1.48$$

After 100.0 mL of 0.100 *M* NaOH has been added (the equivalence point), the pH is 7.00 because the NaOH exactly neutralizes the HCl.

After 110.0 mL of 0.100 *M* NaOH has been added, the pH is determined by the excess NaOH

	HCl	+ NaOH	$\longrightarrow$ NaCl	$+ H_2O$
Start	10.0 mmol	11.0 mmol	0 mmol	
Change	−10.0 mmol	−10.0 mmol	+10.0 mmol	
After rxn	0 mmol	1.0 mmol	10.0 mmol	

$$M_{NaOH} = \frac{1.0\ \text{mmol}}{210.0\ \text{mL}} = 0.0048\ M\ \text{NaOH}$$
$$[OH^-] = 4.8 \times 10^{-3}\ M$$
$$pOH = 2.32 \quad \text{and} \quad pH = 11.68$$

TABLE 16–6 Titration Data for HCl *vs.* NaOH

mL of 0.100 *M* NaOH Added	mmol NaOH Added	mmol Excess Acid or Base	pH
0.0	0.00	10.0 H_3O^+	1.00
20.0	2.00	8.0	1.17
50.0	5.00	5.0	1.48
90.0	9.00	1.0	2.28
99.0	9.90	0.10	3.30
99.5	9.95	0.05	3.60
100.0	10.00	0.00	7.00
100.5	10.05	0.05 OH^-	10.40
110.0	11.00	1.00	11.68
120.0	12.00	2.00	11.96

Table 16–6 displays the data for the titration of 100.0 mL of 0.100 *M* HCl by 0.100 *M* NaOH solution. A few additional points have been included to show the shape of the curve definitively. The data are plotted in Figure 16–2.

Note that the titration curve has a long "vertical section" over which pH changes very rapidly with the addition of very small amounts of base. The pH changes from 3.60 (99.5 mL NaOH added) to 10.40 (100.5 mL of NaOH added) in the vicinity of the equivalence point (100.0 mL NaOH added). The midpoint of the vertical section (pH = 7.00) is the equivalence point.

Indicators with color changes in the range pH 4 to 10 can be used in the titration of strong acids and strong bases, although ideally the color change

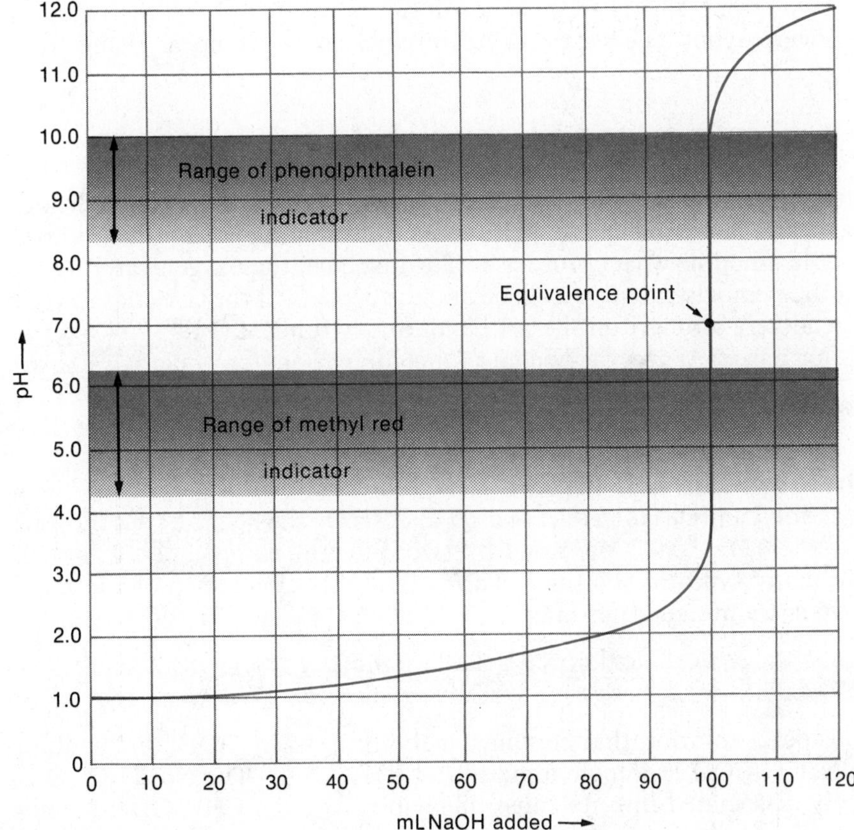

FIGURE 16–2 Titration curve for 100 mL of 0.100 *M* HCl versus 0.100 *M* NaOH. Note that the "vertical section" of the curve is quite long. The titration curves for other strong acids and bases are identical with this one *if* the same concentrations of acid and base are used.

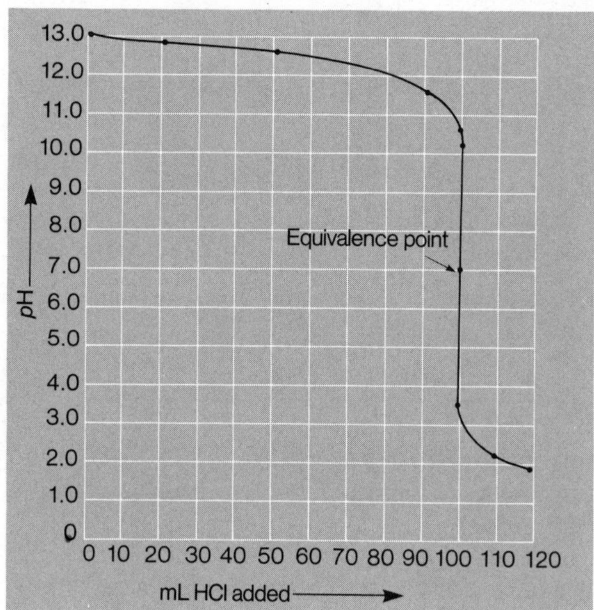

FIGURE 16–3 Titration curve for 100 mL of 0.100 *M* NaOH versus 0.100 *M* HCl. Notice that this titration curve is very similar to that in Figure 16–2, but inverted.

should occur at pH = 7.00. Figure 16–2 shows that the color ranges for methyl red and phenolphthalein, two widely used indicators, both fall within the vertical section of the NaOH/HCl titration.

When a strong acid is added to a solution of a strong base, the titration curve is inverted, but its essential characteristics are the same, as Figure 16–3 shows.

16–7 The Common Ion Effect and Buffer Solutions

It is often useful in laboratory reactions, industrial processes, and bodies of plants and animals to keep the pH nearly constant despite the addition of considerable amounts of acids or bases. For instance, the oxygen-carrying capacity of the hemoglobin in your blood and the activity of the enzymes in your cells depend very strongly on the pH of your body fluids. In order to keep the pH within a narrow range, the body uses a combination of compounds known as a *buffer system,* whose operation depends on the *common ion effect,* a special case of LeChatelier's Principle.

Buffer systems resist changes in pH.

The Common Ion Effect

The term **common ion effect** is used to describe the behavior of solutions in which the same ion is produced by two different compounds. Although there are many types of solutions that exhibit this effect, two of the most frequently encountered kinds are:

1. a solution of a weak acid *plus* a soluble ionic salt of the weak acid;
2. a solution of a weak base *plus* a soluble ionic salt of the weak base.

Consider a solution that contains acetic acid, CH_3COOH, *and* sodium acetate, $NaCH_3COO$, a soluble ionic salt of CH_3COOH. The $NaCH_3COO$ is completely dissociated into its constituent ions, but the CH_3COOH is only slightly ionized, as indicated by the following equations.

$$NaCH_3COO \xrightarrow{H_2O} Na^+ + \boxed{CH_3COO^-}$$

$$CH_3COOH + H_2O \rightleftharpoons \boxed{H_3O^+} + CH_3COO^-$$

In solutions containing both CH_3COOH and $NaCH_3COO$ in reasonable concentrations, both compounds serve as sources of CH_3COO^- ions. The completely dissociated $NaCH_3COO$ provides a high concentration of CH_3COO^- ions, which shifts the equilibrium of the ionization of CH_3COOH far to the left as CH_3COO^- combines with H_3O^+ to form nonionized CH_3COOH and H_2O. This results in a drastic decrease in the concentration of H_3O^+ in the solution. Solutions containing a weak acid *plus a salt of the weak acid* are always less acidic than solutions containing the same concentration of the weak acid alone, as Example 16–16 illustrates.

LeChatelier's Principle (Section 15–5) is applicable to equilibria in aqueous solution.

EXAMPLE 16–16

Calculate the concentration of H_3O^+ and the pH of a solution that is 0.10 M in CH_3COOH and 0.20 M in $NaCH_3COO$.

Solution

The appropriate equations are

$$NaCH_3COO \longrightarrow Na^+ + CH_3COO^-$$
$$CH_3COOH + H_2O \rightleftharpoons H_3O^+ + CH_3COO^-$$

and the ionization constant expression for CH_3COOH is

$$K_a = \frac{[H_3O^+][CH_3COO^-]}{[CH_3COOH]} = 1.8 \times 10^{-5}$$

This ionization constant expression is valid for all solutions that contain CH_3COOH; the fact that the solution also contains $NaCH_3COO$ does not affect its validity. In solutions that contain both CH_3COOH and $NaCH_3COO$, acetate ions are derived from two sources, and the ionization constant is satisfied by the total CH_3COO^- concentration in the solution.

Since $NaCH_3COO$ is completely dissociated, the concentration of CH_3COO^- from $NaCH_3COO$ will be 0.20 mol/L. If we let $x = [H_3O^+]$, then x is also equal to the concentration of CH_3COO^- from CH_3COOH, and the total concentration of CH_3COO^- is $(0.20 + x)$ M. The concentration of nonionized CH_3COOH is $(0.10 - x)$ M. This gives us

$$\underset{0.20\,M}{NaCH_3COO} \longrightarrow \underset{0.20\,M}{Na^+} + \underset{0.20\,M}{CH_3COO^-}$$

$$\underset{(0.10 - x)\,M}{CH_3COOH} + H_2O \rightleftharpoons \underset{x\,M}{H_3O^+} + \underset{x\,M}{CH_3COO^-}$$

Substitution into the ionization constant expression for CH_3COOH gives

$$\frac{(x)(0.20 + x)}{(0.10 - x)} = 1.8 \times 10^{-5}$$

Examination of the above equation indicates that x is a very small number, and we are able to make two simplifying assumptions, namely

$(0.20 + x) = \sim 0.20$ and $(0.10 - x) = \sim 0.10$

Making these assumptions gives

$$\frac{0.20\,x}{0.10} = 1.8 \times 10^{-5} \quad \text{and} \quad x = 9.0 \times 10^{-6}\,M = [H_3O^+]$$

Since $[H_3O^+] = 9.0 \times 10^{-6}\,M$, we find pH = 5.05.

You may verify the validity of the assumptions by substituting the value for x, 9.0×10^{-6}, into the original equation.

To gain an appreciation for just how much the acidity of the 0.10 M CH_3COOH solution is reduced by making the solution 0.20 M in $NaCH_3COO$ also, refer back to Example 16–13, where we found that the H_3O^+ concentration in 0.10 M CH_3COOH is 1.3×10^{-3} mol/L and the pH of the solution is 2.89. Table 16–7 makes the point clearly.

TABLE 16–7 Comparison of $[H_3O^+]$ and pH in Acetic Acid and Sodium Acetate–Acetic Acid Solutions

Solution	$[H_3O^+]$	pH	
0.10 M CH_3COOH	$1.3 \times 10^{-3}\,M$	2.89	ΔpH = 2.16
0.10 M CH_3COOH and 0.20 M $NaCH_3COO$	$9.0 \times 10^{-6}\,M$	5.05	

Comparison of $[H_3O^+]$ in the two solutions shows that $[H_3O^+]$ is 1.4×10^2 times greater in 0.10 M CH_3COOH than in the solution that is both 0.10 M in CH_3COOH and 0.20 M in $NaCH_3COO$, which demonstrates LeChatelier's Principle.

The calculation of $[H_3O^+]$ in solutions containing both a weak acid and a salt of the weak acid can be simplified greatly. If we write the equation for the ionization of a weak monoprotic acid in the following way, a useful relationship is easily evolved

$$\text{H(Anion)} + H_2O \rightleftharpoons H_3O^+ + \text{Anion}^-$$

The ionization constant expression can be written as

$$\frac{[H_3O^+][\text{Anion}^-]}{[\text{H(Anion)}]} = K_a$$

Solving this expression for $[H_3O^+]$ gives

$$[H_3O^+] = \frac{[\text{H(Anion)}]}{[\text{Anion}^-]} \times K_a$$

We now impose two conditions: (1) The concentrations of both the weak acid and its salt are some reasonable value, say greater than 0.050 M; (2) the salt contains a univalent cation. Under these conditions, the concentration of the anion, $[\text{Anion}^-]$, in the solution can be assumed to be the same as the concentration of the salt. With these restrictions, the above expression for $[H_3O^+]$ becomes

$$[H_3O^+] = \frac{[\text{Acid}]}{[\text{Salt}]} \times K_a$$

where [Acid] is the concentration of nonionized acid (for all practical purposes, the total acid concentration) and [Salt] is the concentration of the salt of the weak acid.

If we take the logarithm of the above equation, we obtain

$$\log [H_3O^+] = \log \frac{[\text{Acid}]}{[\text{Salt}]} + \log K_a$$

Multiplying by -1 gives

$$-\log [H_3O^+] = -\log \frac{[\text{Acid}]}{[\text{Salt}]} - \log K_a$$

and rearrangement gives

$$\text{pH} = \text{p}K_a + \log \frac{[\text{Salt}]}{[\text{Acid}]} \qquad \text{where } \text{p}K_a = -\log K_a$$

This relationship is valid only for solutions that contain a weak monoprotic acid and a soluble, ionic salt of the weak acid with a univalent cation, both in reasonable concentrations.

This equation is known as the **Henderson-Hasselbalch equation,** and workers in the biological sciences use it frequently.

Let's consider the second common kind of buffer solution, containing a weak base and its salt. A solution that contains aqueous NH_3 and ammonium chloride, NH_4Cl, a soluble ionic salt of NH_3, is typical. The NH_4Cl is completely dissociated into its constituent ions, but aqueous NH_3 is only slightly ionized, as indicated by the following equations

$$NH_4Cl \xrightarrow{H_2O} NH_4^+ + Cl^-$$

$$NH_3 + H_2O \rightleftharpoons NH_4^+ + OH^-$$

In this solution both NH_4Cl and aqueous NH_3 produce NH_4^+ ions. The completely dissociated NH_4Cl provides a high concentration of NH_4^+ that shifts the equilibrium of the ionization of aqueous NH_3 far to the left, as NH_4^+ ions combine with OH^- ions to form nonionized NH_3 and H_2O. Thus, the concentration of OH^- is decreased significantly. Solutions containing a weak base plus a salt of the weak base are less basic than solutions containing the same concentration of weak base alone, as Example 16–17 illustrates.

EXAMPLE 16–17

Calculate the concentration of OH^- and the pH of a solution that is 0.20 *M* in aqueous NH_3 *and* 0.10 *M* in NH_4Cl.

Solution

The appropriate equations and algebraic representation of concentrations are

$$\begin{array}{lcccc} NH_4Cl & & \longrightarrow NH_4^+ & + & Cl^- \\ 0.10\,M & & 0.10\,M & & 0.10\,M \\ NH_3 & + H_2O & \rightleftharpoons NH_4^+ & + & OH^- \\ (0.20 - x)\,M & & x\,M & & x\,M \end{array}$$

$$\text{Total } [NH_4^+] = (0.10 + x)\,M$$

Substitution into the ionization constant expression for aqueous NH_3 gives

$$K_b = \frac{[NH_4^+][OH^-]}{[NH_3]} = 1.8 \times 10^{-5} = \frac{(0.10 + x)(x)}{(0.20 - x)}$$

Again, we can assume that $(0.10 + x) = \sim 0.10$ and $(0.20 - x) = \sim 0.20$, which gives

$$\frac{0.10x}{0.20} = 1.8 \times 10^{-5} \quad \text{and} \quad x = 3.6 \times 10^{-5}\ M = [OH^-]$$

Since $[OH^-] = 3.6 \times 10^{-5}\ M$, we find pOH = 4.44 and pH = 9.56.

Recall that in Example 16–14 we calculated $[OH^-]$ and pH for 0.20 *M* aqueous NH_3, and comparison of that result with the results obtained here is instructive (Table 16–8).

TABLE 16–8 Comparison of $[OH^-]$ and pH in Ammonia and Ammonium Chloride–Ammonia Solutions

Solution	$[OH^-]$	pH
0.20 *M* aq NH_3	$1.9 \times 10^{-3}\ M$	11.28
0.20 *M* aq NH_3 and 0.10 *M* aq NH_4Cl	$3.6 \times 10^{-5}\ M$	9.56
		$\Delta pH = -1.72$

Note that the concentration of OH^- is 53 times greater in the solution containing only 0.20 *M* aqueous NH_3 than in the solution containing both 0.20 *M* aqueous NH_3 and 0.10 *M* NH_4Cl, another demonstration of LeChatelier's Principle.

We can derive a relationship for concentration of OH^- in solutions containing weak bases *plus* salts of the weak bases, just as we did for weak acids. The equation for the ionization of a monoprotic weak base, in general terms, is

$$\text{Base} + H_2O \rightleftharpoons (\text{Base})H^+ + OH^-$$

The ionization constant expression then becomes

$$\frac{[(\text{Base})H^+][OH^-]}{[\text{Base}]} = K_b$$

Solving this expression for $[OH^-]$ gives

$$[OH^-] = \frac{[\text{Base}]}{[(\text{Base})H^+]} \times K_b$$

Taking the logarithm of both sides of the equation gives

$$\log [OH^-] = \log \frac{[\text{Base}]}{[(\text{Base})H^+]} + \log K_b$$

As long as we consider salts of weak bases that contain univalent anions, $[(\text{Base})H^+] = [\text{Salt}]$. Multiplication by -1 and rearrangement gives the Henderson-Hasselbalch equation for solutions containing a weak base plus a salt of the weak base.

$$\text{pOH} = \text{p}K_b + \log \frac{[\text{Salt}]}{[\text{Base}]} \qquad \text{where } \text{p}K_b = -\log K_b$$

If we consider other salts, such as $(NH_4)_2SO_4$, that contain divalent anions, then $[(Base)H^+] = 2[Salt]$. The above relationship is valid for solutions of weak bases plus salts of weak bases in reasonable concentrations, as Example 16–18 illustrates.

EXAMPLE 16–18

Calculate the concentration of OH^- and the pH of a solution that is 0.15 M in aqueous NH_3 and 0.075 M in $(NH_4)_2SO_4$.

Solution

The equations and algebraic representations are

$$\begin{array}{llll} (NH_4)_2SO_4 & \longrightarrow & 2NH_4^+ & + \; SO_4^{2-} \\ 0.075\,M & & 0.15\,M & 0.075\,M \\ NH_3 \;+\; H_2O & \rightleftharpoons & NH_4^+ & + \; OH^- \\ (0.15 - x)\,M & & x\,M & x\,M \end{array}$$

We can substitute into the ionization constant expression for aqueous NH_3, as we did in Example 16–17

$$K_b = \frac{[NH_4^+][OH^-]}{[NH_3]} = 1.8 \times 10^{-5} = \frac{(0.15 + x)(x)}{(0.15 - x)}$$

which gives $x = 1.8 \times 10^{-5}\,M = [OH^-]$, pOH = 4.74, and pH = 9.26.

Alternatively, we can use the ionization constant expression for aqueous NH_3 in the general form

$$K_b = \frac{[(Base)H^+][OH^-]}{[Base]} = 1.8 \times 10^{-5}$$

$$[OH^-] = \frac{[Base]}{[(Base)H^+]} \times 1.8 \times 10^{-5} = \frac{(0.15)}{(0.15)} \times 1.8 \times 10^{-5}$$

which gives $[OH^-] = 1.8 \times 10^{-5}\,M$, pOH = 4.74, and pH = 9.26.

As another alternative, substitution into the Henderson-Hasselbalch equation gives the same result. $K_b = 1.8 \times 10^{-5}$ so $pK_b = 4.74$.

$$pOH = pK_b + \log \frac{2[Salt]}{[Base]} = 4.74 + \log \frac{0.15}{0.15}$$

$$pOH = 4.74 + 0 = 4.74, \; pH = 9.26$$

16–8 Buffering Action

As we stated earlier, buffer solutions resist changes in pH. By this statement we mean that we can add small amounts of strong base or strong acid to buffer solutions and the pH will change very little. These statements imply that buffer solutions are able to react with both H_3O^+ ions and OH^- ions, and indeed this is the case. The two common kinds of buffer solutions are the ones we have just discussed, namely, solutions containing (1) a weak acid plus a salt of the weak acid and (2) a weak base plus a salt of the weak base.

All buffer solutions contain conjugate acid-base pairs in reasonable concentrations. In all cases *the more acidic component is the one that reacts with (and tends to neutralize) added strong base, and the more basic component reacts with added strong acid.*

1 Solutions of a Weak Acid and a Salt of the Weak Acid

Let us consider a solution containing acetic acid, CH_3COOH, and sodium acetate, $NaCH_3COO$, as a common example of this kind of buffer solution. The more acidic component is CH_3COOH. The more basic component is $NaCH_3COO$ because the acetate ion, CH_3COO^-, is the conjugate base of CH_3COOH. Thus the following reaction tends to neutralize added strong acid, such as HCl

$$HCl + NaCH_3COO \xrightarrow{\sim 100\%} CH_3COOH + NaCl$$

or as a net ionic equation

The net effect is to neutralize most of the H_3O^+ from HCl by forming nonionized CH_3COOH molecules, and to decrease the ratio $[CH_3COO^-]/[CH_3COOH]$, which governs the pH of the solution.

$$\underset{\text{added strong acid}}{H_3O^+} + \underset{\text{base}}{CH_3COO^-} \xrightarrow{\sim 100\%} \underset{\text{weak acid}}{CH_3COOH} + H_2O$$

This reaction goes nearly to completion because CH_3COOH is a *weak* acid; its constituent ions, when mixed from separate sources, have a strong tendency to combine rather than remain separate.

On the other hand, when a strong soluble base, such as NaOH, is added to the $CH_3COOH/NaCH_3COO$ buffer solution it is consumed by the *acidic component*, CH_3COOH

$$\underset{\text{added base}}{NaOH} + \underset{\text{acid}}{CH_3COOH} \xrightarrow{\sim 100\%} \underset{\text{salt}}{NaCH_3COO} + \underset{\text{water}}{H_2O}$$

The net effect is to neutralize most of the OH^- from NaOH, and to increase the ratio $[CH_3COO^-]/[CH_3COOH]$, which governs the pH of the solution.

or as a net ionic equation

$$OH^- + CH_3COOH \xrightarrow{\sim 100\%} CH_3COO^- + H_2O$$

EXAMPLE 16–19

If 0.010 mole of solid NaOH is added to one liter of a buffer solution that is 0.10 *M* in CH_3COOH and 0.10 *M* in $NaCH_3COO$, how much will $[H_3O^+]$ and pH change? Assume no volume change due to the addition of solid NaOH.

Solution

From our earlier discussion, for the 0.10 *M* CH_3COOH and 0.10 *M* $NaCH_3COO$ solution, we can write

$$[H_3O^+] = \frac{[\text{Acid}]}{[\text{Salt}]} \times K_a = \frac{0.10}{0.10} \times 1.8 \times 10^{-5},$$

which gives $[H_3O^+] = 1.8 \times 10^{-5}$ *M* and pH = 4.74 before the NaOH is added. When solid NaOH is added to the solution, the NaOH reacts with CH_3COOH to form more $NaCH_3COO$.

If HCl, rather than NaOH, were added to this buffer solution, it would react with $NaCH_3COO$ to form nonionized CH_3COOH molecules.

	$NaOH$	$+ CH_3COOH$	$\longrightarrow NaCH_3COO$	$+ H_2O$
Start	0.01 mol	0.10 mol	0.10 mol	
Change	−0.01 mol	−0.01 mol	+0.01 mol	
After rxn	0 mol	0.09 mol	0.11 mol	

Since the volume of the solution is 1.0 liter, we now have a solution that is 0.09 *M* in CH_3COOH and 0.11 *M* in $NaCH_3COO$. In this solution

$$[H_3O^+] = \frac{[\text{Acid}]}{[\text{Salt}]} \times K_a = \frac{0.09}{0.11} \times 1.8 \times 10^{-5}$$

which gives $[H_3O^+] = 1.5 \times 10^{-5}\ M$ and pH = 4.82.

The calculation shows that the addition of 0.010 mole of solid NaOH to 1.0 liter of a buffer solution (this is enough NaOH to neutralize 10% of the acid) decreases $[H_3O^+]$ from $1.8 \times 10^{-5}\ M$ to $1.5 \times 10^{-5}\ M$ and increases pH from 4.74 to 4.82, a change of 0.08 pH units. These are very slight changes.

By contrast to the case in Example 16–19, adding 0.010 mole of NaOH to enough pure water to give one liter of solution gives a 0.010 *M* solution of NaOH in which $[OH^-] = 1.0 \times 10^{-2}\ M$, and pOH = 2.00. Since pH + pOH = 14, the pH of this solution is 12.00, an increase of 5 pH units above that of pure water.

Addition of 0.010 mole of solid NaOH to one liter of 0.10 *M* CH_3COOH (pH = 2.89 from Table 16–3) gives a solution that is 0.09 *M* in CH_3COOH and 0.01 *M* in $NaCH_3COO$. The pH of this solution is 3.80, which is 0.91 pH units higher than that of the 0.10 *M* CH_3COOH solution. Table 16–9 summarizes these data.

TABLE 16–9 Changes in pH Caused by Addition of 0.010 Mole of Solid NaOH to One Liter of Solution

Solution	Change in pH
0.10 *M* CH_3COOH } 0.10 *M* $NaCH_3COO$ }	0.08
0.10 *M* CH_3COOH alone	0.91
pure H_2O	5.00

2 Solutions of a Weak Base and a Salt of the Weak Base

A common example of a buffer solution in this category is one containing the weak base aqueous ammonia, NH_3, and its soluble ionic salt ammonium chloride, NH_4Cl. If a strong acid, such as HCl, is added to such a buffer solution, the more basic component, NH_3, neutralizes most of the H_3O^+ from HCl

$$\underset{\text{acid}}{HCl} + \underset{\text{base}}{NH_3} + H_2O \xrightarrow{\sim 100\%} \underset{\text{salt}}{NH_4Cl} + \underset{\text{water}}{H_2O}$$

or in net ionic form

$$H_3O^+ + NH_3 \xrightarrow{\sim 100\%} NH_4^+ + H_2O$$

The net effect is to neutralize most of the H_3O^+ from the HCl and to increase the ratio $[NH_4^+]/[NH_3]$, which governs the pH of the solution.

If a strong soluble base such as NaOH is added to the original buffer solution, it is neutralized by the more acidic component, NH_4Cl, or NH_4^+, which is the conjugate acid of aqueous ammonia

$$NaOH + NH_4Cl \xrightarrow{\sim 100\%} NH_3 + H_2O + NaCl$$

or in net ionic form

$$\underset{\text{added strong base}}{OH^-} + \underset{\text{acid}}{NH_4^+} \longrightarrow \underset{\text{weak base}}{NH_3} + H_2O$$

The net effect is to neutralize most of the OH^- from NaOH, and to decrease the ratio $[NH_4^+]/[NH_3]$, which governs the pH of the solution.

Thus we see that changes in pH, that is, changes in $[H_3O^+]$ and $[OH^-]$, are minimized in buffer solutions because one (basic) component can react with excess H_3O^+ ions and another (acidic) component can react with excess OH^- ions.

16–9 Preparation of Buffer Solutions

Buffer solutions are frequently prepared by mixing other solutions. When solutions are mixed, the volume in which each solute is contained increases and the concentrations of the dissolved species change. This change in concentration must be taken into consideration, as Example 16–20 illustrates.

EXAMPLE 16–20

Calculate the concentration of H_3O^+ in a buffer solution prepared by mixing 200 mL of 0.10 *M* NaF and 100 mL of 0.050 *M* HF. $K_a = 7.2 \times 10^{-4}$ for HF.

Solution

Since two dilute solutions are mixed, we may assume that the volumes are additive, and the volume of the new solution will be 300 mL. Mixing a solution of a weak acid with a solution of a salt of the weak acid does not result in any new chemical species being formed, and we have a straightforward buffer calculation. Therefore, we shall calculate the number of millimoles of each compound and the molarities in the new solution.

$$\underline{?}\text{ mmol NaF} = 200\text{ mL} \times \frac{0.10\text{ mmol NaF}}{\text{mL}} = 20\text{ mmol NaF}$$

$$\underline{?}\text{ mmol HF} = 100\text{ mL} \times \frac{0.050\text{ mmol HF}}{\text{mL}} = 5.0\text{ mmol HF}$$

Since the 300 mL of solution contain 20 mmol of NaF and 5.0 mmol of HF, the molarities of NaF and HF are

$$\frac{20\text{ mmol NaF}}{300\text{ mL}} = 0.067\ M\text{ NaF}$$

and

$$\frac{5.0\text{ mmol HF}}{300\text{ mL}} = 0.017\ M\text{ HF}$$

The appropriate equations and algebraic representations of concentrations are

$$\begin{array}{lcccc} \text{NaF} & \longrightarrow & Na^+ & + & F^- \\ 0.067\ M & & 0.067\ M & & 0.067\ M \end{array}$$

$$\begin{array}{lccccc} \text{HF} & + H_2O & \rightleftharpoons & H_3O^+ & + & F^- \\ (0.017 - x)\ M & & & x\ M & & x\ M \end{array}$$

The ionization constant expression for hydrofluoric acid is

$$K_a = \frac{[H_3O^+][F^-]}{[HF]} = 7.2 \times 10^{-4}$$

Substitution gives

$$\frac{(x)(0.067 + x)}{(0.017 - x)} = 7.2 \times 10^{-4}$$

Some question may arise as to whether we can assume that x is negligible compared to 0.067 and 0.017 in this expression. When in doubt, solve the equation using the simplifying assumption and then determine whether the assumption is valid. Assuming that $(0.067 + x) \approx 0.067$ and that $(0.017 - x) \approx 0.017$ gives

$$\frac{0.067x}{0.017} = 7.2 \times 10^{-4} \qquad x = 1.8 \times 10^{-4}\ M$$

the concentration of H_3O^+ in the solution. Our assumption is valid.

On occasion it is desirable to prepare a buffer solution of a particular pH. Example 16–21 illustrates one method by which such solutions can be prepared. This involves adding a solid salt of a weak base (or weak acid) to a solution of the weak base (or weak acid) itself.

EXAMPLE 16–21

Calculate the number of moles of solid ammonium chloride, NH_4Cl, that must be used to prepare 2.0 liters of a buffer solution that is 0.10 *M* in aqueous ammonia and that has a pH of 9.15.

We could prepare an NH_3/NH_4Cl buffer by mixing aqueous ammonia with less than the stoichiometrically equivalent amount of HCl.

Calculated quantities of *solutions* of weak base (or weak acid) and strong acid (or strong soluble base) also could be mixed as in Example 16–21 to produce a buffer solution of a designated pH.

Solution

Because pH = 9.15, pOH = 14.00 − 9.15 = 4.85

$[OH^-] = \text{antilog}(-4.85) = 10^{-4.85} = 1.4 \times 10^{-5}\ M$

Let's define the necessary molarity of NH_4Cl to be $x\,M$. Since $[OH^-] = 1.4 \times 10^{-5}\ M$, this must be the concentration of OH^- ions produced by ionization of NH_3. The equations and representations of equilibrium concentrations are shown below.

$$NH_4Cl \xrightarrow{100\%} NH_4^+ + Cl^-$$
$$x\,M \qquad x\,M \qquad x\,M$$

$$NH_3 + H_2O \rightleftharpoons NH_4^+ + OH^-$$
$$(0.10 - 1.4 \times 10^{-5})\ M \qquad 1.4 \times 10^{-5}\ M \qquad 1.4 \times 10^{-5}\ M$$

Substitution into the ionization constant expression for aqueous ammonia gives the following

$$K_b = \frac{[NH_4^+][OH^-]}{[NH_3]} = 1.8 \times 10^{-5} = \frac{(x + 1.4 \times 10^{-5})(1.4 \times 10^{-5})}{0.10 - 1.4 \times 10^{-5}}$$

Since NH_4Cl is completely dissociated we may assume that $x \gg 1.4 \times 10^{-5}$ so that $(x + 1.4 \times 10^{-5}) \approx x$. The expression then becomes

$$\frac{(x)(1.4 \times 10^{-5})}{0.10} = 1.8 \times 10^{-5}, \qquad x = 0.13\ M = [NH_4^+] = M_{NH_4Cl}$$

In this case x does *not* represent a *change* in concentration, but rather the initial concentration of NH_4Cl, so we do *not* assume that $x \ll 1.4 \times 10^{-5}$, but rather the converse.

Now we calculate the number of moles of NH_4Cl that must be added to prepare 2.0 L of buffer solution.

$$\underline{?}\ \text{mol}\ NH_4Cl = 2.0\ \text{L} \times \frac{0.13\ \text{mol}\ NH_4Cl}{\text{L}} = 0.26\ \text{mol}\ NH_4Cl$$

This corresponds to 0.26 mol × 53.5 g/mol = 14 g NH_4Cl.

16–10 Polyprotic Acids

Thus far we have restricted our discussion of weak acids to *monoprotic* acids, i.e., acids that furnish only one hydronium ion per molecule. Many weak acids furnish two or more hydronium ions per molecule, and are called **polyprotic** acids. The ionizations of polyprotic acids occur stepwise, that is, one proton at a time. Ionization constant expressions can be written for each step in the ionization of polyprotic acids, as the following example illustrates. Consider phosphoric acid as a typical polyprotic acid. It contains three acidic hydrogen atoms and ionizes in three steps

$$H_3PO_4 + H_2O \rightleftharpoons H_3O^+ + H_2PO_4^-, \quad K_1 = \frac{[H_3O^+][H_2PO_4^-]}{[H_3PO_4]} = 7.5 \times 10^{-3}$$

$$H_2PO_4^- + H_2O \rightleftharpoons H_3O^+ + HPO_4^{2-}, \quad K_2 = \frac{[H_3O^+][HPO_4^{2-}]}{[H_2PO_4^-]} = 6.2 \times 10^{-8}$$

$$HPO_4^{2-} + H_2O \rightleftharpoons H_3O^+ + PO_4^{3-}, \quad K_3 = \frac{[H_3O^+][PO_4^{3-}]}{[HPO_4^{2-}]} = 3.6 \times 10^{-13}$$

Examination of the three ionization constants for H_3PO_4 reveals that K_1 is much larger than K_2, and that K_2 is much larger than K_3. This is generally true for polyprotic *inorganic* acids (refer to Appendix F). Successive ionization constants commonly decrease by a factor of approximately 10^4 to 10^6, although some differences are outside this range. A large decrease in the value of each successive ionization constant means that each step in the ionization of a polyprotic acid occurs to a much lesser extent than the previous step. The extent of ionization in the second and third steps is even less than the values of K_2 and K_3 indicate, because the concentration of H_3O^+ produced in the first step is very large compared to the concentration of H_3O^+ produced in the second and third steps. Since the expressions for K_2 and K_3 include $[H_3O^+]$, they are satisfied by the *total concentration of H_3O^+ in the solution.* As a matter of fact, in all except extremely dilute solutions of H_3PO_4, the concentration of H_3O^+ may be assumed to be that furnished by the first step in the ionization alone.

EXAMPLE 16–22

Calculate the concentrations of all species present in 0.100 *M* H_3PO_4.

Solution

Since K_1 is so much larger than K_2, we assume that the amount of H_3O^+ formed in the second and third steps of the ionization of H_3PO_4 and by the ionization of water is insignificant.

H_3PO_4	$+ H_2O \rightleftharpoons$	H_3O^+	$+ H_2PO_4^-$
$(0.100 - x)\ M$		$x\ M$	$x\ M$

Substitution into the expression for K_1 gives

$$\frac{[H_3O^+][H_2PO_4^-]}{[H_3PO_4]} = \frac{(x)(x)}{(0.100 - x)} = 7.5 \times 10^{-3}$$

This equation must be solved by the quadratic formula. Solving the quadratic equation gives the positive root $x = 2.4 \times 10^{-2}$. Thus, from the first step in the ionization of H_3PO_4

$x = [H_3O^+] = [H_2PO_4^-] = 2.4 \times 10^{-2}\ M$

$(0.100 - x) = [H_3PO_4] = 7.6 \times 10^{-2}\ M$

$x = -3.1 \times 10^{-2}$ is the extraneous root of the quadratic equation, and is discarded because concentrations can't possibly be less than zero!

For the second step in the ionization of H_3PO_4, we use the concentrations of H_3O^+ and $H_2PO_4^-$ obtained from the first step, and let $y = [H_3O^+]$ produced in the second step as well as $[HPO_4^{2-}]$.

$$\begin{array}{llll} H_2PO_4^- & + H_2O \rightleftharpoons & H_3O^+ & + HPO_4^{2-} \\ (2.4 \times 10^{-2} - y)\ M & & (2.4 \times 10^{-2} + y)\ M & y\ M \end{array}$$

(the 2.4×10^{-2} in $[H_3O^+]$ is from 1st step)

Substitution into the expression for K_2 gives

$$\frac{[H_3O^+][HPO_4^{2-}]}{[H_2PO_4^-]} = \frac{(2.4 \times 10^{-2} + y)(y)}{(2.4 \times 10^{-2} - y)} = 6.2 \times 10^{-8}$$

Examination of this equation indicates that 2.4×10^{-2} is much greater than y, and making the usual simplifying assumption gives

$$\frac{(2.4 \times 10^{-2})(y)}{(2.4 \times 10^{-2})} = 6.2 \times 10^{-8}, \qquad y = 6.2 \times 10^{-8}\ M = [HPO_4^{2-}]$$

Also, $y = [H_3O^+]$ produced in the second step. Thus, we find that the concentration of HPO_4^{2-} is equal to K_2.

A general statement can be made for solutions of weak polyprotic acids for which $K_1 \gg K_2$ and that contain no other electrolytes: **the concentration of the anion produced in the second step ionization is always equal to K_2 in solutions of reasonable concentration.** Note that our assumption that the $[H_3O^+]$ produced in the first step ($2.4 \times 10^{-2}\ M$) is much greater than the $[H_3O^+]$ produced in the second step ionization ($6.2 \times 10^{-8}\ M$) is valid.

The pH of most solutions of polyprotic acids is governed by the first step ionization.

We have now calculated the concentrations of all the species in 0.10 M H_3PO_4 solution except PO_4^{3-}, the anion produced in the third step ionization, and OH^-, which is present in all aqueous solutions. To calculate the concentration of PO_4^{3-} we use the equation for the third step ionization and let $z = [PO_4^{3-}]$ and $[H_3O^+]$ produced in the third step.

$$\begin{array}{llll} HPO_4^{2-} & + H_2O \rightleftharpoons & H_3O^+ & + PO_4^{3-} \\ (6.2 \times 10^{-8} - z)\ M & & (2.4 \times 10^{-2} + z)\ M & z\ M \end{array}$$

(the 2.4×10^{-2} in $[H_3O^+]$ is from 1st step)

y was disregarded in step 2 and is also disregarded here.

$$\frac{[H_3O^+][PO_4^{3-}]}{[HPO_4^{2-}]} = \frac{(2.4 \times 10^{-2} + z)(z)}{(6.2 \times 10^{-8} - z)} = 3.6 \times 10^{-13}$$

We can make our simplifying assumption, and find that

$z = 9.3 \times 10^{-19}\ M = [PO_4^{3-}]$.

We have now calculated the concentration of each of the species formed by the ionization of 0.100 M H_3PO_4. For easy comparison these concentrations are listed in Table 16–10. As a point of interest, the concentration of OH^- in

TABLE 16–10 Concentrations of the Species in 0.10 *M* H_3PO_4

Species	Concentration (mol/L)
H_3PO_4	0.076
H_3O^+	0.024
$H_2PO_4^-$	0.024
HPO_4^{2-}	0.000000062
OH^-	0.00000000000042
PO_4^{3-}	0.0000000000000000093

0.100 *M* H_3PO_4 is also included in the tabulation; it is calculated from the known value for $[H_3O^+]$ using the ion product for water, $[H_3O^+][OH^-] = 1.0 \times 10^{-14}$.

Note that nonionized H_3PO_4 is present in greater concentration than any other species in 0.100 *M* H_3PO_4 solution, and that the only other species present in significant concentrations are H_3O^+ and $H_2PO_4^-$. Analogous statements can be made for other weak polyprotic acids for which the last *K* is very small.

Phosphoric acid may be thought of as a typical weak polyprotic acid. Let us now describe solutions of sulfuric acid, a very strong polyprotic acid, and hydrosulfuric acid (commonly called hydrogen sulfide), a very weak polyprotic acid.

EXAMPLE 16–23

Calculate the concentrations of the various species in a 0.10 *M* solution of H_2SO_4, for which $K_2 = 1.2 \times 10^{-2}$.

Solution

The first step ionization of H_2SO_4 is complete, as we have pointed out previously

$$H_2SO_4 + H_2O \longrightarrow H_3O^+ + HSO_4^-$$

and produces 0.10 mol/L of both H_3O^+ and HSO_4^- ions. However, the second step ionization is not complete,

$$HSO_4^- + H_2O \rightleftharpoons H_3O^+ + SO_4^{2-}$$

$$K_2 = \frac{[H_3O^+][SO_4^{2-}]}{[HSO_4^-]} = 1.2 \times 10^{-2}$$

Keeping in mind that the concentration of H_3O^+ is the sum of the concentrations produced in the first and second steps, we may represent the equilibrium concentrations as (where *x* is mol/L of HSO_4^- that ionizes):

$$\underset{(0.10 - x)\,M}{HSO_4^-} + H_2O \rightleftharpoons \underset{(0.10 + x)\,M}{H_3O^+} + \underset{x\,M}{SO_4^{2-}}$$

from 1st step ↗ ↖ from 2nd step

Substitution into the ionization constant expression gives

$$\frac{(0.10 + x)(x)}{0.10 - x} = 1.2 \times 10^{-2}$$

Clearly, x cannot be disregarded, and this quadratic equation must be solved by the quadratic formula (as we did in Example 16–22). This gives $x = 0.010$ and $x = -0.12$, the latter being the extraneous root. We now tabulate the concentrations of the various species in a 0.10 M solution of H_2SO_4

$[H_2SO_4] = \sim 0\ M$

$[H_3O^+] = (0.10 + x)\ M = 0.11\ M$

$[HSO_4^-] = (0.10 - x)\ M = 0.09\ M$

This calculation shows that the extent of the second step ionization in a 0.10 M solution of H_2SO_4 is 10%.

EXAMPLE 16–24

Calculate the concentrations of the various species in a 0.10 M solution of H_2S. (This is a saturated solution.) $K_1 = 1.0 \times 10^{-7}$ and $K_2 = 1.3 \times 10^{-13}$.

Solution

These ionization constants tell us that H_2S is a very weak acid. Let x = mol/L of H_2S that ionize in the first step

$$\underset{(0.10 - x)\ M}{H_2S} + H_2O \rightleftharpoons \underset{x\ M}{H_3O^+} + \underset{x\ M}{HS^-}$$

$$K_1 = \frac{[H_3O^+][HS^-]}{[H_2S]} = \frac{(x)(x)}{(0.10 - x)} = 1.0 \times 10^{-7}$$

We can disregard the x in the denominator, and solving the resulting equation gives $x = 1.0 \times 10^{-4}$. Thus,

$[H_2S] = (0.10 - x) \approx 0.10\ M$

$[HS^-] = [H_3O^+] = x = 1.0 \times 10^{-4}\ M$ (1st step)

The second step ionization involves the anion produced in the first step, and we may represent equilibrium concentrations as (where y = mol/L of HS^- that ionize in the second step):

$$\underset{(1.0 \times 10^{-4} - y)\ M}{HS^-} + H_2O \rightleftharpoons \underset{(1.0 \times 10^{-4} + y)\ M}{H_3O^+} + \underset{y\ M}{S^{2-}}$$

(from 1st step) (from 2nd step)

Substitution into K_2 gives

$$K_2 = \frac{[H_3O^+][S^{2-}]}{[HS^-]} = \frac{(1.0 \times 10^{-4} + y)(y)}{(1.0 \times 10^{-4} - y)} = 1.3 \times 10^{-13}$$

Examination of the quadratic equation suggests that $(1.0 \times 10^{-4} + y) \approx 1.0 \times 10^{-4}$ and $(1.0 \times 10^{-4} - y) \approx 1.0 \times 10^{-4}$, and therefore we have

$$\frac{(1.0 \times 10^{-4})y}{1.0 \times 10^{-4}} = 1.3 \times 10^{-13} \qquad y = 1.3 \times 10^{-13}$$

Clearly, our assumption is valid. The concentration of HS^- that ionizes $(1.3 \times 10^{-13}\ M)$ is very small indeed, but it does give us the equilibrium concentration of sulfide ions.

$[S^{2-}] = y = \underline{1.3 \times 10^{-13}\ M}$

$[HS^-] = (1.0 \times 10^{-4} - y) = (1.0 \times 10^{-4}) - (1.3 \times 10^{-13}) \approx \underline{1.0 \times 10^{-4}\ M}$

$[H_3O^+] = (1.0 \times 10^{-4} + y) = (1.0 \times 10^{-4}) + (1.3 \times 10^{-13}) \approx \underline{1.0 \times 10^{-4}\ M}$

We have already calculated the concentration of nonionized H_2S (~0.10 *M*).

Table 16–11 provides an interesting comparison for 0.10 *M* solutions of the three polyprotic acids we have described in Examples 16–22, 23, and 24. This tabulation indicates the wide range of acidity in 0.10 *M* solutions of polyprotic acids. The concentration of H_3O^+ in 0.10 *M* H_2SO_4 is 1100 times greater than that in 0.10 *M* H_2S!

TABLE 16–11 Comparison of Concentrations of H_3O^+ and Nonionized Acid in 0.10 *M* Solutions of Three Polyprotic Acids

	0.10 *M* H_2SO_4	0.10 *M* H_3PO_4	0.10 *M* H_2S
K_1	very large	7.5×10^{-3}	1.0×10^{-7}
K_2	1.2×10^{-2}	6.2×10^{-8}	1.3×10^{-13}
K_3		3.6×10^{-13}	
$[H_3O^+]$	0.11 *M*	$2.4 \times 10^{-2}\ M$	$1.0 \times 10^{-4}\ M$
[Acid molecules]	~0 *M*	$7.6 \times 10^{-2}\ M$	~0.10 *M*

Key Terms

Amines derivatives of ammonia in which one or more hydrogen atoms have been replaced by organic groups.

Buffer solution solution that resists change in pH; contains either a weak acid and a soluble ionic salt of the acid or a weak base and a soluble ionic salt of the base.

Common ion effect suppression of ionization of a weak electrolyte by the presence in the same solution of a strong electrolyte containing one of the same ions as the weak electrolyte.

End point the point at which an indicator changes color and a titration is stopped.

Equivalence point the point at which chemically equivalent amounts of reactants have reacted.

Indicators for acid-base titrations, organic compounds that exhibit different colors in solutions of different acidities; used to determine the point at which reaction between two solutes is complete.

Ionization constant equilibrium constant for the ionization of a weak electrolyte.

Ion product for water equilibrium constant for the ionization of water, $K_w = [H^+][OH^-] = 1.00 \times 10^{-14}$ at 25°C.

p[ ] the negative logarithm of the concentration (mol/L) of the indicated species.

pH the negative logarithm of the concentration (mol/L) of the H^+ (H_3O^+) ion; scale is commonly used over the range 0 to 14.

Titration procedure in which a solution of known concentration, a standard solution, is added to a solution of unknown concentration until the chemical reaction between the two solutes is complete.

Titration curve for acid-base titration, a plot of pH versus volume of acid or base solution added.

Exercises

Basic Ideas—Strong Electrolytes

1. List names and formulas for:
   (a) the common strong acids
   (b) six weak bases
   (c) the common strong soluble bases
   (d) ten soluble ionic salts
2. Calculate the concentrations of the constituent ions in solutions of the following compounds in the indicated concentrations. The salts are soluble ionic compounds.
   (a) 0.10 *M* HI
   (b) 0.050 *M* KOH
   (c) 0.010 *M* $Sr(OH)_2$
   (d) 0.0020 *M* $Ba(NO_3)_2$
   (e) 0.00030 *M* H_2SO_4
   (f) 0.0035 *M* $Fe_2(SO_4)_3$
3. Calculate the concentrations of the constituent ions in the following solutions. The salts are soluble ionic compounds.
   (a) 4.0 g NaOH in 2.00 liters of solution
   (b) 0.74 g $Ca(OH)_2$ in 200 mL of solution
   (c) 2.64 g $(NH_4)_2SO_4$ in 100 mL of solution
   (d) 3.42 g $Al_2(SO_4)_3$ in 300 mL of solution
   (e) 43.8 g $CaCl_2 \cdot 6H_2O$ in 4.00 liters of solution

The Ionization of Water

4. Why is the ion product for water written in the form $[H_3O^+][OH^-] = 1.0 \times 10^{-14}$?
5. (a) Why is the concentration of H_3O^+ produced by the ionization of water neglected in calculating the concentration of H_3O^+ in a 0.10 *M* solution of HCl?
   (b) Demonstrate that it (H_3O^+ from H_2O) may be neglected.
6. (a) Why is the concentration of OH^- produced by the ionization of water neglected in calculating the concentration of OH^- in a 0.10 *M* solution of NaOH?
   (b) Demonstrate that it (OH^- from H_2O) may be neglected.
7. Calculate and compare the concentrations of OH^- in the solutions described in Exercises 2(a) and 2(e) with the OH^- concentration in pure water.
8. Calculate and compare the concentrations of H_3O^+ in the solutions described in Exercises 2(b), 2(c), 3(a), and 3(b) with the H_3O^+ concentration in pure water.

Logarithms, pH, and pOH

9. Look up and write down the logarithms of the following numbers:
   (a) 0.012
   (b) 0.17
   (c) 0.0036
   (d) 0.00074
   (e) 1.8×10^{-2}
   (f) 1.6×10^{-6}
   (g) 9.3×10^{-12}
   (h) 7.5×10^{-8}
10. Define and illustrate the following terms clearly and concisely: (a) pH, (b) pOH, (c) pCl
11. Refer to Exercise 8 and calculate the pH of each solution.
12. Refer to Exercise 7 and calculate the pOH of each solution.
13. Refer back to Exercise 2(a), 2(b), 2(c), and 2(e) to complete the following table. Is there an obvious relationship between pH and pOH? What is it?

Solution	$[H_3O^+]$	$[OH^-]$	pH	pOH
0.10 *M* HI				
0.050 *M* KOH				
0.010 *M* $Sr(OH)_2$				
0.00030 *M* H_2SO_4				

14. Given one of the following four values, calculate the other three:

	$[H_3O^+]$	$[OH^-]$	pH	pOH
(a) Solution *A*	1.0×10^{-3} *M*			
(b) Solution B			2.000	
(c) Solution *C*		7.4×10^{-6} *M*		

15. Complete the following table by appropriate calculations:

$[H_3O^+]$	pH	$[OH^-]$	pOH
(a) ______	5.27	(c) ______	1.24
(b) ______	12.34	(d) ______	9.67

16. Calculate the following values for each solution:

	$[H_3O^+]$	$[OH^-]$	pH	pOH
(a) 0.020 M NaOH	______	______	______	______
(b) 0.063 M HCl	______	______	______	______
(c) 0.020 M $Ca(OH)_2$	______	______	______	______

17. Given: $A = 2.4 \times 10^{-4}$ and $B = 6.3 \times 10^{-6}$.
(a) What is the ratio of A to B?
(b) A is ______ times greater than B.
(c) What is the ratio of B to A?

18. Given: $C = 1.4 \times 10^{-3}$ and
$D = 4.8 \times 10^{-9}$.
(a) What is the ratio of C to D?
(b) C is ______ times greater than D.
(c) What is the ratio of D to C?

19. Complete the following table by appropriate calculations.

Solution	$[H_3O^+]$	$[OH^-]$	pH	pOH
0.020 M HCl	______	______	______	______
______ M HCl	______	______	2.00	______
0.020 M $Ba(OH)_2$	______	______	______	______
______ M $Ba(OH)_2$	______	1.00×10^{-3} M	______	______

20. In solution E the H_3O^+ concentration is 0.020 mol/L, and in solution F the H_3O^+ concentration is 3.0×10^{-4} mol/L.
(a) What is the ratio of H_3O^+ concentration in solution E to solution F?
(b) What is the ratio of H_3O^+ concentration in solution F to solution E?
(c) Calculate the concentrations of OH^- in solutions E and F.
(d) What is the ratio of OH^- concentration in solution E to solution F?
(e) What is the ratio of OH^- concentration in solution F to solution E?
(f) What is the ratio of H_3O^+ concentration to OH^- concentration in solution E?
(g) What is the ratio of OH^- concentration to H_3O^+ concentration in solution E?
(h) Why are answers (b) and (d) the same while answers (a) and (e) are slightly different?

21. What do we mean when we refer to:
(a) neutral solutions?
(b) acidic solutions?
(c) basic solutions?

Weak Acids and Weak Bases

22. In a solution of a weak acid, $HA + H_2O \rightleftharpoons H_3O^+ + A^-$, the following equilibrium concentrations are found: $[H_3O^+] = 0.0020$ M and $[HA] = 0.098$ M. Calculate the ionization constant for the weak acid, HA.

23. A 0.050 M aqueous solution of a weak, monoprotic acid ($HA + H_2O \rightleftharpoons H_3O^+ + A^-$) is 1.2% ionized. Calculate the ionization constant for HA.

24. In 0.10 M solution a weak acid, $HX + H_2O \rightleftharpoons H_3O^+ + X^-$, is 1.6% ionized. Calculate the ionization constant for the weak acid, HX.

25. A 0.0100 molal solution of acetic acid ($CH_3COOH + H_2O \rightleftharpoons H_3O^+ + CH_3COO^-$) freezes at −0.01938°C. Calculate the ionization constant for acetic acid. A 0.0100 molal solution is sufficiently dilute that it may be assumed to be 0.0100 molar without introducing a significant error.

26. Calculate the concentrations of the various species in 0.10 *M* hypochlorous acid, for which $K_a = 3.5 \times 10^{-8}$.

27. Calculate $[H_3O^+]$, $[OH^-]$, pH, pOH, and % ionization for 0.050 *M* CH_3COOH solution, for which $K_a = 1.8 \times 10^{-5}$.

28. Calculate $[H_3O^+]$, $[OH^-]$, pH, pOH, and % ionization for 0.050 *M* HCN solution, for which $K_a = 4.0 \times 10^{-10}$.

29. (a) In 0.050 *M* CH_3COOH the H_3O^+ concentration is approximately ______ times greater than in 0.050 *M* HCN. Refer to Exercises 27 and 28.
    (b) In 0.050 *M* HCN solution the OH^- concentration is approximately ______ times greater than in 0.050 *M* CH_3COOH solution. Refer to Exercises 27 and 28.
    (c) In 0.050 *M* CH_3COOH solution the % ionization is approximately ______ times greater than in 0.050 *M* HCN solution. Refer to Exercises 27 and 28.

30. Calculate the concentration of H_3O^+, pH, and % ionization for 0.10 *M* solutions of the following weak, monoprotic acids. (a) HCN, (b) HF, (c) HBrO, (d) HN_3

31. The pH of a hydrofluoric acid solution is 3.00. Calculate the molarity of the solution. Can you make a simplifying assumption in this case?

32. A 0.20 *M* solution of a newly discovered monoprotic acid is found to have a pH of 3.22. What is the value of the ionization constant for the acid? Represent the new acid by the general formula HY.

33. Answer the following questions for *0.10 M solutions* of the weak acids listed in Table 16–2.
    (a) Which solution contains
        1. the highest concentration of H_3O^+?
        2. the highest concentration of OH^-?
        3. the lowest concentration of H_3O^+?
        4. the lowest concentration of OH^-?
        5. the highest concentration of nonionized acid molecules?
        6. the lowest concentration of nonionized acid molecules?
    (b) In which solution is
        1. the pH highest?
        2. the pH lowest?
        3. the pOH highest?
        4. the pOH lowest?
    (c) Which solution contains
        1. the highest concentration of the anion of the weak acid?
        2. the lowest concentration of the anion of the weak acid?

34. Calculate $[H_3O^+]$, $[OH^-]$, pH, pOH, and % ionization for 0.050 *M* aqueous ammonia solution, $K_b = 1.8 \times 10^{-5}$. How do these numbers compare with those obtained for 0.050 *M* CH_3COOH solution in Exercise 27? Why?

35. Calculate $[OH^-]$, % ionization, and pH for (a) 0.10 *M* aqueous ammonia, and (b) 0.10 *M* methylamine solution.

36. Refer to Table 16–4 and answer the following questions for *0.10 M solutions* of the weak bases listed there.
    (a) Which solution contains
        1. the highest concentration of OH^-?
        2. the highest concentration of H_3O^+?
        3. the lowest concentration of OH^-?
        4. the lowest concentration of H_3O^+?
        5. the highest concentration of nonionized molecules of weak base?
        6. the lowest concentration of nonionized molecules of weak base?
        7. the highest concentration of the cation of the weak base?
        8. the lowest concentration of the cation of the weak base?
    (b) In which solution is
        1. the pH highest?
        2. the pH lowest?
        3. the pOH highest?
        4. the pOH lowest?

37. Vinegar is a dilute aqueous solution of acetic acid, and is one of the most common acidic materials found in most homes. Household ammonia is a dilute aqueous solution of ammonia that also contains a little

detergent. It is one of the most common basic materials in most homes.

(a) Write the balanced molecular equation for the reaction that occurs when vinegar and household ammonia are mixed.

(b) Would you expect 1.00 L of vinegar that is 5.00% CH_3COOH to neutralize 1.00 L of household ammonia that is 5.00% NH_3? Why?

(c) The density of a sample of household ammonia that is 5.00% NH_3 by mass is 0.980 g/mL. What is the molarity of this solution?

(d) The density of a sample of white vinegar that is 5.00% CH_3COOH by mass is 1.01 g/mL. What is the molarity of this solution?

(e) What volume of the vinegar described in (d) would be required to neutralize 1.00 L of the household ammonia described in (c)?

(f) What volume of the household ammonia would be required to neutralize 1.00 L of the vinegar?

Titration and Titration Curves

38. (a) What is a titration? (b) What are acid-base indicators? (c) What is the essential characteristic of acid-base indicators? (d) What determines the color of an acid-base indicator in an aqueous solution?

39. Define and distinguish between end point and equivalence point in an acid-base titration.

40. What colors do the following indicators exhibit in solutions of the indicated pH?

	pH = 2.00	pH = 11.00
methyl red		
neutral red		
phenol-phthalein		

41. Demonstrate mathematically that bromthymol blue is yellow in solutions of pH 4.00, while it is blue in solutions of pH 9.00. Its ionization constant is 1×10^{-7}.

42. What are titration curves?

43. Construct a table similar to Table 16–6 for the titration of 25.0 mL of 0.100 *M* HNO_3 solution by 0.100 *M* KOH solution. Calculate the pH of the solution before any KOH is added and after the addition of the following amounts of 0.100 *M* KOH: (a) 5.0 mL, (b) 10.0 mL, (c) 12.5 mL, (d) 20.0 mL, (e) 24.0 mL, (f) 24.9 mL, (g) 25.0 mL, (h) 25.1 mL, (i) 27.0 mL, and (j) 30.0 mL. (1) Plot pH *vs.* mL of KOH added. (2) Does this titration curve resemble the curve in Figure 16–2? Why? (3) Over what pH range could indicators change colors for this titration?

44. Construct a table similar to Table 16–6 for the titration of 25.0 mL of 0.100 *M* NaOH solution by 0.100 *M* $HClO_4$ solution. Calculate the pH of the solution before any $HClO_4$ is added and after the addition of the following amounts of 0.100 *M* $HClO_4$: (a) 5.0 mL, (b) 10.0 mL, (c) 12.5 mL, (d) 20.0 mL, (e) 24.0 mL, (f) 24.9 mL, (g) 25.0 mL, (h) 25.1 mL, (i) 27.0 mL, and (j) 30.0 mL. (1) Plot pH *vs.* mL of $HClO_4$ added. (2) Does this titration curve resemble the curve in Figure 16–3? Why? (3) Over what pH range could indicators change colors for this titration?

The Common Ion Effect and Buffer Solutions

45. Define and illustrate the following terms clearly: (a) common ion effect, (b) buffer solution.

46. List two frequently encountered kinds of buffer solutions. List two specific examples of each kind.

47. Consider solutions that contain the indicated concentrations of the following pairs of compounds? Which ones are buffer solutions? Why?
(a) 0.10 *M* HF and 0.10 *M* NaF
(b) 0.10 *M* HCl and 0.10 *M* NaCl
(c) 0.050 *M* HClO and 0.10 *M* NaClO
(d) 0.10 *M* $HClO_4$ and 0.050 *M* $KClO_4$
(e) 0.10 *M* NH_3 and 0.10 *M* NH_4Cl
(f) 0.10 *M* NH_3 and 0.50 *M* NH_4NO_3

48. Calculate the hydronium ion concentration and the pH of the following solutions. After you have done the calculations, refer back to

Exercise 30 and compare $[H_3O^+]$ and pH for solutions containing only the acids.
(a) 0.10 *M* HCN and 0.10 *M* NaCN
(b) 0.10 *M* HF and 0.10 *M* NaF

49. (a) Calculate the ratio of concentrations $CH_3COOH/NaCH_3COO$ that gives solutions with pH = 5.00.
(b) Calculate the ratio of concentrations $CH_3COOH/NaCH_3COO$ that gives solutions with pH = 4.74.

50. Calculate pH for each of the following buffer solutions:
(a) 0.10 *M* HCNO and 0.20 *M* KCNO
(b) 0.050 *M* HCN and 0.025 *M* $Ba(CN)_2$

51. Calculate the concentration of OH^- and the pH of each of the following solutions:
(a) 0.10 *M* aq. NH_3 and 0.10 *M* NH_4NO_3
(b) 0.20 *M* aq. NH_3 and 0.050 *M* $(NH_4)_2SO_4$
(c) 0.10 *M* pyridine and 0.10 *M* pyridinium chloride, C_5H_5NHCl.

52. Calculate the concentration of OH^- and pH for the following buffer solutions:
(a) 0.10 *M* aq. NH_3 and 0.20 *M* NH_4NO_3
(b) 0.10 *M* aq. NH_3 and 0.10 *M* $(NH_4)_2SO_4$

53. Calculate the ratio of concentrations $[NH_3]/[NH_4^+]$ that gives
(a) solutions of pH = 9.00.
(b) solutions of pH = 9.25.

54. Calculate the change in $[H_3O^+]$ and pH for the following buffer solutions:
(a) 0.010 mole of HCl is added to 1.0 liter of a solution that is 0.10 *M* in CH_3COOH and 0.10 *M* in $NaCH_3COO$. Assume no volume change.
(b) 0.010 mole of HNO_3 is added to 500 mL of a solution that is 0.10 *M* in aqueous ammonia and 0.20 *M* in NH_4NO_3. Assume no volume change.
(c) 0.010 mole of NaOH is added to 500 mL of a solution that is 0.10 *M* in aqueous ammonia and 0.20 *M* in NH_4Cl. Assume no volume change.

55. (a) What is the pH of a solution containing 0.20 *M* formic acid, HCO_2H? $K_a = 1.8 \times 10^{-4}$.
(b) What is the pH of a solution containing 0.20 *M* formic acid and 0.10 *M* sodium formate, $NaCO_2H$?
(c) What is the pH of a solution prepared by adding 50 mL of 0.010 *M* NaOH to 100 mL of 0.20 *M* formic acid?
(d) What is the pH of a solution prepared by adding 50 mL of 0.010 *M* NaOH to 100 mL of a solution containing 0.20 *M* HCO_2H and 0.10 *M* $NaCO_2H$?
(e) What is the pH of a solution prepared by adding 50 mL of 0.10 *M* HCl to 100 mL of a solution containing 0.20 *M* HCO_2H and 0.10 *M* $NaCO_2H$?

56. If 100 mL of 0.020 *M* HCl and 100 mL of 0.10 *M* NaCN solutions are mixed, what will the pH of the resulting solution be? Would this be wise? Why?

57. Calculate the pH of each of the following solutions:
(a) 100 mL of 0.100 *M* HCl mixed with 50.0 mL of 0.100 *M* KOH.
(b) 100 mL of 0.100 *M* HCl mixed with 100 mL of 0.100 *M* KOH.
(c) 100 mL of 0.100 *M* HCl mixed with 150 mL of 0.100 *M* KOH.
(d) 100 mL of 0.100 *M* CH_3COOH mixed with 50.0 mL of 0.100 *M* NaOH.

58. Calculate $[H_3O^+]$, $[OH^-]$, pH, and pOH for the following solutions:
(a) 50 mL of 2.0 *M* aqueous NH_3.
(b) 50 mL of a solution containing 2.0 *M* aqueous NH_3 and 1.0 *M* NH_4Cl.
(c) 50 mL of a solution containing 2.0 *M* aqueous NH_3 and 1.0 *M* $(NH_4)_2SO_4$.
(d) A solution prepared by adding 10 mL of 1.0 *M* HNO_3 to the solution in (a).
(e) A solution prepared by adding 10 mL of 1.0 *M* HNO_3 to the solution in (b).
(f) A solution prepared by adding 10 mL of 1.0 *M* KOH to the solution in (c).

59. How many grams of NH_4Cl must be added to 100 mL of 4.0 *M* aqueous NH_3 in order to prepare a buffer solution having a pH of 10.20?

60. How many grams of sodium acetate must be added to 200 mL of 6.0 *M* CH_3COOH to prepare a buffer solution having a pH of 3.62?

61. How much 0.20 *M* NaF would have to be added to 100 mL of 0.10 *M* HF to give a solution of pH = 3.00?

62. How much 0.10 *M* NaCNO would have to be added to 100 mL of 0.10 *M* HCNO to give a solution of pH = 4.00?

63. A buffer was prepared by adding 10 mL of 4.0 *M* $NaCH_3COO$ to 10 mL of 6.0 *M* CH_3COOH and diluting to a total volume of 100 mL. What is the pH of this solution?

Polyprotic Acids

64. Calculate the concentrations of H_3O^+, OH^-, HCO_3^-, and CO_3^{2-} in 0.050 *M* H_2CO_3 solution.

65. Calculate the concentrations of H_3O^+, OH^-, $H(COO)_2^-$, and $(COO^-)_2$ in 0.050 *M* $(COOH)_2$, oxalic acid. Compare the concentrations with those obtained in Exercise 64. How can you explain the difference between the concentrations of HCO_3^- and $H(COO)_2^-$? between CO_3^{2-} and $(COO^-)_2$?

66. Calculate the concentrations of the various species in 0.100 *M* H_3AsO_4 solution. Compare the concentrations with those of the analogous species in 0.100 *M* H_3PO_4 solution (Example 16–22 in the text).

67. Calculate the concentration of each species in each of the following solutions. Include the concentration of OH^-.

(a) 0.020 *M* H_2CO_3
(b) 0.20 *M* H_3AsO_4
(c) 0.15 *M* H_3PO_4
(d) 0.10 *M* H_2S

17 Equilibria in Aqueous Solutions—II

Hydrolysis

Solvolysis describes the reaction of a substance with the solvent in which it is dissolved. We shall consider reactions that occur in aqueous solutions, which are called *hydrolysis* reactions. **Hydrolysis** refers to the reaction of a substance with water or its ions. One common kind of hydrolysis reaction involves the hydrolysis of salts, and particularly the combination of an anion of a weak acid with H_3O^+ from water to form nonionized acid molecules. The removal of H_3O^+ upsets the H_3O^+/OH^- balance in water, and so some salts produce basic solutions

$$\text{Anion}^- + H_3O^+ \rightleftharpoons \underset{\text{weak acid}}{\text{H(Anion)}} + H_2O$$

or the reaction is more commonly represented as

$$\text{Anion}^- + H_2O \rightleftharpoons \text{H(Anion)} + OH^-$$

Recall that in neutral solutions, $[H_3O^+] = [OH^-] = 1.0 \times 10^{-7}\ M$, while in basic solutions $[H_3O^+] < [OH^-]$ or $[OH^-] > 1.0 \times 10^{-7}\ M$ and in acidic solutions $[H_3O^+] > [OH^-]$ or $[H_3O^+] > 1.0 \times 10^{-7}\ M$.

The conjugate base of a strong acid is a very weak base and the conjugate base of a weak acid is a stronger base.

In our earlier discussion of acid-base reactions, we emphasized the fact that in Brønsted-Lowry terminology, anions of strong acids are extremely weak bases while anions of weak acids are stronger bases (Section 11–3). Conversely, cations of strong bases are extremely weak acids and cations of weak bases are stronger acids. To refresh our memories, consider the following examples. Nitric acid, a typical strong acid, is essentially completely ionized in *dilute* aqueous solution, and *dilute* aqueous solutions of nitric acid can be assumed to contain no molecules of nitric acid; only equal numbers of nitrate and hydronium ions as the following equation indicates

$$HNO_3 + H_2O \longrightarrow H_3O^+ + NO_3^-$$

Nitrate ions show almost no tendency to combine with hydronium ions (when mixed from separate sources) to form nonionized nitric acid in dilute aqueous solutions, because the nitrate ion is such a weak base.

On the other hand, acetic acid is a weak acid that is only slightly ionized in dilute aqueous solution

$$CH_3COOH + H_2O \rightleftharpoons H_3O^+ + CH_3COO^-$$

The acetate ion is a much stronger base than the nitrate ion, and acetate ions show a pronounced tendency to combine with H_3O^+ to form nonionized CH_3COOH molecules.

We have examined solutions of strong acids and strong soluble bases in Examples 16–1, 16–2, 16–4, and 16–8. In dilute solutions these compounds are completely ionized or dissociated. Let us now consider dilute aqueous solutions of salts that contain no free acid or base. We have classified acids and bases as strong and weak, and therefore we can identify four distinctly different kinds of salts: (1) salts of strong acids and strong soluble bases, (2) salts of strong soluble bases and weak acids, (3) salts of weak bases and strong acids, and (4) salts of weak bases and weak acids.

17–1 Salts of Strong Acids and Strong Soluble Bases

Solutions of salts derived from strong soluble bases and strong acids are neutral because neither the cation nor the anion of these salts reacts with water. For example, consider an aqueous solution of NaCl, the salt of NaOH and HCl. Sodium chloride is ionic even in the solid state and dissociates into hydrated ions in water. The water also ionizes slightly to produce equal numbers of H_3O^+ and OH^- ions

$$NaCl\ (solid) \longrightarrow Na^+ + Cl^-$$

$$H_2O + H_2O \rightleftharpoons OH^- + H_3O^+$$

Thus, aqueous solutions of NaCl contain four ions, Na^+, Cl^-, H_3O^+ and OH^-. However, the cation of the salt, Na^+, is such a weak acid that it does not react with the anion of water, OH^-. Likewise, the anion of the salt, Cl^-, is such a weak base that it does not react with the cation of water, H_3O^+. Therefore, solutions of salts of strong bases and strong acids are neutral because neither the cation nor the anion of such salts reacts to upset the H_3O^+/OH^- balance in water.

17–2 Salts of Strong Bases and Weak Acids

When salts derived from strong soluble bases and weak acids are dissolved in water, the resulting solutions are always found to be basic. Soluble salts of strong bases and weak acids produce basic solutions because anions of weak acids react with water to form hydroxide ions, as the following example illustrates. Consider a solution of sodium acetate, $NaCH_3COO$, the salt of NaOH and CH_3COOH. It is soluble and dissociates completely in water.

$$NaCH_3COO\text{ (solid)} \xrightarrow{H_2O} Na^+ + CH_3COO^-$$
$$H_2O + H_2O \rightleftharpoons OH^- + H_3O^+$$

Equilibrium is shifted $\longrightarrow$ excess

$$\Updownarrow$$
$$CH_3COOH + H_2O$$

Acetate ion from $NaCH_3COO$ is somewhat basic since it is the conjugate base of a *weak* acid, CH_3COOH. Thus it combines with H_3O^+ and causes more water to ionize. As H_3O^+ is removed from the solution, an excess of OH^- builds up, and so the solution becomes basic. The above equations can be combined into a single equation that represents the reaction much more simply, namely

LeChatelier's Principle applies to equilibria in aqueous solution and it enables us to make accurate predictions.

$$CH_3COO^- + H_2O \rightleftharpoons CH_3COOH + OH^-$$

The equilibrium constant for this reaction is called a (base) hydrolysis constant, or K_b for CH_3COO^-, and the equilibrium expression is written

Similar equations can be written for other salts derived from strong bases and weak acids.

$$K_b = \frac{[CH_3COOH][OH^-]}{[CH_3COO^-]}$$

For the first time, we are able to calculate equilibrium constants using other known expressions. (Previously we have considered equilibrium constants that were determined experimentally, or derived from changes in free energy.) If we multiply the above expression by $[H_3O^+]/[H_3O^+]$, an algebraic manipulation that doesn't change the value of the expression, we have

$$K_b = \frac{[CH_3COOH][OH^-]}{[CH_3COO^-]} \times \frac{[H_3O^+]}{[H_3O^+]} = \frac{[CH_3COOH]}{[H_3O^+][CH_3COO^-]} \times \frac{[H_3O^+][OH^-]}{1}$$

We recognize that

$$K_b = \frac{1}{K_{a\,(CH_3COOH)}} \times \frac{K_w}{1} = \frac{K_w}{K_{a\,(CH_3COOH)}} = \frac{1.0 \times 10^{-14}}{1.8 \times 10^{-5}}$$

or

$$K_b = \frac{[CH_3COOH][OH^-]}{[CH_3COO^-]} = 5.6 \times 10^{-10}$$

Hydrolysis constants, or K_b's for anions of weak acids, can be determined experimentally, and the values obtained from experiments agree with the calculated values. Please note that the K_b we use here refers to a reaction in which the anion of a weak acid acts as a base.

We have calculated the hydrolysis constant for the acetate ion, CH_3COO^-. We can perform the same operations for the anion of any weak monoprotic acid and find that $K_b = K_w/K_a$ where K_a refers to the ionization constant for the weak monoprotic acid from which the anion under consideration is derived, and K_b refers to the hydrolysis constant for the anion.

Note that the equation can be rearranged to

$$K_b \times K_a = K_w$$

a relationship that is valid for *all conjugate acid-base pairs* in aqueous solution. If either K_a or K_b is known, the other can be calculated.

The following examples illustrate the very different base strengths of different anions.

EXAMPLE 17–1

Calculate $[OH^-]$ and pH for 0.10 *M* solutions of (a) $NaCH_3COO$, sodium acetate, and (b) NaCN, sodium cyanide.

Solution

Since we have two distinct problems, let's work them one at a time. (a) The $NaCH_3COOH$ is completely dissociated in 0.10 *M* solution, and the overall equation for its reaction as a base with water is

$$CH_3COO^- + H_2O \rightleftharpoons CH_3COOH + OH^-$$

The equilibrium expression is

$$K_b = \frac{[CH_3COOH][OH^-]}{[CH_3COO^-]} = 5.6 \times 10^{-10}$$

From the balanced equation, we see that $[CH_3COOH] = [OH^-]$, and we represent these concentrations by x.

$$\underset{(0.10 - x)\,M}{CH_3COO^-} + H_2O \rightleftharpoons \underset{x\,M}{CH_3COOH} + \underset{x\,M}{OH^-}$$

Substitution into the equilibrium constant expression gives

$$\frac{[CH_3COOH][OH^-]}{[CH_3COO^-]} = 5.6 \times 10^{-10} = \frac{(x)(x)}{(0.10 - x)}$$

Making our usual simplifying assumption gives $x = 7.5 \times 10^{-6}\,M = [OH^-]$. Since $[OH^-] = 7.5 \times 10^{-6}\,M$, pOH = 5.12 and pH = 8.88. The 0.10 *M* $NaCH_3COO$ solution is distinctly basic.

(b) If we perform the same kind of calculation for 0.10 *M* NaCN, a soluble salt that is completely dissociated in water, we have

$$\underset{(0.10 - y)\,M}{CN^-} + H_2O \rightleftharpoons \underset{y\,M}{HCN} + \underset{y\,M}{OH^-}$$

$$K_b = \frac{[HCN][OH^-]}{[CN^-]} = \frac{K_w}{K_a} = \frac{1.0 \times 10^{-14}}{4.0 \times 10^{-10}} = 2.5 \times 10^{-5}$$

Substitution into this expression gives

$$\frac{(y)(y)}{(0.10 - y)} = 2.5 \times 10^{-5} \quad \text{and} \quad y = 1.6 \times 10^{-3}\,M = [OH^-],$$

from which we find pOH = 2.80 and pH = 11.20.

The 0.10 *M* solution of NaCN is much more basic than the 0.10 *M* solution of $NaCH_3COO$ because CN^- is a much stronger base than CH_3COO^-. This is entirely predictable because HCN is a much weaker acid than CH_3COOH.

To emphasize further the difference in the basicities of CH_3COO^- and CN^-, let's calculate the percentage hydrolysis of these anions in 0.10 *M* solution. For the hydrolysis of 0.10 *M* CH_3COO^- we found that $[OH^-] = 7.5 \times 10^{-6}$ *M*. Because each CH_3COO^- that hydrolyzes produces one OH^-, the concentration of OH^- is the same as the concentration of CH_3COO^- that *hydrolyzed.* Therefore, we can write

$$\% \text{ hydrolysis} = \frac{[CH_3COO^-]_{\text{hydrolyzed}}}{[CH_3COO^-]_{\text{total}}} \times 100\% = \frac{[OH^-]}{[CH_3COO^-]_{\text{total}}} \times 100\%$$

Substituting 7.5×10^{-6} *M* for $[OH^-]$ and 0.10 *M* for $[CH_3COO^-]_{\text{total}}$ gives

$$\frac{7.5 \times 10^{-6}}{0.10} \times 100\% = \underline{0.0075\% \text{ hydrolysis}} \qquad (0.10\ M\ NaCH_3COO)$$

For the 0.10 *M* solution of NaCN, we have an analogous situation in which each CN^- ion that hydrolyzes produces one OH^- ion. Therefore, we can represent the percentage hydrolysis as

$$\% \text{ hydrolysis} = \frac{[CN^-]_{\text{hydrolyzed}}}{[CN^-]_{\text{total}}} \times 100\% = \frac{[OH^-]}{[CN^-]_{\text{total}}} \times 100\%$$

$$= \frac{1.6 \times 10^{-3}}{0.10} \times 100\% = \underline{1.6\% \text{ hydrolysis}} \qquad (0.10\ M\ NaCN)$$

We have just shown that the % hydrolysis for 0.10 *M* NaCN (1.6%) is approximately 213 times greater than the % hydrolysis for 0.10 *M* $NaCH_3COO$ (0.0075%). Table 17–1 summarizes the two calculations. Data for a 0.10 *M* solution of aqueous ammonia are included to provide a reference point. Note that the cyanide ion, CN^-, is a stronger base than aqueous NH_3.

TABLE 17–1 Comparison of 0.10 *M* Solutions of $NaCH_3COO$, NaCN, and NH_3

	0.10 *M* $NaCH_3COO$	0.10 *M* NaCN	0.10 *M* aq NH_3
K_{acid} for parent acid	1.8×10^{-5}	4.0×10^{-10}	
K_b for anion	5.6×10^{-10}	2.5×10^{-5}	K_b for NH_3 is 1.8×10^{-5}
$[OH^-]$	7.5×10^{-6} *M*	1.6×10^{-3} *M*	1.3×10^{-3} *M*
% hydrolysis	0.0075%	1.6%	1.3% ionized
pH	8.88	11.20	11.11

17–3 Salts of Weak Bases and Strong Acids

The second common kind of hydrolysis reaction involves the reaction of the cation (the conjugate acid) of a weak base with OH^- from water to form nonionized molecules of the weak base. The removal of OH^- upsets the H_3O^+/OH^- balance in water, and it produces *acidic* solutions

$$(\text{Base})H^+ + OH^- \longrightarrow \underset{\text{weak base}}{\text{Base}} + H_2O$$

This reaction is more frequently represented as

$$(\text{Base})H^+ + H_2O \longrightarrow \text{Base} + H_3O^+$$

because the removal of OH^- causes more H_2O to ionize to give an excess of H_3O^+.

Aqueous solutions of salts of weak bases and strong acids are acidic because cations derived from weak bases react with water to produce hydronium ions. Consider a solution of ammonium chloride, NH_4Cl, the salt of aqueous ammonia and hydrochloric acid.

Ammonium chloride is an ionic solid that is soluble in water.

$$NH_4Cl\ (\text{solid}) \xrightarrow{H_2O} NH_4^+ + Cl^-$$

$$H_2O + H_2O \rightleftharpoons OH^- + H_3O^+ \text{ (excess)}$$

Equilibrium is shifted →

$$NH_4^+ + OH^- \rightleftharpoons NH_3 + H_2O$$

Ammonium ions, from NH_4Cl, react with OH^- to form nonionized ammonia and water molecules. Since this reaction removes OH^- from the system, it causes more water to ionize and produces an excess of H_3O^+ in the solution. We can conveniently combine the above equations into a single equation that represents the reaction accurately

Analogous reactions can be written for cations derived from other weak bases.

$$NH_4^+ + H_2O \rightleftharpoons NH_3 + H_3O^+$$

The equilibrium constant expression for this reaction is written

$$K_a = \frac{[NH_3][H_3O^+]}{[NH_4^+]}$$

Like hydrolysis constants for anions of weak acids, hydrolysis constants for cations of weak bases can be calculated. If we multiply the above expression by $[OH^-]/[OH^-]$, we have

$$K_a = \frac{[NH_3][H_3O^+]}{[NH_4^+]} \times \frac{[OH^-]}{[OH^-]} = \frac{[NH_3]}{[NH_4^+][OH^-]} \times \frac{[H_3O^+][OH^-]}{1}$$

We recognize that

K_b represents the ionization constant for aq NH_3, the conjugate base of NH_4^+.

$$K_a = \frac{1}{K_{b\,(NH_3)}} \times \frac{K_w}{1} = \frac{1.0 \times 10^{-14}}{1.8 \times 10^{-5}} = 5.6 \times 10^{-10}$$

This manipulation gives the hydrolysis constant

We use K_a to represent the hydrolysis constant for the cation of a weak base.

$K_a = K_{hb}$

$$K_a = \frac{[NH_3][H_3O^+]}{[NH_4^+]} = 5.6 \times 10^{-10}$$

which enables us to calculate pH values for solutions of salts of aqueous ammonia and strong acids.

In calculating the equilibrium constant for NH_4^+, we found that $K_a = K_w/K_b$. The process by which we obtained this hydrolysis constant can be used to obtain equilibrium constants for cations of other weak bases. Alternatively, we can use the relationship

$$K_a \times K_b = K_w \quad \text{(any conjugate acid-base pair)}$$

and find

$$K_a = \frac{K_w}{K_b}$$

the same expression obtained above.

The fact that K_a for the ammonium ion, NH_4^+, is the same as K_b for the acetate ion, CH_3COO^-, should not be surprising. Recall that the ionization constants for CH_3COOH and aqueous NH_3 are equal. Thus, we might expect the anion of CH_3COOH to hydrolyze to exactly the same extent as the cation of aqueous NH_3, and it does.

Example 17–2 illustrates how the appropriate equilibrium constant can be used to calculate concentrations of the various species in solutions containing soluble salts of weak bases and strong acids.

EXAMPLE 17–2

Calculate the pH of a 0.20 *M* solution of ammonium nitrate, NH_4NO_3.

Solution

We recognize that NH_4NO_3 is the salt of aqueous ammonia and nitric acid, a strong acid. Therefore, the cation of the weak base reacts with water

$$\underset{(0.20 - x)\,M}{NH_4^+} + H_2O \rightleftharpoons \underset{x\,M}{NH_3} + \underset{x\,M}{H_3O^+}$$

The equation tells us that $[NH_3] = [H_3O^+]$ and since we don't know either, we represent both as x mol/L. The equilibrium concentration of NH_4^+ is represented as $(0.20 - x)$ mol/L. In our earlier discussion we found that K_a for NH_4^+ is

$$K_a = \frac{[NH_3][H_3O^+]}{[NH_4^+]} = 5.6 \times 10^{-10}$$

Substituting into this expression gives

$$\frac{(x)(x)}{(0.20 - x)} = 5.6 \times 10^{-10}$$

Making the usual simplifying assumption gives $x = 1.1 \times 10^{-5}\,M = [H_3O^+]$, and pH = 4.96. Note that 0.20 *M* NH_4NO_3 solution is distinctly acidic.

Ammonium nitrate is widely used as a fertilizer. As the discussion indicates, it contributes significantly to soil acidity.

17–4 Salts of Weak Bases and Weak Acids

Soluble salts of weak bases and weak acids make up the fourth class of salts we wish to examine. In our earlier discussion we pointed out the fact that anions of weak acids react with water to give basic solutions, and cations of weak bases give acidic solutions. Since salts of weak bases and weak acids contain cations that give acidic solutions and anions that give basic solutions, will solutions of salts of weak bases and weak acids be neutral, basic, or acidic? The answer is that they may be any one of the three, depending on the relative strengths of the weak molecular acid and weak molecular base from which the salt is derived. Thus, salts of weak bases and weak acids may be divided into three types that depend on the values of the ionization constants of their parent weak bases and weak acids.

1 Salts of Weak Bases and Weak Acids for which $K_{base} = K_{acid}$

These K's refer to the parent acid and base from which the salt is derived.

The most common example of a salt of this type is ammonium acetate, NH_4CH_3COO, the salt of aqueous ammonia and acetic acid. Recall that the ionization constants for both aqueous ammonia and CH_3COOH are 1.8×10^{-5}. From our earlier discussion we know that ammonium ions react with water to produce H_3O^+ by the reaction

$$NH_4^+ + H_2O \rightleftharpoons NH_3 + H_3O^+$$

We have shown that K_a for this reaction is

$$K_a = \frac{[NH_3][H_3O^+]}{[NH_4^+]} = 5.6 \times 10^{-10}$$

We also recall that CH_3COO^- reacts with water to produce OH^- by the reaction

$$CH_3COO^- + H_2O \rightleftharpoons CH_3COOH + OH^-$$

K_b for this reaction is

$$K_b = \frac{[CH_3COOH][OH^-]}{[CH_3COO^-]} = 5.6 \times 10^{-10}$$

From the fact that the K_a for NH_4^+ and K_b for CH_3COO^- are equal, we can predict that aqueous solutions of NH_4CH_3COO are neutral. They are neutral because the NH_4^+ produces *exactly* as many H_3O^+ ions as the CH_3COO^- produces OH^- ions in ammonium acetate solutions.

Weak bases and weak acids that have exactly the same ionization constants are very few indeed. Ammonium acetate is the only common example of a salt derived from such a pair. Examples of the next two types of salts are much more common.

2 Salts of Weak Bases and Weak Acids for which $K_{base} > K_{acid}$

Salts of weak bases and weak acids for which K_{base} is greater than K_{acid} are always basic because the anion of the weaker acid hydrolyzes to a greater extent than the cation of the stronger base, as the following example illustrates.

Consider NH_4CN, ammonium cyanide, as a typical salt of a weak base and a weak acid for which the ionization constant of the parent weak base is greater than the ionization constant of the parent weak acid. Since K_a for HCN (4.0×10^{-10}) is much smaller than K_b for NH_3 (1.8×10^{-5}), then K_b for CN^- (2.5×10^{-5}) is much larger than K_a for NH_4^+ (5.6×10^{-10}). This tells us that the CN^- ion hydrolyzes to a much greater extent than NH_4^+ ion, and so solutions of NH_4CN are distinctly basic. Stated differently, CN^- is much stronger as a base than NH_4^+ is as an acid

$$NH_4^+ + H_2O \rightleftharpoons NH_3 + H_3O^+$$
$$CN^- + H_2O \rightleftharpoons HCN + OH^-$$
$$H_3O^+ + OH^- \rightarrow 2H_2O$$

This reaction occurs to greater extent, ∴ solution is basic

3 Salts of Weak Bases and Weak Acids for which $K_{base} < K_{acid}$

Salts of weak bases and weak acids for which $K_{base} < K_{acid}$ are acidic because the cation of the weaker base hydrolyzes to a greater extent than the anion of the stronger acid. Consider ammonium fluoride, NH_4F, as a typical example. It is the salt of aqueous ammonia and hydrofluoric acid.

Recall that K_b for aqueous NH_3 is 1.8×10^{-5} and K_a for HF is 7.2×10^{-4}. The K_a value for NH_4^+ (5.6×10^{-10}) is slightly larger than the K_b value for F^- (1.4×10^{-11}), which tells us that the NH_4^+ ion hydrolyzes to a slightly greater extent than the F^- ion. Or, NH_4^+ is slightly stronger as an acid than F^- is as a base, and so solutions of ammonium fluoride are slightly acidic.

$$NH_4^+ + H_2O \rightleftharpoons NH_3 + H_3O^+$$
$$F^- + H_2O \rightleftharpoons HF + OH^-$$
$$H_3O^+ + OH^- \rightarrow 2H_2O$$

This reaction occurs to greater extent, ∴ solution is acidic

17–5 Titration Curves for Weak Acids and Strong Bases

In Section 16–6 we examined titration curves for strong acids and strong bases. In Figure 16–2 we see that the curve rises very slowly before the equivalence point because there is no buffering action, and then very rapidly near the equivalence point because there is no hydrolysis. The curve becomes almost flat beyond the equivalence point.

When a weak acid is titrated with a strong base, the curve is dramatically different before and at the equivalence point. The solution is buffered before the equivalence point, and it is basic at the equivalence point because salts of weak acids and strong bases hydrolyze to give basic solutions.

Consider the titration of 100.0 mL of 0.100 *M* CH_3COOH with a 0.100 *M* solution of NaOH. (The strong electrolyte is added to the weak electrolyte.) Before any base is added the pH is 2.89 (see Example 16–13).

Titrations are usually done with 50-mL burets. We have chosen to use 100 mL in this example to simplify the arithmetic.

As soon as some NaOH is added, and until the equivalence point is reached, the solution is buffered because it contains both $NaCH_3COO$ and CH_3COOH.

$$\underset{\text{lim amt}}{NaOH} + \underset{\text{excess}}{CH_3COOH} \longrightarrow NaCH_3COO + H_2O$$

After 20.0 mL of NaOH solution has been added, we have

	$NaOH$	$+ \ CH_3COOH$	$\longrightarrow NaCH_3COO + H_2O$
Start	2.00 mmol	10.00 mmol	0
Change	−2.00	−2.00	+2.00
After reaction	0	8.00 mmol	2.00 mmol

These amounts are present in 120 mL of solution, so the concentrations are

$$M_{CH_3COOH} = \frac{8.00 \text{ mmol}}{120 \text{ mL}} = 0.0667\ M\ CH_3COOH$$

$$M_{NaCH_3COO} = \frac{2.00 \text{ mmol}}{120 \text{ mL}} = 0.0167\ M\ NaCH_3COO$$

The pH of this solution is calculated as we did in Example 16–16.

$$\frac{[H^+][CH_3COO^-]}{[CH_3COOH]} = 1.8 \times 10^{-5}$$

$$[H^+] = 1.8 \times 10^{-5} \times \frac{[CH_3COOH]}{[CH_3COO^-]} = 1.8 \times 10^{-5} \times \frac{0.0667}{0.0167}$$

$$[H^+] = \underline{7.2 \times 10^{-5}\ M} \quad \text{and} \quad \text{pH} = \underline{4.14}$$

Just before the equivalence point, the solution contains relatively high concentrations of $NaCH_3COO$ and relatively low concentrations of CH_3COOH. Just after the equivalence point, the solution contains relatively high concentrations of $NaCH_3COO$ and relatively low concentrations of NaOH, both basic components. In both regions our calculations are only approximations. Exact calculations of pH in these regions are beyond the scope of this text.

All points before the equivalence point are calculated in the same way. After some NaOH has been added, the solution contains both $NaCH_3COO$ and CH_3COOH, and so it is buffered until the equivalence point is reached. At the equivalence point, the solution is 0.0500 *M* in $NaCH_3COO$.

$$\begin{array}{lllll} NaOH & + CH_3COOH & \longrightarrow & NaCH_3COO + H_2O \\ 10.0\text{ mmol} & 10.0\text{ mmol} & \longrightarrow & 10.0\text{ mmol} \end{array}$$

$$M_{NaCH_3COO} = \frac{10.0\text{ mmol}}{200\text{ mL}} = 0.0500\ M\ NaCH_3COO$$

The pH of a 0.0500 *M* solution of $NaCH_3COO$ is 8.72. (See Example 17–1 for a similar calculation.) The solution is distinctly basic at the equivalence point, because of the hydrolysis of the acetate ion.

Beyond the equivalence point, the excess NaOH determines the pH of the solution just as it did in the titration of a strong acid (Section 16–6).

Table 17–2 lists several points on the titration curve and Figure 17–1 shows the titration curve for 100.0 mL of 0.100 *M* CH_3COOH titrated with a 0.100 *M* solution of NaOH. This titration curve has a short vertical section (pH = 8–10) and the indicator range is limited. Phenolphthalein is the indicator commonly used when weak acids are titrated with strong bases (see Table 16–5).

The titration curves for weak bases and strong acids are similar to those for weak acids and strong bases except that they are inverted (recall that strong is added to weak). Figure 17–2 displays the titration curve for 100.0 mL of 0.100 *M* aqueous ammonia titrated with 0.100 *M* HCl solution.

When solutions of weak acids are titrated with weak bases, the vertical portion of the curve is much too short for color indicators to be used. The

TABLE 17–2 Titration Data for CH_3COOH vs. NaOH

mL 0.100 *M* NaOH Added	mmol Base Added	mmol excess Acid or Base	pH
0.0 mL	0	10.0 CH_3COOH	2.89
20.0 mL	2.00	8.00	4.14
50.0 mL	5.00	5.00	4.74
75.0 mL	7.50	2.50	5.22
90.0 mL	9.00	1.00	5.70
95.0 mL	9.50	0.50	6.02
99.0 mL	9.90	0.10	6.74
100.0 mL	10.0	0	8.72
101.0 mL	10.1	0.10 OH^-	10.70
110.0 mL	11.0	1.0	11.68
120.0 mL	12.0	2.0	11.96

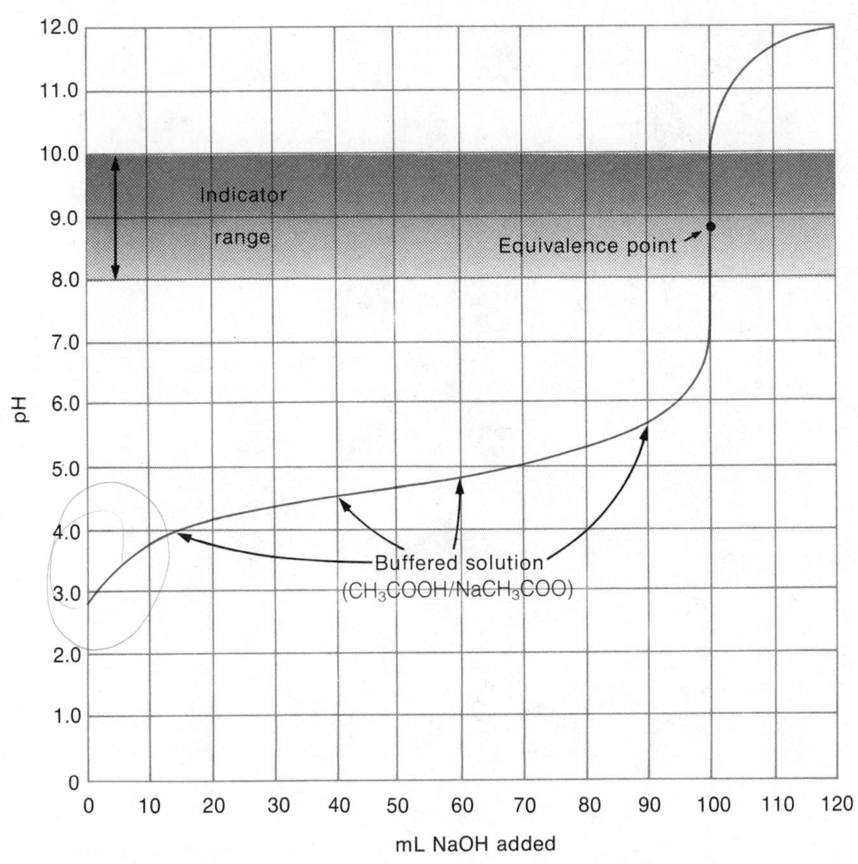

FIGURE 17–1 Titration curve for 100 mL of 0.100 *M* CH_3COOH versus 0.100 *M* NaOH. Notice that the vertical section of this curve is much shorter than the vertical sections in Figures 16–2 and 16–3 because the solution is buffered before the equivalence point.

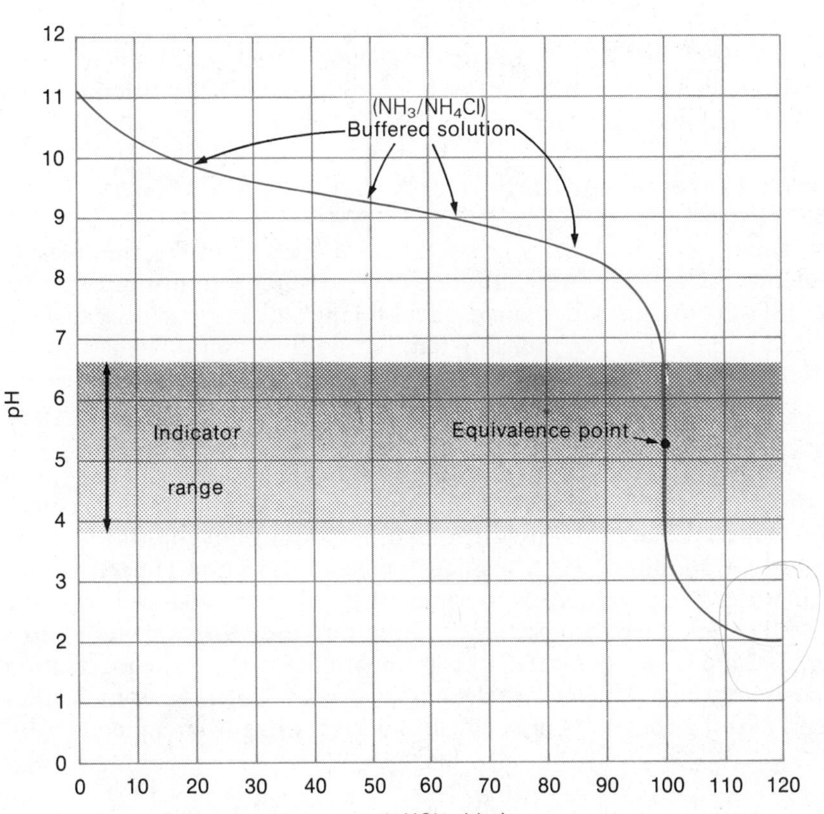

FIGURE 17–2 Titration curve for 100 mL of 0.100 *M* aqueous ammonia versus 0.100 *M* HCl. Notice that the vertical section of the curve is relatively short because the solution is buffered before the equivalence point. Notice also that the curve is very similar to the curve in Figure 17–1, but inverted.

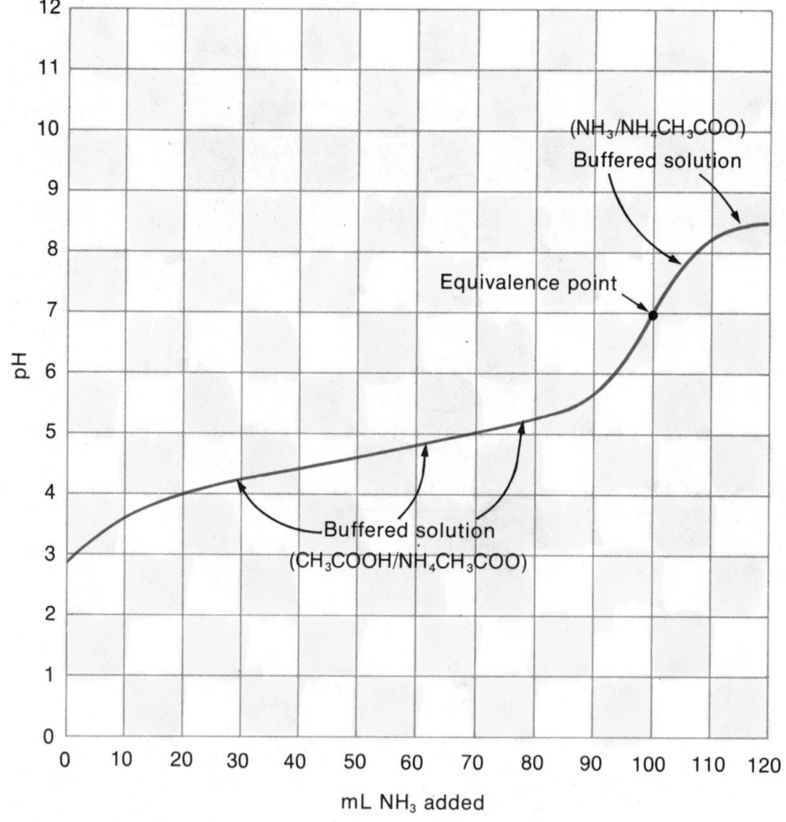

FIGURE 17–3 Titration curve for 100 mL of 0.100 *M* CH_3COOH versus 0.100 *M* aqueous NH_3. Because the solution is buffered before and after the equivalence point, the vertical section of the curve is too short to be noticed. Therefore, color indicators cannot be used in such titrations.

solution is buffered both before and after the equivalence point. Figure 17–3 shows the titration curve for 100.0 mL of 0.100 *M* CH_3COOH solution titrated with 0.100 *M* aqueous ammonia.

17–6 Hydrolysis of Small, Highly Charged Cations

Solutions of many common salts of strong acids are acidic. For this reason, many homeowners apply iron(II) sulfate, $FeSO_4 \cdot H_2O$, or aluminum sulfate, $Al_2(SO_4)_3 \cdot 18H_2O$, to the soil around "acid-loving" plants such as azaleas, camelias, and hollies. You are probably familiar with the sour "acid" taste of alum, $KAl(SO_4)_2 \cdot 12H_2O$, a substance that is frequently added to pickles.

All the common metal ions except those derived from strong soluble bases fall into this category.

Solutions of salts that contain small, highly charged cations together with anions of strong acids are acidic, because such cations hydrolyze to produce excess hydronium ions. Consider aluminum chloride, $AlCl_3$, as a typical example. When solid anhydrous $AlCl_3$ is added to water, the water becomes very warm. In fact, if a large quantity of anhydrous aluminum chloride is added rapidly to a small amount of water, a small "explosion" occurs! The reaction of $AlCl_3$ with water is so highly exothermic that steam is produced, and the escaping steam causes severe spattering of the mixture. As we have pointed out earlier, ions are always hydrated in solution. In many cases the interaction between positively charged ions and the negative end of polar water molecules is sufficiently strong that salts of such cations crystallize from aqueous solution in combination with definite numbers of water molecules. Salts containing Al^{3+}, Fe^{2+}, Fe^{3+}, and Cr^{3+} ions combined with noncoordinating anions usually crystallize from aqueous solutions with six water molecules associated with each metal ion. Many common salts of these cations contain hydrated cations, such as $[Al(OH_2)_6]^{3+}$, $[Fe(OH_2)_6]^{2+}$, $[Fe(OH_2)_6]^{3+}$ and $[Cr(OH_2)_6]^{3+}$ in the solid state. All of these species are octahedral; i.e., the metal ion (M^{n+}) is located at the center of a regular octahedron, and the six water molecules are located at the corners, as illustrated in Figure 17–4. In the metal-oxygen bonds of these hydrated cations, electron density is decreased around the oxygen end of the water molecule by the positively charged metal ions. This decrease in electron density weakens the hydrogen-oxygen bonds in coordinated water molecules relative to the hydrogen-oxygen bonds in noncoordinated water molecules. Consequently, noncoordinated water molecules can "abstract hydrogen ions" from the coordinated water molecules to form hydronium ions and produce acidic solutions as shown in Figure 17–5.

Noncoordinating anions do not form coordinate covalent bonds with metal ions in aqueous solution.

Abstract means "to pull off."

The equation for the reaction is written as

$$[Al(OH_2)_6]^{3+} + H_2O \rightleftharpoons [Al(OH)(OH_2)_5]^{2+} + H_3O^+$$

or even more simply as

$$Al^{3+} + 2H_2O \rightleftharpoons Al(OH)^{2+} + H_3O^+$$

Note that abstraction of a hydrogen ion converts a coordinated water molecule to a coordinated hydroxide ion, and decreases the positive charge on the hydrated species. We write equilibrium constant expressions for the hydrolysis of small, highly charged cations just as we have for other reactions. Thus, the equilibrium constant expression for the hydrolysis of the hydrated aluminum ion is

$$K_a = \frac{[[Al(OH)(OH_2)_5]^{2+}][H_3O^+]}{[[Al(OH_2)_6]^{3+}]} = 1.2 \times 10^{-5}$$

or more simply

$$K_a = \frac{[Al(OH)^{2+}][H_3O^+]}{[Al^{3+}]} = 1.2 \times 10^{-5}$$

Hydrolysis of small, highly charged cations may occur beyond the first step, and in many cases the reactions are quite complex. They may involve two or more cations reacting with each other to form large polymeric species. However, for most common cations, consideration of the first hydrolysis constant is adequate for the calculations we will perform.

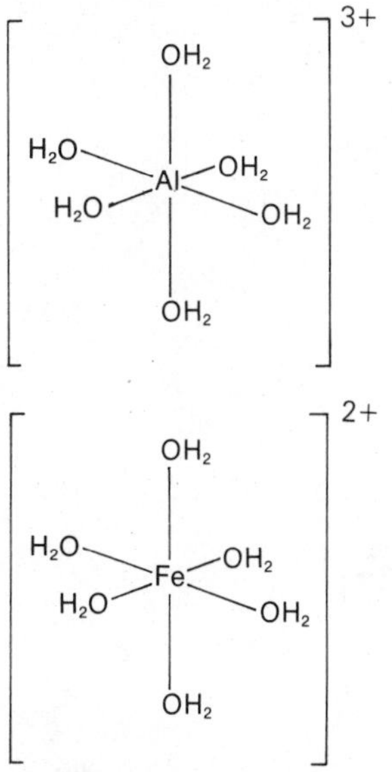

$$\left[\begin{array}{c} Al(OH_2)_6 \end{array}\right]^{3+} \qquad \left[\begin{array}{c} Fe(OH_2)_6 \end{array}\right]^{2+}$$

FIGURE 17–4 Structures of hydrated aluminum and iron(II) ions $[Al(OH_2)_6]^{3+}$ and $[Fe(OH_2)_6]^{2+}$.

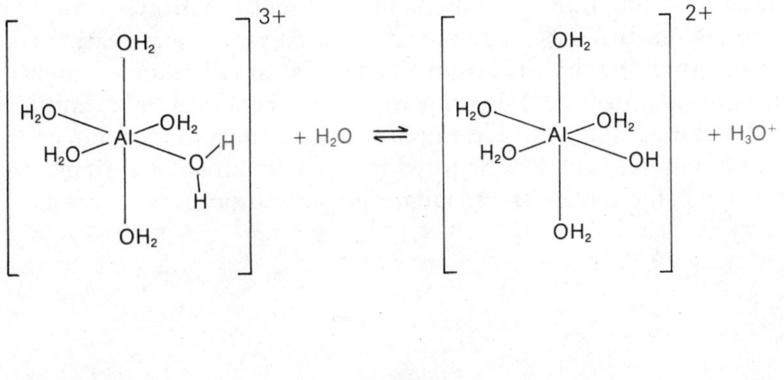

$$[Al(OH_2)_6]^{3+} + H_2O \rightleftharpoons [Al(OH)(OH_2)_5]^{2+} + H_3O^+$$

FIGURE 17–5 Hydrolysis of hydrated aluminum ion to produce H_3O^+, that is, the abstraction of a proton from a coordinated H_2O molecule by a non-coordinated H_2O molecule.

EXAMPLE 17–3

Calculate the pH and percentage hydrolysis in a 0.10 *M* solution of $AlCl_3$.

Solution

As we have just observed, the equation for the reaction can be written as

$$Al^{3+} + 2H_2O \rightleftharpoons Al(OH)^{2+} + H_3O^+$$

and the hydrolysis constant is

$$K_a = \frac{[Al(OH)^{2+}][H_3O^+]}{[Al^{3+}]} = 1.2 \times 10^{-5}$$

The usual algebraic representation of equilibrium concentrations gives

$$\begin{array}{llll} Al^{3+} & + 2H_2O \rightleftharpoons & Al(OH)^{2+} & + H_3O^+ \\ (0.10 - x)\,M & & x\,M & x\,M \end{array}$$

and substitution into the K_a expression gives

$$\frac{(x)(x)}{(0.10 - x)} = 1.2 \times 10^{-5} \quad \text{and} \quad x = 1.1 \times 10^{-3}\,M = [H_3O^+]$$

Since $[H_3O^+] = 1.1 \times 10^{-3}\,M$, pH = 2.96, and we see that the solution is quite acidic. It may be helpful to recall that in 0.10 *M* CH_3COOH, $[H_3O]^+ = 1.3 \times 10^{-3}\,M$, and pH = 2.89. Thus, we see that a 0.10 *M* CH_3COOH solution is only slightly more acidic than a 0.10 *M* $AlCl_3$ solution. To calculate the % hydrolysis, we observe that each Al^{3+} which hydrolyzes produces one H_3O^+. Thus, the equilibrium concentration of H_3O^+ will be equal to the concentration of Al^{3+} which hydrolyzes.

$$\% \text{ hydrolysis} = \frac{[Al^{3+}]_{\text{hydrolyzed}}}{[Al^{3+}]_{\text{total}}} \times 100\% = \frac{[H_3O^+]}{[Al^{3+}]_{\text{total}}} \times 100\%$$

$$= \frac{1.1 \times 10^{-3}}{0.10} \times 100\% = 1.1\% \text{ hydrolyzed}$$

Again, recall that in 0.10 *M* solution, CH_3COOH is 1.3% ionized while in 0.10 *M* solution $AlCl_3$ is 1.1% hydrolyzed, and so the acidities of the two solutions are almost equal.

We might expect that smaller, more highly charged cations are stronger acids than larger, less highly charged cations because the smaller, more highly charged cations interact with coordinated water molecules more strongly. Indeed, this is usually the case, as Table 17–3 shows.

If we compare isoelectronic cations in the same period in the periodic table, the smaller, more highly charged cation is the stronger acid. (Compare K_a's for Li^+ and Be^{2+} and for Na^+, Mg^{2+}, Al^{3+}.) For cations with the same charge from the same group in the periodic table, the smaller cation hydrolyzes to a greater extent. (Compare K_a's for Be^{2+} and Mg^{2+}.) If we compare cations of the same element in different oxidation states, the smaller, more highly charged cation is the stronger acid. (Compare K_a's for Fe^{2+} and Fe^{3+}, and for Co^{2+} and Co^{3+}.) Although there are additional considerations that enable us to predict relatively accurately the extent to which various kinds of cations hydrolyze, the previous statements are adequate for our purposes.

TABLE 17–3 Ionic Radii and Hydrolysis Constants for Some Common Cations

Cation	Ionic Radius (Å)	Hydrated Cation	K_a
Li^+	0.60	$[Li(OH_2)_4]^+$	1×10^{-14}
Be^{2+}	0.31	$[Be(OH_2)_4]^{2+}$	1.0×10^{-5}
Na^+	0.95	$[Na(OH_2)_6]^+$ (?)	10^{-14}
Mg^{2+}	0.65	$[Mg(OH_2)_6]^{2+}$	3.0×10^{-12}
Al^{3+}	0.50	$[Al(OH_2)_6]^{3+}$	1.2×10^{-5}
Fe^{2+}	0.76	$[Fe(OH_2)_6]^{2+}$	3.0×10^{-10}
Fe^{3+}	0.64	$[Fe(OH_2)_6]^{3+}$	4.0×10^{-3}
Co^{2+}	0.74	$[Co(OH_2)_6]^{2+}$	5.0×10^{-10}
Co^{3+}	0.63	$[Co(OH_2)_6]^{3+}$	1.7×10^{-2}
Cu^{2+}	0.96	$[Cu(OH_2)_4]^{2+}$	1.0×10^{-8}
Zn^{2+}	0.74	$[Zn(OH_2)_6]^{2+}$	2.5×10^{-10}
Hg^{2+}	1.10	$[Hg(OH_2)_6]^{2+}$	8.3×10^{-7}
Bi^{3+}	0.74	$[Bi(OH_2)_6]^{3+}$	1.0×10^{-2}

Key Terms

Hydrolysis the reaction of a substance with water or its ions.

Hydrolysis constant an equilibrium constant for a hydrolysis reaction.

Solvolysis the reaction of a substance with the solvent in which it is dissolved.

Exercises

1. Why can we say that 0.10 *M* solutions of $HClO_4$ and HCl contain essentially no molecules of nonionized acid?
2. Why can we say that 0.10 *M* solutions of HF and HNO_2 contain relatively few ions?
3. How may salts be classified conveniently into four classes? For each class, write the name and formula of a salt that fits into that category. Use examples other than those used in illustrations in this chapter.
4. Define and illustrate the following terms clearly and concisely: (a) solvolysis, (b) hydrolysis.
5. Why do salts of strong soluble bases and strong acids give neutral aqueous solutions? Use KNO_3 to illustrate. Write names and formulas for three other salts of strong soluble bases and strong acids.
6. Why do salts of strong soluble bases and weak acids give basic aqueous solutions. Use sodium hypochlorite, NaClO, to illustrate. (Recall that Chlorox®, Purex®, and the other "chlorine bleaches" are 5% NaClO.)
7. Write names and formulas for three salts of strong soluble bases and weak acids other than NaClO.
8. Calculate hydrolysis constants for the following anions of weak acids: (a) F^-, (b) ClO^-, (c) CN^-. Is there a relationship between K_a, the ionization constant for a weak acid, K_b, and the hydrolysis constant for the anion of the weak acid? What is it?
9. Calculate the pH of 0.15 *M* solutions of the following salts: (a) NaF, (b) NaClO, (c) NaCN. (Refer to Exercise 8 for *K*'s.)

10. What is the percent hydrolysis in each of the solutions in Exercise 9?
11. Why do salts of weak bases and strong acids give acidic aqueous solutions? Illustrate with NH_4NO_3, a common fertilizer.
12. Write names and formulas for four salts of weak bases and strong acids.
13. Calculate hydrolysis constants for the following cations of weak bases:
(a) NH_4^+,
(b) $(CH_3)NH_3^+$, methylammonium ion,
(c) $C_5H_5NH^+$, pyridinium ion.
14. Calculate the pH of 0.15 *M* solutions of:
(a) NH_4NO_3,
(b) $(CH_3)NH_3NO_3$,
(c) $C_5H_5NHNO_3$.
15. Can you make a general statement relating parent base strength and extent of hydrolysis of the cations of Exercise 13, by using hydrolysis constants calculated in that exercise?
16. How do pH values for the following pairs of solutions compare? Why?
(a) 0.10 *M* NH_4Cl, ammonium chloride, and 0.10 *M* ammonium nitrate, NH_4NO_3
(b) 0.010 *M* ammonium perchlorate, NH_4ClO_4, and 0.0050 *M* ammonium sulfate, $(NH_4)_2SO_4$
17. Why are some aqueous solutions of weak acids and weak bases neutral, while others are acidic, and still others are basic?
18. Write the name and formula for a salt of a weak acid and a weak base that gives (a) neutral, (b) acidic, and (c) basic aqueous solutions.

Hydrolysis of Small, Highly Charged Cations

19. Why do salts that contain cations related to insoluble bases (metal hydroxides) and anions related to strong acids give acidic aqueous solutions? Use $Cr(NO_3)_3$ to illustrate.
20. Calculate pH and percent hydrolysis for the following:
(a) 0.050 *M* $Al(NO_3)_3$, aluminum nitrate
(b) 0.20 *M* $Co(ClO_4)_2$, cobalt(II) perchlorate
(c) 0.15 *M* $MgCl_2$, magnesium chloride
21. Given pH values for solutions of the following concentrations, calculate hydrolysis constants for the cations:
(a) 0.00050 *M* $CeCl_3$, cerium(III) chloride, pH = 5.99
(b) 0.10 *M* $Cu(NO_3)_2$, copper(II) nitrate, pH = 4.50
(c) 0.10 *M* $Sc(ClO_4)_3$, scandium perchlorate, pH = 3.44
22. (a) Arrange the following Brønsted-Lowry bases in order of increasing base strength: aqueous NH_3, $C_6H_5NH_2$, CN^- and OCN^-.
(b) Suppose you had separate 0.010 *M* solutions of each of the following solutes: NH_3, $C_6H_5NH_2$, NaCN, and NaOCN. Which solution would be most basic and which one would be least basic?

Titration Curves

For Exercises 23–26, calculate $[H_3O^+]$, $[OH^-]$, pH, and pOH at the indicated points. In each case assume that pure acid (or base) is added to exactly one liter of a 0.0100 molar solution of the indicated base (or acid). This simplifies the arithmetic because we may assume that the volume of each solution is constant throughout the titration. Plot each titration curve with pH on the vertical axis and moles of base (or acid) added on the horizontal axis.

23. *Titration Curves—I: HCl* vs. *NaOH*
Solid NaOH is added to one liter of 0.0100 *M* HCl solution. Consult Table 16–5 and list the indicators that could be used satisfactorily in this titration.

Total Moles NaOH Added	$[H^+]$	$[OH^-]$	pH	pOH
none				
0.00100				
0.00300				
0.00500 (50%)				
0.00700				
0.00900 (90%)				
0.00950				
0.0100 (100%)				
0.0105				
0.0120				
0.0150 (50% excess NaOH)				

24. *Titration Curves—II: CH_3COOH* vs. *NaOH*
Solid NaOH is added to one liter of 0.0100 *M* CH_3COOH solution. Consult Table 16–5 and list the indicators that could be used satisfactorily in this titration.

Total Moles NaOH Added	$[H^+]$	$[OH^-]$	pH	pOH
none				
0.00200				
0.00400				
0.00500 (50%)				
0.00700				
0.00900 (90%)				
0.00950				
0.0100 (100%)				
0.0105				
0.0120				
0.0150 (50% excess NaOH)				

25. *Titration Curves—III: HCl* vs. *aqueous NH_3*
Gaseous HCl is added to one liter of 0.0100 *M* aqueous ammonia solution. Consult Table 16–5 and list the indicators that could be used satisfactorily in this titration.

Total Moles HCl Added	$[H^+]$	$[OH^-]$	pH	pOH
none				
0.00100				
0.00300				
0.00500 (50%)				
0.00700				
0.00900 (90%)				
0.00950				
0.0100 (100%)				
0.0105				
0.0120				
0.0150 (50% excess HCl)				

26. *Titration Curves—IV: CH_3COOH vs. NH_3*
Gaseous NH_3 is added to one liter of 0.0100 *M* CH_3COOH solution.

Total Moles NH_3 Added	$[H^+]$	$[OH^-]$	pH	pOH
none				
0.00100				
0.00400				
0.00500 (50%)				
0.00900 (90%)				
0.00950				
0.0100 (100%)				
0.0105				
0.0130				

(a) What is the major difference between the titration curve for the reaction of CH_3COOH and NH_3 and the other curves which you have plotted?

(b) Consult Table 16–5. Can you suggest a satisfactory indicator for this titration? Why?

18 Equilibria in Aqueous Solutions—III

The Solubility Product Principle

In our discussion of ionic equilibrium thus far we have considered only compounds that are readily soluble in water. Many compounds are only slightly soluble in water, and such substances are usually referred to as "insoluble compounds." Nearly all compounds dissolve in water to some extent, however, and we shall now consider those that are only very slightly soluble. As a rough rule-of-thumb, compounds that dissolve in water to the extent of at least 0.020 mole/liter are classified as soluble. Refer to the solubility rules (Section 7–1, part 5) as necessary.

Barium sulfate is the "insoluble" substance taken orally before stomach x-rays are made, because the heavy barium atoms absorb x-rays well. Even though barium ions are quite toxic, barium sulfate can still be taken orally without danger. The compound is so insoluble that most of it passes through the digestive system unchanged, and only an insignificant amount of barium ions dissolves.

18–1 Solubility Product Constants

Let's consider what happens when powdered barium sulfate, $BaSO_4$, is placed in water and stirred vigorously. Careful experiments show that a very small amount of barium sulfate dissolves in water to give a saturated solution, and the $BaSO_4$ that dissolves is completely dissociated into its constituent ions

$$BaSO_4\ (s) \rightleftharpoons Ba^{2+}\ (aq) + SO_4^{2-}\ (aq)$$

The existence of a substance in the solid state is indicated a number of ways. For example, $BaSO_4$ (s), $\underline{BaSO_4}$, and $BaSO_4 \downarrow$ are three different ways of representing solid $BaSO_4$. The last two are frequently used to indicate the formation of solids (precipitates) in reactions in aqueous solution. As a matter of consistency, we shall use the notation in the preceding equation for formulas of solid substances in equilibrium with dissolved species in aqueous solution. The equilibrium expression for the dissolution of $BaSO_4$ in water may be written

$$K_c = \frac{[Ba^{2+}][SO_4{}^{2-}]}{[BaSO_4\ (s)]}$$

Note that this expression contains the concentration of solid $BaSO_4$, $[BaSO_4\ (s)]$. In an equilibrium between a solid substance and its ions in aqueous solution, the "concentration of the solid" has no physical significance, as the following discussion shows. Suppose that one gram of solid $BaSO_4$ is added to 100 mL of water and stirred until the solution is *saturated*. Most of the $BaSO_4$ does not dissolve. In fact, only 0.00025 gram of $BaSO_4$ dissolves in 100 mL of water at 25°C. Even if 10 grams of $BaSO_4$ are added to 100 mL of water, still only 0.00025 gram dissolves at 25°C. No matter how much $BaSO_4$ we add to the 100 mL of water, the same small amount dissolves. The expression "concentration of solid" has no meaning *in equilibria involving solids and their saturated solutions,* since the amount of solid remaining does not affect the concentrations of the ions in solution. In many equilibria involving the dissolution of slightly soluble compounds in water, it is convenient to determine experimentally the concentrations of *dissolved ions* in saturated solutions. Therefore, for such equilibria, we define a new kind of equilibrium constant called a **solubility product constant, K_{sp}**

Strictly speaking, the activity of pure crystalline $BaSO_4$ is the same as its activity in the standard state, that is, $a_{BaSO_4}/a^{\circ}_{BaSO_4} = 1$. Therefore, the denominator of this and other solubility product constants is unity. One may also argue that the concentration of a solid is fixed by its density (Section 15–9).

$$BaSO_4\ (s) \rightleftharpoons Ba^{2+}\ (aq) + SO_4{}^{2-}\ (aq)$$

$$K_{sp} = [Ba^{2+}][SO_4{}^{2-}]$$

The solubility product constant for $BaSO_4$ is the product of the concentrations of the constituent ions.

In general terms, *the* **solubility product expression** *for a compound is the product of the concentrations of the constituent ions, each raised to the power that corresponds to the number of ions in one formula unit of the compound.* This statement is the **solubility product principle.** The following examples illustrate the point.

Consider the dissolution of the slightly soluble compound calcium fluoride, CaF_2, in water

$$CaF_2\ (s) \rightleftharpoons Ca^{2+}\ (aq) + 2F^-\ (aq)$$

Its solubility product expression is $K_{sp} = [Ca^{2+}][F^-]^2$. On the other hand, the dissolution of solid bismuth sulfide, Bi_2S_3, gives two bismuth ions and three sulfide ions per formula unit

$$Bi_2S_3\ (s) \rightleftharpoons 2Bi^{3+}\ (aq) + 3S^{2-}\ (aq)$$

so $K_{sp} = [Bi^{3+}]^2[S^{2-}]^3$. Generally, we may represent the dissolution of a slightly soluble compound as

The units of all solubility product constants are mol/L raised to the appropriate power. Units are commonly omitted from K_{sp}'s.

$$M_yX_z\ (s) \rightleftharpoons y\,M^{z+}\ (aq) + z\,X^{y-}\ (aq)$$

and its solubility product expression is

$$K_{sp} = [M^{z+}]^y[X^{y-}]^z$$

If there are more than two kinds of ions in the formula for a compound, the above generalization still applies. The dissolution of the slightly soluble compound magnesium ammonium phosphate, $MgNH_4PO_4$, in water is represented as

$$MgNH_4PO_4\ (s) \rightleftharpoons Mg^{2+}\ (aq) + NH_4^+\ (aq) + PO_4^{3-}\ (aq)$$

and its solubility product expression is

$$K_{sp} = [Mg^{2+}][NH_4^+][PO_4^{3-}]$$

A word about conventions. We frequently shorten the term "solubility product expression" to "solubility product." Thus, the solubility product for barium sulfate, $BaSO_4$, is written

$$K_{sp} = [Ba^{2+}][SO_4^{2-}] = 1.1 \times 10^{-10}$$

and the solubility product for calcium fluoride, CaF_2, is written

$$K_{sp} = [Ca^{2+}][F^-]^2 = 3.9 \times 10^{-11}$$

Unless otherwise indicated, solubility product constants and solubility data are given for 25°C.

↓ Appendix H

18–2 Experimental Determination of Solubility Product Constants

Solubility product constants can be determined in a number of ways. For example, careful measurements of conductivity show that one liter of a saturated solution of barium sulfate contains 0.0025 gram of dissolved $BaSO_4$. If the solubility of a compound is known, its solubility product can be calculated as Example 18–1 illustrates.

Although we frequently use statements like "the solution contains 0.0025 g of dissolved $BaSO_4$," what we mean is: 0.0025 g of solid $BaSO_4$ dissolves to give a saturated solution that contains equal numbers of Ba^{2+} and SO_4^{2-} ions.

EXAMPLE 18–1

One liter of saturated barium sulfate contains 0.0025 gram of dissolved $BaSO_4$. Calculate the solubility product constant for $BaSO_4$.

Solution

As with previous problems, we write the appropriate (1) chemical equation and (2) equilibrium constant expression. Recall that in a saturated solution, equilibrium exists between solid and dissolved solute.

$$BaSO_4\ (s) \rightleftharpoons Ba^{2+}\ (aq) + SO_4^{2-}\ (aq)$$

$$K_{sp} = [Ba^{2+}][SO_4^{2-}]$$

From the solubility of barium sulfate in water we can calculate its *molar solubility*, the number of moles that dissolve to give one liter of saturated solution.

$$\frac{\underline{?}\ \text{mol } BaSO_4}{L} = \frac{2.5 \times 10^{-3}\ \text{g } BaSO_4}{L} \times \frac{1\ \text{mol } BaSO_4}{233\ \text{g } BaSO_4} = 1.1 \times 10^{-5}\ \text{mol } BaSO_4/L$$

(molar solubility)

We are not suggesting that the concentration of solid $BaSO_4$ is 1.1×10^{-5} *M*, only that its molar solubility is 1.1×10^{-5} *M*.

Now that the molar solubility of $BaSO_4$ is known, reference to the dissolution equation shows each formula unit of $BaSO_4$ dissolves to produce one Ba^{2+} and one SO_4^{2-} ion

$$\begin{array}{lll} BaSO_4\,(s) & \rightleftharpoons Ba^{2+}\,(aq) & + \; SO_4^{2-}\,(aq) \\ 1.1 \times 10^{-5}\,M & 1.1 \times 10^{-5}\,M & 1.1 \times 10^{-5}\,M \end{array}$$

The molar solubility of $BaSO_4$ is 1.1×10^{-5} mol/L, and in a saturated solution $[Ba^{2+}] = [SO_4^{2-}] = 1.1 \times 10^{-5}\,M$. Substituting these values into the solubility product constant expression for $BaSO_4$ gives its solubility product constant

$$K_{sp} = [Ba^{2+}][SO_4^{2-}] = (1.1 \times 10^{-5})(1.1 \times 10^{-5}) = \underline{1.2 \times 10^{-10}}$$

The value calculated here is 1.2×10^{-10}, whereas the accepted value is 1.1×10^{-10}. Round-off error is responsible for the difference.

This expression is particularly useful because it is applicable *to all saturated solutions of* $BaSO_4$ at 25°C. The *origin* of the Ba^{2+} and SO_4^{2-} ions is of no importance. For example, suppose a solution of barium chloride (which contains Ba^{2+} and Cl^- ions) and a solution of sodium sulfate (which contains Na^+ and SO_4^{2-} ions) are mixed at 25°C. (Both of these compounds are very soluble.) As soon as the concentration of Ba^{2+} ions multiplied by the concentration of SO_4^{2-} ions exceeds 1.1×10^{-10}, solid $BaSO_4$ begins to precipitate from the solution. *Or,* if solid $BaSO_4$ is placed in water at 25°C, it dissolves until the concentration of Ba^{2+} multiplied by the concentration of SO_4^{2-} ions just equals 1.1×10^{-10}.

Example 18–2 is similar to Example 18–1, but there are some differences, due to the 2:1 ion stoichiometry of Ag_2CrO_4.

EXAMPLE 18–2

One liter of a saturated solution of silver chromate at 25°C contains 0.0435 g of Ag_2CrO_4. Calculate its solubility product constant.

Solution

The equation for the dissolution of silver chromate in water and its solubility product expression are

$$Ag_2CrO_4\,(s) \rightleftharpoons 2Ag^+\,(aq) + CrO_4^{2-}\,(aq)$$

$$K_{sp} = [Ag^+]^2[CrO_4^{2-}]$$

The molar solubility of silver chromate is calculated as in Example 18–1

$$\frac{?\ \text{mol } Ag_2CrO_4}{\text{L}} = \frac{0.0435\ \text{g } Ag_2CrO_4}{\text{L}} \times \frac{1\ \text{mol } Ag_2CrO_4}{332\ \text{g } Ag_2CrO_4} = 1.31 \times 10^{-4}\ \text{mol/L}$$

molar solubility

We are not suggesting that the concentration of solid Ag_2CrO_4 is $1.31 \times 10^{-4}\,M$, only that 1.31×10^{-4} mol of solid Ag_2CrO_4 dissolves to give a liter of saturated solution.

The equation for dissolution of Ag_2CrO_4 and its molar solubility give the concentrations of Ag^+ and CrO_4^{2-} in a saturated solution

$$\begin{array}{lll} Ag_2CrO_4\,(s) & \rightleftharpoons 2Ag^+\,(aq) & + \; CrO_4^{2-}\,(aq) \\ 1.31 \times 10^{-4}\,M & 2.62 \times 10^{-4}\,M & 1.31 \times 10^{-4}\,M \end{array}$$

Substitution into the solubility product expression for Ag_2CrO_4 gives its solubility product

$$K_{sp} = [Ag^+]^2[CrO_4^{2-}] = (2.62 \times 10^{-4})^2(1.31 \times 10^{-4})$$

$$K_{sp} = \underline{[Ag^+]^2[CrO_4^{2-}] = 9.0 \times 10^{-12}}$$

Solubility products are usually given to only two significant digits.

We have shown that the molar solubility of Ag_2CrO_4 is 1.31×10^{-4} mol/L and its solubility product is 9.0×10^{-12}.

TABLE 18–1 Comparison of Solubilities of $BaSO_4$ and Ag_2CrO_4

Compound	Molar Solubility	K_{sp}
$BaSO_4$	1.1×10^{-5} mol/L	$[Ba^{2+}][SO_4^{2-}] = 1.1 \times 10^{-10}$
Ag_2CrO_4	1.3×10^{-4} mol/L	$[Ag^+]^2[CrO_4^{2-}] = 9.0 \times 10^{-12}$

The values obtained in Examples 18–1 and 18–2 are compared in Table 18–1. These data show that the molar solubility of Ag_2CrO_4 is greater than that of $BaSO_4$. However, the solubility product for $BaSO_4$ is greater than the solubility product for Ag_2CrO_4 because the solubility product expression for Ag_2CrO_4 contains a *squared* term, $[Ag^+]^2$. Thus, we see that care must be exercised in comparing solubility products for different compounds.

Solubility products for some common compounds are tabulated in Appendix H. Refer to this appendix as needed.

18–3 Uses of Solubility Product Constants

If the solubility product for a compound is known, the solubility of the compound in water at 25°C can be calculated conveniently, as Example 18–3 illustrates.

EXAMPLE 18–3

Calculate molar solubilities, concentrations of the constituent ions, and solubilities in g/L for the following compounds: (a) silver chloride, AgCl, and (b) zinc hydroxide, $Zn(OH)_2$.

Solution

(a) The equation for the dissolution of silver chloride and its solubility product expression are

$$AgCl\ (s) \rightleftharpoons Ag^+\ (aq) + Cl^-\ (aq)$$

$$K_{sp} = [Ag^+][Cl^-] = 1.8 \times 10^{-10}$$

The value of K_{sp} is obtained from Appendix H.

Each formula unit of AgCl that dissolves produces one Ag^+ and one Cl^-, and we can let x = the molar solubility, which gives

$$\begin{array}{ccc} AgCl\ (s) & \rightleftharpoons Ag^+\ (aq) & + \ Cl^-\ (aq) \\ x \text{ mol/L} & x\,M & x\,M \end{array}$$

Substitution into the solubility product expression gives

$$[Ag^+][Cl^-] = (x)(x) = 1.8 \times 10^{-10}$$

$$x^2 = 1.8 \times 10^{-10}$$

$$x = 1.3 \times 10^{-5} \text{ mol AgCl/L} \leftarrow \text{molar solubility for AgCl}$$

One liter of saturated AgCl contains 1.3×10^{-5} mole of AgCl at 25°C. Now that we know the value of x, the molar solubility, we also know the concentrations of the constituent ions

$$x = [Ag^+] = [Cl^-] = 1.3 \times 10^{-5} \text{ mol/L}$$

To calculate the mass of AgCl in one liter of saturated solution, we convert 1.3×10^{-5} mol AgCl/L to g AgCl/L

$$\frac{?\text{ g AgCl}}{\text{L}} = \frac{1.3 \times 10^{-5} \text{ mol AgCl}}{\text{L}} \times \frac{143 \text{ g AgCl}}{1 \text{ mol AgCl}} = 1.9 \times 10^{-3} \text{ g AgCl/L}$$

A liter of saturated AgCl solution contains only 0.0019 g of AgCl, and we are justified in referring to AgCl as a slightly soluble compound.

(b) The dissolution of zinc hydroxide, $Zn(OH)_2$, in water and its solubility product expression are represented as

$$Zn(OH)_2\ (s) \rightleftharpoons Zn^{2+}\ (aq) + 2OH^-\ (aq)$$

$$K_{sp} = [Zn^{2+}][OH^-]^2 = 4.5 \times 10^{-17}$$

If, as before, we let x = molar solubility, then $[Zn^{2+}] = x$ and $[OH^-] = 2x$, and we have

$$\underset{x\ mol/L}{Zn(OH)_2\ (s)} \rightleftharpoons \underset{x\ M}{Zn^{2+}\ (aq)} + \underset{2x\ M}{2OH^-\ (aq)}$$

Substitution into the solubility product expression gives

$$[Zn^{2+}][OH^-]^2 = (x)(2x)^2 = 4.5 \times 10^{-17}$$

$$4x^3 = 4.5 \times 10^{-17}$$

$$x^3 = 11 \times 10^{-18}$$

$$x = \underline{2.2 \times 10^{-6}\ mol\ Zn(OH)_2/L} \leftarrow \text{molar solubility of } Zn(OH)_2$$

so

Note: The concentration of OH^- is twice the molar solubility of $Zn(OH)_2$ in pure water because each formula unit of $Zn(OH)_2$ contains two OH^-.

$$x = [Zn^{2+}] = 2.2 \times 10^{-6}\ M$$

$$2x = [OH^-] = 4.4 \times 10^{-6}\ M$$

We can now calculate the mass of $Zn(OH)_2$ in one liter of saturated solution,

$$\frac{?\ g\ Zn(OH)_2}{L} = \frac{2.2 \times 10^{-6}\ mol\ Zn(OH)_2}{L} \times \frac{99\ g\ Zn(OH)_2}{1\ mol\ Zn(OH)_2}$$

$$= \underline{2.2 \times 10^{-4}\ g\ Zn(OH)_2/L}$$

The common ion effect applies to solubility equilibria just as it does to ionic equilibria in general. The solubility of any compound is less in a solution containing an ion common to the compound than in pure water, as long as no other reaction is caused by the presence of the common ion (such as dissolution of an amphoteric metal hydroxide in the presence of a large excess of hydroxide ions). Example 18–4 illustrates the common ion effect.

EXAMPLE 18–4

The molar solubility of magnesium fluoride, MgF_2, in pure water at 25°C is 1.2×10^{-3} mol/L. Calculate the molar solubility of MgF_2 in 0.10 M sodium fluoride, NaF, solution at 25°C. $K_{sp} = 6.4 \times 10^{-9}$ for MgF_2.

Solution

NaF is a soluble, ionic salt, and therefore 0.10 M F^- is produced by

$$\underset{0.10\ M}{NaF\ (s)} \xrightarrow{H_2O} \underset{0.10\ M}{Na^+\ (aq)} + \underset{0.10\ M}{F^-\ (aq)}$$

For MgF_2, a slightly soluble salt,

$$MgF_2\ (s) \rightleftharpoons Mg^{2+}\ (aq) + 2F^-\ (aq)$$

molar solubility = x mol/L $\qquad x\ M \qquad 2x\ M$

$K_{sp} = [Mg^{2+}][F^-]^2 = 6.4 \times 10^{-9}$

$(x)(0.10 + 2x)^2 = 6.4 \times 10^{-9}$

We make the simplifying assumption that $2x \ll 0.10$, so

$(x)(0.10)^2 = 6.4 \times 10^{-9}$

$x = 6.4 \times 10^{-7}\ M$ = molar solubility of MgF_2 in 0.10 M NaF.

(Note that the assumption is valid.) The molar solubility of MgF_2 in 0.10 M NaF ($6.4 \times 10^{-7}\ M$) is much less than in pure water ($1.2 \times 10^{-3}\ M$).

Another application of the solubility product principle is the calculation of concentrations of ions that can exist in solution, and from these the determination of whether or not a precipitate will form in a given solution. This involves comparison of Q_{sp}, the reaction quotient (Section 15–4) applied to solubility equilibria, with K_{sp}.

If $Q_{sp} < K_{sp}$	forward process is favored no precipitation occurs, more solid will dissolve
$Q_{sp} = K_{sp}$	solution is *just* saturated neither forward nor reverse process is favored
$Q_{sp} > K_{sp}$	reverse process is favored, precipitation occurs

Example 18–5 illustrates this idea.

EXAMPLE 18–5

If 100 mL of 0.0010 M sodium sulfate, Na_2SO_4, and 100 mL of 0.010 M barium chloride, $BaCl_2$, solutions are mixed, will a precipitate form?

Solution

First, consider the kinds of compounds mixed and determine whether a reaction *can* occur. Both Na_2SO_4 and $BaCl_2$ are soluble ionic salts; therefore, at the moment of mixing the new solution contains a mixture of Na^+, SO_4^{2-}, Ba^{2+}, and Cl^- ions in the following concentrations (concentrations are halved by doubling the volume of solution):

When dilute aqueous solutions are mixed, their volumes may be added to give the volume of the new solution.

$$Na_2SO_4\ (s) \xrightarrow{H_2O} 2Na^+\ (aq) + SO_4^{2-}\ (aq)$$

0.00050 mol/L $\longrightarrow$ $2(0.00050\ M)$ = 0.0010 M $\qquad$ 0.00050 M

$$BaCl_2\ (s) \xrightarrow{H_2O} Ba^{2+}\ (aq) + 2Cl^-\ (aq)$$

0.0050 mol/L $\longrightarrow$ 0.0050 M $\qquad$ $2(0.0050\ M)$ = 0.010 M

The possibility of forming two new compounds, NaCl and $BaSO_4$, exists. Sodium chloride is a soluble ionic compound and therefore Na^+ and Cl^- do not react in dilute aqueous solutions. However, $BaSO_4$ is only very slightly soluble, and solid $BaSO_4$ will precipitate from the solution *if* the product of the concentrations of Ba^{2+} and SO_4^{2-} exceeds the solubility product for $BaSO_4$. Having determined that a precipitate may form, we perform calculations to determine whether solid $BaSO_4$ precipitates from this particular solution. The solubility product for $BaSO_4$ is 1.1×10^{-10} (Appendix H). If $[Ba^{2+}][SO_4^{2-}]$ exceeds 1.1×10^{-10}, then solid $BaSO_4$ will precipitate. Substituting $[Ba^{2+}] = 0.0050\ M$ and $[SO_4^{2-}] = 0.00050\ M$ into the solubility product expression for $BaSO_4$ we calculate Q_{sp}

Recall that Q has the same form as the equilibrium constant, in this case K_{sp}, but the concentrations are not necessarily equilibrium concentrations.

$$Q_{sp} = [Ba^{2+}][SO_4^{2-}] = (5.0 \times 10^{-3})(5.0 \times 10^{-4}) = 2.5 \times 10^{-6}$$

The product of the actual concentrations of the constituent ions, Q_{sp}, exceeds the solubility product for $BaSO_4$, 1.1×10^{-10}. Since $Q_{sp} > K_{sp}$, solid $BaSO_4$ will precipitate until the concentrations of the constituent ions just satisfy K_{sp} for $BaSO_4$.

Since the actual concentrations of Ba^{2+} and SO_4^{2-} multiplied together give 2.5×10^{-6}, K_{sp} for $BaSO_4$ is exceeded by a factor of approximately 2×10^4, and there will be no difficulty in seeing some precipitated $BaSO_4$.

The human eye is not a particularly sensitive detection device; when two solutions are mixed, enough solid must form so that we can see it if we are to be aware of precipitate formation. As a general rule-of-thumb, a precipitate can be seen with the naked eye if the solubility product of the compound is exceeded by a factor of 1000. If the solubility product is exceeded by a factor less than 1000, the amount of solid will likely be so small that it cannot be seen.

Solubility products enable us to calculate the concentration of an ion necessary to initiate precipitation in a particular solution, as Example 18–6 demonstrates.

EXAMPLE 18–6

What concentration of barium ion (from a soluble compound such as barium chloride) is necessary to start the precipitation of barium sulfate in a solution that is 0.0015 M in sodium sulfate?

Solution

We are dealing with the same compounds we described in Example 18–5. Since Na_2SO_4 is a soluble ionic compound, we know that $[SO_4^{2-}] = 0.0015\ M$. Therefore, we can use K_{sp} for $BaSO_4$ to calculate the concentration of barium ions required to exceed K_{sp}.

$$[Ba^{2+}][SO_4^{2-}] = 1.1 \times 10^{-10}$$

$$[Ba^{2+}] = \frac{1.1 \times 10^{-10}}{[SO_4^{2-}]} = \frac{1.1 \times 10^{-10}}{1.5 \times 10^{-3}} = 7.3 \times 10^{-8}\ M$$

Addition of enough $BaCl_2$ to give a barium ion concentration of $7.3 \times 10^{-8}\ M$ *just satisfies* K_{sp} for $BaSO_4$. Ever-so-slightly more $BaCl_2$ would be required to exceed K_{sp} and initiate precipitation of $BaSO_4$. Therefore, we say

$[Ba^{2+}] \geq 7.3 \times 10^{-8}\ M$ to initiate precipitation of $BaSO_4$

Recall that K_{sp} must be exceeded by a factor of approximately 1000 in order for a precipitate to be visible.

Frequently, it is desirable to remove a particular ion from solution by forming an insoluble compound (as in water purification). Calculations based on solubility products enable us to calculate the concentrations of ions remaining in solution *after* precipitation has occurred.

EXAMPLE 18–7

Suppose we wish to recover silver from an aqueous solution containing a soluble silver compound such as silver nitrate, $AgNO_3$, by precipitating silver ions as the insoluble compound silver chloride, AgCl. What concentration of chloride ion is necessary to reduce the concentration of silver ions to 1.0×10^{-9} *M*? A soluble ionic compound such as NaCl is used as a source of chloride ions.

Solution

The equation for the reaction of interest is

$$Ag^+ (aq) + Cl^- (aq) \rightleftharpoons AgCl (s)$$

and the solubility product for AgCl is

$$[Ag^+][Cl^-] = 1.8 \times 10^{-10}$$

To determine the concentration of Cl^- required to reduce the concentration of Ag^+ to 1.0×10^{-9} *M* we simply solve the solubility product expression for $[Cl^-]$,

$$[Cl^-] = \frac{1.8 \times 10^{-10}}{[Ag^+]} = \frac{1.8 \times 10^{-10}}{1.0 \times 10^{-9}} = 0.18\ M\ Cl^-$$

Thus, to reduce $[Ag^+]$ to 1.0×10^{-9} *M*, NaCl would be added until the final concentration of Cl^- in the solution is 0.18 *M*.

The recovery of silver from solutions used in developing and fixing photographic film and prints presents just such a problem. Silver is an expensive metal, and the recovery is profitable. Moreover, if not recovered, the silver ions would constitute an undesirable pollutant in water supplies.

This is only 0.00000011 g Ag^+/L.

18–4 Fractional Precipitation

On occasion, it is desirable to remove some ions from solution while leaving other ions with very similar properties in solution. The process is called **fractional precipitation.** Consider a solution that contains Cl^-, Br^- and I^- ions. Since these halide ions are derived from elements in the same family in the periodic table, we should expect that there will be similarities in their properties. But we should also expect some differences in properties, and indeed this is what we find when we examine the solubility products for these silver halides.

Compound	Solubility Product
AgCl	1.8×10^{-10}
AgBr	3.3×10^{-13}
AgI	1.5×10^{-16}

The solubility products show that AgI is less soluble than AgBr, and AgBr is less soluble than AgCl. This should not be surprising, since the iodide ion is larger and more polarizable than the bromide ion, and the bromide ion in turn is larger and more polarizable than the chloride ion. As a matter of interest, silver fluoride is quite soluble in water because the small fluoride ion is not easily polarized.

The differences in solubilities of the silver halides are due to increasing polarizability of the anions in going from Cl^- to Br^- to I^- (Section 22–5). This leads to greater covalent character in AgI.

EXAMPLE 18–8

Solid silver nitrate is slowly added to a solution that is 0.0010 *M* each in NaCl, NaBr, and NaI. Calculate the concentration of Ag^+ required to initiate the precipitation of each of these silver halides. (Recall that NaCl, NaBr, NaI, and $AgNO_3$ are soluble compounds that are completely dissociated in dilute aqueous solution. Sodium nitrate, $NaNO_3$, is also a soluble ionic compound so it cannot precipitate from this solution.)

Solution

Let us calculate the concentration of Ag^+ necessary to begin to precipitate each of the halide ions as we did in Example 18–6. The solubility product for AgI is

$$[Ag^+][I^-] = 1.5 \times 10^{-16}$$

Since $[I^-] = 1.0 \times 10^{-3}\ M$, the $[Ag^+]$ that must be exceeded to initiate precipitation of AgI is

$$[Ag^+] = \frac{1.5 \times 10^{-16}}{[I^-]} = \frac{1.5 \times 10^{-16}}{1.0 \times 10^{-3}} = 1.5 \times 10^{-13}\ M$$

Therefore, $[Ag^+] \geq 1.5 \times 10^{-13}\ M$ is required to start precipitation of silver iodide. Repeating this kind of calculation for silver bromide gives

We are not suggesting that a $1.5 \times 10^{-13}\ M$ solution of $AgNO_3$ be added to the solution. We are pointing out the fact that when sufficient $AgNO_3$ has been added to the solution to make $[Ag^+] = 1.5 \times 10^{-13}\ M$, AgI begins to precipitate.

$$[Ag^+] = \frac{3.3 \times 10^{-13}}{[Br^-]} = \frac{3.3 \times 10^{-13}}{1.0 \times 10^{-3}} = 3.3 \times 10^{-10}\ M$$

Thus, $[Ag^+] \geq 3.3 \times 10^{-10}\ M$ is needed to start precipitation of silver bromide. For the precipitation of silver chloride to begin,

$$[Ag^+] = \frac{1.8 \times 10^{-10}}{[Cl^-]} = \frac{1.8 \times 10^{-10}}{1.0 \times 10^{-3}} = 1.8 \times 10^{-7}\ M$$

For silver chloride to precipitate, we must have $[Ag^+] \geq 1.8 \times 10^{-7}\ M$. We have demonstrated that

to precipitate AgI, $[Ag^+] \geq 1.5 \times 10^{-13}\ M$,

to precipitate AgBr, $[Ag^+] \geq 3.3 \times 10^{-10}\ M$,

to precipitate AgCl, $[Ag^+] \geq 1.8 \times 10^{-7}\ M$.

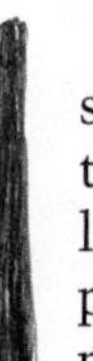

This calculation tells us that when silver nitrate is added slowly to a solution that is 0.0010 *M* in each of NaI, NaBr, and NaCl, silver iodide precipitates first, silver bromide precipitates second, and silver chloride precipitates last. The calculation of the amount of I^- precipitated before Br^- begins to precipitate and the amounts of I^- and Br^- precipitated before Cl^- begins to precipitate is illustrated in Example 18–9.

EXAMPLE 18–9

(a) Calculate the percent of I^- precipitated before AgBr precipitates. Refer to Example 18–8 as necessary for data.

(b) Calculate the percent of I^- and Br^- precipitated before Cl^- precipitates.

Solution

(a) In Example 18–8 we found that $[Ag^+] \geq 3.3 \times 10^{-10}\ M$ is needed to begin precipitation of AgBr. This value for $[Ag^+]$ can be substituted into the solubility product expression for silver iodide to determine $[I^-]$ remaining *in solution* just before AgBr begins to precipitate

$$[Ag^+][I^-] = 1.5 \times 10^{-16}$$

$$[I^-]_{unppt'd} = \frac{1.5 \times 10^{-16}}{[Ag^+]} = \frac{1.5 \times 10^{-16}}{3.3 \times 10^{-10}} = 4.5 \times 10^{-7}\ M$$

The percent of I^- unprecipitated is

$$\%\ I^-_{unppt'd} = \frac{[I^-]_{unppt'd}}{[I^-]_{orig}} \times 100\% = \frac{4.5 \times 10^{-7}}{1.0 \times 10^{-3}} \times 100\%$$

$$= \underline{0.045\%\ I^-\ \text{unprecipitated}}$$

Therefore, 99.955% of the I^- precipitates *before* AgBr begins to precipitate.

(b) Similar calculations show that *just before* AgCl begins to precipitate, $[Ag^+] = 1.8 \times 10^{-7}\ M$ and the $[I^-]$ unprecipitated is calculated as in (a).

In Example 18–7 we did a similar calculation, but stopped short of expressing the result in terms of percent of an ion precipitated.

We have subtracted 0.045% from *exactly* 100%, and therefore we have *not* violated the rules for significant figures.

$$[Ag^+][I^-] = 1.5 \times 10^{-16}$$

$$[I^-]_{unppt'd} = \frac{1.5 \times 10^{-16}}{[Ag^+]} = \frac{1.5 \times 10^{-16}}{1.8 \times 10^{-7}} = 8.3 \times 10^{-10}\ M$$

The percent of I^- unprecipitated just before AgCl precipitates is

$$\%\ I^-_{unppt'd} = \frac{[I^-]_{unppt'd}}{[I^-]_{orig}} \times 100\% = \frac{8.3 \times 10^{-10}}{1.0 \times 10^{-3}} \times 100\%$$

$$= \underline{0.000083\%\ I^-\ \text{unprecipitated}}$$

Therefore, 99.999917% of the I^- precipitates before AgCl begins to precipitate.

A similar calculation for the amount of Br^- precipitated before AgCl begins to precipitate gives

$$[Ag^+][Br^-] = 3.3 \times 10^{-13}$$

$$[Br^-]_{unppt'd} = \frac{3.3 \times 10^{-13}}{[Ag^+]} = \frac{3.3 \times 10^{-13}}{1.8 \times 10^{-7}} = 1.8 \times 10^{-6}\ M$$

$$\%\ Br^-_{unppt'd} = \frac{[Br^-]_{unppt'd}}{[Br^-]_{orig}} \times 100\% = \frac{1.8 \times 10^{-6}}{1.0 \times 10^{-3}} \times 100\%$$

$$= \underline{0.18\%\ Br^-\ \text{unprecipitated}}$$

Thus, 99.82% of the Br^- precipitates before AgCl begins to precipitate.

We have described the series of reactions that occurs when solid $AgNO_3$ is added slowly to a solution that is 0.0010 *M* in Cl^-, Br^-, and I^-. Silver iodide begins to precipitate first, and 99.955% of the I^- precipitates before any solid AgBr is formed. Silver bromide begins to precipitate next, and 99.82% of the Br^- and 99.999917% of the I^- precipitate before any solid AgCl forms.

18–5 Simultaneous Equilibria Involving Slightly Soluble Compounds

Reactions between weak acids or bases and metal ions to form insoluble compounds are very common. Example 18–10 provides insight into the reaction of a metal ion with aqueous ammonia to form an insoluble metal hydroxide.

EXAMPLE 18–10

If a solution is made 0.10 *M* in magnesium nitrate, $Mg(NO_3)_2$, *and* 0.10 *M* in aqueous ammonia, a weak base, will magnesium hydroxide, $Mg(OH)_2$, precipitate? The solubility product for $Mg(OH)_2$ is 1.5×10^{-11} and the ionization constant for aqueous NH_3 is 1.8×10^{-5}.

Solution

Two equilibria must be considered, and we must determine whether or not K_{sp} for $Mg(OH)_2$ is exceeded in the solution. That is, is the ion product in the solution, Q_{sp}, greater than K_{sp}, $[Mg^{2+}][OH^-]^2 = 1.5 \times 10^{-11}$? To answer the question the concentrations of both ions are calculated.

Magnesium nitrate is a soluble ionic compound and $[Mg^{2+}] = 0.10\ M$ in a 0.10 *M* solution of $Mg(NO_3)_2$.

The concentration of OH^- in 0.10 *M* aqueous ammonia is calculated as in Example 16–14.

$$NH_3\ (aq) + H_2O \rightleftharpoons NH_4^+\ (aq) + OH^-\ (aq)$$

$$(0.10 - x)\ M \qquad\qquad x\ M \qquad\qquad x\ M$$

$$\frac{[NH_4^+][OH^-]}{[NH_3]} = 1.8 \times 10^{-5}$$

Substitution into the ionization constant expression for aqueous NH_3 gives

$$\frac{(x)(x)}{(0.10 - x)} = 1.8 \times 10^{-5} \qquad x = 1.3 \times 10^{-3}\ M = [OH^-]$$

Now that the concentrations of both Mg^{2+} and OH^- are known, they can be substituted into the solubility product for $Mg(OH)_2$ to determine whether $Mg(OH)_2$ will precipitate at these concentrations of Mg^{2+} and OH^-.

$$[Mg^{2+}][OH^-]^2 = Q_{sp} \qquad\qquad K_{sp} = [Mg^{2+}][OH^-]^2 = 1.5 \times 10^{-11}$$

$$(0.10)(1.3 \times 10^{-3})^2 = 1.7 \times 10^{-7} = Q_{sp} \qquad\qquad \text{at equilibrium}$$

The ion product, $Q_{sp} = 1.7 \times 10^{-7}$, is greater than K_{sp} for $Mg(OH)_2$, and therefore $Mg(OH)_2$ precipitates until its K_{sp} is just satisfied.

Example 18–11 demonstrates how the concentration of a weak base required to initiate precipitation of an insoluble metal hydroxide may be calculated.

EXAMPLE 18–11

What concentration of aqueous ammonia is necessary to just initiate precipitation of $Mg(OH)_2$ from a 0.10 *M* solution of $Mg(NO_3)_2$? Refer to Example 18–10.

Solution

Two equilibria must be considered:

$$Mg(OH)_2\ (s) \rightleftharpoons Mg^{2+}\ (aq) + 2OH^-\ (aq) \quad K_{sp} = 1.5 \times 10^{-11}$$

and

$$NH_3\ (aq) + H_2O \rightleftharpoons NH_4^+\ (aq) + OH^-\ (aq) \quad K_b = 1.8 \times 10^{-5}$$

First we determine the $[OH^-]$ necessary to initiate precipitation of $Mg(OH)_2$ from 0.10 *M* $Mg(NO_3)_2$. Since $Mg(NO_3)_2$ is completely dissociated, $[Mg^{2+}] = 0.10\ M$.

$$K_{sp} = 1.5 \times 10^{-11} = [Mg^{2+}][OH^-]^2$$

$$[OH^-]^2 = \frac{1.5 \times 10^{-11}}{[Mg^{2+}]} = \frac{1.5 \times 10^{-11}}{0.10} = 1.5 \times 10^{-10}$$

$$[OH^-] = 1.2 \times 10^{-5}\ M$$

Now we determine the concentration of NH_3 that will supply $1.2 \times 10^{-5}\ M\ OH^-$. If we let *x* be the original concentration of NH_3, the equilibrium concentrations are

$$\underset{(x - 1.2 \times 10^{-5})\ M}{NH_3\ (aq)} + H_2O \rightleftharpoons \underset{1.2 \times 10^{-5}\ M}{NH_4^+\ (aq)} + \underset{1.2 \times 10^{-5}\ M}{OH^-\ (aq)}$$

$$K_b = 1.8 \times 10^{-5} = \frac{[NH_4^+][OH^-]}{[NH_3]} = \frac{(1.2 \times 10^{-5})(1.2 \times 10^{-5})}{x - 1.2 \times 10^{-5}}$$

x cannot be neglected because it is the *initial* number of mol/L of NH_3.

$$1.44 \times 10^{-10} = 1.8 \times 10^{-5}\,x - 2.16 \times 10^{-10}$$

$$1.8 \times 10^{-5}\,x = 3.6 \times 10^{-10}$$

$$x = 2.0 \times 10^{-5}\ M = \text{initial } M_{NH_3}$$

Thus, the solution must be ever so slightly greater than $2.0 \times 10^{-5}\ M$ in NH_3 to initiate precipitation of $Mg(OH)_2$ from a 0.10 *M* solution of $Mg(NO_3)_2$.

On occasion, solutions containing a weak base are buffered to decrease their basicity so that a significant concentration of a metal ion that forms an insoluble hydroxide may exist in the solution. Example 18–12 illustrates this.

EXAMPLE 18–12

How many moles of ammonium chloride, NH_4Cl, must be added to 1.0 liter of solution that is to be made 0.10 *M* in $Mg(NO_3)_2$ *and* 0.10 *M* in ammonia to prevent precipitation of $Mg(OH)_2$?

Solution

Note that the concentrations of both $Mg(NO_3)_2$ and NH_3 are the same as in Example 18–10, and we demonstrated that $Mg(OH)_2$ precipitates from that solution. The buffering action of NH_4Cl in the presence of NH_3 decreases the concentration of OH^-. Again, two equilibria must be considered

$$Mg(OH)_2\ (s) \rightleftharpoons Mg^{2+}\ (aq) + 2OH^-\ (aq)$$

$$NH_3\ (aq) + H_2O\ (\ell) \rightleftharpoons NH_4^+\ (aq) + OH^-\ (aq)$$

The concentration of Mg^{2+} cannot be varied, $[Mg^{2+}] = 0.10\ M$. It can be substituted into K_{sp} for $Mg(OH)_2$ and the *maximum concentration of* OH^- *that can exist* without exceeding K_{sp} can be calculated.

$$[Mg^{2+}][OH^-]^2 = 1.5 \times 10^{-11}$$

$$[OH^-]^2 = \frac{1.5 \times 10^{-11}}{[Mg^{2+}]} = \frac{1.5 \times 10^{-11}}{0.10} = 1.5 \times 10^{-10}$$

$$[OH^-] = 1.2 \times 10^{-5}\ M \leftarrow \text{Max. } [OH^-] \text{ possible}$$

You may wish to refer back to Section 16–7 to refresh your memory on buffer solutions.

The maximum $[OH^-]$ has been determined, and K_b for aqueous NH_3 is used to calculate the number of moles of NH_4Cl necessary to buffer 0.10 *M* aqueous NH_3 so that $[OH^-] = 1.2 \times 10^{-5}$ *M*. Let x = number of mol/L of NH_4Cl required

$$\underset{x\,M}{NH_4Cl\,(aq)} \longrightarrow \underset{x\,M}{NH_4^+\,(aq)} + \underset{x\,M}{Cl^-\,(aq)}$$

$$\underset{(0.10 - 1.2 \times 10^{-5})\,M}{NH_3\,(aq)} + H_2O \rightleftharpoons \underset{1.2 \times 10^{-5}\,M}{NH_4^+\,(aq)} + \underset{1.2 \times 10^{-5}\,M}{OH^-\,(aq)}$$

$$K_b = 1.8 \times 10^{-5} = \frac{[NH_4^+][OH^-]}{[NH_3]} = \frac{(x + 1.2 \times 10^{-5})(1.2 \times 10^{-5})}{(0.10 - 1.2 \times 10^{-5})}$$

Two simplifying assumptions, $(x + 1.2 \times 10^{-5}) \approx x$ and $(0.10 - 1.2 \times 10^{-5}) \approx 0.10$, can be made to give

$$\frac{(x)(1.2 \times 10^{-5})}{0.10} = 1.8 \times 10^{-5}$$

$x = 0.15$ mol of NH_4^+ per liter of solution

Addition of 0.15 mol of NH_4Cl to 1.0 L of 0.10 *M* aqueous NH_3 decreases $[OH^-]$ to 1.2×10^{-5} *M*, so that K_{sp} for $Mg(OH)_2$ is not exceeded in this solution.

We have just demonstrated (Examples 18–10, 18–11, 18–12) the fact that when more than one equilibrium must be considered to describe a solution, *all relevant equilibria must be satisfied.* Suppose we are asked to calculate $[H^+]$ and the pH of the solution described in Example 18–12. From the solution of the problem we know

$$[OH^-] = 1.2 \times 10^{-5}\,M$$

We can use the ion-product for water (another equilibrium constant that must be satisfied) to calculate $[H^+]$ and pH.

$$[H^+][OH^-] = 1.0 \times 10^{-14}$$

$$[H^+] = \frac{1.0 \times 10^{-14}}{[OH^-]} = \frac{1.0 \times 10^{-14}}{1.2 \times 10^{-5}} = 8.3 \times 10^{-10}\,M\ H^+$$

Since $[H^+] = 8.3 \times 10^{-10}$ *M*, then pH = 9.08.

18–6 Dissolution of Precipitates

Precipitates can be dissolved by one or more of the three types of reactions outlined below. Dissolution of a precipitate is accomplished by reducing the concentrations of the constituent ions so that K_{sp} is no longer exceeded; i.e., when $Q_{sp} < K_{sp}$ then a precipitate dissolves until the concentrations of the ions just satisfy K_{sp}.

1 Converting an Ion to a Weak Electrolyte

Typical examples are:

a. Acidification of insoluble aluminum hydroxide in contact with its saturated solution converts OH^- ions from its saturated solution to the weak electrolyte, water, shifting the solubility equilibrium to the right until $[Al^{3+}][OH^-]^3 < K_{sp}$, and dissolution occurs.

Recall that solubility products, like other equilibrium constants, are thermodynamic quantities. They tell us nothing about how fast a given reaction occurs, only that it can, or cannot, occur under the specified conditions.

$$Al(OH)_3\ (s) \rightleftharpoons Al^{3+}\ (aq) + 3OH^-\ (aq)$$
$$3H^+\ (aq) + 3OH^-\ (aq) \longrightarrow 3H_2O$$

overall reaction $Al(OH)_3\ (s) + 3H^+\ (aq) \longrightarrow Al^{3+}\ (aq) + 3H_2O$

b. Treatment of magnesium hydroxide with ammonium ion from a salt, such as ammonium chloride, NH_4Cl, converts OH^- ions from its saturated solution to the weak electrolytes, ammonia and water, with the result that $[Mg^{2+}][OH^-]^2 < K_{sp}$, and dissolution occurs.

$$Mg(OH)_2\ (s) \rightleftharpoons Mg^{2+}\ (aq) + 2OH^-\ (aq)$$
$$2NH_4^+\ (aq) + 2OH^-\ (aq) \longrightarrow 2NH_3\ (aq) + 2H_2O$$

overall reaction $Mg(OH)_2\ (s) + 2NH_4^+\ (aq) \longrightarrow Mg^{2+}\ (aq) + 2NH_3\ (aq) + 2H_2O$

Observe that this process, the dissolution of $Mg(OH)_2$ in an NH_4Cl solution, is the reverse of the reaction we considered in Example 18–10, the precipitation of $Mg(OH)_2$ from a solution of aqueous ammonia.

c. Acidification of some metal sulfides, for example manganese(II) sulfide, $K_{sp} = 5.1 \times 10^{-15}$, converts sulfide ions from the saturated solution to hydrogen sulfide, a weak electrolyte, with the result that $[Mn^{2+}][S^{2-}] < K_{sp}$ and MnS dissolves.

$$MnS\ (s) \rightleftharpoons Mn^{2+}\ (aq) + S^{2-}\ (aq)$$
$$2H^+\ (aq) + S^{2-}\ (aq) \longrightarrow H_2S\ (g)$$

overall reaction $MnS\ (s) + 2H^+\ (aq) \longrightarrow Mn^{2+}\ (aq) + H_2S\ (g)$

The latter technique is effective only for the more soluble of the "insoluble" metal sulfides, since the very insoluble sulfides do not produce sufficiently high concentrations of sulfide ions in their saturated solutions to react with even the strongest acids.

2 Converting an Ion to Another Species by Oxidation-Reduction Reaction

Most insoluble metal sulfides can be dissolved in hot dilute nitric acid because nitrate ions oxidize sulfide ions to elemental sulfur, thereby removing sulfide ions from the solution

$$3S^{2-}\ (aq) + 2NO_3^-\ (aq) + 8H^+\ (aq) \longrightarrow 3S\ (s) + 2NO\ (g) + 4H_2O\ (\ell)$$

Consider copper(II) sulfide, CuS, as a typical metal sulfide in equilibrium with its ions. This equilibrium lies far to the left, $K_{sp} = 8.7 \times 10^{-36}$, but removal of the sulfide ions by oxidation to elemental sulfur shifts the equilibrium to the right, and so CuS (s) dissolves in hot dilute HNO_3

$$CuS\ (s) \rightleftharpoons Cu^{2+}\ (aq) + S^{2-}\ (aq)$$

$$3S^{2-}\ (aq) + 2NO_3^-\ (aq) + 8H^+\ (aq) \longrightarrow 3S\ (s) + 2NO\ (g) + 4H_2O\ (\ell)$$

Multiplication of the first equation by three, followed by addition of these two equations and cancellation of like terms, gives the net ionic equation for the dissolution of CuS (s) in hot dilute HNO_3

$$3CuS\ (s) + 2NO_3^-\ (aq) + 8H^+\ (aq) \longrightarrow 3Cu^{2+}\ (aq) + 3S\ (s) + 2NO\ (g) + 4H_2O\ (\ell)$$

3 Complex Ion Formation

Ligand is the name given to an atom or a group of atoms bonded to the central element in complex ions. Ligands are Lewis bases.

Many slightly soluble compounds contain cations that are capable of forming soluble compounds that contain complex ions. They do this by accepting shares in electron pairs from such molecules or ions as NH_3, CN^-, OH^-, SCN^-, F^-, Cl^-, Br^-, I^-, and so on. Such coordinate covalent bonds are formed as these ligands replace water molecules from hydrated metal ions (Section 10–2). Some complex ions and their dissociation constants, K_d, are given in Appendix I.

The smaller K_d is, the more effectively the ligand competes with water for a coordination site on the metal ion; stated differently, the smaller the K_d, the more stable the complex ion is. Many copper(II) compounds react with excess aqueous ammonia to form the deep blue colored complex ion, $Cu(NH_3)_4^{2+}$ (aq).

$$Cu(NH_3)_4^{2+}\ (aq) \rightleftharpoons Cu^{2+}\ (aq) + 4NH_3\ (aq)$$

$$K_d = \frac{[Cu^{2+}][NH_3]^4}{[Cu(NH_3)_4^{2+}]} = 8.5 \times 10^{-13}$$

Recall that Cu^{2+} (aq) is really a hydrated ion, $Cu(OH_2)_4^{2+}$. We could write the above reaction more accurately as

$$Cu(NH_3)_4^{2+} + 4H_2O \rightleftharpoons Cu(OH_2)_4^{2+} + 4NH_3$$

$$K_d = \frac{[Cu(OH_2)_4^{2+}][NH_3]^4}{[Cu(NH_3)_4^{2+}]} = 8.5 \times 10^{-13}$$

EXAMPLE 18–13

What are the concentrations of hydrated Cu^{2+}, NH_3, and $Cu(NH_3)_4^{2+}$ in a 0.20 *M* solution of $Cu(NH_3)_4SO_4$?

Solution

The soluble deep blue complex salt, $Cu(NH_3)_4SO_4$, dissociates completely to produce tetraamminecopper(II) ions and sulfate ions

$$\underset{0.20\ M}{Cu(NH_3)_4SO_4\ (aq)} \xrightarrow{100\%} \underset{0.20\ M}{Cu(NH_3)_4^{2+}\ (aq)} + \underset{0.20\ M}{SO_4^{2-}\ (aq)}$$

Some of the $Cu(NH_3)_4^{2+}$ ions then dissociate. Let x be the concentration of $Cu(NH_3)_4^{2+}$ that dissociates

$$\underset{(0.20 - x)\ M}{Cu(NH_3)_4^{2+}} \rightleftharpoons \underset{x\ M}{Cu^{2+}\ (aq)} + \underset{4x\ M}{4NH_3}$$

$$K_d = \frac{[Cu^{2+}][NH_3]^4}{[Cu(NH_3)_4^{2+}]} = 8.5 \times 10^{-13}$$

Assume $(0.20 - x) \approx 0.20$.

$$8.5 \times 10^{-13} = \frac{x(4x)^4}{0.20 - x} = \frac{256\ x^5}{0.20}$$

$$x^5 = 6.6 \times 10^{-16}$$

Taking the fifth root of both sides of this equation gives

$x = 9.2 \times 10^{-4}\ M = [Cu^{2+}]$

$[NH_3] = 4x = 3.7 \times 10^{-3}\ M$

$[Cu(NH_3)_4{}^{2+}] = (0.20 - x)M = 0.20\ M$

Our assumption, $(0.20 - x) \approx 0.20$, is valid since 9.2×10^{-4} is much less than 0.20.

The dissolution of solid zinc hydroxide in excess sodium hydroxide solution to form the complex ion, $Zn(OH)_4{}^{2-}$, illustrates the amphoteric nature of $Zn(OH)_2$ (Section 11–8)

$$Zn(OH)_2\ (s) + 2OH^- \rightleftharpoons Zn(OH)_4{}^{2-}$$

EXAMPLE 18–14

Some solid $Zn(OH)_2$ is suspended in a saturated solution of $Zn(OH)_2$. A solution of sodium hydroxide is added until all the $Zn(OH)_2$ just dissolves. The pH of the solution is 11.80. What are the concentrations of Zn^{2+} and $Zn(OH)_4{}^{2-}$ in the solution? K_{sp} for $Zn(OH)_2 = 4.5 \times 10^{-17}$, and K_d for $Zn(OH)_4{}^{2-} = 3.5 \times 10^{-16}$.

Solution

The equilibria of interest are

$$Zn(OH)_2\ (s) \rightleftharpoons Zn^{2+}\ (aq) + 2OH^-\ (aq) \qquad K_{sp} = 4.5 \times 10^{-17}$$

$$Zn(OH)_4{}^{2-}\ (aq) \rightleftharpoons Zn^{2+}\ (aq) + 4OH^-\ (aq) \qquad K_d = 3.5 \times 10^{-16}$$

Since we know the pH, we can calculate $[OH^-]$ from pH + pOH = 14.00

$$pOH = 14.00 - pH = 14.00 - 11.80 = 2.20$$

$$[OH^-] = 10^{-pOH} = 10^{-2.20} = 10^{+0.80} \times 10^{-3} = 6.3 \times 10^{-3}\ M = [OH^-]$$

We can use the solubility product relationship to calculate $[Zn^{2+}]$

$$K_{sp} = 4.5 \times 10^{-17} = [Zn^{2+}][OH^-]^2$$

$$4.5 \times 10^{-17} = [Zn^{2+}](6.3 \times 10^{-3})^2$$

$$[Zn^{2+}] = \frac{4.5 \times 10^{-17}}{(6.3 \times 10^{-3})^2} = 1.1 \times 10^{-12}\ M$$

Since both equilibria are established in the same solution, the same $[Zn^{2+}]$ and $[OH^-]$ also satisfy the complex ion dissociation equilibrium

$$K_d = 3.5 \times 10^{-16} = \frac{[Zn^{2+}][OH^-]^4}{[Zn(OH)_4{}^{2-}]}$$

$$[Zn(OH)_4{}^{2-}] = \frac{[Zn^{2+}][OH^-]^4}{3.5 \times 10^{-16}} = \frac{(1.1 \times 10^{-12})(6.3 \times 10^{-3})^4}{3.5 \times 10^{-16}}$$

$$[Zn(OH)_4{}^{2-}] = 5.0 \times 10^{-6}\ M$$

Key Terms

Complex ions ion resulting from the formation of coordinate covalent bonds between simple ions and other ions or molecules.

Dissociation constant equilibrium constant that applies to the dissociation of a complex ion into a simple ion and coordinating species (ligands).

Fractional precipitation removal of some ions from solution by precipitation while leaving other ions with similar properties in solution.

Insoluble compound actually a very slightly soluble compound.

Molar solubility number of moles of a solute that dissolve to produce a liter of saturated solution.

Precipitate solid formed by mixing in solution the constituent ions of an insoluble compound.

Solubility product constant equilibrium constant that applies to the dissolution of a slightly soluble compound.

Solubility product principle the solubility product constant expression for a slightly soluble compound is the product of the concentrations of the constituent ions, each raised to the power that corresponds to the number of ions in one formula unit.

Exercises

1. (a) Are "insoluble" substances really insoluble?
   (b) What do we mean when we refer to insoluble substances?
2. State the solubility product principle. What does it mean?

Solubility Product Constants

3. (a) Why are solubility product constant expressions written as products of concentrations of ions raised to appropriate powers?
   (b) What does the term "concentration of a solid" mean when we discuss an equilibrium involving a slightly soluble compound such as $BaSO_4$ dissolving in water to form a saturated solution?
4. Write an equation for the dissolution of, and the solubility product constant expression for each of the following slightly soluble compounds. (a) AgCl, (b) $CaCO_3$, (c) $AlPO_4$, (d) Ag_2S, (e) BaF_2, (f) $Sn(OH)_2$, (g) $Fe(OH)_3$, (h) Sb_2S_3.
5. What do we mean when we refer to the molar solubility of a compound?

Experimental Determination of K_{sp}'s

6. From the solubility data given for the following compounds, calculate their solubility product constants. Your calculated values may not agree exactly with the solubility products given in Appendix H because round-off errors become large in calculations to two significant figures. Also, there is some disagreement within the scientific community over the exact values for some solubility products. These solubility data were taken from a handbook of chemistry and physics.
   (a) AgBr, silver bromide, 2.7×10^{-6} g/L
   (b) $BaCrO_4$, barium chromate, 0.00041 g/100 mL
   (c) $CaCO_3$, calcium carbonate, 0.0014 g/10 mL
   (d) CaF_2, calcium fluoride, 0.016 g/L
7. Calculate molar solubilities, concentrations of constituent ions, and solubilities in g/L for the following compounds at 25°C. (Refer to Appendix H for solubility product constants.)
   (a) CuI, copper(I) iodide
   (b) BaF_2, barium fluoride
   (c) Ag_2SO_4, silver sulfate
   (d) $Ag_4Fe(CN)_6$, silver hexacyanoferrate(II)
8. Construct a table similar to Table 18–1 for the compounds listed in Exercise 7. Which compound has (a) the highest molar solubility, (b) the lowest molar solubility, (c) the highest solubility expressed in g/L, and (d) the lowest solubility expressed in g/L?

Uses of K_{sp}'s

9. Calculate the number of moles and the mass of:
   (a) CuBr that will dissolve in one liter of 0.010 *M* HBr.
   (b) Ag_2CrO_4 that will dissolve in one liter of 0.10 *M* $AgNO_3$.
   (c) Ag_2CrO_4 that will dissolve in one liter of 0.10 *M* Na_2CrO_4.
10. If the following volumes of the indicated solutions are mixed, will a precipitate form? If so, do you expect to be able to see the solid?
    (a) 100 mL of 0.20 *M* KCl and 100 mL of 0.50 *M* Na_2S
    (b) 100 mL of 0.00050 *M* $AgNO_3$ and 100 mL of 0.0020 *M* NaCl
    (c) 200 mL of 0.015 *M* $Pb(NO_3)_2$ and 100 mL of 0.030 *M* KI
    (d) 20 mL of 0.0015 *M* $AgNO_3$ and 10 mL of 0.0030 *M* Na_3PO_4
11. Suppose you have beakers that contain 100 mL of the following solutions:
    (a) 0.0015 *M* KOH (b) 0.0015 *M* K_2CO_3
    (c) 0.0015 *M* K_2S
    If solid copper(II) nitrate is added (slowly) to each beaker, what concentration of Cu^{2+} is required to initiate precipitation? If solid copper(II) nitrate were added to each beaker until $[Cu^{2+}] = 0.0015$ *M*, what concentrations of OH^-, CO_3^{2-} and S^{2-} would remain in solution, i.e., unprecipitated?
12. Suppose we wish to recover silver from an aqueous solution that contains a soluble silver compound such as $AgNO_3$ by precipitating silver ions in the form of an insoluble compound such as AgCl.
    (a) How much sodium chloride should we add to reduce the silver ion concentration to 1.0×10^{-6} *M*?
    (b) What mass of Ag^+ remains in one liter of solution?
    (c) What volume of the solution contains one gram of Ag^+?
13. If silver ions are to be removed from solution by precipitation of Ag_2S, what final concentration of sulfide ions is required to reduce the Ag^+ concentration to 1.0×10^{-10} molar? What volume of this solution contains 1.0 g of Ag^+?
14. If calcium ions are to be removed from a city water supply by the addition of soda ash, Na_2CO_3, what final concentration of CO_3^{2-} is required to reduce $[Ca^{2+}]$ in the water to 0.000010 *M*? What volume of this solution contains 1.0 g of Ca^{2+}? Assume no dilution due to the addition of solid soda ash.
15. If solid NaOH is added to 0.020 *M* $Mg(NO_3)_2$ until the pH reaches 11.00, what concentration of Mg^{2+} will remain in solution? What % of the Mg^{2+} will be precipitated? Assume no volume change due to the addition of solid NaOH.

Fractional Precipitation

16. What is fractional precipitation?
17. Solid Na_2SO_4 is added slowly to a solution that is 0.10 *M* in $Pb(NO_3)_2$ and 0.10 *M* in $Ba(NO_3)_2$. In what order will solid $PbSO_4$ and $BaSO_4$, form? Calculate the percentage of Ba^{2+} that precipitates just before $PbSO_4$ begins to precipitate.
18. To a solution that is 0.010 *M* in Cu^+, 0.010 *M* Ag^+, and 0.010 *M* in Au^+, *solid* NaCl is added slowly. Assume no volume change due to the addition of solid NaCl.
    (a) Which compound will begin to precipitate first?
    (b) Calculate $[Au^+]$ when AgCl just begins to precipitate. What % of the Au^+ has precipitated at this point?
    (c) Calculate $[Au^+]$ and $[Ag^+]$ when CuCl just begins to precipitate.
19. Solid $Pb(NO_3)_2$ is added slowly to a solution that is 0.010 *M* each in NaOH, K_2CO_3, and Na_2SO_4. In what order will solid $Pb(OH)_2$, $PbCO_3$ and $PbSO_4$ form? Calculate the percentages of OH^- and CO_3^{2-} that precipitate just before $PbSO_4$ begins to precipitate.

Simultaneous Equilibria

20. If a solution is made 0.10 *M* in $Mg(NO_3)_2$, 1.0 *M* in aqueous ammonia, and 3.0 *M* in NH_4NO_3, will $Mg(OH)_2$ precipitate? What is the pH of this solution?
21. If a solution is made 0.10 *M* in $Mg(NO_3)_2$, 1.0 *M* in aqueous ammonia, and 0.10 *M* in NH_4NO_3, will $Mg(OH)_2$ precipitate? What is the pH of this solution?

22. (a) What is the minimum concentration of ammonium nitrate necessary to prevent precipitation of $Mg(OH)_2$ in the solution described in Exercise 21?
(b) What mass of ammonium nitrate is this in one liter of solution?
(c) What is the pH of this solution if it contains the minimum concentration of NH_4NO_3 necessary to prevent the precipitation of $Mg(OH_2)$?
23. If a solution is 0.010 M in manganese(II) nitrate, $Mn(NO_3)_2$, and 0.10 M in aqueous ammonia, will manganese(II) hydroxide, $Mn(OH)_2$, precipitate?
24. If a solution is 1.0×10^{-5} M in $Mn(NO_3)_2$ and 1.0×10^{-3} M in aqueous ammonia, will $Mn(OH)_2$ precipitate?
25. What concentration of NH_4NO_3 is necessary to prevent precipitation of $Mn(OH)_2$ in the solution of Exercise 23.

Dissolution of Precipitates

26. Explain, by writing appropriate equations, how the following insoluble compounds can be dissolved by the addition of a solution of nitric acid. (Carbonates dissolve in strong acids to form carbon dioxide, which is evolved as a gas, and water.) What is the "driving force" for each reaction? (a) $Fe(OH)_2$, (b) $Fe(OH)_3$, (c) $Co(OH)_2$, (d) $PbCO_3$, (e) $(CuOH)_2CO_3$.
27. Explain, by writing equations, how the following insoluble compounds can be dissolved by the addition of a solution of ammonium nitrate or ammonium chloride: (a) $Mg(OH)_2$, (b) $Mn(OH)_2$, (c) $Ni(OH)_2$
28. The following insoluble sulfides can be dissolved in 3 M hydrochloric acid. Explain how this is possible and write the appropriate equations: (a) FeS, (b) ZnS
29. The following sulfides are less soluble than those listed in Exercise 28. These sulfides can be dissolved in hot 6 M nitric acid, an oxidizing acid. Explain how and write the appropriate balanced equations: (a) PbS, (b) CuS, (c) Bi_2S_3
30. Which of the following insoluble metal hydroxides can be dissolved in 6 M sodium hydroxide? (You may wish to refer to Table 11–3.) (a) $Cd(OH)_2$, (b) $Mg(OH)_2$, (c) $Zn(OH)_2$, (d) $Al(OH)_3$, (e) $Cr(OH)_3$, (f) $Ni(OH)_2$, (g) $Cu(OH)_2$.
31. Calculate the concentration of complex ion, metal ion, and ammonia in the following solutions. The complex ions are enclosed in brackets. All these compounds are both soluble and ionic.
(a) 0.10 M $[Ag(NH_3)_2]Cl$
(b) 0.10 M $[Cu(NH_3)_4]SO_4$
(c) 0.10 M $[Co(NH_3)_6](ClO_4)_3$

19 Electrochemistry

Electrochemistry deals with chemical changes produced by an electric current and with the production of electricity by chemical reactions. By its nature, electrochemistry requires some method of introducing a stream of electrons into a reacting chemical system, and some means of withdrawing electrons.

The reacting system is contained in a **cell** and an electric current is led in or out by **electrodes.**

There are two kinds of electrochemical cells, electrolytic cells and voltaic cells.

1. **Electrolytic cells** are those in which electrical energy causes *nonspontaneous* chemical reactions to occur.
2. **Voltaic cells** are those in which *spontaneous* chemical reactions produce electricity and supply it to an external circuit.

The study of electrochemistry has provided much of our knowledge about chemical reactions. The amount of electrical energy consumed or produced can be measured quite accurately. Additionally, the parts of an electrochemical reaction are separated physically, so that one half of a reaction occurs at one electrode while the other half occurs at the other electrode. All electrochemical reactions involve the transfer of electrons and are therefore *oxidation-reduction* reactions.

19–1 Electrical Conduction

Electrical current can be conducted through pure liquid electrolytes or solutions containing electrolytes, and through metal wires or along metallic surfaces. The latter type of conduction is called **metallic conduction.** It involves the flow of electrons along a metal surface with no similar movement of the atoms of the metal and no obvious changes in the metal. **Ionic** or **electrolytic conduction** refers to the conduction of electrical current by the motion of ions through a solution or a pure liquid. The positively charged ions (cations) migrate spontaneously toward the negative electrode, while the negatively charged ions (anions) move toward the positive electrode. Both kinds of conduction, ionic and metallic, occur in electrochemical cells (Figure 19–1).

19–2 Electrodes

Electrodes are usually metal surfaces upon which oxidation or reduction half-reactions occur. They may or may not participate in the reactions. Those that do not react with the reactants or products are called **inert electrodes.** Regardless of the kind of cell, electrolytic or voltaic, the **cathode** is defined as the electrode at which *reduction* occurs as electrons are gained by some species. The **anode** is defined as the electrode at which *oxidation* occurs as electrons are lost by some species.

Electrolytic Cells

Lysis means "splitting apart." In many electrolytic cells, compounds are split into their constituent elements.

Electrolytic cells are electrochemical cells in which *nonspontaneous* chemical reactions are made to occur by the forced input of electrical energy. This process is called **electrolysis.** An electrolytic cell consists of a container for the reaction material, plus electrodes immersed in the reaction material and connected to a source of direct current. Inert electrodes are usually used, so that the electrodes are not involved in the chemical reaction.

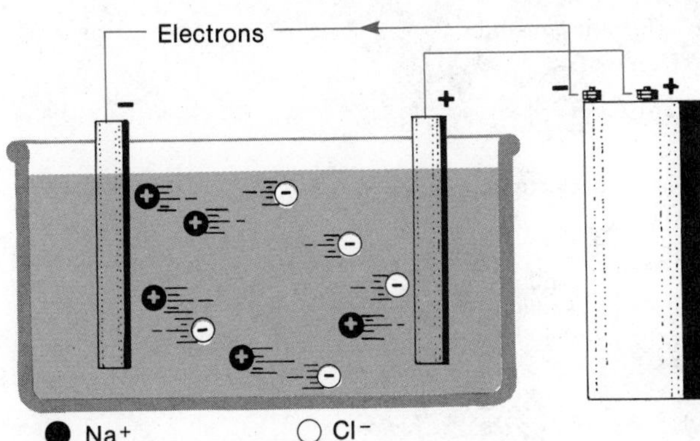

FIGURE 19–1 The motion of ions through a solution is an electric current, and accounts for ionic (electrolytic) conduction. Positively charged ions move toward the negative electrode while negatively charged ions move toward the positive electrode.

We shall discuss several electrochemical cells in this chapter. For most of them, experimental observations will be presented from which we can deduce the electrode reactions and, from them, the overall reactions. We can then construct simplified diagrams of the cells.

19–3 The Electrolysis of Molten Sodium Chloride (The Downs Cell)

Solid sodium chloride does not conduct electricity because, although its ions vibrate about fixed positions, they are not free to move throughout the crystal. However, molten (melted) sodium chloride (melting point, 801°C; a clear, colorless liquid that looks like water) is an excellent conductor because its ions are freely mobile. Consider a cell in which a source of direct current is connected by a wire to two inert graphite electrodes that are immersed in a container of sodium chloride heated above its melting point. When the current is flowing, we observe that

Electrons cannot migrate through the NaCl crystal the way they do in metals (Section 9–17).

1. A pale green gas, which is chlorine, Cl_2, is liberated at one electrode.
2. Molten, silvery-white metallic sodium, Na, forms at the other electrode and rises to the top of the molten sodium chloride.

From this information we can determine the essential features of the cell. Since chlorine is produced, it must be produced by oxidation of chloride ions, and the electrode at which this happens must be the anode. Metallic sodium is produced by reduction of sodium ions at the cathode, where electrons are being forced into the cell. The metal remains liquid because its melting point is only 97.8°C, and it floats because it is less dense than the molten sodium chloride.

$$2Cl^- \longrightarrow Cl_2\ (g) + 2e^- \qquad \text{(oxidation, anode half-reaction)}$$
$$2(Na^+ + e^- \longrightarrow Na\ (\ell)) \qquad \text{(reduction, cathode half-reaction)}$$
$$\underbrace{2Na^+ + 2Cl^-}_{2NaCl\ (\ell)} \longrightarrow 2Na\ (\ell) + Cl_2\ (g) \qquad \text{(overall cell reaction)}$$

The formation of metallic sodium and gaseous chlorine from sodium chloride is nonspontaneous except at temperatures much higher than 801°C, so the dc generator must supply considerable electrical energy to cause this reaction to occur. Electrons are produced at the anode (oxidation) and con-

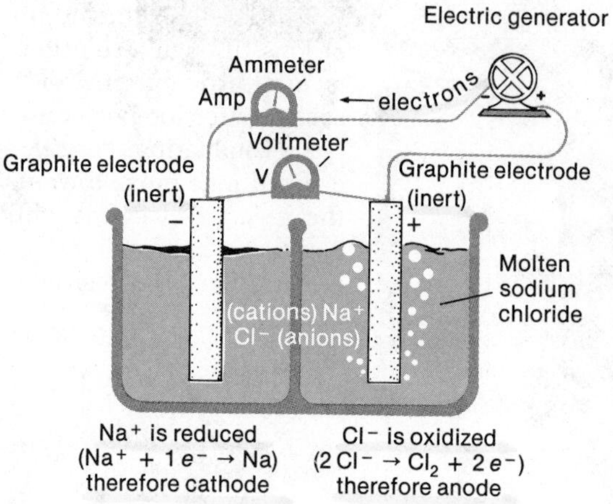

FIGURE 19–2 Apparatus for electrolysis of molten sodium chloride.

sumed at the cathode (reduction) and, therefore, they travel through the wire from *anode to cathode*. Since the reaction is nonspontaneous, the dc source forces electrons to flow, nonspontaneously from the positive electrode to the negative electrode. Thus, the anode is the positive electrode and the cathode the negative electrode *in all electrolytic cells*. Figure 19–2 shows a simplified diagram of the cell.

The direction of spontaneous flow for negatively charged particles is from negative to positive.

The commercial Downs cell is more sophisticated than that depicted in Figure 19–2. Figure 19–3 shows the Downs cell that is used for the industrial electrolysis of sodium chloride. Sodium and chlorine must not be allowed to come in contact with each other because they react spontaneously, rapidly, and explosively to form sodium chloride (ΔG^0 is very negative for this reaction at 801°C). The Downs cell is expensive to use mainly because of the cost of construction, the cost of the electricity, and the cost of heating the sodium chloride to 801°C. However, electrolysis of a molten sodium salt is the only means by which metallic sodium can be obtained, because of its extremely high reactivity. Once liberated by the electrolysis, the liquid sodium is drained off, cooled, and cast into blocks. These must be stored in an inert environment, such as below the surface of mineral oil, to prevent reaction with oxygen or other components of the atmosphere.

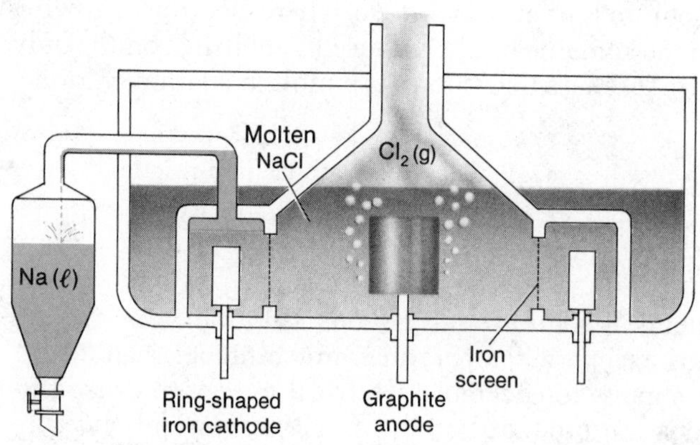

FIGURE 19–3 The Downs cell, the apparatus in which molten sodium chloride is commercially electrolyzed to produce sodium metal and chlorine gas.

Electrolysis of molten compounds is also the common method of obtaining other Group IA and IIA metals (except barium) and aluminum (Chapter 20). Most chlorine is produced by the less expensive electrolysis of aqueous sodium chloride, but that produced in the Downs cell is cooled, compressed, and marketed. This partially offsets the expense of producing metallic sodium.

19–4 The Electrolysis of Aqueous Sodium Chloride

We have indicated many times that all ions in aqueous solutions are hydrated. We will not use the notation that indicates states of substances, (s), (ℓ), (g), and that ions are hydrated in solution, (aq), in the remainder of the text except in those cases where such information is not obvious. This greatly simplifies writing chemical equations.

Consider the electrolysis of a solution of sodium chloride in water, using inert electrodes. The following experimental observations can be made when a sufficiently high voltage is applied across the electrodes of a suitable cell:

1. Gaseous hydrogen is liberated at one electrode, and the solution becomes basic around that electrode.
2. Gaseous chlorine is liberated at the other electrode.

Chloride ions are obviously being oxidized to chlorine in this cell, as they were in the electrolysis of molten sodium chloride. But sodium ions are not reduced to metallic sodium. Instead, gaseous hydrogen and aqueous hydroxide ions are produced by reduction of water molecules at the cathode. It must be that water is more easily reduced than Na^+ ions. This illustrates an important principle of electrochemistry and oxidation-reduction reactions in general:

The most easily oxidized species is oxidized and the most easily reduced species is reduced in preference to other species.

The half-reactions and overall cell reaction for the electrolysis of aqueous sodium chloride solution are

$$\begin{array}{rcll}
2Cl^- & \longrightarrow & Cl_2 + 2e^- & \text{(at anode)} \\
2H_2O + 2e^- & \longrightarrow & 2OH^- + H_2 & \text{(at cathode)} \\
\hline
2H_2O + 2Cl^- & \longrightarrow & 2OH^- + H_2 + Cl_2 & \text{(overall cell reaction)} \\
\underbrace{+\ 2Na^+}_{2NaCl} & & \underbrace{+\ 2Na^+}_{2NaOH} & \text{(spectator ions)}
\end{array}$$

The cell is illustrated in Figure 19–4. As before, the electrons flow from the anode (+) through the wire to the cathode (−) under the influence of the source of electrical energy that causes the nonspontaneous cell reaction to occur.

The overall cell reaction produces gaseous H_2 and Cl_2 and an aqueous solution of sodium hydroxide, sometimes called caustic soda. Solid sodium hydroxide is obtained by evaporation of the residual solution. This process represents the most important commerical preparation of each of these substances, and is much less expensive than the electrolysis of molten sodium chloride since it is not necessary to heat the solution.

Not surprisingly, the fluctuations in the commercial prices of these widely used industrial products nearly always parallel each other.

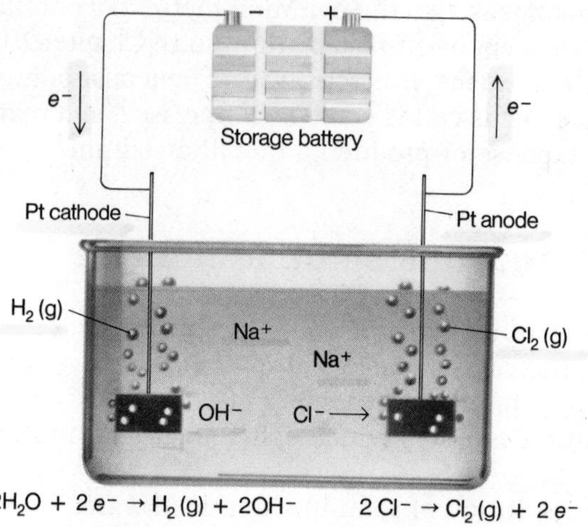

FIGURE 19–4 Electrolysis of aqueous NaCl solution. Although several reactions occur at both the anode and the cathode, the end result is the production of H_2 (g) and NaOH at the cathode and Cl_2 (g) at the anode.

19–5 The Electrolysis of Aqueous Sodium Sulfate

The following observations can be made for the electrolysis of aqueous sodium sulfate using inert electrodes:

1. Gaseous hydrogen is produced at one electrode, and the solution becomes basic around that electrode.
2. Gaseous oxygen is produced at the other electrode, and the solution becomes acidic around that electrode.

As in the previous example, water is reduced in preference to Na^+ at the cathode. Observation 2 suggests that water is also preferentially oxidized, relative to the sulfate ion, SO_4^{2-}, at the anode:

$$2(2H_2O + 2e^- \longrightarrow H_2 + 2OH^-) \qquad \text{(cathode)}$$
$$2H_2O \longrightarrow O_2 + 4H^+ + 4e^- \qquad \text{(anode)}$$
$$6H_2O \longrightarrow 2H_2 + O_2 + \underbrace{4H^+ + 4OH^-}_{4H_2O} \qquad \text{(overall cell reaction)}$$

$$2H_2O \longrightarrow 2H_2 + O_2 \qquad \text{(net reaction)}$$

The net reaction describes the electrolysis of pure water, which happens because water is more easily reduced than Na^+ and more easily oxidized than SO_4^{2-}. The ions of Na_2SO_4 conduct the current through the solution, and take no part in the reaction. The cell is diagrammed in Figure 19–5.

19–6 Faraday's Law of Electrolysis

In 1832–33 Michael Faraday observed that the amount of substance undergoing oxidation or reduction at each electrode during electrolysis is directly proportional to the amount of electricity that passes through the cell. This statement is known as **Faraday's Law of Electrolysis.** The quantitative unit of electricity, now called the **faraday,** is the amount of electricity that reduces one equivalent weight of a substance at the cathode and oxidizes one equiva-

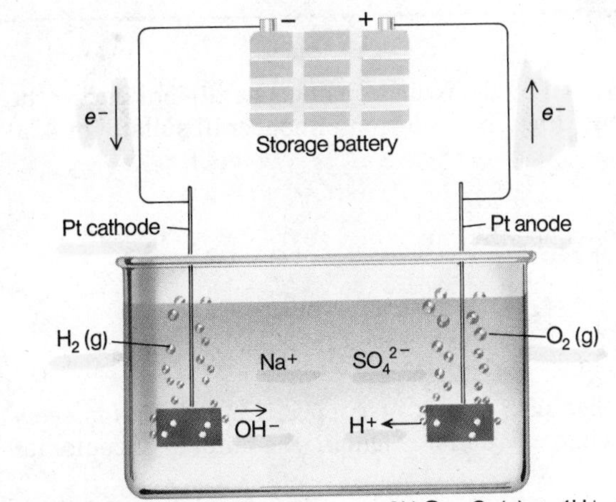

FIGURE 19–5 Electrolysis of aqueous Na_2SO_4 solution to produce H_2 (g) at the cathode and O_2 (g) at the anode.

lent weight of a substance at the anode. This corresponds to the gain or loss, and therefore the passage, of 6.022×10^{23} electrons.

A smaller electrical unit commonly used in physics and electronics is the **coulomb.** One coulomb is formally defined to be the amount of charge that passes a given point when one ampere of electrical current flows for one second, and is also equal to the amount of electricity that will deposit 0.001118 gram of silver at the cathode during the electrolysis of an aqueous solution containing silver ions, Ag^+, without limit of time. One ampere of current equals one coulomb per second. One faraday is found to be equal to 96,487 coulombs of charge.

1 faraday of electricity corresponds to the passage of 1 mole of electrons.

$$1 \text{ ampere} = 1 \frac{\text{coulomb}}{\text{second}}$$

$$1 \text{ faraday} = 6.022 \times 10^{23} e^- = 96{,}487 \text{ coul}$$

We may restate Faraday's Law: During electrolysis one faraday of electricity (96,487 coulombs) reduces and oxidizes, respectively, one equivalent weight of the oxidizing and reducing agents. This corresponds to the passage of 6.022×10^{23} electrons through the cell. Table 19–1 shows the amounts of several elements produced during electrolysis by the passage of one faraday of electricity.

TABLE 19–1 Elements Produced at One Electrode in Electrolysis by One Faraday of Electricity

Half-Reaction	Product (Electrode)	Amount (= 1 Eq Wt)
$Ag^+ (aq) + e^- \longrightarrow Ag (s)$	Ag (cathode)	1 mol = 107.868 g
$2H^+ (aq) + 2e^- \longrightarrow H_2 (g)$	H_2 (cathode)	$\frac{1}{2}$ mol = 1.008 g
$Cu^{2+} (aq) + 2e^- \longrightarrow Cu (s)$	Cu (cathode)	$\frac{1}{2}$ mol = 31.77 g
$Au^{3+} (aq) + 3e^- \longrightarrow Au (s)$	Au (cathode)	$\frac{1}{3}$ mol = 65.6555 g
$2Cl^- \longrightarrow Cl_2 (g) + 2e^-$	Cl_2 (anode)	$\frac{1}{2}$ mol = 35.453 g = 11.2 L_{STP}
$2H_2O (\ell) \longrightarrow O_2 (g) + 4H^+ (aq) + 4e^-$	O_2 (anode)	$\frac{1}{4}$ mol = 7.9997 g = 5.60 L_{STP}

Because most of our calculations will be done to three significant figures, 96,487 coulombs will be rounded off to 96,500 coulombs.

EXAMPLE 19–1

Calculate the mass of copper produced by the reduction of copper(II) ions during the passage of 2.50 amperes of current through a solution of copper(II) sulfate for 45.0 minutes.

Solution

The equation for the reduction of copper(II) ions is

Cu^{2+} +	$2e^-$	$\longrightarrow$	Cu	(at cathode)
1 mol 63.5 g	$2(6.02 \times 10^{23})e^-$ 2(96,500 coul)		1 mol 63.5 g	

From this information we see that 63.5 grams of copper "plate out" for every 2(96,500 coulombs) of electronic charge. We first calculate the number of coulombs passing through the cell.

$$\underline{?}\text{ coul} = 45.0\text{ min} \times \frac{60\text{ s}}{1\text{ min}} \times \frac{2.50\text{ coul}}{\text{s}} = 6.75 \times 10^3\text{ coul}$$

We now calculate the mass of copper produced by the passage of 6.75×10^3 coulombs.

$$\underline{?}\text{ g Cu} = 6750\text{ coul} \times \frac{63.5\text{ g Cu}}{2\text{ mol }e^-\left(96{,}500\dfrac{\text{coul}}{\text{mol }e^-}\right)} = \underline{2.22\text{ g Cu}}$$

Notice how little copper is deposited by this considerable current in 45 minutes.

EXAMPLE 19–2

What volume of oxygen (measured at STP) is produced by the oxidation of water in the electrolysis of copper(II) sulfate in Example 19–1?

Solution

The equation for the oxidation of water, and the equivalence between coulombs and the volume of oxygen produced at STP, are given below

$2H_2O \longrightarrow$	O_2	$+ 4H^+ +$	$4e^-$	(at anode)
	1 mol 22.4 L_{STP}		$4(6.02 \times 10^{23})e^-$ 4(96,500 coul)	

The number of coulombs passing through the cell is the same as in Example 19–1, 6750 coulombs. For every 4(96,500 coulombs) passing through the cell, 22.4 liters of O_2 at STP are produced.

$$\underline{?}\ L_{STP}\ O_2 = 6750\text{ coul} \times \frac{22.4\ L_{STP}\ O_2}{4\text{ mol }e^-\left(96{,}500\dfrac{\text{coul}}{\text{mol }e^-}\right)} = \underline{0.392\ L_{STP}\ O_2}$$

Note again how little oxygen is produced by what seems to be a lot of electricity. This shows why electrolytic production of gases and metals is so expensive.

19–7 Determination of the Charge on an Ion by Electrolysis

Faraday's Law can also be used to determine the charge on an ion. The charge is calculated by electrolyzing a solution containing the ion and relating the amount of the element produced to the number of coulombs of charge passing through the cell.

EXAMPLE 19–3

An aqueous solution of an unknown salt of palladium is electrolyzed by a current of 3.00 amperes passing for 1.00 hour. This results in the reduction of 2.977 grams of palladium ions at the cathode. What is the charge on the palladium ions in this solution?

Solution

As in previous examples, we begin by determining the number of coulombs of charge passing through the cell:

$$\underline{?}\ \text{coul} = 1.00\ \text{hr} \times \frac{60\ \text{min}}{1\ \text{hr}} \times \frac{60\ \text{s}}{1\ \text{min}} \times \frac{3.00\ \text{coul}}{1\ \text{s}} = 1.08 \times 10^4\ \text{coul}$$

Thus 1.08×10^4 coulombs reduce 2.977 grams of palladium ions. The equation for the reduction and the relationship between the number of coulombs and the number of grams of palladium reduced is

$$\begin{array}{lllll} Pd^{n+} & + & ne^- & \longrightarrow & Pd \quad \text{(cathode)} \\ 1\ \text{mol} & & n(6.02 \times 10^{23})e^- & & 1\ \text{mol} \\ 106.4\ \text{g} & & n(96{,}500\ \text{coul}) & & 106.4\ \text{g} \end{array}$$

In the following equation the only unknown is n, the charge on the ion, and we solve for n.

$$\underbrace{2.977\ \text{g Pd}}_{\text{amount of Pd reduced}} = \underbrace{1.08 \times 10^4\ \text{coul}}_{\text{electricity used}} \times \underbrace{\frac{106.4\ \text{g Pd}}{n\ \text{mol}\ e^-\left(96{,}500\ \dfrac{\text{coul}}{\text{mol}\ e^-}\right)}}_{\text{proportionality factor}}$$

$$n = \frac{(1.08 \times 10^4\ \text{coul})(106.4\ \text{g Pd})}{(2.977\ \text{g Pd})(96{,}500\ \text{coul})} = 4$$

The palladium ions are Pd^{4+}. Note that if the charge on an ion is known, then its atomic weight, and therefore its identity, can be determined by the same means.

19–8 Electrolytic Refining and Electroplating of Metals

Electrolytic reduction is the only means by which many active metals can be obtained (Section 20–3). Other less active metals that can be obtained from their ores by less expensive chemical reduction, and metals that occur in nature in the native (uncombined) state, are frequently purified or refined by electrolysis. The electrolytic method used for refining metals is also called *electroplating* when applied to plating a metal onto a surface. For example, impure metallic copper obtained from the chemical reduction of Cu_2S and CuS (Section 20–7) is purified by using an electrolytic cell like the one shown in Figure 19–6. In this cell, thin sheets of very pure copper are made cathodes by connecting them to the negative terminal of a direct-current generator. Impure chunks of copper are connected to the positive terminal and function as anodes. The electrodes are immersed in a solution of copper(II) sulfate and sulfuric acid. When the circuit is closed, copper from the impure anode is oxidized and goes into solution as Cu^{2+} ions, while Cu^{2+} ions from the solution are reduced and plate out as metallic copper on the pure copper cathodes. Other active metals from the impure bars also go into solution after oxidation, but they do not plate out onto the bars of pure copper because of the far greater concentration of the more easily reduced copper ions already in solution. Overall, there is no net reaction, merely a transfer of copper from anode to solution and from solution to cathode

$$\begin{array}{lll} \text{(impure) Cu} \longrightarrow Cu^{2+} + 2e^- & & \text{(anode)} \\ Cu^{2+} + 2e^- \longrightarrow \text{Cu (pure)} & & \text{(cathode)} \\ \hline \text{no net reaction} & & \end{array}$$

Although there is no net reaction, the net effect is that small bars of very pure copper and large bars of impure copper are converted into large bars of very pure copper and small bars of impure copper. The energy provided by the electric generator is used to force a decrease in the entropy of the system by separating the copper from the other ions in the impure bars.

A sludge called anode mud collects under the anodes. It contains such valuable and difficult-to-oxidize elements as gold, platinum, silver, selenium, and tellurium; the separation, purification and sale of these elements reduces the cost of refined copper. Copper can be plated onto other objects by the same mechanism (Figure 19–7).

Examples of metal-plated articles are common in our society. Jewelry and tableware are often plated with silver. Gold is plated on jewelry and electrical contacts. Automobiles have steel bumpers with thin films of chromium. The plating process for a typical bumper requires approximately three seconds of electrolysis, to produce a smooth, shiny surface only 0.0002 mm thick. Rapid plating of metal results in rough, grainy, black surfaces due to the many discontinuities of the metallic lattices produced, because the atoms are deposited

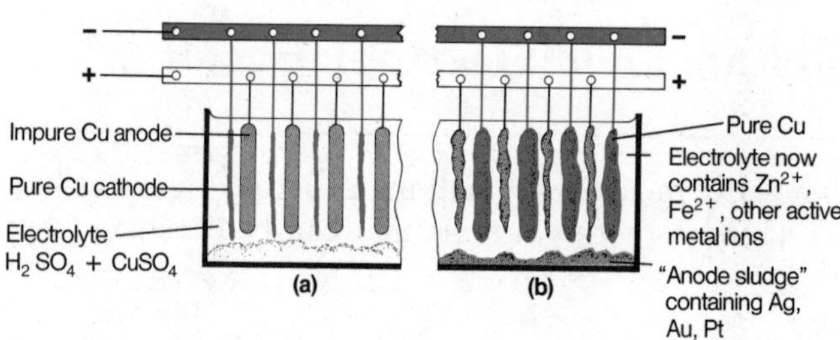

FIGURE 19–6 A schematic of the electrolytic cell used for refining copper. (a) Its appearance before electrolysis; (b) after electrolysis.

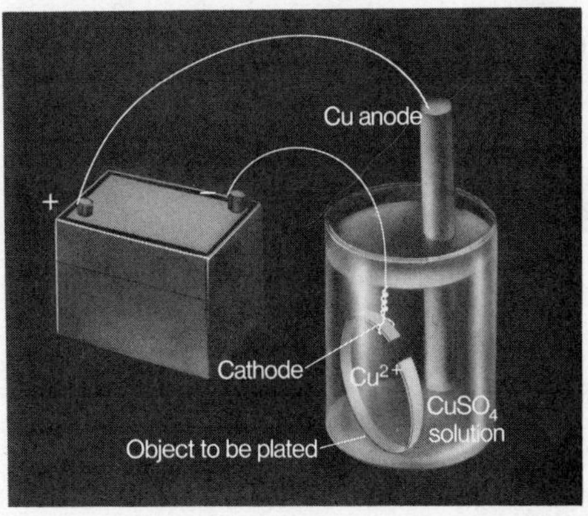

FIGURE 19–7 Electroplating with copper. The anode is made of pure copper, which dissolves during the electroplating process, keeping the concentration of Cu^{2+} constant.

from ions so rapidly that they are not able to form extended lattices. Slower plating produces smooth surfaces. "Tin cans" are steel cans plated electrolytically with tin; these may now be replaced by cans plated in one third of a second with a chromium film only a few atoms thick.

Voltaic or Galvanic Cells

Voltaic or **galvanic cells** are electrochemical cells in which *spontaneous* oxidation-reduction reactions produce electrical energy. The two halves of the redox reaction are separated, so that electron transfer is forced to occur by passage of the electrons through an external circuit. In this way, useful electrical energy is obtained. Everyone is familiar with some voltaic cells. The dry cells commonly used in flashlights, transistor radios, photographic equipment, and many toys and appliances are voltaic cells. Automobile batteries are also voltaic cells. We shall consider first some simple laboratory cells used to measure the potential difference, or voltage, as it is sometimes called, of a reaction under study, and then look at some more common types of galvanic cells.

Named for Alessandro Volta and Luigi Galvani, two Italian physicists of the 18th century.

19–9 The Construction of Simple Voltaic Cells

A **half-cell** contains the oxidized and reduced forms of an element, or other more complex species, in contact with each other. A common kind of half-cell consists of a piece of metal (the electrode) immersed in a solution of its ions. Consider two such half-cells in separate beakers, using two different elements (Figure 19–8). Electrical contact between the two solutions is made by a **salt bridge.** A salt bridge can be any medium through which ions can slowly pass; a most useful one may be prepared by bending a piece of glass tubing into the shape of a "U," filling it with hot saturated potassium chloride/5% agar solution, and allowing it to cool. The cooled mixture "sets" to the consistency of firm gelatin, so the solution does not run out when the tube is inverted (see Figure 19–8). A salt bridge serves two functions: it avoids mixing the electrode solutions, and it makes electrical contact between the two solutions, thereby enabling an electrical circuit to be completed.

Agar is a gelatinous material obtained from algae.

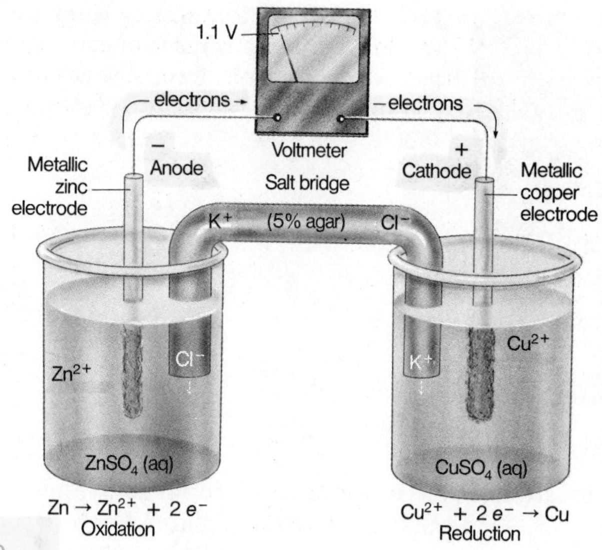

FIGURE 19–8 The zinc/copper voltaic cell utilizes the reaction

$$Zn\ (s) + Cu^{2+}\ (aq) \rightarrow Zn^{2+}\ (aq) + Cu\ (s)$$

The potential of this cell is 1.1 volts.

standard electrode - in TERMS!

A cell in which the reactants are in their thermodynamic standard states (one molar for dissolved ions and one atmosphere partial pressure for gases) is called a **standard cell.**

19–10 The Zinc-Copper Cell

Consider a standard cell that consists of two half-cells, one with a strip of metallic copper immersed in 1.0 *M* copper(II) sulfate solution, and the other with a strip of zinc immersed in 1.0 *M* zinc sulfate solution. The electrodes are connected by a wire, and the solutions by a salt bridge. A voltmeter can be inserted into the circuit to measure the potential difference between the two electrodes, or an ammeter can be inserted to measure the flow of electricity. The electrical current is the direct result of the spontaneous redox reaction that occurs, and we seek to measure the potential difference under standard state conditions.

Neither of these instruments generates electrical energy. •

When the copper-zinc cell is constructed as described, the following experimental observations can be made:

1. The initial voltage is 1.10 volts.
2. The mass of the copper electrode increases, and the concentration of Cu^{2+} decreases in the solution around the copper electrode as the cell operates.
3. The mass of the zinc electrode decreases, and the concentration of Zn^{2+} increases in the solution around the zinc electrode as the cell operates.

From these observations we deduce that the half-reaction at the cathode is the reduction of copper(II) ions to copper metal, which plates out on the copper electrode. The zinc electrode is the anode. It loses mass because the zinc metal is oxidized to zinc ions, which go into solution.

$$Zn \longrightarrow Zn^{2+} + 2e^- \qquad \text{(anode)}$$
$$Cu^{2+} + 2e^- \longrightarrow Cu \qquad \text{(cathode)}$$
$$Cu^{2+} + Zn \longrightarrow Cu + Zn^{2+} \qquad \text{(overall cell reaction)}$$

This is the cell that was shown in Figure 19–8. Electrons are released at the anode and consumed at the cathode. Therefore they flow through the wire from anode to cathode, as is the case in all electrochemical cells. Since the electrons flow spontaneously in all voltaic cells, they go from the negative electrode to the positive electrode. So, in contrast to electrolytic cells, the anode must be negative and the cathode must be positive. In order to maintain electroneutrality and complete the circuit, two Cl^- ions from the salt bridge migrate into the anode solution for every Zn^{2+} ion formed. Simultaneously, two K^+ ions migrate into the cathode solution to replace every Cu^{2+} ion reduced. Neither K^+ nor Cl^- ions is oxidized or reduced in preference to the zinc metal or Cu^{2+} ions.

Compare the +/–, anode/cathode, and oxidation/reduction labels and the direction of electron flow in Figures 19–2 and 19–8.

Voltaic cells are frequently represented in shorthand form, as illustrated below for the cell just considered.

species (and concentrations) in contact with electrode surfaces

salt bridge

$$Zn/Zn^{2+}\ (1.0\ M)\ \|\ Cu^{2+}\ (1.0\ M)/Cu$$

electrode surfaces

The same net reaction occurs when a piece of zinc is dropped into a blue solution of copper(II) sulfate. The zinc dissolves; the blue color of Cu^{2+} disappears; and copper forms on the zinc and then settles to the bottom of the container. But no electricity flows because the two half-reactions are not physically separated as they are in the cell we have just described.

19–11 The Copper-Silver Cell

Let us now consider a similar standard voltaic cell consisting of a strip of copper immersed in 1.0 M copper(II) sulfate and a strip of silver immersed in 1.0 M silver nitrate, $AgNO_3$, solution. A wire and a salt bridge complete the circuit. The following experimental observations can be made:

1. The initial voltage of the cell is 0.46 volt.
2. The mass of the copper electrode decreases, and the Cu^{2+} ion concentration increases in the solution around the copper electrode.
3. The mass of the silver electrode increases, and the Ag^+ ion concentration decreases in the solution around the silver electrode.

We can tell that the copper electrode is the anode in this cell because copper metal is oxidized to Cu^{2+}. The silver electrode is the cathode because Ag^+ ions are reduced to metallic silver.

$$\begin{array}{ll} Cu \longrightarrow Cu^{2+} + 2e^- & \text{(anode)} \\ 2(Ag^+ + e^- \longrightarrow Ag) & \text{(cathode)} \\ \hline Cu + 2Ag^+ \longrightarrow Cu^{2+} + 2Ag & \text{(overall cell reaction)} \end{array}$$

Chloride and potassium ions from the salt bridge migrate into the anode and cathode solutions, respectively, to maintain electroneutrality and to complete

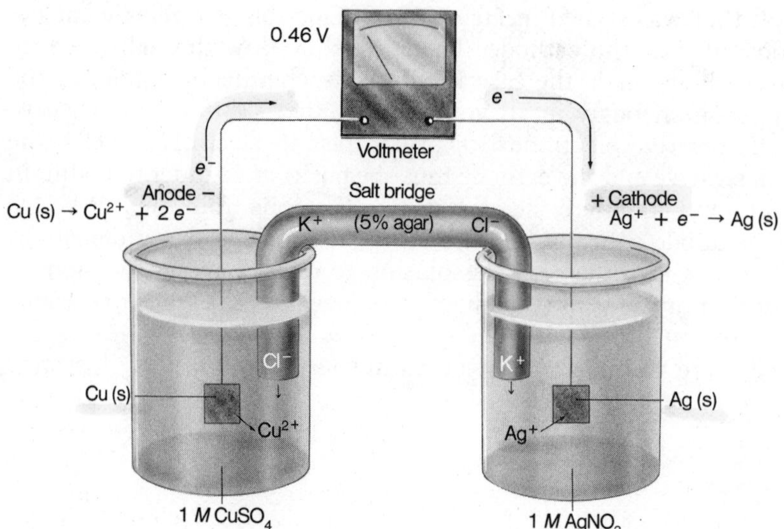

FIGURE 19–9 The copper/silver voltaic cell utilizes the reaction

$$Cu\ (s) + 2Ag^{+}\ (aq) \rightarrow Cu^{2+}\ (aq) + 2Ag\ (s)$$

The potential of this cell is 0.46 volt.

the circuit. Some NO_3^- ions (from the cathode vessel) and some Cu^{2+} ions (from the anode vessel) also migrate into the salt bridge. The cell is diagrammed in Figure 19–9.

Recall that in the zinc-copper cell the *copper electrode* is the *cathode,* while in the copper-silver cell the *copper electrode* is the *anode.* Whether a particular electrode behaves as an anode or a cathode depends on what the other electrode of the cell is. The two cells we have described demonstrate that the Cu^{2+} ion is a stronger oxidizing agent than Zn^{2+}, and Cu^{2+} oxidizes metallic zinc to Zn^{2+}. By contrast, silver ion is a stronger oxidizing agent than Cu^{2+} ion, and Ag^+ oxidizes copper atoms to Cu^{2+}. Conversely, metallic zinc is a stronger reducing agent than metallic copper, and metallic copper is a stronger reducing agent than metallic silver. We can now arrange the species we have studied thus far in order of increasing strength as oxidizing agents and as reducing agents.

Oxidizing Agents	Reducing Agents
$Zn^{2+} < Cu^{2+} < Ag^+$	$Ag < Cu < Zn$
increasing strength →	increasing strength →

Standard Electrode Potentials

The potentials of the standard zinc-copper and copper-silver voltaic cells are 1.10 volts and 0.46 volts, respectively. The magnitude of a cell's potential is a direct measure of the spontaneity of its redox reaction. Higher (more positive) cell potentials indicate greater spontaneity. Under standard conditions the oxidation of metallic zinc by Cu^{2+} ions has a greater tendency to go toward completion than does the oxidation of metallic copper by Ag^+ ions. It is desirable to separate the individual contributions to the total cell potential of the two electrode half-reactions. In this way we can determine the relative tendencies of particular oxidation or reduction half-reactions to occur. Such information gives us a quantitative basis for specifying strengths of oxidizing and reducing agents.

19–12 The Standard Hydrogen Electrode

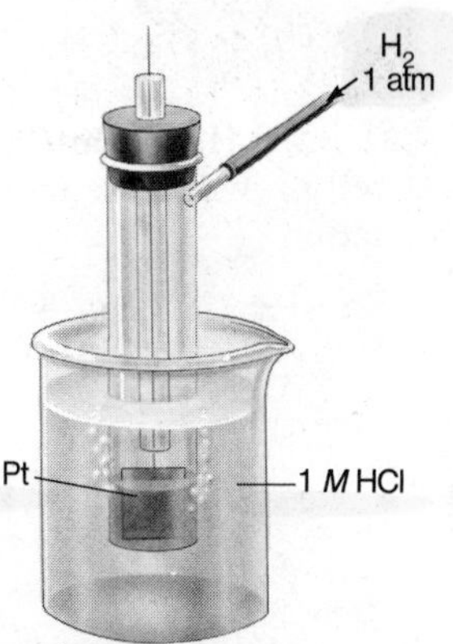

FIGURE 19–10 The standard hydrogen electrode.

It is not possible to determine experimentally the potential of a single electrode, because every oxidation must be accompanied by a reduction (that is, the electrons must have somewhere to go). Therefore, it is necessary to establish some arbitrary standard. By international agreement, the reference electrode is the **standard hydrogen electrode (SHE).** A standard hydrogen electrode contains a piece of metal electrolytically coated with a grainy black surface of inert platinum metal, immersed in a 1.0 *M* H^+ solution. Hydrogen, H_2, is bubbled at one atmosphere pressure through a glass envelope over the platinized electrode (Figure 19–10). By international agreement the standard hydrogen electrode is arbitrarily assigned a potential of exactly 0.0000 . . . volt.

The superscript in E^0 indicates standard conditions.

SHE half-reaction	**E^0 (standard electrode potential)**
$H_2 \longrightarrow 2H^+ + 2e^-$	0.0000 . . . V (SHE as anode)
$2H^+ + 2e^- \longrightarrow H_2$	0.0000 . . . V (SHE as cathode)

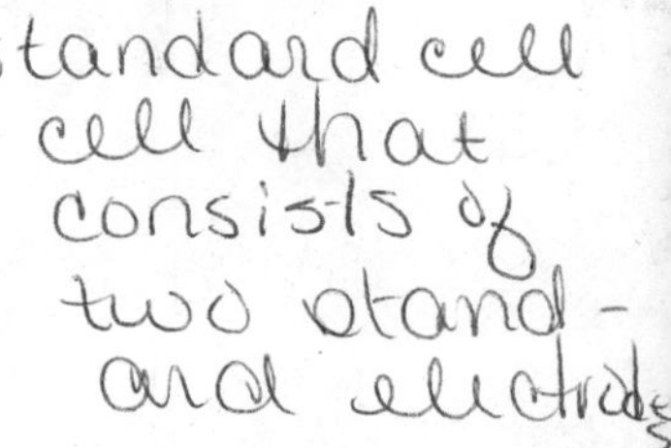

When a cell is constructed of a standard hydrogen electrode and some other standard electrode (half-cell), then the standard cell potential, E^0_{cell}, is arbitrarily taken as the *standard potential of the other electrode.* Let us now consider two cells, one consisting of the standard zinc electrode versus the SHE, and the other consisting of the standard copper electrode versus the SHE.

19–13 The Zinc-SHE Cell

This cell consists of an SHE in one beaker and a strip of zinc immersed in 1.0 *M* zinc sulfate solution in another beaker. A wire and salt bridge complete the circuit. When the circuit is closed, the following observations can be made:

1. As the cell operates, the mass of the zinc electrode decreases and the concentration of Zn^{2+} ion increases in the solution around the zinc electrode.
2. The H^+ concentration decreases in the SHE, and hydrogen is produced.
3. The initial potential of the cell is 0.763 volt.

We can conclude from these observations that the following half-reactions and cell reaction occur

		E^0
(anode)	$Zn \longrightarrow Zn^{2+} + 2e^-$	0.763 V
(cathode)	$2H^+ + 2e^- \longrightarrow H_2$	0.000 V
(cell reaction)	$Zn + 2H^+ \longrightarrow Zn^{2+} + H_2$	$E^0_{cell} = 0.763$ V

The standard potential at the anode *plus* the standard potential at the cathode gives the standard cell potential. Since the standard cell potential is found to be 0.763 volt, and the potential of the SHE is 0.000 volt, the standard potential of the zinc anode is 0.763 volt. The Zn/Zn^{2+} (1.0 *M*) ‖ H^+(1.0 *M*); H_2 (1 atm)/Pt cell is depicted in Figure 19–11.

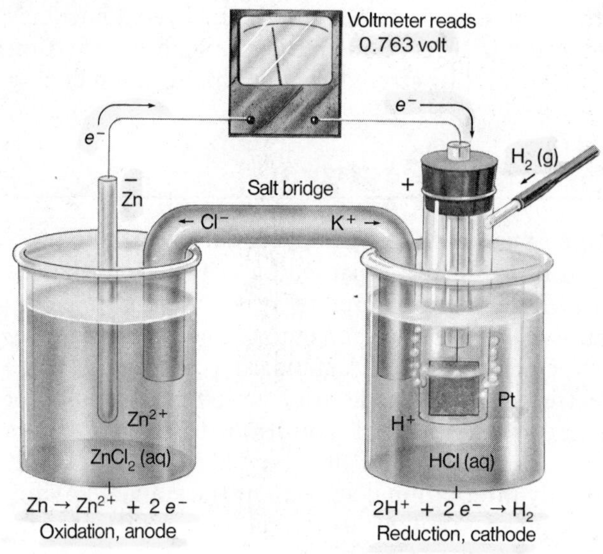

FIGURE 19–11 The Zn/Zn^{2+} (1.0 *M*)||H^+ (1.0 *M*); H_2 (1 atm)/Pt cell in which the reaction

$Zn\ (s) + 2H^+\ (aq) \rightarrow Zn^{2+}\ (aq) + H_2\ (g)$

occurs.

Note that in this cell the SHE is the *cathode,* and metallic zinc reduces H^+ to H_2. The zinc electrode is the anode in this cell as well as in the zinc-copper cell we examined earlier.

19–14 The Copper-SHE Cell

Consider a cell that consists of an SHE in one beaker and a strip of metallic copper immersed in a 1.0 *M* solution of copper(II) sulfate in another beaker. A wire and a salt bridge complete the circuit. The following observations can be made when the cell is in operation:

1. The mass of the copper electrode increases, and the concentration of Cu^{2+} ions decreases in the solution around the copper electrode.
2. Gaseous hydrogen is used up and the H^+ concentration increases in the solution of the SHE.
3. The initial cell potential is 0.337 volt.

Thus the following half-reactions and cell reaction occur.

		E^0
(anode)	$H_2 \longrightarrow 2H^+ + 2e^-$	0.000 V
(cathode)	$Cu^{2+} + 2e^- \longrightarrow Cu$	0.337 V
(cell reaction)	$H_2 + Cu^{2+} \longrightarrow 2H^+ + Cu$	$E^0_{cell} = 0.337$ V

The cell is shown in Figure 19–12.

The SHE functions as the *anode* in this cell, and Cu^{2+} ions oxidize H_2 to H^+ ions. (In the Zn-SHE cell, the SHE functions as the *cathode.*) The standard electrode potential of the copper electrode is 0.337 volt as a *cathode* in the Cu-SHE cell.

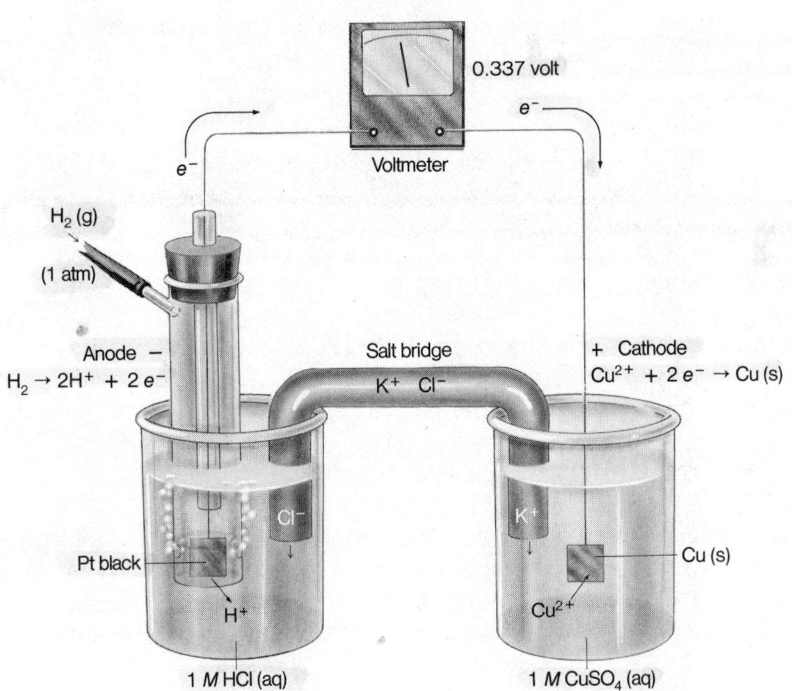

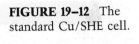

FIGURE 19–12 The standard Cu/SHE cell.

19–15 The Electromotive Series or Activity Series of the Elements

By measuring the potentials of several other standard electrodes versus the SHE in the same manner as described for the standard Zn-SHE and standard Cu-SHE voltaic cells, a series of standard electrode potentials can be established. When the electrodes involve metals or nonmetals in contact with their ions, the resulting series is called the **electromotive series** or **activity series** of the elements. In Section 19–13 we deduced that the standard zinc electrode behaves as the anode versus the SHE and that the standard *oxidation* potential for the zinc electrode is 0.763 volt.

		$E^0_{\text{oxidation}}$
(as anode)	$Zn \longrightarrow Zn^{2+} + 2e^-$	$+0.763$ V
in general:	reduced form $\longrightarrow$ oxidized form + ne^-	standard oxidation potential

Therefore, the *reduction* potential for the standard zinc electrode (to act *as a cathode* relative to the SHE) is the negative of this, or −0.763 volt.

		$E^0_{\text{reduction}}$
(as cathode)	$Zn^{2+} + 2e^- \longrightarrow Zn$	-0.763 V
in general:	oxidized form + ne^- $\longrightarrow$ reduced form	standard reduction potential

By international convention, the standard potentials of electrodes are tabulated for *reduction half-reactions,* indicating the tendencies of the electrodes to behave as cathodes toward the SHE. Those with positive E^0 values for reduction half-reactions do in fact act as cathodes versus the SHE, while those with negative E^0 values for reduction half-reactions behave instead as anodes versus the SHE. Stated differently, *the more positive the* E^0 *value for a half-reaction is, the greater the tendency for the half-reaction to occur in the forward direction as written.* Conversely, the more *negative* the E^0 value for a half-reaction is, the greater the tendency for the half-reaction to occur in the *reverse* direction as written.

The electromotive series is shown in Table 19–2. The more positive the reduction potential is, the stronger the species on the left is as an oxidizing agent and the weaker the species on the right is as a reducing agent.

19–16 Uses of the Electromotive Series

The electromotive series can be used in many ways. One important application is in the determination of the spontaneity of redox reactions. It is not necessary that the series be used only with respect to reactions in electrochemical cells. Standard electrode potentials can be used to determine the spontane-

TABLE 19–2 Standard Aqueous Electrode Potentials at 25°C—The Electromotive Series*

Element	Electrode Reaction		Standard Reduction Potential E^0, volts
Li†	$Li^+ + e^- \rightleftharpoons Li$		−3.045
K	$K^+ + e^- \rightleftharpoons K$		−2.925
Ca	$Ca^{2+} + 2e^- \rightleftharpoons Ca$		−2.87
Na	$Na^+ + e^- \rightleftharpoons Na$		−2.714
Mg	$Mg^{2+} + 2e^- \rightleftharpoons Mg$		−2.37
Al	$Al^{3+} + 3e^- \rightleftharpoons Al$		−1.66
Zn	$Zn^{2+} + 2e^- \rightleftharpoons Zn$		−0.7628
Cr	$Cr^{3+} + 3e^- \rightleftharpoons Cr$		−0.74
Fe	$Fe^{2+} + 2e^- \rightleftharpoons Fe$		−0.44
Cd	$Cd^{2+} + 2e^- \rightleftharpoons Cd$		−0.403
Ni	$Ni^{2+} + 2e^- \rightleftharpoons Ni$		−0.25
Sn	$Sn^{2+} + 2e^- \rightleftharpoons Sn$		−0.14
Pb	$Pb^{2+} + 2e^- \rightleftharpoons Pb$		−0.126
H_2	$2H^+ + 2e^- \rightleftharpoons H_2$	(reference electrode)	0.000
Cu	$Cu^{2+} + 2e^- \rightleftharpoons Cu$		+0.337
I_2	$I_2 + 2e^- \rightleftharpoons 2I^-$		+0.535
Hg	$Hg^{2+} + 2e^- \rightleftharpoons Hg$		+0.789
Ag	$Ag^+ + e^- \rightleftharpoons Ag$		+0.7994
Br_2	$Br_2 + 2e^- \rightleftharpoons 2Br^-$		+1.08
Cl_2	$Cl_2 + 2e^- \rightleftharpoons 2Cl^-$		+1.360
Au	$Au^{3+} + 3e^- \rightleftharpoons Au$		+1.50
F_2	$F_2 + 2e^- \rightleftharpoons 2F^-$		+2.87

Increasing Strength as Oxidizing Agent ↓ (from Li⁺ down to F₂)

Increasing Strength as Reducing Agent ↑ (from F⁻ up to Li)

*Defined in accord with Josiah Gibbs, one of the founders of chemical thermodynamics, and the International Union of Pure and Applied Chemistry, IUPAC-GIBBS STOCKHOLM Sign Convention. Source: "Selected Constants, Oxidation-Reduction Potentials in Aqueous Solution," Gaston Charlot, *IUPAC Supplement,* 1971.

†For pure metals that react violently with water, an amalgam saturated with metal is used as the electrode.

ity of redox reactions in general, and whether or not the half-reactions are physically separated is of no importance. Before this application is illustrated, we must consider two further points about conventions used in the series.

1. The species on the *left* side are all either cations of metals or hydrogen, or elemental nonmetals. These are all *oxidizing agents* (or *oxidized forms* of the elements). Their strengths as oxidizing agents increase from top to bottom, i.e., as the $E^0_{\text{reduction}}$ values become more positive. Fluorine is the strongest oxidizing agent, and Li^+ is a very weak oxidizing agent. (As we shall see in Section 19–17, it is possible to expand a table such as this one to include oxidized and reduced forms, neither of which involve free elements. Such an expanded version is given in Appendix J.)
2. The species on the *right* side are free metals, hydrogen, or anions of nonmetals. These are all reducing agents (reduced forms of the elements). Their strengths as reducing agents increase from bottom to top, i.e., as the $E^0_{\text{reduction}}$ values become more negative. Metallic Li is a very strong reducing agent, and F^- is a very weak reducing agent.

The half-reaction for the standard lithium electrode is

$$Li^+ + e^- \longrightarrow Li \qquad E^0 = -3.045 \text{ V}$$

The very negative E^0 value tells us that this half-reaction does not occur except under extreme circumstances. As a matter of fact, electrolytic methods in nonaqueous cells are required to reduce lithium ions to metallic lithium (Section 19–3). The half-reaction for the standard fluorine electrode is

$$F_2 + 2e^- \longrightarrow 2F^- \qquad E^0 = +2.87 \text{ V}$$

The very positive E^0 value tells us that this half-reaction occurs readily and fluorine, F_2, is the strongest of all common oxidizing agents.

Note that the elements with high ionization energies and highly negative electron affinities have the greatest tendencies to exist as anions. The elements with low ionization energies and positive or slightly negative electron affinities have the greatest tendencies to exist as cations.

The following example will illustrate how E^0 values can be used to predict the spontaneity of a redox reaction.

EXAMPLE 19–4

Will copper(II) ions oxidize metallic zinc to zinc ions, or will Zn^{2+} ions oxidize metallic copper to Cu^{2+}?

Solution

One of the two possible reactions is spontaneous, and the reverse reaction is nonspontaneous. We must determine which one is spontaneous. We already know the answer to this question from experimental results (Section 19–10), but let us demonstrate the procedure for determining the spontaneous reaction.

1. Choose the appropriate half-reactions from a table of standard reduction potentials.
2. Write the equation for the reaction with the more positive E^0 value for reduction first, along with its potential.

3. Then write the equation for the other half-reaction *as an oxidation* and write its *oxidation potential;* that is, reverse the tabulated reduction half-reaction and change the sign of E^0. (Reversing a half-reaction or a complete reaction also changes the sign of its potential.)
4. Balance the electron transfer.
5. Add the reduction and oxidation half-reactions and add the reduction and oxidation potentials. E^0_{cell} *will be positive* for the resulting overall cell reaction. *This indicates that the forward reaction is spontaneous.* (Of course, any overall cell reaction for which E^0_{cell} is negative is nonspontaneous.)

Following these steps, we obtain the equation for the spontaneous reaction

		E^0	
	$1(Cu^{2+} + 2e^- \longrightarrow Cu)$ reduction	+0.337 V	
Zn^{2+}/Zn couple has the less positive reduction potential $\longrightarrow$	$1(Zn \longrightarrow Zn^{2+} + 2e^-)$ oxidation	+0.763 V $\longleftarrow$	oxidation potential
	$Cu^{2+} + Zn \longrightarrow Cu + Zn^{2+}$	$E^0_{cell} = +1.10$ V	
	spontaneous reaction		

The fact that E^0_{cell} is positive tells us that the forward reaction is spontaneous. Reference to Section 19–10 shows that 1.10 volts is, indeed, the potential of the standard zinc-copper voltaic cell and that this is the spontaneous reaction that occurs. Copper(II) ions oxidize metallic zinc to zinc ions and in turn are reduced to metallic copper.

In order to make the reverse reaction (nonspontaneous, negative E^0) occur, electrical energy would have to be supplied with a potential difference greater than 1.10 volts.

$$Cu + Zn^{2+} \longrightarrow Cu^{2+} + Zn \qquad E^0_{cell} = -1.10 \text{ volts}$$

nonspontaneous reaction

EXAMPLE 19–5

Will chromium(III) ions, Cr^{3+}, oxidize metallic copper to copper(II) ions, Cu^{2+}, or will Cu^{2+} oxidize metallic chromium to Cr^{3+} ions?

Solution

As in Example 19–4, we refer to the table of standard reduction potentials and choose the two appropriate half-reactions. The copper half-reaction has the more positive (less negative) reduction potential, so we write it first. Then we write the chromium half-reaction as an oxidation, balance the electron transfer, and add the second half-reaction, and its oxidation potential, to the first. *Note that the potentials are not multiplied by the numbers (2 and 3) used to balance the electron transfer!*

		E^0
$3(Cu^{2+} + 2e^- \longrightarrow Cu)$	reduction	+0.337 V
$2(Cr \longrightarrow Cr^{3+} + 3e^-)$	oxidation	+0.74 V
$2Cr + 3Cu^{2+} \longrightarrow 2Cr^{3+} + 3Cu$		$E^0_{cell} = +1.08$ V

This is the spontaneous reaction because E^0_{cell} is positive. Copper(II) ions spontaneously oxidize metallic chromium to chromium(III) ions and are reduced to metallic copper.

19–17 Electrode Potentials for More Complex Half-Reactions

It is possible to construct half-cells in which both oxidized and reduced species are in solution as ions in contact with inert electrodes. For example, the standard iron(III) ion/iron(II) ion half-cell contains 1.0 *M* concentrations of the two ions and involves the following half-reaction

$$Fe^{3+} + e^- \longrightarrow Fe^{2+} \qquad E^0 = +0.771 \text{ V}$$

Another example is the standard dichromate ($Cr_2O_7{}^{2-}$) ion/chromium(III) ion half-cell, which consists of 1.0 *M* concentrations of each of the two ions in contact with an inert electrode. The balanced half-reaction in acidic solution (1.0 *M* H^+) is

$$Cr_2O_7{}^{2-} + 14H^+ + 6e^- \longrightarrow 2Cr^{3+} + 7H_2O \qquad E^0 = +1.33 \text{ V}$$

Other examples of more complex standard electrode potentials are given in Table 19–3 and in Appendix J.

TABLE 19–3 Standard Electrode Potentials For Selected Half-Cells

Electrode Reaction (Reduction)	Standard Electrode Potential E^0, volts
$Zn(OH)_4{}^{2-} + 2e^- \rightleftharpoons Zn + 4OH^-$	−1.22
$Fe(OH)_2 + 2e^- \rightleftharpoons Fe + 2OH^-$	−0.877
$2H_2O + 2e^- \rightleftharpoons H_2 + 2OH^-$	−0.8277
$PbSO_4 + 2e^- \rightleftharpoons Pb + SO_4{}^{2-}$	−0.356
$NO_3{}^- + H_2O + 2e^- \rightleftharpoons NO_2{}^- + 2OH^-$	+0.01
$Sn^{4+} + 2e^- \rightleftharpoons Sn^{2+}$	+0.15
$AgCl + e^- \rightleftharpoons Ag + Cl^-$	+0.222
$Hg_2Cl_2 + 2e^- \rightleftharpoons 2Hg + 2Cl^-$	+0.27
$O_2 + 2H_2O + 4e^- \rightleftharpoons 4OH^-$	+0.40
$NiO_2 + 2H_2O + 2e^- \rightleftharpoons Ni(OH)_2 + 2OH^-$	+0.49
$Fe^{3+} + e^- \rightleftharpoons Fe^{2+}$	+0.771
$ClO^- + H_2O + 2e^- \rightleftharpoons Cl^- + 2OH^-$	+0.89
$NO_3{}^- + 4H^+ + 3e^- \longrightarrow NO + 2H_2O$	+0.96
$Cr_2O_7{}^{2-} + 14H^+ + 6e^- \rightleftharpoons 2Cr^{3+} + 7H_2O$	+1.33
$MnO_4{}^- + 8H^+ + 5e^- \rightleftharpoons Mn^{2+} + 4H_2O$	+1.51
$PbO_2 + SO_4{}^{2-} + 4H^+ + 2e^- \rightleftharpoons PbSO_4 + 2H_2O$	+1.685

These electrode potentials are analogous to those of the electromotive series, and can be used in similar fashion, as Example 19–6 illustrates.

EXAMPLE 19–6

Will tin(IV) ions, Sn^{4+}, oxidize gaseous nitrogen oxide, NO, to nitrate ions, $NO_3{}^-$, in acidic solution, or will $NO_3{}^-$ oxidize Sn^{2+} to Sn^{4+}?

Solution

The nitrate/nitrogen oxide half-reaction has the more positive E^0 value, so we write the tin(IV)/tin(II) half-reaction as an oxidation. The electron transfer is balanced, and the reduction and oxidation half-reactions are added to obtain the equation for the spontaneous reaction.

	E^0
$2(NO_3^- + 4H^+ + 3e^- \longrightarrow NO + 2H_2O)$	+0.96 V
$3(Sn^{2+} \longrightarrow Sn^{4+} + 2e^-)$	−0.15 V
$2NO_3^- + 8H^+ + 3Sn^{2+} \longrightarrow 2NO + 4H_2O + 3Sn^{4+}$	E^0_{cell} = +0.81 V

Since E^0_{cell} is positive for this reaction, nitrate ions spontaneously oxidize tin(II) ions to tin(IV) ions and are reduced to nitrogen oxide in acidic solution, when all species are present at unit activities.

Now that we are familiar with the use of standard reduction potentials, let us apply them to explain the reaction that occurs in the electrolysis of aqueous solutions of sodium chloride. Recall that the first two electrolytic cells we considered involved molten NaCl and aqueous NaCl. In the first case there was no doubt that metallic sodium would be produced by reduction of Na^+ and gaseous Cl_2 would be produced by oxidation of Cl^-. But in aqueous NaCl we found that H_2O, rather than Na^+, was reduced. This result is consistent with the considerably less negative reduction potential of H_2O as compared to Na^+.

	E^0
$2H_2O + 2e^- \longrightarrow H_2 + 2OH^-$	−0.828 V
$Na^+ + e^- \longrightarrow Na$	−2.714 V

The more easily reduced species, H_2O, is reduced.

19–18 Corrosion

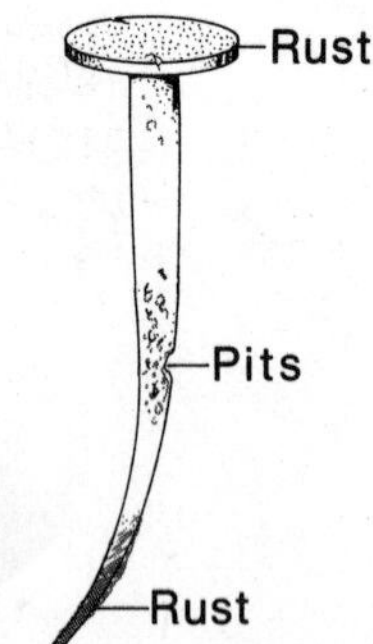

FIGURE 19–13 Corrosion of a bent nail at points of strain and "active" metal atoms.

In the formation of iron rust, Fe_2O_3, iron(II) is ultimately further oxidized to iron(III) ions.

Ordinary **corrosion** is the redox process by which metals are oxidized by oxygen, O_2, in the presence of moisture. There are other kinds, but this is the most common. The problems of corrosion and its prevention are of both theoretical and practical interest, because they are responsible for the loss of billions of dollars annually in metal products. The mechanism of corrosion has been studied extensively, and it is now known that the oxidation of metals occurs most readily at points of strain (where the metals are most "active"). Thus a steel nail, which is mostly iron (Section 20–6), first corrodes at the tip and the head (Figure 19–13). A bent nail is also susceptible to corrosion at the bend.

The end of a steel nail acts as an anode where the iron is oxidized to iron(II) ions

$$Fe \longrightarrow Fe^{2+} + 2e^- \quad \text{(anode)}$$

The electrons produced then flow along the nail to areas containing impurities or weak points, which act as cathodes where oxygen is reduced to hydroxide ions, OH^-, and pits are formed

$$O_2 + 2H_2O + 4e^- \longrightarrow 4OH^- \quad \text{(cathode)}$$

The overall reaction is obtained by balancing the electron transfer and adding the two half-reactions.

$$
\begin{array}{lll}
2(\mathrm{Fe} \longrightarrow \mathrm{Fe}^{2+} + 2e^-) & & \text{(oxidation occurs at anode)} \\
\mathrm{O_2} + 2\mathrm{H_2O} + 4e^- \longrightarrow 4\mathrm{OH}^- & & \text{(reduction occurs at cathode)} \\
\hline
2\mathrm{Fe} + \mathrm{O_2} + 2\mathrm{H_2O} \longrightarrow \underbrace{2\mathrm{Fe}^{2+} + 4\mathrm{OH}^-}_{2\mathrm{Fe(OH)_2}} & & \text{(net reaction)}
\end{array}
$$

The $Fe(OH)_2$ may be dehydrated to iron(II) oxide, FeO, or further oxidized to $Fe(OH)_3$ and then dehydrated to iron rust, Fe_2O_3.

19–19 Corrosion Protection

Several methods for protection of metals against corrosion have been developed. The most widely used methods are (1) plating the metal with a thin layer of a less easily oxidized metal, (2) connecting the metal directly to a "sacrificial anode," a piece of another metal that is more active and therefore preferentially oxidized, (3) allowing a protective film, such as a metal oxide, to form naturally on the surface of the metal, and (4) galvanizing, in which steel is coated with zinc, a more active metal.

The thin layer of tin in tin-plated steel cans is not easily oxidized, and it protects the steel underneath from corrosion. It is deposited by dipping the can into molten tin, or by electroplating. Copper is also less active than iron (see Table 19–2) and is sometimes used for protecting metals when food is not involved. It is deposited by electroplating. Whenever the layer of tin or copper is breached, the iron beneath it corrodes even more rapidly than it would without the coating, because of the adverse electrochemical cell set up.

Figure 19–14 shows an iron or steel pipe connected to a strip of magnesium, a more active metal, to protect the iron from oxidation. The magnesium is preferentially oxidized to magnesium ions and is called a "sacrificial anode." Zinc could be used for the same purpose. Similar methods are used to protect bridges and ships' hulls from corrosion.

Aluminum is a very active metal, so active that it reacts rapidly with oxygen from the air to form a surface layer of aluminum oxide, Al_2O_3, which is so thin that it is transparent. This is a very tough, hard substance that is inert toward oxygen, water, and most other corrosive agents in the environment. For this reason, objects made of aluminum form their own protective layer and need not be treated further to inhibit corrosion.

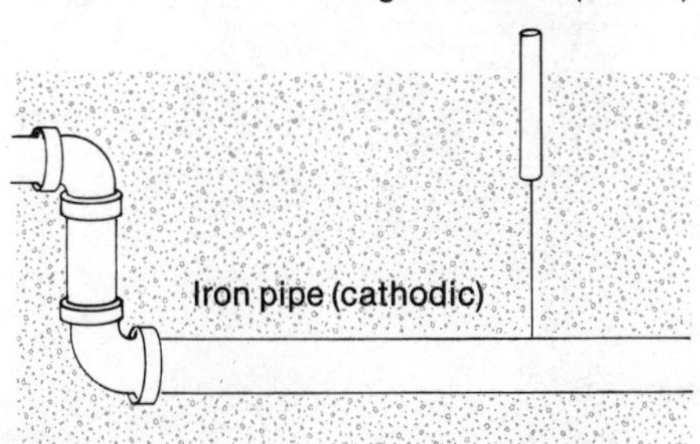

FIGURE 19–14 Cathodic protection of buried iron pipe. A magnesium or zinc bar is oxidized instead of the iron.

Effect of Concentrations (or Partial Pressures) on Electrode Potentials

19–20 The Nernst Equation

If we connect (by a salt bridge and a voltmeter) two half-cells consisting of strips of copper immersed in 1.0 *M* and 0.010 *M* Cu^{2+} ions, we observe a voltage of 0.059 volt. Of course, there would have been a potential of zero if both solutions had been 1.0 *M* in Cu^{2+} ion (or if both had been 0.010 *M*). Deviations of concentrations from 1.0 *M* or of partial pressures from 1.0 atm cause the corresponding potentials to deviate from standard electrode potentials.

Standard electrode potentials, designated E^0, refer to standard state conditions. These standard state conditions are one molar solutions for ions, pressures of one atmosphere for gases, and solids and liquids in their standard states at 25°C. It is important to remember that we refer to *thermodynamic* standard state conditions, and not standard conditions as in gas law calculations.

Developed by Walther Nernst (1864–1941).

The **Nernst equation** was derived to calculate electrode potentials for concentrations and partial pressures other than the standard state values. The Nernst equation is

In this equation, the expression to the right of the minus sign represents the deviation of electrode potential from the E^0 value due to the **nonstandard** conditions.

$$E = E^0 - \frac{2.303\ RT}{nF} \log Q$$

in which E = the electrode potential under the **nonstandard** conditions
E^0 = the **standard** electrode potential
R = the gas constant, 8.314 J/mol · K or 1.987 cal/mol · K
T = the absolute temperature, in K
n = the number of electrons transferred in the half-reaction
F = the faraday, 96,487 coul/mol e^- × 1 J/(V · coul)
= 96,487 J/V · mol e^- or 23,060 cal/(V · mol e^-)
Q = a term analogous to the reaction quotient (Section 15–4) in which dissolved species are expressed in instantaneous molar concentrations and gases in instantaneous partial pressures (in atm) in the same expression

Substituting the values given above into the Nernst equation at 25°C simplifies the equation to

$$E = E^0 - \frac{0.0592}{n} \log Q$$

Recall that this reaction quotient, *Q*, has the same form as the thermodynamic equilibrium constant, *K*, except that the concentrations or partial pressures are not necessarily those at equilibrium concentrations. For the familiar half-reaction involving metallic zinc and zinc ions

$$Zn + 2e^- \rightleftharpoons Zn^{2+} \qquad E^0 = +0.763 \text{ volt}$$

the corresponding Nernst equation is

Since metallic Zn is a solid, its concentration does not appear in *Q*.

$$E = E^0 - \frac{0.0592}{2} \log \frac{[Zn^{2+}]}{1} \qquad \text{(for reduction)}$$

Substituting the E^0 value into the equation

$$E = +0.763 \text{ V} - \frac{0.0592}{2} \log \frac{[Zn^{2+}]}{1}$$

Note that under standard conditions of $[Zn^{2+}] = 1.0$ M, the electrode potential, E, is equal to the standard electrode potential, E^0, because the correction factor is equal to zero

$$E = +0.763 \text{ V} - \frac{0.0592}{2} \log 1.0 \qquad \log(1.0) = 0$$

$$E = +0.763 \text{ V} - 0 \text{ V}$$

$$E = E^0$$

Recall that half-reactions for standard reduction potentials are written

$$\text{Ox} + ne^- \longrightarrow \text{Red}$$

where Red refers to the reduced species and Ox refers to the oxidized species. The Nernst equation for any cathode half-cell, where the reduction half-reaction occurs, is

$$E = E^0 - \frac{0.0592}{n} \log \frac{[\text{Red}]}{[\text{Ox}]}$$

EXAMPLE 19–7

Calculate the (reduction) potential for the Fe^{3+}/Fe^{2+} electrode if the concentration of Fe^{2+} is five times that of Fe^{3+}.

Solution

The reduction half-reaction is

$$Fe^{3+} + e^- \longrightarrow Fe^{2+} \qquad E^0 = +0.771 \text{ V}$$

First we use the given information to calculate the value of Q.

$$Q = \frac{[Fe^{2+}]}{[Fe^{3+}]} = \frac{5[Fe^{3+}]}{[Fe^{3+}]} = 5$$

Now we substitute this into the Nernst equation with $n = 1$

$$E = E^0 - \frac{0.0592}{n} \log Q = +0.771 - \frac{0.0592}{1} \log 5$$

$$E = +0.771 - 0.0592(0.699) = (+0.771 - 0.041) \text{ V}$$

$$E = +0.730 \text{ V}$$

EXAMPLE 19–8

Calculate E for the Fe^{3+}/Fe^{2+} electrode when the Fe^{3+} concentration is five times the Fe^{2+} concentration.

Solution

$$Q = \frac{[Fe^{2+}]}{[Fe^{3+}]} = \frac{1}{5}$$

$$E = +0.771 - \frac{0.0592}{1} \log \left(\frac{1}{5}\right)$$

$$E = +0.771 - 0.0592(-0.699)$$

$$E = (+0.771 + 0.041)\ \text{V}$$

$$E = \underline{+0.812\ \text{V}} \quad \text{(for reduction)}$$

Note that the correction factor, +0.041 volt, differs from that in Example 19–7 only in sign.

EXAMPLE 19–9

Calculate the potential of the chlorine/chloride ion, Cl_2/Cl^-, electrode when the partial pressure of Cl_2 is 10.0 atm and $[Cl^-] = 1.00 \times 10^{-3}\ M$.

Solution

The half-reaction and standard reduction potential are

$$Cl_2 + 2e^- \longrightarrow 2Cl^- \qquad E^0 = +1.360\ \text{volt}$$

Remember that in evaluating Q in the Nernst equation: (1) molar concentrations are used for dissolved species, and (2) partial pressures of gases are expressed in atmospheres.

The appropriate Nernst equation is

$$E = +1.360 - \frac{0.0592}{2} \log \frac{[Cl^-]^2}{P_{Cl_2}}$$

Substituting $[Cl^-] = 1.00 \times 10^{-3}\ M$ and $P_{Cl_2} = 10.0$ atm into the equation,

$$E = +1.360 - \frac{0.0592}{2} \log \frac{(1.00 \times 10^{-3})^2}{10.0}$$

$$E = +1.360 - \frac{0.0592}{2} \log (1.00 \times 10^{-7})$$

$$E = +1.360 - \frac{0.0592}{2} (-7.00) = [+1.360 - (-0.207)]\ \text{V}$$

$$E = \underline{+1.567\ \text{V}}$$

Example 19–10 illustrates how the Nernst equation can be applied to balanced equations for redox reactions, just as we applied it to equations for half-reactions.

EXAMPLE 19–10

Calculate the cell potential for the following cell: one electrode consists of the Fe^{3+}/Fe^{2+} couple in which $[Fe^{3+}] = 1.00\ M$ and $[Fe^{2+}] = 0.100\ M$; the other involves the MnO_4^-/Mn^{2+} couple in acidic solution in which $[MnO_4^-] = 1.00 \times 10^{-2}\ M$, $[Mn^{2+}] = 1.00 \times 10^{-4}\ M$, and $[H^+] = 1.00 \times 10^{-3}\ M$.

Solution

We first determine the reaction that occurs and the standard cell potential, E^0_{cell}, in the usual way

		E^0
$MnO_4^- + 8H^+ + 5e^- \longrightarrow Mn^{2+} + 4H_2O$		+1.51 V
$5(Fe^{2+} \longrightarrow Fe^{3+} + e^-)$		−(+0.771) V
$MnO_4^- + 8H^+ + 5Fe^{2+} \longrightarrow Mn^{2+} + 4H_2O + 5Fe^{3+}$	$E^0_{cell} =$	+0.74 V

Now we apply the Nernst equation to this reaction

$$E_{cell} = E^0_{cell} - \frac{0.0592}{5} \log \frac{[Mn^{2+}][Fe^{3+}]^5}{[MnO_4^-][H^+]^8[Fe^{2+}]^5}$$

Substituting the appropriate values we get

$$E_{cell} = +0.74 \text{ V} - \frac{0.0592}{5} \log \frac{(1.00 \times 10^{-4})(1.00)^5}{(1.00 \times 10^{-2})(1.00 \times 10^{-3})^8(0.100)^5}$$

$$E_{cell} = +0.74 \text{ V} - \frac{0.0592}{5} \log (1.00 \times 10^{+27})$$

$$E_{cell} = +0.74 \text{ V} - \frac{0.0592}{5} (27.0) = (+0.74 - 0.32) \text{ V}$$

$$E_{cell} = +0.42 \text{ V}$$

The reaction shown is spontaneous under the stated conditions, with a potential of +0.42 volt.

19–21 Relationship of E^0_{cell} to ΔG^0 and K

In Section 15–10 the relationship between the standard Gibbs free energy change, ΔG^0, and the thermodynamic equilibrium constant, K, was studied

$$\Delta G^0 = -2.303RT \log K$$

There also exists a simple relationship between ΔG^0 and the standard cell potential, E^0_{cell}, for a redox reaction:

$$\Delta G^0 = -nFE^0_{cell}$$

ΔG^0 can also be thought of as the *maximum electrical work* that can be obtained from a redox reaction. In this equation, n is the number of electrons involved in the overall process, and F is the faraday, 96,487 J/V mol e^- (see Section 19–20).

Combining these two relationships for ΔG^0 enables us to relate E^0_{cell} values to equilibrium constants.

$$\underbrace{-nFE^0_{cell}}_{\Delta G^0} = \underbrace{-2.303RT \log K}_{\Delta G^0}$$

Solving for E^0_{cell}

$$E^0_{cell} = \frac{2.303RT \log K}{nF}$$

or for K

$$\log K = \frac{nFE^0_{cell}}{2.303RT} \qquad K = \text{antilog}\left(\frac{nFE^0_{cell}}{2.303RT}\right)$$

If any one of the three quantities ΔG^0, K, or E^0_{cell} is known, each of the other two can be calculated using one or more of these relationships. You should keep in mind the following correspondences for all redox reactions *under standard state conditions.*

Spontaneity of Forward Reaction	ΔG^0	K	E^0_{cell}	
Spontaneous	$-$	>1	$+$	
At equilibrium	0	1	0	(Standard state conditions)
Nonspontaneous	$+$	<1	$-$	

Recall that ΔG^0 is expressed in joules per mole of occurrences of the entire reaction as *written*. Here, we are asking for the number of joules of free energy change that would accompany the reaction of 2 moles of chromium and 3 moles of tin(IV) ions.

EXAMPLE 19–11

Calculate the standard Gibbs free energy change, ΔG^0, in J/mol at 25°C from standard electrode potentials for the reaction below.

$3Sn^{4+} + 2Cr \longrightarrow 3Sn^{2+} + 2Cr^{3+}$

Solution

To evaluate the potential of this cell, we simply look up the two appropriate half-reactions in a table of standard reduction potentials; then we reverse the one that corresponds to the oxidation and change the sign of its E^0 value. The standard reduction potential for the Sn^{4+}/Sn^{2+} couple is +0.15 volt, while that of the Cr^{3+}/Cr couple is −0.74 volt. Since the equation for the reaction shows Cr being oxidized to Cr^{3+}, the sign of the E^0 value for the Cr^{3+}/Cr couple is reversed. The overall reaction, the sum of the oxidation and reduction half-reactions, has a cell potential equal to the sum of the potentials of the two half-reactions as shown below.

		E^0
$3(Sn^{4+} + 2e^- \longrightarrow Sn^{2+})$		+0.15 V
$2(Cr \longrightarrow Cr^{3+} + 3e^-)$		−(−0.74 V)
$3Sn^{4+} + 2Cr \longrightarrow 3Sn^{2+} + 2Cr^{3+}$	$E^0_{cell} =$	+0.89 V

The positive value of E^0_{cell} indicates that the forward reaction is spontaneous. We now calculate ΔG^0 using the relationship

$$\Delta G^0 = -nFE^0_{cell}$$

$$\Delta G^0 = -(6)\underbrace{\left(\frac{96{,}500 \text{ coul}}{\text{mol}} \times \frac{1 \text{ J}}{\text{V} \cdot \text{coul}}\right)}_{96{,}500 \text{ J/V} \cdot \text{mol}}(+0.89 \text{ V})$$

$$\Delta G^0 = \underline{-5.2 \times 10^5 \text{ J/mol}} \quad \text{or} \quad -5.2 \times 10^2 \text{ kJ/mol}$$

EXAMPLE 19–12

Calculate the equilibrium constant, K, for the reaction in Example 19–11 at 25°C by (a) relating it to ΔG^0, and (b) relating it to E^0_{cell}

Solution

(a) The relationship between ΔG^0 and K is

$$\Delta G^0 = -2.303RT \log K$$

$$-5.2 \times 10^5 \text{ J/mol} = (-2.303)\left(8.314 \frac{\text{J}}{\text{mol} \cdot \text{K}}\right)(298 \text{ K}) \log K$$

The very large value of K is also consistent with the fact that the reaction is spontaneous.

$$\log K = \frac{-5.2 \times 10^5}{(-2.303)(8.314)(298)} = +91$$

$$K = \underline{1 \times 10^{91}}$$

(b) The relationship between E^0_{cell} and K is

$$\log K = \frac{nFE^0_{cell}}{2.303RT}$$

Substituting the known values into the equation, we obtain a numerical expression for K.

$$\log K = \frac{(6)\left(96{,}500\frac{\text{J}}{\text{V}\cdot\text{mol}}\right)(+0.89\text{ V})}{(2.303)\left(8.314\frac{\text{J}}{\text{mol}\cdot\text{K}}\right)(298\text{ K})} = +90$$

$$K = 1 \times 10^{90}$$

This result agrees with that obtained in (a) within round-off error. Both 1×10^{91} and 1×10^{90} are so very large that the difference between the two is insignificant. Therefore, at equilibrium, only very little Sn^{4+} and Cr are present, and we say that for all practical purposes, the reaction goes to completion.

$$K = \frac{[Cr^{3+}]^2[Sn^{2+}]^3}{[Sn^{4+}]^3} = 1 \times 10^{91}$$

Under nonstandard conditions the relationship between the Gibbs free energy change, ΔG, and the cell potential, E_{cell}, is similar to that at standard conditions.

$$\Delta G = -nFE_{cell} \qquad \text{(at nonstandard conditions)}$$

EXAMPLE 19–13

Calculate the equilibrium constant, K, at 25°C for the reaction below.

$$2Cu + PtCl_6^{2-} \longrightarrow 2Cu^+ + PtCl_4^{2-} + 2Cl^-$$

Solution

First we find the appropriate half-reactions. The Cu^+/Cu couple occurs as an oxidation in this reaction, so it is written as an oxidation and the sign of its tabulated E^0 value is changed. The electron transfer is balanced and the oxidation and reduction half-reactions are added. The resulting E^0_{cell} value can be used to calculate the equilibrium constant, K, which *does not change* with changing concentrations.

		E^0
$2(Cu \longrightarrow Cu^+ + e^-)$		$-(+0.521\text{ V})$
$PtCl_6^{2-} + 2e^- \longrightarrow PtCl_4^{2-} + 2Cl^-$		$+0.68\text{ V}$
$2Cu + PtCl_6^{2-} \longrightarrow 2Cu^+ + PtCl_4^{2-} + 2Cl^-$	$E^0_{cell} =$	$+0.16\text{ V}$

The equilibrium constant is calculated using the relationship

$$\log K = \frac{nFE^0_{cell}}{2.303RT}$$

$$\log K = \frac{(2)(96.5 \times 10^3\text{ J/V mol})(+0.16\text{ V})}{(2.303)(8.314\text{ J/mol}\cdot\text{K})(298\text{ K})} = 5.4$$

$$K = 2.5 \times 10^5$$

The equilibrium constant for a reaction does not change w/ concentrations of the reactions. >* To calculate the equilibrium constant we use E^0_{cell} value.

At equilibrium

$$K = \frac{[Cu^+]^2[PtCl_4^{2-}][Cl^-]^2}{[PtCl_6^{2-}]} = 2.5 \times 10^5$$

The forward reaction is spontaneous and the equilibrium lies far to the right.

EXAMPLE 19–14

Calculate ΔG at 25°C for the reaction of Example 19–13 at the concentrations indicated below.

$[PtCl_6^{2-}] = 1.00 \times 10^{-2}\ M$

$[Cu^+] = 1.00 \times 10^{-3}\ M$

$[Cl^-] = 1.00 \times 10^{-3}\ M$

$[PtCl_4^{2-}] = 2.00 \times 10^{-5}\ M$

Solution

The Gibbs free energy change, ΔG, is related to E_{cell} (not E^0_{cell}) by the relationship

$$\Delta G = -nFE_{cell}$$

We must evaluate E_{cell}. For the Cu^+/Cu couple as a reduction the Nernst equation is

Recall that n is the number of electrons transferred in the half-reaction, not in the overall reaction, when we deal with *individual* half-reactions.

$$E = +0.521 - \frac{0.0592}{1}\log\frac{1}{[Cu^+]}$$

$$E = +0.521 - \frac{0.0592}{1}\log\frac{1}{1.00 \times 10^{-3}} = [+0.521 - 0.0592\ (3.00)]\ \text{V}$$

$E = +0.343$ V as a reduction

For the other (reduction) half-reaction

$$E = +0.68 - \frac{0.0592}{2}\log\frac{[PtCl_4^{2-}][Cl^-]^2}{[PtCl_6^{2-}]}$$

$$E = +0.68 - \frac{0.0592}{2}\log\frac{(2.00 \times 10^{-5})(1.00 \times 10^{-3})^2}{1.00 \times 10^{-2}}$$

$$E = +0.68 - \frac{0.0592}{2}\log\ (2.00 \times 10^{-9})$$

$$E = +0.68 - \frac{0.0592}{2}(-8.699) = (+0.68 + 0.26)\ \text{V}$$

$E = +0.94$ V as a reduction

If both potentials are reduction potentials, then

Subtracting a reduction potential is equivalent to adding an oxidation potential.

$$E_{cell} = (E_{PtCl_6^{2-}/PtCl_4^{2-},Cl^-}) - (E_{Cu^+/Cu})$$

$$E_{cell} = (+0.94\ \text{V}) - (+0.343\ \text{V}) = +0.60\ \text{V}$$

We saw earlier that the forward reaction,

$$2Cu + PtCl_6^{2-} \longrightarrow 2Cu^+ + PtCl_4^{2-} + 2Cl^-$$

is spontaneous when all ions are present in 1.0 M concentrations ($E^0_{cell} = +0.16$ V).

Now, we see that it is even more spontaneous ($E_{cell} = +0.60$ V) under the conditions stated. The Gibbs free energy change under these conditions can now be calculated

$$\Delta G = -nFE_{cell}$$

$$\Delta G = -(2)\left(\frac{96.5 \times 10^3 \text{ J}}{\text{V mol}}\right)(0.60 \text{ V})$$

$$\Delta G = -1.2 \times 10^5 \text{ J/mol}$$

Primary Voltaic Cells

Primary voltaic cells are cells in which, once the chemicals have been consumed, further chemical action is not possible. The properties of the electrolytes and/or electrodes are such that they cannot be regenerated by reversing the current flow through the cell using an external direct current source of electrical energy. The most familiar examples of primary voltaic cells are ordinary "dry" cells that are used as energy sources in flashlights and other small appliances.

19–22 The Dry Cell (Leclanché Cell)

The dry cell was patented by Georges Leclanché in 1866 (Figure 19–15). The container of a dry cell is made of zinc, which also serves as one of the electrodes. The other electrode is a carbon rod in the center of the cell. The zinc container is lined with porous paper to separate it from the other materials of the cell. A moist mixture (the cell is not really dry) of ammonium chloride, NH_4Cl, manganese(IV) oxide, MnO_2, zinc chloride, $ZnCl_2$, and a porous, inert filler occupies the space between the paper-lined zinc container and the graphite rod. Dry cells are sealed with wax or other sealant to keep the moisture from evaporating. As the cell operates, after the electrodes are connected, the

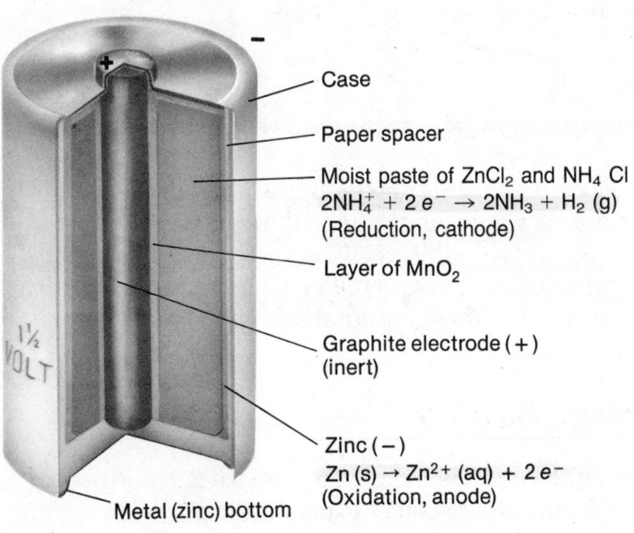

FIGURE 19–15 A representation of a typical dry cell that generates a potential difference of about 1.5 volts.

metallic zinc is oxidized to Zn^{2+} and the electrons liberated are left on the container. Thus the zinc electrode is the negative electrode (anode)

$$Zn \longrightarrow Zn^{2+} + 2e^- \qquad \text{(anode)}$$

The carbon rod is the cathode, at which ammonium ion is reduced

$$2NH_4{}^+ + 2e^- \longrightarrow 2NH_3 + H_2 \qquad \text{(cathode)}$$

Addition of the half-reactions gives the overall cell reaction

$$Zn + 2NH_4{}^+ \longrightarrow Zn^{2+} + 2NH_3 + H_2 \qquad E_{cell} = 1.6\ V$$

As hydrogen is formed, it is oxidized by the manganese(IV) oxide in the cell. This prevents collection of hydrogen on the cathode which would stop the reaction, a situation referred to as **polarization** of the electrode

$$H_2 + 2MnO_2 \longrightarrow 2MnO(OH)$$

The ammonia produced at the cathode combines with zinc ion and forms a soluble compound containing the complex ion, $Zn(NH_3)_4{}^{2+}$

$$Zn^{2+} + 4NH_3 \longrightarrow Zn(NH_3)_4{}^{2+}$$

This reaction prevents polarization due to build-up of ammonia, and it prevents the concentration of Zn^{2+} from increasing substantially, which would decrease the cell potential.

Alkaline dry cells are similar to ordinary dry cells with two exceptions. (1) the electrolyte is basic (alkaline) because it contains potassium hydroxide. (2) The interior surface of the zinc container is rough, which gives a larger surface area than if it were smooth. The reactions that occur in an alkaline dry cell during discharge are:

$$\begin{array}{ll} \text{(anode)} & Zn(s) + 2OH^-\ (aq) \longrightarrow Zn(OH)_2\ (s) + 2e^- \\ \text{(cathode)} & 2MnO_2\ (s) + 2H_2O\ (\ell) + 2e^- \longrightarrow 2MnO(OH)\ (s) + 2OH^-\ (aq) \\ \hline \text{(overall)} & Zn\ (s) + 2MnO_2\ (s) + 2H_2O\ (\ell) \longrightarrow Zn(OH)_2\ (s) + 2MnO(OH)\ (s) \end{array}$$

Alkaline batteries have longer shelf life than ordinary dry cells, and they stand up better under heavy use. The voltage of an alkaline battery is approximately 1.5 V.

Secondary Voltaic Cells

Secondary cells, or *reversible cells,* are cells in which the original reactants can be regenerated by passing a direct current through the cell in the direction opposite to the current flow when the cell is *discharging,* or producing electrical current. This process is referred to as *charging* or recharging a cell or battery. The lead storage battery, used in most automobiles, is the most common example of a secondary voltaic cell.

19–23 The Lead Storage Battery

The lead storage battery is depicted in Figure 19–16. The battery consists of a group of lead plates bearing compressed spongy lead, alternating with a group

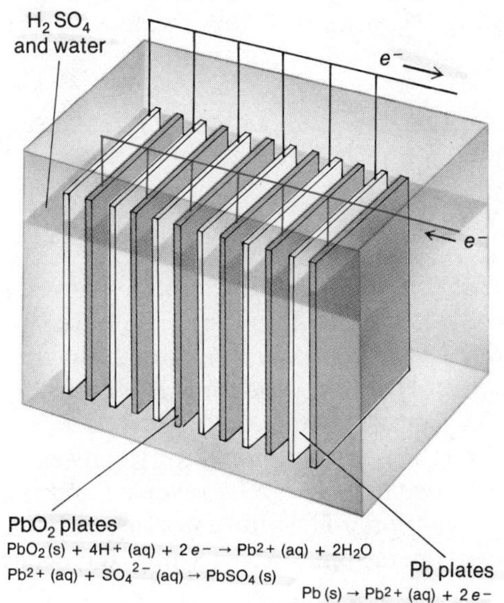

FIGURE 19–16 A schematic representation of a 12-volt lead storage battery. Alternate lead grids are packed with spongy lead and lead(IV) oxide and then connected in series. The grids are immersed in a solution of sulfuric acid, which serves as the electrolyte.

of lead plates bearing lead(IV) oxide, PbO_2. These electrodes are immersed in a solution of about 30% sulfuric acid. When the cell discharges, it operates as a voltaic cell. The spongy lead is oxidized to lead ions and the lead plates accumulate a negative charge

$$Pb \longrightarrow Pb^{2+} + 2e^- \qquad \text{(oxidation)}$$

The lead ions then combine with sulfate ions from the sulfuric acid to form insoluble lead(II) sulfate, which begins to coat the lead electrode

$$Pb^{2+} + SO_4^{2-} \longrightarrow PbSO_4\text{ (s)} \qquad \text{(precipitation)}$$

Thus the net process at the anode *when the cell delivers current* is

$$Pb + SO_4^{2-} \longrightarrow PbSO_4\text{ (s)} + 2e^- \qquad \text{(anode during discharge)}$$

The electrons produced at the lead electrode travel through the external circuit and re-enter the cell at the lead(IV) oxide, PbO_2, electrode, which is the cathode during discharge. Here, in the presence of hydrogen ions, the lead(IV) oxide is reduced to lead(II) ions, Pb^{2+}. These ions, of course, also combine with sulfate ions from the sulfuric acid to form additional insoluble lead sulfate, which coats the lead(IV) oxide electrode.

$$\begin{array}{ll} PbO_2 + 4H^+ + 2e^- \longrightarrow Pb^{2+} + 2H_2O & \text{(reduction)} \\ Pb^{2+} + SO_4^{2-} \longrightarrow PbSO_4\text{ (s)} & \text{(precipitation)} \\ \hline PbO_2 + 4H^+ + SO_4^{2-} + 2e^- \longrightarrow PbSO_4\text{ (s)} + 2H_2O & \text{(cathode during discharge)} \end{array}$$

The net cell reaction and its standard potential during discharge are obtained by adding the net anode and cathode half-reactions and their tabulated potentials. The tabulated E^0 value for the anode half-reaction is reversed in sign since it occurs as oxidation during discharge.

$$
\begin{array}{lr}
 & E^0 \\
Pb + SO_4^{2-} \longrightarrow PbSO_4\ (s) + 2e^- & -(-0.356\ V) \\
PbO_2 + 4H^+ + SO_4^{2-} + 2e^- \longrightarrow PbSO_4\ (s) + 2H_2O & +1.685\ V \\
\hline
Pb + PbO_2 + \underbrace{4H^+ + 2SO_4^{2-}}_{2H_2SO_4} \longrightarrow 2PbSO_4\ (s) + 2H_2O \quad E^0_{cell} = & +2.041\ V
\end{array}
$$

The decrease in the concentration of sulfuric acid provides an easy method for measuring the degree of discharge, because the density of the solution decreases proportionally; just measure the density of the solution with a hydrometer.

One cell creates a potential of about 2 volts. Most automobile batteries are 12-volt batteries that involve six cells connected in series. The potential declines only slightly in use since solid reagents are being consumed. Eventually the sulfuric acid is consumed, leaving only water. (It is for this reason that a discharged lead-acid battery can freeze, rupturing its case.)

When a potential slightly greater than the potential the battery can generate is imposed across the electrodes, the current flow can be reversed and the battery can be recharged by reversal of all reactions. The alternator or generator applies this potential when the engine is in operation. The reactions that occur in a lead storage battery are summarized below.

An *alternator* supplies alternating current, so a rectifier (an electronic device) is required to supply the necessary direct current to the battery. A *generator* supplies direct current.

$$Pb + PbO_2 + 2[2H^+ + SO_4^{2-}] \underset{\text{charge}}{\overset{\text{discharge}}{\rightleftharpoons}} 2PbSO_4\ (s) + 2H_2O$$

Note that sulfuric acid and the electrodes are consumed during discharge and lead sulfate is deposited. After many repeated charge-discharge cycles, some of the lead sulfate falls to the bottom of the container, the sulfuric acid concentration remains correspondingly low, and the battery cannot be recharged fully. It is then traded in on a new one, and the lead is recovered and reused to make new storage batteries.

This is one of the oldest and most successful examples of recycling.

19–24 The Nickel-Cadmium (Nicad) Cell

In recent years a new kind of dry cell, the nickel-cadmium cell, has gained widespread popularity because it can be recharged and thus has a much longer useful life than ordinary (Leclanché) dry cells. Nicad batteries are now used in electronic wristwatches, electronic calculators, and photographic equipment.

The anode is made of cadmium, while the cathode is made of nickel(IV) oxide—the electrolytic solution is basic. The reactions that occur during discharge of a nickel-cadmium battery can be represented as

$$
\begin{array}{ll}
\text{(anode)} & Cd\ (s) + 2OH^-\ (aq) \longrightarrow Cd(OH)_2\ (s) + 2e^- \\
\text{(cathode)} & NiO_2\ (s) + 2H_2O\ (\ell) + 2e^- \longrightarrow Ni(OH)_2\ (s) + 2OH^-\ (aq) \\
\hline
\text{(overall)} & Cd\ (s) + NiO_2\ (s) + 2H_2O\ (\ell) \longrightarrow Cd(OH)_2\ (s) + Ni(OH)_2\ (s)
\end{array}
$$

As in the lead storage battery, the solid reaction product at each electrode adheres to the electrode surface, and so a nicad battery can be recharged by an external source of electricity, that is, the electrode reactions can be reversed. Because no gases are produced by the reactions in a nicad battery, the unit can be sealed. The voltage of a nicad cell is about 1.4 V, slightly less than that of a Leclanché cell.

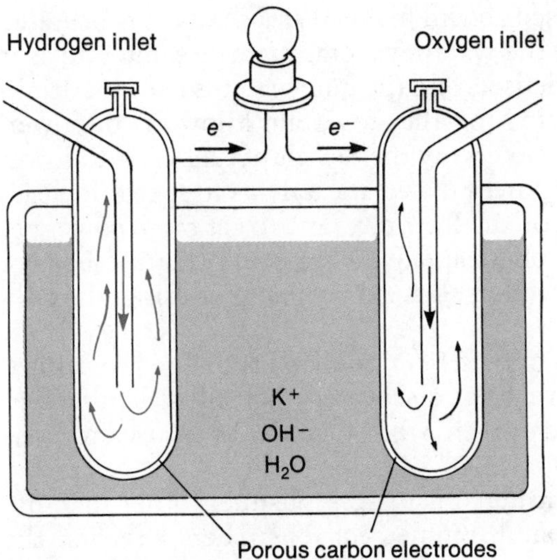

FIGURE 19–17 Schematic drawing of a hydrogen-oxygen fuel cell.

19–25 The Hydrogen-Oxygen Fuel Cell

Fuel cells are voltaic cells in which the reactants are continuously supplied to the cell. The hydrogen-oxygen fuel cell (Figure 19–17) already has many applications and is expected to have even more in the future. It has been widely used in spacecraft to supplement the energy obtained from solar cells. Since liquid hydrogen is carried on board as a propellant, the vapor that ordinarily would be lost is used in a fuel cell to generate electrical power, and incidentally, to provide drinking water.

Hydrogen (the fuel) is supplied to one compartment, the anode, where it is oxidized. Oxygen (the oxidizer) is fed into the cathode compartment, where it is reduced. The diffusion rates of the gases into the cell must be carefully regulated to achieve maximum efficiency. The net reaction is the same as the burning of hydrogen in oxygen to form water, but combustion does not actually occur. Most of the chemical energy from the formation of H—O bonds is converted directly into electrical energy, rather than into heat energy as in combustion.

The efficiency of energy conversion of the fuel cell operation is only 60 to 70% of the theoretical maximum (based on ΔG), but this still represents about twice the efficiency that can be realized from burning hydrogen in a heat engine coupled to a generator.

Oxygen undergoes reduction at the cathode, which consists of porous carbon impregnated with finely divided platinum or palladium catalyst.

$$O_2 + 2H_2O + 4e^- \xrightarrow{Pt} 4OH^- \qquad \text{(cathode)}$$

The hydroxide ions migrate through the electrolyte, an aqueous solution of a base, to the anode. The anode is also porous carbon containing a small amount of catalyst (Pt, Ag, or CoO). Here the hydrogen is oxidized to water

$$H_2 + 2OH^- \longrightarrow 2H_2O + 2e^- \qquad \text{(anode)}$$

The net reaction is obtained by balancing the electron transfer and adding the two half-reactions

$$\begin{array}{ll} O_2 + 2H_2O + 4e^- \longrightarrow 4OH^- & \text{(cathode)} \\ 2(H_2 + 2OH^- \longrightarrow 2H_2O + 2e^-) & \text{(anode)} \\ \hline 2H_2 + O_2 \longrightarrow 2H_2O & \text{(net cell reaction)} \end{array}$$

When the H_2/O_2 fuel cell is used aboard manned spacecraft, it is operated at a high enough temperature that the water evaporates at the same rate as it is produced. The vapor is then condensed, and the pure water is used for drinking. Many cells are usually connected together to obtain kilowatts of power.

Current research is aimed at modifying the design of fuel cells to lower their cost. This includes the development of better catalysts to speed the reactions and allow more rapid generation of electricity, and to get more power per unit volume. The use of fuel cells offers an advantage over other methods of obtaining energy in that the H_2/O_2 cell itself is entirely nonpolluting; the only substance released is water.

A potential future application of the H_2/O_2 fuel cell is tied to the utilization of solar energy. Catalysts have been discovered that allow sunlight to decompose water into hydrogen and oxygen, which could then be used to operate fuel cells.

Fuel cells have also been constructed using gaseous fuels other than hydrogen, such as methane or methanol. Biomedical researchers envision the possibility of using tiny fuel cells to operate heart pacemakers. The disadvantage of other power supplies for pacemakers, which are primary voltaic cells, is that their reactants are eventually consumed so that they require periodic surgical replacement. As long as the fuel and oxidizer are supplied continuously, a fuel cell can, in theory at least, operate forever. It is hoped that, eventually, tiny pacemaker fuel cells can be operated by the oxidation of blood sugar (the fuel) by the body's oxygen at a metal electrode implanted just below the skin.

Key Terms

Activity series the relative order of tendencies for elements and their simple ions to act as oxidizing or reducing agents; also called the electromotive series.

Alkaline battery a dry cell in which the electrolyte contains KOH.

Ampere unit of electrical current; one ampere equals one coulomb per second.

Anode electrode at which oxidation occurs.

Cathode electrode at which reduction occurs.

Cathodic protection protection of a metal (making it a cathode) against corrosion by attaching it to a sacrificial anode of a more easily oxidized metal.

Cell potential potential difference, E_{cell}, between oxidation and reduction half-cells under nonstandard conditions.

Corrosion oxidation of metals in the presence of moisture.

Coulomb unit of electrical charge.

Downs cell electrolytic cell for the commercial electrolysis of molten sodium chloride.

Dry cells ordinary batteries (voltaic cells) for flashlights, radios, etc.; many are Leclanché cells.

Electrochemistry involves chemical changes produced by electrical current and production of electricity by chemical reactions.

Electrodes surfaces upon which oxidation and reduction half-reactions occur in electrochemical cells.

Electrode potentials potentials, E, of half-reactions as reductions versus the standard hydrogen electrode.

Electrolysis process that occurs in electrolytic cells.

Electrolytic cells electrochemical cells in which electrical energy causes nonspontaneous redox reactions to occur.

Electrolytic conduction conduction of electrical current by ions through a solution or pure liquid.

Electromotive series activity series.

Electroplating plating a metal onto a (cathodic) surface by electrolysis.

Faraday one faraday of electricity corresponds to the charge on 6.022×10^{23} electrons, or 96,487 coulombs.

Faraday's Law of Electrolysis one equivalent weight of a substance is produced at each electrode during the passage of 96,487 coulombs of charge through an electrolytic cell.

Fuel cells voltaic cells in which the reactants (usually gases) are supplied continuously.

Galvanic cells see *Voltaic cells*

Half-cell compartment in which the oxidation or reduction half-reaction occurs in a voltaic cell.

Hydrogen-oxygen fuel cell fuel cell in which hydrogen is the fuel (reducing agent) and oxygen is the oxidizing agent.

Hydrometer a device used to measure the densities of liquids and solutions.

Lead storage battery secondary voltaic cell used in most automobiles.

Leclanché cell a dry cell.

Metallic conduction conduction of electrical current through a metal or along a metallic surface.

Nernst equation corrects standard electrode potentials for nonstandard conditions.

Nickel-cadmium cell (battery) a dry cell in which the anode is Cd, the cathode is NiO_2 and the electrolyte is basic.

Polarization of an electrode, build-up of a product of oxidation or reduction at an electrode, preventing further reaction.

Primary voltaic cells voltaic cells in which no further chemical action is possible once the chemicals are consumed; cannot be recharged.

Sacrificial anode a more active metal that is attached to a less active metal to protect the less active metal against corrosion.

Salt bridge "U-shaped" tube containing electrolyte, which connects two half-cells of a voltaic cell.

Secondary voltaic cells voltaic cells that can be recharged; original reactants can be regenerated by reversing the direction of current flow.

Standard cell potential potential difference, E^0_{cell}, between standard reduction and oxidation half-cells.

Standard electrochemical conditions 1.0 M concentration of dissolved ions, 1.0 atm partial pressure of gases, and pure solids and liquids.

Standard electrode potential by convention, potential, E^0, of a half-reaction as a reduction relative to the standard hydrogen electrode when all species are present at unit activity.

Standard electrodes half-cells in which the oxidized and reduced forms of a species are present at unit activity: 1.0 M solutions of dissolved ions, 1.0 atm partial pressure of gases, and pure solids and liquids.

Voltage potential difference between two electrodes; measure of the chemical potential for a redox reaction to occur.

Voltaic cells electrochemical cells in which spontaneous chemical reactions produce electricity; also called galvanic cells.

Exercises

General Concepts

1. Define and distinguish among electrochemical cells, electrolytic cells, voltaic cells, anode, and cathode.
2. Why do all electrochemical cells involve redox reactions?
3. Compare and contrast ionic conduction and metallic conduction.
4. Support or refute the statement that the positive electrode in any electrochemical cell is the one toward which the electrons flow through the wire.

Electrolytic Cells

5. Why do solids like potassium bromide, KBr, and sodium nitrate, $NaNO_3$, not conduct electrical current even though they are ionic? Can these solids be electrolyzed?

6. Why can metallic potassium not be obtained by electrolysis of aqueous potassium chloride, KCl?

7. Support or refute the statement that the Gibbs free energy change, ΔG, for any electrolysis reaction is positive.

8. The following compounds are natural sources of the metals contained within them. For which ones can the pure metal be obtained only by electrolysis of the compound or some other compound prepared from it? Explain. $CaCO_3$, Al_2O_3, NiS, KCl, PbS, Fe_2O_3, $MgCO_3$, Cr_2O_3, Cu_2S.
9. Why must the products of many electrolysis reactions be kept separate once they are produced?
10. Why are there no sodium ions in the overall cell reaction for the electrolysis of aqueous sodium chloride?
11. Consider the electrolysis of molten magnesium chloride with inert electrodes. The following experimental observations can be made when current is supplied.
    (a) Bubbles of pale green chlorine gas, Cl_2, are produced at one electrode.
    (b) Silvery white molten metallic magnesium is produced at the other electrode.

    Diagram the cell, indicating the anode, the cathode, the positive and negative electrodes, the half-reactions occurring at the electrodes, the overall cell reaction, and the direction of electron flow through the wire.
12. Do the same as in Exercise 11 for the electrolysis of molten aluminum oxide, Al_2O_3, dissolved in cryolite, Na_3AlF_6. This is the Hall process for commercial production of aluminum (Section 20–5). The observations are:
    (a) Silvery metallic aluminum is produced at one electrode.
    (b) Oxygen, O_2, bubbles off at the other electrode.
13. Do the same as in Exercise 11 for the electrolysis of aqueous hydrogen chloride, that is, hydrochloric acid, HCl. The observations are:
    (a) Bubbles of gaseous hydrogen are produced at one electrode and the solution becomes less acidic around that electrode.
    (b) Bubbles of chlorine gas are produced at the other electrode.
14. Do the same as in Exercise 11 for the electrolysis of aqueous calcium sulfate, $CaSO_4$. The observations are:
    (a) Bubbles of gaseous hydrogen are produced at one electrode and the solution becomes more basic around that electrode.
    (b) Bubbles of gaseous oxygen are produced at the other electrode and the solution becomes more acidic around that electrode.

Faraday's Law

15. What is (a) a coulomb, (b) electrical current, (c) an ampere, (d) a faraday?
16. If separate solutions of the following salts are electrolyzed with the same current for the same time period, which metal would be produced in the greatest mass? What will be the ratio of numbers of moles of metal plated out? $AgNO_3$, CuCl, $CuCl_2$, $CrCl_3$, $Hg(NO_3)_2$, $Hg_2(NO_3)_2$.
17. Calculate the charge in coulombs on a single electron.
18. Calculate the number of electrons that have a total charge of one coulomb.
19. How many faradays would be required to reduce 1.00 mole of each of the following cations to the free metal?
    (a) Fe^{3+}, (b) Cu^{2+}, (c) Cu^{+}, (d) Ag^{+}, (e) Al^{3+}.
20. How many coulombs would be necessary to plate out 1.00 gram of each of the metals from the ions listed in Exercise 19?
21. Calculate the mass of gold plated out during the electrolysis of gold(III) sulfate with a 0.150 ampere current for 52.0 hours.
22. Refer to Exercise 21. How many grams and how many milliliters of dry O_2, measured at STP, would be produced simultaneously at the anode?
23. How long would electrolysis of aqueous gold(III) sulfate have to occur to produce 5.00 grams of gold at the cathode at a current of 0.150 ampere?
24. How long must the electrolysis of Exercise 21 occur to produce 5.00 liters of dry O_2 measured at STP?
25. How long must the electrolysis of Exercise 21 occur to produce 5.00 liters of O_2, collected over water at 32.0°C and 740 torr pressure? The vapor pressure of water at 32.0°C is 35.7 torr.
26. How many amperes of electrical current would be required for 10.0 hours to produce 1.00 gram of:

(a) Fe from Fe^{2+}? (c) Ag from Ag^{+}?
(b) Cr from Cr^{3+}? (d) Br_2 from Br^{-}?

27. Describe how the charge on an ion may be determined by electrolysis.
28. The total charge of electricity required to plate out 15.54 grams of a metal from a solution of its dipositive ions is 14,475 coulombs. What is the metal?
29. What is the charge on an ion of tin if 14.84 grams of metallic tin are plated out from the passage of 24,125 coulombs through a solution containing the ion?
30. How many grams of H_2 and O_2 are produced during the electrolysis of water under a 1.30 ampere current for 5.00 hours? What volumes of dry gases are produced at STP?
31. How many minutes would be required to:
    (a) produce 3.00 faradays of electricity at a current of 2.00 amperes?
    (b) pass 67,400 coulombs of charge through a cell at a current of 1.50 amperes?
    (c) reduce 0.606 gram of Au^{3+} to metallic gold at a current of 2.50 amperes?
32. What mass of silver could be plated onto a spoon from electrolysis of silver nitrate at a 1.00 ampere current for 10.0 minutes?
33. What mass of platinum could be plated onto a ring from the electrolysis of a platinum(II) salt at a 0.100 ampere current for 30.0 seconds?
34. How many coulombs of electricity would be required to reduce the iron in 36.0 grams of potassium hexacyanoferrate(III), $K_3[Fe(CN)_6]$, to metallic iron?
35. Describe and illustrate the electrolytic purification of impure silver.
36. How does the smoothness or coarseness of a thin film of an electroplated metal depend on the time period of electrolysis?

Voltaic Cells

37. What kind of energy is converted into electrical energy in voltaic cells?
38. What are the functions of a salt bridge?
39. Describe and sketch the following electrodes:
    (a) the standard magnesium electrode, Mg^{2+}/Mg.
    (b) the standard Fe^{3+}/Fe^{2+} electrode.
    (c) the standard chlorine electrode, Cl_2/Cl^{-}.
40. A voltaic cell containing a standard Co^{2+}/Co electrode and a standard Au^{3+}/Au electrode is constructed and the circuit is closed. Without consulting the table of standard reduction potentials, diagram and completely describe the cell from the following experimental observations.
    (a) Metallic gold plates out on one electrode and the gold ion concentration decreases around that electrode.
    (b) The mass of the cobalt electrode decreases, and the cobalt(II) ion concentration increases around the same electrode.
41. Repeat Exercise 40 for a voltaic cell containing standard Fe^{3+}/Fe^{2+} and Ga^{3+}/Ga electrodes. The observations are:
    (a) The mass of the gallium electrode decreases, and the gallium ion concentration around that electrode increases.
    (b) The ferrous ion, Fe^{2+}, concentration increases in the other electrode solution.
42. What are standard electrochemical conditions?
43. What does voltage measure? How does it vary with time in a primary voltaic cell? Why?
44. When metallic copper is placed into aqueous silver nitrate, a spontaneous redox reaction occurs. Why is electricity not produced?
45. Distinguish among (a) primary voltaic cells, (b) secondary voltaic cells, and (c) fuel cells.
46. Sketch and describe the operation of (a) a dry cell, (b) a lead storage battery, and (c) a hydrogen-oxygen fuel cell.

Standard Electrode Potentials

47. Why are we permitted to assign arbitrarily an electrode potential of 0.000 volt to the standard hydrogen electrode?
48. What does the sign of the standard reduction potential of a half-reaction indicate? What does the magnitude indicate?
49. What is the electromotive series? What information does it contain?
50. How was the electromotive series constructed? In general, how do ionization energies and electron affinities of elements vary with their position in the electromotive series?

51. Diagram the following cells. In each case write the balanced equation for the reaction that occurs spontaneously, and calculate the cell potential. Indicate the direction of electron flow, the anode, the cathode, and the polarity (+ or −) of each electrode. In each case assume that the circuit is completed by a wire and a salt bridge.
   (a) A strip of copper is immersed in a solution that is 1.0 *M* in Cu^{2+} and a strip of silver is immersed in a solution that is 1.0 *M* in Ag^+.
   (b) A strip of aluminum is immersed in a solution that is 1.0 *M* in Al^{3+} and a strip of copper is immersed in a solution that is 1.0 *M* in Cu^{2+}.
   (c) A strip of magnesium is immersed in a solution that is 1.0 *M* in Mg^{2+} and a strip of silver is immersed in a solution that is 1.0 *M* in Ag^+.
52. Why are E^0 values for half-reactions *not* multiplied by the factors necesary to balance electron transfer in generating net cell reactions and standard cell potentials, E^0_{cell}?
53. Describe the process of corrosion. How can corrosion of an oxidizable metal be prevented if it must be exposed to the weather?

In answering Exercises 54–61, justify your answer by appropriate calculations.

54. Will Fe^{3+} oxidize Sn^{2+} to Sn^{4+} in acidic solution?
55. Will dichromate ions oxidize fluoride ions (F^-) to free fluorine (F_2) in acidic solution?
56. Will permanganate ions oxidize arsenous acid (H_3AsO_3) to arsenic acid (H_3AsO_4) in acidic solution?
57. Will permanganate ions oxidize hydrogen peroxide (H_2O_2) to free oxygen (O_2) in acidic solution?
58. Will dichromate ions oxidize Mn^{2+} to MnO_4^- in acidic solution?
59. Calculate the standard cell potential, E^0_{cell}, for each of the cells described in Exercises 40 and 41.
60. By consulting a table of standard reduction potentials, determine which of the following reactions are spontaneous under standard electrochemical conditions.
   (a) $Cr^{3+}(aq) + Fe^{2+}(aq) \longrightarrow Cr^{2+}(aq) + Fe^{3+}(aq)$
   (b) $Mn(s) + 2H^+(aq) \longrightarrow H_2(s) + Mn^{2+}(aq)$
   (c) $2Al^{3+}(aq) + 3H_2(g) \longrightarrow 2Al(s) + 6H^+(aq)$
   (d) $H_2(g) \longrightarrow H^+(aq) + H^-(aq)$
   (e) $Cl_2(g) + 2Br^-(aq) \longrightarrow Br_2(\ell) + 2Cl^-(aq)$
61. Which of the following reactions are spontaneous in voltaic cells under standard conditions?
   (a) $Si(s) + 6OH^-(aq) + 2Cu^{2+}(aq) \longrightarrow SiO_3^{2-}(aq) + 3H_2O(\ell) + 2Cu(s)$
   (b) $Zn(s) + 4CN^-(aq) + Ag_2CrO_4(s) \longrightarrow Zn(CN)_4^{2-}(aq) + 2Ag(s) + CrO_4^{2-}(aq)$
   (c) $MnO_2(s) + 4H^+(aq) + Sr(s) \longrightarrow Mn^{2+}(aq) + 2H_2O(\ell) + Sr^{2+}(aq)$
   (d) $2Cr(OH)_3(s) + 6F^-(aq) \longrightarrow 2Cr(s) + 6OH^-(aq) + 3F_2(g)$
   (e) $Cl_2(g) + 2H_2O(\ell) + ZnS(s) \longrightarrow 2HClO(aq) + H_2S(aq) + Zn(s)$
62. Which of each pair is the better oxidizing agent?
   (a) Cd^{2+} or Al^{3+} (c) H^+ or Cr^{2+}
   (b) Sn^{2+} or Sn^{4+} (d) Cl_2 or Br_2
   (e) MnO_4^- in acidic solution or MnO_4^- in basic solution
   (f) F_2 or Pb^{2+}
63. Which of each pair is the better reducing agent?
   (a) Ag or H_2 (b) Pt or Co^{2+}
   (c) Hg or Au
   (d) Cl^- in acidic solution or Cl^- in basic solution
   (e) Ce^{3+} or Cu (f) Rb or H_2
64. Which of each pair is the better oxidizing agent?
   (a) H^+ or Cl_2
   (b) Ni^{2+} or Se in contact with acidic solution
   (c) Fe^{2+} or Zn^{2+}
   (d) Al^{3+} or H^+
   (e) HgS or Au^+ (f) $Cr_2O_7^{2-}$ or Br_2
65. Which of each pair is the better reducing agent?
   (a) H^- or K (b) V or Fe^{2+}
   (c) Zn or H^-
   (d) $Mn(OH)_2$ in basic solution or Mn^{2+} in acidic solution
   (e) H_2S or Zr (f) Ga or Cr

The Nernst Equation

66. How is the Nernst equation of value in electrochemistry?

67. Identify all of the terms in the Nernst equation. What part of the Nernst equation represents the correction factor for nonstandard electrochemical conditions?

68. Calculate the reduction potentials for the following electrodes under the stated conditions. (You will need to write the balanced half-reactions to evaluate Q.)
(a) Fe^{2+}/Fe when $[Fe^{2+}] = 2.00\ M$
(b) $SnF_6^{2-}/Sn, F^-$ when $[SnF_6^{2-}] = 0.500\ M$, and $[F^-] = 1.00 \times 10^{-4}\ M$
(c) $CdS/Cd, S^{2-}$ when $[S^{2-}] = 1.00 \times 10^{-19}\ M$
(d) $Cr(OH)_3/Cr$, OH^- when $[OH^-] = 5.80 \times 10^{-4}\ M$
(e) $TeO_2, H^+/Te$ when $[H^+] = 3.50 \times 10^{-3}\ M$
(f) MnO_4^-, H^+/Mn^{2+} when $[MnO_4^-] = 5.00 \times 10^{-2}\ M$, $[H^+] = 1.00 \times 10^{-3}\ M$, and $[Mn^{2+}] = 6.62 \times 10^{-1}\ M$

69. Calculate the reduction potentials for the following electrodes under the stated conditions. (You will need to write the balanced half-reactions to evaluate Q.)
(a) Cl_2/Cl^- when $P_{Cl_2} = 2.50$ atm and $[Cl^-] = 1.00\ M$
(b) Cl_2/Cl^- when $P_{Cl_2} = 1900$ torr and $[Cl^-] = 1.00\ M$
(c) O_2, H^+/H_2O_2 when $P_{O_2} = 10.0$ atm, $[H^+] = 2.00 \times 10^{-2}\ M$ and $[H_2O_2] = 1.00\ M$
(d) NO_3^-, H^+/NO when $[NO_3^-] = [H^+] = 3.00 \times 10^{-5}\ M$ and $P_{NO} = 0.00350$ atm
(e) H^+/H_2 when $[H^+] = 5.00\ M$ and $P_{H_2} = 1.00 \times 10^{-5}$ atm

70. Write the equation for the spontaneous reaction and calculate the cell potential, E_{cell}, for the following combinations of electrodes from Exercises 68 and 69.
(a) 68(a) and 68(b)
(b) 68(c) and 68(f)
(c) 69(c) and 68(e)
(d) 69(d) and 69(e)

71. Determine the cell potential for each of the following reactions under the specified conditions at 25°C. Concentrations of ions and partial pressures of gases are indicated parenthetically.
(a) $Zn\ (s) + 2H^+\ (1.0 \times 10^{-3}\ M) \longrightarrow Zn^{2+} (3.0\ M) + H_2\ (g)\ (5.0\ atm)$
(b) $Cu\ (s) + 2Ag^+\ (1.0 \times 10^{-6}\ M) \longrightarrow Cu^{2+}\ (5.0 \times 10^{-2}\ M) + 2Ag\ (s)$
(c) $Sn^{2+}\ (0.010\ M) + Ni\ (s)\ (0.010\ M) \longrightarrow Sn\ (s) + Ni^{2+}\ (aq)\ (6.0\ M)$

72. Calculate E_{cell} for each of the following reactions under the specified conditions at 25°C.
(a) $PbO_2\ (s) + SO_4^{2-}\ (1.0 \times 10^{-2}\ M) + 4H^+\ (0.10\ M) + Zn\ (s) \longrightarrow PbSO_4\ (s) + 2H_2O + Zn^{2+}\ (1.0 \times 10^{-5}\ M)$
(b) $2MnO_4^-\ (0.10\ M) + 16H^+\ (3.0\ M) + 10Cl^-\ (3.0\ M) \longrightarrow 2Mn^{2+}\ (1.0 \times 10^{-2}\ M) + 5Cl_2\ (g)\ (0.50\ atm) + 8H_2O$

Relationships Among ΔG^0, E^0_{cell}, and K

73. How are the signs and magnitudes of E^0_{cell}, ΔG^0, and K related for a particular reaction?

74. If E_{cell} is related to ΔG and E^0_{cell} is related to ΔG^0, why is the equilibrium constant K related only to E^0_{cell} and not E_{cell}?

75. In light of your answer to Exercise 52, how do you explain the fact that ΔG^0 for a redox reaction *does* depend on the number of electrons transferred, according to $\Delta G^0 = -nFE^0_{cell}$?

76. Calculate E^0_{cell} from the tabulated standard reduction potentials for each of the following reactions in aqueous solution. Then calculate ΔG^0 and K at 25°C from E^0_{cell}. Which reactions are spontaneous as written?
(a) $Sn^{4+} + 2Fe^{2+} \longrightarrow Sn^{2+} + 2Fe^{3+}$
(b) $2Cu^+ \longrightarrow Cu^{2+} + Cu\ (s)$
(c) $3Zn + 2MnO_4^- + 4H_2O \longrightarrow 2MnO_2\ (s) + 3Zn(OH)_2\ (s) + 2OH^-$

77. Calculate ΔG^0 (overall) and ΔG^0 per mole of metal for each of the following reactions from E^0 values.
(a) Zinc dissolves in dilute hydrochloric acid to produce a solution that contains Zn^{2+}, and hydrogen gas is evolved.
(b) Aluminum dissolves in dilute hydrochloric acid to produce a solution that contains Al^{3+}, and hydrogen gas is evolved.
(c) Silver dissolves in dilute nitric acid to form a solution that contains Ag^+, and NO is liberated as a gas.

(d) Lead dissolves in dilute nitric acid to form a solution that contains Pb^{2+}, and NO is liberated as a gas.

78. Write balanced equations for the following spontaneous reactions. Calculate the standard cell potential for each cell, ΔG^0 (overall), and ΔG^0 per mole of each reactant.
    (a) In acidic solution, dichromate ions (from $K_2Cr_2O_7$) oxidize iodide ions (from NaI) to free iodine.
    (b) In acidic solution, permanganate ions (from $KMnO_4$) oxidize sulfurous acid (H_2SO_3) to sulfate ions.
    (c) In acidic solution, permanganate ions (from $KMnO_4$) oxidize oxalic acid, $(COOH)_2$, to carbon dioxide.

79. Calculate the *standard* Gibbs free energy change, ΔG^0, at 25°C for each of the reactions of Exercise 71.

80. Calculate the thermodynamic equilibrium constant, K, at 25°C for each of the reactions in Exercise 71.

81. Calculate the free energy change, ΔG, for each reaction of Exercise 71 under the stated conditions at 25°C.

82. Given ΔG^0 at 25°C for each of the following reactions, calculate E^0_{cell} and the thermodynamic equilibrium constant, K, at 25°C.
    (a) $MnO_4^- + 8H^+ + 5Fe^{2+} \longrightarrow Mn^{2+} + 5Fe^{3+} + 4H_2O$ $\quad \Delta G^0 = -357$ kJ
    (b) $Cr_2O_7^{2-} + 14H^+ + 6I^- \longrightarrow 2Cr^{3+} + 3I_2\ (s) + 7H_2O$ $\quad \Delta G^0 = -460$ kJ
    (c) $3Zn\ (s) + 8H^+ + 2NO_3^- \longrightarrow 3Zn^{2+} + 2NO\ (g) + 4H_2O$ $\quad \Delta G^0 = -997$ kJ

83. Given the thermodynamic equilibrium constant, K, at 25°C for each reaction below, calculate E^0_{cell} at 25°C.
    (a) $Cl_2\ (g) + 2Br^-\ (aq) \rightleftharpoons Br_2\ (\ell) + 2Cl^-\ (aq)$

$$K = \frac{[Cl^-]^2}{(P_{Cl_2})[Br^-]^2} = 2.9 \times 10^9$$

    (b) $4ClO_3^-\ (aq) \rightleftharpoons Cl^-\ (aq) + 3ClO_4^-\ (aq)$ (in basic solution)

$$K = K_c = \frac{[Cl^-][ClO_4^-]^3}{[ClO_3^-]^4} = 2.4 \times 10^{26}$$

    (c) $Cl_2\ (g) + 2OH^-\ (aq) \rightleftharpoons ClO^-\ (aq) + Cl^-\ (aq) + H_2O\ (\ell)$

$$K = \frac{[ClO^-][Cl^-]}{(P_{Cl_2})[OH^-]^2} = 2.9 \times 10^{32}$$

84. Given the following E^0 values at 25°C, calculate K_{sp} for cadmium sulfide, CdS.

$$Cd^{2+}\ (aq) + 2e^- \longrightarrow Cd\ (s) \qquad E^0 = -0.403\ V$$

$$CdS\ (s) + 2e^- \longrightarrow Cd\ (s) + S^{2-}\ (aq) \qquad E^0 = -1.21\ V$$

85. Refer to Exercise 84. Evaluate ΔG^0 at 25°C for the process
    $CdS\ (s) \rightleftharpoons Cd^{2+}\ (aq) + S^{2-}\ (aq)$

20 Metals and Metallurgy

Metals

We usually think of metals as substances used for structural purposes such as in buildings, trains, railroads, ships, cars, and trucks. Metals also are used for containers and to conduct heat and electricity. In addition, metal ions serve many important biological functions; medical and nutritional research during recent decades has provided most of our insight into these important functions of metals. Sodium, potassium, calcium, and magnesium are all present in substantial quantities in humans.

In this chapter we shall (1) show how small amounts of some metals are important in life processes and mention a few trace nonmetals as well, (2) study the occurrence of metals, (3) examine processes for obtaining some metals from their ores, i.e., metallurgy, and (4) consider some possibilities for conserving metals.

Metals and Life

The major problem associated with investigations of dietary "trace elements" has been the difficulty of measuring the extremely small amounts of these elements present in foods. As an example, the vanadium content of fresh peas is usually less than 4.0×10^{-10} gram per gram of fresh peas. Based on this figure, 2700 tons of fresh peas contain only 1.0 gram of vanadium. Obviously very sensitive techniques are required to detect vanadium (or any other element) in such low concentrations, and sufficiently sensitive techniques have been developed only recently.

In 1681 an English physician, Thomas Sedenham, soaked "iron and steel filings" in cold Rhenish wine and used the resulting solution to treat patients suffering from chlorosis, an iron-deficiency anemia. Iron was the first trace element shown to be essential in the human diet. Around 1850 the French chemist Boussingualt demonstrated that certain salt deposits claimed by South American Indians to cure goiter did, in fact, do so because the salt deposits contained iodine compounds. Iodine is a nonmetal, but it is an essential trace element.

In recent years seven additional trace elements have been found to be essential to human nutrition: copper, manganese, zinc, cobalt, molybdenum, selenium, and chromium. All these elements except selenium are metals. Six other trace elements have been shown to be essential for proper nutrition of various animals. They are the metals tin, vanadium, and nickel, and the nonmetals fluorine, silicon, and arsenic. (We might also mention that the nonmetals carbon, hydrogen, oxygen, nitrogen, phosphorus, and sulfur are all present in large amounts in biological systems.)

At the present time trace elements are divided into three groups, shown in Table 20–1.

A dietary shortage of **iron** is the most common trace element deficiency. Anemia, characterized by a low concentration of hemoglobin in the blood or by a low volume of packed red blood cells, is the common symptom. The recommended dietary allowance for women ages 23 to 50 years is 80% higher than for men in the same age group because of menstrual bleeding.

Iodine (a nonmetal) is needed to prevent goiter due to iodine deficiency, which accounts for approximately 96% of the disease. Iodine is present in two thyroid hormones, thyroxine and triiodothyronine, which increase the metabolic rate and oxygen consumption of cells.

Zinc is present in at least 90 enzymes and in the hormone insulin. Zinc is involved in the functioning of the pituitary and adrenal glands, and in the pancreas and gonads. It plays an important role in the growth processes, including protein synthesis and cell division. Research at the University of Wisconsin demonstrated in 1936 that zinc is essential to animal growth, but it was not until 1964 that human need was demonstrated by research at Wayne State University. Meat and other animal products are the principal dietary sources of zinc for human beings.

TABLE 20–1 Dietary Trace Elements Shown to be Essential in Humans and Animals

Element	Date Discovered to be an Essential Dietary Trace Element in Animals	Dietary Sources	Functions in Humans and/or Animals
Group 1 (Need in Humans Established and Quantified)			
Iron	17th century	Meat, liver, fish, poultry, beans, peas, raisins, prunes	Component of hemoglobin and myoglobin; oxidative enzymes
Iodine*	1850	Iodized table salt, shellfish, kelp	Needed to make the thyroid hormones thyroxine and triiodothyronine; prevents iodine-deficiency goiter
Zinc	1934	Meat, liver, eggs, shellfish	Present in at least 90 enzymes and in the hormone insulin
Group 2 (Need in Humans Established But Not Yet Quantified)			
Copper	1928	Nuts, liver, shellfish	Component of oxidative enzymes; involved in absorption and mobilization of iron needed for making hemoglobin
Manganese	1931	Nuts, fruits, vegetables, whole-grain cereals	Involved in formation of enzymes, bone
Cobalt	1935	Meat, dairy products	Component of vitamin B_{12}
Molybdenum	1953	Organ meats, green leafy vegetables, legumes	Involved in formation of enzymes, proteins
Selenium*	1957	Meat, seafood	Involved in enzyme formation, fat metabolism
Chromium	1959	Meat, beer, unrefined wheat flour	Required for glucose metabolism
Group 3 (Need Established in Animals But Not Yet in Humans)			
Tin	1970	†	Essential for normal growth of rats
Vanadium	1971	†	Needed for optimum growth of chicks and rats
Fluorine*	1972	Fluoridated water	Essential for normal growth of rats
Silicon*	1972	†	Needed for growth and bone development in chicks; needed for growth of rats
Nickel	1974	†	Needed for normal growth and for formation of red blood cells in rats; needed by chicks and swine
Arsenic*	1975	†	Required for normal growth of rats, goats, minipigs; shortage of element can impair reproductive ability of goats, minipigs

*Nonmetals.
†Present in trace quantities in many foods and in the environment.

Copper serves an important function in the body's oxidation processes as a component of several oxidative enzymes. The most widely accepted theory suggests that a copper deficiency causes anemia because copper is needed for the absorption and mobilization of iron required to make hemoglobin. Human need for copper was established in 1928. Nuts, liver, and shellfish are important sources. Copper serves the same oxygen-carrying function in the fluids of certain marine animals that iron serves in higher animals.

Manganese was shown to be essential to animals in 1931. One of its functions is the activation of certain enzymes. **Cobalt** is present in vitamin B_{12}, which prevents pernicious anemia. Its need by humans has been known since 1935. **Molybdenum** was shown to be essential to animals in 1953, and human need has also been established. It is a component of several enzymes and is involved in protein formation. The nonmetal *selenium* is involved in enzyme formation and fat metabolism. Animal need for selenium was demonstrated in 1957.

Chromium, complexed with an unidentified organic molecule, is essential to glucose metabolism. Chromium deficiency has been observed among children with severe protein deficiency in undeveloped countries and among some elderly individuals in the United States.

During the period 1970–1975, the trace elements Sn, V, Ni, F, Si, and As were shown to be essential to certain animals. Their need by humans has not been demonstrated at the present time. However, many scientists believe that these trace elements plus several others are probably essential to human beings. The metals cadmium, lead, and lithium, and the nonmetals bromine and boron, may also be essential trace elements.

Two factors are important in the recent discovery of the role of many trace elements. One is the availability of two highly sensitive analytical techniques, activation analysis and electrothermal atomic absorption spectroscopy, which enable scientists to detect these elements in concentrations of only a few parts per billion. The second is the availability of special isolation chambers that allow scientists to study animals under carefully controlled conditions, free of unwanted contaminants. The diets fed to animals must be carefully purified to keep out even *traces* of unwanted elements. Obviously, metals cannot be used in such cages, and even the air must be carefully purified.

As the results of additional research on the role of trace elements in human nutrition become available, interesting questions arise. For many years some individuals have suggested that drinking a small quantity of boiled seawater daily supplies essential "trace minerals" that have been leached from the soil by eons of rainfall and may no longer be there. In view of the recently established importance of several trace elements that are contained in seawater, it may be that those who advocate drinking small amounts of seawater are on the right track.

Current medical research has shown that low-salt diets are needed for many, if not most, Americans. Although NaCl is required in much greater than trace amounts for metabolism, too much of it and other essential substances can be harmful.

Let us cite one additional bit of evidence on the importance of trace elements in nutrition. Arthur J. Stattelman of the Poultry Disease Research Center at the University of Georgia has demonstrated that when the drinking water of chickens contains 10% seawater, the structure of their feathers is greatly improved. These feathers are up to three times more effective as heat insulators than those of chickens fed the same diet but no seawater. The trace element or elements responsible for improved feather quality have not been identified at this time. How, you may wonder, could this possibly be significant? The cost of producing chickens, as well as other animals used for food, includes the amount of feed required to maintain body temperature at the desired level as well as the amount of feed required for body growth. Chickens with feathers that are more effective heat insulators require less feed to maintain body temperature, and therefore production costs are lower.

20–1 Occurrence of the Metals

In our study of periodicity we learned that metallic character increases toward the left and toward the bottom of the periodic table (Section 4–8) and that oxides of most metals are basic (Section 7–9, part 2). The oxides of some metals (and metalloids) are amphoteric (Section 11–8). We described metallic bonding and related the effectiveness of metallic bonding to the characteristic properties of metals in Section 9–17.

You may wish to review some of these sections.

As we shall see, the properties of the metals are related to the kinds of ores in which the metals are found, and also to the metallurgical processes that must be used to extract the metal from the ore. Metals with negative standard reduction potentials (that is, active metals, those that are easily oxidized) are found only in the combined state in nature. Those with positive reduction potentials, the less active metals, may occur in the uncombined free state or **native ores.** Examples of the latter are copper, silver, gold, and the less abundant platinum, osmium, iridium, ruthenium, rhodium, and palladium. Copper, silver, and gold are also found in the combined state.

Elements that occur uncombined in nature are called **native ores.**

Many rather insoluble compounds of the metals are found in the earth's crust. Solids containing these compounds are the **ores** from which metals are extracted. Ores contain **minerals,** comparatively pure compounds of the metal of interest, mixed with relatively large amounts of sand, soil, clay, rock and other material called **gangue,** a German term meaning mining vein. Water-soluble metal compounds are usually found dissolved in the sea or in salt beds in areas where large bodies of water have evaporated. Metal ores can be classified in terms of the anions with which the metal ion is combined. The most common types are listed with examples in Table 20–2.

TABLE 20–2 Common Classes of Ores

Anion	Examples and Name of Mineral
None (Native ores)	Au, Ag, Pt, Os, Ir, Ru, Rh, Pd, As, Sb, Bi
Oxide	hematite, Fe_2O_3; magnetite, Fe_3O_4; bauxite, Al_2O_3; cassiterite, SnO_2; periclase, MgO; silica, SiO_2
Sulfide	chalcopyrite, $CuFeS_2$; chalcocite, Cu_2S; sphalerite, ZnS; galena, PbS; iron pyrites, FeS_2; cinnabar, HgS
Chloride	rock salt, NaCl; sylvite, KCl; carnallite, $KCl \cdot MgCl_2$
Carbonate	limestone, $CaCO_3$; magnesite, $MgCO_3$; dolomite, $MgCO_3 \cdot CaCO_3$
Sulfate	gypsum, $CaSO_4 \cdot 2H_2O$; epsom salts, $MgSO_4 \cdot 7H_2O$; barite, $BaSO_4$
Silicate	beryl, $Be_3Al_2Si_6O_{18}$; kaolinite, $Al_2(Si_2O_8)(OH)_4$; spodumene, $LiAl(SiO_3)_2$

The most widespread minerals are silicates, but because extraction of metals from silicates is very difficult, metals are obtained from silicate minerals only when there is no other more economical alternative.

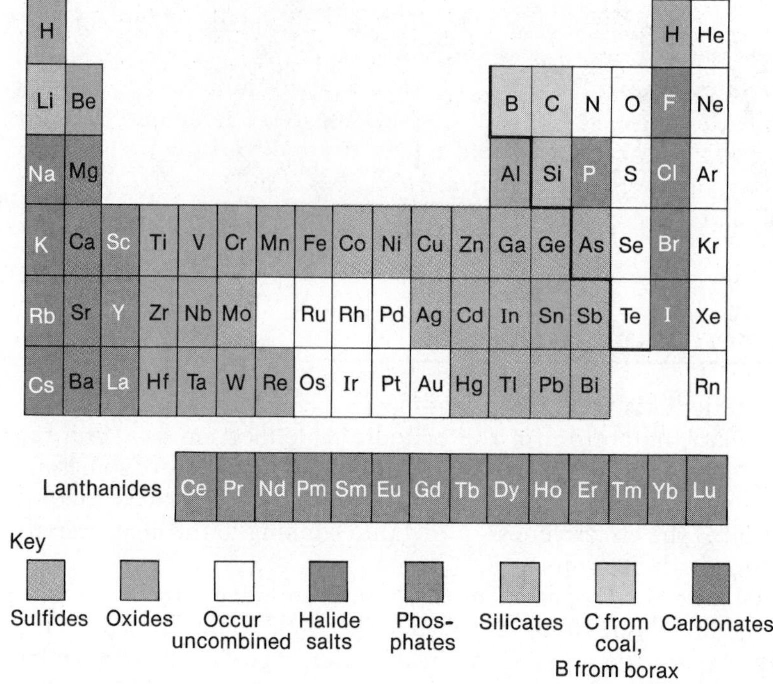

FIGURE 20–1 Natural sources of the elements. The soluble halide salts are found in the ocean or in solid deposits. Most of the noble gases are obtained from air.

Figure 20–1 shows the major natural sources of all the elements with respect to position in the periodic table.

Metallurgy

Metallurgy is the term applied to the commercial extraction of metals from their ores, and preparation of the metals for use. It is usually divided into a sequence of steps: (1) mining, (2) pretreatment of the ore, (3) reduction of the ore to the free metal, (4) refining or purifying the metal, and (5) alloying, or mixing with other appropriate elements, if necessary.

Mixtures of metals frequently have properties that are more desirable for a particular purpose than are those of a free metal. In such cases metals are alloyed.

20–2 Pretreatment of Ores

After an ore is mined, but before it is reduced to a free metal, it must be concentrated by removal of most of the gangue. Most sulfides have relatively high densities and are more dense than gangue. After pulverization, the lighter gangue particles are removed by a variety of methods. One involves blowing the lighter particles away using a cyclone separator (Figure 20–2). In some cases the lighter particles are sifted out through layers of vibrating wire mesh or inclined vibration tables.

The **flotation** method is particularly applicable to sulfides, carbonates, and silicates, which either are not "wet" by water or can be made water-repellent by treatment. Their surfaces are easily covered by layers of oil or other flotation agents. When a stream of air is blown through a swirled suspension of such an ore in water and oil (or other agent), bubbles form on the oil

These are called hydrophobic materials.

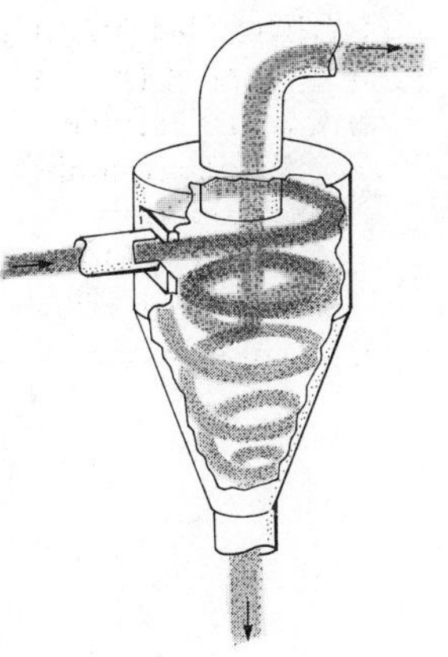

FIGURE 20–2 The cyclone separator, used for enriching metal ores. The crushed ore is blown at high velocity into the separator, where centrifugal force takes the heavier particles, with a high percentage of the desired metal, to the wall of the separator. From there, these particles spiral down to the collection bin at the bottom. Lighter particles, not as rich in the metal, move into the center of the separator, and are carried out the top in the air stream.

surfaces on the mineral particles and cause them to rise to the surface of the suspension. The bubbles are prevented from breaking and escaping by a layer of oil and emulsifying agent at the surface, where a frothy ore concentrate forms. By varying the relative amounts of oil and water, the types of oil additive, the air pressure, and so on, it is even possible to separate one metal sulfide, carbonate, or silicate from another. A flotation apparatus is shown in Figure 20–3.

Another pretreatment process involves chemical modification in order to convert a metal compound to a more easily reduced form. Carbonates and hydroxides may be heated to drive off carbon dioxide and water, respectively, as shown by the following equations

$$CaCO_3\ (s) \xrightarrow{\Delta} CaO\ (s) + CO_2\ (g)$$

$$Mg(OH)_2\ (s) \xrightarrow{\Delta} MgO\ (s) + H_2O\ (g)$$

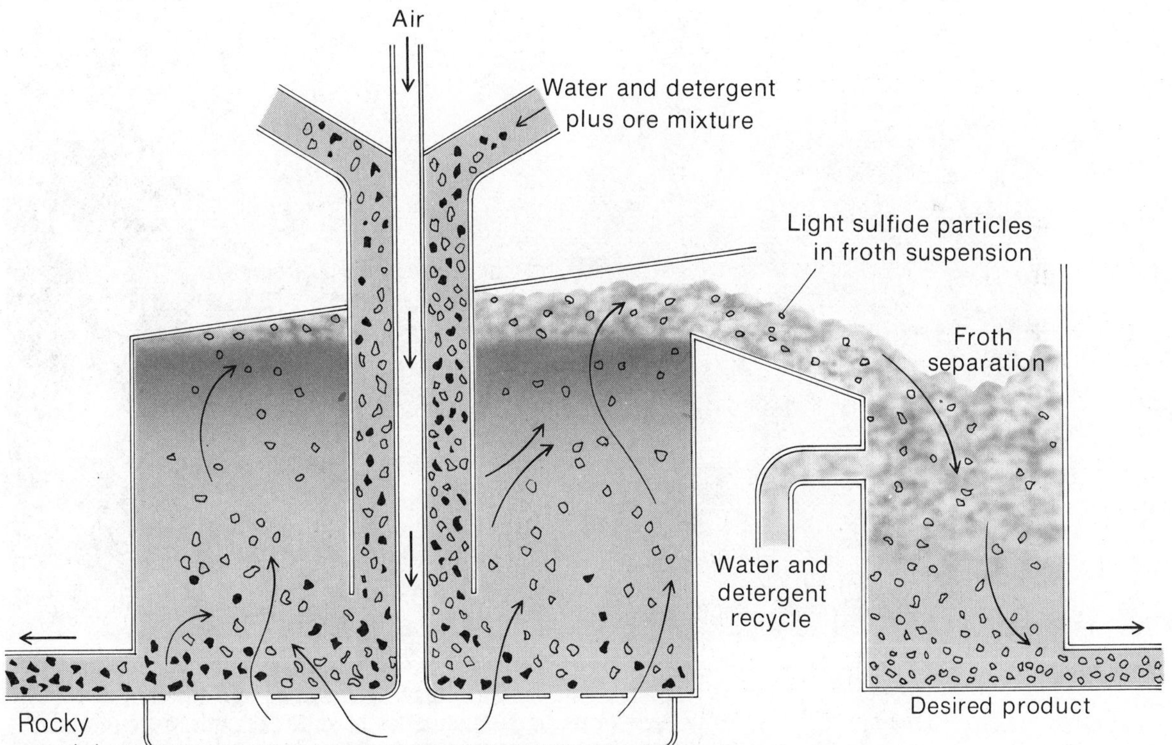

FIGURE 20–3 Flotation process for enrichment of copper sulfide ore. The relatively light sulfide particles are put into suspension in the water-oil-detergent mixture and collected as a froth. The denser material sinks to the bottom of the container.

Some sulfides are converted to oxides by **roasting,** i.e., heating below their melting points in the presence of oxygen from air. For example,

$$2ZnS\ (s) + 3O_2\ (g) \xrightarrow{\Delta} 2ZnO\ (s) + 2SO_2\ (g)$$

Roasting of sulfides has caused problems of air pollution due to the escape of enormous quantities of SO_2 into the atmosphere (Section 7–9). Figure 20–4 shows the damage caused by SO_2 from roasting Cu_2S in Coppertown Basin, Tennessee in the early 1940's. Federal regulations now require limitation of the amount of SO_2 that escapes with stack gases and fuel gases, so most of the SO_2 now is trapped and used in the manufacture of sulfuric acid (Section 23–16).

20–3 Reduction to the Free Metals

The method used for reduction, or smelting, of metal ores to the free metals depends upon the strength of the bonds between metal ions and anions. The stronger the bonds, the more energy is required to break them (and reduce the metal) and, generally, the more expensive the reduction process. The most active metals usually form the strongest bonds.

The least reactive metals can occur in the free state, and therefore require no reduction. Examples are gold, silver, and platinum.

FIGURE 20–4 Coppertown Basin (Ducktown), Tennessee, as photographed in 1943. Copper ore (principally copper sulfide, Cu_2S) had been mined and smelted in this area since 1847. In the early years, large quantities of sulfur dioxide, a by-product, were discharged directly into the atmosphere and killed all vegetation for miles around the smelter. Today the sulfur is reclaimed in the exhaust stacks to make sulfuric acid, but the denuded soil remains a monument to the misuse of the atmosphere.

Weakly active metals, such as mercury, can be obtained from their sulfide ores by roasting alone, which reduces metal ions to the free metal by oxidation of the sulfide ions

$$HgS\ (s) + O_2\ (g) \xrightarrow{\Delta} Hg\ (g) + SO_2\ (g)$$

cinnabar — from air — obtained as vapor; later condensed

Roasting a more active metal sulfide produces a metal oxide, but no free metal

$$2NiS\ (s) + 3O_2\ (g) \xrightarrow{\Delta} 2NiO\ (s) + 2SO_2\ (g)$$

The resulting metal oxides are then reduced to free metals with coke (impure carbon) or carbon monoxide. If carbon must be avoided, other reducing agents such as hydrogen, iron, or aluminum are used

$$SnO_2\ (s) + 2C\ (s) \xrightarrow{\Delta} Sn\ (\ell) + 2CO\ (g)$$

$$WO_3\ (s) + 3H_2\ (g) \xrightarrow{\Delta} W\ (s) + 3H_2O\ (g)$$

The very active metals, such as aluminum and sodium, can be reduced only electrochemically, usually from their anhydrous molten salts. If water is present, it is reduced preferentially. Tables 20–3 and 20–4 summarize reduction processes for some different metal ions.

20–4 Refining or Purification of Metals

Metals obtained from the procedures described above are almost always impure, and further purification (refining) is required in most cases. This may be

TABLE 20–3 Reduction Processes for Some Metals

Metal Ion (Increasing Activity of Metals ↑)	Typical Reduction Process
Lithium, Li^+ Potassium, K^+ Calcium, Ca^{2+} Sodium, Na^+ Magnesium, Mg^{2+} Aluminum, Al^{3+}	Electrolysis of molten salt
Manganese, Mn^{2+} Zinc, Zn^{2+} Chromium, Cr^{2+}, Cr^{3+} Iron, Fe^{2+}, Fe^{3+}	Reaction of oxide with coke (carbon) or carbon monoxide (CO)
Lead, Pb^{2+} Copper, Cu^{2+} Silver, Ag^+ Mercury, Hg^{2+} Platinum, Pt^{2+} Gold, Au^+	Element occurring free, or easily obtained by roasting the sulfide or oxide ore

TABLE 20–4 Some Specific Reduction Processes

Metal	Ore	Reduction Process	Comments
Mercury	HgS, cinnabar	Roast reduction; heat ore in air $HgS + O_2 \xrightarrow{\Delta} Hg + SO_2$	
Copper	Sulfides such as Cu_2S, chalcocite	Blow oxygen through purified molten Cu_2S: $Cu_2S + O_2 \xrightarrow{\Delta} 2Cu + SO_2$	Preliminary ore concentration and purification steps required to remove FeS impurities
Zinc	ZnS, sphalerite	Conversion to oxide and reduction with carbon: $2ZnS + 3O_2 \xrightarrow{\Delta} 2ZnO + 2SO_2$ $ZnO + C \xrightarrow{\Delta} Zn + CO$	Process also used for the production of lead from galena, PbS
Iron	Fe_2O_3, hematite	Reduction with carbon monoxide: $2C\ (coke) + O_2 \xrightarrow{\Delta} 2CO$ $Fe_2O_3 + 3CO \xrightarrow{\Delta} 2Fe + 3CO_2$	
Titanium	TiO_2, rutile	Conversion of oxide to halide salt and reduction with an active metal: $TiO_2 + 2Cl_2 + 2C \xrightarrow{\Delta} TiCl_4 + 2CO$ $TiCl_4 + 2Mg \xrightarrow{\Delta} Ti + 2MgCl_2$	Also used for the reduction of UF_4 obtained from UO_2, pitchblende
Tungsten	$FeWO_4$, wolframite	Reduction with hydrogen: $WO_3 + 3H_2 \xrightarrow{\Delta} W + 3H_2O$	Used also for molybdenum
Aluminum	$Al_2O_3 \cdot nH_2O$, bauxite	Electrolytic reduction (electrolysis) in molten cryolite, Na_3AlF_6, at 1000°C: $2Al_2O_3 \xrightarrow{\Delta} 4Al + 3O_2$	
Sodium	Salt beds, seawater	Electrolysis of molten chlorides: $2NaCl \xrightarrow{\Delta} 2Na + Cl_2$	Also for calcium, magnesium, and other active metals in Groups IA and IIA

accomplished by distillation if the metal is more volatile than its impurities, as in the case of mercury obtained by roasting cinnabar, HgS. Copper, silver, gold and aluminum are among the metals purified electrolytically (Section 19–8). The impure metal is the anode and a small sample of the pure metal is the cathode. Both are immersed in a solution of the desired metal ion.

Zone refining is often used when extremely pure metals are desired for such applications as solar cells and semiconductors (Section 9–17). This method relies on the fact that many impurities will not fit into the lattice of a pure crystal. The process is illustrated in Figure 20–5. An induction heater surrounds a bar of the impure solid and passes slowly from one end to the other. As it passes, it melts portions of the bar, which slowly recrystallize as the heating element moves away. Since the impurity does not fit into the lattice as easily as the element of interest, most of it is carried along in the molten portion until it reaches the end. Repeated passes of the heating element produce a bar of high purity.

The end containing the impurities is sliced off and often recycled.

After purification many metals are alloyed, or mixed with other elements, to change their physical and chemical characteristics. In some cases,

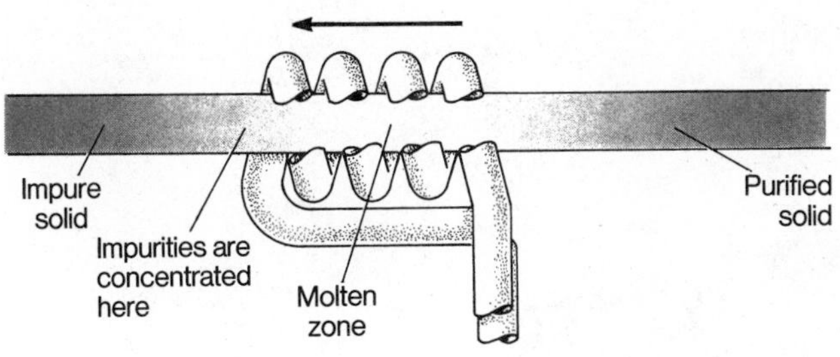

FIGURE 20–5 A zone-refining apparatus.

certain impurities are allowed to remain during refining because their presence improves the desired properties of the metal. For example, the presence of a small amount of carbon in iron greatly enhances its hardness. Some examples of alloys with improved properties are brass, bronze, duralumin, and stainless steel.

The Metallurgies of Specific Metals

The metallurgies of aluminum, iron, copper, gold, and magnesium will now be discussed as specific examples.

20–5 Aluminum

Aluminum is obtained from bauxite, or hydrated aluminum oxide, $Al_2O_3 \cdot xH_2O$. The aluminum ions can be reduced to metallic aluminum only by electrolysis in the absence of water. The crushed bauxite is first purified by dissolving it in a concentrated solution of sodium hydroxide to form soluble $NaAl(OH)_4$, and then precipitating $Al(OH)_3 \cdot xH_2O$ from the filtered solution by blowing in carbon dioxide to neutralize the unreacted sodium hydroxide as well as one OH^- ion per formula unit of $NaAl(OH)_4$. The hydrated compound is dehydrated to Al_2O_3 by heating it. The melting point of Al_2O_3 is 2045°C; electrolysis of pure molten Al_2O_3 would have to be carried out at or above this temperature, with great expense. However, it can be done at a much lower temperature when the Al_2O_3 is mixed with much lower-melting cryolite, a mixture of NaF and AlF_3 often represented as Na_3AlF_6. The molten mixture can be electrolyzed at 1000°C with carbon electrodes. The cell used industrially for this process, called the **Hall Process,** is shown in Figure 20–6.

Recall that Al_2O_3 is amphoteric. Impurities such as oxides of iron, which are not amphoteric, are left behind in the crude ore.

A mixture of compounds typically has a lower melting point than any of the pure compounds.

The inner surface of the cell is plated with carbon or carbonized iron, which functions as the cathode at which aluminum ions are reduced to the free metal. Oxide ions are oxidized to oxygen at the graphite anode, which is consumed almost quantitatively by the oxygen and must be replaced frequently. This represents one of the chief costs of aluminum production.

cathode	$4Al^{3+} + 12e^-$	$\longrightarrow$	$4Al\ (\ell)$
anode	$6O^{2-}$	$\longrightarrow$	$3O_2\ (g) + 12e^-$
net reaction	$4Al^{3+} + 6O^{2-}$	$\longrightarrow$	$4Al\ (\ell) + 3O_2\ (g)$

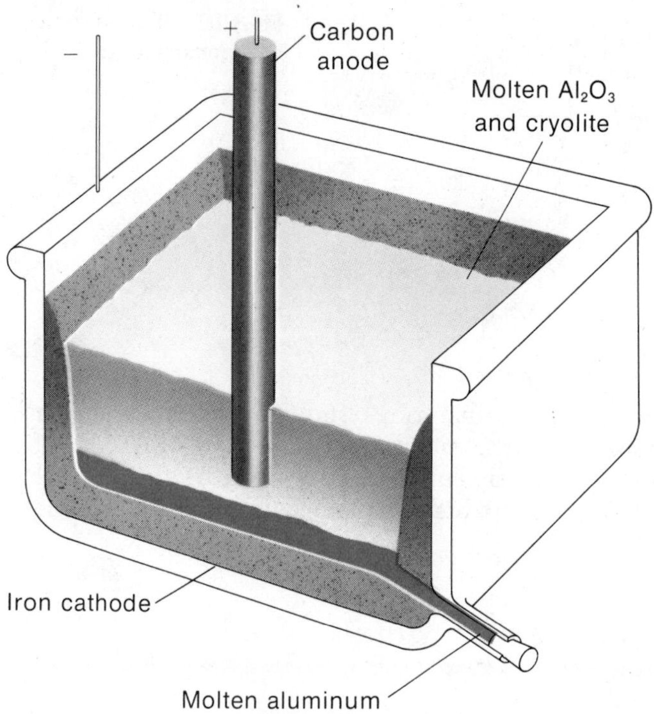

FIGURE 20–6 Schematic drawing of a cell for producing aluminum by electrolysis of a melt of Al_2O_3 in Na_3AlF_6. The molten aluminum collects in the cathode container.

Molten aluminum is more dense than molten cryolite, and it collects in the bottom of the cell until drawn off and cooled to a solid.

Electrolytic cells used in Hall Process. In the foreground, molten aluminum is being poured.

Recently a more economical process has been developed. The anhydrous bauxite is first converted to $AlCl_3$ by reaction with Cl_2 in the presence of carbon. The $AlCl_3$ is then melted and electrolyzed to aluminum and chlorine, using only about 30% as much electricity as in the Hall Process.

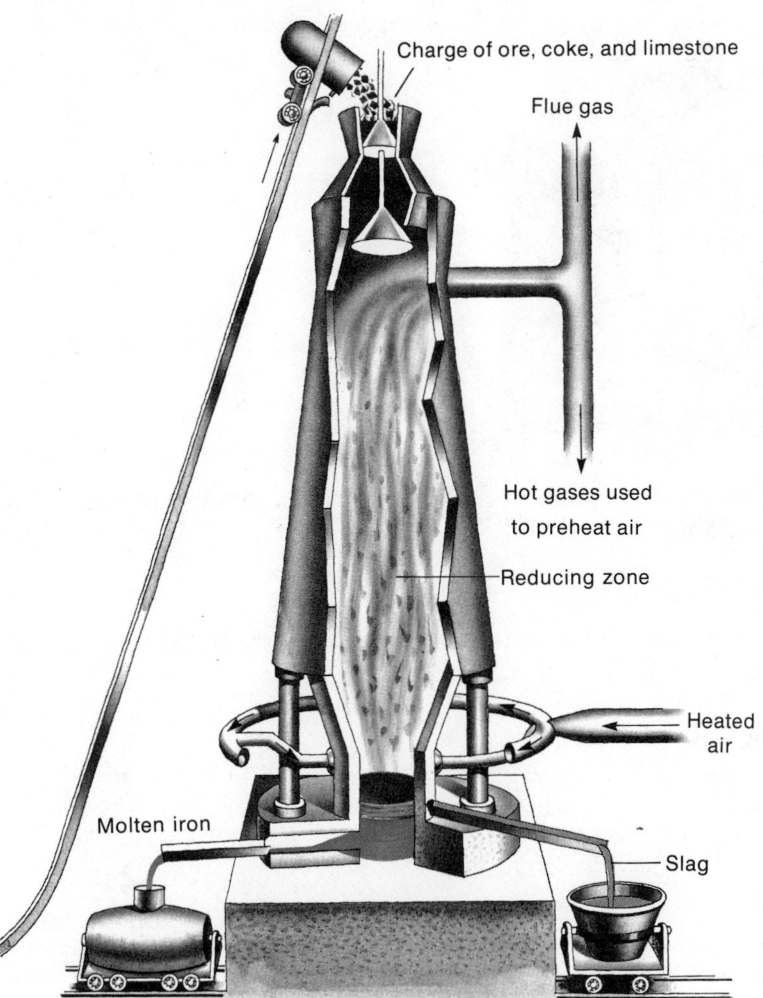

FIGURE 20–7 Blast furnace for reduction of iron ore.

20–6 Iron

The most desirable iron ores contain hematite, Fe_2O_3, or magnetite, Fe_3O_4. As the available supplies of these high grade ores have dwindled, taconite, which is magnetite in a very hard silica rock, has become an important source of iron. The oxide is reduced in blast furnaces (Figure 20–7) by carbon monoxide. Coke mixed with limestone ($CaCO_3$) and crushed ore is admitted at the top of the furnace as the "charge." A blast of hot air from the bottom burns the coke to carbon monoxide with the evolution of more heat

$$2C\ (s) + O_2\ (g) \xrightarrow{\Delta} 2CO\ (g) + \text{heat}$$

Most of the oxide is reduced to molten iron by the carbon monoxide, although some is reduced directly by coke. Several stepwise reductions occur, but the overall reactions for Fe_2O_3 may be summarized as

$$Fe_2O_3\ (s) + 3CO\ (g) \xrightarrow{\Delta} 2Fe\ (\ell) + 3CO_2\ (g) + \text{heat}$$

$$Fe_2O_3\ (s) + 3C\ (s) \xrightarrow{\Delta} 2Fe\ (\ell) + 3CO\ (g) + \text{heat}$$

Iron ore is scooped in an open-pit mine.

Much of the CO_2 reacts with excess coke to produce more CO to reduce the next incoming charge

$$CO_2\ (g) + C\ (s) \xrightarrow{\Delta} 2CO\ (g)$$

The limestone, called a **flux,** is added to react with the silica gangue in the ore to form a molten **slag** of calcium silicate

$$\underset{\text{limestone}}{CaCO_3\ (s)} \xrightarrow{\Delta} CaO\ (s) + CO_2\ (g)$$

$$CaO\ (s) + \underset{\text{gangue}}{SiO_2\ (s)} \xrightarrow{\Delta} \underset{\text{slag}}{CaSiO_3\ (\ell)}$$

The slag is less dense than the molten iron; it floats on the surface of the iron and protects it from atmospheric oxidation. Both are drawn off periodically. Some of the slag is subsequently used in the manufacture of cement.

Molten steel is poured from a basic oxygen furnace.

The iron obtained is still impure and contains carbon, among other things. It is called **pig iron.** If it is remelted, run into molds, and cooled, it becomes **cast iron,** which is brittle because it contains much iron carbide, Fe_3C. If some of the carbon is removed and other metals such as Mn, Cr, Ni, W, Mo, and V are added, the mixture becomes mechanically stronger and is known as **steel.** There are many types of steel containing alloyed metals and other elements in various controlled proportions. Stainless steels show excellent resistance to corrosion and high tensile strength. The most common kind contains 14–18% chromium and 7–9% nickel. If all the carbon is removed, nearly pure iron can be produced. It is silvery in appearance, quite soft and of little use.

Pig iron can also be converted to steel by burning out most of the carbon with oxygen in a basic oxygen furnace, shown in Figure 20–8. Oxygen is blown through a heat-resistant tube inserted below the surface of the molten iron. When the iron is heated to high temperatures, the carbon burns to carbon monoxide, which subsequently escapes and burns to carbon dioxide.

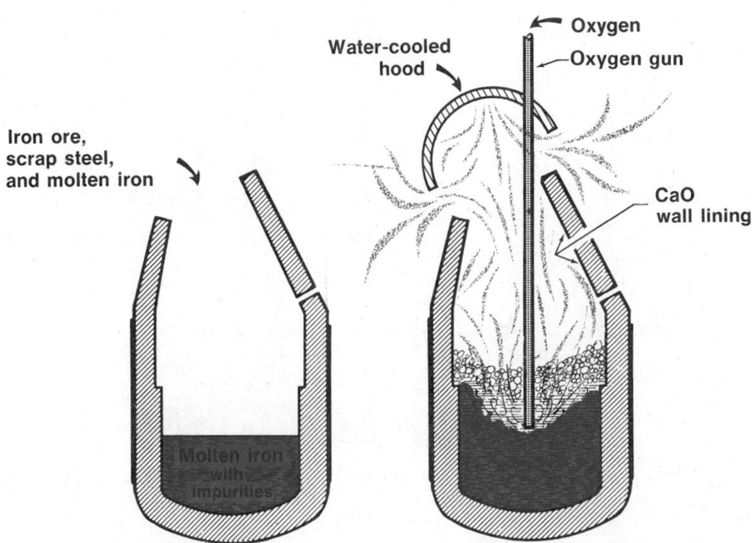

FIGURE 20–8 The basic oxygen process furnace. Much of the steel manufactured today is refined by blowing oxygen through a furnace charged with ore, scrap, and molten iron. The steel industry is one of the nation's largest consumers of oxygen.

20–7 Copper

Copper is so widely used, especially in its alloys such as bronze (copper and tin) and brass (copper and zinc), that it is becoming very scarce. The U.S. Bureau of Mines estimates that the currently known worldwide reserves of copper ore will be exhausted during the first half of the next century. The two principal classes of copper ores are the mixed sulfides of copper and iron, such as chalcopyrite, $CuFeS_2$, and the basic carbonates such as azurite, $Cu_3(CO_3)_2(OH)_2$, and malachite, $Cu_2CO_3(OH)_2$.

It is now profitable to mine ores containing as little as 0.25% copper.

Let us consider $CuFeS_2$ (or $CuS \cdot FeS$) as an example. After the ore is mined, the copper compound is separated by flotation and then roasted to remove volatile impurities. Enough air is used to convert iron(II) sulfide, but not the less active copper(II) sulfide, to the oxide

$$2CuFeS_2\ (s) + 3O_2\ (g) \xrightarrow{\Delta} 2FeO\ (s) + 2CuS\ (s) + 2SO_2\ (g)$$

The roasted ore is then mixed with sand (silica or SiO_2), crushed limestone ($CaCO_3$), and some unroasted ore that contains copper(II) sulfide in a reverberatory furnace at 1100°C (Figure 20–9). Copper(II) sulfide, CuS, is reduced to Cu_2S, which melts. The limestone and silica form a calcium silicate glass, which dissolves the ferrous oxide to form a slag less dense than molten copper(I) sulfide upon which it floats.

$$CaCO_3\ (s) + SiO_2\ (s) \xrightarrow{\Delta} CaSiO_3\ (\ell) + CO_2\ (g)$$

$$CaSiO_3\ (\ell) + FeO\ (s) + SiO_2\ (s) \xrightarrow{\Delta} CaSiO_3 \cdot FeSiO_3\ (\ell)$$

The slag is periodically drained off. The molten copper(I) sulfide is drawn off into a Bessemer converter where it is again heated and air-oxidized to free

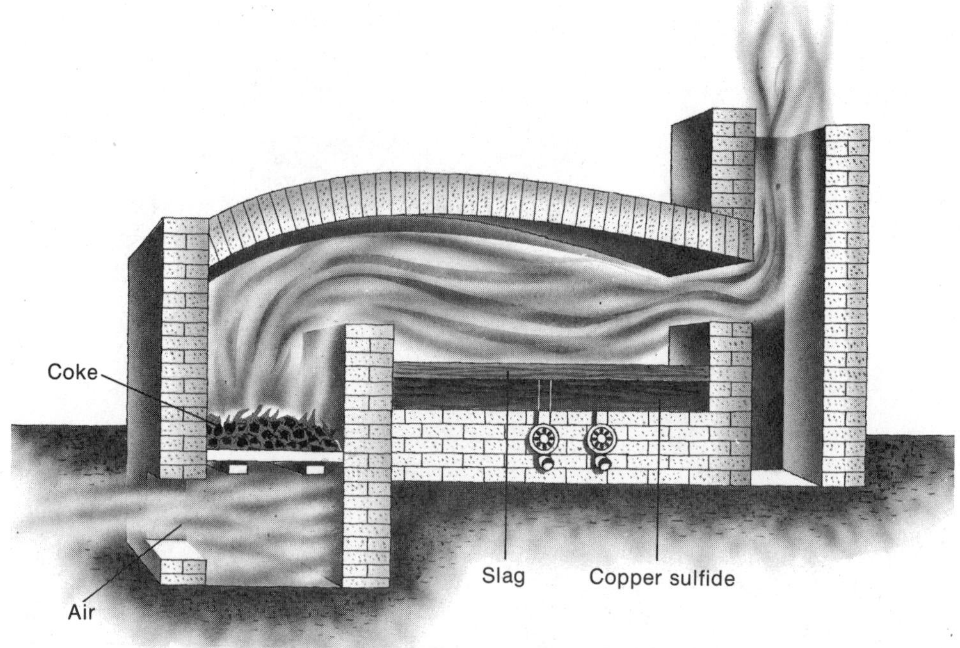

FIGURE 20–9
Reverberatory furnace for the production of copper.

copper and SO_2. This oxidizes the sulfide ions, but not copper(I) ions; they are reduced to metallic copper

$$Cu_2S\ (\ell) + O_2\ (g) \xrightarrow{\Delta} 2Cu\ (\ell) + SO_2\ (g)$$

The impure copper is refined electrolytically as described in detail in Section 19–8.

20–8 Gold

Since gold is an inactive metal it occurs mostly in the native state, although it is sometimes found as gold telluride. Because of its high density, metallic gold can be concentrated by panning. In this operation, gold-bearing sand and gravel are gently swirled with water in a pan. The lighter particles spill over the edge and the denser nuggets of gold remain. Gold is now more efficiently concentrated by sifting crushed gravel in a stream of water on a slightly inclined shaking table containing several low barriers, which impede the descent of the heavier gold particles but allow the lighter particles to pass over. The gold is then amalgamated (alloyed with mercury) and removed, after which the mercury is distilled away, leaving behind the pure gold.

There are about 10 million metric tons of gold in the world's oceans; but since it is dissolved in more than a billion cubic kilometers of seawater at a concentration of 7×10^{-9} g/L, it is unlikely that much of it will ever be reclaimed.

Gold is also recovered from the anode sludge from electrolytic purification of copper (Section 19–8). Since gold is so rare, it is also obtained from very low grade ores by the cyanide process, in which air is bubbled through an agitated slurry of the ore mixed with a solution of sodium cyanide. This causes slow oxidation of the metal and the formation of a soluble complex compound

$$4Au\ (s) + 8CN^-\ (aq) + O_2\ (g) + 2H_2O\ (\ell) \longrightarrow 4Au(CN)_2^-\ (aq) + 4OH^-\ (aq)$$

The free gold can then be regenerated by reduction of $Au(CN)_2^-$ with zinc after filtration, or it may be reduced by electrolysis.

20–9 Magnesium

Although magnesium occurs in carbonate ores in the earth's crust, most of the world's supply of magnesium comes from salt brines and from the sea, which is 0.13% magnesium by mass. Because of its very low density, only 1.74 g/cm^3, it is used in lightweight structural alloys for such things as ladders and aircraft parts.

Magnesium ions are precipitated as the hydroxide by addition of calcium hydroxide (slaked lime) to seawater. The slaked lime is obtained by crushing oyster shells ($CaCO_3$), heating them to produce lime (CaO), and then adding a limited amount of water (slaking).

$$CaCO_3\ (s) \xrightarrow{\Delta} CaO\ (s) + CO_2\ (g)$$

$$CaO\ (s) + H_2O\ (\ell) \longrightarrow Ca(OH)_2\ (s)$$

$$Ca(OH)_2\ (s) + Mg^{2+}\ (aq) \longrightarrow Ca^{2+}\ (aq) + Mg(OH)_2\ (s)$$

The last reaction occurs because K_{sp} for $Mg(OH)_2$, 1.5×10^{-11}, is much less than that for $Ca(OH)_2$, 7.9×10^{-6}. The milky white suspension of magnesium hydroxide is filtered and the solid $Mg(OH)_2$ is then neutralized with hydrochloric acid to produce magnesium chloride solution. Evaporation of the water leaves solid $MgCl_2$ which is then melted and electrolyzed (Figure 20–10) under an inert atmosphere to produce molten magnesium and gaseous chlorine. The products are separated as they are formed to prevent recombination

$$Mg(OH)_2\ (s) + 2[H^+\ (aq) + Cl^-\ (aq)] \longrightarrow [Mg^{2+}\ (aq) + 2Cl^-\ (aq)] + 2H_2O\ (\ell)$$

$$\xrightarrow[\text{and melt}]{\text{evaporate}} MgCl_2\ (\ell) \xrightarrow{\text{electrolysis}} Mg\ (\ell) + Cl_2\ (g)$$

The magnesium is cast into ingots or alloyed with other light metals. If no HCl is available the by-product chlorine is allowed to react with methane, from natural gas, and air to produce more hydrochloric acid for neutralization of

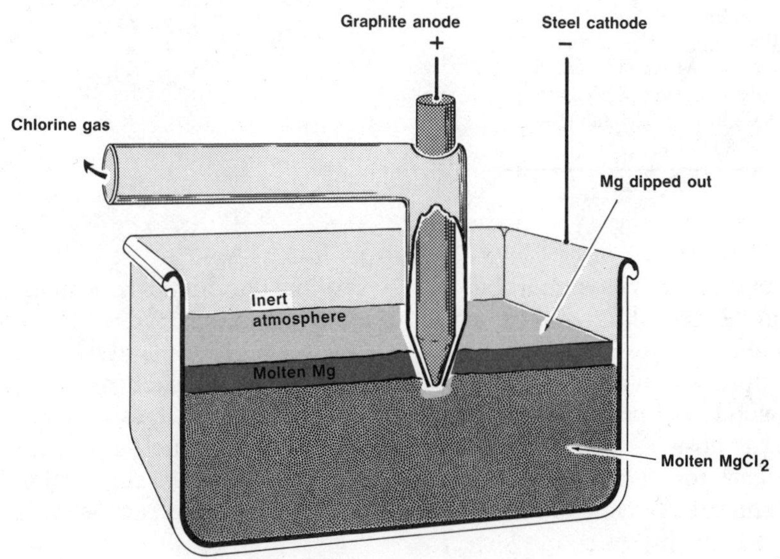

FIGURE 20–10 A cell for electrolyzing molten $MgCl_2$. The magnesium metal is formed on the steel cathode and rises to the top where it is dipped off periodically. Chlorine gas is formed on the graphite anode and is piped off.

magnesium hydroxide. The carbon monoxide from the air–chlorine mixture can be used as a fuel, since it undergoes combustion to carbon dioxide

$$2CH_4 + O_2 + 4Cl_2 \longrightarrow 8HCl + 2CO$$

20–10 Utilization and Conservation of Metal Resources

Fuels and structural materials are the basis for technological societies such as ours. We are already keenly aware of the vulnerable position we have allowed ourselves to get into by placing such an unrealistically low value on nonrenewable fuel resources for so many years. We are also rapidly approaching a similar situation with respect to some of our metal ores, from which we obtain the metals that serve so many structural purposes. The United States is probably the richest country in the world with respect to total resources, but our extravagant use of metal ores has forced us to an ever-increasing dependence on foreign resources. Table 20–5 shows the presently known world reserves of several important metals.

TABLE 20–5 Resources of Metals*

Metal	Location of Reserves (% of Total)
Au	Rep. So. Africa (40)
Hg	Spain (30); Italy (21)
Ag	Communist countries (36); U.S. (24)
Sn	Thailand (33); Malaysia (14)
Zn	U.S. (27); Canada (20)
Pb	U.S. (39)
Cu	U.S. (28); Chile (19)
W	China (73)
Mo	U.S. (58); U.S.S.R. (20)
Mn	Rep. So. Africa (38); U.S.S.R. (25)
Al	Australia (33); Guinea (20)
Co	Rep. Congo (31); Zambia (20)
Pt metals	Rep. So. Africa (47); U.S.S.R. (47)
Ni	Cuba (25); New Caledonia (22); Canada (14)
Fe	U.S.S.R. (33); Canada (14)
Cr	Rep. So. Africa (75)

There are three obvious methods of conserving our metal resources: (1) substitution, (2) recycling, and (3) utilizing lower-grade ores.

Substitution involves replacing some of the metals currently widely used for structural purposes with other more abundant resources, such as glass, granite rock, wood, and perhaps plastics.

It must be kept in mind, though, that most plastics are obtained from by-products of petroleum refining.

Recycling is possible now only for some metal products such as aluminum cans and lead storage batteries (Section 19–23). Unfortunately, the steel in automobiles cannot be recycled economically at the present time because it is present in so many different alloys, all of which have different properties. It is impossible to melt down mixtures of the alloys to produce useful uniform alloys. The presence of the many elements in the mixtures results in steel with very weak structural characteristics. It is also impossible to separate the components of the alloys with any reasonable expenditure of energy. This is why there are so many junkyards in which old automobiles are allowed to rust away.

As automobiles are currently constructed, it would be a tremendously tedious, time-consuming, and therefore costly task to separate all the different alloys from each other.

Copper is one of our most important metals whose supply is running very low. Copper from the wiring in generators, starting motors, and elsewhere in automobiles could be easily recycled if it weren't for the fact that most of the wiring is wound into the frame during assembly. If the automobile industry modified the construction process, the recycling of copper wire would become more economically feasible.

As the sources of ores of a particular metal dwindle, the cost of the metal rises and, consequently, it becomes profitable to extract the metal from **lower-grade ores.** As we have already seen (Section 20–9), magnesium is obtained from seawater, and therefore seawater can be considered an ore. Practically all the known elements can be found in the sea, although most are present in extremely low concentrations. However, as the technology of extraction is improved and the prices of the elements rise sufficiently, we will undoubtedly see other elements being obtained from the sea. The seas have such vast volume that they can be considered to contain virtually an inexhaustible supply of most elements. If we are able to develop a new, cheap and inexhaustible source of energy such as nuclear fusion (Section 28–15), it would become economically feasible to do such things as mine the sea for many elements, separate elements from alloys, and obtain aluminum from the very abundant but intractable aluminosilicates that cover the surface of the earth. Solar energy may be a source of cheap energy in the not too distant future.

In view of the technological advances that would have to occur before we are in a position to approach the extremes discussed in the preceding paragraph, we should concentrate on conserving the resources that are currently available. Figure 20–11 depicts an estimate of the effects of conservation measures on the depletion pattern of a nonrenewable resource. This indicates that if we recycle, improve mining techniques, and reduce per capita use of such a resource, it will be available some time after its supply would run out if we just mine, use, and throw away.

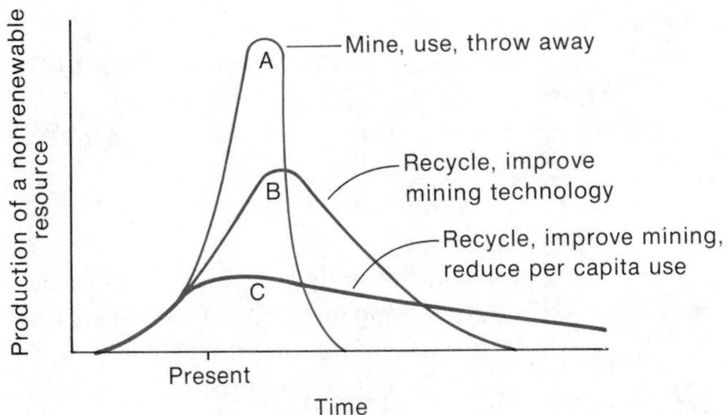

FIGURE 20–11 Alternate depletion patterns for a nonrenewable resource.

Key Terms

Alloying mixing of a metal with other substances (usually other metals) to modify properties.

Charge a sample of crushed ore as it is admitted to a furnace for smelting.

Flotation method by which hydrophobic (water-repelling) particles of an ore are separated from hydrophilic (water-attracting) particles in a metallurgical pretreatment process.

Flux a substance added to react with the charge, or a product of its reduction, in metallurgy; usually added to lower a melting point.

Gangue sand, rock, and other impurities surrounding the mineral of interest in an ore.

Metallurgy refers to the overall processes by which metals are extracted from ores.

Native state refers to the occurrence of an element in an uncombined or free state in nature.

Ore a natural deposit containing a mineral of an element to be extracted.

Refining purifying of a substance.

Roasting heating a compound below its melting point in the presence of air.

Slag unwanted material produced during smelting.

Smelting refers to chemical reduction of a substance at high temperature in metallurgy.

Zone refining method of purifying a bar of metal by passing it through an induction heater; this causes impurities to move along in melted portion.

Exercises

1. What kinds of metals are most apt to occur in the uncombined (native) state in nature?
2. List the six anions (and their formulas) that are most often combined with metals in ores. Give at least one example of an ore of each kind. What anion is the most commonly encountered?
3. Give the five general steps involved in extracting a metal from its ore. Briefly describe the importance of each.
4. Describe the flotation method of ore pretreatment. Are any chemical changes involved?
5. What kinds of ores are roasted? What kinds of compounds are converted to oxides by roasting? What kinds are converted directly to the free metals?
6. Of the following compounds, which would require electrolysis to obtain the free metals: KCl, $Cr_2(SO_4)_3$, Fe_2O_3, Al_2O_3, Ag_2S, $MgSO_4$? Why?
7. At which electrode is the free metal produced in the electrolysis of a metal compound? Why?
8. How are bond strengths related to ease of obtaining free metals from ores? All other factors equal, would it be more economically advantageous to be able to obtain a free metal from its ore by chemical reduction (such as heating a metal oxide with coke, hydrogen, and so on) or by electrolysis? Which method requires the greater energy input? Why?
9. Suggest the best method of obtaining manganese from an ore containing manganese(III) oxide, Mn_2O_3. On what basis do you make the suggestion?
10. What is the purpose of utilizing the basic oxygen furnace after the blast furnace in the production of iron?
11. Describe the metallurgy of (a) copper and (b) magnesium.
12. Describe the metallurgy of (a) aluminum, (b) iron, and (c) gold.
13. What is steel? How does the hardness of iron compare with that of steel?
14. Describe and illustrate the electrolytic refining of copper.
15. (a) Calculate the weight in pounds of sulfur dioxide produced in the roasting of one ton of chalcocite ore containing 6.5% Cu_2S, 0.4% Ag_2S, and no other source of sulfur.

(b) What weight of sulfuric acid can be prepared from the SO_2 generated, assuming 93% of it can be recovered from stack gases and 88% of that recovered can be converted to sulfuric acid?

(c) How many pounds of pure copper can be obtained assuming 75% efficient extraction and purification?

(d) How many pounds of silver can be produced assuming 82% of it can be extracted and purified?

16. Thirty-five pounds of Al_2O_3 obtained from bauxite are mixed with cryolite and electrolyzed. How long would a 0.500 ampere current have to be passed to convert all the Al^{3+} (from Al_2O_3) to aluminum metal? What volume of oxygen collected at 840 torr and 145°C, would be produced in the same period of time?

17. Calculate the percentage of iron in hematite ore containing 62.4% Fe_2O_3 by mass. How many pounds of iron would be contained in one ton of the ore?

18. Find the standard molar enthalpies of formation of Al_2O_3, Fe_2O_3 and HgS in Appendix K. Are the values in line with what might be predicted in view of the methods by which the metal ions are reduced in extractive metallurgy?

19. Using data from Appendix K, calculate ΔG^0_{298} for the following reactions:
(a) $Al_2O_3\ (s) \longrightarrow 2Al\ (s) + \frac{3}{2}O_2\ (g)$
(b) $Fe_2O_3\ (s) \longrightarrow 2Fe\ (s) + \frac{3}{2}O_2\ (g)$
(c) $HgS\ (s) \longrightarrow Hg\ (\ell) + S\ (s)$
Are any of the reactions spontaneous? Are the ΔG^0_{298} values in line with what would be predicted based on the relative activities of the metal ions involved? Do increases in temperature favor these reactions?

21 The Representative Metals

Hydrogen, a *nonmetal*, is included in IA in the periodic table, but is *not* a metal.

In this chapter we shall discuss the representative metals. Recall that the representative elements are those in the A groups of the periodic table and they have valence electrons in their outermost *s* and *p* atomic orbitals. Metallic character increases from top to bottom within groups and from right to left within periods. All the elements in Groups IA (except H) and IIA are metals, as are the lower members of Groups IIIA, IVA, and VA, which are called the post-transition metals. The metals will be covered in this order.

The Alkali Metals (Group IA)

21–1 Properties and Occurrence of the Group IA Metals

See the discussion of electrolysis of sodium chloride in Section 19–3.

Because they are so easily oxidized, the alkali metals are not found free in nature. They are obtained by electrolysis of their molten salts. Sodium (2.6% abundance by mass) and potassium (2.4% abundance) are very common in the earth's crust, but the other IA metals are quite rare. Francium consists only of short-lived radioactive isotopes formed by alpha-particle emission (Section 28–3) from actinium. Both potassium and cesium also have radioisotopes. Potassium-40 is important in the potassium-argon radioactive decay method of dating ancient objects. It has a half-life of 1.28 billion years (Section 28–10). The properties of the alkali metals vary regularly as the group is descended and are summarized in Table 21–1.

The free metals except lithium are soft, silvery corrosive metals that can be cut with a knife; lithium is harder. Cesium is slightly golden and melts in the hand (wrapped in plastic since it is so corrosive). The relatively low melting and boiling points of the alkali metals are consequences of the fairly weak bonding forces, since each atom can furnish only one electron for metallic bonding (Section 9–17). Because their outer electrons are so loosely held, the metals are excellent electrical and thermal conductors. They exhibit the photoelectric effect with low-energy radiation, i.e., they ionize when irradiated with light. These effects become more pronounced with increasing atomic size. For this reason, cesium is used in photoelectric cells.

The ionization energies reveal that the single electron in the outer shell of the IA metals is very easily removed. Cesium and francium are the least electronegative; lithium the most. In all alkali metal compounds the metals exhibit the +1 oxidation state, and virtually all are ionic. The extremely high second ionization energies show that removal of an electron from the next (filled) shell is impossible by ordinary chemical means.

One may expect that the standard reduction potentials of the metal ions would reflect the same trends as the first ionization energies and become more negative descending the group. However, the magnitude of the standard reduction potential of lithium seems unusually large, −3.05 volts. The first ionization energy is the amount of energy absorbed when a mole of *gaseous* atoms ionize, but the standard reduction potential measures the energy released when a mole of *aqueous* ions is reduced to the metal in contact with water.

TABLE 21–1 Properties of the Group IA Metals

Property	Lithium	Sodium	Potassium	Rubidium	Cesium	Francium
Outer electrons	$2s^1$	$3s^1$	$4s^1$	$5s^1$	$6s^1$	$7s^1$
Melting point (°C)	186	97.5	63.6	38.9	28.5	27
Boiling point (°C)	1326	889	774	688	690	677
Density (g/cm^3)	0.534	0.971	0.862	1.53	1.87	—
Atomic radius (Å)	1.52	1.86	2.31	2.44	2.62	—
Ionic radius, M^+ (Å)	0.60	0.95	1.33	1.48	1.69	—
Electronegativity	1.0	1.0	0.9	0.9	0.8	0.8
Standard reduction potential (V)						
$M^+ (aq) + e^- \rightarrow M (s)$	−3.05	−2.71	−2.93	−2.93	−2.92	—
Ionization energy (kJ/mol)						
$M (g) \rightarrow M^+ (g) + e^-$	520	496	419	403	376	—
$M^+ (g) \rightarrow M^{2+} (g) + e^-$	7298	4562	3051	2632	2420	—
Heat of hydration of gaseous ion (kJ/mol)						
$M^+ (g) + xH_2O \rightarrow M^+ (aq)$	−544	−435	−351	−293	−264	—

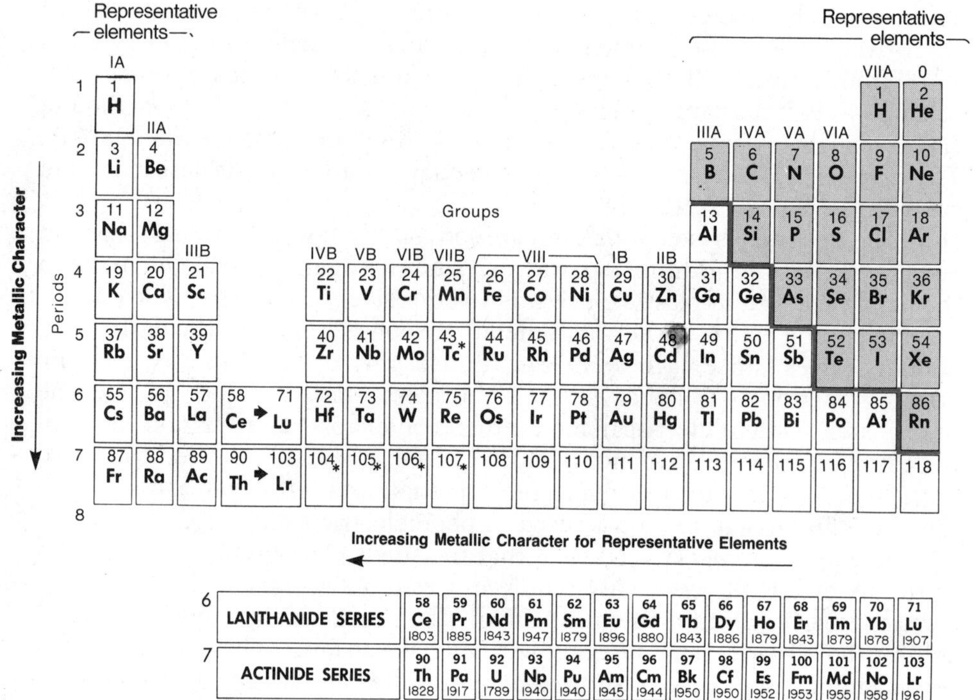

FIGURE 21–1 Periodic table showing the representative metals shaded.

Thus, hydration energies must also be considered. Hydration energy is the energy released when a mole of gaseous ions is hydrated.

$$M^+ (g) + xH_2O (\ell) \longrightarrow M(OH_2)_x^+ + \text{hydration energy}$$

Reference to Table 21–1 shows that the hydration energy of lithium ion is very high, and this accounts for the unusual E^0 value. Since the lithium ion is so small, its charge density (ratio of charge to size, Section 10–2) is very high and it therefore exerts a stronger attraction for polar water molecules than the other IA ions do. Thus water molecules, of course, must be stripped off during the reduction process. The high charge density of lithium ion also accounts for its ability to polarize large anions, which gives rise to a higher degree of covalent character in lithium compounds than in other corresponding alkali metal compounds. For example, LiCl is soluble in ethyl alcohol, a less polar solvent than water; NaCl is not. Salts of the alkali metals with small anions are very soluble in water, but salts with large and complex anions (such as silicates and aluminosilicates) are not soluble to any appreciable extent. Again, there are variations in the Group.

Polarization of an anion refers to distortion of its electron cloud. The ability of a cation to polarize an anion increases with increasing charge density (ratio of charge/size) of the cation.

21–2 Reactions of the IA Metals

Many of the reactions of the alkali metals are summarized in Table 21–2. All are characterized by the loss of one electron per metal atom. These metals are very strong reducing agents. Reactions of the alkali metals with oxygen and hydrogen were discussed in Sections 7–9 and 7–8, those with the halogens in Section 4–10, and those with water in Section 7–3.

TABLE 21–2 Reactions of the IA Metals

Reaction	Remarks
$4M + O_2 \longrightarrow 2M_2O$	limited O_2
$4Li + O_2 \longrightarrow 2Li_2O$	excess O_2 (lithium oxide)
$2Na + O_2 \longrightarrow Na_2O_2$	(sodium peroxide)
$M + O_2 \longrightarrow MO_2$	M = K, Rb, Cs; excess O_2 (superoxides)
$2M + H_2 \longrightarrow 2MH$	molten metals
$6Li + N_2 \longrightarrow 2Li_3N$	at high temperature
$2M + X_2 \longrightarrow 2MX$	X = halogen (Group VIIA)
$2M + S \longrightarrow M_2S$	also with Se, Te of Group VIA
$3M + P \longrightarrow M_3P$	also with As, Sb of Group VA
$2M + 2H_2O \longrightarrow 2MOH + H_2$	K, Rb, and Cs react explosively
$2M + 2NH_3 \longrightarrow 2MNH_2 + H_2$	with NH_3 (ℓ) in presence of catalyst; with NH_3 (g) at high temperature (solutions also contain M^+ + solvated e^-)

The high reactivities of the alkali metals are exemplified by their vigorous reactions with water. Lithium reacts readily, sodium reacts so vigorously that it may ignite, and potassium, rubidium, and cesium burst into flames when dropped into water. The large amounts of heat evolved provide the activation energy to ignite the evolved hydrogen. The elements also react with water vapor in the air, or moisture from the skin.

Alkali metals should *never* be handled with bare hands.

$$2K + 2H_2O \longrightarrow 2[K^+ + OH^-] + H_2 \qquad \Delta H^0 = -390.8 \text{ kJ}$$

For this reason, alkali metals are stored under anhydrous nonpolar liquids such as mineral oil.

As is often true for elements of the second period, lithium differs in many ways from the other members of its family. Its ionic charge density and electronegativity are close to those of magnesium, so lithium compounds resemble those of magnesium in some ways. This is illustrative of the **diagonal similarities** that exist between elements in successive groups near the top of the periodic table.

IA	**IIA**	**IIIA**	**IVA**
Li	Be	B	C
Na	Mg	Al	Si

For example, lithium is the only IA metal that combines with nitrogen to form a nitride, Li_3N. Magnesium readily forms magnesium nitride, Mg_3N_2. Both metals readily combine with carbon to form carbides, whereas the other alkali metals do not react readily with carbon. The solubilities of lithium compounds are closer to those of magnesium compounds than to those of other IA compounds. The fluorides, phosphates, and carbonates of both lithium and magnesium are only slightly soluble, but their chlorides, bromides, and iodides are very soluble. Both lithium and magnesium form normal oxides, Li_2O and MgO, when burned in air at one atmosphere pressure, rather than the peroxides or superoxides formed by other alkali metals.

The IA metal oxides are all basic and react with water to produce strong soluble bases.

$$Na_2O\ (s) + H_2O \longrightarrow 2[Na^+ + OH^-]$$

$$K_2O\ (s) + H_2O \longrightarrow 2[K^+ + OH^-]$$

Since the IA cations are all related to strong soluble bases, they do not hydrolyze in their salt solutions.

21–3 Uses of the Alkali Metals and Their Compounds

Commercially, **sodium** is by far the most widely used alkali metal because it is so abundant. Its salts are absolutely essential for life. The metal itself is used as a reducing agent in the manufacture of drugs and dyes and in the metallurgy of metals such as titanium and zirconium, which are obtained from their chlorides.

$$TiCl_4\ (g) + 4Na\ (\ell) \xrightarrow{\Delta} 4NaCl\ (s) + Ti\ (s)$$

Leaded gasoline is now being phased out of use due to the air pollution problems it causes.

Sodium has been used in the form of sodium-lead alloys in the production of leaded gasoline. The metal is also used as a heat-transfer liquid in some of the fast-breeder reactors of newer nuclear power plants (see Section 28–14). Highway lamps often have sodium filaments, which produce a bright yellow glow. Just a few examples of the wide variety of uses for sodium compounds are: NaOH, called caustic soda or—when in solution—lye or soda lye (production of rayon, cleansers, textiles, soap, paper, petroleum products); Na_2CO_3, soda or soda ash, and $Na_2CO_3 \cdot 10H_2O$, washing soda (and as a substitute for NaOH when a weaker base is required); $NaHCO_3$, baking soda (baking and other household uses); NaCl (table salt and source of all other compounds of sodium and chlorine); $NaNO_3$, Chile saltpeter (a nitrogen fertilizer); $NaHSO_4$ (production of HCl from NaCl); Na_2SO_4, salt cake, a byproduct of HCl manufacture (production of brown wrapping paper and corrugated boxes); and NaH (synthesis of $NaBH_4$, which is used to recover silver and mercury from waste water).

The highly corrosive nature of both lithium and sodium is a major drawback to applications of the pure metals.

Like sodium, metallic **lithium** is used as a heat transfer medium in experimental nuclear reactors because it has the highest heat capacity of any element. Lithium is used as a reducing agent for the synthesis of many organic compounds. Lithium compounds are components of some dry cells and storage batteries as well as some glasses and ceramics. LiCl and LiBr are very hygroscopic and are used in industrial drying processes and air-conditioning. Lithium compounds are also used for the treatment of some types of mental disorders such as manic depression.

Like salts of sodium (and probably lithium), those of **potassium** are also absolutely essential for life. KNO_3, commonly known as nitre or saltpeter, is used as a potassium and nitrogen fertilizer. There are few major industrial uses for potassium that cannot be satisfied as well with the more abundant and cheaper sodium.

There are very few practical uses for the rare metals **rubidium, cesium,** and **francium.** Metallic cesium is, however, used in some photoelectric cells (Section 3–10).

The Alkaline Earth Metals (Group IIA)

21–4 Properties and Occurrence of the Group IIA Metals

The alkaline earth metals are all silvery-white, malleable, ductile, and somewhat harder than their immediate predecessors in Group IA. Activity increases from top to bottom within the group, with calcium, strontium, and barium being considered quite active. They all have two electrons in the highest energy level, which are both lost in ionic compound formation, though not as easily as the outer electron of an alkali metal. Compare their ionization energies in Tables 21–1 and 21–3. While nearly all other IIA compounds are ionic, those of beryllium exhibit a great deal of covalent character. This is due to the extremely high charge density of Be^{2+}. Its 2+ charge is spread over only a very small volume, resulting in high polarizing power. Beryllium compounds therefore resemble those of aluminum in Group IIIA. The IIA elements exhibit the +2 oxidation state in all their compounds. The tendency to react by forming 2+ ions increases from beryllium to radium.

Recall the principle of diagonal similarities mentioned in Section 21–2.

The alkaline earth metals show a wider range of chemical properties than the alkali metals. Although the IIA metals are not quite as reactive as the IA metals, they are much too reactive to occur free in nature, and are obtained by electrolysis of their molten chlorides. In order to increase the electrical conductivity of molten anhydrous beryllium chloride, which is covalent and polymeric

In Section 5–10, we found that gaseous $BeCl_2$ is linear. However, the Be atoms in $BeCl_2$ molecules act as Lewis acids and, in the solid state, accept shares in electron pairs from Cl atoms in other molecules to form polymers.

Cl Cl Cl
Be Be Be
Cl Cl Cl

small amounts of NaCl are added to the melt.

TABLE 21–3 Properties of the Group IIA Metals

Property	Beryllium	Magnesium	Calcium	Strontium	Barium	Radium
Outer electrons	$2s^2$	$3s^2$	$4s^2$	$5s^2$	$6s^2$	$7s^2$
Melting point (°C)	1283	650	845	770	725	700
Boiling point (°C)	2970	1120	1420	1380	1640	1140
Density (g/cm^3)	1.85	1.74	1.55	2.60	3.51	5
Atomic radius (Å)	1.11	1.60	1.97	2.15	2.17	2.20
Ionic radius, M^{2+} (Å)	0.31	0.65	0.99	1.13	1.35	—
Electronegativity	1.5	1.2	1.0	1.0	1.0	1.0
Standard reduction potential (V)						
$M^{2+}(aq) + 2e^- \rightarrow 2M(s)$	−1.85	−2.37	−2.87	−2.89	−2.90	−2.92
Ionization energy (kJ/mol)						
$M(g) \rightarrow M^+(g) + e^-$	899	738	590	549	503	509
$M^+(g) \rightarrow M^{2+}(g) + e^-$	1757	1451	1145	1064	965	(979)
Heat of hydration of gaseous ion (kJ/mol) (approx.)						
$M^{2+}(g) \rightarrow M^{2+}(aq)$	—	−1925	−1650	−1485	−1276	—

TABLE 21–4 Reactions of the Group IIA Metals

Reaction	Remarks
$2M + O_2 \longrightarrow 2MO$	very exothermic (except Be)
$Ba + O_2 \longrightarrow BaO_2$	almost exclusively
$M + H_2 \longrightarrow MH_2$	M = Ca, Sr, Ba at high temperatures
$3M + N_2 \longrightarrow M_3N_2$	at high temperatures
$6M + P_4 \longrightarrow 2M_3P_2$	at high temperatures
$M + X_2 \longrightarrow MX_2$	X = halogen (Group VIIA)
$M + S \longrightarrow MS$	also with Se, Te of Group VIA
$M + 2H_2O \longrightarrow M(OH)_2 + H_2$	M = Ca, Sr, Ba at room temperature; Mg gives MgO at high temperatures
$M + 2NH_3 \longrightarrow M(NH_2)_2 + H_2$	M = Ca, Sr, Ba in NH_3 (ℓ) in presence of catalyst; NH_3 (g) with heat.
$3M + 2NH_3\ (g) \longrightarrow M_3N_2 + 3H_2$	at high temperatures
$Be + 2OH^- + 2H_2O \longrightarrow Be(OH)_4^{2-} + H_2$	only with Be

Calcium and magnesium are quite abundant in the earth's crust, especially as carbonates and sulfates. Beryllium, strontium, and barium are less abundant. All known radium isotopes are radioactive and are extremely rare.

21–5 Reactions of the IIA Metals

Table 21–4 summarizes many of the reactions of the alkaline earth metals which, except for stoichiometry, are quite similar to the corresponding reactions of the alkali metals. Reactions with oxygen and hydrogen were discussed in Sections 7–9 and 7–8, and those with the halogens in Section 4–10.

Except for beryllium, all the alkaline earth metals are oxidized to oxides in air. As is true of the oxides of the alkali metals, the IIA oxides (except BeO) are basic and react with water to give hydroxides. Beryllium hydroxide, $Be(OH)_2$, is quite insoluble in water and is amphoteric. Magnesium hydroxide, $Mg(OH)_2$, is only slightly soluble in water ($K_{sp} = 1.5 \times 10^{-11}$). The hydroxides of calcium, strontium, and barium are considered water-soluble and strong bases.

Amphoterism refers to the ability of a substance to react with both acids and bases.

When one considers the position of beryllium at the top of Group IIA, it is not surprising to find that its oxide is amphoteric while the heavier members of the group have basic oxides. Metallic character increases from top to bottom within a group and from right to left across a period. This results in increasing basicity and decreasing acidity of the oxides in the same directions.

Group IA	Group IIA	Group IIIA
Li_2O basic	BeO amphoteric	B_2O_3 acidic
Na_2O basic	MgO basic	Al_2O_3 amphoteric
K_2O basic	CaO basic	Ga_2O_3 amphoteric
		In_2O_3 basic

↓ Decreasing Acidity of Oxides; Increasing Basicity of Oxides; Increasing Metallic Character of Elements

← Decreasing Acidity of Oxides; Increasing Basicity of Oxides; Increasing Metallic Character of Elements

Calcium, strontium, and barium react with water at room temperature to form their hydroxides and hydrogen (see Table 21–4). Magnesium reacts with steam to produce magnesium oxide and hydrogen. Beryllium does not react with pure water even at red heat. It dissolves in alkali as it reacts to form the complex ion, $Be(OH)_4^{2-}$, and H_2.

Despite the high covalent character of most beryllium compounds, many are soluble in water to give solutions that conduct electricity, because of the high hydration energy of Be^{2+} The tetrahydrated $[Be(OH_2)_4]^{2+}$ ion is formed

$$BeCl_2\ (s) + 4H_2O\ (\ell) \longrightarrow [Be(OH_2)_4]^{2+} + 2Cl^-\ (aq)$$

Group IIA compounds are generally less soluble in water than corresponding IA compounds, but many are quite soluble. All form hydrated ions. Because of the strong attraction of Be^{2+} for water's electrons, the Be—O bonds in $Be(OH_2)_4^{2+}$ are relatively quite strong. Thus the O—H bonds are relatively weak (Section 7–6), and the ion hydrolyzes to produce acidic solutions.

$$[Be(OH_2)_4]^{2+} + H_2O\ (\ell) \longrightarrow [Be(OH_2)_3(OH)]^+ + H_3O^+\ (aq) \qquad K_a = 1.0 \times 10^{-5}$$

The other IIA ions are hexahydrated in most solid compounds. The larger hydrated magnesium ions hydrolyze only slightly

$$[Mg(OH_2)_6]^{2+} + H_2O\ (\ell) \longrightarrow [Mg(OH_2)_5(OH)]^+ + H_3O^+\ (aq) \qquad K_a = 3.0 \times 10^{-12}$$

Hydrated calcium, strontium, and barium ions, like the IA cations, are cations of strong soluble bases and do not hydrolyze.

21–6 Uses of the Alkaline Earth Metals and Their Compounds

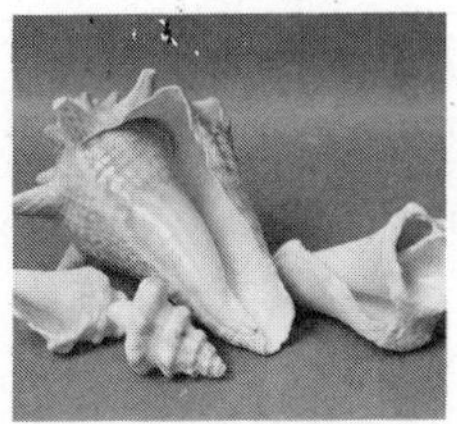

Calcium carbonate and phosphate occur in seashells and animal bones.

Calcium and its compounds are widely used commercially. The element is used as a reducing agent in the metallurgy of uranium, thorium, and other metals, as a scavenger to remove dissolved impurities such as oxygen, sulfur, and carbon in molten metals, and to remove residual gases in vacuum tubes. It is also a component of many alloys. Heating limestone produces *quicklime,* CaO, which can then be treated with water to form *slaked lime,* $Ca(OH)_2$, a cheap base for which industry finds many uses. When slaked lime is mixed with sand and exposed to the CO_2 of the air it hardens to form mortar and, with a binder, lime plaster for coating walls and ceilings. Careful heating of gypsum, $CaSO_4 \cdot 2H_2O$, produces plaster of paris, $2CaSO_4 \cdot H_2O$.

Metallic **magnesium** burns in air with such a brilliant white light that it is used in photographic flash accessories, fireworks, and incendiary bombs. It is very lightweight, and is presently used in many alloys for structural purposes. Given its inexhaustible supply in the oceans, it is likely to find many more structural uses as the reserves of iron ore dwindle and as the energy shortage dictates the construction of automobiles, trucks, and airplanes frcm lighter weight materials. The metal is used as a reagent for many important organic syntheses. When magnesite, $MgCO_3$, is thermally decomposed, *magnesia,* MgO, is produced. Magnesia is an excellent heat insulator which is used in making furnaces, ovens, and crucibles. It can be converted to $Mg(OH)_2$ by reaction with aqueous ammonium salts (it does not slake readily). A milky white aqueous suspension of finely divided $Mg(OH)_2$, called milk of magnesia, is used as a stomach antacid and as a laxative. Anhydrous $MgSO_4$ and $Mg(ClO_4)_2$

are used as drying agents. A number of other magnesium compounds are commercially important.

Because of its rarity **beryllium** has only a few practical uses. It occurs mainly as beryl, $Be_3Al_2Si_6O_{18}$, a gemstone which, with appropriate impurities, may be aquamarine (blue) or emerald (green). The metal itself wasn't readily available for industrial use until 1957. Its properties of very low density and high strength are the basis for its primary use as a structural material. It is also alloyed with copper for use in electrical contacts, springs, and nonsparking tools. Since it is transparent to x-rays, "windows" for x-ray tubes are constructed of beryllium. Beryllium compounds are quite toxic.

Strontium salts are used in fireworks and flares that show the characteristic red glow of strontium in a flame. The metal itself has no practical uses.

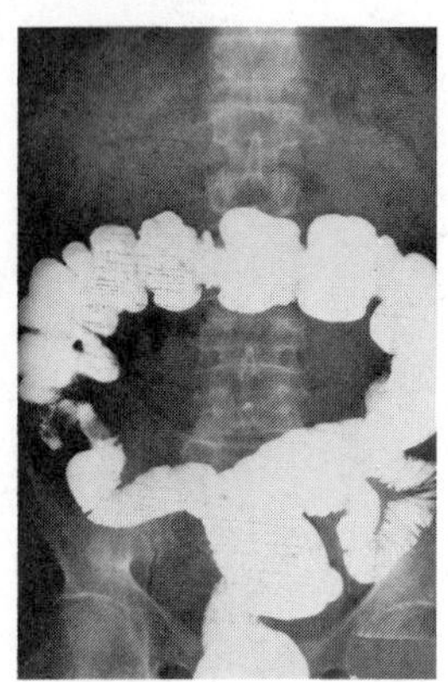

Barium is a constituent of alloys used for spark plugs because of the ease with which it emits electrons when heated. It is used as a degassing agent for vacuum tubes. A slurry of finely divided barium sulfate from barite, $BaSO_4$, is used to coat the gastrointestinal tract for obtaining x-ray photographs (see margin) because it absorbs x-rays so well. It is so insoluble that it is not poisonous; all soluble barium salts are very toxic. A combination of ZnS and $BaSO_4$ (both white) forms a very insoluble bright white paint pigment called lithopone.

The Post-Transition Metals

There are no true metals in Groups VIA, VIIA, and 0.

The properties of Bi (VA) are listed with those of the other Group VA elements in Table 24–1.

The metals below the stepwise division of the periodic table in Groups IIIA through VA are the **post-transition metals.** These include aluminum, gallium, indium, and thallium from Group IIIA; tin and lead from Group IVA; and bismuth from Group VA. Strictly speaking, we earlier (Section 4–8) classified the elements *along* the stepwise division as metalloids; however we shall discuss aluminum here because so many of its properties are characteristic of metals. Aluminum is the only post-transition metal that is considered very reactive.

21–7 Periodic Trends—Groups IIIA and IVA

IIIA	IVA	VA	VIA	VIIA	0
				1 H	2 He
5 B	6 C	7 N	8 O	9 F	10 Ne
13 Al	14 Si	15 P	16 S	17 Cl	18 Ar
31 Ga	32 Ge	33 As	34 Se	35 Br	36 Kr
49 In	50 Sn	51 Sb	52 Te	53 I	54 Xe
81 Tl	82 Pb	83 Bi	84 Po	85 At	86 Rn

Post-transition metals are shaded

The properties of the Groups IIIA and IVA elements, listed in Tables 21–5 and 21–6, respectively, vary less regularly down the groups than those of the IA and IIA metals do.

The Group IIIA elements are all solids. *Boron,* at the top of the group, is a nonmetal. Its melting point, 2300°C, is very high because it crystallizes as a covalent network solid (Sections 9–16 and 25–15). The other elements, aluminum through thallium, crystallize in metallic lattices and consequently have considerably lower melting points.

Gallium is unusual in that it melts when held in the hand, and has the largest liquid-state temperature range of any element. It is used in transistors and high temperature thermometers.

Indium is a soft, bluish metal that is used in some alloys with silver and lead to make good heat conductors. Most indium is used in electronics.

Thallium is a soft, heavy metal that resembles lead. It is quite toxic and is not used in any important practical way.

TABLE 21–5 Properties of the Group IIIA Elements

Property	Boron	Aluminum	Gallium	Indium	Thallium
Outer electrons	$2s^22p^1$	$3s^23p^1$	$4s^24p^1$	$5s^25p^1$	$6s^26p^1$
Physical state (25°C, 1 atm)	solid	solid	solid	solid	solid
Melting point (°C)	2300	660	29.8	156.6	303.5
Boiling point (°C)	2550	2327	2403	2000	1457
Density (g/cm³)	2.34	2.70	5.91	7.31	11.85
Atomic radius (Å)	0.88	1.43	1.22	1.62	1.71
Ionic radius, M^{3+} (Å)	(0.20)*	0.50	0.62	0.81	0.95
Electronegativity	2.0	1.5	1.7	1.6	1.6
Standard reduction potential (V)					
M^{3+} (aq) + $3e^-$ → M (s)	(−0.90)*	−1.66	−0.53	−0.34	(−0.34)†
Oxidation states	−3 to +3	+3	+1, +3	+1, +3	+1, +3
Ionization energy (kJ/mol)					
M (g) → M^+ (g) + e^-	801	578	576	556	586
M^+ (g) → M^{2+} (g) + e^-	2427	1817	1971	1813	1961
M^{2+} (g) → M^{3+} (g) + e^-	3660	2745	2952	2692	2867
Heat of hydration of gaseous ion (kJ/mol)					
M^{3+} (g) + xH_2O → M^{3+} (aq)	—	−4750	−4703	−4159	−4117

* For covalent +3 oxidation state.
† For Tl^+ (aq) + e^- → Tl (s); for Tl^{3+} (aq) + $2e^-$ → Tl^+ (aq), $E^0 = +1.28$V.

TABLE 21–6 Properties of the Group IVA Elements

Properties	Carbon	Silicon	Germanium	Tin	Lead
Outer electrons	$2s^22p^2$	$3s^23p^2$	$4s^24p^2$	$5s^25p^2$	$6s^26p^2$
Physical state (25°C, 1 atm)	solid	solid	solid	solid	solid
Melting point (°C)	3570	1414	937	232	328
Boiling point (°C)	sublimes	2355	2830	2270	1750
Density (g/cm³)	2.25 (graphite)	2.33	5.35	7.30 (white tin)	11.35
Atomic radius (Å)	.017	1.17	1.22	1.40	1.75
Ionic radius, M^{2+} (Å)	—	—	0.73	0.93	1.21
Electronegativity	2.5	1.8	1.9	1.8	1.7
Standard reduction potential (V)					
M^{2+} (aq) + $2e^-$ → M (s)	(+0.39)*	(+0.10)*	−0.3	−0.14	−0.126
Oxidation states	±2, ±4	±4	+2, +4	+2, +4	+2, +4

* For covalent +2 oxidation state.

The elements at the top of Group IVA, carbon, silicon, and germanium, all have relatively high melting points (especially C), like those at the top of Group IIIA, because they crystallize in covalent network lattices, while the metals, tin and lead, have lower melting points.

The atomic radii do not increase regularly upon descending Groups IIIA and IVA. Note that the atomic radius of gallium, 1.22 Å, is *less* than that of aluminum, 1.43 Å, which is directly above gallium in Group IIIA. The intervention of the transition elements between calcium (IIA) and gallium (IIIA), strontium (IIA) and indium (IIIA), and barium (IIA) and thallium (IIIA), as well as intervention of the lanthanides in the latter case, cause the radii of gallium, indium, and thallium to be smaller than would be predicted from the radii of boron and aluminum. The effect of filling the $(n-1)d$ subshell, and the corresponding increase in the nuclear charge over what it would be if there were no intervening transition elements, is the contraction of the size of the atom due to the stronger attraction of outer electrons toward the more highly charged nucleus. Likewise, in Group IVA, the atomic radii of germanium, tin, and lead are contracted for the same reasons.

Compare the radii and densities of these elements with those of the heavier IA and IIA metals.

As we saw in Chapter 4, atomic radii, and therefore atomic volumes, have a strong influence on other properties. For example, note that the densities of Ga, In, and Tl from Group IIIA and Ge, Sn, and Pb from IVA are much higher than those of the elements above them due to the tendency toward contraction of atomic volume.

The smaller than expected atomic volumes of the heavier post-transition elements also result in their valence electrons being quite strongly held. This tends to reduce the reactivities and the metallic character of these elements. Thus several of the oxides and hydroxides of these metals are amphoteric.

The Group IIIA elements all have the ns^2np^1 outer electronic configuration. Aluminum shows only the +3 oxidation state in its compounds, but the heavier metals (Ga, In, Tl) can lose or share either the single *p* valence electron or else the *p* and both *s* electrons to exhibit the +1 and +3 oxidation states, respectively. The IVA elements have the ns^2np^2 configuration. The IVA metals can lose or share the two *p* electrons to exhibit the +2 oxidation state, or they may share all four valence electrons to produce the +4 oxidation state. In general, then, the post-transition metals can exhibit oxidation states of $(g-2)+$ and $g+$ where g = periodic group number. As examples, TlCl and $TlCl_3$ both exist, as do $SnCl_2$ and $SnCl_4$. The stability of the lower state increases upon descending the groups. This is called the **inert *s*-pair effect** since the two *s* electrons remain nonionized or unshared for the $(g-2)+$ oxidation state. To illustrate, $AlCl_3$ exists but not AlCl; TlCl is more stable than $TlCl_3$. Likewise $GeCl_4$ is more stable than $GeCl_2$, but $PbCl_2$ is more stable than $PbCl_4$. However, for the IIIA metals, the +3 compounds are stabilized (relative to +1) in aqueous solution by the very high heats of hydration of the positive ions.

As is generally true, for each pair of compounds covalent character is greater for the higher (more polarizing) oxidation state.

Bi from Group VA exhibits the +3 and +5 oxidation states, but the +5 state is quite rare.

Ga, In, and Tl are all quite rare.

Let us now look closely at some of the more important post-transition metals—aluminum, tin, lead, and to a lesser extent, bismuth.

21–8 Aluminum

Aluminum is the most reactive of the post-transition metals. It is the most abundant metal in the earth's crust (7.5%) and the third most abundant element. Most is contained in aluminosilicate minerals (Section 25–11) such as clays and micas. Unfortunately, it is not economical to extract aluminum from these intractable sources. The metal itself is produced by electrolysis of Al_2O_3 obtained from its ore, *bauxite,* by the Hall process (Section 20–5, Part 1). Aluminum is quite inexpensive compared to most other metals. It is soft and can be extruded into wires or rolled, pressed, or cast into shapes very readily.

Because of its relatively low density, aluminum is often used as a lightweight structural metal. It is often alloyed with Mg and some Cu and Si to increase its strength. The Alcoa building (left) is sheathed in aluminum, which resists corrosion by forming an oxide coating.

Although pure aluminum conducts only about two thirds as much electrical current per unit volume as copper, it is only one third as dense. As a result, aluminum can conduct twice as much current as the same mass of copper. It is now used in electrical transmission lines and has been used in wiring in homes. However, the latter use has been implicated as a fire hazard due to the heat that can be generated during high current flow at the junction of the aluminum wire and fixtures of other metals.

The metal is a good reducing agent

$$Al^{3+}\ (aq) + 3e^- \longrightarrow Al\ (s) \qquad E^0 = -1.66\ V$$

Although aluminum is quite reactive, a thin transparent film of Al_2O_3 forms when aluminum comes into contact with air and this protects it from further oxidation. It is even passive toward nitric acid, HNO_3, a strong oxidizing agent, for this reason. However, if the oxide coating is brushed off with steel wool, aluminum reacts vigorously with HNO_3

Recall the HNO_3 is an oxidizing acid that does not produce hydrogen when it reacts with active metals.

$$Al\ (s) + 4HNO_3\ (aq) \longrightarrow Al(NO_3)_3\ (aq) + NO\ (g) + 2H_2O\ (\ell)$$

The very negative enthalpy of formation of aluminum oxide makes aluminum an especially good reducing agent for other metal oxides. The **thermite reaction** is a spectacular example that generates enough heat to produce molten iron for welding steel.

$$2Al\ (s) + Fe_2O_3\ (s) \longrightarrow 2Fe\ (s) + Al_2O_3\ (s) \qquad \Delta H^0 = -852\ kJ$$

Anhydrous Al_2O_3 occurs naturally as the extremely hard, high-melting mineral *corundum,* which has a network lattice. It is colorless when pure, but becomes colored when a few transition metal ions replace Al^{3+} in the crystal lattice. *Sapphire* is blue and contains some iron and titanium. *Ruby* is red due to the presence of small amounts of chromium.

Aluminum also occurs naturally, and was first discovered, in a series of double salts known as *alums.* Common **alums** are hydrated sulfates of the general formula $M^+M^{3+}(SO_4)_2 \cdot 12H_2O$. The most common alum is $KAl(SO_4)_2 \cdot 12H_2O$. Li^+, Na^+, and NH_4^+ commonly substitute for K^+, and Cr^{3+} and Fe^{3+} can substitute for Al^{3+} to form a large variety of alums. All the alums crystallize in the same kind of lattice and the large, octahedral crystals are easily grown in the laboratory. Alums are used in large quantities in the dye industry and for sizing in the paper industry. Sizing helps paper to retain its shape and to become water-repellent.

A structure of powdered aluminum and iron(III) oxide, called *thermite,* is used to weld steel rods in construction.

The amphoteric nature of aluminum hydroxide, $Al(OH)_3$, has been discussed in Section 11–8. Metallic aluminum, itself, is also amphoteric. Freshly brushed aluminum (brushing removes the protective coating of Al_2O_3) will dissolve in non-oxidizing acids to form aqueous aluminum salts and hydrogen

$$Al\ (s) + 6H^+\ (aq) \longrightarrow Al^{3+}\ (aq) + 3H_2\ (g)$$

and in solutions of strong soluble bases to form aqueous aluminate salts, or more properly tetrahydroxoaluminate salts, and hydrogen.

Most oven cleaners contain NaOH (lye). Can you see why they should not be used to clean aluminum pots and pans?

$$2Al\ (s) + 2OH^-\ (aq) + 6H_2O\ (\ell) \longrightarrow \underset{\text{aluminate ion}}{2Al(OH)_4^-} + 3H_2\ (g)$$

Most anhydrous compounds of aluminum have considerable covalent character, even Al_2O_3. Anhydrous AlF_3 is the only common exception. However, most aluminum salts are ionized in aqueous solution and crystallize from such solutions as hydrated ionic salts because of the high hydration energy of Al^{3+} ion (Table 21–5). For example, aluminum chloride crystallizes from aqueous solution as $AlCl_3 \cdot 6H_2O$, which is $Al(OH_2)_6Cl_3$.

Recall that Al^{3+} (aq) is $Al(OH_2)_6^{3+}$.

By contrast, anhydrous aluminum chloride, like $BeCl_2$, is a three-dimensional polymeric solid in which chlorine atoms form bridges between adjacent aluminum atoms via coordinate covalent bonds. In the vapor phase it exists primarily as **dimers** of $AlCl_3$, that is Al_2Cl_6. In both the solid and vapor phases, the aluminum atoms are tetrahedrally surrounded by chlorine atoms, and are sp^3 hybridized. In this way, the aluminum atoms attain an octet of electrons in their valence shells.

A **dimer** is a molecule formed by the combination of two identical smaller molecules. Compare this with BCl_3 and BF_3 (Section 5–11) which are trigonal planar molecules with sp^2 hybridization at the boron atom.

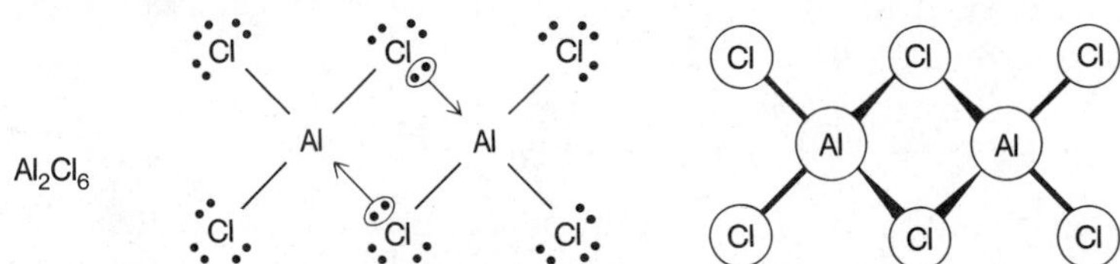

As discussed in Section 17–6, solutions of aluminum salts are acidic because Al^{3+} (aq), a small, highly charged cation, hydrolyzes extensively, with $K_a = 1.2 \times 10^{-5}$

$$Al^{3+} (aq) + H_2O (\ell) \rightleftharpoons Al(OH)^{2+} (aq) + H^+ (aq)$$

or more completely,

$$Al(OH_2)_6^{3+} + H_2O \rightleftharpoons Al(OH_2)_5(OH)^- + H_3O^+.$$

Some antiperspirants contain Al^{3+} salts. The H_3O^+ produced in such hydrolysis reactions kills bacteria in perspiration. You probably have heard of "aluminum chlorhydrate."

21–9 Tin and Lead (Group IVA) and Bismuth (Group VA)

Tin is obtained from cassiterite, SnO_2. It is used in some alloys such as solder (with lead), bronze (with copper), and "white metal." Its widest use is as in pewter and as tin plate, which is produced by dipping sheet iron or steel into molten tin or by electroplating the tin onto the sheet. Tin exists in three allotropic forms, gray tin, malleable tin, and brittle tin. Malleable tin is most common. It is silver-white and resistant to air oxidation.

Lead is bluish-white and malleable, but is more dense than tin. It is most often obtained from galena, PbS. It is used as a protective absorber of x-rays, in battery plates, and in alloys such as solder. Like most heavy metals, lead is toxic. Its use in gasoline as tetraethyllead, $Pb(C_2H_5)_4$, an anti-knock additive, is decreasing since federal regulations now prohibit its use in the United States in newly manufactured cars. Lead "poisons" the catalysts in catalytic converters. $Pb_3(OH)_2(CO_3)_2$ was formerly used as a pigment in white paints. Pb_3O_4 is still used in red corrosion-resistant outdoor paints.

Heavy metals tend to accumulate in the body; they often inhibit enzyme catalyzed reactions in the body.

These Group IVA metals, tin and lead, exhibit the +2 and +4 oxidation states. All four oxides, SnO, SnO_2, PbO, and PbO_2, are amphoteric. PbO_2 is thermally unstable and decomposes to PbO. Tin(II) oxide, SnO, and lead(II) oxide, PbO, are more basic and have greater ionic character than tin (IV) oxide,

SnO_2, and lead(IV) oxide, PbO_2. This is consistent with the general trend that oxides and hydroxides of an element in a higher oxidation state are more acidic and covalent than those of the same element in lower oxidation states.

Both Sn^{2+} and Pb^{2+} ions from *salts* hydrolyze (Sn^{2+} quite extensively) to produce acidic solutions.

Sn(II) = stannous
Sn(IV) = stannic
Pb(II) = plumbous
Pb(IV) = plumbic

$$[Sn(OH_2)_6]^{2+} + H_2O \rightleftharpoons [Sn(OH_2)_5(OH)]^+ + H_3O^+ \qquad K_a \approx 10^{-2}$$

$$[Pb(OH_2)_6]^{2+} + H_2O \rightleftharpoons [Pb(OH_2)_5(OH)]^+ + H_3O^+ \qquad K_a \approx 10^{-8}$$

The first step hydrolysis constant is 1.0×10^{-2}.

Aqueous solutions of Sn^{2+} salts are acidified to inhibit hydrolysis and subsequent precipitation of various basic salts. This, of course, is an application of LeChatelier's Principle.

Lead(IV) compounds can act as oxidizing agents as can be seen by the very positive standard reduction potential

$$Pb^{4+}\ (aq) + 2e^- \longrightarrow Pb^{2+}\ (aq) \qquad E^0 = +1.8\ V$$

Bismuth (Group VA) is a dense metal with a yellowish tinge sometimes found in the uncombined form in nature. It is also found as bismuth(III) oxide, Bi_2O_3, and as bismuth(III) sulfide, Bi_2S_3. It is incorporated into many low-melting alloys to take advantage of its unusual ability to expand upon freezing, thus giving sharp impressions of mold-cast objects. Low-melting bismuth alloys are used in such things as fire alarms, automatic sprinkler systems, and electrical fuses. Bismuth subcarbonate, $(BiO)_2CO_3$, and bismuth subnitrate $(BiO)NO_3$, are used medically in the treatment of stomach and skin disorders.

Bismuth exists primarily in the +3 oxidation state, and only rarely in the +5 oxidation state. The most common oxide, bismuth(III) oxide, Bi_2O_3, is distinctly basic. It dissolves in acids to produce bismuth(III) salts but does not dissolve in bases.

$$Bi_2O_3\ (s) + 6H^+\ (aq) \longrightarrow 2Bi^{3+}\ (aq) + 3H_2O$$

Solutions of bismuth(III) salts hydrolyze so extensively that a precipitate of a basic salt of bismuth forms. For solutions of $BiCl_3$, the reaction is

$$Bi^{3+}\ (aq) + 2H_2O + Cl^- \longrightarrow Bi(OH)_2Cl\ (s) + 2H^+\ (aq)$$

$$Bi(OH)_2Cl\ (s) \longrightarrow BiOCl\ (s) + H_2O$$

Other Bi^{3+} salts react similarly. Hydrolysis can be prevented by acidification.

Bismuth(V) in sodium bismuthate, $NaBiO_3$, is a powerful oxidizing agent. It is often used to oxidize Mn^{2+} to permanganate ion, MnO_4^-, in qualitative analysis

$$NaBiO_3\ (s) + 6H^+\ (aq) + 2e^- \longrightarrow Bi^{3+}\ (aq) + Na^+\ (aq) + 3H_2O \qquad E^0 = \sim 1.6\ V$$

Key Terms

Alkali metals Group IA metals

Alkaline earth metals Group IIA metals

Alums hydrated sulfates of the general formula $M^+M^{3+}(SO_4)_2 \cdot 12H_2O$.

Catenation the bonding together of atoms of the same element to form chains.

Diagonal similarities refers to chemical similarities of elements of Period 2 to elements of Period 3 one group to the right; especially evident toward the left of the periodic table.

Dimer molecule formed by combination of two smaller (identical) molecules.

Double salt solid consisting of two co-crystallized salts.

Inert *s*-pair effect characteristic of the post-transition metals; tendency of the two outermost *s* electrons to remain nonionized or unshared in compounds.

Post-transition metals representative metals in the "*p*-block."

Exercises

General Concepts

1. How do the acidities or basicities of metal oxides vary with oxidation number of the same metal?
2. Discuss the general differences in electronic configurations of (a) representative elements, (b) *d*-transition metals, and (c) *f*-transition metals.
3. Compare the extents to which the properties (in general) of successive elements in the periodic table differ for (a) representative metals, (b) *d*-transition metals, and (c) *f*-transition metals. Explain.
4. How do the physical properties of metals differ from those of nonmetals?
5. Draw electronic configuration diagrams for (a) Ca, (b) Ca^{2+}, (c) Mg^{2+}, (d) Al, (e) Al^{3+}, (f) Ga^{3+}, (g) Co, (h) Ni^{2+}.

Alkali and Alkaline Earth Metals

6. Compare the alkali metals with alkaline earth metals with respect to the following properties: (a) atomic radii, (b) densities, (c) first ionization energies, and (d) second ionization energies. Explain the comparisons.
7. Write general equations for reactions of alkali metals with: (a) hydrogen, (b) sulfur, (c) water, (d) halogens. Represent the metal as M and the halogen as X.
8. Summarize the chemical and physical properties of the alkali metals.
9. Describe the "diagonal relationships" in the periodic table.
10. What is hydration energy? How does it vary for cations of the alkali metals and within the alkaline-earth metals?
11. How do the standard reduction potentials of the alkali metal cations vary? How do those of the alkaline earth cations vary? Why?
12. Why are the standard reduction potentials of lithium and beryllium out of line with respect to group trends?
13. Write general equations for reactions of alkaline earth metals with: (a) hydrogen, (b) chlorine, (c) water, and (d) sulfur.
14. Calculate ΔH^0 values at 25°C for the reactions of one mole quantities of each of the following metals with stoichiometric quantities of water to form metal hydroxides and hydrogen: (a) Li, (b) K, (c) Be, and (d) Mg. What do these values indicate?

Post-Transition Metals

15. Describe the "inert-pair effect" associated with the post-transition metals. What is its cause? What are the most likely oxidation states for post-transition metals in the various groups?
16. Why are M^{3+} ions highly polarizing? What does this mean? Would high polarizing power of the metal lead to high covalent or ionic character of its bonds? For a given oxidation state, are the lighter or heavier members of a family more polarizing? Why? Answer the same questions for IVA and VA elements.
17. Write equations illustrating the amphoterism of two post-transition metal hydroxides.
18. Which hydrolyzes to the greater extent, Sn^{2+} or Pb^{2+}? Why? How can hydrolysis and precipitation of $Sn(OH)_2$ be prevented in stannous salt solutions? Relate this to LeChatelier's Principle.
19. Calculate the pH of a 0.010 *M* aqueous solution of stannous chloride. Assume that K_a for $Sn(OH_2)_6{}^{2+}$ is 1×10^{-2}.

22 The Nonmetallic Elements, Part I: The Noble Gases and Group VIIA

Only about 20% of all the elements are classified as nonmetals. With the exception of hydrogen, they are located in the upper right-hand corner of the periodic table, above the somewhat arbitrary stepwise division line.

In general, metallic character increases from *right to left* within periods and from *top to bottom* within groups. The elements along the stepwise division are called *metalloids* because their properties are intermediate between those of metals and nonmetals. In this chapter we shall consider the chemistry and properties of the noble gases (Group 0) and the halogens (Group VIIA). These two groups best illustrate group trends and individuality of elements within groups. Other nonmetals will be discussed in the next chapters.

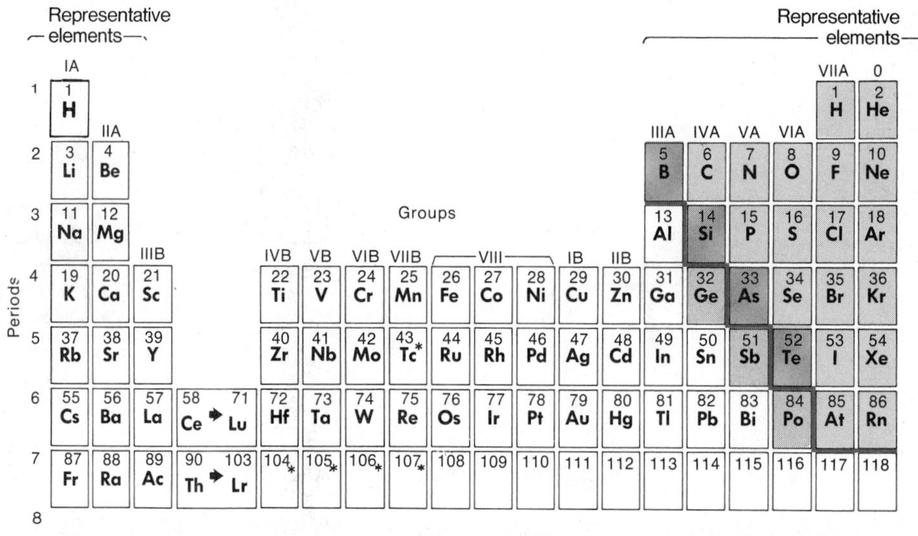

FIGURE 22–1 The periodic table with the nonmetals highlighted. The lanthanides (numbers 58-71) and the actinides (numbers 90-103) are metals, and they have been omitted.

The Noble Gases (Group 0)

22–1 Occurrence, Isolation, and Uses of the Noble Gases

The noble gases are often called the rare gases. They were formerly called the inert gases because it was incorrectly thought that they could not enter into chemical combination.

The noble gases are very low boiling gases that can be isolated by fractional distillation of liquefied air, except for radon, all of whose isotopes are radioactive. Radon must be collected from the radioactive disintegration of radium salts. Table 22–1 gives the percentage of each in the atmosphere.

TABLE 22–1 Percentages of Noble Gases in the Atmosphere (By Volume)

He	Ne	Ar	Kr	Xe	Rn
0.0005%	0.0015%	0.94%	0.00011%	0.000009%	0%

Helium is most economically recovered from helium-containing natural gas fields in the United States. This source was discovered in 1905 by H. P. Cady and D. F. McFarland at the University of Kansas, when they were asked to analyze a nonflammable component of natural gas from a Kansas gas well. The main uses of the noble gases are summarized in Table 22–2.

22–2 Physical Properties of the Noble Gases

The noble gases are colorless, tasteless, and odorless, and all have extremely low melting and boiling points. The only forces of attraction among the atoms in the liquid and solid states are very weak London or van der Waals forces. Since polarizability, and therefore interatomic interaction, increases with in-

TABLE 22–2 Uses of the Noble Gases

Noble Gas	Use	Useful Properties or Reasons
Helium	1. filling of observation balloons and other lighter-than-air craft	nonflammable; 93% of lifting power of flammable hydrogen
	2. He/O_2 mixtures rather than N_2/O_2 for deep sea breathing	He not blood-soluble, prevents nitrogen narcosis and "bends"
	3. diluent for gaseous anesthetics	nonflammable, nonreactive
	4. He/O_2 mixtures for respiratory patients	low density, flows easily through restricted passages
	5. heat transfer medium for nuclear reactors	transfers heat readily; does not become radioactive; chemically inert
	6. industrial applications, such as inert atmosphere for welding easily oxidized metals	chemically inert
	7. liquid He used to maintain very low temperatures in research (cryogenics)	extremely low boiling point
Neon	neon signs	even at low Ne pressure, moderate electric current causes bright orange-red glow; color can be modified by colored glass or mixing with Ar or Hg vapor; flashes with changes in current
Argon	1. inert atmosphere for metallurgical operations such as welding	chemically inert
	2. filling incandescent light bulbs (with Hg)	inert; inhibits vaporization of W and blackening of bulbs
Krypton	used in airport runway and approach lights	gives longer life to incandescent lights than Ar, but more expensive
Xenon	Xe and Kr mixture in high intensity, short exposure photographic flash tubes	both have fast response to electric current
Radon	radiotherapy of cancerous tissues	radioactive

creasing atomic size, the melting and boiling points increase with increasing atomic number. The forces of attraction among helium atoms are so small that helium remains liquid at one atmosphere pressure even at a temperature of 0.001 K. Table 22–3 summarizes some of the physical properties of the noble gases.

A pressure of about 26 atmospheres is required for solidification at this temperature.

TABLE 22–3 Physical Properties of Noble Gases

Property	Helium	Neon	Argon	Krypton	Xenon	Radon
Atomic number	2	10	18	36	54	86
Outer shell e^-	$1s^2$	$2s^22p^6$	$3s^23p^6$	$4s^24p^6$	$5s^25p^6$	$6s^26p^6$
Atomic radius (Å)	0.5	0.70	0.94	1.09	1.30	1.4
Melting point (°C, 1 atm)	−272.2*	−248.6	−189.3	−157	−112	−71
Boiling point (°C, 1 atm)	−268.9	−245.9	−185.6	−152.9	−107.1	−61.8
Density (g/liter at STP)	0.18	0.90	1.78	3.75	5.90	9.73
First ionization energy (kJ/mol)	2372	2081	1521	1351	1170	1037

* at 26 atm.

22–3 Chemical Properties of the Noble Gases

The consensus in the scientific community until the early 1960s was that the Group 0 elements (then called the inert gases) would not combine chemically with any elements. To do so would violate the octet rule. This opinion was strongly held even though there were many known examples of stable substances containing elements with expanded valence shells: SF_6, PCl_5, IF_7, SiF_6^{2-}, and so on.

In 1962, Neil Bartlett and his research group at the University of British Columbia were studying the powerful oxidizing agent, PtF_6. After accidentally preparing and identifying $O_2^+PtF_6^-$ by reaction of oxygen with PtF_6, Bartlett reasoned that xenon also should be oxidized by PtF_6 since the first ionization energy of molecular oxygen is actually slightly larger (1.31×10^3 kJ/mol) than that of xenon (1.17×10^3 kJ/mol). His attempts yielded a red crystalline solid initially believed to be $Xe^+PtF_6^-$ but now known to be more complex.

22–4 Xenon Compounds

Oxygen is second only to fluorine in electronegativity.

Since Bartlett's discovery, many other noble gas compounds have been synthesized. Most are compounds of xenon, and the best characterized compounds are the xenon fluorides, though oxygen compounds are also well known. Reaction of xenon with fluorine, also an extremely strong oxidizing agent, in different stoichiometric ratios produces xenon difluoride, XeF_2, xenon tetrafluoride, XeF_4, and xenon hexafluoride, XeF_6, all colorless crystals, as summarized in Table 22–4.

All the xenon fluorides are formed in exothermic reactions, and are reasonably stable with Xe—F bond energies of about 125 kJ/mol of bonds. For

TABLE 22–4 Xenon Fluorides

Compound	Preparation (Molar ratio Xe:F_2)	Reaction Conditions	Electron Pairs*	Hybridization at Xe	Geometry
XeF_2	1:1–3	400°C or irradiation or elec. discharge	5	sp^3d	Linear
XeF_4	1:5	same as for XeF_2	6	sp^3d^2	Square planar
XeF_6	1:20	300°C and 60 atm or elec. discharge	7	sp^3d^3 (?)	exact geometry undetermined

*Around the Xe atom.

comparison, strong bond energies generally range from about 170 to 500 kJ/mol, whereas bond energies of hydrogen bonds and other weak bonding interactions are typically less than 40 kJ/mol.

The Halogens (Group VIIA)

The elements of Group VIIA are known as **halogens** (Greek: salt formers). The term **halides** is used to describe their binary compounds.

22–5 Properties of the Halogens

The elemental halogens exist as diatomic molecules containing single covalent bonds. Because many of the chemical properties of the halogens are common to the whole group, it is often convenient to let X represent a halogen atom without specifying a particular halogen.

$:\ddot{\underset{..}{X}}:\ddot{\underset{..}{X}}:$ Lewis dot formula for a halogen molecule

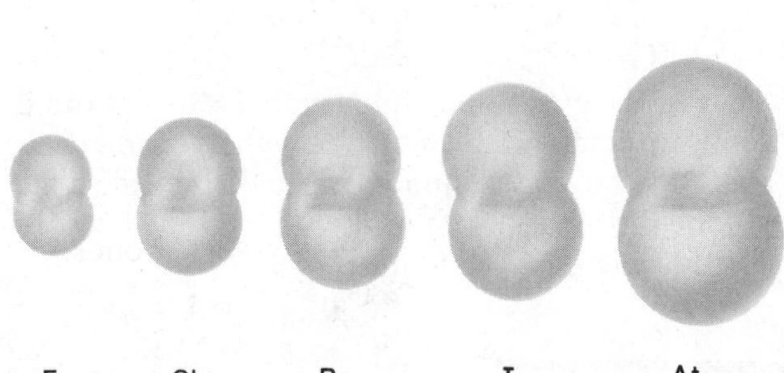

Some of the important properties of the halogens are given in Table 22–5. The properties of the halogens show rather obvious trends. The high electronegativities of the halogens indicate that they attract electrons strongly. Nearly all binary compounds containing a metal and a halogen are ionic.

The high standard reduction potentials indicate that fluorine, chlorine, and bromine are strong oxidizing agents while iodine is a mild oxidizing agent. Conversely, fluoride, chloride, and bromide ions are weak reducing agents while iodide ions are mild reducing agents.

The small fluoride ion (radius = 1.36 Å) is not easily polarized, or distorted by cations, whereas the large iodide ion (radius = 2.16 Å) is. The unit negative charge associated with each halide ion may be thought of as distributed over the entire surface of the ion. Isolated halide ions are spherical, and it is easy to see why there should be significant differences between fluoride and iodide ions. On the average, the single negative charge associated with an iodide ion (surface area = 59 $Å^2$) is spread over a much larger surface area than is the negative charge associated with a fluoride ion (surface area = 23 $Å^2$). Stated another way, the average charge density on the fluoride ion is 2.6 times greater than on the iodide ion. Additionally, the iodide ion has low-lying energy levels that contain large numbers of shielding electrons. Consequently, an I^- ion is

The shapes of monatomic ions are assumed to be spherical. The surface area of a sphere is $4\pi r^2$ or $12.57\ r^2$.

TABLE 22–5 Properties of the Halogens

Properties	Fluorine	Chlorine	Bromine	Iodine	Astatine
Physical state (25°C, 1 atm)	gas	gas	liquid	solid	solid
Color	pale yellow	yellow-green	red-brown	violet (g) black (s)	—
Atomic radius (Å)	0.64	0.99	1.14	1.33	1.40
Ionic radius (X^-)(Å)	1.36	1.81	1.95	2.16	
Outer shell e^-	$2s^22p^5$	$3s^23p^5$	$4s^24p^5$	$5s^25p^5$	$6s^26p^5$
First ionization energy (kJ/mol)	1681	1251	1140	1008	920
Electronegativity	4.0	3.0	2.8	2.5	2.1
Melting point (°C, 1 atm)	−218	−101	−7.1	114	—
Boiling point (°C, 1 atm)	−188	−35	59	184	—
X—X bond energy (kJ/mol)	158	243	192	151	—
Standard reduction potential (V) (aq. soln.)	+2.87	+1.36	+1.08	+0.54	+0.3
Heat of hydration of X^- (kJ/mol)	−510	−372	−339	−301	—
Solubility in water (mol/L, 20°C)	reacts	0.09	0.21	0.0013	—

As a result, compounds containing I^- show greater covalent character.

much more easily polarized (that is, the electron cloud is more easily distorted from spherical geometry) when it interacts with a positively charged ion such as Ag^+. Similar reasoning leads to the conclusion that small fluoride ions should not be easily polarized. The properties of chloride and bromide ions are intermediate between those of fluoride and iodide ions.

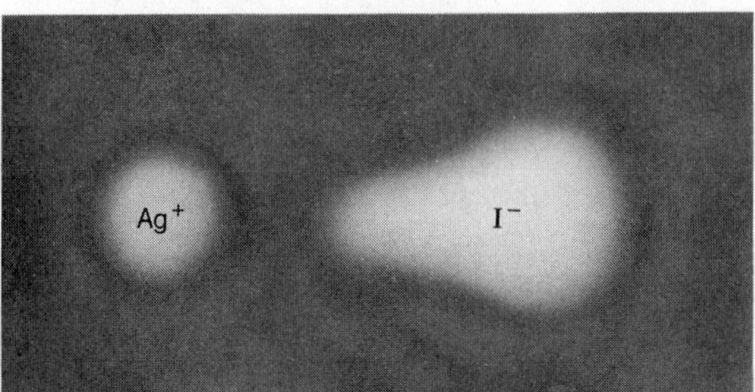

The halogens resemble each other more closely than do elements in any other periodic group, with the exception of the noble gases and possibly the Group IA metals. But their properties do differ significantly. The main reason for the rise in melting and boiling points from fluorine to iodine is the increase in atomic size and the accompanying increase in ease of polarization of outer shell electrons by adjacent nuclei. This results in greater intermolecular attractive forces as described in Section 9–9. All halogens, except the short-lived radioactive element astatine, are decidedly nonmetallic. They show the −1 oxidation number in most of their compounds although, except for fluorine, they also commonly exhibit oxidation numbers of +1, +3, +5, and +7.

22–6 Occurrence, Production, and Uses of the Halogens

The halogens are so reactive that they do not occur free in nature. The most abundant sources of halogens are halide salts, although a primary source of iodine is $NaIO_3$. The halogens are obtained by oxidation of the halide ions

Sodium iodate, $NaIO_3$, occurs as an impurity in the Chilean nitrate deposits, which are primarily $NaNO_3$.

$$2X^- \rightleftharpoons X_2 + 2e^-$$

The order of increasing ease of oxidation of halide ions is $F^- < Cl^- < Br^- < I^- < At^-$.

1 Fluorine

Fluorine occurs in large quantities in the minerals *fluorspar* or *fluorite*, CaF_2, *cryolite*, Na_3AlF_6, and *fluoroapatite*, $Ca_{10}(PO_4)_6F_2$. It also occurs in small amounts in seawater, teeth, bones, and blood. Since F_2 is such a strong oxidizing agent, it cannot be produced by chemical oxidation of fluoride ions. Rather, the pale yellow gas must be prepared by electrolysis of a molten mixture of KF + HF, or KHF_2, in a Monel cell under strictly anhydrous conditions because water is more readily oxidized than F^-.

Monel metal is an alloy of Ni, Cu, Al, and Fe that is resistant to attack by hydrogen fluoride.

$$2KHF_2 \xrightarrow[\text{melt}]{\text{Electrolysis}} \underset{\text{anode}}{F_2\ (g)} + \underset{\text{cathode}}{H_2\ (g)} + 2KF$$

Industrially and in research, fluorine is used in the direct fluorination of inorganic compounds and as a powerful oxidizing agent. Many fluorinated organic compounds, called fluorocarbons, are stable and nonflammable. They have found use as refrigerants, lubricants, plastics, insecticides, coolants, and, until recently, as aerosol propellants. A common refrigerant is freon-12 or CCl_2F_2, and CCl_3F is an insecticide. *Teflon* is a very inert plastic consisting of polymers of $—CF_2—CF_2—$ units, used for nonstick surfaces on cooking utensils and a variety of other articles requiring smooth inert surfaces (such as artificial valves for the heart). Gaseous fluorine is also used in the production of volatile uranium hexafluoride, UF_6, which is used in separating fissionable and nonfissionable uranium isotopes for nuclear reactors. In a very different area, the fluoride ion is effective in preventing dental caries because it encourages formation of fluoroapatite, $Ca_{10}(PO_4)_6F_2$, in teeth instead of the more acid-soluble apatite, $Ca_{10}(PO_4)_6(OH)_2$. The fluoride ion is added to public water supplies in carefully regulated, nontoxic amounts.

Direct reactions of F_2 with other elements or compounds are dangerous because of the vigor with which F_2 oxidizes other substances. They are carried out only rarely and with *extreme* caution.

2 Chlorine

Chlorine (Greek: *chloros*, meaning "green") occurs in abundance in NaCl, KCl, $MgCl_2$, and $CaCl_2$ in salt water and in salt beds. It is also present as HCl in gastric juices. The toxic yellowish-green gas is prepared commercially by electrolysis of concentrated aqueous sodium chloride, in which industrially important hydrogen and caustic soda (NaOH) are also produced. The electrode reactions were given in Section 19–4. A common laboratory method for the preparation of chlorine involves oxidation of HCl with manganese(IV) oxide, MnO_2

21.12 billion pounds of chlorine were produced in the United States in 1981.

Manganese(IV) oxide is often called manganese dioxide.

$$4HCl + MnO_2 \longrightarrow MnCl_2 + Cl_2 + 2H_2O$$

Chlorine is a very important reagent in the production of many widely used and commercially important products. Tremendous amounts are used in extractive metallurgy and in chlorinating hydrocarbons to produce a variety of compounds such as polyvinyl chloride, a plastic. Chlorine is present as Cl_2, $NaClO$, $Ca(ClO)_2$ or $Ca(ClO)Cl$ in household bleaches as well as in bleaches for wood pulp and textiles. Cl_2 itself is used under carefully controlled conditions to kill bacteria in public water supplies.

ClO^- is the hypochlorite ion.

3 Bromine

Bromine (Greek: *bromos,* meaning "stench") is less abundant than fluorine and chlorine. In the elemental form it is a dense, freely flowing, corrosive, dark red liquid with a brick red vapor at room temperature. The element occurs mainly in NaBr, KBr, $MgBr_2$, and $CaBr_2$ in salt water, underground salt brines, and salt beds. The major source for commercial production of bromine is seawater, although its concentration there is only 0.0066%. Seawater is normally alkaline because of the presence of CO_3^{2-} from calcium carbonate. It is first acidified with H_2SO_4 to a pH of 3.5, and chlorine gas is bubbled through it.

Approximately 70 pounds of bromine can be obtained by treatment of a million pounds of seawater.

$$Cl_2\ (g) + 2Br^- \longrightarrow 2Cl^- + Br_2\ (\ell)$$

This displacement reaction occurs because Cl_2 is a stronger oxidizing agent (more active) than Br_2 (Section 7–3).

The bromine vapor is then swept by a stream of air into a saturated alkaline solution of Na_2CO_3 in which it is trapped as a concentrated solution of sodium bromide and sodium bromate. This reaction, the disproportionation of Br_2, occurs readily in basic solution

$$3Br_2\ (g) + 3CO_3^{2-} \longrightarrow 5Br^- + BrO_3^- + 3CO_2\ (g)$$

Bromine vapor can then be obtained conveniently when desired by acidifying this solution with H_2SO_4

$$5Br^- + BrO_3^- + 6H^+ \longrightarrow 3Br_2\ (\ell) + 3H_2O$$

Bromine is used in the production of 1,2-dibromoethane, $C_2H_4Br_2$, a component of leaded gasolines that reacts with lead in automobile engines to form volatile $PbBr_2$. The $PbBr_2$ is exhausted into the atmosphere and prevents buildup of nonvolatile lead compounds inside engines. However, it poses a serious health hazard when present in the air. Its use for this purpose is declining with the phasing out of leaded fuels.

Bromine is also used in the production of silver bromide for light-sensitive eyeglasses and photographic film, in the production of sodium bromide, a mild sedative, and in methyl bromide, CH_3Br, a soil fumigant.

4 Iodine

Iodine (Greek: *iodos,* meaning "purple") is a black crystalline solid with a metallic luster. It exists in equilibrium with a violet vapor at room temperature. The element can be obtained from dried seaweed or shellfish or from $NaIO_3$ impurities in Chilean nitrate ($NaNO_3$) deposits. It is contained in the growth-regulating hormone, thyroxine, produced by the thyroid gland, as described in Chapter 20. About 0.02% of "iodized" table salt is KI, which helps prevent goiter, a condition in which the thyroid enlarges. Iodine is also used as an antiseptic and germicide in the form of tincture of iodine, a solution in alcohol. Silver iodide is used in "cloud seeding" to cause rain.

More recently available is an aqueous solution of an iodine complex of polyvinylpyrolidone or "povidone," which does not sting when applied to open wounds.

The commercial preparation of iodine involves reduction of iodate ion from $NaIO_3$ with sodium hydrogen sulfite, $NaHSO_3$

$$2IO_3^- + 5HSO_3^- \longrightarrow 3HSO_4^- + 2SO_4^{2-} + H_2O + I_2\ (s)$$

Iodine is then purified by sublimation (Figure 9–15). Elemental chlorine or bromine also can be used to displace iodine from iodide salts

$$2I^- + Cl_2 \longrightarrow I_2 + 2Cl^-$$

5 Astatine

Astatine (Greek: "the unstable one") was not isolated until 1940, when Corson, McKenzie, and Segré produced it by bombardment of bismuth targets with alpha particles accelerated in a cyclotron (Section 28–11).

$$^{209}_{83}Bi + \underset{\text{alpha particle}}{^{4}_{2}He} \xrightarrow{\text{cyclotron}} {}^{211}_{85}At + \underset{\text{neutrons}}{2\,^{1}_{0}n}$$

22–7 Reactions of the Free Halogens

The free halogens react directly with most other elements and many compounds. For example, all the Group IA metals react with all the halogens to form simple binary ionic compounds, as described in Section 4–10.

In general, the most vigorous reactions are those of F_2, which usually oxidizes other species to their highest possible oxidation states. On the other hand, I_2 is only a mild oxidizing agent (I^- is a mild reducing agent) and usually does not oxidize substances to high oxidation states. We might also say that F^- stabilizes high oxidation states of cations, while I^- stabilizes low oxidation states. As illustrations, consider the following reactions of halogens with metals that exhibit variable oxidation numbers, iron on the left and copper on the right

$$2Fe + 3F_2 \longrightarrow 2\overset{+3}{Fe}F_3\ \text{(only)}$$

$$2Fe + 3Cl_2\ \text{(excess)} \longrightarrow 2\overset{+3}{Fe}Cl_3$$

$$Fe + Cl_2\ \text{(lim. amt.)} \longrightarrow \overset{+2}{Fe}Cl_2$$

$$2Fe + 3Br_2\ \text{(excess)} \longrightarrow 2\overset{+3}{Fe}Br_3$$

$$Fe + Br_2\ \text{(lim. amt.)} \longrightarrow \overset{+2}{Fe}Br_2$$

$$Fe + I_2 \longrightarrow \overset{+2}{Fe}I_2\ \text{(only)}$$

$$Fe^{3+} + I^- \longrightarrow Fe^{2+} + \tfrac{1}{2}I_2$$

$$Cu + X_2 \longrightarrow \overset{+2}{Cu}X_2 \qquad (X = F, Cl, Br)$$

$$2Cu + I_2 \longrightarrow 2\overset{+1}{Cu}I\ \text{(only)}$$

$$\overset{+2}{Cu}{}^{2+} + 2I^- \longrightarrow \overset{+1}{Cu}I + \tfrac{1}{2}I_2$$

Table 22–6 summarizes some of the reactions of the free halogens.

TABLE 22–6 Some Common Reactions of the Free Halogens

General Reaction	Remarks
$nX_2 + 2M \longrightarrow 2MX_n$	all X_2 with most metals (most vigorous reaction with F_2 and Group IA metals)
$X_2 + nX'_2 \longrightarrow 2XX'_n$	formation of interhalogens
$X_2 + H_2 \longrightarrow 2HX$	
$3X_2 + 2P \longrightarrow 2PX_3$	with all X_2; and with As, Sb, Bi replacing P
$5X_2 + 2P \longrightarrow 2PX_5$	not with I_2; also Sb $\longrightarrow$ SbF_5, $SbCl_5$; As $\longrightarrow$ AsF_5; Bi $\longrightarrow$ BiF_5
$X_2 + H_2S \longrightarrow S + 2HX$	with all X_2
$X'_2 + 2X^- \longrightarrow 2X'^- + X_2$	$F_2 \longrightarrow Cl_2, Br_2, I_2$ $Cl_2 \longrightarrow Br_2, I_2$ $Br_2 \longrightarrow I_2$
$X_2 + C_nH_y \longrightarrow C_nH_{y-1}X + HX$	halogenation; substitution of many saturated hydrocarbons with Cl_2, Br_2
$X_2 + C_nH_{2n} \longrightarrow C_nH_{2n}X_2$	halogenation of hydrocarbon double bonds with Cl_2, Br_2

22–8 Interhalogens

The interhalogens are a class of compounds, XX'_n where X and X′ are different halogens, X being the larger of the two, and n is 1, 3, 5, or 7. They exhibit properties intermediate between those of the elemental halogens. The interhalogens are often made by direct combination in a heated nickel tube. The product depends upon reaction conditions. For example,

Note that in naming the interhalogen compounds, the less electronegative halogen is named first (positive oxidation state) and the more electronegative halogen is named last (negative oxidation state) and given an *-ide* ending. BrF is bromine fluoride, BrF_3 is bromine trifluoride, BrF_5 is bromine pentafluoride, and IF_7 is iodine heptafluoride.

$$Cl_2\,(g) + F_2\,(g) \xrightarrow{200°C} 2ClF\,(g) \qquad :\ddot{\underset{..}{Cl}}:\ddot{\underset{..}{F}}:$$

equal volumes — chlorine fluoride

$$Cl_2\,(g) + 3F_2\,(g) \xrightarrow{280°C} 2ClF_3\,(g)$$

excess — chlorine trifluoride

Lewis structure: :F:Cl:F: with :F: below Cl

Chlorine trifluoride has been described as "fluorine tricked into being a liquid" because it is so readily condensed for storage and transport, yet undergoes the fluorinating reactions of fluorine itself with almost as much release of energy. Table 22–7 shows that several interhalogens are known. Since smaller atoms are grouped around larger atoms, the maximum number of X′ atoms increases as the ratio of the radii of X to X′ increases.

The X—X′ bond energies increase as the differences in electronegativities of X and X′ increase. The interhalogens undergo reactions similar to those of

TABLE 22–7 Interhalogens

		Increasing Oxidation State of X →					
XX'		XX'_3		XX'_5		XX'_7	
ClF	colorless gas	ClF_3	colorless gas	ClF_5	colorless gas	IF_7	colorless gas
BrF	lt. red gas	BrF_3	colorless liq.	BrF_5	colorless liq.		
BrCl	lt. red gas	IF_3	yellow solid	IF_5	colorless liq.		
IF	brown solid	ICl_3	yellow solid				
ICl	dk. red liq.						
IBr	red-brn. solid						

the halogens. The shapes of the molecules are: XX′, linear; XX'_3, T-shaped; XX'_5, square pyramidal; and XX'_7, pentagonal bipyramidal. Examples are shown below.

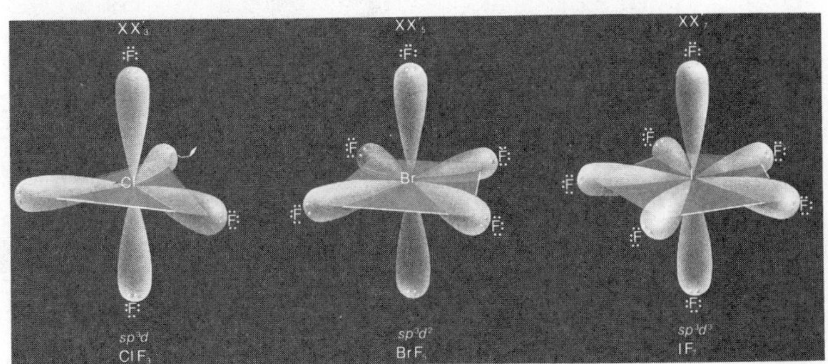

Bromine Chloride as a Disinfectant

For years Cl_2 has been used as a disinfectant for wastewater treatment. Its hydrolysis produces hypochlorous acid, HClO (Section 22–14), which in turn decomposes to produce atomic oxygen, the active disinfectant, and hydrochloric acid. However, Cl_2 may be replaced by the interhalogen bromine chloride, BrCl, since BrCl hydrolyzes exclusively to hypobromous acid, HBrO, and HCl, and it does so at a rate that exceeds that for the hydrolysis of either Cl_2 or Br_2. Its hydrolysis constant is 4000 times greater than that of Br_2. It reacts with ammonia and its organic derivatives, the amines (Section 30–9), in wastewater to form bromamines rather than chloramines, and the bromamines are less toxic to fish. The bromamines are also disinfectants in themselves, and along with the hypobromous acid, are effective over a wider pH range than are chloramines.

Although BrCl itself is more expensive than Cl_2, when all expenses involved are considered it has a lower relative cost, and is less dangerous to handle.

22–9 The Hydrogen Halides

The hydrogen halides (Figure 22–2) are all colorless gases that dissolve in water to give acidic solutions called hydrohalic acids. The gases all have pierc-

For example, aqueous solutions of hydrogen fluoride are called hydrofluoric acid.

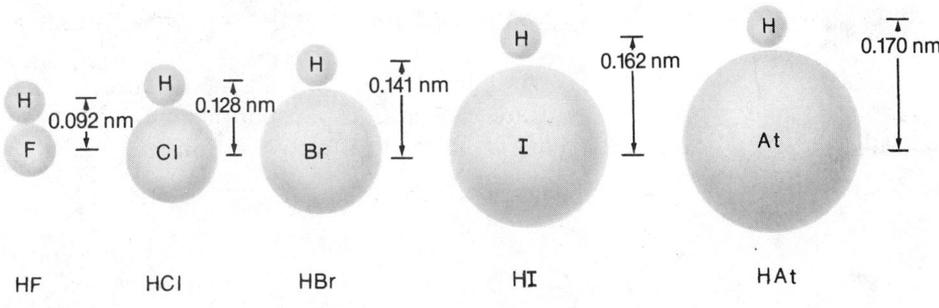

FIGURE 22–2 Relative sizes (approximate) of the hydrogen halides. The distance between the center of the hydrogen atom and the center of the halogen atom (internuclear distance) is indicated for each molecule.

TABLE 22–8 Properties of the Hydrogen Halides

Property	HF	HCl	HBr	HI
Physical state (25°C, 1 atm)	colorless gas	colorless gas	colorless gas	colorless gas
Melting point (°C)	−83.1	−114.8	−86.9	−50.7
Boiling point (°C)	19.54	−84.9	−66.8	−35.4
H—X bond energy (kJ/mol)	569	431	368	297
Apparent % ionization in 0.1 *M* aq. sol'n (18°C)	<10 (wk. acid)	92.6 (st. acid)	93 (st. acid)	95 (st. acid)

ing, irritating odors. Some of the properties of the hydrogen halides are given in Table 22–8.

22–10 Preparation of Hydrogen Halides

All the hydrogen halides may be prepared by direct combination of the elements

$$H_2 + X_2 \longrightarrow 2HX\ (g) \qquad X = F, Cl, Br, I$$

A photochemical reaction is one in which some species (usually a molecule) interacts with radiant energy to produce very reactive species, which then undergo further reaction.

However, the reaction with F_2 to produce HF is explosive and very dangerous under all conditions. The reaction producing HCl does not occur significantly in the dark but occurs rapidly via a photochemical chain reaction when exposed to light. Light energy is absorbed by chlorine molecules, which break apart into very reactive chlorine **radicals** (atoms with unpaired electrons). These subsequently attack hydrogen molecules and produce HCl molecules, leaving hydrogen atoms, also radicals, behind. The hydrogen radicals, in turn, attack chlorine molecules to form HCl molecules and chlorine radicals.

$$Cl_2 \xrightarrow{h\nu} 2{:}\ddot{\underset{..}{Cl}}\cdot \qquad \text{initiation step}$$

$$\left.\begin{array}{l} {:}\ddot{\underset{..}{Cl}}\cdot + H_2 \longrightarrow HCl + H\cdot \\ H\cdot + Cl_2 \longrightarrow HCl + {:}\ddot{\underset{..}{Cl}}\cdot \end{array}\right\} \text{chain propagation steps}$$

This chain reaction continues as long as there is a significant concentration of radicals. The following reaction termination steps eliminate (by combination) two radicals and eventually terminate the reaction.

$$\left.\begin{array}{l} H\cdot + H\cdot \longrightarrow H_2 \\ {:}\ddot{\underset{..}{Cl}}\cdot + {:}\ddot{\underset{..}{Cl}}\cdot \longrightarrow Cl_2 \\ H\cdot + {:}\ddot{\underset{..}{Cl}}\cdot \longrightarrow HCl \end{array}\right\} \text{termination steps}$$

The reaction of H_2 with Br_2 is also a photochemical reaction. The reaction of H_2 with I_2 is very slow, even at high temperatures and with illumination.

A safer way of preparing HF or HCl is the reaction of a metal halide with a nonvolatile acid, such as concentrated sulfuric or phosphoric acid. The more volatile hydrogen halide is evolved from the resulting solution as a gas.

$$\underset{}{NaF\ (s)} + \underset{bp\,=\,338°C}{H_2SO_4\ (\ell)} \longrightarrow NaHSO_4\ (s) + \underset{bp\,=\,20°C}{HF\ (g)}$$

HBr and HI are not prepared from H_2SO_4 because they are oxidized to Br_2 or I_2 by hot concentrated H_2SO_4. H_3PO_4 is a nonoxidizing, nonvolatile acid that can be used to prepare HBr and HI from their salts

$$\underset{\text{bp} = 213°\text{C}}{NaBr\ (s) + H_3PO_4\ (\ell)} \longrightarrow NaH_2PO_4\ (s) + \underset{\text{bp} = -66.8°\text{C}}{HBr\ (g)}$$

Nonmetal halides hydrolyze to produce hydrogen halides and an acid or oxide of the nonmetal

Note that oxidation numbers remain constant in such reactions.

$$BCl_3\ (\ell) + 3H_2O \longrightarrow H_3BO_3\ (aq) + 3HCl\ (g)$$

$$SiCl_4\ (\ell) + 2H_2O \longrightarrow SiO_2\ (s) + 4HCl\ (g)$$

Hydrogen bromide and hydrogen iodide are often prepared by hydrolysis of phosphorus tribromide and phosphorus triiodide, respectively, and are separated from the reaction mixture by distillation

$$PX_3 + 3H_2O \longrightarrow H_3PO_3\ (s) + 3HX\ (g) \qquad X = Cl, Br, I$$

The phosphorus trihalide is usually formed at the time of reaction by mixing red phosphorus with the free halogen.

A commercially important class of reactions is the *halogenation of saturated hydrocarbons.* Hydrogen halides are by-products of such reactions. A general reaction of this type is

$$C_nH_y + X_2 \longrightarrow C_nH_{y-1}X + HX \qquad X = F, Cl, Br, I$$

However, fluorination of hydrocarbons with F_2 is so dangerously explosive that such reactions are not performed. Halogenation of hydrocarbons with I_2 is not economically practical.

22–11 Bonding and Acidity of Anhydrous Hydrogen Halides

The anhydrous hydrogen halides are all covalent molecules. The hydrogen-halogen electronegativity difference (ΔEN) decreases and bond energy decreases with increasing atomic weight (Table 22–8; see also Section 5–1). Thus acidity of the hydrogen halides, which is proportional to the ease of rupture of the H—X bond, increases in the same order.

The hydrogen halides are much less acidic in the pure state than in aqueous solution.

	HF	HCl	HBr	HI
ΔEN:	1.9	0.9	0.7	0.4

increasing acid strength →

The abnormally high melting and boiling points of HF (Table 22–8) are due to very strong hydrogen bonding that is present in solid, liquid, and even gaseous phases. Measurements of electron diffraction and vapor density indicate that a number of different polymeric units, up to $(HF)_5$, exist in gaseous hydrogen fluoride at relatively low temperatures but above its boiling point, 19.5°C. Above 88°C, apparently only HF units are present. The hydrogen bonding is similar to that in H_2O (Section 9–9, part 3). The dashed lines in Figure 22–3 represent hydrogen bonds.

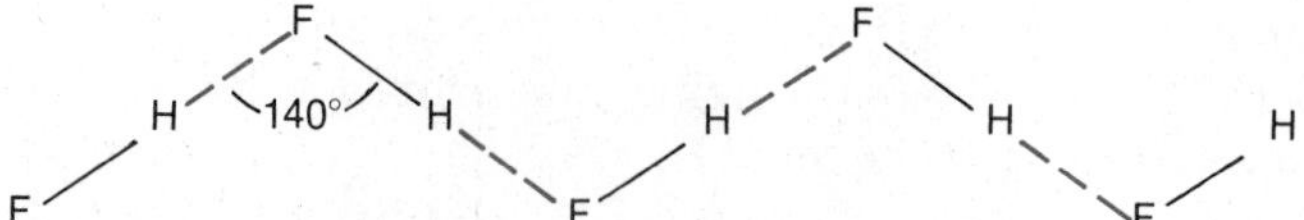

FIGURE 22–3 Hydrogen bonding (dashed lines) in pure, i.e., anhydrous, HF.

22–12 Aqueous Solutions of Hydrogen Halides (Hydrohalic Acids)

The hydrogen halides all dissolve readily in water, producing hydrohalic acids that ionize as shown

$$\mathrm{H{:}\ddot{O}{:}H} + \mathrm{H{:}\ddot{X}{:}} \rightleftharpoons [\mathrm{H{:}\ddot{O}{:}H}]^{+}_{\,\mathrm{H}} + \mathrm{{:}\ddot{X}{:}^{-}}$$

The reaction is essentially complete for dilute HCl (aq), HBr (aq), and HI (aq), which are classified as strong acids. However, because of the strong H—F bonds as well as hydrogen bonding between HF and H_2O (Figure 22–4), dilute HF (aq) is a weak acid with $K_a = 7.2 \times 10^{-4}$. In concentrated solution the more acidic dimeric $(HF)_2$ units are present, and they ionize as shown below.

$$(HF)_2 + H_2O \rightleftharpoons H_3O^+ + HF_2^- \qquad K \approx 5$$

This is the leveling effect (Section 11–5).

Water is a sufficiently strong base that it does not discriminate among the acid strengths of hydrochloric, hydrobromic, and hydroiodic acids. However, reactions in less basic solvents, such as anhydrous acetic acid, indicate that the order of increasing acid strengths of the hydrohalic acids is the same as for the anhydrous hydrogen halides: HF (aq) $\ll$ HCl (aq) $<$ HBr (aq) $<$ HI (aq).

HCl (aq) is known commercially as muriatic acid.

The only acid used to a greater extent in industry than hydrochloric acid is sulfuric acid. Hydrochloric acid is used in the production of metal chlorides, dyes, and many other commercially important products. It is also used on a large scale for dissolving metal oxide coatings from iron and steel prior to galvanizing or enameling.

Hydrofluoric acid is used in the production of fluorine-containing compounds, and for etching glass. The acid reacts with silicates, such as calcium silicate, $CaSiO_3$, in the glass to produce a very volatile and thermodynamically stable compound, silicon tetrafluoride, SiF_4

$$CaSiO_3\ (s) + 6HF\ (aq) \longrightarrow CaF_2\ (s) + SiF_4\ (g) + 3H_2O\ (\ell)$$

Hydrobromic and hydroiodic acids are relatively less important, although they are utilized in the synthesis of bromine-containing and iodine-containing organic compounds and in chemical research.

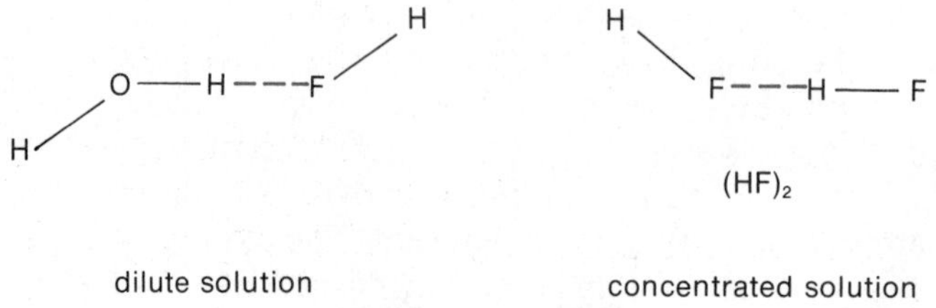

H—O—H- - -F—H

dilute solution

H—F- - -H—F

$(HF)_2$

concentrated solution

FIGURE 22–4 Hydrogen bonding (dashed lines) in dilute and concentrated aqueous solution of hydrofluoric acid.

22–13 Halides of Other Elements

The halogens combine with most other elements to form halides. They range from high melting, water-soluble, ionic electrolytes (metal halides) such as NaCl, KBr, and $CaCl_2$, to volatile, covalent nonelectrolytes (nonmetal halides) like PCl_3 and SF_6. Generally the former properties are those of halides of metals, and the latter properties characterize the halides of nonmetals.

Halides (increasing nonmetallic character of other elements) →

metal halide; ionic lattice; metal ions slightly hydrolyzed if at all; X^- not hydrolyzed except F^-. Examples: NaCl, MgF_2, $NiBr_2$	⟶	intermediate properties	⟶	**nonmetal halide** or complex halide; molecular (covalent) compounds, volatile; completely hydrolyzed. Examples: BF_3, PCl_5, SF_4, $AsCl_3$

22–14 The Oxyacids (Ternary Acids) of the Halogens and Their Salts

Table 22–9 lists the known oxyacids of the halogens and their sodium salts, and some trends in properties.

Only three oxyacids, $HClO_4$, HIO_3, and H_5IO_6, have been isolated in anhydrous form. The others are known only in aqueous solution. In all these acids the hydrogen is bonded through an oxygen.

The Lewis dot formulas and structures of the chlorine oxyanions are shown in Figure 22–5. The four oxyanions of chlorine all have tetrahedral *electronic geometry* but only the perchlorate ion, ClO_4^- has tetrahedral *ionic geometry.* The other three have the ionic geometries that can be predicted by

TABLE 22–9 Oxyacids of the Halogens and Their Salts

Oxidation State	Acid	Name of Acid	Thermal Stability and Acid Strength	Oxidizing Power of Acid	Sodium Salt	Name of Salt	Thermal Stability	Oxidizing Power and Hydrolysis of Anion	Nature of Halogen
+1	HXO	Hypohalous acids	Increases ↓ (from +1 to +7)	Increases ↑ (from +7 to +1)	NaXO	sodium hypohalite	Increases ↓ (from +1 to +7)	Increases ↑ (from +7 to +1)	X = F*, Cl, Br, I
+3	HXO_2	Halous acids			$NaXO_2$	sodium halite			X = Cl, Br (?)
+5	HXO_3	Halic acids			$NaXO_3$	sodium halate			X = Cl, Br, I
+7	HXO_4	Perhalic acids			$NaXO_4$	sodium perhalate			X = Cl, Br, I
+7	H_5XO_6	Paraperhalic acids			several types	sodium paraperhalates			X = I

*The oxidation state of F is −1 in HOF.

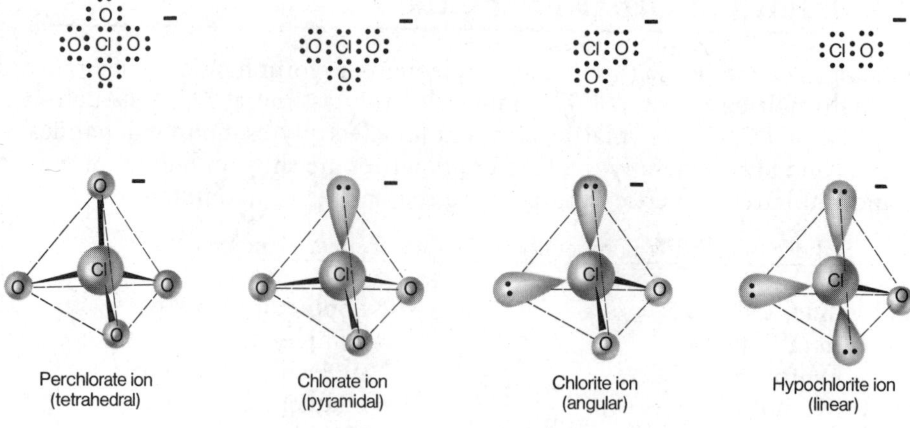

FIGURE 22–5 Lewis dot formulas and structures of the oxyanions of chlorine.

the (imaginary) successive removal of oxygen atoms. The corresponding oxyanions of bromine and iodine have similar structures.

Since oxygen is so electronegative, greater numbers of oxygen atoms around the halogen cause the electron density of the H—O bonds of the acids to be shifted more toward oxygen. This results in an increase in the ease of H—O bond cleavage (Section 11–8); therefore, formation of a hydrogen ion and an anion becomes easier as the number of oxygen atoms increases. Such behavior is observed for the series of oxyacids of any element, *if* all the H atoms are bonded to O atoms. Consistent with this is the observation that the basic strength, or degree of hydrolysis, of the anion of a salt decreases in the same order.

When concentrated, all of these acids are strong oxidizing agents; oxidizing power decreases with increasing number of oxygen atoms and increasing oxidation state of the halogen.

In HOF the oxidation states are: F = −1, H = +1, O = 0.

The only oxyacid of fluorine that has been prepared is hypofluorous acid, HOF. It can be prepared in the vapor phase by passing fluorine rapidly over ice.

$$F_2 + H_2O \xrightarrow{0°C} HOF + HF$$

It is difficult to isolate because HOF reacts with water to produce hydrofluoric acid and hydrogen peroxide.

Aqueous *hypohalous acids* (except HOF) can be prepared by reaction of free halogens (Cl_2, Br_2, I_2) with cold water.

These reactions all involve disproportionation of the halogen.

Hypohalite ions can be represented as either XO^- or OX^-.

$$\overset{0}{X_2} + H_2O \rightleftharpoons \underset{\text{hydrohalic acid}}{\overset{-1}{HX}} + \underset{\text{hypohalous acid}}{\overset{+1}{HOX}} \qquad X = Cl, Br, I$$

The smaller the halogen, the farther to the right the equilibrium lies.

Hypohalite salts can be prepared by reactions of the halogens with *cold* dilute bases.

$$\overset{0}{X_2} + 2NaOH \longrightarrow \underset{\text{sodium halide}}{\overset{-1}{NaX}} + \underset{\text{sodium hypohalite}}{\overset{+1}{NaOX}} + H_2O \qquad X = Cl, Br, I$$

The hypohalites are used as bleaching agents. Sodium hypochlorite is made on a huge commercial scale by electrolyzing aqueous sodium chloride and mixing two of the products, Cl_2 and NaOH (Section 19–4).

Solid household bleaches are usually $Ca(ClO)Cl$, prepared by reaction of Cl_2 with $Ca(OH)_2$.

$$Ca(OH)_2 + Cl_2 \longrightarrow Ca(ClO)Cl + H_2O$$

Sometimes Cl_2 is used as a bleach or as a disinfectant, as in public water supplies, but the above reactions provide the hypochlorite ion when the Cl_2 is introduced into water.

$$\underset{\text{hypohalite ion}}{OX^-} + H_2O \rightleftharpoons \underset{\text{hypohalous acid}}{HOX} + OH^-$$

The hypochlorous acid produced by this hydrolysis then undergoes partial decomposition to HCl and O radicals.

$$HOCl \longrightarrow HCl + :\ddot{O}\cdot$$

These oxygen radicals, which are very strong oxidizing agents, are the effective bleaching and disinfecting agent in aqueous solutions of Cl_2 or hypochlorite salts. Upon heating, the hypohalite ions disproportionate to produce halates and halides.

$$3\overset{+1}{X}O^- \xrightarrow{\Delta} \underset{\text{halate ion}}{\overset{+5}{X}O_3^-} + \underset{\text{halide ion}}{2\overset{-1}{X}^-} \qquad X = Cl, Br, I$$

Anhydrous *halous acids* have not been isolated. Fluorous and iodous acids apparently do not exist, even in aqueous solution, and the existence of bromous acid is questionable. Chlorite ion has long been known, and aqueous solutions of bromite ion have been prepared recently.

Chlorine dioxide, ClO_2, undergoes disproportionation in basic solution to produce chlorite and chlorate salts

$$2\overset{+4}{Cl}O_2 + 2NaOH \longrightarrow \underset{\text{sodium chlorite}}{Na\overset{+3}{Cl}O_2} + \underset{\text{sodium chlorate}}{Na\overset{+5}{Cl}O_3} + H_2O$$

A convenient preparation of aqueous chlorous acid involves reaction of chlorine dioxide with barium hydroxide to produce barium chlorite, a suspension of which is treated with sulfuric acid. The precipitated barium sulfate is removed by filtration, leaving a solution of $HClO_2$

$$\underset{\text{barium chlorite}}{Ba(ClO_2)_2} + H_2SO_4 \longrightarrow BaSO_4 + \underset{\text{chlorous acid}}{2HClO_2}$$

The *halate* salts can be prepared by reaction of free halogens (except F_2) with *hot aqueous* alkali, or by thermal decomposition of hypohalites

$$3\overset{0}{X}_2 + 6NaOH \longrightarrow \underset{\text{sodium halate}}{Na\overset{+5}{X}O_3} + 5Na\overset{-1}{X} + 3H_2O \qquad X = Cl, Br, I$$

Aqueous solutions of chloric acid and bromic acids are prepared by reactions of barium chlorate and bromate with sulfuric acid

$$Ba(XO_3)_2 + H_2SO_4 \longrightarrow \underset{\text{precipitate}}{BaSO_4} + \underset{\text{halic acid}}{2HXO_3} \qquad X = Cl, Br$$

When stoichiometric amounts of reactants are mixed, Ba^{2+} and SO_4^{2-} ions combine to form insoluble $BaSO_4$, which can be separated by filtration. Fairly pure $HClO_3$ and $HBrO_3$ solutions remain.

All *halates* decompose upon heating. Potassium chlorate is used as a convenient laboratory source of small amounts of oxygen, especially when heated with a catalyst like MnO_2, which lowers the temperature of decomposition

$$2K\overset{+5}{Cl}O_3 \xrightarrow[MnO_2]{\Delta} 2K\overset{-1}{Cl} + 3O_2$$

Gentle heating without a catalyst, however, produces potassium perchlorate and potassium chloride by disproportionation:

$$4K\overset{+5}{Cl}O_3 \longrightarrow 3K\overset{+7}{Cl}O_4 + K\overset{-1}{Cl}$$

Neither perfluoric acid nor perfluorate ion is known. The other *perhalic acids* and *perhalates* are known. Anhydrous perchloric acid, $HClO_4$, a colorless oily liquid, distills under reduced pressure (10–20 torr) after reaction of a perchlorate salt with nonvolatile concentrated sulfuric acid

$$NaClO_4 + H_2SO_4 \longrightarrow NaHSO_4 + \underset{\text{perchloric acid distills}}{HClO_4}$$

Treatment of $HClO_4$ with a strong dehydrating agent like P_4O_{10} produces its explosive acid anhydride, dichlorine heptoxide, Cl_2O_7. The oxidation state of Cl is +7 in both compounds.

Hot, concentrated perchloric acid can explode in the presence of reducing agents, especially organic reducing agents. While hot, concentrated $HClO_4$ solutions are very strong oxidizing agents, cold, dilute perchloric acid is only a weak oxidizing agent. In fact, the hydrogen ions in such solutions are reduced to hydrogen with no reduction of perchlorate ions in the presence of fairly strong reducing agents, such as metallic zinc

$$Zn + \underset{\text{cold, dilute}}{2HClO_4} \longrightarrow Zn(ClO_4)_2 + H_2$$

Perchloric acid is the strongest of all common acids with respect to ionization, exceeding even HNO_3 and HCl. It is used commercially as a substitute for sulfuric acid under conditions in which H_2SO_4 could be reduced to SO_2.

22–15 Pseudohalides and Pseudohalogens

Several uni-negative inorganic groups resemble the halide ions in both covalent and ionic compounds (Table 22–10). They are called **pseudohalide** ions (or halogenoid ions). Corresponding dimeric pseudohalogens of four of them have been prepared and characterized.

TABLE 22–10 Pseudohalides and Pseudohalogens

Pseudohalide Ions		Pseudohalogen	
CN^-	cyanide ion	$(CN)_2$	cyanogen
OCN^-	cyanate ion		
ONC^-	isocyanate ion		
SCN^-	thiocyanate ion	$(SCN)_2$	thiocyanogen
$SeCN^-$	selenocyanate ion	$(SeCN)_2$	selenocyanogen
$TeCN^-$	tellurocyanate ion		
N_3^-	azide ion		
$SCSN_3^-$	azidothiocarbonate ion	$(SCSN_3)_2$	azidocarbondisulfide

Among the characteristics that pseudohalogens and halogens have in common are that they:

1. are volatile in the free state
2. are isomorphous (Section 9–15) in free (solid) state
3. combine with metals to form salts; heavy metal salts are insoluble in water
4. form H^+ acids in H_2O (all involving pseudohalogens are weak acids)
5. form interhalogenoids and polyhalogenoid ions
6. form double and complex salts
7. have similar preparative reactions

The following reactions illustrate some of the similarities in reactions

1. $Cl_2 + 2OH^- \longrightarrow Cl^- + OCl^- + H_2O\ (\ell)$ (halogen)
   $(CN)_2 + 2OH^- \longrightarrow CN^- + OCN^- + H_2O\ (\ell)$ (pseudohalogen)
2. $MnO_2\ (s) + 4H^+ + 2Cl^- \longrightarrow Cl_2 + Mn^{2+} + 2H_2O$
   $MnO_2\ (s) + 4H^+ + 2SCN^- \longrightarrow (SCN)_2 + Mn^{2+} + 2H_2O$
3. $PbCl_4 \xrightarrow{\Delta} PbCl_2 + Cl_2$
   $Pb(SCN)_4 \xrightarrow{\Delta} Pb(SCN)_2 + (SCN)_2$

The order of *decreasing oxidizing power* of the halogens and some pseudohalogens has been determined, and is found to be

$$F_2 > Cl_2 > Br_2 > (CN)_2 > (SCN)_2 > I_2 > (SCSN_3)_2 > (SeCN)_2$$

The order of *increasing reducing power* of the halide and pseudohalide anions is

$$F^- < ONC^- < OCN^- < Cl^- < N_3^- < Br^- < CN^- < SCN^- < SCSN_3^- < I^- < SeCN^- < TeCN^-$$

Key Terms

Halogens Group VIIA elements; F, Cl, Br, I, At.

Noble gases Group 0 elements; He, Ne, Ar, Kr, Xe, Rn.

Exercises

The Noble Gases

1. Why are the noble gases so unreactive?
2. Why were the noble gases among the last elements to be discovered?
3. List some of the uses of the noble gases and reasons for the uses.
4. Arrange the noble gases in order of increasing: (a) atomic radii, (b) melting points; (c) boiling points; (d) densities; (e) first ionization energies.
5. Explain the order of increasing melting and boiling points of the noble gases in terms of polarizabilities of the atoms and forces of attraction between them.
6. What gave Neil Bartlett the idea that compounds of xenon could be synthesized?
7. Describe the bonding and geometry in XeF_2, XeF_4, and XeF_6.
8. Compare the bond energies of Xe—F bonds with other typical bonds and with hydrogen bonds.
9. Suggest a reason why neon is not likely to form covalent compounds in light of your response to Exercise 4(e).
10. How many grams of xenon oxide tetrafluoride, $XeOF_4$, and how many liters of HF at STP, could be prepared, assuming complete reaction of 6.50 grams of xenon tetrafluo-

ride, XeF_4, with a stoichiometric quantity of water according to the equation below?

$$6XeF_4\ (s) + 8H_2O\ (\ell) \longrightarrow 2XeOF_4\ (\ell) + 4Xe\ (g) + 16HF\ (g) + 3O_2\ (g)$$

The Halogens

11. Draw Lewis dot formulas for I_2 and Cl_2.
12. Describe the normal physical states and colors of the free halogens.
13. Why do the halogens not occur free in nature?
14. List the halogens in order of increasing: (a) atomic radii, (b) ionic radii, (c) electronegativities, (d) melting points, (e) boiling points, (f) standard reduction potentials.
15. What is the order of increasing X—X (halogen) bond energies? Suggest why the F—F bond energy is less than the Cl—Cl bond energy.
16. Discuss the polarizabilities of halide ions.
17. Give examples of halogen compounds in which a halogen exhibits the +1, +3, +5, and +7 oxidation states. Are these compounds ionic or covalent?
18. Why can fluorine not be prepared by electrolysis of aqueous solutions of fluoride salts?
19. Write the equations describing the half-reactions and net reaction for the electrolysis of molten KF/HF mixtures. At which electrodes are the products formed? What is the purpose of the HF?
20. Write an equation for a common small-scale laboratory preparation of Cl_2 from hydrochloric acid.
21. Discuss the chemistry of the extraction of bromine from seawater.
22. Give two practical uses for compounds of each of the halogens.
23. Write equations describing general reactions of the free halogens, X_2, with (a) Group IA (alkali) metals, (b) Group IIA (alkaline earth) metals, (c) Group IIIA metals. Represent the metals as M.
24. Write equations illustrating the tendency of F^- to stabilize high oxidation states of cations and the tendency of I^- to stabilize low oxidation states. Why is this the case?
25. Why are the free halogens more soluble in water than most nonpolar molecules?
26. The interhalogens are compounds of two different halogens, X and X′, having general formulas XX′, XX'_3, XX'_5, and XX'_7. X is always the larger halogen and the central atom. Draw structures and give hybridizations at the central atom for BrF_3, ClF_5, and IF_7.
27. Ions of the interhalogens (Exercise 26) also exist. An example is the ICl_4^- ion in $KICl_4$. Sketch the ion and indicate the hybridization at I in an ICl_4^- ion.
28. Distinguish between hydrogen bromide and hydrobromic acid.
29. What is the order of decreasing melting and boiling points of the hydrogen halides? Why is the HF out of line?
30. Give a reaction illustrating each of the four general methods for preparation of hydrogen halides.
31. Compare the order of acid strengths of the hydrogen halides with those of the hydrohalic acids.
32. Describe the effect of hydrofluoric acid on glass.
33. Compare the general properties of metal halides with those of nonmetal halides.
34. What is the acid anhydride of perchloric acid?
35. Write the equation for the dehydration of $HClO_4$ with tetraphosphorus decoxide.
36. Name the following compounds: (a) $KBrO_3$, (b) KBrO, (c) $NaClO_4$, (d) $NaClO_2$, (e) HBrO, (f) $HBrO_3$, (g) HIO_3, (h) $HClO_4$.
37. Draw Lewis formulas and structures of the four ternary acids of chlorine.
38. Write equations describing reactions by which the following compounds can be prepared:
    (a) hypohalous acids of Cl, Br, and I (in solution with hydrohalic acids)
    (b) hypohalite salts
    (c) chlorous acid
    (d) halate salts
    (e) a perchlorate salt
    (f) perchloric acid
39. What is the order of increasing acid strength of the ternary chlorine acids? Explain the order.

40. The pseudohalogens are compounds that behave chemically much like the diatomic halogen molecules. Cyanogen, $(CN)_2$ is a pseudohalogen. How would you expect cyanogen to react with ethylene (or ethene),

$$\begin{matrix} H & & & & H \\ & \diagdown & & \diagup & \\ & & C{=}C & & \\ & \diagup & & \diagdown & \\ H & & & & H \end{matrix} \quad ?$$

41. Calculate the surface area ($4\pi r^2$) in square Angstroms (Å) for all the halogen atoms and for the halide ions.
42. Calculate the ratio of rates of diffusion of each of the halogen molecules to that of the fluorine molecule.
43. The volume of a sphere is given by the equation $V = \frac{4}{3}\pi r^3$. Given the radii of the halogen atoms and their atomic weights, calculate the densities of the halogen atoms in grams per milliliter. (Note: These densities do *not* include the empty space between atoms in samples of the elements.)
44. A 0.250 ampere current is applied to molten sodium chloride (see Section 19–3). The Cl_2 collected at 26°C and 774 torr pressure occupies a volume of 32,700 liters. How many moles, grams, and pounds of NaCl must have been electrolyzed? How many pounds of metallic sodium were produced? How long did the cell operate?
45. Assuming that "iodized salt" contains 0.02% potassium iodide and 99.98% sodium chloride, calculate the percentage of the total mass of a sample of "iodized salt" due to iodine alone.
46. What volume of 0.150 normal sodium hydrogen sulfite solution is required to react with sodium iodate to produce 500 grams of solid iodine? What mass of sodium iodate must react?
47. What molarity of hypochlorous acid is produced if 30.0 grams of Cl_2 are bubbled through 500 milliliters of water, and 17.6 grams of Cl_2 escape unreacted? What molarity of hydrochloric acid will have been produced?
48. How many grams and how many milliliters of silicon tetrafluoride are generated at STP by the action of 100 milliliters of 3.00 *M* HF on 65.4 grams of calcium silicate?
49. What volume of oxygen collected at 727 torr and 28.2°C would be generated by the strong heating and complete decomposition of 5.271 grams of potassium chlorate in the laboratory?
50. How many grams of potassium perchlorate could be produced by *gently* heating 5.271 grams of potassium chlorate such that it all disproportionates?
51. Calculate the percentage of chlorine in: (a) Cl_2O_7, (b) $HClO_3$, (c) $HClO_4$, (d) $NaClO_4$.

23 The Nonmetallic Elements, Part II: Group VIA Elements

In this chapter the chemistry of the heavier members of Group VIA will be discussed. **The chemistry of oxygen and its compounds was described in Chapter 7.** The VIA elements show less electronegative character than the halogens. As is true in almost all groups, metallic character increases with increasing atomic weight. Oxygen and sulfur are clearly nonmetallic, but selenium is less so. Tellurium is usually classified as a metalloid and crystallizes in a metal-like crystal lattice, yet its chemistry is mostly nonmetallic. Polonium, all 29 isotopes of which are radioactive, is metallic.

All that is known about polonium has been determined through the use of radioactive tracers.

Irregularities in variations of the properties of elements within a given family increase from the extreme right and left of the periodic table toward the middle, when only the A group elements are considered. Thus, the properties of the Group VIA elements differ more than do the properties of the halogens.

Furthermore, the properties of elements in the second period usually differ significantly from those of other elements in their families, because of the absence of low-energy *d* orbitals. In keeping with this, the properties of oxygen are not very similar to those of the other Group VIA elements.

The outer electronic configuration of the VIA elements is ns^2np^4, and all show a tendency to gain or share two additional electrons in many of their compounds. All form covalent compounds of the type H_2E in which the VIA element (E) exhibits the oxidation number of −2. The maximum number of atoms with which oxygen can bond (coordination number) is four, but sulfur, selenium, tellurium, and probably polonium can bond covalently to as many as six other atoms. This is due to the availability of vacant *d* orbitals in the outer shell of each of the VIA elements except oxygen. One or more of the *d* orbitals can be used to accommodate additional electrons donated by other atoms to form a total of up to six bonds.

The *d* orbitals do not occur until the third shell.

TABLE 23–1 Some Properties of Group VIA Elements

Properties	Oxygen	Sulfur	Selenium	Tellurium	Polonium
Physical state (1 atm, 25°C)	gas	solid	solid	solid	solid
Color	colorless (very pale blue)	yellow	red-gray to black	brass-colored metallic luster	—
Outermost electrons	$2s^22p^4$	$3s^23p^4$	$4s^24p^4$	$5s^25p^4$	$6s^26p^4$
Melting point (1 atm, °C)	−219	112	217	450	254
Boiling point (1 atm, °C)	−183	444	685	990	962
Electronegativity	3.5	2.5	2.4	2.1	1.9
First ionization energy (kJ/mol)	1314	1000	941	869	812
Atomic radius (Å)	0.66	1.04	1.17	1.37	1.4
Ionic (2−) radius (Å)	1.40	1.84	1.98	2.21	—
Common oxidation states	usually −2	−2, +2, +4, +6	−2, +2, +4, +6	−2, +2, +4, +6	−2, +6

Sulfur, Selenium, and Tellurium

23–1 Occurrence, Properties, and Uses of the Elements

1 Sulfur

Sulfur makes up about 0.05% of the earth's crust, and was one of the elements known to the ancients. It was used by the Egyptians as a yellow coloring, and it was burned in some religious ceremonies because of the unusual odor it produced; it is the "brimstone" of the Bible. Alchemists tried to incorporate its "yellowness" into other substances in attempts to produce gold.

Sulfur occurs as the free element, predominantly S_8 molecules, and as metal sulfides such as galena, PbS, iron pyrite, FeS_2, and cinnabar, HgS. To a lesser extent it occurs as metal sulfates such as barite, $BaSO_4$, and gypsum, $CaSO_4 \cdot 2H_2O$, and in volcanic gases as hydrogen sulfide, H_2S, and sulfur dioxide, SO_2.

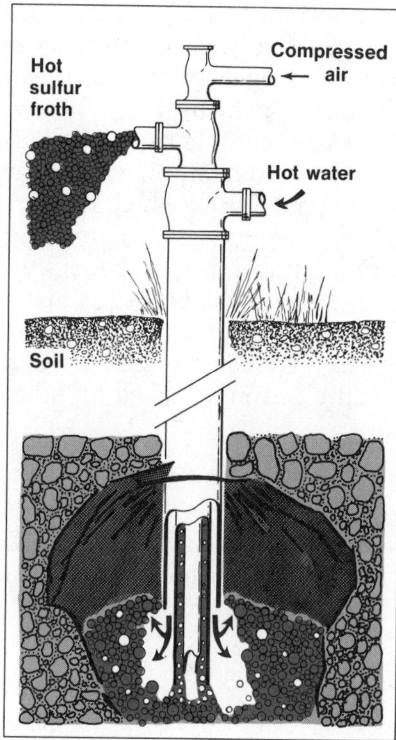

FIGURE 23–1 The Frasch process for sulfur mining. There are three concentric pipes used. Water at a temperature of about 170°C and a pressure of 100 lb/sq in (7 kg/cm^2) is forced down the outermost pipe to melt the sulfur. Hot compressed air is pumped down the innermost pipe, and mixes with the molten sulfur to form a froth that rises through the third.

Sulfur is also a constituent of much naturally occurring organic matter such as petroleum and coal. Its presence in fossil fuels causes environmental and health problems because many sulfur-containing compounds undergo combustion to produce sulfur dioxide (Section 23–9), an air pollutant, when the fuels are burned.

The elemental sulfur mined along the U.S. Gulf Coast is obtained by the **Frasch process,** or "hot water" process, described in Figure 23–1. Most of it is used in the production of sulfuric acid, H_2SO_4, probably the most important of all industrial chemicals. Sulfur is also a component of black gunpowder and is used in the vulcanization of rubber (see Section 29–10) and in the synthesis of many important sulfur-containing organic compounds. Elemental sulfur is a good electrical insulator.

In each of the three physical states, elemental sulfur exists in many forms. While the stable form of oxygen is the diatomic molecule, O_2, the two most stable forms of sulfur, the rhombic (mp 112°C) and monoclinic (mp 119°C) crystalline modifications, consist of S_8 molecules. These are puckered rings containing eight sulfur atoms (Figure 1–12) and all S—S single bonds. The S—S single bond energy (213 kJ/mol) is much greater than that for the O—O single bond (138 kJ/mol), and this accounts for the greater tendency of sulfur atoms to bond to other sulfur atoms and form rings. The rings are arranged differently in the rhombic and monoclinic forms. Below about 150°C liquid sulfur consists mainly of S_8 molecules. Above 150°C it becomes increasingly viscous and darkens as the S_8 rings break apart into chains that interlock with each other through —S—S bonds. The viscosity reaches a maximum at

O_2 is doubly bonded whereas S—S sigma bonds are apparently too long for *pi* bonding to be effective.

180°C, at which point sulfur is dark brown. Above 180°C the liquid thins as the chains are broken down into smaller chains. At 444°C sulfur boils to give a vapor containing S_8, S_6, S_4, and S_2 units.

2 Selenium

Like sulfur, selenium exists in a number of allotropic forms, but only two common crystalline modifications are well-characterized. These are the gray, metal-like hexagonal form and the red, nonmetallic, monoclinic form. Selenium exists mainly as Se_8 molecules in the solid form, but the vapor contains Se_8, Se_6, and Se_2 molecules. There appears to be only one liquid form.

Selenium is quite rare (9×10^{-6} % of the earth's crust) and occurs mainly as an impurity in sulfur, sulfide, and sulfate deposits. It is extracted from the flue dusts resulting from roasting sulfide ores and from the "anode mud" formed in the electrolytic refining of copper (Sections 19–8 and 20–7). It is used as a red coloring in glass. The metal-like form of selenium has an electrical conductivity that is very light-sensitive, so it finds application in photocopying machines and in solar cells. It is also an additive to some types of stainless steel and copper alloys.

3 Tellurium

Tellurium is even less abundant (2×10^{-7}% of the earth's crust) than selenium. It occurs mainly in sulfide ores, especially with copper sulfide, and as the tellurides of gold and silver. Like selenium, it is obtained from the "anode mud" from the electrolytic refining of copper obtained from copper sulfide ores. The element crystallizes as brass-colored metallic hexagonal crystals having low electrical conductivity. It is added to some metals, particularly lead, to increase electrical resistance and improve resistance to heat, corrosion, mechanical shock, and wear. It is also used to impart red, blue, or brown colors to glass. This metalloid also functions as a semiconductor (Section 9–17).

23–2 Reactions of the Nonmetals of Group VIA

Some of the general reactions of the elements of Group VIA are summarized in Table 23–2.

TABLE 23–2 Some Reactions of the VIA Elements

General Equation	Remarks
$xE + yM \longrightarrow M_yE_x$	With many metals
$zE + M_xE_y \longrightarrow M_xE_{y+z}$	Especially with S, Se
$E + H_2 \longrightarrow H_2E$	Decreasingly in the series O_2, S, Se, Te
$E + 3F_2 \longrightarrow EF_6$	With S, Se, Te, and excess F_2
$2E + Cl_2 \longrightarrow E_2Cl_2$*	With S, Se (Te gives $TeCl_2$)
$E_2Cl_2 + Cl_2 \longrightarrow 2ECl_2$*	With S, Se
$E + 2Cl_2 \longrightarrow ECl_4$*	With S, Se, Te, and excess Cl_2
$E + O_2 \longrightarrow EO_2$	With S (with Se, use $O_2 + NO_2$)
$3E + 4HNO_3 \longrightarrow 3EO_2 + 2H_2O + 4NO$	With S, Se, Te
$3E + 6OH^- \longrightarrow EO_3^{2-} + 2E^{2-} + 3H_2O$	With S

* Parallel reactions with Br_2.

23–3 Hydrides of the VIA Elements

HCN is the gas used in gas chambers in some states utilizing capital punishment.

All the VIA elements form covalent compounds of the type H_2E (E = O, S, Se, Te, Po) in which the VIA element is in the −2 oxidation state. While H_2O is a liquid that is absolutely essential for animal and plant life, H_2S, H_2Se, and H_2Te are colorless, noxious, poisonous gases. They are even more toxic than HCN. H_2Po is a liquid. Egg protein contains sulfur and its decomposition forms H_2S, which is responsible for the odor of rotten eggs and for tarnishing silver. H_2Se and H_2Te smell even worse. Fortunately, their odors are usually ample warning of the presence of these poisonous gases. Small amounts of H_2S will cause headaches and nausea. (If these symptoms occur while you are using this common laboratory reagent, you should leave the laboratory and get fresh air immediately.) Exposure to larger amounts can cause fainting and heart or lung failure. H_2S poses an additional problem in that, after long enough exposure, it severely retards the sense of smell so that one is no longer aware of its presence. Some properties of the VIA hydrides are summarized in Table 23–3.

Both the melting point and boiling point of water are very much higher than would be predicted from those of the heavier hydrides (Table 23–3). This is a consequence of hydrogen bonding in ice and liquid water (Section 9–9) caused by the strongly dipolar nature of water molecules. The energy required to overcome hydrogen bonding is reflected in the high melting and boiling points of water. Since the electronegativity differences between hydrogen and the other VIA elements are much smaller than between hydrogen and oxygen, no hydrogen bonding occurs in hydrogen sulfide, selenide, telluride, or polonide. The angular, polar water molecule is described by assuming sp^3 hybridization for oxygen (Section 5–14). The bond angle decreases, as does polarity, as the group is descended. The heavier members, H_2S, H_2Se, and H_2Te, all have bond angles of nearly 90°, and so are best described by assuming that the VIA atom utilizes p^2 bonding.

Water and hydrogen sulfide are produced from their elements in exothermic reactions, whereas the other hydrides are produced by endothermic reactions. Therefore, while water and hydrogen sulfide are stable at room temperature, hydrogen selenide and telluride slowly decompose to the elements. H_2S can be prepared by direct union of the gaseous elements at elevated temperatures on pumice chips

Pumice is a porous solid resulting from solidification of volcanic lava.

$$8H_2 + S_8 \xrightarrow{600°C} 8H_2S$$

TABLE 23–3 Some Properties of the Group VIA Hydrides

Property	H_2O	H_2S	H_2Se	H_2Te
Melting point (°C)	0.00	−85.60	−60.4	−51
Heat of fusion (kJ/mol)	6.01	2.38	—	—
Boiling point (°C)	100	−60.75	−41.5	−1.8
Heat of vaporization (kJ/mol)	40.7	18.7	19.9	24
Density at boiling point (g/mL)	0.958	0.993	2.004	2.650
Heat of formation at 25°C (kJ/mol)	−286	−20.6	29.7	135
Free energy of formation at 25°C (kcal/mol)	−237	−33.6	15.9	130

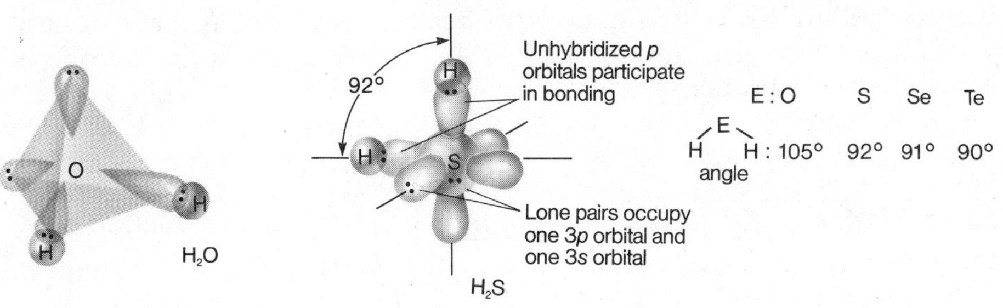

FIGURE 23–2 The group VIA hydrides. The structures of H_2Se and H_2Te are similar to that of H_2S, but the bond angles are slightly smaller because Se and Te atoms are larger than the S atom.

It also can be prepared by reaction between (nonoxidizing) protonic acids and metal sulfides, such as iron(II) sulfide

$$FeS + 2H_2SO_4 \longrightarrow H_2S + Fe(HSO_4)_2$$

H_2S is often prepared in the laboratory by hydrolysis of thioacetamide, a sulfur-containing organic compound. Heating speeds the reaction

$$\underset{\text{thioacetamide}}{CH_3{-}\overset{\overset{\displaystyle S}{\|}}{C}{-}NH_2} + 2H_2O \xrightarrow{\Delta} \underbrace{CH_3{-}\overset{\overset{\displaystyle O}{\|}}{C}{-}O^-NH_4^+}_{\substack{\text{ammonium acetate}\\ NH_4CH_3COO}} + H_2S$$

23–4 Aqueous Solutions of the VIA Hydrides

Aqueous solutions of hydrogen sulfide, selenide, and telluride are acidic, with acid strength increasing upon descending the group: $H_2O < H_2S < H_2Se < H_2Te$. The same trend was observed for increasing acidity of the hydrogen halides (Section 22–12) and will be observed for decreasing basicity of the Group VA hydrides in Chapter 24. This is a consequence of the decrease in average bond energy in the same order. The acid ionization constants are

The solubility of H_2S in water is approximately 0.10 mol/L at 25°C.

		H_2S	H_2Se	H_2Te
$H_2E \rightleftharpoons H^+ + HE^-$	K_1:	1.0×10^{-7}	1.9×10^{-4}	2.3×10^{-3}
$HE^- \rightleftharpoons H^+ + E^{2-}$	K_2:	1.3×10^{-13}	$\sim 10^{-11}$	$\sim 1.6 \times 10^{-11}$

23–5 Reactions of the VIA Hydrides

The Group VIA hydrides are reducing agents, and their reducing strength increases from H_2S to H_2Te. When H_2S acts as a reducing agent it is usually oxidized to elemental sulfur, but strong oxidizing agents may oxidize sulfur to the +4 or +6 oxidation states, for example, SO_2 or SO_3^{2-} and SO_3 or SO_4^{2-}. Gaseous H_2S only slowly reduces moist oxygen, but solutions of H_2S are clouded with elemental sulfur after a matter of days

$$2H_2\overset{-2}{S} + O_2 \xrightarrow{\Delta} 2H_2O + 2\overset{0}{S}$$

Hydrogen telluride solutions undergo an analogous reaction in a few minutes.

Gaseous hydrogen sulfide burns in the air to produce sulfur dioxide and steam when it is heated

$$2H_2\overset{-2}{S} + 3O_2 \xrightarrow{\Delta} 2\overset{+4}{S}O_2 + 2H_2O$$

If excess H_2S is present, elemental sulfur is produced by the reaction of SO_2 with H_2S

$$2H_2\overset{-2}{S} + \overset{+4}{S}O_2 \longrightarrow 3\overset{0}{S} + 2H_2O$$

The same reaction occurs rapidly when both gases are bubbled through a beaker of water.

Many metals will displace hydrogen from hydrogen sulfide. Passing H_2S over the surface of magnesium (free of MgO) results in the formation of magnesium sulfide and hydrogen

$$Mg + H_2S \xrightarrow{\Delta} MgS + H_2$$

The hydrogen sulfide formed in the decomposition of the sulfur-containing protein of eggs tarnishes silverware by a reaction involving reduction of oxygen, oxidation of silver, and formation of black silver sulfide. In this reaction the H_2S is neither reduced nor oxidized, but apparently aids the oxidation of metallic silver

$$4Ag + 2H_2S + O_2 \longrightarrow \underset{\text{black}}{2Ag_2S} + 2H_2O$$

23–6 Metal Sulfides and Analogs

Selenium is often incorporated into glass for coloring purposes in the form of Na_2Se. Subsequent heating causes particles of red colloidal selenium to form.

Two classes of binary metal salts of the VIA elements exist: (1) those containing sulfide, S^{2-}, selenide, Se^{2-}, or telluride, Te^{2-}, ions, and (2) those containing hydrosulfide, HS^-, hydroselenide, HSe^-, or hydrotelluride, HTe^-, ions. These salts may be considered the sulfur, selenium, and tellurium analogs of the oxides, O^{2-}, and hydroxides, OH^-.

Metal sulfides exhibit a wide range of water solubility, and metal ions in aqueous solution are often analytically separated from each other (as in qualitative analysis) with H_2S. Sulfides of some small, highly charged cations are hydrolyzed completely to form insoluble hydroxides and hydrogen sulfide.

$$Al_2S_3 + 6H_2O \longrightarrow 2Al(OH)_3 + 3H_2S$$

23–7 Halides of the VIA Elements

Several halides of the type E_2X_2, EX_2, EX_4, and EX_6 (E = S, Se, Te; X = F, Cl, Br, I) have been prepared. They are listed in Table 23–4.

The large iodine atom combines only with tellurium, the largest of the three VIA elements. The EX_6 species involve only the small fluorine atoms, due partly to the difficulty of fitting six of the larger halogens around the VIA element. This is called *steric* hindrance. The EX_6 species are all octahedral with sp^3d^2 hybridization at the central atom (Section 5–17).

TABLE 23–4 Group VIA Halides

Oxidation State	<+2	+2	+4	+5	+6
Sulfur	S_2F_2	(SF_2)*	SF_4	S_2F_{10}	SF_6
	S_2Cl_2	SCl_2	SCl_4		$SClF_5$
	S_2Br_2				$SBrF_5$
	S_nCl_2				
	S_nBr_2				
Selenium	Se_2Cl_2	$(SeCl_2)$	SeF_4		SeF_6
	Se_2Br_2	$(SeBr_2)$	$SeCl_4$		
			$SeBr_4$		
Tellurium	Te_3Cl_2	$TeCl_2$	TeF_4	Te_2F_{10}	TeF_6
	Te_2Br	$TeBr_2$	$TeCl_4$		
	TeI		$TeBr_4$		
	Te_xI		TeI_4		

* Compounds in parentheses have not been prepared in pure form.

The halides of lower oxidation states, E_2X_2, EX_2, and EX_4, are quite reactive. For example, sulfur tetrafluoride reacts with water readily to produce hydrofluoric acid and sulfur dioxide

$$SF_4\ (g) + 2H_2O\ (\ell) \longrightarrow 4HF\ (g) + SO_2\ (g)$$

However, SF_6 and SeF_6 are stable compounds that are inert toward water. By contrast, TeF_6 is somewhat less stable and decomposes in water to form telluric acid, H_6TeO_6 (Section 23–18), within a day. Because of the unusual stability of gaseous SF_6 it is sometimes inhaled with oxygen for x-ray examinations of the lungs. Certainly this could not be so used if the SF_6 hydrolyzed with water vapor in the lungs to produce HF and SO_3, as might be expected.

Both HF and SO_3, as well as the hydrofluoric and sulfuric acids formed by further reaction with water, destroy lung tissue.

The relative chemical stability of the EX_6 species is due in part to the absence of nonbonding valence electrons around the central atom. Each VIA element has six outer shell electrons, and each is involved in bonding to a halogen atom in EX_6 species. However, the EX_4 species are sp^3d hybridized at the central atom and retain one lone pair of electrons. These nonbonding electrons are susceptible to attack by electron-seeking species such as water to initiate a reaction. The structures of SF_4 and SF_6 are shown below.

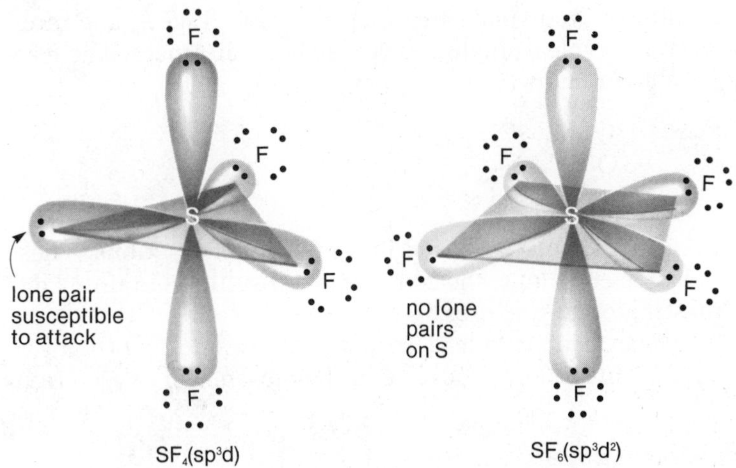

Group VIA Oxides

Although others exist, the most important VIA oxides are the dioxides, which are acid anhydrides of sulfurous, selenous, and tellurous acids, and the trioxides, which are anhydrides of sulfuric, selenic, and telluric acids.

23–8 Physical States of the VIA Dioxides

The tendency toward metallic character upon descending a group is exemplified by the stable forms of the dioxides of S, Se, and Te. The very nonmetallic sulfur is covalently bonded to oxygen in SO_2, as is selenium in polymeric SeO_2, but the more metallic tellurium forms ionic TeO_2. Note the structural similarity of O_3, ozone or "oxygen dioxide," (p. 39) to SO_2.

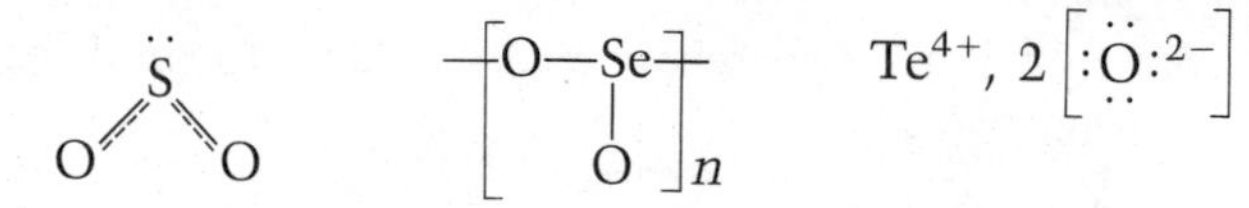

$\ddot{S}$ (O, O) — $-[O-Se(-O)]_n-$ — $Te^{4+}, 2[:\ddot{O}:^{2-}]$

sulfur dioxide molecular gas — selenium dioxide polymeric solid — tellurium dioxide ionic lattice (also PoO_2)

Double bonds are easily formed between atoms like sulfur and oxygen, which have small electronegativity differences and large electronegativity sums. Sulfur and selenium dioxides are quite soluble in water, but tellurium dioxide is only slightly soluble, owing to its high crystal lattice energy.

23–9 Sulfur Dioxide (An Air Pollutant)

Sulfur dioxide is a colorless, poisonous, corrosive gas with a very irritating odor. It causes coughing and nose, throat, and lung irritation when inhaled even in small quantities. It is an angular molecule with trigonal planar electronic geometry involving sp^2 hybridization at the sulfur atom and resonance stabilization. Both sulfur-oxygen bonds are of equal length, 1.43 Å, and intermediate in length between typical single and double bond distances. The resonance structures are shown below.

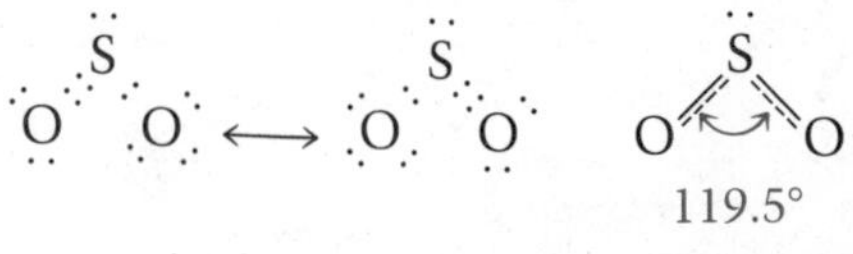

$$\ddot{O}{=}\ddot{S}{-}\ddot{O}: \longleftrightarrow :\ddot{O}{-}\ddot{S}{=}\ddot{O}$$

119.5°

Sulfur dioxide is undesirably produced in a number of reactions. These primarily involve volcanic eruptions, the combustion of sulfur-containing fossil fuels, and the roasting of sulfide ores.

Metals such as copper, zinc, and lead are obtained by smelting their sulfide ores (Section 20–3). This involves heating in the presence of oxygen from the air (roasting)

$$2ZnS + 3O_2 \rightleftharpoons 2ZnO + 2SO_2$$

A waste product of the operation is sulfur dioxide which, in the past, has been almost indiscriminately released into the atmosphere along with some SO_3 produced by its reaction with O_2. However, efforts are now underway to trap gaseous sulfur dioxide and trioxide and use them to produce sulfuric acid. Some coal contains up to 5% sulfur, so both SO_2 and SO_3 are present in flue gases and should be trapped when coal is burned.

If the SO_2 and SO_3 are allowed to escape into the atmosphere they cause highly acidic rain.

No way has been found to remove all the SO_2 from flue gases of power plants, but one way of removing most of it involves the injection of limestone, $CaCO_3$, into the combustion zone of the furnace where it undergoes thermal decomposition to lime, CaO. This then combines with SO_2 to form calcium sulfite, $CaSO_3$, an ionic solid that is then collected

$$CaCO_3 \xrightarrow{\Delta} CaO + CO_2$$

$$CaO + SO_2 \longrightarrow CaSO_3$$

This process is called scrubbing. A disadvantage is that it creates huge quantities of solid waste ($CaSO_3$, unreacted CaO, and other substances) which must be disposed of.

Methods of catalytic oxidation are being used by the smelting industry to convert SO_2 into SO_3 and then into solutions of sulfuric acid, H_2SO_4 (up to 80% by mass). To do so the gases containing SO_2 are passed through a series of condensers containing catalysts to speed up the desired conversions. Again, waste disposal is a major problem. In some cases the impure acid solutions can be used in other operations in the same plant, but in other cases the acid must be sold commercially. This is economically burdensome, because the solutions first must be purified and transportation costs are usually high (since most smelters are quite isolated).

Sulfur dioxide is also an important industrial, commercial, and research chemical. Sulfur dioxide can be prepared in the laboratory by treating sodium sulfite or sodium hydrogen sulfite with a nonvolatile acid such as sulfuric or phosphoric acid.

$$\underset{\text{sodium hydrogen sulfite}}{2NaHSO_3} + H_2SO_4 \longrightarrow Na_2SO_4 + 2SO_2 + 2H_2O$$

It can also be prepared by the action of very weak reducing agents on hot concentrated sulfuric acid. The dissolution of metallic copper is an example

$$Cu + 2H_2SO_4 \longrightarrow CuSO_4 + SO_2 + 2H_2O$$

23–10 Selenium and Tellurium Dioxides

Selenium and tellurium dioxides can be formed by burning the elements

$$Se + O_2 \xrightarrow{\Delta} SeO_2$$

$$Te + O_2 \xrightarrow{\Delta} TeO_2$$

These dioxides can also be prepared by dissolving elemental selenium or tellurium in concentrated nitric acid (they will not dissolve in nonoxidizing acids) and evaporating the resulting solution to dryness. Selenous and tellurous acids are intermediate products that are decomposed to the corresponding dioxides when heated

$$Se + 4HNO_3 \xrightarrow{\Delta} SeO_2 + 2H_2O + 4NO_2$$

$$Te + 4HNO_3 \xrightarrow{\Delta} TeO_2 + 2H_2O + 4NO_2$$

23–11 Sulfur Trioxide

Sulfur trioxide is a liquid that boils at 44.8°C. It is the anhydride of sulfuric acid, and can be formed by the reaction of sulfur dioxide with oxygen. The ordinarily slow but very exothermic reaction is catalyzed commercially in the **contact process** by spongy platinum, silicon dioxide, or vanadium(V) oxide, V_2O_5, at high temperatures (400–700°C).

$$2SO_2\ (g) + O_2\ (g) \underset{\text{catalyst}}{\rightleftharpoons} 2SO_3\ (g) \qquad \Delta H^0 = -197.6\ kJ$$

The high temperature shifts the equilibrium to the left (favors SO_2 and O_2) but allows equilibrium to be achieved so much more rapidly that it is economically advantageous. The SO_3 is then removed from the gaseous reaction mixture by dissolution in concentrated H_2SO_4 (~95% H_2SO_4 by mass) to produce polysulfuric acids, mainly pyrosulfuric acid, $H_2S_2O_7$, which is called oleum or fuming sulfuric acid. Addition of fuming sulfuric acid to water produces commercial H_2SO_4

The term *pyro* means "heat or fire." Pyrosulfuric acid may also be obtained by heating concentrated sulfuric acid, which results in the elimination of one molecule of water per two molecules of sulfuric acid.

$$2H_2SO_4 \longrightarrow H_2S_2O_7 + H_2O$$

$$SO_3 + H_2SO_4 \longrightarrow H_2S_2O_7$$

$$H_2S_2O_7 + H_2O \longrightarrow 2H_2SO_4$$

Alternatively, SO_3 can be obtained by distillation of fuming sulfuric acid. It is used in the preparation of sulfonated oils and sulfonate detergents (Section 10–23).

Sulfur dioxide in polluted air reacts rapidly with oxygen to form sulfur trioxide in the presence of certain catalysts. Particulate matter, or suspended microparticles, such as ammonium nitrate and elemental sulfur, act as efficient catalysts for this oxidation.

Sulfur trioxide contains both SO_3 and S_3O_9 molecules in all three physical states

$$3SO_3 \underset{\text{catalyst}}{\rightleftharpoons} S_3O_9$$

The SO_3 molecule is trigonal planar, containing sp^2 hybridized sulfur, and is represented by the following resonance formulas

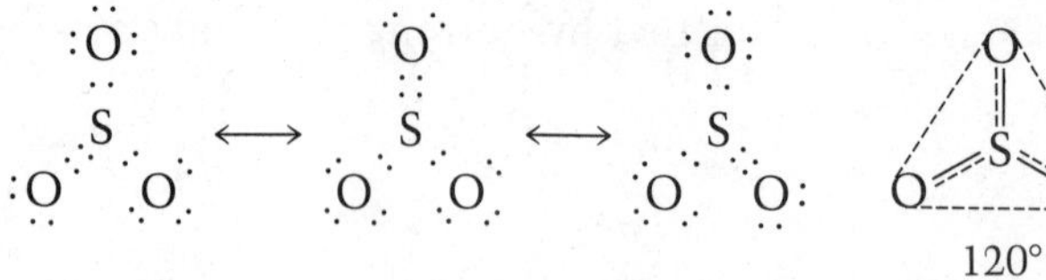

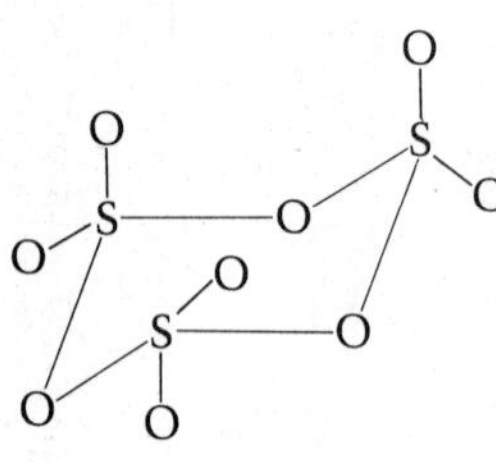

The S_3O_9 molecule is a puckered ring that can be thought of as three distorted SO_4 tetrahedra (sp^3 hybridization at sulfur) joined by common corners, as shown in the margin.

Sulfur trioxide decomposes into sulfur dioxide and oxygen at temperatures above 900°C. It acts as an oxidizing agent in many reactions. As examples, it converts sulfur to sulfur dioxide, SCl_2 to $SOCl_2$ and SO_2Cl_2, phosphorus to P_4O_{10}, and PCl_3 to $POCl_3$. It can also act as a Lewis acid in reactions with Lewis bases such as magnesium oxide.

$$\underbrace{Mg^{2+} + :\ddot{O}:^{2-}}_{\text{magnesium oxide (Lewis base)}} + \underbrace{SO_3}_{\text{sulfur trioxide (Lewis acid)}} \longrightarrow \underbrace{Mg^{2+} + [SO_4]^{2-}}_{\text{magnesium sulfate}}$$

magnesium oxide Lewis base | sulfur trioxide Lewis acid | magnesium sulfate

23–12 Selenium and Tellurium Trioxides

Both SeO_3 and TeO_3 are stronger oxidizing agents than SO_3. They are the anhydrides of selenic and telluric acids, respectively. SeO_3 is prepared by the reaction of potassium selenate and SO_3.

$$\underset{\text{potassium selenate}}{K_2\overset{+6}{Se}O_4} + \overset{+6}{S}O_3 \longrightarrow \underset{\text{potassium sulfate}}{K_2\overset{+6}{S}O_4} + \overset{+6}{Se}O_3$$

Dehydration of telluric acid at 300–600°C produces water-insoluble TeO_3, which decomposes above 400°C to Te_2O_5 and O_2

$$\underset{\text{telluric acid}}{H_6\overset{+6}{Te}O_6} \xrightarrow{\Delta} \overset{+6}{Te}O_3 + 3H_2O$$

Oxyacids of Sulfur, Selenium, and Tellurium

23–13 Sulfurous Acid and Sulfites

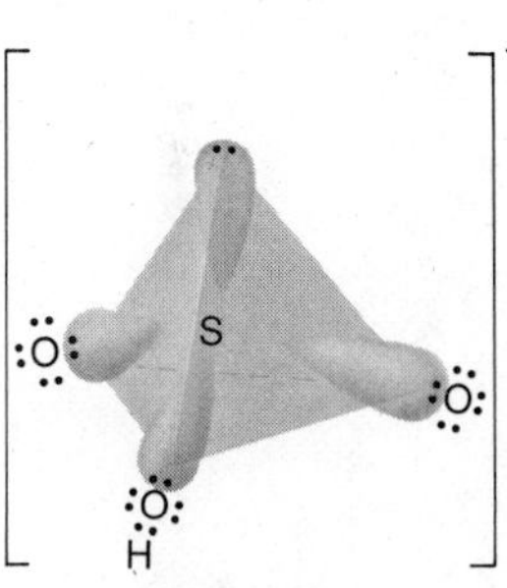

hydrogen sulfite ion

Sulfur dioxide readily dissolves in water to produce solutions of sulfurous acid, H_2SO_3. The acid has not been isolated in anhydrous form

$$H_2O + \overset{+4}{S}O_2 \rightleftharpoons H_2\overset{+4}{S}O_3$$

The acid ionizes in two steps in water

$$\underset{\text{sulfurous acid}}{H_2SO_3} \rightleftharpoons H^+ + \underset{\text{hydrogen sulfite ion (bisulfite ion)}}{HSO_3^-} \qquad K_1 = 1.2 \times 10^{-2}$$

$$HSO_3^- \rightleftharpoons H^+ + \underset{\text{sulfite ion}}{SO_3^{2-}} \qquad K_2 = 6.2 \times 10^{-8}$$

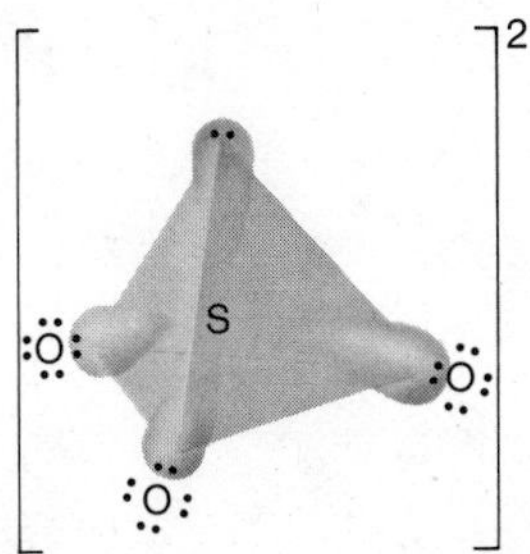

sulfite ion

When SO_2 is bubbled through aqueous NaOH, sodium hydrogen sulfite,

$NaHSO_3$, is produced. This acid salt can be neutralized with additional NaOH or Na_2CO_3 to produce sodium sulfite

$$NaOH + H_2SO_3 \longrightarrow \underset{\text{sodium hydrogen sulfite}}{NaHSO_3} + H_2O$$

$$NaOH + NaHSO_3 \longrightarrow \underset{\text{sodium sulfite}}{Na_2SO_3} + H_2O$$

The sulfite ion is pyramidal and has tetrahedral electronic geometry.

23–14 Selenous Acid and Selenites

Selenous acid, H_2SeO_3, can be prepared by dissolving its anhydride, SeO_2, in water

$$H_2O + \overset{+4}{Se}O_2 \longrightarrow H_2\overset{+4}{Se}O_3$$

The resulting selenous acid is the only one of the *-ous* acids of sulfur, selenium, and tellurium that can be obtained as a solid. H_2SeO_3 is a white hygroscopic solid. It is a diprotic acid only slightly weaker than H_2SO_3. It ionizes in two steps with $K_1 = 2.7 \times 10^{-3}$, and $K_2 = 2.5 \times 10^{-7}$. Neutralization of H_2SeO_3 produces selenite and hydrogen selenite salts.

Hygroscopic substances readily absorb water from the air.

23–15 Tellurous Acid and Tellurites

Tellurous acid, H_2TeO_3, cannot be produced by dissolving TeO_2 in water since TeO_2 is so insoluble. Tellurites are prepared by dissolving TeO_2 in alkaline solutions. Solutions of tellurite salts, upon acidification, presumably produce tellurous acid, which has acid ionization constants of $K_1 = 2 \times 10^{-3}$ and $K_2 = 1 \times 10^{-8}$.

In view of the weakly metallic character of tellurium, it is not surprising that tellurous acid can also ionize slightly as a base in water.

$$H_2TeO_3 \rightleftharpoons TeO(OH)^+ + OH^- \qquad K \approx 10^{-12}$$

23–16 Sulfuric Acid and Sulfates (and Acidic Rain)

More than 40 million tons of sulfuric acid are produced annually worldwide. The contact process (Section 23–11) is used for the commercial production of almost all sulfuric acid and sulfur trioxide. The solution sold commercially as "concentrated sulfuric acid" is 96–98% H_2SO_4 by mass, the rest being water, and is 18 molar H_2SO_4.

Pure sulfuric acid is a colorless oily liquid that freezes at 10.4°C and boils at 290–317°C while partially decomposing to SO_3 and water. Solid and liquid H_2SO_4 are somewhat hydrogen bonded.

The sulfur is sp^3 hybridized in sulfuric acid and in sulfate and hydrogen sulfate ions, resulting in tetrahedral geometry with respect to sulfur.

Tremendous amounts of heat are evolved when concentrated sulfuric acid is diluted. This behavior illustrates the strong affinity of H_2SO_4 for water

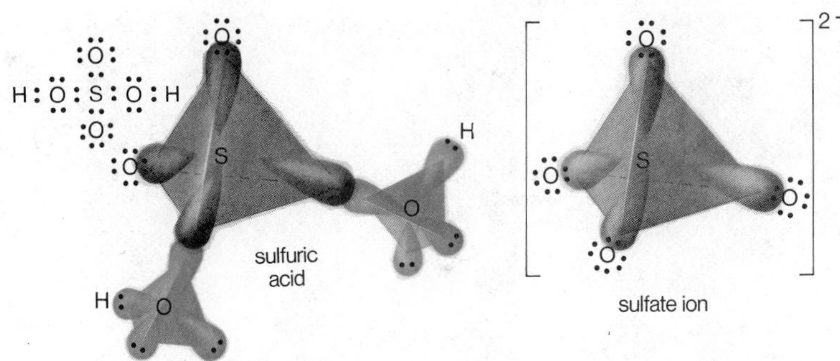

(Section 10–3). Thus, H_2SO_4 is often used as a dehydrating agent. DILUTIONS (AS WITH ALL ACIDS) SHOULD ALWAYS BE PERFORMED BY ADDING THE ACID TO WATER in order to avoid spattering the acid, a phenomenon caused by the heat evolved on mixing.

Sulfuric acid is a strong acid with respect to the first step of its ionization in water. The second ionization occurs to a somewhat lesser extent

$$H_2SO_4 \rightleftharpoons H^+ + \underset{\text{hydrogen sulfate ion}}{HSO_4^-} \qquad K_1 = \text{very large}$$

$$HSO_4^- \rightleftharpoons H^+ + \underset{\text{sulfate ion}}{SO_4^{2-}} \qquad K_2 = 1.2 \times 10^{-2}$$

Hot concentrated sulfuric acid is also a strong oxidizing agent.

The "acidic rain" now experienced especially in highly industrialized areas is due in large part to the presence of sulfuric acid and some sulfurous acid, which are formed by reaction of SO_2 and SO_3 with moisture. The oxides are produced during combustion of sulfur-containing fossil fuels and smelting operations.

It should be pointed out that even rain in "unpolluted" air is slightly acidic because of the presence of carbon dioxide in the atmosphere, which reacts with moisture to produce carbonic acid, H_2CO_3. However, rain in some areas has been found to have a pH of about 4, due mainly to the presence of sulfuric acid.

Sulfuric acid, of course, easily reacts with structural materials such as limestone and marble (both $CaCO_3$)

$$H_2SO_4 + CaCO_3 \longrightarrow CaSO_4 + CO_2 + H_2O$$

The calcium sulfate then washes away in the rain (see Figure 23–3). The acid also breaks down structures composed of metals such as steel (which contains iron) and aluminum (which is always coated with an oxide surface when exposed to air)

$$H_2SO_4 + Fe \longrightarrow FeSO_4 + H_2$$

$$3H_2SO_4 + Al_2O_3 \longrightarrow Al_2(SO_4)_3 + 3H_2O$$

Sulfuric acid also is extremely reactive toward many organic compounds, including those in plants (see Figure 23–4) and human flesh; the lungs are especially susceptible to irritation.

FIGURE 23–3 Two photographs of a statue showing the corrosive power of pollutants causing acidic rain. In 1908, when the picture at the left was taken, the statue at Harten Castle in Westphalia had survived nearly intact for more than two centuries. The photograph at the right, taken 61 years later, in 1969, shows the damage done by increasingly polluted air, and the resulting acid rain.

23–17 Selenic Acid and Selenates

Selenic acid is prepared by oxidation of SeO_2 (or H_2SeO_3) with 30% H_2O_2. Selenium trioxide is not used directly because of its instability. Selenic acid can be obtained in the solid form as colorless hygroscopic crystals. Selenates

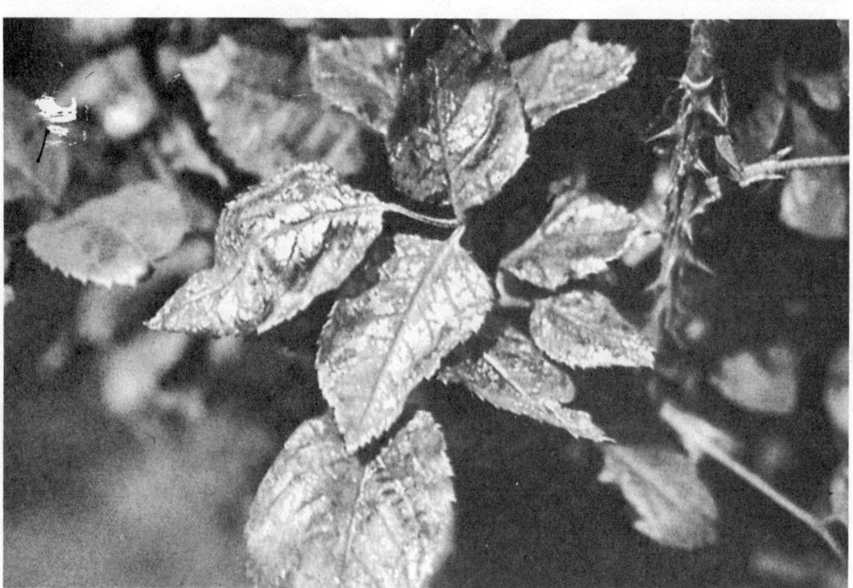

FIGURE 23–4 Rose leaves in Independence, Missouri, show marginal and interveinal necrotic injury from acid rain.

are prepared by neutralization of selenic acid solutions, or by oxidation of selenite salts. The acid is approximately as strong as sulfuric acid and ionizes in two steps

$$H_2SeO_4 \rightleftharpoons H^+ + HSeO_4^- \qquad K_1 = \text{very large}$$

$$HSeO_4^- \rightleftharpoons H^+ + SeO_4^{2-} \qquad K_2 = 1.15 \times 10^{-2}$$

The acid and anions have the same geometries as sulfuric acid and its anions, and are even stronger oxidizing agents.

Interestingly, selenic acid is the only pure acid that will dissolve gold. Gold also can be dissolved in a mixture of concentrated hydrochloric and nitric acids, called "aqua regia."

23–18 Telluric Acid

Telluric acid, H_6TeO_6 or $Te(OH)_6$, is obtained by oxidation of TeO_2 with 30% hydrogen peroxide, H_2O_2; or aqueous chloric acid, $HClO_3$; or a solution of nitric acid, HNO_3, and potassium permanganate, $KMnO_4$. Apparently because of its large size, tellurium is able to form a six-coordinate acid; it does so by undergoing sp^3d^2 hybridization. The H_6TeO_6 molecule is octahedral. The oxidation state of tellurium is +6, as is the case for sulfur and selenium in sulfuric and selenic acids.

Note that H_6TeO_6 has the stoichiometry $H_2TeO_4 \cdot 2H_2O$.

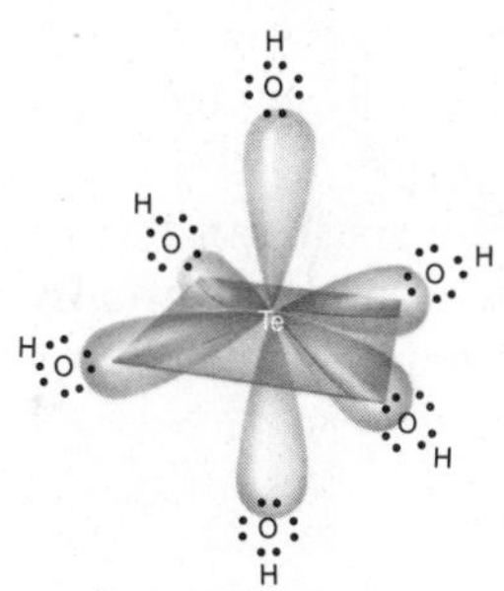

Aqueous solutions of telluric acid undergo ionization in two steps and, in contrast to sulfuric and selenic acids, are only weakly acidic. This behavior is consistent with the more metallic character of tellurium. Salts of the type $MTeO(OH)_5$ and $M_2TeO_2(OH)_4$ are known, as well as Na_6TeO_6 and Ag_6TeO_6.

23–19 Thiosulfuric Acid and Thiosulfates

The prefix *thio-* designates the replacement of an oxygen atom with a sulfur atom. Thus, thiosulfuric acid is $H_2S_2O_3$ and the thiosulfate ion is $S_2O_3^{2-}$. It should not be surprising that the acid is prepared by reacting SO_3 with H_2S (rather than H_2O which would produce H_2SO_4) at low temperatures in ether.

$$H_2S + SO_3 \xrightarrow[\text{ether}]{\text{cold}} \underset{\text{thiosulfuric acid}}{H_2S_2O_3} \qquad \text{(exists only in solution)}$$

The acid is unstable and disproportionates to sulfurous acid and sulfur.

Thiosulfate salts are produced by boiling aqueous sulfite salts with sulfur

$$\underset{\text{sodium sulfite}}{Na_2SO_3} + S \xrightarrow{\Delta} \underset{\text{sodium thiosulfate}}{Na_2S_2O_3}$$

Sodium thiosulfate pentahydrate is known as photographer's "fixer." It dissolves unexposed AgBr in photographic film as it forms soluble $Na_3Ag(S_2O_3)_2$, which contains the complex ion $[Ag(S_2O_3)_2]^{3-}$.

The structures of thiosulfuric acid and the thiosulfate ion are analogous to those of sulfuric acid and the sulfate ion.

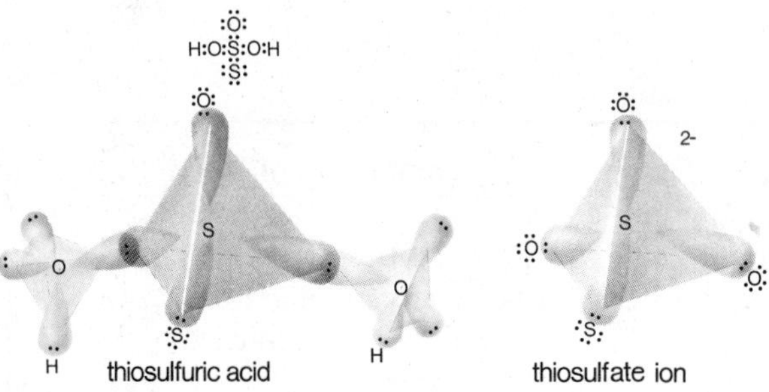

Key Terms

Contact process industrial process by which sulfur trioxide and sulfuric acid are produced from sulfur dioxide.

Frasch process method by which elemental sulfur is mined or extracted, in which it is melted and forced to the surface of the earth as a slurry with superheated water (at 170°C under high pressure).

Particulate matter finely divided solid particles suspended in polluted air.

Polymer a large molecule consisting of chains or rings of linked monomer units usually characterized by high melting and boiling points.

Exercises

1. Write abbreviated electronic configurations for atomic oxygen, sulfur, and tellurium.
2. Write out the electronic configurations of oxide, sulfide, and selenide ions.
3. Characterize the Group VIA elements with respect to color and physical state under normal conditions.
4. The Group VIA elements, except oxygen, can exhibit oxidation states ranging from -2 to $+6$. Why not -3 to $+7$?
5. Give and explain the order of increasing melting and boiling points of the Group VIA elements.
6. Is the order of decreasing first ionization energies of the Group VIA elements consistent with the order of increasing metallic character?
7. Sulfur, selenium, and tellurium are all capable of forming six-coordinate compounds such as SF_6. Give two reasons why oxygen cannot be the central atom in such six-coordinate molecules.
8. Draw diagrams that show the hybridization of atomic orbitals and three-dimensional structures showing all hybridized orbitals and outermost electrons for the following

molecules or ions: (a) H_2S, (b) SF_6, (c) SF_4, (d) SO_2, (e) SO_3.

9. Repeat Exercise 8 for: (a) SeF_6, (b) SO_3^{2-}, (c) SO_4^{2-}, (d) HSO_4^-, (e) thiosulfate ion, $S_2O_3^{2-}$ (one S is central atom).
10. Write equations for the reactions of:
    (a) S, Se, and Te with excess F_2.
    (b) O_2, S, Se, and Te with H_2.
    (c) S, Se, and Te with O_2.
11. Write equations for the reactions of:
    (a) S and Te with HNO_3.
    (b) S and Se with excess Cl_2.
    (c) S and Se with Na, Ca, and Al.
12. What is the order of increasing melting and boiling points and heats of vaporization of the Group VIA hydrides, H_2O, H_2S, H_2Se, H_2Te? What are their physical states under normal conditions? Why is H_2O out of line with the others?
13. What are three general methods for the preparation of hydrogen sulfide?
14. Discuss the acidity of the aqueous Group VIA hydrides including the relative values of acid ionization constants. What is primarily responsible for the order of increasing acidities in this series?
15. Write equations for the reactions of:
    (a) aqueous H_2S with O_2.
    (b) gaseous H_2S with O_2 in the presence of heat.
    (c) excess gaseous H_2S with O_2 in the presence of heat.
    (d) gaseous H_2S with aluminum metal.
    (e) aqueous H_2S with NaOH.
16. Why is sulfur tetrafluoride a potent poison while SF_6 is not? Why is SF_4 so reactive while SF_6 is not?
17. Compare the structures of the dioxides of sulfur, selenium, tellurium, and polonium. How do they relate to the metallic or nonmetallic character of these elements?
18. Write equations for the following reactions:
    (a) the preparation of sulfur dioxide (and copper(II) sulfate + water) by reaction of copper with hot concentrated sulfuric acid.
    (b) the reaction of sulfur dioxide with oxygen.
    (c) the reaction of sulfur dioxide with water.
19. Draw the structure of the S_3O_9 molecule. How are the sulfur atoms hybridized? With what species is this molecule always in equilibrium?
20. Draw a Lewis dash representation of pyrosulfuric acid, $H_2S_2O_7$. Write an equation to show how it is prepared from concentrated sulfuric acid.
21. What are the acid anhydrides of sulfuric acid, selenic acid, and telluric acid?
22. Write equations for reactions of:
    (a) sodium hydroxide with sulfuric acid (1:1 ratio).
    (b) sodium hydroxide with sulfuric acid (2:1 ratio).
    (c) sodium hydroxide with sulfurous acid (1:1 ratio).
    (d) sodium hydroxide with sulfurous acid (2:1 ratio).
    (e) sodium hydroxide with selenic acid, H_2SeO_4, (1:1 ratio).
    (f) sodium hydroxide with selenic acid (2:1 ratio).
    (g) sodium hydroxide with tellurium dioxide (1:1 ratio).
    (h) sodium hydroxide with tellurium dioxide (2:1 ratio).
23. How much sulfur dioxide is produced from the complete combustion of one ton of coal containing 4.4% sulfur impurity?
24. A sterling silver serving piece contains 117 grams of silver. If 0.104 gram of silver sulfide (tarnish) forms by reaction of the silver with H_2S from the decomposition of eggs, how much silver must react? What percentage of the silver tarnishes?
25. Calculate the concentrations of H^+, HSO_3^- and SO_3^{2-} ions present in 0.010 M sulfurous acid, H_2SO_3, solution? $K_1 = 1.2 \times 10^{-2}$ and $K_2 = 6.2 \times 10^{-8}$.
26. Calculate the concentration of OH^- and the pH in 0.10 M aqueous sodium sulfite, Na_2SO_3. See the data in Exercise 25.
27. Write reaction equations for the preparation of (a) thiosulfuric acid and (b) potassium thiosulfate.

24 The Nonmetallic Elements, Part III: Nonmetals of Group VA

In the nitrogen family, nitrogen and phosphorus are nonmetals, arsenic is predominantly nonmetallic, antimony is somewhat more metallic, and bismuth is definitely metallic. Many of the properties of the Group VA elements are listed in Table 24–1.

TABLE 24–1 Properties of the Group VA Elements

Properties	Nitrogen	Phosphorus	Arsenic	Antimony	Bismuth
Physical state (1 atm, 25°C)	gas	solid	solid	solid	solid
Color	colorless	red, white, black	yellow, gray	yellow, gray	gray
Outermost electrons	$2s^22p^3$	$3s^23p^3$	$4s^24p^3$	$5s^25p^3$	$6s^26p^3$
Melting point (°C)	−210	44 (white)	814 (gray)	631 (gray)	271
Boiling point (°C)	−196	280 (white)	sublimes 613	1380	1560
Atomic radius (Å)	0.70	1.10	1.21	1.41	1.46
Electronegativity	3.0	2.1	2.1	1.9	1.8
First ionization energy (kJ/mol)	1402	1012	947	834	703
Oxidation states	−3 to +5	−3 to +5	−3 to +5	−3 to +5	−3 to +5

The possible oxidation states of these elements range from −3 to +5, with emphasis on the odd numbers. The VA elements form only a few monatomic ions. Ions with a charge of 3− occur for nitrogen and phosphorus, as in Mg_3N_2 and Ca_3P_2. Tripositive cations probably exist for antimony and bismuth in such compounds as antimony(III) sulfate, $Sb_2(SO_4)_3$, and bismuth(III) perchlorate pentahydrate, $Bi(ClO_4)_3 \cdot 5H_2O$, although in aqueous solution they are extensively hydrolyzed to SbO^+ or SbOX and BiO^+ or BiOX (X = univalent anion) to give strongly acidic solutions. There are few if any compounds containing ions with 5+ charge, and simple cations of the Group VA elements with this charge do not exist.

All of the Group VA elements exhibit the −3 oxidation state in covalent compounds such as NH_3, PH_3, and AsH_3. The +5 oxidation state is exhibited only in covalent compounds, including phosphorus pentafluoride, PF_5, phosphoric acid, H_3PO_4, and in polyatomic ions such as NO_3^- and PO_4^{3-}. Nitrogen and phosphorus have many oxidation states in their compounds, but the common ones for antimony and bismuth are +3 and +5.

Oxides in which the Group VA elements exhibit the +3 oxidation state are known for all: N_2O_3, P_4O_6, As_4O_6, Sb_4O_6, and Bi_2O_3. The first two are acid anhydrides of nitrous acid, HNO_2, and phosphorous acid, H_3PO_3; both are weak acids. Saturated aqueous solutions of As_4O_6 and Sb_4O_6 are amphoteric, the former having more acidic character and the latter more basic character. Neither oxide dissolves in water to any significant extent. Bismuth(III) oxide, Bi_2O_3, is the anhydride of bismuth(III) hydroxide, $Bi(OH)_3$, an insoluble base. The trend of increasing oxide basicity on descending the group indicates increasingly metallic character of the elements.

Nitrogen

24–1 Occurrence, Properties, and Importance

Nitrogen, N_2, is a colorless, odorless, tasteless gas that makes up about 75% by mass and 78% by volume of the atmosphere. Despite the fact that nitrogen

All proteins contain nitrogen in each of their fundamental amino acid units.

compounds form only a minor portion of the earth's crust, all living matter contains nitrogen. The primary natural inorganic deposits of nitrogen are very localized and consist primarily of potassium and sodium nitrates, KNO_3 and $NaNO_3$. Most sodium nitrate is mined in Chile.

The explanation for the extreme abundance of N_2 in the atmosphere but the low relative abundance of nitrogen compounds elsewhere lies in the chemical inertness of the N_2 molecule. This inertness results from the very high bond energy of the N≡N bond (946 kJ/mol). The Lewis dot formula shows a triple bond and no unpaired electrons, consistent with its experimentally observed diamagnetism in all three physical states.

:N⋮⋮N:

24–2 The Nitrogen Cycle

Although nitrogen molecules are unreactive, nature provides mechanisms by which nitrogen atoms can be incorporated into proteins, nucleic acids, and other nitrogenous compounds. The **nitrogen cycle** is the complex series of reactions by which nitrogen is slowly but continually recycled in the atmosphere, lithosphere (earth), and hydrosphere (water).

When N_2 and O_2 molecules collide in the atmosphere (which is our nitrogen reservoir) in the vicinity of a bolt of lightning, they can absorb enough electrical energy to produce molecules of NO, nitrogen oxide (commonly called nitric oxide). Molecules of NO are quite reactive because they contain one unpaired electron, and NO reacts readily with O_2 to form nitrogen dioxide, NO_2. Most nitrogen dioxide dissolves in rainwater and falls to the earth's surface. Bacterial enzymes reduce the nitrogen in a series of reactions in which amino acids and proteins are produced. These are then used by plants, eaten by animals, metabolized, and excreted as nitrogenous compounds such as urea, $(NH_2)_2CO$, and ammonium salts like $NaNH_4HPO_4$. These products can also be enzymatically converted to ammonia, NH_3, and amino acids, etc.

Another mechanism allows N_2 to be converted directly to ammonia. Members of a certain class of plants called legumes, such as alfalfa and clover, have nodules on their roots; within the nodules live bacteria that contain an enzyme called nitrogenase. These bacteria extract N_2 molecules directly from air trapped in the soil and convert them into ammonia. The ability of nitrogenase to catalyze such a conversion, called **nitrogen fixation,** at usual temperatures and pressures with very high efficiency is a marvel to scientists, who must resort to very extreme and costly conditions to produce NH_3 from nitrogen and hydrogen (Haber process, Sections 15–5 and 24–5). We hope to learn through research how the enzymatic conversion is carried out, so that we can accomplish the transformation much more efficiently.

The suffix *-ase* associated with the name of a biological compound almost invariably signifies that it is an enzyme, or a protein that acts as a catalyst for a specific biological reaction.

Ammonia is the source of nitrogen in many nitrogen-containing fertilizers. Unfortunately, nature does not produce ammonia and related plant nutrient compounds rapidly enough to provide an adequate food supply for the world's growing population. Commercial synthetic fertilizers have helped to lessen this problem; with increasing energy costs, however, the fertilizer shortage is predicted to grow out of hand rapidly in the new few decades unless significant advances are made in the area of agricultural and fertilizer chemistry or in population control.

24–3 Production and Preparation of Nitrogen

Nitrogen is sold commercially in cylinders containing the compressed gas. It is obtained by fractional distillation of air (Section 10–12), and its boiling point is −195.8°C. Tank nitrogen contains traces of oxygen, argon, and water vapor. Further purification can be accomplished by passing the gas over hot copper, which reacts with oxygen to form copper(II) oxide, and then drying it with tetraphosphorus decoxide, P_4O_{10}, a very effective dehydrating agent. The remaining argon is so inert that its presence causes no problem except in the most precise research work.

Small amounts of very pure argon-free nitrogen can be produced by heating pure azides of the Group IA or IIA metals. The reaction produces the metal and pure nitrogen, often explosively

The azide ion is

$:\ddot{N}::N::\ddot{N}:^-$

It is linear. Can you see why?

$$\underset{\text{sodium azide}}{2NaN_3} \xrightarrow{300°C} 2Na + 3N_2$$

24–4 Oxidation States of Nitrogen

No other element exhibits more oxidation states than nitrogen does. Examples are given in Table 24–2.

TABLE 24–2 Oxidation States of Nitrogen and Examples

−3	−2	−1	0	+1	+2	+3	+4	+5
NH_3 ammonia	N_2H_4 hydrazine	NH_2OH hydroxylamine	N_2 nitrogen	N_2O dinitrogen oxide	NO nitrogen oxide	N_2O_3 dinitrogen trioxide	NO_2 nitrogen dioxide	N_2O_5 dinitrogen pentoxide
NH_4^+ ammonium ion		NH_2Cl chloramine		$H_2N_2O_2$ hyponitrous acid		HNO_2 nitrous acid	N_2O_4 dinitrogen tetroxide	HNO_3 nitric acid
NH_2^- amide ion						NO_2^- nitrite ion		NO_3^- nitrate ion

Nitrogen Compounds with Hydrogen

24–5 Ammonia and Ammonium Salts

1 Production

Ammonia is produced commercially by the **Haber process** described in Section 15–5

38 billion pounds of NH_3 were produced in the USA in 1981.

$$N_2\ (g) + 3H_2\ (g) \underset{\Delta,\ \text{pressure}}{\overset{\text{iron}}{\rightleftharpoons}} 2NH_3\ (g) \qquad \Delta H^0 = -92.2\ \text{kJ}$$

Household ammonia is a dilute aqueous solution containing about 5% ammonia and some detergent.

Ammonia can be prepared in the laboratory by treating aqueous ammonia or ammonium salts with excess alkali and distilling. The presence of excess OH^- forces the equilibrium toward the right and NH_3 is evolved as a gas

$$\underset{\text{excess}}{NH_4^+ + OH^-} \xrightleftharpoons{\text{shift}} NH_3 + H_2O$$

2 Properties

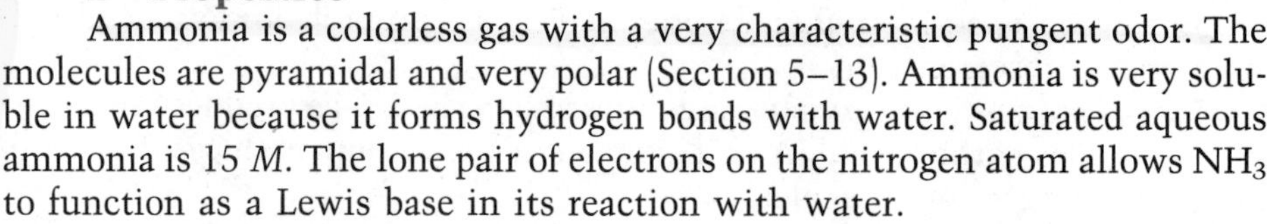

Ammonia is a colorless gas with a very characteristic pungent odor. The molecules are pyramidal and very polar (Section 5–13). Ammonia is very soluble in water because it forms hydrogen bonds with water. Saturated aqueous ammonia is 15 *M*. The lone pair of electrons on the nitrogen atom allows NH_3 to function as a Lewis base in its reaction with water.

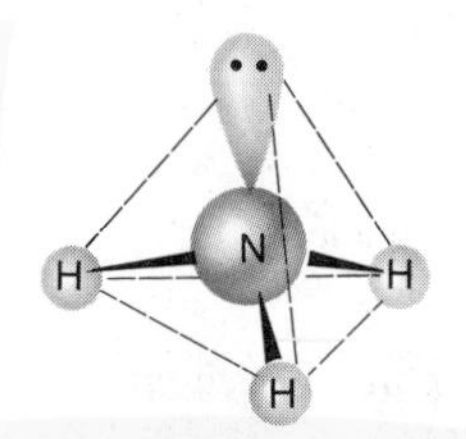

Ammonia, NH_3
mp −77.7°C
bp −33.4°C
polar, pyramidal molecule;
tetrahedral (sp^3) electronic geometry

$$\underbrace{NH_3\,(aq) + H_2O\,(\ell)}_{\text{aqueous ammonia}} \rightleftharpoons NH_4^+ + OH^- \qquad K_b = 1.8 \times 10^{-5}$$

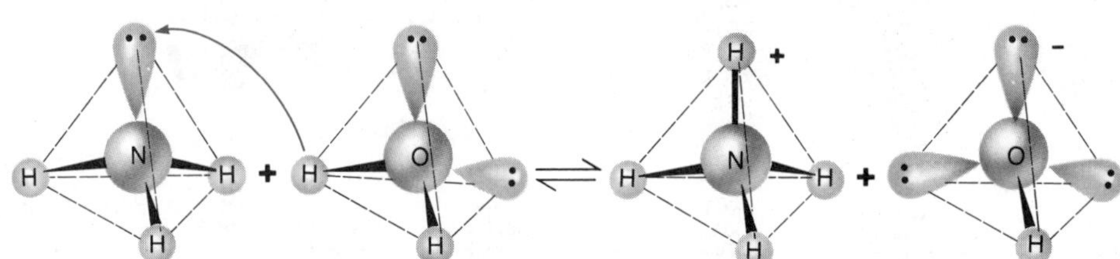

Ammonia acts as a Lewis base in its reaction with metal ions, such as aqueous Co^{2+}, or $[Co(OH_2)_6]^{2+}$, to form ammine complex ions (Sections 18–6 and 27–2), such as $[Co(NH_3)_6]^{2+}$.

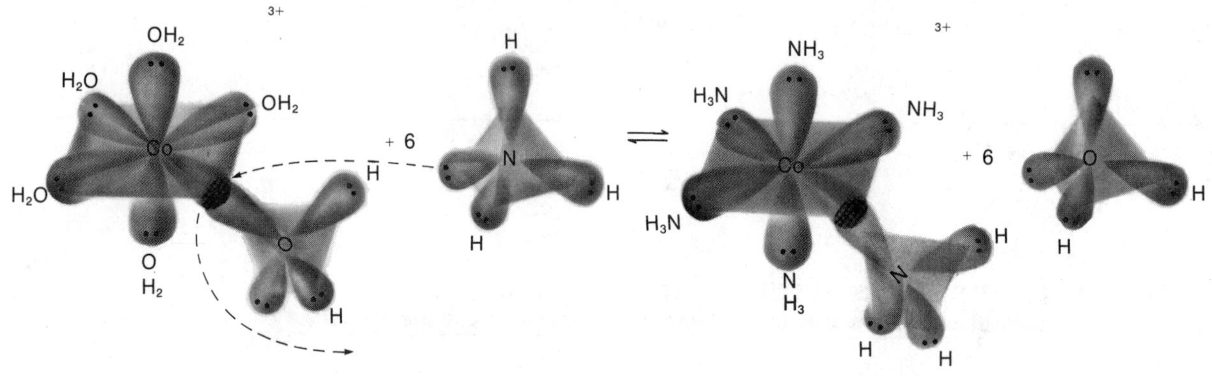

$$\underset{\text{hexaaquacobalt(II) ion}}{[Co(OH_2)_6]^{2+}} + 6NH_3 \xrightarrow{H_2O} \underset{\text{hexaamminecobalt(II) ion}}{[Co(NH_3)_6]^{2+}} + 6H_2O$$

Ammonia also acts as a Lewis base in its reactions with other strong electron acceptors to form "addition compounds." Its reaction with boron trifluoride, BF_3, is a typical example.

$$\underset{(sp^2)}{BF_3} + \underset{(sp^3)}{:NH_3} \longrightarrow \underset{(sp^3)\ (sp^3)}{F_3B:NH_3}$$

Ammonia burns in oxygen when heated to produce nitrogen and water

$$4NH_3\ (g) + 3O_2\ (g) \xrightarrow{\Delta} 2N_2\ (g) + 6H_2O\ (g) \qquad \Delta H^0 = -1.27 \times 10^3\ kJ$$

But in the presence of red-hot platinum, nitrogen oxide, rather than nitrogen, is produced.

$$4NH_3\ (g) + 5O_2\ (g) \xrightarrow[Pt]{\Delta} 4NO\ (g) + 6H_2O\ (g) \qquad \Delta H^0 = -904\ kJ$$

This is an important reaction in the Ostwald process (Section 24–15) for the production of nitric acid.

Liquid ammonia (bp −33.4°C) is used as a solvent for some chemical reactions. It is hydrogen bonded, just as water is, but ammonia is a much more basic solvent. Its weak *autoionization* produces the ammonium ion, NH_4^+, and the amide ion, NH_2^-. Note the similarity to H_2O, which ionizes slightly to produce some H_3O^+ and OH^- ions with $K_w = 10^{-14}$

Liquid NH_3 is a clear, colorless liquid.

$$\underset{base_2}{NH_3\ (\ell)} + \underset{acid_1}{NH_3\ (\ell)} \rightleftharpoons \underset{acid_2}{NH_4^+} + \underset{base_1}{NH_2^-} \qquad K = 10^{-30}$$

The autoionization of NH_3, like that of H_2O, involves the transfer of a proton.

Many ammonium salts are known, and most are very soluble in water. They can be prepared by reactions of ammonia with acids, such as those shown below

$NH_3 + HCN \longrightarrow [NH_4^+ + CN^-]$ ammonium cyanide

$NH_3 + [H^+ + Br^-] \longrightarrow [NH_4^+ + Br^-]$ ammonium bromide

$NH_3 + [H^+ + NO_3^-] \longrightarrow [NH_4^+ + NO_3^-]$ ammonium nitrate

Some ammonium salts contain anions that act as oxidizing agents toward the ammonium ion; when heated, they decompose rapidly and sometimes even explosively

$$\underset{\text{ammonium nitrate}}{2NH_4NO_3\ (s)} \xrightarrow{\Delta} 2N_2\ (g) + 4H_2O\ (g) + O_2\ (g) + \text{Heat}$$

$$\underset{\text{ammonium dichromate}}{(NH_4)_2Cr_2O_7\ (s)} \xrightarrow{\Delta} N_2\ (g) + Cr_2O_3\ (s) + 4H_2O\ (g) + \text{Heat}$$

Ammonium nitrate is often used as a source of nitrogen in fertilizers because of its high nitrogen content (35% by mass) and water solubility.

24–6 Amines

Organic compounds derived from or structurally related to ammonia by replacing one or more hydrogens with organic groups are called **amines.** All involve sp^3 hybridized nitrogen and, like ammonia itself, are weak bases because of the lone pair of electrons. Examples are given below.

$H-\ddot{N}(H)-CH_3$	$H_3C-\ddot{N}(H)-CH_3$	$H_5C_2-\ddot{N}(H)-CH_3$	$H-\ddot{N}(H)-OH$	$H_2\ddot{N}-C(H)(R)-COOH$
methylamine	dimethylamine	ethylmethylamine	hydroxylamine	an amino acid R = organic group

24–7 Hydrazine

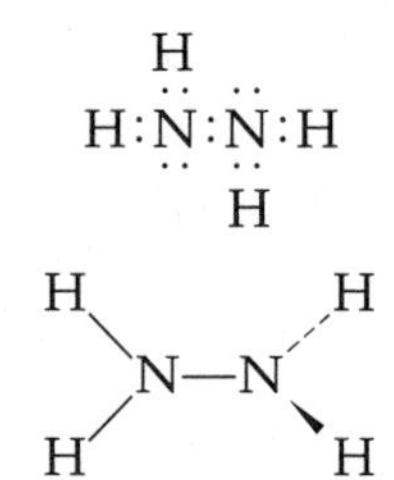

Hydrazine, H_2NNH_2
m.p. 2°C
b.p. 113.5°C

Hydrazine, N_2H_4, a weak base, can be considered a derivative of ammonia produced by the replacement of one hydrogen by an $—NH_2$ group. It is a colorless fuming liquid (bp 114°C) in the anhydrous form, and it can be kept in this condition despite the fact that it is thermodynamically unstable with respect to decomposition ($\Delta H_f^0 = +93.09$ kJ/mol). Its unsymmetrical structure is consistent with its dipole moment of 1.35 D. Hydrazine can be thought of as a nitrogen structural analog of hydrogen peroxide, H_2O_2, but it is a very weak base that ionizes in two steps

$$N_2H_4 + H_2O \rightleftharpoons N_2H_5^+ + OH^- \qquad K_1 = 8.5 \times 10^{-7}$$

$$N_2H_5^+ + H_2O \rightleftharpoons N_2H_6^{2+} + OH^- \qquad K_2 = 8.9 \times 10^{-15}$$

Hydrazine is a powerful reducing agent in basic solution, usually being oxidized to N_2. It is used in conjunction with hydrogen peroxide and other related compounds for rocket fuels. It is also used in the manufacture of certain drugs, industrial chemicals, and dyes.

Nitrogen Oxides

Nitrogen forms several oxides, in which it exhibits positive oxidation states of 1 to 5. (See the right side of Table 24–2.) All the oxides have positive free energies of formation, owing to the high dissociation energy of nitrogen and oxygen molecules. All those listed in Table 24–2 are gases except N_2O_5, a solid that sublimes at only 32.4°C. There is also evidence for the existence of two very unstable oxides, NO_3 and N_2O_6.

24–8 Dinitrogen Oxide (+1 Oxidation State)

Molten ammonium nitrate undergoes internal oxidation-reduction (decomposition) at 170° to 260°C to produce dinitrogen oxide, also called nitrous oxide. Above this temperature range, explosions occur and the products are N_2, O_2, and H_2O

Some dentists use N_2O for its mild anesthetic properties; it is also known as laughing gas because of its side effects.

$$NH_4NO_3 \xrightarrow{\Delta} N_2O + 2H_2O + \text{Heat}$$

Dinitrogen oxide decomposes into nitrogen and oxygen when heated

$$2N_2O \xrightarrow{\Delta} 2N_2 + O_2 + \text{Heat}$$

Dinitrogen oxide supports combustion very well, since it produces oxygen upon heating.

The molecule is linear but nonsymmetric, with a dipole moment of 0.17 D. It is thought to have at least two important contributing resonance structures

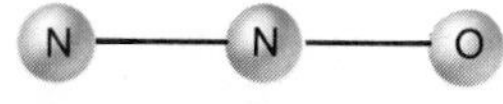

Dinitrogen oxide or nitrous oxide, N_2O
mp −90.8°C
bp −88.8°C

:N::N::O: ⟷ :N:::N:O: N—N—O (1.126 Å, 1.186 Å)

24–9 Nitrogen Oxide (+2 Oxidation State)

The first step of the Ostwald process (Section 24–15) for the production of nitric acid from ammonia is also used for the commercial preparation of nitrogen oxide, NO, also called nitric oxide

$$4NH_3 + 5O_2 \xrightarrow[\Delta]{\text{catalyst}} 4NO + 6H_2O$$

In the laboratory it can be generated by reduction of nitrite ion by iron(II) ions or iodide ions in acidic solution

$$2NaNO_2 + 2FeSO_4 + 3H_2SO_4 \longrightarrow Fe_2(SO_4)_3 + 2NaHSO_4 + 2H_2O + 2NO$$

It is produced in nature by direct reaction of nitrogen with oxygen in the air in electrical storms, but not under usual conditions. (See Examples 13–12, 13–13, 15–10 and 15–12.)

Nitrogen oxide or nitric oxide, NO
mp −163.6°C
bp −151.8°C
one unpaired electron

Nitrogen oxide is a colorless gas that condenses at −152°C to a blue liquid. Gaseous NO is paramagnetic and contains one unpaired electron per molecule.

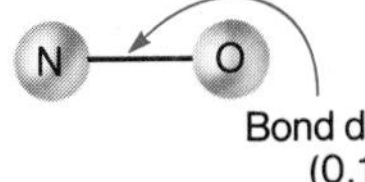

Bond distance (0.115 nm) intermediate between N≡O (0.106 nm) and N=O (0.12 nm)

$$:\dot{N}::\ddot{O}: \longleftrightarrow :\ddot{N}::\dot{O}:$$

Its unpaired electron makes nitric oxide very reactive. Molecules that contain unpaired electrons are called **radicals.** It is somewhat surprising that in the gas phase NO is predominantly monomeric. However, in the solid phase it is dimerized to N_2O_2 and is diamagnetic and polar

N:N with O atoms (Lewis structure)   O=N—N=O

Nitrogen oxide, which is colorless, reacts with oxygen to form nitrogen dioxide, a brown corrosive gas

$$2NO\ (g) + O_2\ (g) \longrightarrow 2NO_2\ (g)$$

There is concern that NO emitted in the exhaust fumes of supersonic transports (SSTs), which fly in the stratosphere, can catalyze the decomposition of ozone, O_3. In the stratosphere, ozone absorbs some of the ultraviolet radiation from the sun before it reaches the surface of the earth. It is thought that the incidence of skin cancer is proportional to exposure to ultraviolet radiation. The atomic oxygen radicals in the following reactions come from the natural radiation-induced breakdown of ozone

$$\begin{array}{l} NO + O_3 \longrightarrow NO_2 + O_2 \\ NO_2 + O \longrightarrow NO + O_2 \\ \hline O_3 + O \longrightarrow 2O_2 \text{ (net reaction)} \end{array}$$

Note that the NO is regenerated in the second step, so it can cause a chain reaction in which a single molecule of NO ultimately is responsible for the decomposition of several molecules of ozone.

24–10 Dinitrogen Trioxide (+3 Oxidation State)

Dinitrogen trioxide is the extremely unstable acid anhydride of nitrous acid, HNO_2. It can be isolated pure only as a solid (mp −103°C). The oxide is prepared as a blue liquid in equilibrium with its reactants by mixing equimolar amounts of nitrogen oxide and nitrogen dioxide at −20°C.

$$NO + NO_2 \rightleftharpoons N_2O_3$$

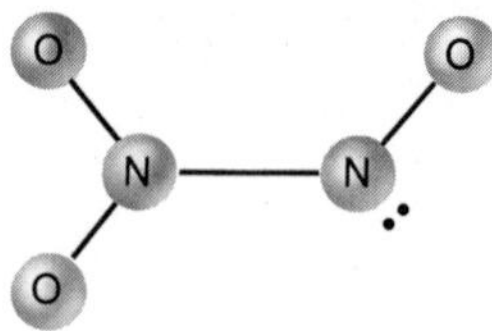

Dinitrogen trioxide, N_2O_3
mp −102°C
bp 3.5°C

Upon boiling (3.5°C) it decomposes completely to a brown gaseous mixture of colorless NO and brown NO_2. Its structure is shown below and in the margin.

:Ö: :Ö:
N:N
:O:

N atoms are sp^2 hybridized

24–11 Nitrogen Dioxide and Dinitrogen Tetroxide (+4 Oxidation State)

Nitrogen oxide is converted into brown nitrogen dioxide by reaction with oxygen (Section 24–9). It is usually prepared in the laboratory by heating heavy metal nitrates such as lead(II) nitrate

$$2Pb(NO_3)_2 \xrightarrow{\Delta} 2PbO + 4NO_2 + O_2$$

or by reduction of concentrated nitric acid (Section 24–15) by a metal.

Nitrogen dioxide molecules contain one unpaired electron each, and they easily dimerize to form colorless, diamagnetic dinitrogen tetroxide, N_2O_4, in a temperature-dependent equilibrium

$$\underset{\text{brown, paramagnetic}}{2NO_2\,(g)} \rightleftharpoons \underset{\text{colorless, diamagnetic}}{N_2O_4\,(g)} \qquad \Delta H^0 = -57.2\ \text{kJ}$$

The formation of N_2O_4 is favored at low temperatures, and NO_2 formation is favored at high temperatures, since the forward reaction is exothermic.

The NO_2 molecule is angular and is represented by resonance structures. The structures of NO_2 and N_2O_4 are represented below and in the margin.

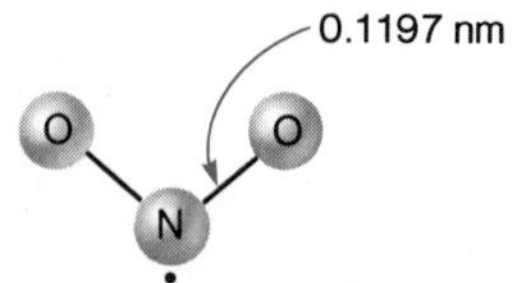

Nitrogen dioxide, NO_2
mp −11.20°C
bp 21.2°C
one unpaired electron
brown gas

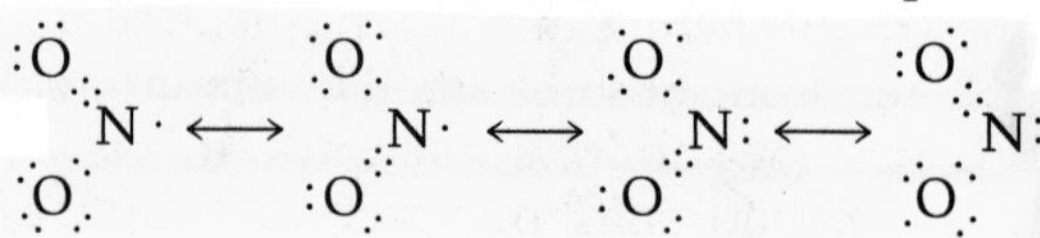

nitrogen dioxide

:O: :O:
N:N ⟷ other resonance structures
:O: :O:

dinitrogen tetroxide

24–12 Dinitrogen Pentoxide (+5 Oxidation State)

Dinitrogen pentoxide, N_2O_5, is the acid anhydride of nitric acid

$$\underset{\text{dinitrogen pentoxide}}{N_2O_5} + H_2O \longrightarrow \underset{\text{nitric acid}}{2HNO_3}$$

The preparation of N_2O_5 is accomplished by dehydration of nitric acid with tetraphosphorus decoxide, P_4O_{10}, a very powerful dehydrating agent

$$4HNO_3 + P_4O_{10} \longrightarrow 4HPO_3 + 2N_2O_5$$

This is called a dehydration because it is equivalent to two occurrences of

$$2HNO_3 \longrightarrow N_2O_5 + H_2O$$

$$H_2O + \tfrac{1}{2}P_4O_{10} \longrightarrow 2HPO_3$$

This is not, however, the actual mechanism of the overall reaction.

The white solid, N_2O_5, is removed by vacuum sublimation at 32.4°C. Below 0°C it is stable, but it decomposes when heated to room temperature at atmospheric pressure. Rapid heating causes it to explode

$$2N_2O_5 \xrightarrow{\Delta} 4NO_2 + O_2 + \text{Heat}$$

In the gaseous state it consists of N_2O_5 molecules, but x-ray diffraction evidence shows that in the solid state it is nitronium nitrate, $NO_2^+NO_3^-$. Dinitrogen pentoxide is a very strong oxidizing agent.

one resonance form of gaseous N_2O_5

solid "N_2O_5": $NO_2^+NO_3^-$, nitronium nitrate

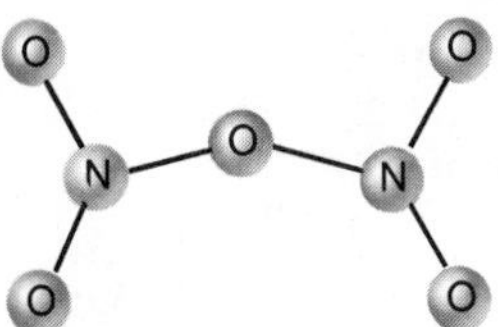

Dinitrogen pentoxide, N_2O_5 (gaseous)
mp 30°C
bp 47°C (dec)

24–13 Nitrogen Oxides and Photochemical Smog

Nitrogen oxides are produced in the atmosphere by natural processes and are important in the nitrogen cycle (Section 24–2). Human activities contribute only about 10% of all the oxides of nitrogen (collectively referred to as NO_x) in the atmosphere, but the human contribution occurs mostly in urban areas where the oxides may be present in concentrations a hundred times greater than in rural areas.

Just as nitrogen oxide is produced naturally by reaction of nitrogen and oxygen in electrical storms, so it is also produced by the same reaction when air is mixed with fossil fuels at high temperatures in internal combustion engines and furnaces.

$$N_2\,(g) + O_2\,(g) \rightleftharpoons 2NO\,(g) \qquad \Delta H^0 = 180\text{ kJ}$$

The reaction does not occur to a significant extent at ordinary temperatures but because it is endothermic it is favored by high temperatures. At the high operating temperatures of internal combustion engines and furnaces, the equilibrium still lies far to the left and only small amounts of NO are produced and released into the atmosphere. However, very small concentrations of nitrogen oxides (in the range of parts per million or parts per billion) are sufficient to cause serious problems.

The nitric oxide radical, NO, reacts with oxygen to produce nitrogen dioxide. Both NO and NO_2 are quite reactive because each possesses an unpaired electron, and they do considerable damage to plants and animals.

Nitrogen dioxide reacts with water vapor in the air to produce corrosive droplets of nitric acid and more NO

$$3\overset{+4}{N}O_2 + H_2O \longrightarrow 2H\overset{+5}{N}O_3 + \overset{+2}{N}O$$

The nitric acid may be washed out of the air by rainwater, or it may react with traces of ammonia in the air to form solid ammonium nitrate, a type of particulate pollutant

$$HNO_3 + NH_3 \longrightarrow NH_4NO_3$$

This situation occurs in all urban areas, but the problem is accentuated in areas with warm, dry climates conducive to light-induced reactions, also called photochemical reactions. Here ultraviolet radiation from the sun produces damaging oxidants. The brownish hazes that often hang over such cities as Los Angeles, Denver, and Mexico City are due to the presence of brown NO_2. Problems begin in the early morning rush hour as relatively high concentrations of NO are exhausted into the air from automobiles and trucks. The NO combines with O_2 to form NO_2. Then, as the sun rises higher into the sky, NO_2 absorbs ultraviolet radiation and breaks down into NO and oxygen radicals

$$NO_2 \xrightarrow{\text{uv}} NO + O$$

The oxygen radicals are extremely reactive, and they combine with oxygen molecules to produce ozone

$$O + O_2 \longrightarrow O_3$$

Ozone is a powerful oxidizing agent that damages rubber, plastic materials, and all plant and animal life. Obviously, the ozone in the upper atmosphere does not pose such a problem, but that generated at the surface of the earth certainly does. It also reacts with hydrocarbons from automobile exhaust and evaporated gasoline to form secondary organic pollutants such as aldehydes and ketones (Section 30–12). The peroxyacyl nitrates (PANs), perhaps the worst of the secondary pollutants, are especially damaging photochemical oxidants that are very irritating to the eyes and throat

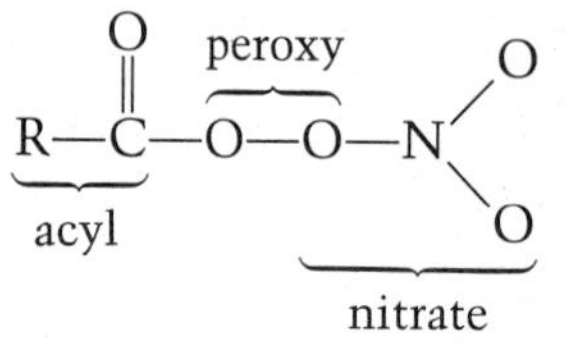

$$\underbrace{R-\overset{\displaystyle O}{\overset{\|}{C}}}_{\text{acyl}}-\overbrace{O-O}^{\text{peroxy}}-\underbrace{N\begin{smallmatrix}\nearrow O \\ \searrow O\end{smallmatrix}}_{\text{nitrate}} \qquad R = \text{hydrocarbon chain or ring}$$

The catalytic converters now being installed in automobile exhaust systems (Section 14–6) are designed to reduce emissions of oxides of nitrogen; unfortunately, evidence indicates that they also increase emission of sulfur oxides and sulfuric acid droplets, also very corrosive pollutants.

Some Oxyacids of Nitrogen and Their Salts

Nitrogen also forms hyponitrous acid, $H_2N_2O_2$, in which N is in the +1 oxidation state, as well as hyponitrite salts such as $Na_2N_2O_2$.

The important oxyacids of nitrogen are nitrous acid, HNO_2, and nitric acid, HNO_3.

24–14 Nitrous Acid and Nitrites (+3 Oxidation State)

Although nitrous acid, HNO_2, is unstable and cannot be isolated in pure form, N_2O_3 can be considered its anhydride. The reaction of N_2O_3 with aqueous alkali produces nitrite salts

$$N_2O_3 + 2[Na^+ + OH^-] \longrightarrow 2[Na^+ + NO_2^-] + H_2O$$

sodium nitrite

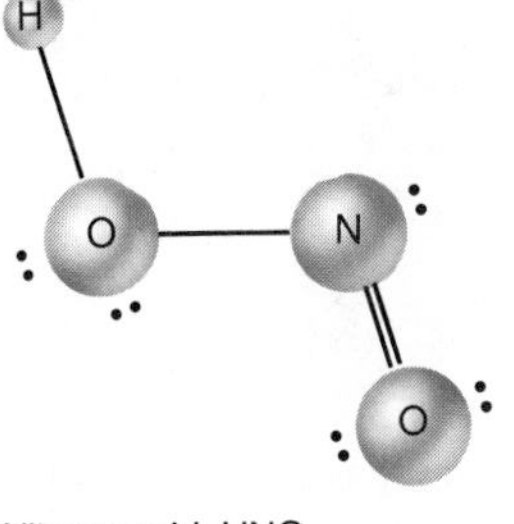

Nitrous acid, HNO_2

The acid is prepared as a pale blue solution when sulfuric acid reacts with cold aqueous sodium nitrite. Nitrous acid is a weak acid ($K_a = 4.5 \times 10^{-4}$). It acts as an oxidizing agent toward strong reducing agents and as a reducing agent toward very strong oxidizing agents.

Dot diagrams for nitrous acid and the nitrite ion are shown below, and their structures are shown in the margin

$$H\!-\!O\!-\!N\!=\!O \qquad [O\!-\!N\!=\!O]^- \longleftrightarrow [O\!=\!N\!-\!O]^-$$

nitrous acid nitrite ion

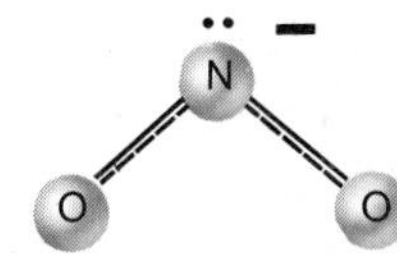

24–15 Nitric Acid and Nitrates (+5 Oxidation State)

Pure nitric acid, HNO_3, is a colorless liquid that boils at 83°C. Light or heat causes it to decompose into NO_2, O_2, and H_2O. The presence of the NO_2 in partially decomposed aqueous nitric acid is responsible for its yellow or brown tinge.

Studies on the vapor phase indicate that nitric acid has the following structure.

$$H\!:\!O\!:\!N(=O)(-O) \longleftrightarrow H\!:\!O\!:\!N(-O)(=O)$$

0.122 nm
115°
130°
115°
0.141 nm
0.122 nm

Nitric acid, HNO_3
mp −42°C
bp 83°C

HNO_3 is commercially prepared by the **Ostwald process.** At high temperatures, ammonia is catalytically converted to nitrogen oxide, which is cooled and then air-oxidized to nitrogen dioxide. Nitrogen dioxide reacts with water to produce nitric acid and some nitrogen oxide. The NO produced in the third step is then recycled into the second step. More than 18,000,000,000 pounds of HNO_3 were produced in the USA in 1981.

$$4NH_3\ (g) + 5O_2\ (g) \xrightarrow[1000°C]{Pt} 4NO\ (g) + 6H_2O\ (g)$$

$$2NO\ (g) + O_2\ (g) \xrightarrow{cool} 2NO_2\ (g)$$

$$3NO_2\ (g) + H_2O\ (\ell) \longrightarrow 2[H^+ + NO_3^-] + NO\ (g)$$

$$4HNO_3 \xrightarrow[\Delta]{uv} 4NO_2 + 2H_2O + O_2$$

decomposition

Nitric acid has many important industrial uses (see Table 11–1), including the production of explosives. It is interesting to note that although the Germans were cut off from their normal supply of nitrates by the allied naval blockade during World War I, they were able to continue the production of munitions. Ammonia was obtained from putrified garbage and then converted into nitric acid, which was used to produce explosives. Thus, the war was extended considerably. (We are indebted to Professor Lloyd Jones for pointing out these facts.)

Nitric acid is very soluble in water (~16 mol/L). It is a strong acid and a strong oxidizing agent. The nature of its reduction products depends on the concentration of the acid used, as indicated below.

1 Action of Nitric Acid on Metals

Acids whose *anions* do not undergo reduction easily are called nonoxidizing acids.

The oxidizing power of nitric acid enables it to dissolve many metals that do not dissolve in nonoxidizing acids. Copper does not dissolve in hydrochloric acid (a very strong but nonoxidizing acid), but dissolves in nitric acid readily. The dissolution of copper in concentrated nitric acid produces nitrogen dioxide

$$Cu + 4[H^+ + NO_3^-] \longrightarrow [Cu^{2+} + 2NO_3^-] + 2NO_2 + 2H_2O$$

while in dilute nitric acid (~3 M) the major reduction product is nitrogen oxide

$$3Cu + 8[H^+ + NO_3^-] \longrightarrow 3[Cu^{2+} + 2NO_3^-] + 2NO + 4H_2O$$

Both reactions probably occur in both cases, but the second predominates in dilute solution and the first in concentrated nitric acid.

When active metals such as zinc are dissolved in nitric acid, a number of reduction products result; the relative amounts of each depend on the acid concentration. In general, the higher the acid concentration, the higher the oxidation state of nitrogen in the reduction product:

The counter ion for Zn^{2+} is NO_3^- in all these reactions.

increasing acid concentration ↑

$$Zn + 2NO_3^- + 4H^+ \longrightarrow Zn^{2+} + 2NO_2 + 2H_2O$$
$$3Zn + 2NO_3^- + 8H^+ \longrightarrow 3Zn^{2+} + 2NO + 4H_2O$$
$$4Zn + 2NO_3^- + 10H^+ \longrightarrow 4Zn^{2+} + N_2O + 5H_2O$$
$$5Zn + 2NO_3^- + 12H^+ \longrightarrow 5Zn^{2+} + N_2 + 6H_2O$$
$$4Zn + NO_3^- + 10H^+ \longrightarrow 4Zn^{2+} + NH_4^+ + 3H_2O$$

↑ increasing oxidation number of nitrogen

2 Action of Nitric Acid on Nonmetals

Many nonmetallic elements dissolve in concentrated nitric acid to form oxides or oxyacids of the nonmetal and nitrogen dioxide. Phosphorus and sulfur are oxidized to phosphoric acid and sulfuric acid, respectively, by hot concentrated nitric acid

$$P_4 + 20[H^+ + NO_3^-] \longrightarrow 4H_3PO_4 + 4H_2O + 20NO_2$$

For simplicity, S is used to indicate sulfur rather than S_8.

$$S + 6[H^+ + NO_3^-] \longrightarrow [2H^+ + SO_4^{2-}] + 2H_2O + 6NO_2$$

3 Nitrates

The nitrate ion, NO_3^-, is a planar, resonance-stabilized ion

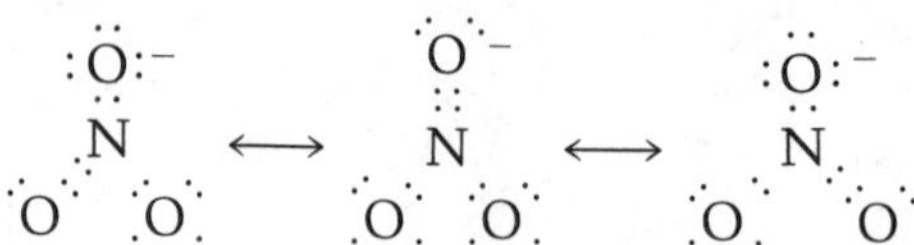

Neutralization of solutions of nitric acid with metal hydroxides, carbonates, or oxides produces water-soluble nitrate salts

$$[Ca^{2+} + 2OH^-] + 2[H^+ + NO_3^-] \longrightarrow [Ca^{2+} + 2NO_3^-] + 2H_2O$$

calcium hydroxide — calcium nitrate

$$CaCO_3 + 2[H^+ + NO_3^-] \longrightarrow [Ca^{2+} + 2NO_3^-] + CO_2 + H_2O$$

calcium carbonate

$$CaO + 2[H^+ + NO_3^-] \longrightarrow [Ca^{2+} + 2NO_3^-] + H_2O$$

calcium oxide

Alkali metal nitrates decompose upon heating to produce oxygen and nitrites

$$2KNO_3 \xrightarrow{\Delta} 2KNO_2 + O_2$$

potassium nitrite

In contrast, heavy metal nitrates undergo thermal decomposition to produce heavy metal oxides, nitrogen dioxide, and oxygen

$$2Pb(NO_3)_2 \xrightarrow{\Delta} 2PbO + 4NO_2 + O_2$$

lead(II) nitrate

Nitrates of very unreactive metals and mercury(II) nitrate, when heated, produce metal oxides that subsequently undergo decomposition to the free metals and oxygen

$$2AgNO_3 \longrightarrow 2Ag + 2NO_2 + O_2$$

24–16 Sodium Nitrite and Sodium Nitrate as Food Additives

Sodium nitrite and sodium nitrate are used as food additives to retard oxidation of meat. In the body, nitrate ion is reduced to nitrite ion, which is then converted to NO. This in turn reacts with the brown oxidized form of the heme in blood. This is the same reaction that keeps meat red longer. The brown color of "old" meat is the result of oxidation of blood, and, though not due to the presence of harmful compounds, is objectionable to many consumers. Nitrites and nitrates retard this oxidation and also prevent growth of botulism bacteria. However, recent controversy has arisen concerning the possibility that nitrites combine with amines under the acidic conditions in the stomach to produce carcinogenic nitrosoamines, which have the following general formula.

Nitrates and nitrites are added during "curing" of meats such as hot dogs, salami, and bacon.

$$\underbrace{R'R}_{\text{amine group}}N{-}\underbrace{N{=}O}_{\text{nitroso group}}$$

R and R′ = organic groups

Phosphorus and Arsenic

24–17 Occurrence, Production, and Uses of Phosphorus

Phosphorus is the only element of Group VA that does not occur uncombined in nature. Like nitrogen, phosphorus is present in all living organisms, as organophosphates and also in calcium phosphates such as hydroxyapatite, $Ca_{10}(PO_4)_6(OH)_2$, and fluoroapatite, $Ca_{10}(PO_4)_6(F)_2$, in bones and teeth. It also occurs in these and related compounds in phosphate minerals, which are mined mostly in Florida and North Africa.

Industrially, the element is obtained from phosphate minerals by heating them to 1200–1500°C in an electric arc furnace with sand (mainly SiO_2) and coke (impure carbon). The reaction is

$$\underset{\substack{\text{calcium phosphate}\\\text{(phosphate rock)}}}{2Ca_3(PO_4)_2} + 6SiO_2 + 10C \xrightarrow{\Delta} \underset{\substack{\text{calcium silicate}\\\text{(slag)}}}{6CaSiO_3} + 10CO + P_4$$

The vaporized phosphorus is condensed to a white solid (mp = 44.2°C; bp = 280.3°C) under water to prevent oxidation. Even when kept under water, white phosphorus slowly converts to the more stable red phosphorus allotrope (mp = 597°C; sublimes at 431°C) (Section 24–19). Red phosphorus and tetraphosphorus trisulfide, P_4S_3, are both used in ordinary matches. They do not ignite spontaneously, yet ignite easily when heated by friction. Both white and red phosphorus are insoluble in water.

The largest single user of phosphorus compounds is the fertilizer industry. Since phosphorus is an essential nutrient and since nature's phosphorus cycle is very slow, owing to the low solubility of most natural phosphates, phosphate fertilizers are absolutely essential. To increase the solubility of the natural phosphates, they are treated with sulfuric acid to produce "superphosphate of lime," a solid consisting of two salts, which is then pulverized and applied as a powder.

This reaction represents the biggest single use of sulfuric acid, the industrial chemical produced in largest quantity.

$$\underset{\text{phosphate rock}}{Ca_3(PO_4)_2} + 2H_2SO_4 + 4H_2O \xrightarrow{\text{evaporate}} \underbrace{[\underset{\substack{\text{calcium}\\\text{dihydrogen}\\\text{phosphate}}}{Ca(H_2PO_4)_2} + \underset{\substack{\text{calcium}\\\text{sulfate}\\\text{dihydrate}}}{2(CaSO_4\cdot 2H_2O)}]}_{\text{superphosphate of lime}}$$

24–18 Arsenic

Small amounts of arsenic occur in free form, but most arsenic is found in the form of yellow arsenic(III) sulfide, As_2S_3. Its natural occurrence in a positive oxidation state (+3) shows the greater metallic character of arsenic compared to nitrogen and phosphorus. It also occurs as arsenopyrite, FeAsS. Roasting of As_2S_3 ore produces As_4O_6, which is separated by sublimation

$$2As_2S_3 + 9O_2 \xrightarrow{\Delta} As_4O_6 + 6SO_2$$

The oxide can be reduced with coke to produce elemental arsenic, As_4

$$As_4O_6 + 6C \longrightarrow 6CO + As_4$$

Arsenopyrite thermally decomposes in the absence of air to iron(II) sulfide and elemental arsenic

$$4FeAsS \xrightarrow[\text{atmos.}]{\overset{\Delta}{\text{inert}}} 4FeS + As_4$$

Pure arsenic occurs in several allotropic forms (below); the most stable, gray arsenic, sublimes at 613°C, and is insoluble in water.

All arsenic compounds are poisonous, and most of their former practical uses are now banned. Arsenic(III) sulfide, As_2S_3, was used in matches. Copper(II) hydrogen arsenite, $CuHAsO_3$, or Scheele's green, was used as a green paint pigment. Lead arsenate, $Pb_3(AsO_4)_2$, and Paris green, $Cu(CH_3COO)_2 \cdot 3Cu(AsO_2)_2$ were used as insecticides; sodium arsenite, Na_3AsO_3, served as a weed killer; and calcium arsenate, $Ca_3(AsO_4)_2$, was used to kill the boll weevil. Although 0.1 g doses of arsenic are lethal to man, it has been demonstrated that minute traces actually stimulate production of red blood cells.

24–19 Allotropes of Phosphorus and Arsenic

Although both phosphorus and arsenic exist in several allotropic forms, there are only two important forms of each. One has a lattice consisting of tetrahedral P_4 or As_4 molecules (white phosphorus and yellow arsenic; see margin). These allotropes are more volatile, less dense, more soluble in organic solvents, more chemically active, and more toxic than the other allotropes. Red phosphorus crystallizes in a polymeric lattice consisting of bonded tetrahedra in which one bond in each P_4 tetrahedron has been broken and replaced by a bond *between* tetrahedra. Gray arsenic consists of puckered six-membered rings within layers.

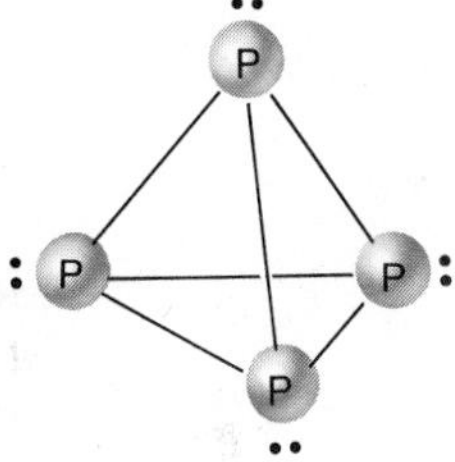

The P_4 molecule

White phosphorus is sometimes called yellow phosphorus because, as its surface layer converts to the more stable red form, it appears to turn yellow and eventually red. White phosphorus was originally used in friction matches, but because of its toxicity it has been replaced by red phosphorus.

At very high temperatures phosphorus and arsenic vapors are thought to contain P_2 and As_2 molecules, analogous to N_2, in addition to P_4 and As_4 molecules.

24–20 Phosphine and Arsine

Phosphorus and arsenic exhibit the −3 oxidation state in phosphine, PH_3, and arsine, AsH_3, both of which have pyramidal molecules with one lone pair of electrons on the central atom, like ammonia. Both are colorless, toxic, ill-smelling gases.

Phosphine is prepared by hydrolysis of calcium phosphide, Ca_3P_2, in a reaction analogous to the hydrolysis of calcium nitride to produce NH_3

$$Ca_3P_2 + 6H_2O \longrightarrow 3[Ca^{2+} + 2OH^-] + 2PH_3$$

Phosphine is unstable, in contrast to ammonia, and reacts instantly with air to give clouds of phosphoric acid. It is a weaker base than ammonia and forms some phosphonium salts (analogous to ammonium salts) such as PH_4I. However, such salts immediately decompose in water and do not ionize to give PH_4^+ ions.

Diphosphine, P_2H_4 or $H_2P—PH_2$ (which is the analog of hydrazine, N_2H_4), occurs as a by-product of the hydrolytic preparation of phosphine from Ca_3P_2. It is a colorless liquid that spontaneously ignites in air; it also disproportionates in light to phosphine and other complex hydrides of greater phosphorus content.

Arsine is even less stable and less basic than phosphine. The reaction of zinc with arsenic compounds such as arsenous acid, H_3AsO_3, in hydrochloric acid solution yields arsine

$$3Zn + H_3AsO_3 + 6[H^+ + Cl^-] \longrightarrow 3[Zn^{2+} + 2Cl^-] + 3H_2O + AsH_3$$

This reaction is the basis for the Marsh test used in research and criminology laboratories as a test for the presence of arsenic in suspected poisoning cases. If the arsine, water vapor, and hydrogen are passed through a heated glass tube, arsine decomposes and a mirror of metal-like arsenic is deposited on the inside of the tube.

24–21 Phosphorus and Arsenic Halides

While there are other phosphorus and arsenic halides, the most common and important ones are the trihalides and pentahalides in which the phosphorus or arsenic exists in the +3 or +5 oxidation state. All eight possible trihalides (PX_3 and AsX_3 for X = F, Cl, Br, I) are known, as well as the pentahalides PF_5, PCl_5, PBr_5, and AsF_5. Mixed halides such as PF_3Cl_2 also exist.

All the trihalides can be prepared by direct union of the elements in stoichiometric ratios

$$P_4 + 6X_2 \longrightarrow 4PX_3$$
$$As_4 + 6X_2 \longrightarrow 4AsX_3 \qquad X = F, Cl, Br, I$$

This method is used for the preparation of PCl_3, PBr_3, and PI_3, but PF_3 is usually made from AsF_3 by reaction with PCl_3, after which the products are separated by fractional distillation

$$\underset{\text{bp } 62.8°C}{AsF_3} + \underset{\text{bp } 76.1°C}{PCl_3} \longrightarrow \underset{\text{bp } 130°C}{AsCl_3} + \underset{\text{bp } -101.2°C}{PF_3}$$

The direct reaction of phosphorus with fluorine can be explosive.

Indirect methods are also used for preparing other trihalides. For example, AsF_3 is usually prepared by distilling it from the reaction mixture of calcium fluoride, tetraarsenic hexoxide, and excess concentrated sulfuric acid

$$6CaF_2 + As_4O_6 + 6H_2SO_4 \longrightarrow 6CaSO_4 + 6H_2O + 4AsF_3$$

The pentahalides can be prepared by reaction of the elements with excess halogen, or of a trihalide with halogen

$$P_4 + 10X_2 \longrightarrow 4PX_5 \qquad X = F, Cl, Br$$
$$As_4 + 10F_2 \longrightarrow 4AsF_5$$
$$PF_3 + F_2 \longrightarrow PF_5$$

The mixed halides are thermally unstable and decompose to simple halides

$$5PF_3Br_2 \xrightarrow{\Delta} 3PF_5 + 2PBr_5$$

All the pentahalides and the phosphorus trihalides hydrolyze rapidly and completely; the arsenic trihalides less so. Their acidic character is obvious from the following reactions

$$\overset{+5}{PX_5} + \underset{\text{excess}}{4H_2O} \longrightarrow \underset{\text{phosphoric acid}}{\overset{+5}{H_3PO_4}} + \underset{\text{hydrohalic acid}}{5HX}$$

$$\underset{\text{excess}}{\overset{+5}{PX_5}} + H_2O \longrightarrow \underset{\text{phosphorus oxyhalide}}{\overset{+5}{POX_3}} + 2HX$$

$$\overset{+3}{PX_3} + 3H_2O \longrightarrow \underset{\text{phosphorous acid}}{\overset{+3}{H_3PO_3}} + 3HX$$

The last reaction is often used to prepare hydrogen bromide and hydrogen iodide.

The covalent trihalides are all pyramidal molecules with tetrahedral electronic geometry. The pentahalides are trigonal bipyramidal in the gaseous state, but X-ray evidence indicates that solid phosphorus pentachloride is ionic and contains $[PCl_4]^+$ tetrahedra and $[PCl_6]^-$ octahedra. Solid phosphorus pentabromide is apparently $[PBr_4]^+Br^-$.

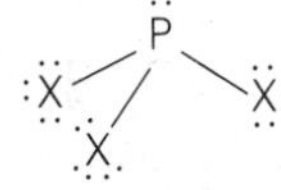

Phosphorus trihalides, PX_3 (X = F, Cl, Br, I)

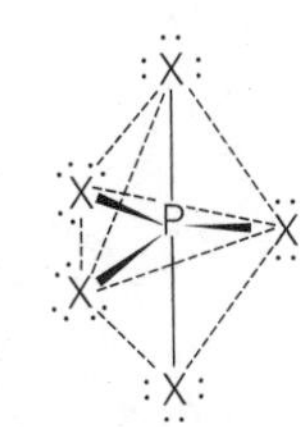

Phosphorus pentahalides, PX_5, in the gas phase (X = F, Cl, Br)

24–22 Oxides of Phosphorus and Arsenic

Several oxides are known, but the most important ones are P_4O_6, P_4O_{10}, As_4O_6, and As_2O_5. Of these, P_4O_6 can be prepared by heating phosphorus under low pressure of oxygen, while high oxygen pressures produce P_4O_{10}. Both are acid anhydrides; as mentioned earlier, P_4O_{10} is a very strong dehydrating agent and will remove H_2O from such compounds as HNO_3. The structure of P_4O_6 is visualized by inserting an oxygen atom between each pair of phosphorus atoms in a P_4 tetrahedron. If an additional oxygen atom is doubly bonded to each phosphorus, the P_4O_{10} structure results (see margin at top of p. 752).

The preparation of As_4O_6, a white solid, is accomplished by roasting As_2S_3, while As_2O_5 is prepared by thermal dehydration of arsenic acid, H_3AsO_4 (Section 24–24). Arsenic acid in turn is prepared by oxidation of As_4O_6 with concentrated nitric acid.

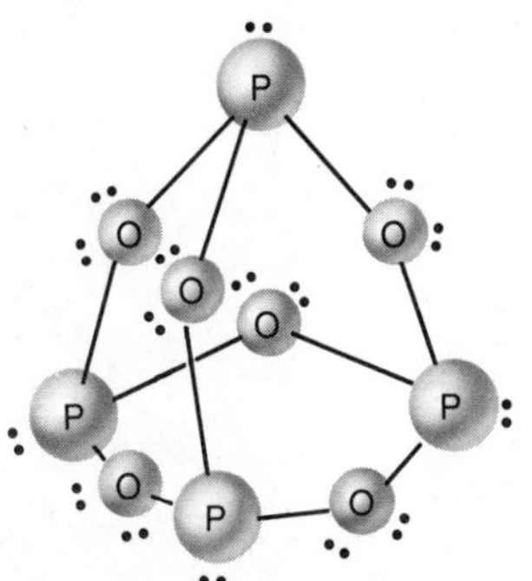

Tetraphosphorus hexoxide or phosphorus (III) oxide, P_4O_6
mp 23.8°C
bp 175.4°C

24–23 Phosphorus Oxyacids and Salts

The oxyacids of phosphorus may contain acidic hydrogens bonded to oxygen

$$\rangle P{-}O{-}H + H_2O \rightleftharpoons \rangle P{-}O^- + H_3O^+$$

or nonacidic hydrogen atoms bonded directly to phosphorus, —P—H. Many

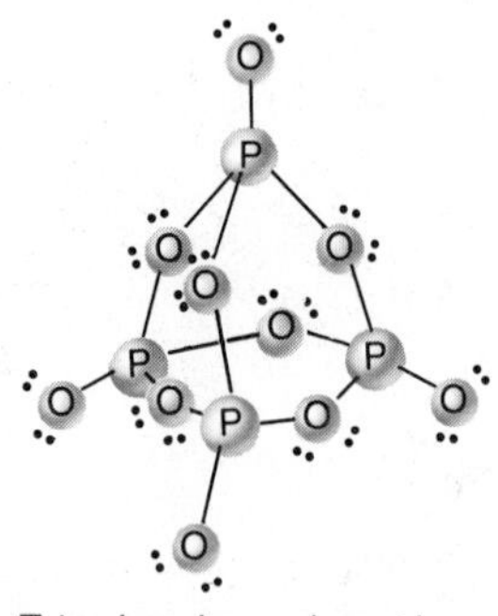

Tetraphosphorus decoxide
or
phosphorus (V) oxide,
P_4O_{10}
sublimes at 358°C

TABLE 24–3 Important Phosphorus Oxyacids

Oxidation State	Formula	Name	No. of Acidic H	Compounds Prepared
+1	H_3PO_2	hypophosphorous acid	1	acid, salts
+3	H_3PO_3	(ortho)phosphorous acid	2	acid, salts
+5	$(HPO_3)_n$	metaphosphoric acids	n	salts (n = 3, 4)
+5	$H_5P_3O_{10}$	triphosphoric acid	3	salts
+5	$H_4P_2O_7$	pyrophosphoric acid	4	acid, salts
+5	H_3PO_4	(ortho)phosphoric acid	3	acid, salts

such oxyacids are known. The names and formulas of the most important ones are given in Table 24–3.

1 Hypophosphorous Acid (+1 Oxidation State)

The barium salt of the monoprotic hypophosphorous acid, H_3PO_2, is prepared by reaction of white phosphorus with aqueous barium hydroxide

$$2P_4 + 3[Ba^{2+} + 2OH^-] + 6H_2O \longrightarrow \underset{\text{barium hypophosphite}}{3Ba(H_2PO_2)_2} + \underset{\text{phosphine}}{2PH_3}$$

The free acid can then be obtained in aqueous solution by reaction of barium hypophosphite with sulfuric acid to form insoluble barium sulfate and hypophosphorous acid

$$Ba(H_2PO_2)_2 + [2H^+ + SO_4^{2-}] \longrightarrow BaSO_4\ (s) + \underset{\text{hypophosphorous acid}}{2H_3PO_2}$$

The free acid is a colorless solid which, in aqueous solution, acts as a comparatively strong monoprotic acid, $K_a \approx 10^{-2}$. The acid and its salts are strong reducing agents. The structures of the acid and hypophosphite ion are shown below.

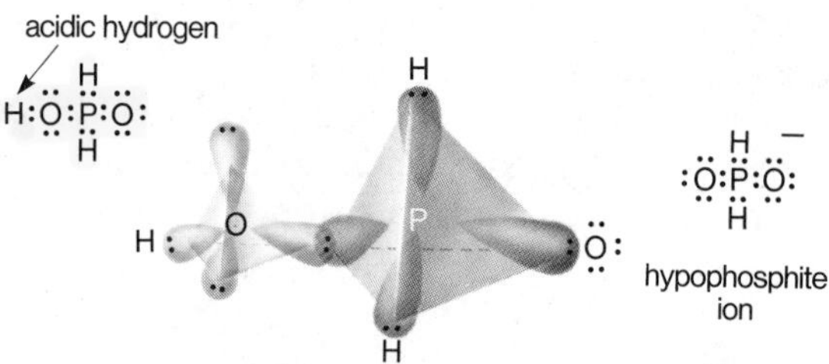

hypophosphorous acid, H_3PO_2

There is some double-bond character between phosphorus and the oxygen atoms to which no hydrogen atoms are bound. This occurs in many other oxyacids in which the central nonmetal atom is from period 3 or lower. The participation of vacant nonmetal d orbitals is involved, as they accept shares of the lone pairs of electrons on the oxygen atoms.

2 Phosphorous Acid (+3 Oxidation State)

Tetraphosphorus hexoxide is the anhydride of (ortho)phosphorous acid, H_3PO_3

$$P_4O_6 + 6H_2O \xrightarrow{\text{cold}} 4H_3PO_3$$

However, the acid is more easily prepared by hydrolysis of PCl_3

$$PCl_3 + 3H_2O \longrightarrow H_3PO_3 + 3HCl$$

The acid is only diprotic ($K_1 = 1.6 \times 10^{-2}$; $K_2 = 7 \times 10^{-7}$) because one hydrogen is directly bonded to phosphorus and will not ionize, as do the two that are bonded to oxygen. Its salts are strong reducing agents in basic solution. Phosphite salts containing either $H_2PO_3^-$ or HPO_3^{2-} ions can be prepared by neutralization of the acid with the appropriate base. The dot diagram and structure of the acid are represented in the margin.

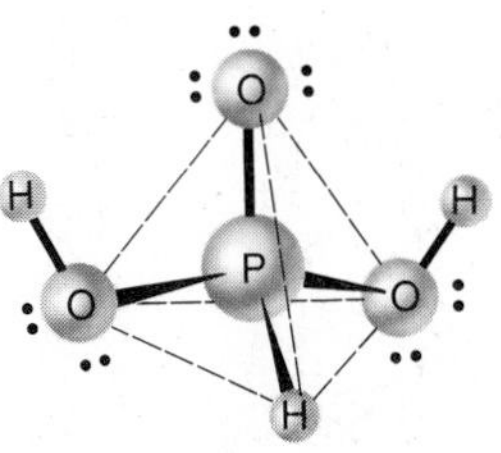

Phosphorous acid, H_3PO_3
mp 73.6°C
bp 200°C (dec)

3 Phosphoric Acid (+5 Oxidation State)

Orthophosphoric acid, or simply phosphoric acid, H_3PO_4, is the most common of the phosphoric acids. Its anhydride is P_4O_{10}

$$P_4O_{10} + 6H_2O \longrightarrow 4H_3PO_4$$

Pure phosphoric acid is a stable, colorless solid. It is a weak triprotic acid

$$H_3PO_4 \rightleftharpoons H^+ + H_2PO_4^- \qquad K_1 = 7.5 \times 10^{-3}$$

$$H_2PO_4^- \rightleftharpoons H^+ + HPO_4^{2-} \qquad K_2 = 6.2 \times 10^{-8}$$

$$HPO_4^{2-} \rightleftharpoons H^+ + PO_4^{3-} \qquad K_3 = 3.6 \times 10^{-13}$$

Phosphoric acid, H_3PO_4
mp 42.35°C

Salts containing each of the three anions can be prepared by stepwise neutralization of the acid. Salts of the dihydrogen phosphate and hydrogen phosphate ions decompose upon heating. For example,

$$\underset{\text{sodium dihydrogen phosphate}}{3NaH_2PO_4\ (s)} \xrightarrow{\Delta} \underset{\text{sodium trimetaphosphate}}{Na_3(PO_3)_3\ (s)} + 3H_2O\ (\ell)$$

The structure of the trimetaphosphate ion is a puckered six-membered ring. Each phosphorus atom is sp^3 hybridized and at the center of a distorted tetrahedron, just as in the simple phosphate ion, PO_4^{3-}.

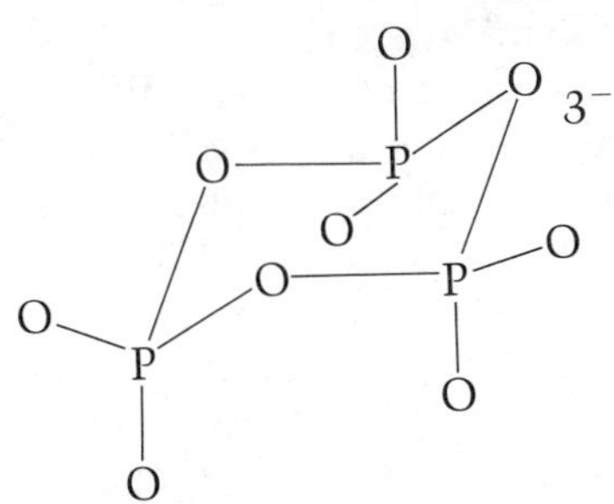

trimetaphosphate ion

Salts containing trimetaphosphate and similar ions have been used in detergents as "builders." They have the ability to complex with (sequester) ions such as Fe^{3+}, Mg^{2+}, and Ca^{2+}, which are responsible for water hardness. However, these "soluble complex phosphate" ions in wastewater flow into rivers and streams in high concentrations when they are so used. The phosphorus, a nutrient, is in a form that is very easily assimilated by plants and algae, which then grow wildly and cause an undesirable condition of vegetation overgrowth called **eutrophication.** When the vegetation dies, its decomposition products are unsightly and have disagreeable odors. The rotting vegetation also uses up the dissolved oxygen supply in the water, so that it can no longer support fish and other aquatic animal life. Detergents cause foaming in flowing waters. For these reasons, the use of phosphate detergents is now discouraged.

By endothermic, we mean a compound whose ΔH_f^0 value is positive.

Organic polyphosphates are very important biological compounds that serve as reservoirs of chemical energy. The greater the number of connected phosphate groups in an organic polyphosphate group, the more "endothermic" the compound is. The energy derived from food is used to make polyphosphate compounds. As such phosphate-phosphate bonds are later broken, energy is released for muscular and other activity. Two important examples in metabolism are adenosine triphosphate (ATP) and adenosine diphosphate (ADP). ATP is hydrolyzed enzymatically to produce ADP and release energy. If R represents the adenosine part of the molecule, the reaction is

$$\underset{\text{ATP}}{R{-}O{-}\overset{\overset{\large O}{\|}}{\underset{\underset{\large OH}{|}}{P}}{-}O{-}\overset{\overset{\large O}{\|}}{\underset{\underset{\large OH}{|}}{P}}{-}O{-}\overset{\overset{\large O}{\|}}{\underset{\underset{\large OH}{|}}{P}}{-}OH} + H_2O \longrightarrow \underset{\text{ADP}}{R{-}O{-}\overset{\overset{\large O}{\|}}{\underset{\underset{\large OH}{|}}{P}}{-}O{-}\overset{\overset{\large O}{\|}}{\underset{\underset{\large OH}{|}}{P}}{-}OH} + \underset{\text{phosphoric acid}}{HO{-}\overset{\overset{\large O}{\|}}{\underset{\underset{\large OH}{|}}{P}}{-}OH} + \text{energy}$$

24–24 Arsenic Oxyacids

The arsenic oxyacids and salts are quite similar to the corresponding phosphorus acids and salts. The acid anhydride of arsenic acid, H_3AsO_4, is As_2O_5, and the anhydride of arsenous acid, H_3AsO_3, is As_4O_6. The oxyacids of arsenic are somewhat weaker than the corresponding phosphorus-containing acids, reflecting the more metallic nature of arsenic.

Key Terms

Haber process a process for the catalyzed industrial production of ammonia from N_2 and H_2 at high temperature and pressure.

Nitrogenases a class of enzymes found in bacteria within root nodules in some plants, which catalyze reactions by which N_2 molecules from the air are converted to ammonia.

Nitrogen cycle the complex series of reactions by which nitrogen is slowly but continually recycled in the atmosphere, lithosphere, and hydrosphere.

Ostwald process a process for the industrial production of nitrogen oxide and nitric acid from ammonia and oxygen.

PANs abbreviation for peroxyacyl nitrates, photochemical oxidants in smog.

Photochemical oxidants photochemically produced oxidizing agents capable of causing severe damage to plants and animals.

Photochemical smog a brownish smog occurring in urban areas receiving large enough amounts of sunlight; caused by photochemical (light-induced) reactions among nitrogen oxides, hydrocarbons, and other components of polluted air which produce photochemical oxidants.

Exercises

1. Characterize each of the Group VA elements with respect to normal physical state and color.
2. Write out complete electron configurations for the Group VA elements as well as for the nitride ion, N^{3-}, and phosphide ion, P^{3-}.
3. Is bismuth classified as a metal or nonmetal? Why? What causes BiO^+ ions to be formed in aqueous bismuth(III) salt solutions such as $Bi(NO_3)_3$? Write an equation for the reaction that occurs. (Refer to Section 21–9.)
4. Phosphorus forms two chlorides: PCl_3 and PCl_5. Classify PCl_3 and PCl_5 as ionic or covalent. Represent their three-dimensional structures.
5. Compare and contrast the properties of (a) N_2 and P_4, (b) HNO_3 and H_3PO_4, (c) N_2O_3, and P_4O_6.
6. Suggest why corresponding phosphorus and arsenic compounds more closely resemble each other than they do the corresponding nitrogen compounds.
7. List natural sources of nitrogen, phosphorus, and arsenic and at least two uses for the first two elements.
8. Describe the natural nitrogen cycle.
9. Discuss the effects of temperature, pressure, and catalysts on the Haber process for the production of ammonia. (You may wish to consult Section 15–5.)
10. Determine the oxidation states of nitrogen in the following molecules: (a) N_2, (b) NO, (c) N_2O_4, (d) HNO_3, (e) HNO_2.
11. Determine the oxidation states of nitrogen in the following species: (a) NO_3^-, (b) NO_2^-, (c) N_2H_4, (d) NH_3, (e) NH_2^-.
12. Calculate the bond energy of the nitrogen—nitrogen triple bond from data in Appendix K. How is this high value related to the general reactivity of N_2?
13. Draw three-dimensional structures showing all outer shell electrons and indicate hybridization (except for N^{3-}) at the central element for the following species: (a) N_2, (b) N^{3-}, (c) NH_3, (d) NH_4^+, (e) NH_2^-, amide ion, (f) N_2H_4.
14. Draw three-dimensional structures showing all outer shell electrons and indicate hybridization at the central element for the following species: (a) NH_2Br, bromamine, (b) HN_3, hydrazoic acid, (c) N_2O_2, (d) $NO_2^+NO_3^-$, solid nitronium nitrate, (e) HNO_3, (f) NO_2^-.
15. Draw three-dimensional structures showing all outer shell electrons for the following species: (a) P_4, (b) P_4O_{10}, (c) As_4O_6, (d) H_3PO_4, (e) AsO_4^{3-}.
16. Write molecular equations for the following reactions:
    (a) thermal decomposition of potassium azide, KN_3
    (b) reaction of gaseous ammonia with gaseous HCl
    (c) reaction of aqueous ammonia with aqueous HCl
    (d) thermal decomposition of ammonium nitrate at temperatures above 260°C.
    (e) reaction of ammonia with oxygen in the presence of red hot platinum catalyst.
    (f) thermal decomposition of nitrous oxide (dinitrogen oxide), N_2O.
    (g) reaction of NO_2 with water
17. Write molecular equations for the following reactions:
    (a) preparation of "superphosphate of lime"
    (b) reaction of phosphorus with limited Cl_2
    (c) reaction of phosphorus with excess Cl_2
    (d) preparation of phosphorous acid, H_3PO_3
    (e) preparation of barium hypophosphite
    (f) preparation of hypophosphorous acid
    (g) preparation of phosphoric acid
18. Write two equations illustrating the ability of ammonia to function as a Lewis base.
19. In liquid ammonia would sodium amide, $NaNH_2$, be acidic, basic, or neutral? Would ammonium chloride, NH_4Cl, be acidic, basic, or neutral? Why?
20. Which of the following molecules have a nonzero dipole moment, i.e., are polar molecules? (a) NH_3, (b) NH_2Cl, (c) NO, (d) N_2H_4, (e) HNO_3, (f) PH_3, (g) As_4O_6.
21. Describe with equations the Ostwald process for the production of nitrogen oxide (nitric oxide), NO, and nitric acid.
22. Why is NO so reactive?
23. Draw a Lewis formula for NO_2. Would you predict that it is very reactive? How about N_2O_4 (or dimerized NO_2)?

24. What are the acid anhydrides of (a) nitric acid, HNO_3, (b) nitrous acid, HNO_2, (c) phosphoric acid, H_3PO_4, (d) phosphorous acid, H_3PO_3?
25. Discuss the problem of NO_x emissions with respect to air pollution. Use equations to illustrate the important reactions.
26. Write equations for reactions of nitric acid with two metals and two nonmetals.
27. Why is nitric acid called an oxidizing acid? How does the concentration of nitric acid affect the oxidation states of nitrogen-containing products of its reactions with metals?
28. Discuss the use of sodium nitrite as a meat preservative.
29. Why is calcium phosphate (phosphate rock) not applied directly as a phosphorus fertilizer?
30. Both sodium nitrate, $NaNO_3$, from Chilean nitrate deposits, and ammonium nitrate, NH_4NO_3, have been used as commercial fertilizers. Both are water-soluble and are easily absorbed by plants.
    (a) Show that ammonium nitrate contains a greater percentage of nitrogen by mass than does sodium nitrate.
    (b) The advantage gained by (a) must be weighed against the disadvantage that the ammonium ion hydrolyzes to produce acidic aqueous solutions, and thus tends to make the soil acidic. This requires application of a basic material such as limestone, $CaCO_3$ with some $MgCO_3$, to neutralize the acidity. Calculate the pH of 0.10 M ammonium nitrate solution, using K_b for $NH_3 = 1.8 \times 10^{-5}$ and $K_w = 1.0 \times 10^{-14}$.
31. Calculate the concentrations of ammonium ions and of amide ions, NH_2^-, in a 5.0×10^{-6} *M* solution of ammonium chloride in liquid ammonia. Assume that NH_4Cl is completely dissociated. *K* for the ionization of liquid ammonia is 1×10^{-30}.
32. Using data in Appendix K, calculate the bond energies of each of the species in the following reaction

    $$N_2\ (g) + O_2\ (g) \rightleftharpoons 2NO\ (g)$$

    Is it surprising that nitrogen and oxygen do not react significantly at room temperature?
33. How much hydrazine could be prepared from the reaction of 100 milliliters of 0.50 *M* aqueous ammonia with 3.00 grams of chloroamine, NH_2Cl, and an excess of sodium hydroxide, assuming 68% yield with respect to the limiting reagent? The equation for the reaction is

    $$NH_3\ (aq) + NH_2Cl + NaOH\ (aq) \longrightarrow N_2H_4 + NaCl\ (aq) + H_2O$$

    The reactants are mixed at low temperature and heated to 80–90°C in the presence of a gelatin catalyst.
34. How many grams of sodium amide must react with excess sodium nitrate at 175°C to produce 10.0 grams of sodium azide, NaN_3, assuming 75% yield?

    $$3NaNH_2 + NaNO_3 \xrightarrow{175°C} NaN_3 + 3NaOH + NH_3$$

35. What volume of dry N_2O would be liberated at STP from the complete thermal decomposition of 50.0 grams of hyponitrous acid, $H_2N_2O_2$? H_2O is the other product. (Write the balanced equation first.)
36. How many grams of "superphosphate of lime" could be prepared by the complete reaction of 22.7 grams of "phosphate rock" containing 62.2% calcium phosphate with an excess of sulfuric acid and water?
37. What will be the molarity of 250 milliliters of a phosphoric acid solution produced by the action of water on 20.0 grams of tetraphosphorus decoxide, P_4O_{10}, assuming complete reaction?
38. Calculate the percentage of phosphorus in P_4O_{10} and in H_3PO_4.
39. If a household detergent contained 25% by mass sodium trimetaphosphate, $Na_3(PO_3)_3$ or $Na_3P_3O_9$, what percentage of phosphorus would it contain? How many grams of phosphorus would be contained in 10.0 grams of the detergent?

25 The Nonmetallic Elements, Part IV: Carbon, Silicon, and Boron

In this chapter we shall examine the remaining nonmetals: carbon, silicon, and boron. Many of the general properties of the elements in Groups IIIA and IVA have been given in Tables 21–5 and 21–6. Carbon is a nonmetal, but silicon and boron are better classified as metalloids. Consistent with their relatively low electronegativities (for nonmetals), neither silicon nor boron forms simple negative ions. Carbon forms the carbide anion, C_2^{2-}, in salt-like compounds with the least electronegative metals, although there is undoubtedly a high degree of covalent character in these carbides. All borides and silicides are classified as covalent.

Many of the chemical characteristics of silicon (Group IVA) are closer to those of boron (IIIA) than to those of carbon (IVA). This tendency for the elements of the second and third periods to exhibit "diagonal relationships" in chemical properties was introduced in Section 21–2. Elements within the same families tend to form compounds with similar stoichiometries, for example, CO_2 and SiO_2; B_2O_3 and Al_2O_3. However, the properties of the compounds of the elements of the second period are often closer to those of the corresponding compounds of the third period element one group to the right. Carbon dioxide is a gas while SiO_2 and B_2O_3 are solids, and both solids are used in making glass. As another example, silicon and boron halides hydrolyze vigorously, whereas halides of carbon are very inert toward water.

Carbon

By virtue of its position in the middle of the representative elements, carbon ($1s^22s^22p^12p^1$) has little tendency to form simple ions. The energy required to remove four outer-shell electrons (sum of the first four ionization energies) is prohibitively large for the existence of C^{4+}, and only the most metallic elements form ionic carbides containing C^{4-} or C_2^{2-} ions. In nearly all of its compounds carbon is bonded covalently.

As a consequence of the high bond energy of C—C single bonds (347 kJ/mol), and of the C—H bond (414 kJ/mol), carbon has a strong tendency to form hydrocarbon chains and rings. The self-linkage of an element to form chains or rings is called **catenation.** Doubly bonded and triply bonded carbon atoms,

$$\rangle C{=}C\langle \quad \text{and} \quad —C{\equiv}C—, \text{ are also common.}$$

The term hydrocarbon refers to a molecule (or portion of a molecule) consisting of only carbon and hydrogen.

Hydrocarbons fall into the domain of organic chemistry, the chemistry of most carbon-containing compounds. Carbon is the only element whose compounds comprise an entire branch of chemistry. The term **organic chemistry** arises from the fact that most compounds essential for plant and animal life contain carbon. The distinction, however, between organic carbon compounds and inorganic carbon compounds is rather arbitrary. In the present chapter we shall consider some of the compounds usually classified as inorganic compounds, most of which contain only one carbon atom per formula unit and no C—H bonds. Chapters 29 and 30 are devoted to an introduction to organic chemistry.

For many years most scientists believed that organic compounds could be obtained only from living organisms. This idea was finally proved false in 1828 when a German chemist, Friedrich Wöhler, synthesized urea (an animal metabolic product) by heating a completely inorganic compound, ammonium cyanate

$$\underset{\substack{\text{ammonium cyanate}\\ \text{"inorganic"}}}{NH_4OCN} \xrightarrow{\Delta} \underset{\substack{\text{urea}\\ \text{"organic"}}}{CO(NH_2)_2}$$

25–1 Occurrence of Carbon

Carbon makes up only about 0.08% of the combined lithosphere, hydrosphere, and atmosphere. It occurs on the crust of the earth mainly in coal and petroleum and in the form of calcium carbonate or magnesium carbonate rocks such as calcite, limestone, dolomite, marble, and chalk. There are also some natural deposits of elemental carbon in the form of diamond and graphite. Carbon dioxide occurs in the atmosphere, which acts as a CO_2 reservoir for photosynthesis in plants.

Carbon dioxide is incorporated by plants and bacteria into the structure of carbohydrates, which are ingested by animals and converted into other biological substances. After plants and animals die, these substances decompose to hydrocarbons and many other organic substances, which may eventually become coal, peat, natural gas, petroleum, and other **fossil fuels.** Both respiration and the complete combustion of fossil fuels produce CO_2 and water to complete the carbon cycle.

1 Allotropes of Carbon

Carbon exists in two crystalline forms, graphite and diamond. The so-called amorphous forms, such as charcoal and carbon black, are really arrangements of graphite crystallites. Carbon black is obtained by burning natural gas in a deficiency of air. Charcoal is the residue from the destructive distillation of wood, in which the wood is heated in the absence of air to drive off volatile substances, and carbonaceous residues are formed at too low a temperature to permit thorough crystallization of the carbon.

Graphite At ordinary temperatures and pressures, graphite is the most stable allotrope of carbon. It is soft, black, and slippery, and its density (2.25 g/cm^3) is less than that of diamond (3.51 g/cm^3). Its properties are closely related to its structure, which consists of sp^2 hybridized carbon atoms arranged in planar layers of six-membered rings (Figure 25–1). Each carbon atom is at the center of a trigonal plane. Layers easily slide horizontally over each other because the van der Waals forces that hold one layer to another are easily overcome. For this reason graphite finds use as a lubricant, as an additive for motor oil, and in pencil "lead." Certain compounds such as CO_2, H_2O, and NH_3 are able to

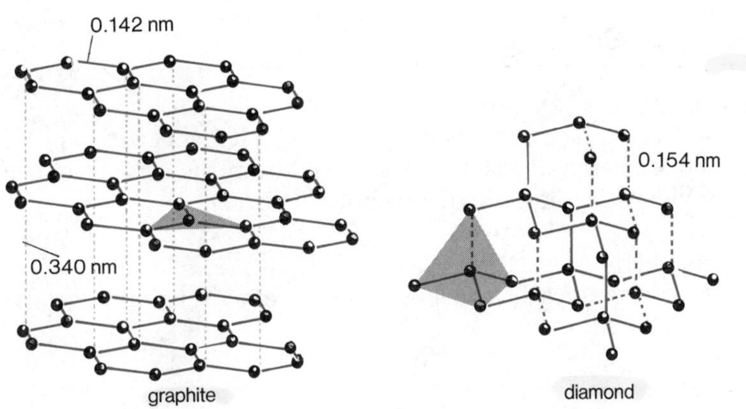

FIGURE 25–1 Allotropes of carbon.

diffuse between the layers and function as cushions, improving the ability of graphite to act as a lubricant. Only three of carbon's four electrons are involved in in-plane bonding. The fourth electron of each is quite mobile and this accounts for graphite's ability to conduct electricity.

Diamond In contrast to graphite, diamond is one of the hardest substances known to man. It is colorless, a nonconductor of electricity, and more dense than graphite. These properties are consistent with the structure of diamond, which consists of a network of tetrahedrally distributed sp^3 hybridized carbon atoms, each separated from its neighbors by only 1.54 Å (as compared with 1.42 Å in-plane and 3.40 Å out-of-plane separations in graphite). This structure (Figure 25–1) involves very strong bonds with no mobile electrons; this accounts for diamond's hardness, the nonconductivity, and the very high melting point of about 3570°C, which is the highest of any element.

2 Coal

Fossil materials are hydrocarbons (Chapter 29) or other carbon-containing material such as oil shale and tar sand. Coal is a fossil material that has an irregular graphite-like framework in which some of the carbon atoms are bonded to hydrogen, making it a hydrocarbon of sorts. It also contains many mineral impurities that remain behind as an ash, as well as other elements such as sulfur that are vaporized as oxides, when the coal is burned.

There is now renewed interest in research aimed at hydrogenation and consequent liquefaction or gasification of coal, in order to use it as a supplement or substitute for our limited supplies of natural gas and petroleum. This is not a new idea, but it is surprising to many that, although Nazi Germany had no oil wells, it managed to fuel a major portion of its mechanized warfare for six years with liquid fuel derived from coal.

The authors are indebted to Professor E. G. Rochow for reminding us of this fact.

3 Coke

Recall that destructive distillation is heating in the absence of air.

When coal is destructively distilled to release valuable volatile hydrocarbons, the residue (which is impure carbon) is called coke. Coke is used as a source of carbon for reduction of metal oxides in metallurgy. It can be converted to graphite for electrodes by the Acheson process, in which a large electric current is passed through a pressed rod of coke for several days to heat it sufficiently for recrystallization of the carbon to occur.

25–2 Reactions of Carbon

1 Reactions with Oxygen

$2C\ (s) + O_2\ (g) \longrightarrow 2CO\ (g)$

$C\ (s) + O_2\ (g) \longrightarrow CO_2\ (g)$

Although carbon is unreactive at room temperature, it reacts with a number of substances when heated. It reacts with oxygen from the air to produce either carbon monoxide or carbon dioxide (Section 7–9, part 2).

2 Reactions with Metal Oxides

Carbon reduces the oxides of the less reactive metals to produce the metal and carbon monoxide

$$Fe_2O_3 + 3C \xrightarrow{\Delta} 2Fe + 3CO$$

This reaction is important in the blast furnace reduction of iron ore with coke (Section 20–6). In such reactions with oxides of the metals, carbides of the metals often are produced. All are very hard, high-melting solids. Examples are beryllium carbide, BeC_2, and aluminum carbide, Al_4C_3. Carbide ions may be either C^{4-} or, more often, C_2^{2-} ($:C:::C:^{2-}$), as in Na_2C_2 and CaC_2. Members of the latter class are also called acetylides; e.g., Na_2C_2 is sodium acetylide. These release acetylene, HC≡CH (Section 29–11), when treated with water or dilute acid.

3 Reactions with Nonmetal Oxides

Carbon reacts with some oxides of nonmetals to produce very hard covalent carbides. Silicon carbide, known by the trade name of carborundum, is one of the hardest substances known. It is produced by reduction of silica (SiO_2) in sand with coke in an electric furnace.

$$SiO_2 + 3C \xrightarrow{3500°C} SiC + 2CO$$

Silicon carbide is industrially important as an abrasive material in sandpaper and grinding wheels. It has the same structure as diamond, except that it contains alternating silicon and carbon atoms in the diamond lattice.

4 Reaction with Steam

Carbon reacts with steam at high temperature in an endothermic reaction to produce "water gas," a mixture of CO and H_2.

$$C + H_2O \xrightarrow{\text{red heat}} CO + H_2$$

Both products burn in air and the mixture constitutes an important industrial fuel.

25–3 Oxides of Carbon

1 Carbon Monoxide

Carbon monoxide is a colorless, odorless, toxic gas that is formed by burning carbon or a hydrocarbon in a deficiency of oxygen (Section 7–9, part 2). Since its carbon–oxygen bond length is intermediate between typical double and triple bond lengths, two resonance forms are thought to contribute to its structure:

An oxide that has the formula C_3O_2 is known, but it is of little importance.

$$:C:::O: \longleftrightarrow :C::\ddot{O}:$$

Carbon monoxide is produced commercially by the "water gas" reaction, above, or by incomplete combustion of hydrocarbons, or by reduction of CO_2 by hot carbon. Pure carbon monoxide can be obtained by dehydrating formic acid with concentrated sulfuric acid

$$\underset{\text{formic acid}}{H{-}\overset{\overset{\displaystyle O}{\|}}{C}{-}O{-}H} + H_2SO_4 \longrightarrow CO + H_2SO_4 \cdot H_2O$$

Carbon monoxide may be considered the anhydride of formic acid, since it dissolves in aqueous alkali to yield salts of formic acid, called formates.

$$CO + [Na^+ + OH^-] \xrightarrow[6\text{ atm}]{200°C} [H{-}\overset{\overset{\displaystyle O}{\|}}{C}{-}O^- + Na^+] \qquad \text{sodium formate}$$

At high temperatures, CO reduces many metal oxides to the free metals, as described under metallurgy in Chapter 20.

$$FeO\ (s) + CO\ (g) \xrightarrow{\Delta} Fe\ (s) + CO_2\ (g)$$

It also disproportionates to more stable carbon and carbon dioxide in a reversible reaction.

$$2CO\ (g) \rightleftharpoons C\ (s) + CO_2\ (g) \qquad \Delta H° = -172\ kJ$$

Carbon monoxide is important industrially not only as a fuel but also as the starting material for the syntheses of many important organic substances. A few examples are

$$CO + Cl_2 \xrightarrow[\text{catalyst}]{\text{uv}} Cl{-}\overset{\overset{\displaystyle O}{\|}}{C}{-}Cl \qquad \text{phosgene or carbonyl chloride}$$

$$CO + 2H_2 \xrightarrow[\Delta,\text{ pressure}]{\text{catalyst}} H{-}\underset{\underset{\displaystyle H}{|}}{\overset{\overset{\displaystyle H}{|}}{C}}{-}O{-}H \qquad \text{methyl alcohol or methanol}$$

$$CO + 3H_2 \xrightarrow[250°C]{\text{catalyst}} H_2O + CH_4 \qquad \text{methane}$$

Carbon monoxide is a poison because of its ability to bind strongly to the iron in the oxygen-carrier protein, hemoglobin, in red blood corpuscles, thus inhibiting the ability of the blood to carry oxygen. The hemoglobin-CO complex is bright red, so often it is not realized that the victim is suffocating. Since CO is odorless and tasteless, the victim is often unaware of danger until it is too late.

Many transition metals react with CO to form coordination complexes (Chapter 27) in which the carbon donates two electrons into vacant *d* orbitals of the metal. (In many such *carbonyl complexes* the metal is formally uncharged and is assigned an oxidation number of zero.) Since they often occur in the same ores, cobalt and nickel are treated with CO to form carbonyls, which can then be separated easily.

Nickel tetracarbonyl is much more toxic than carbon monoxide, and dicobalt octacarbonyl is somewhat less toxic.

$$Ni\ (s) + 4CO\ (g) \longrightarrow Ni(CO)_4\ (\ell) \qquad \text{nickel tetracarbonyl}$$

$$2Co\ (s) + 8CO\ (g) \longrightarrow Co_2(CO)_8\ (s) \qquad \text{dicobalt octacarbonyl}$$

Since nickel tetracarbonyl is more volatile than dicobalt octacarbonyl, it can be distilled from the reaction mixture. Both carbonyls are decomposed to the free metal and carbon monoxide at high temperatures, so the carbon monoxide can be recycled for the same process.

2 Carbon Dioxide (and the Greenhouse Effect)

Carbon dioxide is a colorless, odorless gas. It is the product of complete combustion of elemental carbon and of hydrocarbons. The combustion of gasoline is illustrated by complete oxidation of octane, C_8H_{18}, one of its components.

$$2C_8H_{18}\ (\ell) + 25O_2\ (g) \longrightarrow 16CO_2\ (g) + 18H_2O\ (g) + \text{Heat}$$

Carbon dioxide is produced commercially as a by-product of the fermentation of sugar or starch for production of ethyl alcohol and alcoholic beverages

$$\underset{\substack{\text{glucose}\\ \text{(a simple sugar)}}}{C_6H_{12}O_6\ (s)} \xrightarrow{\text{yeast}} \underset{\substack{\text{ethyl alcohol}\\ \text{or ethanol}}}{2C_2H_5OH\ (\ell)} + 2CO_2\ (g)$$

Photosynthesis in plants involves the conversion of carbon dioxide and water into carbohydrates, such as glucose and other sugars

$$6CO_2 + 6H_2O \xrightarrow{\text{uv}} C_6H_{12}O_6 + 6O_2$$

The CO_2 molecule is linear, with an *sp* hybridized carbon atom and two carbon-oxygen double bonds.

$$:\ddot{O}::C::\ddot{O}:$$

Carbon dioxide is a minor component of the atmosphere, in which it is present in a concentration of about 325 parts per million. Natural processes account for most of the atmospheric CO_2, which strives to achieve or remain in equilibrium with the CO_2 dissolved in natural waters, that contained in the earth's crust (such as limestone, $CaCO_3$), that participating in photosynthesis, and that produced by combustion of fossil fuels.

The great increase in our use of fossil fuels over the past century has caused a significant increase in the concentration of CO_2 in the atmosphere, and this use is accelerating. It is now projected that with the anticipated development and use of synthetic fuels ("synfuels," derived from natural sources such as plants and grain) the concentration of atmospheric CO_2 could double by early in the next century.

This could result in a rise in the average global temperature by 2 to 3°C owing to the **greenhouse effect** caused by the carbon dioxide and water vapor in the atmosphere. Most of the sunlight passing through the atmosphere is in the visible region of the spectrum. Neither carbon dioxide nor water vapor absorbs visible light, so they do not prevent it from reaching the surface of the earth. However, the earth reradiates some of the energy from sunlight in the form of lower energy infrared (heat) radiation, which is readily absorbed by both CO_2 and H_2O (and by the glass or plastic of greenhouses). Thus, some of the heat the earth must lose in order to be in thermal equilibrium is trapped in the atmosphere, and the temperature rises. While a 2 to 3°C increase in the temperature of the atmosphere may not seem like much, it is thought to be sufficient to cause such things as the partial melting of the polar icecaps, a rise in the level of the oceans that would submerge some coastal cities, and a dramatic change in climates.

Ironically, though, there is also evidence to suggest that air pollution in the form of particulate matter may partially counteract this effect, because the particles reflect visible and infrared radiation rather than absorbing it.

25–4 Carbonic Acid and the Carbonates

Carbon dioxide is moderately soluble in water, and a saturated aqueous solution is about 0.033 *M*. Its solubility is increased, in accord with LeChatelier's Principle, by increasing the partial pressure of CO_2. Carbonated beverages contain dissolved CO_2 under pressure. About 0.4% of the dissolved CO_2 reacts with water to form carbonic acid, H_2CO_3, which has not been isolated in pure form

$$CO_2\ (g) + H_2O\ (\ell) \rightleftharpoons H_2CO_3\ (aq)$$

This reaction is easily reversed by heating. The structure of carbonic acid can be represented as follows.

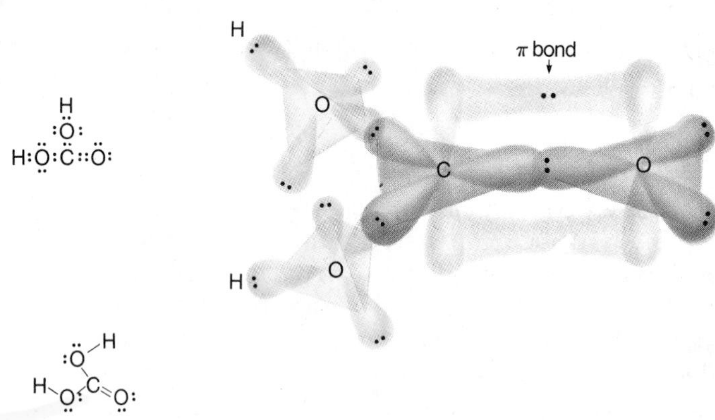

It is a weak diprotic acid, capable of forming both hydrogen carbonate and carbonate salts upon neutralization. Carbonate salts can also be formed when basic solids absorb gaseous carbon dioxide.

$$CO_2\ (g) + 2NaOH\ (s) \longrightarrow Na_2CO_3 \cdot H_2O\ (s)$$

Alkali metal carbonates are soluble in water, but most other metal carbonates are quite insoluble.

Since all natural waters contain dissolved CO_2, it should not be surprising that deposits of $CaCO_3$ and $MgCO_3$ are abundant in areas formerly covered by bodies of water. Limestone caves are formed as ground water containing dissolved CO_2, or H_2CO_3, dissolves $CaCO_3$ and forms soluble $Ca(HCO_3)_2$.

Hydrogen carbonate salts are more soluble than the corresponding carbonates. Hydrogen carbonate ion, HCO_3^-, is often called bicarbonate ion.

$$CaCO_3\ (s) + H_2CO_3\ (aq) \longrightarrow \underset{\text{calcium bicarbonate}}{[Ca^{2+} + 2HCO_3^-]}$$

The carbonates are redeposited as stalactites and stalagmites, composed of $CaCO_3$, as water evaporates and the above reaction is reversed.

$$[Ca^{2+} + 2HCO_3^-] \longrightarrow CaCO_3\ (s) + CO_2\ (g) + H_2O\ (\ell)$$

All metal carbonates decompose upon heating to the oxides and CO_2. The alkali metal carbonates are quite stable and require high temperatures for decomposition

$$Li_2CO_3 \xrightarrow{1310°C} Li_2O + CO_2$$

$$PbCO_3 \xrightarrow{315°C} PbO + CO_2$$

Solutions of alkali metal carbonates are strongly basic because of the hydrolysis of carbonate ion (Section 17–2)

$$CO_3^{2-} + H_2O \rightleftharpoons HCO_3^- + OH^- \qquad K_b = 2.1 \times 10^{-4}$$

Solutions of hydrogen carbonate salts are much less basic because HCO_3^- is a weaker base than CO_3^{2-}

$$HCO_3^- + H_2O \rightleftharpoons H_2CO_3 + OH^- \qquad K_b = 2.4 \times 10^{-8}$$

The carbonate ion is symmetrical and planar, and is represented by three resonance forms. The carbon is sp^2 hybridized and all bond angles are 120°.

25–5 Carbon Tetrahalides

Carbon forms four tetrahalides [CF_4 (gas), CCl_4 (liquid), and CBr_4 and CI_4 (solids)] as well as many chain and ring hydrocarbon compounds with halogens substituted for hydrogen. Of the tetrahalides, only carbon tetrachloride is commercially important. It is prepared by the reaction of carbon disulfide (a liquid with bonding similar to that in CO_2) with Cl_2; iodine, I_2, or antimony pentachloride, $SbCl_5$, is used as a catalyst. The "carbon tet" is separated from the mixture by fractional distillation

$$\underset{}{CS_2 + 3Cl_2} \longrightarrow \underset{\text{bp } 76.7°C}{CCl_4} + \underset{\text{bp } 138°C}{S_2Cl_2}$$

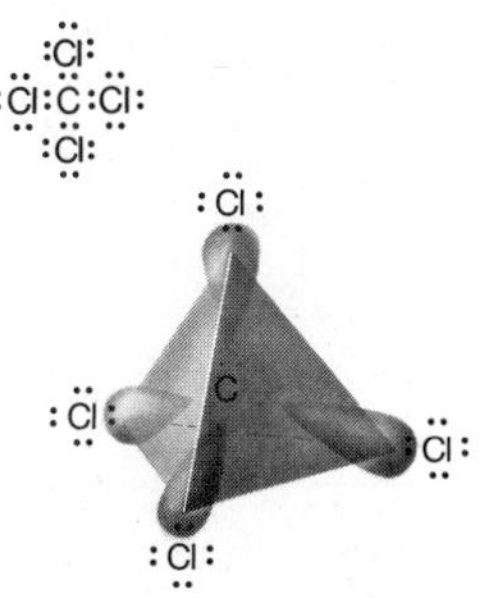

carbon tetrachloride

Carbon tetrachloride formerly was used as a dry cleaning agent because it is an excellent nonpolar solvent for dissolving fats, greases, and oils. It was also used to fight electrical fires because it is so dense (1.58 g/mL), does not conduct electricity, and does not support combustion, and it has been used by stamp collectors in detecting printers' watermarks. Because of its toxicity and high vapor pressure, its use in dry cleaning and stamp collecting is now discouraged. In fact, its use in fire extinguishers and for dry cleaning is now illegal. Used on fires, it produces phosgene, $COCl_2$, which can cause severe liver damage.

Silicon

Silicon is a shiny, blue-gray, high-melting, brittle metalloid. It looks like a metal, but it is chemically more nonmetallic than metallic. It is second only to oxygen in abundance in the earth's crust, about 87% of which is composed of silica (SiO_2) and its derivatives, the silicate minerals (Section 25–11). Silicon itself constitutes 26% of the crust, compared to oxygen's 49.5%. It does not occur free in nature. Pure silicon crystallizes in a diamond-type lattice, but the silicon atoms are less closely packed than carbon. Its density, therefore, is only 2.4 g/cm^3 compared to 3.51 g/cm^3 for diamond.

25–6 Production and Uses of Silicon

Elemental silicon is usually prepared by the high-temperature reduction of silica (silicon dioxide) from sand with coke (impure carbon). An excess of silica prevents the formation of silicon carbide

$$SiO_2\ (\text{excess}) + 2C \xrightarrow{\Delta} Si + 2CO$$

Reduction of a mixture of silicon and iron oxides with coke produces an alloy of iron and silicon known as ferrosilicon, which is used in the production of acid-resistant steel alloys such as "duriron" and in the "deoxidation" of steel. Aluminum alloys for aircraft are strengthened with silicon.

Elemental silicon is also used as the source of silicon in silicone polymers (Section 25–12). Its semiconducting properties (Section 9–17) are used in transistors and solar cells, for which the silicon must be ultrapure. For such applications it is best prepared by reducing a tetrahalide in the vapor phase with an active metal such as sodium or magnesium. The tetrahalide first must be distilled carefully to remove impurities of boron, aluminum, and arsenic halides.

$$SiCl_4 + 4Na \longrightarrow 4NaCl + Si$$

The sodium chloride is then dissolved in hot water, leaving behind quite pure silicon, which is melted and cast into bars. In order to obtain silicon with less than one part per billion impurity, it must be further purified by zone refining (Section 20–4).

25–7 Chemical Properties of Silicon

Si is much larger than C and therefore forms Si–Si bonds that are too long for effective *pi* bonding found in multiple bonds.

The biggest differences between silicon and carbon are that (1) silicon does not form double bonds, (2) it does not readily form Si—Si bonds unless the silicon atoms are bonded to very electronegative elements, and (3) it has vacant 3*d* orbitals in its valence shell with which to accept electrons from donor atoms. The Si—O single bond is the strongest of all silicon bonds and accounts for the stability and prominence of silica and the silicates. Although silicon does possess vacant 3*d* orbitals, it actually does not form many species in which it is bound to more than four atoms. However, the octahedral hexafluorosilicate ion, SiF_6^{2-}, exists, with sp^3d^2 hybridization at Si.

Elemental silicon is quite unreactive. It does not react with solutions of acids except a mixture of nitric acid, an oxidizing acid, and hydrofluoric acid, which dissolves the SiO_2 as it forms to produce hexafluorosilicic acid.

$$Si + 4[H^+ + NO_3^-] + 6HF \longrightarrow \underbrace{[2H^+ + SiF_6^{2-}]}_{\text{hexafluorosilicic acid}} + 4NO_2 + 4H_2O$$

Silicon dissolves in solutions of very strong bases to produce silicate salts

$$Si + 2[K^+ + OH^-] + H_2O \longrightarrow \underbrace{[2K^+ + SiO_3^{2-}]}_{\text{potassium silicate}} + 2H_2$$

25–8 Silicon Halides

All four tetrahalides exist. SiF_4 is a gas, $SiCl_4$ and $SiBr_4$ are volatile liquids, and SiI_4 is a solid. Silicon reacts directly with the halogens to produce tetrahalides. The reaction with F_2 is explosive

$$Si + 2X_2 \longrightarrow SiX_4 \qquad X = F, Cl, Br, I$$

Silicon tetrafluoride is prepared by reaction of concentrated sulfuric acid with a mixture of silica and calcium fluoride

$$SiO_2 + 2H_2SO_4 + 2CaF_2 \longrightarrow 2CaSO_4 + 2H_2O + SiF_4$$

All silicon halides are easily hydrolyzed, in contrast to carbon halides (which do not react with water), although the thermodynamics are similar for both kinds of reactions. Consider CCl_4 and $SiCl_4$ as examples

$$CCl_4\,(\ell) + 2H_2O\,(\ell) \longrightarrow CO_2\,(aq) + 4HCl\,(aq) \qquad \Delta G^0_{298} = -224 \text{ kJ}$$

$$SiCl_4\,(\ell) + 4H_2O\,(\ell) \longrightarrow H_4SiO_4\,(s) + 4HCl\,(aq) \qquad \Delta G^0_{298} = -277 \text{ kJ}$$

Only the latter reaction actually occurs, even though both are thermodynamically spontaneous. The distinguishing factor is a kinetic one. There are vacant *d* orbitals in the outer shell of silicon that can accept an electron pair from an attacking water molecule, whereas the carbon atom has no such orbitals. As a result, the two reactions proceed by different pathways; that of CCl_4 has a much higher activation energy and is so slow that it is not observable. The availability of outer shell *d* orbitals on a central element frequently makes it more reactive than it would be otherwise. The situation in this example is quite similar in that respect to the one described in Section 23–7 for the reactive SF_4 and inert SF_6.

25–9 Silicon Dioxide (Silica)

Silicon dioxide exists in two familiar forms in nature. One is quartz, small chips of which occur in sand. The other is flint (Latin: *silex*), an uncrystallized amorphous type of silica. Silica is properly represented as $(SiO_2)_n$ since it is a polymeric solid of SiO_4 tetrahedra sharing all oxygens among surrounding tetrahedra. Its structure is represented in Figure 25–2. For comparison, solid carbon dioxide or Dry Ice consists of discrete O=C=O molecules, just as does gaseous CO_2.

Some gems and semiprecious stones such as amethyst, opal, agate, and jasper are crystals of quartz with colored impurities. When silica is melted and

FIGURE 25–2 In SiO_2 every silicon atom (small) is bonded tetrahedrally to four oxygen atoms (large).

rapidly cooled, it actually supercools. That is, it cools to a temperature below its normal melting (freezing) point as it forms an amorphous glass called "fused quartz," or more properly "fused silica." Since it does not absorb visible or ultraviolet radiation, fused silica is used in optical instruments and sun lamps. Because it is resistant to most chemical attack and expands or contracts only very slightly with changing temperature, it is used for the manufacture of glassware for use in laboratories. It is, however, rapidly attacked by hydrofluoric acid to form volatile silicon tetrafluoride

$$SiO_2 + 4HF \longrightarrow 2H_2O + SiF_4$$

25–10 Silicates and Silicic Acids

Strong bases react very slowly with silica to form various soluble metal silicates

$$SiO_2 + 2[Na^+ + OH^-] \longrightarrow \underset{\text{sodium metasilicate}}{[2Na^+ + SiO_3^{2-}]} + H_2O$$

$$SiO_2 + 4[Na^+ + OH^-] \longrightarrow \underset{\text{sodium orthosilicate}}{Na_4SiO_4\ (aq)} + 2H_2O$$

$$2SiO_2 + 6[Na^+ + OH^-] \longrightarrow \underset{\text{sodium pyrosilicate}}{Na_6Si_2O_7\ (aq)} + 3H_2O$$

The parent acids of these salts are unstable and cannot be isolated in anhydrous form because they revert to silica. They are metasilicic acid, H_2SiO_3, orthosilicic acid, H_4SiO_4, and pyrosilicic acid or disilicic acid, $H_6Si_2O_7$. Others are thought to exist in solution. Dehydration of any silicic acid produces SiO_2. Silica in a spongy form with 5% water content is called *silica gel.* Because of its tremendous surface area, it adsorbs water and some vapors in relatively large quantities and is used as a drying agent and dehumidifier.

25–11 Natural Silicates

Most of the crust of the earth is made up of silica and silicates. The natural silicates comprise a large variety of compounds, the structures of which are all based on SiO_4 tetrahedra with metal ions occupying spaces between the tetrahedra. The extreme stability of the silicates derives from the donation of extra electrons from oxygen into vacant 3*d* orbitals of silicon. In many common minerals, called aluminosilicates, aluminum atoms replace some silicon atoms with very little structural change. Since an aluminum atom has one less positive charge in its nucleus than silicon does, it is also necessary to introduce another ion, such as K^+ or Na^+, or to substitute, for example, Ca^{2+} for Na^+. The major classes of silicates and their structures are summarized in Table 25–1.

TABLE 25–1 Structures of Silicate Minerals

Structural Arrangement		Anion Unit	Examples
Independent tetrahedra		$(SiO_4)^{4-}$	Forsterite, Mg_2SiO_4 Olivine $(Mg, Fe)_2SiO_4$ Fayalite, Fe_2SiO_4
Two tetrahedra sharing one oxygen		$(Si_2O_7)^{6-}$	Akermanite, $Ca_2MgSi_2O_7$
Closed rings of tetrahedra, each sharing two oxygens		$(Si_3O_9)^{6-}$, $(Si_4O_{12})^{8-}$, $(Si_6O_{18})^{12-}$	Beryl, $Al_2Be_3Si_6O_{18}$
Continuous single chains of tetrahedra, each sharing two oxygens		$(SiO_3)^{2-}$	Pyroxenes: enstatite, $MgSiO_3$, diopside, $CaMg(SiO_3)_2$
Continuous double chains of tetrahedra, sharing two or three oxygens		$(Si_4O_{11})^{6-}$	Amphiboles: anthophyllite, $Mg_7(Si_4O_{11})_2(OH)_2$
Continuous sheets of tetrahedra, each sharing three oxygens		$(Si_4O_{10})^{4-}$	Talc, $Mg_3Si_4O_{10}(OH)_2$ Kaolinite, $Al_4Si_4O_{10}(OH)_8$
Continuous framework of tetrahedra, each sharing all four oxygens	See Figure 25–2	(SiO_2)	Quartz, SiO_2 Albite, $NaAlSi_3O_8$

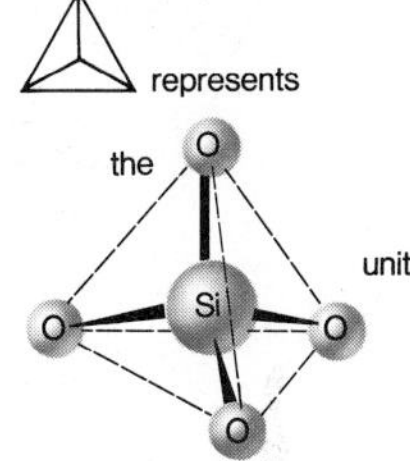

The physical characteristics of the silicates are often suggested by the arrangement of the SiO_4 tetrahedra. A few examples are given in the table. A single-chain silicate, diopside $[CaMg(SiO_3)_2]_n$, and a double-chain silicate, asbestos $[Ca_2Mg_5(Si_4O_{11})_2(OH_2)]_n$, occur as fibrous or needle-like crystals. Talc, $[Mg_3Si_4O_{10}\,(OH)_2]_n$, a silicate with a sheet-like structure, is flaky. Micas are sheet-like aluminosilicates (similar to talc) with about one of every four silicon atoms replaced by aluminum. Muscovite mica is $[KAl_2(AlSi_3O_{10})(OH)_2]_n$. Micas all occur in thin sheets that are easily peeled away from each other.

1 Clays

The clay minerals are also silicates and aluminosilicates with sheet-like structures. They result from the weathering of granite and other rocks. The layers have enormous "inner surfaces" capable of absorbing large quantities of water. Clay minerals often occur as minute platelets with total surface areas greater than those of larger crystals. When wet, the clays are easily shaped. When heated to high temperatures they lose water; when fired in a furnace, they become very rigid. If clays are heated with feldspar, $[KAlSi_3O_8]_n$, and silica, a mixture of crystallites bound by a rigid glass-like matrix results. This is called earthenware, porcelain, or china, according to its composition and properties. Clays are also used in making cement, ceramics, bricks, flowerpots, and other such materials.

2 Glass

Fused sodium silicate, Na_2SiO_3, and calcium silicate, $CaSiO_3$, are the major components of the glass used in such things as drinking glasses, bottles, and window panes. **Glass** is a state of matter, and has no fixed composition or structure. Since it has no regular structure, it does not break evenly along crystal planes but breaks to form rounded surfaces and jagged edges. The basic ingredients are produced by heating a mixture of sodium and calcium carbonates with silica until it melts at about 700°C

$$[CaCO_3 + SiO_2]\ (\ell) \xrightarrow{\Delta} CaSiO_3 + CO_2$$

$$[Na_2CO_3 + SiO_2]\ (\ell) \xrightarrow{\Delta} Na_2SiO_3 + CO_2$$

The resulting "soda-lime" glass is clear and colorless when solidified if all CO_2 bubbles escape and if the amounts of reactants are carefully controlled. Other ingredients may be added to produce certain desired characteristics. Soda-lime glass (soft glass) has a low softening temperature and expands and contracts considerably with changing temperature. Pyrex, a type of borosilicate glass, has some of the sodium and calcium replaced by boron and has a higher percentage of silica. This type of glass is much more resistant to chemical attack and is able to withstand greater thermal and mechanical stress than soda-lime glass. Flint glasses or lead glasses contain potassium and lead, respectively, and have high refractive indices that make them useful for decorative glassware, prisms, and lenses. A variety of substances may be added in small proportions to add color; examples are given in Table 25–2.

TABLE 25–2 Substances Used to Color Glass

Substance	Color	Substance	Color
Calcium fluoride	milky white	iron(II) compounds	green
Copper(I) oxide	red, green, or blue	manganese(IV) oxide	violet
Cobalt(II) oxide	blue	tin(IV) oxide	opaque
Uranium compounds	yellow, green	colloidal selenium	red

25–12 Silicones

Silicones are polymeric organosilicon compounds containing individual or cross-linked Si—O chains or rings in which some of the oxygens of SiO_4 tetrahedra are replaced by such groups as hydroxyl, —OH, methyl, CH_3—, ethyl, C_2H_5—, or phenyl, C_6H_5—. They possess some of the properties of hydrocar-

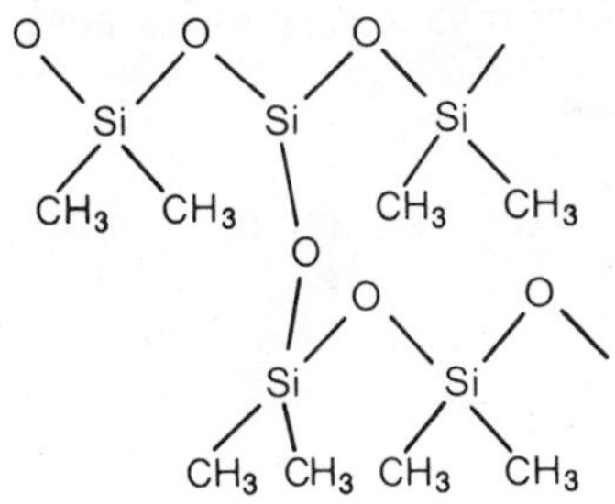

FIGURE 25–3 A cross-linked silicone.

bons and some of the properties of silicon-oxygen compounds. Most are very resistant toward chemical attack and thermal decomposition. The linear, high-molecular-weight polymers are very rubbery. Silicone rubber is not attacked by ozone as ordinary rubber is. Liquid silicones are very fluid even at low temperatures, yet retain viscous fluidity at very high temperatures at which hydrocarbon oils decompose. The properties of a particular silicone, of course, depend upon the substituents and the degree of cross-linking and ring formation. The structure of a typical cross-linked silicone is illustrated in Figure 25–3.

The toys known as "Silly Putty" and "Superball" contain rubbery silicones extended with oil. Other silicones are used as lubricants and waterproof films. Because it does not crack at low temperature, another silicone is used for refrigerator gaskets. Because of their unique combination of properties of chemical inertness and gel-like firmness and pliability, some silicones find application in cosmetic surgery.

Boron

25–13 Occurrence and Uses of Boron

Boron is very rare, constituting only 0.0003% of the earth's crust. Much of it is localized in evaporated lake beds in the southwestern United States in the form of borax, $Na_2B_4O_5(OH)_4 \cdot 8H_2O$, and kernite, $Na_2B_4O_5(OH)_4 \cdot 2H_2O$, hydrated sodium salts of tetraboric acid. Borax is used mainly as a cleansing agent in laundry detergents and as a mild alkali. It is also used in soldering and welding as a flux, since it melts at low temperatures and dissolves metal-oxide films.

25–14 Boric Acids and Boric Oxide

Recrystallized mined borax reacts with sulfuric acid to produce boric acid, H_3BO_3 [or $B(OH)_3$]

$$Na_2B_4O_5(OH)_4 \cdot 8H_2O\ (s) + [2H^+ + SO_4{}^{2-}] \longrightarrow [2Na^+ + SO_4{}^{2-}] + 5H_2O + 4H_3BO_3\ (aq)$$

Boric acid itself is used as a mildly acidic antiseptic. It is a very weak acid

$K_2 = 1.8 \times 10^{-13}$
$K_3 = 1.6 \times 10^{-14}$

$$H_3BO_3\ (aq) \rightleftharpoons H^+ + H_2BO_3^- \qquad K_1 = 7.3 \times 10^{-10}$$

The solid acid melts and loses water at high temperature, producing a colorless, transparent melt of boric oxide, B_2O_3

$$2H_3BO_3 \xrightarrow{\Delta} B_2O_3 + 3H_2O$$

boric oxide

Rapid cooling produces a glass-like solid. Boric oxide is used in the fabrication of the sturdy borosilicate glasses (Section 25–11).

Boric oxide is the anhydride of three common boric acids, two of which are obtained from boric acid (or, more properly, orthoboric acid) by dehydration

$$\underset{\text{orthoboric acid}}{4H_3BO_3} \xrightarrow[-4H_2O]{\Delta} \underset{\text{metaboric acid}}{4HBO_2} \xrightarrow[-H_2O]{\Delta} \underset{\text{tetraboric acid (pyroboric acid)}}{H_2B_4O_7}$$

A number of borate salts containing anions of these acids are known.

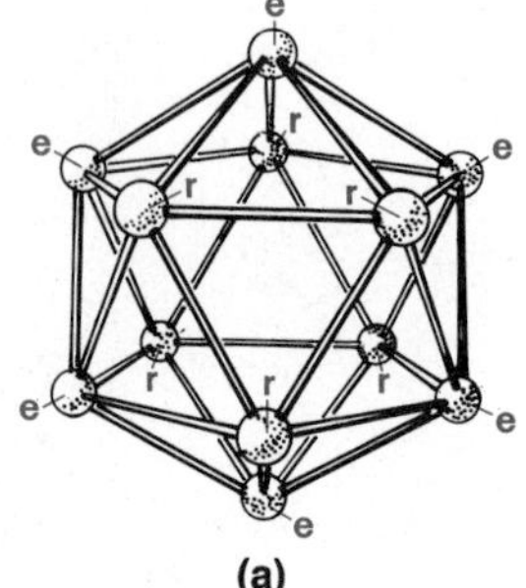

25–15 Elemental Boron

The element can be obtained in low purity in a dark brown amorphous form by the high-temperature reduction of boric oxide with magnesium. The magnesium oxide is dissolved away by hydrochloric acid, leaving the boron behind

$$B_2O_3 + 3Mg \xrightarrow{\Delta} 3MgO + 2B$$

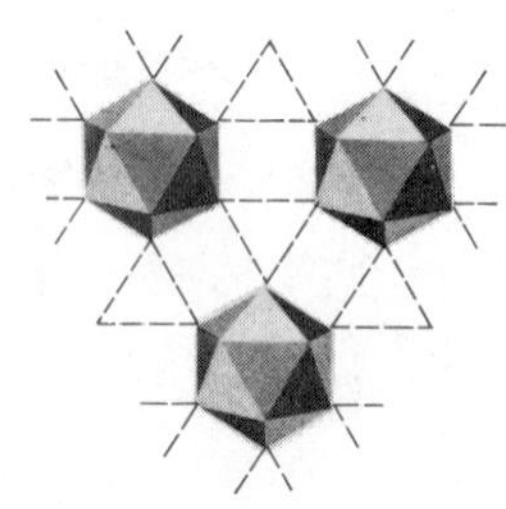

FIGURE 25–4
(a) B_{12}-icosahedron.
(b) Linking of icosahedra in rhombohedral boron.

An *icosahedron* is a regular polyhedron with 20 equilateral triangular faces and 12 vertices.

Boron can be obtained in higher purity by reduction of gaseous boron tribromide or trichloride with hydrogen over a hot tungsten or tantalum filament

$$2BBr_3 + 3H_2 \xrightarrow[1000\text{–}2500°C]{W} 2B + 6HBr$$

Elemental boron exists in several allotropic modifications, of which α-rhombohedral boron is the simplest in structure. All modifications consist of large, complex polyhedral clusters of boron atoms; they are very hard and are semiconductors. The α-rhombohedral form consists of icosahedral B_{12} units. One boron atom occupies each vertex. The structure is shown in Figure 25–4.

25–16 Three-Center Two-Electron Bonds

Three-center two-electron bonds, as well as ordinary two-center bonds, are present in all modifications of elemental boron. In such bonds, boron atoms ("*e*" in Figure 25–4a) from three adjacent icosahedra are held together by two electrons, which are delocalized over a molecular orbital resulting from the overlap of three pseudo-*sp* hybrid orbitals on each boron atom. The pseudo-*sp* hybrids have more *p* character than *s* character. A three-center two-electron bond is represented next.

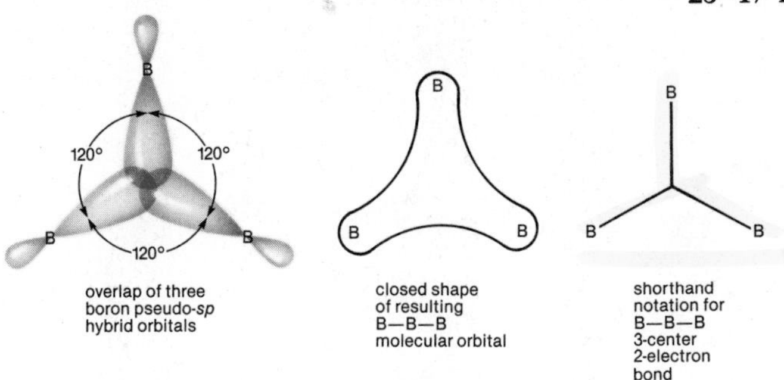

25–17 Boron Hydrides (Boranes)

Boron forms a number of unusual binary compounds with hydrogen called **boron hydrides,** or more commonly **boranes.** They fall into two series, B_nH_{n+4} and the less stable B_nH_{n+6}. Examples are given in Table 25–3, and typical structures are shown in Figure 25–5. A special system of naming is usually applied, in which the prefix indicates the number of boron atoms and the parenthetical number gives the number of hydrogen atoms. The simplest borane is diborane, B_2H_6. No BH_3 exists except as a short-lived reaction intermediate.

The higher boranes are generally prepared by controlled heating of diborane or other boranes

$$2B_2H_6 \xrightarrow{120°C} B_4H_{10} + H_2$$

$$5B_2H_6 \xrightarrow{200\text{–}240°C} 2B_5H_9 + 6H_2$$

$$2B_9H_{15} \xrightarrow{>-30°C} B_8H_{12} + B_{10}H_{14} + 2H_2$$

Acid decomposition of magnesium boride, Mg_3B_2, also produces a number of boranes, but mainly B_2H_6.

Boron trichloride, BCl_3, or boron trifluoride, BF_3, reacts with lithium aluminum hydride, $LiAlH_4$, in ether solution to produce diborane.

$$4BCl_3 + 3LiAlH_4 \xrightarrow{\text{ether}} 2B_2H_6 + 3LiCl + 3AlCl_3$$

TABLE 25–3 Typical Boron Hydrides

B_nH_{n+4}	Name	Phase	B_nH_{n+6}	Name	Phase
B_2H_6	diborane	gas	B_4H_{10}	tetraborane(10)	gas
B_4H_8	tetraborane(8)	gas	B_5H_{11}	pentaborane(11)	liquid
B_5H_9	pentaborane(9)	liquid	B_6H_{12}	hexaborane(12)	liquid
B_8H_{12}	octaborane(12)	solid	$B_{10}H_{16}$	decaborane(16)	solid

FIGURE 25–5 The structures of some boranes, with their numbering conventions and some interatomic distances. The open circles represent boron atoms, and the solid ones represent hydrogen. Note that the thin lines connecting boron atoms are not meant to show bonds, but only define the polyhedral framework.

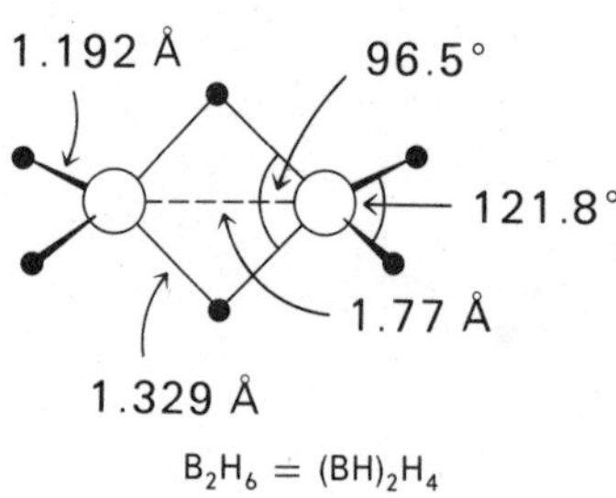

$B_2H_6 = (BH)_2H_4$

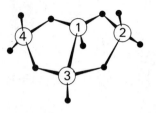

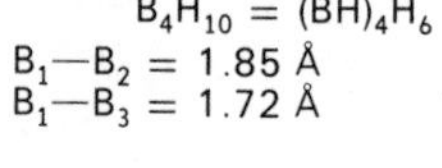

$B_4H_{10} = (BH)_4H_6$
$B_1{-}B_2 = 1.85$ Å
$B_1{-}B_3 = 1.72$ Å

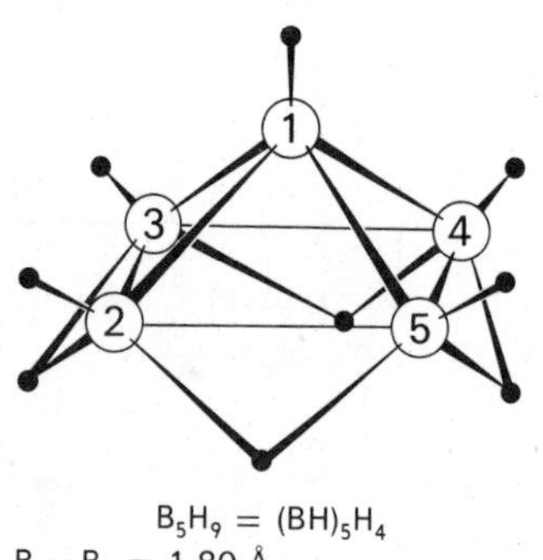

$B_5H_9 = (BH)_5H_4$
$B_3{-}B_4 = 1.80$ Å
$B_1{-}B_4 = 1.69$ Å

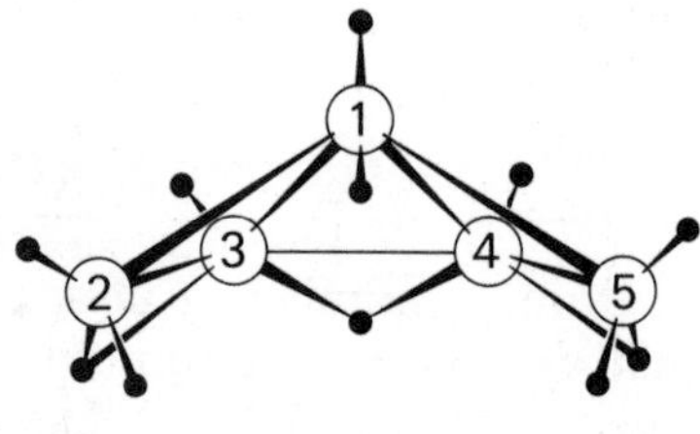

$B_5H_{11} = (BH)_5H_6$
$B_3{-}B_4 = 1.77$ Å $B_1{-}B_5 = 1.87$ Å
$B_1{-}B_4 = 1.72$ Å

Many boranes have structures based on fragments of icosahedra, as shown in Figure 25–5. Each borane contains three-center two-electron bonds. Some are B–B–B bonds (B bridging two B atoms) and others are B–H–B bonds involving "hydrogen bridges," as will be discussed for diborane.

The bonding in B_2H_6 is not well explained by simple Lewis dot formulas. Each of the six hydrogen atoms contributes one electron, and each of the two boron atoms contributes three electrons to give a total of twelve bonding electrons. If two electrons are allotted to each of the four terminal B—H (two-center) bonds, only four electrons remain to form the other bonds. Note also that hypothetical BH_3 would not have a complete octet of bonding electrons. The structure of diborane is:

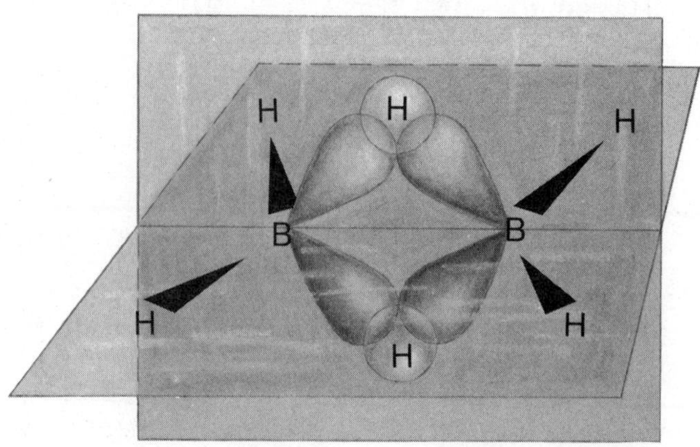

It is thought that the $\mathrm{B}\overset{\mathrm{H}}{\diagup\ \diagdown}\mathrm{B}$ bridges are held together by two three-center two-electron bonds resulting from the overlap of approximately sp^3 hybrid orbitals on each boron atom with the $1s$ orbital of a bridging hydrogen atom. Two electrons occupy each of the resulting pair of three-center molecular orbitals.

```
H   H   H
 :B : B:
H   H   H
```

25–18 Properties of the Boron Hydrides

The boranes are volatile compounds that decompose at red heat to boron and hydrogen. All hydrolyze, but the ease of hydrolysis varies. Halogens, hydrogen halides, or boron trihalides react with boranes to yield halogenated borane derivatives. Diborane itself is a colorless, poisonous gas with an unpleasant odor. All boranes are easily oxidized to B_2O_3 and water, and most ignite spontaneously in air.

The boranes can also react to produce borohydride anions. The simplest of these is tetrahedral, BH_4^-. Examples of syntheses of metal borohydrides are

$$4NaH + B(OCH_3)_3 \xrightarrow{250°C} NaBH_4 + 3NaOCH_3$$

$$2LiH + B_2H_6 \xrightarrow{\text{ether}} 2LiBH_4$$

$$AlCl_3 + 3NaBH_4 \xrightarrow{\Delta} Al(BH_4)_3 + 3NaCl$$

Sodium borohydride, $NaBH_4$, is important in both inorganic and organic chemistry as a reducing agent and a source of hydride ions, H^-.

As a reducing agent, $NaBH_4$ is less vigorous than $LiAlH_4$.

25–19 Boron Halides

All four boron trihalides are known. All are covalent, planar, nonpolar molecules with 120° X—B—X angles, and involve sp^2 hybridization (Section 5–11). Boron trifluoride is a gas, BCl_3 and BBr_3 are liquids, and BI_3 is a solid. By comparison, the more metallic aluminum, which is just below boron in Group IIIA, forms ionic AlF_3, but the other aluminum trihalides are primarily covalent. All BX_3 molecules hydrolyze to produce boric acid and the corresponding hydrogen halide

```
  :F:
:F: B :F:
```

$$BX_3 + 3H_2O \longrightarrow H_3BO_3 + 3HX \qquad X = F, Cl, Br, I$$

The boron trihalides form mixed halides by exchange reactions

$$BBr_3 + BCl_3 \rightleftharpoons BBr_2Cl + BBrCl_2$$

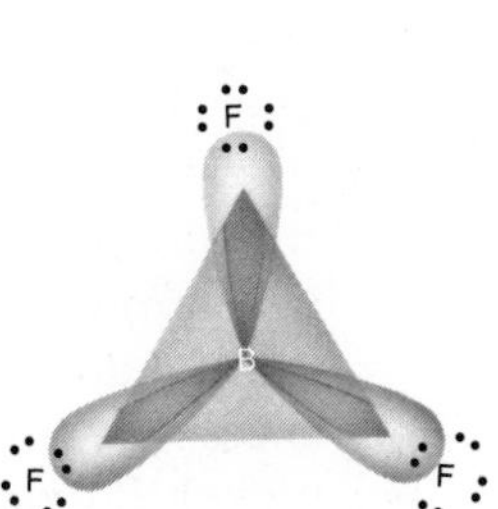

The boron halides are Lewis acids, because the boron atoms lack an octet of electrons in their outer shells in these compounds. For example, BF_3 reacts with ammonia to form an addition compound, $H_3N—BF_3$, and with the fluoride ion to form the tetrafluoroborate ion, BF_4^-.

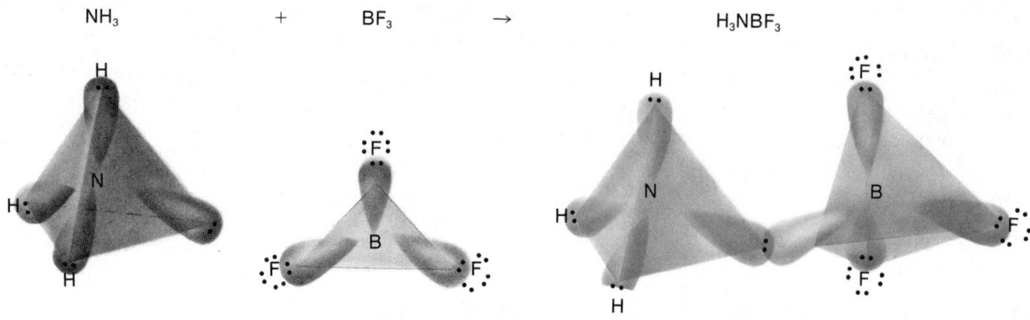

Key Terms

Catenation bonding of atoms of the same element into chains or rings.

Fossil fuels substances consisting largely of hydrocarbons, derived from decay of organic materials under geological conditions of high pressure and temperature; include coal, petroleum, natural gas, peat, and oil shale.

Greenhouse effect trapping of heat on the surface of the earth by carbon dioxide and water vapor in the atmosphere.

Three-center two-electron bond a bond holding three atoms together but involving only two electrons; common in boron compounds.

Exercises

Carbon

1. Why are most compounds of carbon covalent?
2. Compare the properties of CO_2 and SiO_2; B_2O_3 and SiO_2. How do you explain the similarities and differences?
3. Give several natural sources of carbon. Briefly summarize the natural carbon cycle.
4. Sketch the structures of graphite and diamond. Describe their properties and relate them to their structures.
5. What fundamental chemical modification is involved in coal liquefaction and gasification?
6. What is coke? Why is it so important industrially? How can it be converted to graphite?
7. Write equations for the reactions of carbon at high temperatures with:
   (a) limited amounts of O_2,
   (b) excess O_2,
   (c) nickel(II) oxide, NiO,
   (d) silicon dioxide, SiO_2,
   (e) germanium dioxide, GeO_2,
   (f) steam.
8. Draw Lewis formulas and structures for each of the following species. Also indicate the hybridization at each carbon atom. (a) CO_2, (b) H_2CO_3, (c) CO_3^{2-}.
9. Draw Lewis formulas and structures for each of the following species. Also indicate the hybridization at each carbon atom. (a) CCl_4, (b) HCN, (c) C_2H_2.
10. Why are simple (single-element) ions with a charge of 4+ or 4− so rare?
11. Write equations for reactions to show how:
    (a) carbon monoxide can be obtained from formic acid, HCO_2H.
    (b) sodium formate can be produced from CO.
    (c) carbon monoxide disproportionates.
12. Write equations for reactions to show how:
    (a) methanol can be synthesized from CO.

(b) methane can be synthesized from CO.
(c) dicobalt octacarbonyl can be prepared.

13. How does carbon monoxide act as a poison?
14. Describe with equations how stalactites and stalagmites are formed in caves.
15. Why is rainwater generally slightly acidic, even in air that is not polluted with oxides of sulfur and nitrogen?
16. Why is carbon tetrachloride effective in removal of greasy stains from clothing? Why is its use now discouraged?
17. How many liters of methanol (a liquid) at 20°C can be produced from the complete reaction of 500 cubic feet of carbon monoxide, measured at 15 atmospheres pressure and 50°C, with an excess of hydrogen? The specific gravity of methanol is 0.791 at 20°C.
18. A 5.631-gram mixture containing only lithium carbonate and lithium oxide was heated to drive off all the carbon dioxide from the carbonate. A total of 1.434 liters of dry CO_2 was collected at 747 torr and 32°C. What mass of lithium carbonate was contained in the original sample? What percentages of lithium carbonate and lithium oxide made up the sample?

Silicon

19. Why is it logical that if 87% of the earth's crust is composed of silicate minerals, silicates must be unreactive?
20. How is silicon obtained from natural sources? How is it purified for use in semiconductors? How does the process of zone refining work?
21. Why are pure silicon and germanium good semiconducting elements?
22. Write equations for the following reactions:
    (a) silicon with aqueous sodium hydroxide.
    (b) silicon with bromine
    (c) silicon tetrachloride with water (compare with carbon tetrachloride)
23. How does hydrofluoric acid attack glass?
24. Describe the production of glass. How is glass colored? Give three examples of colored glasses and their colors.
25. What is a general characteristic of most silicones? How do the structures of silicones differ from quartz (SiO_2)?
26. Using the data of Appendix K, calculate the ΔG^0 values at 25°C for each of the reactions below for X = F, Cl. Use standard states for each of the species.

    $$Si + 2X_2 \longrightarrow SiX_4$$

    Which reaction is more spontaneous at 25°C?

Boron

27. How can natural borax be converted to boric acid, H_3BO_3?
28. Show how boric oxide, B_2O_3, metaboric acid, HBO_2, and tetraboric acid, $H_2B_4O_7$, can be prepared by dehydration of boric acid.
29. Write equations for the following reactions:
    (a) a preparation of elemental boron.
    (b) the preparation of B_5H_9 from diborane, B_2H_6.
    (c) the preparation of B_4O_{10} from diborane.
    (d) the preparation of diborane from boron trichloride.
    (e) boron trifluoride with water.
30. Write equations for two reactions illustrating the ability of BCl_3 to act as a Lewis acid.
31. Is it surprising that boron participates in three-center two-electron bonds? Why?
32. Describe and illustrate the bonding in two kinds of three-center two-electron bonds.
33. How much borax must react with excess sulfuric acid to produce 100 pounds of boric acid, assuming 95.2% yield based on borax?
34. Calculate the concentrations of all species and the pH in 0.10 M boric acid, H_3BO_3, a triprotic acid. $K_1 = 7.3 \times 10^{-10}$, $K_2 = 1.8 \times 10^{-13}$, $K_3 = 1.6 \times 10^{-14}$.
35. How many grams of diborane, B_2H_6, can be produced from the 88.4% efficient reaction of 10.7 grams of boron trichloride with an excess of lithium aluminum hydride in ether solution?
36. A pure sample of a gaseous boron hydride is found to be 78.4% boron by mass. A 0.2120-gram sample occupies 114.5 mL at STP. What is the molecular formula of the compound?

26 The Transition Metals

The term "transition elements" was originally coined to denote the elements in the middle of the periodic table that provide a transition between the "base-formers" on the left and the "acid-formers" on the right. It actually applies to both the *d*- and *f*-transition elements, all of which are metals, but we commonly use the term "transition metals" to refer to only the more frequently encountered *d*-transition metals. The *f*-transition elements are usually called the rare earths or inner transition elements. In this chapter we shall discuss primarily the *d*-transition metals.

Remember that oxides of most nonmetals are acidic and oxides of most metals (except those having high oxidation states) are basic.

The *d*-Transition Metals

26–1 General Properties

These metals are located between Groups IIA and IIIA in the periodic table. Strictly speaking, in order for an element to be classified a *d*-transition metal,

TABLE 26–1 Properties of the First Transition Series

Properties	Scandium	Titanium	Vanadium	Chromium	Manganese	Iron	Cobalt	Nickel	Copper	Zinc
Melting point (°C)	1541	1660	1890	1900	1244	1535	1495	1453	1083	420
Boiling point (°C)	2831	3287	3380	2672	1962	2750	2870	2732	2567	907
Density (g/cm^3)	3.0	4.51	6.11	7.9	7.2	7.87	8.7	8.91	8.94	7.13
Atomic radius (Å)	1.62	1.47	1.34	1.25	1.29	1.26	1.25	1.24	1.28	1.38
Ionic radius, M^{2+} (Å)	—	0.94	0.88	0.89	0.80	0.74	0.72	0.69	0.70	0.74
Electronegativity	1.3	1.4	1.5	1.6	1.6	1.7	1.8	1.8	1.8	1.6
Std. red. pot. (V) for M^{2+} (aq) + $2e^-$ → M (s)	−2.08*	−1.63	−1.2	−0.91	−1.18	−0.44	−0.28	−0.25	+0.34	−0.76
Ionization energy (kJ/mol)										
first	631	658	650	653	717	759	758	737	745	906
second	1235	1310	1414	1592	1509	1561	1646	1753	1958	1733

* For Sc^{3+} (aq) + $3e^-$ → Sc (s)

it must have a partially filled set of *d* orbitals. Zinc, cadmium, and mercury (the Group IIB elements) and their cations have completely filled sets of *d* orbitals, and therefore are not really *d*-transition metals. However, they are often discussed with *d*-transition metals because of the similarities of their properties to those of the transition metals. All of the other elements in this region have partially filled sets, except copper, silver, and gold (the IB elements) and palladium, all of which have completely filled sets of *d* orbitals. However, some of the cations of these latter elements have only partially filled sets of *d* orbitals.

The following general properties are characteristic of the transition elements:

1. Are all metals.
2. Most are harder, more brittle, and have higher melting points and boiling points and higher heats of vaporization than nontransition metals.
3. Ions and their compounds are often colored.
4. They form many complex ions (see Chapter 27).
5. With few exceptions, they exhibit multiple oxidation states.
6. Many of them are paramagnetic, as are many of their compounds.
7. Many of the metals and their compounds are good catalysts.

Some specific properties of the 3*d*-transition metals are tabulated in Table 26–1.

26–2 Electronic Configurations and Oxidation States

Since some of the valence electrons of the *d*-transition metals are in *d* orbitals one energy level below the highest occupied energy level, the physical and chemical properties of these metals vary less dramatically for consecutive elements than do the properties of representative elements, whose valence electrons are all in the highest occupied energy level. The electronic configurations of the three *d*-transition series are given in Table 26–2. The properties of the transition metals can be correlated roughly with either the total number of *d* electrons or the number of unpaired electrons.

TABLE 26–2 Electronic Configurations of the *d*-Transition Metals

Period 4	Period 5	Period 6
$_{21}Sc$ $[Ar]3d^14s^2$	$_{39}Y$ $[Kr]4d^15s^2$	$_{57}La$ $[Xe]5d^16s^2$
$_{22}Ti$ $[Ar]3d^24s^2$	$_{40}Zr$ $[Kr]4d^25s^2$	$_{72}Hf$ $[Xe]4f^{14}5d^26s^2$
$_{23}V$ $[Ar]3d^34s^2$	$_{41}Nb$ $[Kr]4d^45s^1$	$_{73}Ta$ $[Xe]4f^{14}5d^36s^2$
$_{24}Cr$ $[Ar]3d^54s^1$	$_{42}Mo$ $[Kr]4d^55s^1$	$_{74}W$ $[Xe]4f^{14}5d^46s^2$
$_{25}Mn$ $[Ar]3d^54s^2$	$_{43}Tc$ $[Kr]4d^55s^2$	$_{75}Re$ $[Xe]4f^{14}5d^56s^2$
$_{26}Fe$ $[Ar]3d^64s^2$	$_{44}Ru$ $[Kr]4d^75s^1$	$_{76}Os$ $[Xe]4f^{14}5d^66s^2$
$_{27}Co$ $[Ar]3d^74s^2$	$_{45}Rh$ $[Kr]4d^85s^1$	$_{77}Ir$ $[Xe]4f^{14}5d^76s^2$
$_{28}Ni$ $[Ar]3d^84s^2$	$_{46}Pd$ $[Kr]4d^{10}$	$_{78}Pt$ $[Xe]4f^{14}5d^96s^1$
$_{29}Cu$ $[Ar]3d^{10}4s^1$	$_{47}Ag$ $[Kr]4d^{10}5s^1$	$_{79}Au$ $[Xe]4f^{14}5d^{10}6s^1$
$_{30}Zn$ $[Ar]3d^{10}4s^2$	$_{48}Cd$ $[Kr]4d^{10}5s^2$	$_{80}Hg$ $[Xe]4f^{14}5d^{10}6s^2$

One of the characteristics of most transition metals is the ability to exhibit more than one oxidation state. The *maximum* oxidation state is given by a metal's group number (see Section 26–4), but this is usually not the most stable oxidation state.

Recall that in "building" atoms by the Aufbau Principle, the outer *s* orbitals are filled before the inner *d* orbitals (Section 3–15).

The outer *s* electrons, of course, lie outside the *d* electrons and are always the first ones lost in ionization. All the metals of the first transition series except scandium and zinc exhibit more than one oxidation state. Scandium loses its two 4*s* electrons and its only 3*d* electron to form Sc^{3+}, its single oxidation state, other than zero. Zinc loses its two 4*s* electrons to give only Zn^{2+}.

	3*d*	**4*s***		
$_{21}Sc$ [Ar]	↑ _ _ _ _	⇅	$\xrightarrow{-3e^-}$	$_{21}Sc^{3+}$ [Ar]
$_{30}Zn$ [Ar]	⇅ ⇅ ⇅ ⇅ ⇅	⇅	$\xrightarrow{-2e^-}$	$_{30}Zn^{2+}$ $[Ar]3d^{10}$

All of the other 3*d*-transition metals exhibit at least two oxidation states in their compounds. For example, cobalt can form Co^{2+} and Co^{3+} ions.

	3*d*	**4*s***			**3*d***	**4*s***
$_{27}Co$ [Ar]	⇅ ⇅ ↑ ↑ ↑	⇅	$\xrightarrow{-2e^-}$	$_{27}Co^{2+}$ [Ar]	⇅ ⇅ ↑ ↑ ↑	_
$_{27}Co$ [Ar]	⇅ ⇅ ↑ ↑ ↑	⇅	$\xrightarrow{-3e^-}$	$_{27}Co^{3+}$ [Ar]	⇅ ↑ ↑ ↑ ↑	_

Simple aqueous salts of cobalt(III) oxidize water and are reduced to cobalt(II) salts.

The most common oxidation states of the 3*d*-transition elements are +2, +3, and +4, with +5 and +6 coming next. The elements in the middle of each series exhibit more oxidation states than those at the extreme left and right. As one moves down within a group, higher oxidation states become more stable and more common. This is opposite to the trend observed for representative elements, and occurs because the *d* electrons are more effectively shielded from the nucleus on descending a group and, therefore, are more easily ionized or more available for sharing in forming bonds. For example, cobalt commonly exhibits the +2 and +3 oxidation states, as we have seen. The elements just below cobalt are rhodium and iridium, whose common oxidation states are +3 and +4. The +4 state is slightly more stable for iridium than for the lighter rhodium.

TABLE 26–3 Nonzero Oxidation States of the 3*d*-Transition Metals*

IIIB	IVB	VB	VIB	VIIB	VIIIB			IB	IIB
Sc	Ti	V	Cr	Mn	Fe	Co	Ni	Cu	Zn
		+1 red.	+1 red.	+1 red.		+1 red.	+1 red.	$+\underline{1}$ red.	
	+2 red.	+2 red.	$+\underline{2}$ red.	$+\underline{2}$	$+\underline{2}$ red.	$+\underline{2}$	$+\underline{2}$	$+\underline{2}$	$+\underline{2}$
$+\underline{3}$	+3 red.	$+\underline{3}$ red.	$+\underline{3}$	+3 ox.	$+\underline{3}$	$+\underline{3}$ ox.	+3 ox.	+3 ox.	
	$+\underline{4}$	$+\underline{4}$	+4 ox.	$+\underline{4}$ ox.	+4 ox.	+4 ox.	+4 ox.		
		$+\underline{5}$ ox.	+5 ox.	+5 ox.	+5 ox.				
			$+\underline{6}$ ox.	+6 ox.	+6 ox.				
				$+\underline{7}$ ox.					

* Most common oxidation states underlined; ox. = good oxidizing agent under ordinary conditions; red. = good reducing agent under ordinary conditions.

Despite this divergence in trends between A and B groups, the transition metals do show many of the same trends as the representative elements. The following trends of the transition metals are examples of those that also apply to representative elements. For corresponding compounds involving metals in the same oxidation state from a given B group, the covalent character usually decreases and the ionic character increases as the group is descended. This trend is generally shown by such evidence as increasing conductivity of aqueous solutions and increasing melting and boiling points for the heavier compounds. For example, consider the metal(V) oxides of Group VB.

Oxide	Melting Point	
V_2O_5	690°C	Increasing
Nb_2O_5	1460°C	Ionic
Ta_2O_5	1800°C	↓ Character

For compounds containing the *same elements* in different proportions, the one containing the metal in the lower oxidation state tends to be more ionic. Titanium(II) chloride, $TiCl_2$, and $TiCl_3$ are ionic solids, whereas titanium(IV) chloride, $TiCl_4$, is a molecular liquid.

One further generalization can be made. The oxides and hydroxides of lower oxidation states of a given transition metal are basic, those containing intermediate oxidation states tend to be amphoteric, and those containing high oxidation states tend to be acidic. This will be illustrated in the discussions of the oxides and hydroxides of manganese and chromium that follow in Sections 26–6 and 26–7.

26–3 Color

Most transition metal compounds are colored, a characteristic that distinguishes them from most compounds of the representative elements. What causes the color? In transition metal compounds the *d* orbitals in any one energy level of the metals are not degenerate; that is, they are no longer all of the same energy, in contrast to the situation for isolated atoms. They are com-

TABLE 26–4 Comparison of Colors of Aqueous Solutions of Nitrate Salts of Typical Representative Metal Ions with Those of Typical Transition Metal Ions

Representative Metal Ion	Color of Aq. Solution*	Transition Metal Ion	Color of Aq. Solution*
Na^+	Colorless	Cr^{3+}	Deep blue
Ca^{2+}	Colorless	Mn^{2+}	Pale pink
Mg^{2+}	Colorless	Fe^{2+}	Pale green
Al^{3+}	Colorless	Fe^{3+}	Orchid†
Sn^{2+}	Colorless	Co^{2+}	Pink
Sn^{4+}	Colorless	Ni^{2+}	Green
Pb^{2+}	Colorless	Cu^{2+}	Blue

*This refers to the color of the hydrated metal ions.

†Yellow color comes from colloidal Fe_2O_3 or from Cl^- complexes. It is very difficult to obtain aqueous solutions of Fe^{3+} that are free of Fe_2O_3.

monly split into sets of orbitals separated by energies that correspond to wavelengths of electromagnetic radiation in the visible region (4×10^{-5} cm to 7×10^{-5} cm). The absorption of visible light causes electronic transitions between orbitals in these sets. Our eyes are sensitive to these wavelengths and can distinguish among them quite well. When certain visible wavelengths are absorbed from incoming "white" light, the light not absorbed appears to have the complementary color of the light absorbed. It is this transmitted or reflected light that we see. For instance, a compound that absorbs yellow light (5.5×10^{-5} cm) appears blue. Table 26–4 compares the colors of transition metal nitrates in aqueous solution with those of representative metal nitrates. The colors of transition metal compounds will be discussed further in the chapter on coordination compounds.

Complementary color pairs are: blue and yellow, red and cyan (blue-green), and green and magenta (blue-red).

26–4 Classification into Subgroups

The transition metals and zinc, cadmium, and mercury are subdivided into eight groups designated by a Roman numeral (I to VIII) followed by "B." The Roman numeral designates the maximum oxidation number exhibited by members of the group. This does not mean that simple ions with these charges exist; no simple ions of these elements possess a charge greater than 3+. The elements in corresponding A and B groups form many compounds of similar stoichiometry, as illustrated in Table 26–5.

TABLE 26–5 Typical Compounds and Ions of the A and B Groups

IA	IIA	IIIA	IVA	VA	VIA	VIIA
$NaCl$	$MgBr_2$	$Al(NO_3)_3$	CCl_4	$POCl_3$	SO_4^{2-}	Cl_2O_7
KNO_3	$CaCl_2$	$Ga(OH)_3$	PbO_2	PO_4^{3-}	$H_2S_2O_7$	$HClO_4$
IB	**IIB**	**IIIB**	**IVB**	**VB**	**VIB**	**VIIB**
$CuCl$	$ZnBr_2$	$Sc(NO_3)_3$	$TiCl_4$	$VOCl_3$	CrO_4^{2-}	Mn_2O_7
$AgNO_3$	$CdCl_2$	$Y(OH)_3$	ZrO_2	VO_4^{3-}	$H_2Cr_2O_7$	$HMnO_4$

However, while the compositions of corresponding compounds of A and B group elements are often similar, their chemical properties are usually dissimilar.

Group VIIIB consits of three columns of three metals each, which have no counterparts among the representative elements. Each horizontal row is called a **triad** and is named after the best-known metal of the row. The triad of the fourth period is called the *iron triad*. The other two are called the *palladium triad* and the *platinum triad*.

VIIIB		
Fe	Co	Ni
Ru	Rh	Pd
Os	Ir	Pt

26–5 Generalizations on Periodic Trends in Physical Properties

The properties of the transition metals vary somewhat more irregularly with respect to position in the periodic table than do those of the representative elements. The irregularity is due to a combination of factors and a detailed discussion is beyond the level of the text. However, we will state some rough generalizations that are valid, for the most part.

It should be mentioned that zinc, cadmium, and mercury show properties that are generally inconsistent with the gross trends. Each of these metals has a pseudo-noble gas electronic configuration, in that each has completely filled sets of *d* orbitals and *s* orbitals, with a noble gas core. The reasons for such behavior are related to the unusual stability of electronic configurations involving only completely filled sets of orbitals. Copper, palladium, silver, and gold, which also have filled sets of *d* orbitals, behave in the same way. Let us now consider trends in atomic radii, densities, magnetism and melting points.

The noble gases themselves exhibit properties that are out of line with the general periodic trends of the representative elements (Chapter 4).

1 Atomic Radii

The atomic radii of the *d*-transition (and IIB) metals are given in Table 26–6. Like the representative elements, the *d*-transition metals generally decrease in atomic radii from left to right across a given period as the nuclear charge (and therefore the nuclear attraction for electrons in the outer shell) increases. But these decreases are not as consistent as for the representative elements. The radii of elements at the end of a transition series increase because the increases in effective nuclear charge are outweighed by greater repulsions among *d* electrons in a given set of orbitals.

The radii of representative elements increase substantially upon descending a group, as electrons occupy energy levels farther from the nucleus. Likewise, the transition metals of the fifth period have larger radii than those of the fourth period. But the transition metals of the sixth period have nearly the

TABLE 26–6 Atomic Radii of the *d*-Transition Metals (Å)

Period 4	Sc	Ti	V	Cr	Mn	Fe	Co	Ni	Cu	Zn
(3*d*)	1.62	1.47	1.34	1.25	1.29	1.26	1.25	1.24	1.28	1.38
Period 5	Y	Zr	Nb	Mo	Tc	Ru	Rh	Pd	Ag	Cd
(4*d*)	1.80	1.60	1.46	1.39	1.36	1.34	1.34	1.37	1.44	1.54
Period 6	La	Hf	Ta	W	Re	Os	Ir	Pt	Au	Hg
(5*d*)	1.87	1.58	1.46	1.39	1.37	1.35	1.36	1.38	1.44	1.57

The lanthanide contraction has important effects on the properties of the sixth period transition and post-transition elements, as we saw in Section 21–7.

The f electrons shield outer electrons less effectively than do s or p electrons.

same radii as the metals above them. This phenomenon is known as the **lanthanide contraction,** following lanthanum (atomic number 57). The insertion of 14 lanthanides, with 14 poorly shielding f electrons two shells inside the outermost shell (and 14 positive charges in the nucleus), preceding hafnium (atomic number 72) results in higher effective nuclear charges felt by the outermost electrons than would be predicted if f electrons were good charge shielders. The higher effective nuclear charges tend to pull the outer electrons closer to the nucleus, which produces smaller radii than would be anticipated otherwise. The radii of the d-transition metals are plotted against periodic position in Figure 26–1.

2 Densities

Osmium and iridium, with densities of 22.6 g/cm^3, are the densest elements known.

As expected, densities vary inversely with the atomic radii (Tables 26–1 and 26–6, and Figures 26–1 and 26–2). Densities increase as the radii decrease within given periods and increase upon descending a group. In going from the fifth period to the sixth period, the metals have considerably more protons and neutrons but nearly the same atomic volume. Thus, the sixth-period transition metals are very dense because of the lanthanide contraction.

3 Magnetism

Many transition metals and metal ions have one or more unpaired electrons and are paramagnetic (Section 3–15). Figure 26–3 shows the good agreement between atomic theory and experiment in correlating the number of unpaired electrons with the observed degree of paramagnetism. Magnetic measurements are also important in explaining the colors and bonding of transition metal compounds, as will be seen in the next chapter.

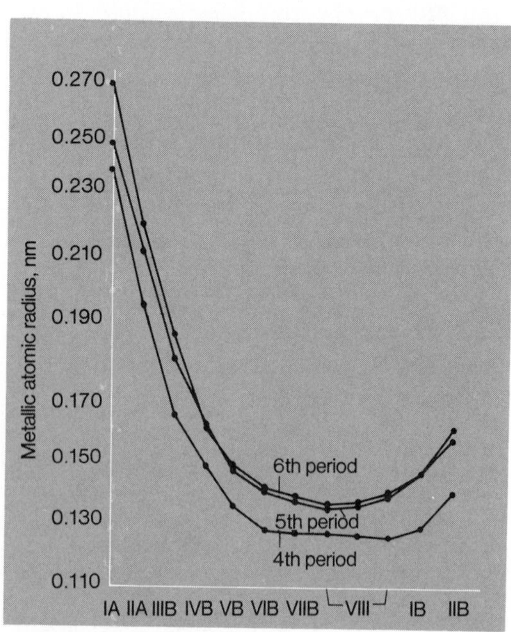

FIGURE 26–1 Metallic atomic radii of alkali, alkaline earth, and transition metals of the *4th*, *5th*, and *6th* periods.

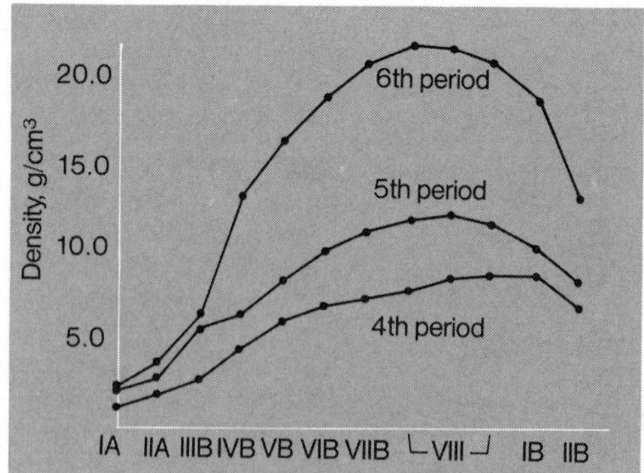

FIGURE 26–2 Densities of the alkali, alkaline earth, and transition metals of the *4th*, *5th*, and *6th* periods.

The metals of the iron triad (iron, cobalt, and nickel) are the only elements in the entire periodic table to exhibit **ferromagnetism** in the uncombined state. This property, which is stronger than paramagnetism, allows a substance to become permanently magnetized when placed in a magnetic field, as randomly oriented electron spins align themselves with an applied field. In order to exhibit ferromagnetism, the atoms of the substance must be within the proper range of sizes to allow the unpaired electrons on adjacent atoms to interact cooperatively with each other, but not to the extent that they pair. Experimental evidence suggests that in ferromagnets, atoms cluster together into **domains** that contain large numbers of atoms in fairly small volumes. The atoms within each domain interact cooperatively with each other.

Certain alloys, not necessarily containing one of these metals, also exhibit ferromagnetism.

4 Melting Points

Other properties of transition metals that are roughly related to the number of unpaired electrons are melting point and boiling point. The transition metals, of course, crystallize in metallic lattices. The strength of metallic bonding increases with the availability of electrons to participate in the bonding by delocalization over the lattice (Section 9–17). Thus, note the similarity of the shapes of the plots in Figures 26–3 and 26–4. Since the alkali and alkaline earth metals have only one or two outer electrons available, their melting points are relatively low in comparison to other solids. But the increasing availability of *d* electrons, especially unpaired electrons, causes an increase in melting point of the transition metals from the beginning toward the middle of

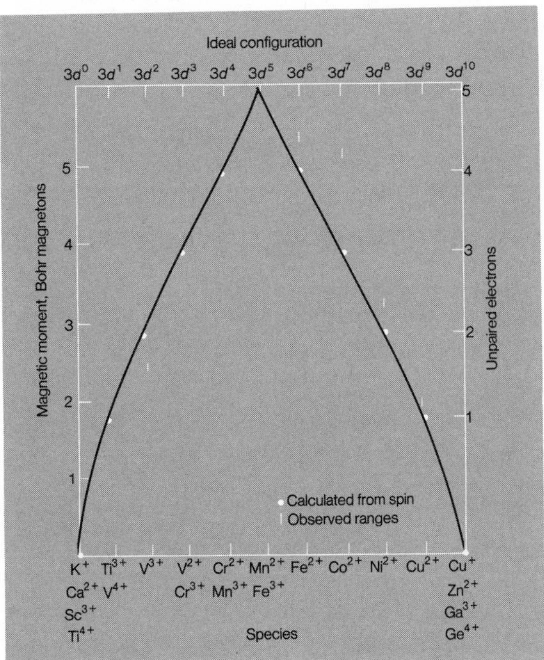

FIGURE 26–3 Magnetic properties of ions and elements in the first transition series and a few ions of representative elements.

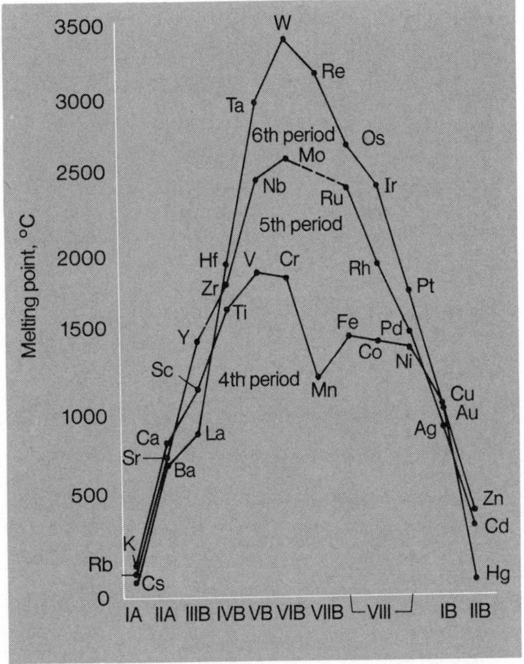

FIGURE 26–4 Melting points of alkali, alkaline earth, and transition metals of the 4*th*, 5*th*, and 6*th* periods.

each transition series. As electron pairing increases toward the right of the transition series, melting points decrease.

Although most transition metals are hard with fairly high melting points, the elements of Group IIB [zinc, cadmium, and mercury (a liquid)] are quite soft with low melting points. This might be expected, since they have pseudo-noble gas electron configurations and no unpaired electrons available for metallic bonding.

We have now studied the variations in some *physical* properties of the transition metals with position in the periodic table. In the next two sections we shall describe the variations in *chemical* properties of the oxides and hydroxides of manganese and chromium as the oxidation state of the metal increases. These variations are typical of those observed for the oxides and hydroxides of many other transition metals.

26–6 Manganese Oxides and Hydroxides

Manganese exhibits more oxidation states than any other 3*d*-transition metal, with the +2 oxidation state being most stable under ordinary conditions. Potassium permanganate, in which manganese is in the +7 oxidation state, is stable; it is a very strong oxidizing agent and common laboratory reagent. Table 26–7 shows some compounds of manganese in different oxidation states, and some properties of each.

Just as the nonmetallic hydroxides, or more commonly oxyacids, increase in acidity as the oxidation number of the central nonmetal increases, so do the transition metal hydroxides.

Compare, for example, the "hydroxides" of chlorine of Group VIIA (Chapter 22) with those of manganese from Group VIIB in Table 26–8. Permanganic and perchloric acids are both very strong acids and very strong oxidizing agents. Permanganic acid, which exists only in solution, can be neutralized

TABLE 26–7 Some Compounds of Manganese

Oxidation State	Oxide	BP (°C) of Oxide*	Hydroxide	Acid or Base Character of Oxide	Related Salt	Name of Salt
+2	MnO green	1785	$Mn(OH)_2$	basic	$MnSO_4$ reddish	manganese(II) sulfate
+3	Mn_2O_3 brown or black	d > 940	$Mn(OH)_3$	weakly basic	$Mn_2(SO_4)_3$ green	manganese(III) sulfate
+4	MnO_2 gray to black	d > 535	H_2MnO_3 $[MnO(OH)_2]$	amphoteric	$CaMnO_3$† brown	calcium manganite
+6	MnO_3 (?) reddish	d	H_2MnO_4 $[MnO_2(OH)_2]$	acidic	K_2MnO_4 green	potassium manganate
+7	Mn_2O_7 green-brown	d > 55	$HMnO_4$	strongly acidic	$KMnO_4$ purple	potassium permanganate

* The symbol d indicates decomposition above the specified temperature.

† Simplified representation; manganite salts usually exist in condensed forms such as $CaO \cdot 2MnO_2 = CaMn_2O_5$ and $CaO \cdot 5MnO_2 = CaMn_5O_{11}$.

TABLE 26-8 Variation of Acidity of Hydroxides with Oxidation States

Oxidation State						
+1	+2	+3	+4	+5	+6	+7
$ClOH$ [$HClO$] hypochlorous acid		$ClO(OH)$ [$HClO_2$] chlorous acid		$ClO_2(OH)$ [$HClO_3$] chloric acid		$ClO_3(OH)$ [$HClO_4$] perchloric acid
	$Mn(OH)_2$ manganese(II) hydroxide	$Mn(OH)_3$ manganese(III) hydroxide	$MnO(OH)_2$ [H_2MnO_3] manganous acid		$MnO_2(OH)_2$ [H_2MnO_4] manganic acid	$MnO_3(OH)$ [$HMnO_4$] permanganic acid

Increasing Covalent Character →
Increasing Acid Strength

with alkali to produce salts such as potassium permanganate, $KMnO_4$, which are also very strong oxidizing agents.

$$MnO_4^- + 8H^+ + 5e^- \longrightarrow Mn^{2+} + 4H_2O \qquad E^0 = +1.51 \text{ volts}$$

Manganeous and manganic acids, H_2MnO_3 and H_2MnO_4, are also unstable and can be isolated only in the form of their salts, the manganites and manganates. Manganese(II) and manganese(III) hydroxides are relatively stable, although $Mn(OH)_2$ is air-oxidized to manganese(III) oxyhydroxide, $MnO(OH)$, which reacts with water to produce $Mn(OH)_3$. Dimanganese heptoxide, Mn_2O_7, is a dark brown, explosive liquid, which is the acid anhydride of permanganic acid. Likewise dichlorine heptoxide, Cl_2O_7, a colorless explosive liquid, is the acid anhydride of perchloric acid.

Although the properties of several of the compounds of manganese and chlorine in high oxidation states are very much alike, it should be stated that the chemical properties are usually not as similar for other compounds involving corresponding A and B Group elements. Remember that manganese is a metal and chlorine is a nonmetal. However, the trends regarding acidity and basicity as a function of oxidation state are generally applicable.

The other elements of Group VIIB form many compounds of stoichiometry similar to those of manganese. Technetium (element 43) consists only of radioisotopes, but its chemistry has been studied. Rhenium (element 75) was discovered only in 1925, but it has been investigated quite thoroughly. The heptoxides, Tc_2O_7 and Re_2O_7, dissolve in water to give strongly acidic solutions, as does Mn_2O_7.

$$Mn_2O_7 + 3H_2O \longrightarrow \underbrace{2[H_3O^+ + MnO_4^-]}_{\text{permanganic acid}}$$

$$Tc_2O_7 + 3H_2O \longrightarrow \underbrace{2[H_3O^+ + TcO_4^-]}_{\text{pertechnetic acid}}$$

$$Re_2O_7 + 3H_2O \longrightarrow \underbrace{2[H_3O^+ + ReO_4^-]}_{\text{perrhenic acid}}$$

While Mn_2O_7 is dangerously explosive, Tc_2O_7 and Re_2O_7 are stable enough to be sublimed. This is evidence of the greater stability of the higher oxidation states as a transition metal group is descended. Perrhenate salts such as potassium perrhenate, $KReO_4$, are also more thermally stable than permanganates. For example, $KMnO_4$ decomposes to evolve oxygen at 200°C, but $KReO_4$ distills at 1370° and 1 atmosphere with no decomposition. As we would predict from its greater stability, $KReO_4$ is not as strong an oxidizing agent as $KMnO_4$.

26–7 Chromium Oxides and Hydroxides

Typical of the metals near the middle of each transition series, chromium shows a number of oxidation states, of which +2, +3, and +6 are the most common. Table 26–9 lists some of the common chromium compounds.

1 Oxidation-Reduction

The most stable oxidation state of chromium is +3. Solutions of blue chromium(II) salts are readily air-oxidized to chromium(III), so Cr^{2+} is a good reducing agent

$$Cr^{3+} + e^- \longrightarrow Cr^{2+} \qquad E^0 = -0.41 \text{ volt}$$

Chromium(VI) species are oxidizing agents. Basic solutions containing the chromate ion, $CrO_4{}^{2-}$, are weakly oxidizing; but acidification produces the dichromate ion, $Cr_2O_7{}^{2-}$ (also chromium(VI) oxide), a powerful oxidizing agent

$$Cr_2O_7{}^{2-} + 14H^+ + 6e^- \longrightarrow 2Cr^{3+} + 7H_2O \qquad E^0 = +1.33 \text{ volts}$$

Cr^{2+} is chromous ion; Cr^{2+} (aq) is $Cr(OH_2)_6^{2+}$. Cr^{3+} is chromic ion; Cr^{3+} (aq) is $Cr(OH_2)_6^{3+}$.

2 Chromate-Dichromate Equilibrium

Red chromium(VI) oxide, CrO_3, is the acid anhydride of two acids, chromic acid, H_2CrO_4, and dichromic acid, $H_2Cr_2O_7$. Neither acid has ever been

TABLE 26–9 Some Common Chromium Compounds

Oxidation State	Oxide	Hydroxide	Name	Acid or Base Character of Oxide	Related Salt	Name
+2	CrO black	$Cr(OH)_2$	chromous or chromium(II) hydroxide	basic	$CrCl_2$ anhydrous, colorless; aqueous, lt. blue	chromous or chromium(II) chloride
+3	Cr_2O_3 green	$Cr(OH)_3$	chromic or chromium(III) hydroxide	amphoteric	$CrCl_3$ anhydrous, violet; aqueous, green	chromic or chromium(III) chloride (acid solution)
					$KCrO_2$ green	potassium chromite (basic solution)
+6	CrO_3 dk. red	H_2CrO_4 or $[CrO_2(OH)_2]$	chromic acid	weakly acidic	K_2CrO_4 yellow	potassium chromate
		$H_2Cr_2O_7$ or $[Cr_2O_5(OH)_2]$	dichromic acid	acidic	$K_2Cr_2O_7$ orange	potassium dichromate

isolated in the pure form, although chromate and dichromate salts are common. CrO_3 dissolves in water to produce a strongly acidic solution containing hydrogen ions and (predominantly) dichromate ions

$$2CrO_3 + H_2O \longrightarrow \underbrace{[2H^+ + Cr_2O_7^{2-}]}_{\substack{\text{dichromic acid} \\ \text{(red-orange)}}}$$

From such solutions orange dichromate salts can be crystallized after adding a stoichiometric amount of base. Addition of excess base produces a yellow solution from which only yellow chromate salts can be obtained. The two Cr(VI) anions exist in solution in a pH-dependent equilibrium as follows

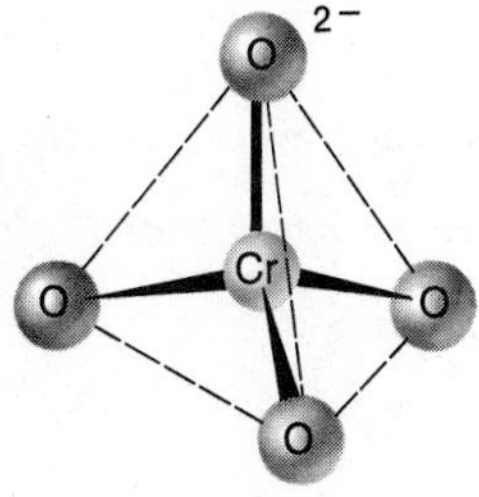

Chromate ions, CrO_4^{2-}

$$\underset{\text{yellow}}{2CrO_4^{2-} + 2H^+} \rightleftharpoons \underset{\text{orange}}{Cr_2O_7^{2-} + H_2O}$$

$$K_c = \frac{[Cr_2O_7^{2-}]}{[CrO_4^{2-}]^2[H^+]^2} = 4.2 \times 10^{14}$$

We see that decreasing the concentration of H^+ by addition of OH^- shifts the equilibrium to the left and produces more CrO_4^{2-}, while acidification shifts the equilibrium in favor of $Cr_2O_7^{2-}$.

The dichromate ion and dimanganese heptoxide are isoelectronic, and have the same geometry as dichlorine heptoxide.

All three species involve two tetrahedra sharing a corner.

```
 :Ö:   :Ö:            ┌   :Ö:  :Ö:   ┐2−         :Ö:  :Ö:
:Ö:Mn:Ö:Mn:Ö:         │ :Ö:Cr:Ö:Cr:Ö: │          :Ö:Cl:Ö:Cl:Ö:
 :Ö:   :Ö:            └   :Ö:  :Ö:   ┘           :Ö:  :Ö:

 O    O    O       ┌ O    O    O ┐2−     O    O    O
  Mn    Mn         │   Cr   Cr   │         Cl   Cl
O  |     |  O      │O  |     |  O│       O  |    |  O
   O     O         └   O     O   ┘          O    O
```

Dehydration of a chromate or dichromate salt with concentrated sulfuric acid, an excellent dehydrating agent, produces chromium(VI) oxide, CrO_3, a strong oxidizing agent. A powerful "cleaning solution" used for removing greasy stains and coatings from laboratory glassware is made by adding concentrated H_2SO_4 to a concentrated solution of potassium dichromate. The active ingredients are CrO_3, an oxidizing agent, and sulfuric acid, an excellent solvent.

Be *extremely* careful whenever using this cleaning solution. It can be very dangerous due to the tremendous amounts of heat liberated when it contacts glassware coated with organic solvents.

We might expect that since acidification of a solution of a salt containing the CrO_4^{2-} ion produces $Cr_2O_7^{2-}$ ion (and CrO_3), treatment of a solution of another strong oxidizing agent, a permanganate salt, MnO_4^-, with H_2SO_4 would produce Mn_2O_7, and that the resulting solution could be used as a cleaning solution. Indeed, Mn_2O_7 is produced from strongly acidic solution; but such a solution *should not be prepared* in the laboratory as a cleaning solution because Mn_2O_7 is so dangerously explosive.

The amphoterism of chromium(III) hydroxide is demonstrated by its reactions with both acid and base.

$$Cr(OH)_3 + 3H^+ \longrightarrow Cr^{3+} + 3H_2O$$

$$Cr(OH)_3 + OH^- \longrightarrow Cr(OH)_4^- \qquad (\text{or } CrO_2^- \cdot 2H_2O)$$

Molybdenum (element 42) and tungsten (element 74) are just below chromium in Group VIB. Molybdenum(VI) oxide, MoO_3, and WO_3 are both more thermally stable, less acidic, and weaker oxidizing agents than CrO_3.

26–8 Transition Metals as Catalysts

Homogeneous reactions occur in only one phase. Heterogeneous reactions involve more than one phase.

Transition metals and their compounds function as effective catalysts in many reactions, both homogeneous and heterogeneous. The unreactive (noble) metals such as platinum, palladium, and gold are sometimes used in the finely divided solid state to provide surfaces upon which heterogeneous reactions can occur. Other transition metals are thought to be effective homogeneous catalysts because their *d* orbital vacancies can accept electrons from reactants to form intermediates that subsequently decompose to products. The following reactions are typical of those catalyzed by transition metals or metal ions.

1. Haber process for the production of ammonia (Sections 15–5 and 24–5)

$$N_2 + 3H_2 \xrightarrow[\substack{500°C \\ 400 \text{ atm}}]{Fe_2O_3} 2NH_3$$

2. The reaction between sodium and liquid ammonia to produce sodium amide and hydrogen

$$2Na + 2NH_3\ (\ell) \xrightarrow[\text{cold}]{Fe} \underbrace{2[Na^+ + NH_2^-]}_{\text{sodium amide}} + H_2$$

3. Contact process for the production of sulfur trioxide in the manufacture of sulfuric acid (Section 23–16)

$$2SO_2 + O_2 \xrightarrow[400°C]{V_2O_5} 2SO_3$$

4. Oxidation of thallium(I) ions by cerium(IV) ions in aqueous solution

$$2Ce^{4+} + Tl^+ \xrightarrow{Mn^{2+}} 2Ce^{3+} + Tl^{3+}$$

5. Bromination of benzene

$$\underset{\text{benzene}}{C_6H_6} + Br_2 \xrightarrow{FeBr_3} \underset{\text{bromobenzene}}{C_6H_5Br} + HBr$$

6. Hydrogenation of an olefin (unsaturated hydrocarbon)

$$\underset{\text{an olefin}}{RCH{=}CH_2} + H_2 \xrightarrow{Pt} RCH_2CH_3 \qquad R = \text{organic group}$$

	IA	IIA	IIIB		IVB	VB	VIB	VIIB	VIII	VIII	VIII	IB	IIB	IIIA	IVA	VA	VIA	VIIA	0
1	1 H																	1 H	2 He
2	3 Li	4 Be												5 B	6 C	7 N	8 O	9 F	10 Ne
3	11 Na	12 Mg												13 Al	14 Si	15 P	16 S	17 Cl	18 Ar
4	19 K	20 Ca	21 Sc		22 Ti	23 V	24 Cr	25 Mn	26 Fe	27 Co	28 Ni	29 Cu	30 Zn	31 Ga	32 Ge	33 As	34 Se	35 Br	36 Kr
5	37 Rb	38 Sr	39 Y		40 Zr	41 Nb	42 Mo	43 Tc*	44 Ru	45 Rh	46 Pd	47 Ag	48 Cd	49 In	50 Sn	51 Sb	52 Te	53 I	54 Xe
6	55 Cs	56 Ba	57 La	58 Ce → 71 Lu	72 Hf	73 Ta	74 W	75 Re	76 Os	77 Ir	78 Pt	79 Au	80 Hg	81 Tl	82 Pb	83 Bi	84 Po	85 At	86 Rn
7	87 Fr	88 Ra	89 Ac	90 Th → 103 Lr	104 Ung*	105 Unp*	106 Unh*	107 Uns*	108	109	110	111	112	113	114	115	116	117	118

Groups; Periods

6	LANTHANIDE SERIES	58 Ce	59 Pr	60 Nd	61 Pm*	62 Sm	63 Eu	64 Gd	65 Tb	66 Dy	67 Ho	68 Er	69 Tm	70 Yb	71 Lu
7	ACTINIDE SERIES	90 Th	91 Pa	92 U	93 Np*	94 Pu*	95 Am*	96 Cm*	97 Bk*	98 Cf*	99 Es*	100 Fm*	101 Md*	102 No*	103 Lr*

FIGURE 26–5 The periodic table with the *f*-transition metals (the lanthanides and actinides) highlighted.

Transition metal ions are also present at the active sites of many important biological catalysts called enzymes. Some examples are cited in Chapter 27.

The Rare Earths (*f*-Transition Metals)

All the rare earth elements are metals in which an *f* sublevel *two* shells inside the outermost occupied shell is being filled. Chemical properties of elements are governed largely by the electrons in the outermost shells. All the rare earth elements (atomic numbers 58–71 and 90–103, shaded in Figure 26–5) have two *s* electrons in their outermost shell and either eight or nine electrons (two *s*, six *p*, and zero or one *d*) in the next shell inward (Table 26–10). This accounts for the chemical similarity of these elements. Most exhibit the ionic +3 oxidation state by ionization of the outer two *s* electrons and either the *d* electron one shell inward or else one of the *f* electrons two shells inward. While the *d*-transition metals show a strong tendency to form coordination compounds, the *f*-transition metals show markedly less tendency in this direction.

Other oxidation states do exist.

The lanthanides are really not as rare as the name *rare earths* implies. Mixtures of lanthanides and actinides occur in minerals such as monazite and

TABLE 26–10 Electronic Configurations of Lanthanides and Formulas of Their Chlorides

Atomic Number	Name	Symbol	Electronic Configuration of Neutral Element	Formula of Chloride	Oxidation State of Lanthanide
57	Lanthanum*	La	(Xe) $5d^1 6s^2$	$LaCl_3$	+3
58	Cerium	Ce	(Xe) $4f^1 5d^1 6s^2$	$CeCl_3$, $CeCl_4$	+3, +4
59	Praseodymium	Pr	(Xe) $4f^3 6s^2$	$PrCl_3$, $PrCl_4$	+3, +4
60	Neodymium	Nd	(Xe) $4f^4 6s^2$	$NdCl_3$	+3
61	Promethium	Pm	(Xe) $4f^5 6s^2$	$PmCl_3$	+3
62	Samarium	Sm	(Xe) $4f^6 6s^2$	$SmCl_2$, $SmCl_3$	+2, +3
63	Europium	Eu	(Xe) $4f^7 6s^2$	$EuCl_2$, $EuCl_3$	+2, +3
64	Gadolinium	Gd	(Xe) $4f^7 5d^1 6s^2$	$GdCl_3$	+3
65	Terbium	Tb	(Xe) $4f^9 6s^2$	$TbCl_3$, $TbCl_4$	+3, +4
66	Dysprosium	Dy	(Xe) $4f^{10} 6s^2$	$DyCl_3$	+3
67	Holmium	Ho	(Xe) $4f^{11} 6s^2$	$HoCl_3$	+3
68	Erbium	Er	(Xe) $4f^{12} 6s^2$	$ErCl_3$	+3
69	Thulium	Tm	(Xe) $4f^{13} 6s^2$	$TmCl_3$	+3
70	Ytterbium	Yb	(Xe) $4f^{14} 6s^2$	$YbCl_2$, $YbCl_3$	+2, +3
71	Lutetium	Lu	(Xe) $4f^{14} 5d^1 6s^2$	$LuCl_3$	+3

*Lanthanum is properly a *d*-transition metal.

gadolinite. However, these elements are difficult to obtain in a pure form because of their chemical similarities. Historically, the lanthanides and some actinides have been tediously separated from each other after extraction from their minerals only by hundreds of fractional recrystallizations of mixtures of their compounds. Recently they have been separated more efficiently by chromatography. **Chromatography** is a separation technique that relies upon the differences in attraction of similar compounds in solution for ionic sites on insoluble resins.

The actinides are all radioactive. Most do not occur naturally and have been prepared only since 1940 by nuclear reactions. Uranium (number 92) is the best-known naturally occurring actinide. It has been known for over 200 years, but has achieved major use and notoriety since 1939, when Otto Hahn and Lise Meitner discovered nuclear fission in uranium. Its oxide ores are treated with nitric acid; the nitrate that is formed is decomposed to pure oxide, which can be reduced to the metal with calcium. To enrich uranium in its fissionable ^{235}U isotope for use in nuclear power plants (Section 28–14), the oxide is converted to UF_4 with HF, and this is oxidized to the volatile covalent UF_6 by fluorine. The vapor of UF_6, which contains both $^{238}UF_6$ and $^{235}UF_6$, is then subjected to repeated diffusion through porous barriers to concentrate the $^{235}UF_6$, in accordance with Graham's law. More recently, gas centrifuges are being used for the concentrating process.

Several of the rare earths have some commercial importance. Cerium alloyed with iron is used in "flints" for cigarette lighters; praseodymium and neodymium are used in sunglasses and glassblowers' goggles; and carefully controlled mixtures of rare earths are used in the fluorescent screens of color televisions. Thorium nitrate has been used for more than a century in gas mantles in lanterns and street lamps.

Key Terms

Domain a cluster of atoms in a ferromagnetic substance, all of which align in the same direction in the presence of an external magnetic field.

Ferromagnetism the ability of a substance to become permanently magnetized by exposure to an external magnetic field.

Lanthanide contraction a decrease in the radii of the elements following the lanthanides relative to what would be expected if there were no *f*-transition metals.

Triad a horizontal row of three elements in Group VIIIB.

Exercises

1. How are the *d*-transition metals distinguished from other elements?
2. What are the general properties of the *d*-transition metals?
3. Why are trends in variations of properties of successive *d*-transition metals less regular than trends among successive representative elements?
4. Write out the entire electronic configurations for the following species: (a) V, (b) Fe, (c) Cu, (d) Zn, (e) Fe^{3+}, (f) Ni^{2+}, (g) Ag, (h) Ag^+.
5. Why do copper and chromium atoms have "unexpected" electronic configurations?
6. Why are most transition metal compounds colored?
7. Discuss the similarities and differences among elements of corresponding A and B groups of the periodic table, IIIA and IIIB for example.
8. What are the three triads of the *d*-transition metals?
9. Why do we say that zinc, cadmium, and mercury have pseudo-noble gas configurations?
10. What is the lanthanide contraction? Which transition elements does it affect? How does it affect radii and densities?
11. Copper exists in the +1, +2, and +3 oxidation states. Which is the most stable? Which is a good oxidizing agent and which a good reducing agent?
12. For a given transition metal in different oxidation states, how does the acidic character of its oxides increase? How do ionic and covalent character vary? Characterize a series of metal oxides as examples.
13. For different transition metals in the same oxidation state in the same group (vertical column) of the periodic table, how do covalent character and acidic character of their oxides vary? Why? Cite evidence for the trends.
14. Chromium(VI) oxide is the acid anhydride of what two acids? What is the oxidation state of the chromium in the acids?
15. How is the number of unpaired electrons per atom or ion experimentally determined?
16. What is ferromagnetism? What are domains? Which pure elements are ferromagnetic?
17. What single characteristic seems to be the most important in determining melting points of the *d*-transition metals?
18. Compare the structures and properties of Mn_2O_7, $Cr_2O_7^{2-}$, and Cl_2O_7.
19. Give four examples of reactions in which *d*-transition metals act as catalysts. What common characteristic of the *d*-transition metals makes them effective as catalysts?
20. How many grams of Co_3O_4 (a mixed oxide, $CoO \cdot Co_2O_3$) must react with excess aluminum to produce 300 g of metallic cobalt, assuming 72% yield?

$$3Co_3O_4 + 8Al \xrightarrow{\Delta} 9Co + 4Al_2O_3 + 4.03 \times 10^3 \text{ kJ}$$

21. What is the ratio of $[Cr_2O_7^{2-}]$ to $[CrO_4^{2-}]$ at 25°C in a solution prepared by dissolving 1.0×10^{-3} mol of sodium dichromate, $Na_2Cr_2O_7$, in enough of an aqueous solution buffered at pH = 3.00 to produce 200 mL of solution?

22. Answer Exercise 21 for a solution buffered at pH = 11.00.
23. Which would you expect to be the stronger oxidizing agent, Co^{3+} or Cr^{3+}? Why?
24. Classify the following substances as acidic or basic: MnO, Mn_2O_3, and Mn_2O_7. Explain.
25. Arrange the substances of Exercise 24 in order of increasing covalent character.
26. Calculate ΔG^0 for the oxidation of ferrous ion to ferric ion by O_2 in acidic solution.

$$4Fe^{2+}\ (aq) + 4H^+\ (aq) + O_2\ (g) \longrightarrow 4Fe^{3+}\ (aq) + 2H_2O\ (\ell) \qquad E^0 = +0.47\ V$$

27. Without consulting a table of electrode potentials, indicate which of the following substances should be strong oxidizing agents: Cr, $Co(OH)_2$, NiO(OH), Cu_2O, CrO_3. Justify your choices.
28. Scandium is quite an active metal. What volume of hydrogen, measured at STP, is produced by the reaction of 6.00 g of scandium with excess hydrochloric acid?

27 Coordination Compounds

27–1 Coordination Compounds

In Chapter 11 we discussed Lewis acid-base reactions. According to this view a *base* makes available a share in an electron pair and an *acid* accepts a share in an electron pair, to form a **coordinate covalent bond.** Such bonds are frequently represented by arrows, as shown below.

Covalent bonds in which the shared electron pair is provided by one atom are known as coordinate covalent bonds.

$$H_3N: + BF_3 \longrightarrow H_3N \rightarrow BF_3$$

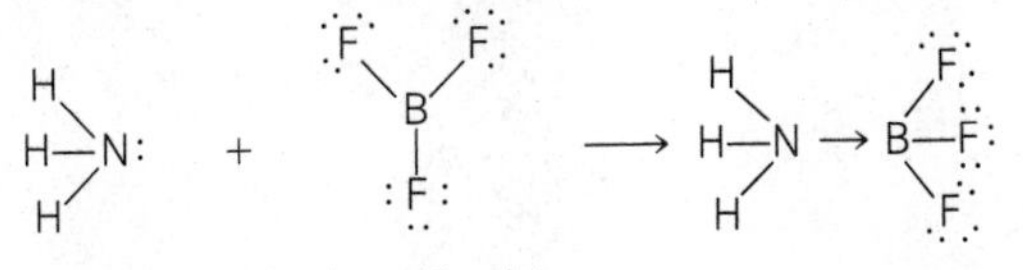

ammonia, a Lewis base; boron trifluoride, a Lewis acid

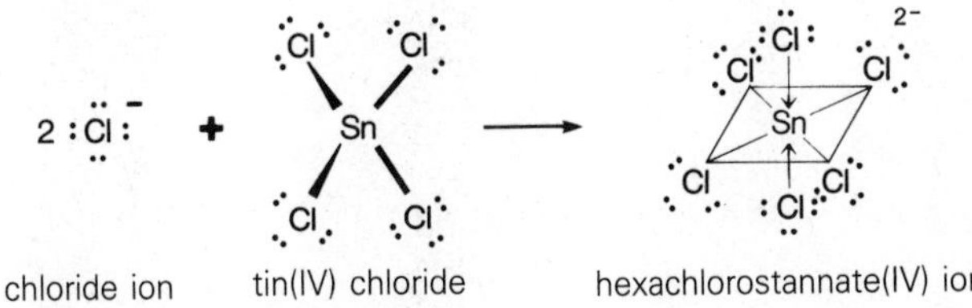

$$2\,:\ddot{Cl}:^- + SnCl_4 \longrightarrow [SnCl_6]^{2-}$$

chloride ion, a Lewis base; tin(IV) chloride, a Lewis acid; hexachlorostannate(IV) ion

In the case of this equation, the arrows do not imply that two Sn—Cl bonds are any different from the others. Once formed, all the Sn—Cl bonds in the $SnCl_6^{2-}$ ion are indistinguishable.

Most d-transition metal ions have vacant d orbitals that can accommodate shares in electron pairs, and they show a marked tendency to act as Lewis acids in forming coordinate covalent bonds in **coordination compounds (coordination complexes** or **complex ions).** Examples of transition metal ions or molecules showing coordinate covalence are $[Cr(OH_2)_6]^{3+}$, $[Co(NH_3)_6]^{3+}$, $[Ni(CN)_4]^{2-}$, $[Fe(CO)_5]$ and $[Ag(NH_3)_2]^+$. Many such complexes are very stable, as indicated by their low dissociation constants, K_d. See Section 18–6, part 3 and Appendix I.

Many biologically important substances are d-transition metal coordination compounds in which large, complicated organic species are bound to a metal ion through coordinate covalent bonds. One example is hemoglobin (Figure 27–1), a protein in blood that contains iron bound to a large, nearly planar organic group called a porphyrin ring; another is vitamin B-12, a large complex of cobalt. There are many practical applications of coordination compounds in such areas as water treatment, soil and plant treatment, protection of metal surfaces, analysis of trace amounts of metals, electroplating, and textile dyeing.

The bonding in transition metal complexes was not well understood until the pioneering research of Alfred Werner, a Swiss chemist of the 1890s and early 1900s who won the Nobel Prize in Chemistry in 1913. Although tremendous advances have been made in the field of coordination chemistry, especially in the last quarter century, Werner's classic work still remains the most important contribution by any single researcher.

Double salts are ionic solids resulting from the crystallization into a single lattice of two salts from the same solution. In the example given, the solid is produced from an aqueous solution of iron(II) sulfate, $FeSO_4$, and ammonium sulfate $(NH_4)_2SO_4$.

Prior to Werner's work the formulas of transition metal complexes were written with dots, $CrCl_3 \cdot 6H_2O$, $AgCl \cdot 2NH_3$, just like double salts such as iron(II) ammonium sulfate hexahydrate, $FeSO_4 \cdot (NH_4)_2SO_4 \cdot 6H_2O$. The properties of solutions of double salts are the properties expected for solutions made by mixing the individual salts. However, a solution of $AgCl \cdot 2NH_3$, or more appropriately $[Ag(NH_3)_2]Cl$, behaves entirely differently from either a solution of (very insoluble) silver chloride or a solution of ammonia. The dots have been called "dots of ignorance," since they signified that the mode of bonding was unknown. The table on the next page illustrates the types of experiments Werner performed and interpreted to lay the foundations for modern coordination theory.

Werner was able to isolate platinum(IV) compounds with the formulas given in the first column of Table 27–1. He added excess silver nitrate to solutions of carefully weighed amounts of each of the five salts. The precipi-

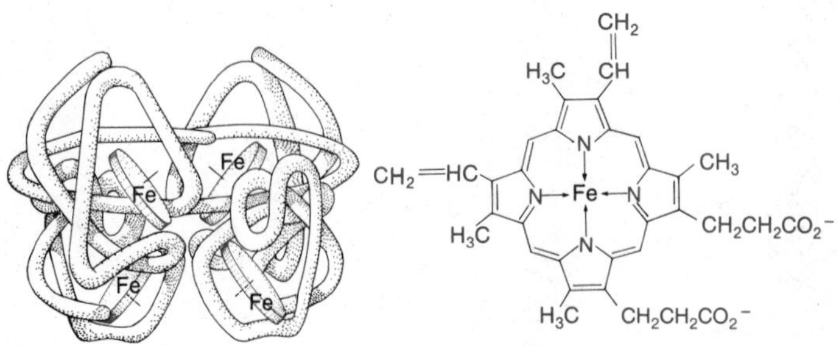

FIGURE 27–1 The shape of the tetrameric hemoglobin molecule (mol. wt.— 64,500) Individual atoms not shown. The four heme groups are represented by the disks (left).

TABLE 27–1 Interpretation of Experimental Data by Werner

Formula	Moles AgCl Precipitated per Formula Unit of Complex by Excess $AgNO_3$	Number of Ions per Formula Unit Indicated by Conductance	True Formula
$PtCl_4 \cdot 6NH_3$	4	5	$[Pt(NH_3)_6]Cl_4$
$PtCl_4 \cdot 5NH_3$	3	4	$[Pt(NH_3)_5Cl]Cl_3$
$PtCl_4 \cdot 4NH_3$	2	3	$[Pt(NH_3)_4Cl_2]Cl_2$
$PtCl_4 \cdot 3NH_3$	1	2	$[Pt(NH_3)_3Cl_3]Cl$
$PtCl_4 \cdot 2NH_3$	0	0	$[Pt(NH_3)_2Cl_4]$

Here the square brackets do not imply molar concentrations; instead they enclose the coordination sphere (Section 27–3).

tated silver chloride was collected by filtration, dried, and weighed to determine the number of moles of silver chloride produced, and therefore the number of chloride ions precipitated per formula unit. The results appear in the second column. Werner reasoned that the precipitated chloride ions must be free (uncoordinated), while the unprecipitated chloride ions must be directly bound to platinum so they could not be removed by silver ions. He also measured the equivalent conductances of salt solutions of known concentrations to determine the number of ions per formula unit. The results are given in the third column. Piecing the evidence together, he concluded that the correct formulations are the ones listed in the last column, in which the following Pt(IV) complex ions or molecules are present (from top to bottom): $[Pt(NH_3)_6]^{4+}$, $[Pt(NH_3)_5Cl]^{3+}$, $[Pt(NH_3)_4Cl_2]^{2+}$, $[Pt(NH_3)_3Cl_3]^+$ and $[Pt(NH_3)_2Cl_4]$. The species within the square brackets, NH_3 and Cl^-, are bonded by coordinate covalent bonds to the Lewis acid, Pt(IV) ion. The charges on the complex ions are the sums of the constituent charges.

27–2 The Ammine Complexes

The **ammine complexes** are complex species that contain ammonia molecules with coordinate covalent bonds to metal ions. Because the ammine complexes are important compounds that have been known for a long time, let us describe them briefly.

Since most metal hydroxides are insoluble in water, aqueous ammonia reacts with nearly all metal ions to form insoluble metal hydroxides, or hydrated oxides. The exceptions are the cations of the strong soluble bases, i.e., cations of the Group IA metals and the heavier members of Group IIA (Ca^{2+}, Sr^{2+}, Ba^{2+}). You have probably observed some examples in your laboratory work.

$$Cu^{2+} + 2NH_3 + 2H_2O \longrightarrow Cu(OH)_2\ (s) + 2NH_4^+$$

$$Cr^{3+} + 3NH_3 + 3H_2O \longrightarrow Cr(OH)_3\ (s) + 3NH_4^+$$

In general terms, we may represent this reaction as

$$M^{n+} + nNH_3 + nH_2O \longrightarrow M(OH)_n\ (s) + nNH_4^+$$

In Section 11–8 we pointed out the fact that the hydroxides of some metals and some metalloids are amphoteric, i.e., they dissolve in an excess of a strong soluble base. Aqueous ammonia is such a weak base ($K_b = 1.8 \times 10^{-5}$)

TABLE 27–2 Common Metal Ions that Form Soluble Complexes with an Excess of Aqueous Ammonia*

Metal Ion	Insoluble Hydroxide Formed by Limited Aq. NH_3	Complex Ion Formed by Excess Aq. NH_3
Co^{2+}	$Co(OH)_2$	$[Co(NH_3)_6]^{2+}$
Co^{3+}	$Co(OH)_3$	$[Co(NH_3)_6]^{3+}$
Ni^{2+}	$Ni(OH)_2$	$[Ni(NH_3)_6]^{2+}$
Cu^+	$CuOH \rightarrow \frac{1}{2}Cu_2O$†	$[Cu(NH_3)_2]^+$
Cu^{2+}	$Cu(OH)_2$	$[Cu(NH_3)_4]^{2+}$
Ag^+	$AgOH \rightarrow \frac{1}{2}Ag_2O$†	$[Ag(NH_3)_2]^+$
Zn^{2+}	$Zn(OH)_2$	$[Zn(NH_3)_4]^{2+}$
Cd^{2+}	$Cd(OH)_2$	$[Cd(NH_3)_4]^{2+}$
Hg^{2+}	$Hg(OH)_2$	$[Hg(NH_3)_4]^{2+}$

* The ions of Rh, Ir, Pd, Pt, and Au show similar behavior, but since these ions are seldom encountered in elementary texts, they are not included here.
† CuOH and AgOH are unstable and decompose to the corresponding oxides.

that the concentration of OH^- is not high enough to dissolve the amphoteric hydroxides and form hydroxo complexes.

However, several metal hydroxides do dissolve in an excess of aqueous ammonia to form ammine complexes. For example, the hydroxides of copper and cobalt are readily soluble in an excess of aqueous ammonia solution

$$Cu(OH)_2\ (s) + 4NH_3 \rightleftharpoons [Cu(NH_3)_4]^{2+} + 2OH^-$$

$$Co(OH)_2\ (s) + 6NH_3 \rightleftharpoons [Co(NH_3)_6]^{2+} + 2OH^-$$

Interestingly, all metal hydroxides that exhibit this behavior are derived from the 12 metals of the cobalt, nickel, copper, and zinc families. All the common cations of these metals except $Hg_2{}^{2+}$ (which disproportionates) form soluble complexes in the presence of excess aqueous ammonia; the common ones are listed in Table 27–2.

27–3 Important Terms in Coordination Chemistry

Before we go further, let us define and illustrate a few terms that will make our discussion of coordination compounds easier. The Lewis bases in coordination compounds may be molecules, anions, or (rarely) cations, and are called **ligands** (Latin, *ligare:* to bind). The **donor atoms** of the ligands are the atoms that actually donate electron pairs to metals. In some instances it is not possible to identify donor atoms, since the bonding electrons are not localized on specific atoms. For example, some small organic molecules such as ethylene, $H_2C{=}CH_2$, can bond to a transition metal through the electrons associated with their double bonds. Examples of typical simple ligands are listed in Table 27–3.

Ligands that can bond to a metal through only one donor atom at a time, such as those in Table 27–3, are **unidentate** (Latin, *dent:* tooth). Many ligands can bond simultaneously through more than one donor atom and are called **polydentate.** Of the polydentates, those that bond through two, three, four, five, or six donor atoms are called bidentates, tridentates, quadridentates, quinquedentates, and sexidentates, respectively. The resulting complexes,

TABLE 27–3 Typical Simple Ligands*

Molecule	Name	Name as Ligand	Ion	Name	Name as Ligand
$:\underline{N}H_3$	ammonia	ammine	$:\ddot{\underline{Cl}}:^-$	chloride	chloro
$:\ddot{\underline{O}}H_2$	water	aqua	$:\ddot{\underline{F}}:^-$	fluoride	fluoro
$:\underline{C}{\equiv}O:$	carbon monoxide	carbonyl	$:\underline{C}{\equiv}N:^-$	cyanide	cyano
$:\underline{P}H_3$	phosphine	phosphine	$:\ddot{\underline{O}}H^-$	hydroxide	hydroxo
$:\underline{N}{=}\ddot{O}$	nitrogen oxide	nitrosyl	$:\underline{N}(O)O^-$	nitrite	nitro†

*Donor atoms are underlined.

†Oxygen atoms can also function as donor atoms, in which case the ligand name is "nitrito."

consisting of the metal atom or ion and polydentate ligands, are called **chelate complexes** (Greek, **chele:** claw).

The **coordination number** of a metal atom or ion in a complex is the number of donor atoms to which it is coordinated, not necessarily the number of ligands. The **coordination sphere** includes the metal or metal ion and its ligands, but no uncoordinated counter-ions. For example, the coordination sphere of hexaamminecobalt(III) chloride, $[Co(NH_3)_6]Cl_3$, is the hexaamminecobalt(III) ion, $[Co(NH_3)_6]^{3+}$. Illustrations of these terms are given in Table 27–4. The naming of complexes will be discussed in the following section.

27–4 Naming Coordination Compounds

Because many thousands of rather complicated coordination compounds are known and thousands of new ones will be discovered during the next decade, the International Union of Pure and Applied Chemistry (IUPAC) has adopted a set of rules for naming such compounds. The rules are based on those originally devised by Werner. They are summarized below.

1. Consistent with the rules of nomenclature for compounds of the representative elements, cations are named before anions.
2. In naming the coordination sphere, ligands are named in alphabetical order. The prefixes di = 2, tri = 3, tetra = 4, penta = 5, hexa = 6, etc., are used to specify the number of a particular kind of coordinated ligand. For example, in dichloro, the "di" indicates that two Cl^- act as ligands, and this prefix is not considered in alphabetizing. However, when the prefix denotes the number of substituents on a single ligand, as in dimethylamine, $NH(CH_3)_2$, it *is* used to alphabetize ligands. For complicated ligands (usually chelating agents), other prefixes such as bis = 2, tris = 3, tetrakis = 4, pentakis = 5, and hexakis = 6 indicate the number of ligands, whose names are enclosed in parentheses.
3. The names of anionic ligands end in the suffix -o. Examples are: F^-, fluoro; OH^-, hydroxo; O^{2-}, oxo; S^{2-}, sulfido; $CO_3{}^{2-}$, carbonato; CN^-, cyano; $SO_4{}^{2-}$, sulfato; $NO_3{}^-$, nitrato; $S_2O_3{}^{2-}$, thiosulfato.

TABLE 27–4 Ligands and Coordination Complexes

Ligand(s)	Classification	Complex	Oxidation Number of M	Coordination Number of M
NH_3 ammine	unidentate	$[Co(NH_3)_6]^{3+}$ hexaamminecobalt(III)	+3	6
$H_2N—CH_2—CH_2—NH_2$ (or N N) ethylenediamine (en)	bidentate	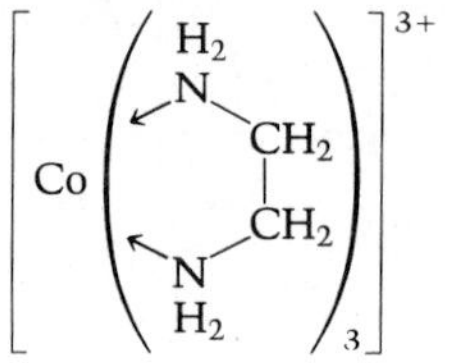 $[Co(en)_3]^{3+}$ tris(ethylenediamine)cobalt(III) ion	+3	6
Br^- bromo; $H_2N—CH_2—CH_2—NH_2$ ethylenediamine (en)	unidentate; bidentate	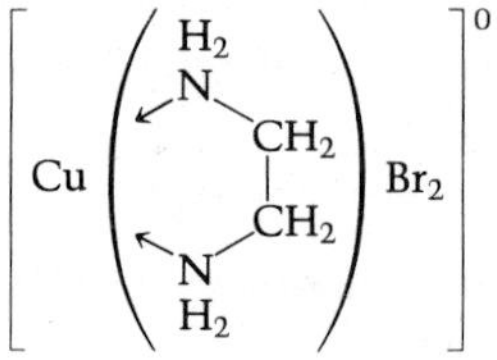 $[Cu(en)\,Br_2]$ dibromoethylenediaminecopper(II) ion	+2	4
$H_2N—CH_2—CH_2—N(H)—CH_2—CH_2—NH_2$ (or N N N) diethylenetriamine (dien)	tridentate	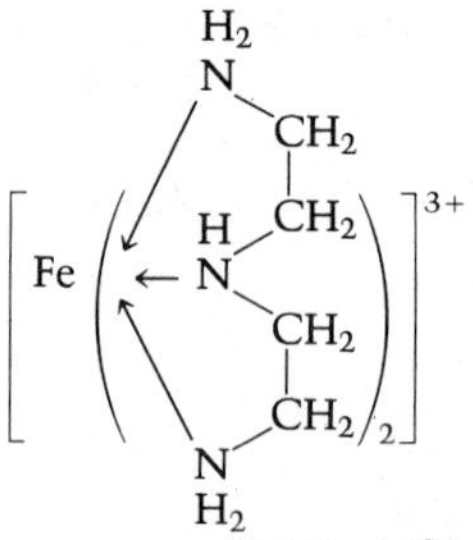 $[Fe(dien)_2]^{3+}$ bis(diethylenetriamine)iron(III) ion	+3	6
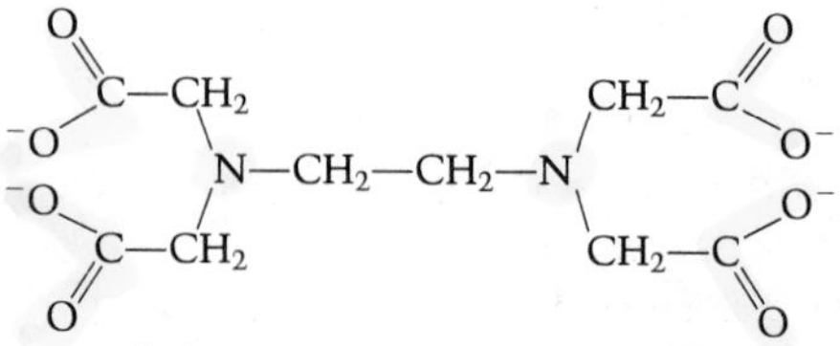 $(^-O_2C—CH_2)_2N—CH_2—CH_2—N(CH_2—CO_2^-)_2$ ethylenediaminetetraacetato (edta)	sexidentate	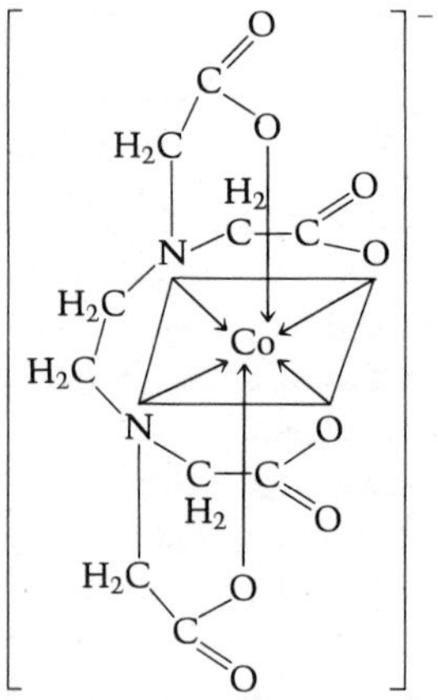 $[Co(edta)]^-$ ethylenediaminetetraacetatocobaltate(III) ion	+3	6

4. The names of neutral ligands are usually unchanged. Some important exceptions are NH_3, ammine; H_2O, aqua; CO, carbonyl; NO, nitrosyl.
5. If the metal exhibits variable oxidation states, the oxidation number of the metal ion is designated by a Roman numeral in parentheses following the name of the complex ion or molecule.
6. The suffix "ate" at the end of the name of the complex signifies that it is an anion. If the complex is neutral or cationic, no suffix is used. The English stem is usually used for the metal, but where the naming of an anion is awkward, the Latin stem is substituted. For example, ferrate is used rather than ironate, and plumbate rather than leadate.

The term *ammine* (two m's), which signifies the presence of ammonia as a ligand, is distinctly different from the term *amine* (one m), which describes a class of organic compounds (Section 30–9) that can be considered to be derived from ammonia.

Several examples are given to illustrate the rules.

$K_2[Cu(CN)_4]$	potassium tetracyanocuprate(II)
$[Ag(NH_3)_2]Cl$	diamminesilver chloride
$[Cr(OH_2)_6](NO_3)_3$	hexaaquachromium(III) nitrate
$[Co(en)_2Br_2]Cl$	dibromobis(ethylenediamine)cobalt(III) chloride
$[Ni(CO)_4]$	tetracarbonylnickel(0)
$[Pt(NH_3)_4][PtCl_6]$	tetraammineplatinum(II) hexachloroplatinate(IV)
$[Fe(OH_2)_5(NCS)]SO_4$	pentaaquathiocyanatoiron(III) sulfate
$[Cu(NH_3)_2(en)]Br_2$	diammine(ethylenediamine)copper(II) bromide
$Na[Al(OH)_4]$	sodium tetrahydroxoaluminate
$Na_2[CrOF_4]$	sodium tetrafluorooxochromate(IV)
$Na_2[Sn(OH)_6]$	sodium hexahydroxostannate(IV)
$[Cu(NH_3)(OH_2)Br_2]$	ammineaquadibromocopper(II)
$[Co(en)_3](NO_3)_3$	tris(ethylenediamine)cobalt(III) nitrate
$K_4[Ni(CN)_2(ox)_2]$	potassium dicyanobis(oxalato)nickelate(II)
$[Co(NH_3)_4(OH_2)Cl]Cl_2$	tetraammineaquachlorocobalt(III) chloride
$[RuCl_3\{P(C_6H_5)_3\}_2]$	trichlorobis(triphenylphosphine)ruthenium(III)

Water is written OH_2 rather than H_2O to emphasize that oxygen is the donor atom.

The oxidation state of aluminum is not given since it is always +3.

The abbreviation ox represents the oxalate ion $(COO)_2^{2-}$.

27–5 Structures of Coordination Compounds

The structures of coordination compounds are governed largely by the coordination number of the metal, and can be predicted quite accurately for most of them by application of the VSEPR method (Chapter 5). In most cases, lone pairs of electrons in *d* orbitals have only minimal influences on geometry since they are not in the outer shell. Table 27–5 summarizes the geometries and hybridizations associated with common coordination numbers.

Transition metal complexes with coordination numbers as high as seven, eight, and nine are known, but they are very rare. For coordination number five, the trigonal bipyramidal structure generally predominates over the square pyramid, although the energies associated with the two geometries are very close. Both tetrahedral and square planar geometries are quite common for complexes with coordination number four. Note that the tabulated geometries are ideal geometries. The actual structures are sometimes distorted, especially if ligands of more than one type are present. The distortions are due to compensations for the unequal electric fields generated by the different ligands at the coordination sites.

TABLE 27–5 Geometries and Hybridizations for Various Coordination Numbers

Coordination Number	Geometry	Hybridization	Example
2	 linear	*sp*	$[Ag(NH_3)_2]^+$ $[Cu(CN)_2]^-$
4	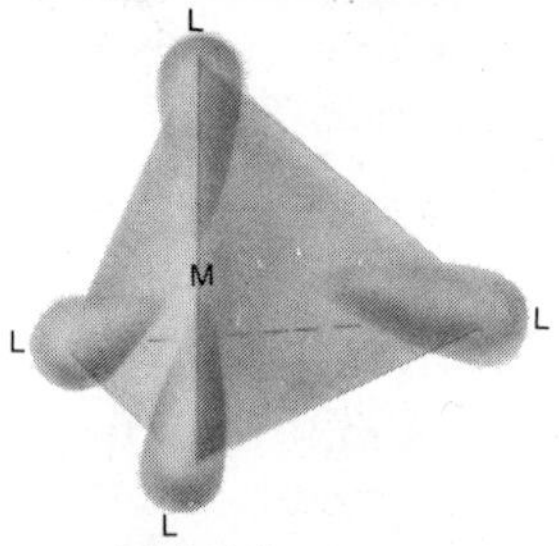 tetrahedral	sp^3	$[Zn(CN)_4]^{2-}$ $[Cd(NH_3)_4]^{2+}$
4	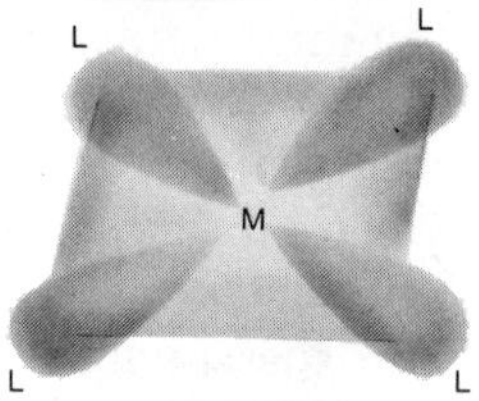 square planar	dsp^2 or sp^2d	$[Cu(OH_2)_4]^{2+}$ $[Pt(NH_3)_2Cl_2]$
5	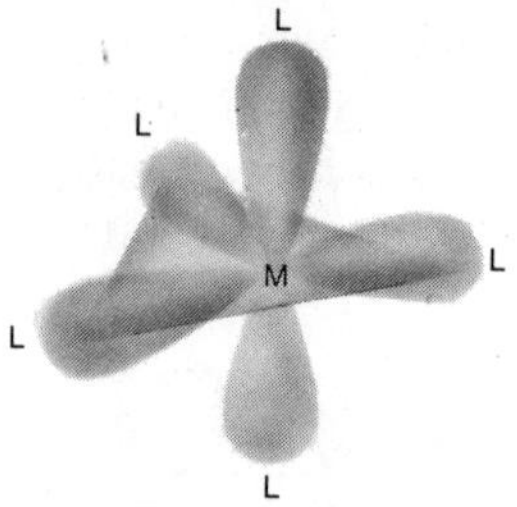 trigonal bipyramid	dsp^3	$[Fe(CO)_5]$ $[CuCl_5]^{3-}$
5	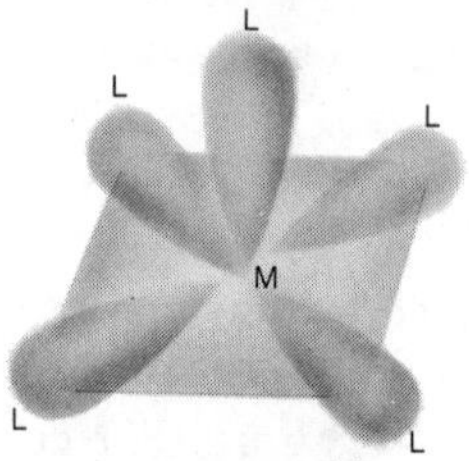 square pyramid	d^2sp^2	$[NiBr_3\{P(C_2H_5)_3\}_2]$ $[RuCl_3\{P(C_6H_5)_3\}_2]$
6	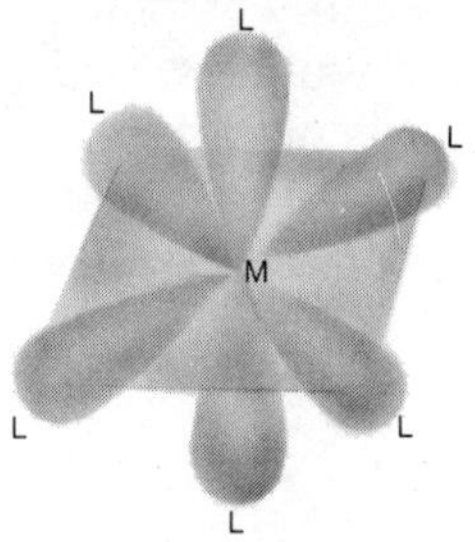 octahedral	d^2sp^3 or sp^3d^2	$[Fe(CN)_6]^{4-}$ $[Fe(OH_2)_6]^{2+}$

Isomerism in Coordination Compounds

Isomers are substances that have the same number and kinds of atoms arranged differently. *Because their structures are different, isomers have different properties.* Isomers can be broadly classed into two major categories, structural isomers and stereoisomers, each of which can be further subdivided as shown below.

The term *isomers* comes from the Greek word meaning "equal parts."

Structural Isomers	**Stereoisomers**
1. ionization isomers	1. geometric (position) isomers
2. hydrate isomers	2. optical isomers
3. coordination isomers	
4. linkage isomers	

Of the two broad classes, stereoisomers are more important and noteworthy. Before considering stereoisomers, we shall give examples of the four types of structural isomers. While distinctions between stereoisomers involve only one coordination sphere, and the same ligands and donor atoms, the differences between **structural isomers** involve either more than one coordination sphere or else different donor atoms on the same ligand. They contain different *atom-to-atom bonding sequences.*

27–6 Structural Isomers

1 Ionization Isomers

These isomers result from the interchange of ions inside and outside the coordination sphere. For example, red-violet $[Co(NH_3)_5Br]SO_4$ and red $[Co(NH_3)_5SO_4]Br$ are ionization isomers.

Isomers such as those shown here may *or may not* exist in the same solution in equilibrium. This depends on the thermodynamic and kinetic properties of the reaction mechanism, if any, that interconverts a pair of isomers. Generally such isomers are formed by *different* reactions.

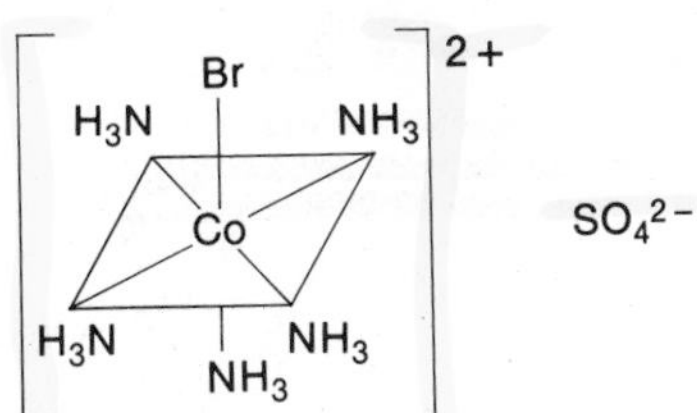

SO_4^{2-}

$[Co(NH_3)_5Br]SO_4$
pentaamminebromocobalt(III) sulfate

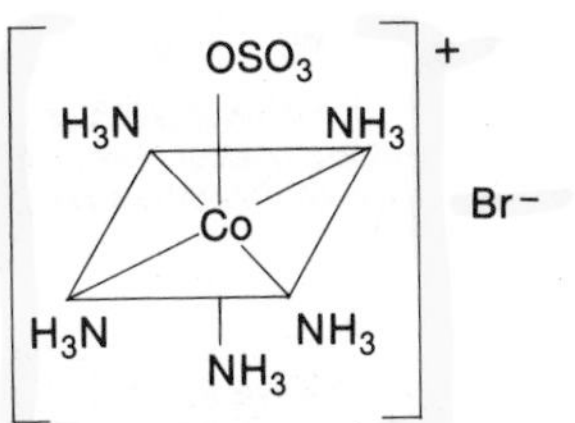

Br^-

$[Co(NH_3)_5SO_4]Br$
pentaamminesulfatocobalt(III) bromide

A solution of the sulfate reacts with a solution of barium chloride to precipitate white barium sulfate, but a solution of the bromide does not. Equimolar solutions of the two also have different electrical conductivities. The sulfate solution conducts the electric current better because its ions have 2+ and 2− charges rather than 1+ and 1−. Other examples of this type of isomerism include:

$[Pt(NH_3)_4Cl_2]Br_2$	and	$[Pt(NH_3)_4Br_2]Cl_2$
$[Pt(NH_3)_4SO_4](OH)_2$	and	$[Pt(NH_3)_4(OH)_2]SO_4$
$[Co(NH_3)_5NO_2]SO_4$	and	$[Co(NH_3)_5SO_4]NO_2$
$[Cr(NH_3)_5SO_4]Br$	and	$[Cr(NH_3)_5Br]SO_4$

2 Hydrate Isomers

Hydration isomerism and ionization isomerism are quite similar. In some crystalline complexes, water can occur in more than one way, inside and outside the coordination sphere. For example, solutions of the three hydrate isomers given below yield three, two, and one mole of silver chloride precipitate, respectively, per mole of complex when treated with excess silver nitrate.

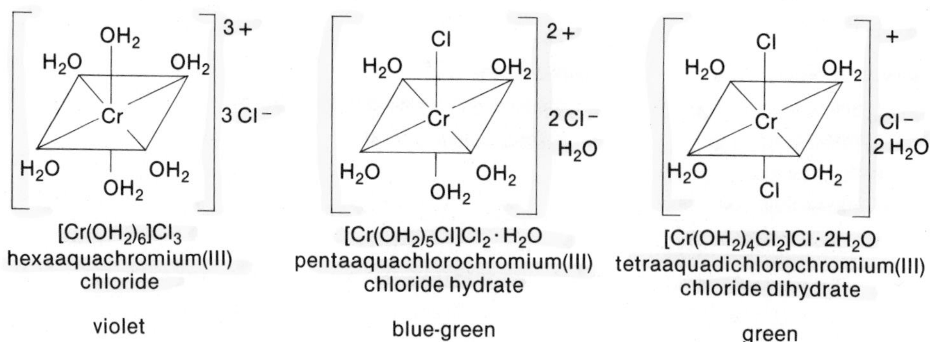

3 Coordination Isomers

Coordination isomerism can occur in compounds containing both complex cations and complex anions. Such isomers involve exchange of ligands between cation and anion, i.e., between coordination spheres.

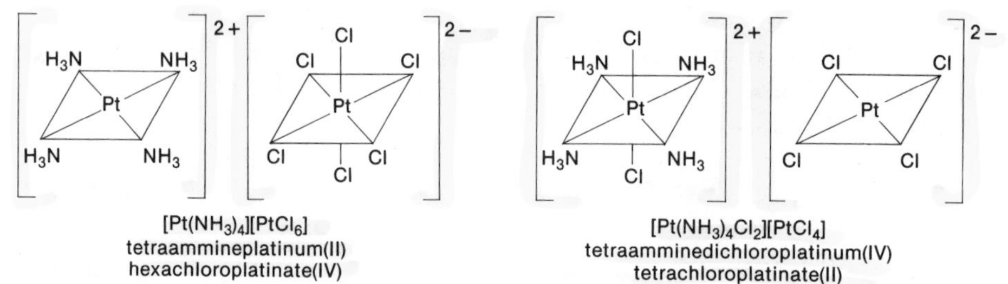

4 Linkage Isomers

Certain ligands can bind to a metal ion in more than one way. Examples of such ligands are cyano, —CN^-, and isocyano, —NC^-; nitro, —NO_2^-, and nitrito, —ONO^-. The donor atoms are on the left in these representations. Examples of linkage isomers are given below.

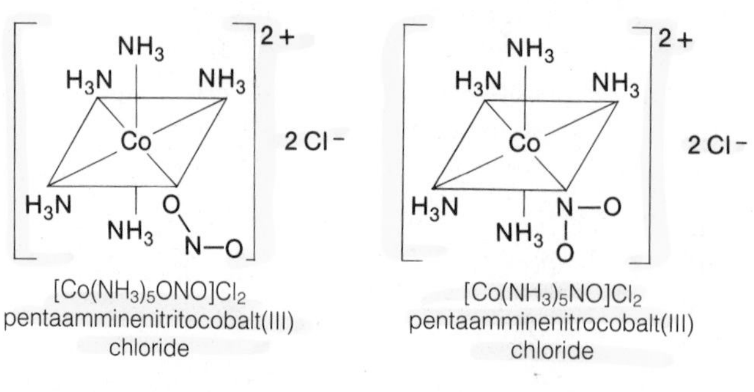

27–7 Stereoisomers

Compounds that contain the same atoms and the same atom-to-atom bonding sequences, but that differ only in the spatial arrangements of the atoms relative to the central atom, are **stereoisomers.** Stereoisomerism can exist only in complexes with coordination number four or greater. Since the most common coordination numbers among coordination complexes are four and six, they will be used to illustrate stereoisomerism.

Because of symmetry, a complex with coordination number 2 or 3 can have only one spatial arrangement; all apparent "isomers" are equivalent to just turning the complex around. Try building models to see this.

1 Geometrical Isomers

Stereoisomers that are not optical isomers (see p. 806) are **geometrical isomers** or position isomers. *Cis-trans* isomerism is one kind of geometrical isomerism. *Cis* means "adjacent to" and *trans* means "on the opposite side of." *Cis-* and *trans*-diamminedichloroplatinum(II) are shown below.

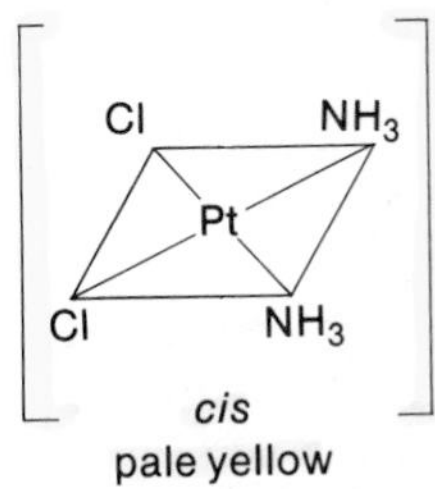

cis
pale yellow

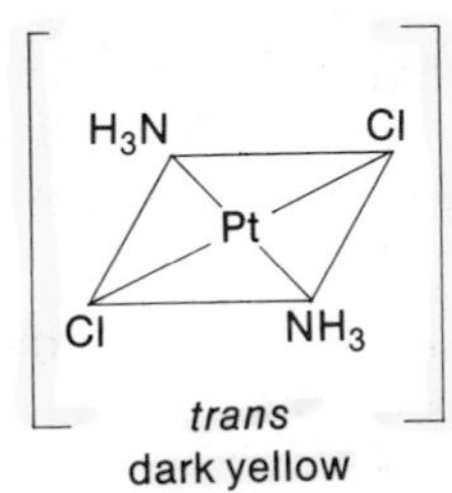

trans
dark yellow

In the *cis* complex, the chloro groups are closer to each other (on the same side of the square) than they are in the *trans* complex. The ammine groups are also closer together in the *cis* complex.

Since all ligands in tetrahedral complexes of the type ML_4 are equidistant from each other, they do not exhibit geometrical isomerism.

Several types of isomerism are possible for octahedral complexes. For example, complexes of the type $MA_2B_2C_2$ can exhibit several isomeric forms. Consider as an example $[Cr(OH_2)_2(NH_3)_2Br_2]^+$. Each of the like ligands may be *trans* to each other or *cis* to each other.

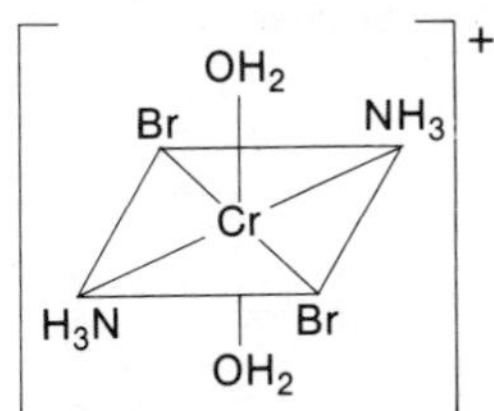

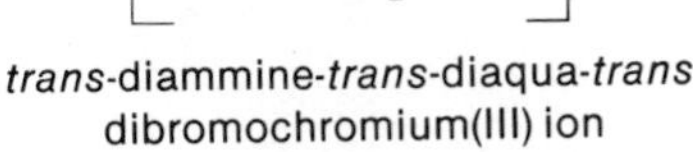
trans-diammine-*trans*-diaqua-*trans*-dibromochromium(III) ion

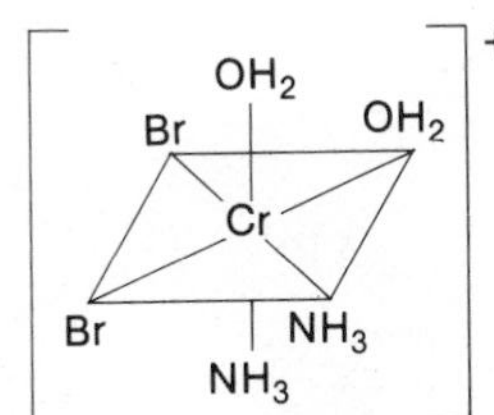

cis-diammine-*cis*-diaqua-*cis*-dibromochromium(III) ion

One pair of ligands may be *trans* to each other, but the others *cis*.

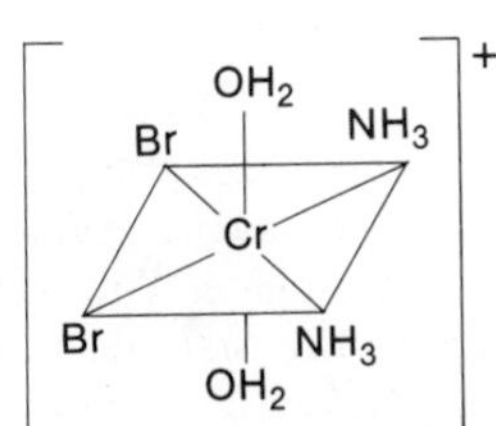

cis-diammine-*trans*-diaqua-*cis*-dibromochromium(III) ion

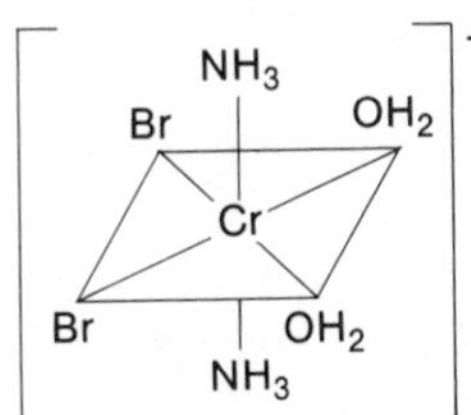

trans-diammine-*cis*-diaqua-*cis*-dibromochromium(III) ion

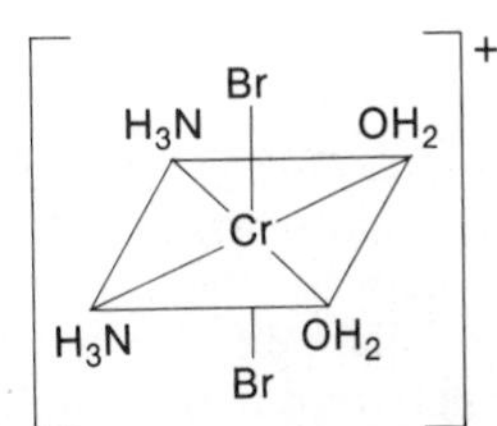

cis-diammine-*cis*-diaqua-*trans*-dibromochromium(III) ion

See whether you can discover why there is no *trans-trans-cis* isomer.

Interchanging the positions of the ligands further produces no new geometric isomers. However, one of the five geometric isomers can exist in two distinct forms called optical isomers.

2 Optical Isomers

The *cis*-diammine-*cis*-diaqua-*cis*-dibromochromium(III) geometric isomer exists in two forms that bear the same relationship to each other as left and right hands. They are *nonsuperimposable* mirror images of each other and are called **optical isomers** or **enantiomers.**

Alfred Werner was also the first person to demonstrate optical activity in an inorganic compound (not of biological origin). This demonstration silenced critics of his theory of coordination compounds. In his own opinion this was his greatest achievement, his other contributions notwithstanding. Louis Pasteur had demonstrated the phenomenon of optical activity many years earlier in compounds of biological origin.

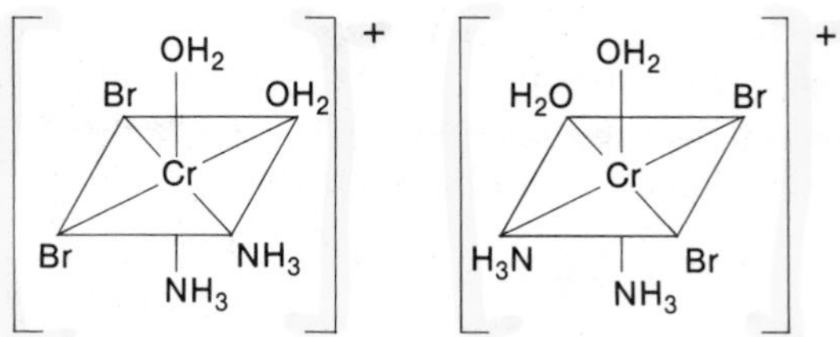

Optical isomers of *cis*-diammine-*cis*-diaqua-*cis*-dibromochromium(III) ion

Optical isomers have identical physical and chemical properties except that they interact with polarized light in different ways. Separate equimolar solutions of the two will rotate a plane of polarized light (see Figures 27–2 and 27–3) by equal amounts but in opposite directions. One solution is **dextrorotatory** (rotates to the *right*) and the other is **levorotatory** (rotates to the *left*).

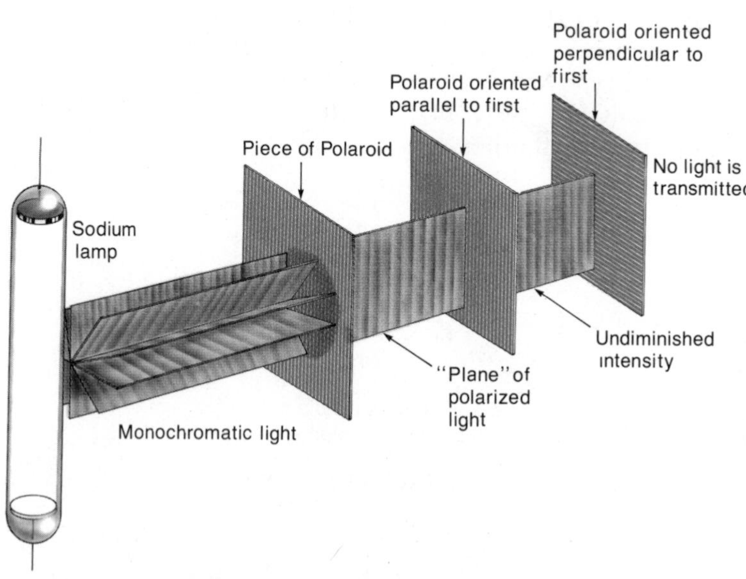

FIGURE 27–2 Light from a lamp or from the sun consists of electromagnetic waves that vibrate in all directions perpendicular to the direction of travel. "Polaroid" material absorbs all waves except those that vibrate in a single plane. The third Polaroid sheet, with a plane of polarization at right angles to the first, absorbs the polarized light completely.

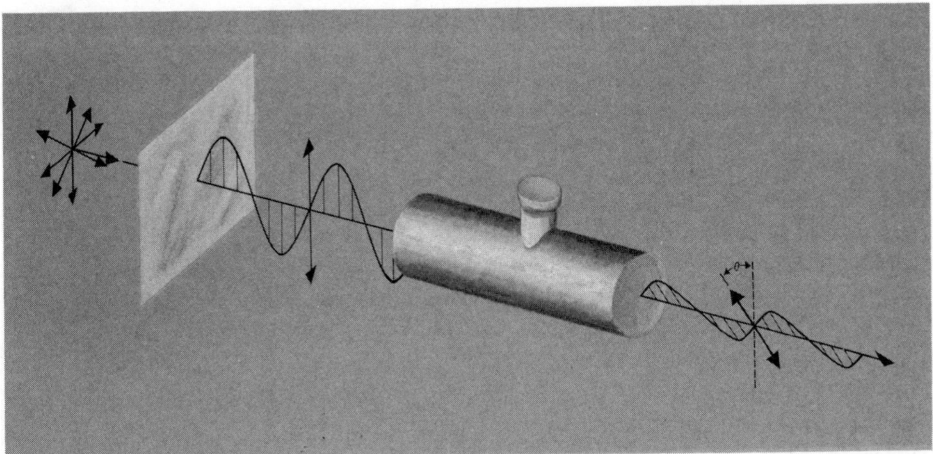

FIGURE 27–3 The plane of polarization of plane polarized light is rotated through an angle (θ) as it passes through an optically active medium. Species that rotate the plane to the right (clockwise) are dextrorotatory and those that rotate to the left are levorotatory.

The optical isomers are thus called *dextro* and *levo* isomers. The phenomenon by which a plane of polarized light is rotated is called **optical activity** and can be measured by a device called a polarimeter (see Figure 27–3) or by more sophisticated instruments. A single solution containing equal amounts of the two isomers, called a **racemic mixture,** does not rotate a plane of polarized light because the equal and opposite effects of the two isomers exactly cancel. Thus, in order to exhibit optical activity, the *dextro* and *levo* isomers (sometimes designated as delta, Δ, and lambda, Λ, isomers) must be separated from each other by one of a number of chemical or physical processes broadly called optical resolution. Another example of a pair of optical isomers is shown below. Note that they both contain ethylenediamine, a bidentate ligand.

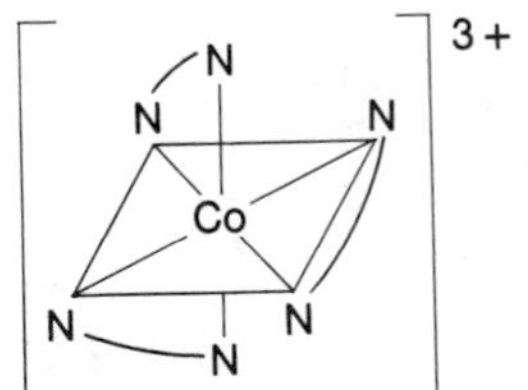

Λ-tris(ethylenediamine)cobalt(III) ion

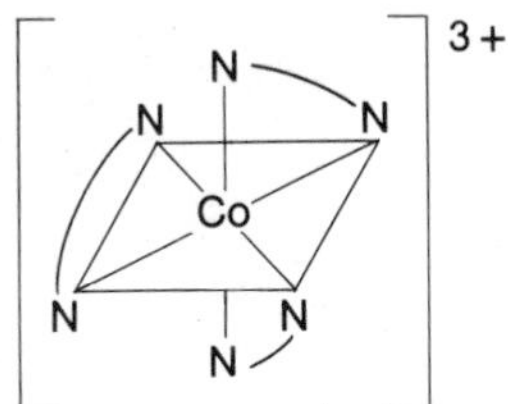

Δ-tris(ethylenediamine)cobalt(III) ion

Bonding in Coordination Compounds

Bonding theories for coordination compounds should be able to account for structural features, colors, and magnetic properties. The earliest accepted theory was the **Valence Bond** theory, which can account for structural and magnetic properties, but offers no explanation for the wide range of colors of coordination compounds. However, it has the advantage of being a simple treatment of bonding using the classical picture of the chemical bond. The **Crystal Field** theory gives quite satisfactory explanations of color as well as structure and magnetic properties for many coordination compounds. We shall discuss the Valence Bond and Crystal Field theories.

27–8 Valence Bond Theory

A basic assumption of the Valence Bond theory applied to coordination compounds is that the coordinate bonds between ligands and metal are entirely covalent. Of course, this is not really true. All bonds have some degree, however slight, of both covalent and ionic character.

1 Coordination Number 6

The shapes of the d orbitals of a metal atom are depicted in Figure 3–20. The lobes of the $d_{x^2-y^2}$ and d_{z^2} orbitals are directed along the x, y, and z axes. The lobes of the d_{xy}, d_{yz}, and d_{xz} orbitals bisect the axes. The six ligand donor atoms in an octahedral complex are located at the corners of an octahedron, two along each of the three axes. Valence Bond theory postulates that they must donate electrons into a set of octahedrally hybridized d^2sp^3 or sp^3d^2 metal orbitals. We shall distinguish between d^2sp^3 and sp^3d^2 hybridization very shortly. Thus the two metal d orbitals used in the hybridization must be the $d_{x^2-y^2}$ and d_{z^2} orbitals, since they are the only ones directed along the x, y, and z axes. This is illustrated in Figure 27–4.

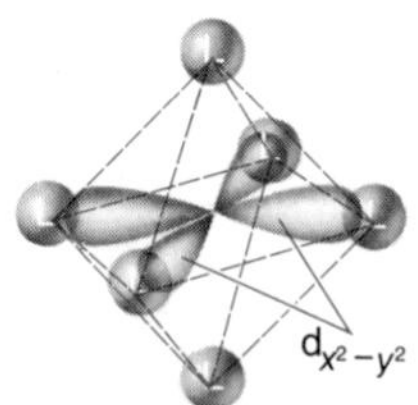

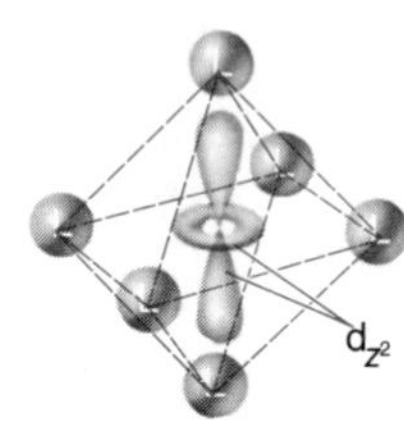

FIGURE 27–4 Orientation of $d_{x^2-y^2}$ and d_{z^2} orbitals relative to the ligands in an octahedral complex.

Let us take one example and see how the Valence Bond theory accounts for its properties. Hexaaquairon(III) perchlorate, $[Fe(OH_2)_6](ClO_4)_3$, is known to be paramagnetic, with a magnetic moment corresponding to five unpaired electrons per iron atom. The valence bond description of the bonding in the $[Fe(OH_2)_6]^{3+}$ ion is as follows. Atomic iron has the electronic configuration

	3d	4s
Fe [Ar]	⇅ ↑ ↑ ↑ ↑	⇅

The Fe^{3+} ion is formed upon ionization of the 4s electrons and one of the 3d electrons

	3d	4s
Fe^{3+} [Ar]	↑ ↑ ↑ ↑ ↑	_

This is in agreement with Hund's Rule.

To account for the experimentally observed fact that $[Fe(OH_2)_6]^{3+}$ has five unpaired electrons, each 3d orbital is assumed to have one unpaired electron in the complex. The vacant $4d_{x^2-y^2}$ and $4d_{z^2}$ orbitals are hybridized with the vacant 4s and 4p orbitals.

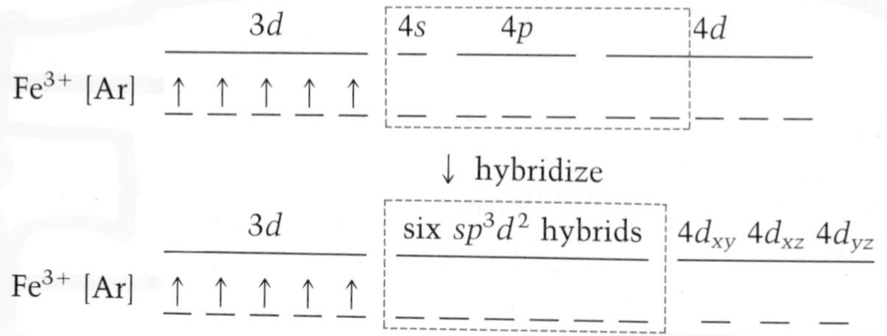

Each of the six water ligands now donates two electrons into one of the six sp^3d^2 orbitals, forming the six coordinate covalent bonds. The electrons originally on the oxygen atoms are represented as "×" rather than "↑," even though, in actuality, electrons are indistinguishable. Note that these are the only *bonding* electrons, and that none of the 3*d* electrons of iron is involved in bonding.

		3*d*	sp^3d^2	$4d_{xy}$ $4d_{xz}$ $4d_{yz}$
$[Fe(OH_2)_6]^{3+}$	[Ar]	↑ ↑ ↑ ↑ ↑	×× ×× ×× ×× ×× ××	_ _ _

Since a set of outer *d* orbitals is involved in the hybridization, $[Fe(OH_2)_6]^{3+}$ is called an **outer orbital complex.**

The hexacyanoferrate(III) ion (also known commonly as the ferricyanide ion), $[Fe(CN)_6]^{3-}$, also involves Fe^{3+} (a d^5 ion), but its magnetic moment indicates only one unpaired electron per iron atom. The $3d_{x^2-y^2}$ and $3d_{z^2}$ orbitals (rather than those in the 4*d* shell) are involved in d^2sp^3 hybridization. This forces pairing of all but one of the nonbonding 3*d* electrons of Fe^{3+}.

A convenient way to describe *d*-transition metal ions is to indicate the number of nonbonding electrons in *d* orbitals.

		3*d*		4*s*	4*p*
Fe^{3+}	[Ar]	↑ ↑ ↑	↑ ↑	_	_ _ _

↓ hybridize

		$3d_{xy}$ $3d_{xz}$ $3d_{yz}$	six d^2sp^3 hybrids
$[Fe(CN)_6]^{3-}$	[Ar]	⇅ ⇅ ↑	×× ×× ×× ×× ×× ××

Since only inner *d* orbitals are used in hybridization, $[Fe(CN)_6]^{3-}$ is called an **inner orbital complex.** As in the previous case, none of the original 3*d* electrons of the Fe^{3+} ion is involved in bonding.

In deciding whether the hybridization at the central metal ion of octahedral complexes is sp^3d^2 (outer orbital) or d^2sp^3 (inner orbital), we must know the results of magnetic measurements that indicate the number of unpaired electrons.

To account for the single unpaired electron in $[Co(NH_3)_6]^{2+}$, derived from a d^7 ion, by Valence Bond theory, it is necessary to postulate the *promotion* of a 3*d* electron to a 5*s* orbital as d^2sp^3 hybridization occurs.

		3*d*		4*s*	4*p*	5*s*
Co^{2+}	[Ar]	⇅ ⇅ ↑	↑ ↑	_	_ _ _	_

↓ hybridize

		$3d_{xy}$ $3d_{xz}$ $3d_{yz}$	d^2sp^3	5*s*
$[Co(NH_3)_6]^{2+}$	[Ar]	⇅ ⇅ ⇅	×× ×× ×× ×× ×× ××	↑

This structure for $[Co(NH_3)_6]^{2+}$ is consistent with the fact that $[Co(NH_3)_6]^{2+}$ is easily oxidized to $[Co(NH_3)_6]^{3+}$, an extremely stable complex ion.

The electronic configurations and hybridizations of some octahedral complexes are shown in Table 27–6.

TABLE 27–6 Bonding and Hybridization in Some Octahedral Complexes

Metal Ion	Outer Electron Configuration				Complex Ion	Type	Outer Electron Configuration		
	3d	*4s*	*4p*	*4d*			*3d*	hybrid orbitals	*4d*
Co^{3+}	↑↓ ↑ ↑ ↑ ↑	—	— — —	— — — — —	$Co(NH_3)_6^{3+}$	inner; diamagnetic	↑↓ ↑↓ ↑↓	d^2sp^3: xx xx xx xx xx xx	— — — — —
					CoF_6^{3-}	outer; paramagnetic	↑↓ ↑ ↑ ↑ ↑	sp^3d^2: xx xx xx xx xx xx	— — —
Co^{2+}	↑↓ ↑↓ ↑ ↑ ↑	—	— — —	— — — — —	$Co(OH_2)_6^{2+}$	outer; paramagnetic	↑↓ ↑↓ ↑ ↑ ↑	sp^3d^2: xx xx xx xx xx xx	— — —
Mn^{2+}	↑ ↑ ↑ ↑ ↑	—	— — —	— — — — —	$Mn(CN)_6^{4-}$	inner; paramagnetic	↑↓ ↑↓ ↑	d^2sp^3: xx xx xx xx xx xx	— — — — —
Ni^{2+}	↑↓ ↑↓ ↑↓ ↑ ↑	—	— — —	— — — — —	$Ni(OH_2)_6^{2+}$	outer; paramagnetic	↑↓ ↑↓ ↑↓ ↑ ↑	sp^3d^2: xx xx xx xx xx xx	— — —
Cr^{3+}	↑ ↑ ↑ — —	—	— — —	— — — — —	$Cr(NH_3)_6^{3+}$	inner; paramagnetic	↑ ↑ ↑	d^2sp^3: xx xx xx xx xx xx	— — — — —
Fe^{2+} and Co^{3+}	↑↓ ↑ ↑ ↑ ↑	—	— — —	— — — — —	$Fe(CN)_6^{4-}$ and $Co(CN)_6^{3-}$	inner; diamagnetic	↑↓ ↑↓ ↑↓	d^2sp^3: xx xx xx xx xx xx	— — — — —

2 Coordination Number 4

Most complexes with coordination number four are either tetrahedral or square planar. Tetrahedral geometry results from sp^3 hybridization. The IIB metals zinc, cadmium, and mercury in their +2 oxidation states are the B group metals *commonly* involved in tetrahedral complexes. Since they all have d^{10} configurations, their inner d orbitals cannot participate in hybridization and it is not surprising that they exhibit sp^3 hybridization. Tetraamminezinc ion, for instance, is diamagnetic and tetrahedral.

There are a few tetrahedral complexes of other B group elements.

	3d					4s	4p		
Zn^{2+} [Ar]	⇅	⇅	⇅	⇅	⇅	_	_	_	_

	3d					sp^3			
$[Zn(NH_3)_4]^{2+}$ [Ar]	⇅	⇅	⇅	⇅	⇅	xx	xx	xx	xx

Because Cu^+ is also a d^{10} ion, this also represents the electronic configuration of diamagnetic, tetrahedral $[Cu(CN)_4]^{3-}$.

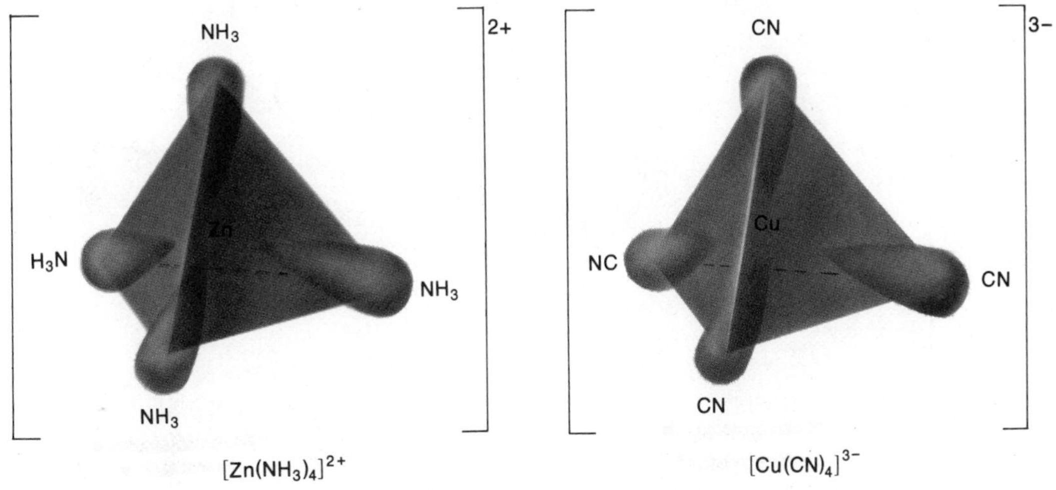

$[Zn(NH_3)_4]^{2+}$ $[Cu(CN)_4]^{3-}$

Although they are not as common, complexes of a few other transition metal ions are also tetrahedral. For example, $[NiCl_4]^{2-}$ is paramagnetic with two unpaired electrons, and is tetrahedral.

	3d					4s	4p		
Ni^{2+} [Ar]	⇅	⇅	⇅	↑	↑	_	_	_	_

	3d					sp^3			
$[NiCl_4]^{2-}$ [Ar]	⇅	⇅	⇅	↑	↑	xx	xx	xx	xx

Most four-coordinate transition metal complexes are square planar and dsp^2 hybridized (utilizing the $d_{x^2-y^2}$ orbital) at the central metal ion. The metals most commonly forming square planar complexes are Cu(II), Au(III), Co(II), Ni(II), Pt(II), and Pd(II). For example, the tetracyanonickelate(II) ion, $[Ni(CN)_4]^{2-}$, is square planar and diamagnetic.

	3d					4s	4p		
Ni^{2+} [Ar]	⇅	⇅	⇅	↑	↑	_	_	_	_

	3d				dsp^2				4p
$[Ni(CN)_4]^{2-}$ [Ar]	⇅	⇅	⇅	⇅	xx	xx	xx	xx	_

We see that Valence Bond theory predicts that paramagnetic four-coordinate Ni(II) complexes are tetrahedral, and diamagnetic four-coordinate Ni(II) complexes are square planar.

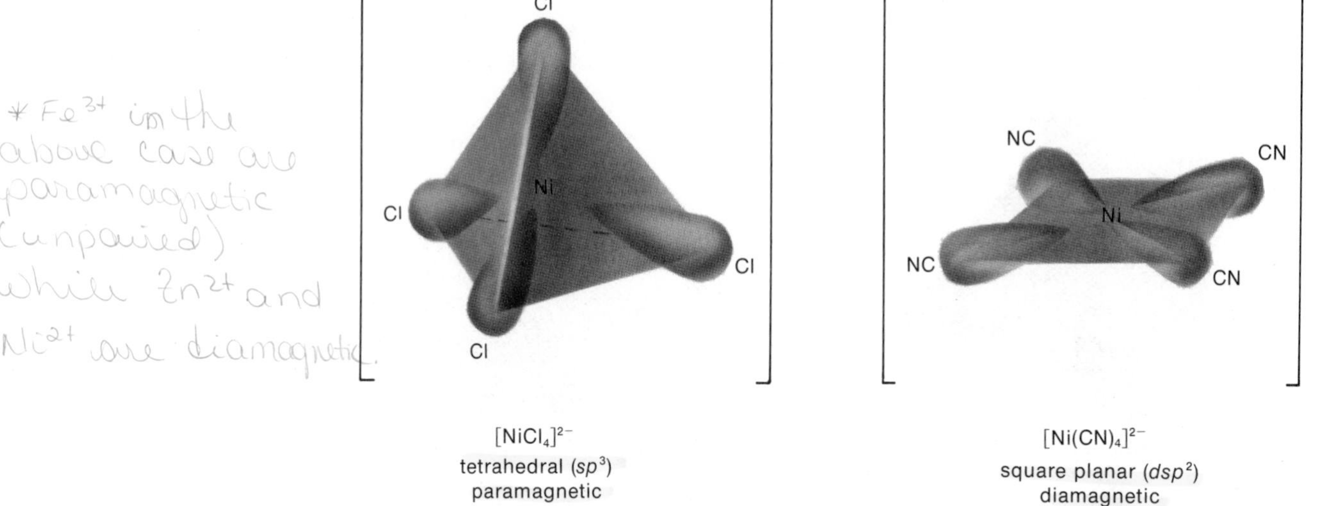

$[NiCl_4]^{2-}$
tetrahedral (sp^3)
paramagnetic

$[Ni(CN)_4]^{2-}$
square planar (dsp^2)
diamagnetic

However, the recent discovery of paramagnetic Ni(II) complexes that are square planar suggests hybridization involving the $4d_{x^2-y^2}$, $4p_x$, $4p_y$, and $4s$ orbitals to give sp^2d (outer orbital) hybrids at the Ni(II). This leaves two unpaired electrons in the $3d$ orbitals of Ni(II). Thus we see that, although Valence Bond theory accurately predicts most hybridizations from structures, it does have some flaws.

27–9 Crystal Field Theory

Hans Bethe and J. H. van Vleck originally developed the Crystal Field theory between 1919 and the early 1930's, but it was not widely used until the 1950s. In its pure form it assumes that the bonds between ligand and metal ion are completely ionic. Both ligand and metal ion are treated as infinitesimally small, nonpolarizable point charges. Recall that Valence Bond theory assumes complete covalence. The outgrowth of Crystal Field theory, Ligand Field theory, attributes partial covalent and partial ionic character to the bonds.

1 Coordination Number 6

As we have seen, the $d_{x^2-y^2}$ and d_{z^2} orbitals are directed along a set of mutually perpendicular *x*, *y*, and *z* axes. As a group these orbitals are called the $\boldsymbol{e_g}$ orbitals. The d_{xy}, d_{yz}, and d_{xz} orbitals, or collectively the t_{2g} orbitals, bisect the axes. Since the ligand donor atoms approach the metal ion along the axes during the formation of octahedral complexes, there are greater repulsions among ligand electrons and metal ion electrons in the e_g orbitals than among ligand electrons and those in t_{2g} orbitals. Crystal Field theory proposes that the approach of the six donor atoms (point charges) along the axes sets up an electric field (the crystal field) that removes the degeneracy of the set of *d* orbitals and splits them into two sets, the t_{2g} set at lower energy and the e_g set at higher energy.

Recall that degenerate orbitals are orbitals of equal energy.

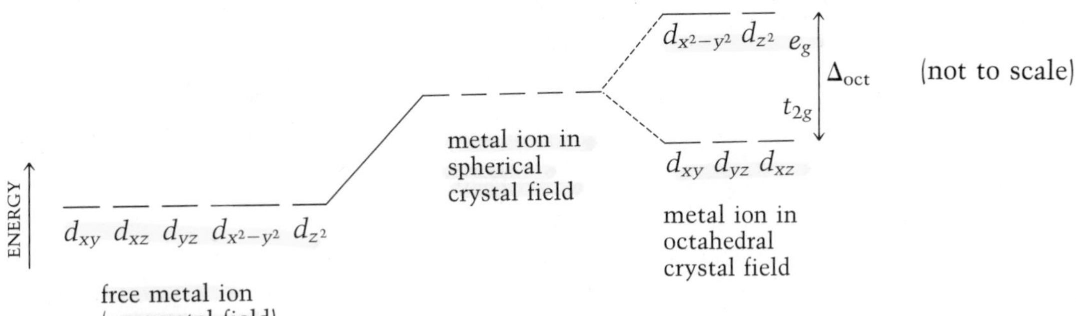

The energy separation between the two sets, called $\Delta_{octahedral}$ or Δ_{oct}, is proportional to the *crystal field strength* of the ligands, that is, how strongly the ligand electrons repel the metal electrons. (See Section 27–10.)

Δ_{oct} is sometimes called $10Dq$. Its typical values are between 100 and 400 kJ/mol.

The *d* electrons on the metal ion occupy the t_{2g} set in preference to the higher-energy e_g set. In fact, the t_{2g} orbitals are called **nonbonding orbitals** in octahedral complexes, but the e_g orbitals are called **antibonding orbitals** because electrons that are forced to occupy these orbitals are quite strongly repelled by the relatively close approach of ligand electrons, and tend to destabilize the octahedral complex. This will be discussed further under the section entitled Crystal Field Stabilization Energy.

Let us now consider the cases of the hexafluorocobaltate(III) ion, $[CoF_6]^{3-}$, and the hexaamminecobalt(III) ion, $[Co(NH_3)_6]^{3+}$. Both contain a d^6 ion and already have been treated in terms of Valence Bond theory (Table 27–6). The former is a paramagnetic outer orbital complex; the latter a diamagnetic inner orbital complex. We shall focus our attention on the *d* electrons only.

In the free ion, or in the absence of ligands, Co^{3+} has six electrons (four unpaired) in its 3*d* orbitals

$$\begin{array}{ll} & 3d \\ Co^{3+}\ [Ar] & \underline{\uparrow\downarrow}\ \underline{\uparrow}\ \underline{\uparrow}\ \underline{\uparrow}\ \underline{\uparrow} \end{array}$$

This is a high spin complex; see below.

Since its magnetic moment indicates that $[CoF_6]^{3-}$ also has four unpaired electrons per ion, the electrons must be arranged with four in t_{2g} orbitals and two in e_g orbitals

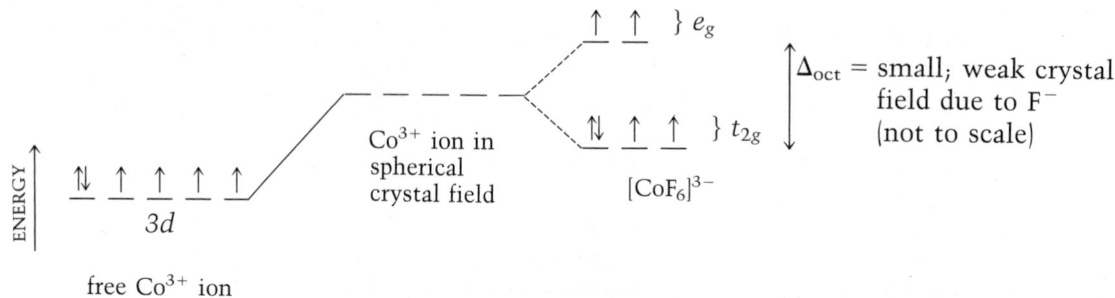

This is a low-spin complex; see p. 815.

On the other hand, $[Co(NH_3)_6]^{3+}$ is diamagnetic, and all six d electrons must be paired in the t_{2g} orbitals

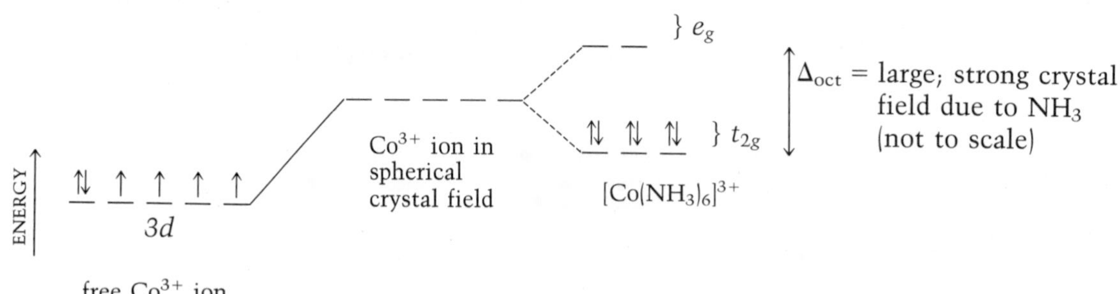

The difference in configurations is a consequence of the relative magnitudes of the crystal field splitting, Δ_{oct}, due to the different crystal field strengths of fluoride and ammonia ligands. The ammonia molecule more readily donates electrons into vacant metal orbitals than does the very electronegative fluoride ion, which holds its electrons very strongly. As a result the crystal field splitting generated by the close approach of six ammonia molecules to the metal ion is greater than that produced by the approach of six fluoride ions. That is,

$$\Delta_{oct} \text{ for } [Co(NH_3)_6]^{3+} > \Delta_{oct} \text{ for } [CoF_6]^{3-}$$

The crystal field splitting for $[CoF_6]^{3-}$ is so small that an energetically more favorable situation results if two electrons remain unpaired in the antibonding e_g orbitals rather than pairing with electrons in the lower energy, nonbonding t_{2g} orbitals. Recall that Hund's Rule requires that electrons singly occupy a set of degenerate orbitals before pairing in any one of the set, in order to avoid the unnecessary expenditure of energy in pairing electrons, or bringing two negatively charged particles into the same region of space. If the approach of a set of ligands removes the d orbital degeneracy but causes a Δ_{oct} less than the electron pairing energy, P, then the electrons will continue to occupy singly the resulting nondegenerate orbitals rather than pair. After all d orbitals are half-filled, additional electrons will pair with electrons in the t_{2g} set. This is the case for $[CoF_6]^{3-}$, which is called a **high spin complex.**

The spin pairing energy is "high" compared to the crystal field splitting energy.

For $[CoF_6]^{3-}$: F^- is weak field ligand so $P > \Delta_{oct}$, thus, high spin complex

(P = electron pairing energy)

A high spin complex in the crystal field treatment corresponds to an outer orbital complex by the valence bond treatment.

In contrast, the $[Co(NH_3)_6]^{3+}$ ion is a **low spin complex.** The Δ_{oct} generated by the strong field ligand, ammonia, is greater than the electron pairing energy, so electrons become paired in t_{2g} orbitals before any occupy the antibonding e_g orbitals.

The spin pairing energy here is "low" compared to the crystal field splitting energy.

For $[Co(NH_3)_6]^{3+}$: NH_3 is strong field ligand so $\Delta_{oct} > P$, thus, low spin complex

A low spin complex corresponds to an inner orbital complex by the valence bond treatment.

Low spin configurations exist only for octahedral complexes having metal ions with d^4, d^5, d^6, and d^7 configurations. For d^1–d^3 and d^8–d^{10} ions only one possibility exists. In each case the configuration is designated as high spin. All d^n possibilities are shown in Table 27–7.

2 Coordination Number 4

Figure 27–5 shows that tetrahedrally distributed ligands are closer to and interact more strongly with the t_{2g} orbitals than with the e_g orbitals. Conse-

TABLE 27–7 High and Low Spin Octahedral Configurations

d^n	Examples	High Spin	Low Spin
d^1	Ti^{3+}	_ _ e_g ↑ _ _ t_{2g}	same as high spin
d^2	Ti^{2+}, V^{3+}, Zr^{2+}	_ _ e_g ↑ ↑ _ t_{2g}	same as high spin
d^3	V^{2+}, Cr^{3+}	_ _ e_g ↑ ↑ ↑ t_{2g}	same as high spin
d^4	Mn^{3+}, Re^{3+}	↑ _ e_g ↑ ↑ ↑ t_{2g}	_ _ e_g ↑↓ ↑ ↑ t_{2g}
d^5	Mn^{2+}, Fe^{3+}, Ru^{3+}	↑ ↑ e_g ↑ ↑ ↑ t_{2g}	_ _ e_g ↑↓ ↑↓ ↑ t_{2g}
d^6	Fe^{2+}, Ru^{2+}, Pd^{4+}, Rh^{3+}, Co^{3+}	↑ ↑ e_g ↑↓ ↑ ↑ t_{2g}	_ _ e_g ↑↓ ↑↓ ↑↓ t_{2g}
d^7	Co^{2+}, Rh^{2+}	↑ ↑ e_g ↑↓ ↑↓ ↑ t_{2g}	↑ _ e_g ↑↓ ↑↓ ↑↓ t_{2g}
d^8	Ni^{2+}, Pt^{2+}, Au^{3+}	↑ ↑ e_g ↑↓ ↑↓ ↑↓ t_{2g}	same as high spin
d^9	Cu^{2+}	↑↓ ↑ e_g ↑↓ ↑↓ ↑↓ t_{2g}	same as high spin
d^{10}	Zn^{2+}, Ag^{+}, Hg^{2+}	↑↓ ↑↓ e_g ↑↓ ↑↓ ↑↓ t_{2g}	same as high spin

FIGURE 27–5 The five *d* orbitals in a tetrahedral environment of ligands, L.

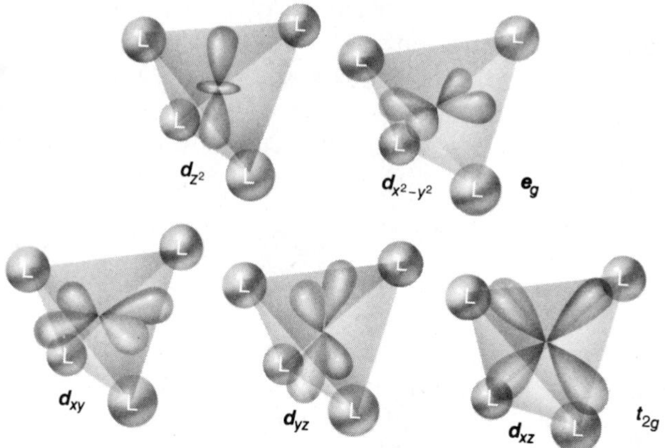

quently, a tetrahedral environment splits a set of *d* orbitals so that the e_g orbitals are lower in energy than the t_{2g} orbitals, just the reverse of the octahedral splitting.

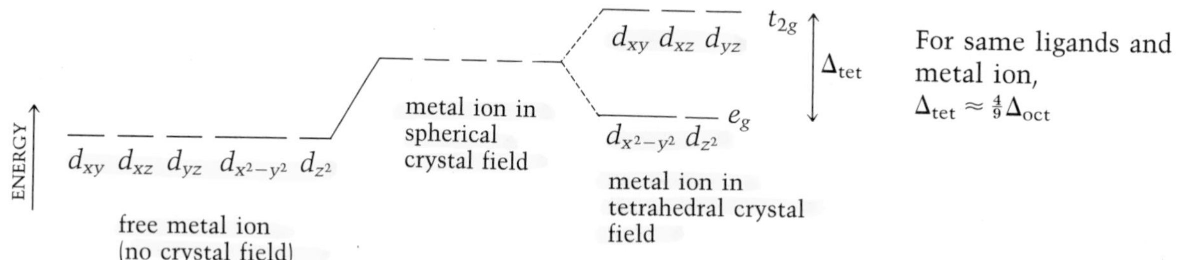

Because an octahedral splitting is caused by six ligands, while a tetrahedral splitting is caused by only four, the tetrahedral splitting caused by a given set of four ligands is smaller than the octahedral splitting caused by six of the same ligands. An approximate relationship is that $\Delta_{tet} \approx \frac{4}{9}\Delta_{oct}$. Since the tetrahedral splittings are so small, even strong field ligands like CN^- are unable to cause low spin tetrahedral complexes to form. That is, $\Delta_{tet} < P$. All tetrahedral complexes found to date are high spin.

Figure 27–6 shows the spatial relationship of a set of *d* orbitals to an environment of square planar ligands. Since the four ligands are located along the *x* and *y* axes, the $d_{x^2-y^2}$ orbital interacts most strongly with the ligand orbitals and the d_{z^2} the least strongly. Intermediate are the d_{xy}, d_{yz}, and d_{xz} orbitals; the one having both *x* and *y* components, the d_{xy} orbital, has the highest energy of the three. We may think of a square planar complex as derived from an octahedral complex by removal of the two ligands along the *z* axis. The square planar splitting of orbitals is depicted below.

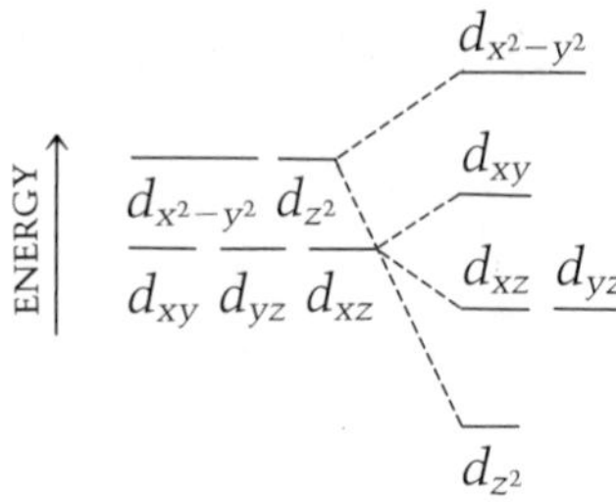

octahedral ⟶ square planar

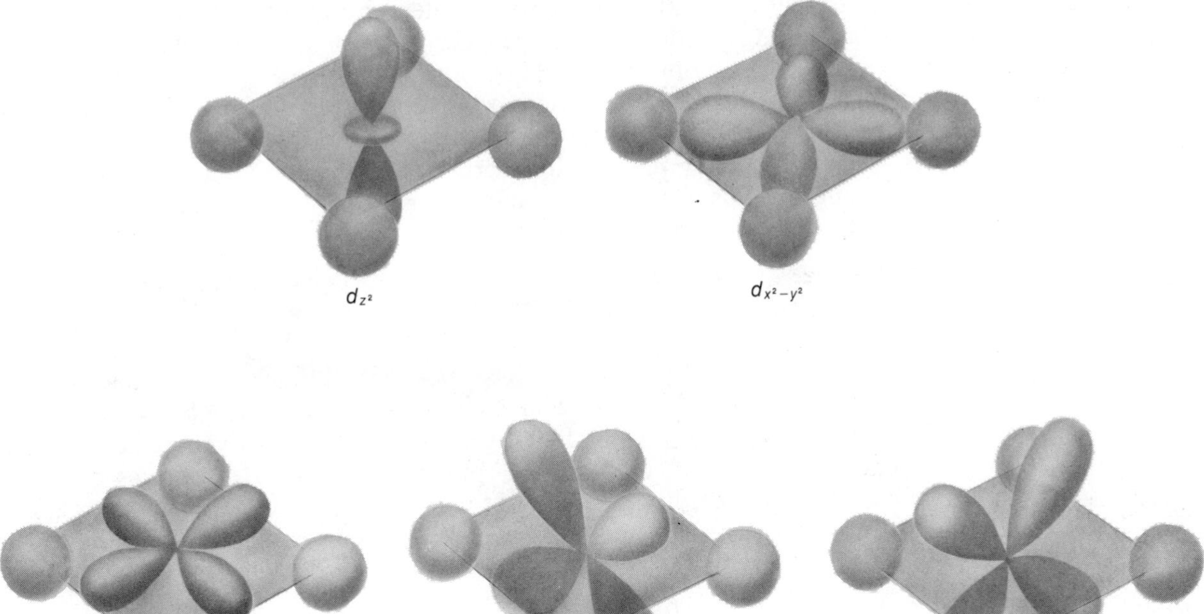

FIGURE 27–6 The five *d* orbitals in a square planar environment.

We would predict that square planar $[Cu(NH_3)_4]^{2+}$, a d^9 case, would have its only unpaired electron in the $d_{x^2-y^2}$ orbital, since it is highest in energy. Experimentally it is found that the ion is paramagnetic, with only one unpaired electron which is postulated to be in the $d_{x^2-y^2}$ orbital. Likewise $Ni(CN)_4^{2-}$, a d^8 case, is diamagnetic with a vacant $d_{x^2-y^2}$ orbital.

ENERGY ↑

↑ $d_{x^2-y^2}$
⇅ d_{xy}
d_{xz} ⇅ ⇅ d_{yz}
⇅ d_{z^2}

$[Cu(NH_3)_4]^{2+}$
square planar, d^9

ENERGY ↑

— $d_{x^2-y^2}$
⇅ d_{xy}
d_{xz} ⇅ ⇅ d_{yz}
⇅ d_{z^2}

$[Ni(CN)_4]^{2-}$
square planar, d^8

27–10 Color and the Spectrochemical Series

The colors of transition metal complexes arise from the absorption of visible light as electrons undergo transitions from one *d* orbital to another. (See Section 26–3.) One transition of a high spin octahedral Co(III) complex is depicted below.

Planck's constant is $h = 6.63 \times 10^{-34}$ J·s

Ground State		**Excited State**	**Energy of light absorbed**
e_g ↑ ↑		⇅ ↑	
t_{2g} ⇅ ↑ ↑	Absorption of light →	↑ ↑ ↑	$E = h\nu = \Delta_{oct}$

The frequency (ν), and therefore the wavelength and color, of the light absorbed is related to Δ_{oct}.* This, in turn, depends upon the crystal field strength of the ligands. So the colors and visible absorption spectra of transition metal complexes, as well as their magnetic properties, all yield information regarding the strengths of the ligand-metal interactions.

By interpreting the visible spectra of numerous complexes, it is possible to arrange ligands in the order of increasing crystal field strengths. Some common ligands are

$$I^- < Br^- < Cl^- < F^- < OH^- < \underrightarrow{H_2O < (COO)_2^{2-} < NH_3 < en < NO_2^-} < CN^-$$

increasing crystal field strength

Such an arrangement is called a **spectrochemical series.** Strong field ligands, such as CN^-, usually produce low spin complexes where possible, and high crystal field splittings. Weak field ligands, like Cl^-, usually produce high spin complexes and small crystal field splittings. Low spin complexes generally absorb higher energy (shorter wavelength) light than do high spin complexes. The colors of several six-coordinate Cr(III) complexes are listed in Table 27–8.

TABLE 27–8 Colors of Some Chromium(III) Complexes

$[Cr(OH_2)_4Br_2]Br$	green	$[Cr(CON_2H_4)_6][SiF_6]_3$	green
$[Cr(OH_2)_6]Br_3$	bluish-grey	$[Cr(NH_3)_5Cl]Cl_2$	purple
$[Cr(OH_2)_4Cl_2]Cl$	green	$[Cr(NH_3)_4Cl_2]Cl$	violet
$[Cr(OH_2)_6]Cl_3$	violet	$[Cr(NH_3)_6]Cl_3$	yellow

In $[Cr(NH_3)_6]Cl_3$, the Cr(III) is bonded to six strong-field ammonia ligands, which produce a relatively high value of Δ_{oct} and cause the $[Cr(NH_3)_6]^{3+}$ ion to absorb relatively high-energy visible light in the blue and violet regions. The color that we see when we look at a sample of the compound (either in solid form or in solution) is the light that is *not* absorbed. Thus, we see a yellowish color, which is the complementary color of blue.

We see the light that is transmitted (passes through the sample) or that is reflected by the sample.

Water is a weaker field ligand than ammonia, and therefore Δ_{oct} is less for $[Cr(OH_2)_6]^{3+}$ than for $[Cr(NH_3)_6]^{3+}$. As a result, $[Cr(OH_2)_6]Br_3$ absorbs lower energy (longer wavelength) light. This causes the reflected and transmitted light to be higher-energy bluish-gray, the color we ascribe to $[Cr(OH_2)_6]Br_3$.

27–11 Crystal Field Stabilization Energy

We have seen that transition metals have a strong tendency to form stable coordination complexes, particularly those with coordination number six. One of the major factors contributing to the stability of complex ions relative to free, or uncomplexed, ions is the **Crystal Field Stabilization Energy** (CFSE)

* The numerical relationship between Δ_{oct} and the wavelength, λ, of the absorbed light is found by combining the expressions $E = h\nu$ and $\nu = c/\lambda$, where c is the speed of light

$$\Delta_{oct} = EN_A = \frac{hcN_A}{\lambda} \quad \text{where } N_A \text{ is Avogadro's number}$$

associated with complexes containing metal ions having certain electronic configurations. Here we will consider the CFSE's of various types of octahedral complexes. For a given set of six ligands, the t_{2g} and e_g orbitals are split by a certain amount of energy, Δ_{oct}. Each of the t_{2g} orbitals is $\frac{2}{5}\Delta_{oct}$ *below* the energy of the set of degenerate *d* orbitals in a spherical (homogeneous) field, and each of the e_g orbitals is $\frac{3}{5}\Delta_{oct}$ *above* the energy of the unsplit *d* orbitals.

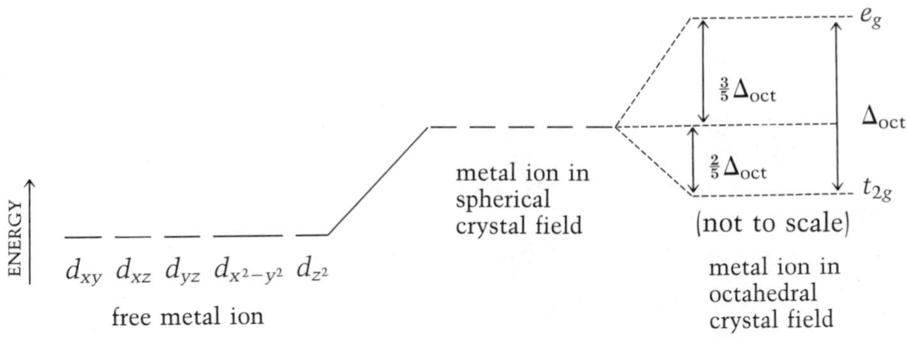

There are three t_{2g} orbitals and two e_g orbitals, and the total energy of the sets of t_{2g} and e_g orbitals is the same as the energy of the set of degenerate *d* orbitals in a spherical field. That is, the change in total energy of the system as a result of the approach of the six ligands along the *x*-, *y*-, and *z*-axes is zero.

$$\Delta E = \underset{(t_{2g})}{3(-\tfrac{2}{5}\Delta_{oct})} + \underset{(e_g)}{2(\tfrac{3}{5}\Delta_{oct})} = 0$$

But electrons occupying the lower energy (nonbonding) t_{2g} orbitals of an octahedral complex are lower in energy than they would be if they occupied orbitals of the metal ion in a spherical crystal field. Electrons occupying the (antibonding) e_g orbitals are higher in energy than if they occupied *d* orbitals of the metal ion in a spherical field. The lower the total energy of a system, the more stable it is. The CFSE of a complex is a measure of the net energy of stabilization (relative to the ion in a spherical field) of a metal ion's electrons as the complex forms.

Table 27–7 shows the electronic configurations of the common high spin and low spin octahedral complexes of *d*-transition metals. If an energy change of $-\frac{2}{5}\Delta_{oct}$ (stabilization) is assigned each t_{2g} electron and an energy change of $+\frac{3}{5}\Delta_{oct}$ (destabilization) is assigned each e_g electron, the CFSE is the sum of the energies of all the electrons in the metal atom or ion.

For example, for a high spin d^7 complex such as $[Co(OH_2)_6]^{2+}$, the CFSE is $5(-\frac{2}{5}\Delta_{oct}) + 2(+\frac{3}{5}\Delta_{oct}) = -\frac{4}{5}\Delta_{oct}$. That is, the *energy released* by the octahedral splitting of the occupied orbitals in formation of the complex ion is $\frac{4}{5}\Delta_{oct}$. The results of similar calculations for d^n configurations are given in Table 27–9. Those configurations with the most negative CFSE's are generally the ones for which large numbers of stable octahedral complexes are known. Note that no configuration can produce a CFSE greater than zero. That is, no *d*-transition metal ions should be less stable in an octahedral ligand environment than in a spherical crystal field.

Most magnetic and spectral properties of complex species of the *d*-transition elements can be accounted for by the Crystal Field theory, and this fact is

TABLE 27–9 Crystal Field Stabilization Energies for Octahedral d^n Complexes

d^n	High Spin	Low Spin	d^n	High Spin	Low Spin
d^0	0	same as high spin			
d^1	$-\frac{2}{5}\Delta_{oct}$	same as high spin	d^6	$-\frac{2}{5}\Delta_{oct}$	$-\frac{12}{5}\Delta_{oct}$
d^2	$-\frac{4}{5}\Delta_{oct}$	same as high spin	d^7	$-\frac{4}{5}\Delta_{oct}$	$-\frac{9}{5}\Delta_{oct}$
d^3	$-\frac{6}{5}\Delta_{oct}$	same as high spin	d^8	$-\frac{6}{5}\Delta_{oct}$	same as high spin
d^4	$-\frac{3}{5}\Delta_{oct}$	$-\frac{8}{5}\Delta_{oct}$	d^9	$-\frac{3}{5}\Delta_{oct}$	same as high spin
d^5	0	$-\frac{10}{5}\Delta_{oct}$	d^{10}	0	same as high spin

strong evidence for the general validity of the Crystal Field theory. Consider the heats (enthalpies) of hydration for the series of 2+ ions shown in Figure 27–7. All these ions form octahedral, weak field complexes with water $[M(OH_2)_6]^{2+}$. Recall that $\Delta H_{hydration}$ is the amount of energy absorbed when one mole of gaseous ions forms hydrated ions

$$M^{2+}(g) + 6H_2O \longrightarrow [M(OH_2)_6]^{2+}(aq)$$

The experimental values for Ca^{2+} (d^0), Mn^{2+} (d^5), and Zn^{2+} (d^{10}) show no CFSE because CFSE is zero for octahedral d^0, d^5, and d^{10} weak field (high spin) complexes. (See Table 27–9.)

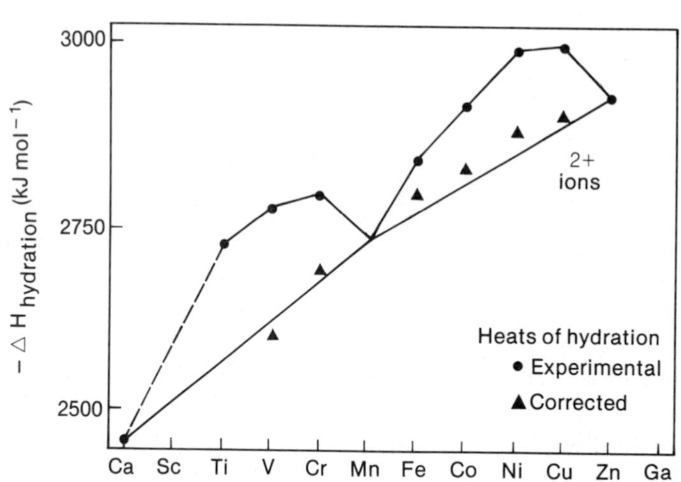

FIGURE 27–7 Heats of hydration of transition metal ions. When values of CFSE from spectroscopic Δ_{oct}'s are subtracted from experimental values of $\Delta H_{hydration}$ (●), a plot of the "corrected" values of $\Delta H_{hydration}$ (▲) versus atomic number is very nearly a straight line.

Key Terms

Ammine complexes complex species that contain ammonia molecules bonded to metal ions.

Chelate a ligand that utilizes two or more donor atoms in binding to metals.

***cis-trans* isomerism** a type of geometrical isomerism related to the angles between like ligands.

Coordinate covalent bond a covalent bond in which both shared electrons are donated by the same atom; a bond between a Lewis base and a Lewis acid.

Coordination compound or complex a compound containing coordinate covalent bonds.

Coordination isomers isomers involving exchange of ligands between complex cation and complex anion of the same compound.

Coordination number the number of donor atoms coordinated to a metal.

Coordination sphere the metal ion and its coordinated ligands but not any uncoordinated counter-ions.

Crystal Field Stabilization Energy a measure of the net energy of stabilization gained by a metal ion's nonbonding *d* electrons as result of complex formation.

Crystal Field theory theory of bonding in transition metal complexes in which ligands and metal ions are treated as point charges; a purely ionic model; ligand point charges represent the crystal (electrical) field perturbing the metal's *d* orbitals containing nonbonding electrons.

Δ_{oct} energy separation between e_g and t_{2g} sets of metal *d* orbitals caused by octahedral complexation of ligands; sometimes called $10Dq$.

Δ_{tet} energy separation between t_{2g} and e_g sets of metal *d* orbitals caused by tetrahedral complexation of ligands.

Dextrorotatory refers to an optically active substance that rotates the plane of plane polarized light clockwise; also called dextro.

Donor atom a ligand atom whose electrons are shared with a Lewis acid.

e_g orbitals set of $d_{x^2-y^2}$ and d_{z^2} orbitals; those *d* orbitals within a set with lobes directed along the *x*, *y*, and *z* axes.

Enantiomers optical isomers.

Geometrical isomers stereoisomers that are not mirror images of each other; also known as position isomers.

High spin complex crystal field designation for an outer orbital complex; no electrons are paired in either t_{2g} or e_g orbitals before all orbitals are singly occupied.

Hydrate isomers isomers of crystalline complexes that differ in whether water is present inside or outside the coordination sphere.

Inner orbital complex valence bond designation for a complex in which the metal ion utilizes *d* orbitals one shell inside the outermost occupied shell in its hybridization.

Ionization isomers isomers that result from interchange of ions inside and outside the coordination sphere.

Isomers different substances that have the same formula.

Levorotatory refers to an optically active substance that rotates the plane of plane polarized light counterclockwise; also called levo.

Ligand a Lewis base in a coordination compound.

Linkage isomers isomers in which a particular ligand bonds to a metal ion through different donor atoms.

Low spin complex crystal field designation for an inner orbital complex; contains electrons paired in t_{2g} orbitals before *eg* orbitals are occupied in octahedral complexes.

Optical activity the rotation of plane polarized light by one of a pair of optical isomers.

Optical isomers stereoisomers that differ only by being nonsuperimposable mirror images of each other, like left and right hands; also called enantiomers.

Outer orbital complex valance bond designation for a complex in which the metal ion utilizes *d* orbitals in the outermost (occupied) shell in hybridization.

Pairing energy energy required to pair two electrons in the same orbital.

Plane polarized light light waves in which all the electric vectors are oscillating in one plane.

Polarimeter a device used to measure optical activity.

Polydentate refers to ligands with more than one donor atom.

Racemic mixture an equimolar mixture of dextro and levo optical isomers that is, therefore, optically inactive.

Spectrochemical series arrangement of ligands in order of increasing ligand field strength.

Square planar complex complex in which the metal is in the center of a square plane, with ligand donor atoms at each of the four corners.

Stereoisomers isomers that differ only in the way that atoms are oriented in space; consist of geometrical and optical isomers.

Strong field ligand ligand that exerts a strong crystal or ligand electrical field and generally forms low spin complexes with metals when possible.

Structural isomers applied to coordination compounds; isomers whose differences involve more than a single coordination sphere or else different donor atoms; include ionization isomers, hydrate isomers, coordination isomers, and linkage isomers.

t_{2g} orbitals set of d_{xy}, d_{yz}, and d_{xz} orbitals; those d orbitals within a set with lobes bisecting the x, y, and z axes.

Weak field ligand ligand that exerts a weak crystal or ligand field and generally forms high spin complexes with metals.

Exercises

Basic Concepts

1. What property of transition metals allows them to form coordination compounds easily?
2. Suggest more appropriate designations for $NiSO_4 \cdot 6H_2O$, $Cu(NO_3)_2 \cdot 4NH_3$, and $Ni(NO_3)_2 \cdot 6NH_3$.
3. Describe the experiments of Alfred Werner on the compounds of the general formula $PtCl_4 \cdot nNH_3$ where $n = 2, 3, 4, 5, 6$.
4. For each of the compounds of Exercise 3, write formulas indicating the species within the coordination sphere. Also indicate the charges on the complex ions and provide a name for each compound.
5. Distinguish among the terms ligands, donor atoms, and chelates.

Naming Coordination Compounds

6. Give systematic names for the following compounds:
   (a) $[Ni(CO)_4]$,
   (b) $Na_2[Co(OH_2)_2(OH)_4]$,
   (c) $[Ag(NH_3)_2]Br$,
   (d) $[Cr(en)_3](NO_3)_3$,
   (e) $[Pt(NH_3)_4(NO_2)_2]\ (NO_3)_2$,
   (f) $K_2[Cu(CN)_4]$.
7. Name the following compounds systematically:
   (a) $K_4[NiF_6]$,
   (b) $[Mo(NCS)_2(en)_2]ClO_4$,
   (c) $[Mn(NH_3)_2(OH_2)_3(OH)]SO_4$,
   (d) $[Co(en)_2Cl_2]I$,
   (e) $(NH_4)_3[CuCl_5]$,
   (f) $[Co(NH_3)_5SO_4]NO_2$.
8. Write formulas for the following compounds:
   (a) sodium tetracyanocadmate
   (b) hexaamminecobalt(III) chloride
   (c) diaquadicyanocopper(II)
   (d) potassium hexachloropalladate(IV)
   (e) *cis*-diaquabis(ethylenediamine)cobalt(III) perchlorate
   (f) ammonium *trans*-dibromobis(oxalato)-chromate(III)
9. Write formulas for the following compounds:
   (a) *trans*-diamminedinitroplatinum(II)
   (b) rubidium tetracyanozincate
   (c) triaqua-*cis*-dibromochlorochromium(III)
   (d) pentacarbonyliron(0)
   (e) sodium pentacyanocobaltate(II)
   (f) hexammineruthenium(III) tetrachloronickelate(II)

Ammine Complexes

10. Which of the following insoluble metal hydroxides will dissolve in an excess of aqueous ammonia? (a) $Zn(OH)_2$, (b) $Cr(OH)_3$, (c) $Fe(OH)_2$, (d) $Ni(OH)_2$, (e) $Cd(OH)_2$.
11. Write net ionic equations for the reactions of Exercise 10 that occur.
12. Write net ionic equations for reactions of solutions of the following transition metal salts in water with a *limited amount* of aqueous ammonia: (It is not necessary to show the ions as hydrated.) (a) $CuCl_2$, (b) $Zn(NO_3)_2$, (c) $Fe(NO_3)_3$, (d) $Hg(NO_3)_2$, (e) $MnCl_3$.
13. Write *net ionic* equations for the reactions of the insoluble products of Exercise 12 with an *excess* of aqueous ammonia, if such a reaction occurs.

Structures of Coordination Compounds

14. Write formulas and provide names for three complex cations in each of the following categories:
    (a) cations coordinated to only unidentate ligands
    (b) cations coordinated to only bidentate ligands
    (c) cations coordinated to two bidentate and two unidentate ligands
    (d) cations coordinated to one tridentate ligand, one bidentate ligand, and one unidentate ligand
    (e) cations coordinated to one tridentate ligand and three unidentate ligands

15. Provide formulas and names for three complex anions fitting each of the descriptions given in Exercise 14.

Isomerism in Coordination Compounds

16. Distinguish between structural isomers and stereoisomers.

17. Distinguish between an optically active complex and a racemic mixture.

18. Write the formula for a potential ionization isomer of each of the following compounds. Name each one.
    (a) $[Cr(NH_3)_4I_2]Br$,
    (b) $[Ni(en)_2(NO_2)_2]Cl_2$,
    (c) $[Fe(NH_3)_5CN]SO_4$.

19. Write the formula for a potential hydrate isomer of each of the following compounds. Name each one. (a) $[Cu(OH_2)_4]Cl_2$ and (b) $[Ni(OH_2)_5Br]Br \cdot H_2O$.

20. Write the formula for a potential coordination isomer of each of the following compounds. Name each one.
    (a) $[Co(NH_3)_6][Cr(CN)_6]$ and
    (b) $[Ni(en)_3][Cu(CN)_4]$

21. Write the formula for a potential linkage isomer of each of the following compounds. Name each one. (a) $[Co(en)_2(NO_2)_2]Cl$ and (b) $[Cr(NH_3)_5(CN)](CN)_2$.

22. How many geometrical isomers (or forms) can exist in theory for the following species?
    (a) $[Pt(NH_3)_2Cl_2]^-$ (square planar)
    (b) $[Pt(NH_3)Cl_3]^-$ (square planar)
    (c) $[Zn(NH_3)_2Cl_2]$ (tetrahedral)
    Sketch each form and name each one.

23. The ion $[Co(en)_2Cl_2]^+$ exists as two geometrical isomers, one of which occurs as two optical isomers. Sketch and name all three geometrical forms. Label each according to the above description.

24. How many *geometrical* and *optical* isomers can exist in theory for the following species?
    (a) $[Co(NH_3)_2Cl_4]^-$
    (b) $[Co(NH_3)_3Cl_3]$
    (c) $[Co(NH_3)(en)Cl_3]$
    Sketch and name them all. (The N atoms of en can bond only at adjacent positions.)

25. How many *geometrical* and *optical* isomers can exist, in theory, for the following species?
    (a) $[CoCl_2Br_2I_2]^{4-}$
    (b) $[Cr(ox)_3]^{3-}$, in which ox = oxalato,

    $$\left[\begin{array}{c} O \qquad\quad O \\ \;\; C-C \\ {}^-O \qquad\quad O^- \end{array}\right].$$

    Represent ox as $[O\frown O^{2-}]$.

Valence Bond Theory

26. Give the hybridizations and sketch the structures of each of the complexes of Exercises 6 and 8. It is not necessary to distinguish between d^2sp^3 and sp^3d^2 hybridization. $[Ni(CO)_4]$ and $[Cd(CN)_4]^2$ are tetrahedral; the other four-coordinate complexes are square planar.

27. Repeat Exercise 26 for the complexes of Exercises 7 and 9. $[NiCl_4]^{2-}$ and $[Zn(CN)_4]^{2-}$ are tetrahedral; $[Pt(NH_3)_2(NO_2)_2]$ is square planar.

28. Why are the d_{z^2} and $d_{x^2-y^2}$ orbitals utilized in d^2sp^3 and sp^3d^2 hybridization rather than two of the d_{xy}, d_{yz}, and d_{xz} orbitals?

29. On the basis of the spectrochemical series, determine whether the following complexes are inner orbital or outer orbital, and diamagnetic or paramagnetic:
    (a) $[Cu(OH_2)_6]^{2+}$, (e) $[CrCl_4Br_2]^{3-}$,
    (b) $[MnF_6]^{3-}$, (f) $[Co(en)_3]^{3+}$,
    (c) $[Co(CN)_6]^{3-}$, (g) $[Fe(OH)_6]^{3-}$,
    (d) $[Cr(NH_3)_6]^{3+}$, (h) $[Fe(NO_2)_6]^{3-}$.

30. Draw diagrams showing outer electron configurations and hybridizations at the metal ions for each of the complexes in Exercise 29.
31. Draw diagrams showing outer electronic configurations and hybridizations for the following four-coordinate complexes:
(a) $[Cd(NH_3)_4]^{2+}$, tetrahedral,
(b) $[Cu(NH_3)_4]^{2+}$, square planar,
(c) $[Zn(CN)_4]^{2-}$, tetrahedral,
(d) $[Pt(NH_3)_4]^{2+}$, square planar.

Crystal Field Theory

32. Describe clearly what Δ is. Why is Δ for a tetrahedral ligand field always less than Δ for an octahedral field? How is Δ actually measured experimentally? How is it related to the spectrochemical series?
33. Δ_{oct} has been referred to as 10 Dq in the past. Suggest a reason for calling Δ_{oct} 10 units Dq, or anything else, for that matter.
34. On the basis of the spectrochemical series, determine whether the complexes of Exercise 29 are low spin or high spin complexes.
35. Write out the electron distribution in t_{2g} and e_g orbitals for the following in an octahedral field.

Metal ions	Ligand Field Strength
V^{2+}	weak
Mn^{2+}	strong
Mn^{2+}	weak
Ni^{2+}	weak
Cu^{2+}	weak
Fe^{3+}	strong
Cu^{+}	weak
Ru^{3+}	strong

36. Write formulas for two complex ions that would fit into each of the categories of Exercise 35. Name the complex ions you list.
37. Describe the relationship among Δ, the electron pairing energy, and whether or not a complex is high spin or low spin. Illustrate the relationship with Fe^{2+} in strong and weak octahedral fields.
38. What is crystal field stabilization energy? Given the following spectrophotometrically measured values of Δ_{oct}, calculate the CFSE's for the ions. Remember to determine first whether the complex is high spin or low spin.

Complex ion	Δ_{oct}*
(a) $[Co(NH_3)_6]^{3+}$	22,900 cm^{-1}
(b) $[Ti(OH_2)_6]^{2+}$	20,300 cm^{-1}
(c) $[Cr(OH_2)_6]^{3+}$	17,600 cm^{-1}
(d) $[Co(CN)_6]^{3-}$	33,500 cm^{-1}
(e) $[Co(OH_2)_6]^{2+}$	10,000 cm^{-1}
(f) $[Cr(en)_3]^{3+}$	21,900 cm^{-1}
(g) $[Cu(OH_2)_6]^{2+}$	13,000 cm^{-1}
(h) $[V(OH_2)_6]^{3+}$	18,000 cm^{-1}

*The cm^{-1} is an energy unit; 1 cm^{-1} = 11.96 J/mol.

Equilibria Involving Coordination Compounds

39. Calculate (a) the molar solubility of $Zn(OH)_2$ in pure water, (b) the molar solubility of $Zn(OH)_2$ in 0.10 *M* NaOH solution, and (c) the concentration of $[Zn(OH)_4]^{2-}$ ions in the solution of (b). K_{sp} for $Zn(OH)_2 = 4.5 \times 10^{-17}$ and K_d for $[Zn(OH)_4]^{2-} = 3.5 \times 10^{-16}$.
40. Calculate the pH of a solution prepared by dissolving 0.10 mol of tetramminecopper(II) chloride, $[Cu(NH_3)_4]Cl_2$, in 1.0 L of solution. K_d for $[Cu(NH_3)_4]^{2+} = 8.5 \times 10^{-13}$; K_b for aqueous $NH_3 = 1.8 \times 10^{-5}$. Ignore hydrolysis of Cu^{2+}.
41. Would a solution of 0.10 mol of tetramminecopper(II) sulfate, $[Cu(NH_3)_4]SO_4$, in 1.0 L of aqueous solution have higher, lower, or the same pH as the solution of Exercise 40? Why? (No calculation is necessary.)

28 Nuclear Chemistry

Chemical properties are determined by electronic distributions and are only indirectly influenced by atomic nuclei. In previous chapters we have been concerned with ordinary chemical reactions and have focused attention on electronic configurations. Nuclear reactions involve changes in the composition of nuclei, and are extraordinary in the sense that they are often accompanied by the release of tremendous amounts of energy and by transmutations of elements. Some of the characteristics that differentiate between nuclear reactions and ordinary chemical reactions are summarized below.

Nuclear Reaction	Ordinary Chemical Reaction
1. Elements may be converted from one to another.	1. No new elements can be produced.

2. Particles within the nucleus are involved.	2. Usually only outermost electrons participate.
3. Often accompanied by release or absorption of tremendous amounts of energy.	3. Accompanied by release or absorption of relatively small amounts of energy.
4. Rate of reaction is independent of factors such as concentration, temperature, catalyst, and pressure.	4. Rate of reaction is influenced by external factors.

Many ancient alchemists spent their lives attempting to convert other heavy metals into gold without success. Years of failure and the acceptance of Dalton's atomic theory early in the nineteenth century convinced the scientific community that elements are not interconvertible. However, Henri Becquerel's discovery of "radioactive rays" emanating from a uranium compound in 1896, and the subsequent determination of the nature of radioactive rays by Ernest Rutherford, showed that atoms of one element may indeed be converted into atoms of other elements by spontaneous nuclear disintegrations, or **natural radioactivity.** (As shown many years later, deliberate nuclear reactions initiated by bombardment of nuclei with accelerated subatomic particles or other nuclei can also transform one element into another.)

The stimulus of Becquerel's discovery led other researchers, including Marie and Pierre Curie, to discover and study new radioactive elements. Many radioactive isotopes, or **radioisotopes,** now have important medical, agricultural, and industrial uses.

Nuclear fission is the splitting of a heavy nucleus into lighter nuclei. **Nuclear fusion** is the combination of light nuclei to produce a heavier nucleus. The huge amounts of energy released per unit mass of nuclear fuels give controlled nuclear fission (Sections 28–12, 28–14) and nuclear fusion (Section 28–15) great promise for supplying a large portion of our future energy demands. Research is currently aimed at surmounting some of the technological problems associated with safe and efficient use of nuclear fission reactors and with the development of continuously controlled fusion reactors.

28–1 The Nucleus

If enough nuclei were gathered together to occupy one cubic centimeter, the total weight would be about 250 million tons!

The nucleus comprises only a minute fraction of the total volume of an atom, yet nearly all the mass of an atom resides in the nucleus. Thus nuclei are extremely dense. It has been experimentally shown that nuclei of all elements have approximately the same density, 2.44×10^{14} g/cm^3.

From an electrostatic point of view it is amazing that positively charged protons (and uncharged neutrons) can be packed so closely together. Yet nonradioactive nuclei do not spontaneously decompose, so they must be stable. In the early twentieth century, when Rutherford postulated the nuclear model of the atom, scientists were puzzled by such a situation. In attempts to explain the phenomenon, physicists have since detected many very short-lived subatomic particles (in addition to protons, neutrons, and electrons) as products of nuclear reactions. Well over a hundred have been identified, and several more are found every year. Their functions are not entirely understood, but it is now thought that they play important roles in overcoming the coulombic proton–

The elucidation of nuclear structure has become a fascinating and very complex problem.

TABLE 28–1 Abundance of Naturally Occurring Nuclides

Protons	even	even	odd	odd
Neutrons	even	odd	even	odd
Number of nuclides	157	52	50	5

proton repulsions and in binding nuclear particles **(nucleons)** together. The attractive forces among nucleons appear to operate over only extremely small distances, about 10^{-13} cm.

28–2 Neutron-Proton Ratio and Nuclear Stability

Most naturally occurring **nuclides** have even numbers of protons and even numbers of neutrons; 157 nuclides fall into this category. Nuclides with odd numbers of both are least common (there are only five), and those with odd-even combinations are intermediate in abundance (Table 28–1).

The term "isotope" applies only to different forms of the same element. The term "nuclide" is used to refer to different atomic forms of all elements.

In addition to the seeming even-even preference, there seem to be certain "magic numbers" of protons and neutrons that impart exceptional stability to nuclides, reminiscent of the 2, 8, 18, and 32 electrons that result in very stable filled electronic shells. Nuclides with a number of protons *or* a number of neutrons *or* a sum of these two equal to 2, 8, 20, 28, 50, 82, and 126 have unusual stability. Examples are $^{4}_{2}He$, $^{16}_{8}O$, $^{42}_{20}Ca$, $^{88}_{38}Sr$, and $^{208}_{82}Pb$. This suggests a shell model for the nucleus similar to the shell model of electronic configurations.

Figure 28–1 shows a plot of the number of neutrons (*N*) versus number of protons (*Z*) for the stable nuclides. For low atomic numbers, below about 20,

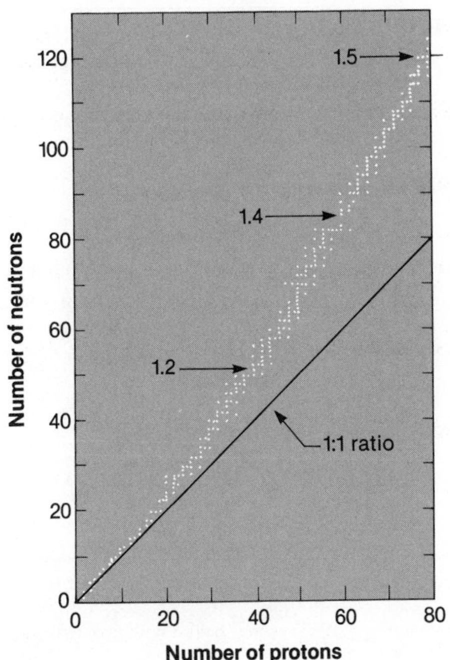

FIGURE 28–1 A plot of the number of neutrons versus the number of protons in stable nuclei. As the atomic number increases, the neutron-to-proton ratio of the stable nuclei increases. The stable nuclei are located in an area of the graph known as the band of stability. The majority of radioactive nuclei occur outside this band.

the most stable nuclides have equal numbers of protons and neutrons ($N = Z$). Above atomic number 20 the most stable nuclides have more neutrons than protons. Each dot in the band represents a stable nuclide. Careful examination reveals a vague stepwise shape associated with the plot, owing to the stability of nuclides with even numbers of nucleons.

28–3 Radioactive Decay

Nuclei whose neutron-to-proton ratios lie outside the stable region undergo spontaneous radioactive decay by emitting one or more particles and/or electromagnetic rays. The type of decay that occurs usually depends upon whether the nucleus is above, below, or to the right of the band of stability. The most common types of radiation emitted in decay processes and their properties are summarized in Table 28–2.

The particles can be emitted at different kinetic energies equal to the energy equivalent of the mass loss of the products relative to reactants (see discussion of mass defect and binding energy in Section 3–8) minus the energy associated with subsequently emitted gamma rays. Frequently radioactive decay leaves a nucleus in an excited (high energy) state. In such cases the decay is accompanied by subsequent gamma-ray emission in which the energy of the gamma ray ($h\nu$) is equal to the energy difference between the ground and excited nuclear states. This is analogous to the emissions of lower energy electromagnetic radiation that occur as atoms in excited electronic states return to their ground states (Section 3–11). Determinations of the energies of gamma

TABLE 28–2 Common Types of Radioactive Emissions

Type and Symbol*	Identity	Relative Mass (amu)	Charge	Velocity	Penetration
Beta (β^-, ${}_{-1}^{0}\beta$, ${}_{-1}^{0}e$)	electron	0.00055	1−	≤90% of speed of light	low to moderate, depending on energy
Positron† (${}_{+1}^{0}\beta$, ${}_{+1}^{0}e$)	positively charged electron	0.00055	1+	≤90% of speed of light	low to moderate, depending on energy
Alpha (α, ${}_{2}^{4}\alpha$, ${}_{2}^{4}He$)	helium nucleus	4.0026	2+	≤10% of speed of light	low
Proton (${}_{1}^{1}p$, ${}_{1}^{1}H$)	proton, hydrogen nucleus	1.0073	1+	≤10% of speed of light	low to moderate, depending on energy
Neutron (${}_{0}^{1}n$)	neutron	1.0087	0	≤10% of speed of light	very high
Gamma (${}_{0}^{0}\gamma$) ray	high-energy electro-magnetic radiation like x-rays	0	0	speed of light	high

* The number at the upper left of the symbol is the number of nucleons, and the number at the lower left is the number of positive charges.

† On the average, a positron exists for only about a nanosecond (1×10^{-9} second) before colliding with an electron and being converted into the corresponding amount of energy.

radiation from decay processes strongly suggest that nuclear energy levels are quantized just as are electronic energy levels, adding further support for a shell model for the nucleus.

$$ {}^{M}_{Z}E^{\star} \longrightarrow {}^{M}_{Z}E + {}^{0}_{0}\gamma $$

excited nucleus

Recall that the energy of electromagnetic radiation is $E = h\nu$, where h is Planck's constant and ν is the frequency.

The penetrating abilities of the particles or rays are proportional to their energies. Beta particles and positrons are about 100 times more penetrating than the heavier and slower-moving alpha particles. They can be stopped by a one-eighth inch (0.3 cm) thick aluminum plate. They can burn skin severely but cannot reach internal organs. Alpha particles have low penetrating ability and cannot damage or penetrate skin. However, they can damage sensitive internal tissue if inhaled. The high-energy gamma rays have great penetrating power and severely damage both skin and internal organs. They travel at the speed of light and can be stopped by only thick layers of concrete or lead.

28–4 Nuclei Above the Band of Stability

Nuclei in this region have too high a ratio of neutrons to protons, and they undergo decays that decrease the ratio. The most common methods of accomplishing this are **beta emission** or, less commonly, **neutron emission.** A beta particle is really an electron ejected from the nucleus as a neutron is converted into a proton

$$ {}^{1}_{0}n \longrightarrow {}^{1}_{1}p + {}^{0}_{-1}\beta $$

Beta emission results in a simultaneous increase in the number of protons and decrease in the number of neutrons by one. Examples of beta particle emission are

$$ {}^{228}_{88}\mathrm{Ra} \longrightarrow {}^{228}_{89}\mathrm{Ac} + {}^{0}_{-1}\beta $$

$$ {}^{234}_{90}\mathrm{Th} \longrightarrow {}^{234}_{91}\mathrm{Pa} + {}^{0}_{-1}\beta $$

$$ {}^{14}_{6}\mathrm{C} \longrightarrow {}^{14}_{7}\mathrm{N} + {}^{0}_{-1}\beta $$

Neutron emissions simply decrease the number of neutrons by one, without changing the atomic number. A lighter isotope is formed.

$$ {}^{137}_{53}\mathrm{I} \longrightarrow {}^{136}_{53}\mathrm{I} + {}^{1}_{0}n $$

$$ {}^{17}_{7}\mathrm{N} \longrightarrow {}^{16}_{7}\mathrm{N} + {}^{1}_{0}n $$

Note that in all equations for nuclear reactions, *the sums of the mass numbers and atomic numbers* (positive nuclear charges) *of the reactants equal the same sums for the products.*

28–5 Nuclei Below the Band of Stability

These nuclei can increase their neutron-to-proton ratios by undergoing **positron emission,** or **electron capture** (K capture).

A positron has the mass of an electron but a positive charge. Positrons are emitted as protons are converted to neutrons

$$ {}^{1}_{1}p \longrightarrow {}^{1}_{0}n + {}^{0}_{+1}\beta $$

By transforming a proton into a neutron, positron emission results in a *decrease* in atomic number and an increase in the number of neutrons by one, with no change in mass number.

$$^{38}_{19}K \longrightarrow {}^{38}_{18}Ar + {}^{0}_{+1}\beta$$

$$^{15}_{8}O \longrightarrow {}^{15}_{7}N + {}^{0}_{+1}\beta$$

The same effect can be accomplished by electron capture (K capture), in which an electron from the K shell ($n = 1$) is captured by the nucleus.

$$^{106}_{47}Ag + {}^{0}_{-1}e \longrightarrow {}^{106}_{46}Pd$$

$$^{37}_{18}Ar + {}^{0}_{-1}e \longrightarrow {}^{37}_{17}Cl$$

Some nuclides undergo both electron capture and positron emission. Sodium-22 is an example.

$$^{22}_{11}Na + {}^{0}_{-1}e \longrightarrow {}^{22}_{10}Ne \qquad (3\%)$$

$$^{22}_{11}Na \longrightarrow {}^{22}_{10}Ne + {}^{0}_{+1}\beta \qquad (97\%)$$

In addition some nuclei, especially heavier ones, undergo **alpha emission.** Alpha particles are actually helium nuclei, $^{4}_{2}He$, two protons and two neutrons. They carry a double positive charge, but charge is usually not shown on atomic species in nuclear reactions. An example is the alpha emission of lead-204. Of course, this also results in an increase of the neutron-to-proton ratio.

$$^{204}_{82}Pb \longrightarrow {}^{200}_{80}Hg + {}^{4}_{2}\alpha$$

28–6 Nuclei with Atomic Number Greater than 82

All nuclides with atomic number greater than 82 are radioactive and are located beyond the band of stability. Many such nuclides decay by alpha emission.

$$^{226}_{88}Ra \longrightarrow {}^{222}_{86}Rn + {}^{4}_{2}\alpha$$

$$^{210}_{84}Po \longrightarrow {}^{206}_{82}Pb + {}^{4}_{2}\alpha$$

The first of the decays shown, that of radium-226, was originally reported in 1902 by Rutherford and Soddy and was the first transmutation of an element ever observed. Heavy nuclides also decay less commonly by beta emission, positron emission, and electron capture.

Some of the isotopes of uranium ($Z = 92$) and elements of higher atomic number, the **transuranium elements,** also disintegrate via spontaneous nuclear fission, in which a heavy nuclide splits into nuclides of intermediate masses and neutrons.

$$^{252}_{98}Cf \longrightarrow {}^{142}_{56}Ba + {}^{106}_{42}Mo + 4\,{}^{1}_{0}n$$

28–7 Detection of Radiations

1 Photographic Detection

Emanations from radioactive substances affect photographic plates just as ordinary visible light does. Becquerel's discovery of radioactivity resulted from the unexpected exposure of a photographic plate wrapped in black paper by a nearby enclosed sample of a uranium-containing compound, potassium uranyl sulfate. After developing and fixing, the intensity of the spot is related to the amount of radiation that struck the plate, but detection of radiation on a quantitative basis by this method is difficult and tedious.

2 Detection by Fluorescence

This method is based upon the ability of **fluorescent** substances to absorb high-energy radiation such as gamma rays and subsequently emit visible light. As the gamma radiation is absorbed, the absorbing atoms jump to excited electronic states. In returning to their ground states, the excited electrons go through a series of transitions, at least some of which result in the emission of light of visible wavelengths. This method may be used for the quantitative detection of radiation, using an instrument called a **scintillation counter.**

The phenomenon of fluorescence can be utilized for other purposes. A familiar application is in fluorescent clock dials. A typical dial is coated with fluorescent zinc sulfide, ZnS, containing about one part per 100,000 of radioactive emitter, such as tritium. The constantly decaying tritium excites the ZnS so that it glows continuously.

3 Cloud Chambers

The original cloud chamber was devised by C. T. R. Wilson in 1911. The chamber contains air saturated with water vapor. The withdrawal of a piston expands and cools the air, causing droplets of water vapor to condense on particles emitted from a radioactive substance. The paths of the particles, which in some ways reflect their properties, can be followed by observing the fog-like tracks produced. The tracks may be photographed and studied in detail. Figures 28–2 and 28–3 show a cloud chamber and a cloud chamber photograph.

Vapors other than water can be used to saturate the air.

4 Gas Ionization Counters

A common gas ionization counter is the **Geiger-Müller** counter, which is depicted in Figure 28–4. Radiation enters the tube through a thin window. Windows of different stopping powers can be used to let only radiation of certain penetrating powers pass through. The tube is filled with a gas, usually argon, some molecules of which become ionized when struck by the radiation. A potential difference of about 1200 volts is applied between the inner wire and the tube shell, which causes a current of electrons to flow toward the (+) wire and Ar^+ ions to flow toward the (−) tube shell. Amplification of the pulse from each ionization causes a clicking sound in a loudspeaker, or activates a digital counter that registers the amount of radiation entering the tube. A chart recorder may also be attached to the circuit to give a graphic display of the intensity of radiation versus time.

28–8 Rates of Decay and Half-Life

The rates of all radioactive decays are independent of temperature and obey first-order kinetics. All radionuclides have different stabilities and decay at different rates. Some samples decay nearly completely in fractions of a second and others only after millions of years or more. In Section 14–8 we found that *first-order* processes occur at rates proportional only to the concentration of one reacting substance.

$$\text{Rate of decay} = k[A]$$

$$\log\left(\frac{A_0}{A}\right) = \frac{kt}{2.303}$$

The 2.303 is a conversion factor from natural to base-10 logarithms.

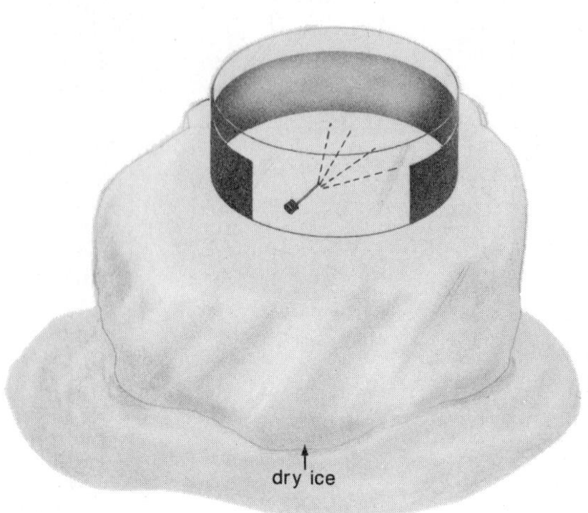

FIGURE 28–2 A cloud chamber. The emitter is glued onto a pin stuck into a stopper that is mounted onto the chamber wall. The chamber has some methyl alcohol in the bottom and rests on dry ice. The cool air near the bottom becomes supersaturated with methyl alcohol vapor. When an emission speeds through this supersaturated vapor, ions are produced that serve as "seeds" about which the vapor condenses, forming tiny droplets, or fog.

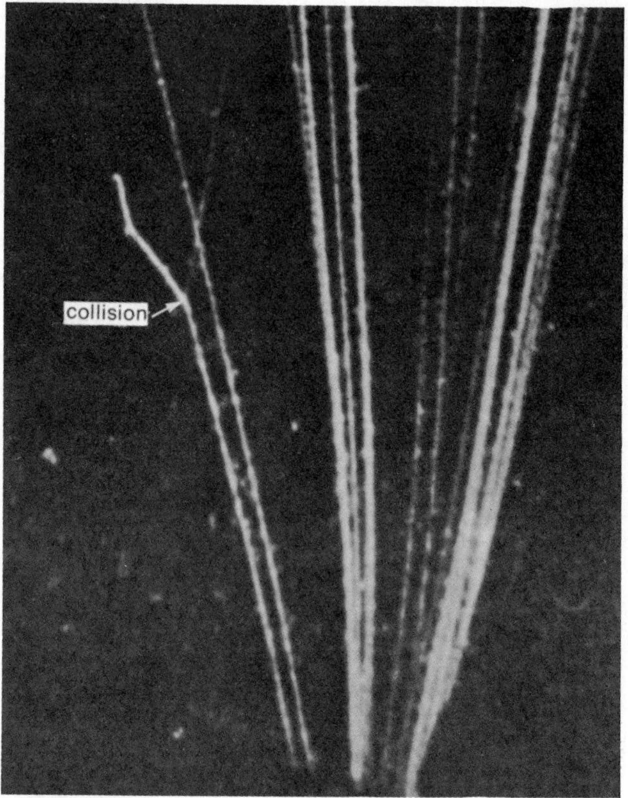

FIGURE 28–3 A historic cloud chamber photograph of alpha tracks in nitrogen gas. The forked track was shown to be due to a speeding proton (going off to the left) and an isotope of oxygen (going off to the right). It is assumed that the alpha particle struck the nucleus of a nitrogen atom at the point where the track forks.

Here A represents the amount of decaying radionuclide of interest remaining at some time t, and A_0 is the amount of the nuclide present at the beginning of the decay process. The k is the specific rate constant, which is different for each radioactive nuclide. If N represents the number of disintegrations per unit time, a similar relationship holds

$$\log\left(\frac{N_0}{N}\right) = \frac{kt}{2.303}$$

We also found that the half-life, $t_{1/2}$, for a first order reaction, or the amount of time required for half the original reacting (decaying) sample to react (time after which $A = \frac{1}{2}A_0$ or $N = \frac{1}{2}N_0$), is given by the equation

$$t_{1/2} = \frac{0.693}{k}$$

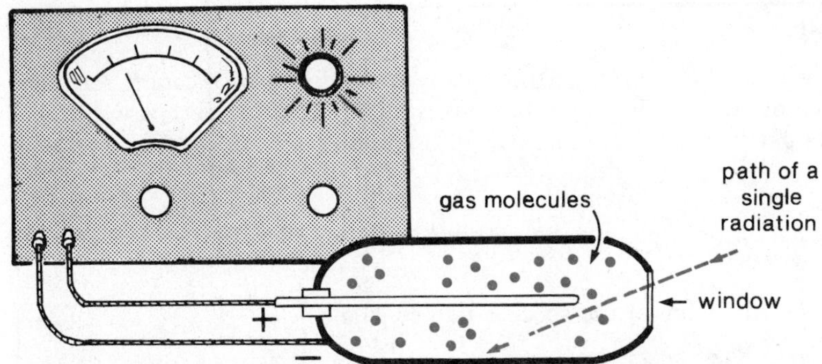

FIGURE 28–4 Principle of operation of a gas ionization counter. The center wire is positively charged, and the shell of the tube is negatively charged. When radiation enters through the window, it ionizes one or more gas atoms. The electrons are attracted to the central wire, and the positive ions are drawn to the shell. This constitutes a pulse of electric current, which is amplified and displayed on the meter or other readout.

Strontium-90 is an isotope that was introduced into the atmosphere by the atmospheric testing of nuclear weapons. Because of the chemical similarity of strontium to calcium, it now occurs with calcium in measurable quantities in milk, bones, and teeth as a result of its presence in food and water supplies. It is a radionuclide that undergoes beta emission with a half-life of 28 years, and it may cause leukemia, bone cancer, or other related disorders. If we begin with a 16 μg sample of $^{90}_{38}Sr$, 8 μg will remain after one half-life of 28 years. After 56 years 4 μg remain; after 84 years, 2 μg; after 112 years 1 μg; after 140 years 0.5 μg remain, and so on. The situation is depicted graphically in Figure 28–5.

In 1963 a treaty was signed by the United States, the Soviet Union, and the United Kingdom prohibiting the further testing of nuclear weapons in the atmosphere. Since then, strontium-90 has been disappearing from the air, water, and soil according to the curve in Figure 28–5, so the treaty has accomplished its aim to the present time (1983).

Similar plots for other radionuclides all show the same shape of **exponential decay curve.** Approximately ten half-lives must pass for radionuclides to lose 99.9% of their radioactivity, or 280 years for $^{90}_{38}Sr$.

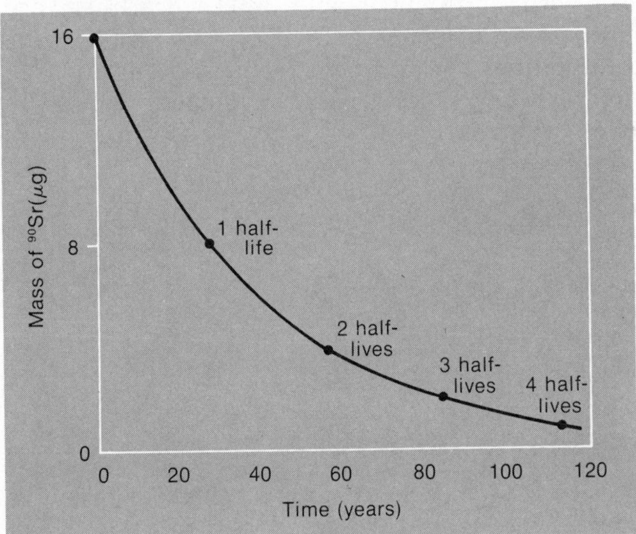

FIGURE 28–5 The decay of a 16 μg sample of $^{90}_{38}Sr$.

Gamma rays will destroy both cancerous and normal cells, so the beams of gamma rays must be directed as nearly as possible toward only cancerous tissue and concentrated there.

EXAMPLE 28–1

The "cobalt treatments" used in medicine to arrest certain types of cancer rely on the ability of gamma rays to destroy cancerous tissues. Cobalt-60 decays with the emission of beta particles and gamma rays, with a half-life of 5.27 years.

$${}^{60}_{27}Co \longrightarrow {}^{60}_{28}Ni + {}^{0}_{-1}\beta + {}^{0}_{0}\gamma$$

How much of a 3.42 μg sample of cobalt-60 remains after 30 years?

Solution

We first determine the value of the specific rate constant.

$$t_{1/2} = \frac{0.693}{k}$$

$$k = \frac{0.693}{t_{1/2}} = \frac{0.693}{5.27 \text{ yr}} = 0.131 \text{ yr}^{-1}$$

This value can now be used to determine the ratio of A_0 to A after 30 years

$$\log\left(\frac{A_0}{A}\right) = \frac{kt}{2.303} = \frac{0.131 \text{ yr}^{-1} (30 \text{ yr})}{2.303} = 1.71$$

Taking the antilog of both sides, $\frac{A_0}{A}$ = antilog 1.71 = $10^{1.71}$ = 51.3

Since A_0 = 3.42 μg,

$$A = \frac{A_0}{51.3} = \frac{3.42 \ \mu\text{g}}{51.3}$$

$$= 0.0667 \ \mu\text{g} \ {}^{60}_{27}\text{Co remains after 30 years.}$$

28–9 Disintegration Series

Many radionuclides cannot attain nuclear stability by only one nuclear reaction, so they decay in a series of disintegrations. A few such series are known to occur in nature. Two begin with isotopes of uranium, ^{238}U and ^{235}U, and one begins with ^{232}Th. All end with a stable isotope of lead (Z = 82). Uranium-238 decays by alpha emission to thorium-234 in the first step of one series. Thorium-234 subsequently emits a beta particle to produce protactinium-234 in the second step of the series, which can be summarized as shown. Emitted particles are shown over the arrows.

$${}^{238}_{92}U \xrightarrow{\alpha} {}^{234}_{90}Th \xrightarrow{\beta} {}^{234}_{91}Pa \xrightarrow{\beta} {}^{234}_{92}U \xrightarrow{\alpha} {}^{230}_{90}Th \xrightarrow{\alpha} {}^{226}_{88}Ra \xrightarrow{\alpha}$$

$${}^{222}_{86}Rn \xrightarrow{\alpha} {}^{218}_{84}Po \xrightarrow{\alpha} {}^{214}_{82}Pb \xrightarrow{\beta} {}^{214}_{83}Bi \xrightarrow{\beta} {}^{214}_{84}Po \xrightarrow{\alpha}$$

$${}^{210}_{82}Pb \xrightarrow{\beta} {}^{210}_{83}Bi \xrightarrow{\beta} {}^{210}_{84}Po \xrightarrow{\alpha} {}^{206}_{82}Pb$$

The net reaction is

$${}^{238}_{92}U \longrightarrow {}^{206}_{82}Pb + 8\ {}^{4}_{2}He + 6\ {}^{0}_{-1}\beta$$

There are "branchings" possible in the chain at various points. That is, two successive decays may be replaced by alternate decays, but they always result

TABLE 28–3 Emissions and Half-Lives of Members of Natural Radioactive Series*

^{238}U Series	^{235}U Series	^{232}Th Series
$^{238}_{92}U \to \alpha$; 4.51×10^{9} y	$^{235}_{92}U \to \alpha$; 7.1×10^{8} y	$^{232}_{90}Th \to \alpha$; 1.41×10^{10} y
$^{234}_{90}Th \to \beta$; 24.1 d	$^{231}_{90}Th \to \beta$; 25.5 h	$^{228}_{88}Ra \to \beta$; 6.7 y
$^{234}_{91}Pa \to \beta$; 6.75 h	$^{231}_{91}Pa \to \alpha$; 3.25×10^{4} y	$^{228}_{89}Ac \to \beta$; 6.13 h
$^{234}_{92}U \to \alpha$; 2.47×10^{5} y	$\beta \leftarrow {}^{227}_{89}Ac \to \alpha$; 98.8% ($\beta$), 21.6 y	$^{228}_{90}Th \to \alpha$; 1.91 y
$^{230}_{90}Th \to \alpha$; 8.0×10^{4} y	$\alpha \leftarrow {}^{227}_{90}Th$, 18.2 d; $^{223}_{87}Fr \to \beta$, 22 m	$^{224}_{88}Ra \to \alpha$; 3.64 d
$^{226}_{88}Ra \to \alpha$; 1.60×10^{3} y	$^{223}_{88}Ra \to \alpha$; 11.4 d	$^{220}_{86}Rn \to \alpha$; 55.3 s
$^{222}_{86}Rn \to \alpha$; 3.82 d	$^{219}_{86}Rn \to \alpha$; 4.00 s	$\beta \leftarrow {}^{216}_{84}Po \to \alpha$; 0.014% ($\beta$), 0.14 s
$\beta \leftarrow {}^{218}_{84}Po \to \alpha$; 0.04% ($\beta$), 3.05 m	$\beta \leftarrow {}^{215}_{84}Po \to \alpha$; 5×10^{-4}% (β), 1.78×10^{-3} s	$\alpha \leftarrow {}^{216}_{85}At$, 3×10^{-4} s; $^{212}_{82}Pb \to \beta$, 10.6 h
$\alpha \leftarrow {}^{218}_{85}At$, 2 s; $^{214}_{82}Pb \to \beta$, 26.8 m	$\alpha \leftarrow {}^{215}_{85}At$, 10^{-4} s; $^{211}_{82}Pb \to \beta$, 36.1 m	$\beta \leftarrow {}^{212}_{83}Bi \to \alpha$; 66.3% ($\beta$), 60.6 m
$\beta \leftarrow {}^{214}_{83}Bi \to \alpha$; 99.96% ($\beta$), 19.7 m	$\beta \leftarrow {}^{211}_{83}Bi \to \alpha$; 99.7% ($\beta$), 2.16 m	$\alpha \leftarrow {}^{212}_{84}Po$, 3.0×10^{-7} s; $^{208}_{81}Tl \to \beta$, 3.10 m
$\alpha \leftarrow {}^{214}_{84}Po$, 1.6×10^{-4} s; $^{210}_{81}Tl \to \beta$, 1.32 m	$\alpha \leftarrow {}^{211}_{84}Po$, 0.52 s; $^{207}_{81}Tl \to \beta$, 4.79 m	$^{208}_{82}Pb$
$^{210}_{82}Pb \to \beta$; 20.4 y	$^{207}_{82}Pb$	
$\beta \leftarrow {}^{210}_{83}Bi \to \alpha$; ~100% ($\beta$), 5.01 d		
$\alpha \leftarrow {}^{210}_{84}Po$, 138 d; $^{206}_{81}Tl \to \beta$, 4.19 m		
$^{206}_{82}Pb$		

* The abbreviations are y, year; d, day; m, minute; and s, second.

in the same final product. Only the most prevalent decays were given above for the uranium-238 series. Table 28–3 outlines in detail the ^{238}U, ^{235}U, and ^{232}Th disintegration series, showing half-lives and alternate pathways. The less prevalent decays are denoted by broken arrows. For any particular decay, the decaying nuclide is called the **mother** nuclide and the product nuclide is the **daughter.** There are also decay series of varying lengths starting with some of the man-made radionuclides (Section 28–11).

28–10 Uses of Radionuclides

Radionuclides find many practical applications either because they decay at known rates or, in some cases, simply because they emit radiation continuously.

1 Radioactive Dating

The ages of articles of organic origin can be estimated by **radiocarbon dating.** The radioisotope carbon-14 is produced continuously in the upper atmosphere, as nitrogen atoms capture cosmic-ray neutrons

$$^{14}_{7}N + ^{1}_{0}n \longrightarrow ^{14}_{6}C + ^{1}_{1}H$$

The carbon-14 nuclei react with oxygen molecules to form $^{14}CO_2$. This process continually supplies the atmosphere with radioactive $^{14}CO_2$, which is removed from the atmosphere by photosynthesis. Cosmic-ray intensity is related to the sun's activity. As long as cosmic-ray intensity remains constant, the amount of $^{14}CO_2$ in the atmosphere remains constant. Since it is incorporated into living plants and organisms just as ordinary $^{12}CO_2$ is, a certain fraction of all carbon atoms in living substances is carbon-14, which decays via beta emission with a half-life of 5730 years.

In recent decades, atmospheric testing of nuclear warheads has also caused fluctuations in the natural abundance of ^{14}C.

$$^{14}_{6}C \longrightarrow ^{14}_{7}N + ^{0}_{-1}\beta$$

After death the plant no longer participates in photosynthesis and no longer takes up $^{14}CO_2$. Other organisms that consume plants for food stop doing so at death. Thus, the emissions from the ^{14}C in dead tissue decrease with the passage of time, and the activity per gram of carbon is a measure of the length of time elapsed since death. Comparison of ages of ancient trees calculated from carbon-14 activity with those determined by counting rings indicates that cosmic ray intensity has varied somewhat throughout history. As a result, the calculated ages must be corrected for these variations, but this can be done. The carbon-14 technique is useful only for dating objects less than 50,000 years old. Older objects possess too little activity to be dated accurately.

Methods used for dating older objects are the **potassium-argon** and **uranium-lead methods.** Potassium-40 is present in all living organisms. It decays by electron capture to argon-40, with a half-life of 1.3 billion years.

$$^{40}_{19}K + ^{0}_{-1}e \longrightarrow ^{40}_{18}Ar$$

Gaseous argon is easily lost from minerals, and therefore, measurements based on the $^{40}K/^{40}Ar$ method may or may not be reliable.

Because of its longer half-life, potassium-40 can be used to date objects up to 1 million years old by determination of the ratio of $^{40}_{19}K$ to $^{40}_{18}Ar$ in the sample. The uranium-lead method is based on the natural uranium-238 decay series, which ends with the production of stable lead-206. This method can be used for dating uranium-containing minerals several billion years old. All the lead-206 in such minerals is *assumed* to have come from uranium-238. Because of the very long half-life of $^{238}_{92}U$, 4.5 billion years, the amounts of intermediate nuclei can be neglected. The oldest object found to date is a meteorite 4.6 billion years old, which fell in Mexico in 1969. Results of $^{238}U/^{206}Pb$ studies support the "big bang" theory, which indicates that our solar system was born about 6 billion years ago.

EXAMPLE 28–2

A sample of uranium ore is found to contain 4.64 mg of ^{238}U and 1.22 mg of ^{206}Pb. Estimate the age of the ore. The half-life of ^{238}U is 4.51×10^9 years.

Solution

First we calculate the amount of ^{238}U that must have decayed to produce 1.22 mg of ^{206}Pb, using the isotopic masses.

$$\underline{?}\text{ mg } ^{238}U = 1.22 \text{ mg } ^{206}Pb \times \frac{238 \text{ mg } ^{238}U}{206 \text{ mg } ^{206}Pb} = 1.41 \text{ mg } ^{238}U$$

Thus the sample must have originally contained 4.64 mg + 1.41 mg = 6.05 mg of ^{238}U.

We must next evaluate the specific rate (disintegration) constant, k, which will be used in the final calculation

$$t_{1/2} = \frac{0.693}{k}$$

$$k = \frac{0.693}{t_{1/2}} = \frac{0.693}{4.51 \times 10^9 \text{ yr}} = 1.54 \times 10^{-10} \text{ yr}^{-1}$$

Now we calculate the age of the sample, t

$$\log\left(\frac{A_0}{A}\right) = \frac{kt}{2.303}$$

$$\log\left(\frac{6.05 \text{ mg}}{4.64 \text{ mg}}\right) = \frac{(1.54 \times 10^{-10} \text{ yr}^{-1})t}{2.303}$$

$$\log 1.30 = (6.69 \times 10^{-11} \text{ yr}^{-1})t$$

$$\frac{0.114}{6.69 \times 10^{-11} \text{ yr}^{-1}} = t$$

$$t = 1.70 \times 10^9 \text{ years}$$

The ore is approximately 1.7 billion years old.

2 Medical Uses of Radionuclides

The use of cobalt radiation treatments for cancerous tumors has already been described in Example 28–1. Several other nuclides find uses as **radioactive tracers** in medicine. Since radioisotopes of an element have the same chemical properties as stable isotopes of the same element, they can be used to "label" an element in compounds. A radiation detector can be used to follow the path of the element throughout the body. Salt solutions containing ^{24}Na can be injected into the bloodstream in order to follow the flow of blood and locate obstructions in the circulatory system. Thallium-201 tends to concentrate in healthy heart muscle tissue, while technetium-99 concentrates in abnormal heart tissue. The two can be used in a complementary fashion in surveying the extent of damage from heart disease. Iodine-131 concentrates in the thyroid gland, the liver, and certain parts of the brain, and is used to monitor goiter and other thyroid problems, as well as liver and brain tumors. The heat generated by the decay of plutonium-238 can be converted into electrical energy in heart pacemakers. The relatively long half-life of the isotope allows the device to be used for ten years before replacement.

3 Research Applications for Radionuclides

The pathways of chemical reactions can also be investigated using radioactive tracers. If radioactive $^{35}S^{2-}$ ion is added to a saturated solution of cobalt sulfide in equilibrium with solid cobalt sulfide, the solid becomes radioactive. This is taken as evidence that sulfide ion exchange occurs between solid and solution in the solubility equilibrium

$$\text{CoS (s)} \rightleftharpoons \text{Co}^{2+} \text{ (aq)} + \text{S}^{2-} \text{ (aq)} \qquad K_{sp} = 8.7 \times 10^{-23}$$

Photosynthesis is the process by which the carbon atoms of carbon dioxide are incorporated into glucose, $C_6H_{12}O_6$, in the presence of chlorophyll in green plants during daylight hours

$$6CO_2 + 6H_2O \xrightarrow[\text{chlorophyll}]{\text{sunlight}} C_6H_{12}O_6 + 6O_2$$

The process is not really as simple as the net equation implies; it actually occurs in a series of steps producing a number of intermediate products. By studying the reaction with labeled $^{14}CO_2$, the intermediate molecules that are produced can be identified as the ones containing the radioactive ^{14}C atoms.

Tetrahydrocannabinol, shown below, provides another example. It is the active ingredient in marijuana and hashish. By synthesizing different samples in which different positions in the molecule are labeled with ^{14}C, determinations can be made as to which metabolic products of tetrahydrocannabinol are derived from which parts of the molecule. This gives some insight as to possible physiological effects of its use. By injecting doses of the labeled samples into the bloodstream and ultimately isolating the metabolic products from the urine, it is found that some unmetabolized THC remains after three days, and its metabolic products are present for eight days.

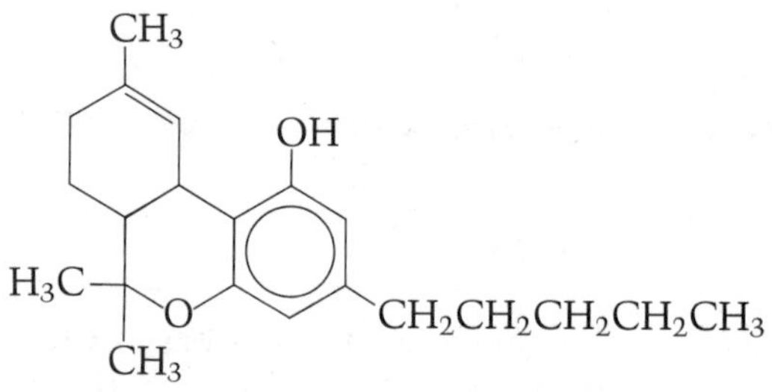

CH_3

OH

H_3C O $CH_2CH_2CH_2CH_2CH_3$

CH_3

4 Agricultural Uses of Radionuclides

DDT is a polychlorinated hydrocarbon whose use as a pesticide has been criticized because of the possible toxicity to humans and animals associated with continuous exposure to it. It decomposes only very slowly, persisting in the environment for a long time, and it tends to concentrate in fatty tissues.

The use of DDT in controlling the screw-worm fly has been replaced by a radiologic technique. The irradiation of the male flies with gamma rays alters their reproductive cells so as to render them sterile. When large numbers of sterilized males are released in an infested area, they mate with females who produce no offspring, resulting in reduction and eventual disappearance of the population. The procedure works because females of this species mate only once. When an area is highly populated with sterile males, the probability of a "productive" mating is very small.

Labeled samples of fertilizers can also be used to investigate nutrient uptake by plants, and to study the growth of crops. Gamma irradiation of some foods allows them to be stored for longer periods. For example, it retards the sprouting of potatoes and onions.

5 Industrial Uses of Radionuclides

There are many applications of radiochemistry in industry and engineering, two of which will be mentioned here. In instances where great precision is required in the manufacture of strips or sheets of metal of definite thicknesses, the penetrating powers of various kinds of radioactive emissions can be utilized. The thickness of the metal can be correlated with the intensity of radiation passing through it. The flow of a liquid or gas through an underground pipeline can be monitored with a Geiger counter by injecting a sample containing a radioactive substance. Leaks in pipelines can also be detected in this way.

28–11 Artificial Transmutations of Elements (Nuclear Bombardment)

The first artificially induced nuclear reaction was performed by Rutherford in 1915. He bombarded nitrogen-14 with alpha particles to produce an isotope of oxygen and a proton.

$$^{14}_{7}N + ^{4}_{2}He \longrightarrow ^{1}_{1}H + ^{17}_{8}O$$

Such reactions are often indicated in abbreviated form, with the bombarding particle and emitted subsidiary particles shown parenthetically between the mother and daughter nuclei.

$$^{14}_{7}N\ (^{4}_{2}\alpha,\ ^{1}_{1}p)\ ^{17}_{8}O$$

Several thousand artificially induced reactions have been carried out with such bombarding projectiles as neutrons, protons, deuterons ($^{2}_{1}H$), alpha particles, and other small nuclei.

1 Bombardment with Positive Ions

A problem arises with the use of positively charged nuclei as projectiles. In order to cause a nuclear reaction to occur, they must actually collide with the target nuclei, which are also positively charged. Coulombic repulsion prevents collisions unless the projectiles have sufficient kinetic energy to overcome it. The required kinetic energies become greater with increasing atomic numbers of the target and of the bombarding particle. Of course, this problem does not arise when neutrons are used.

Particle accelerators called **cyclotrons** (atom smashers) and **linear accelerators** have surmounted the problem of repulsion. Figure 28–6 depicts a cyclotron. A cyclotron consists of two hollow D-shaped electrodes called "dees," both of which are in an evacuated enclosure between the poles of an electromagnet. The particles to be accelerated are introduced at the center in the gap between the dees. The dees are connected to a source of high frequency alternating current which keeps them oppositely charged. At first, the positively charged particles are attracted toward the negative electrode. As the charge is reversed on the dees, the particles are repelled by the first dee (now positive) and attracted to the other. As this process repeats itself in synchronism with the motion of the particles, they accelerate along a spiral path and eventually emerge through an exit hole oriented in such a way that the beam of particles hits the target atoms (Figure 28–7).

The first cyclotron was constructed by E. O. Lawrence and M. S. Livingston at the University of California in 1930.

The path of the particle is initially circular because of the interaction of the particle's charge with the electromagnet's field; as the particle gains energy, the path widens into a spiral.

In a linear accelerator, the particles are accelerated through a series of tubes within an evacuated chamber (Figure 28–8). The odd-numbered tubes are at first negatively charged, and the even ones positively charged. The positively charged particle is attracted toward the first tube. As it passes through the tube, the charges on the tubes are reversed so that the particle is repelled

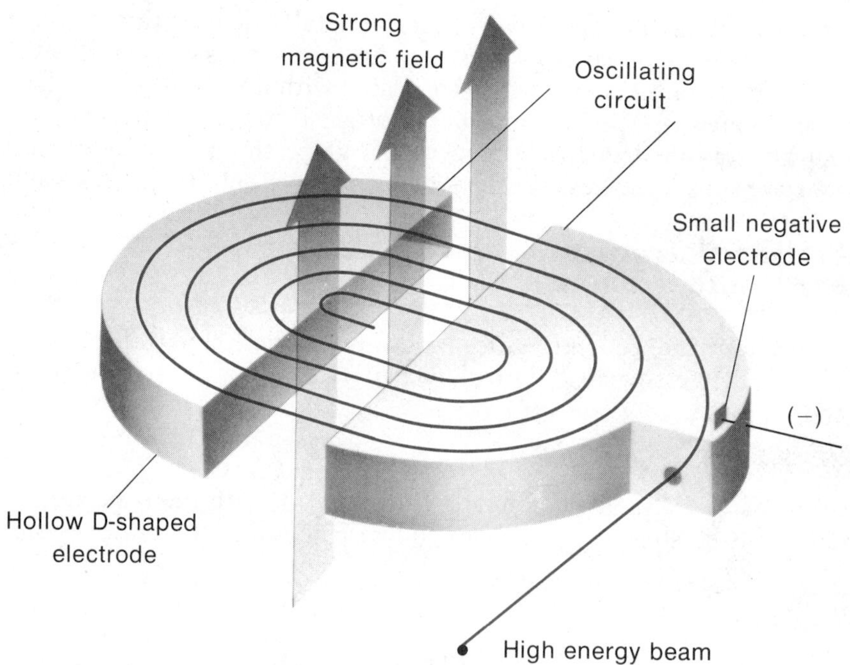

FIGURE 28–6 Schematic representation of a cyclotron.

The first linear accelerator was built in 1928 by a German physicist, Rolf Wideroe.

One gigaelectron volt (GeV) = 1×10^9 eV = 1.60×10^{-10} J. This is sometimes called 1 billion electron volts (BeV) in the United States.

out of the first tube (now positive) and toward the second (negative) tube. As the particle nears the end of the second tube the charges are again reversed, and as this process is repeated the particle is accelerated to very high velocities. Since the polarity is changed at constant frequency, the lengths of subsequent tubes increase to accommodate the increased distance traveled by the accelerating particle per unit time. The bombardment target is located outside the last tube. If the initial polarities are reversed, negatively charged particles can also be accelerated. The longest linear accelerator is approximately two miles long and was completed in 1966 at Stanford University. It is capable of accelerating electrons to energies of nearly 20 GeV.

Many nuclear reactions have been induced by bombardment with positively charged particles. At the time of development of particle accelerators

FIGURE 28–7 A beam of deuterons (bright area) from a cyclotron at the University of California. Nuclear reactions take place when deuterons and other atomic particles strike the nuclei of atoms.

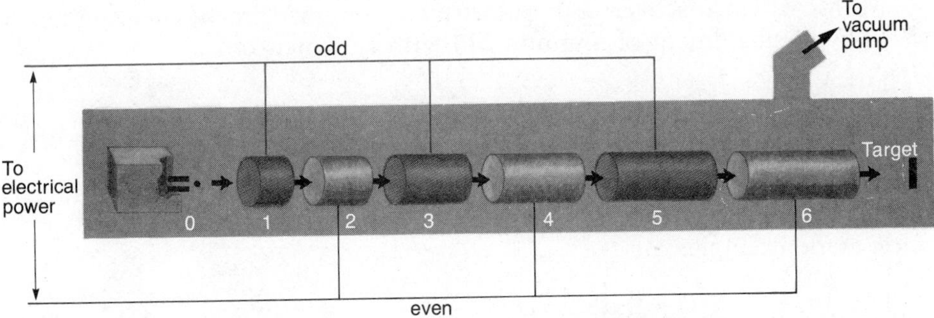

FIGURE 28–8 Diagram of an early type of linear accelerator. An alpha emitter is placed in the container at the left; only those alpha particles can escape that happen to be emitted in line with the series of accelerating tubes.

there were a few gaps among the first 92 elements in the periodic table. Particle accelerators were used between 1937 and 1941 to synthesize three of the four "missing" elements: number 43 (technetium), number 85 (astatine), and number 87 (francium).

$${}^{96}_{42}\text{Mo} + {}^{2}_{1}\text{H} \longrightarrow {}^{97}_{43}\text{Tc} + {}^{1}_{0}n$$

$${}^{209}_{83}\text{Bi} + {}^{4}_{2}\text{He} \longrightarrow {}^{210}_{85}\text{At} + 3\ {}^{1}_{0}n$$

$${}^{230}_{90}\text{Th} + {}^{1}_{1}\text{H} \longrightarrow {}^{223}_{87}\text{Fr} + 2\ {}^{4}_{2}\text{He}$$

Many hitherto unknown, unstable, artificial isotopes of known elements were also synthesized so that their nuclear structures and behavior could be studied.

2 Neutron Bombardment

Since neutrons bear no charge, they are not repelled by nuclei as are positively charged projectiles, and so need not be accelerated to produce bombardment reactions. Neutrons can be generated in several ways. A frequently used method involves bombardment of beryllium-9 with alpha particles.

$${}^{9}_{4}\text{Be} + {}^{4}_{2}\text{He} \longrightarrow {}^{12}_{6}\text{C} + {}^{1}_{0}n$$

Nuclear reactors (Section 28–14) are also used as neutron sources. Neutrons ejected in nuclear reactions usually possess high kinetic energies and are called **fast neutrons.** When they are used as projectiles they cause reactions, such as (n, p) or (n, α) reactions, in which subsidiary particles are ejected. **Slow neutrons** are produced when fast neutrons collide with **moderators** such as hydrogen, deuterium, oxygen, or the carbon atoms in paraffin, and these neutrons are more likely to be captured by the target nuclei. Bombardments with slow neutrons cause mostly neutron-capture (n, γ) reactions

Fast neutrons are moving so rapidly that they are likely to pass right through a target nucleus without reacting. Hence the probability of a reaction is low, even though the reactions may be very energetic.

$${}^{200}_{80}\text{Hg} + {}^{1}_{0}n \longrightarrow {}^{201}_{80}\text{Hg} + {}^{0}_{0}\gamma$$

The fourth "missing" element, number 61 (promethium), was synthesized by fast neutron bombardment of neodymium-142.

$${}^{142}_{60}\text{Nd} + {}^{1}_{0}n \longrightarrow {}^{143}_{61}\text{Pm} + {}^{0}_{-1}\beta$$

E. M. McMillan discovered the first transuranium element, neptunium, in 1940 by bombardment of uranium-238 with slow neutrons.

$${}^{238}_{92}U + {}^{1}_{0}n \longrightarrow {}^{239}_{92}U + {}^{0}_{0}\gamma$$

$${}^{239}_{92}U \longrightarrow {}^{239}_{93}Np + {}^{0}_{-1}\beta$$

Fifteen additional elements have since been prepared by neutron bombardment or by bombardment of nuclei so produced with positively charged particles. Some examples are

$$\left.\begin{array}{l} {}^{238}_{92}U + {}^{1}_{0}n \longrightarrow {}^{239}_{92}U + {}^{0}_{0}\gamma \\ {}^{239}_{92}U \longrightarrow {}^{239}_{93}Np + {}^{0}_{-1}\beta \\ {}^{239}_{93}Np \longrightarrow {}^{239}_{94}Pu + {}^{0}_{-1}\beta \end{array}\right\} \text{plutonium}$$

$${}^{239}_{94}Pu + {}^{4}_{2}He \longrightarrow {}^{242}_{96}Cm + {}^{1}_{0}n \qquad \text{curium}$$

$${}^{246}_{96}Cm + {}^{12}_{6}C \longrightarrow {}^{254}_{102}No + 4\,{}^{1}_{0}n \qquad \text{nobelium}$$

28–12 Nuclear Fission and Fusion

Isotopes of some elements with atomic numbers above 80 are capable of undergoing fission in which they split into nuclei of intermediate masses and emit one or more neutrons. Some fissions are spontaneous; others require that the activation energy be supplied by bombardment. A given nucleus can split in many different ways, liberating enormous amounts of energy. Some of the possible fissions resulting from bombardment of fissionable uranium-235 with fast neutrons are shown below. The uranium-236 is a short-lived intermediate.

$${}^{235}_{92}U + {}^{1}_{0}n \longrightarrow [{}^{236}_{92}U] \longrightarrow \begin{cases} {}^{160}_{62}Sm + {}^{72}_{30}Zn + 4\,{}^{1}_{0}n + \text{energy} \\ {}^{146}_{57}La + {}^{87}_{35}Br + 3\,{}^{1}_{0}n + \text{energy} \\ {}^{140}_{56}Ba + {}^{93}_{36}Kr + 3\,{}^{1}_{0}n + \text{energy} \\ {}^{144}_{55}Cs + {}^{90}_{37}Rb + 2\,{}^{1}_{0}n + \text{energy} \\ {}^{144}_{54}Xe + {}^{90}_{38}Sr + 2\,{}^{1}_{0}n + \text{energy} \end{cases}$$

In dealing with atomic structure (Section 3–8) we discussed mass defects and binding energies. Recall that the binding energy of an atom is the amount of energy released in forming the atom from its constituent subatomic particles, and is related by the Einstein equation, $E = mc^2$, to its mass defect, m. We could say that the binding energy represents the amount of energy that must be supplied to the atom to break it apart into subatomic particles.

The term "nucleon" refers to a nuclear particle, either a neutron or a proton.

Reference to Figure 28–9, a plot of binding energy per nucleon versus mass number, shows that the highest binding energies per nucleon, and therefore the greatest stabilities, are associated with atoms of intermediate mass number. The most stable atom is ${}^{56}_{26}Fe$, with a binding energy per nucleon of 8.80 MeV. Thus, fission is an energetically favorable process for heavy atoms, since atoms with intermediate masses and greater binding energies per nucleon are formed.

Which isotopes of which elements undergo fission? Experiments with particle accelerators have revealed that all elements with an atomic number of

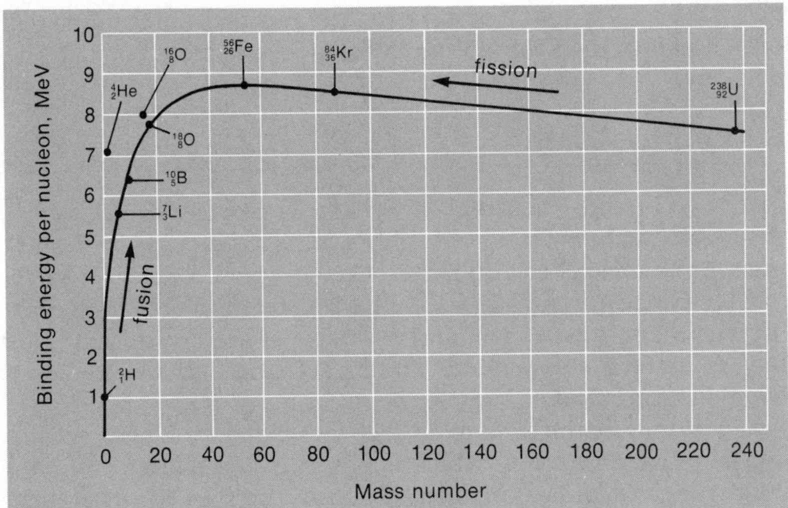

FIGURE 28–9 Variation in nuclear binding energy with atomic mass.

1 MeV = 1.60×10^{-13} J.

80 or more have one or more isotopes capable of undergoing fission provided they are bombarded at the right energy. Nuclei with atomic numbers between 89 and 98 fission spontaneously with long half-lives: 10^{17} yr to 10^{4} yr. Nuclei with atomic numbers of 98 or more spontaneously fission with shorter half-lives from 60.5 days to a few milliseconds. One of the *natural* decay modes of the transuranium elements is via spontaneous fission. In fact, nuclides with *mass numbers* greater than 250 do this because they are just too big to be stable. Nuclides with mass numbers between 225 and 250 will not undergo fission spontaneously (except for a few with extremely long half-lives) but can be induced to do so when bombarded with particles of relatively low kinetic energies. Particles that can supply the appropriate activation energy include slow (and fast) neutrons, protons, and alpha particles, as well as fast electrons. As we proceed toward nuclei lighter than mass 225 the activation energy required to induce fission rises very rapidly.

In Section 28–2 we discussed the preference of nuclei to have even numbers of protons and even numbers of neutrons. Thus when we consider the fission of nuclides with mass numbers of 225 to 250, the isotopes of uranium for example, we should not be surprised to learn that both ^{233}U and ^{235}U can be excited to a fissionable state by slow neutrons much more easily than ^{238}U, since they are less stable. In fact, it is so difficult to cause fission in ^{238}U that this isotope is said to be "nonfissionable."

Fusion, the joining of light nuclei to form heavier nuclei, is favorable for the very light atoms. In both fission and fusion the energy liberated is due to the loss of mass that accompanies the reactions. Much larger amounts of energy per unit mass of reacting atoms are produced in fusion than in fission.

Spectroscopic evidence indicates that the sun is a tremendous fusion reactor consisting of 73% hydrogen, 26% helium, and 1% other elements. The major fusion reaction that occurs is thought to involve the combination of a deuteron, $^{2}_{1}$H, and a triton, $^{3}_{1}$H, at tremendously high temperatures to form a helium nucleus and a neutron

$$^{2}_{1}\text{H} + ^{3}_{1}\text{H} \longrightarrow ^{4}_{2}\text{He} + ^{1}_{0}n + \text{energy}$$

The deuteron and triton are the nuclei of two isotopes of hydrogen, called deuterium and tritium. Deuterium occurs naturally in water. When the D_2O is purified as "heavy water," it can be used for several types of chemical analysis.

Thus solar energy is actually a form of fusion energy, which is, in fact, the only unlimited kind of energy available to us. Even our fossil fuels are just "left over" fusion energy.

28–13 The Atomic Bomb (Fission)

On the average, two or three neutrons are produced per fission reaction. If sufficient fissionable material, the **critical mass,** is contained in a small enough volume, an uncontrolled explosive chain reaction can result as the neutrons ejected from the initial fission collide with other fissionable atoms to sustain and expand the process. If too few fissionable atoms are present, most of the neutrons escape the mass and no chain reaction occurs. Figure 28–10 depicts a fission chain reaction.

One type of atomic bomb (Figure 28–11) contains two subcritical portions of fissionable material, one of which is driven into the other to form a supercritical mass by an ordinary chemical explosive such as trinitrotoluene (TNT). An uncontrolled nuclear fission explosion results. Tremendous amounts of heat energy are released, as well as many radionuclides whose effects are devastating to life and the environment. The radioactive dust and debris is called **fallout.**

28–14 Nuclear Reactors (Fission)

Controlled fission reactions in nuclear reactors are of great use and even greater potential. No possibility of nuclear explosion exists because the fuel elements of a nuclear reactor have neither the composition nor the extremely compact arrangement of the critical mass of a bomb at the instant of explosion. However, certain dangers are associated with nuclear energy generation. There is the possibility of "meltdown" if the cooling system is not utilized properly. This will be discussed with respect to cooling systems in light water reactors. Proper shielding precautions also must be taken to insure that the radionu-

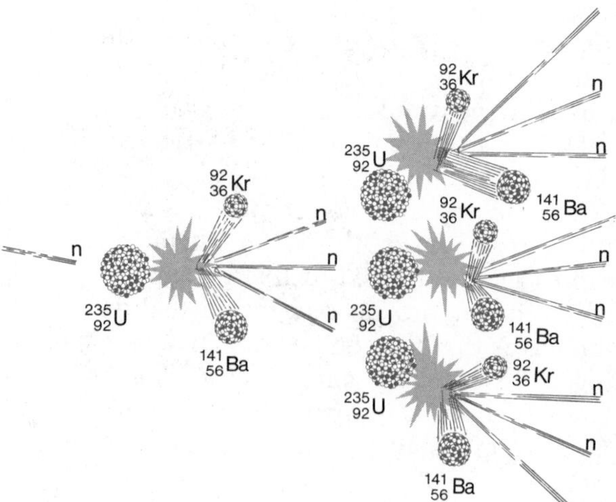

FIGURE 28–10 Self-propagating nuclear chain reaction. A stray neutron induces a single fission, liberating three more neutrons. Each of these induces another fission, each of which is accompanied by release of two or three neutrons. The chain continues to branch in this way, finally resulting in an explosive rate of fission.

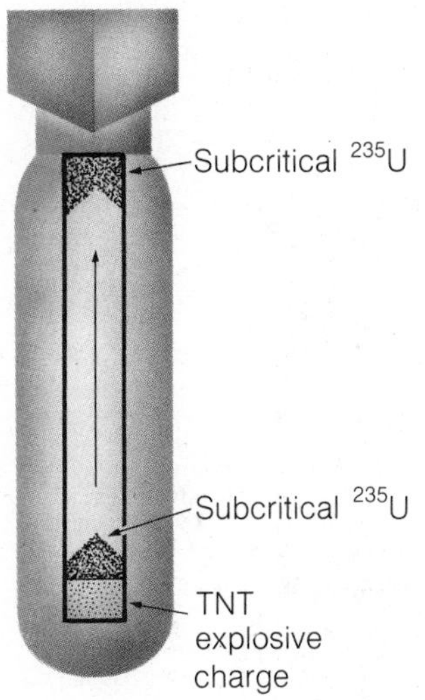

FIGURE 28–11 One design used in atomic bombs. A conventional explosive is used to bring two subcritical masses together to form a supercritical mass.

clides produced are always contained within vessels from which neither they nor their radiations can escape. Long-lived radionuclides from spent fuel must be stored underground in heavy, shock-resistant containers until they have decayed to the point that they are no longer biologically harmful. As examples, strontium-90 ($t_{1/2}$ = 28 years) and plutonium-239 ($t_{1/2}$ = 24,000 years) must be stored for 280 years and 240,000 years, respectively, before they lose 99.9% of their activities. Critics of nuclear energy contend that the containers could fail over such long periods, or as a result of earth tremors, and that transportation and reprocessing accidents could cause environmental contamination with radionuclides. They claim that river water used for cooling is returned to the rivers with too high a heat content (thermal pollution), thus disrupting marine life. It should be noted, though, that fossil fuel electric power plants also cause the same thermal pollution, kilowatt for kilowatt. The problem of espionage also presents itself. Plutonium-239, a fissionable material, could be stolen from reprocessing plants and used to construct atomic weapons.

Proponents of the development of nuclear energy argue that the potential risks are far outweighed by the advantages. Nuclear energy plants do not pollute the air with oxides of sulfur, nitrogen, carbon, and particulate matter as do fossil fuel electric power plants. The big advantage of nuclear fuels, though, is the enormous amount of energy liberated per unit mass of fuel. With rapidly declining fossil fuel reserves, it seems likely that nuclear energy and solar energy will become increasingly important. At present, nuclear reactors provide nearly 10% of the electrical energy consumed in the United States.

Most commercial nuclear power plants in the United States are "light water" reactors, moderated and cooled by ordinary water, but the greatest potential exists for **breeder reactors,** which actually produce more fissionable material than they consume, as we shall see. Let us look at the components and operation of light water reactors first.

The Soviet Union has placed especially high priority on the development and use of breeder reactors.

1 Light Water Reactors

A schematic diagram of a light water reactor plant is shown in Figure 28–12. The main difference between this and a fossil fuel plant is that the

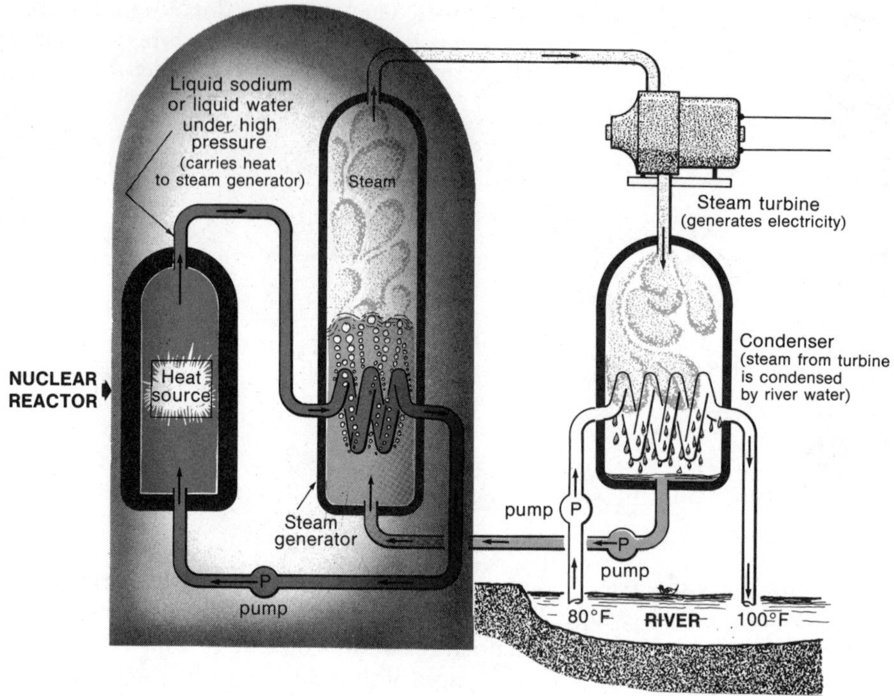

FIGURE 28–12 A schematic diagram of a light water reactor plant.

reactor core at the left replaces the usual furnace in which coal, oil, or natural gas is burned. Such a fission reactor consists of five main components: (1) fuel, (2) moderator, (3) control rods, (4) cooling system, and (5) shielding.

Fuel Rods of U_3O_8 enriched in uranium-235 serve as the fuel. Unfortunately, uranium ores contain only about 0.7% uranium-235. Most of the rest is nonfissionable uranium-238. The enrichment is accomplished in processing and reprocessing plants by separating $^{235}UF_6$ from $^{238}UF_6$, prepared from the ore as described at the end of Chapter 26. Separation by diffusion is based upon Graham's Law, which relates rate of diffusion to the molecular weights of gases (Section 8–14). Another separation procedure depends upon the ultracentrifuge.

A potentially more efficient method of enrichment involves the use of sophisticated tunable lasers to ionize uranium-235 selectively and not uranium-238. The ionized uranium-235 could then be made to react with negative ions to form another compound, easily separated from the mixture. The success of this method will lie in our ability to construct lasers capable of producing radiation monochromatic enough to excite one isotope and not the other, a very difficult challenge.

Moderator The most efficient fission reactions occur with slow neutrons. Thus the fast neutrons ejected during fission must be slowed by collisions with atoms of comparable mass which do not absorb them. Such materials are called **moderators.** The most commonly used moderator is ordinary water, although graphite is sometimes used. The most efficient moderator is helium, which slows neutrons but does not absorb them all. The next most efficient one is "heavy water" (deuterium oxide, $^2_1H_2O$ or $^2_1D_2O$), but this is so expensive that it has been used chiefly in research reactors. However, there is a Canadian-designed power reactor that uses heavy water and is correspondingly more neutron-efficient.

Control Rods Control of the rate of a fission reaction is achieved by using movable control rods, usually of cadmium or boron steel. They are inserted or removed automatically from spaces between the fuel rods. Cadmium and boron are good neutron absorbers

$$^{10}_{5}B + ^{1}_{0}n \longrightarrow ^{7}_{3}Li + ^{4}_{2}\alpha$$

The more neutrons absorbed by the control rods, the fewer fissions occur, and the less heat is produced. Hence the heat output is governed by the control system that operates the rods.

Cooling System Actually two cooling systems are needed. First, the moderator itself serves as a coolant for the reactor. It transfers the fission-generated heat to a second system called a steam generator, which converts water to steam. The steam then goes to the turbines to drive the generator to produce electricity. Another coolant (river water, seawater, or recirculated water) condenses the steam from the turbine, and the condensate is then recycled into the steam generator.

Note that the water that actually comes in contact with the nuclear fuel flows in a closed system, and is not released to the environment.

The danger of meltdown arises if the reactor is shut down quickly. The disintegration of radioactive fission products still goes on at a furious rate, fast enough to overheat the fuel elements and to melt them. So it isn't enough to shut down the fission reaction; efficient cooling must be continued until the short-lived isotopes are gone and the heat from their disintegration is dissipated. Only then can the circulation of cooling water be stopped.

The accident that occurred in 1979 at Three Mile Island, near Harrisburg, Pennsylvania, was the result of stopping the water pumps too soon *and* the inoperability of the emergency pumps. So a combination of mechanical malfunctions, errors, and carelessness produced the overheating that damaged the fuel assembly. It did not and *could not explode,* although some melting of core material did occur.

Shielding It is essential that the people and surrounding countryside be adequately shielded from possible exposure to radioactive nuclides. Thus the entire reactor is enclosed in a steel containment vessel, which is housed in a thick-walled concrete building. The operating personnel are protected further by a so-called biological shield, a thick layer of organic material made of compressed wood fibers. This absorbs the neutrons and beta and gamma rays that would be absorbed in the human body.

The neutrons are the worst radiation problem. The human body contains a high percentage of H_2O, which absorbs neutrons very efficiently.

2 Breeder Reactors (Fission)

It is predicted that our limited supply of uranium-235 will only last another 50 years. However, nonfissionable uranium-238 is about 100 times more plentiful and can be converted into fissionable plutonium-239. In fact this does take place to some extent in light water reactors.

$$^{238}_{92}U + ^{1}_{0}n \longrightarrow ^{239}_{94}Pu + 2\ ^{0}_{-1}\beta$$

Fissionable uranium-233 can also be produced by neutron bombardment of thorium-232

$$^{232}_{90}Th + ^{1}_{0}n \longrightarrow ^{233}_{92}U + 2\ ^{0}_{-1}\beta$$

It is possible to build reactors, called *breeder reactors,* that not only generate large quantities of heat from fission, but also generate more fuel than they use because neutrons are purposely absorbed in a thorium or uranium "blanket" to cause the above reactions to occur. However, the design of the breeder reactor has several difficulties associated with it. This type of reactor requires the use of fast neutrons, so no moderator is needed, but control is more difficult. It also must operate at higher temperatures than light water reactors. Thus water cannot be used as a coolant, and liquid sodium, which is not a neutron moderator, is used instead. Sodium is very reactive, particularly so at high temperatures, and has a tendency to attack the walls of its container. Heat must be transferred very efficiently because plutonium-239 melts at a relatively low temperature of 640°C.

An additional problem is that plutonium-239 is one of the most toxic substances known, so any release would be disastrous.

28–15 Nuclear Fusion (Thermonuclear) Energy

As mentioned earlier, fusion reactions are accompanied by even greater energy production per unit weight of reacting atoms than fission reactions. However, they can be initiated only by extremely high temperatures. The fusion of $^{2}_{1}H$ and $^{3}_{1}H$ occurs at the lowest temperature of any fusion reaction known, but even this is 40,000,000 K! Such temperatures exist in places like the sun and other stars, but they are next to impossible to achieve and contain on earth. **Thermonuclear** bombs (called fusion bombs, or hydrogen bombs) of incredible energy have been detonated in tests but, thankfully, never in war. In them the necessary activation energy is supplied by the explosion of a fission bomb.

There is hope that fusion reactions can be harnessed for generation of energy for domestic power. Because of the tremendously high temperatures required, no currently known structural material can confine the reaction. At such high temperatures all molecules dissociate and most atoms ionize, resulting in the formation of a new state of matter called a **plasma.** A very high temperature plasma is so hot that it melts and decomposes anything it touches, including, of course, structural components of a reactor. The technological progress required to build a viable fusion reactor probably represents the greatest challenge ever faced by the scientific and engineering community.

Recent attempts at the containment of lower-temperature plasmas by external magnetic fields have been successful, and they lend encouragement to our hopes. However, fusion as a practical energy source lies far into the future at best. The biggest advantages of its use would be that (1) the deuterium fuel can be found in a virtually inexhaustible supply, the oceans; and (2) fusion reactions would produce only radionuclides of very short half-life, primarily tritium ($t_{1/2}$ = 12.3 years), so there would be no long-term waste-disposal problem. If controlled fusion can be brought about, it could free our culture of all its dependence on uranium and fossil fuels.

Key Terms

Alpha particle (α) a helium nucleus.

Artificial transmutation an artificially induced nuclear reaction caused by bombardment of a nucleus with subatomic particles or small nuclei.

Band of stability band containing nonradioactive nuclides in a plot of number of neutrons versus atomic number.

Beta particle (β) electron emitted from the nucleus when a neutron decays to a proton and an electron.

Breeder reactor a nuclear reactor that produces more fissionable nuclear fuel than it consumes.

Chain reaction a reaction which, once initiated, sustains itself and expands.

Cloud chamber a device for observing the paths of speeding particles as vapor molecules condense on them to form foglike tracks.

Control rods rods of materials such as cadmium or boron steel that act as neutron absorbers (not merely moderators) used in nuclear reactors to control neutron fluxes and therefore rates of fission.

Critical mass the minimum mass of a particular fissionable nuclide in a given volume required to sustain a nuclear chain reaction.

Cyclotron a device for accelerating charged particles along a spiral path.

Daughter nuclide nuclide that is produced in a nuclear decay.

Fast neutron a neutron ejected at high kinetic energy in a nuclear reaction.

Fluorescence absorption by a substance of high energy radiation and subsequent emission of visible light.

Gamma ray (γ) high energy electromagnetic radiation.

Half-life of a radionuclide, the time required for half of a given sample to undergo radioactive decay.

Heavy water water containing deuterium, a heavy isotope of hydrogen, ^2_1H.

***K* capture** absorption of a *K* shell ($n = 1$) electron by a proton as it is converted to a neutron.

Linear accelerator a device used for accelerating charged particles along a straight line.

Moderator a substance such as hydrogen, deuterium, oxygen, or paraffin capable of slowing fast neutrons upon collision.

Mother nuclide nuclide that undergoes nuclear decay.

Nuclear fission the process in which a heavy nucleus splits into nuclei of intermediate masses and one or more protons are emitted.

Nuclear fusion the combination of light nuclei to produce a heavier nucleus.

Nuclear reaction involves a change in the composition of a nucleus and can evolve or absorb an extraordinarily large amount of energy.

Nuclear reactor a system in which controlled nuclear fission reactions generate heat energy on a large scale, which is subsequently converted into electrical energy.

Nucleons particles comprising the nucleus.

Nuclides refer to different atomic forms of all elements in contrast to "isotopes" which refer only to different atomic forms of a single element.

Plasma a physical state of matter which exists at extremely high temperatures in which all molecules are dissociated and most atoms are ionized.

Positron nuclear particle with the mass of an electron but opposite charge.

Radiation high energy particles or rays emitted in nuclear decay processes.

Radioactive dating method of dating ancient objects by determining the ratio of amounts of mother and daughter nuclides present in an object and relating the ratio to the object's age via half-life calculations.

Radioactive tracer a small amount of radioisotope replacing a nonradioactive isotope of the element in a compound whose path (for example, in the body) or whose decomposition products are to be monitored by detection of radioactivity; also called a radioactive label.

Radioactivity the spontaneous disintegration of atomic nuclei.

Radioisotope a radioactive isotope of an element.

Radionuclide a radioactive nuclide.

Scintillation counter device used for the quantitative detection of radiation.

Slow neutron a fast neutron slowed by collision with a moderator.

Thermonuclear energy energy from nuclear fusion reactions.

Transuranium elements the elements with atomic number greater than 92 (uranium); none occur naturally and all must be prepared by nuclear bombardment of other elements.

Exercises

1. How do nuclear reactions differ from ordinary chemical reactions?
2. How do nuclear fission and nuclear fusion differ? How are they similar?
3. Describe what is meant by "magic numbers" of nucleons.
4. Explain the fact that a plot of number of neutrons versus atomic number for all nuclei shows a band of stable nuclei with a somewhat step-like shape.
5. Describe the characteristics of alpha particles, beta particles, and gamma rays.
6. Fill in the missing symbols in the following nuclear reactions:
   (a) $^{23}_{11}Na + ? \longrightarrow ^{23}_{12}Mg + ^{1}_{0}n$
   (b) $^{96}_{42}Mo + ^{4}_{2}He \longrightarrow ^{100}_{43}Tc + ?$
   (c) $^{232}_{90}Th + ? \longrightarrow ^{240}_{96}Cm + 4\,^{1}_{0}n$
   (d) $? + ^{1}_{1}H \longrightarrow ^{29}_{14}Si + ^{0}_{0}\gamma$
   (e) $^{209}_{83}Bi + ? \longrightarrow ^{210}_{84}Po + ^{1}_{0}n$
   (f) $^{238}_{92}U + ^{16}_{8}O \longrightarrow ? + 5\,^{1}_{0}n$
7. Write the symbols for the daughter nuclei in the following radioactive decays:
   (a) $^{237}_{92}U \xrightarrow{-\beta}$
   (b) $^{13}C \xrightarrow{-n}$
   (c) $^{11}B \xrightarrow{-\gamma}$
   (d) $^{224}Ra \xrightarrow{-\alpha}$
   (e) $^{18}F \xrightarrow{-^{1}_{1}p}$
   (f) $^{40}_{19}K \xrightarrow{+\beta}$ (*K* capture)
8. Write the symbols for the daughter nuclei in the following nuclear reactions:
   (a) $^{60}_{28}Ni(n,p)$
   (b) $^{98}_{42}Mo(^{1}_{0}n,\beta)$
   (c) $^{35}_{17}Cl(p,\alpha)$
   (d) $^{20}_{10}Ne(\alpha,\gamma)$
   (e) $^{15}_{7}N(p,\alpha)$
   (f) $^{10}_{5}B(n,\alpha)$
9. Predict the kind of decays you would expect for the following radionuclides:
   (a) $^{60}_{27}Co$ (n/p ratio too high)
   (b) $^{20}_{11}Na$ (n/p ratio too low)
   (c) $^{224}_{88}Rn$
   (d) $^{64}_{29}Cu$ (n/p ratio too low)
   (e) $^{238}_{92}U$
   (f) $^{11}_{6}C$
10. Name and describe four methods for detection of radiation.
11. Why must all radioactive decays be first order?
12. Describe the process by which steady-state (constant) ratios of carbon-14 to (nonradioactive) carbon-12 are attained in living plants and organisms. Describe the method of radiocarbon dating.
13. Name some radionuclides that have medical uses and give the uses.
14. Describe how radionuclides can be used in: (a) research, (b) agriculture, (c) industry.
15. Describe how (a) cyclotrons and (b) linear accelerators work.
16. Why do both fission and fusion reactions release energy?
17. Summarize how an atomic bomb works, including how the nuclear explosion is initiated.
18. Discuss the pros and cons of the use of nuclear energy instead of other more conventional types of energy based on fossil fuels.
19. Describe and illustrate the essential features of a light water fission reactor.
20. How is fissionable uranium-235 separated from nonfissionable uranium-238?
21. Distinguish between moderators and control rods of nuclear reactors.

22. What are the major advantages and disadvantages of fusion as a potential energy source, as compared to fission?

23. **Particle** **Mass**

Particle	Mass
neutron	1.0087 amu
proton	1.0073 amu
electron	0.00055 amu

Given the information above and any information found in the appendices calculate the following for $^{65}_{29}Cu$ (actual mass = 64.9278 amu):
(a) mass defect in amu/atom
(b) mass defect in g/mol
(c) binding energy in ergs/atom
(d) binding energy in J/atom
(e) binding energy in kcal/mol
(f) binding energy in kJ/mol

24. Francium-223 undergoes β decay with a half-life of 22 minutes. How much of a 26 μg sample remains after 2.0 hours?

25. Strontium-90 is one of the harmful radionuclides resulting from nuclear fission explosions. It decays by beta emission with a half-life of 28 years. How long would it take for 99.99% of a given sample released in an atmospheric test of an atomic bomb to disintegrate?

26. Carbon-14 decays by beta emission with a half-life of 5730 years. Assuming a particular object originally contained 6.50 μg of carbon-14 and now contains 0.76 μg of carbon-14, how old is the object?

27. Plutonium-239 decays by alpha particle and gamma ray emission with a half-life of 24,000 years.
(a) Write the balanced equation for the decay.
(b) Calculate the specific rate constant, k, for the decay.
(c) Is ^{239}Pu or ^{235}U ($t_{1/2}$ = 710 million years) more stable?

28. Potassium-42 undergoes beta decay and γ emission. If, after 50 hours, 57 minutes, a 2.000-gram sample of ^{42}K has decayed to only 0.116 g, what is the half-life of ^{42}K?

29 Organic Chemistry I: Hydrocarbons

Urea, $H_2N—CO—NH_2$, is the principal end product of metabolism of nitrogen-containing compounds in mammals. It is eliminated in the urine. An adult man excretes about 30 g of urea in 24 hours.

Organic chemistry may be described as the chemistry of substances that contain carbon-hydrogen bonds. Originally the term "organic" was used to describe compounds of plant or animal origin. However, in 1828 Friedrich Wöhler synthesized urea by boiling ammonium cyanate with water and overthrew the "vital force" theory, which held that organic compounds could be made only by living things.

$$\underset{\text{(an inorganic compound)}}{\underset{\text{ammonium cyanate}}{NH_4OCN}} \xrightarrow[\text{boil}]{H_2O} \underset{\text{(an organic compound)}}{\underset{\text{urea}}{H_2N—\overset{\overset{\displaystyle O}{\|}}{C}—NH_2}}$$

Today it is commonplace to manufacture organic compounds from inorganic materials.

Carbon is the first element in Group IVA in the periodic table, which means that it has four electrons in its outermost shell and can be expected to form single bonds with four other atoms (or with other carbon atoms, for that matter).

Of the Group IVA elements, carbon is a nonmetal, silicon and germanium are metalloids, and tin and lead are metallic elements.

Although carbon constitutes only about 0.027% of the earth's crust, it is unique among the elements in that it bonds to itself to form millions of different compounds. The ability of an element to bond to itself is known as **catenation.** Several elements exhibit catenation, but carbon does so to a much greater extent than any other element. Carbon atoms bond to each other to form long chains, branched chains, and rings that may also have chains attached to them.

Saturated Hydrocarbons

Saturated hydrocarbons, those containing only single bonds, compose the major portion of petroleum and natural gas. Although we normally think of petroleum and natural gas as fuel sources, we should keep in mind that most synthetic organic materials are derived from these two sources. Well over half of the top 50 commercial chemicals are organic compounds derived in this way. Organic compounds are used for a variety of purposes such as clothing, fuel, plastics, paints, medicine, and foodstuffs. Many plastics and other synthetic materials find increasing uses as structural materials.

Recall that hydrocarbons contain only C and H.

As the use of coal increases, it is likely that more commercially important compounds will be derived from coal.

29–1 The Alkanes

The hydrocarbons are among the simplest and most common compounds of carbon. The *saturated hydrocarbons,* or **alkanes,** are compounds in which each carbon atom is bonded to four other atoms, and each hydrogen atom is, of course, bonded to only one carbon atom. The term saturated means that only single covalent bonds are present.

The term "saturated" comes from early studies in which chemists tried to add hydrogen to various organic substances. Those to which no more hydrogen could be added were called saturated, by analogy with saturated solutions.

In Section 5–12 we examined the structure of the simplest alkane, *methane,* CH_4, and found that methane molecules are tetrahedral, with sp^3 hybridization at carbon (Figure 29–1).

Ethane, C_2H_6, is the next simplest saturated hydrocarbon, and its structure is quite similar to that of methane. Two carbon atoms share a pair of electrons, and each carbon atom shares an electron with each of three hydrogen atoms. Both carbon atoms are sp^3 hybridized (Figure 29–2). One may visualize the formation of an ethane molecule from two methane molecules by removing one H atom (and its electron) from each CH_4 molecule and then joining the fragments. *Propane,* C_3H_8, is the next member of the family (Figure 29–3). Again, it can be considered as the result of removing a hydrogen atom each from a methane molecule and an ethane molecule, and joining the fragments.

There are two compounds that have the formula C_4H_{10}. These molecules are structural **isomers** (Section 27–6). The structures of these two isomeric

FIGURE 29–1 A molecule of methane, CH_4, formed by overlap of the four sp^3 carbon orbitals with the *s* orbitals of four hydrogen atoms. The resulting molecule is tetrahedral. See Section 5–12 for more detail.

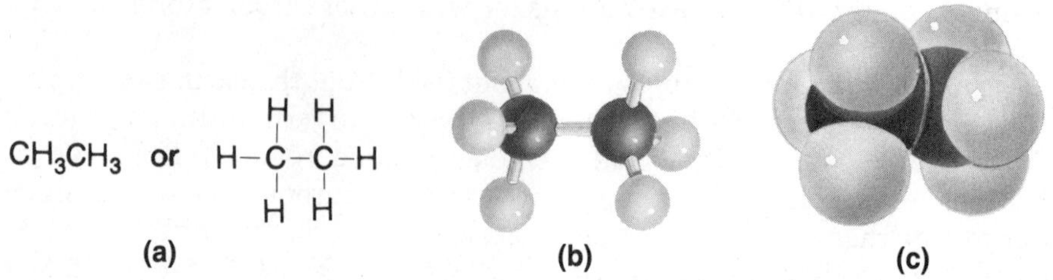

CH_3CH_3 or
```
   H H
   | |
 H-C-C-H
   | |
   H H
```
(a) (b) (c)

FIGURE 29–2 Models of ethane, C_2H_6. (a) The condensed and line formulas for ethane. (b) Ball-and-stick model and (c) a space-filling model of ethane.

butanes are shown in Figure 29–4. These two structures correspond to the two ways in which a hydrogen atom could be removed from a propane molecule and replaced by a $—CH_3$ group. If the hydrogen were removed from either of the end carbon atoms and replaced by a $—CH_3$, the result is *normal butane,* abbreviated as *n*-butane. "Normal," or "straight chain," refers to hydrocarbon chains in which there is no branching. If the hydrogen were removed from the central carbon atom of propane, however, the result of replacing it with a $—CH_3$ group would be 2-methylpropane, or *isobutane,* the simplest *branched-chain* hydrocarbon.

Examination of the formulas of the saturated hydrocarbons discussed to this point indicates that their formulas can be written in general terms as C_nH_{2n+2}, where n refers to the number of carbon atoms per molecule. The first five members of the series are

	CH_4	C_2H_6	C_3H_8	C_4H_{10}	C_5H_{12}
n =	1	2	3	4	5
$2n + 2$ =	4	6	8	10	12

Each saturated hydrocarbon differs from the next by CH_2, a **methylene group.** A series of compounds in which each member differs from the next member by

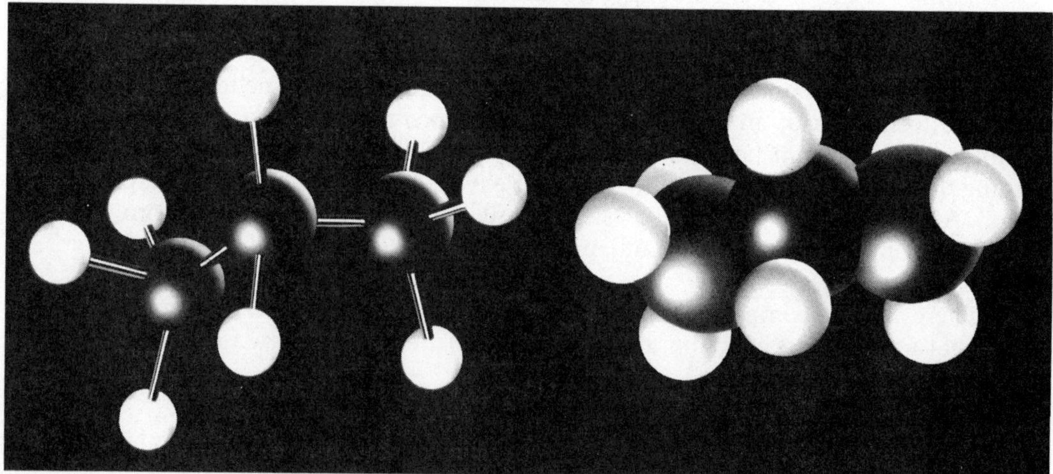

FIGURE 29–3 Ball-and-stick and space-filling models of propane, C_3H_8.

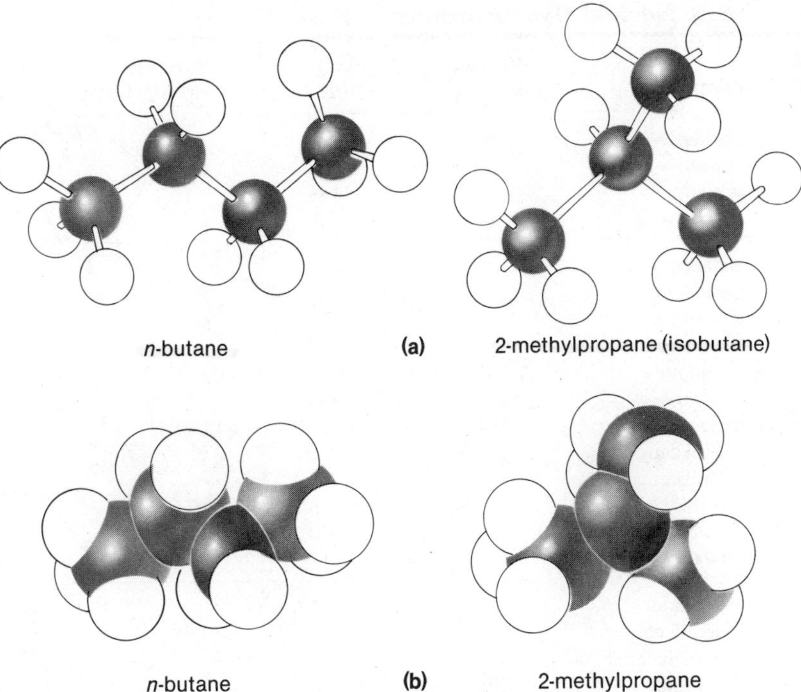

FIGURE 29–4 Models of n-butane, $CH_3CH_2CH_2CH_3$, and isobutane, $CH_3CH(CH_3)CH_3$: (a) Ball-and-stick and (b) space-filling models of the two isomeric butanes.

Dark spheres represent C atoms, and light spheres represent H atoms.

a specific number and kind of atoms is referred to as a **homologous series.** The chemical and physical properties of members of a homologous series are usually closely related. For example, the boiling points of the lighter members of the saturated hydrocarbon series are shown in Figure 29–5. Note that there is a regular increase in boiling points as the molecular weights of the normal hydrocarbons increase. This is due to the increase in effectiveness of London forces (Section 9–9, part 4).

Because there are so many straight-chain hydrocarbons, some method for naming them systematically is necessary. The names of the first 20 members

"Straight-chain" does not mean that the molecules are linear. It means that no carbon atom is bonded to more than two other carbon atoms. The bond angles around each carbon atom are approximately 109°28′.

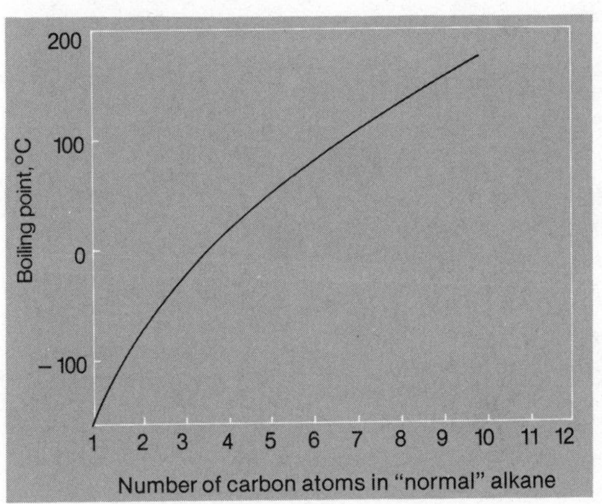

FIGURE 29–5 A plot of boiling point versus the number of carbon atoms in normal, i.e., straight-chain, saturated hydrocarbons.

TABLE 29–1 Some Normal Hydrocarbons (Alkanes)

Molecular Formula	Name	Boiling Point (°C)	Melting Point (°C)	State at Room Temp.
CH_4	methane	−161	−184	
C_2H_6	ethane	−88	−183	Gas
C_3H_8	propane	−42	−188	
C_4H_{10}	*n*-butane	+0.6	−138	
C_5H_{12}	*n*-pentane	36	−130	
C_6H_{14}	*n*-hexane	69	−94	
C_7H_{16}	*n*-heptane	98	−91	
C_8H_{18}	*n*-octane	126	−57	
C_9H_{20}	*n*-nonane	150	−54	
$C_{10}H_{22}$	*n*-decane	174	−30	
$C_{11}H_{24}$	*n*-undecane	194.5	−25.6	Liquid
$C_{12}H_{26}$	*n*-dodecane	214.5	−9.6	
$C_{13}H_{28}$	*n*-tridecane	234	−6.2	
$C_{14}H_{30}$	*n*-tetradecane	252.5	+5.5	
$C_{15}H_{32}$	*n*-pentadecane	270.5	10	
$C_{16}H_{34}$	*n*-hexadecane	287.5	18	
$C_{17}H_{36}$	*n*-heptadecane	303	22.5	
$C_{18}H_{38}$	*n*-octadecane	317	28	
$C_{19}H_{40}$	*n*-nonadecane	330	32	
$C_{20}H_{42}$	*n*-eicosane	205 (at 15 torr)	36.7	Solid

of the family, as prescribed by the International Union of Pure and Applied Chemistry (IUPAC), are listed in Table 29–1. Note that the first four members' names must be memorized, but the names of the later members have prefixes (from Greek) that give the number of carbon atoms in the molecules. All names have the *-ane* ending, characteristic of all alkanes.

Branched-chain alkanes will be named in the next section.

We have already noted that there are two saturated hydrocarbons that have the formula C_4H_{10}. When we come to the C_5 hydrocarbons, we find there are three possible arrangements of the atoms, and indeed three different *pentanes* are known (Figure 29–6).

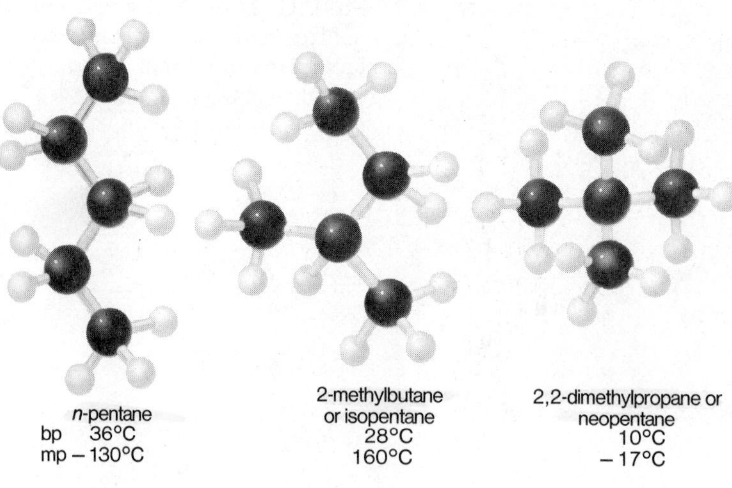

FIGURE 29–6 Three-dimensional models of the three isomeric pentanes, each containing five C atoms and twelve H atoms arranged in a different way to give three isomeric structures.

TABLE 29–2 Isomeric Hexanes, C_6H_{14}

Name	Formula	Boiling Point (°C)	Melting Point (°C)
n-hexane	$CH_3CH_2CH_2CH_2CH_2CH_3$	68.7	−94
2-methylpentane	$CH_3CH_2CH_2CHCH_3$ | CH_3	60.3	−153.7
3-methylpentane	$CH_3CH_2CHCH_2CH_3$ | CH_3	63.3	−118
2,2-dimethylbutane	CH_3 | $CH_3CH_2CCH_3$ | CH_3	49.7	−99.7
2,3-dimethylbutane	$CH_3CH—CHCH_3$ | | CH_3 CH_3	58.0	−128.4

Note that the boiling points of the pentanes decrease as branching increases (Figure 29–6). For isomers of a given molecular weight, boiling points and melting points generally decrease with increased branching. Properties such as boiling point depend on the forces between molecules, which in turn depend on the "contact area" presented by each molecule to its neighbors. As the degree of branching increases for a series of molecules of the same molecular weight, the molecules become more compact (as seen in Figure 26–6) and the "contact area" becomes smaller.

The number of structural isomers increases rapidly as the number of carbon atoms in saturated hydrocarbons increases. There are five isomeric *hexanes* (Table 29–2).

Table 29–3 displays the number of isomers of some saturated hydrocarbons (alkanes) containing increasing numbers of carbon atoms. Clearly most of the isomers have not been, and probably never will be, prepared and isolated.

TABLE 29–3 Number of Possible Isomers of Alkanes

Carbon Content	Isomers
C_6	18
C_9	35
C_{10}	75
C_{11}	159
C_{12}	355
C_{13}	802
C_{14}	1,858
C_{15}	4,347
C_{20}	366,319
C_{25}	36,797,588
C_{30}	4,111,846,763

29–2 Nomenclature of Saturated Hydrocarbons

The names of the lower "straight-chain" alkanes were listed in Table 29–1. You may wish to refer to these to refresh your memory. Branched-chain hydrocarbons are named by naming the longest continuous chain of carbon atoms and indicating the position and kind of substituent attached to the chain. **Alkyl group substituents** attached to the longest chain are thought of as fragments of hydrocarbon molecules obtained by the removal of one hydrogen atom. They are given names related to the parent hydrocarbons from which they are derived, as shown in the following list. Other alkyl groups are named similarly.

Parent Hydrocarbon	**Alkyl Group**
methane CH_4	methyl $CH_3—$
H | H—C—H | H	H | H—C— | H

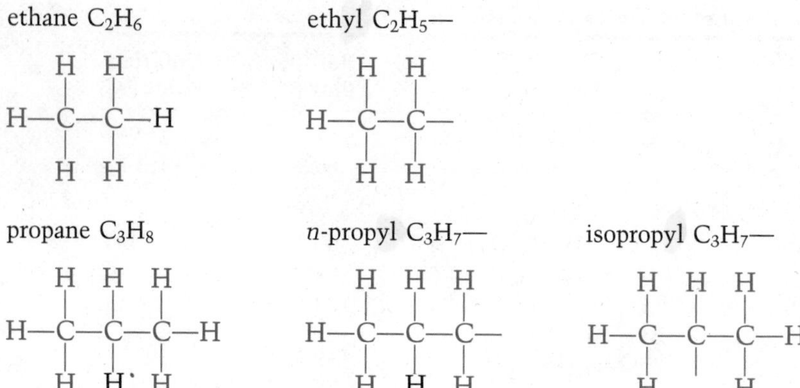

Let us name some saturated branched hydrocarbons. The rules are: (1) Find the longest continuous chain of carbon atoms; (2) number the carbon atoms in the chain beginning at the end nearest the branching; (3) assign the name and (position) numbers that indicate the substituents; (4) the longest continuous chain is always numbered so that the substituents have the lowest possible numbers.

Name the following compound.

Parentheses are used to conserve space. Formulas written with parentheses must indicate unambiguously the structure of the compound. The parentheses here indicate that the CH_3 group is attached to the C that precedes it.

```
             H
             |
  H   H—C—H   H   H   H   H
  |          |          |   |   |   |
H—C————C————C—C—C—C—H
  |          |          |   |   |   |
  H          H          H   H   H   H
```

or $CH_3CH(CH_3)CH_2CH_2CH_2CH_3$

Following rules 1 and 2 allows us to number the carbon atoms in the longest chain.

```
        CH3
        |
CH3—CH—CH2—CH2—CH2—CH3
 1    2    3    4    5    6
```

It is incorrect to name the compound as 5-methylhexane because that violates rules 2 and 4.

The methyl group is attached to the *second* carbon atom in a *six-carbon* chain and so the compound is named 2-methylhexane.

The following examples illustrate further the rules of nomenclature.

```
       CH3
 1    2|   3     4     5
CH3—C—CH2—CH2—CH3
       |
       CH3
```

or $CH_3C(CH_3)_2(CH_2)_2CH_3$

2,2-dimethylpentane

Remember that line (dash) formulas indicate atoms that are bonded to each other. They do *not* show molecular geometry.

```
 1     2     3
CH3—CH2—CH—CH3
             |4
             CH2
             |5
             CH3
```

3-methylpentane

```
                 CH3
 1     2     3|    4     5
CH3—CH2—CH—CH2—CH3
```

better written so as to emphasize the 5-C chain

In general terms, we use the symbol R *to represent any alkyl group.* For example, the saturated hydrocarbons may be represented by the general formula R—H.

29–3 Conformations

A **conformation** is one specific geometry of a molecule. Conformations of compounds differ from one another by the *extent of rotation about a single bond.* The distance between two singly bonded carbon atoms is relatively independent of the structure of the rest of the molecule and is very nearly equal to 1.54 Å. Rotation about single carbon-carbon bonds is possible; in fact, it occurs rapidly. It might appear that there is an infinite number of kinds of ethane molecules, depending on the rotation of one carbon atom with respect to the other. However, at room temperature ethane molecules possess sufficient thermal energy to cause rapid rotation about the single carbon-carbon bond from one conformation to another. Therefore there is only one kind of ethane molecule, and no isomers are known. There is, however, a small preference for the staggered conformation over the eclipsed conformation (see Figure 29–7) because there is less repulsive interaction between hydrogen atoms on adjacent carbon atoms in the staggered conformation.

Rotation is not possible around carbon-carbon double bonds at room temperature; see Section 29–8.

Let us consider two conformations of *n*-butane (Figure 29–8). Again, there is a slight preference for the second (staggered) conformation, but at room temperature many conformations are present in a sample of pure *n*-butane.

29–4 Chemical Properties of the Alkanes

The saturated hydrocarbons are chemically relatively inert materials. For many years they were known as paraffin hydrocarbons because they undergo relatively few reactions. The alkanes do not react with such powerful oxidizing agents as potassium permanganate and potassium dichromate. However, they do react with the halogens, with oxygen when ignited, and with concentrated nitric acid. As expected, the members of a homologous series show similar chemical properties. If we study the chemistry of one of these compounds, we can make predictions about the others with a fairly high degree of certainty.

Paraffin is a mixture of high-molecular-weight alkanes, and is chemically quite unreactive.

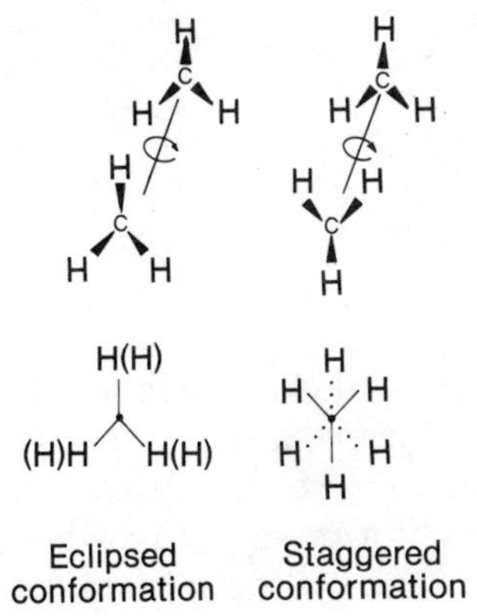

H(H)

(H)H H(H)

Eclipsed conformation

Staggered conformation

FIGURE 29–7 Two possible conformations of ethane. Rotation from one to the other is easily possible, as shown by the curved arrows. The lower formulas show how the hydrogens would be arranged if the molecule were viewed end on along the carbon-carbon bond axis.

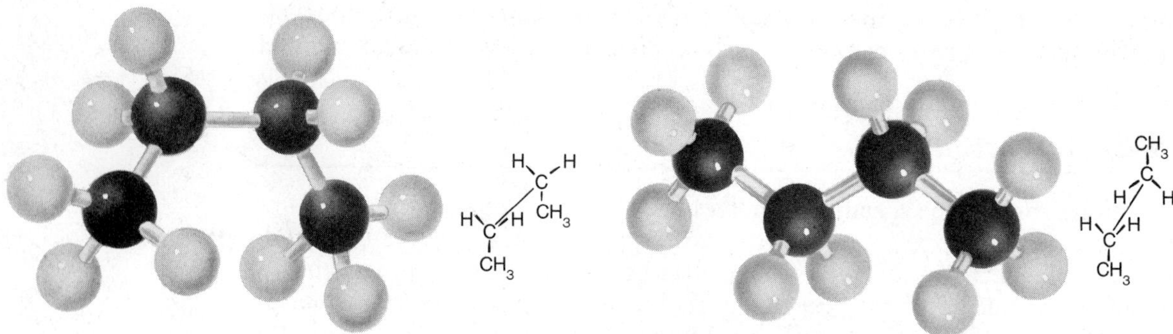

FIGURE 29–8 Two conformations of n-butane, n-C_4H_{10}.

29–5 Substitution Reactions

A reaction in which an atom (or group of atoms) replaces another atom (or group of atoms) in an organic compound is a *substitution reaction.*

Since the saturated hydrocarbons contain only single covalent bonds, they can react without a big disruption of the molecular structure only by *displacement or substitution of one atom for another.* At room temperature, chlorine and bromine react very slowly with saturated straight-chain hydrocarbons. At higher temperatures, and particularly in the presence of sunlight or other source of ultraviolet light, hydrogen atoms in the hydrocarbon can be replaced easily by halogen atoms. The reactions are called **halogenation** reactions. The mechanism of reaction of chlorine with methane may be represented as

$$:\ddot{Cl}:\ddot{Cl}: \xrightarrow[\text{sunlight}]{\text{heat or}} 2:\ddot{Cl}\cdot$$

chlorine molecule — chlorine atoms (free radicals)

A free radical is an atom or group of atoms that contains at least one unpaired electron. Some carry an electrical charge, others do not.

$$H:\underset{H}{\overset{H}{\ddot{C}}}:H + :\ddot{Cl}\cdot \longrightarrow H:\underset{H}{\overset{H}{\ddot{C}}}\cdot + H:\ddot{Cl}:$$

methane — chlorine atom — methyl radical — hydrogen chloride

$$H:\underset{H}{\overset{H}{\ddot{C}}}\cdot + :\ddot{Cl}:\ddot{Cl}: \longrightarrow H:\underset{H}{\overset{H}{\ddot{C}}}:\ddot{Cl}: + :\ddot{Cl}\cdot, \text{ etc.}$$

methyl radical — chlorine molecule — methyl chloride — chlorine atom

The reaction of chlorine with methane is called a **free radical chain reaction.** It is quite similar to the reaction of chlorine with hydrogen (Section 22–10). Chlorine molecules absorb light energy and split into very reactive chlorine atoms. Some chlorine atoms may recombine to form chlorine molecules, but others attack a hydrocarbon molecule, removing one of the hydrogen atoms to form hydrogen chloride. The methyl radical, in turn, reacts with chlorine molecules to form methyl chloride. The overall reaction is usually represented as

Note that only one-half of the chlorine atoms occurs in the organic product. The other half forms hydrogen chloride, a commercially valuable compound.

$$\underset{\text{chlorine}}{Cl{-}Cl} + \underset{\text{methane}}{H{-}\overset{\displaystyle H}{\underset{\displaystyle H}{\overset{|}{\underset{|}{C}}}}{-}H} \xrightarrow[\text{uv}]{\text{heat or}} \underset{\substack{\text{chloromethane, bp = 23.8°C} \\ \text{(methyl chloride)}}}{H{-}\overset{\displaystyle H}{\underset{\displaystyle H}{\overset{|}{\underset{|}{C}}}}{-}Cl} + HCl$$

Many organic reactions are complex, and produce more than a single product. For example, the chlorination of methane may produce several other products in addition to methyl chloride, or chloromethane, as the following equations show

$$Cl{-}Cl + H{-}\overset{\displaystyle H}{\underset{\displaystyle H}{\overset{|}{\underset{|}{C}}}}{-}Cl \longrightarrow \underset{\substack{\text{dichloromethane, bp = 40.2°C} \\ \text{(methylene chloride)}}}{Cl{-}\overset{\displaystyle H}{\underset{\displaystyle H}{\overset{|}{\underset{|}{C}}}}{-}Cl} + HCl$$

$$Cl{-}Cl + Cl{-}\overset{\displaystyle H}{\underset{\displaystyle H}{\overset{|}{\underset{|}{C}}}}{-}Cl \longrightarrow \underset{\substack{\text{trichloromethane, bp = 61°C} \\ \text{(chloroform)}}}{Cl{-}\overset{\displaystyle H}{\underset{\displaystyle Cl}{\overset{|}{\underset{|}{C}}}}{-}Cl} + HCl$$

$$Cl{-}Cl + Cl{-}\overset{\displaystyle H}{\underset{\displaystyle Cl}{\overset{|}{\underset{|}{C}}}}{-}Cl \longrightarrow \underset{\substack{\text{tetrachloromethane, bp = 76.8°C} \\ \text{(carbon tetrachloride)}}}{Cl{-}\overset{\displaystyle Cl}{\underset{\displaystyle Cl}{\overset{|}{\underset{|}{C}}}}{-}Cl} + HCl$$

The mixture of products formed in an organic reaction can be separated by physical methods such as fractional distillation, which relies on differences in the boiling points of different compounds.

When a hydrocarbon has more than one carbon atom, the reaction with chlorine is considerably more complex than the chlorination of methane. The first step in the chlorination of ethane gives a product that contains one chlorine atom per molecule.

Ethyl chloride is widely used by athletic trainers as a spray-on pain killer.

$$Cl{-}Cl + \underset{\text{ethane}}{H{-}\overset{\displaystyle H}{\underset{\displaystyle H}{\overset{|}{\underset{|}{C}}}}{-}\overset{\displaystyle H}{\underset{\displaystyle H}{\overset{|}{\underset{|}{C}}}}{-}H} \xrightarrow[\text{uv}]{\text{heat or}} \underset{\substack{\text{chloroethane, bp = 13.1°C} \\ \text{(ethyl chloride)}}}{H{-}\overset{\displaystyle H}{\underset{\displaystyle H}{\overset{|}{\underset{|}{C}}}}{-}\overset{\displaystyle H}{\underset{\displaystyle H}{\overset{|}{\underset{|}{C}}}}{-}Cl} + HCl$$

When a second hydrogen atom is replaced, two products are possible, and indeed a mixture of both is obtained.

The product mixture does not contain equal numbers of moles of the dichloroethanes, so we do not show a stoichiometrically balanced equation. Since reactions of saturated hydrocarbons with chlorine produce a multiplicity of products, the reactions are not always as useful as might be desired.

$$\text{Cl—Cl} + \text{H—}\underset{\text{H}}{\overset{\text{H}}{\text{C}}}\text{—}\underset{\text{H}}{\overset{\text{H}}{\text{C}}}\text{—Cl} \xrightarrow[\text{uv}]{\text{heat or}} \left\{ \begin{array}{c} \text{H—}\underset{\text{H}}{\overset{\text{H}}{\text{C}}}\text{—}\underset{\text{H}}{\overset{\text{Cl}}{\text{C}}}\text{—Cl} \\ \text{1,1-dichloroethane} \\ \text{bp} = 57°\text{C} \\ \text{Cl—}\underset{\text{H}}{\overset{\text{H}}{\text{C}}}\text{—}\underset{\text{H}}{\overset{\text{H}}{\text{C}}}\text{—Cl} \\ \text{1,2-dichloroethane} \\ \text{bp} = 84°\text{C} \end{array} \right\} + \text{HCl}$$

The cyclic saturated hydrocarbons, or **cycloalkanes,** are more symmetrical and can be converted into single monosubstituted products in good yield. The first four cycloalkanes are indicated below. Note that the general formula for cycloalkanes is C_nH_{2n}.

CH_2 / $H_2C—CH_2$ (ring)	$CH_2—CH_2$ / $CH_2—CH_2$ (ring)	CH_2, CH_2, CH_2, $CH_2—CH_2$ (ring)	CH_2, CH_2, CH_2, CH_2, CH_2, CH_2 (ring)
cyclopropane	cyclobutane	cyclopentane	cyclohexane

The reaction of cyclopentane with chlorine produces only a single product, cyclopentyl chloride or chlorocyclopentane.

$$\text{cyclo-}(CH_2)_5 + Cl_2 \xrightarrow[\text{uv}]{\text{heat or}} \text{cyclo-}(CHCl)(CH_2)_4 + HCl$$

cyclopentane → cyclopentyl chloride

29–6 Oxidation of Alkanes

The principal use of hydrocarbons is as fuels such as gasoline, diesel fuel, and the heating oil and "natural gas" that are used extensively in both residential heating and industry. Hydrocarbons burn in excess oxygen to produce carbon dioxide and water in highly exothermic processes (Section 7–9, parts 3 and 4).

methane $CH_4 + 2O_2 \longrightarrow CO_2 + 2H_2O + 891\ kJ$

n-octane $2C_8H_{18} + 25O_2 \longrightarrow 16CO_2 + 18H_2O + 1.090 \times 10^4\ kJ$

These reactions are the basis for the important uses of the hydrocarbons as fuels. The **heat of combustion** is the amount of energy *liberated* per mole of hydrocarbon burned. Heats of combustion are assigned positive values (Table 29–4), and are therefore equal in magnitude but opposite in sign to ΔH^0 values for combustion reactions. The combustion of hydrocarbons produces large volumes of gases in addition to large amounts of heat. The sudden expansion of these gases drives the pistons or turbine blades in internal combustion engines.

Recall that ΔH^0 is negative for an exothermic process.

TABLE 29–4 Heats of Combustion of Some Alkanes

Hydrocarbon		Heat of Combustion kJ/mol	Heat of Combustion J/g
methane	CH_4	891	55.7
propane	C_3H_8	2220	50.5
n-pentane	*n*-C_5H_{12}	3507	48.7
n-octane	*n*-C_8H_{18}	5450	47.8
n-decane	*n*-$C_{10}H_{22}$	6737	47.4
ethanol*	C_2H_5OH	1372	29.8

* Not an alkane; included for comparison only.

The heat of combustion for ethanol, which is low compared to those of the hydrocarbons, is included in Table 29–4 for comparison with common hydrocarbons. Even when converted to a per-gram basis, the value for ethanol is low compared with those for saturated hydrocarbons. Clearly, the complete combustion of a given mass of ethanol produces considerably less energy than complete burning of the same mass of hydrocarbons in fuels. Ethanol is usually produced by fermentation of grain and other agricultural products. Considerable amounts of petroleum products are required to produce fertilizers to grow crops and to produce fuels to operate farm equipment. Additionally, energy is required to distill alcohol out of the fermentation mixture. Grain that is converted to alcohol must be taken from the world's supply of food. Until other sources of alcohol are found, it is not likely to become an important fuel.

In the absence of sufficient oxygen, partial combustion of hydrocarbons occurs. The products may be carbon monoxide (a very poisonous gas) or carbon (the source of carbon deposits on spark plugs, in the cylinder head, and on the pistons of automobile engines). All hydrocarbons undergo similar reactions. The reactions of methane with insufficient oxygen are

$$2CH_4 + 3O_2 \longrightarrow 2CO + 4H_2O$$

$$CH_4 + O_2 \longrightarrow C + 2H_2O$$

TABLE 29–5 Petroleum Fractions

Fraction*	Principal Composition	Distillation Range
Natural gas	C_1–C_4	below 20°C
Bottled gas	C_5–C_6	20–60°
Gasoline	C_4–C_{12}	40–200°
Kerosene	C_{10}–C_{16}	175–275°
Fuel oil, diesel oil	C_{15}–C_{20}	250–400°
Lubricating oils	C_{18}–C_{22}	above 300°
Paraffin	C_{23}–C_{29}	mp 50–60°
Asphalt		viscous liquid ("bottoms fraction")
Coke		solid

* Other descriptions and distillation ranges have been used, but all are similar.

29–7 Petroleum

Petroleum or crude oil consists mainly of hydrocarbons, but small amounts of organic compounds containing nitrogen, sulfur, and oxygen are present. Each oil field produces petroleum with a particular set of characteristics. Distillation of petroleum produces several fractions, as shown in Table 29–5.

The process is called **thermal cracking.**

Since gasoline is so much in demand, higher hydrocarbons are "cracked" to increase the amount of gasoline that can be made from a barrel of petroleum. Higher-molecular-weight hydrocarbons (C_{15} to C_{18}) are heated, in the absence of air and in the presence of a catalyst, to produce a mixture of lower-molecular-weight hydrocarbons (C_5 to C_{12}) that can be used in gasoline.

Engine knock is caused by premature detonation of fuel in the combustion chamber.

The **octane number** (rating) of a gasoline refers to how smoothly the gasoline burns and how much engine "knock" it produces. Isoöctane, 2,2,4-trimethylpentane, has excellent combustion properties and was assigned an arbitrary octane number of 100. Normal heptane, $CH_3—(CH_2)_5—CH_3$, has very poor combustion properties and was assigned an octane number of zero.

$$CH_3—C(CH_3)_2—CH_2—CH(CH_3)—CH_3 \qquad CH_3—CH_2—CH_2—CH_2—CH_2—CH_2—CH_3$$

isoöctane
octane number = 100

n-heptane
octane number = 0

Mixtures of isoöctane and *n*-heptane were prepared and burned in test engines to establish the "octane" scale. The octane number of such a mixture is the percentage of isoöctane in it. When gasolines are burned in standard test engines, they can be assigned octane numbers based on the compression ratio at which they begin to knock. A 90-octane gasoline produces the same amount of knock as the 90% isoöctane–10% *n*-heptane mixture. Branched-chain compounds produce less knock than straight-chain compounds. The octane numbers of two isomeric hexanes illustrate the point.

$$CH_3—(CH_2)_4—CH_3 \qquad (CH_3)_3C—CH_2—CH_3$$

n-hexane
octane number = 25

2,2-dimethylbutane
octane number = 92

For many years tetraethyllead, $Pb(C_2H_5)_4$, was used as a gasoline additive to improve smoothness of burning and increase octane rating. However, the discharge of lead into the atmosphere is believed to be hazardous to health, and the lead compounds "poison" the catalyst in catalytic converters. Therefore, "ethyl" gasoline is being phased out of use.

Gasoline containing tetraethyllead is known as "ethyl" gasoline or "leaded" gasoline.

Unsaturated Hydrocarbons

There are three classes of unsaturated hydrocarbons: (1) the alkenes and their cyclic counterparts, the cycloalkenes, (2) the alkynes and the cycloalkynes, and (3) the aromatic hydrocarbons.

29–8 The Alkenes (Also Called Olefins)

The general formula for the simplest **alkenes** is C_nH_{2n} because these compounds contain one carbon-carbon double bond per molecule. The roots for their names are derived from the alkanes having the same number of carbon atoms as the longest chain containing the double bond. In the common (sometimes called trivial) system of nomenclature, the suffix *-ylene* is added to the characteristic root. In systematic (IUPAC) nomenclature, the suffix *-ene* is added to the characteristic root. In chains of four or more carbon atoms, the position of the double bond is indicated by a numerical prefix that shows the *lowest numbered* doubly bonded carbon atom.

Recall that the cycloalkanes may also be represented by the general formula C_nH_{2n}.

There is only one position possible for a double bond in chains of two or three carbon atoms.

	$CH_2{=}CH_2$	$CH_3{-}CH{=}CH_2$	$\overset{4}{C}H_3\overset{3}{C}H_2{-}\overset{2}{C}H{=}\overset{1}{C}H_2$
trivial	ethylene	propylene	*n*-butylene
systematic	ethene	propene	1-butene

$\overset{1}{C}H_3\overset{2}{C}H{=}\overset{3}{C}H\overset{4}{C}H_3$ — 2-butene

$CH_3{-}\underset{\displaystyle CH_3}{\underset{|}{C}}{=}CH_2$ — isobutylene, methylpropene

In naming more complex alkenes, the double bond takes (positional) preference over substituents on the carbon chain and is always assigned the lowest possible number.

$\overset{4}{C}H_3{-}\overset{3}{C}H_2{-}\underset{\displaystyle CH_3}{\underset{|}{\overset{2}{C}}}{=}\overset{1}{C}H_2$ — 2-methyl-1-butene

$\overset{1}{C}H_3{-}\overset{2}{C}H{=}\overset{3}{C}H{-}\underset{\displaystyle CH_3}{\underset{|}{\overset{4}{C}H}}{-}\overset{5}{C}H_3$ — 4-methyl-2-pentene

numbers are separated by commas, & the number and names are separated by the "hyphen"

Some alkenes have two or more carbon-carbon double bonds per molecule. The suffixes -adiene, -atriene, etc. are used to indicate the number of (C=C) double bonds in a molecule.

$\overset{1}{C}H_2{=}\overset{2}{C}H{-}\overset{3}{C}H{=}\overset{4}{C}H_2$ — 1,3-butadiene

$\overset{4}{C}H_3{-}\overset{3}{C}H{=}\overset{2}{C}{=}\overset{1}{C}H_2$ — 1,2-butadiene

1,3-Butadiene and similar molecules that contain alternating single and double bonds are described as having **conjugated double bonds.** Such compounds are of special interest because of the polymerization reactions they undergo (Section 29–9).

The cycloalkenes are represented by the general formula C_nH_{2n-2}

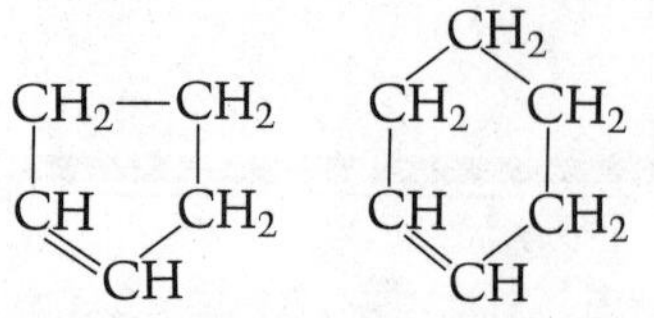

```
CH2—CH2          CH2
|    |         /    \
CH   CH2     CH2    CH2
 \\ /         |      |
  CH          CH    CH2
               \\  /
                CH
```

cyclopentene   cyclohexene

The bonding in ethylene was described in detail in Section 5–18. The hybridization (sp^2) and bonding at other double-bonded carbon atoms is similar. Recall that both carbon atoms in C_2H_4 are located at the centers of trigonal planes.

Rotation about C=C double bonds is *not* possible. (In Section 29–3, we pointed out that rotation about C—C single bonds is possible.) Therefore, compounds that have the general formulas (XY)C=C(XY) exist as a pair of *cis-trans* isomers. Figure 29–9 shows the *cis-trans* isomers of 1,2-dichloroethene.

Carbon-carbon double bonds are shorter than carbon-carbon single bonds, 1.34 Å versus 1.54 Å, because the two shared electron pairs draw the carbon nuclei closer than does a single electron pair. Although the physical properties of the alkenes are similar to those of the alkanes, their chemical properties are quite different.

29–9 Reactions of the Alkenes

Most reactions of the *alkanes* that do not disrupt the carbon skeleton are **substitution reactions,** but the *alkenes* are characterized by **additions** to the double bond. Consider the reactions of ethane and ethene with chlorine

ethane $CH_3—CH_3 + Cl_2 \longrightarrow CH_3—CH_2Cl + HCl$ (substitution)

ethene $CH_2=CH_2 + Cl_2 \longrightarrow \underset{Cl}{CH_2}—\underset{Cl}{CH_2}$ (addition)

More functional groups will be considered later.

Carbon-carbon double bonds represent **reaction sites,** and so are called **functional groups.** Most addition reactions proceed rapidly at room tempera-

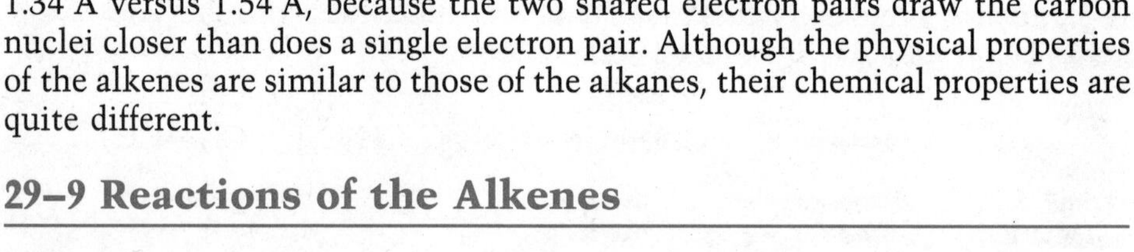

FIGURE 29–9 Two isomers of 1,2-dichloroethene are possible because rotation around the double bond is restricted. This is an example of geometric isomerism.

cis-1,2-dichloroethene
m.p. −80.5°C, b.p. 60.3°C

trans-1,2-dichloroethene
m.p. −50°C, b.p. 47.5°C

ture. By contrast, the substitution reactions of the alkanes usually require catalysts and elevated temperatures.

Bromine adds readily to the alkenes to give dibromides. The reaction with ethene is

$$Br_2 + CH_2{=}CH_2 \longrightarrow \underset{\displaystyle Br\qquad Br}{CH_2{-}CH_2}$$

1,2-dibromoethane
(ethylene dibromide)

Ethylene dibromide is added to gasoline that contains tetraethyllead, $Pb(C_2H_5)_4$, to prevent the accumulation of lead in engines. It forms $PbBr_2$, which is volatile at engine-operating temperatures and is expelled through the exhaust system. The lead bromide "poisons" the catalytic converters in some modern automobiles, so the use of lead-containing gasoline in these automobiles is prohibited by law.

The addition of bromine to alkenes is used as a simple qualitative test for unsaturation. Bromine, a dark red liquid, is dissolved in a nonpolar solvent such as methylene chloride, CH_2Cl_2. When an alkene is added, the solution becomes colorless as the bromine reacts with the alkene to form a colorless compound. This reaction may be used to distinguish between alkanes and alkenes.

Hydrogenation is an extremely important reaction of the alkenes. At elevated temperatures, under high pressures, and in the presence of an appropriate catalyst (finely divided Pt, Pd, or Ni), hydrogen adds across double bonds.

Note that hydrogenation corresponds to reduction.

$$CH_2{=}CH_2 + H_2 \xrightarrow[\text{heat}]{\text{catalyst}} CH_3{-}CH_3$$

ethene ethane

Unsaturated hydrocarbons are converted to saturated hydrocarbons in the manufacture of high octane gasoline and aviation fuels. Unsaturated vegetable oils may also be converted to solid cooking fats (shortening) by hydrogenation (Figure 29–10).

$$\begin{array}{l} CH_2OC(=O)(CH_2)_7CH{=}CH(CH_2)_7CH_3 \\ | \\ CHOC(=O)(CH_2)_7CH{=}CH(CH_2)_7CH_3 \\ | \\ CH_2OC(=O)(CH_2)_7CH{=}CH(CH_2)_7CH_3 \end{array} \xrightarrow[\text{Ni catalyst, heat}]{3H_2} \begin{array}{l} CH_2OC(=O)(CH_2)_{16}CH_3 \\ | \\ CHOC(=O)(CH_2)_{16}CH_3 \\ | \\ CH_2OC(=O)(CH_2)_{16}CH_3 \end{array}$$

olein, an oil, liquid — stearin, a fat, solid

Polymerization, the combination of many small molecules to form large molecules (polymers), is another important reaction of the alkenes. Polyethylene provides a common example. In the presence of appropriate catalysts (a

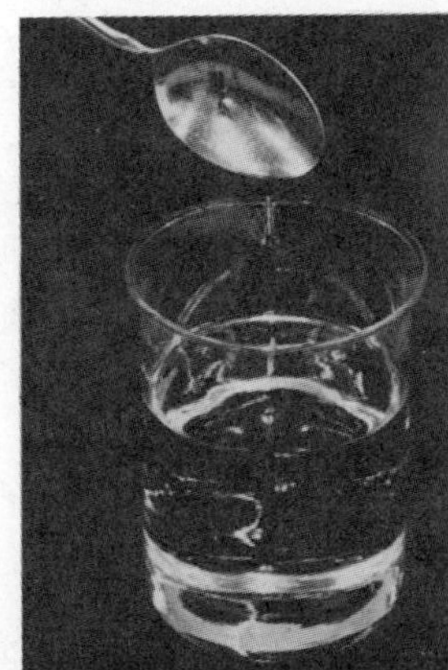

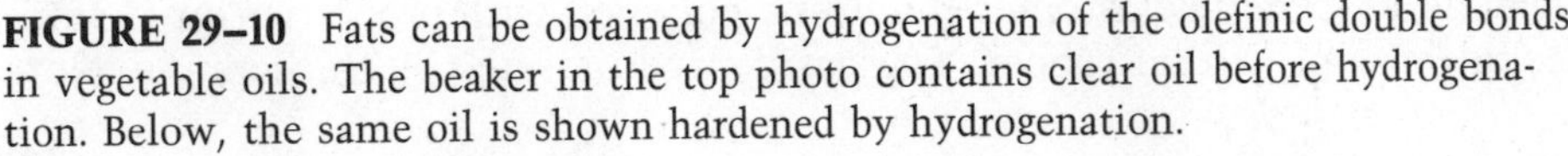

FIGURE 29–10 Fats can be obtained by hydrogenation of the olefinic double bonds in vegetable oils. The beaker in the top photo contains clear oil before hydrogenation. Below, the same oil is shown hardened by hydrogenation.

mixture of aluminum trialkyls, R_3Al, and titanium tetrachloride, $TiCl_4$), ethylene polymerizes into chains containing 800 or more carbon atoms.

$$nCH_2{=}CH_2 \xrightarrow{\text{catalyst}} \!\!+\!CH_2{-}CH_2\!+\!\!_n$$

ethylene — polyethylene

The polymer may be represented as $CH_3(CH_2{-}CH_2)_nCH_3$ where n is approximately 400. Polyethylene is a tough, flexible plastic widely used as an electrical insulator and for the fabrication of such things as unbreakable refrigerator dishes, plastic cups, and squeeze bottles. Polypropylene is made by polymerizing propylene, $CH_3{-}CH{=}CH_2$, in much the same way that polyethylene is produced.

Teflon® is made by polymerizing tetrafluoroethylene in a reaction that is similar to the formation of polyethylene.

$$nCF_2{=}CF_2 \xrightarrow[\text{heat}]{\text{catalyst}} \!\!+\!CF_2{-}CF_2\!+\!\!_n$$

tetrafluoroethylene — "Teflon"

The molecular weight of Teflon is between 1×10^6 and 2×10^6; i.e., approximately 20,000 $CF_2{=}CF_2$ molecules polymerize to form a single giant molecule. Teflon is a very useful polymer in that it does *not* react with concentrated acids and bases, or with most oxidizing agents, nor does it dissolve in most organic solvents.

Natural rubber is obtained from the sap of the rubber tree, a sticky liquid called latex. Rubber is a polymeric hydrocarbon formed (in the sap) by the combination of about 2000 molecules of 2-methyl-1,3-butadiene, commonly called isoprene. The molecular weight of rubber is about 136,000.

$$2nCH_2{=}\overset{\overset{\displaystyle CH_3}{|}}{C}{-}CH{=}CH_2 \longrightarrow \!+\!CH_2{-}\overset{\overset{\displaystyle CH_3}{|}}{C}{=}CH{-}CH_2{-}CH_2{-}\overset{\overset{\displaystyle CH_3}{|}}{C}{=}CH{-}CH_2\!+\!\!_n$$

isoprene

When natural rubber becomes warm, it flows and becomes sticky. To eliminate this problem, **vulcanization** is used. This is the process in which sulfur is added to rubber and the mixture is heated to approximately 140°C. Sulfur atoms combine with some of the double bonds in the linear polymer molecules to form bridges that bond one rubber molecule to another. This cross-linking by sulfur atoms converts the linear polymer into a three-dimensional polymer. Fillers and reinforcing agents are added during the mixing process to improve the wearing qualities of rubber and to form colored rubber. Carbon black is the most common reinforcing agent, and zinc oxide, barium sulfate, titanium dioxide, and antimony(V) sulfide are common fillers.

Numerous other polymers are elastic enough to come under the generic name of rubber.

Some synthetic rubbers are superior to natural rubber in some ways. Neoprene is a synthetic elastomer (an elastic polymer) with properties quite similar to those of natural rubber. The basic structural unit is chloroprene. Note that chloroprene differs from isoprene in having a chlorine atom rather than a methyl group as a substituent on the 1,3-butadiene chain.

$$n\text{CH}_2\!=\!\text{CH}-\underset{\displaystyle\text{C}}{\overset{\displaystyle\text{Cl}}{|}}\!=\!\text{CH}_2 \xrightarrow{\text{polymerization}} \text{+CH}_2-\text{CH}\!=\!\underset{\displaystyle\text{C}}{\overset{\displaystyle\text{Cl}}{|}}-\text{CH}_2\text{+}_n$$

chloroprene — neoprene

Neoprene is less affected by gasoline and oil, and is more elastic than natural rubber. It resists abrasion well, and is not swollen or dissolved by hydrocarbons. It is widely used to make hoses for oil and gasoline, electrical insulation, and automobile and refrigerator parts.

29–10 Oxidation of Alkenes

The alkenes, like the alkanes, burn in excess oxygen to form carbon dioxide and water in exothermic reactions.

$$CH_2{=}CH_2 + 3O_2 \text{ (excess)} \longrightarrow 2CO_2 + 2H_2O + 1387 \text{ kJ}$$

When an alkene (or any other unsaturated organic compound) is burned in air, a yellow luminous flame is observed and considerable amounts of soot, unburned carbon, are obtained. This reaction provides a qualitative test for unsaturation, because saturated hydrocarbons burn in air without the production of significant amounts of soot.

Another qualitative test for unsaturation involves the reaction of alkenes (and, as we shall see presently, the alkynes) with cold dilute aqueous solutions of potassium permanganate, $KMnO_4$. Potassium permanganate solutions are purple to pink, depending on their concentrations. The addition of an unsaturated hydrocarbon results in the disappearance of the characteristic permanganate color and the formation of manganese dioxide, MnO_2, a brown solid. With ethene the reaction is

$$3CH_2{=}CH_2 + 2KMnO_4 + 4H_2O \longrightarrow 3\underset{\displaystyle\text{OH}}{\underset{|}{\text{CH}_2}}-\underset{\displaystyle\text{OH}}{\underset{|}{\text{CH}_2}} + 2MnO_2 + 2KOH$$

ethylene glycol

Other alkenes undergo similar reactions. Hot or concentrated $KMnO_4$ solutions oxidize ethane all the way to carbon dioxide and water.

29–11 The Alkynes

The **alkynes,** or acetylenic hydrocarbons, contain carbon-carbon triple bonds. Those with one triple bond per molecule have the general formula C_nH_{2n-2}. They are named like the alkenes except that the suffix -yne is added to the characteristic root. The first member of the series is commonly called acetylene, and experiments show that its molecular formula is C_2H_2. It may be transformed into ethene and then to ethane by the addition of hydrogen. These reactions suggest that the formula for acetylene is H—C≡C—H.

$CH\equiv CH$	$CH_3—C\equiv CH$	$CH_3—CH_2—C\equiv CH$	$CH_3—C\equiv C—CH_3$	$CH_3—CH(CH_3)—C\equiv CH$
ethyne acetylene	propyne	1-butyne	2-butyne	3-methyl-1-butyne

The triple bond takes positional preference over substituents on the carbon chain and is assigned the lowest possible number in naming.

The bonding in alkynes is similar to that described specifically for acetylene in Section 5–19. The triply bonded carbon atoms are all *sp* hybridized, and they and the adjacent atoms lie on a straight line (Figure 29–11).

$$CaC_2 + 2H_2O \longrightarrow CH\equiv CH + Ca(OH)_2$$

Acetylene lamps are charged with calcium carbide. Very slow addition of water produces acetylene, which is burned as it is produced. Acetylene is also used in the oxyacetylene torch for welding and cutting metals. When it is burned with oxygen, the flame reaches temperatures of 3000°C.

$$2CH\equiv CH + 5O_2 \longrightarrow 4CO_2 + 2H_2O + 2611 \text{ kJ}$$

Since the alkynes contain two pi bonds, both of which are concentrated sources of electrons, they are more reactive than the alkenes. The most common reaction of the alkynes is addition across the triple bond. The reactions with hydrogen and with bromine are typical.

$$H—C\equiv C—H \xrightarrow{H_2} H_2C=CH_2 \xrightarrow{H_2} H_3C—CH_3$$

$$H—C\equiv C—H \xrightarrow{Br_2} \underset{\text{1,2-dibromoethene}}{HBrC=CBrH} \xrightarrow{Br_2} \underset{\text{1,1,2,2-tetrabromoethane}}{H—CBr_2—CBr_2—H}$$

These reactions are thought of as stepwise reactions, but stopping them after the first step is difficult because the products still contain a reactive double bond.

Aromatic Hydrocarbons

Originally the word **aromatic** was applied to pleasant-smelling substances. We now use the word to describe benzene, its derivatives, and certain other compounds that exhibit similar chemical properties. Some have very foul odors because of substituents on the benzene ring. On the other hand, many fragrant compounds do not contain benzene rings.

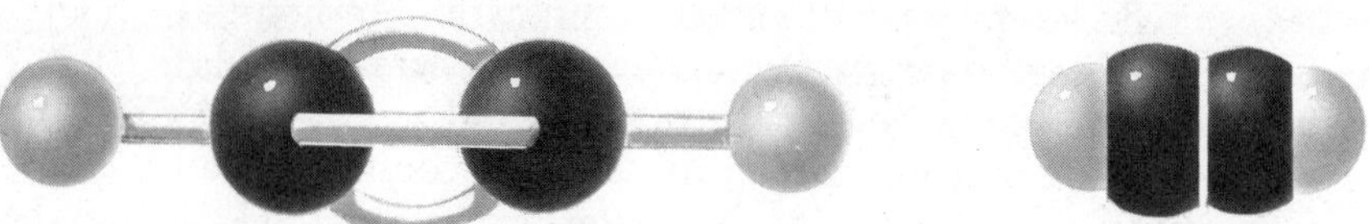

FIGURE 29–11 Models of acetylene, H—C≡C—H.

In the nineteenth century the development of the coal tar industry in Germany provided an enormous stimulus to the systematic study of organic chemistry. The production of steel requires large amounts of coke, which is prepared by heating bituminous coal to high temperatures in the absence of air. The production of coke also results in the formation of *coal gas* and *coal tar;* the latter serves as a source of aromatic compounds.

Distillation of coal tar produces a variety of aromatic compounds, as we shall see shortly.

When a ton of coal is converted to coke, about 5 kg of aromatic compounds can be obtained from it. Because of the enormous amount of coal converted to coke, coal tar is produced in large quantities.

The principal components of coal gas are hydrogen (~50%) and methane (~30%). Small amounts of NH_3, N_2, CO, CO_2, H_2O, and some low-molecular-weight hydrocarbons are also present.

29–12 Benzene

Benzene is the simplest aromatic hydrocarbon, and by studying its reactions we can learn a great deal about aromatic hydrocarbons. Benzene was discovered in 1825 by Michael Faraday when he fractionally distilled a by-product oil obtained in the manufacture of illuminating gas from whale oil.

Elemental analysis and determination of its molecular weight show that the true formula for benzene is C_6H_6. The formula suggests that it is a highly unsaturated substance, but its properties are quite different from those of open-chain unsaturated hydrocarbons. It reacts with neither bromine water (aqueous Br_2) nor solutions of potassium permanganate, reagents that invariably react with unsaturated *aliphatic* hydrocarbons.

Aliphatic hydrocarbons contain no aromatic rings.

The facts that only one monosubstitution product is obtained in numerous reactions and that no addition products can be prepared indicate conclusively that benzene has a *symmetrical ring structure.* Stated differently, every hydrogen atom is equivalent to every other hydrogen atom, and this is possible only in a symmetrical ring structure:

```
      H
      C
  HC     CH
  |      |        (skeleton)
  HC     CH
      C
      H
```

Only the *positions* of the atoms are indicated in this structure. The debate over the structure and bonding in benzene raged for about 30 years. In 1865, Friedrich Kekulé suggested that the structure of benzene was intermediate between two structures that we now call resonance structures. He actually proposed an ill-defined equilibrium between them

```
      H                   H
      C                   C
  HC//   CH          HC     \\CH
  |      ||   <-->   ||       |
  HC\\   CH          HC     //CH
      C                   C
      H                   H
```

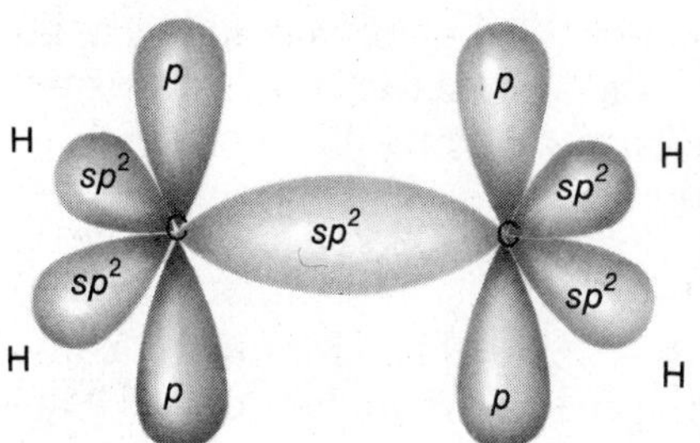

FIGURE 29–12 sp^2 hybridized carbons with *p* orbitals parallel.

We now describe benzene as having a structure intermediate between the two resonance structures shown, and represent it as

H
C
HC CH
HC CH
C
H

or more simply as

Benzene molecules are planar (all 12 atoms lie in a plane). This suggests sp^2 hybridization of each carbon, with one electron being promoted to the unhybridized *p* orbital (Figure 29–12), as in ethene (Section 5–18).

The structure of benzene is described in detail in Section 6–6, part 2 in terms of MO theory.

The six sp^2 hybridized carbon atoms lie in a plane, and the unhybridized *p* orbitals extend above and below the plane. Side-by-side overlap of the *p* orbitals leads them to merge into pi orbitals (Figure 6–12).

The electrons associated with the pi bonds are *delocalized* over the entire benzene ring (recall that all C's are chemically equivalent, and so are all the H's); this is depicted in Figure 29–13.

Thus, the available data support writing the formula for benzene as the intermediate form shown. Scale models of benzene and toluene (methylbenzene) are shown in Figures 29–13 and 29–14.

29–13 Other Aromatic Hydrocarbons

The distillation of coal tar provides four volatile fractions, as well as pitch that is used for surfacing roads and in the manufacture of "asphalt" roofing (Figure 29–15).

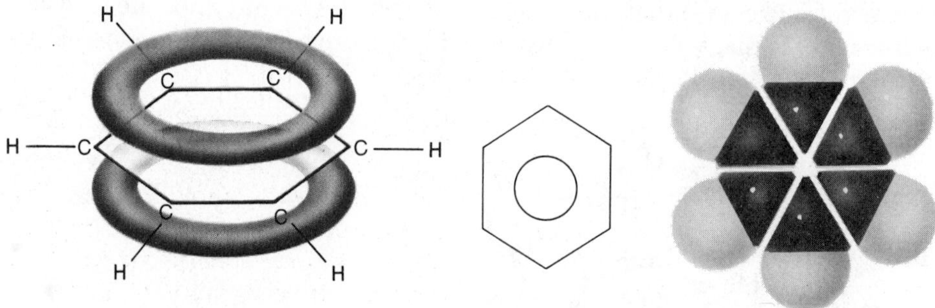

FIGURE 29–13 The electron distribution and models of the benzene, C_6H_6, molecule.

Eight aromatic hydrocarbons are obtained in significant amounts by efficient fractional distillation of the "light oil" fraction. They are called "coal-tar crudes" (Table 29–6).

The structures for benzene and toluene were given earlier. Note that three isomeric **xylenes** are listed in Table 29–6. The *o-*, *m-*, and *p-* refer to relative positions of substituents on the benzene ring. The xylenes are dimethylbenzenes, of which there are three distinctly different compounds that have the formula $C_6H_4(CH_3)_2$. The *ortho-* prefix refers to two substituents located on *adjacent* carbon atoms; i.e., 1,2-dimethylbenzene is *o*-xylene. The *meta-* prefix refers to substituents located on carbon atoms 1 and 3. Thus, 1,3-dimethylbenzene is *m*-xylene. The *para-* prefix refers to substituents located on carbon atoms 1 and 4, and 1,4-dimethylbenzene is *p*-xylene.

The structures of naphthalene, anthracene, and phenanthrene are

naphthalene, $C_{10}H_8$ anthracene, $C_{14}H_{10}$ phenanthrene, $C_{14}H_{10}$

These compounds are examples of "condensed" or "fused" ring systems. Note that there are no hydrogen atoms attached to the carbon atoms that are involved in fusion of *aromatic* rings. Many naturally occurring compounds contain fused rings. For example, fused rings occur in many hormones such as estradiol and testosterone, the major female and male sex hormones, respectively.

estradiol testosterone

ortho-xylene
bp = 144°C
mp = −27°C

meta-xylene
bp = 139°C
mp = −54°C

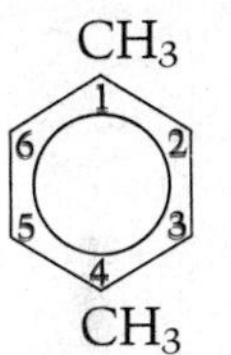

para-xylene
bp = 138°C
mp = 13°C

Note that the rings in the major sex hormones are mostly cycloalkane type rings rather than aromatic rings. Structural formulas imply the presence of H atoms to complete the bonding at each C.

29–14 Reactions of Aromatic Hydrocarbons

Early research on the reactions of the aromatic hydrocarbons was richly rewarded as chemists learned to prepare a great variety of dyes, drugs, flavors and perfumes, and explosives. More recently, large numbers of polymeric materials such as plastics and fabrics have been prepared from these "coal-tar crudes."

The most common kind of reaction of the aromatic ring is substitution, i.e., replacement of a hydrogen atom by another atom or group of atoms.

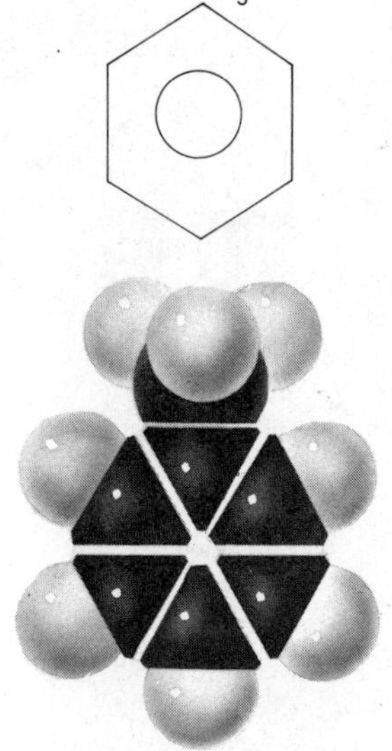

FIGURE 29–14 Models of toluene, $C_6H_5CH_3$, which may be thought of as a derivative of benzene in which one H atom has been replaced by a $—CH_3$ group.

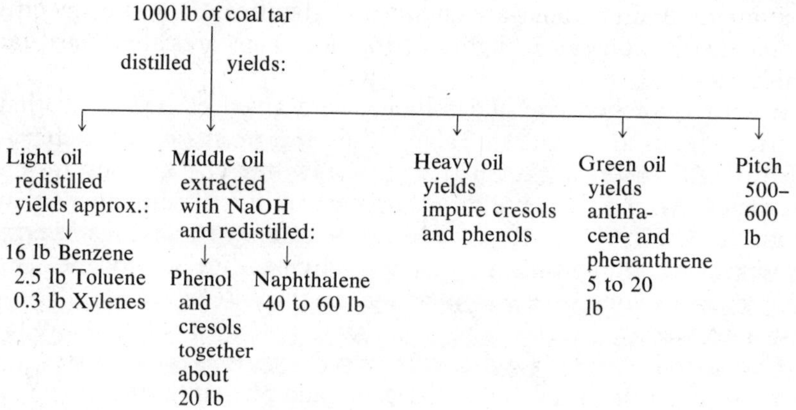

1000 lb of coal tar

distilled | yields:

- **Light oil** redistilled yields approx.: ↓ 16 lb Benzene, 2.5 lb Toluene, 0.3 lb Xylenes
- **Middle oil** extracted with NaOH and redistilled: ↓ Phenol and cresols together about 20 lb; ↓ Naphthalene 40 to 60 lb
- **Heavy oil** yields impure cresols and phenols
- **Green oil** yields anthracene and phenanthrene 5 to 20 lb
- **Pitch** 500–600 lb

FIGURE 29–15 Fractions obtained from coal tar.

Halogenation, with chlorine or bromine, occurs readily in the presence of iron or anhydrous iron(III) chloride (a Lewis acid) catalyst.

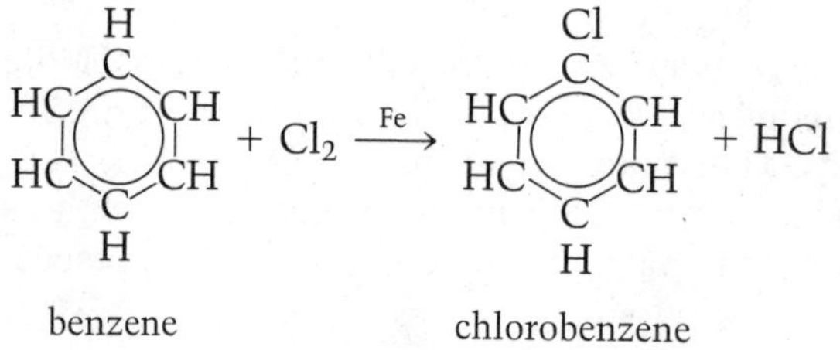

$$C_6H_6 + Cl_2 \xrightarrow{Fe} C_6H_5Cl + HCl$$

benzene   chlorobenzene

When iron is used as a catalyst, it reacts with chlorine to form iron(III) chloride.

The equation is usually written in condensed form as

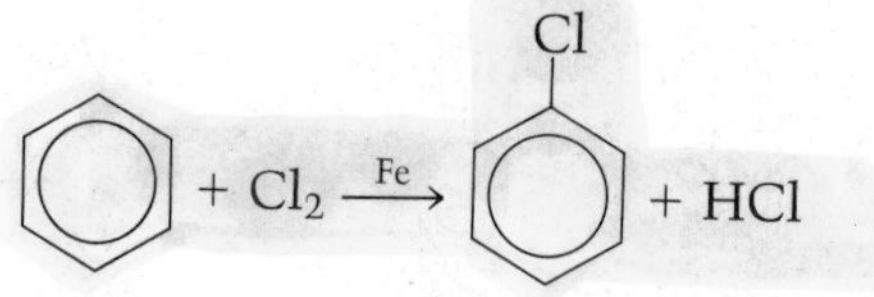

$$C_6H_6 + Cl_2 \xrightarrow{Fe} C_6H_5Cl + HCl$$

TABLE 29–6 The Coal-Tar Crudes

Name	Formula	Boiling Point °C	Melting Point °C	Solubility
benzene	C_6H_6	80	+6	
toluene	$C_6H_5CH_3$	111	−95	
o-xylene	$C_6H_4(CH_3)_2$	144	−27	All
m-xylene	$C_6H_4(CH_3)_2$	139	−54	insoluble
p-xylene	$C_6H_4(CH_3)_2$	138	+13	in
naphthalene	$C_{10}H_8$	218	+80	water
anthracene	$C_{14}H_{10}$	342	+218	
phenanthrene	$C_{14}H_{10}$	340	+101	

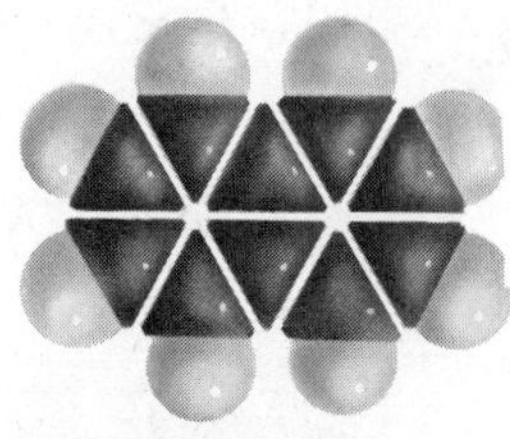

FIGURE 29–16 A scale model of naphthalene, $C_{10}H_8$, the simplest "condensed" aromatic ring system.

Aromatic rings can undergo *nitration* in a mixture of concentrated nitric and sulfuric acids at low temperatures. The sulfuric acid is a catalyst and a dehydrating agent.

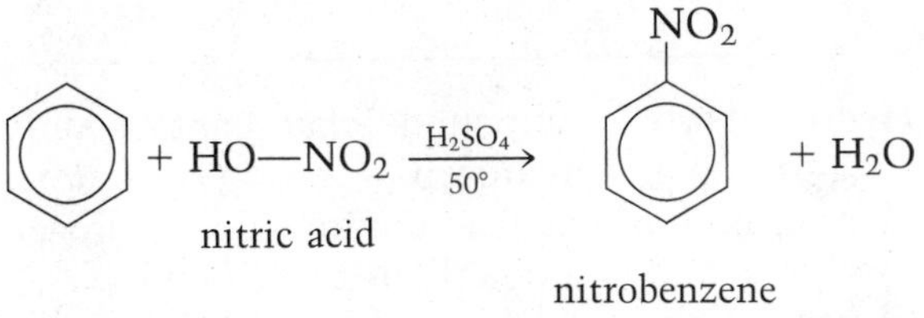

$$C_6H_6 + HO{-}NO_2 \xrightarrow[50°]{H_2SO_4} C_6H_5NO_2 + H_2O$$

nitric acid — nitrobenzene

isomers:
1. positional
2. geometric
3. optical
4. functional.

TNT (2,4,6-trinitrotoluene) is manufactured by the nitration of toluene in several steps

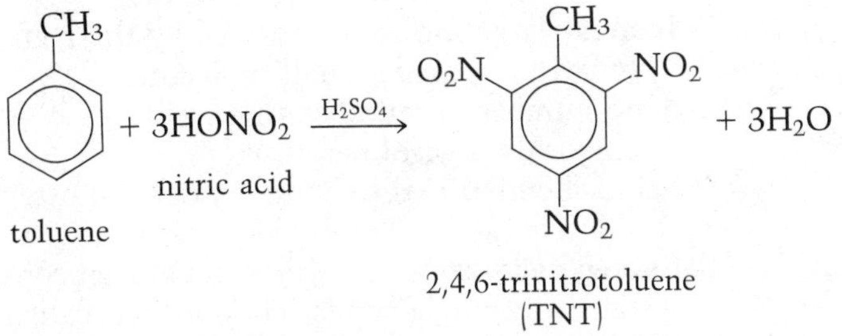

$$C_6H_5CH_3 + 3HONO_2 \xrightarrow{H_2SO_4} CH_3C_6H_2(NO_2)_3 + 3H_2O$$

toluene — nitric acid — 2,4,6-trinitrotoluene (TNT)

Like other hydrocarbons, aromatic hydrocarbons burn in air to release large amounts of energy

$$2C_6H_6 + 15O_2 \longrightarrow 12CO_2 + 6H_2O + 6548\ kJ$$

benzene

The reactions of strong oxidizing agents with alkylbenzenes, which are benzene molecules containing alkyl side chains, illustrate strikingly the stability of the benzene ring system. Heating toluene with a basic solution of potassium permanganate results in a nearly quantitative yield of benzoic acid. The ring itself remains intact; only the nonaromatic portion of the molecule is oxidized

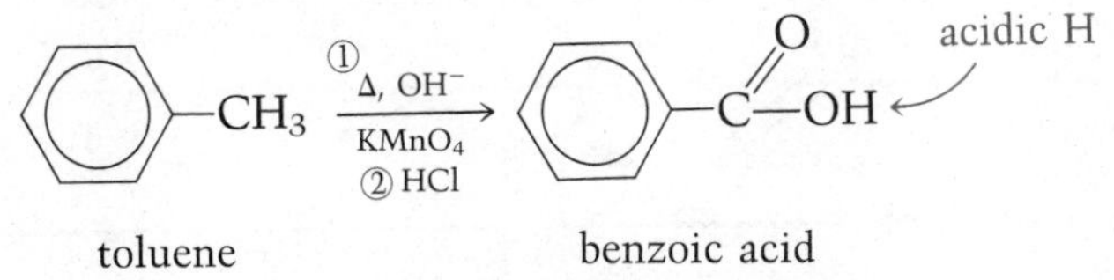

$$C_6H_5{-}CH_3 \xrightarrow[KMnO_4\ ②\ HCl]{①\ \Delta,\ OH^-} C_6H_5{-}\overset{O}{\overset{\|}{C}}{-}OH$$

toluene — benzoic acid (acidic H)

Frequently only the product of interest is shown in organic reactions. The reduction product is MnO_2, which is removed by filtration of the basic solution. Acidification with a strong inorganic acid results in the precipitation of benzoic acid, an insoluble weak acid.

Two or more alkyl groups on an aromatic ring are oxidized to yield a diprotic acid, as the following example illustrates.

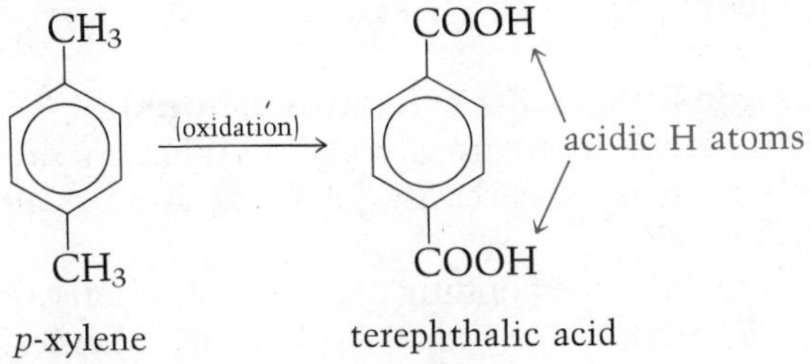

$$CH_3C_6H_4CH_3 \xrightarrow{(oxidation)} HOOC{-}C_6H_4{-}COOH$$

p-xylene — terephthalic acid (acidic H atoms)

Organic reactions are sometimes written in extremely abbreviated form. This is often the case when a variety of common oxidizing agents will accomplish the desired conversion.

As we shall see in the next chapter, terephthalic acid is used to make "polyesters," an important class of polymers.

Key Terms

Addition reaction a reaction in which two atoms or groups of atoms are added to a molecule, one on each side of a double or triple bond.

Alkanes see *Saturated hydrocarbons*

Alkenes (olefins) unsaturated hydrocarbons that contain one or more carbon-carbon double bonds.

Alkyl group a group of atoms derived from an alkane by the removal of one hydrogen atom.

Alkylbenzene a compound containing an alkyl group bonded to a benzene ring.

Alkynes unsaturated hydrocarbons that contain one or more carbon-carbon triple bonds.

Aromatic hydrocarbons benzene and its derivatives.

Catenation the ability of an element to bond to itself.

Conformations structures of a compound that differ by the extent of rotation about a single bond.

Conjugated double bonds double bonds that are separated from each other by one single bond, as in C=C—C=C.

Cycloalkanes cyclic saturated hydrocarbons.

Homologous series a series of compounds in which each member differs from the next by a specific number and kind of atoms.

Hydrocarbons compounds that contain only carbon and hydrogen.

Hydrogenation the reaction in which hydrogen adds across a double or triple bond.

Pi bonds chemical bonds formed by the side-to-side overlap of atomic orbitals.

Polymerization the combination of many small molecules to form large molecules.

Polymers large molecules formed by the combination of many small molecules.

Octane number a number that indicates how smoothly a gasoline burns.

Organic chemistry the chemistry of substances that contain carbon-hydrogen bonds.

Saturated hydrocarbons hydrocarbons that contain only single bonds. They are also called *alkanes* or *paraffin hydrocarbons.*

Sigma bonds chemical bonds formed by the end-to-end overlap of atomic orbitals.

Structural isomers compounds that contain the same number of the same kinds of atoms in different geometric arrangements.

Substitution reaction a reaction in which an atom or a group of atoms is replaced by another atom or group of atoms.

Vulcanization the process in which sulfur is added to rubber and heated to 140°C.

Unsaturated hydrocarbons hydrocarbons that contain double or triple carbon-carbon bonds.

Exercises

Basic Ideas

1. (a) What is organic chemistry? (b) What was the "vital force" theory? (c) What happened to the "vital force" theory?
2. (a) What is catenation? (b) How is carbon unique among the elements?
3. How many "everyday" uses of organic compounds can you think of? List them.
4. (a) What are the principal sources of organic compounds? (b) Some chemists argue that the ultimate source of all naturally occurring organic compounds is carbon dioxide. Could this be possible? Hint: Think about the origin of coal, natural gas, and petroleum.

Alkanes (Saturated Hydrocarbons)

5. (a) What are hydrocarbons? (b) What are saturated hydrocarbons? (c) What are the alkanes?
6. Describe the bonding in and the geometry of molecules of the following alkanes.

(a) methane, (b) ethane, (c) propane, (d) *n*-butane. How are the formulas for these compounds similar? Different?

7. (a) What are "normal" hydrocarbons? (b) What are branched-chain hydrocarbons? (c) Cite three examples of each.
8. (a) What is a homologous series? (b) Provide specific examples of compounds that are members of a homologous series. (c) What is a methylene group? (d) How does each member of a homologous series differ from compounds that come before and after it in the series?
9. (a) How do the melting points and boiling points of the normal alkanes vary with molecular weight? (b) Do you expect them to vary in this order? (c) Why?

Isomerism and Nomenclature

10. (a) What are structural isomers? (b) Draw all possible structural isomers for compounds that have the following formulas: C_4H_{10}, C_5H_{12}, C_6H_{14}. (c) Why are there no structural isomers for C_3H_8?
11. Draw structural formulas for and write the names of the first ten straight-chain alkanes.
12. (a) What are alkyl groups? (b) Draw structures for and write the names of five alkyl groups. (c) What is the origin of the name for an alkyl group?
13. Write the name for each structural isomer you drew in answering Exercise 10.
14. What are conformations?

Reactions of the Alkanes

15. Why are the alkanes also called paraffin hydrocarbons?
16. (a) What is a substitution reaction? (b) What is a halogenation reaction?
17. (a) Describe the reaction of methane with chlorine in the presence of ultraviolet light. (b) Write equations that show formulas for all compounds that can be formed by reaction (a). (c) Write names for all compounds in these equations.
18. (a) Describe the reaction of ethane with chlorine in the presence of ultraviolet light. (b) Write equations that show formulas for all compounds that can be formed by reaction (a). (c) Write names for all compounds in these equations.
19. Why are the halogenation reactions of the alkanes of limited value?
20. What is a free radical chain reaction?
21. (a) What are the cycloalkanes? (b) How do they differ from other alkanes?
22. Why does halogenation of cycloalkanes give a single product, whereas halogenation of normal alkanes gives a multiplicity of products?

Oxidation of Alkanes

23. (a) What are the principal uses of hydrocarbons? (b) Why are hydrocarbons used for these purposes? (c) What does heat of combustion mean? Illustrate.
24. How does the heat of combustion of ethyl alcohol compare with the heats of combustion of saturated hydrocarbons on a per mole as well as a per gram basis?

Petroleum

25. (a) What is petroleum? (b) What are the principal fractions obtained by the distillation of petroleum? (c) For what is each fraction used?
26. (a) What do we mean when we say that higher molecular weight hydrocarbons can be cracked? (b) Why is "cracking" important?
27. (a) To what does the octane rating of a gasoline refer? (b) Why is octane rating important?
28. What kinds of hydrocarbons have low octane ratings? High octane ratings?

Unsaturated Hydrocarbons

29. (a) What are alkenes? (b) What other names are used to describe alkenes?
30. (a) How does the general formula for the alkenes differ from the general formula for the alkanes? (b) Why are the general formulas identical for alkenes and cycloalkanes that contain the same number of carbon atoms?
31. Write names and formulas for six alkenes.
32. (a) What are dienes? (b) Provide some examples.

33. (a) What are cycloalkenes? (b) What is their general formula? (c) Provide three examples.

Bonding in the Alkenes

34. Describe the bonding at each carbon atom in (a) ethene, (b) propene, (c) 1-butene, and (d) 2-butene.
35. (a) What are geometric (structural) isomers? (b) Why is rotation around a double bond not possible at room temperature? (c) What do *cis* and *trans* mean? (d) Draw structures for *cis-* and *trans-*1,2-dichloroethene. How do their melting and boiling points compare?
36. Distinguish between conformations and isomers.
37. How do carbon-carbon single bond lengths and carbon-carbon double bond lengths compare? Why?
38. Most reactions of the alkanes that do not disrupt the carbon skeleton are substitution reactions, while the alkenes are characterized by addition to the double bond. What does this statement mean?
39. (a) What are functional groups? (b) What is the functional group that occurs in all alkenes?
40. Write equations for two reactions in which halogens undergo addition reactions with alkenes. Name all compounds.
41. How can bromination be used to distinguish between alkenes and alkanes?
42. (a) What is hydrogenation? (b) Why is it important? (c) Write equations for two reactions that involve hydrogenation of alkenes. (d) Name all compounds in (c).
43. What is the difference between a vegetable oil and a shortening? Illustrate.

Polymers

44. (a) What is polymerization? (b) Write equations for three polymerization reactions.
45. (a) What is rubber? (b) What is vulcanization? (c) What is the purpose of vulcanizing rubber? (d) What are fillers and reinforcing agents? (e) What is their purpose?
46. (a) What is an elastomer? (b) Cite a specific example. (c) What are some of the advantages of neoprene compared to natural rubber?

Oxidation of Alkenes

47. (a) Describe two qualitative tests that can be used to distinguish between alkenes and alkanes? (b) Cite some specific examples. (c) What functional group is the basis for the qualitative distinction between alkanes and alkenes?

The Alkynes

48. (a) What are alkynes? (b) What other name is used to describe them? (c) What is the general formula for alkynes? (d) How does the general formula for alkynes compare with the general formula for cycloalkenes? Why?
49. (a) Write names and formulas for four alkynes. (b) What is the functional group in alkynes?

Bonding in the Alkynes

50. Describe the bonding in and the geometry of the alkynes you listed in Exercise 49. (Section 5–19)
51. (a) What is the most familiar alkyne? (b) What is its common name? (c) For what is it commonly used? (d) How is it prepared?

Reactions of the Alkynes

52. (a) Why are alkynes more reactive than alkenes? (b) What is the most common kind of reaction that alkynes undergo? (c) Write equations for four such reactions. (d) Name all compounds in part (c).

Aromatic Hydrocarbons

53. (a) What are aromatic hydrocarbons? (b) What is the principal source of aromatic hydrocarbons?
54. (a) What is the most common aromatic hydrocarbon? (b) From what source was it first isolated? When?
55. What is coal tar? coal gas?

Bonding in Benzene

56. (a) Describe the structure of and the bonding in benzene. (b) What evidence indicates conclusively that benzene has a symmetrical ring structure?
57. (a) Draw resonance structures for benzene. (b) What are resonance structures? (c) What

do we mean when we say that the electrons associated with the pi bonds in benzene are delocalized over the entire ring?

Other Aromatic Hydrocarbons

58. (a) Write names and draw structural formulas for seven aromatic hydrocarbons (other than benzene). (b) How is each structurally related to benzene?
59. What are coal-tar crudes?
60. How are the prefixes *ortho-*, *meta-*, and *para-* used in naming aromatic compounds?

Reactions of Aromatic Hydrocarbons

61. (a) What is the most common kind of reaction that the benzene ring undergoes? (b) Write equations for the reaction of benzene with chlorine in the presence of an iron catalyst and for the analogous reaction with bromine.
62. Write equations to illustrate both aromatic and aliphatic substitution reactions of toluene using (a) chlorine and (b) bromine.
63. Write equations to illustrate the nitration of benzene and toluene.
64. (a) Do you expect aromatic hydrocarbons to produce soot as they burn? Why? (b) Would you expect the flames to be blue or yellow?
65. Write equations to illustrate the oxidation of the following aromatic hydrocarbons by potassium permanganate in basic solution. (a) toluene, (b) ethylbenzene, (c) 1,2-dimethylbenzene.

30 Organic Chemistry II: Functional Groups

The study of organic chemistry is greatly simplified by assuming that hydrocarbons represent parent compounds, and that other compounds are derived from them. **Functional groups** are groups of atoms that represent potential reaction sites in organic compounds. Because the reactions of a given functional group are similar in most compounds, we are able to systematize large amounts of information easily by studying functional groups and their characteristic reactions.

We shall introduce several common classes of organic compounds and study a few of their important reactions.

Alcohols and Phenols

Alcohols and phenols contain the hydroxyl group (—O—H) as their functional group. **Alcohols** may be considered to be derived from saturated or unsaturated hydrocarbons by replacing at least one hydrogen atom by a hydroxyl group. The properties of alcohols are the properties of a hydroxyl group attached to an *aliphatic* carbon atom, —C—O—H. Ethanol, or ethyl alcohol, is the most common example.

Originally the term *aliphatic* meant "fat-like." Present use of the term is: the parent hydrocarbon was an alkane, an alkene, an alkyne, or one of their cyclic counterparts.

When a hydrogen atom attached to an aromatic ring is replaced by a hydroxyl group, the resulting compound is known as a **phenol.** Such a compound behaves more like an acid than an alcohol.

Alternatively, we may view alcohols and phenols as derivatives of water in which one H atom has been replaced by an organic group

H—O—H (water)

H—C(H)(H)—C(H)(H)—O—H (ethanol)

C_6H_5—O—H (phenol)

The simplest phenol is called phenol. The most common member of a class of compounds is frequently called by the class name. Salt, sugar, alcohol, and phenol are common examples.

Indeed, this is by far the better view. The structure of water was discussed in detail in Section 5–14. The hydroxyl group in an alcohol or a phenol is covalently bonded to a carbon atom, but recall that the O—H bond is quite polar. The oxygen atom retains its two unshared electron pairs, and the C—O—H bond angle is nearly 104.5°.

Clearly the presence of a bonded alkyl or aryl group changes the properties of the —O—H group. *Alcohols* are so very weakly acidic that they are thought of as neutral compounds. *Phenols* are weakly acidic. For phenol, $K_a = 10^{-10}$.

A group derived from an *aromatic* hydrocarbon by removing a hydrogen atom is called an aryl (Ar) group.

Many reactions of alcohols depend on whether the hydroxyl group is attached to a carbon that is bonded to one, two, or three other carbon atoms. If we represent alkyl groups as R, we can illustrate the three classes of alcohols.

R—C(H)(H)—OH (a primary (1°) alcohol)

R—C(R)(H)—O—H (a secondary (2°) alcohol)

R—C(R)(R)—O—H (a tertiary (3°) alcohol)

Recall that R represents an alkyl group.

FIGURE 30–1 Models of ethanol (also called ethyl alcohol or grain alcohol), CH_3CH_2OH.

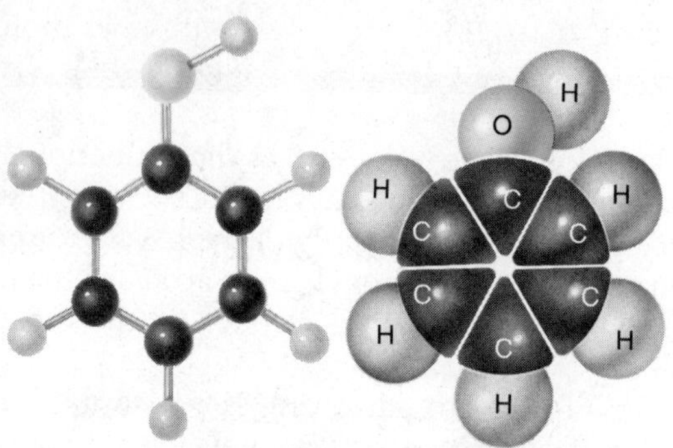

FIGURE 30–2 Models of phenol, C_6H_5OH.

Note that primary alcohols contain one R group, secondary alcohols contain two R groups, and tertiary alcohols contain three R groups bonded to the carbon atom to which the —OH group is attached. The R groups may be the same or different.

30–1 Nomenclature of Alcohols and Phenols

The systematic names of alcohols consist of the characteristic stem plus an -ol ending. A numeric prefix indicates the position of the —OH group in chains of three or more carbon atoms.

$CH_3—OH$	$CH_3—CH_2—OH$	$CH_3—CH_2—CH_2—OH$	$CH_3—CH(OH)CH_3$
methanol methyl alcohol wood alcohol	ethanol ethyl alcohol grain alcohol	1-propanol *n*-propyl alcohol a primary alcohol	2-propanol isopropyl alcohol a secondary alcohol rubbing alcohol

There are four four-carbon alcohols containing one —OH per molecule

$CH_3CH_2CH_2CH_2OH$

1-butanol **1°**
normal butyl alcohol

$$CH_3—\underset{\underset{CH_3}{|}}{CH}—CH_2OH$$

2-methyl-1-propanol **1°**
isobutyl alcohol

$$CH_3CH_2\underset{\underset{OH}{|}}{CH}CH_3$$

2-butanol **2°**
secondary butyl alcohol

$$CH_3—\overset{\overset{CH_3}{|}}{\underset{\underset{CH_3}{|}}{C}}—OH$$

2-methyl-2-propanol **3°**
tertiary butyl alcohol

There are eight saturated five-carbon alcohols containing one —OH per molecule. They are often called "amyl" alcohols. Two examples are

$CH_3CH_2CH_2CH_2CH_2OH$

1-pentanol
n-pentyl alcohol
n-amyl alcohol

$(CH_3)_2CHCH_2CH_2OH$

3-methyl-1-butanol
isopentyl alcohol
isoamyl alcohol

The **polyhydric alcohols** contain more than one —OH group per molecule. Those containing two OH groups per molecule are called **glycols.** Important examples of polyhydric alcohols include

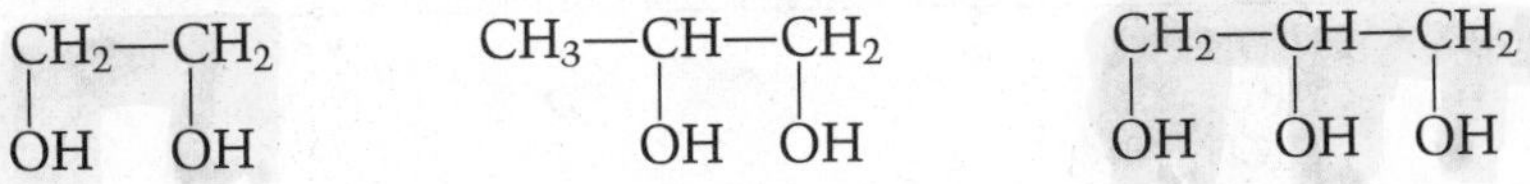

$$\underset{\text{OH}}{\text{CH}_2}-\underset{\text{OH}}{\text{CH}_2} \qquad \text{CH}_3-\underset{\text{OH}}{\text{CH}}-\underset{\text{OH}}{\text{CH}_2} \qquad \underset{\text{OH}}{\text{CH}_2}-\underset{\text{OH}}{\text{CH}}-\underset{\text{OH}}{\text{CH}_2}$$

1,2-ethanediol
ethylene glycol
(permanent antifreeze)

1,2-propanediol
propylene glycol

1,2,3-propanetriol
glycerine or glycerol
(the moisturizer in cosmetics)

Phenols are nearly always referred to by their common names. Some examples are

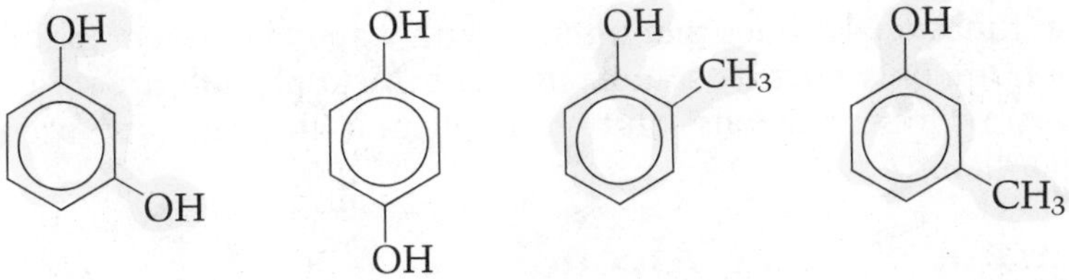

resorcinol (benzene ring with OH at positions 1 and 3)

hydroquinone (benzene ring with OH at positions 1 and 4)

o-cresol (benzene ring with OH and CH_3 on adjacent carbons)

m-cresol (benzene ring with OH and CH_3 at positions 1 and 3)

As you might guess, the cresols occur in "creosote," a wood preservative that has been widely used. There is a third isomer, *p*-cresol, in which the —OH and —CH_3 groups occur on opposite corners of the aromatic ring.

30–2 Physical Properties of Alcohols and Phenols

Since the hydroxyl group is quite polar while alkyl groups are nonpolar, the properties of alcohols depend on two factors: (1) the number of hydroxyl groups per molecule, and (2) the size of the organic portion of the molecule. We may think of alcohols as consisting of a polar part, —OH, and a nonpolar part, R.

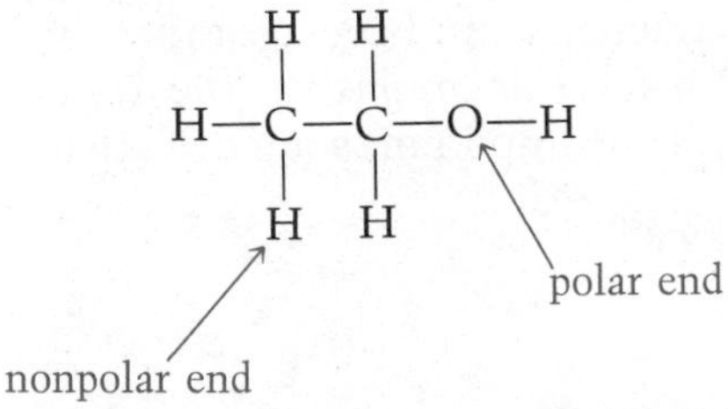

$$\text{H}-\overset{\text{H}}{\underset{\text{H}}{\text{C}}}-\overset{\text{H}}{\underset{\text{H}}{\text{C}}}-\text{O}-\text{H}$$

nonpolar end

polar end

The low-molecular-weight monohydric alcohols are miscible with water in all proportions. Beginning with the four butyl alcohols, solubility in water decreases rapidly with increasing molecular weight because the nonpolar part of the molecules becomes larger and larger. Many polyhydric alcohols are very soluble in water because they contain two or more polar hydroxyl groups per molecule.

Table 30–1 shows that the boiling points of normal primary alcohols increase, while their solubilities in water decrease, with increasing molecular weight. The boiling points of the alcohols are much higher than those of the

TABLE 30–1 Physical Properties of Normal Primary Alcohols

Name	Formula	BP, °C	Solubility in H_2O, g/100 g at 20°C
methanol	CH_3OH	65	Completely miscible
ethanol	CH_3CH_2OH	78.5	Completely miscible
1-propanol	$CH_3CH_2CH_2OH$	97	Completely miscible
1-butanol	$CH_3CH_2CH_2CH_2OH$	117.7	7.9
1-pentanol	$CH_3CH_2CH_2CH_2CH_2OH$	137.9	2.7
1-hexanol	$CH_3CH_2CH_2CH_2CH_2CH_2OH$	155.8	0.59

corresponding alkanes (Table 29–1) because of the hydrogen bonding of the hydroxyl groups.

Unless other functional groups that interact with water are present in the molecules, phenols are only slightly soluble in water. Most phenols are solids at room temperature. Dilute aqueous solutions of phenols are frequently used as antiseptics and disinfectants.

30–3 Preparation of Some Alcohols

Methanol, or methyl alcohol, was produced for many years by the destructive distillation of wood, and it is often called wood alcohol. Most methanol is now prepared from carbon monoxide and hydrogen at high temperatures and pressures in the presence of a mixed oxide catalyst (oxides of Zn, Cu, Cr)

$$CO + 2H_2 \xrightarrow[\substack{400°C \\ \text{cat.}}]{150\text{ atm}} CH_3OH$$

Methanol is used as a temporary antifreeze (bp = 65°C), as a solvent for varnishes and shellacs, and as the starting material in the manufacture of formaldehyde (Section 30–13). It is very toxic and causes permanent blindness when taken internally.

Fermentation is an enzymatic process carried out by certain kinds of bacteria.

Ethanol, or ethyl alcohol, was first prepared by fermentation *a long time ago*—the most ancient literature contains references to beverages that were obviously alcoholic! The fermentation of blackstrap molasses, the residue from the purification of cane sugar, sucrose, is one important source of ethanol

$$\underset{\text{cane sugar}}{C_{12}H_{22}O_{11}} + H_2O \xrightarrow{\text{yeast}} \underset{\text{ethanol}}{4CH_3CH_2OH} + 4CO_2$$

The starch in grain, potatoes, and similar substances can be converted into sugar by malt; this is followed by fermentation to produce ethanol, which is the most important industrial alcohol. The industrial preparation involves the hydration of ethene from petroleum, using H_2SO_4 as a catalyst.

$$\underset{\text{ethene}}{H-\overset{H}{\overset{|}{C}}=\overset{H}{\overset{|}{C}}-H} + \underset{\text{sulfuric acid}}{H^{+}\,{}^{-}OSO_3H} \xrightarrow{\text{cold}} \underset{\text{ethyl hydrogen sulfate}}{H-\underset{H}{\underset{|}{\overset{H}{\overset{|}{C}}}}-\underset{H}{\underset{|}{\overset{H}{\overset{|}{C}}}}-OSO_3H}$$

$$H-\overset{H}{\underset{H}{C}}-\overset{H}{\underset{H}{C}}-OSO_3H + H-OH \longrightarrow H-\overset{H}{\underset{H}{C}}-\overset{H}{\underset{H}{C}}-OH + HOSO_3H$$

steam ethanol

Note that these reactions amount to the addition of water to ethene, since H_2SO_4 is regenerated in the second reaction.

Ethylene glycol is prepared by the reaction of ethylene with the hypochlorous acid present in chlorine water, and the subsequent hydrolysis of the product in an aqueous solution of sodium carbonate.

$$CH_2{=}CH_2 + HOCl \longrightarrow \underset{OH}{CH_2}-\underset{Cl}{CH_2} \xrightarrow[2Na^+CO_3^{2-}]{H_2O} \underset{OH}{CH_2}-\underset{OH}{CH_2}$$

hypochlorous acid (chlorine + water) — ethylene chlorohydrin — ethylene glycol (1,2-ethanediol)

Ethylene glycol is completely miscible with water and is widely used as a permanent antifreeze (bp = 197°C).

30–4 Reactions of Alcohols and Phenols

The alcohols are *very weakly acidic* compounds. However, they do not react with strong soluble bases. They are much weaker acids than water (see Table 30–4), but some of their reactions are analogous to those of water.

The very reactive metals react with alcohols to form **alkoxides** with the liberation of hydrogen

$$2CH_3-OH + 2Na \longrightarrow 2Na^{+}\,{}^{-}OCH_3 + H_2$$

sodium methoxide (an alkoxide)

$$2CH_3CH_2-OH + 2Na \longrightarrow 2Na^{+}\,{}^{-}OCH_2CH_3 + H_2$$

sodium ethoxide (an alkoxide)

Note the similarity between these reactions and the reaction of water with active metals.

$$2H-OH + 2Na \longrightarrow 2Na^+OH^- + H_2 \quad \text{(can occur explosively)}$$

The alkoxides of low molecular weight are strong bases that react with water (i.e., hydrolyze) to form the parent alcohol and a strong soluble base.

$$Na^{+}\,{}^{-}OCH_2CH_3 + H-OH \longrightarrow CH_3CH_2OH + Na^+OH^-$$

sodium ethoxide — ethyl alcohol

Phenols also react with metallic sodium to produce **phenoxides** in reactions that are analogous to those of alcohols. Since phenols are more acidic than alcohols, their reactions are more vigorous.

$$2\,C_6H_5-O-H + 2Na \longrightarrow 2\,C_6H_5-O^-Na^+ + H_2$$

phenol — sodium phenoxide

The hydroxyl groups of alcohols can be replaced by atoms or by groups of atoms. The hydrohalic acids react with alcohols in the presence of certain Lewis acid catalysts to form **alkyl halides.** Concentrated hydrochloric acid reacts with primary alcohols very slowly at elevated temperatures

$$H^+Cl^- + CH_3CH_2OH \xrightarrow[\Delta]{ZnCl_2} \underset{\text{ethyl chloride}}{CH_3CH_2Cl} + H_2O$$

Simple inorganic esters may be thought of as compounds that contain one or more alkyl groups covalently bonded to the anion of a *ternary* inorganic acid. Unless specifically indicated, the term *ester* refers to organic esters.

Alcohols react with common inorganic oxyacids to produce **inorganic esters.** For instance, nitric acid reacts with alcohols to produce nitrates.

$$CH_3CH_2OH + \underset{\text{nitric acid}}{HONO_2} \longrightarrow \underset{\text{ethyl nitrate}}{CH_3CH_2{-}O{-}NO_2} + H_2O$$

The reaction of nitric acid with glycerol produces the principal explosive ingredient of dynamite.

Interestingly, nitroglycerine is used as a vasodilator (it dilates the blood vessels) to decrease arterial tension in persons who have heart disease. Perhaps you know someone who takes "nitro" pills.

$$\underset{\text{glycerol}}{\begin{array}{l} CH_2{-}OH \\ | \\ CH{-}OH \\ | \\ CH_2{-}OH \end{array}} + 3HONO_2 \xrightarrow{H_2SO_4} \underset{\substack{\text{glyceryl trinitrate} \\ \text{(nitroglycerine)}}}{\begin{array}{l} CH_2{-}ONO_2 \\ | \\ CH{-}ONO_2 \\ | \\ CH_2{-}ONO_2 \end{array}} + 3H_2O$$

Cold concentrated sulfuric acid reacts with alcohols to form **alkyl hydrogen sulfates.** The reaction with lauryl alcohol is an important industrial reaction

$$\underset{\substack{\text{lauryl alcohol} \\ \text{1-dodecanol}}}{CH_3CH_2CH_2CH_2CH_2CH_2CH_2CH_2CH_2CH_2CH_2CH_2{-}OH} + H_2SO_4 \longrightarrow$$

$$\underset{\text{lauryl hydrogen sulfate}}{CH_3(CH_2)_{10}CH_2{-}O{-}SO_3H} + H_2O$$

The reaction of an alkyl hydrogen sulfate with sodium hydroxide produces the sodium salt of the alkyl hydrogen sulfate

You might check a detergent box at the next opportunity to see whether it contains sodium lauryl sulfate, also called sodium dodecyl sulfate.

$$CH_3(CH_2)_{10}CH_2{-}O{-}SO_3H + Na^+OH^- \longrightarrow$$

$$\underset{\substack{\text{sodium lauryl sulfate} \\ \text{a detergent}}}{CH_3(CH_2)_{10}CH_2{-}OSO_3^-Na^+} + H_2O$$

Sodium salts of the alkyl hydrogen sulfates that contain approximately 12 carbon atoms are excellent detergents, and they are also biodegradable. Soaps and detergents were discussed in Section 10–23.

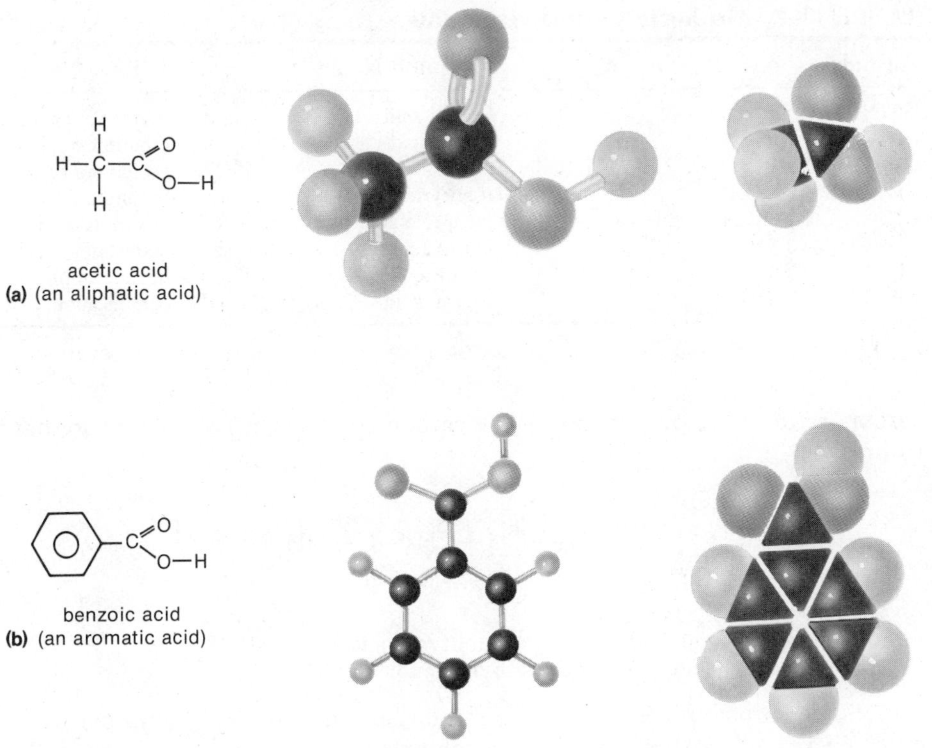

acetic acid
(a) (an aliphatic acid)

benzoic acid
(b) (an aromatic acid)

FIGURE 30–3 (a) Models of acetic acid, (b) models of benzoic acid.

Carboxylic Acids

30–5 Nomenclature, Structure, and Acid Strength of Carboxylic Acids

Compounds that contain the **carboxyl group,** $-\overset{\displaystyle O}{\overset{\|}{C}}-O-H$, are found to be acidic, and they are called **carboxylic acids.** The general formula is R—COOH, and most are *weak acids* compared with strong inorganic acids such as HCl. However, they are much stronger acids than most phenols. These acids are named systematically by dropping the terminal *-e* from the name of the parent hydrocarbon and adding *-oic acid.* However, old habits are hard to break, and organic acids are usually called by common names. In aromatic acids the carboxyl group is attached to the aromatic ring as shown in Figure 30–3.

Aliphatic carboxylic acids are sometimes referred to as *fatty acids* because many have been obtained from animal fats.

Organic acids occur widely in natural products, and many have been known since ancient times. As Table 30–2 indicates, the common names are often derived from a Greek or Latin word that indicates the original source.

The names of derivatives of carboxylic acids are often derived from the trivial names of the acids. Positions of substituents may be indicated by lower-case Greek letters beginning with the carbon *adjacent* to the carboxyl

Formic acid was obtained by distillation of ants (L: *formica*, ant); acetic acid occurs in vinegar (L: *acetum*, vinegar); butyric acid in rancid butter (L: *butyrum*, butter); stearic acid in animal fats (Gr: *stear*, beef suet).

n-Caproic acid is one of the so-called "goat acids." Its odor is responsible for the name.

TABLE 30–2 Aliphatic Carboxylic Acids

Formula	Common Name	IUPAC Name
$HCOOH$	formic acid	methanoic acid
CH_3COOH	acetic acid	ethanoic acid
CH_3CH_2COOH	propionic acid	propanoic acid
$CH_3CH_2CH_2COOH$	*n*-butyric acid	butanoic acid
$CH_3CH_2CH_2CH_2CH_2COOH$	*n*-caproic acid	hexanoic acid
$CH_3(CH_2)_{10}COOH$	lauric acid	dodecanoic acid
$CH_3(CH_2)_{14}COOH$	palmitic acid	hexadecanoic acid
$CH_3(CH_2)_{16}COOH$	stearic acid	octadecanoic acid

carbon, rather than by numbering the carbon chain (again, old habits are hard to break).

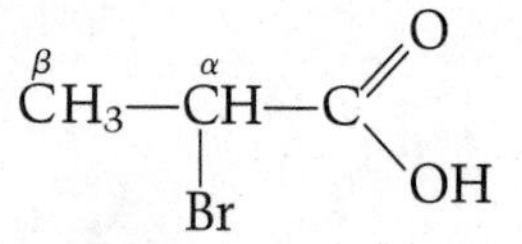

$$\overset{\beta}{C}H_3-\overset{\alpha}{C}H(Br)-C(=O)OH \qquad \overset{\gamma}{C}H_3-\overset{\beta}{C}H(CH_3)-\overset{\alpha}{C}H_2-C(=O)-OH$$

α-bromopropionic acid (2-bromopropanoic acid)

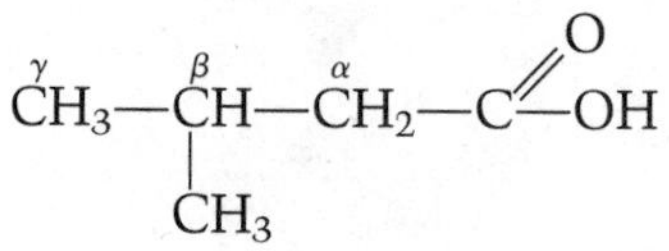

β-methylbutyric acid (3-methylbutanoic acid)

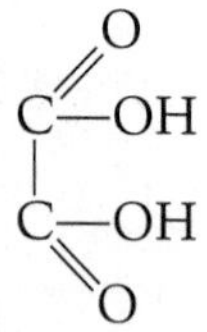

$$HO-C(=O)-C(=O)-OH$$

oxalic acid (an aliphatic acid)

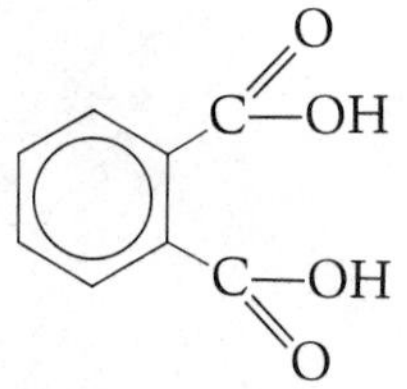

Some carboxylic acids contain more than one carboxyl group per molecule. These acids are known almost exclusively by their common names. Oxalic acid is an aliphatic **dicarboxylic acid,** while phthalic acid is a typical aromatic dicarboxylic acid.

Aromatic acids are called by their common names or named as derivatives of benzoic acid, which is considered the "parent" aromatic acid.

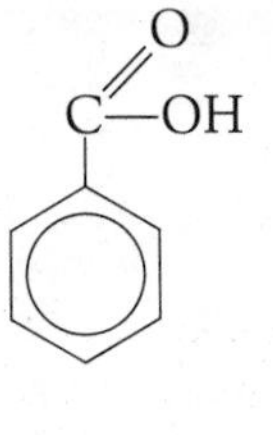

$$C_6H_4(COOH)_2$$

phthalic acid (an aromatic acid)

$$C_6H_5-C(=O)-OH \qquad Cl-C_6H_4-C(=O)-OH \qquad CH_3-C_6H_4-C(=O)-OH$$

benzoic acid 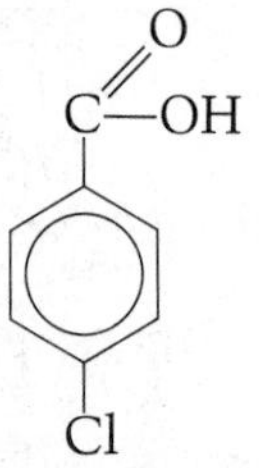 *p*-chlorobenzoic acid 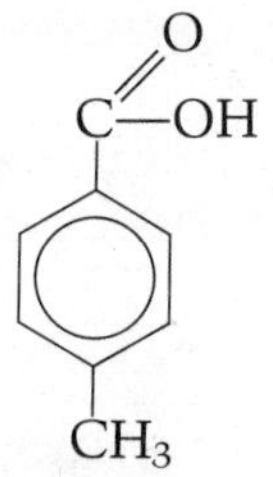 *p*-toluic acid

Since many reactions of carboxylic acids involve displacement of the —OH group by another atom or group of atoms, we find it useful to name the

TABLE 30–3 Aliphatic Dicarboxylic Acids

Formula	Name
$HOOC-COOH$	oxalic acid
$HOOC-CH_2-COOH$	malonic acid
$HOOC-CH_2CH_2-COOH$	succinic acid
$HOOC-CH_2CH_2CH_2-COOH$	glutaric acid
$HOOC-CH_2CH_2CH_2CH_2-COOH$	adipic acid

non—OH portion of acid molecules because they occur in numerous compounds. These compounds are thought of as derivatives of carboxylic acids.

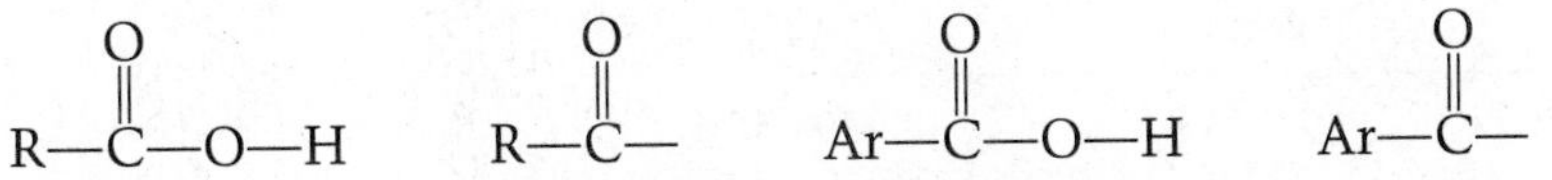

$$\underset{\text{an aliphatic carboxylic acid}}{R{-}\overset{O}{\overset{\|}{C}}{-}O{-}H} \qquad \underset{\text{an aliphatic acyl group}}{R{-}\overset{O}{\overset{\|}{C}}{-}} \qquad \underset{\text{an aromatic carboxylic acid}}{Ar{-}\overset{O}{\overset{\|}{C}}{-}O{-}H} \qquad \underset{\text{an aromatic acyl group}}{Ar{-}\overset{O}{\overset{\|}{C}}{-}}$$

Acyl groups are named as derivatives of the parent acid by dropping *-ic acid* and adding -yl to the characteristic stem. Some examples are

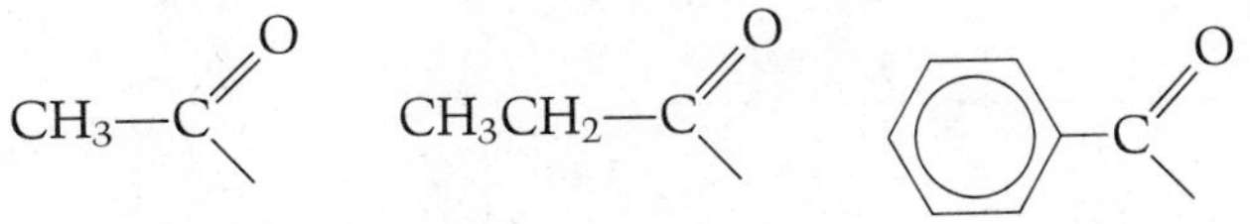

$CH_3{-}C(=O){-}$ acetyl group

$CH_3CH_2{-}C(=O){-}$ propionyl group

$C_6H_5{-}C(=O){-}$ benzoyl group

Four important classes of acid derivatives are formed by the displacement of the hydroxyl group by another atom or group of atoms to form compounds that contain an acyl group

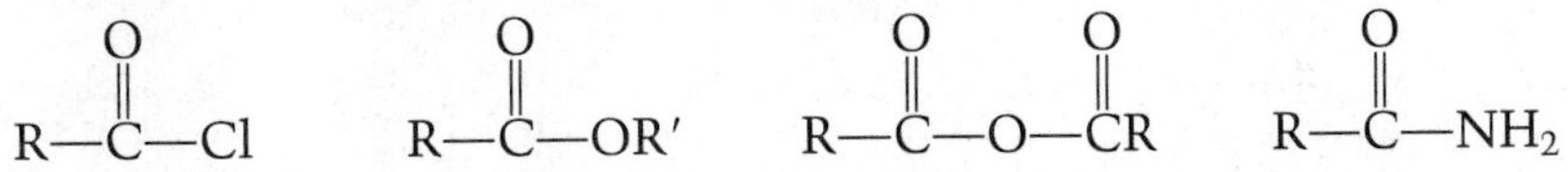

$$\underset{\text{an acyl chloride (an acid chloride)}}{R{-}\overset{O}{\overset{\|}{C}}{-}Cl} \qquad \underset{\text{an ester}}{R{-}\overset{O}{\overset{\|}{C}}{-}OR'} \qquad \underset{\text{an acid anhydride}}{R{-}\overset{O}{\overset{\|}{C}}{-}O{-}\overset{O}{\overset{\|}{C}}R} \qquad \underset{\text{an amide}}{R{-}\overset{O}{\overset{\|}{C}}{-}NH_2}$$

Aromatic compounds of the same types are encountered frequently.

In Section 16–4 we discussed the extent of ionization of acetic acid.

$$CH_3COOH \rightleftharpoons H^+ + CH_3COO^-$$

$$K_a = \frac{[H^+][CH_3COO^-]}{[CH_3COOH]} = 1.8 \times 10^{-5}$$

It is 1.3% ionized in 0.10 *M* solution. The acid strengths of the monocarboxylic acids are approximately the same, regardless of the length of the chain. Their acid strength increases dramatically when highly electronegative substituents are present on the α-carbon atom, i.e., the carbon attached to the —COOH group. (See acetic acid and the three substituted acetic acids listed in Table 30–4.)

Derivatives of Carboxylic Acids

30–6 The Acyl Halides (Acid Halides)

The **acyl halides,** sometimes called **acid halides,** are much more reactive than their parent acids, and consequently they are often used in reactions to intro-

TABLE 30–4 Ionization Constants of Some Carboxylic Acids

Name	Formula	K_a
formic acid	$HCOOH$	1.8×10^{-4}
acetic acid	CH_3COOH	1.8×10^{-5}
propionic acid	CH_3CH_2COOH	1.4×10^{-5}
monochloroacetic acid	$ClCH_2COOH$	1.5×10^{-3}
dichloroacetic acid	$Cl_2CHCOOH$	5.0×10^{-2}
trichloroacetic acid	Cl_3CCOOH	2.0×10^{-1}
benzoic acid	C_6H_5COOH	6.3×10^{-5}
phenol	C_6H_5OH	1.3×10^{-10}
ethanol	CH_3CH_2OH	$\sim 10^{-18}$

Phenol and ethanol are not carboxylic acids. They are included in Table 30–4 to show that they are only weakly acidic when compared to carboxylic acids.

duce an acyl group into another molecule. They are usually prepared by treating acids with PCl_3, PCl_5, or $SOCl_2$ (thionyl chloride).

$$\underset{\text{acetic acid}}{CH_3-\overset{\overset{\displaystyle O}{\|}}{C}-O-H} + \underset{\text{phosphorus pentachloride}}{PCl_5} \longrightarrow \underset{\text{acetyl chloride (an acyl chloride or an acid chloride)}}{CH_3-\overset{\overset{\displaystyle O}{\|}}{C}-Cl} + HCl\,(g) + \underset{\text{phosphorus oxychloride}}{POCl_3}$$

In general terms, the reaction of acids with PCl_5 may be represented as

$$\underset{\text{acid}}{R-\overset{\overset{\displaystyle O}{\|}}{C}-OH} + PCl_5 \longrightarrow \underset{\text{an acyl chloride or an acid chloride}}{R-\overset{\overset{\displaystyle O}{\|}}{C}-Cl} + HCl\,(g) + POCl_3$$

30–7 Esters

When an organic acid is heated with an alcohol, an equilibrium is established with the resulting **ester** and water. The reaction is catalyzed by traces of strong inorganic acids, such as a few drops of concentrated H_2SO_4

$$\underset{\text{acetic acid}}{CH_3-\overset{\overset{\displaystyle O}{\|}}{C}-OH} + \underset{\text{ethyl alcohol}}{CH_3CH_2-OH} \xrightleftharpoons{H^+,\ \Delta} \underset{\text{ethyl acetate, an ester}}{CH_3-\overset{\overset{\displaystyle O}{\|}}{C}-O-CH_2CH_3} + H_2O$$

In general terms the reaction (where R and R′ may be the same or different alkyl groups) may be represented as:

Numerous experiments have demonstrated conclusively that the —OH group from the acid and —H from the alcohol are the atoms that form water molecules.

$$\underset{\text{acid}}{R-\overset{\overset{\displaystyle O}{\|}}{C}-O-H} + \underset{\text{alcohol}}{R'-O-H} \rightleftharpoons \underset{\text{ester}}{R-\overset{\overset{\displaystyle O}{\|}}{C}-O-R'} + H_2O$$

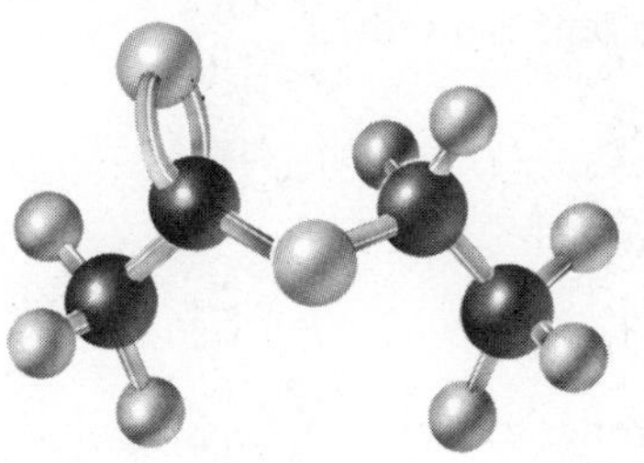
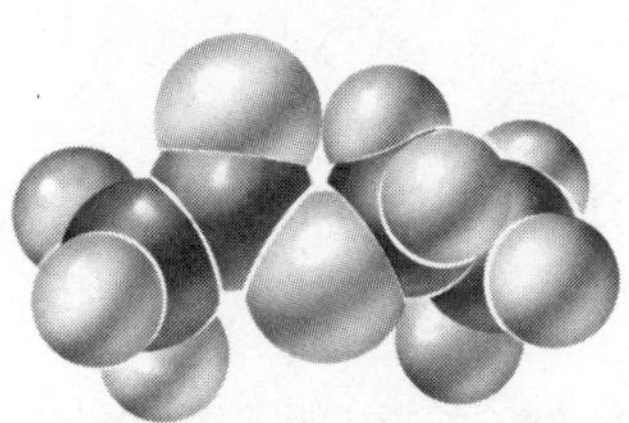

FIGURE 30–4 Models of ethyl acetate,

$$CH_3-\overset{\overset{\displaystyle O}{\|}}{C}-O-CH_2-CH_3,$$

an ester. The $CH_3-\overset{\overset{\displaystyle O}{\|}}{C}-$ fragment is derived from acetic acid, the parent acid, while the $-O-CH_2-CH_3$ fragment is derived from ethanol, the parent alcohol.

Esters are nearly always called by their common names, which consist of the name of the alkyl group in the alcohol first, then the name of the anion derived from the acid. This statement does *not* imply that both oxygen atoms come from the acid; they don't.

Reactions between acids and alcohols are usually quite slow and require prolonged boiling (refluxing). However, the reactions between most acyl halides and most alcohols occur very rapidly.

$$\underset{\text{acetyl chloride}}{CH_3-\overset{\overset{\displaystyle O}{\|}}{C}-Cl} + \underset{\text{ethyl alcohol}}{CH_3-CH_2-O-H} \longrightarrow \underset{\text{ethyl acetate}}{CH_3-\overset{\overset{\displaystyle O}{\|}}{C}-O-CH_2CH_3} + HCl$$

Most simple esters are pleasant-smelling substances. They are responsible for the flavors and fragrances of most fruits and flowers as well as many of the artificial fruit flavors that are used in cakes, candies, and ice cream (Table 30–5).

Esters of low molecular weight are excellent solvents for nonpolar compounds. Many nail polish removers contain ethyl acetate, an excellent solvent that also gives nail polish removers their characteristic odor.

Most esters are not very reactive, and strong reagents are required for many of their reactions. Esters can be hydrolyzed by refluxing with solutions of strong bases.

$$\underset{\text{ethyl acetate}}{CH_3-\overset{\overset{\displaystyle O}{\|}}{C}-O-CH_2-CH_3} + Na^+OH^- \xrightarrow{\Delta} \underset{\text{sodium acetate}}{CH_3\overset{\overset{\displaystyle O}{\|}}{C}-O^-Na^+} + \underset{\text{ethanol}}{CH_3CH_2OH}$$

TABLE 30–5 Some Common Esters

Ester	Formula	Odor of
n-butyl acetate	$CH_3COOC_4H_9$	bananas
ethyl butyrate	$C_3H_7COOC_2H_5$	pineapples
n-amyl butyrate	$C_3H_7COOC_5H_{11}$	apricots
n-octyl acetate	$CH_3COOC_8H_{17}$	oranges
isoamyl isovalerate	$C_4H_9COOC_5H_{11}$	apples
methyl salicylate	$C_6H_4(OH)(COOCH_3)$	oil of wintergreen
methyl anthranilate	$C_6H_4(NH_2)(COOCH_3)$	grapes

In general terms the hydrolysis of esters may be represented as

$$\underset{\text{ester}}{R{-}\overset{\overset{\displaystyle O}{\|}}{C}{-}O{-}R'} + Na^+OH^- \xrightarrow{\Delta} \underset{\text{salt of an acid}}{R{-}\overset{\overset{\displaystyle O}{\|}}{C}{-}O^-Na^+} + \underset{\text{alcohol}}{R'OH}$$

The hydrolysis of esters in the presence of strong soluble bases is called **saponification** because the hydrolysis of fats and oils produces soaps.

Recall that glycerol is

$$\begin{array}{l} CH_2{-}OH \\ | \\ CH{-}OH \\ | \\ CH_2{-}OH \end{array}$$

Fats (solids) and **oils** (liquids) are esters of glycerol and aliphatic acids of high molecular weight. "Fatty acids" are any and all organic acids that occur in fats and oils (as esters).

Fats and oils may be represented by the general formula

$$\begin{array}{l} CH_2{-}O{-}\overset{\overset{\displaystyle O}{\|}}{C}{-}R \\ | \\ CH{-}O{-}\overset{\overset{\displaystyle O}{\|}}{C}{-}R \\ | \\ CH_2{-}O{-}\overset{\overset{\displaystyle O}{\|}}{C}{-}R \end{array}$$

where the R's may be the same or different groups, and the fatty acid portions

$$({-}\overset{\overset{\displaystyle O}{\|}}{C}{-}R)$$

may be saturated or unsaturated.

Most fatty acids contain even numbers of carbon atoms because they are synthesized in the body from two-carbon acetyl groups.

Fats are solid esters of glycerol and (mostly) saturated acids. Oils are liquid esters that are derived primarily from unsaturated acids and glycerol. The acid portions of fats almost always contain an even number of carbon atoms, usually 16 or 18. The acids that occur in fats and oils most frequently are

Butyric	$CH_3CH_2CH_2COOH$
Lauric	$CH_3(CH_2)_{10}COOH$
Myristic	$CH_3(CH_2)_{12}COOH$
Palmitic	$CH_3(CH_2)_{14}COOH$
Stearic	$CH_3(CH_2)_{16}COOH$
Oleic	$CH_3(CH_2)_7CH{=}CH(CH_2)_7COOH$
Linolenic	$CH_3CH_2CH{=}CHCH_2CH{=}CHCH_2CH{=}CH(CH_2)_7COOH$
Ricinoleic	$CH_3(CH_2)_5CHOHCH_2CH{=}CH(CH_2)_7COOH$

Figure 30–5 is a scale model of stearic acid, a long-chain saturated fatty acid.

Naturally occurring fats and oils are mixtures of many different esters. Milk fat, lard, and tallow are familiar important fats. Soybean oil, cottonseed oil, linseed oil, palm oil, and coconut oil are familiar examples of important oils.

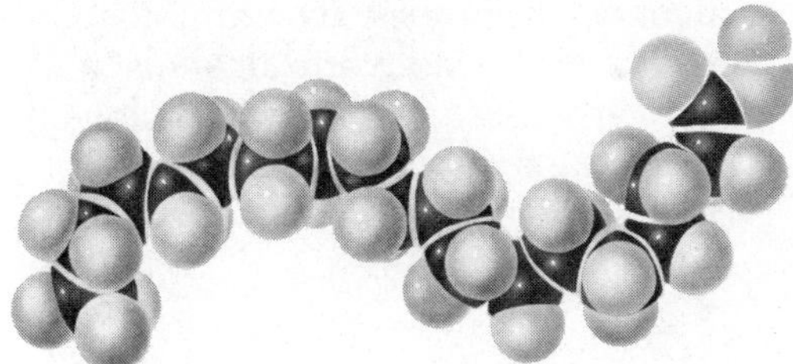

FIGURE 30–5 Scale model of a long chain fatty acid, stearic acid, $CH_3(CH_2)_{16}COOH$.

The triesters of glycerol are called glycerides. Simple glycerides are esters in which all three R groups are identical. Two examples are

$$\begin{array}{ll}
CH_2OC(=O)(CH_2)_{14}CH_3 & CH_2OC(=O)(CH_2)_{16}CH_3 \\
| & | \\
CHOC(=O)(CH_2)_{14}CH_3 & CHOC(=O)(CH_2)_{16}CH_3 \\
| & | \\
CH_2OC(=O)(CH_2)_{14}CH_3 & CH_2OC(=O)(CH_2)_{16}CH_3 \\
\text{glyceryl tripalmitate} & \text{glyceryl tristearate} \\
\text{(palmitin)} & \text{(stearin)}
\end{array}$$

Glycerides are frequently called by their common names, which are indicated in parentheses in the examples above. The common name is the characteristic stem for the parent acid plus an *-in* ending.

Like other esters, fats and oils can be hydrolyzed in strongly basic solution to produce salts of the acids and the alcohol, glycerol. The resulting sodium salts of long-chain fatty acids are soaps.

$$\begin{array}{l}
CH_2{-}O{-}C(=O){-}(CH_2)_{16}CH_3 \\
| \\
CH{-}O{-}C(=O){-}(CH_2)_{16}CH_3 + 3Na^+OH^- \xrightarrow{\Delta} \underset{\text{sodium stearate, a soap}}{3CH_3{-}(CH_2)_{16}C(=O){-}O^-Na^+} + \underset{\text{glycerol}}{\begin{array}{l} CH_2{-}OH \\ | \\ CH{-}OH \\ | \\ CH_2{-}OH \end{array}} \\
| \\
CH_2{-}O{-}C(=O){-}(CH_2)_{16}CH_3 \\
\text{glyceryl tristearate, a fat}
\end{array}$$

In Section 10–23 we described the cleansing action of soaps and detergents. In Section 29–9 we mentioned the hydrogenation of oils to convert them into fats.

Waxes are esters of fatty acids and alcohols other than glycerol. Most are derived from long-chain fatty acids and long-chain monohydric alcohols, both of which usually contain even numbers of carbon atoms. Beeswax is largely $C_{15}H_{31}COOC_{30}H_{61}$, while carnauba wax contains $C_{25}H_{51}COOC_{30}H_{61}$. Both are esters of myricyl alcohol, $C_{30}H_{61}OH$.

30–8 Polyesters

Dihydric alcohols contain two —OH groups per molecule.

When *dihydric alcohols* react with *dicarboxylic acids,* ester linkages may be formed at each end of each molecule to build up large molecules containing many ester linkages. The resulting *polymeric esters* are called **polyesters.** The familiar synthetic fiber Dacron is a polyester prepared from ethylene glycol and terephthalic acid.

$$HO-\overset{O}{\overset{\|}{C}}-C_6H_4-\overset{O}{\overset{\|}{C}}-OH \quad H-OCH_2CH_2O-H \quad HO-\overset{O}{\overset{\|}{C}}-C_6H_4-\overset{O}{\overset{\|}{C}}-OH \quad H-OCH_2CH_2O-H$$

terephthalic acid — ethylene glycol — terephthalic acid — ethylene glycol

↓ remove water

$$HOCH_2CH_2O-\left[\overset{O}{\overset{\|}{C}}-C_6H_4-\overset{O}{\overset{\|}{C}}-OCH_2CH_2O\right]_n-H$$

polyethylene terephthalate (Dacron)

Dacron, the fiber produced from this polyester, absorbs very little moisture, and its properties are very nearly the same when wet or dry. Additionally, it possesses exceptional elastic recovery properties, so it is used to make "no-iron" or "permanent press" fabrics. This polyester can also be made into films, such as Mylar, of great strength. There are many other polyesters.

Amines

The **amines** are considered to be derivatives of ammonia in which one or more hydrogen atoms have been replaced by alkyl or aryl groups. Amines are basic compounds, and their basicity depends on the nature of the organic substituents (Table 30–6).

The common names of tetramethylenediamine, *putrescine,* and pentamethylenediamine, *cadaverine,* are indeed suggestive.

The odors of amines are quite unpleasant; many of the malodorous compounds released as fish decays are simple amines. Amines of high molecular weight are nonvolatile, and they have little odor. One of the raw materials used to manufacture nylon, hexamethylenediamine, is an aliphatic amine. Many aromatic amines are used to prepare organic dyes that are widely used in indus-

TABLE 30–6 Some Properties of Amines

Name	Formula	Boiling Point °C	Ionization Constant, K_b
ammonia	NH_3	−33.4	1.8×10^{-5}
methylamine	CH_3NH_2	−6.5	50×10^{-5}
dimethylamine	$(CH_3)_2NH$	7.4	74×10^{-5}
trimethylamine	$(CH_3)_3N$	3.5	7.4×10^{-5}
ethylamine	$CH_3CH_2NH_2$	16.6	47×10^{-5}
aniline	$C_6H_5NH_2$	184	4.2×10^{-10}
ethylenediamine	$H_2NCH_2CH_2NH_2$	116.5	8.5×10^{-5}
pyridine	C_5H_5N	115.3	15×10^{-10}

trial societies. Amines are also used to produce many medicinal products, including local anesthetics and sulfa drugs.

30–9 Structure and Nomenclature of Amines

Recall that ammonia is a Lewis base because there is one unshared pair of electrons on the nitrogen atom (Section 11–12). As a reference point, a 0.10 *M* solution of aqueous ammonia is 1.3% ionized. (See Section 16–4.)

There are three classes of amines, depending on whether one, two, or three hydrogen atoms have been replaced by organic groups. They are called primary, secondary, and tertiary amines, respectively.

NH_3	RNH_2	R_2NH	R_3N
H—N(—H)—H	CH_3—N(—H)—H	CH_3—N(—H)—CH_3	CH_3—N(—CH_3)—CH_3
ammonia	methylamine (primary)	dimethylamine (secondary)	trimethylamine (tertiary)

Models for ammonia, methylamine, dimethylamine, and trimethylamine are shown in Figure 30–6.

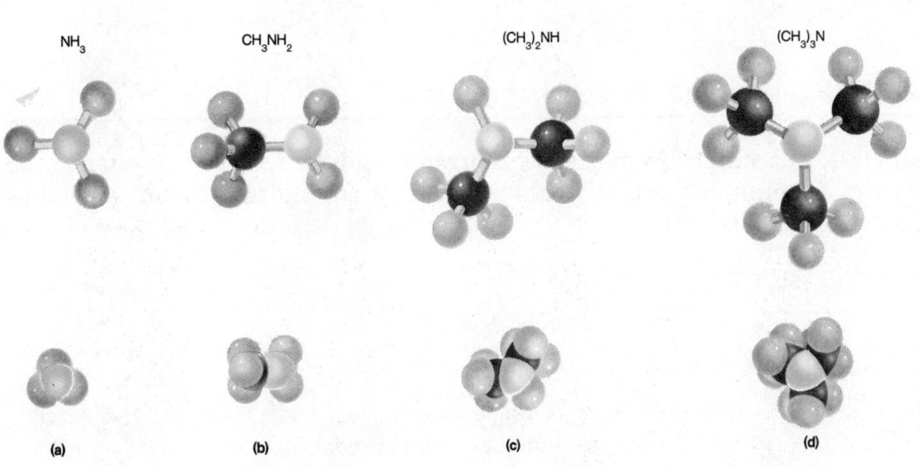

FIGURE 30–6 Models of (a) ammonia, (b) methylamine (c) dimethylamine, and (d) trimethylamine.

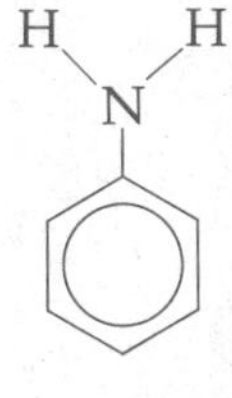

H H N (benzene ring)

aniline (primary)

Aniline is the simplest aromatic amine, and many aromatic amines are named as derivatives of aniline.

Heterocyclic amines contain nitrogen as a part of the ring, bound to two carbon atoms. Many heterocyclic amines are found in coal tar and a variety of natural products

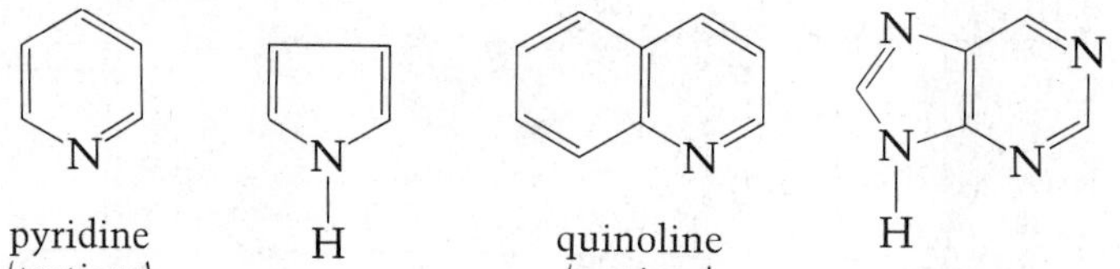

pyridine (tertiary)

pyrrole (secondary)

quinoline (tertiary)

purine

Reference to Table 30–6 shows that aliphatic amines are much stronger bases than aromatic and heterocyclic amines. Most low molecular weight aliphatic amines are somewhat stronger bases than ammonia.

The aliphatic amines of low molecular weight are soluble in water. Even aliphatic diamines of fairly high molecular weight are soluble in water, because each molecule contains two highly polar $—NH_2$ groups. The $—NH_2$ group is called the **amino** group in amines and amino acids.

Reactions of amines with water are similar to that of ammonia with water (see Section 16–4)

$$NH_3 + H_2O \rightleftharpoons \underset{\text{ammonium ion}}{NH_4^+} + OH^-$$

$$CH_3NH_2 + H_2O \rightleftharpoons \underset{\text{methylammonium ion}}{CH_3NH_3^+} + OH^-$$

$$(CH_3)_2NH + H_2O \rightleftharpoons \underset{\text{dimethylammonium ion}}{(CH_3)_2NH_2^+} + OH^-$$

There are no general methods by which all types of amines can be prepared. We shall leave the preparation of amines for the organic chemistry course.

30–10 Reactions of Amines

All classes of amines form salts of inorganic acids in reactions similar to the reaction of ammonia and hydrochloric acid (or any inorganic acid, as well as many organic acids).

$$NH_3 + H^+Cl^- \longrightarrow \underset{\text{ammonium chloride}}{NH_4^+Cl^-}$$

$$CH_3NH_2 + H^+Cl^- \longrightarrow CH_3NH_3^+Cl^-$$

methylammonium chloride
(methylamine hydrochloride)

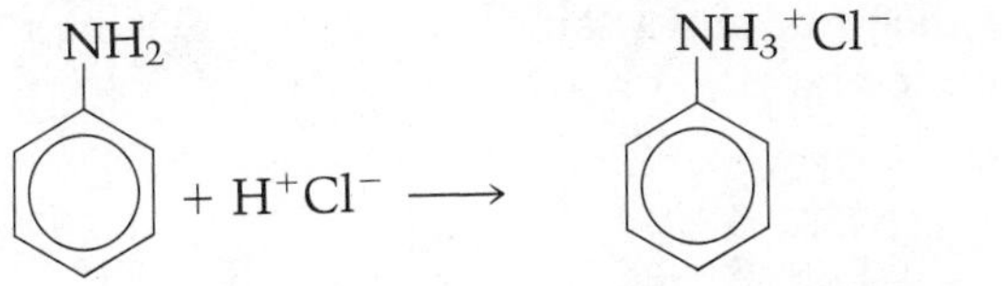

aniline — anilinium chloride (aniline hydrochloride)

Most salts of amines are soluble in water, but insoluble in hydrocarbon solvents. As the above examples illustrate, the common names for salts of amines consist of (name of amine) + hydro + (name of anion of acid).

Just as the reaction between ammonium salts and strong bases liberates ammonia, the reaction of strong bases with salts of amines liberates the free amine.

$$NH_4{}^+Cl^- + Na^+OH^- \longrightarrow NH_3\ (g) + Na^+Cl^- + H_2O$$

$$CH_3NH_3{}^+Cl^- + Na^+OH^- \longrightarrow \underset{\text{methylamine}}{CH_3NH_2} + Na^+Cl^- + H_2O$$

The fact that amines react with acids to form salts is frequently used to extract amines (for example, strychnine and nicotine) from their natural sources by treatment with acids. The resulting solutions are then made basic with sodium hydroxide, which liberates the free amine.

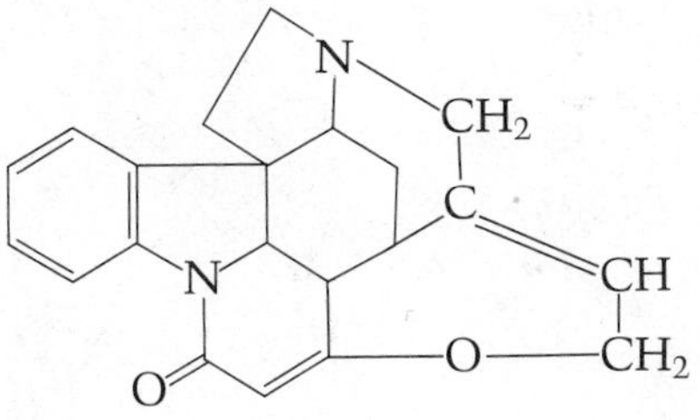

strychnine

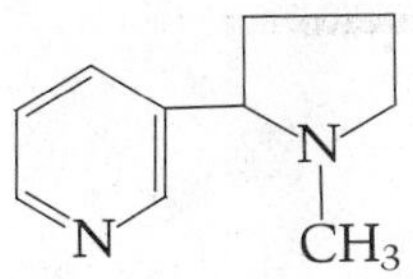

nicotine

Amides

30–11 Preparation, Structure, and Nomenclature of Amides

Amides are thought of as derivatives of primary or secondary amines and organic acids. Amides contain the $-\overset{\overset{\displaystyle O}{\|}}{C}-N\langle$ grouping of atoms. However, they are usually *not* prepared by the reaction of an amine with an organic acid. Acyl halides (acid halides) and acid anhydrides react with primary and secondary amines to produce amides readily. The reaction of an acyl halide with a primary or secondary amine produces an amide and a salt of the amine. One half of the amine is converted to an amide and the other half to a salt.

$$2CH_3NH_2 + CH_3{-}\overset{\overset{\displaystyle O}{\|}}{C}{-}Cl \longrightarrow CH_3{-}\overset{\overset{\displaystyle O}{\|}}{C}{-}N\!\!\begin{array}{l}{}^{\diagup}H\\ {}_{\diagdown}CH_3\end{array} + CH_3NH_3{}^+Cl^-$$

methylamine a primary amine — acetyl chloride an acyl halide — N-methylacetamide (an amide) — methylammonium chloride (a salt)

The reaction of acid anhydrides (see colored box) with primary or secondary amines produces an amide and a salt of the amine. Note that only one half of the amine is converted to an amide and the other half to a salt.

$$2(CH_3)_2NH + CH_3\overset{\overset{\displaystyle O}{\|}}{C}{-}O{-}\overset{\overset{\displaystyle O}{\|}}{C}{-}CH_3 \longrightarrow CH_3{-}\overset{\overset{\displaystyle O}{\|}}{C}{-}N\!\!\begin{array}{l}{}^{\diagup}CH_3\\ {}_{\diagdown}CH_3\end{array} + CH_3{-}\overset{\overset{\displaystyle O}{\|}}{C}{-}O^- \;{}^+H_2N(CH_3)_2$$

dimethylamine a secondary amine — acetic anhydride an acid anhydride — N,N-dimethylacetamide — dimethylammonium acetate (a salt)

Acid anhydrides are usually prepared by indirect methods, but this illustration shows the structural relationship that is important for our purposes.

The relationship between monocarboxylic acids and their anhydrides is

$$\begin{array}{l} CH_3{-}\overset{\overset{\displaystyle O}{\|}}{C}{-}O{-}H \\ CH_3{-}\overset{\overset{\displaystyle O}{\|}}{C}{-}OH \end{array} \longrightarrow \begin{array}{l} CH_3{-}\overset{\overset{\displaystyle O}{\|}}{C}\diagdown \\ \qquad\quad O + H_2O \\ CH_3{-}\underset{\underset{\displaystyle O}{\|}}{C}\diagup \end{array}$$

two molecules of acetic acid — one molecule of acetic anhydride

Acetanilide, the amide of acetic acid and aniline (which is sometimes called antifebrin), is used to treat headaches, neuralgia, and mild fevers. N,N-diethyl-*m*-toluamide, the amide of metatoluic acid and N,N-diethylamine, is the active ingredient in some insect repellents.

$$CH_3\overset{\overset{\displaystyle O}{\|}}{C}{-}NH{-}C_6H_5 \qquad m\text{-}CH_3C_6H_4{-}\overset{\overset{\displaystyle O}{\|}}{C}{-}N(CH_2CH_3)_2$$

acetanilide — N,N-diethyl-*m*-toluamide

The polymeric amides are an especially important class of compounds, also called **polyamides. Nylon** is the best known polymeric amide. It is prepared by heating anhydrous hexamethylenediamine with anhydrous adipic acid, a dibasic acid.

$$\underset{\text{adipic acid}}{\text{HO—}\overset{\overset{\displaystyle O}{\|}}{C}\text{—(CH}_2)_4\text{—}\overset{\overset{\displaystyle O}{\|}}{C}\text{—OH}} + \underset{\text{hexamethylenediamine}}{\text{H}_2\text{N—(CH}_2)_6\text{—NH}_2} \xrightarrow[-\text{H}_2\text{O}]{\text{heat}}$$

$$\underset{\substack{\text{Nylon 6-6} \\ \text{(a polyamide)}}}{\text{—NH}\left(\overset{\overset{\displaystyle O}{\|}}{C}\text{—(CH}_2)_4\text{—}\overset{\overset{\displaystyle O}{\|}}{C}\text{—NH—(CH}_2)_6\text{—NH}\right)_n\overset{\overset{\displaystyle O}{\|}}{C}\text{—}}$$

This substance is known as Nylon 6-6 because the parent diamine and dicarboxylic acid each contain six carbon atoms.

The molecular weight of the polymer varies from about 10,000 to about 25,000. It melts at approximately 260° to 270°C.

Molten Nylon is drawn into threads that, after cooling to room temperature, can be stretched to about four times their original length (Figure 30–7). The "cold drawing" process orients the polymer molecules so that their long axes are parallel to the fiber axis. At regular intervals there are N—H---O hydrogen bonds that cross-link oxygen and nitrogen atoms on adjacent chains to give strength to the fiber.

We might note in passing that petroleum is the ultimate source of both adipic acid and hexamethylenediamine. This statement does not imply that these compounds are present in petroleum, only that they are made from petroleum. Similar statements can be made for many industrial chemicals. The price of petroleum is an important factor in our economy because so many different products are derived from petroleum.

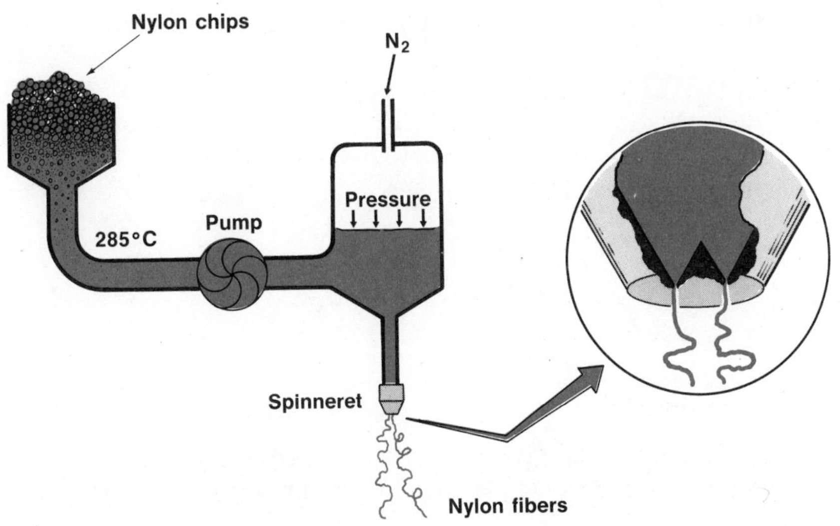

FIGURE 30–7 Fibers of synthetic polymers are made by extrusion of the molten material through tiny holes, called spinnerets. After cooling, they are stretched to about four times their original length.

Aldehydes and Ketones

30–12 Structure and Nomenclature of Aldehydes and Ketones

Aldehydes and ketones contain the carbonyl group, $\rangle C{=}O$. **Aldehydes** are compounds in which one alkyl or aryl group and one hydrogen atom are bonded to the carbonyl group. **Ketones** have two alkyl or aryl groups bonded to a carbonyl group.

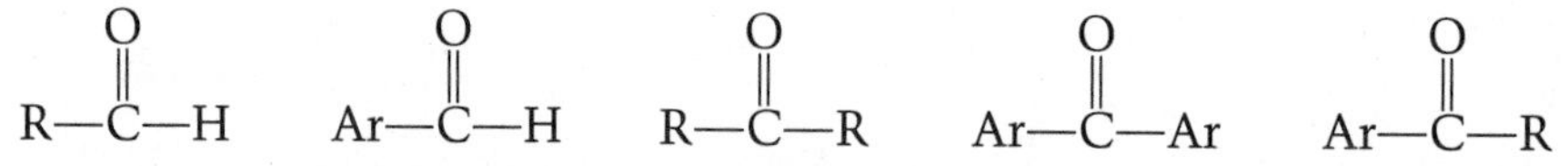

$R-\overset{O}{\overset{\|}{C}}-H$	$Ar-\overset{O}{\overset{\|}{C}}-H$	$R-\overset{O}{\overset{\|}{C}}-R$	$Ar-\overset{O}{\overset{\|}{C}}-Ar$	$Ar-\overset{O}{\overset{\|}{C}}-R$
aliphatic aldehyde	aromatic aldehyde	aliphatic ketone	aromatic ketone	mixed ketone

Models of formaldehyde, the simplest aldehyde, and acetone, the simplest ketone, are shown in Figure 30–8.

Aldehydes are usually called by their common names, which are derived from the name of the acid formed when the aldehyde is oxidized (Table 30–7). The systematic name is derived from the name of the parent hydrocarbon. The suffix *-al* is added to the characteristic stem. The carbonyl group takes positional preference over other substituents.

Formaldehyde has been used as a disinfectant and as a preservative for biological specimens (this includes embalming fluid) for many years. Its most important use is in the production of certain plastics and in binders for plywood. Many important natural substances are aldehydes and ketones. Examples include sex hormones, some vitamins, camphor, and the flavors from almonds and cinnamon. Since aldehydes contain a carbon-oxygen double bond, they are very reactive compounds. Reaction usually occurs by addition to the carbonyl group.

Simple, commonly encountered ketones are usually called by their common names, which are derived by naming the alkyl or aryl groups attached to the carbonyl group

$CH_3\overset{O}{\overset{\|}{C}}CH_3$	$CH_3\overset{O}{\overset{\|}{C}}CH_2CH_3$	$CH_3CH_2\overset{O}{\overset{\|}{C}}CH_2CH_3$
acetone	methyl ethyl ketone	diethyl ketone

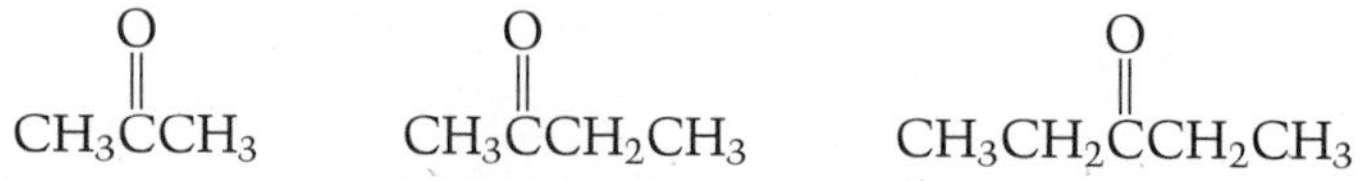

When the benzene ring (C_6H_5—) is a substituent, it can be called a phenyl group.

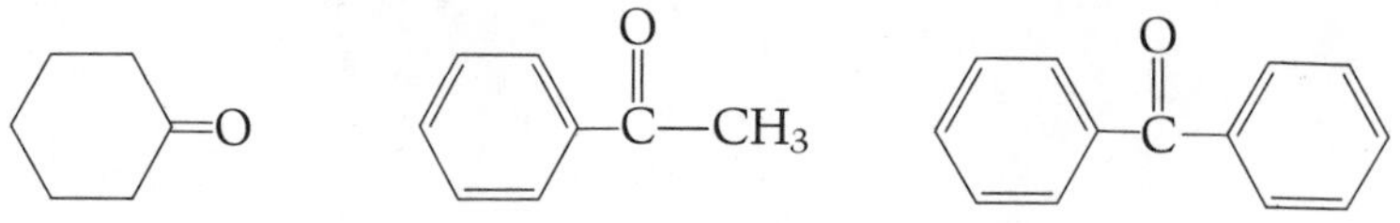

$C_6H_{10}{=}O$	$C_6H_5-\overset{O}{\overset{\|}{C}}-CH_3$	$C_6H_5-\overset{O}{\overset{\|}{C}}-C_6H_5$
cyclohexanone	acetophenone (methyl phenyl ketone)	benzophenone (diphenyl ketone)

In some cases trivial names that give few clues about structure are used. Acetone, acetophenone, and benzophenone are three examples given above.

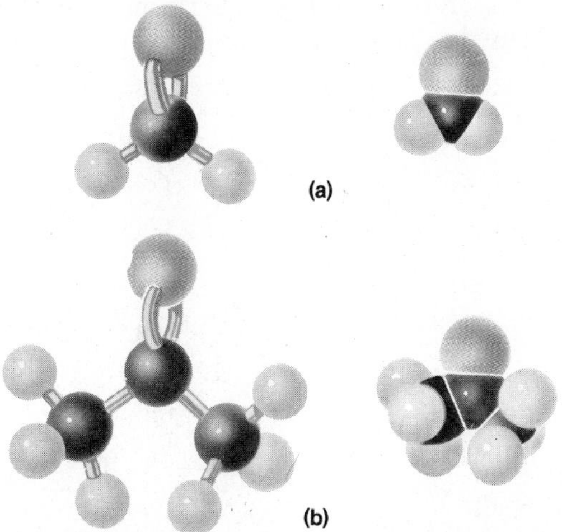

FIGURE 30–8 (a) Models of formaldehyde, HCHO, the simplest aldehyde, and (b) models of acetone, $CH_3—CO—CH_3$, the simplest ketone.

The systematic names for ketones are derived from their parent hydrocarbons. The suffix *-one* is added to the characteristic stem

$$\overset{1}{CH_3}-\overset{\overset{O}{\|}}{\underset{2}{C}}-\overset{3}{CH_2}-\overset{4}{CH_3} \qquad \overset{1}{CH_3}-\overset{2}{CH_2}-\overset{\overset{O}{\|}}{\underset{3}{C}}-\overset{\overset{CH_3}{|}}{\underset{4}{CH}}-\overset{5}{CH_2}-\overset{6}{CH_3}$$

2-butanone 4-methyl-3-hexanone

The ketones are excellent solvents, and acetone is particularly useful because it dissolves most organic compounds yet is completely miscible with water. Acetone is widely used as a solvent in the manufacture of lacquers, paint removers, explosives, plastics, drugs, and disinfectants. Some ketones of

TABLE 30–7 Properties of Some Simple Aldehydes

Common Name (IUPAC Name)	Formula	Boiling Point (°C)
formaldehyde (methanal)	$H-\overset{\overset{O}{\|}}{C}-H$	−21
acetaldehyde (ethanal)	$CH_3-\overset{\overset{O}{\|}}{C}-H$	20.2
propionaldehyde (propanal)	$CH_3CH_2\overset{\overset{O}{\|}}{C}-H$	48.8
benzaldehyde	$C_6H_5-\overset{\overset{O}{\|}}{C}-H$	179.5

high molecular weight are used extensively in the blending of perfumes. Structures of some naturally occurring aldehydes and ketones are given below.

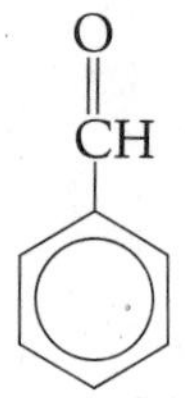

benzaldehyde (almonds)

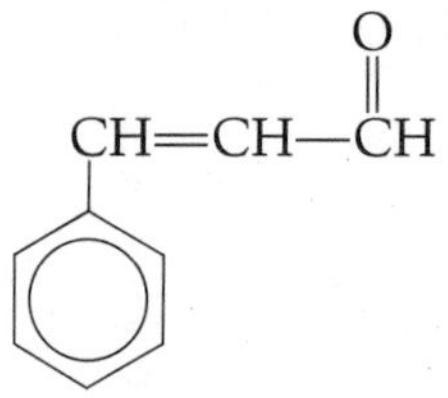

cinnamaldehyde (cinnamon)

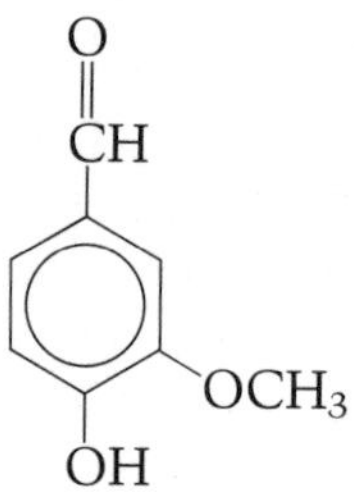

vanillin (vanilla)

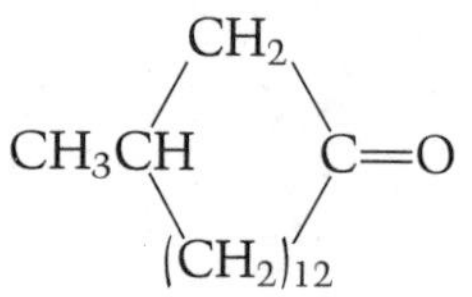

muscone (musk deer, used in perfumes)

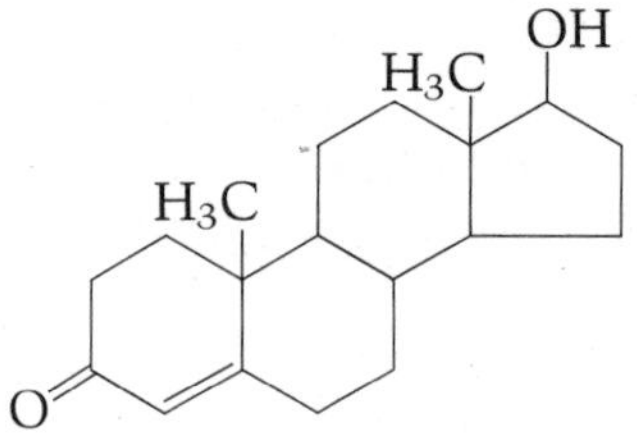

testosterone (male sex hormone)

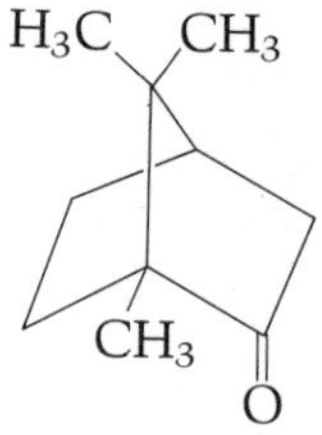

camphor

30–13 Preparation of Aldehydes and Ketones

Since aldehydes are easily oxidized to acids, they must be removed from the reaction mixture as rapidly as they are formed. Aldehydes have lower boiling points than the alcohols from which they are formed.

Aldehydes may be prepared by the oxidation of *primary* alcohols. The reaction mixture is heated to a temperature slightly above the boiling point of the aldehyde so that it distills out as soon as it is formed. Potassium dichromate, in the presence of dilute sulfuric acid, is the common oxidizing agent.

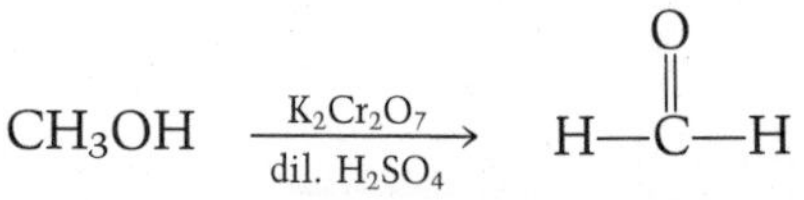

$$CH_3OH \xrightarrow[\text{dil. } H_2SO_4]{K_2Cr_2O_7} H\text{—}\overset{\overset{\large O}{\|}}{C}\text{—}H$$

methanol, bp = 65°C → methanal (formaldehyde), bp = −21°C

Ketones may be prepared by the oxidation of *secondary* alcohols. Ketones are not as susceptible to oxidation as are aldehydes, and alkaline solutions of potassium permanganate may be used as the oxidizing agent.

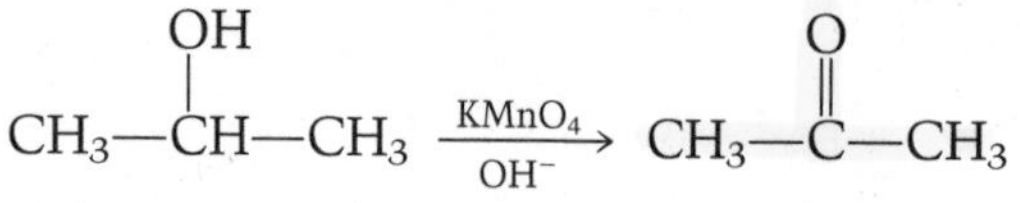

$$CH_3\text{—}\overset{\overset{\large OH}{|}}{C}H\text{—}CH_3 \xrightarrow[OH^-]{KMnO_4} CH_3\text{—}\overset{\overset{\large O}{\|}}{C}\text{—}CH_3$$

isopropyl alcohol, 2-propanol → acetone

$$\underset{\text{2-butanol}}{CH_3{-}CH_2{-}\overset{\overset{\displaystyle OH}{|}}{C}H{-}CH_3} \xrightarrow[OH^-]{KMnO_4} \underset{\substack{\text{2-butanone}\\ \text{methyl ethyl ketone}}}{CH_3{-}CH_2{-}\overset{\overset{\displaystyle O}{\|}}{C}{-}CH_3}$$

Aldehydes and ketones may be prepared commercially by a catalytic process that involves passing alcohol vapors and air over a copper gauze or powder at approximately 300°C.

$$\underset{\text{methanol}}{2CH_3OH} + O_2 \xrightarrow[300°C]{Cu} \underset{\text{formaldehyde}}{2H{-}\overset{\overset{\displaystyle O}{\|}}{C}{-}H} + 2H_2O$$

Formaldehyde is quite soluble in water; the gaseous compound may be dissolved in water to give a 40% solution.

Acetaldehyde may be prepared by the similar oxidation of ethanol.

$$2CH_3CH_2OH + O_2 \xrightarrow[300°C]{Cu} 2CH_3{-}\overset{\overset{\displaystyle O}{\|}}{C}{-}H + 2H_2O$$

Ethers

When the word ether is mentioned, most people think of the well-known anesthetic, diethyl ether. There are many ethers, and they are used for a variety of other purposes such as artificial flavors, refrigerants, and as a very important class of solvents. **Ethers** are compounds in which an oxygen atom is bonded to two organic groups.

$$-\overset{|}{\underset{|}{C}}-O-\overset{|}{\underset{|}{C}}-$$

Note the difference between an ether and an ester. An ether does not have a carbon-oxygen double bond.

Alcohols are considered derivatives of water in which one hydrogen atom has been replaced by an organic group. Ethers may be considered derivatives of water in which both hydrogen atoms have been replaced by organic groups (Figure 30–9). However, the similarity is only structural because ethers are not very polar and are chemically quite unreactive.

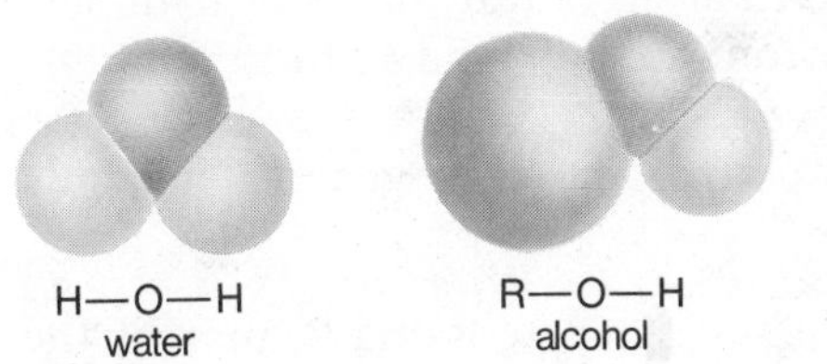

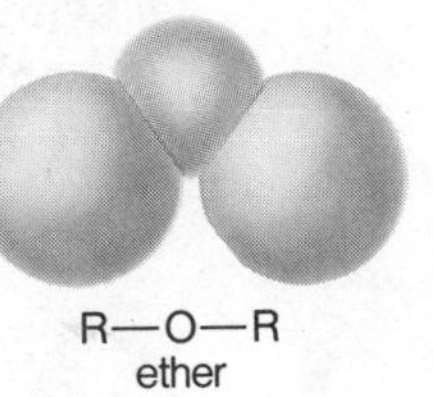

R—O—R
ether

FIGURE 30–9 Models showing the structural relationship among water, alcohol, and ethers.

Three kinds of ethers are known: (1) aliphatic, (2) aromatic, and (3) mixed ethers. Common names are used for ethers in most cases

$CH_3—O—CH_3$	$CH_3—O—CH_2CH_3$	$C_6H_5—O—CH_3$	$C_6H_5—O—C_6H_5$
methoxymethane dimethyl ether (an aliphatic ether)	methoxyethane methyl ethyl ether (an aliphatic ether)	methoxybenzene methyl phenyl ether anisole (a mixed ether)	phenoxybenzene diphenyl ether (an aromatic ether)

Ethers are quite unreactive, and we shall not discuss their reactions here. Diethyl ether is a very low boiling liquid, bp = 35°C. Dimethyl ether is a gas that is used as a refrigerant. The aliphatic ethers of higher molecular weights are liquids, and the aromatic ethers are liquids or solids.

Even ethers of low molecular weight are only slightly soluble in water. Diethyl ether is an excellent solvent for organic compounds. It is widely used to extract organic compounds from plants and other natural sources.

Ethers burn readily, and care must be exercised to avoid fires when ethers are used in laboratories or anywhere else. Diethyl ether is also oxidized at room temperature to a nonvolatile, explosive peroxide by oxygen in the air. For this reason, ethereal solutions should never be evaporated to dryness because of the danger of peroxide explosions, unless proper precautionary steps have been taken to destroy all peroxides in advance.

Key Terms

Acid anhydride compound produced by dehydration of a carboxylic acid; general formula is $R—\overset{O}{\overset{\|}{C}}—O—\overset{O}{\overset{\|}{C}}—R$.

Acyl group group of atoms remaining after removal of an —OH group of a carboxylic acid.

Alcohol hydrocarbon derivative containing an —OH group attached to a carbon atom not in an aromatic ring.

Aldehyde compound in which an alkyl or aryl group and a hydrogen atom are attached to a carbonyl group; general formula, $R—\overset{O}{\overset{\|}{C}}—H$.

Amide compound containing the $—\overset{O}{\overset{\|}{C}}—N\langle$ group.

Amine compound that can be considered a derivative of ammonia in which one or more hydrogens are replaced by alkyl or aryl groups.

Amino acid compound containing both an amino group and a carboxylic acid group.

Amino group the $—NH_2$ group.

Aryl group group of atoms remaining after a hydrogen atom is removed from an aromatic system.

Carbonyl group the $—\overset{O}{\overset{\|}{C}}—$ group.

Carboxylic acid compound containing a $—\overset{O}{\overset{\|}{C}}—O—H$ group.

Dehydrogenation removal of two hydrogen atoms from a compound to form a double bond.

Ester compound of the general formula $R—\overset{O}{\overset{\|}{C}}—O—R'$ where R and R′ may be the same or different, and may be either aliphatic or aromatic.

Ether compound in which an oxygen atom is bonded to two alkyl or two aryl groups, or one alkyl and one aryl group.

Fat solid triester of glycerol and (mostly) saturated fatty acids.

Fatty acid an aliphatic acid; many can be obtained from animal fats.

Functional group generally, a group of atoms that represents a potential reaction site in an organic compound.

Glyceride triester of glycerol.

Heterocyclic amine amine in which the nitrogen is part of a ring.

Ketone compound in which a carbonyl group is bound to two alkyl or two aryl groups, or to one alkyl and one aryl group.

Phenol hydrocarbon derivative containing an —OH group bound to an aromatic ring.

Polyamide a polymeric amide.

Polyester polymeric ester.

Polyhydric alcohol alcohol containing more than one —OH group.

Oil liquid triester of glycerol and unsaturated fatty acids.

Saponification hydrolysis of esters in the presence of strong soluble bases.

Soap sodium salt of long chain fatty acid.

Exercises

Alcohols and Phenols—Structure and Nomenclature

1. What is the functional group of alcohols and phenols?
2. (a) What are alcohols and phenols?
   (b) How do they differ?
   (c) Why can alcohols and phenols be viewed as derivatives of hydrocarbons? As derivatives of water?
3. Distinguish between alkyl and aryl groups.
4. Draw structural formulas and write names for six alcohols and three phenols.
5. (a) Distinguish among primary, secondary, and tertiary alcohols. (b) Write names and formulas for three alcohols of each type.
6. (a) Draw structural formulas for and write the names of the four (saturated) alcohols that contain four carbon atoms and one —OH group per molecule. (b) Draw structural formulas for and write the names of the eight (saturated) alcohols that contain five carbon atoms and one —OH group per molecule. Which ones may be classified as primary alcohols? Secondary alcohols? Tertiary alcohols?
7. What are glycols?

Physical Properties of Alcohols and Phenols

8. Refer to Table 30–1 and explain the trends in boiling points and solubilities of alcohols in water listed there.
9. Why are glycols more soluble in water than monohydric alcohols that contain the same number of carbon atoms?
10. Why are most phenols only slightly soluble in water?

Preparation of Alcohols

11. Why are methyl alcohol and ethyl alcohol called wood alcohol and grain alcohol respectively?
12. (a) How is methyl alcohol prepared commercially? (b) List some uses for methyl alcohol.
13. (a) What is the important source of ethyl alcohol obtained by fermentation? (b) Write the overall equation for this reaction. (c) What is blackstrap molasses?
14. Describe the industrial preparation of ethyl alcohol and write balanced equations for the reactions.
15. Describe the industrial preparation of ethylene glycol and write balanced equations for the reactions.

Reactions of Alcohols

16. (a) Write equations for the reactions of three alcohols with metallic sodium. (b) Name all compounds in these equations. (c) Are these reactions similar to the reaction of metallic sodium with water? How?
17. (a) What are alkoxides? (b) What do we mean when we say that the low molecular weight alkoxides are strong bases?

18. (a) Write equations for the reaction of nitric acid with the following alcohols, methanol, ethanol, *n*-propanol, *n*-butanol.
    (b) Name the inorganic ester formed in each case.
    (c) What are inorganic esters?
19. (a) What is nitroglycerine? (b) How is it produced? (c) List two important uses for nitroglycerine. Are they similar?
20. (a) What are alkyl hydrogen sulfates? (b) How are they prepared? (c) Can alkyl hydrogen sulfates be classified as acids? Why? (d) What are detergents? (e) What is sodium lauryl sulfate? (f) What is the common use of sodium lauryl sulfate?

Carboxylic Acids—Names and Formulas

21. (a) What are carboxylic acids? (b) Draw structural formulas for and write the names of seven carboxylic acids. (c) Why are aliphatic carboxylic acids sometimes called fatty acids? Cite two examples.

Reactions of Carboxylic Acids

22. (a) What are acyl chlorides or acid chlorides? (b) How are they prepared? (c) Write equations for the preparation of four acid chlorides. Name all compounds.
23. (a) What are esters? (b) Draw structures for four esters and write their names.
24. Write equations for the formation of three different esters starting with an acid and an alcohol in each example. Name all compounds.
25. Write equations for the formation of three esters starting with an acid chloride and an alcohol in each case. Name all compounds.

Occurrence and Hydrolysis of Esters

26. List six naturally occurring esters and their sources.
27. (a) What is saponification? (b) Why is this kind of reaction called saponification?
28. Write equations for the hydrolysis of (a) methyl acetate, (b) ethyl formate, (c) *n*-butyl acetate, and (d) *n*-octyl acetate. Name all products.
29. (a) What are fats? Oils? (b) Write the general formula for fats and oils.
30. What are glycerides? Distinguish between simple glycerides and mixed glycerides.
31. Write the names and formulas of some acids that occur in fats and oils (as esters, of course).
32. Write names and formulas for three simple glycerides.
33. Write equations for the hydrolysis (saponification) of glyceryl tristearate and glyceryl tripalmitate.
34. (a) What are soaps? (b) What is the difference between a soap and a detergent?
35. What are waxes?

Polyesters

36. (a) What are polyesters? (b) What is Dacron? (c) How is Dacron prepared?
37. Is it reasonable to assume that a polyester can be made from propylene glycol and terephthalic acid? If so, sketch its structure.

Amines

38. (a) What are amines? (b) Why are amines thought of as derivatives of ammonia?
39. (a) Distinguish among primary, secondary, and tertiary amines. (b) Write names and formulas for three amines in each of the following classes without using any compound more than once: (1) aliphatic amines, (2) aromatic amines, (3) primary aliphatic amines, (4) secondary aliphatic amines, and (5) tertiary aliphatic amines.
40. Why are aqueous solutions of amines basic?
41. Show that the reactions of amines with inorganic acids such as HCl are similar to the reactions of ammonia with inorganic acids.

Amides

42. (a) What are amides? (b) Draw structural formulas for and write the names of three amides.
43. Write equations for the preparation of an amide starting with
    (a) a primary amine and an acid chloride
    (b) a secondary amine and an acid chloride
    (c) a primary amine and an acid anhydride
    (d) a secondary amine and an acid anhydride

44. What are polyamides? Why are they an important class of compounds?
45. (a) What is Nylon? (b) How is it prepared?
46. Common Nylon is called Nylon 6-6. (a) What does this mean? (b) Can you visualize other Nylons?

Aldehydes and Ketones

47. (a) Distinguish between aldehydes and ketones. (b) Cite three examples (each) of aliphatic and aromatic aldehydes and ketones by drawing structural formulas and naming the compounds.
48. (a) List several naturally occurring aldehydes and ketones. (b) What are their sources? (c) What are some uses of these compounds?
49. Describe the preparation of three aldehydes from alcohols and write appropriate equations. Name all reactants and products.
50. Describe the preparation of three ketones from alcohols and write appropriate equations. Name all reactants and products.
51. (a) Describe the commercial preparation of aldehydes and ketones. (b) Write equations for the preparation of two aldehydes and two ketones. (c) Name all reactants and products.

Ethers

52. (a) What are ethers? (b) Draw structural formulas for four ethers. Name them.
53. Distinguish between aliphatic ethers and mixed ethers.
54. List some uses for some ethers.

Appendices

APPENDIX A

Some Mathematical Operations

We frequently use very large or very small numbers in chemistry, and such numbers are conveniently expressed in *scientific* or *exponential notation.*

A.1 Scientific Notation

In scientific notation a number is expressed as the *product of two numbers.* By convention, the first number, called the digit term, is between 1 and 10. The second number, called the *exponential term,* is an integer power of 10. Some examples follow.

$10000 = 1 \times 10^4$	$24327 = 2.4327 \times 10^4$
$1000 = 1 \times 10^3$	$7958 = 7.958 \times 10^3$
$100 = 1 \times 10^2$	$594 = 5.94 \times 10^2$
$10 = 1 \times 10^1$	$98 = 9.8 \times 10^1$
$1 = 1 \times 10^0$	Recall, $(\text{any base})^0 = 1$ (by definition)
$1/10 = 0.1 = 1 \times 10^{-1}$	$0.32 = 3.2 \times 10^{-1}$
$1/100 = 0.01 = 1 \times 10^{-2}$	$0.067 = 6.7 \times 10^{-2}$
$1/1000 = 0.001 = 1 \times 10^{-3}$	$0.0049 = 4.9 \times 10^{-3}$
$1/10000 = 0.0001 = 1 \times 10^{-4}$	$0.00017 = 1.7 \times 10^{-4}$

The digit term of the numerator is divided by the digit term of the de-
places the decimal point must be shifted to give the number in long form. A *positive exponent* indicates that the decimal point is *shifted right* that number of places to give the long form, and a *negative exponent* indicates that the decimal point is *shifted left,* as the following examples illustrate.

$$7.3 \times 10^3 = 73 \times 10^2 = 730 \times 10^1 = 7300$$

$$8.6 \times 10^{-3} = 86 \times 10^{-4} = 860 \times 10^{-5}$$

$$0.0086 = 0.086 \times 10^{-1} = 0.86 \times 10^{-2} = 8.6 \times 10^{-3}$$

When numbers are written in *standard scientific notation* there is one nonzero digit to the left of the decimal point.

1 Addition and Subtraction

In addition and subtraction, all numbers are converted to the same power of 10 and the digit terms are added or subtracted.

$$(4.21 \times 10^{-3}) + (1.4 \times 10^{-4}) = (4.21 \times 10^{-3}) + (0.14 \times 10^{-3}) = \underline{4.35 \times 10^{-3}}$$

$$(8.97 \times 10^{4}) - (2.31 \times 10^{3}) = (8.97 \times 10^{4}) - (0.231 \times 10^{4}) = \underline{8.74 \times 10^{4}}$$

2 Multiplication

The digit terms are multiplied in the usual way, the exponents are added algebraically, and the product is written with one nonzero digit to the left of the decimal.

$$(4.7 \times 10^{7})(1.6 \times 10^{2}) = (4.7)(1.6) \times 10^{7+2} = 7.52 \times 10^{9} = \underline{7.5 \times 10^{9}} \quad \text{(two sig. figs.)}$$

$$(8.3 \times 10^{4})(9.3 \times 10^{-9}) = (8.3)(9.3) \times 10^{4-9} = 77.19 \times 10^{-5} = \underline{7.7 \times 10^{-4}} \quad \text{(two sig. figs.)}$$

3 Division

The digit term of the numerator is divided by the digit term of the denominator, the exponents are subtracted algebraically, and the quotient is written with one nonzero digit to the left of the decimal.

$$\frac{8.4 \times 10^{7}}{2.0 \times 10^{3}} = \frac{8.4}{2.0} \times 10^{7-3} = \underline{4.2 \times 10^{4}}$$

$$\frac{3.9 \times 10^9}{8.4 \times 10^{-3}} = \frac{3.9}{8.4} \times 10^{9-(-3)}$$
$$= 0.4643 \times 10^{12} = \underline{4.6 \times 10^{11}} \quad \text{(two sig. figs.)}$$

4 Powers of Exponentials

The digit term is raised to the indicated power in the usual way, and the exponent is multiplied by the number that indicates the power.

$$(1.2 \times 10^3)^2 = (1.2)^2 \times 10^{3 \times 2}$$
$$= 1.44 \times 10^6 = \underline{1.4 \times 10^6} \quad \text{(two sig. figs.)}$$

$$(3.0 \times 10^{-3})^4 = (3.0)^4 \times 10^{-3 \times 4}$$
$$= 81 \times 10^{-12} = \underline{8.1 \times 10^{-11}}$$

Electronic calculators: To square a number: (1) enter the number, (2) touch the (x^2) button.

These instructions are applicable to the Texas Instruments series. If your calculator has other notation, consult your instruction booklet.

$(7.3)^2 = 53.29 = \underline{53}$ (two sig. figs.)

To raise a number to a higher power: (1) enter the number, (2) touch the (y^x) button, (3) enter the power, and (4) touch the (=) button.

$(7.3)^4 = 2839.8241 = \underline{2.8 \times 10^3}$ (two sig. figs.)

$(7.30 \times 10^2)^5 = 2.0730716 \times 10^{14} = \underline{2.07 \times 10^{14}}$ (three sig. figs.)

5 Roots of Exponentials

The exponent must be divisible by the desired root if a calculator is not used. The root of the digit term is extracted in the usual way and the exponent is divided by the desired root.

$$\sqrt{2.5 \times 10^5} = \sqrt{25 \times 10^4} = \sqrt{25} \times \sqrt{10^4} = \underline{5.0 \times 10^2}$$

$$\sqrt[3]{2.7 \times 10^{-8}} = \sqrt[3]{27 \times 10^{-9}} = \sqrt[3]{27} \times \sqrt[3]{10^{-9}} = \underline{3.0 \times 10^{-3}}$$

Electronic calculators: To extract the square root of a number: (1) enter the number, and (2) touch the ($\sqrt{x}$) button.

$\sqrt{23} = 4.7958315 = 4.8$ (two sig. figs.)

To extract a higher order root: (1) enter the number, (2) touch the $\sqrt[x]{y}$) button, (3) enter the root to be extracted, and (4) touch the (=) button.

On some models, this function is performed by the combination of the (INV) and (y^x) buttons.

$\sqrt[3]{12.0} = 2.2894285 = \underline{2.29}$ (three sig. figs.)

$\sqrt[4]{1.2 \times 10^{-9}} = 5.8856619 \times 10^{-3} = \underline{5.9 \times 10^{-3}}$ (two sig. figs.)

A.2 Logarithms

The power to which the base 10 must be raised to equal a number is called the *logarithm* (log) of the number. In the simplest view, a logarithm is an exponent. The number 10 must be raised to the third power to equal 1000. Therefore the logarithm of 1000 is 3, written as log 1000 = 3. Some examples are listed below.

Number	Exponential Expression	Logarithm
1000	10^3	3
100	10^2	2
10	10^1	1
1	10^0	0
1/10 = 0.1	10^{-1}	−1
1/100 = 0.01	10^{-2}	−2
1/1000 = 0.001	10^{-3}	−3

Most numbers are neither exact multiples nor exact fractions of 10. A table of common (base 10) logarithms is a tabulation of *fractional powers of the base 10.* These fractional powers are called *mantissas.* The four-digit numbers in Appendix B are mantissas or fractional powers of 10.

The following examples are illustrative:

$10^{0.0414} = 1.1$, so $\log 1.1 = 0.0414$ $\qquad$ $10^{0.3222} = 2.1$, so $\log 2.1 = 0.3222$

$10^{0.6435} = 4.4$, so $\log 4.4 = 0.6435$ $\qquad$ $10^{0.9956} = 9.9$, so $\log 9.9 = 0.9956$

Note that the first two digits of a number are tabulated in the left-hand vertical column, and the third digit is listed in the top horizontal row. For example, we find the log of 3.33 by locating 3.3 in the first vertical column and moving right until we come to the mantissa under the 3 column, where we find .5224. Thus,

$$\log 3.33 = 0.5224 \qquad \text{or} \qquad 10^{0.5224} = 3.33$$

The above procedure can be reversed. Suppose that we know that $\log x = 0.7259$. How do we obtain x? First, the terminology. The number x is called the *antilogarithm* of 0.7259. We locate .7259 in table of mantissas, look to the left and see that .7259 is located in row that starts with 5.3. Then we look to the top of column and see that .7259 is located in the column that is headed by 2. Therefore,

$$\text{antilog } 0.7259 = 5.32 \qquad \text{or} \qquad 10^{0.7259} = 5.32$$

For numbers much larger or much smaller than those in these examples, logarithms may be obtained by expressing the number in exponential form. Remember: logarithms are powers of 10 and they are treated like all other exponents, i.e., when two numbers are multiplied together their exponents are added.

$$\log 37 = \log (3.7 \times 10^1) = \log 3.7 + \log 10^1 = 0.5682 + 1 = \underline{1.5682}$$

so

$$\log 37 = 1.5682 \qquad \text{or} \qquad 10^{1.5682} = 37$$

Recall that $\log 10 = 1$ and $\log 100 = 2$, so log 37 must be greater than 1 but less than 2.

$$\log 0.037 = \log (3.7 \times 10^{-2}) = \log 3.7 + \log 10^{-2} = 0.5682 + (-2) = \underline{-1.4318}$$

so

$$\log 0.037 = -1.4318 \qquad \text{or} \qquad 10^{-1.4318} = 0.037$$

Recall that $\log 0.01 = -2$ and $\log 0.1 = -1$, so log 0.037 must be greater than -2 but less than -1.

Electronic calculators: To obtain the logarithm of a number from an electronic calculator, (1) enter the number, and (2) touch the (log x) button.

log 7.4 = 0.86923172 = 0.869 (three sig. figs.)

log $7.4 \times 10^{-3} = -2.1307683 = 2.13$ (three sig. figs.)

To obtain an antilogarithm, (1) enter the logarithm, (2) touch the (INV) button, and then (3) touch the (log x) button.

Antilog 1.301 = 19.998619 = 20.0 (three sig. figs.)

1 Multiplication Using Logarithms $\log xy = \log x + \log y$

To multiply two numbers, we (1) add their logarithms and then (2) find the antilogarithm (number) that corresponds to the logarithm of the product. Suppose we wish to multiply 213 by 42.6.

$$\begin{array}{r} \log\ 213 = 2.3284 \\ +\log 42.6 = 1.6294 \\ \hline \log \text{product} = 3.9578 \approx 3 + 0.9576 \end{array}$$ (nearest mantissa listed in Appendix B)

antilog $3.9576 = 9.07 \times 10^3$

We are justified in reporting only three significant figures, since we have not interpolated. We've simply taken the antilog of the mantissa nearest to 0.9578.

Interpolation is a process for approximating values of mantissas between those listed in the table. If you need to use it, you can find an explanation in any common algebra book.

Thus, we write $(213)(42.6) = 9.07 \times 10^3$ (three sig. figs.)

2 Division Using Logarithms $\log \frac{x}{y} = \log x - \log y$

Division is just the reverse of multiplication. Suppose we wish to divide 213 by 42.6.

$$\begin{array}{r} \log\ 213 = 2.3284 \\ -\log 42.6 = 1.6294 \\ \hline \log \text{quotient} = 0.6990 \end{array}$$

antilog 0.6990 = 5.00

Or, we may write 213/42.6 = 5.00

3 Extraction of Higher Order Roots Using Logarithms

Although most electronic calculators have the capability of extracting higher order roots (p. A.3), your instructor may insist that you use logarithms to demonstrate that you understand the process.

Recall that $\sqrt[5]{y} = (y)^{1/5}$ or in general terms $\sqrt[n]{y} = (y)^{1/n}$. To evaluate the fifth root of an expression such as

$(2x)^2(3x)^3 = 1.6 \times 10^{-72}$

we perform the indicated multiplication and obtain:

$108\,x^5 = 1.6 \times 10^{-72}$ or $x^5 = 1.48 \times 10^{-74}$

Since we wish to extract the fifth root, the exponent must be divisible by 5 if a calculator is not used. Therefore, we decrease the exponent to −75 by shifting the decimal point one place to the right.

$$x^5 = 14.8 \times 10^{-75}$$

and extract the fifth root of the exponential term.

$$x = \sqrt[5]{14.8 \times 10^{-75}} = \sqrt[5]{14.8} \times \sqrt[5]{10^{-75}} = \sqrt[5]{14.8} \times 10^{-15}$$

To obtain the fifth root of 14.8, we (1) look up its logarithm, (2) divide the logarithm by five, and (3) take the antilog of the logarithm obtained in step (2).

(1) $\log 14.8 = 1.1703$

(2) $\dfrac{\log 14.8}{5} = \dfrac{1.1703}{5} = 0.2341$

(3) antilog $0.2341 = 1.71$

Thus,

$$\sqrt[5]{14.8} = 1.71 \qquad \text{and} \qquad \sqrt[5]{14.8 \times 10^{-75}} = 1.71 \times 10^{-15} = x,$$

which should be reported as

$$\underline{x = 1.7 \times 10^{-15}}$$

because the original equation, $108\,x^5 = 1.6 \times 10^{-72}$, contains only two significant figures on the right side.

A.3 Quadratic Equations

Algebraic expressions of the form

$$ax^2 + bx + c = 0$$

are called **quadratic equations.** Each of the constant terms (a, b, and c) may be either positive or negative. All quadratic equations may be solved by the **quadratic formula:**

$$x = \frac{-b \pm \sqrt{b^2 - 4ac}}{2a}$$

If we wish to solve the quadratic equation $3x^2 - 4x - 8 = 0$, we observe that $a = 3$, $b = -4$, and $c = -8$. Substitution of these values into the quadratic formula gives

$$x = \frac{-(-4) \pm \sqrt{(-4)^2 - 4(3)(-8)}}{2(3)} = \frac{4 \pm \sqrt{16 + 96}}{6} = \frac{4 \pm \sqrt{112}}{6} = \frac{4 \pm 10.6}{6}$$

The two roots of this quadratic equation are

$$\underline{x = 2.4} \qquad \text{and} \qquad \underline{x = -1.1}$$

As you construct and solve quadratic equations based on the observed behavior of matter, you must decide which root has physical significance. Examination of the *equation that defines* x always gives clues about possible values for x. In this way you can tell which is extraneous (has no physical significance). Negative roots are *usually* extraneous.

When you have solved a quadratic equation you should always check the values you obtain by substitution into the original equation. In the above example we obtained $x = 2.4$ and $x = -1.1$. Substitution of these values into the original equation, $3x^2 - 4x - 8 = 0$, shows that both roots are correct. Note, however, that such substitutions often don't give a perfect check because some round-off error has been introduced.

Four-Place Table of Logarithms

	0	1	2	3	4	5	6	7	8	9
1.0	.0000	.0043	.0086	.0128	.0170	.0212	.0253	.0294	.0334	.0374
1.1	.0414	.0453	.0492	.0531	.0569	.0607	.0645	.0682	.0719	.0755
1.2	.0792	.0828	.0864	.0899	.0934	.0969	.1004	.1038	.1072	.1106
1.3	.1139	.1173	.1206	.1239	.1271	.1303	.1335	.1367	.1399	.1430
1.4	.1461	.1492	.1523	.1553	.1584	.1614	.1644	.1673	.1703	.1732
1.5	.1761	.1790	.1818	.1847	.1875	.1903	.1931	.1959	.1987	.2014
1.6	.2041	.2068	.2095	.2122	.2148	.2175	.2201	.2227	.2253	.2279
1.7	.2304	.2330	.2355	.2380	.2405	.2430	.2455	.2480	.2504	.2529
1.8	.2553	.2577	.2601	.2625	.2648	.2672	.2695	.2718	.2742	.2765
1.9	.2788	.2810	.2833	.2856	.2878	.2900	.2923	.2945	.2967	.2989
2.0	.3010	.3032	.3054	.3075	.3096	.3118	.3139	.3160	.3181	.3201
2.1	.3222	.3243	.3263	.3284	.3304	.3324	.3345	.3365	.3385	.3404
2.2	.3424	.3444	.3464	.3483	.3502	.3522	.3541	.3560	.3579	.3598
2.3	.3617	.3636	.3655	.3674	.3692	.3711	.3729	.3747	.3766	.3784
2.4	.3802	.3820	.3838	.3856	.3874	.3892	.3909	.3927	.3945	.3962
2.5	.3979	.3997	.4014	.4031	.4048	.4065	.4082	.4099	.4116	.4133
2.6	.4150	.4166	.4183	.4200	.4216	.4232	.4249	.4265	.4281	.4298
2.7	.4314	.4330	.4346	.4362	.4378	.4393	.4409	.4425	.4440	.4456
2.8	.4472	.4487	.4502	.4518	.4533	.4548	.4564	.4579	.4594	.4609
2.9	.4624	.4639	.4654	.4669	.4683	.4698	.4713	.4728	.4742	.4757
3.0	.4771	.4786	.4800	.4814	.4829	.4843	.4857	.4871	.4886	.4900
3.1	.4914	.4928	.4942	.4955	.4969	.4983	.4997	.5011	.5024	.5038
3.2	.5051	.5065	.5079	.5092	.5105	.5119	.5132	.5145	.5159	.5172
3.3	.5185	.5198	.5211	.5224	.5237	.5250	.5263	.5276	.5289	.5302
3.4	.5315	.5328	.5340	.5353	.5366	.5378	.5391	.5403	.5416	.5428
3.5	.5441	.5453	.5465	.5478	.5490	.5502	.5514	.5527	.5539	.5551
3.6	.5563	.5575	.5587	.5599	.5611	.5623	.5635	.5647	.5658	.5670
3.7	.5682	.5694	.5705	.5717	.5729	.5740	.5752	.5763	.5775	.5786
3.8	.5798	.5809	.5821	.5832	.5843	.5855	.5866	.5877	.5888	.5899
3.9	.5911	.5922	.5933	.5944	.5955	.5966	.5977	.5988	.5999	.6010
4.0	.6021	.6031	.6042	.6053	.6064	.6075	.6085	.6096	.6107	.6117
4.1	.6128	.6138	.6149	.6160	.6170	.6180	.6191	.6201	.6212	.6222
4.2	.6232	.6243	.6253	.6263	.6274	.6284	.6294	.6304	.6314	.6325
4.3	.6335	.6345	.6355	.6365	.6375	.6385	.6395	.6405	.6415	.6425
4.4	.6435	.6444	.6454	.6464	.6474	.6484	.6493	.6503	.6513	.6522
4.5	.6532	.6542	.6551	.6561	.6571	.6580	.6590	.6599	.6609	.6618
4.6	.6628	.6637	.6646	.6656	.6665	.6675	.6684	.6693	.6702	.6712
4.7	.6721	.6730	.6739	.6749	.6758	.6767	.6776	.6785	.6794	.6803
4.8	.6812	.6821	.6830	.6839	.6848	.6857	.6866	.6875	.6884	.6893
4.9	.6902	.6911	.6920	.6928	.6937	.6946	.6955	.6964	.6972	.6981
5.0	.6990	.6998	.7007	.7016	.7024	.7033	.7042	.7050	.7059	.7067
5.1	.7076	.7084	.7093	.7101	.7110	.7118	.7126	.7135	.7143	.7152
5.2	.7160	.7168	.7177	.7185	.7193	.7202	.7210	.7218	.7226	.7235
5.3	.7243	.7251	.7259	.7267	.7275	.7284	.7292	.7300	.7308	.7316
5.4	.7324	.7332	.7340	.7348	.7356	.7364	.7372	.7380	.7388	.7396
5.5	.7404	.7412	.7419	.7427	.7435	.7443	.7451	.7459	.7466	.7474
5.6	.7482	.7490	.7497	.7505	.7513	.7520	.7528	.7536	.7543	.7551
5.7	.7559	.7566	.7574	.7582	.7589	.7597	.7604	.7612	.7619	.7627
5.8	.7634	.7642	.7649	.7657	.7664	.7672	.7679	.7686	.7694	.7701
5.9	.7709	.7716	.7723	.7731	.7738	.7745	.7752	.7760	.7767	.7774

Four-Place Table of Logarithms

	0	1	2	3	4	5	6	7	8	9
6.0	.7782	.7789	.7796	.7803	.7810	.7818	.7825	.7832	.7839	.7846
6.1	.7853	.7860	.7868	.7875	.7882	.7889	.7896	.7903	.7910	.7917
6.2	.7924	.7931	.7938	.7945	.7952	.7959	.7966	.7973	.7980	.7987
6.3	.7993	.8000	.8007	.8014	.8021	.8028	.8035	.8041	.8048	.8055
6.4	.8062	.8069	.8075	.8082	.8089	.8096	.8102	.8109	.8116	.8122
6.5	.8129	.8136	.8142	.8149	.8156	.8162	.8169	.8176	.8182	.8189
6.6	.8195	.8202	.8209	.8215	.8222	.8228	.8235	.8241	.8248	.8254
6.7	.8261	.8267	.8274	.8280	.8287	.8293	.8299	.8306	.8312	.8319
6.8	.8325	.8331	.8338	.8344	.8351	.8357	.8363	.8370	.8376	.8382
6.9	.8388	.8395	.8401	.8407	.8414	.8420	.8426	.8432	.8439	.8445
7.0	.8451	.8457	.8463	.8470	.8476	.8482	.8488	.8494	.8500	.8506
7.1	.8513	.8519	.8525	.8531	.8537	.8543	.8549	.8555	.8561	.8567
7.2	.8573	.8579	.8585	.8591	.8597	.8603	.8609	.8615	.8621	.8627
7.3	.8633	.8639	.8645	.8651	.8657	.8663	.8669	.8675	.8681	.8686
7.4	.8692	.8698	.8704	.8710	.8716	.8722	.8727	.8733	.8739	.8745
7.5	.8751	.8756	.8762	.8768	.8774	.8779	.8785	.8791	.8797	.8802
7.6	.8808	.8814	.8820	.8825	.8831	.8837	.8842	.8848	.8854	.8859
7.7	.8865	.8871	.8876	.8882	.8887	.8893	.8899	.8904	.8910	.8915
7.8	.8921	.8927	.8932	.8938	.8943	.8949	.8954	.8960	.8965	.8971
7.9	.8976	.8982	.8987	.8993	.8998	.9004	.9009	.9015	.9020	.9026
8.0	.9031	.9036	.9042	.9047	.9053	.9058	.9063	.9069	.9074	.9079
8.1	.9085	.9090	.9096	.9101	.9106	.9112	.9117	.9122	.9128	.9133
8.2	.9138	.9143	.9149	.9154	.9159	.9165	.9170	.9175	.9180	.9186
8.3	.9191	.9196	.9201	.9206	.9212	.9217	.9222	.9227	.9232	.9238
8.4	.9243	.9248	.9253	.9258	.9263	.9269	.9274	.9279	.9284	.9289
8.5	.9294	.9299	.9304	.9309	.9315	.9320	.9325	.9330	.9335	.9340
8.6	.9345	.9350	.9355	.9360	.9365	.9370	.9375	.9380	.9385	.9390
8.7	.9395	.9400	.9405	.9410	.9415	.9420	.9425	.9430	.9435	.9440
8.8	.9445	.9450	.9455	.9460	.9465	.9469	.9474	.9479	.9484	.9489
8.9	.9494	.9499	.9504	.9509	.9513	.9518	.9523	.9528	.9533	.9538
9.0	.9542	.9547	.9552	.9557	.9562	.9566	.9571	.9576	.9581	.9586
9.1	.9590	.9595	.9600	.9605	.9609	.9614	.9619	.9624	.9628	.9633
9.2	.9638	.9643	.9647	.9652	.9657	.9661	.9666	.9671	.9675	.9680
9.3	.9685	.9689	.9694	.9699	.9703	.9708	.9713	.9717	.9722	.9727
9.4	.9731	.9736	.9741	.9745	.9750	.9754	.9759	.9763	.9768	.9773
9.5	.9777	.9782	.9786	.9791	.9795	.9800	.9805	.9809	.9814	.9818
9.6	.9823	.9827	.9832	.9836	.9841	.9845	.9850	.9854	.9859	.9863
9.7	.9868	.9872	.9877	.9881	.9886	.9890	.9894	.9899	.9903	.9908
9.8	.9912	.9917	.9921	.9926	.9930	.9934	.9939	.9943	.9948	.9952
9.9	.9956	.9961	.9965	.9969	.9974	.9978	.9983	.9987	.9991	.9996

APPENDIX C

Common Units, Equivalences, and Conversion Factors

Fundamental Units of the SI System

The metric system was begun by the French National Assembly in 1790 and has undergone many modifications. The International System of Units or *Système International* (SI), which represents an extension of the metric system, was adopted by the 11th General Conference of Weights and Measures in 1960. It is constructed from seven base units, each of which represents a particular physical quantity (Table I).

TABLE I SI Fundamental Units

Physical Quantity	Name of Unit	Symbol
Length	metre	m
Mass	kilogram	kg
Time	second	s
Temperature	kelvin	K
Amount of substance	mole	mol
Electric current	ampere	A
Luminous intensity	candela	cd

The first five units listed in Table I are particularly useful in general chemistry. They are defined as follows.

1. The *metre* was redefined in 1960 to be equal to 1,650,763.73 wavelengths of a certain line in the emission spectrum of krypton-86.
2. The *kilogram* represents the mass of a platinum-iridium block kept at the International Bureau of Weights and Measures at Sèvres, France.
3. The *second* was redefined in 1967 as the duration of 9,192,631,770 periods of a certain line in the microwave spectrum of cesium-133.
4. The *kelvin* is 1/273.16 of the temperature interval between absolute zero and the triple point of water.
5. The *mole* is the amount of substance that contains as many entities as there are atoms in exactly 0.012 kg of carbon-12 (12 g of ^{12}C atoms).

Prefixes Used with Traditional Metric Units and SI Units

Decimal fractions and multiples of metric and SI units are designated by using the prefixes listed in Table II. Those most commonly used in general chemistry are underlined.

TABLE II Traditional Metric and SI Prefixes

Factor	Prefix	Symbol	Factor	Prefix	Symbol
10^{12}	tera	T	10^{-1}	deci	d
10^{9}	giga	G	10^{-2}	centi	c
10^{6}	mega	M	10^{-3}	milli	m
10^{3}	kilo	k	10^{-6}	micro	μ
10^{2}	hecto	h	10^{-9}	nano	n
10^{1}	deka	da	10^{-12}	pico	p
			10^{-15}	femto	f
			10^{-18}	atto	a

Derived SI Units

In the International System of Units, all physical quantities are represented by appropriate combinations of the base units listed in Table I. A list of the derived units frequently used in general chemistry is given in Table III.

TABLE III Derived SI Units

Physical Quantity	Name of Unit	Symbol	Definition
Area	square metre	m^2	
Volume	cubic metre	m^3	
Density	kilogram per cubic metre	kg/m^3	
Force	newton	N	$kg\ m/s^2$
Pressure	pascal	Pa	N/m^2
Energy	joule	J	$kg\ m^2/s^2$
Electric charge	coulomb	C	A s
Electric potential difference	volt	V	J/(A s)

Common Units of Mass and Weight

1 pound = 453.59 grams

1 pound = 453.59 grams = 0.45359 kilogram
1 kilogram = 1000 grams = 2.205 pounds
1 gram = 10 decigrams = 100 centigrams = 1000 milligrams
1 gram = 6.022×10^{23} atomic mass units
1 atomic mass unit = 1.6606×10^{-24} gram
1 short ton = 2000 pounds = 907.2 kilograms
1 long ton = 2240 pounds
1 metric tonne = 1000 kilograms = 2205 pounds

Common Units of Length

1 inch = 2.54 centimeters (exactly)

1 mile = 5280 feet = 1.609 kilometers
1 yard = 36 inches = 0.9144 meter
1 meter = 100 centimeters = 39.37 inches = 3.281 feet = 1.094 yards
1 kilometer = 1000 meters = 1094 yards = 0.6215 mile
1 Ångstrom = 1.0×10^{-8} centimeter = 0.10 nanometer = 1.0×10^{-10} meter = 3.937×10^{-9} inch

Common Units of Volume

1 quart = 0.9463 liter
1 liter = 1.056 quarts

1 liter = 1 cubic decimeter = 1000 cubic centimeters
= 0.001 cubic meter
1 milliliter = 1 cubic centimeter = 0.001 liter
= 1.056×10^{-3} quart
1 cubic foot = 28.316 liters = 29.902 quarts = 7.475 gallons

Common Units of Force* and Pressure

1 atmosphere = 760 millimeters of mercury
= 1.013×10^5 pascals
= 14.70 pounds per square inch
1 bar = 10^5 pascals
1 torr = 1 millimeter of mercury
1 pascal = 1 kg/m s^2 = 1 N/m^2

* Force: 1 newton (N) = 1 kg m/s^2, i.e., the force that when applied for 1 second gives a 1 kilogram mass a velocity of 1 meter per second.

Common Units of Energy

1 joule = 1×10^7 ergs

1 thermochemical calorie* = 4.184 joules = 4.184×10^7 ergs = 4.129×10^{-2} liter-atmospheres
= 2.612×10^{19} electron volts
1 erg = 1×10^{-7} joule = 2.3901×10^{-8} calorie
1 electron volt = 1.6022×10^{-19} joule = 1.6022×10^{-12} erg = 96.487 kJ/mol†
1 liter-atmosphere = 24.217 calories = 101.32 joules = 1.0132×10^9 ergs
1 British thermal unit = 1055.06 joules = 1.05506×10^{10} ergs = 252.2 calories

* The amount of heat required to raise the temperature of one gram of water from 14.5°C to 15.5°C.
† Note that the other units are per particle and must be multiplied by 6.022×10^{23} to be strictly comparable.

APPENDIX D

Physical Constants

Quantity	Symbol	Traditional Units	SI Units
Acceleration of gravity	g	980.6 cm/s	9.806 m/s
Atomic mass unit (1/12 the mass of ^{12}C atom)	amu or u	1.6606×10^{-24} g	1.6606×10^{-27} kg
Avogadro's number	N	6.022×10^{23} particles/mol	6.022×10^{23} particles/mol
Bohr radius	a_0	0.52918 Å 5.2918×10^{-9} cm	5.2918×10^{-11} m
Boltzmann constant	k	1.3807×10^{-16} erg/K	1.3807×10^{-23} J/K
Charge-to-mass ratio of electron	e/m	1.7588×10^{8} coulomb/g	1.7588×10^{11} C/kg
Electronic charge	e	1.6022×10^{-19} coulomb 4.8033×10^{-10} esu	1.6022×10^{-19} C
Electron rest mass	m_e	9.1095×10^{-28} g 0.00054859 amu	9.1095×10^{-31} kg
Faraday constant	F	96,487 coulombs/eq 23.06 kcal/volt eq	96,487 C/mol e^- 96,487 J/V mol e^-
Gas constant	R	$0.08206 \frac{\text{L atm}}{\text{mol K}}$ $1.987 \frac{\text{cal}}{\text{mol K}}$	$8.3145 \frac{\text{Pa dm}^3}{\text{mol K}}$ 8.3145 J/mol K
Molar volume (STP)	V_m	22.414 L/mol	22.414×10^{-3} m^3/mol 22.414 dm^3/mol
Neutron rest mass	m_n	1.67495×10^{-24} g 1.008665 amu	1.67495×10^{-27} kg
Planck's constant	h	6.6262×10^{-27} erg s	6.6262×10^{-34} J s
Proton rest mass	m_p	1.6726×10^{-24} g 1.007277 amu	1.6726×10^{-27} kg
Rydberg constant	R_∞	3.289×10^{15} cycles/s 2.1799×10^{-11} erg	1.0974×10^{7} m^{-1} 2.1799×10^{-18} J
Velocity of light (in a vacuum)	c	2.9979×10^{10} cm/s (186,281 miles/second)	2.9979×10^{8} m/s

$\pi = 3.1416$
$e = 2.7183$
$\ln X = 2.303 \log X$

$2.303\,R = 4.576$ cal/mol · K = 19.15 J/mol · K
$2.303\,RT$ (at 25°C) = 1364 cal/mol = 5709 J/mol

APPENDIX E

Some Physical Constants for Water and a Few Common Substances

Vapor Pressure of Water at Various Temperatures

Temperature °C	Vapor Pressure torr	Temperature °C	Vapor Pressure torr	Temperature °C	Vapor Pressure torr	Temperature °C	Vapor Pressure torr
−10	2.1	21	18.7	51	97.2	81	369.7
−9	2.3	22	19.8	52	102.1	82	384.9
−8	2.5	23	21.1	53	107.2	83	400.6
−7	2.7	24	22.4	54	112.5	84	416.8
−6	2.9	25	23.8	55	118.0	85	433.6
−5	3.2	26	25.2	56	123.8	86	450.9
−4	3.4	27	26.7	57	129.8	87	468.7
−3	3.7	28	28.3	58	136.1	88	487.1
−2	4.0	29	30.0	59	142.6	89	506.1
−1	4.3	30	31.8	60	149.4	90	525.8
0	4.6	31	33.7	61	156.4	91	546.1
1	4.9	32	35.7	62	163.8	92	567.0
2	5.3	33	37.7	63	171.4	93	588.6
3	5.7	34	39.9	64	179.3	94	610.9
4	6.1	35	42.2	65	187.5	95	633.9
5	6.5	36	44.6	66	196.1	96	657.6
6	7.0	37	47.1	67	205.0	97	682.1
7	7.5	38	49.7	68	214.2	98	707.3
8	8.0	39	52.4	69	223.7	99	733.2
9	8.6	40	55.3	70	233.7	100	760.0
10	9.2	41	58.3	71	243.9	101	787.6
11	9.8	42	61.5	72	254.6	102	815.9
12	10.5	43	64.8	73	265.7	103	845.1
13	11.2	44	68.3	74	277.2	104	875.1
14	12.0	45	71.9	75	289.1	105	906.1
15	12.8	46	75.7	76	301.4	106	937.9
16	13.6	47	79.6	77	314.1	107	970.6
17	14.5	48	83.7	78	327.3	108	1004.4
18	15.5	49	88.0	79	341.0	109	1038.9
19	16.5	50	92.5	80	355.1	110	1074.6
20	17.5						

Specific Heats and Heat Capacities for Some Common Substances

Substance	Specific Heat		Heat Capacity	
	cal/g·°C	J/g·°C	cal/mol·°C	J/mol·°C
Al (s)	0.215	0.900	5.81	24.3
Ca (s)	0.156	0.653	6.25	26.2
Cu (s)	0.092	0.385	5.85	24.5
Fe (s)	0.106	0.444	5.92	24.8
Hg (ℓ)	0.0331	0.138	6.62	27.7
H_2O (s), ice	0.500	2.09	9.00	37.7
H_2O (ℓ), water	1.00	4.18	18.0	75.3
H_2O (g), steam	0.484	2.03	8.71	36.4
C_6H_6 (ℓ), benzene	0.415	1.74	32.4	136
C_6H_6 (g), benzene	0.249	1.04	19.5	81.6
C_2H_5OH (ℓ), ethanol	0.587	2.46	27.0	113
C_2H_5OH (g), ethanol	0.228	0.954	10.5	420
$(C_2H_5)_2O$ (ℓ), diethyl ether	0.893	3.74	41.1	172
$(C_2H_5)_2O$ (g), diethyl ether	0.561	2.35	25.8	108

Heats of Transformation and Transformation Temperatures of Several Substances

Substance	MP	Heat of Fusion		ΔH_{fus}		BP	Heat of Vaporization		ΔH_{vap}	
	°C	cal/g	J/g	kcal/mol	kJ/mol	°C	cal/g	J/g	kcal/mol	kJ/mol
Al	658	94.5	395	2.54	10.6	2467	2515	10520	67.9	284
Ca	851	55.7	233	2.23	9.33	1487	963	4030	38.6	162
Cu	1083	49.0	205	3.11	13.0	2595	1146	4790	72.8	305
H_2O	0.0	79.8	333	1.44	6.02	100	540	2260	9.73	40.7
Fe	1530	63.7	267	3.56	14.9	2735	1515	6340	84.6	354
Hg	−39	2.7	11	5.57	23.3	357	69.8	292	14.0	58.6
CH_4	−182	14.0	58.6	0.22	0.92	−164	—	—	—	—
C_2H_5OH	−117	26.1	109	1.20	5.02	78.0	204	855	9.39	39.3
C_6H_6	5.48	30.4	127	2.37	9.92	80.1	94.3	395	7.36	30.8
$(C_2H_5)_2O$	−116	23.4	97.9	1.83	7.66	35	83.9	351	6.21	26.0

APPENDIX F

Ionization Constants for Weak Acids at 25°C

Acid	Formula and Ionization Equation	K_a
Acetic	$CH_3COOH \rightleftharpoons H^+ + CH_3COO^-$	1.8×10^{-5}
Arsenic	$H_3AsO_4 \rightleftharpoons H^+ + H_2AsO_4^-$	$K_1 = 2.5 \times 10^{-4}$
	$H_2AsO_4^- \rightleftharpoons H^+ + HAsO_4^{2-}$	$K_2 = 5.6 \times 10^{-8}$
	$HAsO_4^{2-} \rightleftharpoons H^+ + AsO_4^{3-}$	$K_3 = 3.0 \times 10^{-13}$
Arsenous	$H_3AsO_3 \rightleftharpoons H^+ + H_2AsO_3^-$	$K_1 = 6.0 \times 10^{-10}$
	$H_2AsO_3^- \rightleftharpoons H^+ + HAsO_3^{2-}$	$K_2 = 3.0 \times 10^{-14}$
Benzoic	$C_6H_5COOH \rightleftharpoons H^+ + C_6H_5COO^-$	6.3×10^{-5}
Boric	$H_3BO_3 \rightleftharpoons H^+ + H_2BO_3^-$	$K_1 = 7.3 \times 10^{-10}$
	$H_2BO_3^- \rightleftharpoons H^+ + HBO_3^{2-}$	$K_2 = 1.8 \times 10^{-13}$
	$HBO_3^{2-} \rightleftharpoons H^+ + BO_3^{3-}$	$K_3 = 1.6 \times 10^{-14}$
Carbonic	$H_2CO_3 \rightleftharpoons H^+ + HCO_3^-$	$K_1 = 4.2 \times 10^{-7}$
	$HCO_3^- \rightleftharpoons H^+ + CO_3^{2-}$	$K_2 = 4.8 \times 10^{-11}$
Citric	$H_3C_6H_5O_7 \rightleftharpoons H^+ + H_2C_6H_5O_7^-$	$K_1 = 7.4 \times 10^{-3}$
	$H_2C_6H_5O_7^- \rightleftharpoons H^+ + HC_6H_5O_7^{2-}$	$K_2 = 1.7 \times 10^{-5}$
	$HC_6H_5O_7^{2-} \rightleftharpoons H^+ + C_6H_5O_7^{3-}$	$K_3 = 4.0 \times 10^{-7}$
Cyanic	$HOCN \rightleftharpoons H^+ + OCN^-$	3.5×10^{-4}
Formic	$HCOOH \rightleftharpoons H^+ + HCOO^-$	1.8×10^{-4}
Hydrazoic	$HN_3 \rightleftharpoons H^+ + N_3^-$	1.9×10^{-5}
Hydrocyanic	$HCN \rightleftharpoons H^+ + CN^-$	4.0×10^{-10}
Hydrofluoric	$HF \rightleftharpoons H^+ + F^-$	7.2×10^{-4}
Hydrogen peroxide	$H_2O_2 \rightleftharpoons H^+ + HO_2^-$	2.4×10^{-12}
Hydrosulfuric	$H_2S \rightleftharpoons H^+ + HS^-$	$K_1 = 1.0 \times 10^{-7}$
	$HS^- \rightleftharpoons H^+ + S^{2-}$	$K_2 = 1.3 \times 10^{-13}$
Hypobromous	$HOBr \rightleftharpoons H^+ + OBr^-$	2.5×10^{-9}
Hypochlorous	$HOCl \rightleftharpoons H^+ + OCl^-$	3.5×10^{-8}
Nitrous	$HNO_2 \rightleftharpoons H^+ + NO_2^-$	4.5×10^{-4}
Oxalic	$H_2C_2O_4 \rightleftharpoons H^+ + HC_2O_4^-$	$K_1 = 5.9 \times 10^{-2}$
	$HC_2O_4^- \rightleftharpoons H^+ + C_2O_4^{2-}$	$K_2 = 6.4 \times 10^{-5}$
Phenol	$HC_6H_5O \rightleftharpoons H^+ + C_6H_5O^-$	1.3×10^{-10}
Phosphoric	$H_3PO_4 \rightleftharpoons H^+ + H_2PO_4^-$	$K_1 = 7.5 \times 10^{-3}$
	$H_2PO_4^- \rightleftharpoons H^+ + HPO_4^{2-}$	$K_2 = 6.2 \times 10^{-8}$
	$HPO_4^{2-} \rightleftharpoons H^+ + PO_4^{3-}$	$K_3 = 3.6 \times 10^{-13}$
Phosphorous	$H_3PO_3 \rightleftharpoons H^+ + H_2PO_3^-$	$K_1 = 1.6 \times 10^{-2}$
	$H_2PO_3^- \rightleftharpoons H^+ + HPO_3^{2-}$	$K_2 = 7.0 \times 10^{-7}$
Selenic	$H_2SeO_4 \rightleftharpoons H^+ + HSeO_4^-$	K_1 = very large
	$HSeO_4^- \rightleftharpoons H^+ + SeO_4^{2-}$	$K_2 = 1.2 \times 10^{-2}$
Selenous	$H_2SeO_3 \rightleftharpoons H^+ + HSeO_3^-$	$K_1 = 2.7 \times 10^{-3}$
	$HSeO_3^- \rightleftharpoons H^+ + SeO_3^{2-}$	$K_2 = 2.5 \times 10^{-7}$
Sulfuric	$H_2SO_4 \rightleftharpoons H^+ + HSO_4^-$	K_1 = very large
	$HSO_4^- \rightleftharpoons H^+ + SO_4^{2-}$	$K_2 = 1.2 \times 10^{-2}$
Sulfurous	$H_2SO_3 \rightleftharpoons H^+ + HSO_3^-$	$K_1 = 1.2 \times 10^{-2}$
	$HSO_3^- \rightleftharpoons H^+ + SO_3^{2-}$	$K_2 = 6.2 \times 10^{-8}$
Tellurous	$H_2TeO_3 \rightleftharpoons H^+ + HTeO_3^-$	$K_1 = 2 \times 10^{-3}$
	$HTeO_3^- \rightleftharpoons H^+ + TeO_3^{2-}$	$K_2 = 1 \times 10^{-8}$

APPENDIX G

Ionization Constants for Weak Bases at 25°C

Base	Formula and Ionization Equation	K_b
Ammonia	$NH_3 + H_2O \rightleftharpoons NH_4^+ + OH^-$	1.8×10^{-5}
Aniline	$C_6H_5NH_2 + H_2O \rightleftharpoons C_6H_5NH_3^+ + OH^-$	4.2×10^{-10}
Dimethylamine	$(CH_3)_2NH + H_2O \rightleftharpoons (CH_3)_2NH_2^+ + OH^-$	7.4×10^{-4}
Ethylenediamine	$(CH_2)_2(NH_2)_2 + H_2O \rightleftharpoons (CH_2)_2(NH_2)_2H^+ + OH^-$	$K_1 = 8.5 \times 10^{-5}$
	$(CH_2)_2(NH_2)_2H^+ + H_2O \rightleftharpoons (CH_2)_2(NH_2)_2H_2^{2+} + OH^-$	$K_2 = 2.7 \times 10^{-8}$
Hydrazine	$N_2H_4 + H_2O \rightleftharpoons N_2H_5^+ + OH^-$	$K_1 = 8.5 \times 10^{-7}$
	$N_2H_5^+ + H_2O \rightleftharpoons N_2H_6^{2+} + OH^-$	$K_2 = 8.9 \times 10^{-16}$
Hydroxylamine	$NH_2OH + H_2O \rightleftharpoons NH_3OH^+ + OH^-$	6.6×10^{-9}
Methylamine	$CH_3NH_2 + H_2O \rightleftharpoons CH_3NH_3^+ + OH^-$	5.0×10^{-4}
Pyridine	$C_5H_5N + H_2O \rightleftharpoons C_5H_5NH^+ + OH^-$	1.5×10^{-9}
Trimethylamine	$(CH_3)_3N + H_2O \rightleftharpoons (CH_3)_3NH^+ + OH^-$	7.4×10^{-5}

APPENDIX H

Solubility Product Constants for Some Inorganic Compounds at 25°C

Substance	K_{sp}
Aluminum compounds	
$AlAsO_4$	1.6×10^{-16}
$Al(OH)_3$	1.9×10^{-33}
$AlPO_4$	1.3×10^{-20}
Antimony compounds	
Sb_2S_3	1.6×10^{-93}
Barium compounds	
$Ba_3(AsO_4)_2$	1.1×10^{-13}
$BaCO_3$	8.1×10^{-9}
$BaC_2O_4 \cdot 2H_2O^*$	1.1×10^{-7}
$BaCrO_4$	2.0×10^{-10}
BaF_2	1.7×10^{-6}
$Ba(OH)_2 \cdot 8H_2O^*$	5.0×10^{-3}
$Ba_3(PO_4)_2$	1.3×10^{-29}
$BaSeO_4$	2.8×10^{-11}
$BaSO_3$	8.0×10^{-7}
$BaSO_4$	1.1×10^{-10}
Bismuth compounds	
$BiOCl$	7.0×10^{-9}
$BiO(OH)$	1.0×10^{-12}
$Bi(OH)_3$	3.2×10^{-40}
BiI_3	8.1×10^{-19}
$BiPO_4$	1.3×10^{-23}
Bi_2S_3	1.6×10^{-72}
Cadmium compounds	
$Cd_3(AsO_4)_2$	2.2×10^{-32}
$CdCO_3$	2.5×10^{-14}
$Cd(CN)_2$	1.0×10^{-8}
$Cd_2[Fe(CN)_6]$	3.2×10^{-17}
$Cd(OH)_2$	1.2×10^{-14}
Cds	3.6×10^{-29}
Calcium compounds	
$Ca_3(AsO_4)_2$	6.8×10^{-19}
$CaCO_3$	4.8×10^{-9}
$CaCrO_4$	7.1×10^{-4}
$CaC_2O_4 \cdot H_2O^*$	2.3×10^{-9}
CaF_2	3.9×10^{-11}
$Ca(OH)_2$	7.9×10^{-6}
$CaHPO_4$	2.7×10^{-7}
$Ca(H_2PO_4)_2$	1.0×10^{-3}
$Ca_3(PO_4)_2$	1.0×10^{-25}
$CaSO_3 \cdot 2H_2O^*$	1.3×10^{-8}
$CaSO_4 \cdot 2H_2O^*$	2.4×10^{-5}
Chromium compounds	
$CrAsO_4$	7.8×10^{-21}
$Cr(OH)_3$	6.7×10^{-31}
$CrPO_4$	2.4×10^{-23}
Cobalt compounds	
$Co_3(AsO_4)_2$	7.6×10^{-29}
$CoCO_3$	8.0×10^{-13}
$Co(OH)_2$	2.5×10^{-16}
CoS (α)	5.9×10^{-21}
CoS (β)	8.7×10^{-23}
$Co(OH)_3$	4.0×10^{-45}
Co_2S_3	2.6×10^{-124}
Copper compounds	
$CuBr$	5.3×10^{-9}
$CuCl$	1.9×10^{-7}
$CuCN$	3.2×10^{-20}
Cu_2O ($Cu^+ + OH^-$)†	1.0×10^{-14}
CuI	5.1×10^{-12}
Cu_2S	1.6×10^{-48}
$CuSCN$	1.6×10^{-11}
$Cu_3(AsO_4)_2$	7.6×10^{-36}
$CuCO_3$	2.5×10^{-10}
$Cu_2[Fe(CN)_6]$	1.3×10^{-16}
$Cu(OH)_2$	1.6×10^{-19}
CuS	8.7×10^{-36}
Gold compounds	
$AuBr$	5.0×10^{-17}
$AuCl$	2.0×10^{-13}
AuI	1.6×10^{-23}
$AuBr_3$	4.0×10^{-36}
$AuCl_3$	3.2×10^{-25}
$Au(OH)_3$	1×10^{-53}
AuI_3	1.0×10^{-46}

APPENDIX H (*Continued*)

Solubility Product Constants for Some Inorganic Compounds at 25°C

Substance	K_{sp}	Substance	K_{sp}
Iron compounds		**Mercury compounds**	
$FeCO_3$	3.5×10^{-11}	Hg_2Br_2	1.3×10^{-22}
$Fe(OH)_2$	7.9×10^{-15}	Hg_2CO_3	8.9×10^{-17}
FeS	4.9×10^{-18}	Hg_2Cl_2	1.1×10^{-18}
$Fe_4[Fe(CN)_6]_3$	3.0×10^{-41}	Hg_2CrO_4	5.0×10^{-9}
$Fe(OH)_3$	6.3×10^{-38}	Hg_2I_2	4.5×10^{-29}
Fe_2S_3	1.4×10^{-88}	$Hg_2O \cdot H_2O$ (Hg_2^{2+} + $2OH^-$)†	1.6×10^{-23}
		Hg_2SO_4	6.8×10^{-7}
		Hg_2S	5.8×10^{-44}
		$Hg(CN)_2$	3.0×10^{-23}
Lead compounds		$Hg(OH)_2$	2.5×10^{-26}
$Pb_3(AsO_4)_2$	4.1×10^{-36}	HgI_2	4.0×10^{-29}
$PbBr_2$	6.3×10^{-6}	HgS	3.0×10^{-53}
$PbCO_3$	1.5×10^{-13}		
$PbCl_2$	1.7×10^{-5}		
$PbCrO_4$	1.8×10^{-14}		
PbF_2	3.7×10^{-8}		
$Pb(OH)_2$	2.8×10^{-16}	**Nickel compounds**	
PbI_2	8.7×10^{-9}	$Ni_3(AsO_4)_2$	1.9×10^{-26}
$Pb_3(PO_4)_2$	3.0×10^{-44}	$NiCO_3$	6.6×10^{-9}
$PbSeO_4$	1.5×10^{-7}	$Ni(CN)_2$	3.0×10^{-23}
$PbSO_4$	1.8×10^{-8}	$Ni(OH)_2$	2.8×10^{-16}
PbS	8.4×10^{-28}	NiS (α)	3.0×10^{-21}
		NiS (β)	1.0×10^{-26}
		NiS (γ)	2.0×10^{-28}
Magnesium compounds			
$Mg_3(AsO_4)_2$	2.1×10^{-20}		
$MgCO_3 \cdot 3H_2O$*	4.0×10^{-5}	**Silver compounds**	
MgC_2O_4	8.6×10^{-5}	Ag_3AsO_4	1.1×10^{-20}
MgF_2	6.4×10^{-9}	$AgBr$	3.3×10^{-13}
$Mg(OH)_2$	1.5×10^{-11}	Ag_2CO_3	8.1×10^{-12}
$MgNH_4PO_4$	2.5×10^{-12}	$AgCl$	1.8×10^{-10}
		Ag_2CrO_4	9.0×10^{-12}
		$AgCN$	1.2×10^{-16}
		$Ag_4[Fe(CN)_6]$	1.6×10^{-41}
		Ag_2O (Ag^+ + OH^-)†	2.0×10^{-8}
Manganese compounds		AgI	1.5×10^{-16}
$Mn_3(AsO_4)_2$	1.9×10^{-11}	Ag_3PO_4	1.3×10^{-20}
$MnCO_3$	1.8×10^{-11}	Ag_2SO_3	1.5×10^{-14}
$Mn(OH)_2$	4.6×10^{-14}	Ag_2SO_4	1.7×10^{-5}
MnS	5.1×10^{-15}	Ag_2S	1.0×10^{-49}
$Mn(OH)_3$	$\sim 1 \times 10^{-36}$	$AgSCN$	1.0×10^{-12}

APPENDIX H *(Continued)*

Solubility Product Constants for Some Inorganic Compounds at 25°C

Substance	K_{sp}	Substance	K_{sp}
Strontium compounds		**Zinc compounds**	
$Sr_3(AsO_4)_2$	1.3×10^{-18}	$Zn_3(AsO_4)_2$	1.1×10^{-27}
$SrCO_3$	9.4×10^{-10}	$ZnCO_3$	1.5×10^{-11}
$SrC_2O_4 \cdot 2H_2O$*	5.6×10^{-8}	$Zn(CN)_2$	8.0×10^{-12}
$SrCrO_4$	3.6×10^{-5}	$Zn_2[Fe(CN)_6]$	4.1×10^{-16}
$Sr(OH)_2 \cdot 8H_2O$*	3.2×10^{-4}	$Zn(OH)_2$	4.5×10^{-17}
$Sr_3(PO_4)_2$	1.0×10^{-31}	$Zn_3(PO_4)_2$	9.1×10^{-33}
$SrSO_3$	4.0×10^{-8}	ZnS	1.1×10^{-21}
$SrSO_4$	2.8×10^{-7}		
Tin compounds			
$Sn(OH)_2$	2.0×10^{-26}		
SnI_2	1.0×10^{-4}		
SnS	1.0×10^{-28}		
$Sn(OH)_4$	1×10^{-57}		
SnS_2	1×10^{-70}		

* Since $[H_2O]$ does not appear in equilibrium constants for equilibria in aqueous solution in general, it does *not* appear in the K_{sp} expressions for hydrated solids.

† Very small amounts of oxides dissolve in water to give the ions indicated in parentheses. Solid hydroxides are unstable and decompose to oxides as rapidly as they are formed.

APPENDIX I

Dissociation Constants for Complex Ions

Dissociation Equilibrium	K_d
$[AgBr_2]^- \rightleftharpoons Ag^+ + 2Br^-$	7.8×10^{-8}
$[AgCl_2]^- \rightleftharpoons Ag^+ + 2Cl^-$	4.0×10^{-6}
$[Ag(CN)_2]^- \rightleftharpoons Ag^+ + 2CN^-$	1.8×10^{-19}
$[Ag(S_2O_3)_2]^{3-} \rightleftharpoons Ag^+ + 2S_2O_3{}^{2-}$	5.0×10^{-14}
$[Ag(NH_3)_2]^+ \rightleftharpoons Ag^+ + 2NH_3$	6.3×10^{-8}
$[Ag(en)]^+ \rightleftharpoons Ag^+ + en^*$	1.0×10^{-5}
$[AlF_6]^{3-} \rightleftharpoons Al^{3+} + 6F^-$	2.0×10^{-24}
$[Al(OH)_4]^- \rightleftharpoons Al^{3+} + 4OH^-$	1.3×10^{-34}
$[Au(CN)_2]^- \rightleftharpoons Au^+ + 2CN^-$	5.0×10^{-39}
$[Cd(CN)_4]^{2-} \rightleftharpoons Cd^{2+} + 4CN^-$	7.8×10^{-18}
$[CdCl_4]^{2-} \rightleftharpoons Cd^{2+} + 4Cl^-$	1.0×10^{-4}
$[Cd(NH_3)_4]^{2+} \rightleftharpoons Cd^{2+} + 4NH_3$	1.0×10^{-7}
$[Co(NH_3)_6]^{2+} \rightleftharpoons Co^{2+} + 6NH_3$	1.3×10^{-5}
$[Co(NH_3)_6]^{3+} \rightleftharpoons Co^{3+} + 6NH_3$	2.2×10^{-34}
$[Co(en)_3]^{2+} \rightleftharpoons Co^{2+} + 3en^*$	1.5×10^{-14}
$[Co(en)_3]^{3+} \rightleftharpoons Co^{3+} + 3en^*$	2.0×10^{-49}
$[Cu(CN)_2]^- \rightleftharpoons Cu^+ + 2CN^-$	1.0×10^{-16}
$[CuCl_2]^- \rightleftharpoons Cu^+ + 2Cl^-$	1.0×10^{-5}
$[Cu(NH_3)_2]^+ \rightleftharpoons Cu^+ + 2NH_3$	1.4×10^{-11}
$[Cu(NH_3)_4]^{2+} \rightleftharpoons Cu^{2+} + 4NH_3$	8.5×10^{-13}
$[Fe(CN)_6]^{4-} \rightleftharpoons Fe^{2+} + 6CN^-$	1.3×10^{-37}
$[Fe(CN)_6]^{3-} \rightleftharpoons Fe^{3+} + 6CN^-$	1.3×10^{-44}
$[HgCl_4]^{2-} \rightleftharpoons Hg^{2+} + 4Cl^-$	8.3×10^{-16}
$[Ni(CN)_4]^{2-} \rightleftharpoons Ni^{2+} + 4CN^-$	1.0×10^{-31}
$[Ni(NH_3)_6]^{2+} \rightleftharpoons Ni^{2+} + 6NH_3$	1.8×10^{-9}
$[Zn(OH)_4]^{2-} \rightleftharpoons Zn^{2+} + 4OH^-$	3.5×10^{-16}
$[Zn(NH_3)_4]^{2+} \rightleftharpoons Zn^{2+} + 4NH_3$	3.4×10^{-10}

* en represents ethylenediamine, $H_2NCH_2CH_2NH_2$

APPENDIX J

Standard Reduction Potentials in Aqueous Solution at 25°C

Acidic Solution	Standard Reduction Potential, E^0 (volts)
Li^+ (aq) + e^- ⟶ Li (s)	−3.045
K^+ (aq) + e^- ⟶ K (s)	−2.925
Rb^+ (aq) + e^- ⟶ Rb (s)	−2.925
Ba^{2+} (aq) + $2e^-$ ⟶ Ba (s)	−2.90
Sr^{2+} (aq) + $2e^-$ ⟶ Sr (s)	−2.89
Ca^{2+} (aq) + $2e^-$ ⟶ Ca (s)	−2.87
Na^+ (aq) + e^- ⟶ Na (s)	−2.714
Mg^{2+} (aq) + $2e^-$ ⟶ Mg (s)	−2.37
H_2 (g) + $2e^-$ ⟶ $2H^-$ (aq)	−2.25
Al^{3+} (aq) + $3e^-$ ⟶ Al (s)	−1.66
Zr^{4+} (aq) + $4e^-$ ⟶ Zr (s)	−1.53
ZnS (s) + $2e^-$ ⟶ Zn (s) + S^{2-} (aq)	−1.44
CdS (s) + $2e^-$ ⟶ Cd (s) + S^{2-} (aq)	−1.21
V^{2+} (aq) + $2e^-$ ⟶ V (s)	−1.18
Mn^{2+} (aq) + $2e^-$ ⟶ Mn (s)	−1.18
FeS (s) + $2e^-$ ⟶ Fe (s) + S^{2-} (aq)	−1.01
Cr^{2+} (aq) + $2e^-$ ⟶ Cr (s)	−0.91
Zn^{2+} (aq) + $2e^-$ ⟶ Zn (s)	−0.763
Cr^{3+} (aq) + $3e^-$ ⟶ Cr (s)	−0.74
HgS (s) + $2H^+$ (aq) + $2e^-$ ⟶ Hg (ℓ) + H_2S (g)	−0.72
Ga^{3+} (aq) + $3e^-$ ⟶ Ga (s)	−0.53
$2CO_2$ (g) + $2H^+$ (aq) + $2e^-$ ⟶ $(COOH)_2$ (aq)	−0.49
Fe^{2+} (aq) + $2e^-$ ⟶ Fe (s)	−0.44
Cr^{3+} (aq) + e^- ⟶ Cr^{2+} (aq)	−0.41
Cd^{2+} (aq) + $2e^-$ ⟶ Cd (s)	−0.403
Se (s) + $2H^+$ (aq) + $2e^-$ ⟶ H_2Se (aq)	−0.40
$PbSO_4$ (s) + $2e^-$ ⟶ Pb (s) + SO_4^{2-} (aq)	−0.356
Tl^+ (aq) + e^- ⟶ Tl (s)	−0.34
Co^{2+} (aq) + $2e^-$ ⟶ Co (s)	−0.28
Ni^{2+} (aq) + $2e^-$ ⟶ Ni (s)	−0.25
$[SnF_6]^{2-}$ (aq) + $4e^-$ ⟶ Sn (s) + $6F^-$ (aq)	−0.25
AgI (s) + e^- ⟶ Ag (s) + I^- (aq)	−0.15
Sn^{2+} (aq) + $2e^-$ ⟶ Sn (s)	−0.14
Pb^{2+} (aq) + $2e^-$ ⟶ Pb (s)	−0.126
N_2O (g) + $6H^+$ (aq) + H_2O + $4e^-$ ⟶ $2NH_3OH^+$ (aq)	−0.05
$2H^+$ (aq) + $2e^-$ ⟶ H_2 (g) **(reference electrode)**	0.0000
AgBr (s) + e^- ⟶ Ag (s) + Br^- (aq)	0.10
S (s) + $2H^+$ (aq) + $2e^-$ ⟶ H_2S (aq)	0.14
Sn^{4+} (aq) + $2e^-$ ⟶ Sn^{2+} (aq)	0.15
Cu^{2+} (aq) + e^- ⟶ Cu^+ (aq)	0.153
SO_4^{2-} (aq) + $4H^+$ (aq) + $2e^-$ ⟶ H_2SO_3 (aq) + H_2O	0.17
SO_4^{2-} (aq) + $4H^+$ (aq) + $2e^-$ ⟶ SO_2 (g) + $2H_2O$	0.20
AgCl (s) + e^- ⟶ Ag (s) + Cl^- (aq)	0.222
Hg_2Cl_2 (s) + $2e^-$ ⟶ 2Hg (ℓ) + $2Cl^-$ (aq)	0.27
Cu^{2+} (aq) + $2e^-$ ⟶ Cu (s)	0.337
$[RhCl_6]^{3-}$ (aq) + $3e^-$ ⟶ Rh (s) + $6Cl^-$ (aq)	0.44

APPENDIX J *(Continued)*

Standard Reduction Potentials in Aqueous Solution at 25°C

Acidic Solution	Standard Reduction Potential, E^0 (volts)
Cu^+ (aq) + e^- $\longrightarrow$ Cu (s)	0.521
TeO_2 (s) + $4H^+$ (aq) + $4e^-$ $\longrightarrow$ Te (s) + $2H_2O$	0.529
I_2 (s) + $2e^-$ $\longrightarrow$ $2I^-$ (aq)	0.535
H_3AsO_4 (aq) + $2H^+$ (aq) + $2e^-$ $\longrightarrow$ H_3AsO_3 (aq) + H_2O	0.58
$[PtCl_6]^{2-}$ (aq) + $2e^-$ $\longrightarrow$ $[PtCl_4]^{2-}$ (aq) + $2Cl^-$ (aq)	0.68
O_2 (g) + $2H^+$ (aq) + $2e^-$ $\longrightarrow$ H_2O_2 (aq)	0.682
$[PtCl_4]^{2-}$ (aq) + $2e^-$ $\longrightarrow$ Pt (s) + $4Cl^-$ (aq)	0.73
$SbCl_6^-$ (aq) + $2e^-$ $\longrightarrow$ $SbCl_4^-$ (aq) + $2Cl^-$ (aq)	0.75
Fe^{3+} (aq) + e^- $\longrightarrow$ Fe^{2+} (aq)	0.771
Hg_2^{2+} (aq) + $2e^-$ $\longrightarrow$ 2Hg (ℓ)	0.789
Ag^+ (aq) + e^- $\longrightarrow$ Ag (s)	0.7994
Hg^{2+} (aq) + $2e^-$ $\longrightarrow$ Hg (ℓ)	0.855
$2Hg^{2+}$ (aq) + $2e^-$ $\longrightarrow$ Hg_2^{2+} (aq)	0.920
NO_3^- (aq) + $3H^+$ (aq) + $2e^-$ $\longrightarrow$ HNO_2 (aq) + H_2O	0.94
NO_3^- (aq) + $4H^+$ (aq) + $3e^-$ $\longrightarrow$ NO (g) + $2H_2O$	0.96
Pd^{2+} (aq) + $2e^-$ $\longrightarrow$ Pd (s)	0.987
$AuCl_4^-$ (aq) + $3e^-$ $\longrightarrow$ Au (s) + $4Cl^-$ (aq)	1.00
Br_2 (ℓ) + $2e^-$ $\longrightarrow$ $2Br^-$ (aq)	1.08
ClO_4^- (aq) + $2H^+$ (aq) + $2e^-$ $\longrightarrow$ ClO_3^- (aq) + H_2O	1.19
IO_3^- (aq) + $6H^+$ (aq) + $5e^-$ $\longrightarrow$ $\frac{1}{2}I_2$ (aq) + $3H_2O$	1.195
Pt^{2+} (aq) + $2e^-$ $\longrightarrow$ Pt (s)	1.2
O_2 (g) + $4H^+$ (aq) + $4e^-$ $\longrightarrow$ $2H_2O$	1.229
MnO_2 (s) + $4H^+$ (aq) + $2e^-$ $\longrightarrow$ Mn^{2+} (aq) + $2H_2O$	1.23
$N_2H_5^+$ (aq) + $3H^+$ (aq) + $2e^-$ $\longrightarrow$ $2NH_4^+$ (aq)	1.24
$Cr_2O_7^{2-}$ (aq) + $14H^+$ (aq) + $6e^-$ $\longrightarrow$ $2Cr^{3+}$ (aq) + $7H_2O$	1.33
Cl_2 (g) + $2e^-$ $\longrightarrow$ $2Cl^-$ (aq)	1.360
BrO_3^- (aq) + $6H^+$ (aq) + $6e^-$ $\longrightarrow$ Br^- (aq) + $3H_2O$	1.44
ClO_3^- (aq) + $6H^+$ (aq) + $5e^-$ $\longrightarrow$ $\frac{1}{2}Cl_2$ (g) + $3H_2O$	1.47
Au^{3+} (aq) + $3e^-$ $\longrightarrow$ Au (s)	1.50
MnO_4^- (aq) + $8H^+$ (aq) + $5e^-$ $\longrightarrow$ Mn^{2+} (aq) + $4H_2O$	1.51
$NaBiO_3$ (s) + $6H^+$ (aq) + $2e^-$ $\longrightarrow$ Bi^{3+} (aq) + Na^+ (aq) + $3H_2O$	~1.6
Ce^{4+} (aq) + e^- $\longrightarrow$ Ce^{3+} (aq)	1.61
2HClO (aq) + $2H^+$ (aq) + $2e^-$ $\longrightarrow$ Cl_2 (g) + $2H_2O$	1.63
Au^+ (aq) + e^- $\longrightarrow$ Au (s)	1.68
PbO_2 (s) + SO_4^{2-} (aq) + $4H^+$ (aq) + $2e^-$ $\longrightarrow$ $PbSO_4$ (s) + $2H_2O$	1.685
NiO_2 (s) + $4H^+$ (aq) + $2e^-$ $\longrightarrow$ Ni^{2+} (aq) + $2H_2O$	1.7
H_2O_2 (aq) + $2H^+$ (aq) + $2e^-$ $\longrightarrow$ $2H_2O$	1.77
Pb^{4+} (aq) + $2e^-$ $\longrightarrow$ Pb^{2+} (aq)	1.8
Co^{3+} (aq) + e^- $\longrightarrow$ Co^{2+} (aq)	1.82
F_2 (g) + $2e^-$ $\longrightarrow$ $2F^-$ (aq)	2.87

Basic Solution

Basic Solution	Standard Reduction Potential, E^0 (volts)
SiO_3^{2-} (aq) + $3H_2O$ + $4e^-$ $\longrightarrow$ Si (s) + $6OH^-$ (aq)	−1.70
$Cr(OH)_3$ (s) + $3e^-$ $\longrightarrow$ Cr (s) + $3OH^-$ (aq)	−1.30
$[Zn(CN)_4]^{2-}$ (aq) + $2e^-$ $\longrightarrow$ Zn (s) + $4CN^-$ (aq)	−1.26
$Zn(OH)_2$ (s) + $2e^-$ $\longrightarrow$ Zn (s) + $2OH^-$ (aq)	−1.245

APPENDIX J (*Continued*)

Standard Reduction Potentials in Aqueous Solution at 25°C

Basic Solution	Standard Reduction Potential, E^0 (volts)
$[Zn(OH)_4]^{2-}$ (aq) + $2e^-$ $\longrightarrow$ Zn (s) + $4OH^-$ (aq)	−1.22
N_2 (g) + $4H_2O$ + $4e^-$ $\longrightarrow$ N_2H_4 (aq) + $4OH^-$ (aq)	−1.15
SO_4^{2-} (aq) + H_2O + $2e^-$ $\longrightarrow$ SO_3^{2-} (aq) + $2OH^-$ (aq)	−0.93
$Fe(OH)_2$ (s) + $2e^-$ $\longrightarrow$ Fe (s) + $2OH^-$ (aq)	−0.877
$2NO_3^-$ (aq) + $2H_2O$ + $2e^-$ $\longrightarrow$ N_2O_4 (g) + $4OH^-$ (aq)	−0.85
$2H_2O$ + $2e^-$ $\longrightarrow$ H_2 (g) + $2OH^-$ (aq)	−0.8277
$Fe(OH)_3$ (s) + e^- $\longrightarrow$ $Fe(OH)_2$ (s) + OH^- (aq)	−0.56
S (s) + $2e^-$ $\longrightarrow$ S^{2-} (aq)	−0.48
$Cu(OH)_2$ (s) + $2e^-$ $\longrightarrow$ Cu (s) + $2OH^-$ (aq)	−0.36
CrO_4^{2-} (aq) + $4H_2O$ + $3e^-$ $\longrightarrow$ $Cr(OH)_3$ (s) + $5OH^-$ (aq)	−0.12
MnO_2 (s) + $2H_2O$ + $2e^-$ $\longrightarrow$ $Mn(OH)_2$ (s) + $2OH^-$ (aq)	−0.05
NO_3^- (aq) + H_2O + $2e^-$ $\longrightarrow$ NO_2^- (aq) + $2OH^-$ (aq)	0.01
O_2 (g) + H_2O + $2e^-$ $\longrightarrow$ OOH^- (aq) + OH^- (aq)	0.076
HgO (s) + H_2O + $2e^-$ $\longrightarrow$ Hg (ℓ) + $2OH^-$ (aq)	0.0984
$[Co(NH_3)_6]^{3+}$ (aq) + e^- $\longrightarrow$ $[Co(NH_3)_6]^{2+}$ (aq)	0.10
N_2H_4 (aq) + $2H_2O$ + $2e^-$ $\longrightarrow$ $2NH_3$ (aq) + $2OH^-$ (aq)	0.10
$2NO_2^-$ (aq) + $3H_2O$ + $4e^-$ $\longrightarrow$ N_2O (g) + $6OH^-$ (aq)	0.15
Ag_2O (s) + H_2O + $2e^-$ $\longrightarrow$ 2Ag (s) + $2OH^-$ (aq)	0.34
ClO_4^- (aq) + H_2O + $2e^-$ $\longrightarrow$ ClO_3^- (aq) + $2OH^-$ (aq)	0.36
O_2 (g) + $2H_2O$ + $4e^-$ $\longrightarrow$ $4OH^-$ (aq)	0.40
Ag_2CrO_4 (s) + $2e^-$ $\longrightarrow$ 2Ag (s) + CrO_4^{2-} (aq)	0.446
NiO_2 (s) + $2H_2O$ + $2e^-$ $\longrightarrow$ $Ni(OH)_2$ (s) + $2OH^-$ (aq)	0.49
MnO_4^- (aq) + e^- $\longrightarrow$ MnO_4^{2-} (aq)	0.564
MnO_4^- (aq) + $2H_2O$ + $3e^-$ $\longrightarrow$ MnO_2 (s) + $4OH^-$ (aq)	0.588
ClO_3^- (aq) + $3H_2O$ + $6e^-$ $\longrightarrow$ Cl^- (aq) + $6OH^-$ (aq)	0.62
$2NH_2OH$ (aq) + $2e^-$ $\longrightarrow$ N_2H_4 (aq) + $2OH^-$ (aq)	0.74
OOH^- (aq) + H_2O + $2e^-$ $\longrightarrow$ $3OH^-$ (aq)	0.88
ClO^- (aq) + H_2O + $2e^-$ $\longrightarrow$ Cl^- (aq) + $2OH^-$ (aq)	0.89

APPENDIX K

Selected Thermodynamic Values

Species	$\Delta H^0_{f298.15}$ kcal/mol	$\Delta H^0_{f298.15}$ kJ/mol	$S^0_{298.15}$ cal/mol·K	$S^0_{298.15}$ J/mol·K	$\Delta G^0_{f298.15}$ kcal/mol	$\Delta G^0_{f298.15}$ kJ/mol
Aluminum						
Al (s)	0	0	6.77	28.3	0	0
$AlCl_3$ (s)	−168.3	−704.2	26.45	110.7	−150.3	−628.9
Al_2O_3 (s)	−400.5	−1676	12.17	50.92	−378.2	−1582
Barium						
$BaCl_2$ (s)	−205.56	−860.1	30	126	−193.8	−810.9
$BaSO_4$ (s)	−350.2	−1465	31.6	132	−323.4	−1353
Beryllium						
Be (s)	0	0	2.28	9.54	0	0
$Be(OH)_2$ (s)	−216.8	−907.1	—	—	—	—
Bromine						
Br (g)	26.74	111.8	41.80	174.9	19.70	82.4
Br_2 (ℓ)	0	0	36.384	152.23	0	0
Br_2 (g)	7.387	30.91	58.641	245.4	0.751	3.14
BrF_3 (g)	−61.09	−255.6	69.89	292.4	−54.84	−229.5
HBr (g)	−8.70	−36.4	47.463	198.59	−12.77	−53.43
Calcium						
Ca (s)	0	0	9.95	41.6	0	0
Ca (g)	46.04	192.6	36.993	154.8	37.98	158.9
Ca^{2+} (g)	459.0	1920	—	—	—	—
CaC_2 (s)	−15.0	−62.8	16.8	70.3	−16.2	−67.8
$CaCO_3$ (s)	−288.45	−1207	22.2	92.9	−269.78	−1129
$CaCl_2$ (s)	−190.0	−795.0	27.2	114	−179.3	−750.2
CaF_2 (s)	−290.3	−1215	16.46	68.87	−277.7	−1162
CaH_2 (s)	−45.1	−189	10	42	−35.8	−150
CaO (s)	−151.9	−635.5	9.5	40	−144.4	−604.2
CaS (s)	−115.3	−482.4	13.5	56.5	−114.1	−477.4
$Ca(OH)_2$ (s)	−235.80	−986.6	18.2	76.1	−214.33	−896.8
$Ca(OH)_2$ (aq)	−239.68	−1002.8	18.2	76.15	−207.37	−867.6
$CaSO_4$ (s)	−342.42	−1433	25.5	107	−315.56	−1320
Carbon						
C (s, graphite)	0	0	1.372	5.740	0	0
C (s, diamond)	0.4533	1.897	0.568	2.38	0.6930	2.900
C (g)	171.291	716.7	37.7597	158.0	160.442	671.3
CCl_4 (ℓ)	−32.37	−135.4	51.72	216.4	−15.60	−65.27
CCl_4 (g)	−24.6	103	74.03	309.7	−14.49	−60.63
$CHCl_3$ (ℓ)	−32.14	−134.5	48.2	202	−17.62	−73.72
$CHCl_3$ (g)	−24.65	−103.1	70.65	295.6	−16.82	−70.37
CH_4 (g)	−17.88	−74.81	44.492	186.2	−12.13	−50.75
C_2H_2 (g)	54.19	226.7	48.00	200.8	50.00	209.2
C_2H_4 (g)	12.49	52.26	52.45	209.5	16.28	68.12
C_2H_6 (g)	−20.24	−84.86	54.84	229.5	−7.86	−32.9

APPENDIX K (*Continued*)

Selected Thermodynamic Values

Species	$\Delta H^0_{f\,298.15}$ kcal/mol	$\Delta H^0_{f\,298.15}$ kJ/mol	$S^0_{298.15}$ cal/mol · K	$S^0_{298.15}$ J/mol · K	$\Delta G^0_{f\,298.15}$ kcal/mol	$\Delta G^0_{f\,298.15}$ kJ/mol
C_3H_8 (g)	−24.820	−103.8	64.51	269.9	−5.614	−23.49
C_6H_6 (ℓ)	11.718	49.03	41.30	172.8	29.756	124.5
C_8H_{18} (ℓ)	−64.23	−268.8	—	—	—	—
C_2H_5OH (ℓ)	−66.37	−227.7	38.4	161	−41.80	−174.9
C_2H_5OH (g)	−56.19	−235.1	67.54	282.6	−40.29	−168.6
CO (g)	−26.416	−110.5	47.219	197.6	−32.780	−137.2
CO_2 (g)	−94.051	−393.5	51.06	213.6	−94.254	−394.4
CS_2 (g)	28.05	117.4	56.82	237.7	16.05	67.15
$COCl_2$ (g)	−53.30	−223.0	69.13	289.2	−50.31	−210.5
Cesium						
Cs^+ (aq)	−59.2	−248	31.8	133	−67.41	−282.0
CsF (aq)	−135.9	−568.6	29.5	123	−133.49	−558.5
Chlorine						
Cl (g)	29.082	121.7	39.457	165.1	25.262	105.7
Cl^- (g)	−54	−226	—	—	—	—
Cl_2 (g)	0	0	53.288	223.0	0	0
HCl (g)	−22.062	−92.31	44.646	186.8	−22.777	−95.30
HCl (aq)	−40.02	−167.4	13.17	55.10	−31.35	−131.2
Chromium						
Cr (s)	0	0	5.68	23.8	0	0
$(NH_4)_2Cr_2O_7$ (s)	−431.8	−1807	—	—	—	—
Copper						
Cu (s)	0	0	7.923	33.15	0	0
CuO (s)	−37.6	−157	10.19	42.63	−31.0	−130
Fluorine						
F^- (g)	−77.0	−322	—	—	—	—
F^- (aq)	−78.66	−329.1	—	—	66.08	276.5
F (g)	18.88	78.99	37.917	158.6	14.80	61.92
F_2 (g)	0	0	48.44	202.7	0	0
HF (g)	−64.8	−271	41.508	173.7	−65.3	−273
HF (aq)	−78.66	−329.1	—	—	−66.08	−276.5
Hydrogen						
H (g)	52.095	218.0	27.391	114.6	48.581	203.3
H_2 (g)	0	0	31.208	130.6	0	0
H_2O (ℓ)	−68.315	−285.8	16.71	69.91	−56.687	−237.2
H_2O (g)	−57.796	−241.8	45.104	188.7	−54.634	−228.6
H_2O_2 (ℓ)	−44.88	−187.8	26.2	109.6	−28.78	−120.4
Iodine						
I (g)	25.5	106.6	43.18	180.66	16.77	70.16
I_2 (s)	0	0	27.757	116.1	0	0
I_2 (g)	14.923	62.44	62.28	260.6	4.627	19.36
ICl (g)	4.25	17.78	59.14	247.4	−1.32	−5.52

APPENDIX K (*Continued*)

Selected Thermodynamic Values

	$\Delta H^0_{f\,298.15}$		$S^0_{298.15}$		$\Delta G^0_{f\,298.15}$	
Species	**kcal/mol**	**kJ/mol**	**cal/mol·K**	**J/mol·K**	**kcal/mol**	**kJ/mol**
Iron						
Fe (s)	0	0	6.52	27.3	0	0
FeO (s)	−65.0	−272	—	—	—	—
Fe_2O_3 (s)	−197.0	−824.2	20.89	87.40	−177.4	−742.2
Fe_3O_4 (s)	−267.3	−1118	35.0	146	−242.7	−1015
FeS_2 (s)	−42.42	−177.5	29.202	122.2	−39.84	−166.7
$Fe(CO)_5$ (ℓ)	−185.0	−774.0	80.8	338	−168.6	−705.4
$Fe(CO)_5$ (g)	−175.4	−733.8	106.4	445.2	−166.65	−697.3
Lead						
Pb (s)	0	0	15.49	64.81	0	0
$PbCl_2$ (s)	−85.90	−359.4	32.5	136	−75.08	−314.1
PbO (s, yellow)	−51.94	−217.3	16.42	68.70	−44.91	−187.9
$Pb(OH)_2$ (s)	−123.3	−515.9	21	88	−100.6	−420.9
PbS (s)	−24.0	−100.4	21.8	91.2	−23.6	−98.7
Lithium						
Li (s)	0	0	6.70	28.0	0	0
LiOH (s)	−116.45	−487.23	12	50	−106.1	−443.9
LiOH (aq)	−121.51	−508.4	0.9	4	−107.82	−451.1
Magnesium						
Mg (s)	0	0	7.77	32.5	0	0
$MgCl_2$ (s)	−153.40	−641.8	21.4	89.5	−141.57	−592.3
MgO (s)	−143.84	−601.8	6.4	27	−136.13	−569.6
$Mg(OH)_2$ (s)	−221.00	−924.7	15.09	63.14	−199.27	−833.7
MgS (s)	−83.0	−347	—	—	—	—
Mercury						
Hg (ℓ)	0	0	18.17	76.02	0	0
$HgCl_2$ (s)	−53.6	−224	34.9	146	−42.7	−179
HgO (s, red)	−21.71	−90.83	16.80	70.29	−13.995	−58.56
HgS (s, red)	−13.9	−58.2	19.7	82.4	−12.1	−50.6
Nickel						
Ni (s)	0	0	7.20	30.1	0	0
$Ni(CO)_4$ (g)	−144.1	−602.9	98.09	410.4	−140.4	−587.3
NiO (s)	−58.4	−244	9.22	38.6	−51.7	−216
Nitrogen						
N_2 (g)	0	0	45.77	191.5	0	0
N (g)	112.979	472.704	36.613	153.19	108.886	455.579
NH_3 (g)	−11.02	−46.11	45.97	192.3	−3.94	−16.5
N_2H_4 (ℓ)	12.10	50.63	28.97	121.2	35.67	149.2
$(NH_4)_3AsO_4$ (aq)	−303	−1268	—	—	—	—
NH_4Cl (s)	−75.15	−314.4	22.6	94.6	−48.15	−201.5
NH_4Cl (aq)	−71.76	−300.2	—	—	—	—
NH_4I (s)	−48.14	−201.4	28.0	117	−26.9	−113
NH_4NO_3 (s)	−87.37	−365.6	36.11	151.1	−43.98	−184.0
NO (g)	21.57	90.25	50.347	210.7	20.69	86.57

APPENDIX K (*Continued*)

Selected Thermodynamic Values

Species	$\Delta H^0_{f\,298.15}$ kcal/mol	$\Delta H^0_{f\,298.15}$ kJ/mol	$S^0_{298.15}$ cal/mol·K	$S^0_{298.15}$ J/mol·K	$\Delta G^0_{f\,298.15}$ kcal/mol	$\Delta G^0_{f\,298.15}$ kJ/mol
NO_2 (g)	7.93	33.2	57.35	240.0	12.26	51.30
N_2O (g)	19.61	82.05	52.52	219.7	24.90	104.2
N_2O_4 (g)	2.19	9.16	72.70	304.2	23.38	97.82
N_2O_5 (g)	2.7	11	85.0	356	27.5	115
N_2O_5 (s)	−10.3	−43.1	42.5	178	27.2	114
NOCl (g)	12.57	52.59	63.0	264	15.86	66.36
HNO_3 (ℓ)	−41.61	−174.1	37.19	155.6	−19.31	−80.79
HNO_3 (g)	−32.28	−135.1	63.63	266.2	−17.87	−74.77
HNO_3 (aq)	−49.37	−206.6	35.0	146	−26.41	−110.5
Oxygen						
O (g)	59.553	249.2	38.467	161.0	55.389	231.8
O_2 (g)	0	0	49.003	205.0	0	0
O_3 (g)	34.1	143	57.08	238.8	39.0	163
OF_2 (g)	5.5	23	58.95	246.6	9.7	41
Phosphorus						
P (g)	14.08	58.91	66.89	279.9	5.85	24.5
P_4 (s, white)	0	0	42.4	177	0	0
P_4 (s, red)	−17.6	−73.6	21.80	91.2	−11.59	−48.5
PCl_3 (g)	−73.22	−306.4	74.49	311.7	−68.42	−286.3
PCl_5 (g)	−95.35	−398.9	84.3	353	−77.59	−324.6
PH_3 (g)	1.3	5.4	50.22	210.1	3.2	13
P_4O_{10} (s)	−713.2	−2984	54.70	228.9	−644.8	−2698
H_3PO_4 (s)	−306.2	−1281	26.41	110.5	−267.5	−1119
Potassium						
K (s)	0	0	15.2	63.6	0	0
KCl (s)	−104.33	−436.5	19.74	82.6	−97.70	−408.8
$KClO_3$ (s)	−93.50	−391.2	34.20	143.1	−69.29	−289.9
KI (s)	−78.37	−327.9	25.43	106.4	−77.20	−323.0
KOH (s)	−101.51	−424.7	18.86	78.91	−90.57	−378.9
KOH (aq)	−115.0	−481.2	22.0	92.0	−105.06	−439.6
Silicon						
Si (s)	0	0	4.50	18.8	0	0
$SiBr_4$ (ℓ)	−95.1	−398	—	—	—	—
SiC (s)	−15.6	−65.3	3.97	16.6	−15.0	−62.8
$SiCl_4$ (g)	−157.03	−657.0	79.02	330.6	−147.47	−617.0
SiH_4 (g)	8.2	34	48.88	204.5	13.6	56.9
SiF_4 (g)	−385.98	−1615	67.49	282.4	−375.88	−1573
SiI_4 (g)	−31.6	−132	—	—	—	—
SiO_2 (s)	−217.72	−910.9	10.00	41.84	−204.75	−856.7
H_2SiO_3 (s)	−284.1	−1189	32	134	−261.1	−1092
Na_2SiO_3 (s)	−258	−1079	—	—	—	—
H_2SiF_6 (aq)	−557.2	−2331	—	—	—	—

APPENDIX K (*Continued*)

Selected Thermodynamic Values

Species	$\Delta H^0_{f\,298.15}$ kcal/mol	$\Delta H^0_{f\,298.15}$ kJ/mol	$S^0_{298.15}$ cal/mol·K	$S^0_{298.15}$ J/mol·K	$\Delta G^0_{f\,298.15}$ kcal/mol	$\Delta G^0_{f\,298.15}$ kJ/mol
Silver						
Ag (s)	0	0	10.17	42.55	0	0
Sodium						
Na (s)	0	0	12.2	51.0	0	0
Na (g)	25.98	108.7	36.715	153.6	18.67	78.11
Na^+ (g)	144	601	—	—	—	—
NaBr (s)	−86.03	−359.9	—	—	—	—
NaCl (s)	−98.232	−411.0	17.30	72.38	−91.785	−384
NaCl (aq)	−97.302	−407.1	27.6	115.5	−93.94	−393.0
Na_2CO_3 (s)	−270.3	−1131	32.5	136	−250.4	−1048
NaOH (s)	−101.99	−426.7	—	—	—	—
NaOH (aq)	−112.24	−469.6	11.9	49.8	−100.184	−419.2
Sulfur						
S (s, rhombic)	0	0	7.60	31.8	0	0
S (g)	66.636	278.8	40.094	167.8	56.949	238.3
S_2Cl_2 (g)	−4.4	−18	79.2	331	−7.6	−31.8
SF_6 (g)	−289	−1209	69.72	291.7	−264.2	−1105
H_2S (g)	−4.93	−20.6	49.16	205.7	−8.02	−33.6
SO_2 (g)	−70.944	−296.8	59.30	248.1	−71.748	−300.2
SO_3 (g)	−94.58	−395.6	61.34	256.6	−88.69	−371.1
$SOCl_2$ (ℓ)	−49.2	−206	—	—	—	—
SO_2Cl_2 (ℓ)	−93.0	−389	—	—	—	—
H_2SO_4 (ℓ)	−194.548	−814.0	37.501	156.9	−164.938	−690.1
H_2SO_4 (aq)	−216.90	−907.5	4.1	17	−177.34	−742.0
Tin						
Sn (s)	0	0	12.32	51.55	0	0
$SnCl_2$ (s)	−83.6	−350	—	—	—	—
$SnCl_4$ (ℓ)	−122.2	−511.3	61.8	258.6	−105.2	−440.2
$SnCl_4$ (g)	−112.7	−471.5	87.4	366	−103.3	−432.2
SnO_2 (s)	−138.8	−580.7	12.5	52.3	−124.2	−519.7
Titanium						
$TiCl_4$ (ℓ)	−192.2	−804.2	60.31	252.3	−176.2	−737.2
$TiCl_4$ (g)	−182.4	−763.2	84.8	354.8	−173.7	−726.8
Tungsten						
W (s)	0	0	7.80	32.6	0	0
WO_3 (s)	−201.45	−842.9	18.14	75.90	−182.62	−764.1
Zinc						
ZnO (s)	−83.24	−348.3	10.43	43.64	−76.08	−318.3
ZnS (s)	−49.23	−205.6	13.8	57.7	−48.11	−201.3

APPENDIX L

Answers to Even-Numbered Exercises

Chapter 1

18. (a) 4.32×10^3 (b) 6.87×10^3 (c) 0.174 (d) 7.89 (e) 9.24×10^{-3} (f) 3.00×10^{-2} 20. (a) 3 sig fig (b) 2 sig fig (c) 3 sig fig (d) 4 sig fig
22. (a) 1.0×10^3 (b) 4.39×10^4 (c) 2.86×10^{-4} (d) 9.8765×10^{-5} (e) 1.0000×10^4 (5 sig fig), 1.000×10^4 (4 sig fig), 1.00×10^4 (3 sig fig), 1.0×10^4 (2 sig fig), 1×10^4 (1 sig fig)
24. (a) 19.79 (b) 0.305 (c) 0.49
26. (a) 3.65×10^3 (b) 1.75×10^3 (c) 5.28 (d) 1.08×10^3 28. (a) kilo (b) milli (c) mega (d) deci (e) centi (f) deci (g) milli (h) micro
32. (a) 7.58×10^3 m (b) 7.58×10^4 cm (c) 0.478 kg (d) 9.87×10^3 g (e) 1.386 L (f) 3.692×10^3 mL (g) 1.126 L (h) 786 cm^3 34. 91.4 m
36. 88 km/hr 38. (a) 3.78×10^3 mL (b) 131 mL, 2.00 in (c) 3.06×10^6 cm^3 (d) 61.0 in^3
40. \$1.51/gal 42. (a) 1.00×10^3 g, 1.00 kg (b) 4.25×10^3 g, 4.25 kg (c) 1.4×10^3 g, 1.4 kg (d) 113 g, 0.113 kg 44. (a) 45.4 kg (b) 220 lb (c) 2.27×10^4 cg (d) 2.02×10^9 mg
48. 8.88×10^7 Al atoms 54. 7.20 g/cm^3
56. 13.6 g Hg, 7.20 g Cr, 1.89 g Hg/g Cr
60. 29 mL 62. 66.3 g 64. 0.48 kg for 25 mL
66. (a) 7.8 cm^3 (b) 7.9 g/cm^3
68. (a) 2.35×10^3 cm^3 (b) 13.3 cm, 5.24 in
70. (a) 37.778°C (b) 93.333°C (c) −173.2°C, −279.8°F (d) 37.8°C, 311 K 72. Answers for vertical columns: 4.92×10^3 ft, 1.48×10^4 ft, 3.00×10^4 ft; 305 m, 3.05×10^3 m, 6.00×10^3 m, 1.08×10^4 m; 41°F, 23°F, −15°F, −69°F; 13°C, −15°C, −44°C 74. −3.14 kJ, −750 cal 76. 43.8°C
78. (a) 897 J (b) 2.72×10^3 J (c) 30.2
80. 300 g 82. 0.487 J/g·°C

Chapter 2

6. Answers for vertical columns: S, Ag; 10.81 amu, 55.847 amu, 107.868 amu; 10.81 g, 32.06 g, 55.847 g 8. (a) 5.50 mol (b) 3.0×10^{23} atoms (c) 9.03×10^{23} atoms (d) 8.67×10^{22} atoms (e) 2.53×10^{21} atoms (f) 34.3:1 or 103:3
10. (a) 0.0833 mol (b) 1.66×10^{-24} mol (c) 9.40×10^{-4} mol 14. (a) 60.3% Mg, 39.7% O (b) 69.9% Fe, 30.1% O (c) 32.4% Na, 22.6% S, 45.0% O (d) 29.2% N, 8.3% H, 12.5% C, 50.0% O (e) 8.1% Al, 14.5% S, 72.0% O, 5.4% H
16. TiO_2 18. Fe_2O_3 20. 0.669 g 0/1.00 g Ti and 0.502 g 0/1.00 g Ti yield an oxygen mass ratio of 4:3 per 1.00 g Ti, 0.430 g 0/1.00 g Fe and 0.382 g 0/1.00 g Fe yield an oxygen mass ratio of 9:8 per 1.00 g Fe 22. $MgSO_3$ 24. MgS_2O_3
26. (a) C_2H_4O (b) $C_4H_8O_2$ 28. 2.5 mol
30. 3.00×10^{24} atoms 32. 1.93×10^{24} atoms
34. $C_6H_{14}O_2N_2$ 36. 13.33 g, 17.33 g, 24.67 g
38. (a) 82.9 g $CaCl_2$ (b) 52.9 g Cl_2 40. (a) All P is consumed (b) 3.5 g P (c) 11.5 g compound
42. 10.0 g + 5.0 g = 15.0 g = 11.5 g + 3.5 g
52. 112 g 54. 125 g 56. 12.8 g
58. 45.3 g, 27.2 g 60. 14.4%
62. 292 g 64. 10 g 66. (a) 11.0 g CO_2, 32.0 g SO_2, 43.0 g total (b) 15.2 g 68. C_6H_7N
70. 0.59 g $CaCO_3$, 0.43 g $MgCO_3$ 72. 1.950 g
74. (a) 5.60 g (b) 5.60 g (c) 6.6 g (d) 19.2 g
76. 27.4 g $K_2Cr_2O_7$, 373 g H_2O 78. 109 g soln,

9.05 g NH_4Cl 80. 39.0 g 82. 641 mL 84. 2.50 *M* 86. 0.0410 *M* 88. 250 g 90. 50.9 mL 92. 39.9 mL 94. 863 mL 96. 40.0 mL 98. 150 mL 100. 192 mL 102. 4.06 *M*

Chapter 3

18. nuclear density = 1.45×10^{13} g/cm^3, atomic density = 14.5 g/cm^3 26. 79.907 amu 28. 7.49% ^{6}Li 30. 10.8 amu, 35.5 amu 36. 0.751 amu/atom, 0.751 g/mol 38. 0.439 amu/atom, 0.439 g/mol, 3.95×10^{10} kJ/mol 42. (a) 3.00×10^{14} Hz (b) 5.77×10^{14} Hz (c) 5.66×10^{14} Hz (d) 3.48×10^{8} Hz (e) 1.50×10^{17} Hz 46. 4.3 light years 54. 2.179×10^{-18} J 58. 239 kJ/mol, 57.1 kcal/mol 62. 1.3×10^{-31} nm

Chapter 4

80. (a) −3 (b) 0 (c) +3 (d) +1 (e) +5 (f) +4 (g) +5

Chapters 5, 6, 7

None

Chapter 8

14. 7.73 L 16. 345 torr, 0.454 atm, 4.60×10^4 Pa, 46.0 kPa 18. 2, 100% 24. 462 mL, (a) Celsius scale: 2.0, 100%, Kelvin scale: 1.15, 15.5% 26. 212 K 28. (b) 25°C, 298 K, 77°F (c) 760 torr, 760 mm Hg, 1 atm, 1.013×10^5 Pa, 1013 kPa 32. 1.86×10^3 K, 1.59×10^{3}°C 34. to 3 atm 38. 1.43 g/L, 1.43×10^{-3} g/mL 40. (a) 3.74 mol (b) 120 g 42. 20.0 g/mol 44. 320 g 48. 65.7 L 50. 0.0200 mol N_2, 2.41×10^{22} atoms 52. 148 L 54. 29.6 g/mol, 1.33% error 58. C_4H_{10} 62. 1.75 atm 64. 84 g 66. 0.0594 g 74. $R_{HI}/R_{CH_3OH} = 0.500$ 76. 0.250 mi/s 82. 15.4 atm, 15.3 atm, 0.7% difference 86. (a) 100 L (b) 150 L 88. 6.72 L 90. 82.0% 92. 5.16 L 94. 4.48 L 96. 17.9 L H_2, 31.0 L C_2H_2

Chapter 9

26. 46.7°C 28. 7.51 kJ 30. -2.59×10^4 J 32. 58.7°C 34. 24.6°C 50. $4Na^+$:$4Cl^-$ = NaCl 52. (a) $1Cs^+$:$1Cl^-$ = CsCl (b) $4Zn^{2+}$:$4S^{2-}$ = ZnS 60. 8.28 cm^3/mol 62. 12.0 g/cm^3 64. 24.3 g/mol, Mg 66. 68.0% occupied, 32.0% empty space 68. 1.246 Å, 8.101 Å^3 70. 2.3490 Å

Chapter 10

18. 32 g 20. 90.0 g 24. 0.20 *M* 26. 0.50 *M* 28. 50 g 30. 123 g 32. 6.00 *M* 34. 0.222 36. 111 g 40. (a) 0.306 torr (b) 23.45 torr 42. 0.328 *m* 44. 7.41 *m* 46. 25.6 g 52. −3.72°C, 101.0°C 54. 0.36°C, 82.63°C 56. 40 amu 58. 106.4°C, 223.5°F 62. −0.21°C 64. 87% 66. 61.7% 72. 3.78×10^{-3} atm, 2.87 torr 74. 1.46×10^3 g/mol

Chapter 11

60. (a) 0.442 *M* (b) 0.122 *M* (c) 0.0666 *M* 62. 2.0×10^3 g 64. 2.32 *M* 66. 0.667 *M* 68. 0.500 *M* 70. 3.00 L NaOH, 2.00 L H_3PO_4 72. 618 mL 74. 100 mL 82. 160 mL 84. 30 mL 86. 25.2% pure 88. 0.09227 *M* 90. 0.03773 *M* 92. 0.393 g 94. 0.65 tablet, 65% 96. 1.50 *N* 98. 0.333 *M*, 1.00 *N* 100. 0.050 *M* 102. 0.0901 *N* NaOH, 0.100 N HCl 104. 0.200 *N*, 0.200 *M* 106. 61.3% pure

Chapter 12

2. (a) +2, +3, +4, −3, −2, −1, +5 (b) +2, +4, 0, −2, −2, +3, +4 (c) 0, −2, +4, +6, +4, +6, +6, 4. (a) −2, +4, +6, +2, +2.5 (b) +3, +3, +6, +6 (c) +3, +3, +3 18. 4.0 mL 20. 10 mL 24. 59.3% 26. 0.200 *M*, 1.00 *N* 28. 0.533 *M*, 0.533 *N* 30. Answers for vertical columns, formula weights: 158 g, 294 g, 248 g, 64.1 g, 37.0 g; equiv per mole: 5, 6, 1, 2, 2; equiv wt: 31.6 g, 49.0 g, 248 g, 32.0 g, 18.5 g; mass of 0.1500 equiv: 4.74 g, 7.35 g, 37.2 g, 4.80 g, 2.78 g 32. 90.9 g/equiv 34. 3.2 g 36. (b) 0.0400 *M* (c) 0.240 *N* (d) 0.160 *N* (e) 0.0800 *M* (f) 36.8 g

Chapter 13

16. $q = 854$ J, w = 63.3 J, $\Delta E = 791$ J 18. -4.58×10^5 J/mol Mg 22. (b) −41.85 kJ/g, −3264 kJ/mol 24. $\Delta H = 6.02$ kJ, $\Delta E = 6.18$ kJ 26. (a) −3267 kJ (b) −3267 kJ vs. −3263 kJ 34. (a) 459.4 kJ (b) 1707 kJ (c) −1103 kJ 36. −90.6 kJ, exothermic 38. −112 kJ 40. −1042 kJ 42. −375 kJ 44. (a) −9.32 kJ (b) -8.46×10^6 kJ 46. −64.9 kJ 48. −114 kJ 50. −24.8 kJ 52. −509 kJ 54. −128 kJ 56. −399 kJ 58. −65.3 kJ 66. 391 kJ/mol bonds 68. 434 kJ/mol bonds 70. −93 kJ 72. 294 kJ/mol bonds 74. −649 kJ 76. −2463 kJ 96. (a) $\Delta H^0 = -277.8$ kJ, $\Delta S^0 = 0.0214$ kJ/K, $\Delta G^0 = -284.2$ kJ (b) $\Delta H^0 = 25.2$ kJ, $\Delta S^0 = 0.3069$ kJ/K, $\Delta G^0 = -66.3$ kJ 100. (a) below 3.27×10^3 K 102. (a) 370 K or 97°C 104. 432 K or 159°C 110. (a) −959.3 kJ, −70.54 kJ, 8.3 kJ (b) -2.05×10^3 kJ 112. (a) −268 J/K (b) 1.2×10^2 J/K (c) 67.1 J/K (d) −167 J/K (e) 32 J/K 114. (a) −86.7 kJ (b) −75.1 kJ (c) −64.1 kJ (d) 500°C (Note: Reaction is gas phase reaction at 500°C and 1000°C but not 25°C.)

Chapter 14

12. (a) time^{-1} (b) $M^{-1}\cdot$time^{-1} or L/mol·time (c) $M^{-2}\cdot$time^{-1} or L^2/mol$^2\cdot$time (d) $M^{-0.5}\cdot$time^{-1} or L$^{0.5}$/mol$^{0.5}\cdot$time 14. Rate = $k[A][B][C]^\circ = (2.0 \times 10^{-2}\ M^{-1}\cdot\text{min}^{-1})[A][B]$ 16. first order 18. increase by factor of $3 \times 3^2 = 27$ 20. Rate = $k[A][B]^2 = (2.5 \times 10^{-3}\ M^{-1}\cdot\text{s}^{-1})[A][B]^2$ 22. Rate = $k[A]^2[B]^3 = (20\ M^{-4}\cdot\text{s}^{-1})[A]^2[B]^3$ 24. $k = 0.5\ M^{-1}\cdot\text{s}^{-1}$, Rate$_2$ = 0.20 $M\cdot$s^{-1}, Rate$_3$ = 0.80 $M\cdot$s^{-1} 26. $k_2 = 3.65(k_1)$ 34. (a) 2.5×10^6 s (b) 1.8×10^6 s or 21 days (c) 0.23 g (d) 0.42 g 36. (a) 92 yr (b) 1.2 *M* NO_2, 55 g NO_2 (c) 1.3 *M* NO

Chapter 15

6. $b < e < a < d < c$ 8. $Q = K$, equilibrium; $Q < K$, net forward reaction; $Q > K$, net reverse reaction 10. $K_c = 2.5 \times 10^{-3}$ 12. 56% 14. $[COCl_2] = 0.041\ M$, $[CO] = [Cl_2] = 0.059\ M$ 16. $[COCl_2] = 0.0415\ M$, $[CO] = [Cl_2] = 0.0585\ M$ 18. $K_c = 2.51 \times 10^{-2}$ 20. $[SbCl_5] = 7.40 \times 10^{-4}\ M$, $[SbCl_3] = 6.99 \times 10^{-3}\ M$, $[Cl_2] = 2.60 \times 10^{-3}\ M$ 22. 68.5% 28. $P_{N_2O_4} = 0.00212$ atm, $P_{NO_2} = 0.0158$ atm, $P_{total} = 0.0179$ atm 30. (a) $P_{Cl_2} = 0.24$ atm (b) $[Cl_2] = 9.8 \times 10^{-3}\ M$ (c) 159 g ICl 32. (a) $K_p = K_c = 50$ (b) $P_{H_2} = P_{I_2} = 0.7$ atm, $P_{HI} = 4.6$ atm, $P_{total} = 6.0$ atm (c) 0.12 mol I_2 30 g I_2 40. $Q = 1.9$, net reverse reaction will occur 42. (a) $K_c = 64$ (b) [A] = 0.22 *M* 44. (a) $K_c = 0.45$ (b) [A] = 0.07 *M*, [B] = [C] = 0.18 *M* (c) [A] = 0.52 *M*, [B] = [C] = 0.48 *M* 46. (a) $K_p = 1.7$ (b) $P_{PCl_5} = 24$ atm, $P_{PCl_3} = P_{Cl_2} = 6.4$ atm (c) $P_{total} = 37$ atm (d) $P_{PCl_5} = 30$ atm 48. (a) $[N_2O_4] = 0.0102\ M$, $[NO_2] = 0.00688\ M$ (b) $[N_2O_4] = 4.51 \times 10^{-3}\ M$, $[NO_2] = 4.59 \times 10^{-3}\ M$ (c) $[N_2O_4] = 0.0222\ M$, $[NO_2] = 0.0102\ M$ 50. (a) $K \gg 1$, ΔG^0 is neg, net forward reaction favors equilibrium (b) $K = 1$, $\Delta G^0 = 0$, equilibrium (c) $K << 1$, ΔG^0 is positive, net reverse reaction favors equilibrium 52. At 500°C, $K_p = 0.11$ and $K_c = 1.7 \times 10^{-3}$. At 1000°C, $K_p = 1.3 \times 10^2$ and $K_c = 1.2$. 54. (a) $\Delta G^0 = -38$ kJ (b) $K = 1.2 \times 10^{-3}$ (c) $\Delta G^0 = 54$ kJ

Chapter 16

2. (a) $[H^+] = [I^-] = 0.10\ M$ (b) $[K^+] = [OH^-] = 0.050\ M$ (c) $[Sr^{2+}] = 0.010\ M$, $[OH^-] = 0.020\ M$ (d) $[Ba^{2+}] = 0.0020\ M$, $[NO_3^-] = 0.0040\ M$ (e) $[H^+] = 0.00060\ M$, $[SO_4^{2-}] = 0.00030\ M$ (f) $[Fe^{3+}] = 0.0070\ M$, $[SO_4^{2-}] = 0.010\ M$ 8. 0.050 M KOH, $[H_3O^+] = 2.0 \times 10^{-13}\ M$; 0.010 M $Sr(OH)_2$, $[H_3O^+] = 5.0 \times 10^{-13}\ M$; 0.050 M NaOH, $[H_3O^+] = 2.0 \times 10^{-13}\ M$; 0.050 M $Ca(OH)_2$, $[H_3O^+] = 1.0 \times 10^{-13}\ M$ 12. 0.10 M HI, pOH = 13.00; 0.00030 M H_2SO_4, pOH = 10.76 14. (a) $[H_3O^+] = 1.0 \times 10^{-3}\ M$, $[OH^-] = 1.0 \times 10^{-11}\ M$, pH = 3.00, pOH = 11.00 (b) $[H_3O^+] = 1.00 \times 10^{-2}\ M$, $[OH^-] = 1.00 \times 10^{-12}\ M$, pH = 2.000, pOH = 12.000 (c) $[H_3O^+] = 1.4 \times 10^{-9}\ M$, $[OH^-] = 7.4 \times 10^{-6}\ M$, pH = 8.87, pOH = 5.13 16. (a) $[H_3O^+] = 5.0 \times 10^{-13}\ M$, $[OH^-] = 0.020\ M$, pH = 12.30, pOH = 1.70 (b) $[H_3O^+] = 6.3 \times 10^{-2}\ M$, $[OH^-] = 1.6 \times 10^{-13}\ M$, pH = 1.20, pOH = 12.80 (c) $[H_3O^+] = 2.5 \times 10^{-13}\ M$, $[OH^-] = 0.040\ M$, pH = 12.60, pOH = 1.40 18. (a) C/D = 2.9×10^5 (b) C = (2.9×10^5)D (c) D/C = 3.4×10^{-6} 20. (a) 67 (b) 1.5×10^{-2} (c) In E, $[OH^-] = 5.0 \times 10^{-13}\ M$, In F, $[OH^-] = 3.3 \times 10^{-11}\ M$ (d) 1.5×10^{-2} (e) 66 (f) In E, $[H_3O^+]/[OH^-] = 4.0 \times 10^{10}$ (g) In E, $[OH^-]/[H_3O^+] = 2.5 \times 10^{-11}$ (h) round-off errors 22. $K_a = 4.1 \times 10^{-5}$ 24. $K_a = 2.6 \times 10^{-5}$ 26. $[HOCl] = 0.20\ M$, $[H_3O^+] = [OCl^-] = 8.4 \times 10^{-5}\ M$ 28. $[HCN] = 0.050\ M$, $[H_3O^+] = [CN^-] = 4.5 \times 10^{-6}\ M$, $[OH^-] = 2.2 \times 10^{-9}\ M$, pH = 5.35, 9.0×10^{-3}% ionization 30. (a) $[HCN] = 0.10\ M$, $[H_3O^+] = [CN^-] = 6.3 \times 10^{-6}\ M$, pH = 5.20, 6.3×10^{-3}% ionization (b) $[HF] = 0.09\ M$, $[H_3O^+] = [F^-] = 8.1 \times 10^{-3}\ M$, pH = 2.09, 8.1% ionization (c) $[HOBr] = 0.10\ M$, $[H_3O^+] = [OBr^-] = 1.6 \times 10^{-5}\ M$, pH = 4.80, 0.016% ionization (d) $[HN_3] = 0.10$ M, $[H_3O^+] = [N_3^-] = 1.4 \times 10^{-3}\ M$, pH = 2.85, 1.4% ionization 32. $K_a = 3.6 \times 10^{-6}$ 34. $[NH_3] = 0.50\ M$, $[NH_4^+] = [OH^-] = 9.5 \times 10^{-4}\ M$, $[H_3O^+] = 1.0 \times 10^{-11}\ M$, pH = 10.98, pOH = 3.02, 1.9% ionization

40.

indicator	pH = 2.00	pH = 11.00
methyl red	red	yellow
neutral red	red	yellow
phenolphthalein	colorless	red

44.

mL of 0.100 M $HClO_4$ added	mmol acid	mmol excess base or acid	pH
0.0	0	2.50 OH^-	13.00
5.0	0.50	2.00	12.82
10.0	1.00	1.50	12.63
12.5	1.25	1.25	12.52
20.0	2.00	0.50	12.05
24.0	2.40	0.10	11.31
24.9	2.49	0.01	10.30
25.0	2.50	0.00	7.00
25.1	2.51	0.01 H^+	3.70
27.0	2.70	0.20	2.41
30.0	3.00	0.50	2.04

48. (a) $[H_3O^+] = 4.0 \times 10^{-10}\ M$ and pH = 9.40 compared to $[H_3O^+] = 6.3 \times 10^{-6}\ M$ and pH = 5.20 for 0.10 M HCN. (b) $[H_3O^+] = 7.2 \times 10^{-4}\ M$ and pH = 3.14 compared to $[H_3O^+] = 8.1 \times 10^{-3}\ M$ and pH = 2.09 for 0.10 M HF. 50. (a) pH = 3.76 (b) pH = 9.40 52. (a) pOH = 5.04, pH = 8.96, $[OH^-] = 9.1 \times 10^{-6}\ M$ (b) pOH = 5.04, pH = 8.96, $[OH^-] = 9.1 \times 10^{-6}\ M$ 54. (a) $[H_3O^+]$ goes from $1.8 \times 10^{-5}\ M$ to $2.2 \times 10^{-5}\ M$ and pH from 4.74 to 4.66 (b) $[OH^-]$ goes from $9.0 \times 10^{-6}\ M$ to $6.5 \times 10^{-6}\ M$ and pH from 8.95 to 8.81 (c) $[OH^-]$ goes from $9.0 \times 10^{-6}\ M$ to $1.2 \times 10^{-5}\ M$ and pH from 8.95 to 9.08 56. $[H_3O^+] = 1.0 \times 10^{-10}\ M$, pH = 10.00 58. (a) $[H_3O^+] = 1.7 \times 10^{-12}\ M$, $[OH^-] = 6.0 \times 10^{-3}\ M$, pH = 11.78, pOH = 2.22 (b) $[H_3O^+] = 2.8 \times 10^{-10}\ M$, $[OH^-] = 3.6 \times 10^{-5}\ M$, pH = 9.56, pOH = 4.44 (c) $[H_3O^+] = 5.6 \times 10^{-10}\ M$, $[OH^-] = 1.8 \times 10^{-5}\ M$, pH = 9.26, pOH = 4.74 (d) $[H_3O^+] = 6.2 \times 10^{-11}\ M$, $[OH^-] = 1.6 \times 10^{-4}\ M$, pH = 10.20, pOH = 3.80 (e) $[H_3O^+] = 3.7 \times 10^{-10}\ M$, $[OH^-] = 2.7 \times 10^{-5}\ M$, pH = 9.43, pOH = 4.57 (f) $[H_3O^+] = 4.5 \times 10^{-10}\ M$, $[OH^-] = 2.2 \times 10^{-5}\ M$, pH = 9.34, pOH = 4.66 60. 7.4 g $NaCH_3COO$ 62. 3.5×10^2 mL 0.10 M NaCNO 64. $[H_3O^+] = [HCO_3^-] = 1.4 \times 10^{-4}\ M$, $[H_2CO_3] = 0.050\ M$, $[CO_3^{2-}] = 4.8 \times 10^{-11}\ M$, $[OH^-] = 7.1 \times 10^{-11}\ M$

66. **Concentrations of Species in 0.100 M H_3XO_4**

Species	Concentration (mol/L)	Species	Concentration (mol/L)
H_3PO_4	0.076	H_3AsO_4	0.095
H_3O^+	0.024	H_3O^+	0.0049
$H_2PO_4^-$	0.024	$H_2AsO_4^-$	0.0049
HPO_4^{2-}	6.2×10^{-8}	$HAsO_4^{2-}$	$5.6 \times \times 10^{-8}$
OH^-	4.2×10^{-13}	OH^-	2.0×10^{-12}
PO_4^{3-}	9.3×10^{-19}	AsO_4^{3-}	3.4×10^{-18}

Chapter 17

8. (a) $K_b = 1.4 \times 10^{-11}$ (b) $K_b = 2.9 \times 10^{-7}$ (c) $K_b = 2.5 \times 10^{-5}$ 10. (a) 9.3×10^{-4}% hydrolysis (b) 0.14% hydrolysis (c) 1.3% hydrolysis 14. (a) $[H_3O^+] = 9.2 \times 10^{-6}\ M$, pH = 5.04 (b) $[H_3O^+] = 1.7 \times 10^{-6}\ M$, pH = 5.77 (c) $[H_3O^+] = 1.0 \times 10^{-3}\ M$, pH = 3.00 20. (a) pH = 3.11, 1.5% hydrolysis (b) pH = 5.00, 5.0×10^{-3}% hydrolysis (c) pH = 6.17, 4.5×10^{-4}% hydrolysis 22. (a) NH_3, $K_b = 1.8 \times 10^{-5}$; $C_6H_5NH_2$, $K_b = 4.2 \times 10^{-10}$; CN^-, $K_b = 2.5 \times 10^{-5}$; OCN^-, $K_b = 2.9 \times 10^{-11}$ (b) $NaOCN < C_6H_5NH_2 < NH_3 < NaCN$
24. Some extra points have been added to give a better titration curve.

mol NaOH	[H⁺]	[OH⁻]	pH	pOH
none	$4.2 \times 10^{-4}\ M$	$2.4 \times 10^{-11}\ M$	3.38	10.62
0.00200	7.2×10^{-5}	1.4×10^{-10}	4.14	9.86
0.00300	4.2×10^{-5}	2.4×10^{-10}	4.38	9.62
0.00400	2.8×10^{-5}	3.6×10^{-10}	4.56	9.44
0.00500	1.8×10^{-5}	5.5×10^{-10}	4.74	9.26
0.00600	1.2×10^{-5}	8.3×10^{-10}	4.92	9.08
0.00700	7.7×10^{-6}	1.3×10^{-9}	5.11	8.89
0.00800	4.6×10^{-6}	2.2×10^{-9}	5.34	8.66
0.00900	2.0×10^{-6}	4.9×10^{-9}	5.69	8.31
0.00950	9.5×10^{-7}	1.1×10^{-8}	6.02	7.98
0.0100	4.2×10^{-9}	2.4×10^{-6}	8.37	5.63
0.0105	2.0×10^{-11}	5.0×10^{-4}	10.70	3.30
0.0110	1.0×10^{-11}	1.0×10^{-3}	11.00	3.00
0.0120	5.0×10^{-12}	2.0×10^{-3}	11.30	2.70
0.0130	3.3×10^{-12}	3.0×10^{-3}	11.48	2.52
0.0150	2.0×10^{-12}	5.0×10^{-3}	11.70	2.30
0.0300	5.0×10^{-13}	2.0×10^{-2}	12.30	1.70

26. Some extra points have been added to give a better titration curve.

mol NH_3	[H⁺]	[OH⁻]	pH	pOH
none	$4.2 \times 10^{-4}\ M$	$2.4 \times 10^{-11}\ M$	3.38	10.62
0.00100	1.6×10^{-4}	6.2×10^{-11}	3.79	10.21
0.00200	7.2×10^{-5}	1.4×10^{-10}	4.14	9.86
0.00300	4.2×10^{-5}	2.4×10^{-10}	4.38	9.62
0.00400	2.7×10^{-5}	3.7×10^{-10}	4.57	9.43
0.00500	1.8×10^{-5}	5.6×10^{-10}	4.74	9.26
0.00700	7.7×10^{-6}	1.3×10^{-9}	5.11	8.89
0.00800	4.5×10^{-6}	2.2×10^{-9}	5.35	8.65
0.00900	2.0×10^{-6}	5.0×10^{-9}	5.70	8.30
0.00950	9.5×10^{-7}	1.1×10^{-8}	6.02	7.98
0.0100	1.0×10^{-7}	1.0×10^{-7}	7.00	7.00
0.0105	1.1×10^{-8}	9.0×10^{-7}	7.95	6.05
0.0110	5.6×10^{-9}	1.8×10^{-6}	8.26	5.74
0.00115	3.7×10^{-9}	2.7×10^{-6}	8.43	5.57
0.00130	1.8×10^{-9}	5.4×10^{-6}	8.74	5.26

Chapter 18

6. (a) $K_{sp} = 2.0 \times 10^{-16}$ (b) $K_{sp} = 2.6 \times 10^{-10}$ (c) $K_{sp} = 2.0 \times 10^{-6}$ (d) $K_{sp} = 3.7 \times 10^{-11}$ 10. (b) $Q_{sp} = 2.5 \times 10^{-7}$ is > $(1000)(1.8 \times 10^{-10})$, precipitate will form that can be seen (c) $Q_{sp} = 1.0 \times 10^{-6}$ is > 8.7×10^{-9} but < $(1000)(8.7 \times 10^{-9})$, precipitate will be formed but probably not be seen (d) $Q_{sp} = 1.0 \times 10^{-12}$ is > $(1000)(1.3 \times 10^{-20})$, precipitate will form that can be seen 12. (a) $[Cl^-] = 1.8 \times 10^{-4}\ M$ (b) 1.1×10^{-4} g Ag^+/L (c) 9.1×10^3 L 14. $[CO_3^{2-}] = 4.8 \times 10^{-4}\ M$, 2.5×10^3 L 18. (a) AuCl (b) $[Au^+] = 1.1 \times 10^{-5}\ M$, 99.89% Au^+ pptd (c) $[Au^+] = 1.1 \times 10^{-8}\ M$, $[Ag^+] = 9.5 \times 10^{-6}\ M$ 20. $Q_{sp} = 3.6 \times 10^{-12}$ is < 1.5×10^{-11}, $Mg(OH)_2$ will not precipitate, pH = 8.78 22. (a) 1.5 M NH_4NO_3 (b) 120 g NH_4NO_3 (c) pH = 9.08 24. $Q_{sp} = 1.7 \times 10^{-13}$ is < $(1000)(4.6 \times 10^{-14})$, precipitate will form that probably cannot be seen

Chapter 19

18. $6.241 \times 10^{18}\ e^-$ 20. (a) 5.19×10^3 coul (b) 3.04×10^3 coul (c) 1.52×10^3 coul (d) 893 coul (e) 1.07×10^4 coul 22. 2.33 g, 1.63×10^3 mL 24. 5.74×10^5 s or 6.65 days 26. (a) 0.0961 amp (b) 0.155 amp (c) 0.0248 amp (d) 0.0335 amp 28. 207 g/mol, Pb^{2+}

30. 0.242 g H_2, 1.94 g O_2, 2.72 L H_2, 1.36 L O_2 32. 0.672 g Ag 34. 3.17×10^4 coul 54. $E^0 = +0.621$ V, spontaneous 56. $E^0 = +0.93$ V, spontaneous 58. $E^0 = -0.18$ V, nonspontaneous 60. (a) $E^0 = -1.18$ V, nonspontaneous (b) $E^0 = +1.18$ V, spontaneous, (c) $E^0 = -1.66$ V, nonspontaneous (d) $E^0 = -2.25$ V, nonspontaneous (e) $E^0 = +0.28$ V, spontaneous 62. (a) Cd^{2+} (b) Sn^{4+} (c) H^+ (d) Cl_2 (e) MnO_4^-, acidic (f) F_2 64. (a) H_2 (b) Pt (c) Hg (d) Cl^-, basic (e) Cu (f) Rb 68. (a) E = −0.43 V (b) $E = +0.10$ V (c) $E = -0.65$ V (d) $E = -1.11$ V (e) $E = +0.384$ V (f) E = +1.21 V 70. (a) $SnF_6^{2-} + 2Fe(s) \longrightarrow Sn(s) + 6F^- + 2Fe^{2+}$, $E = +0.53$ V (b) $2MnO_4^- + 16H^+ + 5Cd(s) + 5S^{2-} \longrightarrow 2Mn^{2+} + 8H_2O + 5CdS(s)$, $E = +1.86$ V (c) $2O_2(g) + Te(s) + 2H_2O(\ell) \longrightarrow 2H_2O_2 + TeO_2(s)$, $E = +0.227$ V (d) $2NO_3^- + 2H^+ + 3H_2(g) \longrightarrow 2NO(g) + 4H_2O(\ell)$, $E = +0.37$ V 72. (a) $E = 2.418$ V, spontaneous (b) $E = 0.24$ V, spontaneous 76. (a) $E^0 = -0.62$ V, $\Delta G^0 = 1.2 \times 10^2$ kJ, $K = 1.1 \times 10^{-21}$ (b) $E^0 = +0.35$ V, $\Delta G^0 = -34$ kJ, $K = 8.3 \times 10^5$ (c) $E^0 = +1.833$ V, $\Delta G^0 = -1.061 \times 10^3$ kJ, $K = 1.008 \times 10^{186}$ 78. (a) $E^0 = +0.80$ V, $\Delta G^0 = -4.6 \times 10^2$ kJ, $\Delta G^0 = -4.6 \times 10^2$ kJ/mol $K_2Cr_2O_7$, $\Delta G^0 = -77$ kJ/mol NaI (b) $E^0 = +1.34$ V, $\Delta G^0 = -1.29 \times 10^3$ kJ, $\Delta G^0 = -6.45 \times 10^2$ kJ/mol $KMnO_4$, $\Delta G^0 = -2.58 \times 10^2$ kJ/mol H_2SO_3 (c) $E^0 = +2.00$ V, $\Delta G^0 = -1.93 \times 10^3$ kJ, $\Delta G^0 = -965$ kJ/mol $KMnO_4$, $\Delta G^0 = -386$ kJ/mol $(COOH)_2$ 80. (a) $K = 6.43 \times 10^{25}$ (b) $K = 4.24 \times 10^{15}$ (c) $K = 5.2 \times 10^3$ 82. (a) $E^0 = +0.740$ V, $K = 3.77 \times 10^{62}$ (b) $E^0 = +0.794$ V, $K = 3.72 \times 10^{80}$ (c) $E^0 = +1.72$ V, $K = 3.44 \times 10^{174}$ 84. $\Delta G^0 = 15.6 \times 10^4$ J, $K_{sp} = 4.57 \times 10^{-28}$

Chapter 20

16. 1.8×10^8 s or 2.1×10^3 days, 7.2×10^3 L O_2

Chapter 21

41. (a) −446.2 kJ (b) −390.8 kJ (c) −431.2 kJ

Chapter 22

10. 2.33 g $XeOF_4$, 1.88 L HF 42. $Cl_2/F_2 = 0.732$, $Br_2/F_2 = 0.488$, $I_2/F_2 = 0.387$, $At_2/F_2 = 0.301$ 44. 1.36×10^3 mol Cl_2, 2.72×10^3 mol NaCl, 1.59×10^5 g NaCl, 350 lb NaCl, 138 lb Na, 2.92×10^5 hr 46. 131 L $NaHSO_3$ solution; 8.70 g $NaIO_3$ 48. 5.20 g SiF_4, 1.12 L SiF_4 50. 4.469 g $KClO_4$

Chapter 23

24. 0.0906 g 0.0774% Ag tarnished 26. $[OH^-] = 1.3 \times 10^{-4}$ *M*, pOH = 3.89, pH = 10.11

Chapter 24

10. (a) 0 (b) +2 (c) +4 (d) +5 (e) +3 12. 946 kJ/mol bonds 30. (a) 16.5% N in $NaNO_3$, 35.0% N in NH_4NO_3 (b) $[H_3O^+] = 7.5 \times 10^{-6}$ *M*, pH = 5.12 32. 946 kJ/mol N≡N bonds, 498.4 kJ/mol O═O bonds, 631.7 kJ/mol N═O bonds 34. 23.9 g $NaNH_2$ 36. 26.3 g superphosphate 38. 43.7% P in P_4O_{10}, 31.6% P in H_3PO_4

Chapter 25

18. 0.0563 mol CO_2, 4.16 g Li_2CO_3, 73.9% Li_2CO_3, 26.1% LiCl 26. For *X* = F, $\Delta G^0 = \Delta G_f^0$ for SiF_4 (g) = −1615 kJ and *X* = Cl, $\Delta G^0 = \Delta G_f^0$ for $SiCl_4$ (g) = −657 kJ. Formation of SiF_4 (g) is the more spontaneous process. 34. $[H_3BO_3] = 0.10$ *M*, $[H_3O^+] = [H_2BO_3^-] = 8.5 \times 10^{-6}$ *M*, $[HBO_3^{2-}] = 1.8 \times 10^{-13}$ *M*, $[BO_3^{3-}] = 3.4 \times 10^{-22}$ *M*, pH = 5.07 36. BH_3 simplest formula, 41.5 g/mol, B_3H_9 molecular formula

Chapter 26

20. 569 g Co_3O_4 22. $[Cr_2O_7^{2-}]/[CrO_4^{2-}] = 4.2 \times 10^{-8}$ 26. -1.8×10^2 kJ 28. 4.48 L H_2

Chapter 27

38. (a) −657 kJ/mol + 2P (b) −194 kJ/mol (c) −253 kJ/mol (d) −900 kJ/mol + 2P (e) −95.7 kJ/mol (f) −314 kJ/mol (g) −93.3 kJ/mol (h) −172 kJ/mol 40. $[NH_3] = 6.7 \times 10^{-4}$ *M*, $[NH_4^+] = [OH^-] = 1.3 \times 10^{-4}$ *M*, pOH = 3.9, pH = 10.1

Chapter 28

24. $k = 0.032$ min^{-1}, 0.56 μg Fr remain
26. $k = 1.21 \times 10^{-4}$ yr^{-1}, 1.8×10^4 yr
28. $k = 5.60 \times 10^{-2}$ hr^{-1}, $t_{1/2} = 12.4$ hr

Chapters 29, 30

None

Illustration and Table Credits

Chapter 1

Figure 1–1, photo by Jim Mergenthaler. Figure 1–6, from the CHEM Study film, "Chemical Families." Figure 1–14, courtesy of U.S. National Bureau of Standards. Figure 1–15, courtesy of Arthur H. Thomas Company. Figure 1–18, from F. Brescia, S. Mehlman, F. C. Pellegrini, and S. Stambler, *Chemistry: A Modern Introduction,* 2nd ed., Saunders College Publishing, 1978. Figure 1–19, photo of Nalgene® graduated cylinder and buret supplied by Nalge Company, Division of Sybron Corporation.

Chapter 2

Figure 2–5, from W. L. Masterton, E. J. Slowinski, and C. L. Stanitski, *Chemical Principles,* 5th ed., Saunders College Publishing, 1981.

Chapter 3

Figures 3–1, 3–3, from Brescia, Mehlman, Pellegrini, and Stambler. Figures 3–2, 3–9, 3–11, 3–20, from Masterton, Slowinski, and Stanitski. Figures 3–4, 3–5, from E. I. Peters, *Introduction to Chemical Principles,* 3rd ed., Saunders College Publishing, 1982. Figure 3–7, adapted from D. B. Murphy, V. Rousseau, and W. F. Kieffer, *Foundations of College Chemistry,* 1st ed., John Wiley & Sons, Inc., 1969.

Chapter 4

Figure 4–1, from *Annalen der Chemie und Pharmacie,* VIII, Supplementary Volume for 1872, p. 151. Figures 4–2, 4–5, 4–6, from Masterton, Slowinski, and Stanitski. Tables 4–2 and 4–7, adapted from Brescia, Mehlman, Pellegrini, and Stambler. Tables 4–4, 4–5, adapted from K. F. Purcell and J. C. Kotz, *Inorganic Chemistry,* Saunders College Publishing, 1977.

Chapter 5

Figures 5–2, 5–3, from Masterton, Slowinski, and Stanitski.

Chapter 7

Figure 7–1, from Brescia, Mehlman, Pellegrini, and Stambler. Figures 7–2, 7–4, 7–6, 7–8, 7–10, 7–12, 7–13, 7–14, from H. C. Metcalfe, J. E. Williams, and J. F. Castka, *Modern Chemistry,* Holt, Rinehart, and Winston, 1982. Figure 7–3, from P. P. Berlow, D. J. Burton, and J. I. Routh, *Introduction to the Chemistry of Life,* Saunders College Publishing, 1982. Figures 7–7, 7–9, from M. M. Jones, D. O. Johnston, J. T. Netterville, and J. L. Wood, *Chemistry, Man and Society,* 4th ed., Saunders College Publishing, 1983. Figure 7–15, courtesy of the Los Angeles County Air Pollution Control District. Table 7–6, from W. L. Masterton and E. J. Slowinski, *Chemical Principles,* 2nd ed., Saunders College Publishing, 1969.

Chapter 8

Figure 8–1 (right) from P. Highsmith, *Physics, Energy and Our World,* Saunders College Publishing, 1975. Figures 8–2, 8–7, 8–11, from Masterton, Slowinski, and Stanitski. Figures 8–3, 8–6, 8–8, 8–12, 8–13, from Brescia, Mehlman, Pellegrini, and Stambler. Figures 8–10, 8–14, from W. L. Masterton, E. J. Slowinski, and E. T. Walford, *Chemistry,* Holt, Rinehart and Winston, 1980. Table 8–4, from J. A. Campbell, *Chemical Systems—Energetics, Dynamics and Structure,* W. H. Freeman and Co., 1970, courtesy of the author.

Chapter 9

Figure 9–17(a), courtesy of Eastman Kodak Company.

Chapter 10

Figure 10–3, after Peters. Figure 10–6, courtesy of Arthur H. Thomas Company. Figure 10–15, reprinted by permission, AMERICAN SCIENTIST 61:280 (1973), "Desalination," by Ronald F. Probstein. Figures 10–16, 10–19, from Brescia, Mehlman, Pellegrini, and Stambler.

Chapter 11

Figure 11–1(d), from Berlow, Burton, and Routh.

Chapter 12

Figure 12–1, from Masterton, Slowinski, and Stanitski.

Chapter 13

Figures 13–1, 13–4, from Masterton, Slowinski, and Stanitski.

Chapter 14

Figure 14–10, from Masterton, Slowinski, and Stanitski.

Chapter 16

Figure 16–1, courtesy of Arthur H. Thomas Company. Table 16–5, from Brescia, Mehlman, Pellegrini, and Stambler.

Chapter 19

Figures 19–1, 19–2, from Brescia, Mehlman, Pellegrini, and Stambler. Figures 19–3, 19–14, 19–16, 19–17, from Masterton, Slowinski, and Stanitski. Figures 19–4, 19–5, 19–15, from Masterton, Slowinski, and Walford. Table 19–2, from G. Charlot, *IUPAC Supplement,* "Selected Constants, Oxidation-Reduction Potentials in Aqueous Solution," 1971.

Chapter 20

Figure on p. 670, from Metcalfe, Williams, and Castka. Figures on p. 672, courtesy of American Iron and Steel Institute. Figure 20–1, from Masterton, Slowinski, and Stanitski. Figures 20–3, 20–6, from Masterton, Slowinski, and Walford. Figures 20–4, 20–8, 20–10, from Jones, Johnston, Netterville, and Wood. Figures 20–7, 20–9, from E. G. Rochow, *Modern Descriptive Chemistry*, Saunders College Publishing, 1977. Table 20–1, reprinted with permission from *Chemical and Engineering News*, 26 March 1979, copyright 1979 American Chemical Society. Table 20–3, from *Chemistry: A Contemporary Approach*, by G. T. Miller, Jr., © 1976, by Wadsworth Publishing Company, Inc., Belmont, California 94002. Reprinted with permission. Table 20–4, from R. G. Gymer, *Chemistry in the Natural World*, 1976, D. C. Heath & Co. Table 20–5, after D. H. Meadows, et al., *Limits to Growth*, Universe Books, New York, 1972.

Chapter 21

Figure on p. 688, from Berlow, Burton, and Routh. Figure on p. 690, courtesy of Aluminum Company of America. Figure on p. 691, courtesy of ORGOTHERM, Inc. Tables 21–1, 21–3, adapted from Rochow and by permission of the publisher from *Inorganic Chemistry* by Jacob Kleinberg, William J. Argersinger, Jr., and Ernest Griswold (Lexington, MA: D. C. Heath and Company, 1960). Tables 21–2, 21–4, adapted with permission of the publisher from *Inorganic Chemistry* by Jacob Kleinberg, William J. Argersinger, Jr., and Ernest Griswold (Lexington, MA: D. C. Heath and Company, 1960) and from T. Moeller, *Advanced Inorganic Chemistry*, John Wiley & Sons, Inc., 1952.

Chapter 23

Figure 23–1, from Masterton, Slowinski, and Walford. Figure 23–4, from E. M. Winkler, *Stone*, Springer-Verlag, 1973. Figure 23–5, from National Air Pollution Control Administration, HEW Public Health Service, 1970, Publication No. AP-71.

Chapter 25

Figure 25–2, from Masterton, Slowinski, and Walford. Figure 25–5, from Purcell and Kotz. Table 25–1, adapted from B. Mason, *Principles of Geochemistry*, 3rd ed., John Wiley & Sons, Inc., 1966.

Chapter 27

Figure 27–1, from M. F. Perutz et al., reprinted by permission from *Nature*, vol. 185, p. 416, copyright © 1960 Macmillan Journals, Limited. Figure 27–7, adapted from P. George and D. S. McClure, *Progress in Inorganic Chemistry*, vol. 1, p. 381, 1959, F. A. Cotton, ed., Wiley (Interscience).

Chapter 28

Figure 28–1, from Masterton, Slowinski, and Walford. Figures 28–3, 28–10, 28–12, from Brescia, Mehlman, Pellegrini, and Stambler. Figures 28–6, 28–11, from M. Merken, *Physical Science with Modern Applications*, 2nd ed., Saunders College Publishing, 1980. Figure 28–7, courtesy of the Lawrence Berkeley Laboratory, University of California, Berkeley. Table 28–3, adapted with permission from R. C. Weast, ed., *CRC Handbook of Chemistry,* 53rd ed., copyright CRC Press, Inc., Boca Raton, FL, 1973; and from J. A. Dean, ed., *Lange's Handbook of Chemistry,* 11th ed., copyright McGraw-Hill Book Company, 1973.

Chapter 29

Figure 29–10, courtesy of Proctor & Gamble Company. Figure 29–12, from Masterton, Slowinski, and Stanitski. Figure 29–15 and Table 29–6, reprinted with permission of Macmillan Publishing Company from Conant and Blatt, *The Chemistry of Organic Compounds,* 5th ed., copyright © 1959 by Macmillan Publishing Company.

Chapter 30

Figure 30–7, from Masterton, Slowinski, and Walford. Table 30–1, from H. Hart and R. D. Schuetz, *Organic Chemistry,* 3rd ed., p. 101, copyright © 1966 by Houghton Mifflin Company; used by permission.

Index

Entries in *italics* indicate illustrations; page numbers followed by *t* indicate tables; page numbers followed by *m* indicate margin notes, by *fn*, footnotes.

D

E

F

G

P

Q

R

The Electronic Configurations of the Atoms of the Elements

Element	Atomic Number	1*s*	2*s*	2*p*	3*s*	3*p*	3*d*	4*s*	4*p*	4*d*	4*f*	5*s*
		Populations of Subshells										
H	1	1										
He	2	2										
Li	3	2	1									
Be	4	2	2									
B	5	2	2	1								
C	6	2	2	2								
N	7	2	2	3								
O	8	2	2	4								
F	9	2	2	5								
Ne	10	2	2	6								
Na	11				1							
Mg	12				2							
Al	13	Neon core			2	1						
Si	14				2	2						
P	15				2	3						
S	16				2	4						
Cl	17				2	5						
Ar	18	2	2	6	2	6						
K	19							1				
Ca	20							2				
Sc	21						1	2				
Ti	22						2	2				
V	23						3	2				
Cr	24						5	1				
Mn	25	Argon core					5	2				
Fe	26						6	2				
Co	27						7	2				
Ni	28						8	2				
Cu	29						10	1				
Zn	30						10	2				
Ga	31						10	2	1			
Ge	32						10	2	2			
As	33						10	2	3			
Se	34						10	2	4			
Br	35						10	2	5			
Kr	36	2	2	6	2	6	10	2	6			
Rb	37											1
Sr	38											2
Y	39									1		2
Zr	40									2		2
Nb	41	Krypton core								4		1
Mo	42									5		1
Tc	43									5	2	2
Ru	44									7	1	1
Rh	45									8		
Pd	46									10		
Ag	47									10		1
Cd	48									10		2